# 2nd International Conference on Engineering and Technology for Sustainable Development (ICET4SD 2017)



MATEC Web of Conferences Volume 154 (2018)

Yogyakarta, Indonesia
13 – 14 September 2017

**Editors:**

**S. Ma'mun**
**H. Tamura**
**M.R.A. Purnomo**

ISBN: 978-1-5108-5910-4

**Printed from e-media with permission by:**

Curran Associates, Inc.
57 Morehouse Lane
Red Hook, NY 12571

**Some format issues inherent in the e-media version may also appear in this print version.**



Printed by Curran Associates, Inc. (2018)

For additional information, please contact EDP Sciences – Web of Conferences at the address below.

EDP Sciences – Web of Conferences
17, Avenue du Hoggar
Parc d'Activité de Courtabœuf
BP 112
F-91944 Les Ulis Cedex A
France

Phone: +33 (0) 1 69 18 75 75
Fax: +33 (0) 1 69 28 84 91

contact-edps@webofconferences.org

**Additional copies of this publication are available from:**

Curran Associates, Inc.
57 Morehouse Lane
Red Hook, NY 12571 USA
Phone: 845-758-0400
Fax: 845-758-2633
Email: curran@proceedings.com
Web: www.proceedings.com

# TABLE OF CONTENTS

## Preface

The 2[nd] International Conference on Engineering and Technology for Sustainable Development (ICET4SD) 2017 with the theme of *Sustainable Energy for a Better Future* is an international conference organized by the Faculty of Industrial Technology, Universitas Islam Indonesia (UII) in collaboration with UII's strategic partners Kansai University Japan, Sirindhorn International Institute of Technology (SIIT) Thailand, Universiti Teknologi MARA (UiTM) Malaysia and Center for Energy Studies – Universitas Gadjah Mada (UGM) Indonesia.

As an international academic forum, the conference aims to exchange ideas, knowledge, and expertise in various fields such as chemical engineering, electrical engineering, industrial engineering and management system, informatics, mechanical engineering, textile engineering, and other engineering disciplines through various sessions consisting of keynote lectures, oral and poster presentations, and discussions. The conference is also expected to foster networking, collaboration and joint effort among the conference participants to advance the theory and practice as well as to identify major trends and critical issues in those fields. This biannual conference is a continuation of the 1[st] ICET4SD that was held in 2015.

All submitted articles to the 2[nd] ICET4SD 2017 were gone through a double-blind peer review by competent reviewers. There were 200 articles from 300 submissions that were accepted to be presented both oral and poster presentations. The forum was attended by more than 300 participants from 4 countries and from 46 reputable affiliations.

We are proud to have a partnership with MATEC Web of Conferences in order to make the 2[nd] ICET4SD 2017 conference proceedings widely usable by the scientific community. After a thorough consideration based on the quality of accepted papers, the 2[nd] ICET4SD Scientific Committee has selected 149 articles to be published in MATEC Web of Conferences.

We express our deepest gratitude to all members of the committee, reviewers, keynote speakers, paper contributors and participants for their valuable contributions. We do hope this international conference will sustainably be hosted and develop the society continuously.

**Dr. Sholeh Ma'mun**
Chairman of the 2[nd] ICET4SD 2017
The 2[nd] International Conference on Engineering and Technology for Sustainable Development
Yogyakarta, Indonesia, February 2018

edp sciences

Web of Conferences

# Statement of Peer review

In submitting conference proceedings to *Web of Conferences*, the editors of the proceedings certify to the Publisher that

1. They adhere to its **Policy on Publishing Integrity** in order to safeguard good scientific practice in publishing.
2. All articles have been subjected to peer review administered by the proceedings editors.
3. Reviews have been conducted by expert referees, who have been requested to provide unbiased and constructive comments aimed, whenever possible, at improving the work.
4. Proceedings editors have taken all reasonable steps to ensure the quality of the materials they publish and their decision to accept or reject a paper for publication has been based only on the merits of the work and the relevance to the journal.

Title, date and place of the conference

The 2nd International Conference on Engineering and Technology for Sustainable Development (ICET4SD 2017)
13 - 14 September 2017
Yogyakarta, Indonesia

Proceedings editor(s):

Dr. Sholeh Ma'mun
Prof. Hiroshi Tamura
Dr. M. Ridwan Andi Purnomo

Date and editor's signature

22 January 2018

# Modification of growth medium of mixed-culture species of microalgae isolated from southern java coastal region

*Hanifrahmawan* Sudibyo[1], *Yuni* Purwanti[2], *Yano Surya* Pradana[1], *Thoriq Teja* Samudra[2], *Arief* Budiman[1], *Eko Agus* Suyono[2,*]

[1]Chemical Engineering Department., Universitas Gadjah Mada, Jalan Grafika No. 2, Yogyakarta, Indonesia, 55281
[2]Faculty of Biology, Universitas Gadjah Mada, Jalan Teknika Selatan, Yogyakarta, Indonesia

**Abstract.** Globally, there is growing interest in microalgae as production organisms. Microalgae contain lipids (oil), proteins and carbohydrates (sugars), and, especially marine algae have been used as food and feed for centuries. Recently, production cost reduction related to the supply of growth nutrients is necessary to make it profitable. Therefore, utilization of molasses, a byproduct of sugar production, as the natural carbon, macronutrients, and micronutrients sources can be helpful. The analysis showed that the content of sucrose, glucose, fructose, potassium, zinc, and magnesium was 68.4% w/w, 18.5% w/w, and 13.1% w/w, 5.5% w/w, 3.91 ppm, and 1,370 ppm respectively. This work aimed to determine the effect of molasses addition to the physio-chemical properties of multi-culture species of microalgae isolated from southern Java coastal region in Indonesia grown under mixotrophic culture. The cultivation in this work used medium which was self-formulated by the authors consisting of $NaNO_3$ (5 mL/L), $H_3BO_3$ (1 mL/L), EDTA (1 mL/L), $N_2H_2PO_4$ (5 mL/L), $FeSO_4$ (1 mL/L), $MgSO_4$ (1 mL/L), NaCl (1 mL/L), micronutrients (1 mL/L), vitamin $B_1$ (1 mL/L), and vitamin $B_{12}$ (1 mL/L) in 500 mL of water. The medium will be treated to have molasses concentration of 0.05% v/v, 0.15% v/v, 0.25% v/v, 0.35% v/v, and 0.45% v/v.

## 1 Introduction

Many species of microalgae naturally has high nutrient content. To increase it, optimization could be made by modifying the growth determining factors such as nitrogen levels, light intensity, temperature, concentration of dissolved $CO_2$, cultivation procedures, and salinity [1]. Environmental stress and limitations of nutrients (nitrogen for green microalgae and silica for diatoms) causes an accumulation of certain nutrient inside the cell. Under certain conditions, usually triggered by stress or atrophy, the nutrient content can be increased up to more than a half of the dry weight of the cell.

Current situation for efficient microalgae cultivation used mixotrophic system as the way. In the mixotrophic system, cell growth is not completely dependent on photosynthesis as the light energy is not the only factor restricting the growth. Both light and organic carbon substrate take part. The mixotrophic system allows the integration of both components of photosynthesis and heterotrophic during the diurnal cycle. This system will reduce the impact of shortage of biomass for dark respiration and lower the amount of organic substances which are used during cell growth [2].

To cultivate microalgae in the mixotrophic system, utilization of low cost carbon source from agroindustry waste which is molasses is studied. Previous study by O'Hara and Mundree [4], molasses possessed pH level of about 5 and contained carbon, nitrogen, inorganic salts such as P, Na, K, Ca, and other metals such as Cu, Zn, Mn, Fe, and Mg as well as vitamins [5]. All those minerals are necessary for better growth of microalgae. The other way to enhance the efficiency of microalgae cultivation is by using mixed culture grown microalgae [5]. Mixed culture condition enabled the self-defense of the microalgae species against "predators" like other microbes or bacteria. Therefore, this study aims to determine effect of molasses addition to the cell growth profile represented by cell amount and biomass dry weight and to the nutrients profile (lipid and chlorophyll)

## 2 Methodology

### 2.1 Materials and apparatus

Mixed cultures-grown microalgae were isolated from South Coast of Yogyakarta, Glagah, Indonesia. The growth medium used was supplied by Merck and using the formula shown by Table 1 and Table 2 [6] as well as molasses medium with the volume percentage of 0.05%, 0.15%, 0.25%, 0.35%, and 0.45%

* Corresponding author: eko_suyono@ugm.ac.id

**Table 1.** Growth medium formula of the macronutrients

| Component | Stock Solution, g/L | Amount taken from stock solution, mL stock solution /L medium |
|---|---|---|
| | Macronutrients | |
| $NaNO_3$ | 250 | 5 |
| $H_3BO_3$ | 33.6 | 1 |
| EDTA.2Na | 45 | 1 |
| $NaH_2PO_4$ | 175 | 1 |
| $FeSO_4$ | 5 | 1 |
| $MgSO_4.7H_2O$ | 50 | 1 |
| NaCl | 5 | 1 |

**Table 2.** Growth medium formula of the micronutrients and vitamins

| Component | Stock Solution, g/L | Amount taken from stock solution, mL stock solution /L medium |
|---|---|---|
| | Micronutrients | |
| $MnCl.4H_2O$ | 0.36 | 1 |
| $ZnSO_4$ | 0.021 | 1 |
| $CoCl_2.6H_2O$ | 0.02 | 1 |
| $(NH_4)_2Mo_7O_{24}.4H_2O$ | 0.0093 | 1 |
| $CuSO_4.5H_2O$ | 0.02 | 1 |
| | Vitamin | |
| Vitamin B1 | 15.25 mg/100 mL | 1 |
| Vitamin B2 | 12.5 mg/100 mL | 1 |

For analytical routines, there were ethanol 70% (CV. General Labora), chloroform (merck), methanol (Merck), distilled water, and acetone. In this study, the molasses was obtained from PT. Madu Baru, Bantul, D.I. Yogyakarta. The laboratory analysis from PT. Madu Baru gave composition of molasses as shown by Table 3. The analysis of carbon source content was performed using an HPLC system (Knauer smart line RI detector 2300, Germany) with a column of Aminex HPX-87C 300 x 7.8 mm (Bio-Rad, USA) at 85°C. Injection volume is 20 μL and mobile phase was deionized water at flow rate of 0.6 mL/min. The analysis of element inside molasses was performed using Atomic Absorption Spectroscopy (AAS).

The experimental apparatus used in this study were spectrophotometer (Genesys 10 UV-VIS Thermoscientific), cuvette, micropipette, pipette tip, oven (ESCO Isoterm), analytical balance, pH meter, autoclave (ES-215 TOMY), digital lux meter LX-1010B, centrifuge (UNIVERSAL 320 R HettichC), microscope, optilab, and haemocytometer

## 2.2 Cultivation Methods

The molasses was diluted to eliminate water-insoluble material in accordance with the prescribed concentration. Liquid left for 24 hours in room temperature and then the supernatant is filtered with a 0.5-μm filter paper and conditioned at a pH in accordance with the culture [7]. To prepare the growth medium, the chemicals were added to 1,000 mL of distilled water and were then homogenized. The pH level of the medium was set in between 7 – 8. The growth medium was then sterilized by autoclaving for 15 minutes at the pressure of 15 psi and temperature of $121^0$C. Afterward, mixed-culture microalgae isolate of 250 mL were mixed in 250 mL growth medium with variation of volume percentage of molasses as many as 0.05%, 0.15%, 0.25%, 0.35%, and 0.45%. Cultures were incubated for 10 days with continuous aeration and lighting

**Table 3.** Analysis of components inside molasses

| Molasses | | Carbon Source inside Molasses | |
|---|---|---|---|
| Components | Content | Components | Content |
| Carbon | 61.44% | Sucrose | 68.4% |
| Nitrogen | 6.02% | | |
| Phosphor | 0.29% | Glucose | 18.5% |
| Sodium | 0.35% | | |
| Potassium | 5.3% | Fructose | 13.1% |
| Calcium | 0.75% | | |
| Copper | 2.21 ppm | Oxalic acid | 4.10% |
| Zinc | 3.89 ppm | | |
| Manganese | 4.69 ppm | Citric acid | 3.22% |
| Iron | 78.26 ppm | | |
| Magnesium | 1,373 ppm | | |

### 2.3 Parameters measurement

Cell counting method follow *Haemocytometer* method as explained by [8]. Calculations done by counting the cells in the second chamber at the top and bottom *Haemocytometer*.

Productivity of the biomass is evaluated by measuring the biomass dry weight using gravimetric method by taking 5 mL sample and by filtering it with filter paper. Biomass was then wrapped in filter paper and put in the microwave for 15 minutes at 34°C until the constant weight was achieved.

Lipid data retrieval is conducted by following the method of Bligh and Dryer [9]. To measure the chlorophyll content, microalgae samples as many as 5 mL was taken and put inside the tube. Afterward, it was centrifuged at rotational speed of 3,300 rpm for 15 minutes. The supernatant was taken and allowed to remain at the bottom of the tube and then 2 mL of acetone was added to be re-centrifuged at 1,800 rpm for 10 minutes. Samples were transferred into a glass cuvette and placed in a spectrophotometer with absorbance at a wavelength of 645 and 662 nm. Based on the absorbance of the spectrophotometer, the concentrations of chlorophyll A and B were calculated using the following equation [10]

$$\text{number of cell}/_{\text{mL}} = \frac{\text{counted cell}}{5} \times \text{dilution factor} \times 25 \times 10^4 \quad (1)$$

$$\text{Chlorophyll A (mg/L)} = 11.75 \text{ x (Abs 662)} - 2.35 \text{ x (Abs 645)} \quad (2)$$

$$\text{Chlorophyll B (mg/L)} = 18.61 \text{ x (Abs 645)} - 3.96 \text{ x (Abs 662)} \quad (3)$$

## 3 Results and Discussion

Based on Fig. 1 (a), the lag phase occurs until day 1. At this phase, microalgae tried to adapt with the medium and prepare energy for cell division so that the growth rate was still very slow. From day 1 – day 8, the log phase occurred. At this stage, there was an increase into very rapid growth rate of actively dividing cells of microalgae. After the $8^{th}$ day, the the phase of the microalgae growth seems to directly enter the death phase.

The highest density in the control was 2.4 million cells/mL while on the molasses medium with 0.45% molasses achieved the highest density as many as 4.2 million cells / mL. This indicated that the content of molasses in the medium can increase the growth rate of microalgae. According to study by Kosakowska et al. [11], high concentrations of $Fe^{2+}$ in the growth medium increased the absorption rate of nitrate and chlorophyll concentration during the cultivation period which led to the increasing of the rate of photosynthesis. The energy resulted from photosynthesis was used for growth. Iron deficiency can reduce the efficiency of photosynthesis of microalgae.

From Table 3, it was shown that molasses also contained high magnesium content. Magnesium was the main cation cell and the base material of chlorophyll. Primary cation cell and inorganic cofactor were useful for many enzymatic reactions functioning in the unification of the substrate and enzyme. Magnesium was also an activator for *ribulosebisphosphate carboxylase* and *phosphoenolpyruvate carboxylase* which were two essential enzymes in carbon fixation during photosynthesis. Therefore, the higher the magnesium contents in the medium, the more active the photosynthesis activity in the cell.

Sucrose is also instrumental in the microalgae growth by participating in the metabolic processes. In addition to cell division, a carbon source in the form of sucrose is also used for cell expansion. The more nutrients are available, the more the cellulose produced as primary metabolites. The cellulose binds to form solid and compact bonding. At the moment of microalgae cells were dried in the oven, the measured dry weight was the cell wall. In the bulk of the cell wall, cellulose was the major composing component.

Fig. 1 (b) showed that the highest dry weight of biomass in control was 1.35 mg/mL while on the treatment with the molasses addition to the growth medium with molasses concentration of 0.45% gave the highest dry weight as many as 1.46 mg/mL. Fig. 2 showed that the dry weight per cell density increased along the time. This occurred because the nutrients contained in the molasses tend to be used for growth activity than for the formation of cell walls.

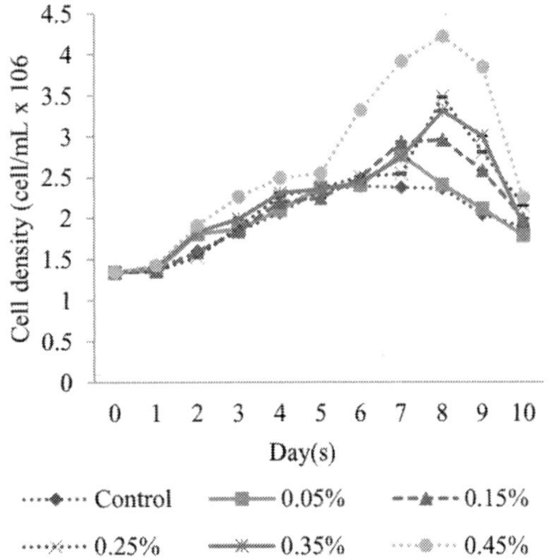


(a)

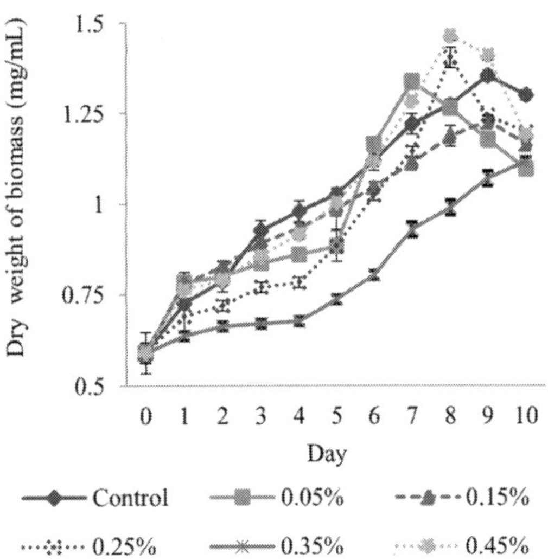


(b)

**Fig. 1.** (a) The cell density profile versus time (b) Dry weight of microalgae biomass profile versus time during the cultivation of microalgae consortium of Glagah isolate

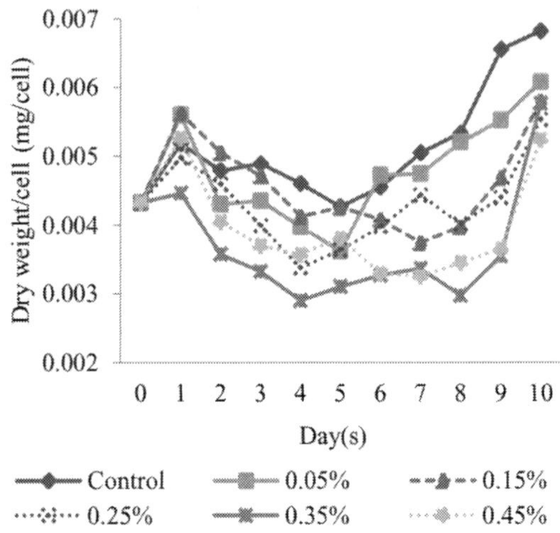


**Fig. 2.** Profile of biomass dry weight/cell density of microalgae consortium of Glagah isolates

Fig.s 3 showed that the highest content of chlorophyll A was in the medium with molasses concentration of 0.25% reaching up to 1,794 mg/mL, much higher than the control (only 1,287 mg/mL). On the other hand, the content of chlorophyll B in medium with 0.25%, 0.35%, 0.45% molasses concentration were relatively the same but were still much higher than the control chlorophyll B production.

On the first day, microalgae were still in the lag phase where they adapt with the new conditions so that the chlorophyll content increased slightly. On day 2-7, the chlorophyll content increased significantly because microalgae had adapted with the new environment and begin to increase their metabolic activities. As the consequence of the abundant availability of sufficient nutrients, the growth of microalgae and chlorophyll synthesis became faster. On day 8, chlorophyll content in microalgae started to decline because the nutrient content in the medium started running out [12,13].

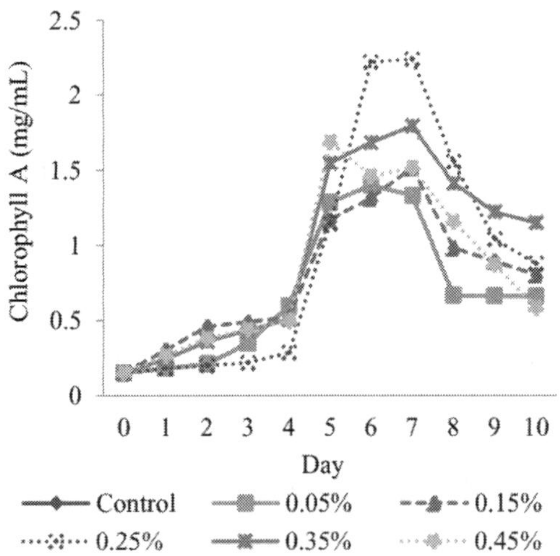


(a)

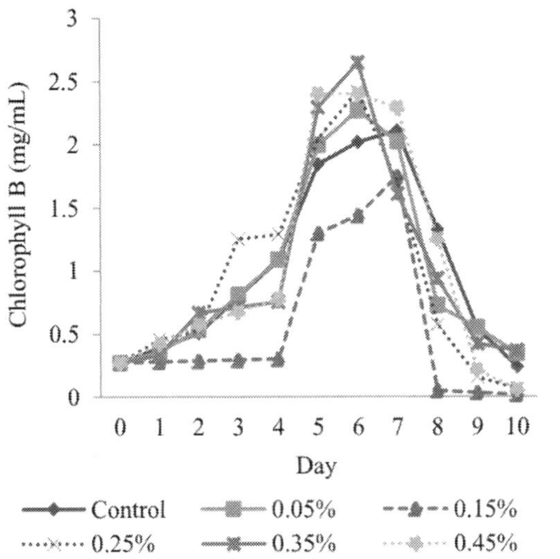


(b)

**Fig. 3.** Profile of (a) Chlorophyll A and (b) Chlorophyll B content

Chlorophyll content in molasses medium was higher than the control because of the magnesium contained in the molasses. According to Chen and Shetty [14], magnesium functioned as the unification of the substrate and enzyme through many enzymatic reactions. It is important for reactions involving ATP because magnesium served to connect the ATP molecule to the active site of the enzyme. The more magnesium content in the medium, the higher the rate of photosynthesis in the cells was. Magnesium also acts as an activator for *ribulosebisphosphate carboxylase* and *phosphoenolpyruvate carboxylase* [2], two essential enzymes in carbon fixation during photosynthesis. Magnesium contained in the addition of molasses will stimulate metabolic activity, so that microalgae can grow better and produce more chlorophyll [13].

In Fig. 4, the comparison between chlorophyll A and B show the positive trend on daily basis for 10 days of cultivation. This phenomenon showed that the molasses addition made microalgae adapted through capturing more lights since the main function in capturing light for photosynthesis was performed by chlorophyll A, while chlorophyll B functioned to extend the range of wavelengths capture capability of chlorophyll A.

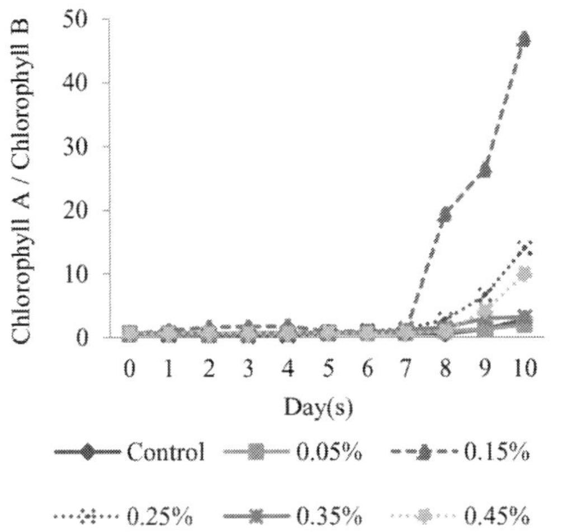


**Fig. 4.** Chlorophyll A/B profile of cultivation with and without molasses addition

In Fig. 3 and Fig. 5, on day 8, lipid content increased significantly as the chlorophyll content in microalgae decreased. The decrease in chlorophyll content was stress expression of microalgae due to lack of nitrogen supply after several days of cultivation so that the remaining carbon source was converted into lipids for their survival.

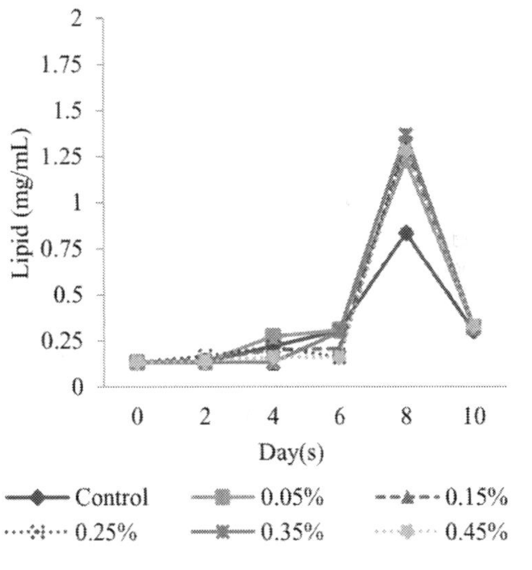


(a)

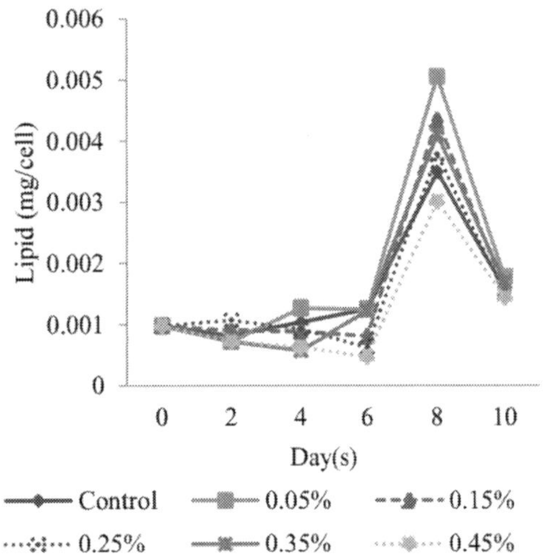


(b)

**Fig. 5.** Profile of a). lipid content of microalgae consortium of Glagah isolate b). lipid/cell density profile after 10 days of microalgae cultivation

## 4 Conclusions

The addition of molasses to the growth medium of microalgae had positive impact due to better nutrient content inside the microalgae. However, in the case of increasing the chlorophyll A content in microalgae, there was optimum concentration of molasses (asymptote point) which was 0.25%. Meanwhile, for the other parameters, the higher the molasses concentration, the better the nutrients content in the microalgae.

The authors are grateful for the financial support from Indonesian General Directorate of Higher Education (DIKTI) in the form of PUPT research grant.

## References

1. H. Sudibyo, Y.S. Pradana, T.T. Samudra, A. Budiman, Indarto, E.A. Suyono, Energy Procedia, *Study of cultivation under different colors of light and growth kinetic study of Chlorella zofingiensis Dönz for biofuel production,* **105**, 270-276, 2017.
2. A. Karemore, R. Pal, R. Sen, Algal Res., *Strategic enhancement of algal biomass and lipid in Chlorococcum infusionum as bioenergy feedstock*, **2**, 113–121, 2013
3. I. O'Hara, S. Mundree, *Sugarcane-based Biofuels and Bioproducts.* pp. 117-119, New York, USA: John Wiley & Sons, 2016.
4. N.S. El-Gendy, H.R. Madian, S.S. Abu, Int. J. Microbiol., *Design and Optimization of a Process for Sugarcane Molasses Fermentation by Saccharomyces cerevisiae Using Response Surface Methodology*, **18**, 2, 2013.
5. Y.S. Pradana, H. Sudibyo, E.A. Suyono, A. Budiman, Energy Procedia, *Oil Algae Extraction of Selected Microalgae Species Grown in Monoculture and Mixed Cultures for Biodiesel Production,* **105**, 277-282, 2017.
6. E.A. Suyono, M. Nuhamunada, N. Ramadhani, Ramdhaniyah, Asian Jr. of Microbiol. Biotech. Env. Sc., *Lipid Content from Monoculture of Micro Algae Chlorella zofingiensis Dönz and mixed culture of Glagah isolate in laboratory scale and raceway pond for biodiesel production,* **18 (1)**, 95-100, 2016.
7. Y. Goksungur, F. Mantzouridou, T. Roukas, J. Chem. Technol. Biotechnol., *Optimization of the production of β-carotene from molasses by Blakeslea trispora: a statistical approach*, **77**, 933–943, 2002.
8. N.A. Punchard, *Haemocytometer Instruction Sheet* (for improved Neubauer *Haemocytometer*), London, UK: University of East London, 2001.
9. Y. Han, Q. Wen, Z. Chen, P. Li, Energy Procedia, *Review of Methods Used for Microalgal Lipid-Content Analysis*, **12**, 944-950, 2011.
10. S. Dere, T. Gunes, R. Sivaci, Turk. J. Bot., *Spectrophotometric Determination of Chlorophyll - A, B and Total Carotenoid Contents of Some Algae Species Using Different Solvents,* **22 (1)**, 13-17, 1998
11. A. Kosakowska, J. Lewandowska, J. Ston, K. Burkiewicz, Biometals, *Qualitative and quantitative composition of pigments in Phaeodactylum tricornutum (Bacillariophyceae) stressed by iron,* **17**, 45–52, 2004.
12. A.G. Moat, J.W. Foster, M.P. Spector, *Microbial Physiology $4^{th}$ ed.* New York, USA: John Wiley and Sons Inc., 2002.
13. L. Taiz, E. Zeiger, *Plant Physiology $3^{th}$ ed.* Sunderland, England: Sinauer Associates, 2002.
14. J. Chen, H.P.C. Shetty, Culture of Marine Feed Organisms, p. 38, Bangkok, Thailand: National Inland Institute Kasetsart University Campus, 1991

# $CO_2$ gasification of microalgae (*N. Oculata*) – A thermodynamic study

*Muflih Arisa* Adnan[1,2*] and *Mohammad Mozahar* Hossain[2]

[1]Department of Chemical Engineering, Islamic University of Indonesia, Yogyakarta 55584, Indonesia
[2]Department of Chemical Engineering, King Fahd University of Petroleum & Minerals, Dhahran 31261, Saudi Arabia

**Abstract.** A new model of $CO_2$ gasification has been developed in the Aspen Plus. The potential of microalgae (*N. oculata)* for $CO_2$ gasification also has been investigated. The present gasification process utilizes the $CO_2$ at atmospheric pressure as the gasifying agent. The steam is also injected to the gasification to enhance the $H_2$ production. The composition of the producer gas and gasification system efficiency (GSE) are used for performance evaluation. It is found that the $CO_2$ gasification of microalgae produces a producer gas with a high concentration of CO and $H_2$. The GSE indicates that the process works at high performance.

## 1 Introduction

The increase of the fossil fuel consumption during the last decades causes a global warming. The renewable fuel has been considered as a potential solution of the corresponding problem. Among the available renewable resources, biomass is considered as a potential alternative resources with minimum negative impact due to its ability to reduce the $CO_2$ from the atmosphere and its low sulfur content [1].

Gasification is a potential biomass conversion process since it produces a producer gas which mainly consists of $H_2$, CO, $CO_2$ and $CH_4$ [2]. The gasifying agent has a considerable effect on the producer gas composition. [3], [4] and [5, 6] reported the injection of steam to the gasifier enhanced the $H_2$ production in the producer gas. The use of $CO_2$ as a gasifying agent attracts the attention of some researchers. For instance, [7] investigated the gasification of pine sawdust using $CO_2$ over Ni/Al precipitation catalyst. [8] studied the gasification of beech sawdust in the entrained flow reactor. [9] developed a model to investigate the performance of biomass gasification with $CO_2$ using fluidized bed gasifier. Recently, some research have focused on microalgae cultivation and processing because it offers positive environmental impacts [10-12]. To the best of our knowledge, only one literature discussing the $CO_2$ gasification of algae. [13] carried out $CO_2$ gasification of microalgae char using high-pressure thermogravimetric analyzer (HP-TGA).

## 2 Methodology

The minimization of Gibbs free energy is used for thermodynamic modeling of gasification process which involves solid, liquid and gas phases [14]. The process flow diagram of the gasification system can be seen in Fig. 1. The biomass and the gasifying agents ($CO_2$ and $H_2O$) are fed to the gasifier separately. It is worth noting that the gasifier is modelled using two blocks: the DECOM block (RYield) and the Reduction block (RGibbs). A Fortran code is embedded in the DECOM block to convert the microalgae feed stock (non-conventional component) into conventional components. While the gasification reactions occur based on the method of Minimization of Gibbs Free Energy in the Reduction block. It is worth mentioning that during gasification process the following consecutive reactions occur in the gasifier:

| | | |
|---|---|---|
| Boudouard reaction | $C + CO_2 \leftrightarrow 2CO \quad \Delta H^0_{298} = +172$ MJ/kmol | (1) |
| Steam reforming | $C + H_2O \leftrightarrow CO + H_2 \quad \Delta H^0_{298} = +131$ MJ/kmol | (2) |
| Methane formation | $C + 2H_2 \leftrightarrow CH_4 \quad \Delta H^0_{298} = -74$MJ/kmol | (3) |
| Water-gas shift reaction | $CO + H_2O \leftrightarrow CO_2 + H_2 \quad \Delta H^0_{298} = -42$ MJ/kmol | (4) |
| Methane reforming | $CH_4 + H_2O \leftrightarrow CO + 3H_2 \quad \Delta H^0_{298} = +206$ MJ/kmol | (5) |
| $CO_2$ reforming | $CH_4 + CO_2 \leftrightarrow 2CO + 2H_2 \quad \Delta H^0_{298} = +247$ MJ/kmol | (6) |

The gasification products flow to the cyclone to separate the gaseous products ($H_2$, CO, $H_2O$, and $CO_2$) and solid products (ash and unconverted char). The gaseous products are fed to the reformer. The reformer products is sent to the $CO_2$ absorber to capture the $CO_2$ in the reformer products. The pure $CO_2$ from the absorber is then cooled and compressed to 523 K and 80 bar, respectively. The producer gas is expanded and cooled to 3 bar and 298 K, respectively. The ultimate and proximate analysis of the biomass, including the heating value of biomass are summarized in Table 1. The operating conditions of the gasification process are summarized in Table 2.

* Corresponding author: muflih.adnan@uii.ac.id

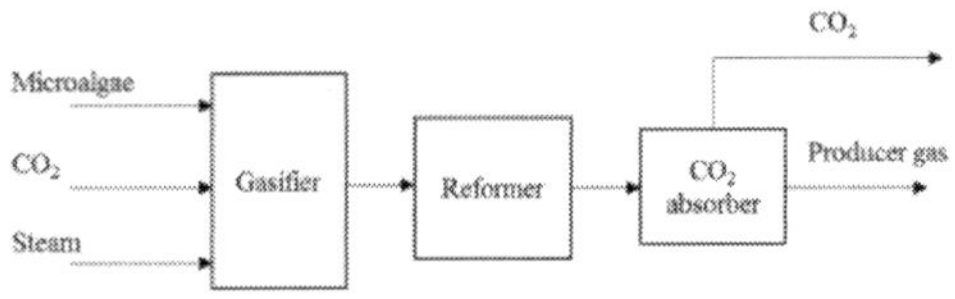


**Fig. 1.** Block diagram of the gasification process.

The performance of the gasification is evaluated in term of the dry composition of producer gas and gasification system efficiency (GSE) (Eq. (7)). One should note that the energy required for $CO_2$ absorption using amine is 3 MJ/kg $CO_2$ absorbed [15].

$$GSE = \frac{m_{pg} \times LHV_{pg} + E_{gen}}{m_{ma} \times LHV_{ma} + E_{req}} \quad (7)$$

Where $m$ and $LHV$ are mass flowrate and low heating value, respectively, while subscript $pg$ and $ma$ refer to the producer gas and the microalgae, respectively. The $E_{gen}$ represents the total energy generated from the process while the $E_{req}$ represents the total energy consumed by the process.

**Table 1.** Proximate and ultimate analysis of microalgae.

| | Microalgae (*N. oculata*) [11] |
|---|---|
| Proximate (wt.%) | |
| Moisture | 6.71 |
| Volatile matters | 78.94 |
| Fixed carbon | 7.95 |
| Ash | 6.4 |
| Ultimate analysis (wt.%) | |
| C | 47.50 |
| H | 6.15 |
| O | 46.35 |
| N | n.r |
| S | n.r |
| Cl | n.r |
| High heating value (MJ/kg) | 15.07 |

n.r : not reported

**Table 2.** Operating conditions in the simulation.

| | |
|---|---|
| Inlet temperature of biomass, boiler feed water, and $CO_2$ | 25°C |
| Temperature of steam entering gasifier | 327°C |
| Temperature of gasifier | 70°C |
| Temperature of reformer | 800°C |
| Pressure in the all blocks | 2000 kPa |
| Stream class | MIXCINC |
| Thermodynamic package | Peng-Robinson |

It is worth noticing that the model validation of the gasification block is conducted using the same feedstock and operating conditions with the ones employed by [16]. The composition of the producer gas from the gasifier is used as the baseline for model validation. It is clearly shown in Table 3 that the results of our simulation agree closely with the results reported by other author [16]. Indeed the relative error for $H_2$ and CO as the major producer gas component was less than 10%.

**Table 3.** Model validation. (Biomass $CH_{1.4}O_{0.6}$, $CO_2/C$ = 0.5, T = 800 °C and P = 1 atm).

| | Our simulation | Chaiwatanodom [16] | %Error |
|---|---|---|---|
| $H_2$ | 30.45% | 30.98 % | 1.75% |
| CO | 63.13% | 59.78 % | 5.31% |
| $CO_2$ | 6.04% | 9.01 % | n.r |
| $CH_4$ | 0.38% | 4.30 % | n.r |

## 3 Results and discussion

A microalgae feedstock with mass flow rate equal to 100 kg/h is fed to the gasifier. In addition, a $CO_2$ and a steam with mass flow rate of 151 kg/h, and 62 kg/h, respectively, are selected as the gasifying agents. It is worth noticing that the gasification system consists of the gasifier, the reformer and the $CO_2$ absorber, as shown in Fig. 1.

It is clearly shown in Fig. 2a that the microalgae disappears while the amount of $H_2$, CO, $CO_2$, and $CH_4$ significantly increase in the product stream of the gasifier. This indicates that the main task of the gasifier is the conversion of the solid feed stock (e.g., microalgae) into the gaseous products ($H_2$, CO, $CO_2$, $CH_4$ and $H_2O$) with the help of the gasifying agents ($H_2O$ and $CO_2$). The decrease of the amount of microalgae and $H_2O$ indicates that the microalgae is decomposed into gaseous products through steam reforming reaction (Eq. (2)). This is also confirmed by the increase of the CO and $H_2$ in the gasifier product. One should notice that the methane formation reaction (Eq. (3)) also takes place during $CO_2$ gasification of microalgae since the $CH_4$ presents in the gasifier product. When one look at Fig. 2a, the amount of $CO_2$ in the gasifier products is higher than that of in the gasifier feed. This indicates that water-gas shift reaction (Eq. (4)) occurs in the gasifier. Indeed, this also can be attributed to the decrease of the $H_2O$ amount in the gasifier product.

In the reformer unit, the gasifier products is further reacted to produce higher concentration of combustible gases (CO and $H_2$). Fig. 2b indicates that the amount of $CO_2$ and $CH_4$ in the reformer feed stream are higher than their counterparts in the reformer product stream. This can be attributed to the methane reforming reaction (Eq. (5)) and $CO_2$ reforming reaction (Eq. (6)). This is confirmed by the increase of the amount of CO and $H_2$ in the stream of the reformer product.

The amounts of the components in the feed of $CO_2$ absorber and in the producer gas stream are depicted in Fig. 2c. The amount of the $CO_2$ in the producer gas stream is lower than its counterpart in the stream of

$CO_2$ absorber feed. This indicates that the $CO_2$ absorber significantly reduces the amount of non-combustible gas in the producer gas stream. It is worth mentioning that the producer gas consists of 45% $H_2$, 45% CO, 6% $CO_2$ and 4% $CH_4$ (dry basis), while the GSE of the process is 68%.

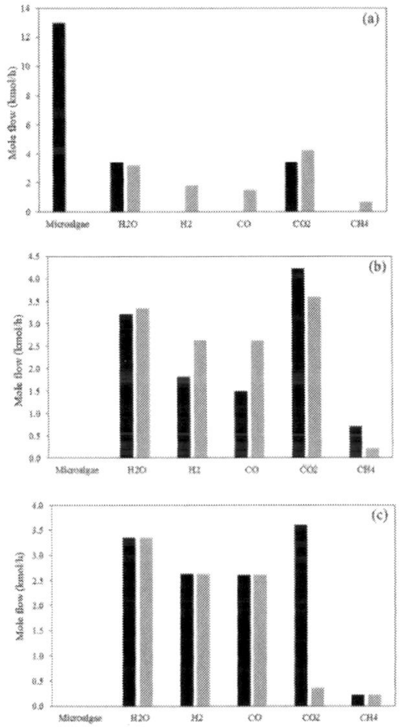


**Fig. 2.** Mole flow of the feed and product of (a) the gasifier, (b) the reformer, and (c) the $CO_2$ absorber. (feed: black; product: grey)

## 4 Conclusion

The production of producer gas with a high concentration of CO and $H_2$ via the $CO_2$ gasification of microalgae is simulated using Aspen Plus. The results show that the $CO_2$ gasification of microalgae can provide the producer gas with a high concentration of $H_2$ and CO. It is also reported that the GSE of the process is relatively high. In addition, high purity of $CO_2$ is also produced as the side-product of the process. Although the present model is developed only for the purpose of the preliminary study of the $CO_2$ gasification of algae. The results indicate that the $CO_2$ gasification of algae can potentially be a promising alternative of green technology with further optimization.

The research team acknowledges the financial support provided by King Abdul Aziz City for Science and Technology (KACST) to this research under KACST-TIC for CCS project no 03. The team also thanks the facilities and support provided by KFUPM.

## References

1. de Lasa, H., et al., *Catalytic Steam Gasification of Biomass: Catalysts, Thermodynamics and Kinetics.* Chemical Reviews, 2011. **111**(9): p. 5404-5433.
2. Balu, E., U. Lee, and J.N. Chung, *High temperature steam gasification of woody biomass – A combined experimental and mathematical modeling approach.* International Journal of Hydrogen Energy, 2015. **40**(41): p. 14104-14115.
3. Naqvi, M., et al., *An experimental study on hydrogen enriched gas with reduced tar formation using pre-treated olivine in dual bed steam gasification of mixed biomass compost.* International Journal of Hydrogen Energy, 2016. **41**(25): p. 10608-10618.
4. Guan, Y., et al., *Steam catalytic gasification of municipal solid waste for producing tar-free fuel gas.* International Journal of Hydrogen Energy, 2009. **34**(23): p. 9341-9346.
5. Adnan, M.A., et al., *Enhancement of hydrogen production in a modified moving bed downdraft gasifier – A thermodynamic study by including tar.* International Journal of Hydrogen Energy, 2017. **42**(16): p. 10971-10985.
6. Adnan, M.A., et al., *Feed compositions and gasification potential of several biomasses including a microalgae: A thermodynamic modeling approach.* International Journal of Hydrogen Energy, 2017. **42**(27): p. 17009-17019.
7. Garcia, L., et al., *CO2 as a gasifying agent for gas production from pine sawdust at low temperatures using a Ni/Al coprecipitated catalyst.* Fuel Processing Technology, 2001. **69**(2): p. 157-174.
8. Billaud, J., et al., *Influence of H2O, CO2 and O2 addition on biomass gasification in entrained flow reactor conditions: Experiments and modelling.* Fuel, 2016. **166**: p. 166-178.
9. Cheng, Y., Z. Thow, and C.-H. Wang, *Biomass gasification with CO2 in a fluidized bed.* Powder Technology, 2016. **296**: p. 87-101.
10. López-González, D., et al., *Kinetic analysis and thermal characterization of the microalgae combustion process by thermal analysis coupled to mass spectrometry.* Applied Energy, 2014. **114**: p. 227-237.
11. Ali, S.A.M., S.A. Razzak, and M.M. Hossain, *Apparent kinetics of high temperature oxidative decomposition of microalgal biomass.* Bioresource Technology, 2015. **175**: p. 569-577.
12. Kebelmann, K., et al., *Intermediate pyrolysis and product identification by TGA and Py-GC/MS of green microalgae and their extracted protein and lipid components.* Biomass and Bioenergy, 2013. **49**: p. 38-48.
13. Soreanu, G., et al., *CO2 gasification process performance for energetic valorization of microalgae.* Energy, 2017. **119**: p. 37-43.

14. Shabbar, S. and I. Janajreh, *Thermodynamic equilibrium analysis of coal gasification using Gibbs energy minimization method.* Energy Conversion and Management, 2013. **65**: p. 755-763.
15. Peters, L., et al., *CO2 removal from natural gas by employing amine absorption and membrane technology—A technical and economical analysis.* Chemical Engineering Journal, 2011. **172**(2–3): p. 952-960.
16. Chaiwatanodom, P., S. Vivanpatarakij, and S. Assabumrungrat, *Thermodynamic analysis of biomass gasification with CO2 recycle for synthesis gas production.* Applied Energy, 2014. **114**: p. 10-17.

MATEC Web of Conferences **154**, 01003 (2018) https://doi.org/10.1051/matecconf/201815401003
*ICET4SD 2017*

# Multiple linear regression for the analysis of the parameters used in dyes decolourisation by ozonation techniques

*Siti* Fatimah[1,*], and *Wiharto* Wiharto[2]

[1]Department of Chemical Engineering, Muhammadiyah University of Surakarta, Indonesia.
[2]Department of Informatic, Sebelas Maret University, Surakarta, Indonesia

**Abstract.** The ozonation process of dye Acid Orange 7 (AO7), Acid Yellow 19 (AY19), and Acid Black 1 (AB1) have been performed. The experimental results predicted the magnitude of the influence of each variable by using Multiple Linear Regression (MLR). This process produces predictive modeling of the variables studied. The variables studied in this ozonation process are independent variables consisting of ozone concentration (mg/L), concentration of dye (mg/L), pH and temperature ($^0$C). The dependent variables studied were the percentage of dye decolorization. The feasibility of the prediction model used has also been tested using the t-test. Based on the prediction model, $R^2$ values for AO7, AY19, and AB1 dyes are 0.84. 0.87, and 0.93 respectively. The analysis of the influence of the independent variables on the percentage of dye decolorization predicted that the concentration of ozone, dyestuff concentration and pH, significantly influenced the ozonation process of AO7, AY19, and AB1. In the ozonation process the temperature variables have a significant effect on AY19 and AB1, but stronger influence on AO7 ozonation process. Based on the predictions of the influence of each variable, the process in the laboratory can be optimized more effectively and efficiently, so that the cost and time factor can be reduced.

## 1 Introduction

The industrial world produces liquid waste resulting from its industrial activities. Paper, pulp, cosmetics, paint, printing, leather, pharmaceuticals, beauty and plastics industries are some of the industries that use synthetic dyes [1]. The use of dye in the process will result in hazardous waste which can endanger life. The dyes used in industrial processes can be classified into two groups, i.e. azo and non-azo groups, where the differences between the two is the presence of the azo group (-N = N-). An azo dye has azo groups that can be more than one. While non-azo dye does not have any azo group. Mostly, azo dyestuffs are more widely used than non-azo groups. This is because the azo dye has a complex structure, so that stabilize the staining process, and the product does not quickly fade [2]. Some commonly used azo dyestuffs include Acid Black 1 (AB1), Acid Yellow 19 (AY19), and Acid Orange 7 (AO7) [3]. The characteristics of the three dyestuffs are presented in Table 1.

Through simple handling, azo dyes contained in textile industry waste are very difficult to process, so that appropriate techniques are needed. Currently various techniques for handling textile waste are being developed so that the waste that comes out of the industry is not harmful to the life of water biota. Azo dye waste discharged into rivers or aquatic ecosystems is capable of influencing water transparency, which affects the penetration of sunlight from rivers. Furthermore, they can also be toxic and mutagenic to organisms or aquatic biota.

**Table 1**. Dye Characteristics

| Dyes | Structure | Formula | Mono/diazo | $\lambda_{max}$ (nm) | $M_w$ (g/mol) |
|---|---|---|---|---|---|
| Acid Orange 7 | 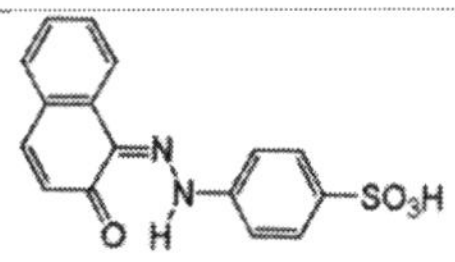 | $C_{16}H_{11}N_2NaO_4S$ | Monoazo | 484 | 350,32 |
| Acid Yellow 19 | 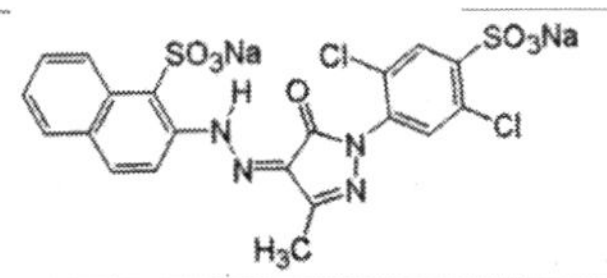 | $C_{20}H_{12}C_{12}N_4Na_2O_7S_2$ | Monoazo | 401 | 551,29 |

* Corresponding author: sf120@ums.ac.id

| | | | | | |
|---|---|---|---|---|---|
| Acid Black 1 | 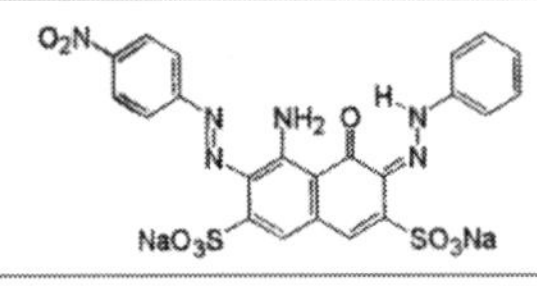 | $C_{22}H_{14}N_6Na_2O_9S_2$ | Diazo | 618 | 616,49 |

To overcome the problem, an effective and an efficient technique are needed. One of the techniques currently under development is the use of Advance Oxidizing Process [1]. The Advanced Oxidation Process (AOPs) method has the main advantage of being able to degrade / completely separate nonbiodegradable hazardous compounds in waste through oxidation [2]. This technique uses chemicals such as oxidizing agents or reducing agents as a treatment based on the formation of hydroxyl radicals produced in the oxidation process. Some examples of research that have used this technique are the use of ozone [3], fenton and sono-fenton [4], persulfate-hydrogen peroxide [1], combination of ozone and electrolysis [5], peroxydisulfate [2], non-thermal AOPs (Advanced Oxidation Process) [6], photocatalyst-zerovalent/copper [7], UV/$H_2O_2$ [8]. The AOPs technology is a combination of several process to produce free hydroxyl radicals (•OH). These hydroxyl radicals are selective against pollutants so that they are effectively used to handle textile industry waste. Ozone has a high oxidation rate of 2.08 V so that it can produce free hydroxyl radicals (•OH). The hydroxyl radicals is considerable powerful to oxidize both organic and non-organic compounds [1].

Some of the variables that affect the degradation efficiency using ozonation include pH, initial concentration of dyestuff, ozone concentration, and temperature. The dye AO7, AY19 and AB1 have different amount of azo bonds, AB1 has two azo bonds while AO7 and AY19 have one azo bond. These differences in the number of azo groups will affect the degradation efficiency of the dyestuff under the same treatment conditions. It has also been investigated using artificial neural network modeling (ANN) by Fatimah et.al [9] that the difference in the number of azo bonds present in the dye affects the percentage of decolorization. In acidic conditions, the ozone molecule is highly selective in attacking the active group of the dyestuff, such as the presence of double bonds or aromatic groups present in the dyestuff. In the basic conditions of hydroxyl radicals are produced from the process of decomposition or decomposition of the dyestuff itself. Under alkaline pH conditions up to neutral pH, there may be reactions which cannot produce radical reactions of hydroxyl formation. The pH value of the solution will decrease with the formation of the hydroxyl compound from the ozonation process. The reaction mechanism of the ozonation process is that the ozone reacts directly with the dyestuff through the ozonolysis process and through the formation of a radical chain. The reaction between ozone and anion hydroxide to intermediate radicals and hydroxyl radicals is essential in the oxidation process of this dye [10].

The degradation efficiency of the ozonation process can be done systematically by testing each variable by making the other variable fixed. This is done repeatedly for several times to produce maximum efficiency. Repetition of this experiment will require time and it can be very costly. The prediction on the strength of the influence of the variables of each experiment can be determined so that it can optimize the existence of each variable. To analyze the performance of the degradation process through ozonation, method of analysis of Multiple Linear Regresion (MLR) can be used. MLR analysis investigated a linear relationship between two or more independent variables ($X_1$, $X_2$, ... .$X_n$) with the dependent variable (Y). This analysis is to reveal the direction of the relationship between independent variables and dependent variables, whether each independent variable has positive or negative effects, and to predict the value of the dependent variable if the value of the independent variable increases or decreases [11]. The data used are usually scaled intervals or ratios. This method involved the development of a single or simple regression. This study used independent variables consisting of 4 parameters namely the concentration of ozone (mg/L), concentration of dye (mg/L), pH, and temperature ($^0$C). The dependent variable consists of one parameter only that was the percentage of decolourization (%).

The MLR model of dyestuff decolorization by using adsorption has been performed by Ho et.al [12]. This study only analyzed the data set by using Chi square so that the power between variables were not explained in detail. Another study by Muthukumar et.al [13] describes the effect of active groups on some dyes on COD values using UV/$H_2O_2$ modeled by statistical tests. To determine the effect of the dye strength on the fabric have been modeled with MLR also by Timofei et.al [14] and Wang et.al [15]. The MLR application is also used by Yao et.al [16] to find out how much the molecular structure affects molecular activity.

Referring to the previous research, it was proposed to use MLR to find out how much influence of each of the variables in the process of dye decolourization in ozonation. From the results of this analysis, it could be known which variables were the most dominant influence so that it would become possible to manage the time and costs in waste processing.

## 2 Material and Method

### 2.1 Material

In this study the data used were secondary data obtained from research that has been done by Kasiri et.al [17]. The parameters used in analyzing this model consisted of 4 parameters as the independent variable and 1 parameter as the dependent variable. Independent variables consisted of ozone concentration (mg/L), concentration of dye (mg/L), pH, and temperature (°C). The dependent variable consisted of the percentage value of decolorization.

### 2.2 Experimental design

Secondary data obtained was processed by using software SPSS version 23. In this research Multiple Linear Regression analysis (MLR) was used. The advantages of using the MLR was that we are able to determine how much influence of independent variables to dependent variables. MLR model can be written in Equation 1. In the design of this study each variable is encoded into a symbol that it was so easy to interpret.

$$Y = \alpha_0 + \alpha_1 X_1 + \alpha_1 X_2 + \alpha_2 X_3 + \alpha_3 X_4 \quad (1)$$

Explain:

| | |
|---|---|
| Y | : Percentage of decolorisation (%) |
| $\alpha_0$ | : Constanta or Determination Coefficients |
| $\alpha_1$ | : Coefficient regression for ozone concentration |
| $\alpha_2$ | : Coefficient regression for dyes concentration |
| $\alpha_3$ | : Coefficient regression for pH |
| $\alpha_4$ | : Coefficient regression for temperature |
| $X_1$ | : ozone concentration (mg/L) |
| $X_2$ | : dyes concentration (mg/L) |
| $X_3$ | : pH |
| $X_4$ | : temperature (°C) |

### 2.3 Parameter analysed

The determination analysis in multiple linear regression is used to know the percentage of contribution of independent variables ($X_1, X_2, X_3$, .................. $X_n$) simultaneously to the dependent variable (Y). This coefficient shows how much percentage of variation of independent variable used in the model was able to explain variation of dependent variable. The value of $R^2$ equals 0, so there was no percentage contribution of the influence given by the independent variable to the dependent variable, or the variation of the independent variable used in the model did not explain the slightest variation of the dependent variable. On the contrary, when $R^2$ equals 1, the percentage of contribution given by the independent variable to the dependent variable is perfect, or the variation of the independent variable used in the model explains 100% variation of the dependent variable. This experiment also analyzed ANOVA from each of the equations obtained. If the significance value of ANOVA <0.05 then the regression equation used is feasible. To find out which variable is most influential and the possibility of unused variable then used enter method, which putting all the variables into the analysis at once [11].

## 3 Result and Discussion

The AOPs (Advance Oxidation Process) technique involves oxidation reactions sourced from ozone ($O_3$) in solution to produce free hydroxyl radical •OH. This hydroxyl radical has a very large role in oxidizing the dye so that it can decrease the presence of the dyestuff. Ozone has an unstable oxygen element, so it is very reactive. This can be explained in the following reaction:

$$O_2 + UV \text{ or } h\upsilon < 310\ nm \rightarrow \bullet O \quad (2)$$

$$\bullet O + O_2 \rightarrow O_3 \quad (3)$$

Species of •O is radical so that when it collided with water, it will form hydroxyl ions ($OH^-$), thus forming a radical •OH, which in turn will play a role in overhauling the bonds of the dyestuff compounds.

Parameters used in this ozonation process are ozone concentration (mg/L), concentration of dye (mg/L), pH and temperature (°C). Each variable parameter has an influence with different dominance. To know how was the big influence of each variable, this was done by MLR analysis. Based on the MLR equation using the enter method, the data obtained as shown in Table 2. As shown on Table 2, the obtained values of $R^2$ for AO7, AY19, AB1 are equal to 0.84, 0.87, and 0.93. This shows that the effect of free variable to the percentage value of dye decolorization are 84%, 87%, and 93% respectively. The rest of the whole 100% value are the influence of other than ozone concentration, dye concentration, pH, and temperature. The p-value of the ANOVA test of each dye is <0.005. This indicates that this regression equation could be used to analyze the influence of independent variable to decolorization percentage. The regression equation for each dye is shown in Table 2.

**Table 2.** The Result of Statistical MLR Equations

| Dyes | $R^2$ | Sig.ANV | Regresion equation |
|---|---|---|---|
| AO7 | 0,84 | 0,000 | $Y=23,328-0,001X_1-0,650X_2+4,556X_3+6,50X_4$ |
| AY19 | 0,87 | 0,000 | $Y=49,473+0,005X_1-0,708X_2+4,167X_3+0,46X_4$ |
| AB1 | 0,93 | 0,000 | $Y=73,546+0,006X_1-0,775X_2+7X_3-0,483X_4$ |

Based on the correlation significance test in Table 3, it was found that in the ozonation process of AO7, the influence of concentration $O_3$ exert a relatively small effect on this process. The most influential variables in this process are the concentration of AO7, pH and temperature. It can be explained that for the addition of 1 mg/L of dye concentration, with increased the pH conditions and increased in temperature, will increase the percentage of decolorization. The resulting hydroxyl radical can increase the effectiveness of decolorization. In the process of analysis of the effect of independent variables on ozonation AY19, it was obtained that the concentration variable AY19 and pH is the variable that gave the most significant influence. With the increase in pH then the percentage of decolorization will be greater. It can be explained that in AY19 there are two sulphonic acid groups and the presence of chlorine and methyl groups. The presence of these constituents will affect the formation of hydroxyl radicals. In the process of ozonation AB1 the most influential variable is the concentration of AB1, pH, and temperature. AB1 structure contains two sulphonic acid groups, one nitro and amine group. This, of course, affects the presence of hydroxyl radicals resulting from the ozonation process. The more the amount of substituents in the dye causes the decolorization process to become more chalenging and require the other variable to be more effective.

Table 3. The Significance of Correlation

| Dye | Paramaters | Correlation | | | | |
|---|---|---|---|---|---|---|
| | | $[O_3]$ | [dye] | pH | T | Constanta |
| AO7 | p-value (Sig.) | 0,845 | 0,000 | 0,000 | 0,016 | 0,55 |
| | t | -0,197 | -5,133 | 5,396 | 2,567 | 2,013 |
| AY19 | p-value (Sig.) | 0,403 | 0,000 | 0,000 | 0,112 | 0,000 |
| | t | 0,850 | -5,272 | 7,143 | -1,644 | 5,466 |
| AB1 | p-value (Sig.) | 0,164 | 0,000 | 0,000 | 0,005 | 0,000 |
| | t | 1,434 | -9,373 | 8,271 | 3,088 | 7,513 |

The resulting analysis showed that the pH factor played a most significant role in the ozonation processes of AO7, AY19, and AB1. In acidic pH the ozone is available as molecular ozone and in alkaline pH it decomposes into secondary free radicals such as $\bullet OH$ and $\bullet HO_2$[18].

$$\bullet O_2^- + H^+ \rightarrow \bullet HO_2$$
$$2\bullet HO_2 \rightarrow H_2O_2 + O_2$$
$$H_2O_2 + \bullet O_2^- \rightarrow \bullet HO + O_2 + OH^-$$
$$H_2O_2 + h\upsilon \rightarrow 2\bullet HO_2$$

Even though hydroxyl radical has a higher oxidation potential than ozone its selectivity is much less. The oxidizing potential of ozone decrease from 2,8V in acidic pH to 1,4V in alkaline solution. This indicates that the ozone stability decrease with increasing pH, it results in the generation of secondary free radicals.

## 4 Conclusion

Analysis of the effect of free variables on dye decolorization of AO7, AY19, and AB1 can use Multiple Linear Regression method. This MLR equation is effective in determining the most significant independent variables affecting the dye decolorization process. Based on the analysis, the variable of dyestuff concentration and pH is the most influential factor on this process. Temperature variables significantly influence the decolorization of AY19 and AB1, whereas at AO7 the temperature variables have more significant effect. This equation can be conditioned for laboratory research so that it can be regulated by technical procedures that affect time and cost.

## References

1. S. Yang *et al.*, "Degradation efficiencies of azo dye Acid Orange 7 by the interaction of heat, UV and anions with common oxidants: Persulfate, peroxymonosulfate and hydrogen peroxide," *J. Hazard. Mater.*, vol. **179**, no. 1–3, pp. 552–558, Jul. 2010.
2. H. Lin, H. Zhang, and L. Hou, "Degradation of C. I. Acid Orange 7 in aqueous solution by a novel electro/Fe3O4/PDS process," *J. Hazard. Mater.*, vol. **276**, pp. 182–191, Jul. 2014.
3. M. B. Kasiri, N. Modirshahla, and H. Mansouri, "Decolorization of organic dye solution by ozonation; Optimization with response surface methodology," *Int. J. Ind. Chem.*, vol. **4**, no. 1, p. 3, 2013.
4. E. Basturk and M. Karatas, "Advanced oxidation of Reactive Blue 181 solution: A comparison between Fenton and Sono-Fenton Process," *Ultrason. Sonochem.*, vol. **21**, no. 5, pp. 1881–1885, Sep. 2014.
5. X.-C. Ruan, M.-Y. Liu, Q.-F. Zeng, and Y.-H. Ding, "Degradation and decolorization of reactive red X-3B aqueous solution by ozone integrated

* Corresponding author: sf120@ums.ac.id

with internal micro-electrolysis," *Sep. Purif. Technol.*, vol. **74**, no. 2, pp. 195–201, Aug. 2010.

6. B. Jiang, J. Zheng, Q. Liu, and M. Wu, "Degradation of azo dye using non-thermal plasma advanced oxidation process in a circulatory airtight reactor system," *Chem. Eng. J.*, vol. **204–206**, pp. 32–39, Sep. 2012.
7. Y. Yuan *et al.*, "Removal of High-Concentration C.I. Acid Orange 7 from Aqueous Solution by Zerovalent Iron/Copper (Fe/Cu) Bimetallic Particles," *Ind. Eng. Chem. Res.*, vol. **53**, no. 7, pp. 2605–2613, Feb. 2014.
8. J. Sun *et al.*, "Degradation of azo dye Acid black 1 using low concentration iron of Fenton process facilitated by ultrasonic irradiation," *Ultrason. Sonochem.*, vol. **14**, no. 6, pp. 761–766, Sep. 2007.
9. S. Fatimah and W. Wiharto, "The use of artificial neural network for modeling the decolourization of acid orange 7 solution of industrial by ozonation process," *IOP Conf. Ser. Mater. Sci. Eng.*, vol. **172**, p. 12052, Feb. 2017.
10. J. Hrubec, Ed., *Quality and Treatment of Drinking Water II*, vol. **5 / 5C**. Berlin, Heidelberg: Springer Berlin Heidelberg, 1998.
11. W. Wiharto, H. Kusnanto, and H. Herianto, "Hybrid System of Tiered Multivariate Analysis and Artificial Neural Network for Coronary Heart Disease Diagnosis," *Int. J. Electr. Comput. Eng. IJECE*, vol. **7**, no. 2, pp. 1023–1031, 2017.
12. Y.-S. Ho, W.-T. Chiu, and C.-C. Wang, "Regression analysis for the sorption isotherms of basic dyes on sugarcane dust," *Bioresour. Technol.*, vol. **96**, no. 11, pp. 1285–1291, Jul. 2005.
13. M. Muthukumar, D. Sargunamani, and N. Selvakumar, "Statistical analysis of the effect of aromatic, azo and sulphonic acid groups on decolouration of acid dye effluents using advanced oxidation processes," *Dyes Pigments*, vol. **65**, no. 2, pp. 151–158, May 2005.
14. S. Timofei, L. Kurunczi., and T. Suzuki, "Multiple Linear Regression (MLR) and Neural Network (NN) calculations of some disazo dye adsorption on cellulose," *Dyes Pigments*, vol. **34**, no. 3, pp. 181–193, Jul. 1997.
15. X. Wang *et al.*, "Quantitative structure–affinity relationship study of azo dyes for cellulose fibers by multiple linear regression and artificial neural network," *Chemom. Intell. Lab. Syst.*, vol. **134**, pp. 1–9, May 2014.
16. X. J. Yao *et al.*, "Comparative Study of QSAR/QSPR Correlations Using Support Vector Machines, Radial Basis Function Neural Networks, and Multiple Linear Regression," *J. Chem. Inf. Comput. Sci.*, vol. **44**, no. 4, pp. 1257–1266, Jul. 2004.
17. M. B. Kasiri, N. Modirshahla, and H. Mansouri, "Decolorization of organic dye solution by ozonation; Optimization with response surface methodology," *Int. J. Ind. Chem.*, vol. **4**, no. 1, pp. 1–10, 2013.
18. T. Setyaningtyas and D. W. Dwiasi, "Degradasi Zat Warna Azo Tartrazin Pada Limbah Cair Mie Dengan Metode AOPs (Advanced Oxidation Processes)," *Molekul*, vol. **7**, no. 2, pp. 153–162, 2012.

 https://doi.org/10.1051/matecconf/201815401004

# Utilization of robusta coffee waste as a renewable energy material - bioetanol

Sutarno*, *Abdul Malik* Kholiq

*Department of Chemical Engineering Faculty of Industrial Technology Indonesian Islamic University*
*Jl. Kaliurang Km 14.5 Sleman – Yogyakarta, Indonesia*

**Abstract.** A research on robusta coffee waste has been conducted as a renewable energy material - Bioethanol. This research was carried out by hydrolysis and fermentation process using Zymomonasmobilis and Saccharomyces cerevisiae (Zymomonasmobilis) bacteria to obtain the best catalyst type in the process of hydrolysis of coffee skin to glucose and the effect of fermentation time on bioethanol content produced. This research was conducted by varying the fermentation time of 7 days; 8 days; 9 days and 10 days. The fermentation fluid was then distilled and tested for bioethanol using a refractometer. Furthermore, bioethanol concentration in the analysis using.

## 1 Preface

Currently high energy consumption is not comparable with the availability of fossil energy sources (especially petroleum) is increasingly depleted. Sooner or later the world's oil reserves will be exhausted. This is due to its limited availability and non-renewable. Therefore, it takes alternative energy that can be used as a substitute for fossil energy. Research on renewable energy continues to be developed, even becoming one of the important programs of various countries to reduce dependence on oil energy that its availability continues to decline. One product that allows to be developed is bioethanol that can be used as fuel for solving energy problems today.

Bioethanol has several advantages over other renewable energies. Among them have higher oxygen content (35%) so that it burns more perfectly, octane is higher (118), and more environmentally friendly because it contains lower exhaust emissions 19-25%. In addition, the substrate for bioethanol production is quite abundant in Indonesia. This product is expected to be able to replace the fuel oil of motor vehicles and industrial machinery. In addition, bioethanol can be produced in microorganisms continuously [2].

One example of material that can be used to produce ethanol is robusta coffee waste that is widely available in rural Indonesia. The production of coffee processing involves a number of byproducts after the coffee beans undergo the grinding process using a traditional machine or tool. In the process, between the skin of the shell and the coffee beans will be separated. If the mill uses the machine and is done in the factory, it is generally not used and accumulates as a by-product of the grinding process. This skin is mostly waste and will be disposed of or used as animal feed so it takes effort to process this waste into bioethanol as renewable energy.

*Corresponding Author: 925210102@uii.ac.id

## 2 Research Methods

The material used is robusta coffee leather waste obtained from farmers of Cangkringan village slopes of Mount Merapi in Sleman district of Yogyakarta. Aquadest, HCl, NaOH, sugar, Saccharomyces cereviseae, and Zymomonas mobilis purchased at UGM Faculty of Biology Laboratory Yogyakarta and Urea purchased in one shop in Yogyakarta city. The tools used are hydrolysis, fermentation and distillation apparatus

### 2.1 Raw Material Preparation

Clean the coffee peel from the dirt, dry it with oven at 100 °C for 2 hours. Then break the skin of the coffee by blending / milling to form a powder, sift the skin of coffee on the sieve 80 mesh and analysis of cellulosanya content with spectrophotometer.

### 2.2 Hydrolysis process

Weigh the powder of 100 grams of coffee skin. Add aquadest and acid catalyst according to the variable that is run $H_2SO_4$ and HCl with volume ratio 10, 20, 30% v/v until total of 1 liter solution. Enter into the hydrolysis flask and hydrolysis with a temperature of 100 °C for 4 hours. Filter the solution of hydrolysis and filtrate taken to be analyzed its glucose level with spectrophotometer.

### 2.3 Fermentation Process

Take the filtrate from the hydrolysis process of 500 ml and add NaOH 1 N to pH = 6. Sterilize in autoclave at 120 °C for 15 minutes. Chill to room temperature. Enter starter

Zymomonasmobilis with variable: starter volume 9, 10, 11% v/v and shuffled. Cover the fermented bottle until tightly and the gas is flowed into another bottle containing the water. Fermentation according to fermentation time variables are 2, 3, 4, 5, 6 and 7 days with fermentation temperature 30 °C. Filter and take filtrate for distillation process.

### 2.4 Distillation Process

The fermented filtrate was distilled at 80 °C to obtain higher levels as desired and then analyzed the ethanol content.

## 3 Results and Discussion

### 3.1 Hydrolysis process

Hydrolysis process is the process of solving complex compounds into simple compounds with the help of water. The first treatment of hydrolysis of coffee skin powder is done with acid, ie HCl 1 N solution. HCl which is a strong acid will tend to give protons if dissolved in water, so this acid will change entirely into base pair / conjugate. HCl solution will hydrolyze the coffee skin powder through a chain-cutting process which results in cellulose/glucose.

This process was performed using 4 (four) samples each weighing 100 grams, using a varied HCl 1 N catalyst that is 250 ml, 300 ml, 350 ml and 400 ml. This hydrolysis process is carried out for 4 hours of each sample at a temperature of 100 °C. The temperature selected in this study is relatively high, because in the process of hydrolysis using acid catalysts require very high temperatures for the hydrolysis process can occur.

Selected HCl catalysts due to HCl produce higher glucose when compared with $H_2SO_4$. This is because $H_2SO_4$ is cellulose while HCl is not, so the resulting glucose levels are less.

Hydrolysis by using acid will produce a more tenuous structure, so water is more volatile at the time of drying. The powdered structure of the coffee powder is densely packed to be higher in water holding capacity and there is a breaking of the hydrogen bonding bonds in the linear chain.

Variations in the number of HCl 1 N catalysts used were intended to compare the cellulose content of coffee skin powder. The best catalyst was used in the concentration of catalyst HCl 1 N 20% (v/v).

In the process of acid hydrolysis, the H + group of HCl will convert the cellulose group from the coffee skin powder to the free radical group. The free radical group will then bind to OH from water and will react to produce glucose. In this study, the higher the concentration of HCl used with the same hydrolysis time is 4 hours, then the resulting glucose level higher. This happens because the free radical groups of cellulose powder of coffee skin produced more and more, so that glucose produced maximum. However, in this study the results obtained glucose is not too high due to several factors, namely the presence of lignin in the coffee, the hirolisa temperature Unstable and calculation of tools and materials used.

Thus, hydrolysis of coffee skin powder with optimum results using HCl 25% for 4 hours that produce glucose as much as 11.02%. The results of each sample will then be fermented to ensure that the filtrate can be used as a fermentation substrate which, when fermented, can produce ethanol.

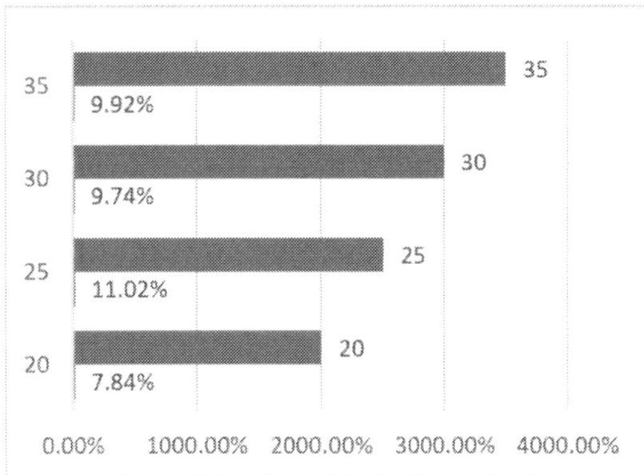


**Fig. 1.** Analysis of glucose levels in Robusta coffee

### 3.2 Fermentation Process

The process of fermentation of sugar from coffee skin hydrolysis to bioethanol using Zymomonasmobilis bacteria and *Saccharomyces cerevisiae*. The optimum temperature of the fermentation process using Zymomomobilis is in the pH range of 4-7.

The purpose of this research is to find the best catalyst type in the process of hydrolyzing the coffee skin into glucose and to find the fermentation time in the process of fermentation of glucose from coffee skin to ethanol.

The fermentation process takes place between 7 days; 8 days; 9 days and 10 days. The experimental results of the effect of fermentation time on bioethanol concentration are shown in Fig. 1. Without additional yeast support, fermentation is able to change the sugar content present in coffee coffee leaf waste. However, the 24-hour fermentation will not be perfect. Imperfect fermentation without yeast, caused by 2 things i.e.; First, in the open air there are spores of yeast Saccharomyces cerevisiae. But the population, certainly not as much when specifically mixed in the results of pulping the coffee fruit. Secondly, in the open air there is also Acetobacter aceti bacteria that will convert sugar into acetic acid. With the active bacteria Acetobacter aceti, the yeast Saccharomyces cerevisiae will be pressed and not developed so that fermentation is not running perfectly. With the help of yeast, precisely Acetobacter bacteria are pressed, and not developed. Because the rise of one yeast population, will inhibit the growth of other types of bacteria (fermentation process in coffee beans, 2011). Fermentation with the help of yeast will shorten the time. Research conducted by Siswati on coffee skin waste has a completely different result with the results of this study. The longer the fermentation time the higher the resulting ethanol content [7].

The relationship between the length of time of fermentation with the ethanol content produced is also linear, the longer the fermentation time is used then the ethanol content produced is also linear, since the fermentation time is influenced by the sugar concentration, the culture used and the fermentation temperature [6]. The best fermentation time used was for 7 days resulting in ethanol content of 67.99%, and on days 8 to 10 showed a decrease in ethanol content of 44.85%, 38.96% san 43.30%. The imperfect fermentation results due to the spores of yeast Saccaromyces cereviseae and Acetobacter aceti bacteria that will convert sugar to acetic acid. With the Acetobacter aceti bacteria, the Asacharomyces cereviseae bacteria will be pushed and not fully developed.

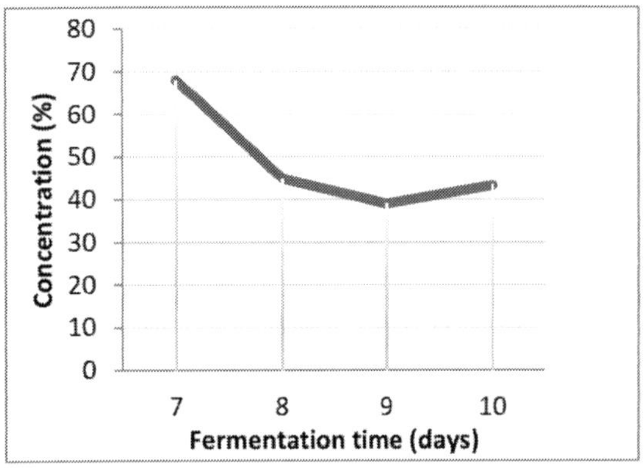


**Fig. 2.** Graph of the relationship between fermentation time and ethanol content produced

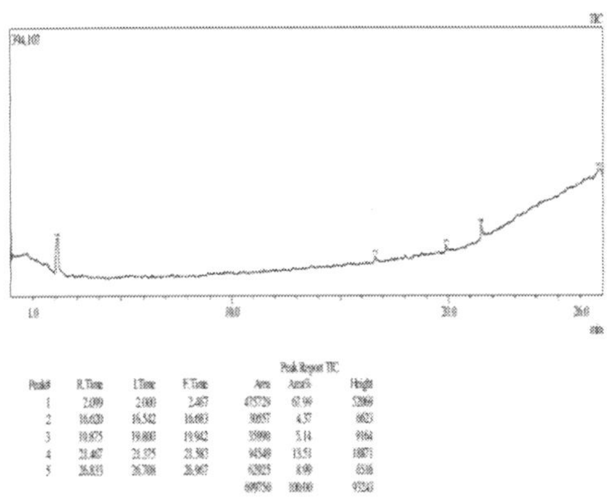

**Fig. 3.** Graph of bioethanol obtained from GCMC analysis

The result of GCMC analysis in Figure 3 shows that the bioethanol obtained at Retention Time (RT) 2.099 minutes is 67.99%.

### 3.3 Distillation Process

The result of fermentation of each filtered and filtrate sample is not through the next step, that is the distillation process to get the ethanol produced because most likely the ethanol produced from the fermentation process will evaporate and run out if heated.

### 3.4 Bioethanol produced

Based on fermentation process for 7 days resulted bioetnol with ethanol content equal to 67,99%, and on day 8 until 10 showed decrease of ethanol content equal to 44,85%, 38,96% and 43,30%.

## 4 Conclusion

The results of this study can be concluded that the skin of coffee can be used as an alternative raw material of bioethanol production by hydrolysis and fermentation process. Coffee skin contains 50.03% cellulose, after hydrolyzed using HCl 1 N with varying volume ratio yielding the highest glucose level of 11.02% with 25% HCl level.

In this study, in the fermentation process using 2 (two) bacteria, the bacteria Saccharomyces cereviseae and zymomonas mobilis. The use of these two bacteria to make the fermentation process faster.

Result of ethanol analysis from fermentation filtrate, yielding ethanol level of 67,99% at 7th day fermetation and on day 8 until 10 showed decrease of ethanol content equal to 44,85%, 38,96% san 4330%. The imperfect fermentation results due to the spores of yeast Saccaromyces cereviseae and Acetobacter aceti bacteria that will convert sugar to acetic acid. With the Acetobacter aceti bacteria the Asacharomyces cereviseae bacteria will be pushed and not developed perfectly.

The study was assisted by Eka Riska Khailani, Eida Ocvita, and Fappy Sofhia Ratnasari, published in laboratory research. All three are alumni of Chemical Engineering UII Yogyakarta. In connection with that, the authors say thank you.

## References

1. N. Azizah, et al. *The Effect of Old Fermentation on Alcohol Levels, Ph, And Gas Production In The Fermentation Process Of Bioethanol Whey With Pineapple Skin Substitution*. J. Appl. FT, **1**, 2, (2012).
2. Z. Efendi, L. Treasures, *Nutritional content of fruit coffee fermentation : Case Study of Water Village Meles Beb Sub Curup Timur Subdistrict*, J. BPTP Bkl, (2014).
3. G. Arasyid et al (2016). *Hydrolysis Processes for Ethanol from Lignosellulosic Material : A Review*, Bioresources **2**, 3, 472-499, (2016).
4. Hapsari, M. Amalia, A. Pramashinta. *Making Bioethanol from Cassava Rubber (Manihot glaziovii) for Household Stove Fuel as Efforts to Accelerate Conversion of Kerosene to Biofuels*. J. CIT, **2**, 2. (2013).
5. Y. Indartono, *Bioethanol, Alternative Renewable Energy: Achievement Study Machinery and Implementation in the field Physics*, LIPI., (2005)
6. M. Judoamidjoyo, A.a. Darwis and E.G.Said, *Fermentation Technology*. Jakarta: Rajawali Press. (1982).

7. N.D. Siswati, Yatim, M. Hidayanto, R, Bioetanol from Waste Leather Coffee with S.M. Lee, Y.M. Koo, *Pilot scale production of cellulose using Trichoderma reesei* Rut C-30 in fed-batch *mode. J. M B.***11**: 229-233 (2001).
8. J. Mohammad, Taherzadeh, K. Karimi. *Enzyme-based Hydrolysis Processes for Ethanol from Lignocellulosic Materials : A review.* (2007).
9. Widyotomo, Sukrisno. *Potential And Technology Of Coffee Waste Diversification Become Quality And Value Added Products.* Review of Coffee and Cocoa Research **1** (1) 2013, pp. 63-80. (2012).
10. Wignyanto, et al. *Effect of Sugar Reducer Pineapple Reduction Concentration and Saccharomyces Cerevisiae Inoculum On Ethanol Fermentation.* J. AT, **2**, 1, pp. 68-77 (2001).
11. S. Onuki. *Bioethanol : Industrial Production Process and Recent Studies.* Iowa State University. Department of Agricultural and Biosystems Engineering, (2007).
12. S.M. Lee, Y.M. Koo. *Pilot scale production of cellulose using Trichoderma reesei Rut C-30 in fed-batch mode.* J. MB..11: 229-233. (2001).
13. Komarayati, Sri, Gusmailina. *Bioethanol Prospect as a Substitute Kerosene.* Bogor: Center for Research and Development of Forest Products. (2010).
14. Raudah, Ernawati. *The Use Of Arabica Coffee Skin From The Pulping Process For Making Bioethanol.* J. S T, Department of Chemical Engineering, State Polytechnic of Lhokseumawe **10.** 21, (2011).

# Production of antipollutan mask based activated carbon from wasted coconut shell

*Fauzan* Nazif, and *Mahmud* Sudibandriyo*

Departemen Teknik Kimia, Universitas Indonesia, Depok, Indonesia

**Abstract**. Indonesia is one of the countries with the highest levels of air pollution in the world. Air pollution in Indonesia, especially in Jakarta due to the number of private vehicles increased at least 10% every year. This air pollution can have an impact on public health. One effort to do as a protection of people health is to use a mask. Activated carbon can be coated to mask in order to improve the effectiveness in reducing the pollutants. One good material used as material for activated carbon is coconut shell. Selection of coconut shell as the raw material of activated carbon is also based on cellulose content of 26.06%, hemicellulose content 27.07% and a lignin content of 29.40% in the dry state. This research was done in some variation such as activation methods, activated carbon mass, and adhesive material types. Based on pollutants adsorption test, mask with 6 grams of activated carbon, chemically activated, and used *TEOS* as adhesive is the best variation that able to adsorb as much 76,25% of $CO_2$ Pollutants. Mask made in this research, has saturation time as long as 4 hours under high $CO_2$ concentration.

## 1 Introduction

The use of motor vehicles have an impact on public health as motor vehicles emit harmful substances such as lead / pb, suspended particulate matter (SPM), nitrogen oxide ($NO_x$), hydrocarbons (HC), carbon monoxide (CO), carbon Dioxide ($CO_2$) and sulfur oxide ($SO_x$). Among of these hazardous substances, carbon dioxide gas ($CO_2$) has the greatest composition on motor vehicle emissions of 14% [5]. In addition, carbon dioxide ($CO_2$) levels in 2014 have reached the highest record for 30 years since 1984 at 400 ppm [2]. Based on previous research, carbon dioxide gas levels will continue to increase as human consumption of energy increases [5].

Along with the increasing emission of air pollutant gas, it is necessary to prevent the bad aspects on health. Prevention widely used today is the use of it is relatively cheap, but their effectiveness still questionable. The commonly widespread mask can only adsorb airborne particulates, whereas harmful gases such as $NO_x$, $SO_x$, CO, and $CO_2$ cannot be adsorbed [1]. The current mask development is to use activated carbon as a coating material. In this case, coconut shell is used to produce activated carbon. The use of coconut shells is also useful to reduce coconut waste. Need to be studied further to evaluate its effectiveness in reducing the pollutants. However, coating coconut shell activated carbon with dip coating method on a mask is a new way to decreasing the concentration of various air pollutants, especially $CO_2$ which has the highest concentration among other harmful emission gas.

## 2 Methodology

Methods applied in this study was started by activated carbon preparation and characterization then continued by adsorption test. Some variations were done such as activation methods, activated carbon mass, and types of adhesive used in purpose to get the best variation to be coated on mask. Adsorbent capacity to adsorb pollutant gas that contains carbon dioxide and pressurized air was examined. Moreover, the saturation time of activated carbon mask was also studied by using the best mask variation.

### 2.1 Adsorbent Preparation

Coconut shells were washed by water to release dirt then dried in oven at 120°C for 5 hours to reduce moisture content. The dried samples were then crushed and sieved to a particle size of 150 mesh. Carbons were activated by two methods. Physical activation was done on reactor by flowing $CO_2$ at 200 ml/minutes and increasing the temperature from room temperature to 850°C. Chemical activation was done by impregnating materials in 5M $ZnCl_2$ solution with impregnating ratio 3/1, then pirolized materials in reactor under the stream of $N_2$ at 200 ml/minutes and increasing the temperature from room temperature to 650°C [6,7]. Those temperature was held for an hour and then the reactor was cooled down to 30°C. The resulting product from the reactor was rinsed with distilled water until the pH was around 7. The activated carbon then was dried at 120°C for an hour to release the moisture content. Activated carbon was dissolved on solution containing water and adhesive for coating

* Corresponding author: msudib@che.ui.ac.id

process. Two types of adhesive used, were *Sodium Silicate* and *TEOS*. Coating solution was stirred by hot plate magnetic stirrer at 85°C for 30 minutes. Flannel was soaked in coating solution for 10 minutes, then being dried on vacuum oven at 120°C for 4 hours before added to mask.

### 2.2 Adsorbent Characterization

The characterizations were done to determine surface area and iodine number of resulting coconut shell based activated carbon. BET method was used to measure the surface area using 0.3 gram of activated carbon samples. Iodine characterization was also done to determine the iodine number, a characterization that indicates ability to adsorbed amount of iodine per gram activated carbon (mg iodine/g activated carbon). 0.1 gram of sample was stirred and heated with 10 ml of iodine solution for 1 hour. After that, 5 ml of the mixed solution was titrated by $Na_2S_2O_3$. The volume of $Na_2S_2O_3$ used for titration then was inputted to iodine number calculation.

### 2.3 Adsorption Test

Adsorption test was carried on modified compartments for 1 hour for each mask variations. Carbon dioxide concentration were collected every 10 minutes. Adsorption test also used on determine the best variations of mask. There are five adsorption tests done in order to determine the effect of methods used in carbon activation process, the effect of activated carbon mass on mask, the effect of adhesive being used, the effect of initial $CO_2$ concentration, and measurement of the saturation time from the best mask variation.

The experimental equipment is schematically shown on Figure 1. The inlet gas, carbon dioxide and pressurized air was controlled by check valve, and flowmeter. Flowmeters were adjusted until reached the desired initial concentration. Initial $CO_2$ concentration ($C_{in}$) could be measured on section A by read $CO_2$ concentration (ppm) from $CO_2$ detector. The gas flows through section B because of driving the existing of exhaust fan. Between section A and B activated carbon mask were placed, and final concentration of $CO_2$ ($C_{out}$) was measured by $CO_2$ detector in section B.

The amount of adsorbed $CO_2$ was determined by

$$\% Adsorption = \frac{C_{in} - C_{out}}{C_{in}} \times 100\% \quad (1)$$

Where $C_{in}$ = initial concentration of $CO_2$, and $C_{out}$ is final concentration of $CO_2$

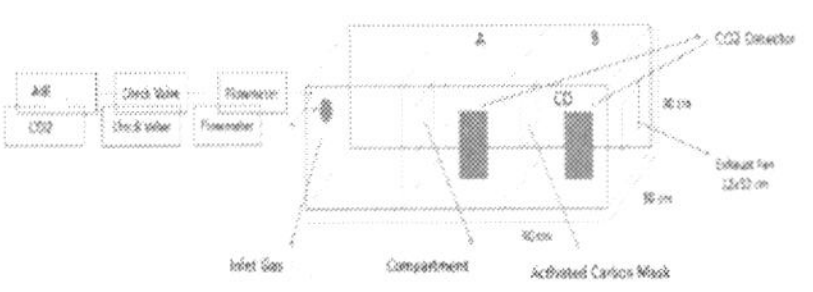

**Fig. 1.** Schematic Diagram for Adsorption Test

## 3 Results and Discussion

Adsorption test occurred in modified compartments by using $CO_2$, $N_2$ and $O_2$ as inlet gas to simulating real atmosphere condition. The operation conditions are 1 bar, 25°C, and flow rate (L/min) depends on initial concentration. Therefore, based on operational condition adsorption that happened is physical adsorption.

Results of five adsorption test is graphically shown by Figures 2 – 6.

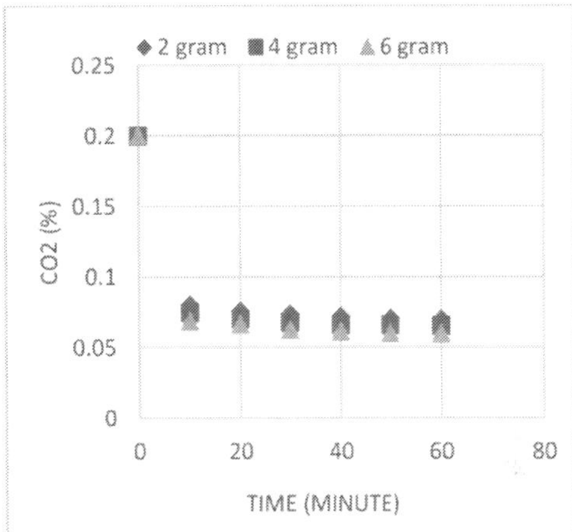


**Fig. 2.** Effect of activated carbon mass

Based on the graph in Figure 2 it has been shown that 6 gram activated carbon on mask is capable of reducing $CO_2$ by 70.00% (from 2000 ppm to 600 ppm), 4 gram activated carbon on mask is capable of reducing $CO_2$ by 67.30% (from 2000 ppm to 654 ppm), 2 gram activated carbon on mask is capable of reducing $CO_2$ by 65.25% (from 2000 ppm to 695 ppm). This indicates that higher amount of activated carbon coated into the mask will increases the $CO_2$ adsorption capacity because more adsorbent can adsorb more adsorbate at the same time unit and volume of space.

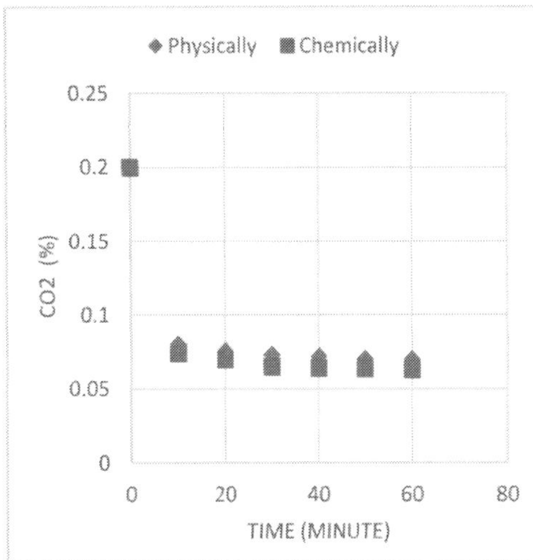


**Fig. 3.** Effect of activation method

Based on the graph in Figure 3 it has been shown that activated carbon mask which chemically activated is capable of reducing $CO_2$ by 68.50% (from 2000 ppm to become 630 ppm). Activated carbon mask that physically activated is capable of reducing $CO_2$ by 65,25% (from 2000 ppm to become 695 ppm). Result indicates that activated carbon mask which chemically activated give higher adsorption capacity than physically activated. This is because in chemical activation, $ZnCl_2$ compounds remove the impurity components that block the pores and pyrolysis using $N_2$ gas which will open more pores than with physically activation. In addition, BET test results also show that the activated carbon that chemically activated has a surface area of 432.26 $m^2$ / g. This value is greater than the activated carbon that physically activated with a surface area of 323.57 $m^2$ / g. The larger surface area of active carbon will increase the adsorption capacity of the activated carbon.

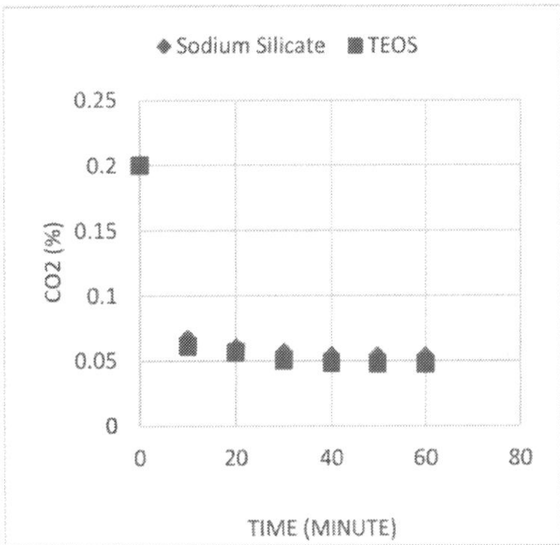


**Fig. 3.** Effect of adhesive type

Figure 4 been shown that activated carbon mask which used *TEOS* as adhesive is capable of reducing $CO_2$ by 76.25% (from 2000 ppm to 475 ppm). Activated carbon mask which used *Sodium Silicate* as adhesive is capable of reducing $CO_2$ by 73.40% (from 2000 ppm to 532 ppm). These results show that *TEOS* is more suitable to applied as an adhesive in activated carbon coating process than *Sodium Silicate*. This is because *Sodium Silicate* has a greater viscosity than *TEOS*, so the addition of *Sodium Silicate* will potentially block the pores of the activated carbon resulting in reduced adsorption capacity. *Sodium Silicate* actually has a stronger adhesion but adhesion is more suitable when applied to harder, larger and non microporous materials. In this experiment the activated carbon used is a powder type that has micropores so application of *Sodium Silicate* is not suitable. Based on previous research, the application of *TEOS* in nano sized particles can increase the surface area [3]. So the application of *TEOS* in this research is appropriate because *TEOS* has the ability as to increase the surface area besides as an adhesive.

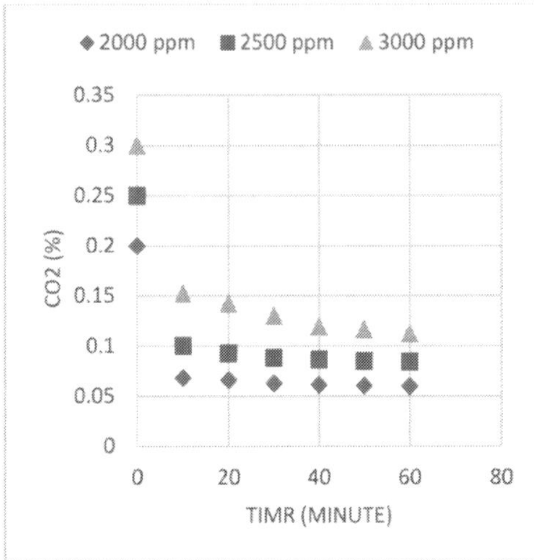


**Fig. 4.** Effect of increased inlet

Figure 5 indicates that the activated carbon mask remains capable of adsorbing $CO_2$ despite an increase in initial $CO_2$ concentration. At an initial concentration of 2000 ppm, the activated carbon mask has adsorption capacity of 70.00% with a final concentration of 600 ppm. At an initial concentration of 2500 ppm active carbon mask has adsorption capacity of 66.20% with final concentration of 845 ppm. At an initial concentration of 3000 ppm the activated carbon mask has adsorption capacity of 62.53% with a final concentration of 1124 ppm. From these results it appears that the final concentration of $CO_2$ increases with increasing initial $CO_2$ concentration. This is because the activated carbon mask has the same adsorption capacity so the more $CO_2$ adsorbed at one time will have certain limitations. In addition, an increase in the initial concentration of $CO_2$ decreases the adsorption capacity because an initial increase in $CO_2$ concentration requires a greater flow rate as well. Based on previous research, the greater flow rate in the adsorption process the shorter contact time between adsorbate with adsorbent, so that adsorption process becomes unoptimal [4]. It is proven in this research, at the initial $CO_2$ concentration of 2000 ppm adsorption capacity of activated carbon mask is 70.00% while at concentration of 3000 ppm adsorption capacity decreased to 62.53%.

https://doi.org/10.1051/matecconf/201815401005

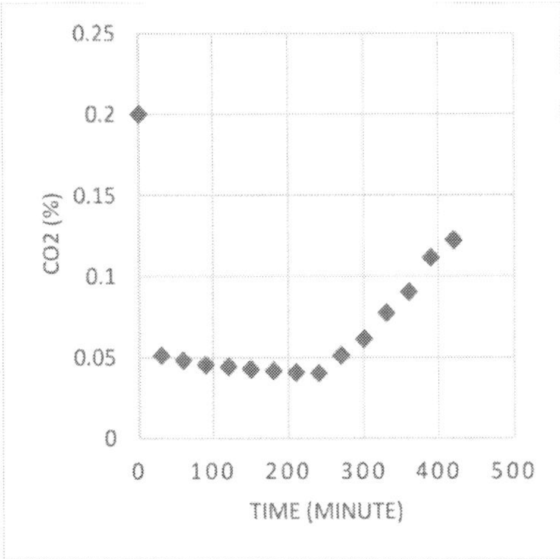


**Figure 5** Saturation Time Measurement

Figure 6 been shown that the activated carbon mask becomes saturated over a span of time between 240 - 270 minutes. This means that the saturated time of activated carbon is approximately 4 hours. This results was under high $CO_2$ concentration of 2000 ppm. A high concentration will speed up the saturated process of activated carbon as more pores of activated carbon adsorb $CO_2$. In addition, the amount of activated carbon coated also affects the saturation time. The amount of activated carbon and TEOS attached to this activated carbon mask is 0.99 grams, a relatively small amount thus accelerating the saturated process of activated carbon

The author would like to thank all parties that have supported this study, especially Universitas Indonesia via Grant of International Publication Indexed for Thesis of UI's Students or known as PITTA.

## 4 Conclusion

- Coconut shell activated carbon has surface area of 432.26 $m^2/g$ for chemically activated, and 323.57 $m^2/g$ for physically activated
- Antipollutan mask that contain 6 gram of chemically activated carbon, and using TEOS as adhesive has the best adsorption rate of 76.25%
- Activated carbon in the mask has saturated time of 4 hours

## References

1. A.R., Hidayu, and N. Muda, *Preparation and Characterization Of Impregnated Activated Carbon From Palm Kernel Shell And Coconut Shell For $CO_2$ Capture*. Science Direct **148**,106 – 113 (2016)
2. Girrard, J.P, *Technology of Meat and Meat Production* (Elis Hoorwood, New York, 1992)
3. Kobayashi, Akira, *et al*, *Effect of Surface Treatment by Using TEOS on Adhesive Force for Spherical Fused Silica* (Powder Tech, Japan, 2011)
4. Mistar, Eka, *et al*, *Effect of Flow Rate on Kinetics Adsorption of Methylene Blue with Activated Activated Coconut Coconut NAOH* (Universitas Serambi Mekkah, 2017)
5. Samiaji, Toni, *CO2 Gas in the Indonesia's Territory*. (LAPAN, 2011)
6. Truong, Xuan, *et al*, *A Potential natural energy absorption material – coconut mesocarp*. International Journal of Mechanical Science **115 – 116**, 564 -573 (2016)
7. Yang, Kunbin, *et al*, *Preparation of high surface area activated carbon from coconut shell using microwave heating*. Bioresource Technology **101**,6163 – 6169 (2010)

# Manufacture of fiber composite materials *musa acuminate l.* prepared by the randomized position with polymer matrix resin

*Lailan* Ni'mah [1,*], *M. Rizka* Akbari [2], *Ferdi Akhmad* Khan [3], and *M. Afief* Ma'ruf[4]

[1,2,3] Lambung Mangkurat University, Chemical Engineering, Faculty of Engineering, Jl. A. Yani KM 36 Banjarbaru, South Kalimantan, Indonesia
[4] Lambung Mangkurat University, Civil Engineering, Faculty of Engineering, Jl. A. Yani KM 36 Banjarbaru, South Kalimantan

**Abstract.** Composite is a system composed by mixing two or more different materials, in the form and composition of materials that do not dissolve each other. In general, the composite material is a material that has some properties that may not be owned by each component. This study aims to determine the effect of immersion banana fiber in NaOH solution, the composition ratio of the difference between the banana fiber to resin used and find the resistance of composite products produced against the pressure test and tensile test. This study begins with taking fiber banana followed by delignification process to remove lignin in the raw material, then molding composites made with a composition that has been set. The resulting composite will be tested durability to perform pressure tests (ASTM D 695) and a tensile test (ASTM D 638). To test tap on the position of the randomized to the concentration of NaOH 0%, 3%, 5% and 7% respectively, each for 53.716 N/mm 2, 43.184 N/mm 2, 81.724 N/mm 2 and 30.032 N/mm 2. For tensile test on an randomized position with NaOH concentration of 0%, 3%, 5% and 7% respectively, each for 21.519 M.Pa, 18.359 M.Pa, 31.033 M.Pa and 21.064 M.Pa.

## 1 Introduction

Bananas as the leading commodity, contributing to the national fruit production reached 34% (angka tetap Tahun 2004, Ditjen. Horti, 2005) is 4,874,439 tonnes of 14,348,456 tons of national fruit production. Distribution of banana production areas in almost all regions in Indonesia, with distribution of the highest production was in Java (West Java, East Java and Central Java) in the amount of 3,108,377 tons or 63.7% of the total national production of bananas, whereas in other areas such as Sumatra (Lampung, North Sumatra and South Sumatra) amounted to 940 390 tonnes, or 19.3%, Sulawesi (South Sulawesi, Central Sulawesi and North Sulawesi) by 6%, the rest of the Nusa Tenggara, Bali and Kalimantan. Based on the Census of Agriculture 2003 (SP 2003, BPS 2004), households of farmers involved in the cultivation of bananas in Indonesia as many as 16 million or 30.3% of farm households as a whole, amounting to 52.9 million farm households. This illustrates that every 10 farmers, three of whom grow bananas well as garden plants as well as plant garden / farm. Banana production in the province of South Kalimantan, according to fixed numbers in 2004, the directorate general of horticulture in 2005, and the directorate general of fruit crops in 2006 reached 67 362 tons of harvested area of 1,873 Ha.

Banana tree is a tree is familiar to the people of Indonesia. Banana production is still focused on the fruit as a commodity in trade. Until now, the stem of a banana tree mostly used by farmers in small amounts, especially as tobacco wrapping materials and materials straps on food wrappers. The potential for a large banana tree frond has not been used for materials in engineering, particularly the field of engineering, especially in the field of composites. If this technology can be developed that will enhance the use value of banana waste, so the waste banana not just be a waste of plantations but can be harnessed into something more economical.

Manufacture of composite materials have previously had to go through a treatment for the process of making the fiber because if the banana is used without treatment will cause the fiber surfaces become dirty and formed wax that can cause the fiber surface becomes softer, so that the bonding fibers to the matrix becoming weaker and lower tensile strength [1].

### 1.1 Composite

Composite is a system composed by mixing two or more different materials, in the form and composition of materials that do not dissolve each other. In general, the composite material is a material that has some properties that may not be owned by each component. In this sense it is certainly the combination is not limited to material matriknya [2].

The nature and characteristics of the composite [3] are determined by:

1. The constituent composite material. Characteristics of the composite. Determined based on the characteristics of the constituent material, according to

* Corresponding author: lailan.nimah@gmail.com

the rule of mixture so that the results will be inversely proportional.

2. Form and preparation of structural composites. Form and way of preparation of composites will affect the characteristics of the composite.
3. Interaction between constituent. When there is interaction between authors will improve the properties of the composites.

Composite materials are generally composed of two elements, namely fiber (fiber) as a filler, and binder fiber is called the matrix. Inside the main element is a composite fiber, while the bonding material using a polymer material that is easily formed and has a high binding power. The main use of its own fiber to determine characteristics of composite materials, such as stiffness, strength and mechanical properties of the other. As fillers, fibers used to resist most of the forces acting on the composite material, the matrix itself has the function of protecting and binding fibers in order to work well against the forces that occur. Therefore, the material used for the fiber material is strong, rigid and brittle, while the matrix material chosen materials are soft and resistant to chemical treatment.

Natural fibers have a strength ranging from 220 MPa (palm fruit fiber) to 1500 MPa (flax fiber) and a Young's modulus between 6 GPa (palm fruit fiber) up to 80 GPa (flax), as well as its density ranges from 1.25 g / cm $^3$ up to 1.5 g / cm $^3$. While the glass fiber type E has a strength 2200 MPa and a Young's modulus of 73 GPa, and a density of 2.55 g / cm $^3$, so for some natural fibers such as flax, hemp, jute and sisal have specific modulus competitive with glass fiber [4]. In the merger between the fiber and resin, fiber will serve as an amplifier *(reinforcemenfiller)* which typically have high strength and stiffness, while functioning as an adhesive or a resin matrix to maintain the position of the fiber, transmits shear forces and also functions as a coating fibers. Predominantly fiber will determine the strength and stiffness of the composite. The smaller the size of the fiber, it will provide adhesion and strength is getting better, because the ratio between the surface and the larger the fiber volume [5]. The mechanical properties of the composite is strongly influenced by the orientation of the fibers, the composite can be *quasi-isotropic* when used short fibers that are oriented randomly, *anisotropic* when used long fibers are oriented in multiple directions, or *orthotropic* when used long fibers that are oriented primarily in the direction perpendicular [6].

The need for the placement of fiber and different fiber direction in Figure 1 to make fiber reinforced composites differentiated into several sections including [7]:

a. Continuous fiber composite (fiber-reinforced composites to continue)
*Continuous or uni-directional,* fiber arrangement has long and straight, forming lamina between matriknya. Composite type is most commonly used. This type has a weakness on the separation between the layers. This is because the strength between layers is affected by the matrix.
b. Woven fiber composite (composite reinforced with fibers woven).
The composite is not easily influenced by the separation between the layers because of the composition of the fiber also binds between the layers. But the arrangement of fibers that are not so straight prolongation result will weaken the strength and stiffness. Composite consists of matrix layer followed by a layer of woven fiber arrangement.
c. Discontinuous fiber composite
Composites with shorter fibers, random type often used in the production of large volumes due to lower manufacturing costs. Disadvantages of this type of random fibers is still below the mechanical properties of the fiber reinforcement with a straight on the same type of fiber.
d. Hybrid fiber composite
Composite is a combination of the type of straight fibers with random fibers. This type is used in order to change their shortcomings properties of both types and can combine strengths.

### 1.2 Fiber Of *Musa Acuminate* L.

South Kalimantan is one banana production centers mainly bananas Kepok. This variety is a type of processed bananas which have advantages in flavor and relatively more resistant to impact, and storability. Kepok banana production areas are in Hulu Sungai Tengah, Banjar, Tapin and Kotabaru. Production Distribution of banana varieties in the central areas of major banana in South Kalimantan.

Banana fiber obtained from kepok banana tree (Musa paradisiaca) is a fiber that has good mechanical properties. The mechanical properties of banana fiber has a density of 1.35 g / cm3, the content of 63-64% cellulose, hemicellulose (20%), lignin content of 5%, the average tensile strength of 600 MPa, the average tensile modulus 17.85 GPa and the length of 3,36% [8].

### 1.3 Polymer Resin

Unsaturated polyester resin or polyester is often called the matrix of the composite. These resins include also the thermoset resin, the liquid resin thermoset polymer is converted into hard and brittle solids that formed by chemical crosslinking polymer chains that form very strong. Thermoset resins do not melt due to heating. At the time of printing, the resin is not necessary given the pressure, because when the liquid still has a relatively low viscosity, hardens at room temperature with the use of a catalyst without generating gas (unlike other thermoset resins). In general, polyester resin strong against acid except oxidizing acids, but has a low resistance to bases. If the resin is put into boiling water for 300 hours it will be broken and cracked. Polyester is widely used in the form of composite materials [9].

https://doi.org/10.1051/matecconf/201815401006

# 2 Material And Methodology

## 2.1 The course of study

Raw material such as banana bark is cleaned and dried for 1 week, then dikerus with a wire brush and drawn fiber. After the obtained fiber, the fiber is soaked in a solution of NaOH with each concentration of 3%, 5% and 7% for 2 hours. Then the fiber is washed with distilled water and drained.

## 2.2 Manufacture of composites

Banana fibers are arranged in a mold, then doused with a resin matrix in accordance with a predefined composition.

## 2.3 Analysis

Analysis of the results of the composite is done with 2 test sample, namely the pressure test (ASTM D 695) and a tensile test (ASTM D 638).

# 3 Results and Discussion

## 3.1 Compressive Strength Testing

Testing the compressive strength of the composite obtained from the maximum suppressive force divided by the cross-sectional area is pressed, with volume 99% polyester, catalyst 1% and 2,500 mm wide field press $^{2.}$ Data from the measurement of compressive strength of the composite polyester are shown in Table 1.

**Table 1.** Results of testing the compressive strength of the composite polyester.

| NaOH (%) | Number of Fibers | Fc (N) | Ac ($mm^2$) | Compressive Strength ($N/mm^2$) |
|---|---|---|---|---|
| 0 | 30 | 134.290 | 2.500 | 53,716 |
| 3 | 30 | 107.960 | 2.500 | 43,184 |
| 5 | 30 | 204.310 | 2.500 | 81,724 |
| 7 | 30 | 75.080 | 2.500 | 30,032 |

Variation randomized position obtained graph the relationship between the compressive strength and NaOH involves the variation shown in Figure 1.

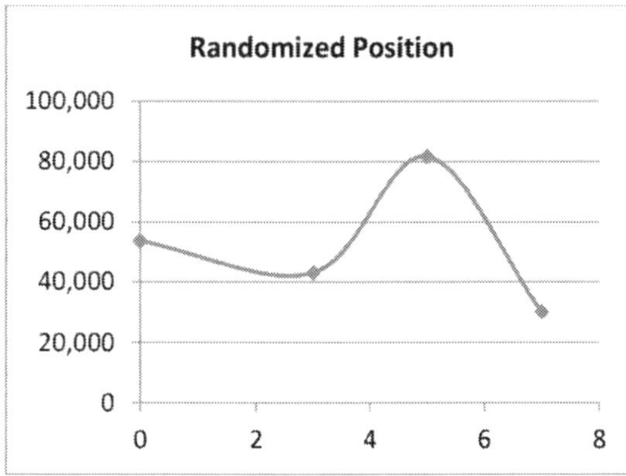


**Fig. 1.** Graph compressive strength of the composite polyester with treatment variations NaOH.

Based on Figure 1 compressive strength values obtained for randomized sequentially position each of 53.716 N / mm $^{2,}$ 43.184 N / mm $^{2,}$ 81.724 N / mm $^{2}$ and 30.032 N / mm $^{2}$ Seen that the compressive strength of greatest value to a position held by the variation randomized treatment of NaOH 5% of 81.724 N / mm $^{2.}$ The compressive strength of the composite material is strongly influenced by the composition of the constituent materials, the bond between peguat and filler composite. One of the physical properties of composite boards are very influential on other mechanical properties are therefore increasingly meeting the density of a filler so it's better compressive strength value[10].

The compressive strength of the composite material in this study was obtained from 8 specimens with a variation of the position of the fiber that is randomized and NaOH treatment variations respectively 0%, 3%, 5% and 7% cuboid measuring 5 x 5 x 5 cm. Based. The compressive strength of the composite material in the randomized position with variation of NaOH 3% and 7% decline in value. For a variation of 3% NaOH was influenced by the decrease in the incidence of *voids* in the composite, *voids* or air bubbles affect the bond between the matrix and fiber. The gap in the fiber or fibers less than perfect shape that causes the matrix will not be able to fill the empty space in the mold. If the composite is receiving load, the voltage area will move to the *void* that will reduce the strength of the composite. Composite strength is inversely proportional to *the void* that is more *voids,* the composite will be more fragile and *void* if its less then the composite will be stronger. As for the variation of NaOH 7% decline is not only influenced by the emergence of *a void,* but also due to the variation of 7% NaOH treatment causes the fiber diameter will be enlarged so that the composite strength will be reduced. Composite strength is inversely proportional to the diameter of the fiber used, the smaller the diameter of the fibers will produce a higher strength of the composite[11]. Increased compressive strength is inversely proportional to the percentage of the composite matrix on the board, this is due to better mechanical properties than the matrix filler, it is also due to voids generated is directly proportional to the percentage of the board matrix composites[10].

The composition of 99% polyester and 1% of the variation catalyst NaOH treatment banana fiber obtained greatest compressive strength value obtained in NaOH 5% variation in the randomized position of 81.724 N / mm [2]. Judging from the treatment variations NaOH 5%, the highest value is obtained because when fibers treated with NaOH 5% strength of the fiber is optimal, wherein the fiber has a small diameter and the pores are narrow so it does not easily absorb water, it is this which makes composites stronger than other NaOH treatment[3]. Jam, when the direction of each fiber spread, its power will be spread in all directions, the composite strength will increase[11].

### 3.2 Tensile Strength Testing

Tensile strength of the composite is strongly influenced by its matrix and *filler*. Namely binder or matrix composite protectors while yaiu filler *filler* composite (fiber). Here is a powerful data tensile polyester composite with a variation of the position and NaOH treatment variations on banana fiber are shown in Table 2.

**Table 1.** Results of testing the tensile strength of the composite polyester.

| NaOH (%) | Number of Fibers | Tensile Strength M.Pa |
|---|---|---|
| 0 | 10 | 21.519 |
| 3 | 10 | 18.359 |
| 5 | 10 | 31.033 |
| 7 | 10 | 21.064 |

For a variety of randomized position obtained graph of tensile strength and NaOH involves the variation shown in Figure 2.

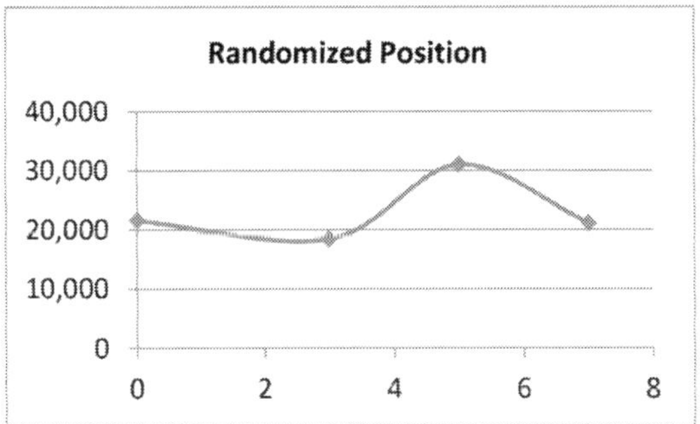


**Fig. 2.** Graph tensile strength of polyester composite with NaOH treatment variations.

Based on Figure 2 tensile strength values obtained for randomized positions sequentially, each for 21.519 M.Pa, M.Pa 18.359, 31.033 and 21.064 M.Pa M.Pa, it appears that the greatest tensile strength values for the position held by the variation randomized treatment of NaOH 5% amounting to 31.033 M.Pa. Tensile strength of the composite material is strongly influenced by the composition of the constituent materials, the bond between peguat and filler composite.

Tensile strength composite material obtained from eight specimens with a variation of the position of the fiber that is randomized and NaOH treatment variations respectively 0%, 3%, 5% and 7%. Tensile strength of the composite material in the randomized position with variation of NaOH 3% and 7% decline in value. At variations NaOH 3% on the position of the randomized of this decline is affected by the onset of *voids* in the composite, *voids* or air bubbles affect the bond between the matrix and fiber, the gap in the fiber or fiber form which is less than perfect is causing the matrix will not be able to fill the empty space in the mold , thereby reducing the kekuatankomposit. As for the variation of NaOH 7% in the second position of the decline is not only influenced by the emergence of *a void,* but also due to the variation of 7% NaOH treatment causes the fiber diameter will be enlarged so that the composite strength will be reduced.

The composition of 99% polyester and 1% of the variation catalyst NaOH treatment banana fiber tensile strength values obtained the largest obtained in NaOH 5% variation in the randomized position of 31.033 M.Pa. Judging from the treatment variations NaOH 5%, the highest value is obtained because when fibers treated with NaOH 5% strength of the fiber is optimal, wherein the fiber has a small diameter and the pores are narrow so it does not easily absorb water, it is this which makes composites stronger than other NaOH treatment [3]. Meanwhile, if viewed from a position that is used is an randomized position, the value of higher strength than weaving position, this is because in the randomized position of the mechanical properties of fiber in 1 direction will be weakened, if the direction of each fiber spread, its power will be spread in all directions, the composite strength will increase[11].

## 4 Conclusion

The comparison between the best concentration of NaOH concentration on composite from banana stem is 5% at 31.033 M.Pa, while for the best press test is 5% NaOH that is equal to 81.724 N / mm$^2$.

Factors affecting the value of compressive strength and tensile strength value on the composite are the compositions of the constituents, namely the bond between the composite strength and filler, and the presence of voids (air bubbles) in the composite.

Thanks to Kemenristekdikti by means of SIMLITABMAS for supporting grants at Penelitian Dosen Pemula..

## References

1. Nurudin, A., Karakterisasi Kekuatan Mekanik Komposit Berpenguat Serat Kulit Waru (Hibiscus Tiliaceus) Kontinyu Laminat Dengan Perlakuan Alkali Bermatriks Polyester.(2011)
2. Surdia,T., Saito, S., , Pengetahuan Bahan Teknik, PT. Dainippon Gitakarya Printing, Indonesia (1985)

3. Pramono, Catur, 2005, Pengaruh Perlakuan Alkali Kadar 5% dengan Lama Perendaman 0 jam, 2 jam, 4 jam, dan 6 jam terhadap Sifat Tarik Serat Pelepah Pisang Kepok (2005)
4. Mueller, D. H., Krobjilowski, A. New Discovery in The Properties of Composites Reinforced With Natural Fiber, Journal of Industrial Textiles, Vol. 33, No. 2-October 2003, pp.111-130 (2003)
5. Riedel, U., Natural fibre reinforced biopolymers as construction materials – new discoveries, 2nd Int Wood and Natural Fibre Composites Symposium, Kassel, Germany (1999)
6. Jamasri, Prospek Pengembangan Komposit Serat Alam Di Indonesia, Pengukuhan Guru Besar Fakultas Teknik Universitas Gajah Mada, pp. 3-19 (2008)
7. Jacobs J.A., Kilduft T.K., Engineering Material Technology Structure, Processing, Property and Selection 2th, Prentice Hall, Inc. A Simon Schuster Company, USA. (1994)
8. Lokantara, P., Analisis Kekuatan Impact Komposit Polyester-Serat Tapis Kelapa Dengan Variasi Panjang Dan Fraksi Volume Serat Yang Diberi Perlakuan NaOH, Fakultas Teknik, Universitas Udayana, Indonesia (2012)
9. Mubarak, A., Karakterisasi Sifat Mekanis Material Biokomposit Unidirectional Laminae Serat Heliconia-Resin Poliester, Jurusan Fisika, IPB (2006)
10. Ach. Kusairi, S., Ni'mah, L., Utilization Fibers and Palm Kernel Shells and Tapioca Adhesive as Matrix in the Manufacture of Composite Boards as an Alternative Raw Material in Furniture Industry, International Journal of ChemTech Research, CODEN (USA): IJCRGG, ISSN: 0974-4290, Vol. 8.,No.4, pp. 1645-1655 (2015)
11. Porwanto, Andri Daniel, Karakterisasi Komposit Berpenguat Serat Bambu dan Serat Gelas Sebagai Alternatif Bahan Baku Industri (2008)
12. Ni'mah, L., Sutomo, E.W., Simbolon, R.J., Manufacture Of Gypsum Board From Eggshell Waste Material, ARPN Journal of Engineering and Applied Sciences, ISSN 1819-6608, VOL. 11, NO. 16, pp. 9933-9940 (2016)

# Simulation of biodiesel production using hydro-esterification process from wet microalgae

*Yano Surya* Pradana*, *Ardian* Fauzi, *Sangga Hadi* Pratama, and *Hanifrahmawan* Sudibyo

Chemical Engineering Department, Faculty of Engineering, Universitas Gadjah Mada, Jalan Grafika No. 2, Yogyakarta 55281, Indonesia

**Abstract.** Recently, algae have received a lot of attention as a new biomass source for the production of renewable energy, such as biodiesel. Conventionally, biodiesel is made through esterification or transesterification of oils where the process involves a catalyst and alcohol to be reacted in a reactor. However, this process is energy intensive for drying and extraction step. To overcome this situation, we studied simulation of a new route of hydro-esterification process which is combine hydrolysis and esterification processes for biodiesel production from wet microalgae. Firstly, wet microalgae treated by hydrolyzer to produce fatty acids (FAs), separated with separator, and then mixed with methanol and esterified at subcritical condition to produce fatty acid methyl esters (FAMEs) while $H_2SO_4$ conducted as the catalyst. Energy and material balance of conventional and hydrolysis-esterification process was evaluated by Aspen Plus. Simulation result indicated that conventional route is energy demanding process, requiring 4.40 MJ/L biodiesel produced. In contrast, the total energy consumption of hydrolysis-esterification method can be reduced significantly into 2.43 MJ/L biodiesel. Based on the energy consumption comparison, hydro-esterification process is less costly than conventional process for biodiesel production.

## 1 Introduction

Today, population and industrial growth trigger the high energy requirement, especially for electricity and transportation as human life support [1]. From this case, renewable energy has an important role in the fulfillment of future energy demand in the world. It is intended to substitute fossil energy which is limited, unstable prices and causing greenhouse effect [2, 3]. One of promising alternative energy to be developed is bio-based diesel fuel, called biodiesel.

Biodiesel can be produced from various oil feedstocks, which is divided into three generations. First generation of biodiesel feedstocks (G-1) is coming from edible vegetable oils, such as palm oil, soybean oil and sunflower oil. These feedstocks have high price and contradictive with food sector, although they generate high quality biodiesel. In order to improve economic value, biodiesel is initiated from second generation of biodiesel feedstocks (G-2), such as used cooking oil and palm fatty acid distillate [4]. These low-cost feedstocks produce medium-to-low quality biodiesel due to their impurities.

Recently, algae have received a lot of attention as a third generation (G-3) source for biodiesel production [5]. Indonesia, as tropical country, which gets sun a year around, has big potential for cultivating microalgae in large scale due to the need of sunlight for growing. Some of the main characteristics which set algae apart from other biomass are higher biomass production, higher oil or starch content, and do not require agricultural land [6]. Moreover, cultivation of microalgae can be utilized as $CO_2$ removal and wastewater treatment medium in industry [7].

Generally, processes including in biodiesel production from microalgae are cultivation, harvesting, drying, oil extraction, oil transesterification/esterification [8]. Cost required for biodiesel production through these processes is high, which is less competitive than petroleum-based diesel. One of solution to overcome this problem is drying process elimination that requires high energy supply. In this case, oil extraction process is run with wet microalgae as fresh feed.

In oil extraction process, hydrolysis is proposed to convert wet microalgae into fatty acids in order to reduce energy supply compared to solvent extraction. By this process, water content in microalgae is very important to be used as hydrolysis medium. The next process is esterification to react fatty acids with alcohol producing biodiesel [9]. Both processes can be accelerated by using acid catalyst [10], which does not require catalyst separation between these two processes.

In this study, we studied simulation of a new route of hydro-esterification process which is combine hydrolysis and esterification processes for biodiesel production from wet microalgae. This route was expected to reduce energy supply which is effected on production cost. Wet microalgae treated by hydrolyzer to produce fatty acids and then

*Corresponding author: yanopradana@ugm.ac.id

esterified with methanol at subcritical condition to produce fatty acid methyl esters while $H_2SO_4$ conducted as the catalyst. Material and energy balance of hydrolysis-esterification process was evaluated by Aspen Plus and compared with conventional process.

## 2 Methods

### 2.1 Materials

Wet microalgae was represented by *Chlorella sp.*, assumed as lipid (2%w/w), carbohydrate (1.5%w/w), protein (5.5%w/w) and water (91%w/w), algal oil was taken to be consist of fatty acids, represented as oleic acid, and biodiesel was represented as methyl oleate [10]. Supporting materials required to carry out hydro-esterification process were methanol and sulfuric acid. Thermodynamic properties of major components used in this simulation were available on Table 1.

**Table 1.** Thermodynamic properties of components in biodiesel production from wet microalgae

| Properties | Oleic acid | Methyl oleate | Methanol |
|---|---|---|---|
| Chemical Formula | $C_{18}H_{34}O_2$ | $C_{19}H_{36}O_2$ | $CH_3OH$ |
| Molar mass (kg/kmol) | 282.46 | 296.55 | 32.04 |
| Freezing point (°C) | 13.0 | -20.0 | -97.6 |
| Flash point (°C) | - | 92.4 | 11.0 |
| Density (kg/m³) | 895.0 | 873.0 | 792.0 |
| Boiling point (°C) | 360.0 | 351.4 | 64.8 |
| Critical temp. (°C) | 507.85 | 490.85 | 239.43 |
| Critical pressure (kPa) | 1390.0 | 1280.0 | 8095.9 |
| $H^o_{f298,L}$ (kcal/mol) | -194.82 | -173.11 | -56.97 |

### 2.2 Procedures

A new route of hydro-esterification process for biodiesel production from wet microalgae was simulated to observe the conversion profile and energy requirement in Aspen Plus. Simulation process mainly involved chemical components definition, thermodynamic model selection, capacity determination, and proper operating condition and unit selection. Non Random Two Liquid (NRTL) model was used to simulate the whole biodiesel production process [11].

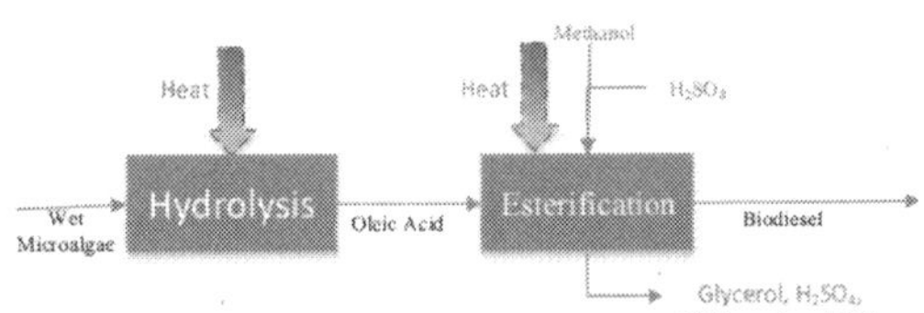


**Fig. 1.** Diagram of hydro-esterification process

Microalgae with total flow rate of 1000 kmol/hr was used as simulation basis. Hydrolyzer was operated at temperature of 260 °C and pressure of 5050 kPa, as known as subcritical condition. Esterification process was carried out in optimum temperature (90-110 °C), pressurized condition, mass fraction of sulfuric acid of 1.0 %w/w microalgae and methanol-oleic acid ratio of 9:1. The route of these processes was shown as block diagram in Fig. 1. Kinetics model of oleic acid esterification was developed by using observation data as shown in Table 2.

**Table 2.** Yield of biodiesel after 80 min of reaction [12]

| Temperature [°C] | Catalyst [% w/w] | Alcohol to oleic acid ratio | Yield of Biodiesel [%] |
|---|---|---|---|
| 90 | 1.0 | 9:1 | 87.0 ± 0.3 |
| 110 | 1.0 | 9:1 | 88.2 ± 0.3 |

Homogeneous model was considered in terms of a concentration in a batch reactor, which follows Eq. 1.

$$-\frac{dC_A}{dt} = k_A.C_A.C_M - k_A^{-1}.C_F.C_W \quad (1)$$

where $k_A$ and $k_A^{-1}$ were the second order kinetic constant (forward and reverse reaction, respectively). Equation 1 can be changed to be reaction equilibrium as follows:

$$-\frac{dC_A}{dt} = k_A.\left(C_A.C_M - \frac{C_F.C_W}{K}\right) \quad (2)$$

where, $C_A$, $C_M$, $C_F$ and $C_W$ were oleic acid, methanol, FAME, and water concentration (mol/L), respectively. Furthermore, $C_A$, $C_M$, $C_F$ and $C_W$ were expressed as a function of the FFAs conversion (X):

$$C_A = C_{A0}.(1 - X) \quad (3)$$

$$C_M = C_{M0} - C_{A0}.X \quad (4)$$

$$C_F = C_W = C_{A0}.X \quad (5)$$

where $C_{A0}$ and $C_{M0}$ was initial concentration of oleic acid and methanol (mol/L), respectively.

Optimization of kinetic and equilibrium constant was carried out in MATLAB, following Arrhenius and Van't Hoff equation, as shown in Eq. 6 and 7:

$$k_A = A\exp(-\frac{E}{RT}) \quad (6)$$

$$\frac{d\ln K}{dT} = -\frac{\Delta H_R{}^o}{RT^2} \quad (7)$$

where A was collision factor; E was activation energy; R was ideal gas constant; T was temperature; K was equilibrium constant; $\Delta H_R{}^o$ was standard enthalpy change of reaction.

For further, optimized kinetic model was used to analyze the sensitivity biodiesel conversion over different operating condition and simulate mass and energy balance of hydro-esterification process by using Aspen Plus. To simplify the complexity of simulation, the following assumptions were taken. First, the mode of all heat exchangers were counter-current. Second, the minimum temperature approached in all heat exchangers was set at 10 °C. Third, the isentropic efficiency of compressors and pumps was assumed to be 80%. The last one, the heat and pressure loss during the biodiesel production process were assumed to be negligible. The result of hydro-esterification simulation was then compared to conventional biodiesel production, as shown in Fig. 2, reported by [8].

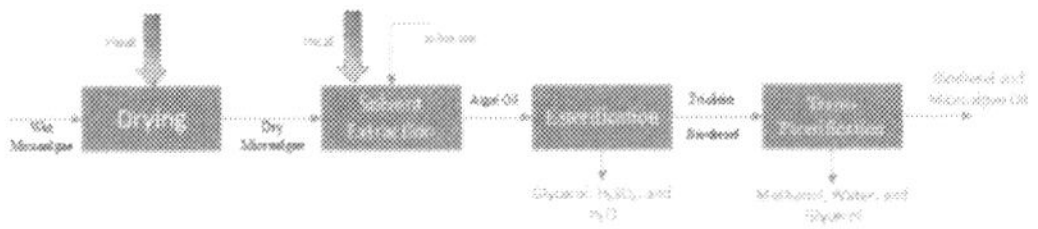


**Fig. 2.** Diagram of conventional biodiesel production from wet microalgae

## 3 Results and Discussion

Simulation of hydro-esterification in Aspen Plus was started by simulating hydrolysis process of wet microalgae. It utilized remaining water in harvested microalgae to produce fatty acids, represented as oleic acid, and glycerol. Simulation of this step applied NRTL thermodynamic model to calculate the mixture properties based on functional-groups that depict the components. Exception for triolein, the boiling point, density, pressure, critical temperature, and volume were determined by drawing the components inside the program.

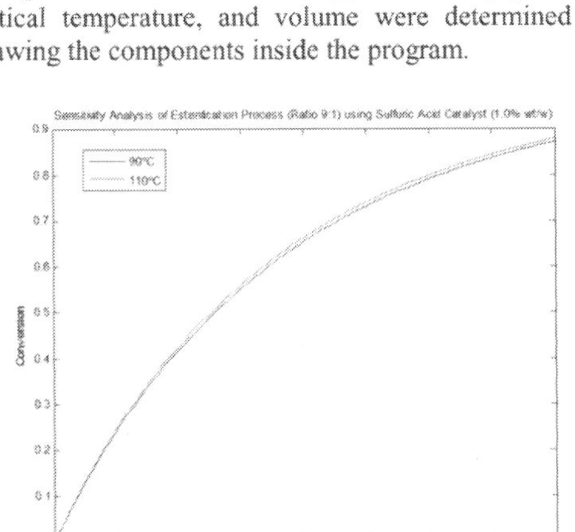


**Fig. 3.** Optimized kinetic model (2D) of oleic acid esterification (methanol-oleic acid ratio of 9:1 and sulfuric acid percentage of 1.0 %w/w)

The oleic acid was then put into esterification reactor to convert oleic acid into methyl oleate. The kinetic model of this step was predicted by optimizing kinetic data in Table 2, reported by [12] in Fig. 3 and Fig. 4 shows the trend of optimized kinetic model which was influenced by reaction temperature at methanol-oleic acid ratio of 9:1 and sulphuric acid percentage of 1%w/w.

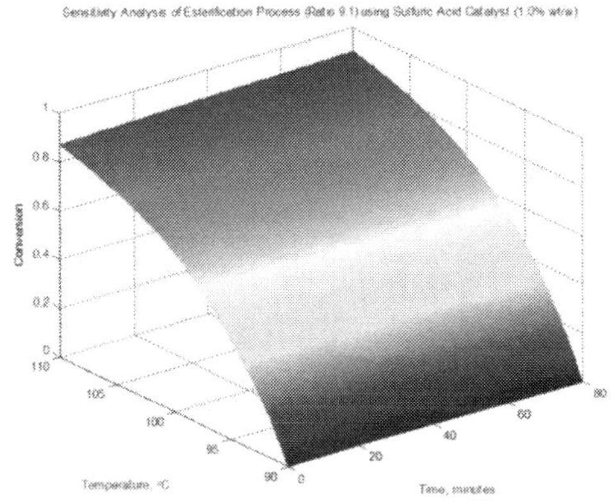


**Fig. 4.** Optimized kinetic model (3D) of oleic acid esterification (methanol-oleic acid ratio of 9:1 and sulfuric acid percentage of 1.0 %w/w)

The change of reaction temperature (90-110 °C) did not increase biodiesel conversion significantly. It can be seen the slight differences of optimized kinetic constants in reaction temperature of 90 and 110 °C, as shown in Table 3. Thus, the simulation of oleic acid esterification used reversible kinetic model at reaction temperature of 90 °C, methanol-oleic acid ratio of 9:1 and sulphuric acid percentage of 1 %w/w.

For those conditions, simulation of hydro-esterification process resulted material and heat configuration, as shown in Table 4. In hydrolysis stage, the energy consumption is used to hydrolyze wet microalgae in order to avoid energy-intensive units, i.e. drying and extraction. In this simulation, the sensible heat of glycerol is recovered by heat exchanger to heat wet microalgae as heat integration effort. As a result, 56.31 kmol/hr of oleic acid could be produced with total energy consumption of hydrolysis process is 9.59 MW, equally 2.07 MJ/L biodiesel. In esterification stage, the energy-intensive units are in heating and pumping process of methanol-$H_2SO_4$ mixture into 90 °C and 3 bar (1.05 MW). As a result, 48.92 kmol/hour of methyl oleate could be produced with total energy consumption is 1.67 MW, equally 0.36 MJ/L biodiesel.

**Table 3.** Optimized kinetic and equilibrium constant

| Temperature [°C] | Catalyst [% w/w] | Methanol-Oleic acid ratio | $k_A$ [L/mol/min] | K | E/R [1/°C] |
|---|---|---|---|---|---|
| 90 | 1.0 | 9:1 | 0.0031 | 0.0051 | 308.0269 |
| 110 | 1.0 | 9:1 | 0.0032 | 0.0032 | |

**Table 4.** Material and heat configuration in hydro-esterification simulation from wet microalgae

| Stream specification | Hydrolysis | | Esterification | |
|---|---|---|---|---|
| | Input | Output | Input | Output |
| Temperature, °C | 25 | 260 | 90 | 50 |
| Pressure, bar | 101 | 5050 | 303 | 101 |
| Total mass flow, kg/hr | 48252.98 | 48252.98 | 33120.13 | 33120.13 |
| Total mole flow, kmol/hr | 1000.00 | 1096.58 | 573.01 | 573.01 |
| Volume flow, $m^3$/hr | 66.07 | 65.21 | 39.94 | 40.71 |
| Component mole fraction, (%) | | | | |
| Triolein | 0.0190 | 0.0010 | - | - |
| Oleic acid | 0.0010 | 0.0510 | 0.0980 | 0.0130 |
| Phenylalanine | 0.0550 | - | - | - |
| Sucrose | 0.0150 | - | - | - |
| Methanol | - | - | 0.8840 | 0.7990 |
| $H_2SO_4$ | - | - | 0.0170 | 0.0170 |
| Water | 0.9100 | 0.7900 | - | 0.0850 |
| Glycerol | - | 0.1580 | - | - |
| Methyl Oleate | - | - | - | 0.0850 |
| Enthalpy, GJ/hr | -83.4810 | -75.2270 | -41.3290 | -39.8880 |

Fig. 5 shows compared energy consumed for biodiesel production using conventional and hydro-esterification process. From this figure, conventional biodiesel production needs more energy to carry out all of processes. Total energy consumed by this process is 4.40 MJ/L biodiesel while hydro-esterification process is 2.43 MJ/L biodiesel. The highest contribution of energy consumption is drying process to remove water content [8], requiring 2.37 MJ/L biodiesel. In hydro-esterification process, drying process is removed and hydrolysis is proposed to substitute solvent extraction, utilizing remaining water as agent.

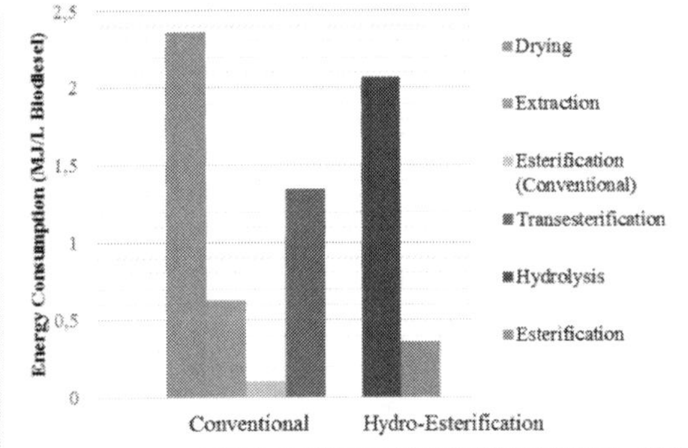


**Fig. 5.** Compared energy consumption of biodiesel production

## 4 Conclusion

In this study, the feasibility analysis of hydro-esterification biodiesel production process from microalgae was investigated and compared with conventional route. Simulation results indicated that conventional route was energy demanding process, requiring 4.40 MJ/L for biodiesel production. In contrast, the total energy consumption of hydrolysis-esterification method could be reduced significantly into 2.43 MJ/L biodiesel. Based on the energy consumption comparison, hydro-esterification process is less costly than conventional process for biodiesel production.

The authors are thankful to Universitas Gadjah Mada, Indonesia for financial support through Research Grant No. 365/DIT.LIT/2017.

## References

1. M.K. Lam, K.T. Lee. Biotechnology Advances **30**, 673-690 (2012)
2. M. Rahimi, B. Aghel, M.Alitabar, A. Sepahvand, H.R. Ghasempour, Energy Convers. Manag **79**, 599-605 (2014)

3. Y.S. Pradana, A. Prasetya, *AIP Conference Proceedings* **1823**, 020069 (2017)
4. D.R. Sawitri, Sutijan, A. Budiman, ARPN Journal of Engineering and Applied Sciences, **11** (16), 9951-9957.
5. H. Sudibyo, Y.S. Pradana, T.T. Samudra, A. Budiman, Indarto, E.A. Suyono, , Energy Procedia **105**, 270-276 (2017)
6. S.S. Oncel, Renewable and Sustainable Energy Reviews **26**, 241-264 (2013)
7. Y.S. Pradana, Y. Kusumastuti, F.N. Rahma, N. Effendy, Chemical Engineering Transactions **56**, 1549-1554 (2017)
8. C. Song. Q. Liu, N. Ji, S. Deng , J. Zhao, S. Li, Y. Kitamura, Bioresource Technology **214**, 747-754 (2016)
9. R.D. Kusumaningtyas, S. Purwono, R. Rochmadi, A. Budiman, Int. J. Exergy **15 (4)**, 447-467 (2014)
10. G.D. Machado, T.L. de Souza, D.A.G. Aranda, F.L.P. Pessoa, M. Castier, V.F. Cabral, L. Cardozo-Filho, Chemical Engineering and Processing: Process Intensification **103**, 37-45 (2016)
11. Y. Zhang, M.A. Dube, D.D. McLean, M. Kates, Bioresource Technology **89**, 1–16 (2003)
12. I.L. Lucena, G.F. Silva, F.A.N. Fernandes, Industrial and Engineering Chemistry Research **47**, 6885-6889 (2008)

# Non-square open-loop dynamic model of methyl acetate production process by using reactive distillation column

*Ahmad Misfa* Kurniawan[1], *Renanto* Handogo[1,*], *Hao-Yeh* Lee[2], and *Juwari Purwo* Sutikno[1]

[1]Department of Chemical Engineering, Institut Teknologi Sepuluh Nopember, 60111 Surabaya, Indonesia
[2]Department of Chemical Engineering, National Taiwan University of Science and Technology, 10607 Taipei, Taiwan

**Abstract.** The development of mathematics model especially non-square open-loop dynamic model has grown rapidly since four decades. Because of the need of simplicity and accuracy, industrial world is demanding a dynamic mathematical model as simple as possible but it still can represent the whole real process. Therefore, to meet those needs, an open-loop dynamic model of methyl acetate production process by using reactive distillation column has been built. Feed ratio, reflux ratio and reboiler duty were chosen to be the manipulated variables to make a better performance rather than only reflux flowrate and reboiler duty. Methyl acetate fraction in distillate and water fraction in bottom were still be selected as controlled variables, so a non-square open-loop dynamic model was formed.

## 1 Introduction

In the development of chemical industry on a large scale, it is necessary to understand the whole process behavior. This is an important issue because it can be used as a guidance to get the desired production result and to keep the operating conditions as safe as possible. The behavior of a process plant can be represented by the mathematical model of its process [1]. Therefore, the development of a mathematical model is very effective in the observation of industrial processes that have a large production scale [2]. There are many way to build a mathematic model of a process plant including derivation of mass and energy balance. In 1973, Wood and Berry introduced a mathematical model in the form of transfer function matrix of the distillation column. This matrix represents dynamic process of a distillation column in the form of 2x2 MIMO which has two controlled variables and two manipulated variables. Those two controlled variables are distillate fraction and bottom fraction. While those two manipulated variables are reflux ratio and steam flowrate [3]. Then in 1978, a 3x3 MIMO separation process between benzene and toluene by using a distillation column was proposed by Doukas and Luyben. Where there are three controlled variables of toluene fraction in bottom column, toluene fraction in top column, and benzene fraction in side draw. And there are three manipulated variables of reflux ratio, side draw flowrate, and reboiler duty [4]. Then in 1987, a non-square MIMO matrix was introduced by Prett and Morari. This mathematic model was referred to as shell standard control problem where it has two controlled variables and three manipulated variables [5]. Considering the importance of the development of the mathematical model, it is necessary to have a model as simple as possible to facilitate the simulation process but it can represent the reality of the process being modeled. In this paper, the development of mathematical model which can represent RDC (Reactive Distillation Column) process as a whole in the form of non-square MIMO with FOPTD (first order plus time delay) transfer function will be put forward.

## 2 Process Description

The process as shown in Figure 1 was suggested by Tang, et al. (2005) as the most sufficient and efficient configuration for production of methyl acetate by using reactive distillation column.[6] In reactive distillation column configuration, reaction and separation process are taking place only in one column. Rather than using reactor as reaction device and distillation column as separation device, reactive distillation configuration is more economic because it can minimize total annual cost compared to the conventional configuration [7].

### 2.1 Production of Methyl Acetate by Using Reactive Distillation Column

In this process, each reactant (methanol and acetic acid) are entering reactive distillation column in equimolar amount at the same temperature and pressure condition. Top product of that distillation column is methyl acetate with water and remaining reactant (methanol and acetic acid) as the impurities. Whereas the bottom product of the distillation column is containing water with a little bit of remaining reactant and methyl acetate [6].

* Corresponding author: renanto@chem-eng.its.ac.id

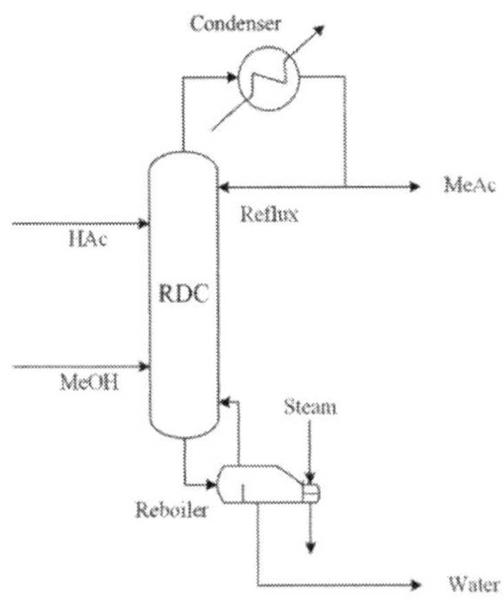


**Fig. 1.** Process scheme of the methyl acetate production.

Operating condition for the reactive distillation process are shown below:

- There are 38 total stage including condenser and reboiler.
- Column diameter is equal to 1.0327 meter.
- Reaction zone is in stage-3 until stage-36.
- First feed is pure acetic acid which is entering column at stage-4 in the amount of 50 kmol/hr with 2 atm pressure as liquid.
- Second feed is pure methanol which is entering column at stage-28 in the amount of 50 kmol/hr with 2 atm pressure as liquid.
- Top column pressure is equal to 1.1 atm and condenser pressure is equal to 1 atm.
- The purity of the methyl acetate as top product of the distillation process has to greater than 98%, while the purity of the water as bottom product of the distillation process has to greater than 98%.
- Catalyst density is equal to 770 kg/m$^3$ with assumption of 50% reactive liquid hold up (m$^3$)

It is known that reaction equation of methyl acetate is shown in equation (1).

$$r = m_{cat} k_f \alpha_{HAc} \alpha_{MeOH} - k_r \alpha_{MeAc} \alpha_{H_2O} \quad (1)$$

where,

$$k_f = 2.961 \times 10^4 e^{\frac{-49190}{RT}} \left[ \frac{kmol}{kg \cdot s} \right]$$

$$k_r = 1.348 \times 10^6 e^{\frac{-69230}{RT}} \left[ \frac{kmol}{kg \cdot s} \right]$$

UNIQUAC thermodynamic model for methyl acetate production process is shown in Table 1 [6].

**Table 1.** UNIQUAC Thermodynamic Model for Esterification Process of Methyl Acetate.

| Comp. i | HAc | HAc | HAc | MeOH | MeOH | MeAc |
|---|---|---|---|---|---|---|
| Comp. j | MeOH | MeAc | $H_2O$ | MeAc | $H_2O$ | $H_2O$ |
| Temperature unit | K | K | K | K | K | K |
| $a_{ij}$ | -0.97039 | 0.43637 | 0.051007 | 0.71011 | -3.1453 | -0.01014 |
| $a_{ji}$ | 2.0346 | -1.1162 | 0.29355 | -0.72476 | 2.0585 | -0.96295 |
| $b_{ij}$ | -390.26 | 62.186 | -422.38 | -62.972 | 575.68 | -593.7 |
| $b_{ji}$ | -65.245 | -81.848 | 98.12 | -326.2 | -219.04 | 265.83 |
| $c_{ij}$ | 0 | 0 | 0 | 0 | 0 | 0 |
| $c_{ji}$ | 0 | 0 | 0 | 0 | 0 | 0 |
| $d_{ij}$ | 0.003061 | -0.00027 | 0.00024 | -0.00117 | 0.006071 | 0.002161 |
| $d_{ji}$ | -0.00316 | 0.001331 | 7.67E-05 | 0.002355 | -0.00701 | -0.0002 |

## 3 Process Simulation

Process simulation of reactive distillation column was built by using Aspen Plus 7.3. To get the mathematic model of its control system, dynamic conversion was needed so that the step test method can be applied to that control system. This step test method was based on the process reaction curve found by Ziegler and Nichols in 1942. A matrix of process transfer function can be found by using that method [8].

In the dynamic simulation built by using Aspen Plus Dynamic 7.3, two PI controllers had been placed to maintain condenser pressure and condenser level at certain value by manipulating condenser duty and distillate flowrate. These two controllers are in auto mode (closed-loop) to maintain the stability of the process while step test is taking place. Control scheme of methyl acetate production process during the step test is shown in Figure 2.

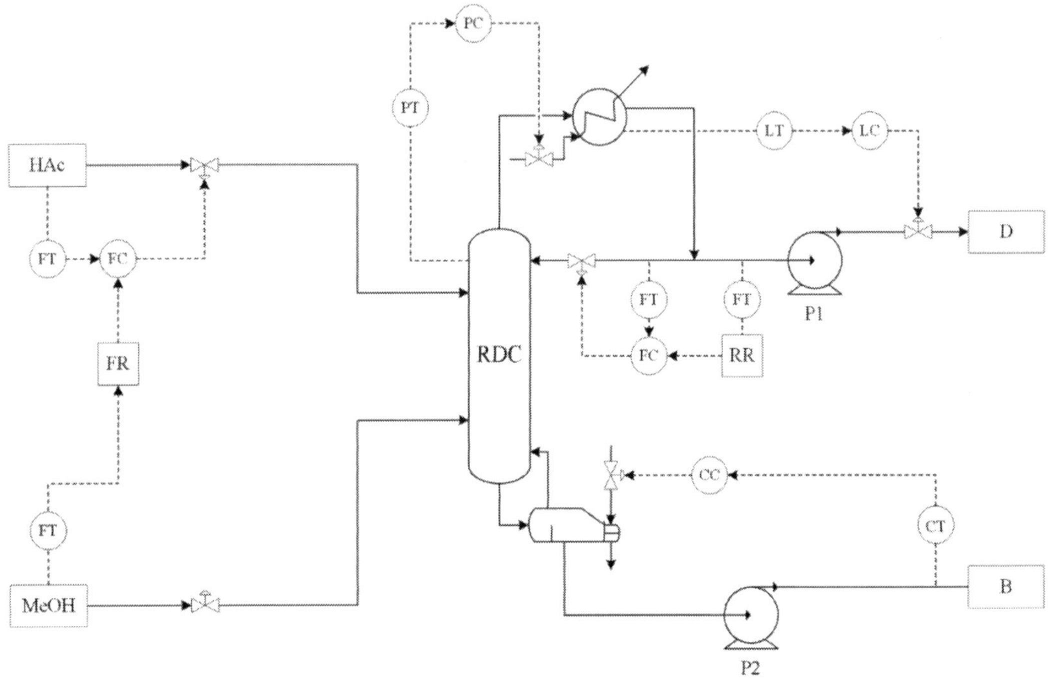


**Fig. 2.** Scheme of methyl acetate production process during the step test.

There are three manipulated variables of feed ratio (FR), reflux ratio (RR), and reboiler duty (QR) and two controlled variables of methyl acetate fraction in distillate ($X_D$) and water fraction in bottom ($X_B$). Those two controlled variables were chosen based on the requirement of product purity (methyl acetate in distillate) that has to be maintained equal or greater than 98% and water purity in bottom is maintained equal or greater than 98% too. Applying reflux ratio as a manipulated variable can be done by placing multiplier as a ratio station to adjust reflux flowrate with considering distillate flowrate. The magnitude of this multiplier is the reflux ratio its self. The next manipulated variable was reboiler duty. To adjust reboiler duty, a controller was used to connect water fraction in bottom as the controlled variable and reboiler duty as the manipulated variable. But, it was set in manual mode (open-loop) so that a disturbance changes can be given to reboiler duty manually. The last manipulated variable was feed ratio. This manipulated variable was chosen because ratio of acetic acid and methanol feed entering the column must be able to adapt to the reaction needs in the reactive distillation column.

## 4 Step Test

Step test was done by giving a step disturbance changes in one manipulated variable while maintaining the other manipulated variables constant and then recording response of all controlled variables from the beginning (when the disturbance is given) until a new steady state value of each controlled variables attained. This step was repeated to the other manipulated variables.

### 4.1. Step Test Result

Calculation based on graphical perusal was the method to determine every parameters of the process transfer function. To compensate error from the graphical perusal, aspen calculation was used to validate graphical method calculation. Response of every controlled variables on the changes of every manipulated variables can be shown in Figure 3.

* Corresponding author: renanto@chem-eng.its.ac.id

https://doi.org/10.1051/matecconf/201815401008

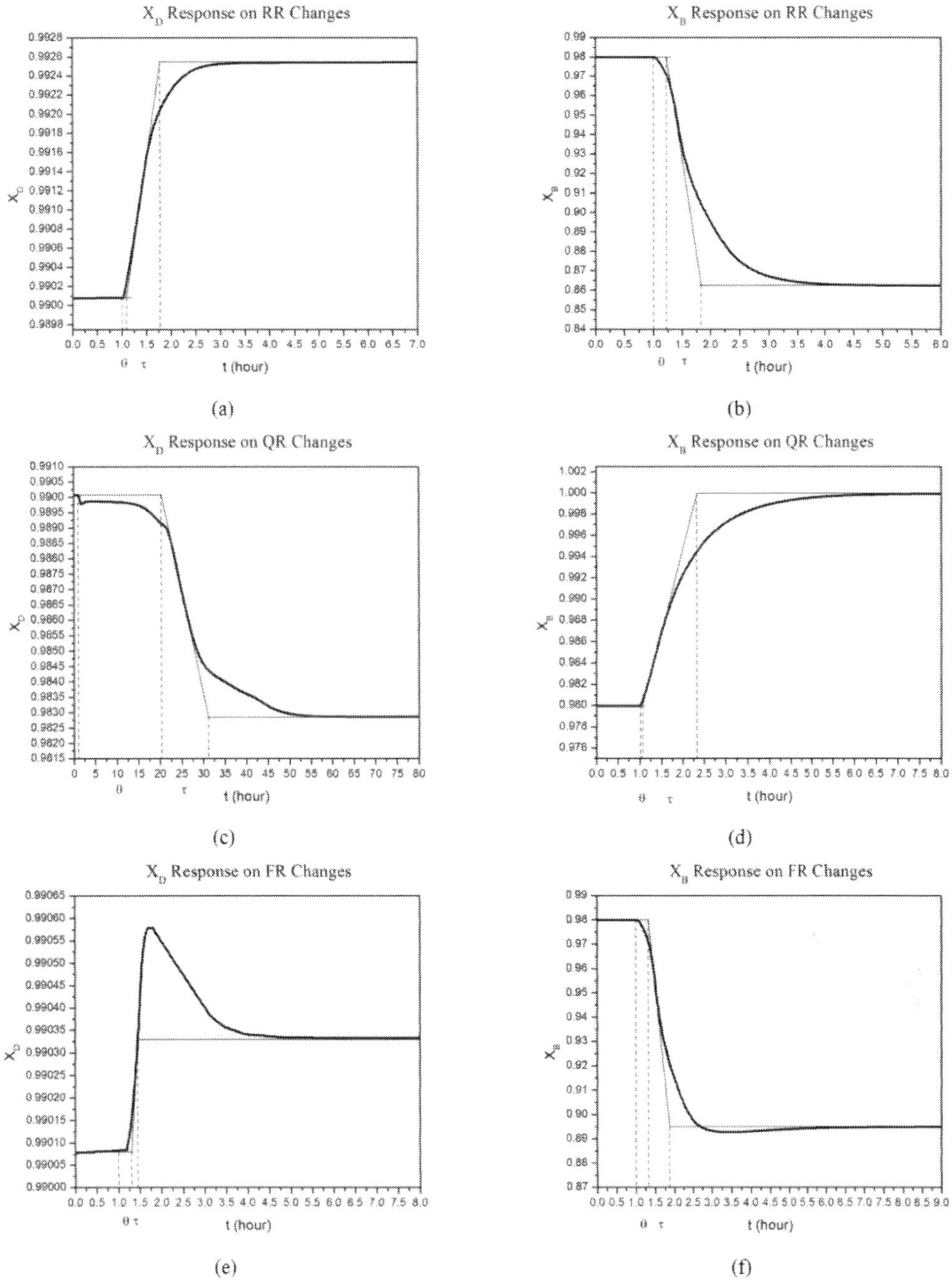


**Fig. 3.** Response from all controlled variables on the changes of every manipulated variables.

The first manipulated variable that was changed was reflux ratio. Magnitude of the reflux ratio was changed 10% from initial value. The initial reflux ratio was 1.98616 and then it was changed to 2.18478 after 1 hour. Figure 3.a and 3.b show $X_D$ and $X_B$ responses to the RR changes sequentially. Then, in the changes of reboiler duty as the second manipulated variables, reboiler duty was changed 2% from the initial value. The initial reboiler duty was 0.879685 MMkcal/hr and then it was changed to 0.897488 MMkcal/hr after 1 hour. Figure 3.c and 3.d show $X_D$ and $X_B$ responses to the QR changes sequentially. The last manipulated variable that was changed was feed ratio. Magnitude of the feed ratio was changed 10% from initial value. The initial feed ratio was 1 and then it was changed to 1.1 after 1 hour. Figure 3.e and 3.f show $X_D$ and $X_B$ responses to the FR changes sequentially. After recording all of the responses, graphical calculation based on the process reaction curve found by Ziegler and Nichols was done.[8] All parameters of process transfer function that were obtained from graphical calculation can be seen in Table 2. To verify that value, a calculation utilized a toolbox from Aspen has been done. Those results can be seen in Table 3.

It is shown from Table 3 that the graphical calculation only slightly different with Aspen calculation and it is tolerable. It should be noted that there is no delay caused from measuring element in the Aspen simulation process. The simplified open-loop dynamic model was expressed as shown in equation (2).

$$\begin{bmatrix} X_D \\ X_B \end{bmatrix} = \begin{bmatrix} \frac{0.0249e^{-0.07s}}{0.7s+1} & \frac{-0.3679e^{-19.36s}}{10.53s+1} & \frac{0.0025e^{-0.32s}}{0.13s+1} \\ \frac{-1.1867e^{-0.23s}}{0.62s+1} & \frac{1.0185e^{-0.05s}}{1.27s+1} & \frac{-0.8658e^{-0.34s}}{0.53s+1} \end{bmatrix} \begin{bmatrix} RR \\ QR \\ FR \end{bmatrix} \quad (2)$$

**Table 2.** All parameters of process transfer function obtained from graphical calculation.

| | RR | | | QR | | | FR | | |
|---|---|---|---|---|---|---|---|---|---|
| | K | τ | θ | K | τ | θ | K | τ | θ |
| $X_D$ | 0.02474 | 0.69 | 0.09 | -0.36105 | 10.8 | 19.3 | 0.00254 | 0.14 | 0.31 |
| $X_B$ | -1.17497 | 0.61 | 0.22 | 0.99805 | 1.27 | 0.05 | -0.84848 | 0.54 | 0.33 |

**Table 3.** Comparison between Aspen and graphical calculation.

| | | RR | | QR | | FR | |
|---|---|---|---|---|---|---|---|
| | | Graphic | Aspen | Graphic | Aspen | Graphic | Aspen |
| $X_D$ | K (%/%) | 0.02474 | 0.024892 | -0.36105 | -0.36791 | 0.00254 | 0.00251 |
| | τ (hr) | 0.69 | 0.705931 | 10.8 | 10.53324 | 0.14 | 0.132899 |
| | θ (hr) | 0.09 | 0.070644 | 19.3 | 19.36572 | 0.31 | 0.324612 |
| $X_B$ | K (%/%) | -1.17497 | -1.18674 | 0.99805 | 1.018478 | -0.84848 | -0.86583 |
| | τ (hr) | 0.61 | 0.624313 | 1.27 | 1.267701 | 0.54 | 0.533481 |
| | θ (hr) | 0.22 | 0.232339 | 0.05 | 0.054459 | 0.33 | 0.33636 |

## 5 Conclusion

An open-loop dynamic model that represent methyl acetate production process by using reactive distillation column has been built based on process reaction curve of Ziegler and Nichols. The open-loop dynamic model as shown in Figure 2 is a 2x3 MIMO process containing a simple transfer function model FOPTD (first order plus time delay).

The authors would like to acknowledge the financial support of DIKTI - PMDSU research program on a contract of 003246.280/IT2.11/PN.08/2015 and PUPT (Penelitian Terapan Unggulan Perguruan Tinggi) Project No. 582/PKS/ITS/2017.

## References

1. R. Aris, Chem. Eng. Sci. **46**, 7 (1991)
2. S.S.E.H. Elnashaie, F.M. Alhabdan, Math. Comp. Mod. **17**, 3 (1993)
3. R.K. Wood, M.W. Berry, Chem. Eng. Sci. **28**, 1707-1717 (1973)
4. N. Doukas, W. Luyben, Anal. Instrum. **16**, 51-58 (1978)
5. D.M. Prett, M. Morari, *The Shell Process Control Workshop* (Butterworths, Stoneham, USA, 1987)
6. Y.T. Tang, Y.W. Chen, H.P. Huang, C.C. Yu, AIChE J. **51**, 6 (2005)
7. H.S. Fogler, *Elements of Chemical Reaction Engineering 3rd ed* (Prentice Hall of India, India, 2004)
8. J.G. Ziegler, N.B. Nichols, Trans. of the ASME **64**, 759-768 (1942)

# Increased lipids production of *Nannochloropsis oculata* and *Chlorella vulgaris* for biodiesel synthesis through the optimization of growth medium composition arrangement by using bicarbonate addition

*Dianursanti**, *Zenitha Lintang* Agustin, and *Dwini Normayulisa* Putri

Chemical Engineering Department, Faculty of Engineering, University of Indonesia, Kampus Baru UI, Depok, West Java, 16424

**Abstract.** *Chlorella vulgaris* and *Nannochloropsis oculata* are a highly potential microalgae to be used in pilot-scale of biodiesel synthesis. The essential content from these microalgae is the fatty acid of lipid which is the main target for the feed and biodiesel industries. One of the key factor in improving lipid microalgae are the arrangemment of nutrients in the growth medium. Research on the regulation of nutrients using bicarbonate ($HCO_3^-$) as an additional inorganic carbon source has been done by many studies, but the yield of lipids obtained has not been much. The aim of the study was to improve the lipid yield of *Chlorella vulgaris* and *Nannochloropsis oculata*. Variation of [$HCO_3^-$] which added to Walne medium were 25 ppm and 75 ppm, while the Walne medium without the addition of bicarbonate acts as control. The results showed that [$HCO_3^-$] 75 ppm could increase *Chlorella vulgaris* biomass by 0.9162 g/l with 17.0% wt, while *Nannochloropsis oculata* produced the greatest lipid content in [$HCO_3^-$] 25 ppm of 20.3% wt and the largest biomass on [$HCO_3^-$] 75 ppm of 1.7233 g/l.

## 1 Introduction

Transportation and electricity are the most crucial daily needs and must be met to support human welfare. On the other hand, the increase population can causes an increasing of these sector demand and raising various problems in the energy field. To respond and address these issues, researchers are always looking for and developing renewable energy sources. The best candidate of this potential source is microalgae because it can produce a large number of lipid in a small area everyday, so it better than crop plants [1].

The most commonly used microalgae as a renewable energy sources for biodiesel synthesis are *Nannochloropsis* and *Chlorella* [2]. Both microalgae can produce a high lipid content in its cells, i.e 22.7-29.7% of the *Nannochloropsis* dry weight and 5-58% of the *Chlorella* dry weight [3]. One of the most common problems is the acquisition of biomass and lipid microalgae that do not meet production targets for feed and biodiesel industries.

Based on the potential of these microalgae, University of Indonesia is actively researching and developing *Nannochloropsis* and *Chlorella* to contribute in the production of biodiesel, especially in Depok, West Java. One of effort to increase the lipid microalgae production is re-optimize the regulation of growth medium using bicarbonate based on previous studies, so that the University of Indonesia has valuable assets to play a role in biodiesel synthesis.

The selection of bicarbonate is considered to be an important point for increasing the lipid production of Nannochloropsis and Chlorella in the upstream biodiesel synthesis study. According to Devgoswami, the solubility of bicarbonate ($HCO_3^-$) is higher than carbon dioxide ($CO_2$), so it can be directly absorbed by the microalgae [4]. In addition, bicarbonate acts as a carbon source that is closely related to photosynthesis and the accumulation of lipids in cells [5].

## 2 Materials and Methods

### 2.1 Microalgae and Growth Medium Compositions

The microorganisms used during this study were *Nannochloropsis oculata* and *Chlorella vulgaris* microalgae. The isolate is cultivated in the Laboratory of Bioprocess Engineering, Department of Chemical Engineering, Faculty of Engineering, University of Indonesia, Depok, West Java. Isolates of both types microalgae are always maintained and reproduced in the range of logarithmic-stationary growth phases.

The growth medium of *Nannochloropsis* and *Chlorella* used in this study was the Walne medium in seawater. Walne medium composition can be seen in Table 1 [6].

* Corresponding author: danti@che.ui.ac.id

**Table 1.** Walne Medium Composition

| Nutrient solution | | Trace element solution | | Vitamin solution | |
|---|---|---|---|---|---|
| **Materials** | **per litre** | **Materials** | **per litre** | **Materials** | **per litre** |
| $NaNO_3$ | 100,0 g | $ZnCl_2$ | 21,0 g | Thiamine.HCl (vit. B1) | 1,0 g |
| $H_3BO_3$ | 33,6 g | $CoCl_2.6H_2O$ | 20,0 g | Cyanocobalamin (vit. B12) | 0,05 g |
| $Na_2EDTA$ | 45,0 g | $(NH_4)_6Mo_7O_{24}.4H_2O$ | 9,0 g | | |
| $NaH_2PO_4.H_2O$ | 20,0 g | $CuSO_4.5H_2O$ | 20,0 g | | |
| $FeCl_3.6H_2O$ | 1,3 g | | | | |
| $MnCl_2.4H_2O$ | 0,36 g | | | | |
| trace element solution | 1 ml | | | | |
| vitamin solution | 100 µl | | | | |

## 2.2 Stock Solution of $NaHCO_3$

The $NaHCO_3$ solution is used as an additional carbon source solution in the Walne medium. The purpose of making $NaHCO_3$ stock solution in this study was to provide homogeneous conditions on each treatment and to obtain significant results in the work culture. The $NaHCO_3$ stock solution made for this study was 300 ppm which can be prepared by dissolving 300 mg of $NaHCO_3$ in 1000 ml of distilled water. Dilution for working culture using a mole ratio between $NaHCO_3$ and $HCO_3^-$.

## 2.3 Working Culture and Culture Condition

The working culture of *Nannochloropsis* and *Chlorella* uses glass-reactor bottles with a capacity of 1500 ml. Variations of $[HCO_3^-]$ which added to Walne medium were 25 ppm and 75 ppm, whereas treatment without addition of $[HCO_3^-]$ was used as a control. The composition of the working culture in this study can be seen in Table 2.

**Table 2.** Working Culture Compositions

| $[HCO_3^-]$ treatment (ppm) | *Nannochloropsis oculata* | | | *Chlorella vulgaris* | | |
|---|---|---|---|---|---|---|
| | **0** | **25** | **75** | **0** | **25** | **75** |
| **Inoculum (ml)** | 500 | 500 | 500 | 500 | 500 | 500 |
| **$NaHCO_3$ solution addition (ml)** | 0 | 125 | 375 | 0 | 125 | 375 |
| **Walne medium (ml)** | 1000 | 875 | 625 | 1000 | 875 | 625 |

With a working culture composition that looks like above, the *Nannochloropsis* and *Chlorella* culture are set with initial optical density (OD) of ±0.4. The set of reactor then placed near the white lamp with light intensity that has been set for each microalga, i.e 3000--3500 lux for *Nannochloropsis* and 2000 lux for *Chlorella*. The both culture then harvested for lipid extraction after reaching the stationary phase.

## 2.4 Harvesting Biomass

After reaching the stationary phase, *Nannochloropsis oculata* and *Chlorella vulgaris* cultures are harvested. The working culture is harvested by centrifugation at 4000 rpm for 15 minutes to obtain biomass [7]. The precipitated biomass then separated from the supernatant and transferred to a container. Container with wet biomass then dried naturally. The dry weight of biomass (X) can be determined by the following equation:

$$X\ (g/l) = \frac{weight\ final\ (g) - weight\ initial\ (g)}{sample\ volume\ (l)} \quad (1)$$

## 2.5 Lipid Extraction

The lipid extraction method used in this study is a modification of the Bligh & Dyer method [7]. Bligh & Dyer modification method can be done by means of wet biomass mixed with chloroform: methanol (2:4), then sonicated for 20 minutes. Then, the mixture was added with chloroform:aquadest (2:2) and re-sonicated for 20 minutes. The mixture then centrifuged for 15 minutes to form three layers. The bottom layer is the extracted lipid. The extracted lipid then placed into a vial, which has been known the empty weight, for the solvent evaporation. After all the solvent has evaporated, the vial is then weight. Microalgal lipid ($Y_L$) can be determined by the following equation:

$$Y_L\ (\%) = \frac{weight\ final\ (g) - weight\ initial\ (g)}{sample\ biomass\ (g)}\ x\ 100\% \quad (2)$$

# 3 Results and Discussions

## 3.1 Results

### *3.1.1 Growth and Biomass Production*

* Corresponding author: danti@che.ui.ac.id

The cultivation of *Nannochloropsis oculata* and *Chlorella vulgaris* with variations of [$HCO_3^-$] addition in the Walne medium reached the stationary phase at 288-336 hours. Growth of both microalgae can be seen in Figure 1. Based on Figure 1, the control culture of *Nannochloropsis oculata* and *Chlorella vulgaris* experienced the lowest growth, while the [$HCO_3^-$] 75 ppm experienced the highest growth.

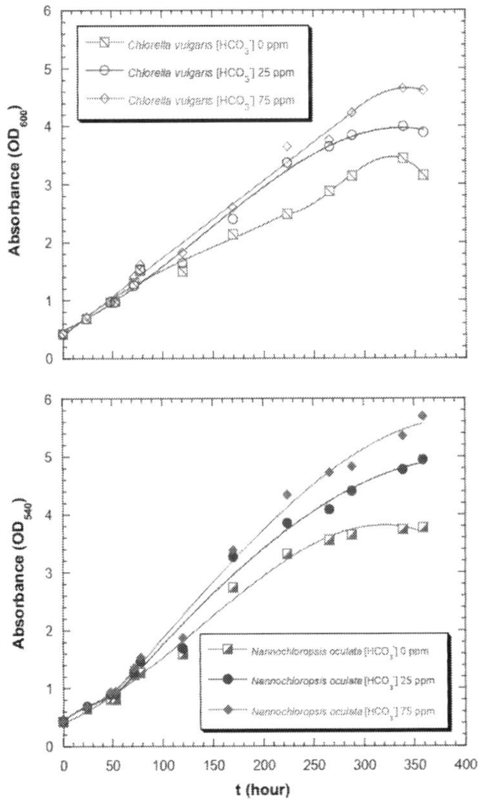


**Fig. 1.** Growth of *Nannochloropsis oculata* and *Chlorella vulgaris*in [$HCO_3^-$] treatments

Dry biomass of *Nannochloropsis oculata* and *Chlorella vulgaris* can be seen in Figure 2. According to Figure 2, the control treatment from *Nannochloropsis oculata* and *Chlorella vulgaris* also had the lowest biomass of 0.8954 g/l and 0.743 g/l, whereas the highest biomass was owned by [$HCO_3^-$] 75 ppm treatment of 1.7233 g/l and 0.9162 g/l. Increased growth and biomass on the addition of [$HCO_3^-$] into the culture medium has also been demonstrated in Devgoswami et al. [4], Ibrahim [8], Lin et al. [9], White et al. [10], and Agustin [11].

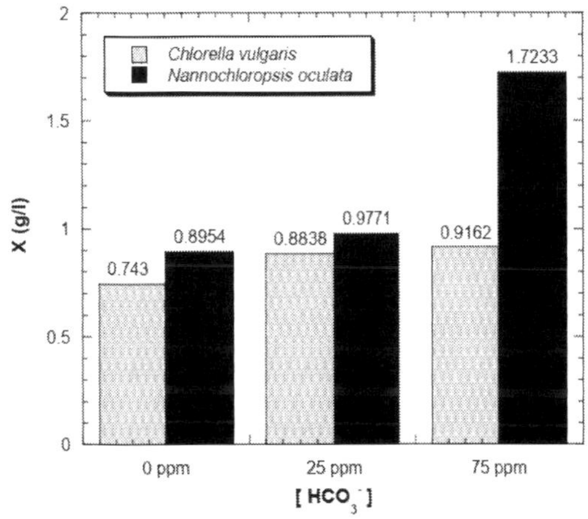


**Fig. 2.** Dry biomass of *Nannochloropsis oculata* and *Chlorella vulgaris*in [$HCO_3^-$] treatments

### 3.1.2 Lipid Production

Yield lipid of *Nannochloropsis oculata* and *Chlorella vulgaris* can be seen in Figure 3. According to Figure 3, the lowest lipid is controlled by the control treatment of both cultures, i.e. 12.1% dry weight of *Nannochloropsis oculata* and 11.8% dry weight of *Chlorella vulgaris*. The highest lipid was obtained from cultures of *Nannochloropsis oculata* with [$HCO_3^-$] 25 ppm of 20.3% dry weight and *Chlorella vulgaris* with [$HCO_3^-$] 75 ppm of 17% dry weight. Increased lipid in cultures that given of [$HCO_3^-$] also been demonstrated in Devgoswami et al. [4] and Agustin [11].

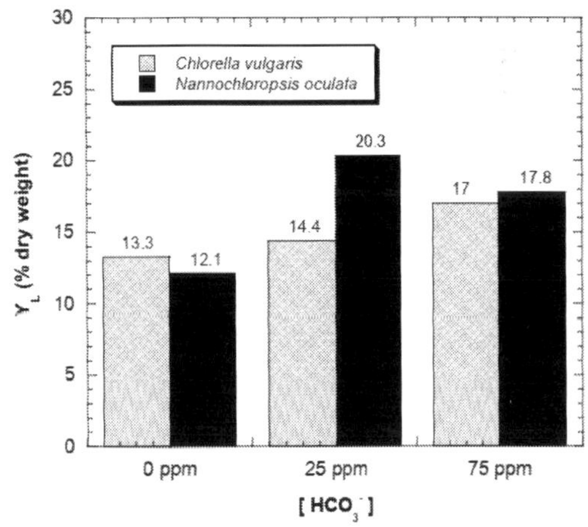


**Fig. 3.** Yield lipid of *Nannochloropsis oculata* and *Chlorella vulgaris*in [$HCO_3^-$] treatments

## 3.2 Discussions

The lowest growth in control culture was caused by no addition of carbon source that can enhance the process of photosynthesis, so cultures rely only on air derived from aeration. This is in accordance with Gardner et al., that culture with the availability of low carbon sources will lead to low cell growth because photosynthesis is inhibited [12]. When [$HCO_3^-$] is added to the cultivation medium, the bicarbonate ($HCO_3^-$) will induce the process of photosynthesis. The induction occurs in photosystems II (PS II) of light reactions and resulting in a large number of NADPH and ATP [5].

In addition, bicarbonate ($HCO_3^-$) also serves as a carbon to replace source of $CO_2$ in the medium and in the dark reaction of photosynthesis because the solubility of bicarbonate ($HCO_3^-$) in water is greater than that of $CO_2$. When [$HCO_3^-$] is added to the cultivation medium, the carbon source will increase to bind to the Rubisco enzyme (Ribulose-1,5-biphosphate), resulting in G3P synthesis (glyceraldehyde-3-phosphate or triose-3-phospate) at The Calvin cycle will increase. Increased of G3P will increase the formation of glucose and lipids, because 28 G3P molecules will produce 14 molecules of glucose and 1 molecule TAG C18 [13]. According to Agustin, glucose is used as a source of energy for cell growth, so that biomass increases [11].

Furthermore, the remaining bicarbonate ($HCO_3^-$) in the medium still serves to induce lipid synthesis to form energy reserves because the nutrient conditions reduced when culture entered the stationary phase. Bicarbonate ($HCO_3^-$) will induce the formation of malonil-CoA which acts as a substrate in lipid synthesis. Carbohydrates will be broken down into acetyl-CoA which is a precursor of respiration and lipid synthesis. Acetyl-CoA then interacts with bicarbonate ($HCO_3^-$) to form malonil-CoA. According to Foster, increased production of malonil-CoA will cause acetyl-CoA for the Krebs cycle then diverted to lipid synthesis, since the enzyme activity in the Krebs cycle is inhibited [14].

## 4. Conclusion

The results showed that [$HCO_3^-$] 75 ppm treatment could increase *Chlorella vulgaris* biomass by 0.9162 g/l with 17.0% wt, while *Nannochloropsis oculata* produced the largest biomass on [$HCO_3^-$] 75 ppm treatment of 1.7233 g/l and the greatest yield lipid content in [$HCO_3^-$] 25 ppm treatment of 20.3% wt. The addition of bicarbonate into the Walne medium with the composition are set in such way on this study has been able to induce the production of biomass and lipids of *Nannochloropsis oculata* and *Chlorella vulgaris* significantly compared with control treatments and prior studies.

The authors would like to thank the financial support provided by Universitas Indonesia through the PITTA 2016-2017 funding scheme under Grant managed by the Directorate for Research and Public Services (DRPM) Universitas Indonesia, West Java, Indonesia.

## References

1. A. Demirbas, M.F. Demirbas. *Algae energy: Algae as a new source of biodiesel*. Springer-Verlag (2010)
2. Y. Chisti. Biotechnol. Adv., **25**, 3 (2007)
3. T.M. Mata, A.A. Martins, N.S. Caetano. Renew. Sust. Energ. Rev., **14**, 1 (2010)
4. Ch.R. Devgoswami, M.C. Kalita, J. Telukdar, R. Bora, P. Sharma. Afr. J. Biotechnol., **10**, 61 (2011)
5. W.G. Hopkins, N.P.A. Huner. *Introduction to plant physiology, 4th ed.* John Wiley & Sons (2009)
6. R.A. Andersen. *Alga Culturing Techniques*. Elsevier Academic Press (2005)
7. D.N. Putri. *Sintesis Biodiesel dari Mikroalga Nannochloropsis sp. melalui Reaksi Transesterifikasi menggunakan Katalis Heterogen CuO/Zeolit*. Undergraduate Thesis (2016)
8. A.I. Ibrahim. *Pertumbuhan dan produksi lemak selular Scenedesmus sp.* Undergraduate Thesis (2008)
9. Q. Lin, N. Gu, G. Li, J. Lin, L. Huang, L. Tan. Bioresour. Technol., **111** (2012)
10. D.A. White, A. Pagarette, P. Rooks, S.T. Ali. J. Appl. Phycol., **25** (2013)
11. Z.L. Agustin. *Pengaruh penambahan variasi konsentrasi $HCO_3^-$ terhadap produksi lemak Nannochloropsis*. Undergraduate Thesis (2014)
12. R.D. Gardner, E. Lohman, R. Gerlach, K.E Cooksey, B.M. Peyton. Biotechnol. Bioeng., **10** (2013)
13. L. de Jaeger, R.E.M. Verbeek, R.B. Draaisma, D.E. Martens, J. Springer, G. Eggink, R.H. Wijffels. Biotechnol. Biofuels, **7**, 1 (2014)
14. D.W. Foster. J. Clin. Invest., **122**, 6 (2012)

# Simulation study of the biodiesel production from palm fatty acid distillate using palm empty fruit bunch catalyst in reactive distillation column

*Arif* Hidayat* and *Bachrun* Sutrisno

Chemical Engineering Department, Universitas Islam Indonesia, Indonesia

**Abstract.** In this paper, the simulation study of biodiesel production from Palm Fatty Acid Distillate (PFAD) using Palm Empty Fruit Bunch (PEFB) catalyst as the heterogeneous acid catalyst in reactive distillation (RD) column was studied. The main parameters of the basic RD column such as the number of stages, molar ratio of feed, total number of stages, and reboiler heat duty were employed in the simulation study. The equilibrium stage and rate-based models were used for the RD simulation studies. The simulation studies show that the equilibrium stage model could describe the column profiles only qualitatively while the rate-based model was capable in describing the column profiles quantitatively and qualitatively. From simulation results, it can be concluded that the reactive distillation process with the Palm Empty Fruit Bunch catalyst offered advantages over the conventional process because it could eliminate the requirement of post-processing separation and purification at cost-effective column design and operating conditions. The optimum condition for producing biodiesel by reactive distillation using PEFB catalyst was found as follows: 4:1 of methanol to PFAD molar feed ratio, reflux ratio of 1.5, reboiler heat duty of 2.107 kJ/h and 3 reactive stages. This condition provided a FFA conversion of 97 wt% and biodiesel purity of 97.5%.

## 1 Introduction

Biodiesel is a mono alkyl ester of fatty acid derived from vegetable oils and animal fats. Biodiesel is conventionally produced via transesterification of triglycerides, which is the reaction of vegetable oils with alcohol to produce alkyl esters and glycerol, in the presence of an alkaline catalyst. Biodiesel can also be produced by esterification of free fatty acid (FFA) with alcohol to fattyacid alkyl ester (biodiesel) and water (by-product). The main advantages of using biodiesel are, it provide lower hydrocarbon and carbon monoxide emissions, higher cetane numbers, less smoke and particulates, provide engine lubricity to low sulfur diesel fuels and are biodegradable and nontoxic.

One of the major concern for biodiesel production is economic feasibility. The overall biodiesel cost consists of raw material (production and processing), catalyst, biodiesel processing (energy consumables and labour), transportation (raw materials and final products) and taxes [1]. To date, most biodiesel plants are using refined vegetable oils as their main feedstock. However, these feedstock is impractical and uneconomical due to high feedstock cost and priority as food resources. Therefore, the cost of refined vegetable oils contributed nearly 80% of the overall biodiesel production cost. The use of low cost feedstocks such as Palm Fatty Acid Distillate (PFAD) can reduce biodiesel production cost and increase supply while avoiding the food versus fuel problem. Meanwhile, carbon-based solid acids are considered ideal catalysts for many reactions due to their chemical inertness, stable under acidic basic conditions, have very high surface area, renewable biomass sources, and good mechanical and thermal stability [2].

Biodiesel is currently produced in batch processes [3]. However, using batch process for biodiesel production have many major drawbacks, such as: high alcohol demand required to shift equilibrium towards fatty esters, advanced process for catalyst neutralization which produce a salt product waste, high separation costs of fatty ester products from the reaction products, high costs caused by complex processes including multiple reactors and separation units. Therefore, to solve these problems, in recent years several researchers have developed a sustainable biodiesel production process based on reactive distillation [4].

Reactive Distillation is the process in which chemical reaction and distillation separation are carried out in one equipment. The reactive distillation process offers several benefits over conventional processes in which the reaction and the product separation are done in series, especially for reactions limited by equilibrium constraints such as esterification reaction (Equation 1).

Under these above conditions, the reaction was assumed to be pseudohomogeneous, second-order in the forward and reverse direction. The reaction rate can be described as:

$$R_1\text{–COOH} + \text{R–OH} \Leftrightarrow R_1\text{–COO–R} + H_2O \qquad (1)$$

* Corresponding author: arif.hidayat@uii.ac.id

From this reaction, fatty acid is reacted with methanol to produce biodiesel and water and the reaction is reversible. Therefore, conversion of the reaction is limited by the concentration of products. In order to control the reaction in forward direction, removal of biodiesel and water from the system is important. As the biodiesel and water in reactive distillation are continuously separated from the reaction zone, no limiting chemical equilibrium can be established and thus the reaction velocity is maintained at a high rate, resulting in greater yields. Other benefits of reactive distillation can include the minimization of side reactions and the utilization of the heat of reaction for the mass transfer within the same column. Therefore the capital investment and operating costs are significantly lower with reactive distillation than for conventional processes. Examples for successful applications of reactive distillation have, among others, been reported for esterifications, etherifications, alkylations and isomerizations [5].

In recent literatures, many works about reactive distillation can be found. He et al. (2006) explored the biodiesel production by reactive distillation from canola oil and methanol using potassium hydroxide as catalyst. Kumar and Mahajani (2007) studied the production of biodiesel by esterification of lactic acid with n-butanol by reactive distillation. Bock et al. (1997) conducted experimental production of myristic acid with isopropanol by using homogenous acid-based catalysts such as sulfuric acid in a two-column facility. Steinigeweg and Gmehling (2003) developed a technically optimized reactive distillation process for the production of decanoic acid methyl esters by esterification of decanoic acid and methanol using Amberlystas heterogeneous catalysts. Conceptual design of RD column to produce 2-ethylhexyl dodecanoate using sulfated zirconia as a solid acid catalyst was investigated by Omota et al. (2003). Bhatia et al. (2006) carried out esterification of palmitic acid (PA) with isopropanol to make isopropyl palmitate (IPA) in a pilot plant scale obtaining high conversion. Singh et al. (2004) investigated biodiesel production from canola oil and methanol using potassium hydroxide as catalyst by reactive distillation.

To best our knowledge, no reports were found on biodiesel production from PFAD by using PEFB catalyst in a reactive distillation column. The present work focused on the operating parameters that affected esterification of PFAD with methanol in a continuous reactive distillation column. The aim of this study was to understand the relationship between operating conditions and the response FFA conversion. Furthermore, the effect of various operating conditions on the performance of the reactive distillation for biodiesel production is investigated.

## 2 Simulation Study

Figure 1 presents the process flowsheet. The flowsheet was simulated by means of a chemical process simulator (ASPEN plus V 7.4). The key equipment is the reactive distillation (RD) column. This contains in principle three zones, rectifying, reaction and stripping zone. The RD column was simulated using the RADFRAC module of Aspen Plus, which relies on the equilibrium stage model. The RADFRAC is a physical equilibrium stage model able to solve the set of equations that describe each equilibrium stage: material balances, enthalpy balances and phase equilibrium relations between the vapour and the liquid leaving the stage.The RADFAC model has been extensively used in RD and model is based on assumption that the vapor from the stage below and liquid from the stage above are brought into intimate and well-mixed contact on the stage. The vapor and liquid streams leaving the stage are assumed to be in equilibrium with each other and the overall reactive distillation process is modeled as a sequence of these equilibrium stages.

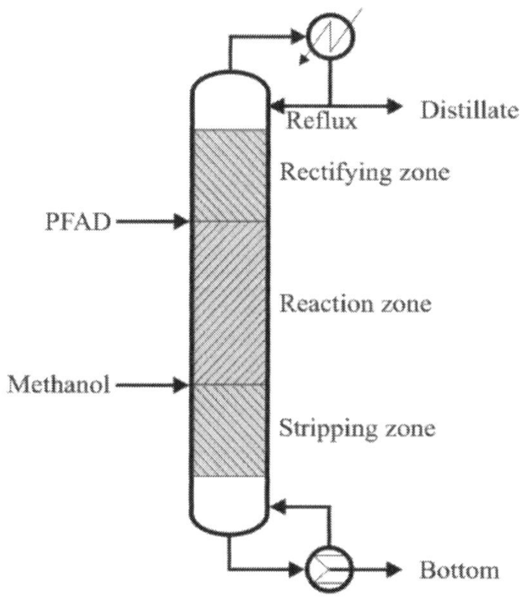


**Fig. 1.** Process flowsheet for RD column of free fatty acid esterification.

The catalyst used in the model column is PEFB catalyst and kinetics data of esterification reaction of FFA with methanol used in the model is applied from Hidayat's report [2]. The kinetic model of this work was built on the following assumptions: (a) the rate of the non-catalyzed reactions is negligible compared with the catalyzed reactions; (b) the catalytic activity of all sites on the catalyst surface is the same; (c) the reaction mixture can be considered as a pseudo-homogeneous system where there is no mass transfer limitation. This can be expected due to high agitation intensity and a high dissolving power of alcohol in oils; (d) the overall process kinetics is chemically controlled; (e) the reaction mixture is perfectly mixed, ensuring its uniform composition. The power law kinetics model is used to represent the reaction in Equation (1) with k (kinetics factor) and E (activation energy) of 1.432×1011 $kmol.m^{-3}.s^{-1}$ and 5.188×104 $kJ.kmol^{-1}$, respectively. The thermodynamic model was the non-random two-liquid (NRTL) model. The temperature of the reactive zone was limited by the boiling temperature of the methanol (64.7 °C). Details about the RD column design are given in Table 1.

**Table 1.** Rate constant at different temperatures.

| Parameter | Value |
|---|---|
| Total number of stages | 20 |
| Number of reactive stages | 3 (from 3 to 5) |
| PFAD feed stage | Stage 5 at 60°C |
| Methanol feed stage | Stage 3 |
| Methanol to PFAD molar ratio | 4 : 1 |
| Total feed of reactants | 50 mol/h |
| Distillate to feed ratio | 0.75 |
| Type of condenser (stage 1) | Total |
| Type of reboiler (stage 10) | Kettle |

## 3 Results and Discussion

### 3.1. Basic RD

At the standard condition, the feed streams consisting of PFAD and methanol at the flow rates of 10 and 40 kmol/h, respectively, are separately fed to a RD column. The column is operated at pressure of 1 atm with the reflux ratio of 3, and feed temperature of 60 °C. The residence time in the column is ~10 min, much less than the retention times of 60-90 min in the conventional processes reported [13, 14]. To reduce the amount of PFAD in the final product, the fatty acid is fed above and methanol below the reactive zone, respectively. The composition and temperature profiles in the basic RD column are shown in Figure 2.There is a large amount of methanol and water at the top of the column, whereas less biodiesel are observed. The bottom products consist of mostly fatty acid methyl ester (biodiesel product). Under the basic RD column, the conversion of FFA is 91.22% and the purity of biodiesel is 99.16%. This was an indication that effective reaction conversion and separation were achieved in the column. The unreacted methanol, the lowest-boiling point substance, is withdrawn from the reactive distillation column as a distillate stream. As can be seen from the temperature profile shown in Figure 2, the temperature of the stage near the PFAD feed section was found to be low. This was due to the reflux flow effects. The temperature gradient was increased from tray 2 to the reboiler section.

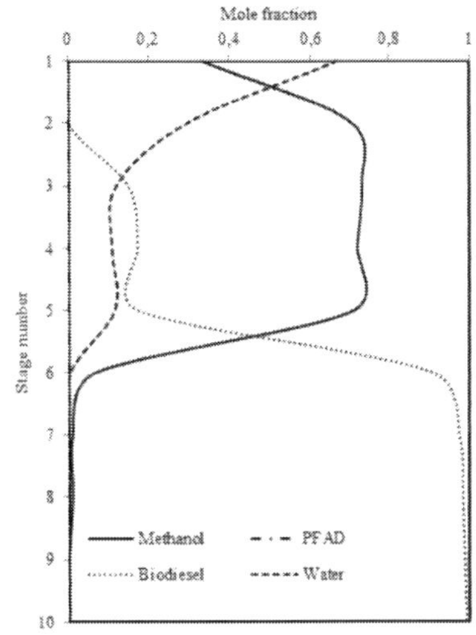


(a)

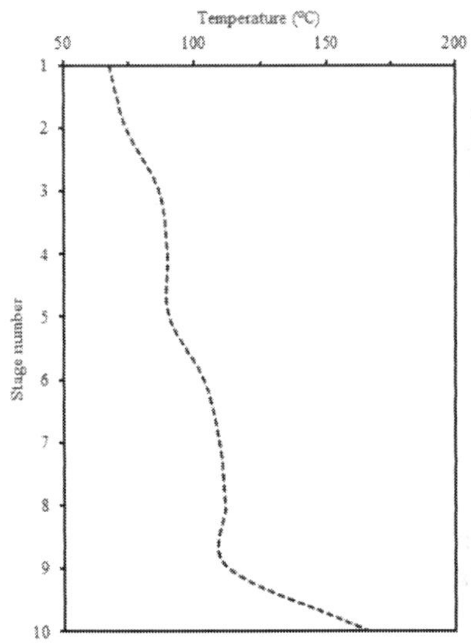


(b)

**Fig. 2.** Liquid compositons (a) and temperature (b) profiles along the basic RD unit.

### 3.2 Effect of Molar Feed Ratio of Methanol and PFAD

The molar feed ratio of methanol and PFAD is one of the key parameters on the performance of the RD column. Figure 3 shows the effect of molar feed ratio of methanol to PFAD on the conversion of FFA. Figure 3 shows the effect of molar feed ratio methanol to PFAD ratio on FFA conversion in the product. Increase methanol to PFAD molar ratio from 3 : 1 to 4 : 1 increases the FFA conversion in the product from 76.67% to 81.22% and increase from 82.67% to 82.91% when molar feed ratio methanol to PFAD molar ratio increases from 5 : 1 to 6 : 1. The simulation results demonstrate that the conversion of FFA increase with increasing feed molar ratio. Methanol to PFAD feed molar ratio of 4 is therefore suggested because too low amount of methanol to PFAD feed molar ratio would cause biodiesel product to be operated under severe conditions and even of specifications below the standards. However, increasing the effect of molar feed ratio of methanol to PFAD increased the energy consumption because of higher condenser and reboiler duties. Yield and purity of biodiesel were found to increase with increasing methanol to PFAD feed molar ratio. Total energy usage in the process increased with methanol to PFAD feed molar ratio due to additional energy consumed to heat methanol. Methanol to PFAD feed molar ratio of 4 was found suitable condition tomeet biodiesel standard values (>96.5%) at reduced energy onsumption [15].

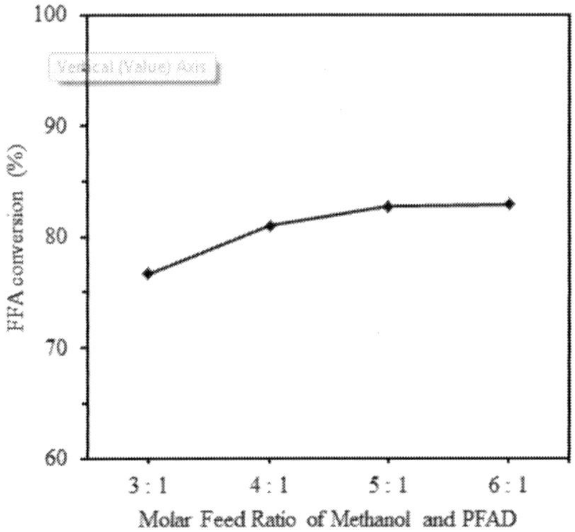


**Fig. 3.** Effect of molar feed ratio of methanol and oil on conversion of FFA.

### 3.3 Effect of a Number of Stages

Figure 4 demonstrates the effect of changing the number of stages on the reactive distillation performance when the feed ratio of methanol and PFAD is 4: 1. It is noted that the esterification reaction is carried out at reaction zone (stage 3 to 5 of the column). The results indicate that adding more stages improves the system performance. The biodiesel purity reached 97.34% at number of 10thstages. When the number of stages increase at 15 and 20, the biodiesel purity were obtained at 98.67% and 98.93%, respectively. Increasing the stages causes the separation of product from unreacted reactant more complete, thereby enhancing the biodiesel purity. Biodiesel yield and purity significantly increased by adding more number of stages owing to more contacting area and residence times in the separation zone [16]. It is noticed that the performance of the reactive distillation is slightly improved when a number of the stages are higher than 15.

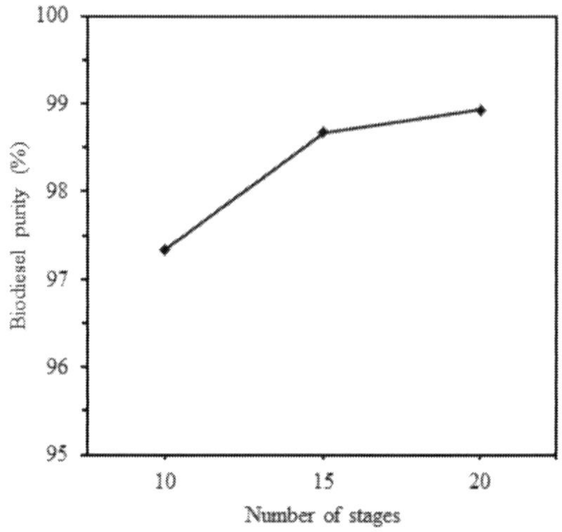


**Fig. 4.** Effect of Number of Stages on Biodiesel Purity.

### 3.4 Effect of Feed Location of Methanol

The location of a feed stream is an important design parameter having an effect on a reactive distillation performance. In general, a lighter reactant should be fed on the bottom stage of the reactive zone, whereas a heavier reactant is fed on the top stage of the reactive zone. When moving the feed stage of methanol, the number of reactive zone would be change. Due to large differences in the boiling temperatures between methanol and other components of the reaction mixture therefore small rectifying stage is required in order to obtain pure methanol as distillate. From the simulation results, moving the feed stage of methanol down to the bottom of the reactive distillation column decreases the conversion of FFA as shown in Figure 5.

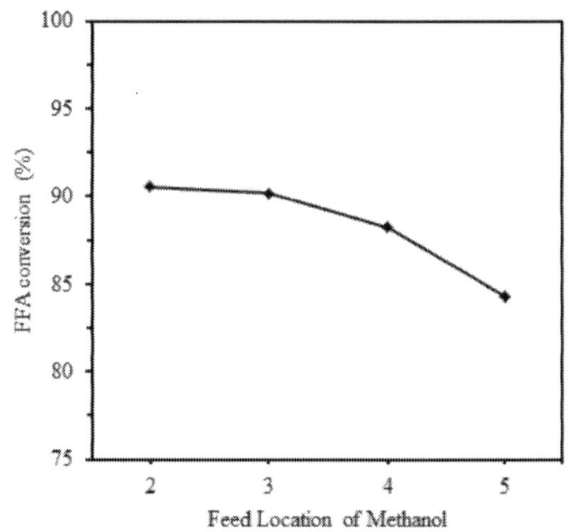


**Fig. 5.** Effect of Feed Location of Methanol on Conversion of FFA.

### 3.5 Effect of reflux ratio and reboiler heat duty

Increasing the reboiler duty from 1 to 1.5.107 kJ/h increases the FFA conversion due to vaporization of methanol in the column, thereby lowering methanol availability in the liquid phase of the reaction zone. Figure 6 shows that the maximum FFA conversion of 97.81% was achieved with the reboiler duty set on 2 kJ/h. The purity of biodiesel, as shown in Figure 6, increased with the reboiler duty owing to lower liquid holdups of methanol in the bottom streams. Nevertheless, the reboiler duty obviously increased the energy consumption in the process directly as seen in Figure 6. The operation of reactive distillation at reboiler heat duty of 2 kJ/h has given high biodiesel purity, assuming no thermal decomposition of biodiesel. This is due to the large amount of methanol in the distillate stream being removed from the column.

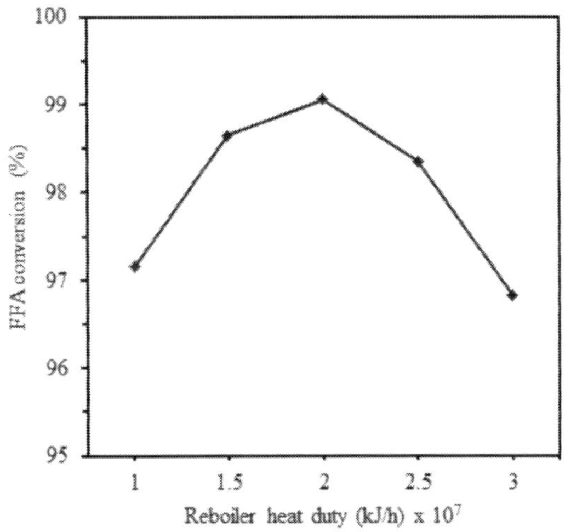


**Fig. 6.** Effect of reboiler heat duty on conversion of FFA.

## 4 Conclusions

Biodiesel production from esterification of FFA on PFAD with methanol by using reactive distillation (RD) was investigated. The use of RD along with PEFB catalyst was found beneficial over the conventional process of sequential esterification and separation. The optimum condition for producing biodiesel by reactive distillation using PEFB catalyst was found as follows: 4:1 of methanol to PFAD molar feed ratio, reflux ratio of 1.5, reboiler heat duty of 2.107 kJ/h and 3 reactive stages. This condition provided a FFA conversion of 97 wt% and biodiesel purity of 97.81%. From simulation results, it can be concluded that the reactive distillation process with the heterogeneous catalyst offered advantages over the conventional process because it could eliminate the requirement of post-processing separation and purification at cost-effective column design and operating conditions. The reactive distillation process provided higher yield and purity of biodiesel than the conventional process and required much less energy consumption. The use of reactive distillation with heterogeneous catalysts eliminated many operation steps such as neutralization, washing, separation, recovery, and waste disposal operations.

The adsorption experiments indicated that modified coal fly ash was effective in removing of Remazol Blue. The percentage removal of dyes increased while the modified fly ash dosage increased. The percentage removal of dyes increased with decreased initial

## References

1. A. Hidayat, B. Sutrisno, AIP Conference Proceedings, **1840**, 050001, (2017)
2. A. Hidayat, Rochmadi, K. Wijaya, A. Budiman, AIP Conference Proceedings, **1699**, 050005 (2015)
3. A. Hidayat, Rochmadi, K. Wijaya, A. Budiman,, IOP Conference Series: Materials Science and Engineering, **105** (1), 012026
4. Z. Mufrodi, Rochmadi, Sutijan, A. Budiman, Engineering Journal, **19** (2), 29-39 (2013)
5. R.D. Kusumaningtyas, Rochmadi, S. Purwono, A. Budiman, Int. J. Exergy, **15** (4), 447-467 (2014)
6. B.B. He, A.P. Singh, J.C.A. Thompson, Trans ASAE, **49**, 107-112 (2006)
7. R. Kumar, S.M. Mahajani, Ind. Eng. Chem. Res., **46**, 6873-6882 (2007)
8. H. Bock, G. Wozny, B. Gutsche, Chem. Eng. Process.: Process Intensif, **36**, 101-109 (1997)
9. S. Steinigeweg, J. Gmehling, Ind. Eng. Chem. Res., **42** (15), 3612-3619 (2003)
10. F. Omota, A.C. Dimian, A. Bliek, Chem. Eng. Sci., **58**, 3159-3174 (2003)
11. S. Bhatia, A.L. Ahmad, A.R. Mohamed, S.Y.Chin, Chem. Eng. Sci., **61**, 7436-7477 (2006)
12. A.P. Singh, B.B. He, J.C. Thompson, Proceedings of the ASAE/CSAE annual international meeting, Ottawa, Canada (2004)
13. D. Darnoko, M. Cheryan, J Am Oil Chem Soc, **77**, 1269-1272 (2000)
14. H. Noureddini, D. Harkey, V. Medikonduru, J Am Oil Chem Soc, **75**, 1775-1783 (1998)
15. N. Boon-anuwat, W. Kiatkittipong, F. Aiouache, S. Assabumrungrat, Chem. Eng. Process.: Process Intensif, **92**, 33-44 (2015)
16. L. Simasatitkul, P.Siricharnsakunchai, Y. Patcharavorachot, S. Assabumrungrat, A. Arpornwichanop, Korean J Chem Eng, **28** (3), 649-655 (2011)

# Catalytic performance of surface modified-niobic acid

*Adid Adep* Dwiatmoko*

Clean Energy and Chemical Engineering, University of Science and Technology, Daejeon, South Korea
(#Current affiliation: Research Center for Chemistry, Indonesian Institute of Sciences, 15314, Tangerang Selatan, Indonesia)

**Abstract.** Surface modification of niobic acid with phosphoric acid and its catalytic performance in the hydrolysis reaction has been studied. A series of catalysts, surface-modified with different concentration of phosphoric acid, have been prepared and characterized. It was found that the catalyst selectivity to glucose increased with the increase of acid concentration used for the surface modification. The extent of selectivity enhancement of the modified catalysts showed a good relationship with the atomic ratio of phosphorous and niobium on the surface.

## 1 Introduction

Since the first report by Tanabe [1], niobic acid (NA) has been recognized to be an effective catalyst for several reactions involved with participation or liberation of water molecules [2, 3]. It is an insoluble polymeric oxide with different amounts of hydrated water depending on a preparation method [1]. The surface of niobic acid generally exhibits a high acid strength, even in the presence of water, as long as it remains in an amorphous form [4]. Moreover, calcination or acid treatment makes its acid properties to be tuned in a certain manner [5]. Especially, the treatment with mineral acids has been reported to be effective in suppressing the crystallization of niobic acid and increasing its surface acidity as well [6]. More recently, we have synthesized phosphoric acid-treated niobic acid catalyst exhibited high activities for hydrolysis of disaccharides to glucose in water, where the conversion and glucose selectivity were 52% and 82%, respectively [7]. However, discussion on the relationship between its catalytic activity and acid property was not fully made in the above report.

Herein we report the investigation into catalytic behaviors of niobic acid, modified with different concentrations of phosphoric acid on the hydrolysis of cellobiose. Thus, expected products formed from cellobiose would be glucose, fructose, HMF and organic acids (levulinic acid and formic acid). Aside from these products identified, some unidentified products could be produced. Since one of major concerns in this case is a higher selectivity of a certain product, this work was focused on finding an active and selective phosphoric acid-treated niobic acid catalyst for cellobiose conversion. From the result that glucose, instead of HMF or organic acids, was favorably produced using the catalyst; an in-depth explanation was formulated for unveiling the effect of phosphoric acid treatment on enhancing the selectivity of glucose.

## 2 Experimental

### 2.1 Catalyst preparation

Niobic acid was modified with different concentration of phosphoric acid (x), denoted as P–NA(x) in which x is the acid concentration (0.5, 1, and 2) in a unit of mol/L, following the method described elsewhere [7, 8]. Briefly, niobic acid (1 g), utilized as an as-received form, was immersed in 20 mL of phosphoric acid solution with a specified concentration and stirred at room temperature for 72 h. The suspension was then filtered and thoroughly washed with large amounts of distilled water until a neutral pH was achieved. The resulting material was subsequently dried in the oven at 100 °C overnight.

### 2.2 Catalyst characterization

The phase structure of prepared catalysts was observed by a powder X-ray diffraction (XRD) analysis using a Rigaku X-ray diffractometer with Cu Kα($\lambda$= 0.15406 nm) radiation source operated at 40 kV and 30 mA, where 2θ value was in the range of 5–90°. The surface P and Nb species were characterized by X-ray photoelectron spectroscopy (XPS) using a PHI 5000 VersaProbe (Ulvac-PHI) analytical instrument with a monochromator Al KαX-ray source (1486.6 eV) operated at 25 W and 15 kV with the spot size of 100 μm by 100 μm. The binding energies of the XPS data were referenced to the C 1s peak at 284.6 eV with an accuracy of 0.1 eV and charge neutralization was employed.

### 2.3 Catalytic Reaction

In order to evaluate the catalytic performances of prepared catalysts, a mixture containing cellobiose (0.2 g), distilled water (20 mL), catalyst (0.1 g) was prepared and heated

* Corresponding author: adid.adep.dwiatmoko@lipi.go.id

with stirring in a stainless-steel autoclave under a $N_2$ atmosphere at 140 °C. After the reaction for 4 h, the mixture was cooled to room temperature and then filtered for analysis of products. The final concentrations of identified products were quantitatively determined by a high-performance liquid chromatography using a Bio-Rad HPX–87H column (7.8 × 300 mm) at 35 °C equipped with a refractive index detector (Young Lin YL 9170 RID), where the mobile phase was 0.005 M sulfuric acid and the flow rate was 0.5 mL/min.

## 3 Results and Discussion

The activity of the prepared catalysts was evaluated in cellobiose hydrolysis reaction. Cellobiose, a model substrate of cellulose, is converted to glucose through hydrolysis reaction, followed by dehydration of water to form HMF in acid-catalyzed reaction. Furthermore, in the presence of acid species, HMF can be converted to organic acids, such as levulinic acid and formic acid. Polymer products and insoluble humins can be generated from these products by recombination with sugars or oligosaccharides via aldol condensation [9].

The effect of phosphoric acid concentration employed for surface modification of niobic acid was examined at a concentration ranging from 0.5 to 2 mol/L. As shown in Figure 1, the activity of the catalysts was dependent on the concentration of phosphoric acid utilized. When the concentration of phosphoric acid was 0.5 mol/L, the cellobiose conversion was 20%. Furthermore, at a higher concentration of 1 mol/L, the cellobiose conversion increase up to 39.6%. Further increase of the concentration of phosphoric acid up to 2 mol/L did not lead to the increase of the catalyst activity. Interestingly, the selectivity of glucose showed a similar tendency with the trend of catalyst activity. In the acid concentration of 0.5 mol/L, the selectivity of glucose was 36.1%, and it significantly improved up to 82.7% by increasing acid concentration to 1 mol/L. This selectivity does not significantly change by further increase of phosphoric acid concentration. In addition, the formation of HMF were not observed, which was somewhat inhibited by the surface modification of phosphoric acid on the catalysts.

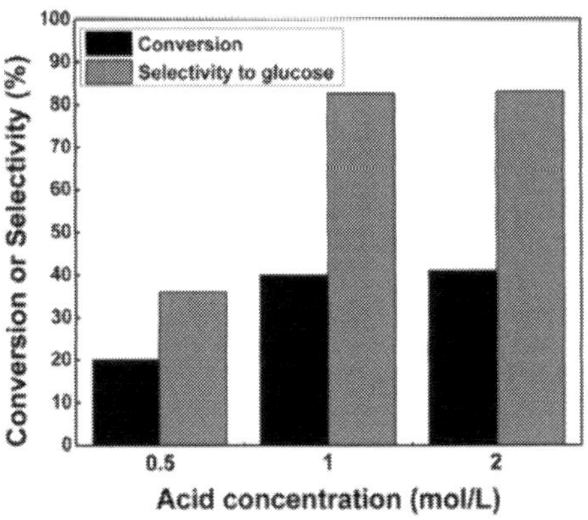


**Fig. 1.** Effect of phosphoric acid concentration on the catalytic activity of surface-modified niobic acid catalysts in the hydrolysis of cellobiose.

XRD analysis was performed to examine the effect of the surface modification on the structural change of the catalysts. Figure 2 shows the XRD patterns of the prepared catalysts, P–NA(x). Niobic acid (NA), which has an amorphous structure, shows only broad diffraction peaks at 2θ = 14.5°, 28.7°, 49.1° and 57.7°. When phosphoric acid was applied for modification with concentration of 1 mol/L or higher, these peaks was either weaken or relieved, and then, clear diffractions at 2θ = 11.8°, 14.5°, 19.6°, 22.9°, 27.8°, 27.8°, 39.7°, 44.4°, 57.2°, 60.9°, and 64.7° started to appear, indicating the formation of a crystalline phase of niobium phosphate structure occurred [6]. This result suggested that niobium phosphate was formed on the surface of niobic acid during introduction with phosphoric acid at concentration of equal to or higher than 1 mol/L.

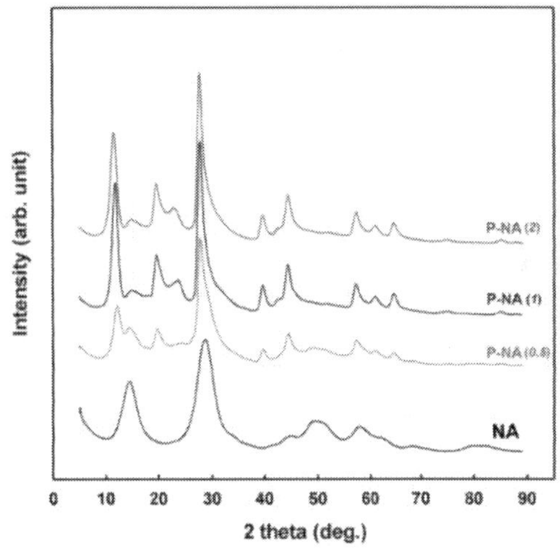


**Fig. 2.** XRD patterns of the prepared catalysts.

Figure 3 shows the XPS patterns of phosphorus species on the catalysts surface. No binding energy peak of P2p was observed on the original niobic acid (NA) due to the absence of the surface modification with phosphoric acid. XPS spectras of P2p for the surface-modified catalysts show that with an increase in concentration of phosphoric acid, the XPS intensity of the P2p peak on the modified catalysts increases. The P2p XPS spectra of the catalysts showed the presence of phosphate ions of $H_2PO_4^-$, which corresponds to a signal centered at around 133.0±1.0 eV with a full width at half maximum of 2.1 eV [10, 11], consistent with the result from XRD analysis. A small peak at the binding energy around 133.0 eV can be detected for the P-NA(0.5) catalyst. This binding energy was gradually shifted to a higher binding energy with an increase in concentration of phosphoric acid employed. A high-intensity peak was recorded at the binding energy of 133.5 eV for the P-NA(2) catalyst.

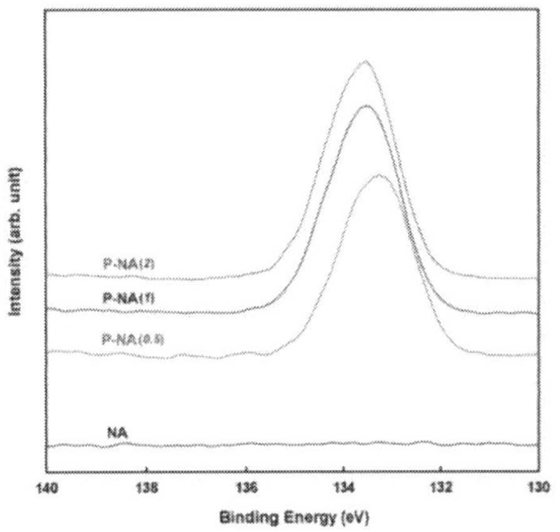


**Fig. 3.** XPS spectra characterizing P2p on the surface of the catalysts.

Furthermore, as shown in Table 1, the atomic concentration of niobium, oxygen and phosphorous on the catalysts surface as a function of phosphoric acid employed during modification was measured by XPS. The atomic ratio of phosphorous and niobium (P/Nb) was then calculated. In the absence of the surface modification, the phosphorous was not detected, so that the ratio of P/Nb was not applicable. As the concentration of phosphoric acid increased from 0.5 mol·L-1 to 1 mol·L-1, the ratio of P/Nb increased from 0.89 to 1.11. Further increase of phosphoric acid concentration up to 2 mol·L-1, did not lead to the increase of the P/Nb ratio. Based on this observation, it is suggested that the interaction of phosphate group on the catalyst surface was saturated at 1 mol·L-1 of phosphoric acid employed during the surface modification. Increasing the concentration of phosphoric acid, would not affect the number of phosphate groups attached on the catalyst surface.

**Table 1. Surface** concentration of the catalysts.

| Catalyst | Atomic concentration (%) | | | P/Nb | O/Nb |
|---|---|---|---|---|---|
| | Nb | O | P | | |
| NA | 14.76 | 39.99 | - | - | 2.71 |
| P-NA(0.5) | 10.38 | 62.37 | 9.21 | 0.89 | 6.01 |
| P-NA(1) | 9.15 | 60.50 | 10.16 | 1.11 | 6.61 |
| P-NA(2) | 10.09 | 65.93 | 10.96 | 1.09 | 6.53 |

The above mentioned results clearly demonstrated that the surface modification of niobic acid catalyst with mineral acids enhanced the catalyst selectivity to glucose. It was found that the selectivity of phosphoric acid-modified niobic acid was significantly higher than that of obtained with sulfuric acid-modified catalyst. HMF and organic acids were still obtained during the reactions. It is well-known that the acidic property of the catalyst could be modified by introducing mineral acids onto its surface. Thus, the possible explanation of the different results obtained using $H_3PO_4$–$Nb_2O_5$ and $H_2SO_4$–$Nb_2O_5$ could be related to its acid property. It was suggested that the formation of further products, such as HMF and organic acids could be related to the difference in the surface acidity of the catalysts. Based on that, we determined the acid strength of the modified catalysts by Hammett indicator. After being modified with 1 mol·L$^{-1}$ phosphoric acid, its $H_o$ value was $-5.6 < H_o \leq -3.3$. Interestingly, the $H_o$ values of the modified catalyst with 1 mol·L$^{-1}$ $H_3SO_4$ was found to be $-8.2 < Ho \leq -5.6$. This shows that the $H_2SO_4$–$Nb_2O_5$ has a stronger acid sites compared to the $H_3PO_4$–$Nb_2O_5$, which lead to the formation of other products from cellobiose hydrolysis, aside from glucose. This was also supported by the fact that sulfuric acid ($pK_a = -3$) is a stronger mineral acid than phosphoric acid ($pK_a = 2.12$).

Taking into account that the XRD analysis shows that the niobium phosphate was formed on the catalyst surface, the surface reaction which occurs between phosphoric acid and niobic acid could be described as depicted in Figure 4. This attached ions on the niobic acid surface acts as additional active and selective sites to convert cellobiose to glucose.

**Fig. 4.** Interaction between phosphoric acid and catalyst surface.

Another issue in the heterogeneous catalytic system is in regards to the catalyst stability. Since the stability of the catalyst is also an important issue in heterogeneous catalysis, we did carry out the recyclability test of the modified catalyst in our catalytic reaction. Figure 5 shows the experimental results of the recyclability of the $H_3PO_4$–$Nb_2O_5$(*2*) catalyst in the hydrolysis of cellobiose five runs. After each run, the catalyst was separated using centrifuge, and subsequently used without any treatment before being applied for the next catalytic reaction. As shown in the figure, the average cellobiose conversion was about 31.3% with the average glucose yield of 25.3%. Even if the active species might leach out of the catalyst surface, there was no significant change in the activity of the catalyst after being used for five runs. Obviously, the catalyst is a sufficiently stable and highly selective for the conversion of cellobiose under our reaction conditions.

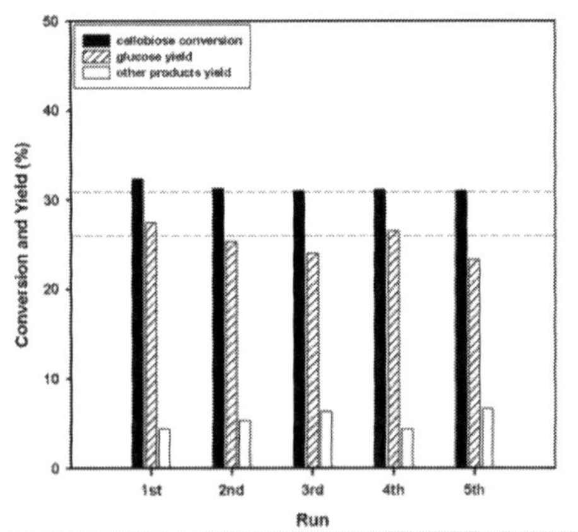


**Fig. 5.** Stability test of the prepared catalyst.

## 4 Conclusions

We have demonstrated in this work, that the surface modification of the niobic acid catalyst ($Nb_2O_5 \cdot xH_2O$) with phosphoric acid has improved the selectivity in the formation of glucose from cellobiose hydrolysis. Selectivity of glucose increased up to 82.7% with the increase of phosphoric acid concentration used for surface modification up to 1 mol·L$^{-1}$. Further increase of phosphoric acid concentration did not lead to substantial changes in both the glucose yield and selectivity. The extent of selectivity enhancement of the modified catalysts showed a good relationship with the atomic ratio of phosphorous and niobium on the surface measured by XPS. It is believed that the principal contributor to the selectivity enhancement is the presence of dihydrogen-phosphate ($H_2PO_4^-$) as an active species on the catalysts surface formed from the reaction which occurs between phosphoric acid and niobic acid, confirmed by XPS. This study shows the surface modification of niobic acid catalysts promotes the formation of valuable products from cellobiose and, further, lignocellulosic materials.

This work was done by the financial support of Korea Institute of Science and Technology at Korea Institute of Science and Technology when the author studied. Guidance and supervision of Prof. Young-Woong Suh and Prof. Dong Jin Suh are greatly acknowledged. The results have been presented by Prof. Dong Jin Suh in Europacat X.

## References

1. Tanabe, K., *Niobic acid as an unusual acidic solid material.* Materials chemistry and physics, 1987. **17**(1-2): p. 217-225.
2. Doiseau, A., F. Rataboul, and N. Essayem. *Xylose dehydration in furfural using a continuous fix-bed reactor: Insights on phosphate addition to niobic acid performances.*
3. Reche, M.T., et al., *Niobic acid nanoparticle catalysts for the aqueous phase transformation of glucose and fructose to 5-hydroxymethylfurfural.* Catalysis Science & Technology, 2016. **6**(19): p. 7334-7341.
4. Chen, Z.-h., T. Iizuka, and K. Tanabe, *Niobic acid as an efficient catalyst for vapor phase esterification of ethyl alcohol with acetic acid.* Chemistry Letters, 1984. **13**(7): p. 1085-1088.
5. Shishido, T., et al., *Brønsted acid generation over alumina-supported niobia by calcination at 1173 K.* Catalysis letters, 2009. **129**(3-4): p. 383-386.
6. Okazaki, S., et al., *The effect of phosphoric acid treatment on the catalytic property of niobic acid.* Bulletin of the Chemical Society of Japan, 1987. **60**(1): p. 37-41.
7. Dwiatmoko, A.A., Rinaldi, N., *Preparasi dan Karakterisasi Katalis Asam Padat Berbasis Niobia untuk Produksi Senyawa Glukosa.* Research Journal of Industrial Pollution Prevention Technology, 2017. **Submitted**.
8. Okazaki, S. and A. Kurosaki, *Acidic properties and catalytic activities of niobic acid treated with phosphoric acid.* Catalysis Today, 1990. **8**(1): p. 113-122.
9. Shimizu, K.-i., et al., *Effects of Brønsted and Lewis acidities on activity and selectivity of heteropolyacid-based catalysts for hydrolysis of cellobiose and cellulose.* Green Chemistry, 2009. **11**(10): p. 1627-1632.
10. Yang, F., et al., *Conversion of biomass into 5-hydroxymethylfurfural using solid acid catalyst.* Bioresource technology, 2011. **102**(3): p. 3424-3429.
11. de Pietre, M.K., et al., *H3PO4-and H2SO4-treated niobic acid as heterogeneous catalyst for methyl ester production.* Reaction Kinetics, Mechanisms and Catalysis, 2010. **99**(2): p. 269-280.

# Bioadsorbent based on water hyacinth modified with citric acid for adsorption of methylene blue in water

*Eko* Siswoyo[1,*], *Anandhitya Rhezsa* Adrian[1], and *Shunitz* Tanaka[2]

[1]Environmental Engineering Program, Faculty of Civil Engineering and Planning, Islamic University of Indonesia (UII), Jl. Kaliurang Km 14.5 Yogyakarta 55584, Indonesia
[2]Graduate School of Environmental Science, Hokkaido University, Sapporo Kitaku North 10 West 5, Hokkaido 060-0810, Japan

**Abstract.** Recently, textile wastewater contamination has been a common problem in Indonesia because many textile industries disposed their wastewater directly to the river without proper treatment. The purpose of this study was to investigate the ability of a low-cost adsorbent based on water hyacinth. The adsorbent is prepared from water hyacinth powder (WHP) and water hyacinth modified with citric acid (WHMCA) for removal of dyes methylene blue (MB) in water. In order to know the properties of the adsorbent, WHP and WHMCA were characterized by FTIR and SEM, respectively. The effect of dose of adsorbent (10 – 100 mg), pH condition (2 – 10), contact time (30 min – 24 h) and initial concentration of methylene blue (100 – 600 mg/L) were investigated. Based on Langmuir isotherm adsorption model, it was found that the adsorption capacity of the adsorbent before and after modification with citric acid for MB was 261 mg/g and 320 mg/g, respectively and the optimum contact time was 2 h. The finding of the present study proved that WHMCA should be considered as a promising adsorbent in the near future.

## 1 Introduction

In the recent year, textile industries grow very fast, especially for some cities in Indonesia which known as batik industry. Indonesia has many textile industries from micro-scale to macro-scale, and each industry produces textile wastewater consisting of various dyes. This problem will not be solved in short period, because of increasing demand is always followed by the increase in production. There are many ways to treat textile wastewater, however, high cost and limit of technology make producers ignore to treat their wastewater. Disposal of wastewater containing dyes compounds into water sources has caused serious environmental impact because dyeing effluent will deplete the dissolved oxygen content in water, inhibit sunlight from reaching to the water sources and also contaminate the living organism in the water [1].

In this study, a low-cost adsorbent was developed based on Water hyacinth to adsorb one of the basic dyes, methylene blue (MB) in water. In recent years, there is only few researcher concerning on adsorption with water hyacinth. In many places, it is very easy to find water hyacinth and people usually assume that water hyacinth is a disturber plant, therefore it is just usually disposed of as waste. There are some studies on the using of water hyacinth as an adsorbent, such as to adsorb copper [2], oil [3] and ammonia [4].

In the present study, water hyacinth will be modified with citric acid. Cellulose cell in water hyacinth will bind carboxyl groups of the citric acid. The low-cost citric acid has been used as a modification agent in order to attach free carboxyl groups onto the hydroxyl of polysaccharide matrix of cellulosic waste materials [8].

In this present research, the water hyacinth was modified with citric acid in order to improve its adsorption capacity for cationic dyes [1]. Water hyacinth was dried and crushed into powder, named water hyacinth powder (WHP). After that WHP was modified by citric acid, named water hyacinth modified citric acid (WHMCA). By this modification, some functional groups of citric acid such as carboxylic acid is expected can increase the adsorption capacity of the adsorbent [5], [9] and [10].

In this study, WHMCA is used for treat methylene blue. Further, WHMCA can be used for treat others pollutant which has positive charge. The negative charge of WHMCA can be used to attract cationic particles. It is possible that WHMCA can treat heavy metal which has positive charge because WHMCA has carboxyl groups.

## 2 Materials and Methods

### 2.1. Preparation of adsorbent

Water hyacinth was dried and crushed into powder, named water hyacinth powder (WHP). Then, WHP was soaked into 1,3 M citric acid solution (4 ml citric acid for 1 g WHP) and stirred for 2 h following [5]. The slurry was dried in oven for 24 h. After that, it was dried on 150°C by using a furnace for 30 minutes, neutralized with

* Corresponding author: eko_siswoyo@uii.ac.id

distilled water and NaOH until pH becomes 7. After neutralization, the slurry was dried for 24 h and crushed into powder, named water hyacinth modified by citric acid (WHMCA).

WHMCA was characterized by FTIR to know the amount of the functional groups and to know the modified was successfully conducted or not. Beside FTIR, WHMCA was also characterized by SEM to know the visualization of the pore and the surface of the adsorbent.

### 2.1. Process of adsorption

The effect of dosage of the adsorbent, pH condition, contact time and initial concentration of methylene blue was investigated in this section. Firstly, the various dosage is done to obtain the optimum amount of adsorbent. Various 10, 25, 50, 75 and 100 mg of WHMCA was tried to adsorb 500 ppm methylene blue in pH 6 with 2 hours contact time. Second, various pH condition (2, 4, 6, 8, and 10) was studied to adsorb 500 ppm of methylene with the optimum dosage in pH 6 and 2 hours contact time.

Third, various of contact time from 30 min, 1 hour, 2 hours, 4 hours, 6 hours, 15 hours, 18 hours and 24 hours was tried to adsorb 500 ppm of methylene blue with optimum dosage and optimum pH condition. Finally, several kinds of initial concentration of methylene blue were investigated in order to get the maximum adsorption capacity. The initial concentrations of methylene blue were 100, 200, 300, 400, 500 and 600 ppm. The spectrophotometer devices were utilized with 665 nm in order to know the concentration of methylene blue in the water.

## 3 Results and Discussion

### 3.1 Properties of adsorbent

The result of SEM image of the adsorbents is shown in Figure 3.1. It is very difficult to identify the effect of citric acid activation on adsorbent by using SEM image, however, it can be seen from the figure that surface area of the adsorbent is probably higher after activation process.

In order to know the presence of functional groups in the adsorbent, FTIR analysis was utilized and the result is shown in Figure 3.2. WHMCA and WHP were analyzed by FTIR, Jasco FT/IR-4100, and SEM, JEOL JSM-636OLA. The FTIR spectra indicate that there are new functional groups in WHMCA. There are two new peaks in 1717 cm-1 and 1620 cm-1, this data shows that carboxyl groups appear after modification and indication that the modification has succeed. Different with FTIR spectra, the result of SEM analysis cannot be analyzed because of the irregular form of the adsorbent. This condition makes the adsorbent cannot be compared withother adsorbents.

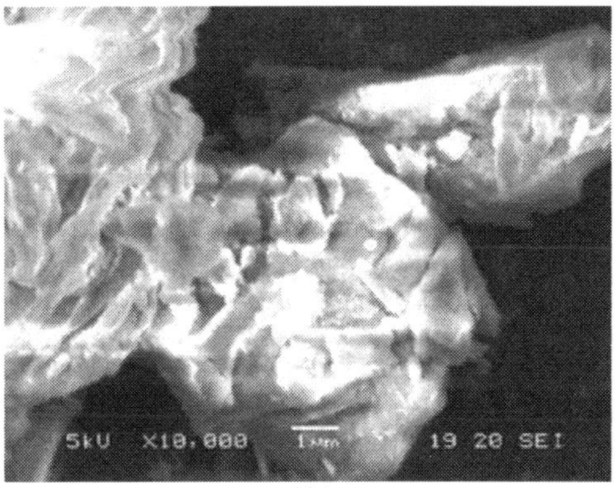


(A)

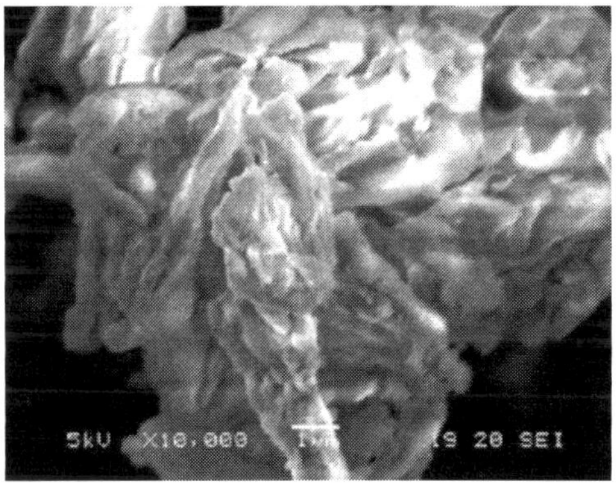


(B)

**Figure 3.1** SEM image of adsorbent (A. WHP and B. WHMCA)

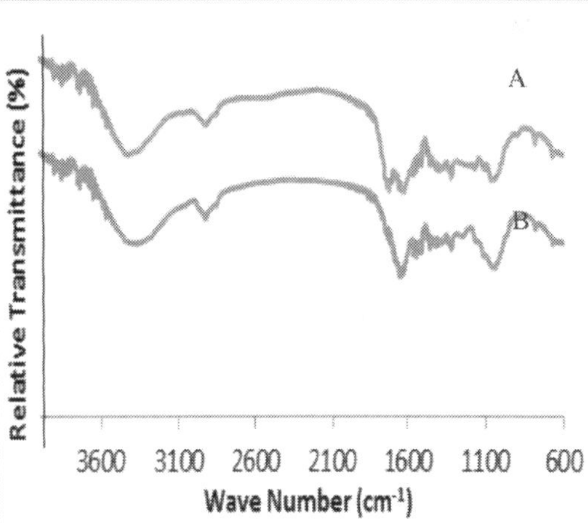


**Figure 3.2** FTIR spectra of adsorbents (A) WHMCA and (B) WHP

### 3.2 Effect of adsorbent mass on the adsorption of methylene blue

Effect of adsorbent dosage was studied to find out the optimum adsorption capacity for the further experimental test. The removal rate of methylene blue is shown in Fig. 3.3. It was clear that removal rate increased for both initial concentrations (250 and 500 ppm) from 10 to 50 mg. After 50 mg of adsorbent, there are no changes in removal rate. Moreover, WHMCA has high capability to remove methylene blue because it has 94% of removal for 500 ppm initial concentration of methylene blue.

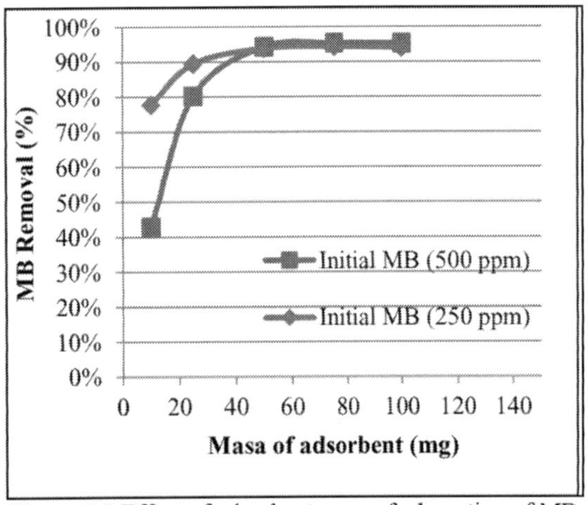


**Figure 3.3** Effect of adsorbent mass of adsorption of MB

### 3.3 Effect of stirring time on the adsorption of methylene blue

It was clear from Fig. 3.4 that optimum contact time for removal of 500 ppm of methylene blue was 30 minutes for WHMCA with 94% of removal efficiency. After 30 minutes of shaking, the increase of methylene blue removal is not significant. This result indicated that the adsorbent reached equilibrium condition in about 30 minutes which is similar to some other adsorbent materials [12] and [13]. Therefore, this adsorbent is suitable for the big scale of wastewater treatment plant considering the energy consumption [5].

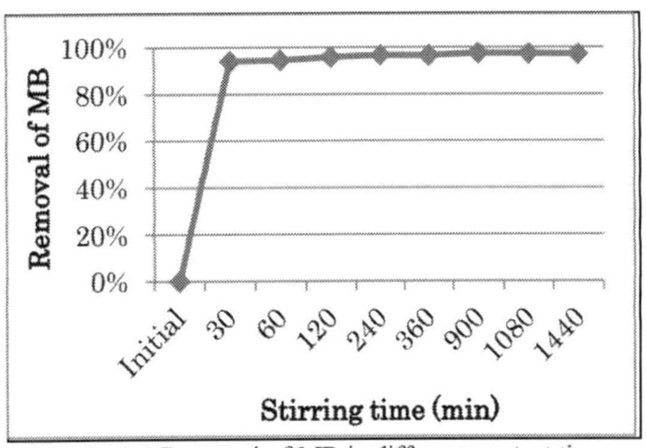


**Figure 3.4** Removal of MB in different contact time

### 3.4. Effect of pH on the adsorption of methylene blue

In adsorption process, pH is one of the important factors. The pH of the solution affected the surface charge of the adsorbent, the dissociation of functional groups on the active sites of the adsorbent, and also the structure of the dye molecule [1] and [11]. Fig. 3.5 shows that for methylene blue adsorption, pH condition does not have a significant influence. This evidence is similar to the previous study [6] where the adsorption of dyes for different pH resulted in similar adsorption capacity. The lowest removal is at pH 2 because in low pH, the surface of adsorbent is surrounded by hydrogen ion and disturb methylene blue adsorption process [7]. Moreover, the main functional groups in WHMCA used to adsorb methylene blue is carboxyl (COO-). In low pH, carboxyl is protonated into COOH that have weak adsorption to methylene blue.

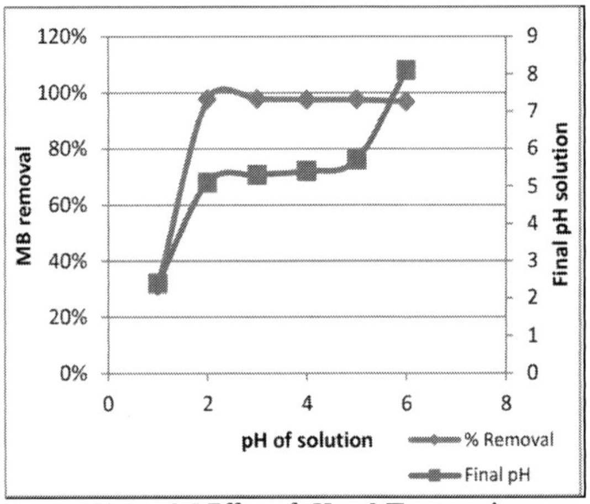


**Figure 3.5** Effect of pH on MB removal

### 3.5 Adsorption isotherm

By various initial concentrations, it could be determined isotherm model for methylene blue adsorption. Langmuir and Freundlich models were applied in order to know adsorption process of methylene blue and the result can be seen from Table 3.1.

**Table 3.1.** Isoterm Models Between WHP and WHMCA

| Model Isoterm | Constants | WHP | WHMCA |
|---|---|---|---|
| Langmuir | $K_L$ | 0,0704 | 0,081 |
| | $q_m$ (mg/g) | 261 | 320 |
| | $R^2$ | 0,996 | 0,919 |
| Freundlich | $K_F$ | 37,5 | 42,8 |
| | N | 2,414 | 2,068 |
| | $R^2$ | 0,877 | 0,747 |

Data shows that the adsorbent is fit to Langmuir model. Negative charge from citric acid in WHMCA will bind positive charge of methylene blue, and principally it will make monolayer because of electrostatic repulsion between methylene blue particles and make electrostatic attraction between methylene blue particles and WHMCA. From Table 3.1, it was clear that the maximum capacity of WHMCA is higher than water hyacinth powder without modification (WHP). Data of adsorption capacity (qm) proved that the modification was successfully conducted to increase adsorption capacity.

## 4 Conclusion

Bio adsorbent based on Water hyacinth was successfully modified by citric acid to remove methylene blue in the water. WHMCA showed high adsorption capacity towards methylene blue, with maximum capacity up to 320 mg/g and fitted with Langmuir model. The presence of some functional groups in the adsorbent is considered as an important factor in the adsorption process. This study proved that citric acid modification in water hyacinth could improve the adsorption capacity of methylene blue in water, therefore the bio adsorbent based on Water hyacinth could be considered as an alternative adsorbent instead of activated carbon.

## References

1. Guo, J., Li, B., Liu, L., and Lv, Kangle. 2014. *Removal of Methylene Blue From Aqueous Solution by Chemically Modified Bamboo*. Chemosphere 111, pp. 225–231.
2. Zheng, J. C., 2009. Feng, H. M., Lam, M. H. W., Lam, P. K. S., Ding, Y. W., and Yu, H. Q. *Removal of Cu(II) in Aqueous Media by Biosorption Using Water Hyacinth Roots as a Biosorbent Material*. Journal of Hazardous Materials 171, pp. 780–785.
3. Faisol. A., Afrizal, R. and Sari, S.R. 2008. Preparation of Oil Adsorbent based on Water Hyacinth (Pembuatan Oil Adsorbant dari Eceng Gondok). Jurnal Teknik Kimia, No. 4, Vol. 15. pp. 44-49.
4. Zaman, B. and Sutrisno, E. 2006. Absorption of Amonia by Water Hyacinth in Medical Wastewater (case study of Panti Wilasa Hospital, Semarang) *(Kemampuan Penyerapan Eceng Gondok terhadap Amoniak dalam Limbah Rumah Sakit Berdasarkan Umur dan Lama Kontak (Studi Kasus : RS. Panti Wilasa, Semarang)*. Jurnal PRESIPITASI Vol.1 No.1 pp. 49-54.
5. Siswoyo E, Endo N, Mihara Y and Tanaka S., 2014. *Agar-encapsulated adsorbent based on leaf of Platanus sp. to adsorb cadmium ion in water*. Water Science and Technology, 70.1, pp. 89–94.
6. Sarioglu, M. and Atay, U. A. 2006. *Removal of Methylene Blue by Using Biosolid*. Global NEST Journal, Vol. 8, No. 2, pp. 113-120.
7. Han, Runping. 2006. *Removal of methylene blue from aqueous solution by chaff in batch mode*. Journal of Hazardous Materials B137, pp. 550–557.
8. Reddy, D. H. K., Seshaiah, K., Reddy A. V. R. and Lee, S. M. 2012. *Optimization of Cd(II), Cu(III) and Ni(II) Biosorption by Chemically Modified Moringa oleifera Leaves Powder*. Carbohydrate Polymers 88, pp. 1077-1086.
9. Li, Z., Teng, T. T., Alkarkhi, A. F. M., Rafatullah, M. and Low, L. W. 2013. Chemical modification of Imperata cylindricaleaf powder for heavy metal ion adsorption. Water Air Soil Pollution 224:1505, pp. 1–14.
10. Low, K. S., Lee, C. K. and Mak, S. M. 2004. Sorption of copper and lead by citric acid modified wood. Wood Science and Technology 38, pp. 629–640.
11. Siswoyo, E. and Tanaka, S. 2013. Development of eco-adsorbent based on solid waste of paper industry to adsorb cadmium ion in water. Journal of Clean Energy Technology 1 (3), pp. 198–201.
12. Siswoyo, E., Mihara, Y. and Tanaka, S. 2014. Determination of key components and adsorption capacity of a low cost adsorbent based on sludge of drinking water treatment plant to adsorb cadmium ion in water. Applied Clay Science Vol. 97-98. pp. 146-152.
13. Siswoyo, E., Firachmatika, A. and Kautsar, R.B. 2016. Removal of Cu (II) in Water by Using Adsorbent Based on Volcanic Ash of Mount Kelud in Indonesia. International Journal of Environmental Science and Development, Vol. 7 No. 9. pp. 657-660.

# Wet impregnation of silver oxide on Lampung natural zeolite as an adsorbent to produce oxygen-enriched air using PSA technique

*Adenia Gita* Dianty, *Andrea Rizky Sabrina* Harahap, *Claudia* Harfian, *Setijo* Bismo*

*Chemical Engineering Department, Universitas Indonesia, Kampus Baru UI Depok, Depok 16424, Indonesia*

**Abstract.** High purity oxygen can be used for various things. Oxygen purification method to be applied on this research is the Pressure Swing Adsorption (PSA) technique. The adsorbent that would be used is a natural zeolite, namely ZAL (Zeolite Alam Lampung). Natural zeolite has non-polar properties, so it will adsorb gas with high quadruple moment, which is nitrogen. The varied variable is the size of adsorbent and the concentration of $H_2SO_4$. The sizes are 18-35 mesh, 35-60 mesh, and 60-100 mesh. While the $H_2SO_4$ concentration are 1M, 2M and 3M. Size of ZAL, and the concentration of sulfuric acid ($H_2SO_4$) were varied to get the optimum value. The adsorbent will be activated in aqua demine, $H_2SO_4$, NaOH, and through a calcination process. The purpose of this pre-treatment was to remove the impurities and to increase the specific surface area of zeolite. Moreover, ZAL will also be modified by wet impregnation technique using $AgNO_3$ solution with 1%-wt loading nominal. The morphology, composition, and crystal phase were characterized by BET, scanning electron microscopy with energy dispersive X-ray spectroscopy (SEM-EDX). The result of the adsorption process was analyzed by using GC (Gas Chromatograph). ZAL 35-60 mesh 1M showed the best performance on adsorbing nitrogen, with its lowest peak at 60356 μV. The result of this research suggested that ZAL itself has the main role on adsorbing nitrogen, while $Ag_xO$ did not give any significant effect

## 1 Introduction

Oxygen is a vital chemical element that has an important role, either for a human metabolic system as well as for other purposes. The utilization of oxygen will increase for high purity oxygen. High purity oxygen can naturally be used widely, such as in chemical processes, medical applications, combustion processes, bleaching in the paper industry, wastewater treatment, fuel cells, and also to produce ozone. Oxygen can be obtained from air. However, air contains a variety of gases which is, oxygen, nitrogen, argon, and others. In this study, the process of purification of oxygen from air using PSA (Pressure Swing Adsorption) technique will be examined. PSA technique produced high purity molecules with relatively low cost. The adsorption process that applies on PSA technique is based on the pressure different, in which the gas molecules will be bound to the adsorbent. The use of PSA to separate oxygen from air was designed so that the nitrogen is adsorbed, while oxygen passes. The process of separating oxygen from the air PSA techniques will rely on the use of mineral aluminosilicate, known as zeolite to be a selective adsorbent of nitrogen [6].

Zeolite is a mineral that has a very unique crystal structure. Zeolite have a wide range of uses, as catalysts, ion exchangers and as absorbers of organic compounds. Natural zeolite as an adsorbent capable of capturing nitrogen to enrich oxygen content in air. Therefore, this study will focus on the preparation of natural zeolite as well as its activation especially ZAL (Natural Zeolite containing Clinoptilolite as major substituent, from Sidomulyo, Lampung–Indonesia) as an adsorbent. The unique crystal structure of the zeolite allows for the modification of zeolites as needed. Natural zeolite from Lampung, Indonesia has wide surface area and good thermal stability, and it is abundant [7]. ZAL will be modified with silver oxide ($Ag_xO$) to improve the properties of ZAL as an adsorbent so that it will increase adsorption and produce high-purity oxygen [8]. The modification method that used is wet impregnation. Wet impregnation is a preparation technique with the objection to filled ZAL pores with $Ag_xO$ to broaden the adsorption surface area of ZAL, by soaking the ZAL with aqueous solution of $AgNO_3$ [3].

## 2 Methodology and Results

### 2.1 Methodology/Experimental

Three sizes of ZAL was used, which was the 18 – 35 mesh, 35 – 60 mesh, and 60 – 100 mesh. Beforehand, the ZAL sizes were uniformed by milling and shaker. First, the ZAL was washed thoroughly with aqua demine water, then dried in oven at 110°C for 2 hours. Second, the ZAL was pre-treated with 1 mole/L, 2 mole/L and 3 mole/L of $H_2SO_4$ and 1 mole/L NaOH for respectively an hour to remove impurities, enlarge pore volume and surface area

* Corresponding author: sbismo@che.ui.ac.id

https://doi.org/10.1051/matecconf/201815401013

of ZAL [2]. Before the impregnated process, ZAL were washed again with aqua demine water until the pH was neutral, then dried in the oven for 2 hours at 110°C. Then, the treated ZAL was impregnated in aqueous solution of $AgNO_3$ in an agitator for 12 hours with the loading percentage of 1%-wt. Lastly, $Ag_xO$ /ZAL go through a calcination process at 350°C for 1.5 hours to release the $NO_2$.

Specific surface area and pore volume was determined by BET method using nitrogen adsorption at the boiling temperature of liquid nitrogen. ASAP 2020 by Micromeritics was used for BET instrument. Before measurements, samples were degassed at 300 °C. Surface morphology of $Ag_xO$ /ZAL were observed by scanning electron microscopy (SEM) and surface composition analysis was carried out by energy dispersive X-ray spectroscopy (EDX). The magnitude for SEM was x5,000 and the voltage was 20 kV. The accelerating voltage for EDX was 20 kV and the energy range for EDX was between 0-20 keV.

The adsorption activity test of ZAL was investigated using PSA technique, on a packed bed reactor (PBR) made out of stainless steel. Air from the environment was compressed with a compressor through a heater with the temperature of 75°C – 95°C, after that the air went through a silica gel bed to remove its water content. Then, the PSA process occur for about 30 minutes with the pressure of 4 bar. The gas sample was taken at every 5 minutes for each running process, in which it was then analyzed by Gas Chromatography (GC) to know the composition and the intensity of every component that was contained within.

### 2.2 Results and Discussions

The BET characterization was done to three ZAL samples that has the best result after GC analysis, which was ZAL 35 – 60 mesh 1 mole/L, ZAL 60 – 100 mesh 1 mole/L, and ZAL 60 – 100 mesh 2 mole/L. The purpose of the BET characterization was to identify the pore volume, so the amount of solid $AgNO_3$ that needed to be used can be known to obtain the 1%-wt impregnation. The Table 1 below, shows the result of BET characterization and the difference of pores volume and specific surface area before and after the pre-treatment process for the 60 – 100 mesh ZAL. As we can see on the Table 1 below, the pore diameter, surface area and pore volume of each ZAL increased. The increase was caused by the removal of impurities within ZAL pores because of the activation by aqua demine, sulfuric acid, and also natrium hydroxide. The increase on pore diameter, pore volume and surface area will also increase ZAL capability on adsorbing nitrogen and produce oxygen enriched-air.

**Table 1.** BET Characterization of ZAL

| Paramet-er | ZAL 35-60 mesh 1 mole/L | ZAL 60-100 mesh 1 mole/L | | ZAL 60-100 mesh 2 mole/L | |
|---|---|---|---|---|---|
| | After | Before | After | Before | After |
| Pore diameter (nm) | 11.047 | 9.105 | 11.000 | 9.105 | 11.138 |
| Surface area ($m^2/g$) | 62.849 | 52.938 | 51.850 | 52.938 | 64.373 |
| Pore volume ($cm^3/g$) | 0.117 | 0.091 | 0.099 | 0.091 | 0.127 |

The impregnation was done to the three selected ZAL in agitator for 12 hours with a respectable amount of $AgNO_3$ used. The wet impregnation turned ZAL color from pure white into golden yellow [4]. The color change in zeolite after the impregnation process can be seen on the Table 2 below. This was caused by the formation of silver clusters within ZAL pores and surfaces [5]. The amount of $Ag_xO$ that impregnated within the zeolite pores and surface correspond to the color change. The more AgxO amount that got impregnated, will caused the zeolite to turn into dark yellow [4]. Based on Table 2, we can see that $Ag_xO$/ZAL 60 – 100 mesh 1 mole/L has the darkest color of all, then it followed by $Ag_xO$/ZAL 35 – 60 mesh 1 mole/L, and after that $Ag_xO$/ZAL 60 – 100 mesh 2 mole/L. A calcination was done to remove water molecules, impurities, and to form a bond between the silver oxide and ZAL. Then, to know the success of the impregnation process, EDX, and XRF characterization will be done to know the detailed composition of the ZAL samples.

**Table 2.** Color Changes of ZAL after Impregnation

| ZAL | ZAL 35-60 mesh 1 mole/L | ZAL 60-100 mesh 1 mole/L | ZAL 60-100 mesh 2 mole/L |
|---|---|---|---|
| Before Impregnation |  |  | 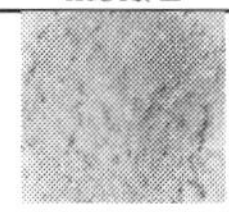 |
| After Impregnation |  |  |  |

**Table 3.** EDX characterization of AgO/ZAL

| Composition | %-atom | | |
|---|---|---|---|
| | After Impregnation | | |
| | 35 - 60 mesh 1 mole/L | 60 - 100 mesh 1 mole/L | 60 - 100 mesh 2 mole/L |
| O | 56.57 | 61.85 | 42.82 |
| Na | 1.60 | 1.37 | 1.31 |
| Al | 5.38 | 5.64 | 6.50 |
| Si | 34.90 | 29.83 | 45.33 |
| K | 1.22 | 1.01 | 2.07 |
| Fe | - | - | 1.50 |
| Ag | 0.33 | 0.30 | 0.47 |

| Composition | %-atom | | |
|---|---|---|---|
| | After Impregnation | | |
| | 35 - 60 mesh 1 mole/L | 60 - 100 mesh 1 mole/L | 60 - 100 mesh 2 mole/L |
| O | 56.57 | 61.85 | 42.82 |
| Na | 1.60 | 1.37 | 1.31 |
| Al | 5.38 | 5.64 | 6.50 |
| Si | 34.90 | 29.83 | 45.33 |
| K | 1.22 | 1.01 | 2.07 |
| Fe | - | - | 1.50 |
| Ag | 0.33 | 0.30 | 0.47 |

Surface morphology and composition of ZAL for before and after impregnation process. As seen on both Table 3 and Table 5, there were many impurities in the surface area of ZAL before impregnation and calcination. Based on the analysis and characterization of ZAL before and after impregnation, it showed that the ratio for Si/Al was increased. Siloxane is an organosilicon compound which has Si-O-Si chain. Si-O itself is a destructed framework from acid cleaning phase [1]. The increased ratio of Si/Al was caused by dealumination process, that made the ZAL became hydrophobic, more stable under high temperature and high acid environment and can adsorb non-polar molecules. $Ag_xO$ /ZAL will have a high selectivity to nitrogen because nitrogen is a non-polar molecule that has high dipole moment. As we can see on Table 5, by comparing ZAL images from SEM characterization, we can see that the zeolite crystals look clearer and more define. This due to the calcination process after impregnation, that makes all of the impurities that were trapped inside the zeolite pores removed.

**Table 4.** Si/Al of ZAL samples

| Sample | | Si/Al Ratio | %Change |
|---|---|---|---|
| 35 - 60 mesh 1 mole/L | Activated | 6.584 | 1.494 |
| | Impregnated | 6.486 | |
| 60 - 100 mesh 1 mole/L | Activated | 6.111 | 13.491 |
| | Impregnated | 5.286 | |
| 60 - 100 mesh 2 mole/L | Activated | 6.086 | 14.567 |
| | Impregnated | 6.973 | |

**Table 5.** SEM characterization of ZAL samples

| ZAL | ZAL 35-60 mesh 1 mole/L | ZAL 60-100 mesh 1 mole/L | ZAL 60-100 mesh 2 mole/L |
|---|---|---|---|
| Before Impregnation | 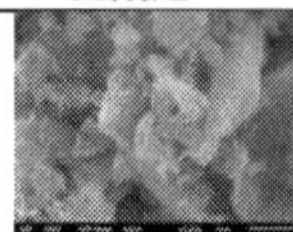 | 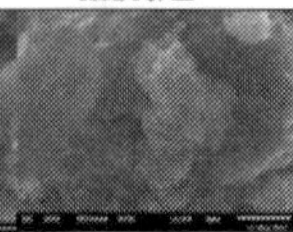 | 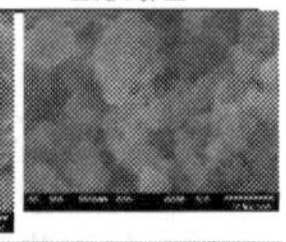 |
| After Impregnation |  |  | 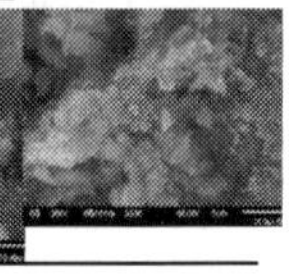 |

Table 6 showed the loading percentage of $Ag_xO$ for three samples of ZAL after the impregnation process. The values are only reaching half of the designated values, which is only 0.5 %-wt $Ag_xO$. This may be happened because not all of the Ag latch through the ZAL pores and surfaces. While the Si/Al ratio showed that the impregnated 35 – 60 mesh 1 mole/L ZAL has the highest value, and this is quite a different result from the EDX characterization one. The XRF characterization take measurement from 5 grams of ZAL, and average it, while EDX characterization only take measurement from only a small portion of ZAL sample. So it is more accurate to look at XRF result of characterization. The high ratio of Si/Al will make ZAL become more responsive to nitrogen and help the adsorption process become better. Based on the XRF characterization as seen on table 6, the impregnated ZAL sample that have the highest Si/Al ratio is the 35 – 60 mesh 1 mole/L. This impregnated ZAL samples will have a better absorbency level rather than the other two samples. The GC analysis result will be show on Figure 2, and will be explain afterwards. The ZAL sample that has the highest amount of $Ag_xO$ impregnated is the 60 – 100 mesh 1 mole/L sample, this support the statements that the more $Ag_xO$ impregnated in the zeolite pores or surface will make the color of the zeolite become darker.

**Table 6.** XRF characterization of ZAL samples

| Element | Parameter | | |
|---|---|---|---|
| | 35 - 60 mesh 1 mole/L | 60 - 100 mesh 1 mole/L | 60 - 100 mesh 2 mole/L |
| Ag (%-weight) | 0.25 | 0.39 | 0.30 |
| Si (%-weight) | 83.7 ±0.3 | 80.7 ±0.4 | 81.3 ±0.1 |
| Al (%-weight) | 10.2 ±0.3 | 10.6 ±0.4 | 10.2 ±0.5 |
| Si/Al Ratio | 8.20 | 7.61 | 7.97 |

The Figure 1 showed the graph that correlating the time of taking the sample versus the peak height that represents the intensity of nitrogen component analyzed in the GC. The ZAL that were used to adsorbed nitrogen were the 35 – 60 mesh 1 mole/L, ZAL 60 – 100 mesh 1 mole/L, and 60 – 100 mesh 2 mole/L. ZAL that had been pre-treated had the capability to adsorb nitrogen, which shown on the graph that for some sample the peak height has a slight decreased. The reason on why the graph tends to decreased to a low point and increased again was because the ZAL was already saturated with nitrogen molecules, in which the adsorption rate became lower. As shown on the Figure 2, the best ZAL adsorbent that has high stability of adsorbing nitrogen was the 60 – 100 mesh 2 mole/L [4]. While the others still tend to fluctuates slightly. But, ZAL 35 – 60 mesh 1 mole/L has the lowest peak number that correlates that the adsorbent has the best capability to adsorb nitrogen.

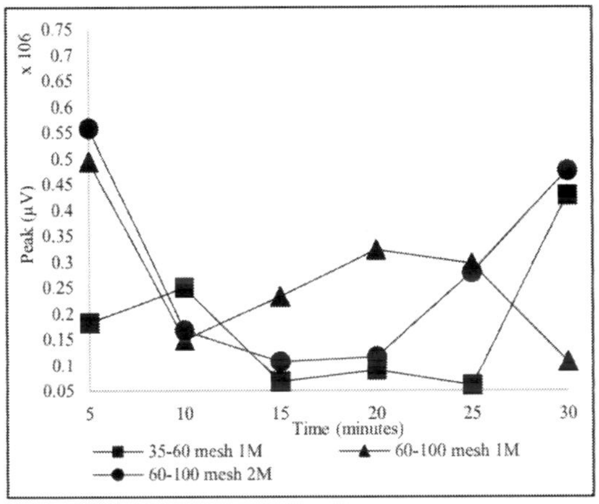


**Figure 1.** Time vs Peak for Pre-treated ZAL

The GC analysis was also done to impregnated $Ag_xO$ /ZAL, and can be seen on Figure 2. The figures showed graphs that compared the ZAL before and after impregnated by $Ag_xO$. The $Ag_xO$/ZAL that have the highest performance on adsorbing nitrogen is the 35 – 50 mesh 1 mole/L. This may be caused by the low Ag impregnated within the zeolite sample, and also the high level of Si/Al ratio. Si/Al ratio have a significant effect on zeolite as explained before. The $Ag_xO$ /ZAL adsorbent shown a low performance to adsorb nitrogen, as the peak are higher rather than the not impregnated ZAL. This phenomenon not only shown on $Ag_xO$ /ZAL 35 – 60 mesh, but also with the 60 – 100 mesh. The increased graph maybe caused by the impregnation of $Ag_xO$, and it may plug the pores of ZAL that making it hard for nitrogen molecules to be adsorbed. Based on this, the impregnation process of $Ag_xO$, decreased the performance of ZAL. The fluctuate graph in almost every end of the data was caused by the saturated condition on the adsorbent, which means that the adsorbent can no longer adsorb as much as nitrogen as before. As can be seen on Figure 4, the best adsorbent is ZAL 35 – 60 mesh 1M. That was caused by its pore volume. The size of ZAL also corresponding to how better the adsorbent works. The smaller ZAL size has the biggest pore volume and also specific surface area, which helped the $Ag_xO$ to stick to the pore volume and making it a better adsorbent.

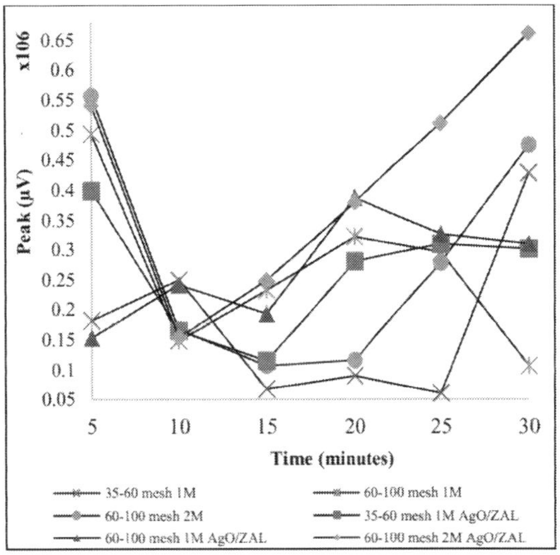


**Figure 2.** Time vs Peak for Impregnated and Un-impregnated ZAL

The explanation above also supported by the result of BET characterization on Table 7. As can be seen on Table 7, the pore diameter, pore volume and surface area of the impregnated $Ag_xO$/ZAL decreased. This mainly caused by the impregnation process itself, in which the ZAL pores are plugged by the silver oxide and make it harder for nitrogen molecules to be adsorb within the pores or surfaces of the adsorbent.

**Table 7.** BET Characterization for Impregnated and Un-impregnated ZAL

| Parameter | 35 – 60 mesh 1 mole/L | | 60 – 100 mesh 2 mole/L | |
|---|---|---|---|---|
| | Activate | Impreg. | Activate | Impreg. |
| Pore diameter (nm) | 11.047 | 10.952 | 11.138 | 11.019 |
| Surface area ($m^2$/g) | 62.849 | 60.455 | 64.373 | 55.412 |
| Pore volume ($cm^3$/g) | 0.117 | 0.111 | 0.127 | 0.102 |

The authors acknowledge Universitas Indonesia and PITTA program for the financial support. The authors also thankful to Badan Pengkajian dan Penerapan Teknologi (BPPT) on PUSPIPTEK, Serpong, Fakultas Matematika dan Ilmu Pengetahuan Alam (FMIPA) and Fakultas Teknik of Universitas Negeri Jakarta for the instruments to characterize the adsorbents.

## References

1. Dianursanti, M. Delaamira, S. Bismo, Y. Muharam. *Effect of Reaction Temperature on Biodiesel Production from Chlorella vulgaris using CuO/Zeolite as Heterogeneous Catalyst.* 2nd International Conference on Tropical and Coastal Region Eco Development, Bali.(2016)

2. Ertan, A., and Ozkan, F. *$CO_2$ and $N_2$ Adsorption on the Acid (HCl, $HNO_3$, $H_2SO_4$, and $H_3PO_4$) Treated Zeolite*. Adsorption, ***11***, 151-156. (2005)
3. Haber, J., Derewinski, M., Dwigaji, S. & Ritter, G. *Proc Int. Symp. on Zeolite Catalyst,* 1995: Acta Chim et hys, Szeged. {1995)
4. Jacobs, P., Uytterhoeven, J. & Beyer, H. *Chem. Soc.*, s.l.: Faraday Trans. (1979)
5. Mara, A., Wijaya, K., Trisunaryati, W. & Mudasir, M. *Effect of Sulfuric Acid Treatment and Calcination on Natural Zeolites of Indonesia,* Yogyakarta: Universitas Gajah Mada. (2016)
6. Scott, P. *Oxygen – Pressure Swing Adsorption,* US: IChem Ltd. (2000)
7. Slamet, Ellyana, M., Bismo, S. *Modifikasi Zeolit Alam Lampung dengan Fotokatalis $TiO_2$ melalui Metode Sol Gel dan Aplikasinya untuk Penyisihan Fenol.* Jurnal Teknologi, **1**, 59-68. (2008)
8. Syafii, F., Sugiarti, S. and Charlena. *Modifikasi Zeolit Melalui Interaksi dengan $Fe(OH)_3$ untuk Meningkatkan Kapasitas Tukar Anion,* Bogor: Institut Pertanian Bogor. (2010)

# Shrinking core model of demineralization of chitin isolation from shrimp shell

*Ahmad* Fadli[1,*], *Subkhan* Maulana[1], and Drastinawati[1]

[1] Department of Chemical Engineering, Universitas Riau, Pekanbaru, Indonesia

**Abstract.** Chitin is a naturally abundant polymer and most of it was used as a surfactant and cosmetic raw materials. Chitin be able to derive from natural sources like shrimp shell waste. The aim of this research was to study the kinetic model of chitin demineralization with approach to shrinking core model (SCM) with variation of hydrochloric acid concentration. There models were used: Fluid Film Layer Diffusion, Ash Layer Diffusion and Chemical Reaction. It was began with deproteination at first with NaOH 3,5 % and continued with demineralization using hydrochloric acid at concentration 0,3 N; 0,6 N; 0,9 N and 1,2 N with variation of reaction time at 15, 30 and 15 minute. The calcium content in the product was analyzed after it was dried in the air oven. The results showed calcium concentration decrease along with increasing of hydrochloric acid concentration. It get kinetic model at concentration 0,3 N of hydrochloric acid usage with form:

$$1-\left(\frac{r_c}{R}\right)^3 = X_B$$

with the highest value of $R^2$ is 0,9555 and the smallest percentage error is 5,92%. Meanwhile with usage of hydrochloric acid at concentration 0,6 N; 0,9 N and 1,2 N were get kinetic model with form:

$$\frac{t}{\tau} = 1-3(1-X_B)^{2/3}+2(1-X_B)$$

with the highest value of $R^2$ is 0,9794.

## 1 Introduction

Potential source of fishery industry such as shrimp shell waste has been limited to feed and a lot of it has caused environmental problem. Utilization of waste for added value features two strategic approaches; both provide benefit to the producers and solve environmental problems by carrying out the principle of "Zero Waste Products" [10]. Chitin is abundant compound in nature estimated annual production reaches $10^{10}$-$10^{12}$ tons [8]. Chitin content in shrimp shell waste is about 20-50% dry weight [4]. Chitin properties are stable against chemical reactions, non-toxic and biodegradable. As the second most abundant biopolymer in the world, chitin can be found mainly in invertebrates, insects, marine diatoms, algae, fungi, and yeasts. Chitin has widely used such as mobilization of enzymes, metal adsorbents, films and fibers, anti-radar material and anti-cancer drugs. Conventionally, chitin was isolated from crustacean shells by deproteination in alkaline solution and followed by demineralization in acid solution [2]. One source of chitin is shrimp shell waste. Shrimp shell waste can be obtained from shrimp processing industry in Indragiri Hilir regency with the amount of waste about 1-3 tons/month. The main raw material of dried shrimp processing business is red shrimp (*Metapenaeus* Sp).

Demineralization is the process of removing minerals contained in a material by the addition of a low concentration of hydrochloric acid solution. This process aims to eliminate minerals contained in shrimp shell [3]. Shrimp shell generally contains 20-50% of minerals depending on the species [7]. Calcium carbonate ($CaCO_3$) is the main composition of minerals contained in shrimp shells, wherein the calcium concentration be able to determine by the complexometric method [5]. Calcium content reduction occurring in shrimp waste particle can be described using a heterogeneous kinetic model with the shrinking core model approach [9]. Wherein, the film layer and ash layer were hydrochloric acid and protein respectively. The utilization of shrimp shell waste into chitin, chitosan and its derivative products is highly potential for wider applications and improves its added value [8]. The wide range of chitin applications makes it useful to be studied for both isolation and kinetic data. Chitin isolation requires an appropriate reactor to produce high quality chitin. Reactor design also required kinetics data at any concentration, temperature and reaction time for better control and optimization of the process. Based on the description above, kinetic data which can be obtained should be used to design and optimize the demineralization reactor to get good quality chitin.

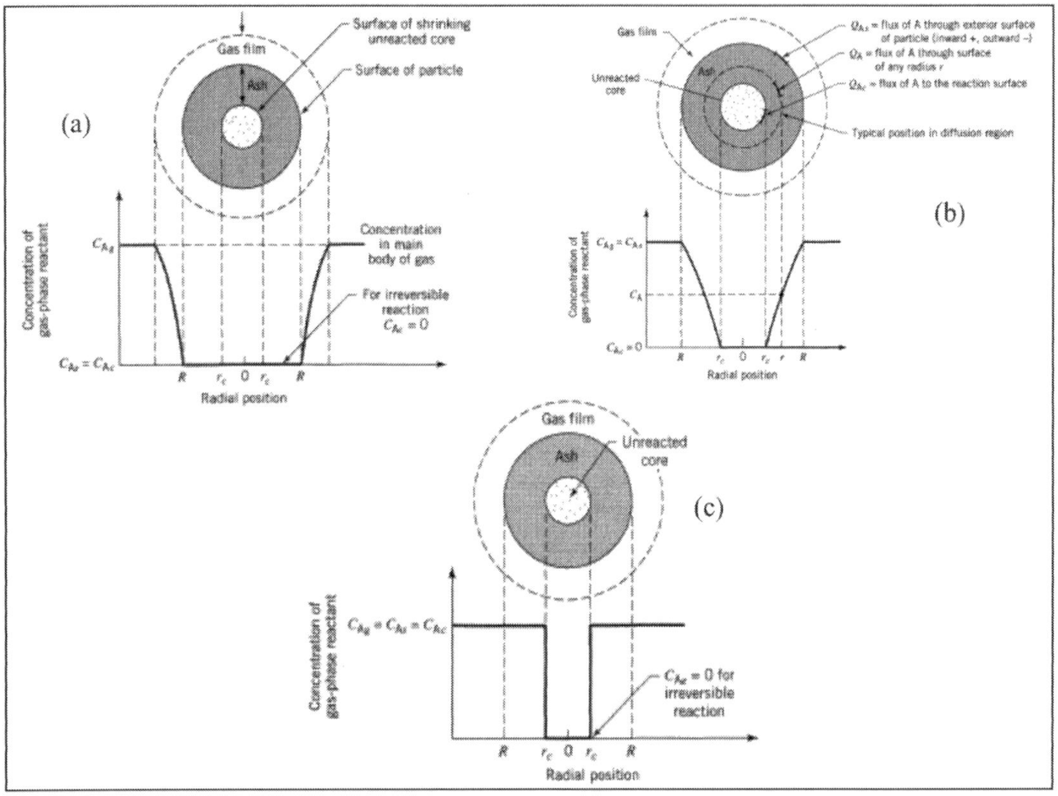


**Fig. 1.** (a) Fluid Film Controls (Model 1) (b) Ash Layer Diffusion Controls (Model 2) (c) Chemical Reaction Controls (Model 3) [6].

## 2 Materials and Methods

The shrimp shell waste was obtained from local dried shrimp industry in Riau, Indonesia. The exoskeleton of the shrimp was manually washed, dried, and sieved at 20 dan 35 mesh. Hydrochloric acid pro analysis (pa) product grade from Merck was procured and prepared into four concentrations (0.3, 0.6, 0.9, and 1.2 N). Materials for analysis using complexometric titration such as solid EDTA, EBT indicator, buffer pH 10, $HNO_3$ and $H_2SO_4$ were purchased and 0,01 M EDTA solution was prepared by dissolving 3.7724 g EDTA ($Na_2H_2Y.{}_2H_2O$) in 50 ml deionized water with constant stirring for 15 min in a 250 ml volumetric flask. When the solid crystals of EDTA were completely dissolved, the volume was made up to the mark.

## 3 Experimentation

### 3.1 Deproteination

100 g of shrimp shell was soaked and stirred in 3.5% solution of NaOH in the mass to volume ratio of 1:10 for 240 min by boiling at 65°C for deproteination process. After a given time interval, the content of the conical flask was quickly filtered using filter paper. The solid residue then washed severally with distilled water, until neutrality and its alkalinity determined by the pH-indicator paper and a pH meter (Kent EIL 7055). The deproteinized samples were dried and weighed and referred to as demineralized sample for 0 minute.

### 3.2 Demineralization

Various concentrations of hydrochloric acid solution (0.3, 0.6, 0.9, and 1.2 N) were prepared by dissolving hydrochloric acid in distilled water. The prepared shrimp shells placed in a 250 ml conical flask and quickly mixed with 0.3M hydrochloric acid solution in the mass to volume ratio of 1:10. The content of the conical flask was maintained for 15 minutes, with stirring using a magnetic stirrer. After a given time interval, the content of the conical flask was quickly injected with syringe and filtered using a filter paper. The solid residue then washed severally with distilled water, until neutrality and its acidity determined by pH-indicator paper and a pH meter (Kent EIL 7055). The demineralized samples were dried and weighed.

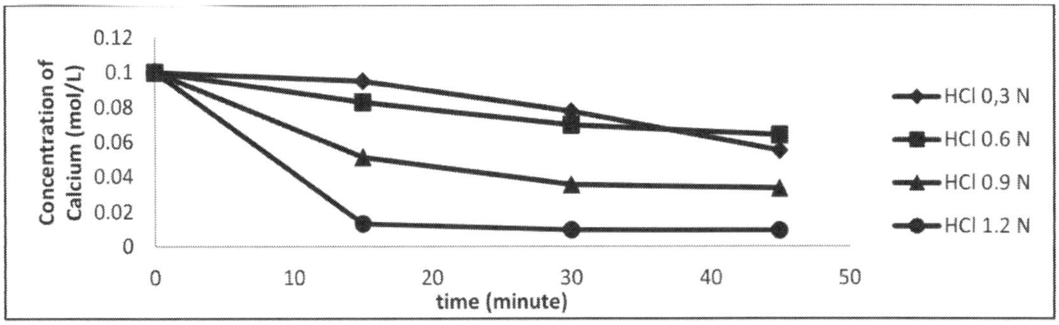


**Fig. 2.** Reduction of calcium content of demineralized shrimp shell due to hydrochloric acid treatment

These steps were then repeated in turns for other reaction times of 30 and 45 minutes. The entire procedure was then repeated for the other hydrochloric acid concentrations of 0.6, 0.9 and 1.2 N.

### 3.3 Analysis

For the determination of calcium content, 0.5 g of sample was mixed into 5 mL of sulfuric acid and 5 mL of nitric acid. In the pipette a 1 mL aliquot of the sample solution into a conical flask with added 100 mL of distilled water. In the pipette a 1 mL aliquot of the sample solution into another conical flask and add 0.1 g of EBT indicator with 1 ml of buffer solution (pH-10) and swirl the solution to dissolve the indicator. The chitin solution was titrated with the EDTA solution. The endpoint is a color change from pink/red to blue.

## 4 Kinetics modeling

The shrinking core model (SCM) was considered as it models fluid-particle reactions [20]. The models for the various SCM control mechanism are:

1. For fluid-film diffusion control (FFDC)

$$\frac{t}{\tau} = X_B \quad (1)$$

2. For ash layer diffusion control (ALDC)

$$\frac{t}{\tau} = 1 - 3(1 - X_B)^{2/3} + 2(1 - X_B) = M_2 \quad (2)$$

3. For chemical reaction control (CRC)

$$\frac{t}{\tau} = 1 - (1 - X_B)^{\frac{1}{3}} = M_3 \quad (3)$$

Where τ: time for complete reaction, $X_B$: conversion of calcium The right hand side of equation 1, 2 and 3 were plotted against t and the fit $R^2$ were computed.

## 5 Result and Discussion

Figure 1 presents the progression in the concentration of calcium in the demineralized shrimp shells with time for the various acid concentrations considered. Figure 2 indicates a progressive drop in the concentration of the calcium for all hydrochloric acid concentrations along with time. Reduction of calcium content and increment of conversion or amount of calcium removed directly proportional due to increasing of hydrochloric acid concentration. For instance, after 45 minutes of demineralization the conversion of calcium for 0.3N and 1.2 N hydrochloric acid solutions were 45% and 90,74% respectively. Figures 3-6 below present the fitting of the kinetic data to the shrinking core model.

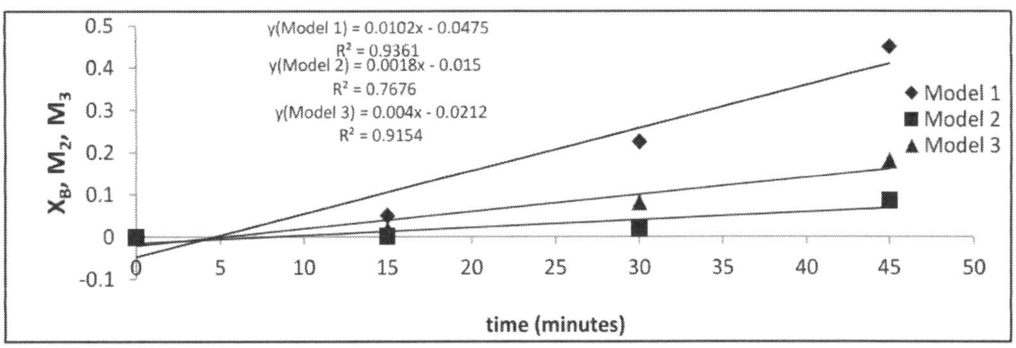


**Fig. 3.** SCM for shrimp shell demineralization using 0.3N hydrochloric acid solution

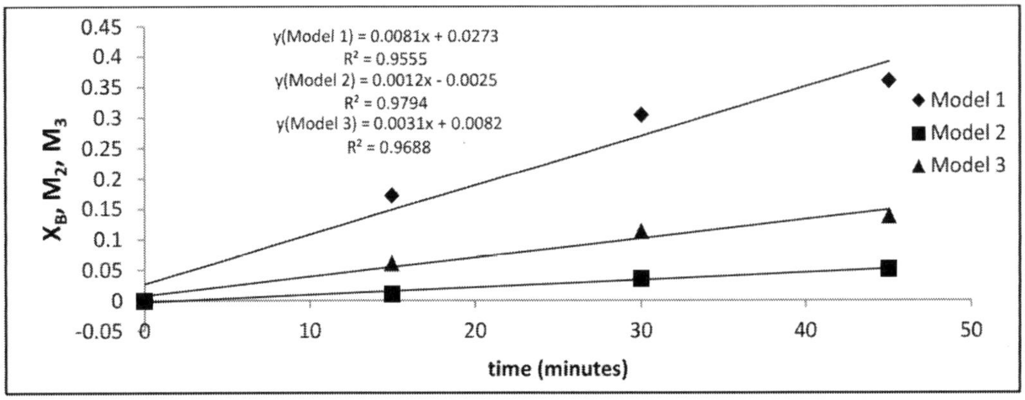


**Fig. 4.** SCM for shrimp shell demineralization using 0.6N hydrochloric acid solution

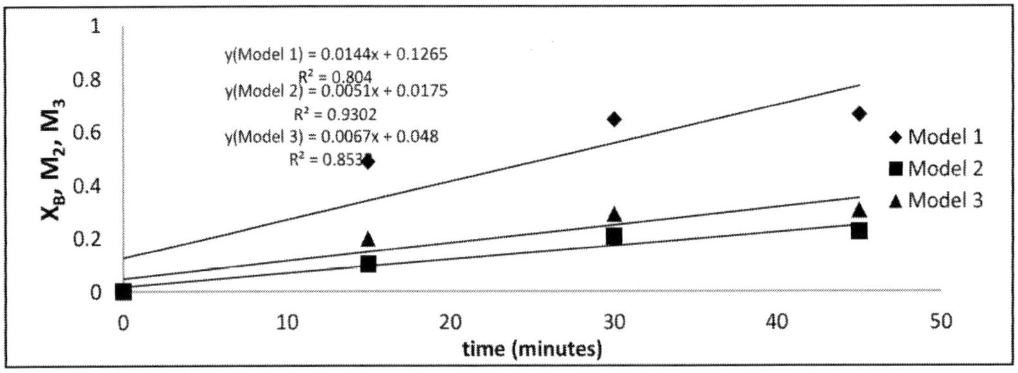


**Fig. 5.** SCM for shrimp shell demineralization using 0.9M hydrochloric acid solution

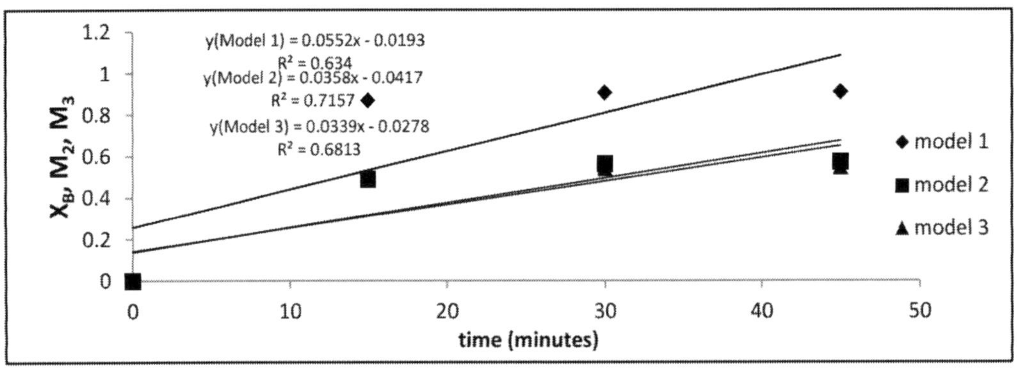


**Fig. 6.** SCM for shrimp demineralization using 1.2 N hydrochloric acid solution

Figures 3-6 denoted all of kinetics modeling equations $X_B$, $M_2$, $M_3$ are plot against time to find out the highest determination coefficient ($R^2$) value. Figures 3 showed that the fluid film diffusion control (model 1) gave a better approximation of the demineralization process compared to ash layer diffusion control mechanism (model 2) and chemical reaction control (model 3). The highest coefficient determination ($R^2$) value is fluid film diffusion model (model 1). This kinetic model occurs when the reactant has a sufficient driving force to diffuse in the film layer. Liquid films in shrimp shell particles was formed when the Reynold number (Re)>1000 [7] which Re in this experiment is 1524. The laminar boundary layer enveloping the particles made low concentration reactant such as hydrochloric acid 0.3N still in diffuse stages through the film layer to the shrimp shell particle. Figures 4-6 showed that the ash layer diffusion control mechanism (model 2) gave a better approximation of the demineralization process compared to fluid film diffusion control (model 2) or chemical reaction control (model 3). The relative magnitudes of the $R^2$ values are indicative of this: for hydrochloric acid at 0.3N, 0.9N and 1.2N considered, the highest of $R^2$ values were for ash layer diffusion model. There is a difference of results between demineralization using hydrochloric acid concentration of 0.3N and 0.6N, 0.9N, 1.2N.

Demineralization using hydrochloric acid 0.3N gave a better approximation of kinetic model of film diffusion controls (model 1). Demineralization reaction using 0.6 N HCl; 0.9 N and 1,2 N showed that the ash layer diffusion (model 2) gave a better approximation compared to the fluid film diffusion (model 1) and chemical reaction controls (model 3).

This is due to the increasing of acid concentration as well as driving force on this reaction so that the diffusion can take place more optimal. One of the most important factor of the kinetic model is the concentration of reactant that affect the intensity of diffusion. Due to increasing of concentration of reactant then it brings hydrochloric acid with greater driving force become easier to diffuse through the film layer and follow by diffuse through the ash layer.

For ash layer diffusion model (model 2) reaction it is visualized that a wall of ash (nonreactive materials) prevents the fluid from moving freely to the zone of reaction (unreacted core). Demineralization of shrimp shells made the other unreactive materials such as protein and chitin should be a resistance. The results of this investigation are widely different from that was reported by Ameh *et al.* [1] which it was reported that the chemical reaction controls (model 3) gave the best prediction. Ameh *et al.* [1] also reported the XRF machine being unfamiliar with biological materials (matrix). This result showed that calcium content analysis with complexometric has widely difference result against Ameh *et al.* [1] which was used XRF for analysis. On the other hand it was showed that SCM give a better step and approximation according to increment of acid concentration.

## 6 Conclusion

Shrimp shell was demineralized using hydrochloric acid at concentration 0.3N, 0.6N, 0.9N, and 1.2N. The concentration of hydrochloric acid affected the conversion of calcium and its kinetic modeling. Analysis of the kinetic data using shrinking core model denoted different results. Demineralization used hydrochloric acid at concentration 0.3N got film diffusion control (model 1) and at concentration 0.6N, 0.9N, 1.2N indicated ash layer diffusion control (model 2) as the best approximation.

## References

1. Ameh, A.O., M.T. Isa, T.J. Adeleye, K.K. Adama, Kinetics of demineralization of shrimp exoskeleton in chitin and chitosan synthesis, Leonardo Electronic Journal of Practices and Technologies. **4**(3) pp. 32-37 (2013)
2. Ameh A.O., D. Abutu, M. T. Isa, U. Rabiu, Kinetics of demineralization of shrimp shell using lactic acid, Leonardo Electronic Journal of Practices and Technologies. **24** pp. 13-22 (2014)
3. Ameh A.O., M.T. Isa, D. Abutu, A. Danlami, Kinetic modelling of the demineralization of shrimp exoskeleton using citric acid, Leonardo Electronic Journal of Practices and Technologies. **25** pp. 99-108 (2014)
4. Horton, R.H., Moran, L.A., Ochs, R.S., Rawn. J.D. and Scrimgeour, K.G, *Principles of Biochemistry*, Third edition, Prentice-Hall, Inc., (New York, 2002)
5. Hussain, Z., A. Nazir, U. Shafique, M. Salman, Comparative study for the determination of metals in milk samples using flame-AAS and EDTA Complexometric Titration, Journal of Scientific Research **40** pp. 9-14 (2010)
6. Levenspiel, O, *Chemical Reaction Engineering*, 3rded, John Willey and Sons Inc., (Singapore, 1999)
7. Munson, B., D.F. Young, T.H. Okiishi, *Fluid Mechanic*, 4th ed, Erlangga., (Jakarta, 2004)
8. Ravichandran, S., G. Rameshkumardan A.R. Prince, Biochemical Composition of Shell and Flesh of the Indian White Shrimp *Penaeus Indicus* (*H.Milne* Edwards 1837), Journal of Scientific Research., **4**(3) pp. 191-194 (2009)
9. Roberts, G.A.F. *Chitin Chemistry*, 1st ed, MacMillan., (London 1992)
10. Wibowo, S, *Research on utilization of shrimp fishery waste for chitosan production and its application to support food industry,* Applied Research Incentive Program, Research and Development Agency of Marine and Fisheries., (Jakarta, 2008)

# Supercritical $CO_2$ extraction of red butterfly wing leaves: process parametric study towards extraction yield

*Mohd Azahar* Mohd Ariff[1,*], *Zulkifli* Abdul Ghani[1], *Noor 'Aina* Abdul Razak[1], *Mohamed Syazwan* Osman[1], and *Jefri* Jaafar[2]

[1]Faculty of Chemical Engineering, Universiti Teknologi MARA, 13500 Permatang Pauh, Penang, Malaysia
[2]Faculty of Chemical Engineering, Universiti Teknologi MARA, 40450 Shah Alam, Selangor, Malaysia

**Abstract.** This research illustrates the parametric studies for determining the relationship between the operating parameters used in the supercritical fluid extraction (SFE) and the extraction yield of Red Butterfly Wing (RBW) leaves. In this study, SFE used carbon dioxide ($CO_2$) as the main solvent and ethanol as the co-solvent. SFE was operated by manipulating few parameters such as temperature, pressure and also particle size in order to determine the extraction yield. The RBW leaves were purchased from the local supplier and then undergone cleaning, drying, grinding as well as sieving processes. 5 grams of grinded leaves was then run through SFE machine. For the temperature, it was observed that the extraction yield increased starting from 40°C to 50°C and dropped after that. The optimum temperature was 50°C with extraction yield of 2.94%. For the pressure, the trend of the extraction yield was directly proportional to the pressure. As the pressure increased, the extraction yield also increased. The optimum extraction yield was 2.45 % at 375 bar. Meanwhile, the particle size was inversely proportional to the extraction yield. The optimum extraction yield was 3.26 % at 63 µm.

## 1 Introduction

Red Butterfly wing (RBW) plant has gained a huge popularity lately. RBW plant is an ornamental plant Fabaceae, or known scientifically as *Mariposa Christia Vespertilionis* [1], others as red butterfly wing, island peak, mariposa (butterfly in Spanish) or rerama in Malaysia [2]. It is commonly known as "Mariposa" or "Red butterfly wing" because of the similarity of its leaves with the color and shape of a butterfly [3]. In a study by Upadhyay *et al.* [4], RBW plant traditionally can treat snake bites, tuberculosis, heal bone fracture, increase blood circulation bronchitis and cold. Other than that, the crushed leaves of RBW can be applied on body parts as it provides a cure for scabies [4]. Furthermore, an early study had been conducted to highlight the potential of *Mariposa Christia Vespertilionis (MCV)*, focusing on its pharmacology properties. The whole plant was extracted using cyclohexane and was reported to have an anti-plasmodial activity towards Plasmodium falciparum FCB1 with IC50 value of 10.8 µg/ml [5]. The method used by Upadhyay *et al.* [4] had been conducted on MCV as anti-plasmodial agent, in-vitro and in-vivo and bioassay-guided studies were applied to characterize the active compounds. In 2013, Hofer *et al.* [6] conducted a study to find out the effect of MCV on human medullary thyroid carcinoma and human intestinal neuroendocrine tumors. They found that ethyl acetate extracted from MCV plant had high inhibition of cancer cells whilst not affecting normal human fibroblasts. Moreover, the extracts caused a change of gene expression in both the carcinomas and tumors. From the results, it implied a great potential of using MCV against cancer.

SFE uses $CO_2$ as a medium for the extraction which is faster than the conventional method that resulted in no residual solvent in the final extract because $CO_2$ is a gas under ambient condition. The physical-chemical properties of supercritical $CO_2$ possesses a higher diffusivity and lower viscosity than conventional liquid solvents [7]. However, pure $CO_2$ does not have sufficient solvation power for polar sativoside and needs to add a polar co-solvent. Experimental co-solvents were water, methanol, ethanol, and mixtures of these solvents [8]. Extraction by conventional methods present some problems because the use of flammable or toxic solvents. At certain temperature and pressure condition, liquid and vapor phases of substance become indistinguishable or known as critical condition. The study will use the SFE to extract the RBW leaves. Throughout the study, the effect of manipulating the chosen operating parameters or variables such as temperature, pressure, and particle sample size will be conducted. From the manipulations, different extracted samples will be collected. Supercritical $CO_2$ offers many advantages over traditional solvents and it can achieve the same results without having the negative side effects of large waste streams. The problem arise when it comes to the yield of extracted sample. The extracted sample could be small in quantity and low in quality. For different methods, the quantity and the quality differed greatly.

* Corresponding author: azahar.ariff@ppinang.uitm.edu.my

 https://doi.org/10.1051/matecconf/201815401015

## 2 Material and methods

In this research, the chemical used was liquid $CO_2$ as the main solvent and ethanol as a co-solvent in SFE. The RBW leaves were purchased from EES Biotech Herbs Sdn. Bhd.

### 2.1 Methodology

The RBW leaves were purchased and taken to the Chemical Engineering Laboratory of UiTM Permatang Pauh, Penang for further process. Before drying process, the RBW leaves were cleaned and washed carefully. After washing the leaves, they were undergone drying process. In the oven, the leaves were dried at 35°C constantly throughout the day. The drying process took 48 hours to ensure that the leaves dried completely. After 48 hours, the leaves and stems were separated. The leaves were grinded and sieved into five different sizes which were 2mm, 1mm, 500μm, 250μm, and 63μm. Following the grinding and sieving procedure, the RBW were run in the SFE machine. 5g of the sample was weighed by using analytical balance model ATY224. After that, the sample was covered in thin cloth and put into the vessel of SFE machine. The extraction process occur in the SFE machine. The extracting parameters such as temperature, pressure and particle size were manipulated at the SFE machine. The temperature used in this study were 40°C, 45°C, 50°C, 55°C and 60°C. The pressure used were 275 bar, 300 bar, 325 bar, 350 bar, and 375 bar. The particle size used was the same as the grinding size used. The extraction time was within one hour with 10 minutes time interval. After that, rotary evaporator was used to get the real mass of the extracted sample by separating the ethanol from the extracted. In order to get the extraction yield, the equation below was used:

*Yield, y (%) = [(Final weight – Initial weight) / Sample weight] x 100.*

## 3 Results and discussion

### 3.1 Effect of temperature

The study of temperature effect on the extraction yield had been carried out in the range of 40°C to 60°C while the pressure and particle size were kept constant. The pressure and particle applied were 250 bar and 2 mm respectively. The details for the experimental results obtained are presented in Figure 1.

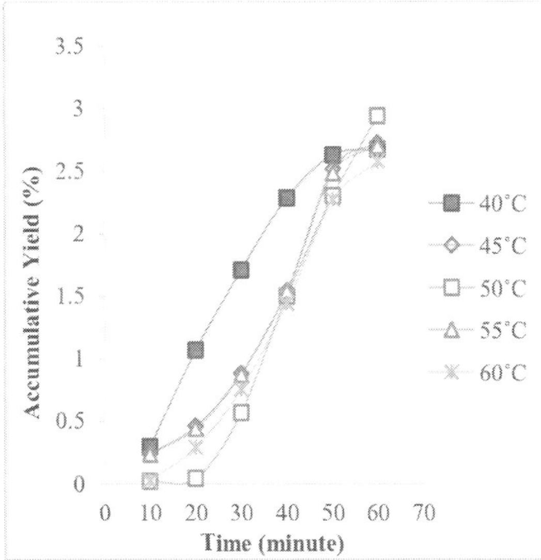


**Figure 1.** Graph for the extracted RBW by manipulating the temperature

From the results, it shows that when the temperature increases from 40°C up to 50°C, it results in the increment of extraction yield. Other studies also had shown similar observations for the extraction yield increment as the temperature increased [9, 10]. After 50°C, the extraction yield is slightly decreases from 2.94 % at 50°C to 2.70 % and 2.58 % at 55°C and 60°C respectively. The same results also had been made on the supercritical $CO_2$ extraction of *Hedyotis diffusa* [11]. Meanwhile, the study of temperature effect had also shown that at 50°C the optimum extraction yield was 2.94 %. The temperature actually has two competing functions which are solute sublimation and supercritical $CO_2$ solvent density [12]. By increasing the temperature, the solute sublimation will be increased while the supercritical $CO_2$ density decreased. It will result in extracting more soluble. The enhancement of solute sublimation may be more dominant than the decrement in density which the extraction yield appears to increase from 40°C to 50°C. It is also reported that the extraction yield would be decreased when the temperature decreased at low pressure, but the extraction yield would be increased when the pressure increased at high temperature [13].

### 3.2 Effect of pressure

The effect of pressure on the extraction yield had been carried out within 60 minutes under five different pressure which the range were from 275 bar to 375 bar while keeping the temperature and particle size at constant which were 40°C and 2 mm respectively. The extraction yields of RBW are presented in Figure 2.

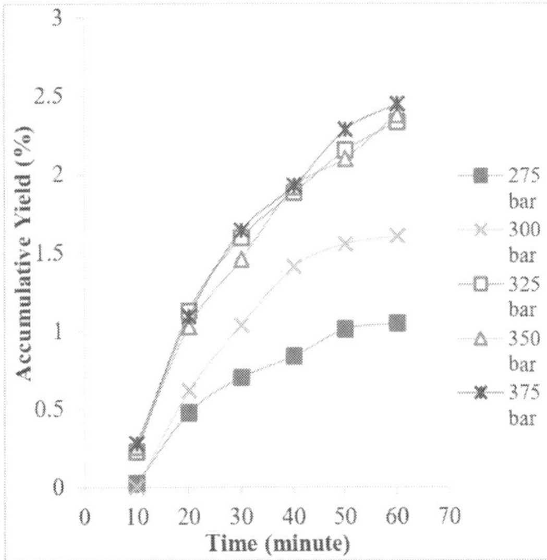


**Figure 2.** Graph for the extracted RBW by manipulating the pressure

The accumulative yield values obtained under the selected pressures, 275 bar, 300 bar, 325 bar, 350 bar, and 375 bar were 1.05 %, 1.60 %, 2.33 %, 2.38 %, and 2.45 % respectively. The highest total extraction yield was recorded at pressure 375 bar. It is shown from the results that the total extraction yield within 275 bar up to 375 bar had an increasing pattern. It indicates that when the pressure increases, the extraction yield will be increased too. This can be explained by the dependence of supercritical $CO_2$ density on pressure [14]. When the set pressure is increased, the supercritical $CO_2$ will bear high density resulting in its solvating power in order to solve the solute in the sample will be higher. The relationship between the density and solvating power will affect the extraction yield. Hence, the higher the solvating power due to high density will resulted in high extraction yield because more extract can be flushed out. Other studies also showed that the extraction yield increased as the pressure increased [14, 15, 16]. For an example, Lisichkov *et al.* [17] had observed the extraction of Cyprinus carpio L. using SFE which the yield increased as the pressure increased. Zhao & Zhang [13, 18] had also stated that the oil yield from both Moringa oleifera seeds and Eucalyptus leaves were significantly increased as the pressure increased. The results actually indicates that the pressure is highly dependent on the dissolving behavior of the extracted sample in supercritical $CO_2$, which will give more soluble extract to have a higher extraction yield.

### 3.3 Effect of particle size

To study the effect of particle size, five sizes had been chosen which were 2 mm, 1 mm, 500 µm, 250 µm, and 63µm. Other operating parameters such as temperature and pressure were kept constant at 50°C and 300 bar. The results are presented in Figure 3.

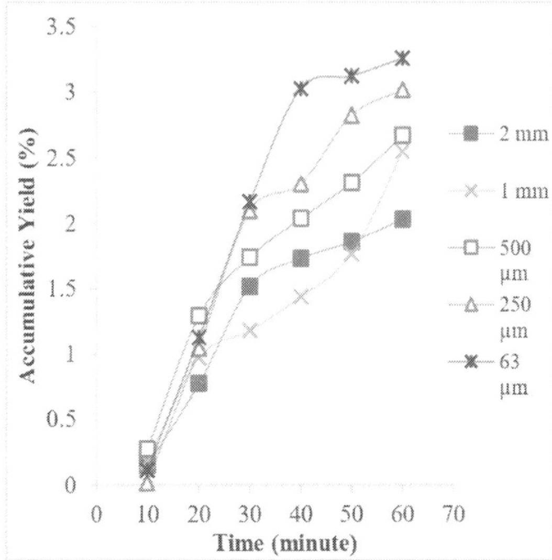


**Figure 3.** Graph for the extracted RBW by manipulating the particle size

The accumulative extraction yield for samples with size of 2 mm, 1 mm, 500 µm, 250 µm, and 63 µm were 2.03 %, 2.55 %, 2.67 %, 3.02 %, and 3.26 % respectively. From the result obtained, it can be seen that a reduction in particle size had resulted in slightly extraction yield increment. The positive effect of a reduction particle size is that the internal mass transfer resistance in the RBW becomes lower and thus leading to the extraction yield increment [14]. As can be seen from Figure 3, the extraction yield steadily increases with the extraction time, and will slow down and then level off because of extracted component exhaustion. It is similar to the study of temperature and pressure variations. Smaller particle size can produce higher extraction yield compared to larger size because small size of particle can give more solute available at the particle surface [14]. In other words, the particle size actually has a direct impact or effect on the extraction yield because smaller particle can enhance the internal diffusion process.

## 4 Conclusion

In conclusion, the effect of SFE operating parameters such as temperature, pressure and particle size towards extraction yield had been observed. As the temperature increased with constant pressure and particle size, the total extraction yield of RBW sample would be increased and after 50°C the extraction yield was decreased from 2.94 % to 2.70 % and 2.58 % at 55°C and 60°C respectively. The optimum extraction yield was 2.94 % at 50°C. The effect of pressure showed that when the pressure increased while the temperature and particle size kept constant, the extraction yield would be increased. The optimum yield obtained was 2.45 % at pressure 375 bar. Meanwhile, the effect of particle size showed that when the particle size decreased while other parameters were kept constant, the total extraction yield would be increased. The optimum yield was 3.26 % recorded at particle size of 63μm.

The authors would like to acknowledge all the staffs of the Faculty of Chemical Engineering laboratory UiTM Penang Branch for their support and insights on using the SCFE equipment. This work is being funded by UiTM under the Internal Research Acculturation Grant Scheme (iRAGS) (600-RMI/DANA 5/3/IRAGS (43/2015)).

## References

1. G.K. Dash, *An appraisal of Christia vespertilionis (L. F.) bakh. F.: A promising medicinal plant*, Int. J. of Pharmacognosie and Phytochem Research **8(6)**, 1037–1039 (2016)
2. H. Bunawan, S.N. Bunawan, S.N. Baharum, *The Red Butterfly Wing (Christia vespertilionis): A promising cancer cure in Malaysia*, Int. J. Pharm Sci. **7(8)**, (2015)
3. P.A. Whiting, *Commercial production of Christia Subcordata Moench by establishing cultural practices and by applying plant growth regulators* (A Master Thesis, Graduate Faculty of the University Of Georgia, Athens, Georgia, USA), (2007)
4. H.C. Upadhyay, B.S. Sisodia, H.S. Cheema, J. Agrawal, A. Pal, M.P. Darokar, S.K. Srivastava, *Novel anti-plasmodial agents from Christia vespertilionis.* Natural product communications **8**, 1591-1594 (2013)
5. J. Nguyen-Pouplin, H. Tran, T.A. Phan, C. Dolecek, J. Farrar, T.H. Tran, P. Caron, B. Bodo, P. Grellier, *Antimalarial and cytotoxic activities of ethno pharmacologically selected medicinal plants from South Vietnam*, J. Ethnopharmacol **109**, 417-427 (2007)
6. D. Hofer, G. Schwach, N. Ghaffari Tabrizi-Wizsy, A. Sadjak, S. Sturm, H. Stuppner, R. Pfragner, *Christia vespertilionis plant extracts as novel anti-proliferative agent against human neuroendocrine tumor cells*, Oncology reports **29**, 2219-2226 (2013)
7. A. Erkucuk, I. Akgun, O. Yesil-Celiktas, *Supercritical $CO_2$ extraction of glycosides from Stevia rebaudiana leaves: Identification and optimization*, J. Supercrit. Fluids **51**, 29-35 (2009)
8. J. Pól, E.V. Ostrá, P. Karásek, M. Roth, K. Benešová, P. Kotlaříková, J. Čáslavský, *Comparison of two different solvents employed for pressurized fluid extraction of stevioside from Stevia rebaudiana: methanol versus water*, J. ABC. **388**, 1847-1857 (2007)
9. Z. Huang, M-J. Yang, S-F. Liu, Q. Ma, *Supercritical carbon dioxide extraction of Baizhu: Experiments and modelling*. The J. of Supercritical Fluids **58**, 31-39 (2011)
10. Rai, A, Mohanty, B., and Bhargava, R. (2015). Modelling and response surface analysis of supercritical extraction of watermelon seed oil using carbon dioxide. Separation and Purification Technology, 141, 354-365.
11. M-C. Wei, Y.C. Yang, *Kinetic studies for ultrasound-assisted supercritical carbon dioxide extraction of triterpenic acids from healthy tea ingredient Hedyotis diffusa and Hedyotis corymbosa*, Separation and Purification Technology **142**, 316-325 (2015)
12. L.T. Taylor, *Supercritical fluid extraction vol. 4*, Wiley-Interscience (1996)
13. S. Zhao, D. Zhang, *Supercritical CO2 extraction of Eucalyptus leaves oil and comparison with Soxhlet extraction and hydro-distillation methods*, Separation and Purification Technology **133**, 443-451 (2014)
14. Z. Huang, M-J. Yang, Q. Ma, S-F. Liu, *Supercritical $CO_2$ extraction of Chinese lantern: Experimental and OEC modelling*. Separation and Purification Technology **159**, 23-34 (2016)
15. G. Wejnerowska, P. Heinrich, J. Gaca, *Separation of squalene and oil from Amaranthus seeds by supercritical carbon dioxide*, Separation and Purification Technology **110**, 39-43 (2013)
16. K. Tomita, S. Machmudah, R. Fukuzato, H. Kanda, A.T. Quitain, M. Sasaki, M. Goto, *Extraction of rice bran oil by supercritical carbon dioxide and solubility consideration*, Separation and Purification Technology **125**, 319-325 (2014)
17. K. Lisichkov, S. Kuvendziev, Z. Zeković, M. Marinkovski, *Influence of operating parameters on the supercritical carbon dioxide extraction of bioactive components from common carp (Cyprinus carpio L.) viscera*. Separation and Purification Technology **138**, 191-197 (2014)
18. S. Zhao, D. Zhang, *A parametric study of supercritical carbon dioxide extraction of oil from Moringa oleifera seeds using a response surface methodology*, Separation and Purification Technology **113**, 9-17 (2013)

# Droplet size distribution measurement of water in kerosene by the buoyancy weighing-bar method

*Rondang* Tambun*, *Piety* Sibagariang, *Brigitta* Alphantaria, and *Ahmad Mulia* Rambe

Department of Chemical Engineering, Universitas Sumatera Utara, Padang Bulan, Medan 20155, Indonesia

**Abstract.** The buoyancy weighing-bar method (BWM) is a method that requires a simple apparatus set up yet produce the high accuracy result in particle size distribution measurement both settling and floating particles. The principle of this experiment that measurement the density change in a suspension due to particle migration is measured by weighing buoyancy against a weighing bar hung in the suspension, and the particle size distribution is calculated using the length of the weighing bar and the time–course change in the apparent mass of the weighing bar. In this study, the BWM was applied to determine the separation time and droplet size distribution for liquid-liquid systems with the different density. The mixture of 99 % of kerosene and 1 % of water were used as samples. The data obtained of separation time of water-kerosene mixture by the BWM were analyzed by the gas chromatography, and the droplet size distributions were compared to Coulter counter method. Based on the data obtained, the BWM could be used to determine the optimal time of water-kerosene separation. The BWM also could measure the droplet size distribution of water in kerosene and the data obtained were comparable to that measured by Coulter counter method. Hence, the BWM is a novel method in separation time determination of water-kerosene mixture and droplet size distribution measurement of water in kerosene.

## 1 Introduction

The buoyancy weighing-bar method (BWM) is a method that requires a simple apparatus set up yet produce the high accuracy result in particle size distribution measurement, for both settling and floating particle. The principle of this method that measurement the density change in a suspension due to particle migration is measured by weighing buoyancy against a weighing bar hung in the suspension, and the particle size distribution (PSD) is calculated using the length of the weighing bar and the time–course change in the apparent mass of the weighing bar. This apparatus consists of an analytical balance with a hook for under-floor weighing, a personal computer, and a weighing bar, which is used to detect the density change of suspension [1-4]. The BWM also could be used to estimate the size distribution measurement of fine particles by using combination with the Rosin-Rammler equation [5]. Obata et al. have reported the measurements of PSD by fluidization in the laminar to turbulent flow region, and one of them can be obtained by a graphical measurement using a fluidization curve [6-8]. The fluidization curve is constructed using the same algorithm as the sedimentation balance method [9].

In this work, the droplet size distribution (DSD) for liquid-liquid systems with the different density would be determined. Some studies have investigated the DSD by using the other methods, such as light scattering method [10], microscope [11], nuclear magnetic resonance [12], etc. These methods could produce highly accurate results within a shorter time, but they require extremely expensive equipment. Hence, a simple and cost effective method to determine the DSD is highly demanded, that is BWM. Beside the DSD, the BWM also could be used to estimate the separation time of liquid-liquid systems with the different density. The principle of BWM in this experiment is analogous to distributions of the settling particles and floating particles. In this method, the initial buoyant mass of the weighing bar W0 depends on the droplets between the top and bottom of the weighing bar in a suspension. The initial density of suspension $\rho_{S0}$, initial buoyant mass of the weighing bar $W_0$, and initial apparent mass of the weighing bar $G_0$ in a suspension at t = 0 are given by the following equations.

$$\rho_{S0} = \rho_L + \frac{C_0}{\rho_P}(\rho_P - \rho_L) \tag{1}$$

$$W_0 = V_B \rho_{S0} \tag{2}$$

$$G_0 = V_B \rho_B - W_0 = V_B(\rho_B - \rho_{S0}) \tag{3}$$

where $\rho_L$ and $\rho_P$ are the liquid density and droplet density, respectively. $C_0$ is the initial droplet mass

* Corresponding author: rondang@usu.ac.id

concentration of suspension, $\rho_B$ is the density of the weighing bar in suspension, and $V_B$ is the volume of the weighing bar.

The droplet mass concentration of suspension C(t) decreases with time because the large droplets settle. The density of suspension $\rho_S$, buoyant mass of the weighing bar W, and apparent mass of the weighing bar G in a suspension at time t are expressed as

1

$$\rho_S(t) = \rho_L + \frac{C(t)}{\rho_P}(\rho_P - \rho_L) \quad (4)$$

$$W(t) = V_B \rho_S(t) \quad (5)$$

$$G(t) = V_B \rho_B - W(t) = V_B(\rho_B - \rho_S(t)) \quad (6)$$

The droplet mass concentration of suspension $C(t)$ becomes zero once all the small droplets also settle. The final density of suspension $\rho_{S\infty}$, final buoyant mass of the weighing bar $W_\infty$, and final apparent mass of the weighing bar $G_\infty$ in a suspension at $t = \infty$ are given by the following equations.

$$\rho_{S\infty} = \rho_L \quad (7)$$

$$W_\infty = V_B \rho_L \quad (8)$$

$$G_\infty = V_B \rho_B - W_\infty = V_B(\rho_B - \rho_L) \quad (9)$$

Equation (10) shows the mass balance of droplets [13].

$$C_0 - C(t) = C_0 \int_{x_i}^{x_{max}} f(x)\mathrm{d}x + C_0 \int_{x_{min}}^{x_i} \frac{v(x)t}{h} f(x)\mathrm{d}x \quad (10)$$

The left side in Eq. (10) is the quantity of droplets that move onto the bottom side of the weighing bar. The first term on the right side represents the mass of droplets larger than droplet size $x_i$ among the droplets that move, while the second term on the right side is the mass of droplets smaller than droplets size $x_i$ among the droplets that move. In the sedimentation balance theory [9], the mass frequency function of a given sample droplets is $f(x)$ where $x$ is the Stokes diameter. From Eqs. (2), (5), (8), and (10), Eq. (11) describes the mass balance of particles in a suspension [13].

$$W_0 - W(t) = (W_0 - W_\infty)\int_{x_i}^{x_{max}} f(x)\mathrm{d}x + (W_0 - W_\infty)\int_{x_{min}}^{x_i} \frac{v(x)t}{h} f(x)\mathrm{d}x \quad (11)$$

where $v(x)$ is the settling velocity of the droplet and $f(x)$ is the mass frequency of the droplet size $x$. Differentiating Eq. (11) with respect to the time $t$ gives

$$-\frac{dW}{dt} = (W_0 - W_\infty)\int_{x_{min}}^{x_i} \frac{v(x)}{h} f(x)dx \quad (12)$$

From Eqs. (11) and (12),

$$W = W_R + \left(\frac{dW}{dt}\right)t \quad (13$$

The apparent mass of the weighing bar, which is given by Eq. (6), gradually increases from $G_0$ to $G_\infty$. The volume and density of the weighing bar are constant values. Differentiating Eq. (6) with respect to time $t$ gives

$$\frac{dG}{dt} = -\frac{dW}{dt} \quad (14)$$

Therefore, according to Eqs. (6), (13), and (14),

$$G = V_B \rho_B - W_R + \left(\frac{dG}{dt}\right)t = G_R + \left(\frac{dG}{dt}\right)t \quad (15)$$

where $G_R = V_B\rho_B - W_R$. The value of $G_R$ is calculated from the tangent line based on Eq. (15). The cumulative mass oversize is

$$R = 100\int_{x}^{x_{max}} f(x)dx = \frac{G_0 - G_R}{G_0 - G_\infty} \times 100 = 100 - D \quad (16)$$

where $D$ is the cumulative mass undersize.

Droplets size $x$ is given by Stokes formula,

$$x = \sqrt{\frac{18\mu_L v(x)}{g(\rho_L - \rho_P)}} \quad (17)$$

The settling velocity of the droplets $v(x)$ is calculated using Eq. (18)

$$v(x) = \frac{h}{t} \quad (18)$$

where $h$ is the length of the weighing bar and $t$ is the time lapse. From Eqs. (17), time $t$ is an inverse function of droplet size $x$. The DSD of the suspended droplets is calculated using the droplet size at each time and then plotting the corresponding cumulative mass undersize $D$. According to equations above, this study will try to investigate the DSD of water in kerosene by using BWM.

## 2 Materials and Methods

Figure 1 schematically illustrates this experiment. The weighing–tools was weighing–bar ( $\phi$ 10.0 mm x 210.0 mm, submerged length: 200.0 mm) which was composed of alumunium (density: 2700 kg/m$^3$). The samples consist of kerosene ($\rho$ = 0.810 g/cm$^3$) and water ($\rho$= 0.99708 g/cm$^3$) were placed in a 1000 ml of graduated cylinder with the ratio 99 % of kerosene and 1 % of water. In this experiment, the influence of graduated cylinder diameter also was investigated. The graduated cylinder used in this experiment were diameter of 65.0 mm and diameter of 60.0 mm. The analytical balance (minimum readout mass 0.1 mg) had a below–balance–weighing hook for hanging measurement. Using a hanging wire, which did not extend due to the weight of the weighing bar, the weighing bar was hung from the analytical balance. The room temperature was approximately 298 K. After thoroughly stirring the suspension using an agitator, the weighing bar was set with the balance. The measuring data, which consist of time t and the corresponding mass of the bar $G_B$, were recorded. The measuring time was 1 hours and the data were collected in 1–second interval. After the measurement, the DSD was estimated based on the above–described theory. The Coulter counter method (Counter LS 100) was used as comparison method to measure DSD, and the gas chromatography was used to determine the purity of kerosene.

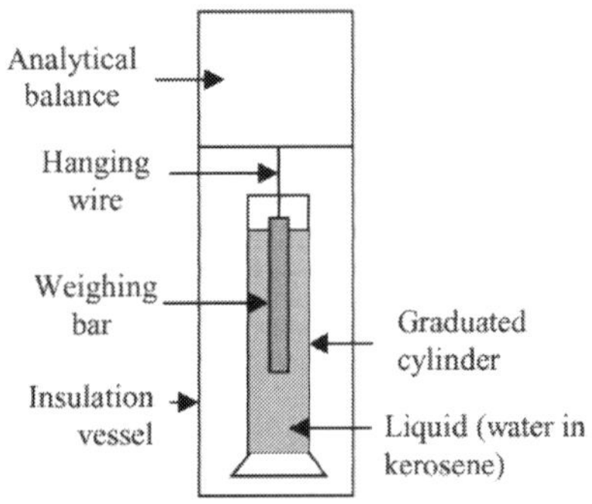


**Fig. 1**. Schematic diagram of experimental apparatus

## 3 Results and Discussions

Figure 2 showed the change with time in the apparent mass of weighing-bar $G_B$ when ratio 99% : 1% of kerosene and water was used.

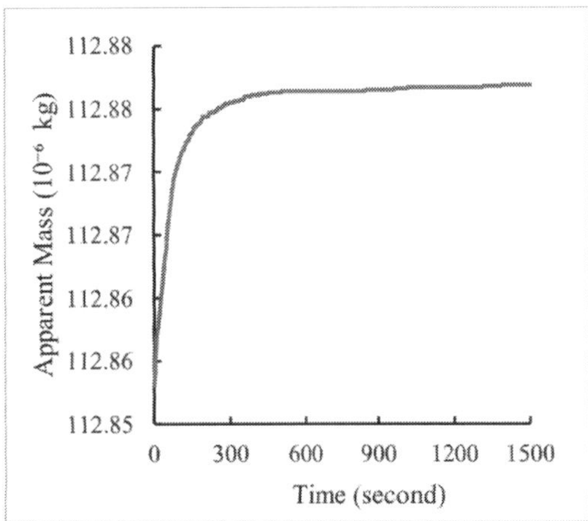


**Fig. 2**. Apparent mass of the weighing-bar as a function of time when ratio of kerosene and water was 99 % : 1%

The graduated cylinder diameter of 65 mm was used in this experiment. The figure showed that the apparent mass of the weighing-bar increased until all the water settled below the lower end of the weighing-bar, and then the apparent mass of the weighing-bar became constant. The change in the apparent mass was due to the change in the buoyant mass against the weighing-bar as well as water settling [14]. We could see at figure 2 that the weighing-bar started to become constant at 300 seconds of experiment when the larger droplets had settled, and constant at 1500 seconds when the middle and small droplets also settled below the weighing-bar. By using the gas chromatography, the concentrations of kerosene were about 89.3359 % and 98.9741 % at 300 seconds and 1500 seconds, respectively. Thus, figure 2 showed us that the separation time of water-kerosene mixture could be detected by using BWM.

Figure 3 showed the DSD of water in kerosene by BWM when ratio 99% : 1% of kerosene and water was used. These DSD were estimated by Stokes formula. The result obtained by using BWM gave the similar result to that measured by Coulter counter method. Hence, BWM could measure DSD of water in kerosene by using Stokes formula when ratio 99% : 1% of kerosene and water was used.

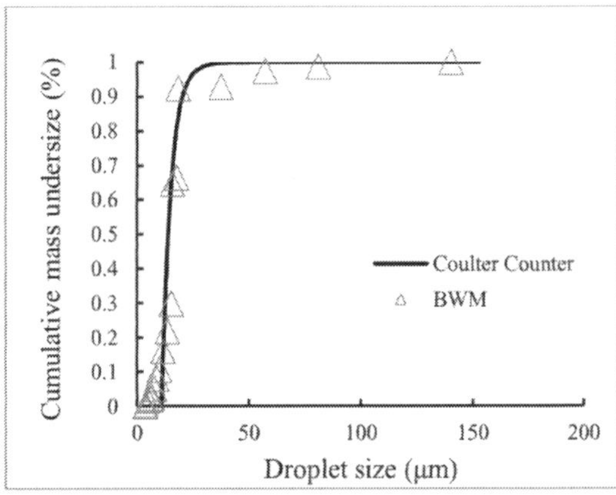


**Fig. 3**. The DSD of water in kerosene by using BWM when ratio of kerosene and water was 99 % : 1%

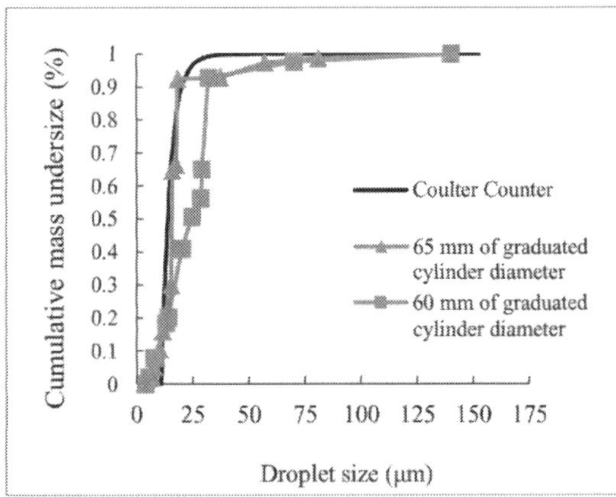


**Fig. 4**. Influence of graduated cylinder diameter in DSD determination of water in kerosene by using BWM when ratio of kerosene and water was 99 % : 1 %

Figure 4 showed the influence of graduated cylinder diameter in DSD determination of water in kerosene by using BWM when ratio of kerosene and water was 99 % : 1 %. The DSD obtained by using graduated cylinder diameter of 65.0 mm gave the closer result to that measured by a Coulter counter method than by using graduated cylinder diameter of 60.0 mm. It was caused the wall effect and sectional area ratio of weighing bar and graduated cylinder influenced the migration velocity of droplet [15].

## 4 Conclusions

We have experimentally investigated the application of BWM to measure the DSD and optimal time of water-kerosene separation. From this study, the following results were obtained:

1. The BWM could determine the separation time of kerosene-water mixture when ratio 99 % : 1 % of kerosene and water was used.
2. The BWM could be used to measure the DSD of water in kerosene when ratio 99 % : 1 % of kerosene and water was used, and the precision of the DSD by using BWM is comparable to that obtained by a Coulter counter method.
3. The BWM used graduated cylinder diameter of 65.0 mm gave the closer result to that measured by Coulter counter method than using graduated cylinder diameter of 60.0 mm.

The authors would like to thank Universitas Sumatera Utara for Research Grant, under TALENTA Fundamental Research Scheme 2017.

## References

1. E. Obata, Y. Ohira, M. Ohta, Powder Technol. **196**, 2 (2009)
2. T. Motoi, Y. Ohira, E. Obata, Powder Technol. **201**, 3 (2010)
3. Y. Ohira, K. Furukawa, R. Tambun, M. Shimadzu, E. Obata, J. Sed. Soc. Jpn. **69**, 1 (2010)
4. R. Tambun, T. Motoi, M. Shimadzu, Y. Ohira, E. Obata, Adv. Powder Technol. **22**, 4 (2011)
5. R. Tambun, K. Furukawa, M. Hirayama, M. Shimadzu, S. Yamanaka, Y. Ohira, J. Chem. Eng. Jpn. **49**, 2 (2016)
6. E. Obata, H. Watanabe, K. Mukaida, M. Akiyoshi, K. Ando, Kagaku Kougaku Ronbun. **12,** 619–621 (1986)
7. N.G Stanly-Wood, E. Obata, H. Takahashi, K. Ando, Powder Technol. **60**, 1 (1990)
8. E. Obata, K. Ando, Encyclopedia of Fluid Mechanics, Supplement 2, Gulf Publishing, Houston, (1993)
9. S. Odén, Soil Sci. **19,** 1–35 (1925)
10. Y. Mao, L. Yong, H. Tao, W. Shimin, X. Yiqian, Wuhan Uni. J. of Nat. Sci. **3**, 4 (1998)
11. J. A. Boxall, C.A. Koh, E.D. Sloan, A.K. Sum, D.T. Wu, Ind. Eng. Chem. Res **49**, 3 (2010)
12. E.O. Fridjonsson, B.F. Graham, M. Akhfash, E.F. May, M.L. Johns, Energy Fuels **28**, 3 (2014)
13. T. Allen, Particle size measurement, 4th edition, Chapman and Hall, London, (1990)
14. R. Tambun, M. Shimadzu, Y. Ohira E., Obata, J. Chem. Eng. Jpn. **45**, 4 (2012)
15. R. Tambun, K. Nakano, M. Shimadzu, Y. Ohira, E. Obata, Adv. Powder Technol. **23**, 6 (2012)

# Modification process in nixtamalization of folic acid-rich dent corn (*Zea mays identata*) and its identification as smart food fortificant

*Agustine* Susilowati*, *Puspa Dewi* Lotulung, *Yati* Maryati, and *Aspiyanto*

Research Center for Chemistry – Indonesian Institute of Sciences, Kawasan PUSPIPTEK, Serpong 15314, South Tangerang, Banten, Indonesia.

**Abstract.** A modification on nixtamalization process of dent corn (*Zea mays identata*) was conducted in order to recover natural folic acid-rich corn. Nixtamalization process on varieties of white dent corn and yellow dent corn subsequently were performed by steeping solution of $Ca(OH)_2$ at concentrations of 0, 10, 20 and 30 % (w/w corn dissolved protein) for 18 hours, and boiling at 90 °C for 15, 30, 45 and 60 minutes. Result of research showed that concentration of $Ca(OH)_2$ solution becoming more and more high and long boiling time increased both folic acid and reducing sugar, dropped total solids and total sugar, and fluctuated dissolved protein for both types of corn. Nixtamalization optimalization of white dent corn and yellow dent corn were achieved at combination of $Ca(OH)_2$ 20 % (w/w corn dissolved protein) for 60 minutes of boiling and $Ca(OH)_2$ 30 % for 30 minutes of boiling and gave folic acid of 466.81 and 506.74 µg/mL, respectively. In this condition, it is occurred an increase of folic acid 192.3 % (1.9 folds) and 139.89 % (1.4 folds) when compared to initial material of corn. Identification on folic acid monomer and glutamic acid monomer of both nixtamalized dent corn and yellow dent corn at optimum operation condition displayed domination of folic acid monomer with molecular weight (MW) 442.56 Dalton (Da.) with relative intensity 25.51 %, and 441.73 Da. with relative intensity 100 %, while glutamic acid monomer of nixtamalized yellow dent corn and nixtamalized white dent corn were dominated by monomer with MWs of 148.27 Da. and 148.32 Da., and relative intensity 3.73 and 1.8 %.

## 1 Introduction

Cereal grains have been the staple component of human diet for thousands of years. Corn, wheat and rice together accounted for about 89 % of total cereal production worldwide. Today, cereal grains are still the most important source of calories to the majority of the world population. In developing countries, about 60 % of calories are derived from cereals, whereas in the developed world it is only 30 % [1-3]. Corn (*Zea mays identata*) is one of the best cereal grains as potential source of polysaccharides to produce energy and to build body, which is supported by functional properties as sufficient high source of protein (7 – 9 %). Due to high protein concentration, corn has the most important source of folic acid, in which folic acid is composed by glutamic acid expressed as dissolved protein [4]. In order to achieve corn products as natural source of folic acid for smart food and to diversify a wide variety and attractive products, it can be conducted through nixtamalization process.

Nixtamalization process on corn is generally performed to achieve corn flours with proved protein quality, increase calcium content and better physical properties in application on food products. Nixtamalization using suspension of calcium hydroxide will inhibit its occurrence of fractionation of amylose and amylopectin chain, so that they are binding and inhibiting its occurrence of retrogradation (snack products with high crispy level) or inhibiting its occurrence of syneresis (sauce and pudding products). Nixtamalization on corn is a process, in which corn is boiled directly in water solution of $Ca(OH)_2$ without filtering, steeped overnight, and washed and cooled at room temperature (26 – 28 °C) in order to obtain and improve nixtamal product with better physical property (texture, color, flavor, aroma, taste, shelf life), maintain functional property (minimize loss of folic acid), ease starch gelatinization, increase nutritional value, adsorb water mass, and drain a portion of its skin and pericarp from corn grains [5, 6]. Nixtamalization process on corn is done to soften corn grain and facilite absorbing $Ca(OH)_2$ so that nixtamalized product undergoing a series of significant changes on the physical and chemical properties (color, aroma), and the morphological and rheological properties of starch will be generated. Softening and swelling on the optimum corn starch gel (65 – 85 °C) is able to enlarge volume (22 folds) when compared to corn starch granula at 95 °C. High content of folic acid in corn (kernel) (26 µg) can be kept to select type

* Corresponding author: agustine_1408@yahoo.co.id

of process and facilite sineresis and retrogradation on corn starch [6-9].

Folic acid, (2*S*)-2-[4-[(2-amino-4-oxo-1H-pteridin-6-yl) methylamino]benzoyl] mino]pentanedioic acid, N-[4-[(2-amino-1,4-dihydro-4-oxo- 6-pteridinyl)methyl] amino] benzoyl]- L-glutamic acid, Pteroylglutamic acid, Vitamin $B_9$, Vitamin M, Folacin ($C_{19}H_{19}N_7O_6$) has sensitive properties on sun's light, oxygen, heat and high temperature (particularly boiling) so that to isolate and keep its presence of folic acid in enriched food is important process [10]. Folic acid in enriched food (smart food) is considered as a shape of real food with appropriate nutritional value. It consist of components giving contribution on smart brain, and very important to be consumed before and during early pregnancy to prevent emergence of serious birth defects of the brain (anencephaly) as Neural Tube Defects (NTDs), spine (spina bifida), low-birth-weight baby and malformation in the general population [11, 12].

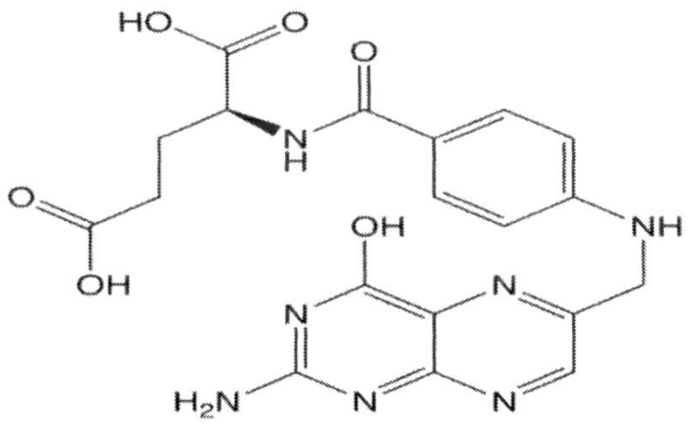

**Fig. 1.** Structure of folic acid.

Effect of all processes on status of folic acid and glutamic acid become a reference guide how far folic acid monomer is achieved by Liquid Chromatography-Mass Spectrometry (LC-MS) instrument. Identification on folic acid and glutamic acid through LC-MS instrument is operated to monitor MW-monomer domination according to intensity in order to display nixtamal characteristic. This matter is related to sensitive of folic acid and various shapes and folic acid structure to reach operation condition effect of nixtamalization. By using chromatography, it will be separated a molecular mixture based on difference in migration rate and molecular distribution in stationary phase (adsorben) and mobile phase (eluen), while mass spectrometry will ionize analit according to electro spray ionization (ESI) principle to gas phase (fine aerosol) [13, 14]. The goal of this experimental activity was to find out characteristic of the best nixtamalized yellow dent corn and white dent corn by different time in boiling and concentration of solution of $Ca(OH)_2$, and its identification on folic acid monomer.

## 2 Experimental design

### 2.1 Materials and Equipments

Main materials used in this experimental activity were dry yellow dent corn and dry white dent corn from Tangerang, standard folic acid, standard glutamic acid, all the chemicals used were reagent grade. Process equipments used in this experimental works were nixtamalization process system, grinder (local), sieve of 60 mesh (Retsch, Germany) and Dead-End Stirred Ultrafiltration Cell (SUFC) (Model 8200, Amicon Bioseparation, MILLIPORE, U.S.A.) for preparation analyses of Liquid Chromatography-Mass Spectrometry (LC-MS). While, main analysis instrument was LC-MS. Mariner Biospectrometry) equipped with LC (Hitachi L 6200).

Experimental works were subsequently performed by selecting corn grains, steeping whole corn in solution of $Ca(OH)_2$ with concentrations of 0, 10, 20, 30 and 40 % (w/w corn dissolved protein) for 18 hours, respectively, followed then by boiling at 90 °C for 15, 30, 45 and 60 minutes, draining, drying and reducing size at room temperature. Analyses were performed on dissolved protein (Lowry method) [15], total solids (Gravimetric method), reducing sugar (Somogyi-Nelson), total sugars (Phenol Sulphate method) [16], folic acid (UV-vis Spectrophotometer) [17]. Identification on folic acid and glutamic acid were conducted by means of LC-MS equipped with LC (Hitachi L 6200) [18]. Process and analysis were conducted in duplicate. Data processed in this description were based on the result of average analysis.

### 2.2 Procedure

#### *2.2.1 Nixtamalization process*

A corn nixtamalization process consist of washing and steeping in water in a 1 : 4 mixture with water. On raw corn steeping was then added by $Ca(OH)_2$ with concentrations of 0, 10, 20, 30 and 40 % (w/w corn dissolved protein) for 18 hours of steeping time. For each treatment was performed a boiling at 90 °C for 0, 10, 20 , 30, 40, 50, and 60 minutes, cooling overnight, and washing to remove a portion of its skin or the remaining pericarp and excess lime, and draining so that it is produced steeped corn (nixtamal). Steeped corn (nixtamal) was ground into masa, grinded to reduce size, and sieved through a sieve of 60 mesh to produce nixtamalized corn flours for further process.

#### 2.2.2 *Identification of folic acid through LC-MS*

Samples of extract as a result of purification result was conducted by using nixtamal corn from the best process condition through microfiltration (MF) membrane (0.15 µm) [19], standard folic acid and standard glutamic acid. Analyses of monomers folic acid and glutamic acid were performed by LC-MS instrument. The LC-MS system consisted of an ion trap Mass Spectrometer and a Liquid Chromatography. LC system is integrated with Q-tof Mass Spectrometer by Electro Spray Ionization (ESI), in which scan mode is performed in range of *m/z* 100 – 1200 at 140 °C. LC (Hitachi L 6200) uses a C18-18 RP (5 µm particle size, 15 cm x 1 mm i.d.) column from Supelco (Bellefonte, PA). Type of solvent used ismethanol at a flowrate of 0.1mL/min. And injection volume was 2uL [18].

https://doi.org/10.1051/matecconf/201815401017

## 3.Results and Discussion

### 3.1 Characteristic of corn.

Selection on dent corns are based on a facility to achieve and prepare them broader when compared to others varieties. Both types of corn show difference in colour, size and uniform of grains, as shown in Figure 2, and composition of dent corns are showed in Table 1. White dent corn displays high concentration of folic acid (211.24 ug/mL) when compared to yellow dent corn (159.7 ug/mL). This matter is appropriate with higher concentration of dissolved protein (0.07 mg/mL) than that yellow dent corn (0.05 mg/mL). It had been known that folic acid (pteroyl-L-Glutamic acid) is a derivative glutamic acids so that it is soluble in water relating with amino acids in general [9].

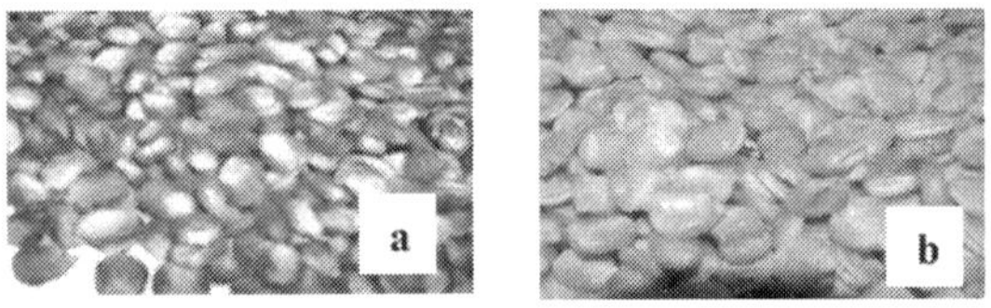


**Fig. 2.** (a) Yellow dent corn and (b) white dent corn (*Zea mays identata*).

Meanwhile, sufficient high concentration of corn protein (± 9 %), higher when compared to other cereals (rice), however under soys (± 14 %) so that corn protein has potential use as source of protein and energy (carbohydrate). The whole compositions related to types and varietis of corn, particularly flavonoid phenolic pigment, such as beta-caroten, lutein, xanthins and cryptoxanthin pigment with vitamin A in yellow dent corn which do not contained in white dent corn. This matter causes both stronger aroma and taste of yellow dent corn. Different trend is seemed on total sugars, in which white dent corn yields lower content of total sugars (208.38 mg/mL) compared with yellow dent corn (310.22 mg/mL), on the other hand, higher content of reducing sugars (982.16 mg/mL) compared with yellow dent corn (640.18 mg/mL). Total sugars relating with high content of starch (72 %) is main component in composing corn carbohydrate. This matter is related with types, varieties of corns and post harvesting processes. Reducing sugars are relating with boiling process of corn on difference in boiling time and alkali concentration before and after nixtamalization. Composing simple sugars with reduction property during nixtamalization has a role in bonding folic acid of nixtamalized corn. Similar trend seems at fat concentration, in which yellow dent corn shows higher concentration of fat (4.65 %) when compared to white dent corn (3.81 %). Besides, due to variety of corn, this matter is also effected by post harvesting drying processes and fat content in initial material when harvesting. Corn fat is enabled to have an important role in emulsification process through homogenation process under both optimal rotation speed and time prior to drying process performed in order to produce nixtamalized flour with better oil absorber in application products.

**Table 1.** Composition of yellow dent corn and white dent corn (*Zea mays identata*)

| Kind of dent corn | Component | | | | | |
|---|---|---|---|---|---|---|
| | Folic acid (μg/mL) | Total solid (%) | Dissolved protein (mg/mL) | Total sugar (μg/mL) | Reducing sugar (mg/mL) | Fat (%) |
| Yellow | 87.52 | 159.70 | 0.05 | 310.22 | 12.30 | 4.65 |
| White | 93.86 | 211.24 | 0.07 | 208.38 | 9.95 | 3.81 |

### 3.2 Effect of nixtamalization process on composition

Nixtamalization process is conducted by adding solution of $Ca(OH)_2$ of 0, 10, 20 and 30 % (w/w, dissolved protein of corn) on yellow dent corn and white dent corn, and boiling for 15, 30, 45 and 60 minutes, respectively.

#### *3.2.1.Folic acid*

Nixtamalization process on yellow dent corn increases folic acid until ability level of binding (20 %) followed by dropping folic acid for all boiling time, as showed Figure 3. Optimization of folic acid was achieved at solution of $Ca(OH)_2$ 20 % (w/w corn dissolved protein) for 60 minutes boiling and gave 466.81 ug/mL. Compared with folic acid of initial material (159.70 ug/mL), the optimal process condition increases folic acid from 159.70 ug/mL to 466.81 ug/mL (192.3 % or 1.9 times). This increase of folic acid is caused by its occurrence of starch gelatinization on corn seed link. In same time, solution of $Ca(OH)_2$ will be absorbed causing its linked components, such as folic acid. While, decrease of folic acid is possibility caused by excess in solution of $Ca(OH)_2$ so that it is occurred coagulation of protein and lysis on further processes (filtration, cooling, draining, reducing size). Properties of folic acid (pteroyl-L-glutamic acid) is dissolved in water [9] coagulated by alkali or $Ca(OH)_2$ solution and sensitive on light, oxigent, high temperature and process treatments using pressure [20, 21].

Different trend seems on white dent corn, in which increase of folic acid is still running to the highest concentration of $Ca(OH)_2$ solution (30%), although is fluctuating, as displayed in Figure 4. To achieve the optimal folic acid, nixtamalization rate demonstrated at higher concentration of $Ca(OH)_2$ solution will need shorter cooking time. In other words, adding $Ca(OH)_2$ solution enables to strengthen corn starch gel by binding amylose chain during boiling process causing stronger structure of corn starch, in which folic acid will be binded strong at starch gel at same time. It had been known that corn has high content amylose of 28 % and amylopectin of 72 %, so that they are easy to fulfil sineresis and retrogradation [22]. Optimization of folic acid at nixtamalized white dent corn is obtained at $Ca(OH)_2$ solution of 30 % for 30 minutes of boiling with result of 506.74 ug/mL. Compared with folic acid in initial material of white dent corn (211.24 ug/mL), the optimum process condition raises folic acid from 211.24 ug/mL to 506.74 ug/mL (139.89 % or 1.4 times). While, decrease of folic acid is possibility caused by excess in solution of $Ca(OH)_2$ so that it is occurred coagulation of protein and lysis on further processes (filtration, cooling, draining, reducing size). Properties of folic acid (pteroyl-L-glutamic acid) is dissolved in water (Belitz, *et al*, 2009) coagulated by alkali or $Ca(OH)_2$ solution and sensitive on light, oxigent, high temperature and process treatments using pressure [20, 21]. Different trend seems on white dent corn, in which increase of folic acid is still running to the highest concentration of $Ca(OH)_2$ solution (30 %), although is fluctuating, as displayed in Figure 4.

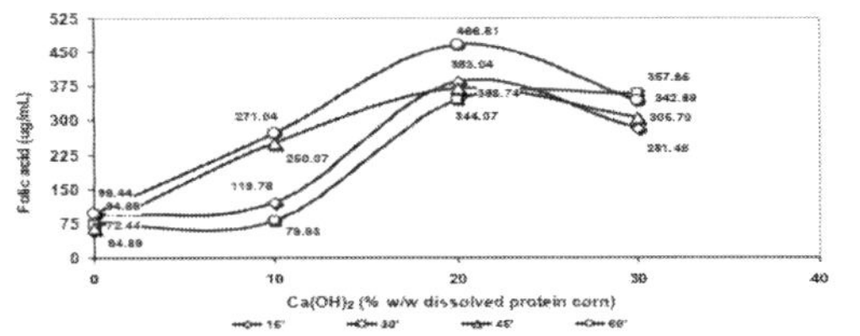


**Fig. 3**. Relationship between $Ca(OH)_2$ concentration and boiling time at 90 °C on folic acid in nixtamalized yellow dent corn.

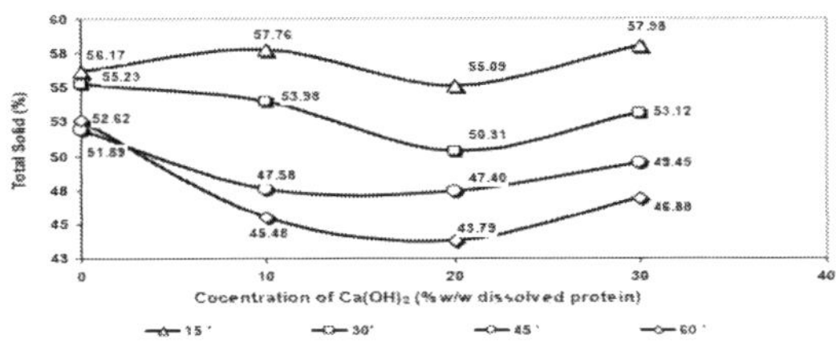


**Fig. 4**. Relationship between $Ca(OH)_2$ concentration and boiling time at 90 °C on folic acid in nixtamalized white dent corn.

To achieve the optimal folic acid, nixtamalization rate demonstrated at higher concentration of $Ca(OH)_2$ solution will need shorter cooking time. In other words, adding $Ca(OH)_2$ solution enables to strengthen corn starch gel by binding amylose chain during boiling process causing stronger structure of corn starch, in which folic acid will be binded strong at starch gel at same time. It had been known that corn has high content amylose of 28 % and amylopectin of 72 %, so that they are easy to fulfil sineresis and retrogradation [22]. Optimization of folic acid at nixtamalized white dent corn is obtained at result of 506.74 ug/mL. Compared with folic acid in Compared with folic acid in initial material of white dent corn (211.24 ug/mL), the optimum process condition raises folic acid from 211.24 ug/mL to 506.74 ug/mL(139.89 % or1.4times).

### 3.2.2 Total solids

Total solids is an accumulation for the whole components in nixtamalized corn both soluble and insoluble extracted according to Gravimetric method [16]. Nixtamalization process on yellow dent corn gave fluctuated total solids and was tend to drop, as demonstrated in Figure 5. Dropping total solids was possiblity caused by its occurrence of water mass adsorption at net of corn grains causing swelling starch gel to decrease total solids. Optimization of total solids reached at treatment of $Ca(OH)_2$ solution of 30 % (w/w corn soluble protein) for 15 minutes of boiling gave 57.98 %. Compared with initial material of corn (87.52 %), the optimum process condition declined total solids to 57.98 % (33.75 %).

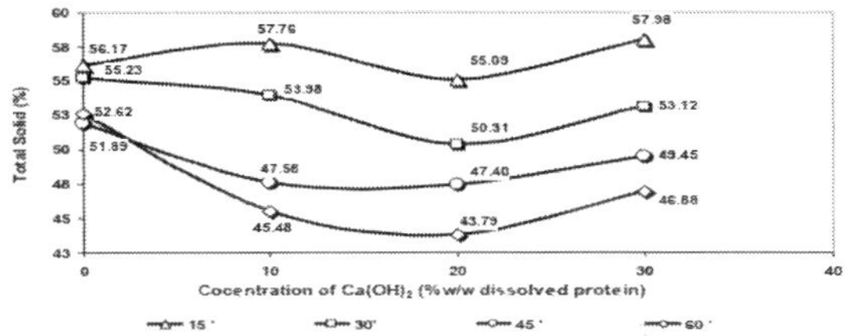


**Fig. 5**. Relationship between $Ca(OH)_2$ concentration and boiling time at 90 °C on recovery of total solids in nixtamalized yellow dent corn.

Different trend seemed at nixtamalized white dent corn, in which raising total solids is occurred at treatment of $Ca(OH)_2$ solution of 10% (w/w corn soluble protein) and almost all boiling time followed by decreasing fluctuation relating with increasing $Ca(OH)_2$ concentration, as showed in Figure 6.

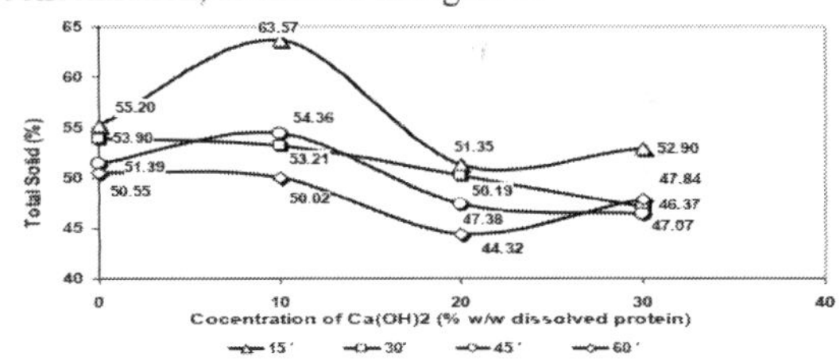


**Fig. 6**. Relationship between $Ca(OH)_2$ concentration and boiling time at 90 °C on recovery of total solids in nixtamalized white dent corn.

Optimization of total solids (63.57 %) was achieved at treatment combination of $Ca(OH)_2$ solution of 10 % for 15 minutes of boiling, and became more and more low relating with increasing $Ca(OH)_2$ solution concentration. When compared to total solids of initial corn material (93.86 %), the optimum process condition dropped total solids from 93.86 % to 63.57 % (32.37 %). This matter displayed that niztamalization rate was tend to lower total solids relating with long cooking time, although adding solution of $Ca(OH)_2$ became more and more much amount. Excess in concentration of $Ca(OH)_2$ solution of 10 %(w/w corn soluble protein), boiling process system was not able to retain gelatinization in order to dissolve components of corn, so that is dropped total solids.

### *3.2.3 Dissolved protein*

Dissolved protein of nixtamalized corn is all amino acids, peptides, and protein derivative, which is soluble in water and is binded by other components as a result of nixtamalization process. Dissolved protein of nixtamalized corn becomes a main process reference guide because of folic acid. Folic acid is protein derivative owning chemical structure of combination from pteridin heterocyclic, para-aminobenzoat acid (PABA) and pteroyl-L-glutamic acid. Its presence of high protein of corn (7 – 10 %) becomes reference guide to select this cereals, besides high content of folic acid (26 µg) and its potential use as source of energy [9]. Nixtamalization process on yellow dent corn gave optimization (0.038 mg/mL) achieved at treatment combination of concentration of $Ca(OH)_2$ of 10 % (w/w corn dissolved protein) and boiling time of 15 minutes, as displayed in Figure 7.

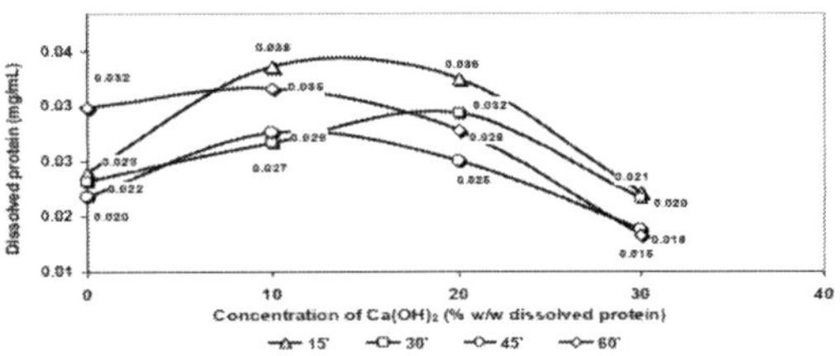


**Fig. 7**. Relationship between $Ca(OH)_2$ concentration and boiling time at 90 °C on recovery of dissolved protein in nixtamalized yellow dent corn.

Compared with dissolved protein of initial corn material, the optimum process condition lowered dissolved protein from 0.05 mg/mL to 0.038 mg/mL (24 % or 0.24 times). Simultaneously decrease of dissolved protein is taken place at concentration $Ca(OH)_2$ solution of 30 %(w/w corn dissolved protein). This matter is possibility caused by coagulation to bind $Ca(OH)_2$ solution, in which long boiling time will become more and more low dissolved protein. This matter seems for the longest boiling time (60 minutes). Other possibilities were caused by processes relating with after boiling process, such as homogenization (8000 rpm, 30 minutes) or size reducing. Similar trend seemed at nixtamalized white dent corn, in which rate of nixtamalization reaction was fluctuative and was tend to raise relating with high concentration of $Ca(OH)_2$ solution. Increasing is reverse proportional with boiling time, in which boiling time becoming more and more long will drop dissolved protein, as demonstrated in Figure 8.

Optimization process was achieved at concentration of $Ca(OH)_2$ solution of 30 % (w/w corn dissolved protein) for 15 minutes of boiling time and gave soluble protein 0.041 mg/mL followed by lowering (0.033 mg/mL) relating with long boiling time for 60 minutes. When compared to dissolved protein of initial white dent corn material (0.07 mg/mL), the optimum process condition drops dissolved protein from 0.07 mg/mL to 0.041 mg/mL (41.43 % or 0.41 times).

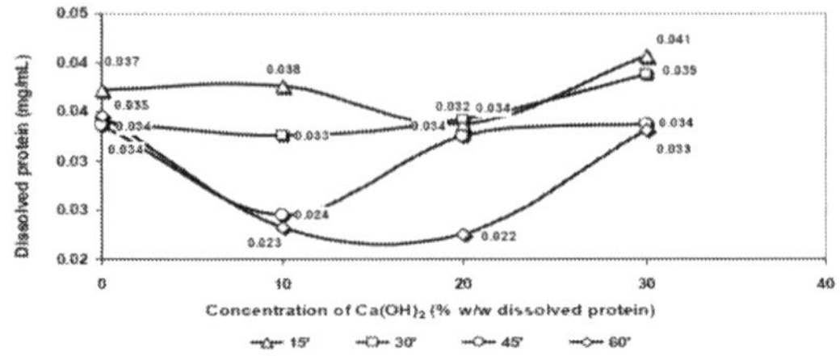


**Fig. 8**. Relationship between $Ca(OH)_2$ concentration and boiling time at 90 °C on recovery of dissolved protein in nixtamalized white dent corn.

### 3.2.4 *Total sugar*

Total sugars on nixtamalized corn is a guidelines from all carbohydrates according to Phenol – Sulphate metode [16]. In boiling process of corn (90 °C), it is happened fractionation of amylose (27 %) and amylopectin (72 %) contained in corn to form gel at corn granular net and dissolve glucose expressed as total sugars. Treatment on yellow dent corn showed fluctuated nixtamalization process rate relating with increase of solution of $Ca(OH)_2$. Treatment of boiling time of 15 minutes raises total sugars until maximum concentration of $Ca(OH)_2$ (20 % w/w corn dissolved protein) to produce higher concentration of total sugar when compared to all treatments followed by dropping to maximum concentration of $Ca(OH)_2$ (30 % w/w corn dissolved protein). In other processes displays an increase of total sugars, although maximum concentration of $Ca(OH)_2$ (30 % w/w corn dissolved protein) and boiling time longer than 15 minutes, as shown in Figure 9.

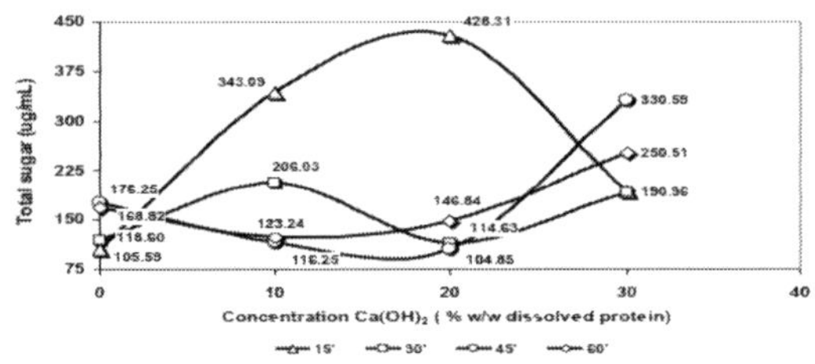


**Fig. 9**. Relationship between $Ca(OH)_2$ concentration and boiling time at 90 °C on recovery of total sugars in nixtamalized yellow dent corn.

Optimization process was achieved by adding solution of $Ca(OH)_2$ of 20% (w/w corn dissolved protein) for 15 minutes of boiling and gave total sugars concentration of 428.31 ug/mL. In this condition, it showed an increase of total sugars concentration from 428.31 ug/mL to 310.22 ug/mL (initial material of yellow dent corn 38.07 % (0.38 times). Whereas, white dent corn demonstrated a different trend of reaction rate, in which declining total sugars is taken place simultaneously at all process treatments, as showed in Figure 10.

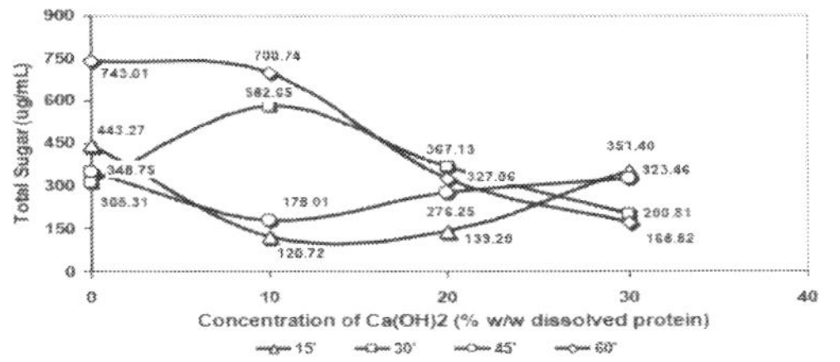


**Fig. 10**. Relationship between $Ca(OH)_2$ concentration and boiling time at 90 °C on recovery of total sugars in nixtamalized white dent corn.

Process optimization was obtained without adding solution of Ca $(OH)_2$ for 60 minutes of boiling and gave total sugars of 743.01 ug/mL. When compared to total sugars of initial corn material, optimum process condition increased total sugars from 208.38 ug/mL to 743.01 ug/mL (256.56 % or 2.56 times). This matter is possibility occurred autohydrolysis on corn granules netting like heat reaction on polysaccharides208.38 ug/mL become 743.01 ug/mL.

### *3.2.5 Reducing sugar*

Reducing sugar at nixtamalized corn is parameter used to degrade polysaccharides (starch) of corn through both heat reaction and natural enzymatic by bacteria during steeping process in water and boiling with or without adding solution of $Ca(OH)_2$. These processes are enabled to form reducing sugars, such as glucose and fructose according to Nelson-Somogyi method [16]. High content of amylose and amylopectin in corn causes retrogradation and syneresis easily. Adding solution of $Ca(OH)_2$ enables non-reactive fractionation process so that formation of monosaccharide with reduction property can be controlled. Nixtamalization process of yellow dent corn and white dent corn demonstrates a similar pattern, in which process rate displayes an fluctuated increase of reducing sugars relating with raising solution of $Ca(OH)_2$, however it is reverse proportional with boiling time. In other words, high concentration of $Ca(OH)_2$ solution and short boiling time will increase reducing sugar, as displayed in Figures 11 and 12.

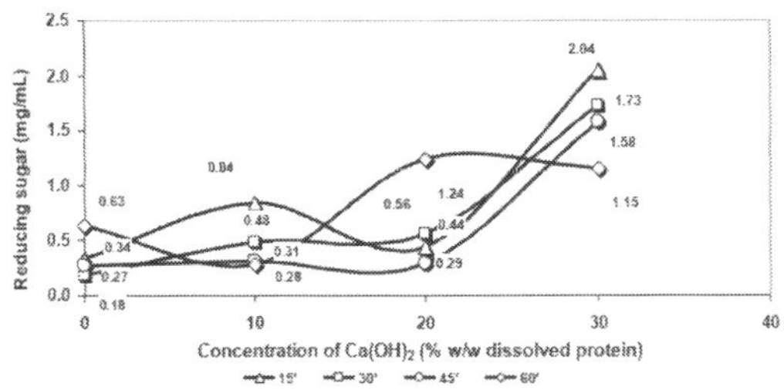


**Fig. 11**. Relationship between $Ca(OH)_2$ concentration and boiling time at 90 °C on recovery of reducing sugar in nixtamalized yellow dent corn.

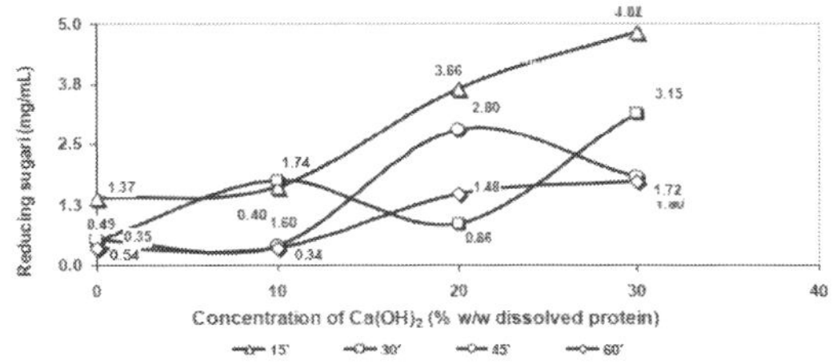


**Fig. 12**. Relationship between $Ca(OH)_2$ concentration and boiling time at 90 °C on recovery of reducing sugar in nixtamalized white dent corn.

Optimization of both nixtamalized yellow dent corn and white dent corn were achieved at concentration of $Ca(OH)_2$ of 30 % (w/w corn protein) for 15 minutes yielded reducing sugars of 2.04 mg/mL and 4.82 mg/mL, respectively. Compared with both reducing sugars of initial material of yellow dent corn (12.30 mg/mL) and white dent corn (9.95 mg/mL), optimum process condition declined reducing sugars of 83.41 % and 51.56 %, respectively. This matter showed that solution of $Ca(OH)_2$ will increase starch gelatinization. On the other hand, at similar period, heat effect with long boiling time will retain gelatinization in order to drop reducing sugars.

From description above, it had been known that based on the highest concentration of folic acid both yellow dent corn and white dent corn, optimization of nixtamalization process were achieved at treatment combination of concentration of $Ca(OH)_2$ solution of 20 % (w/w corn dissolved protein) for 60 minutes of boiling time and concentration of $Ca(OH)_2$ solution of 30 % (w/w corn dissolved protein) for 30 minutes of boiling time, and gave nixtamalized corn with folic acid concentration of 466.80 ug/mL and 506.74 ug/mL, as shown in Figures 13a and 13b.

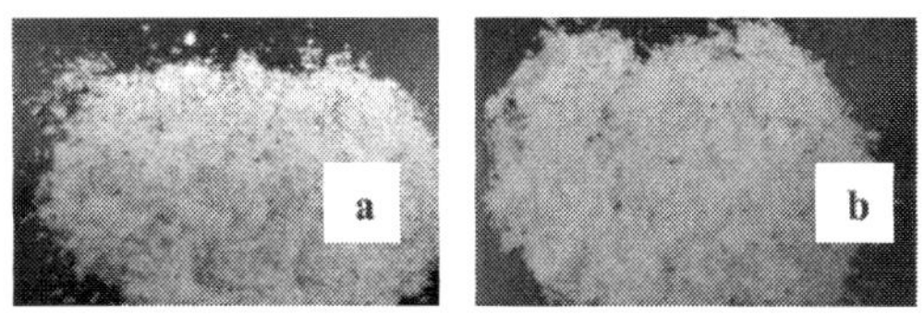


**Fig. 13.** (a) Nixtamalized yellow dent corn and (b) nixtamalized white dent corn at optimum process condition.

## 3.2 Identification of folic acid and glutamic acid monomer in nixtamalized corn

Identification on monomer of nixtamalized corn was performed on folic acid and glutamic acid as a part of folic acid. Identification result on standard glutamic acid was achieved 1 peak (T3.0) at retention time of 0 – 10 minutes with relative intensity of 100 %, in which in mass spectra *m/z* 111 – 784 from T3.0 showed domination of compounds with molecular weight (MW) of 148.1479 Da. (100 %), as shown in Figures 14 and 15.

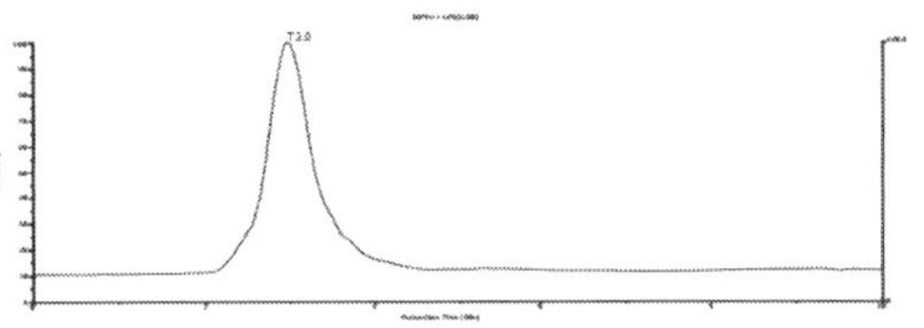

**Fig. 14.** Chromatogram of standard glutamic acid with retention time (T)3.0.

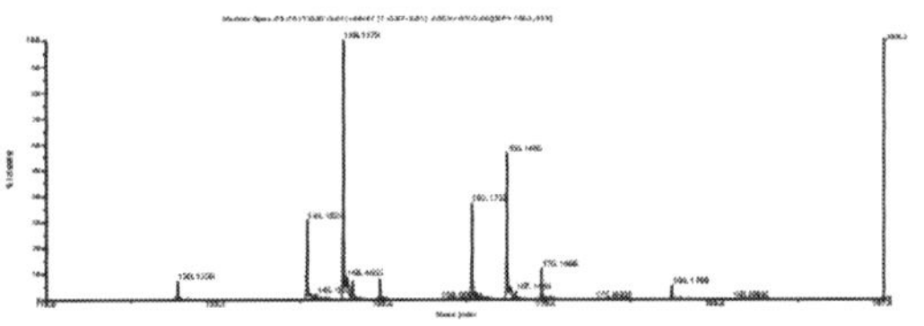

**Fig. 15.** Mass spectra in T 3.0 at *m/z* 128 – 207 at standard glutamic acid.

Folic acid is combination between pteridin heterocyclic, para-aminobenzoate acid (PABA) and glutamic acid [9]. In standard folic acid was reached 1 peak (T1.7) at retention time of 0 – 10 minutes with relative intensity of 100 %, in which in mass spectra *m/z* 425 – 498 from T1.8 displayed domination of compounds with MW of 442.76 Da. (100 %), 443.7 (25 %), and 441.48 (5 %), as showed in Figures 16 and 17.

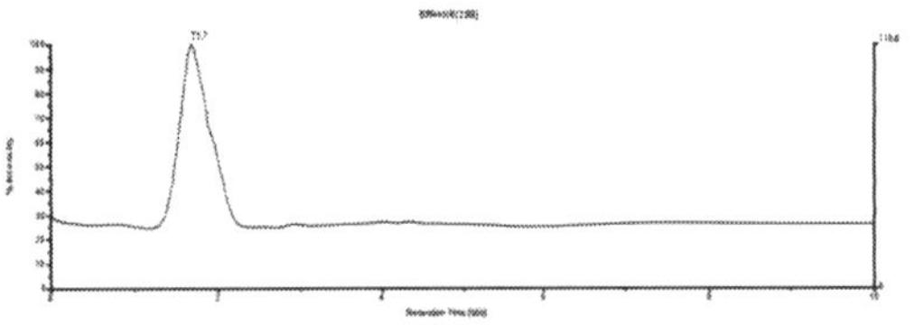

**Fig. 16.** Chromatogram of standard folic acid with retention time (T)3.0.

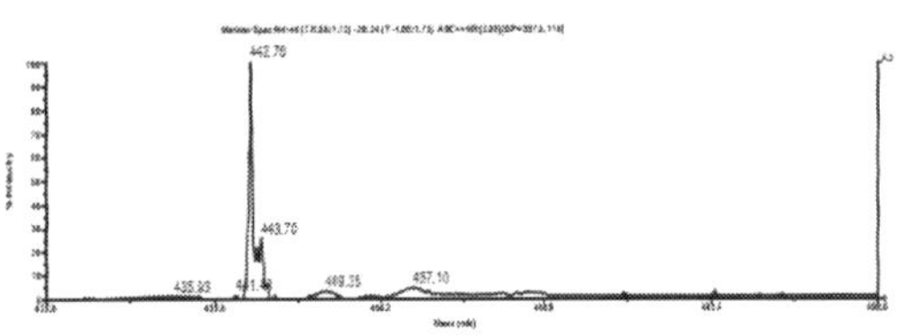

**Fig. 17.** Mass spectra in T3.0 at *m/z* 425 – 498 at standard folic acid.

Folic acid has MW of 441 Da. [9], however, due to LC-MS system (type of solvent, injection concentration, flow rate), it enables to degrade folic acid [18]. By using LC-MS method had been known that a compound indicated difference in MW, in which its possibility is as $M^+$, $M^+$ $Na^+$, $2M^{++}$ or $2M^+$, $Na^+$. This matter is caused by its presence of ionization as a consequence of sensitivity of LC-MS instrument relating to eluent used.

### *3.3.1 Nixtamalized yellow dent corn*

Identification result on the best treatment of nixtamalized yellow dent corn (boiling time of 60 minutes, concentration of $Ca(OH)_2$ solution of 20 % (w/w corn dissolved protein) was obtained chromatogram with 1 peak (T1.5) at retention time 0 – 10 minutes, as displayed in Figure 18.

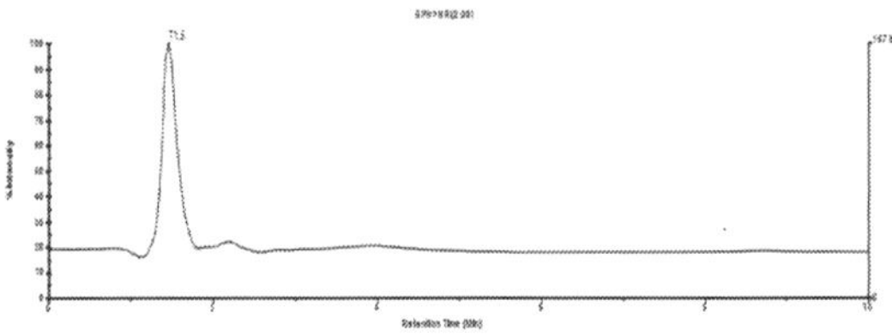

**Fig. 18.** Chromatogram of nixtamalized yellow dent corn from optimum process condition with retention time 0 – 10 minutes (T)1.5.

In mass spectra from T1.39 ranging *m/z* 99 – 1200 (Figure 19) did not shows glutamic acid monomer, although it had been known that relative intensity of glutamic acid was 1.41 %, while at this mass spectra range did not resulted folic acid. Glutamic acid monomer has been monitored at mass spectra ranging *m/z* 147.43 – 149 with MW between 147.94 – 148.93 Da. and relative intensity from 1.18 to 0.81 %. This mass spectra was dominated by monomer with MW of 148.27 Da. and relative intensity of 3.73 %, as demonstrated in Figure 20. Similar matter seemed on folic acid monomer, in which at mass spectra ranging *m/z* 440 – 443 demonstrates folic acid monomer with MW ranging 441.02 – 443.51 Da. and relative intensity of 4.93 – 11.82 %. This mass spectra was dominated by monomer with MW of 442.56 Da. and relative intensity of 25.51%, as shown in Figure 21.

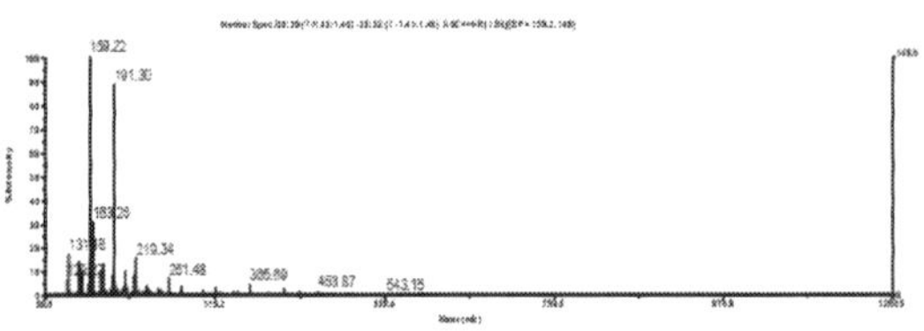

**Fig. 19.** Mass spectra of nixtamalized yellow dent corn from optimum process condition ranging *m/z* 99 – 1200.

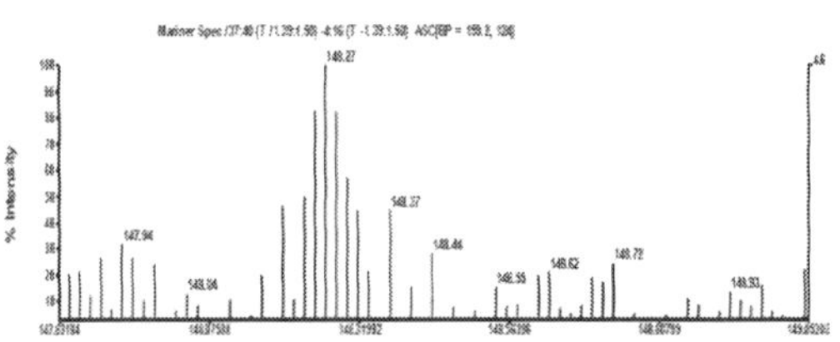

**Fig. 20.** Glutamic acid monomer in nixtamalized yellow dent corn from optimum process condition at mass spectra ranging *m/z* 147.43 – 149 in T1,5.

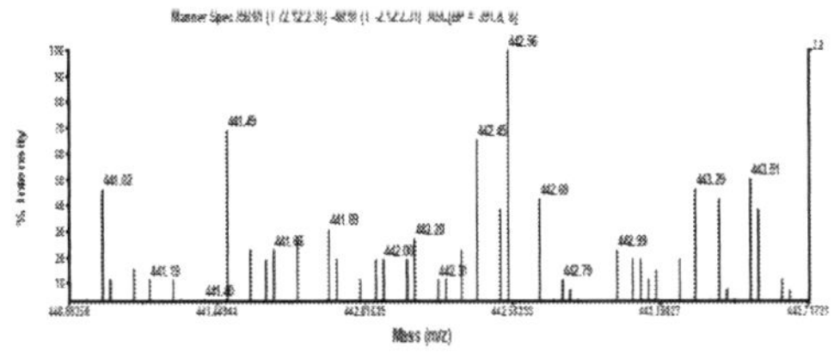

**Fig. 21.** Folic acid monomer in nixtamalized yellow dent corn from optimum process condition at mass spectra ranging *m/z* 440 – 443 in T 1,5.

### *3.3.2 Nixtamalized white dent corn*

Identification result on nixtamalized white dent corn from the best treatments (boiling of 30 minutes, concentration of Ca(OH)2 solution of 30 % (w/w corn dissolved protein) was achieved chromatogram with 1 peak (T1.5) at retention time of 0 – 10 minutes, as shown in Figure 22. Mass spectra from T1.5 ranging *m/z* 99 – 1200 (Figure 23) did not demonstrate glutamic acid monomer, although it is monitored clearly at mass spectra ranging *m/z* 147.39 – 148.96 with MW ranging 147.92 – 148.86 Da. and relative intensity from 0.71 to 1.8 %. This mass spectra was dominated by MW of 148.32 Da. and relative intensiy of 1.8 %, as showed in Figure 24. Folic acid monomer with MW of 441.94 Da. was reached with relative intensity of 0.12 % at ranging *m/z* 99 – 1200 (Figure 23), however at mass spectra ranging *m/z* 440.81 – 443.38 with MW of 440.92 – 443.23 Da., and relative intensity from 0.43 – 2.29 % displayed clearer.

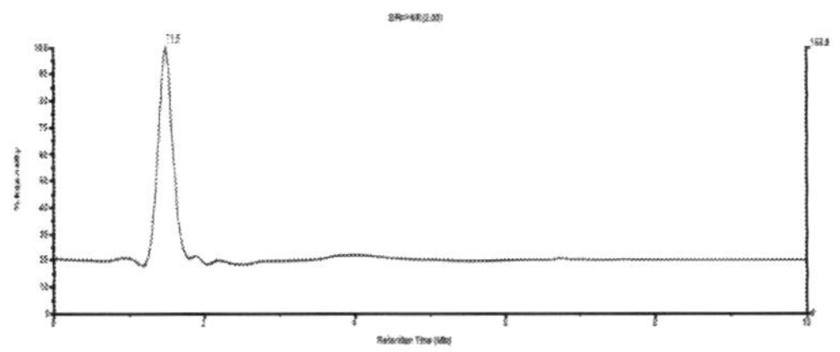

**Fig. 22.** Chromatogram of nixtamalized white dent corn from optimum process condition with retention time 0 – 10 minutes (T)1.5.

**Fig. 23.** Mass spectra of nixtamalized white dent corn from optimum process condition ranging *m/z* 99 – 1200.

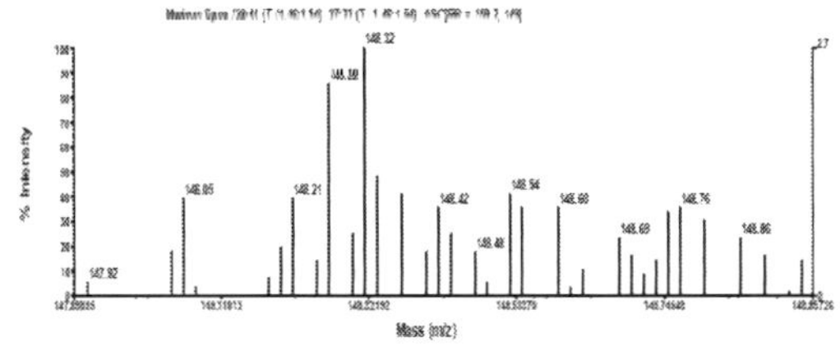

**Fig. 24.** Glutamic acid monomer at mass spectra ranging *m/z* 147.39 – 148.96 in T1,5 nixtamalized white dent corn from optimum process condition .

This mass spectra was dominated by a monomer with MW 441.73 Da. and relative intensity of 100 %, as displayed in Figure 25. On these whole identifications had showed that boiling (90 °C) and further processes (boiling, reducing size) remains still folic acid and glutamic acid by bonding of $Ca(OH)_2$ on nixtamal. Glutamic acid is combination between pteridin heterocylic, para-aminobenzoat (PABA) and glutamic acid [9].

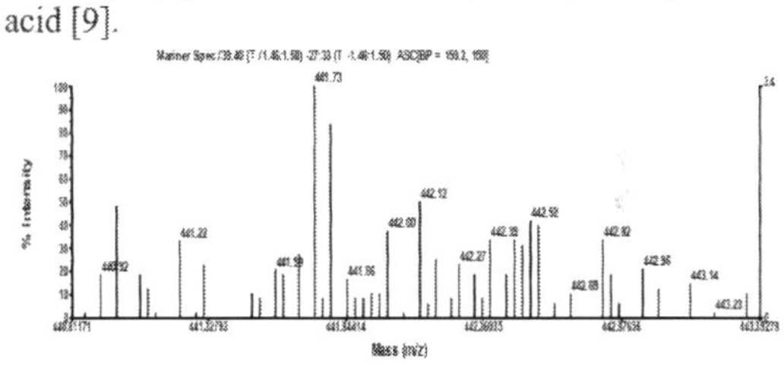

**Fig. 25.** Folic acid monomer in nixtamalized white dent corn from optimum process condition at mass spectra *m/z* 440.81 – 443.38 in T 1,5.

## 4 Conclusions

Nixtamalization process with difference concentration of solution of $Ca(OH)_2$ and boiling time of corn affected on the overall nixtamalization compositions. $Ca(OH)_2$ solution concentration becoming more and more high increases concentrations of folic acid and reducing sugar, declines total solids and total sugars, and tends to fluctuate and drop dissolved protein both two types of corn. Based on the highest folic acid, optimization of nixtamalization treatments at yellow dent corn and white dent corn were achieved at combination between concentration Ca(OH)2 solution of 20 % (w/w corn dissolved protein) for 60 minutes of boiling and concentration Ca(OH)2 solution of 30 % (w/w corn dissolved protein) for 30 minutes of boiling. In these optimum conditions, it is produced both nixtamalized flours of yellow dent corn and white dent corn with compositions of folic acid of 466.81 and 506.74 μg/mL, total solids of 43.78 and 47.06 %, dissolved protein of 0.029 and 0.039 mg/mL, total sugars of 146.840 and 200.809 ug/mL, and reducing sugar of 1.236 and 3.151 mg/mL, respectively, and give an increase on folic acid 192.3 % (1.9 folds) and 139.89 % (1.4 folds) compared with initial materials of yellow dent corn and white dent corn. Identification on folic acid and glutamic acid monomers in both nixtamalized yellow dent corn and nixtamalized white dent corn at optimum operation condition demonstrated domination of folic acid monomer of MW of 442.56 and 441.73 Da. with relative intensity of 25.51 and 100 %, respectively. While, glutamic acid monomer in both nixtamalized yellow dent corn and nixtamalized white dent corn were dominated by monomers with MW of 148.27 and 148.32 Da., and relative intensities of 3.73 and 1.8 %, respectively.

## References


1. Anonymous. Recommendations for the use of folic acid to reduce the number of cases of spina bifida and other neural tube defects. MMWR Recomm. Rep., **41** (No. RR.14) (1992).
2. De Wals, P., Rusen, I. D., Lee, N. S., Morin, P., and Niyonsenga, T. Trend in prevalence of neural tube defects in Quebec. Birth Defects Res A Clin Mol Teratol., 67, p.919 – 23 (2003).
3. Anonymous. EUROCAT. Special Report : Prevention of Neural Tube Defects by Periconceptional Folic Acid Supplementation
4. Wronkowska, M. Wet-Milling of Cereals, Journal of Food Processing and Preservation ISSN 1745-4549, **40**, p.572 – 580. doi:10.1111/jfpp.12626 (2016).
5. Rooney, L.W. and Serna-Salvidar, S.O. *Food uses whole corn and dry milled factions*. in Corn : Chemistry and Technology (Watson, S. A. and Ramsted, P. E., ed.). ACCC. St. Paul, M. N. (1987), F. (ed.), Molecular Structure and Function or Food Carbohydrate. *Appl. Sci. Publ. Ltd.*, London (1973).
6. Argun, M. S. and Dogan, I. S. Effects of varying nixtamalization conditions on the calcium absorption and pasting properties of dent and flint corn flours, Journal of Food Process Engineering, ISSN 1745–4530, **40** : e12436. Wiley Periodicals, Inc., p.1 – 7. doi:10.1111/jfpe.12436 (2017).
7. Banks, W., Greenwood, C. T. and Muir, D. D. The Structure of Starch in Birch, G. G. and Green, L
8. Nilsen, R. M., Leoncini, E., Gastaldi, P., Allegri, V., Agostino, R., Faravelli, F., Ferrazzoli, F., Finale, E., Ghirri, P., Scarano, G., and Mastroiacovo, P. Prevalence and determinants of preconception folic acid use : an Italian multicenter survey, Italian Journal of Pediatrics, Published online July 13, **42** : 65, p.1 – 10. DOI 10.1186/s13052-016-0278-z. Open Access. CrossMark (2016).
9. Belitz, H. D., Grosch, W., and Schieberle, P. *Food Chemistry*. 4th revised and extended edition, Springer-Verlag, Berlin Heidelberg. ISBN 3-540-15043-9 (2009).
10. Anonymous. *Folic acid, Semolus*. Accessed at 8 December 2014 (2008).
11. Anonymous. FDA approves folic acid fortification of corn masa flour, U. S. Food and Drug Administration News Release, April 14 (2016).
12. Hamner, H. C. and Tinker, S. C. Fortification of corn masa flour with folic acid in the United States: an overview of the evidence, ANNALS OF THE NEW YORK ACADEMY OF SCIENCES. ISSN 0077-8923. **1312**, p.8 – 14. New York Academy of Sciences (2014).
13. Onggo, D. General Principles in Electrospray Mass Spectrometry : A New Technique in Mass Spectral Analysis. *Journal of Mass Spectrum*, Vol. **3**, No. 2, pp.115 – 131, Wikipedia. Accessed at 3 February 2013 (1998).
14. *Dass, Chreh. Dasar-dasar Spektrometri Massa Kontemporer. John Wiley & Sons, Inc., pp.151 – 194.* doi : 10.1002/9780470118498.ch5. ISBN 9780470118498. Accessed at 20 April 2017 (2007).
15. *Lowry, O. H., Rosebrough, N. J., Farr, A. L., Randall, R. J. Protein measurement with Folin phenol reagent. J. Biol. Chem. 193 (1), pp.265 – 276.* PMID 14907713 *(1951).*
16. AOAC. *Official Methods of Analysis*. Association of Official Analytical Chemists International, 20th. Edition, Washington D. C. (2016).
17. Ruengsitagoon, W. and Hattanat, N. Simple spectrophotometric method for determination of folic acid. *The 4th Annual Northeast Pharmacy Research Conference*, Thailand (2012).
18. Eichhorn, P. and Knepper, T. P. Electrospray ionization mass spectrometric studies on the amphoteric surfactant cocamidopropylbetaine. *Journal of Mass Spectrum*, June : **36** (6), pp.677 – 684. DOI : 10.1002/jms.170 (2001).
19. Susilowati, A., Iskandar, Y. M., Melanie, H., Maryati, Y., Lotulung, P. D., Aryani, D. G. *Pengembangan konsentrat sayuran hijau dan kacang-kacangan terfermentasi pada jagung (Zea mays L.) pramasak sebagai sumber asam folat untuk formula pangan pintar*. Laporan Hasil Penelitian, Program Tematik, Kedeputian IPT, Tahun Anggaran 2015, Pusat Penelitian Kimia – LIPI, Kawasan PUSPIPTEK, Serpong, Tangerang Selatan (2015).
20. Verlinde, P., Oey, I., Deborggraeve, W., Me, J., Hendrickx, M. E., Van Loey, A. M. Mechanism and related kinetics of 5-methyltetrahydrofolic acid degradation during combined high hydrostatic pressure–thermal treatments. *Journal of Agricultural and Food Chemistry*, **57** *(15) :* 6803 – 14. Accessed at 25 April 2017 (2009).
21. Delchier, N., Ringling, C., Maingonnat, J. F., Rychlik, M., Renard, C. M. G. C. Mechanisms of folate losses during processing: diffusion vs heat degradation. *Food Chem.*, 157: pp.439 – 47. Accessed at 25 April 2017 (2014).
22. Fleche, G. *Chemical Modification and Degradation of Starch* in G.MA. Van Beynum and J. A. Rolls (eds.), Marcel Dekker Inc., New York, U. S. A. (1985).

MATEC Web of Conferences **154**, 01018 (2018) https://doi.org/10.1051/matecconf/201815401018
*ICET4SD 2017*

# The effect of treatment with activated carbon on the metal content in reuse of lubricating oil waste

Riyanto*, *Tri Astuti* Agussetianingrum, and *Dwi Putri* Juliantydjawi

*Department of Chemistry, Faculty of Mathematics and Natural Science, Islamic University of Indonesia, Jalan Kaliurang KM 14,5, Sleman Yogyakarta Indonesia 55584*

**Abstract.** The effect of treatment with activated carbon to the metal content in the reuse of lubricating oil waste has been done. Waste lubricating oil has treatment using 1-butanol and KOH as a solvent and coagulant. The research studies of the effect of activated carbon to the metals are Pb, Cr, Ca, Mg and Fe in lubrication oil waste after treatment. Waste lubricating oil treatment by the adsorption method using activated carbon with various weight are 0,5; 1,0; and 1,5 g. The metal concentrate was analysis after and before treatment using Atomic Adsorption Spectrophotometer (AAS). The analysis result shown the best weight of activated carbon for decreased of metals contains in lubricating oil waste is 1.5 g. Metal concentration of Pb, Cr, Ca, Mg and Fe in lubricating oil waste before treatment are 181.0002, 10.7198, 1019.0220, 325.8788 and 365.1329 mg/L (ppm), respectively. Metal concentration of Pb, Cr, Ca, Mg and Fe in lubricating oil waste after treatment are 47.5670, not detection, 2.6871, 44.3251 and 222.043 mg/L (ppm), respectively. Base on ASTM D5185 standard shown the Ca, Mg and Cr metals concentration according to the new lubricating oil quality standard. Pb and Fe metals concentration after process are still above the new lubricating oil quality standard. As a conclusion is activated carbon is a good material for treatment of waste lubricating oil, especially to reduce metal concentrations.

## 1 Introduction

Waste lubricating oil is hazardous and toxic waste according to Government Regulation No.101, the year 2014. Waste lubricating oil can hazards to humans, the environment, and other living creatures so that there is a need for B3 waste management processes including waste oil. Waste lubricating oil waste contains contaminants such as water, salt, impurities, metals and others [1]. The metal on used oil is derived from engine wear. On the move of the engine is turned on, the moving machine will occur friction on the metal which will cause the release of particles or metal molecules from the event [2].

Heavy metals are very toxic and will harm the environment. The properties of heavy metals are difficult to degrade, can accumulate in organisms including shellfish and fish, and endanger human health, easily accumulate in sediments, so their concentration is higher than the concentration of metals in water. The waste lubricating oil contains heavy metals are Fe, Ca and Mg which are semi toxic metals and toxic metals are Pb and Cr.

According to Ali et al. [1], the potential of waste lubricating oil is very extraordinary. So many industries have done waste oil waste management into new oil that can be used again. Methods used such as the method of vacuum and hydrotreating distillation. This method is difficult to be a low investment model in small scale industries, as well as with acid clay method. The acid clay method has low investment but is not environmentally friendly with its tar products containing sulfuric acid and is very dangerous.

Several studies of waste lubricating oil processing into new oil have been widely practiced, for example, Mohammed et al. [3] using solvent extraction and adsorption methods. In the extraction process varied solvents are n-hexane, 1-butanol, ether, 1-hexanol, carbon tetra chloride and acetone. Udonne [4] has been to recycle waste lubricating oil using four methods are adding acid, vacuum distillation, acid addition and addition of activated charcoal. The quality of the oil is tested at the point of flame, pour point, type density, metal content, viscosity, and sulfur content. The extraction and adsorption of lubricants have been done by Kamal and Khan [5]. In this study, the best solvent was 1-butanol produced by the maximum sediment followed by methyl ethyl ketone, 1-hexanol, and 2-butanol. Hussein, et al. [6] has been researched is oil refining may be performed by solvent extraction. The solvents used are alcohols (methanol, 1-propanol, butyl alcohol and isobutanol), ketones (acetone and methyl ethyl ketones) and hydrocarbon solvents ie n-hexane.

Reduction of the concentration of heavy metal content in waste water using activated carbon has been done by Dinesh et al. [7]. Activated carbon has a good ability to absorption of heavy metals. Several researchers has been used activated carbon for absorption $Pb^{2+}$ [8], absorption of heavy metals [9] and removal of heavy metals from

*Corresponding author: riyanto@uii.ac.id

industry waste water [10]. Heavy metals analyzed in this research are metal Pb, Fe, Cu, Cd. Maoura, et al. [11] in the study explains that the oil waste has a substance that is carcinogenic such as PAH so that the research aims to reduce the content of PAH in used oil. The best method of waste lubricating oil treatment has been done by Mohammed et al. [3]. Waste lubricating oil has been treatment using 1-butanol dissolution with a 3: 1 ratio with oil, coagulation using KOH and adsorption using acid-activated clay. The method uses fewer organic solvents, uses most of the processes at room temperature and reduces the effects of global warming and has a lower investment. Fuadi et al. [12] has been done using activated carbon as the heavy metal adsorbent in leachate water is able to absorb heavy metals in the waste.

The weakness of some researchers conducted is not yet report the effect of treatment on the metal content and conformity with ASTM standard. In this paper will deliver the results of the study the effect of treatment with activated carbon to the metal content and organic compounds in the reuse of lubricating oil waste. The aim of this study was to the effect of treatment of waste lubricating oil using activated carbon to the metals concentrations (Pb, Cr, Ca, Mg, Fe). The quality of new lubricating oil or waste lubricating oil after treatment was compared with ASTM standard.

## 2 Experimental Sections

### 2.1 Materials and Instrumentations

This study uses chemicals such as waste lubricating oil from collected from a motor vehicle repair shop, potassium hydroxide (KOH), 1-butanol, activated carbon, hydrochloride acid (HCl), nitric acid ($HNO_3$). All chemicals used in this research are of pro analysis grade and manufactured by Merck. All the solutions in the research have been prepared using distilled water. This research was used Atomic Absorption Spectrophotometer (AAS) from Perkin Elmer PinAAcle 900T for determination of heavy metals and Gas Chromatography-Mass Spectrometry (GC-MS) from Shimadzu QP 2010 SE was used for determination of organic compounds in before and after waste lubricating oil treatment.

### 2.2 Waste lubricating oil treatment

Waste lubricating oil and 1-butanol is inserted into 100 mL beaker glass with a ratio of 1: 3. The mixture is added 2 g of potassium hydroxide (KOH). The mixture is stirred for 30 minutes or until all of the mixture dissolves at 350 rpm. After stirring process is allowed to stand for 24 hours at room temperature. Then the solution is filtered with Whatman filter paper 4.2 with the help of the vacuum pump. The filtrate is distilled to separate 1-butanol with oil. The end result of this step is the new lubricating oil. The activated carbon is inserted into the porcelain cup, then activated with the oven at 105 ° C for 1 hour. Activated carbon with a variation of 0.5, 1.0 and 1.5 g. The activated carbon is introduced into the sample solution. The mixture was stirred with the stirrer for 60 minutes or until all mixed and heating 60 °C. The mixture is allowed to stand for 24 hours, then filtered. The result is new lubricating oil and analysis using AAS and GC-MS.

### 2.3 Preparation of sample for analysis heavy metals using AAS

Waste lubricating oil before and after treatment was weighed 1.68 g in the porcelain cup, and then heated with the oven at 100 °C for 1 hour. The sample was heated for 1 hour with a temperature of 600 °C using the furnace. The result of this process is ash. The ash was added $HNO_3$ and HCl each of 5 mL, then heated until the vapor is lost. The sample solution is included in a 100 mL volumetric flask and diluted with distillate water. The solution was analyzed by AAS to determine the heavy metal content of Mg, Fe, Pb, Ca, and Cr.

### 2.4 Preparation of sample for analysis organic compounds using GC-MS

Waste lubricating oil before and after treatment was taken as much as 15 mL and added with n-hexane as much as 60 mL and acetonitrile as much as 40 mL. The sample is inserted in a separating funnel, then shaken for 1 hour and allowed to stand for 30 minutes until two layers are formed. The upper layer is collected and then evaporated using rotary evaporator at 74°C and analyzed by GC-MS.

## 3 Result and Discussion

### 3.1 Waste lubricating oil before and after treatment with activated carbon

Figure 1 shows the color changes of waste lubricating oil before treatment (1A) and wastes lubricating oil before treatment (1B). Waste lubricating oil after treatment is obtained oil with the same color as new oil or lubricating oil before use. Waste lubricating oil has a black color and wastes lubricating oil after treatment has a brownish yellow color. Waste lubricating oil that has been the treatment to produce oil that has the same color with the new oil.

### 3.2 Chromatogram GC-MS waste lubricating oil before and after treatment

Figure 2 and Table 1 shown the waste lubricating oil has been chemicals compounds are 3-penten-1-ol, n-Hexane, methyl cyclopentane, 1,2,4-trimethyl benzene, cyclopentane, cycloheptane, 2-methyl naphthalene, hexadecanoic acid methyl ester, nonyl-diphenylamine, octyl-diphenylamine, 1,2-benzene dicarboxylic acid, dioctyl ester, 1,4-dibutoxybutane. The 3-penten-1-ol and n-Hexane in the waste lubricating oil have the highest levels compared to other compounds. The compounds of n-hexane have been derived from the solvent used in the extraction process. Carboxylic compounds are produced from the oxidation process of the compound with the OH group.

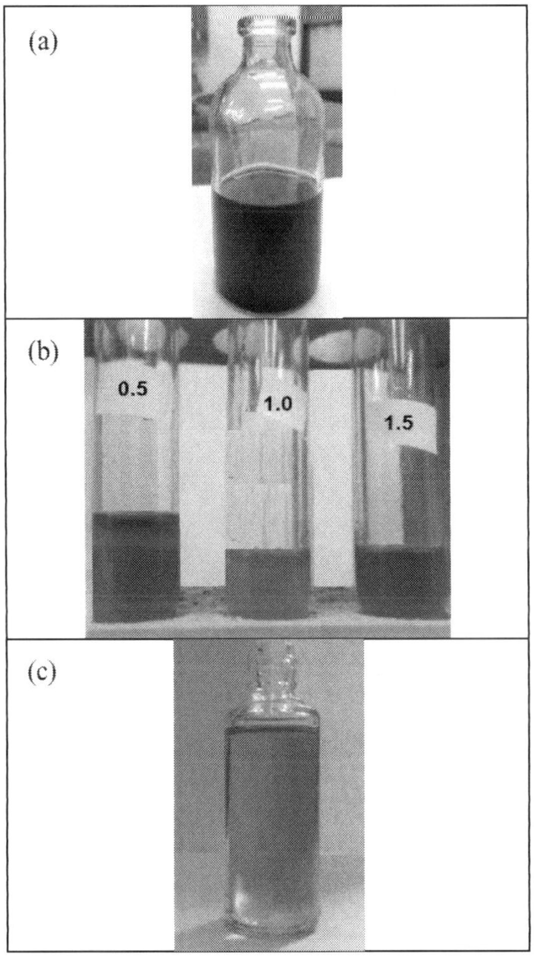


**Figure 1.** Image of waste lubricating oil before treatment (A) after treatment with activated carbon 0.5, 1.0 and 1.5 g (B) and new lubricating oil (C)

The results of this study are different from the results of the study Khalaf et al. [13]. Khalaf et al. [13] has been done separation and identification of organic compounds in lubricating oil additives using GC-MS. Lubricating oil have compounds are 2-ethyl-1-hexanol, 2-(t-butyl)-phenol, 2,4-bis-(t-butyl)-1-methoxy benzene, 2,6-bis-(t-butyl)-2,5-cyclohexadiene-1,4-dione, 2,4-bis-(t-butyl)-phenol, 3,5-bis-(t-butyl)-phenol, decanoic acid methyl ester, tridecanol, 1,1-diphenyl hydrazine; 1,2-benzene dicarboxylic acid dibutyle ester, S-triazol 1,5-A-pyridine 8-amino 2-phenyl, phenol-2,6-bis-(t-butyl), phenol-2,5-bis-(tbutyl), phenol - 2,4,6-Tris-(t-butyl).

Waste lubricating oil after treatment using butanol 45 and 60 mL has been the result the chemicals compounds is di-9-octadecenoyl-glycerol. This chemicals compound is the main compound in the lubricating oil. The n-butanol compound in the waste lubricating oil is from the solution was used treatment.

Table 2 shown is the effect of treatment of waste lubricating oil using activated carbon to the concentration of Pb, Cr, Ca, Mg and Fe metals. Waste lubricating oil to contain the metals of Pb, Cr, Ca, Mg and Fe is 181.00, 10.72, 1019.02, 325.88, and 365.13 mg/kg, respectively. When the values obtained in the waste lubricating oil after treatment with activated carbon samples were compared with the standard values, it has been determined that the metals have reached to standard levels for the new lubricating oil (Table 2). The Pb, Cr, Ca, Mg, Fe concentrations in waste lubricating oil before and after treatment were compared with international standards. Pb and Fe values were higher than standard values of ASTM D5185. The obtained results showed that, with the exception of Pb and Fe, the metal concentrations in water did not exceed ASTM D5185.

A significant decrease in concentration occurred in Cr metal. Cr metal concentration on waste lubricating oil of 10.2 mg/kg drops to very low concentration so it is not detected with AAS (Table 2). Cr metal is the heavy metal so its presence in oil is very dangerous. Cr metal is very low concentrations indicate excellent oil quality. Zajac et al. [14] has been the determination of metals content in used oil using XRF method. The contents of iron, and copper also were high, at Fe 16.5-267 mg/kg. Low manganese, chromium, and lead contents were found in the ranges Mn 0.41-4.27 mg/kg, Cr 0.36-10.66 mg/kg, and Pb 0.9-13.71 mg/kg. The difference in the results is due to the difference in the sample of used oil analyzed.

Pb and Cr metals are classified as hazardous, toxic and carcinogenic substances. Pb and Cr, when introduced into water reservoirs, is significantly absorbed by sediments. However, it also is subject to strong bioaccumulation, particularly in phytoplankton, which effects its fast incorporation into the food chain. In lubricating oils metals mainly comes from the piston rings and the shaft.

## 3 Conclusions

Waste lubricating oil after treatment has color same new lubricating oil. The color is brownish yellow produced by using adsorbed by activated carbon with weight as 1.5 g. The resulting color is brownish yellow, while the new lubricating oil color is crystal clear yellow. The effect of adsorbed by activated carbon with weight as 1.5 g to metal contained in waste lubricating oil is the decrease in Ca and Pb metals. The best amount of activated carbon used for metal adsorption in waste lubricating oil treatment is 1.5 g. This indicates that the increasing amount of activated carbon, the increase of absorption capacity of the metal in the waste lubricating oil. As a conclusion is activated carbon is a good material for treatment of waste lubricating oil, especially to reduce metal concentrations.

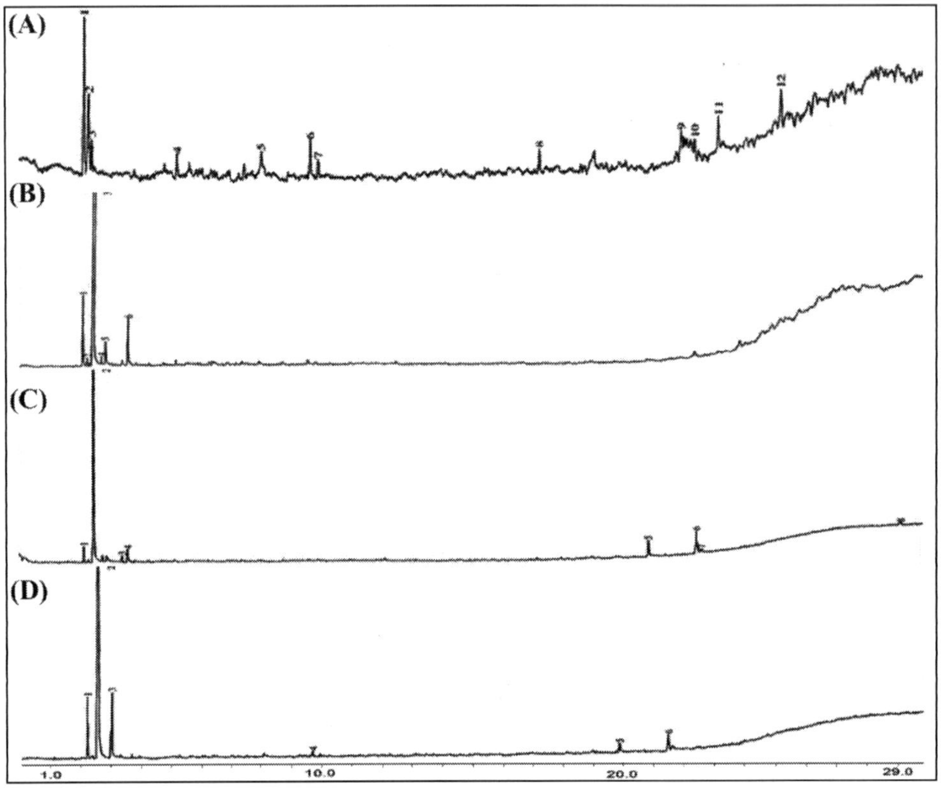


**Figure 2.** Chromatogram from analysis waste lubricating oil (A) after treatment using 30 (B) 45 (C) and 60 mL (D) n-butanol using GC-MS

**Table 1.** Chemical composition of waste lubricating oil before and after treatment using 30, 45 and 60 mL butanol

| Retention time | Name of compounds | Waste lubricating oil % area | After treatment with Butanol in mL (% area) | | |
|---|---|---|---|---|---|
| | | | 30 | 45 | 60 |
| 2.192 | 3-penten-1-ol | 28.69 | - | - | - |
| 2.367 | n-Hexane | 19.86 | - | - | - |
| 2.467 | Methylcyclopentane | 4.80 | - | - | - |
| 5.267 | 1,2,4-Trimethylbenzene | 4.59 | - | - | - |
| 8.100 | Cyclopentacycloheptane | 3.01 | - | - | - |
| 9.725 | 2-Methylnaphthalene | 7.87 | - | - | - |
| 9.988 | 2-methyl-naphthalene | 3.42 | - | - | - |
| 17.283 | Hexadecanoic acid methyl ester | 4.40 | - | - | - |
| 21.967 | Nonyl-diphenylamine | 5.72 | - | - | - |
| 22.408 | Octyl-diphenylamine | 3.16 | - | - | - |
| 23.183 | 1,2-Benzene dicarboxylic acid, dioctyl ester | 6.65 | - | - | - |
| 25.200 | 1,4-Dibutoxybutane | 7.83 | - | - | - |
| 2.201 | 3-Pentyn-1-ol | - | 2.11 | 0.94 | 1.21 |
| 2.367 | 2-methyl-butanal | - | 0.31 | - | - |
| 2.531 | Butyl alcohol-N | - | 95.50 | 92.77 | 96.20 |
| 2.831 | 2-methyl-1-butanol | - | 0.32 | - | - |
| 2.962 | Acetamide | - | 0.71 | - | 1.85 |
| 3.660 | 3-methyl-butanoic acid | - | 1.06 | 0.71 | - |
| 3.482 | 2-methyl-3-hexanone | - | - | 0.45 | - |
| 20.905 | 2-hydroxy-1,3propanediyl ester-hexadecanoic acid | - | - | 1.56 | - |
| 22.492 | Di-9-octadecenoyl-glycerol | - | - | 2.42 | 0.46 |
| 22.667 | 2-hydroxy-1,3propanediyl ester-Octadecanoic acid | - | - | 0.42 | - |
| 29.225 | 2-hexadecanol | - | - | 0.73 | - |
| 9.716 | 1-methyl-naphthalene | - | - | - | 0.10 |
| 19.857 | Butyl ester hexadecanoic acid | - | - | - | 0.18 |

 https://doi.org/10.1051/matecconf/201815401018

**Table 2.** The effect of treatment using activated carbon of waste lubricating oil to concentration of metal

| Sample | Concentration of metal (mg/kg) | | | | |
|---|---|---|---|---|---|
| | Pb | Cr | Ca | Mg | Fe |
| Waste lubricating oil | 181.00 | 10.72 | 1019.02 | 325.88 | 365.13 |
| 0.5 g activated carbon | 158.25 | nd* | 41.34 | 15.31 | 46.70 |
| 1.0 g activated carbon | 51.33 | nd* | 15.34 | 21.18 | 90.58 |
| 1.5 g activated carbon | 47.57 | nd* | 2.69 | 44.33 | 222.04 |
| ASTM D5185 standard | 16 | 0.12 | 1103 | 870.80 | 2.76 |

nd* = not detection

## References

1. S.M.W. Ali, A. Ripin, A. Ahmad, J. Appl. Sci. 10, 21, 2688 (2010)
2. S.M.W. Ali, A. Ripin, A. Ahmad, J. Appl. Sci. 10, 21, 2688 (2010)
3. M.S. Al-Nozili, F.A. Abeed, M.M. Ahmed, Int. J. Emerging Technol. Adv. Eng. 4,10, 304 (2014)
4. J.D. Udonne, J. Petroleum Gas Eng.2, 2, 12 (2010)
5. A. Kamal, F. Khan, Oil Gas Sci. Technol.-Rev. 64, 2, 191 (2009)
6. M. Hussein, A.A. Amer, A.S. Gassaberah, Am. J. Env. Eng. Sci. 1, 3, 44 (2014)
7. M. Dinesh, K.P. Singh, V.K. Singh, J. Hazard. Mater. 152, 1045 (2008)
8. Y. Kikuchi, Q. Qian, M.J. Machida, J. Carbon. 44, 195 (2006)
9. A. Kassim, C.G. Joseph, Z. Zainal, M.Z. Hussein, M.D.J. Haron, A.H. Abdullah, J. Indian Chem. Soc. 81, 946 (2004)
10. K. Kadirvelu, K. Thamaraiselvi, C. Namasiyayam, Bioresource Tech. 76, 63 (2001)
11. H.I. Khalaf, M.J.M. Hassan, O.A. Hassan, J. Al-Nahrain University. 15, 3, 62 (2012)
12. N.A.B. Fuadi, A.S. Ibrahim, K.N. Ismail, J. Purity, Util. React. Environ. 1, 5, 252 (2012)
13. L.G.M. Maoura, Assuncao, J.L. Filho, A.C.S. Ramos, Braz. J. Petrol.Gas, 4, 3, 91 (2010)
14. G. Zajac, J.S. Barglowicz, T. Slowik, A. Kuranc, A. Kaminska, Pol. J. Environ. Stud. 24, 5, 2277 (2015)

# Acid modified jackfruit wood sawdust as biosorbent for the removal of Fe(II) from aqueous solutions

*Tintin* Mutiara*, *Primadonna Ciko* Karisa, and *Intan* Mujahidah

Department of Chemical Engineering, Islamic University of Indonesia, Jl. Kaliurang Km.14.5 Sleman, 55584 Yogyakarta, Indonesia

**Abstract.** The adsorption of Fe(II) from aqueous solutions by acid modified jackfruit wood sawdust biosorbent was investigated in a batch adsorption system. The experiment conducted at room temperature and pH 5 resulted adsorption behaviour of chemical modified Jackfruit wood sawdust biosorbent. The removal of Fe(II) has been found to be the function of initial Fe(II) concentration, acid type and acid concentration. The best removal was at low Fe(II) initial concentration, the higher initial concentration resulted on lower adsorption. Untreated jackfruit wood sawdust has the highest adsorption efficiency, followed by acetic acid, citric acid and tartaric acid modified jackfruit wood sawdust. The maximum uptake capacity of Fe (II) ion was 1.05 mg/g by untreated jackfruit wood sawdust.

## 1 Introduction

Environment contamination by heavy metals drawn concern in the last few decades. Certain concentration of those heavy metals become severe health hazard on human and animals. Heavy metals are actually needed in industrial processes such as electroplating, aerospace, and metal industries [1]. Disposal streams from industries such as mining, fertilizer, tannery, pulp industries, contain various kind of heavy metals for example Cd, Pb, Ni, Cr, As, Cu, Fe and so on [2]. If those streams disposed to environment continuously without proper treatment, it could damage environment. Metals are non-biodegradable compound and tend to accumulate in organism which generate various diseases and disorders [3].

Various methods, such as precipitation, ion exchange, membrane processes, adsorption, coagulation, flotation, etc., have been applied for the removal of heavy metals from aqueous solution [4-7]. Some of those processes are often restricted because of technical or economic constraints [8]. Therefore, it is important to study and find more efficient and cheaper new process.

In the last few years, biosorption become popular because it has been successfully eliminate heavy metals from aqueous solutions. Several biosorbents have been tested to remove heavy metal from aqueous solutions, some of them are sugarcane bagasse [9], bagasse fly ash [10], coconut husk [11], Saw dust and neem bark [12], Chinese ephedra residue [13], sawdust white pine [14], activated carbon [15], etc. wood sawdust is biomass which mainly consist of cellulose (40 - 50%), hemicellulose (20 - 40%) and lignin (20 - 40%) [16]. Wood sawdust is biowaste from wood industry and its accumulation generate pollution and health problem [17]. In Indonesia, the yield of sawmill industry waste is around 50 – 60 %, and about 15 - 20% consist of wood sawdust. It was estimated that sawdust waste in Indonesia reached 0.78 million $m^3$/year [18] and it is still not optimally utilized.

Recent studies focus on a series of batch adsorption experiments to develop the potential of wood sawdust on removal metal ion from aqueous solutions. Salazar and Leyva [14] studied the adsorption of Pb (II) on acid modified and natural white pine sawdust. Citric acid modification at 25 °C and pH 5 had maximum adsorption capacity towards Pb (II) of 304 mg/g. Jeon and Kim [19] investigated adsorption capacity of modified oak sawdust with phosphoric acid on Pb (II). At pH 4 its maximum adsorption capacity of 192 mg/g. Naiya, et al. [12] studied the dosage level and equilibrium contact time of saw dust and neem bark on the removal of Zn (II) and Cd (II). Meena, et al. [20] treated acacia arabica sawdust to remove lignin which cause poor adsorption and red colour of the sawdust. The adsorption capacity of treated sawdust was 111.61 mg/g on Cr (VI), 52.38 mg/g on Pb (II), 20.62 mg/g Hg (II), and 5.64 mg/g on Cu (II). Untreated Chinese ephedra residue as adsorbent for the removal of heavy metal ion was studied by Feng and Zhang [13] resulted maximum uptake capacity for Pb (II) ions was 1.0 mmol/g at 298K.

This study deals with batch adsorption experiment determines the adsorption capacity of acid modified jackfruit wood sawdust for removal of Fe (II) from aqueous solutions. The effects of acid type, acid concentration, and initial Fe (II) concentration on the adsorption capacity were investigated.

* Corresponding author: tintin.mutiara@uii.ac.id

## 2 Materials and Methods

### 2.1. Materials

The sawdust used throughout this work was from wood of jackfruit (Artocarpus heterophyllus Lam.) obtained from local park near Islamic University of Indonesia, Sleman, Yogyakarta. The wood was milled and washed with deionized water several times then dried at 80 °C for 24 hours and sieved to 80 mesh of size.

Fe (II) stock solution of 1000 mg/L was prepared by dissolving particular mass of $FeSO_4$ in deionized water. The stock solution was diluted in deionized water to desired concentration.

### 2.2 Methods

#### *2.2.1 Biosorbent preparation*

Jackfruit sawdust was modified by acetic acid, citric acid and tartaric acid. 20 gram of jackfruit wood sawdust was soaked and mixed in 200 ml of acid solution for 2 hours at 60 °C. The concentration of acid were 0.5 M, 1.0 M, and 1.5 M. Afterwards, the mixture was allowed to cool down to room temperature, then the biosorbent was separated from solution by filtration with filter paper. The biosorbent washed with deionized water for several times until the pH of washing solution was constant, then oven dried at 50 °C.

#### *2.2.2 Batch adsorption*

A gram of biosorbent was equilibrated with 50 mL of the Fe(II) solution of known concentration. A 10 mL sample was taken and analysed to determine the concentration.

The following mathematic equation to calculate the Fe(II) adsorbed in biosorbent.

$$q_e = \frac{(C_o - C_e)V}{W} \quad (1)$$

The removal percentage (R%) or adsorption efficiency was calculated with following equation.

$$R\% = \left(\frac{C_o - C_e}{C_o}\right) x100 \quad (2)$$

Where $q_e$ (mg/g) is the mass of Fe(II) adsorbed in a gram of biosorbent at equilibrium, Co (mg/L) and Ce (mg/L) are concentration of Fe(II) at initial condition and at equilibrium, respectively. V (L) is the volume of Fe(II) solution and W(g) is the mass of biosorbent.

#### *2.2.3 Characterization of biosorbent*

The structure and surface functional groups were identified by Fourier Transform Infra Red (FTIR) spectroscopy. The FTIR spectra of modified biosorbent were recorded between 400 and 4000 $cm^{-1}$ in Shimadzu IR Prestige 21 Fourier Transform Infrared Spectroscopy.

## 3 Results and Discussions

### 3.1 Infra Red Spectra

In this work FTIR was used to collect information about the chemical structure and functional groups of the modified biosorbent. The FTIR spectra is shown in Fig.1, it is quite similar to mainly cellulose materials such as pistachio-nut shell, olive-waste, rockrose and Neem sawdust [21, 15, 22, 23].

Fig. 1 shows a number of peaks, indicating the complex nature of jackfruit sawdust. It showed strong and broad O-H stretching band centered at 3390 $cm^{-1}$, it also indicated $-NH_2$ group. The band located at 2903 $cm^{-1}$ assigned to the C–H group function in methyl and methylene groups. The peak at 1739 $cm^{-1}$ indicated to –C=O, band of carboxylic group. The peak at 1427 $cm^{-1}$ indicated the bending of C–H. The bands in the range of 1300–1000 $cm^{-1}$ could be assigned to the C–O stretching vibration of alcohols and carboxylic acids. The functional groups that important in sorption are –OH, –NH, carbonyl, and carboxylic groups [24]. Based on FTIR data, this modified biosorbent has potential to be adsorbent for the treatment of industrial waste effluents.

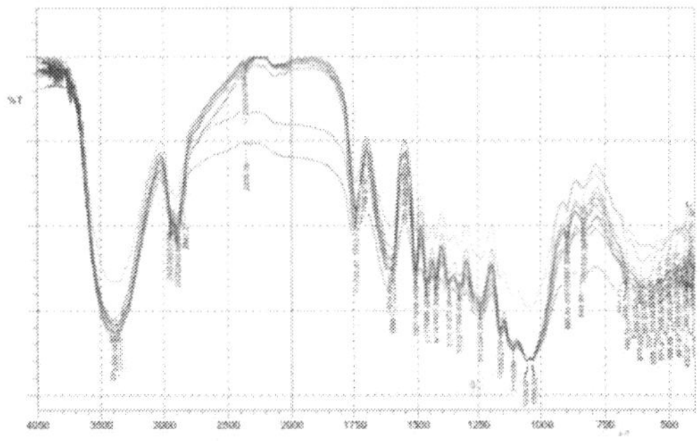

**Fig. 1.** FTIR spectra of prepared biosorbent.

### 3.2 Effect of initial concentration of Fe (II)

The experiment was conducted to investigate the effect of initial concentration of Fe (II) on adsorption capacity of modified biosorbent. The variation of initial concentration was at the range of 5 ppm to 40 ppm, room temperature and pH 5.

Fig. 2 showed the adsorption capacity per gram biosorbent increased at lower initial concentration. The adsorption capacity reached its peak around initial concentration of 24 ppm and then gradually decreased. The maximum uptake capacity of Fe (II) ion was 1.05 mg/g by untreated jackfruit wood sawdust.

https://doi.org/10.1051/matecconf/201815401019

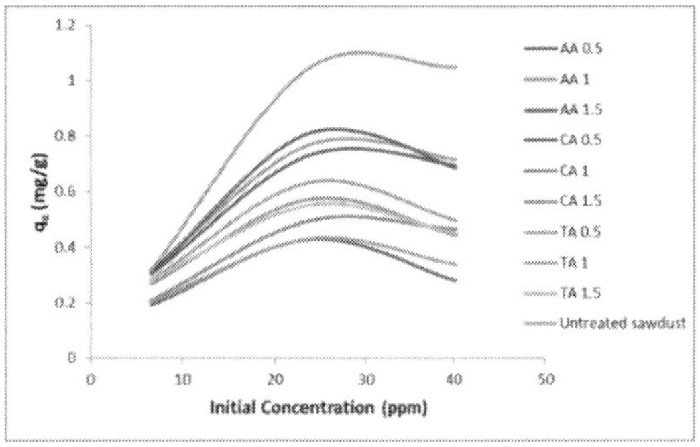


**Fig. 2.** Mass of Fe(II) adsorbed in every gram of acetic acid (AA), citric acid (CA), tartaric acid (TA) modified biosorbent and untreated sawdust

### 3.3. Effect of acid modification on Fe(II) adsorption capacity

The effect of chemical modification on percentage of the removal is illustrated in Fig.3. In this experiment, ratio between biosorbent and Fe(II) solution was 1 gram biosorbent to 50 ml Fe (II) solution. As seen in Fig.3, the best removal was at low initial concentration, the higher initial concentration resulting on lower adsorption. Sufficient adsorption sites are available for adsorption at low initial concentration. However, at higher concentrations the numbers of heavy metal ions are relatively higher compared to availability of adsorption sites [20]. Untreated jackfruit wood sawdust has the highest adsorption efficiency and acetic acid modified biosorbent has better adsorption efficiency compared to citric acid and tartaric acid modified biosorbent.

It is needed further study on adsorption behaviour of jackfruit wood sawdust for the development of cheap technology for the removal of heavy metal ion from aqueous solutions.

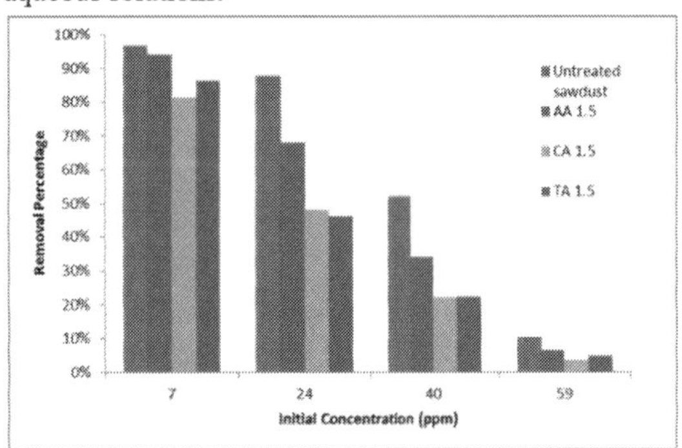


**Fig. 3.** Variation removal percentage of Fe(II) by modified biosorbent

## 4 Conclusion

Adsorption of heavy metal by modified Jackfruit wood sawdust biosorbent is a process related to initial concentration of Fe (II) solutions and also related to chemical modification. The best removal was at low initial concentration, the higher initial concentration resulting on lower adsorption. This study showed that untreated jackfruit wood sawdust has higher adsorption efficiency compared to the ones with acid modification and this can be a start to lead further study on adsorption behaviour of jackfruit wood sawdust for the development of cheap technology for the removal of heavy metal ions from aqueous solutions.

Acknowledgement is made to the Department of Chemical Engineering, Islamic University of Indonesia for the support of this research fund.

## References

1. Benhima, H., Chiban, M., Sinan, F., Seta, P., Persin, M. Removal of lead and cadmium ions from aqueous solution by adsorption onto micro-particles of dry plants. Colloids and Surfaces B: Biointerfaces, **61** 10-16. (2008).
2. Barakat, M. A. New trends in removing heavy metals from industrial wastewater. Arabian J. Chem. **4** 361–377. (2011).
3. Manzoor, Q., Nadeem, R., Iqbal, M., Saeed, R., Ansari, T.M. Organic acids pretreatment effect on Rosa bourbonia phyto-biomass for removal of Pb(II) and Cu (II) from aqueous media. Bioresour. Technol. **132** 446–452. (2013).
4. Esalah, J.O., Weber, M.E., Vera, J.H. Removal of lead, cadmium and zinc from aqueous solutions by precipitation with sodium di-(n-octyl) phosphinate. Can. J. Chem. Eng. **78** 948–954. (2000).
5. Zouboulis, A.I., Matis, K.A., Lanara, B.G., Neskovic, C.L. Removal of cadmium from dilute solutions by hydroxyapatite. II. Flotation studies, Sep. Sci. Technol. **32** 1755–1767. (1997).
6. Ravindran, V., Stevens, M.R., Badriyha, B.N., Pirbazari, M. Modeling the sorption of toxic metals on chelant-impregnated adsorbent. AIChE J. **45** 1135–1146. (1999).
7. Canet, L.,Lipide, M., Seta, P. Efficient facilitated transport of lead, cadmium, zinc and silver across a flat sheet-supported liquid membrane mediated by lasalocid A, Sep. Sci. Technol. **37** 1851–1860. (2002).
8. Zouboulis A I, Loukidou M X, Matis K A. Biosorption of toxic metals from aqueous solutions by bacteria strains isolated from metalpolluted soils. Process Biochem; **39**: 909–916. (2004)
9. Demirbas, A. Heavy metal adsorption onto agro-based waste materials: a review. J. Hazard. Mater **157**, 220-229. (2008).
10. Srivastava, V.M., Mall, I.D., Mishra, I.M. Equilibrium modeling of single and binary adsorption of cadmium and nickel onto bagasse fly ash. Chem. Eng. J. **117**, 79-91. (2006).
11. Hasan, S.M. and Ahmad, R. The potential of cost-effective coconut husk for the removal of toxic metal ions for environmental protection. J. Environ. Manag. **81**, 286-295. (2006).

12. Naiya, T.K., Chowdhury, P., Bhattacharya, A.K., Das, S.K. Sawdust and neem bark as low-cost natural biosorbent for adsorptive removal of Zn(II) and Cd(II) ions from aqueous solutions. Chemical Engineering Journal **148** 68–79. (2009).
13. Feng, N. and Zhang, F. *Untreated Chinese ephedra residue as biosorbents for the removal of Pb2+ ions from aqueous solutions*. Procedia Environmental Sciences **18** 794 – 799. (2013).
14. Salazar-Rabago, J.J. and Leyva-Ramos, R. Novel biosorbent with high adsorption capacity prepared by chemical modification of white pine (Pinus durangensis) sawdust. Adsorption of Pb(II) from aqueous solutions. Journal of Environmental Management **169** 303-312 . (2016).
15. Baccar, R., Bouzid, J., Feki, M., Montiel, A. Preparation of activated carbon from Tunisian olive-waste cakes and its application for adsorption of heavy metal ions. Journal of Hazardous Materials **162** 1522–1529. (2009).
16. Lee, J.M. *Biochemical Engineering*. Prentice Hall, New Jersey. (1992).
17. Akunwa, N.K., Muhammad, M.N., Akunna, J.C. Treatment of metalcontaminated wastewater: a comparison of low-cost biosorbents. Journal Environment Management **146**, 517-523. (2014).
18. Pari, G. and Roliadi, H. *Alternative Technology for the Utilization of Biomass Waste from Wood Industries*. Proceeding of the International Workshop on Better Utilization of Forest Biomass for Local Community and Environments, Research and Development Center for Forest Products Technology, Bogor. (2004).
19. Jeon, C. and Kim, J.H. Removal of lead using phosphorylated sawdust. J. Ind. Eng. Chem. **15**, 910-913. (2009).
20. Meena, A.K., Kadirvelu, K., Mishra, G.K., Rajagopal, C., Nagar, P.N. Adsorptive removal of heavy metals from aqueous solution by treated sawdust (Acacia arabica). Journal of Hazardous Materials **150** 604–611. (2008).
21. Gomez-Serrano, V., Pastor-Villegas, J., Duran-Valle, C.J., Valenzuela-Calahorro, C. Heat treatment of rockrose char in air. Effect on surface chemistry and porous texture. Carbon **34** 533–538. (1996).
22. Pastor-Villegas, J., Valenzuela-Calahorro, C., Bernalte-Garcia, A., Gomez-Serrano, V. Charcterisation study of char and activated carbon prepared from raw and extracted rockrose, Carbon **31** 1061–1069. (1993).
23. Vinodhini, V. and Das, N.. Mechanism of Cr(VI) Biosorption by Neem Sawdust, American-Eurasian Journal of Scientific Research 4 (4): 324-329. (2009).
24. Gupta, H, dan Gogate, P.R. Intensified removal of copper from waste water using activated watermelon based biosorbent in the presence of ultrasound, Ultrason. Sonochem. http://dx.doi.org/10.1016/j.ultsonch. (2015)

# Removal of calcium and magnesium ions from hard water using modified Amorphophallus *campanulatus* skin as a low cost adsorbent

*Ajeng Yulianti Dwi* Lestari*, *Abdul* Malik, Sukirman, *Muhammad Irfan* Ilmi, and *Mahathir* Sidiq

Chemical Engineering Department, Faculty of Industrial Technology, Universitas Islam Indonesia, Jalan Kaliurang KM 14,5 Besi Sleman 55584, Yogyakarta, Indonesia

**Abstract.** Low cost adsorbent from *Amorphophallus campanulatus* skin has successfully synthesized to remove calcium and magnesium ions in the syntetic hard water. *A. campanulatus* skin were dried, crushed into powder form and modified by acid modification. A batch experiment with various parameters was used in this research. Various isotherm models were applied to fit the experimental data. Adsorption capacity of Ca and Mg on KB and KM adsorbents in 100 ppm solution respectively 10,85 mg/g, 27,64 mg/g, 1,79 mg/g and 20,1 mg/g. It was found out that the adsorption behavior of hard mineral ions by adsorbents match well with the Dubinin Radushkevich isotherm model. Based on the result, it can be concluded that a acid modified *A. campanulatus* skin is quite potential as a new low cost adsorbent which is expected to be applied to Indonesian groundwater which have high degree of hardness.

## 1 Introduction

Hard water is water that has high mineral content mainly calcium and magnesium ions. Water hardness causes a variety of problems in either house hold water supply or industrial water systems, such as the buildup of lime-scale that fouls plumbing and promotes galvanic corrosion, and the formation of membrane scaling in seawater desalination. Water softening is a process to remove the water hardness from various water streams and the conventional methods include the lime-soda ash treatment and ion exchange resin [1]. Permanent hardness is usually caused by the presence of calcium and magnesium sulfates in water which become more soluble as temperature rises. Therefore, permanent hardness cannot be removed by boiling. It can be removed by using a water softener which works on the principle of ion exchange in which calcium and magnesium ions are exchanged with sodium or potassium ions, reducing the concentration of hardness minerals to tolerable levels and thus making the water softer and giving it a smoother feeling [2].

Adsorption is the process that can separate objects which neither impossible to be applied nor impracticed by conventional techniques [3]. Zeolite, activated carbon, activated alumina are used as adsorbent. But recently there are researches that concern on making of the novel adsorbent. This novel adsorbent is made not only from unusual sources but also made from the polisaccarides. Polisaccarides which is choosen as adsorbents due to their easily to synthesize, have large adsorption capacity and cheap [4]. Elephant foot yam contains polisaccarides that not explore before. So this project focused on evaluation of the characteristics of the acid modification of the elephant foot yam skin adsorbents and their ability to adsorp both ion $Ca^{2+}$ and $Mg^{2+}$ in aqueous solution.

## 2 Material and methods

The main materials of this subject are *A. campanulatus* skin, pure distilled water, Merck's hydrochloric acid and Merck's sodium hydroxide. The selected fresh foot yam tubers is obtain from the farmers at Ngawi, East Java, Indonesia which has the itchy yellowish tuber.

### 2.1 Pretreatment of *A. campanulatus* skin

*A. campanulatus* fresh tuber firstly cleaned then peeled the skin. The skin then dried in the oven at 50°C for 24 hours. The dried skin then grinded and sieved on 200 mesh sieve after that the sieved skin has been stored into a sterile plastic bag

### 2.2 Modification of A. campanulatus skin

Sieved *A. campanulatus* skin then modified with acid to improve its adsorption ability. Five hundred grams of sieved *A. campanulatus* skin then dissolved into 2 litres of HCl solution for 3 hours. This acid mixture the neutralized with NaOH solution until pH 7 was reached. The neutral mixture then separated with filter paper. The distilate was dried in oven at 60°C for 1-2 days. The modified skin then stored into a sterile plastic bag.

*Corresponding author: aydlestari@uii.ac.id

### 2.3 Characterization of the adsorbents

There are two characterization of this project. They are suface morphology characterization and chemical bonding characterization. The adsorbents surface morphology was visualized by FEI Inspect S50 Scanning Electron Microscope (SEM). The chemical bonding was determined with Thermo Nicolet Avatar 360 Fourier Transform Infrared Spectroscopy (FTIR).

### 2.4 Determination adsorption properties of adsorbent

Isotherm adsorption was determined by mixing 1.5 g of sample adsorbents (KB and KM) with 25 mL differential concentration of Ca and Mg artificial hard water (100, 200, 300, 400, 500 [ppm]). Mixture then mixed well for about 30 minutes. After the mixing the adsorbent separated from the filtrate by the Whatmann 40 ashless filter paper, the final concentration of the mixture then analized using Perkin Elmer PinAAcle 900T Atomic Absoprtion Spectrofotometer (AAS). Experimental data then fitted with Langmuir, Freundlich, Temkin and Dubinin Radushkevich isotherm model and studied which the appropriate model(s).

## 3 Result and discussion

### 3.1 Effect of acid modification

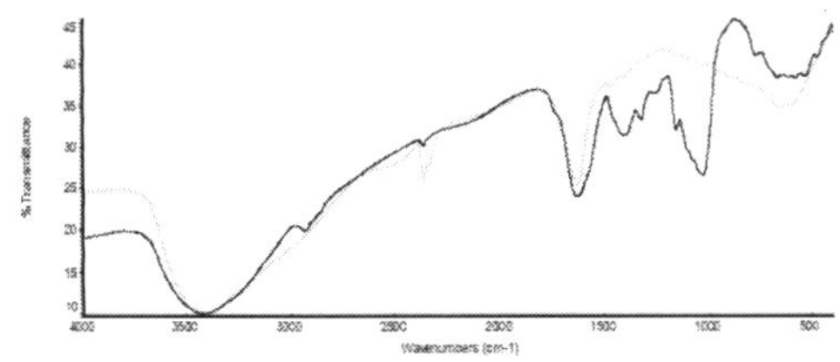

**Figure 0.1** FTIR Spectrums of KB and KM

Figure 3.1 represented FTIR spectra of the chemical bonding that occurred in the adsorbent KB and KM. The black line represented KB spectrum and the light blue line represented KM spectrum. Spectra of KB and KM showed the -OH bonds on wavelength around 3.400/cm, the C=O bonds on wavelength around 1.800/cm, the C-O bonds on wavelength 1.000/cm. After the modification, both KB and KM's -OH bonds was tend to be more symmetrical. It can be said that the process of modification with hydrochloric acid may alter-OH bond.

### 3.2 Characterization of the adsorbents

SEM photograph showed in Figure 3.2 below with 5.000x magnification of KB and 500x magnification of KM. Figure KB that represent the pure dried *A. campanulatus* skin and figure KM represent the modified dried *A. campanulatus* skin. It showed that KM has larger pore than KB due to the diference of magnification. SEM showed that the pore diameter of KB is 412,6 nm and the pore diameter of KM is 64,69 µm. Modification process influences the morpfology and chemical bonding of adsorbents. Based on those characteristic, the KM sample assessed to have a better adsorption ability.

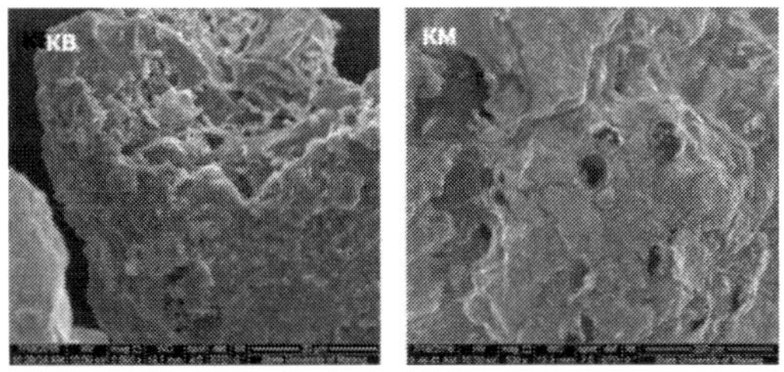


**Figure 0.2** SEM Graphs of KB and KM

### 3.3 Effect of initial concentration

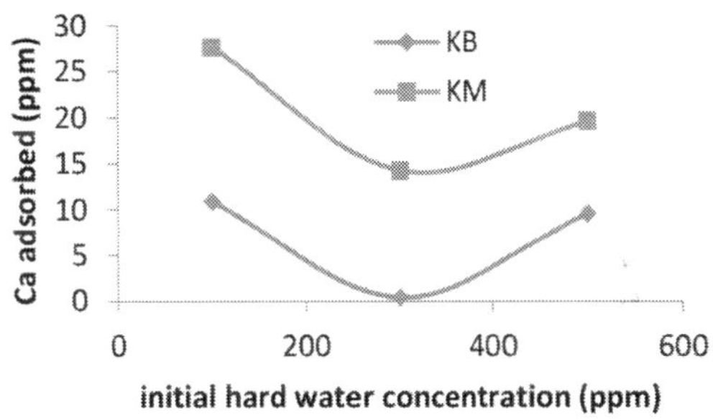


**Figure 0.3** Calcium adsorption capacity on KB and KM in various concentration

The relationship between the initial Ca and Mg concentration and the adsorption capacities on KB and KM was studied. As shown in Figure 3.3 above, the adsorption capacities of KB and also KM for Ca was correlated with the initial Ca ion concentration because the process was depend on the concentration. When the concentration rose from 100 ppm to 500 ppm, the adsorption capacities of KB and KM increased from 10.85 to 27.64 mg/g in 100 pmm solution, 0.42 to 14.24 mg/g in 300 ppm solution and 9.57 to 19.56 mg/g in 500 ppm solution. Figure 3.4 below represents the correlation between Mg initial concentration and Mg adsorbed both on KB and KM surface. When the concentration rose from 100 ppm to 500 ppm, the adsorption capacities of KB and KM increased from 1.79 to 20.1 mg/g in 100 pmm solution, 11.33 to 33.66 mg/g in 300 ppm solution and 10.92 to 21.84 mg/g in 500 ppm solution.

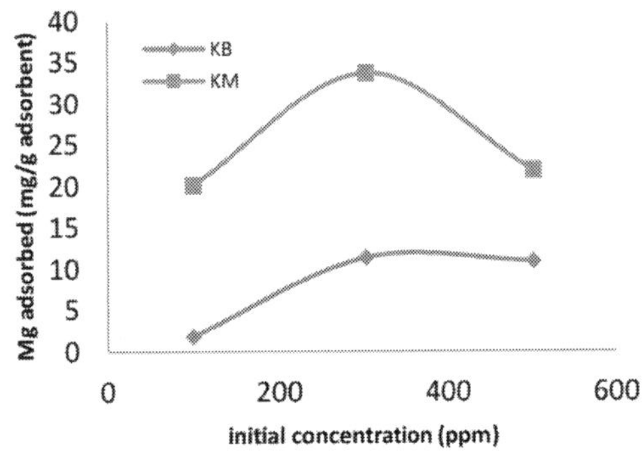


**Figure 0.4** Magnesium adsorption capacity on KB and KM in various concentration

### 3.4 Adsorption isoterm of calcium and magnesium ion into the adsorbents

The adsorption isotherms showed the relations between the concentration of adsorbat and its degree of accumulation of Ca and Mg onto surface of adsorbent at room temperature. Several models of adsorption isotherm have been used to fit to the experimental data. Fitting model used to evaluate isotherm performances for water adsorption. These isotherm models are the Freundlich model, Langmuir model, Temkin model and Dubnin Radushkevich model. Figure 3.5 and Figure 3.6 showed the plot of data isoterm of Ca adsorption onto the KB and KM adsorbents. They showed that the Ca adsorption onto KB and KM are suitable with Dubinin isotherm. Figure 3.7 and Figure 3.8 showed the plot of data isoterm of Mg adsorption onto the KB and KM adsorbents. They showed that the Mg adsorption onto KB and KM are also suitable with Dubinin isotherm.

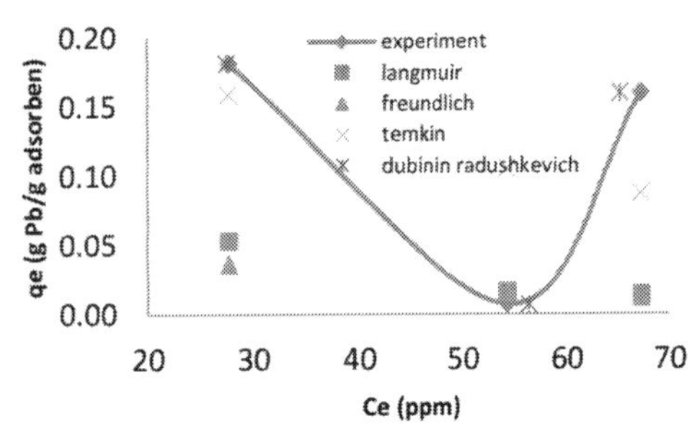


**Figure 0.5** Ca onto KB

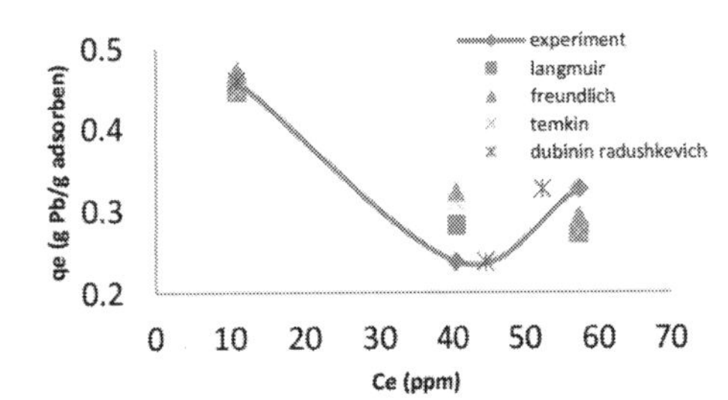


**Figure 0.6** Ca onto KM

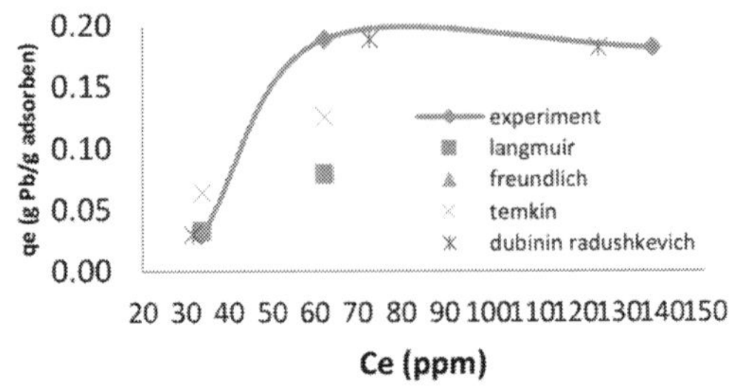


**Figure 0.7** Mg Sorption on KB

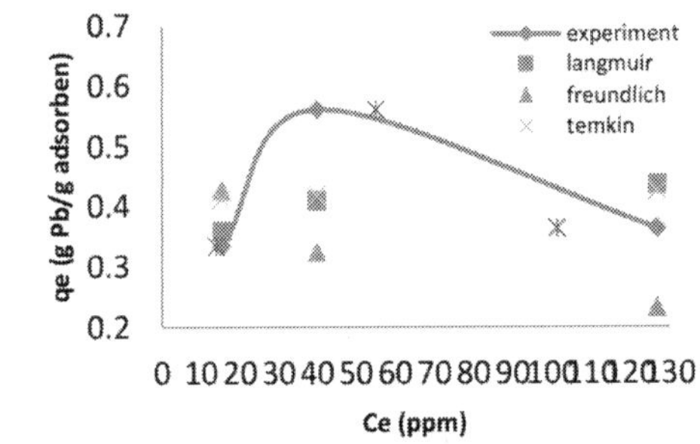


**Figure 0.8** Mg Sorption on KM

Table 3.1 showed the details of Ca adsorption isotherm parameters onto KB and KM adsorbents.

**Table 0.1** Calcium Isothem Parameters onto KB and KM

| | Langmuir | | | Freundlich | | |
|---|---|---|---|---|---|---|
| | $K_L$ | $q_0$ | $R^2$ | $K_F$ | n | $R^2$ |
| KB | 0,044 | 0,009 | 0,131 | 2,418 | 0,789 | 0,101 |
| KM | 0,207 | 0,249 | 0,601 | 0,946 | 3,448 | 0,968 |
| | Temkin | | | Dubinin Radushkevich | | |
| | $A_t$ | B | $R^2$ | $K_{ad}$ | $q_s$ | $R^2$ |
| KB | 0,005 | 0,080 | 0,153 | 1,674 | 120,542 | 0,994 |
| KM | 0,001 | 0,105 | 0,676 | 0,902 | 77,246 | 0,988 |

Table 3.2 showed the details of Mg adsorption isotherm parameters onto KB and KM adsorbents.

**Table 0.2** Magnesium Isotherm Parameters on KB and KM

| | Langmuir | | | Freundlich | | |
|---|---|---|---|---|---|---|
| | $K_L$ | $q_0$ | $R^2$ | $K_F$ | n | $R^2$ |
| KB | 0,006 | 0,117 | 0,843 | 0,040 | 0,820 | 0,667 |
| KM | 0,254 | 0,451 | 0,161 | 0,946 | 3,448 | 0,586 |
| | Temkin | | | Dubinin Radushkevich | | |
| | $A_t$ | B | $R^2$ | $K_{ad}$ | $q_s$ | $R^2$ |
| KB | 0,055 | 0,102 | 0,647 | 2,512 | 196,960 | 0,962 |
| KM | 0,005 | 0,007 | 0,003 | 1,448 | 134,424 | 0,926 |

### 3.5 Comparison of KB and KM with other Ca adsorbent

Table 3 showed the list of adsorption capacity of Ca and Mg by some synthetic adsorbents not only conventional adsorbent but also low cost natural based adsorbent. This study said that the KM has the adsorption ability better than KB for bith ions. Another advantage of KM especially compared to adsorbent conventional is the availability of raw materials for the manufacture of adsorbent abundant sources of biomass, easily regenerated, adsorbent regeneration after use more environmental friendly.

**Table 0.3** Mg and Ca adsorption capacity on various adsorbents

| Adsorbent | Adsorption Capacity (mg/g) | | Initial concentration (mg/L) | | Reference |
|---|---|---|---|---|---|
| | $Ca^{2+}$ | $Mg^{2+}$ | $Ca^{2+}$ | $Mg^{2+}$ | |
| KB | 10.85 | 1.79 | 100 | 100 | This study |
| KM | 27.64 | 20.10 | 100 | 100 | This study |
| MPC | 85% removed | 85% removed | 208 | 147 | [5] |
| Quartz sand | 60.00 | | 450 | | [6] |
| SDBS-SMB | 29.27 | 29.27 | 120 | 120 | [7] |
| Kaolin based geopolimer | 76.00 | 40.00 | | | [8] |

## 4 Conclusion

Novel adsorbent was succesfully synthesized from *Amorphophallus campanulatus* skin modified by hydrochloric acid. The experiment focused on investigation of the effect of modification and the adsorption ability for Ca and Mg ions in aquoeus solution. Modification caused KB more porous and the changed the infrared spectrum of –OH bonding. The adsorption behaviour is dependent on the initial concentration of Ca and Mg ion. Adsorption capacity Ca and Mg on KB and KM in 100 ppm aqueous are 10.85 mg/g, 27.64 mg/g, 1.79 mg/g and 20.1 mg/g. The adsorption follows the Dubinin isotherm. The study also evaluates that the KM has better adsorption ability than KB. So hydrochloric acid modification on the *Amorphophallus campanulatus* skin gave a potential application as another low cost natural based adsorbent.

Authors would say thank to Directorate of Research and Community Service Islamic University of Indonesia for the financial support. We also thank Rudi Syah Putra, head of Integrated Laboratory Islamic University of Indonesia and Abdulloh Fuad, head of Central Laboratory of Mineral & Advanced Materials, Malang State University for their guidance with the experimental element of this research.

## References

1. C. Liu, L. Shi, R. Wang. Jour *of Memb Sci*, **486,** 169-176. (2015)
2. MA Tofighy, T Mohammadi. *Desal.,* **268,** 208-213. (2011)
3. KS Knaebel. *Adsorption Research Inc's publication*: 1-21. (2009)
4. DW O'Connell, C Birkinshaw, TF O'Dwyer. *Bior Tech*, **99,** 6709–6724. (2008)
5. HS Altundogan, A Topdemir, M Cakmak, N Bahar. *Jour of the Tai Inst of Chem Eng*. 1-7. (2015)
6. Y Chen, R Fan, D An, Y Cheng, H Tan. (2016). *Jour of Env Sci, article in press* (2016).
7. NNA Kadir, M Shahadat, S Ismaii. *App Clay Sci*, **137**, 168-175. (2017)
8. M Naghsh, K Shams. *App Clay Sci,* **146,** 238-245. (2017)

# Preparation of activated carbon from banana peel waste for reducing air pollutant from motorcycle muffler

*Achmad* Chafidz[1,*], *Widi* Astuti[2], *Dhoni* Hartanto[2], *Aulia Septiani* Mutia[2], *Purtiah Rantau* Sari[2]

[1]Department of Chemical Engineering, Universitas Islam Indonesia, Yogyakarta 55584, Indonesia
[2]Department of Chemical Engineering, Universitas Negeri Semarang, Semarang 50229, Indonesia

**Abstract.** The exhaust gas or emission from motor vehicle contains various pollutants and some of them are toxic and very harmful for human health. In the present work, we prepared an activated carbon to reduce the toxic pollutants (via adsorption process) from the exhaust gas of the motor vehicle (particularly motorcycle in this work). The activated carbon was prepared from local banana peel which considered as an agricultural waste without economic value. To prepare the activated carbon, banana peel was carbonized using furnace at 500°C. Then the resulted carbon was activated using chemical activation with $ZnCl_2$ and physical activation with either microwave (300 W) or furnace (700°C). The prepared activated carbon was then characterized using Scanning Electron Microscope (SEM) and Fourier Transform Infrared (FT-IR) analysis. The SEM result showed that the pore of microwave induced $ZnCl_2$ activated carbon ($AC_M$) was larger than the pore of virgin char/carbon and furnace induced $ZnCl_2$ activataed carbon ($AC_F$). The FT-IR spectrum of $AC_M$ showed some peaks at 3408.59 $cm^{-1}$, 1589.25 $cm^{-1}$, and 1093.63 $cm^{-1}$. They are assigned to O-H stretching, C=O stretching, and C-OH stretching. Additionally, the adsorption performance of the prepared activation carbon to reduce the pollutants concentration (i.e. CO and CH) from the exhaust gas of motor vehicle was investigated. The results showed that the activated carbon induced by microwave and $ZnCl_2$ activation ($AC_M$) gave the best result. The adsorption or removal efficiency of gas CO reached approximately 97.64 %vol.

## 1 Introduction

Nowdays, air pollution resulted from the exhaust emission of motor vehicle engine is a topic of concern, due to the significant number of motor vehicles in the world, especially in the big cities. Exhaust emission of motorcycle can contribute about 70-80 percent of air pollution. The exhaust gas or emission of motor vehicle contains various pollutants such as carbon monoxide (CO), carbon dioxide ($CO_2$), hydrocarbon (HC), nitrogen oxide ($NO_x$), sulfur oxide ($SO_x$), particulate, etc. Some of the pollutants are toxic and very harmful for human health. One of ways to reduce the air pollution is via adsorption process by using an activated carbon. Activated carbon is a porous material that behaves as a powerful adsorbent because of high surface area, adequate pore size distribution, and variable characteristics of surface chemistry, low ash content, and relatively high mechanical strength. The activated carbon are used in many application such as removal or adsorb of harmful gasses like CO and HC gas. The synthesis of activated carbon can be carried out through two step process that is carbonization and activation [1]. Activation step can be performed by physical or chemical process. In physical process, the precursor is heated at 700-1100°C in presence of oxidizing agents [1] while in chemical process, the precursor is reacted with chemical agent including $H_3PO_4$ [2], KOH [3] and $ZnCl_2$ [4].

On the other hand, agricultural biomass waste is a renewable resource that can be further utilized to provide a broad range of useful chemicals, materials and energy, which have economic values [5]. Banana peel is one of agricultural biomass waste that has potential to be used as a raw material for the production of activated carbon [6]. It is because banana peel has high lignocellulose content, approximately 90.3% (dry weight based) [7]. Activated carbon can be prepared by chemical activation process using $ZnCl_2$. The use of $ZnCl_2$ as an activating reagent is because it can make high specific surface area [8]. After chemical activation, the next process is physical activation using heat either by using microwave or furnace. In term of microwave heating, the energy is freely transformed into heat inside the particles by ionic conduction and dipole rotation. The microwave induced reaction is more effective and faster than a conventional furnace [9].

## 2 Experimental

### 2.1 Preparation of activated carbon

Banana peels were collected as wastes from the local fruit markets in Semarang city, Indonesia. They were cut into small pieces of about 2 cm x 2 cm, washed with water to remove surface dirt then sun dried in 3 days and dried

* Corresponding author: achmad.chafidz@uii.ac.id

using oven at 110°C until the weight is constant. The dried banana peels were then crushed into powder form. The powdered banana peel was then carbonized using furnace at 500°C for 1 hour. The carbonized material (char) was soaked in zink chloride ($ZnCl_2$) solution with an impregnation ratio ($ZnCl_2$:char) of 3:1 (w/w) in 10 ml aquades for 14 hours and then dehydrated in an oven at 110°C for 6 hours. The sample was then physically activated using either microwave and furnace. The sample that was activated by microwave (called $AC_M$) was heated at 300 W for 5 min. Whereas, the sample that was activated by furnace (called $AC_F$) was heated at 700°C for 1 hour. The resulted activated carbon (AC) were washed with HCl 0.2 N after cooling to dissolve and remove any residual ash, and subsequently followed by washing with hot distillled water until pH reached 7.0. Finally, the activated carbons were dried in an oven at 110°C until the weight is constant. In summary, there will be three samples of activated carbon, which are: banana peel char/carbon (without activation), activated carbon by microwave ($AC_M$), and activated carbon by furnace ($AC_F$).

## 2.2 Characterization of activated carbon

The morphology of char and activated carbon (i.e. $AC_M$ and $AC_F$) were investigated by scanning electron microscope (SEM). Then the functional groups present in the raw material (banana peel), $AC_M$ before and after gas emission adsorption of motor vehicle were evaluated from Fourier Transform Infrared Spectrophotometer (FT-IR) spectra, which were recorded between 4000 and 450 $cm^{-1}$.

## 2.3 Muffler design and adsorption test

The performance of three activated carbon samples (i.e. char/without activation, $AC_M$, and $AC_F$) was measured by analyzing the exhaust emission from the motorcycle muffler by using gas analyzer, STARGAS 898 (Tecnotest, Italy). The motorcycle used in this work was a 4-stroke engine motorcycle, Mio brand 2012 (Honda, Indonesia). Prior to the tests, the samples were put inside the motorcycle muffler, as shown in Figure 1. The muffler used in this study has a little different design than the muffler in general. In this study, each of the activated carbon samples with a weight of 10 g was placed in an adsorbent holder/container made of stainless steel tube with 5 cm long and diameter fit to the channel of muffler. The stainless steel material was selected due to its superior corrosion resistance. The adsorbent container was placed on the channel after the gas discharge from the engine. At both ends of the tube were closed with a mesh (i.e. 150 mesh), so that the activated carbon granular will not come out of the adsorbent tube due to the high pressure of the exhaust gas. The muffler design with the adsorbent (i.e. activated carbon) container is presented in Fig. 1.

The working principles of the test are as follow: First, the exhaust gas from the combustion process of the motor engine enters into the adsorbent tube. Then the gas makes a contact with both carbon/char (without activation) and the activated carbon (either $AC_M$ or $AC_F$), and subsequently the gas pollutants (e.g. CO and CH) are absorbed by the activated carbon. Expectedly, the gas coming out of the adsorbent tube has low CO and HC gas concentration, and then flows into the exhaust drain and

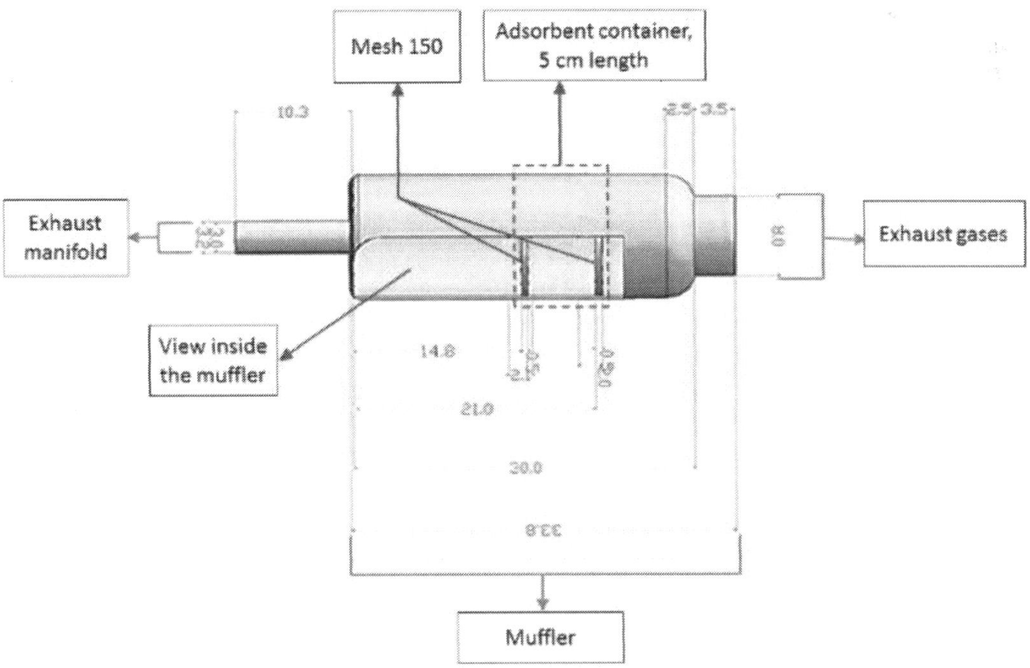


**Fig. 1.** Muffler design used for the gas emission test

released to the environment. The muffler has a simple design, because it can be dismantled and the activated carbon on the tube adsorbent can be easily replaced when it is saturated. Additionally, this design is also cheap, because the adsorbent tube used is not too long and thus, only need small amount of adsorbent/activated carbon. From this test results, the adsorption efficiency of each activated carbon samples were then compared to determine the best activated carbon in reducing gas pollutant (i.e. CO and HC) in the gas emission of motorcycle.

## 3 Results and discussion

### 3.1 Morphological analysis

Scanning electron microscopy (SEM) technique was used to observe the surface morphology of the activated carbon samples. The surface morphology of char (carbon without activation), $AC_M$, and $AC_F$ were evaluated by the SEM at 1000X magnification and the obtained images are shown in Fig. 2.

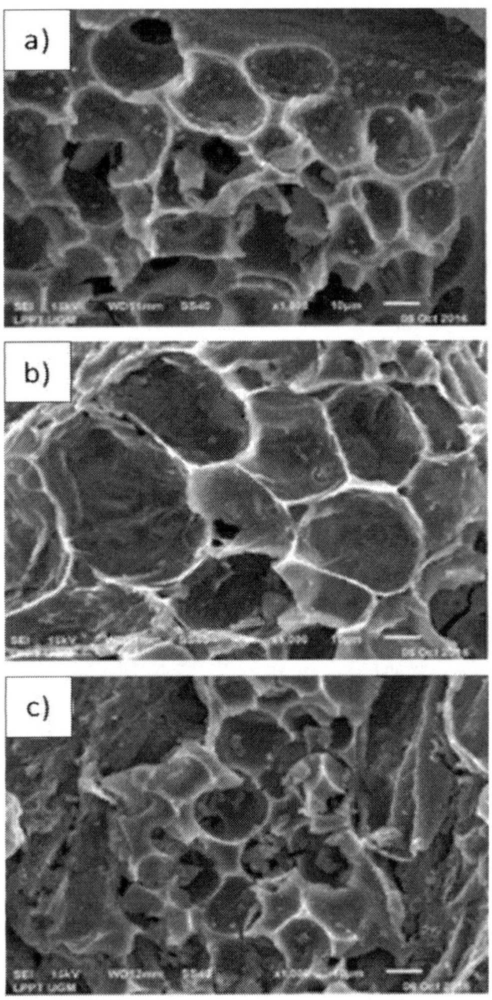


**Fig. 2.** SEM micrograph of (a) Char; (b) $AC_M$; and (c) $AC_F$.

### 3.2 Fourier Transform InfraRed (FT-IR) analysis

The FT-IR analysis was carried out to determine the availability of functional groups. Functional group analysis was performed on banana peel carbon powder (i.e. char); $ZnCl_2$ activated carbon and microwave ($AC_M$), the $AC_M$ before and after adsorption test. The functional group analysis was performed at wavelength range of 4000 - 370 cm-1. The FT-IR test results of banana peel carbon/char and banana peel based activated carbon (i.e. $AC_M$) are presented in Fig. 3. Whereas, the FT-IR test results of $AC_M$ before and after the adsorption are depicted in Fig. 4.

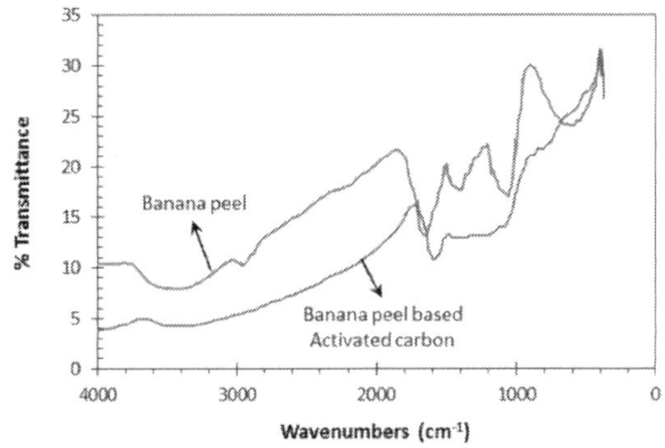


**Fig. 3.** FT-IR spectrums of banana peel and banana peel carbon based activated carbon by $ZnCl_2$ + Microwave (i.e. $AC_M$)

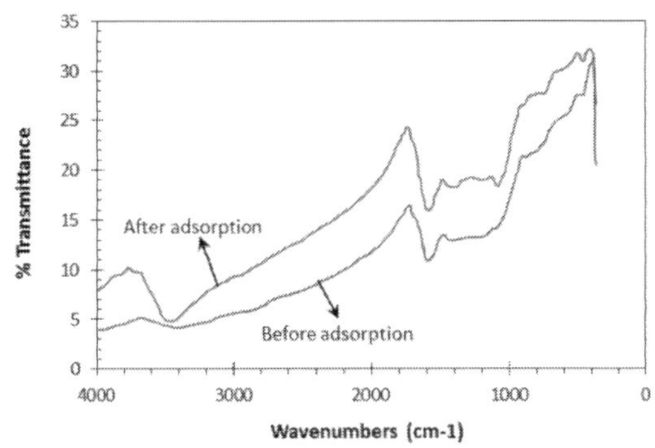


**Fig. 4.** FT-IR spectrums of banana peel carbon based activated carbon by $ZnCl_2$ + Microwave (i.e. $AC_M$); before and after the adsorption tests

As seen in Fig. 3, the banana peel spectra shows the absorption peak at 3430.59 $cm^{-1}$ indicating the presence of an O-H functional group [11]. The absorption peak at 1642.28 $cm^{-1}$ shows the presence of C=C groups of aromatic compounds [11]. While the presence of C-O-C of an ether group was indicated by uptake of 1037.03 $cm^{-1}$ [11]. The peak at 577 $cm^{-1}$ wavelength indicates the presence of an C-H group. Whereas, for the spectra of banana peel based activated carbon which activated using $ZnCl_2$ and microwave (i.e. $AC_M$) shows an absorption peak at 3408.59 $cm^{-1}$ indicating the presence of O-H group which may contain water, alcohol and phenol components [11]. The absorption peak at 1589.25 $cm^{-1}$ shows the presence of C=C groups of aromatic compounds, while the C-O-C group is indicated at the wave number 1093.63 cm$^{-1}$ [11]. Additionally, as seen in the figure, the FT-IR spectra of banana peel after carbonized and activated using $ZnCl_2$ and microwave undergo some changes in the position and intensity if compared with the initial banana peel spectra. From the figure, it can be seen that the C=C group of aromatic compounds was shifting and getting shallower, meaning that the aromatic group decreased after being activated. It is because the lignin compound which contains aromatic group degraded and re-structured into another compound. The wide absorption peak at 577 cm$^{-1}$ after activation was lost. It is because this peak is O-H deformation group in which the group will undergo a change in shape to another compound.

Additionally, as seen in Fig. 4, there was a comparison between FT-IR spectra of activated carbon ($AC_M$) before and after adsorption test. There was a significant change in the absorption peak of 3438.64 cm$^{-1}$ after adsorption test with the gas emission of motorcycle. This absorption peak indicates the presence of an O-H group that allows the presence of water, alcohol, and phenol components. The odds of the O-H group multiply after adsorption was likely due to reaction between hydrocarbons based fuel in motorcycles with the oxygen to produce $CO_2$ and $H_2O$. The resulting $H_2O$ might be adsorbed on the activated carbon, and thus increasing the O-H group on the activated carbon spectra after adsorption test.

### 3.3 Exhaust emission analysis

The analysis of reduction of CO and HC gas in the exhaust emission of motorcycle (Mio 2012, Honda) was carried out by using a gas analyzer instrument, STARGAS 898 (Technotest, Italy). The analysis was divided into 4 parts, which are: 1) direct gas emission measurements (without activated carbon); 2) using carbon/char, before activation; 3) using activated carbon activated by $ZnCl_2$ and microwave ($AC_M$); and 4) using activated carbon activated by $ZnCl_2$ and furnace ($AC_F$). The objective of the analysis is to find out the gas emission levels (especially CO and HC) of various variables tested. The data collection of gas emission analysis was carried out for 8 minutes with constant motor speed of 60 km/hour. Every minute the results were printed out. The first gas emission analysis results (without using activated carbon) has resulted in CO emission level of 7.558 %vol and HC gas of 470 ppm volume. Whereas, the next analysis results were presented in term of gas CO and HC removal efficiency (compared to the first analysis), which are shown in Fig. 5 and Fig. 6 respectively.

As seen in both Figs. 5 and 6, the concentration of gas emissions from the muffler that did not use activated carbon was higher than by using activated carbon. The activation of banana peel based carbon is believed can improve the adsorption of gas CO and HC. Additionally, the adsorption efficiency of the activated carbon, $AC_M$ was higher than the char (without activation) and $AC_F$. The gas CO emission levels before and after the use of activated carbon ($AC_M$) were 7.558% and 0.174% (%Vol). Whereas, the gas CO emission standard for motorcycle production in 2010 and above was 4.5%, which is below 7.558%.

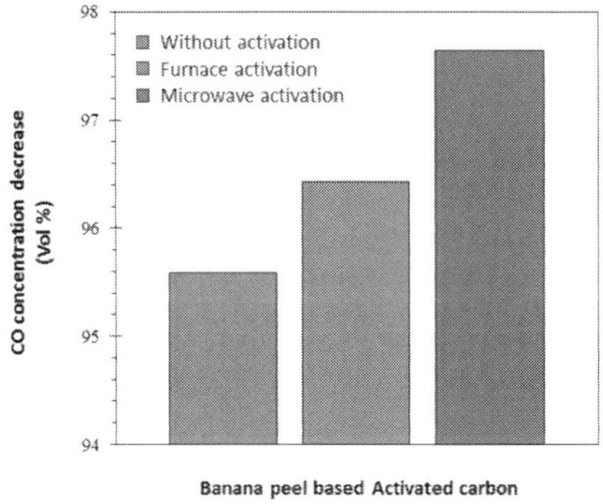


**Fig. 5.** Gas CO removal efficiency

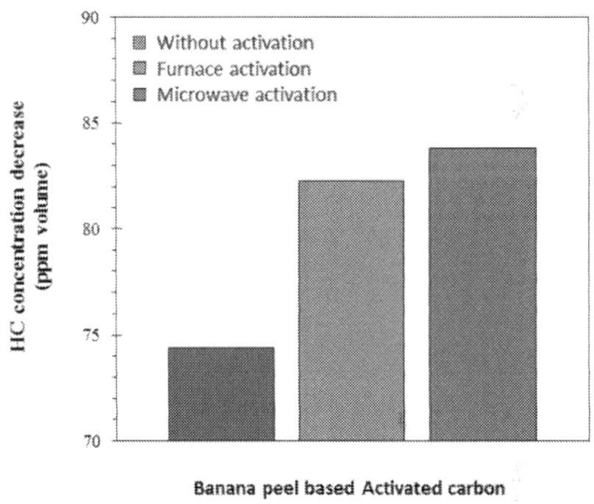


**Fig. 6.** Gas HC removal efficiency

It can be concluded that the motorcycle being tested without using activated carbon did not meet the gas emission standard set by the government. In other hand, by using activated carbon, the gas CO emission level met the standard, which is much less than 4.5%. Additionally, the gas HC emission levels before and after the use of activated carbon ($AC_M$) were 470 ppm and 58 ppm (ppm Vol.). Whereas, the gas HC emission standard for motorcycle production in 2010 and above was 2000 ppm. This indicates that even without the use of activated carbon, the exhaust emission from motorcycles have met the standard. Nevertheless, by using activated carbon (either $AC_M$ or $AC_F$), the gas HC emission level was much less than that without the use activated carbon.

## 4 Conclusion

In the present work, activated carbon was prepared from local banana peel which considered as an agricultural waste without economic value. Three samples of activated carbon were prepared which are: banana peel char/carbon (without activation), activated carbon by microwave ($AC_M$), and activated carbon by furnace ($AC_F$). The SEM result showed that the pore of microwave induced $ZnCl_2$ activated carbon ($AC_M$) was larger than the pore of virgin char/carbon and furnace induced $ZnCl_2$ activated carbon ($AC_F$). The FT-IR spectrum of $AC_M$ showed some peaks at 3408.59 $cm^{-1}$, 1589.25 $cm^{-1}$, and 1093.63 $cm^{-1}$. They are assigned to O-H stretching, C=O stretching, and C-OH stretching. Additionally, the adsorption performance of the prepared activation carbon to reduce the pollutants concentration (i.e. CO and CH) from the exhaust gas of motor vehicle was investigated. The results showed that the activated carbon induced by microwave and $ZnCl_2$ activation ($AC_M$) gave the best result. The adsorption or removal efficiency of gas CO reached approximately 97.64 %vol.

## References

1. O. Pezoti Jr., A.L. Cazetta, I.P.A.F. Souza, K.C. Bedin, A.C. Martins, T.L. Silva and V.C. Almaida, J. Ind. Eng. Chem, **20** (2014) 4401-4407.
2. W. Astuti, R.A. Hermawan, H. Mukti and N.R. Sugiyono, AIP Conf. Proc., **1788** (2017) 030082-1–030082-5.
3. J.S. Nazzal, W. Kaminska, B. Michalkiewicz and Z.C. Koren, Industrial Crops and Products, **47** (2013) 153-159.
4. M. Acikyildiz, A. Gurses and S. Karaca, Microporous Mesoporous Mater., **198** (2014) 45-49.
5. W. Astuti, T. Sulistyaningsih and M. Maksiola, Asian J. Chem., **29** (2017) 133-138.
6. L.C.A. Oliveira, E. Pereira, I.R. Guimaraes, A. Vallone, M. Pereira, J.P. Mesquita and K. Sapag, J. Hazard. Mater, **165** (2009) 87-94
7. H. Tibolla, F.M. Pelissari, M.I. Rodrigues and F.C. Menegalli, Industrial Crops and Products, (2016) 1-11.
8. L. H. Tzong, Chem. Eng. J., **158** (2010) 129-142.
9. D. Liu, J. Taiwan Inst. Chem. Eng., **59** (2016) 563-568.
10. K. Mohanty, D. Das and M.N. Biswas, Chem. Eng. J., **115** (2005) 121-131.
11. W. Astuti, T. Sulistyaningsih and M. Maksiola, Jurnal Teknologi, **79** (2016) 35-42.

# The effect of organosolv pretreatment on optimization of hydrolysis process to produce the reducing sugar

*Fibrillian Zata* Lini, *Tri* Widjaja*, *Nuniek* Hendrianie, *Ali* Altway, *Siti* Nurkhamidah, and *Yumarta* Tansil

Chemical Engineering Department, 60111 Institut Teknologi Sepuluh Nopember, Indonesia

**Abstract.** As the fossil energy decrease such as petroleum and natural gas, that are encourages a lot of research to develop new sources of energy from renewable raw materials. One of the source is through reducing sugar (glucose and xylose) obtained from coffee pulp waste; this is due to abundant production of coffee pulp every year reaching 743 kg/ha. In addition, this waste has not been used optimally and the cellulose and hemicellulose content of the coffee is high. The purpose of this study is to get the optimal operating condition for reducing sugar production from coffee pulp waste. The method used for optimization is Response Surface Methodology with Central Composite Design. The optimum operation condition obtained was pH 4.63 at 34°C for 16.29 hours of hydrolysis. As a result, the predicted yield gained was 0.147 grams of reducing sugars / gram of cellulose+hemicellulose. The result indicates the gained yield was 0.137 grams of reducing sugars / gram of cellulose+hemicellulose.

## 1 Introduction

Due to the improvements of technologies, the fossil based energy source will always decreasing, therefore the invention of renewable alternative energy sources must be developed to fulfill the energy needs (Zaldivar et al., 2001). Bioethanol is a choice of alternative energy source to replace the function of fossil based energy source. It can reduce both fossil based energy source and greenhouse effect that become main environment issue nowadays. Bioethanol itself is a renewable fuel source that will not produce extra carbon dioxide from its usage and can be categorized as clean alternative energy that can be renewed (Widjaja et al., 2014).

The most common method of bioethanol production is using sugar or starch containing materials as the raw material. Those kind of materials (sugar or starch) are commonly used either as main food source or livestock feed (Hahn-Hagerdal et al., 2006). Sugar or starch based materials for bioethanol production will make another problem, which is the food and livestock feed source will decrease. Another material can be used for bioethanol production, especially from agricultural waste that cannot be processed again such as rice straw, sorghum stem, coconut coir dust, or coffee pulp waste (Widjaja et al., 2015). By using those materials, the agricultural waste issue can be solved.

Agricultural waste consists of lignocellulosic biomass that can be converted into reducing sugar which will be fermented to produce bioethanol. The usage of lignocellulosic biomass has a very high potential to be developed due to the availability of that kind of material and lignocellulosic biomass cannot be used for either food source or main livestock feed (Geng et al., 2012). Coffee pulp waste is a lignocellulosic biomass that is very abundant in Indonesia, which can reach 743 kilograms each hectare of coffee farm. Coffee pulp waste has a high potential to be raw material to produce reducing sugar which will be fermented to produce bioethanol. Besides the biomass availability, the composition of coffee pulp consists of high cellulose and hemicellulose content which can be converted to reducing sugar. In previous study, the coffee pulp waste consist of cellulose (58.36%); hemicellulose (21.8%); lignin (5.05%); protein (0.81%); tannin (3.11%); and pectin (2.16% ) (Widjaja et al., 2017).

Commonly, bioethanol production process from lignocellulose biomass consists of biomass pretreatment process, hydrolysis, reducing sugar fermentation process, and then continued to separation and purification process (Cardona and Sanchez, 2007). There are commonly need three methods for pretreatment process which are organosolv, hydrothermal, and ammonium methods.

The hydrotermal pretreatment has major advantages due to no addition of chemicals and no requirement of corrosion resistant materials for hydrolysis reactors. It must be operated in high temperature and short residence time, the increasing of operating temperature causes higher saturated water vapour pressure, which results in faster hydrolysis of volatile matter contents. However, it increases the energy consumption in the pretreatment process. Furthermore, higher residence time causes pyrolysis, which results in charring of biomass along with higher energy requirement (Chandra et al., 2012).

In the previsious study (Domanski et al., 2016), the use of ammonia for pretreatment of lignocellulosic materials results in biomass swelling and increasing of biomass porosity, reduction in the degree of polymerization and the crystallinity of cellulose and an

* Corresponding author: papatri2003@yahoo.com

effective delignification. Ammonia is thought to act selectively with lignin bonds, especially C-O-C bonds, as well as ester and ether bonds, which selectively reduces lignin in the lignocellulose structure. Aqueous ammonia as a chemical has several desirable traits that make it a good candidate for an industrial platform. Additionally, ammonia is relatively inexpensive, and less harmful byproducts that could inhibit enzymatic hydrolysis created from the interaction of aqueous ammonia with lignocellulosic biomass. However, most of the ammonium pretreatments require relatively high temperatures (over 100 $^0$C) and high pressure.

Organosolv pretreatment is a separation process of fiber using organic solvent with or without acid or base catalyst to hydrolyze lignin in biomass. This pretreatment can increase the cellulose which is effective to degrade hemicellulose or lignin and to decrease the cyrztalline. The greater the solvent concentration, the greater the reduction of lignin which results from dissolving small part of hemicellulose and increasing the cellulose. Organosolv pretreatment is carried out at 100-250°C with acidic catalyst. With the addition of acidic catalyst, high temperature will decompose the xylose to be furfural, hydroxylemthyl furfural and polyphenol which will inhibit the pretreatment. Another disadvantage of this method is the requirement of high amount of solvent and high energy recovery consumption. Ethanol is used as the solvent because of its cheap price. Compared to another chemical pretreatment, the advantages of Oganosolv pretreatment are a relatively pure lignin which will be produced and less energy consumption. From environmental point of view, the separation of lignin will decrease waste water treatment problem. The increasing of cellulose from the removal of lignin will increase the maximum glucose concentration (Widjaja et al., 2016).

The hydrolysis step consists of two techniques: enzymatic hydrolysis and acidic hydrolysis. Enzymatic hydrolysis process is chosen due to lower energy consumption in the process because the process is done in mild condition and its substrate specific and rarely formed byproduct from its process (Sun and Cheng, 2002). Cellulose is the main composition that will be processed to be reducing sugar and this component is very abundant in environment. Based on this reason, cellulase enzyme commonly used to hydrolyze cellulose to be glucose as reducing sugar. Beside cellulose, hemicellulose can be utilized to produce reducing sugar, especially in coffee pulp waste. Due to this reason, the combination of cellulase and xylanase enzyme in hydrolysis process must be developed to reach better reducing sugar production.

Enzymatic hydrolysis is affected by some factors which are the variety of substrate, enzyme activity and operation condition (temperature, pH, hydrolysis time, etc.). There were many studies that focused on the optimization of production yield and rate in enzymatic hydrolysis process(Sun and Cheng, 2002). Operating condition must be optimized due to the economic terms from the intense usage of enzyme. Response surface methodology (RSM) is a statistical method that commonly used to generate optimum value from modelling and optimize variables that affect the model. This method commonly used in enzymatic hydrolysis optimization (Ferreira et al., 2009).

This study was focused on the optimization of enzymatic hydrolysis of organosolv pretreated coffee pulp using response surface methodology (RSM) with central composite design. The enzymatic hydrolysis was done using cellulase and xylanase enzyme mixture to obtain higher reducing sugar yield.

## 2 Experimental

### 2.1 Coffee Pulp Pretreatment Process

In this study, coffee pulp waste as the substrate must be pretreated before hydrolyzed in enzymatic process. The pretreatment consisted of two processes, mechanical and chemical pretreatment. In mechanical pretreatment, coffee pulp waste was dried under the sunlight for 12 hours and grinded to reach 100-120 mesh size. Grinded coffee pulp pretreated chemically using organosolv method with 50% ethanol as the solvent [10]. Coffee pulp was mixed with citric acid with 1:16 (w/w) ratio and added by NaOH 4N solution until pH 3 was reached. The coffee pulp and citric acid solution then was heated at 80°C for 75 minutes with stirring speed of 600 rpm. After this process, the solution was separated between solid and liquid. The solid obtained was treated by 100 mL of 50% ethanol solution and heated at 50°C in a magnetic stirrer with stirrer speed of 500 rpm for 2 hours. The solid form after the addition of ethanol solution must be washed until the pH was neutral and then dried at 60°C until reach constant weight. The crystallinity index (CI) of celluloses and hemicellulose was measured using XRD (X'Pert PRO, PANalytical, Departement of materials and metallurgy, ITS, INDONESIA).

### 2.2 Enzyme Preparation

The enzyme used in this study was combination of pure cellulase enzyme and pure xylanase enzyme. Pure cellulase enzyme was obtained from *Aspergillus niger* Sigma Aldrich C1184- 25KU whereas pre-xylanase enzyme was obtained from *Trichoderma longibrachiatum* Sigma Aldrich X2629-100G. One gram of each enzyme was diluted with citric buffer until 100 mL of volume.

### 2.3 Enzyme Hydrolysis

One gram of pretreated coffee pulp was added by 18.6 U of pure cellulase enzyme solution and 18.6 U of pure xylanase enzyme solution. 18.6 U was obtained activity enzyme.

**Table 1.** Initial Activity Enzyme Of The Cellulase And Xylanase Used In This Work.

| pH | Activity Enzyme (U/mL) | | Enzyme required/ one gram of coffee pulp |
|---|---|---|---|
| | Cellulase Enzyme | Xylanase Enzyme | |
| 3 | 1.241 | 3.123 | 20.994 |
| 5 | 2.523 | 3.901 | 12.141 |
| 7 | 2.775 | 4.079 | 11.262 |

Citric buffer was added until it reached 30 mL of its hydrolysis volume. The enzymatic hydrolysis was done in a shaking incubator with 125 rpm shaking speed in various operating conditions that optimized using central composite design method.

### 2.4 Optimization Technique

Response surface methodology (RSM) is a mathematical and statistical method used for modelling purpose and problems analysis in which the variables used have direct effect on the response. This methodology was conducted to get optimum value not only for the response but also for the variables (Mesa et al., 2011).

Based on previous study, enzymatic hydrolysis was generally affected by hydrolysis temperature, pH, and hydrolysis time (Hamada et al., 2013). In this study, these factors were used as the main factors that affect the enzymatic hydrolysis result significantly. A model with $2^3$ factorial design in face centered central composite design with 6 star points (a=1) and 5 times center point replication ($n_0$=5) resulted 19 experimental variables that must be done to obtain optimum operational enzymatic hydrolysis value. This experiment was done in 2 time replication (duplo).

**Table 2.** Factors and code values of factorial design from Design Expert® 10.

| Factor | Variable | Unit | Coded Factor | | |
|---|---|---|---|---|---|
| | | | -1 | 0 | +1 |
| A | pH | - | 3 | 5 | 7 |
| B | Temperature | °C | 30 | 40 | 50 |
| C | Hydrolysis Time | Hours | 12 | 18 | 24 |

The comparability between the model and experiments was done using Design Expert® 10 by looking the R-square and adjusted R-square that gathered from the model and experiment. The validation was done by comparing the optimum value of reducing sugar yield from experiment to the predicted value of it.

## 3 Result and Discussion

### 3.1 Coffee Pulp Pretreatment

In this study, coffee pulp waste used for the main substrate was converted to reduce sugar. This waste was chosen due to the high content of cellulose and hemicellulose. Pretreatment process was done to break the solid structure lignin content inside coffee pulp waste, which could inhibit enzymatic hydrolysis process. Mechanical pretreatment was done to make the contact area of the coffee pulp waste wider and made it easier to be pretreated and hydrolyzed. The next step was chemical pretreatment process using organosolv method with citric acid and ethanol as a solvent. The coffee pulp and citric acid solution in a solid:liquid ratio was 1:16 w:w. The additions was obtained from previous studies (Pratiwi et al., 2010) using the best condition of cacao pulp. To improves the ethanol selectivity with respect to lignin and to improve the ability of ethanol delignifying, in this study, the addition of NaOH 4N as catalyst agen during organosolv ethanol pretreatment (Mesa et al., 2011) was used. And then, 50% ethanol was added to degrade lignin. The addition of citric acid aims to reduce the content of pectin in the coffee pulp. Based on previous study by (Mesa et al., 2011), sugarcane bagasse was conducted to a dilute-acid pretreatment (0.2M $H_2SO_4$ solution in a solid:liquid ratio of 1:5 w:w, at 120°C for 40 min). Then, it was followed by organosolv treatment, which was carried out in a 1000ml Parr reactor. Reactions were performed using the acid pretreated sugarcane bagasse and the ethanol and NaOH solution in a solid:liquid ratio of 1:7 w:w. In order to alleviate the corrosions caused by mineral acids in the acid-catalyzed ethanol pretreatment, in the previous study by (Zhao et al., 2009) was using acetic acid as catalyst instead of sulfuric acid.

**Table 3.** The Result of Pretreatment Process using 50% Ethanol

| Substrate | % Cellulose (w/w) | % Hemicellulose (w/w) | % Lignin (w/w) |
|---|---|---|---|
| Untreated | 58.36 | 21,80 | 5,05 |
| Pretreated | 66.82 | 25,08 | 0,12 |

Table 3 shows that organosolv pretreatment method using 50% ethanol as the solvent able to degrade lignin content in coffee pulp waste significantly. It shows a significant effect, the lignin content degraded from 5.05% to 0.12%. Lignin was able to inhibit enzyme's activity during enzymatic hydrolysis. Based on previous study (Mesa et al., 2011), organosolv pretreatment process was able to degrade hemicellulose inside the substrate. However, the hemicellulose content was increasing, instead of degrading. Hemicellulose content in coffee pulp waste was not degraded due to the pretreatment temperature, which at 80$^{0}$C, therefore the hemicellulose content did not degrade (Widjaja et al., 2016).

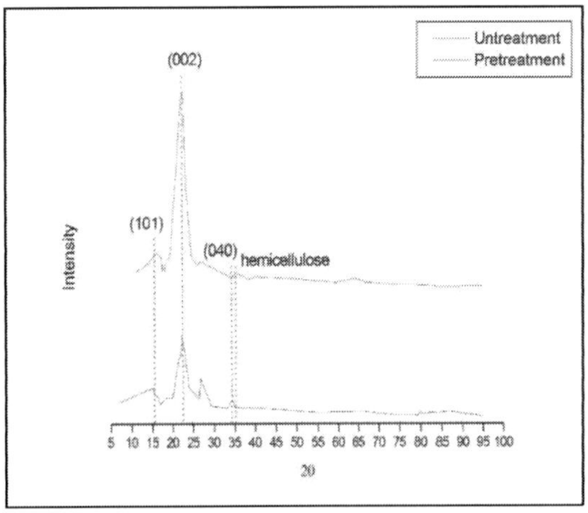


**Fig. 1.** XRD Analysis for Untreated and Pretreated Coffee Pulp Waste. Below the figure

In Figure 1, untreated coffee pulp waste shows diffraction peaks at 2θ= 16$^{0}$; 22$^{0}$; and 34$^{0}$ which represent reflector plane (101); (002); and (040). The diffraction peaks shown in Figure 1 was similar to that from the former study done by Park et. al. in 2010, in which those peaks were common characteristic for original (untreated) cellulose crystalline (Park et al., 2010). After coffee pulp waste was treated using organosolv method with 50% ethanol as the solvent, the reflector plane of (101) and (002) shows increasing in their intensity. The result of pretreated coffee pulp waste indicates the characteristics of cellulose II, that is commonly found in pretreated lignocellulosic biomass substrate using solvation and regeneration, in which the area of reflector plane (101) and (002) became wider (Xiao et al., 2012).

The increasing of cellulose crystalline after being pretreated using organosolv method made the reflector plane (101) and (002) areas became wider. The increasing of reflector areas implies that lignin content in coffee pulp waste was degraded (Sangian et al., 2015). 2θ= 35$^{0}$ reflector plane in untreated coffee pulp waste did not show any diffraction peak, however after the pretreatment process, it shows a little diffraction peak, which was hemicellulose reflector plane (Hamada et al., 2013). This means that after pretreatment process, hemicellulose content was increasing.

## 3.2 Enzymatic Hydrolysis Optimization

Optimization of enzymatic hydrolysis was done to get optimum operation condition in the enzymatic hydrolysis of pretreated coffee pulp waste. The aim of this optimization is to gather efficient operating condition in reducing sugar production process using enzymatic hydrolysis with pure cellulose enzyme nd pure xylanase enzyme combination. Enzymatic hydrolysis was the core process in bioethanol production process from lignocellulosic biomass. By optimizing hydrolysis process, the reducing sugar production yield would increase.

In this study, the optimization technique was carried out using response surface methodology (RSM). There were many kinds of variables that influence on the sugar production processes reduction, therefore central composite design (CCD) was used to determine the effect of each variable on the model of enzymatic hydrolysis process using coffee pulp waste as the substrate.

Based on ANOVA analysis in Table 4, the model used had significant effect of this experiment. This was shown by the value of F-value 78.81 and p-value (Prob>F) which was less than 0.05. The significance of each variable can be determined by the p-value (Prob>F). If this value was less than 0.05 (<0.05), it means that the variable used in this process has significant effect in enzymatic hydrolysis result. In this experiment, A, B, AB, AC, $A^2$, ABC, $A^2B$, and $AB^2$ were shown that those had significant effect to the model. The other variables did not show significant effect because p-value (Prob>F) bigger than 0.1 (>0.1).

**Table 4.** ANOVA Result to the Response in Enzymatic Hydrolysis Process of Coffee Pulp Waste.

| Source | Sum of Squares | Degree of Freedo | Mean Square | F Value | p-Value (Prob>F) |
|---|---|---|---|---|---|
| Model | 0.055 | 13 | $4.224 \times 10^{-3}$ | 78.81 | <0.0001 |
| A: pH | $4.410 \times 10^{-4}$ | 1 | $4.410 \times 10^{-4}$ | 8.23 | 0.0085 |
| B: Temperature | $8.07 \times 10^{-4}$ | 1 | $8.074 \times 10^{-4}$ | 15.06 | 0.0007 |
| C: Hydrolysis Time | $4.134 \times 10^{-5}$ | 1 | $4.134 \times 10^{-5}$ | 0.77 | 0.3885 |
| AB | 0.017 | 1 | 0.017 | 309.28 | <0.0001 |
| AC | $5.531 \times 10^{-4}$ | 1 | $5.531 \times 10^{-4}$ | 10.32 | 0.0037 |
| BC | $3.525 \times 10^{-5}$ | 1 | $3.525 \times 10^{-5}$ | 0.66 | 0.4253 |
| $A^2$ | $2.758 \times 10^{-3}$ | 1 | $2.758 \times 10^{-3}$ | 51.45 | <0.0001 |
| $B^2$ | $7.875 \times 10^{-5}$ | 1 | $7.875 \times 10^{-5}$ | 1.47 | 0.2373 |

| Source | Sum of Squares | Degree of Freedo | Mean Square | F Value | p-Value (Prob>F) |
|---|---|---|---|---|---|
| $C^2$ | $1.456 \times 10^{-4}$ | 1 | $1.456 \times 10^{-4}$ | 2.72 | 0.1124 |
| ABC | $1.329 \times 10^{-3}$ | 1 | $1.329 \times 10^{-3}$ | 24.80 | <0.0001 |
| $A^2B$ | $1.338 \times 10^{-3}$ | 1 | $1.338 \times 10^{-3}$ | 24.95 | <0.0001 |
| $A^2C$ | $2.102 \times 10^{-4}$ | 1 | $2.102 \times 10^{-4}$ | 3.92 | 0.0593 |
| $AB^2$ | $3.369 \times 10^{-3}$ | 1 | $3.369 \times 10^{-3}$ | 62.85 | <0.0001 |
| $AC^2$ | 0 | 0 | | | |
| $B^2C$ | 0 | 0 | | | |
| $BC^2$ | 0 | 0 | | | |
| $A^3$ | 0 | 0 | | | |
| $B^3$ | 0 | 0 | | | |
| $C^3$ | 0 | 0 | | | |
| Residual | $1.286 \times 10^{-3}$ | 24 | | | |
| Lack of Fit | $5.708 \times 10^{-5}$ | 1 | | 1.07 | 0.3122 |
| Pure Error | $1.229 \times 10^{-3}$ | 23 | | | |
| Total | 0.056 | 37 | | | |

$R^2 = 0.9771$; adjusted $R^2 = 0.9647$

Lack of fit value of 1.07 indicates that the lack of fit in this model did not have significant effect on the pure error. This means that the model can be used for the experiment. Besides the determination coefficient, $R^2$ value of 0.9771 reveals that the model was significant and was able to be used in this experiment. $R^2$ value which was bigger than 0.75 (>0.75) implies that the model can be used properly.

In this experiment, the variables that have significant effect on enzymatic hydrolysis process of coffee pulp waste were pH and hydrolysis temperature based on the p-value (Prob>F) of 0.0085 and 0.0007 respectively. This result was in a good agreement with the previous study that shows that pH and hydrolysis temperature had significant effect on enzymatic hydrolysis process of lignocellulosic biomass (Ferreira et al., 2009). It was due to substrate and enzyme bonding role during hydrolysis process and the movement of substrate to active side of the enzyme which was affected by buffer type. In this process, citric acid buffer had better effect than acetate buffer for enzymatic hydrolysis process (Peña-Farfal et al., 2004). Temperature had significant effect on enzymatic hydrolysis process, in which enzymatic hydrolysis process in mild temperature (45-50$^{o}$C) would obtain the best result of reducing sugar (Duff and Murray, 1996).

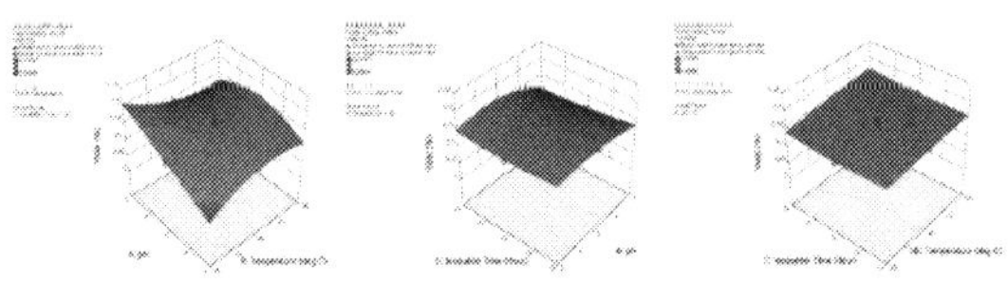

**Fig. 2.** Depicts Response Surface with Central Composite Design Plot in the Optimization of Enzymatic Hydrolysis to the Yield of produced Reducing Sugar which shows: (a) pH against Temperature, (b) Hydrolysis Time against pH, dan (c) Hydrolysis Time against Temperature. Below the figure

The equation gathered from response surface methodologywas the connection among the manipulated variables that was used during enzymatic hydrolysis process, which was the response yield of produced reducing sugar. The equation was:

$$Yield = 1912 + 7.595 x 10^{-5} abc - 5 x 10^{-4} a^2 b - 3.377 x 10^{-4} a^2 c - 1.6 x 10^{-4} ab^2 + 0.021 a^2 + 7.620 x 10^{-4} b^2 + 1.434 x 10^{-4} c^2 + 0.018 ab - 1.513 x 10^{-4} ac - 3.550 x 10^{-4} bc - 0.468 a - 0.074 b + 3.583 x 10^{-3} c \quad (1)$$

a, b, and c values were actual value in various pH, temperature, and hydrolysis time condition. The equation was then solved to get optimal value of each variable to the response. The optimization process was using Design Expert® 10 and this is shown in Table 5.

Based on Table 5, optimum reducing sugar yield from experiments was 0.137 gram of reducing sugar / gram (cellulose + hemicellulose) with error value of 0.065. The experiment value was in a good agreement compared to predicted yield value of 0.147 gram of reducing sugar / gram (cellulose + hemicellulose) with reducing sugar yield range of 0.03955 – 0.23389 gram of reducing sugar / gram (cellulose + hemicellulose). From the experimental result, it can be concluded that the model was in a good agreement to the experiment.

**Table 5.** Predicted and Actual Yield with Optimum Value of Each Variable.

| Variables | | Predicted Yield | Actual Yield | Error |
|---|---|---|---|---|
| pH | 4.63 | 0.147 | 0.137 | 0.065 |
| Temperature | 34$^{o}$C | | | |
| Hydrolysis Time | 16.29 hours | | | |

From the previous study, the optimum condition of enzymatic hydrolysis was at pH 4.5- 5.0 with hydrolysis temperature at 40-50$^{o}$C. Enzymatic hydrolysis had several benefits compared to other hydrolysis methods which had mild condition of hydrolysis, did not have corrosiveness and less utility needs. In addition, the hydrolysate that was produced contained less toxic agents or components that were able to inhibit fermentation process. For those reasons, enzymatic hydrolysis was widely used compared to other hydrolysis method. Meanwhile, the pH value of 4.63 in this study was in a good agreement compared to that in the previous study. However, the optimized value of hydrolysis temperature (34°C) was a little bit lower compared to that in the previous study (40–50°C). The difference happened due to the different substrate that was used in enzymatic hydrolysis process and the combination of enzyme type used. Optimum temperature was affected by hydrolyzed substrate, enzyme type as the hydrolysis agent, and hydrolysis time (Palmqvist and Hahn-Hägerdal, 2000). Optimum hydrolysis time in enzymatic hydrolysis for coffee pulp waste was 16.29 hours. From previous experiments, the substrate was still able to be hydrolyzed after 16.29 hours of hydrolysis time, but this result would be another problem for further process because the reducing sugars can be inhibitor for enzymatic hydrolysis process (Linde et al., 2007). In short, it can be concluded that optimum value of hydrolysis time in enzymatic hydrolysis was 16.29 hours.

## 4 Conclusion

From the experiments and analysis, it can be concluded that organosolv method pretreatment using 50% ethanol as the solvent was able to increase both cellulose and hemicellulose content to 66.82% (w/w) and 25.05% (w/w), respectively. This process has significant effect in lignin degradation in coffee pulp waste that was shown by the decreasing of lignin content from 5.05% (w/w) to 0.12% (w/w).

The optimization of enzymatic hydrolysis process using combination of pure cellulose and xylanase enzyme using response surface methodology producing pH value of 4.63, hydrolysis temperature of 34°C, and hydrolysis time for 16.29 hours with predicted reducing sugar yield of 0.147 gram reducing sugar / gram (cellulose + hemicellulose). From ANOVA analysis, it shows that pH and hydrolysis temperature had significant effect on enzymatic hydrolysis process of coffee pulp waste.

The authors would like to express their deepest gratitude to Ministry of Research, Technology and Higher Education of the Republic of Indonesia for providing financial support for carrying out this research with research's contract number: 179/SP2H/LT/DRPM/III/2016.

## References

1. Cardona, C.A., Sánchez, Ó.J., 2007. Fuel ethanol production: Process design trends and integration opportunities. Bioresour. Technol. 98, 2415–2457.
2. Chandra, R., Takeuchi, H., Hasegawa, T., 2012. Hydrothermal pretreatment of rice straw biomass : A potential and promising method for enhanced methane production. Appl. Energy 94, 129–140.
3. Domanski, J., Borowski, S., Marchut-Mikolajczyk, O., Kubacki, P., 2016. Pretreatment of rye straw with aqueous ammonia for conversion to fermentable sugars as a potential substrates in biotechnological processes. Biomass and Bioenergy 91, 91–97.
4. Duff, S.J.B., Murray, W.D., 1996. Bioconversion of forest products industry waste cellulosics to fuel ethanol: A review. Bioresour. Technol. 55, 1–33.
5. Ferreira, S., Duarte, A.P., Ribeiro, M.H.L., Queiroz, J.A., Domingues, F.C., 2009. Response surface optimization of enzymatic hydrolysis of Cistus ladanifer and Cytisus striatus for bioethanol production. Biochem. Eng. J. 45, 192–200.
6. Geng, A., Xin, F., Ip, J. yu, 2012. Ethanol production from horticultural waste treated by a modified organosolv method. Bioresour. Technol. 104, 715–721.
7. Hahn-Hägerdal, B., Galbe, M., Gorwa-Grauslund, M.F., Lidén, G., Zacchi, G., 2006. Bio- ethanol - the fuel of tomorrow from the residues of today. Trends Biotechnol. 24, 549– 556.
8. Hamada, Y., Yoshida, K., Asai, R., Hayase, S., Nokami, T., Izumi, S., Itoh, T., 2013. A possible means of realizing a sacrifice-free three component separation of lignocellulose from wood biomass using an amino acid ionic liquid. Green Chem. 15, 1863.
9. Linde, M., Galbe, M., Zacchi, G., 2007. Simultaneous saccharification and fermentation of steam-pretreated barley straw at low enzyme loadings and low yeast concentration.
10. Enzyme Microb. Technol. 40, 1100–1107.
11. Mesa, L., González, E., Cara, C., González, M., Castro, E., Mussatto, S.I., 2011. The effect of organosolv pretreatment variables on enzymatic hydrolysis of sugarcane bagasse. Chem. Eng. J. 168, 1157–1162.
12. Palmqvist, E., Hahn-Hägerdal, B., 2000. Fermentation of lignocellulosic hydrolysates. I: Inhibition and detoxification. Bioresour. Technol. 74, 17–24.
13. Park, S., Baker, J.O., Himmel, M.E., Parilla, P. a, Johnson, D.K., 2010. Cellulose crystallinity index: measurement techniques and their impact on interpreting cellulase performance. Biotechnol. Biofuels 3, 10.
14. Peña-Farfal, C., Moreda-Piñeiro, A., Bermejo-Barrera, A., Bermejo-Barrera, P., Pinochet- Cancino, H., De Gregori-Henríquez, I., 2004. Use of enzymatic hydrolysis for the multi- element determination in mussel soft tissue by inductively coupled plasma-atomic emission spectrometry. Talanta 64, 671–681.
15. Pratiwi, Eka, Yatim, Endahwati, L., 2010. Pemanfaatan limbah kulit buah cokelat sebagai bioetanol. Makal. Semin. Nas. Tek. Kim. Soebardjo Brotohardjono "Ketahanan Pangan dan Energi" 1–10.
16. Sangian, H.F., Kristian, J., Rahma, S., Agnesty, S.Y., Gunawan, S., Widjaja, A., 2015.
17. Comparative Study of the Preparation of Reducing Sugars Hydrolyzed from High-Lignin Lignocellulose Pretreated with Ionic Liquid, Alkaline Solution and

Their Combination. J. Eng. Technol. Sci. 47, 137–148.

18. Sun, Y., Cheng, J., 2002. Hydrolysis of lignocellulosic materials for ethanol production : a review q. Bioresour. Technol. 83, 1–11.
19. Widjaja, T., Altway, A., Ni'Mah, H., Tedji, N., Rofiqah, U., 2015. Technique of ethanol food grade production with batch distillation and dehydration using starch-based adsorbent. AIP Conf. Proc. 1699.
20. Widjaja, T., Altway, A., Nurkhamidah, S., Edahwati, L., Lini, F.Z., Oktafia, F., 2016. The Effect of Pretreatment and Variety of Microorganisms to The Production of Ethanol from Coffee Pulp 11, 1056–1060.
21. Widjaja, T., Altway, A., Permanasari, A.R., Gunawan, S., 2014. Production of Ethanol As A Renewable Energy by Extractive Fermentation. Appl. Mech. Mater. 493, 300–305.
22. Widjaja, T., Iswanto, T., Altway, A., Shovitri, M., Rachmania Sri, J., 2017. Methane Production from Coffee Pulp by Microorganism of Rumen Fluid and Cow Dung in Co- digestion. Chem. Eng. Trans. 56, 1465–1470.
23. Xiao, W., Yin, W., Xia, S., Ma, P., 2012. The study of factors affecting the enzymatic hydrolysis of cellulose after ionic liquid pretreatment. Carbohydr. Polym. 87, 2019– 2023.
24. Zaldivar, J., Nielsen, J., Olsson, L., 2001. Fuel ethanol production from lignocellulose: A challenge for metabolic engineering and process integration. Appl. Microbiol.
25. Biotechnol. 56, 17–34.
26. Zhao, X., Cheng, K., Liu, D., 2009. Organosolv pretreatment of lignocellulosic biomass for enzymatic hydrolysis. Appl. Microbiol. Biotechnol. 82, 815–827

 https://doi.org/10.1051/matecconf/201815401023

# Comparison of *Nannochloropsis sp.* cells disruption between hydrodynamic cavitation and conventional extraction

*Martomo* Setyawan[1, 2], *Panut* Mulyono[1], Sutijan[1], and *Arief* Budiman[1, 3, *]

[1] Chemical Engineering Department, Gadjah Mada University, Jalan Grafika 2 Yogyakarta, Indonesia
[2] Faculty of Industrial Engineering, Ahmad Dahlan University, Jalan Prof DR Soepomo SH Yogyakarta, Indonesia
[3] Center for Energy Studies, Gadjah Mada University, Sekip K IA, Yogyakarta, Indonesia

**Abstract.** Biodiesel production from microalgae is one of the solution of the future energy problem, but its production cost is still high. One of the costly stages of this process is the lipid extraction process. It can be reduced by microalgae cell disruption. One of the mechanical method to cell disruption with the lowest energy requirement is hydrodynamic cavitation. This aim of this study is to evaluate the distribution coefficient and the mass transfer coefficient value of lipid extraction of *Nannochloropsis sp.* assisted by hydrodynamic cavitation and compare with conventional extraction. The hydrodynamic cavitation extraction was done at 34 °C, 1 atm. The conventional extraction was done at 34 °C, 1 atm with stirring speed 260 and 1000 rpm. The experimental result shows that the distribution coefficient dependent on the temperature with the values for 50, 44, 38 and 34 °C were 0.502, 0.394, 0.349, and 0.314 respectively. And it was according to Van' Hoff equation with the values of ΔH° was 20.718 kJ/mol and ΔS° was 58.05 J/mol/K. The hydrodynamic cavitation extraction was faster than conventional. The mass transfer coefficient values for hydrodynamic cavitation, conventional 260 rpm and 1000 rpm were 7.373, 0.534 and 0.121 1/s respectively.

## 1 Introduction

The increasing energy demand in the recent years and the decreasing fossil fuel stock will induce the energy problem related to the supply of raw material [1]. One of the solutions is to find the alternative renewable energy sources. Some renewable energy sources have been investigated by some researchers. The large stock of renewable energy source that already exists is palm oil to make biodiesel, but producing biodiesel from palm oil will make a threat to the food security [2]. Non-edible energy sources have been investigated to develop the second generation of biodiesel. These sources were like palm fatty acid distillate [3], oil palm empty fruit bunches (EFB)[4], wood waste [5], and non-edible seeds such as Jatropha seed [6]. However these non-edible energy sources are limited inventory. At the last recent time, some researchers give their attention to make the third generation of biodiesel from microalgae lipids [7].

Microalgae have been proved as a candidate for fuel production (algae fuel) with several benefits. The benefits of microalgae are their higher photosynthetic efficiency, higher biomass production, and faster growth than to the other energy crops [8]. Microalgae can be cultivated in large scale on non-arable lands and do not need potable water to grow. So, there is not threatening to the food production [9]. The important problem associated with using fuel is the increasing carbon dioxide emission as a pollutant that causes global warming. Another benefit from microalgae is to prevent global warming caused by carbon dioxide emission into the air, since microalgae need carbon dioxide to grow. Carbon dioxide as a burning product can be absorbed by microalgae are used for photosynthesis to product lipids [10].

The processes to produce biodiesel from microalgae involve cultivation, harvesting, lipid extraction and transesterification. Intensive research in this field over several years are still hindered by high overall production costs. It makes uncompetitive in the market. So, finding the processes with low production costs become preponderant [11]. Among of these steps of processes, the lipid extraction process is the most costly process with the portion of 70 - 80% of total cost [12]. Lipid extractions were reported in some literatures [9, 10, 11]. It was starting from conventional to the advanced methods of extraction. The conventional extraction just using chemical solvent and stirring, this process has low efficiency and need a long time of extraction [13]. The advanced methods of extraction were done by assisted cell disruption, this processes got higher efficiency than the conventional extraction. So mechanical treatment, such as mechanical disruption is needed to ensure the complete extraction of lipids.

* Corresponding author: abudiman@ugm.ac.id

https://doi.org/10.1051/matecconf/201815401023

Lipid extractions assisted by mechanical cell disruption have been published. Autoclaving, bead-beating, high-pressure homogenization, and ultrasonification were commonly practiced methods [14, 15, 16]. The hydrodynamic cavitation is one of the mechanical cell disruptions that needs the smallest amount of energy [17]. Hydrodynamic cavitation that was used to cell disruption technique was investigated, with the level of cell disruption was determined by lipids extracted and chlorophyll released. It was found that for lipids extraction, hydrodynamic cavitation technique was requiring energy as big as 3 MJ/kg with the microalgae concentration 1.5-2% w/w [18]. Hydrodynamic cavitation to assist lipid extraction simultaneously from *Nannochloropsis Salina sp.* was reported with variables of amount of energy input, time of extraction as compared to ultrasonic cavitation. This hydrodynamic cavitation was resulting in higher extraction yield at the same energy input with the microalgae concentration was used 2% w/vol [19].

However to scale up lipid extraction assisted by hydrodynamic cavitation need the value of mass transfer coefficient. The aim of this study is to investigate the value of mass transfer coefficient. To evaluate this coefficient, the mathematical models are proposed and the experimental data are used to verify the model. The mathematical models are including equilibrium and mass balance models.

## 2 Materials and Methods

### 2.1 Materials

Dry microalgae *Nannochloropsis sp.*was purchased from *Balai Budidaya Air Payau* in Situbondo East Java Indonesia. N-Hexane technical grade (Density of n-hexane = 0.66 kg/cc and normal boiling point = 68 °C) was purchased from PT. Brataco Chemica, a local chemical store in Yogyakarta Indonesia. Methanol technical grade (Density of methanol = 0.791 kg/cc, molecule weight of methanol = 32.034 g/mol and normal boiling point = 64.7 °C) was purchased from CV. Multi Kimia, a local chemical store in Yogyakarta Indonesia.

### 2.2 Equipment

Equilibrium experiment for conventional extraction was carried out in the 150 ml erlenmeyer and the magnetic stirrer with heater. To ensure there was no solvent vaporization this Erlenmeyer was equipped with the condenser pipe. The experiments of hydrodynamic cavitation were carried out on batch extraction cavitation (Figure 1). The equipments were configured of a compressor to provide the driving compressed air to drive the mixture of solvent and microalga. Venturi was used to generate cavitation. Sample and product chamber were to place the sample and collect raw product respectively. The separation equipment were thickener and evaporator, thickener was used to separate the fluid and solid product by gravitational force, and the evaporator was used to evaporate the solvent, and lipid can be found as a residue.

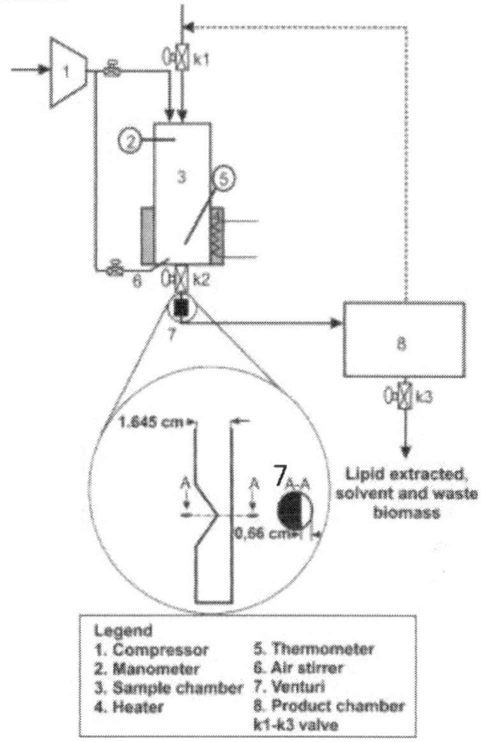


**Fig 1.** The scheme of hydrodynamic cavitation equipment

### 2.3 Experimental Procedures

The equilibrium experiments were done by mixing dry *Nannochloropsis sp.* with the mixture solvent methanol and hexane with the various ratios at the erlenmeyer. The mixture then was mixed for 8 hours at the 300 rpm mixing and then solid and fluid should be separated using a centrifuge. To determine lipids concentration in the fluid phase, measured liquid volume and weight, and lipids content in the fluid phase were determined by gravimeter after all the solvent were evaporated.

The hydrodynamic cavitation extraction experiments were done with 10 grams dry weight of *Nannochloropsis sp.* and solvent consists of 95 ml of hexane and 41 ml of methanol. All of the material were inputted into the sample chamber and then were flowed through the venturi with the pressure sample chamber 6.8 atm. Repeat this procedure for several rounds (2, 3, 4, etc.) as variable extraction. After the extraction processes have been finished, the fluid and solid phase were separated by sedimentation process. The remaining solids were washed with 25 ml of methanol and 25 ml of hexane separately. Washing methanol and hexane were combined with solvents that have been separated in the first step. The next step was separating lipid and solvent by distillation to obtain a residue with a constant weight. The residue was weighed and recorded as $w_1$. It was washed with hexane 5 ml and the remaining solids were dried until the weight remains constant and recorded as $w_2$ and lipid weight ($w_p$) obtained from microalgae was calculated using

$$w_p = w_1 - w_2 \quad (1)$$

Extraction yield is the weight of extracted lipids which is compared with the weight of dry microalgae in the beginning.

$$y = \frac{w_p}{w_{mi}} \quad (2)$$

where y is extraction yield, $w_{mi}$ is weight of dry microalgae at the beginning.

### 2.4 Mathematical model

In the solid-fluid extraction at the initial condition concentration lipids in the solvent (y) is zero, the changing value of y as function of the time is equal with the number lipids release from the solid ($j_T$), so it can be written as the equation (3)

$$m_f \frac{\partial y}{dt} = j_T \quad (3)$$

where $m_f$ represents the mass of fluid phase, t represents time. Lipids concentration in the solid can be calculated by lipids mass balance in the solid that can be written as

$$m_s \frac{dx}{dt} = -j_T \quad (4)$$

where $m_s$ represent the dry microalgae mass. The amount lipids release are equal with the mass transfer coefficient multiply with the concentration gradient between microalgae surface and in the bulk of liquid, and it can be written as

$$j_T = k_T a_T \rho_f (y^* - y) \quad (5)$$

where $k_T a_T$ represent volumetric mass transfer coefficient, $\rho_f$ represent fluid density, and $y^*$ represent lipids concentration at the microalgae surface. Value of $y^*$ can be predicted by equations

$$y^* = K.x \quad (6)$$

where K represents distribution coefficient, x represents lipid concentration in the solid phase. The value of K can be calculated from data of equilibrium experiment with the equation (7).

$$K = \frac{C_\infty}{C_s} \quad (7)$$

where $C_\infty$ is the lipid concentration in the solvent (g oil/ solvent volume) and $C_s$ is the lipid concentration in the solid phase (g oil/g dry microalga). The value of $C_\infty$ was taken from experiment and $C_s$ was calculated from initial concentration reduced by lipid released divided by the weight of dry microalgae.

The thermodynamic parameter of enthalpy change $\Delta H^o$ (kJ/mol) and entropy change $\Delta S^o$ (J/mol/K) are evaluated by Van't Hoff equation [20]:

$$\ln K = -\frac{\Delta H^o}{RT} + \frac{\Delta S^o}{R} \quad (8)$$

where R is the universal gas constant (J/mol/K). The other parameter Gibbs free energy change $\Delta G$ (kJ/mol) is determined by equation

$$\Delta G = \Delta H^o - T\Delta S^o \quad (9)$$

Model evaluation was done by calculating determination coefficient ($R^2$).

## 3 Results and discussions

The dry *Nannochloropsis sp.* microalgae was determined the total lipid content using stirring extraction method. 5 grams of dry *Nannochloropsis sp.* was extracted using 47.5 ml hexane and 20.5 ml methanol then stirred at speed of stirring 1100 rpm for 2 hours operation every stage. This extraction processes were done 3 stages at the room temperature (34 °C). The total lipid content of dry *Nannochloropsis sp.* that was used as the sample was 10.46% gram lipids/gram of dry *Nannochloropsis sp.* microalgae. The lipids composition were determined by Gas Chromatography-mass spectrometry. And the lipids were composed of fatty acids with C20 and C17 atoms chain with a composition of 69.81% and 14.63% respectively. The content of fatty acids in these lipids indicate that lipids from microalgae *Nannochloropsis sp.* potentially to be processed into biodiesel.

### 3.1 Equilibrium experiment

The yield of extraction lipid from 5 grams dry *Nannochloropsis sp.* using mixer solvents hexane (23 ml) and methanol (10 ml) as function of time extraction. The extraction processes was done by conventional methods with stirring speed 300 rpm at the room temperature (32 °C). For extraction time under 8 hours the extraction processes have not reached equilibrium state because there is increasing yields. At this case for 5 hours extraction time the yield is 5.25% and for 8 hours the yield is 6.76% and then after 8 hours the yield relatively constant. It is mean after 8 hours the extraction processes have reached the equilibrium state.

Figure 2 shows the various relationship between concentration lipids in the solids phase (x) and the fluid phase (y) at the equilibrium state. The variation of concentration was setting by different number of ratio dry microalgae and the mixture solvent, but the solvent volume ratio between hexane and methanol was fixed 2.3:1. The experiments were done at the temperatures 34, 38, 44 and 50 °C. From Figure 2 shown that the temperature took effect to the distribution coefficient (K) value. To determine the K value, it was approached by linier regression for each temperature data group.

The linier regression results and the determine coefficients ($R^2$) and distribution coefficients values are listed in the Table 1.

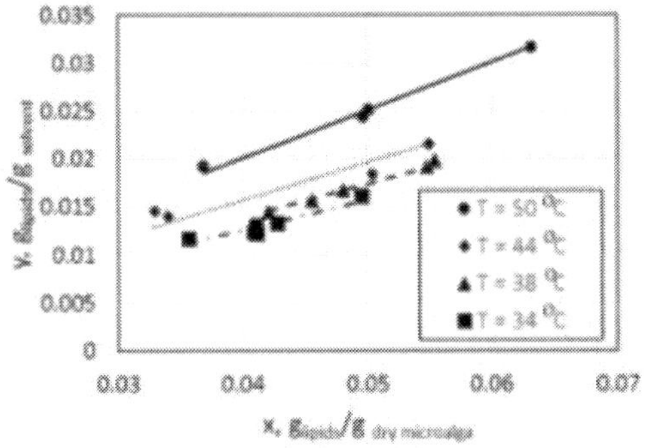


**Fig 2.** Variation ratio lipids concentration in the solid (x) and in the fluid (y) at different temperature

## 3.2 Enthalpy change, entropy change and Gibbs free energy change

Table 1 shows K values for different temperatures, the highest temperature in this experiment give the largest K value, and decreasing temperatures tend to decrease K values. The relation of K as function of the temperature is approaching by Van't Hoff equation (Equation (6)), using this equation the enthalpy change and entropy change can be predicted. The approaching results using Van't Hoff equation are shown in the Figure 3.

**Table 1.** Values of distribution coefficient, enthalpy change and entropy change

| T, °C | K | $R^2$ | ΔH°, J/mol | ΔS°, J/mol/K |
|---|---|---|---|---|
| 50 | 0.502 | 0.990 | 20718.59 | 58.05099 |
| 44 | 0.394 | 0.858 | | |
| 38 | 0.349 | 0.987 | | |
| 32 | 0.314 | 0.891 | | |

Figure 3 shows plotting data and approaching line K value as function of temperatures. The approaching line is well represent the data with the $R^2$ value is 0.946, with the enthalpy change and entropy change value are 20718.59 J/mol and 58.05 J/mol/K. The value of Gibbs free energy change as function of temperatures are calculated using equation (9). Values of Gibbs free energy for various temperatures are listed in the Table 2.

Enthalpy change (ΔH°) value is 20.72 kJ/mol it is mean that the extraction lipid from *Nannochloropsis sp.* is endothermic process. The ΔH° of this process is larger than the range of ΔH° of oil extraction process that was agreement between 4 – 13.5 kJ/mol [21]. It is mean that to extract lipid from *Nannochloropsis sp.* need more energy than another oil plant sources. The positive value of ΔS° is indicating this process is irreversible.

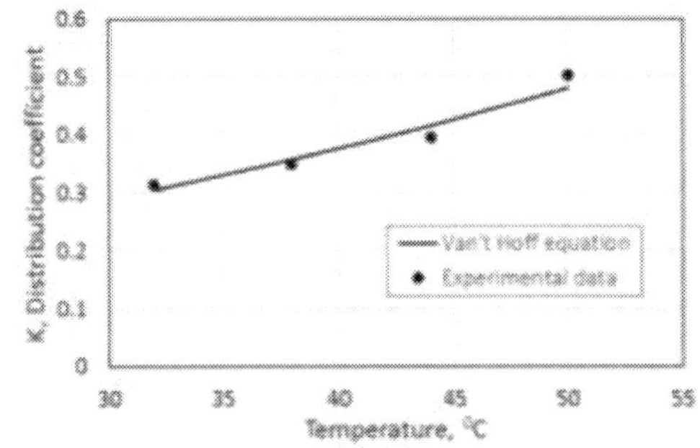


**Fig 3.** K values as function temperatures and the approaching line

**Table 2.** Values of Gibbs free energy change for some temperatures

| T, °C | ΔG, J/mol | $R^2$ |
|---|---|---|
| 50 | 1969.312 | 0.946 |
| 44 | 2317.355 | |
| 38 | 2665.399 | |
| 32 | 3013.442 | |

The value of Gibbs free energy for temperature 50 °C and below are positive, this is indicated that the extraction lipid from *Nannochloropsis sp.*for this temperature range is not spontaneous. Based on the thermodynamic evaluation, the lipid extraction from *Nannochloropsis sp.* need more energy than another oil-plant sources. This energy is used to cell disruption.

## 3.3 Comparison of hydrodynamic cavitation and conventional lipid extraction

Figure 4 shows comparison phenomena of lipid extraction as function of extraction time among hydrodynamic cavitation with pressure booster 6.8 atm, conventional with 260 rpm stirring speed and conventional with 1000 rpm stirring speed. All of the experiment were done at the 34 °C and pressure 1 atm. The amount of dry *Nannochloropsis sp.*, hexane and methanol for all experiment were constant. The weight of dry *Nannochloropsis sp.* was 5 grams, volume of hexane was 47.5 ml and volume methanol was 20.5 ml.

From Figure 4, we may see that the phenomena of lipid extraction assisted by hydrodynamic cavitation is the fastest methods as compared to the conventional methods with stirring speed 260 and 1000 rpm. The time to get 8.9% yield by hydrodynamic cavitation extraction need time less than 10 minutes, while using conventional methods for extraction time 60 minutes the yields 7.3% for 1000 rpm stirring speed and 5.44% for 260 rpm stirring speed . This comparison is indicate that extraction process assisted by hydrodynamic cavitation is very effective methods. And if it is compared to another cell disruption methods the hydrodynamic cavitation is most effective [19].

### 3.4. Volumetric mass transfer coefficient

Volumetric mass transfer coefficient of *Nannochloropsis sp.* lipid extraction was evaluated using equations (3), (4), (5) and (6). The experimental data of equilibrium were used to calculate K by equation (7). Equations (3) and (4) were solved using Runge Kutta Method and the value of volumetric mass transfer coefficient was evaluated using one variable minimization Golden Section Method. The model calculation results represent in the Figure 4. It shows phenomena of lipid extraction from *Nannochloropsis sp.* using three different methods there were hydrodynamic cavitation (HC), extraction with 260 rpm stirring speed and 1000 rpm stirring speed. The value of volumetric mass transfer coefficient of three different methods are listed in the Table 3.

**Table 3.** Volumetric mass transfer coefficient of three different methods

| $k_T a_T$, 1/s | | |
|---|---|---|
| HC | 1000 rpm stirring | 260 rpm stirring |
| 7.375 | 0.534 | 0.121 |

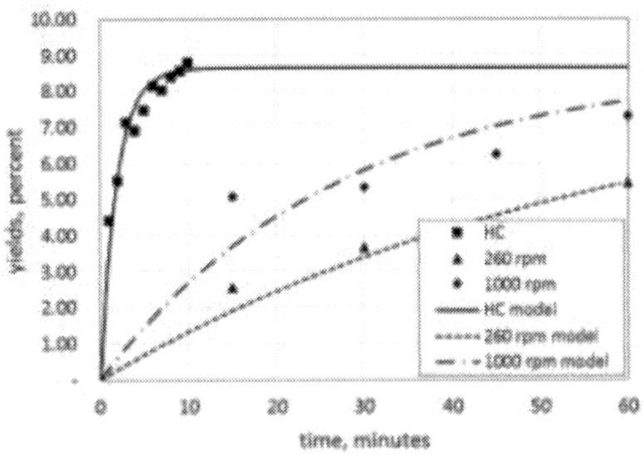


**Fig 4**. Experimental and simulated model of three different methods

## 4 Conclusions

The distribution coefficient of lipid extraction process of *Nannochloropsis sp.* using methanol and hexane solvents are affected by temperature with the values for temperatures 34, 38, 44 and 50 °C are 0.314, 0.349, 0.394 and 0.502, respectively. The value of thermodynamic parameter ΔH° and ΔS° are 20.72 kJ/mol and 58.05 J/mol/K respectively. The Gibbs free energy values are affected by temperatures. These values for temperatures 34, 38, 44 and 50 °C are 3.013, 2.665, 2.317 and 1.969 kJ/mol respectively. Based on the thermodynamic parameter values lipid extraction of *Nannochloropsis sp.* process is endothermic, irreversible and not spontaneous.

Lipid extraction by hydrodynamic cavitation is the fastest process as compared to the conventional methods. The volumetric mass transfer coefficient value of hydrodynamic cavitation, conventional methods 1000 rpm and 260 rpm are 7.373, 0.534 and 0.121 1/s respectively.

We would like to thank Ahmad Dahlan University, Yogyakarta, Indonesia which has provided research fund and Process System Engineering Research Group, Chemical Engineering Department, Gadjah Mada University, Yogyakarta, Indonesia which has provided research facilities.

## References

1. R. D. Kusumaningtyas, S. Purwono, Rocmadi, A. Budiman, IJEX, **15**, 447-467. (2014)
2. T. Suganya, M. Varman, H.H. Masjuki, S. Renganathan, Renew. Sust. Energ. Rev., **55**, 909–941 (2016).
3. D.R. Sawitri, Sutijan, A. Budiman, JEAS, **11**, 9951-9957 (2016)
4. Sunarno, Rochmadi, P. Mulyono, A. Budiman. AIP Con. Proc., **1737**, 060008 (2016).
5. D.R.Wicakso,., Sutijan, Rochmadi, A. Budiman, yst, AIP Con. Proc., **1737**, 060009 (2016)
6. M.D. Supardan, Fahrizal, R. Moulana, D. Safrida, Satriana, W.A.W. Mustapha, JESTEC, **12**, 847-859 (2017)
7. H. Sudibyo, Y.S. Pradana, T.T. Samudra, A. Budiman, Indarto, E.A. Suyono, Energy Procedia, **105**, 270-276 (2017)
8. T.M. Mata, A. Martins, N.S. Caetano, Renew. Sust. Energ. Rev., **14**, 217–232 (2010)
9. J. Singh, S. Gu, Renew. Sust. Energ. Rev., **14**, 2596–2610 (2010)
10. E. Sualli, R. Sarbatly, Renew. Sust. Energ. Rev., **16**, 4316– 4342 (2012)
11. I. Hariskos, C. Posten, Biotechnol. J., **9**, 739–752 (2014)
12. M.A. Islam, R.J. Brown, I. O'Hara, M. Kent, K Heimann, Energy Convers. Manage., **88**, 307–316 (2014)
13. Y.S. Pradana, H. Sudibyo, E.A. Suyono, Indarto, A. Budiman, Energy Procedia, **105**, 277-282 (2017)
14. R. Halim, M.K. Danquah, P.A. Webley, Biotechnol. Adv., **30** (3), 709–732 (2012)
15. A.R. Byreddy, A. Gupta, C.J. Barrow, M. Puri, Marine Drugs, **13**, 5111-5127 (2015)
16. J.Y. Lee, C. Yoo, S.Y. Jun, C.Y. Ahn, H.M. Oh, Bioresour. Technol., **101**, s75–s77 (2010)
17. A.K. Lee, D.M. Lewis, P.J. Ashman, Bioresour. Technol., **128**, 199–206 (2013)
18. A.K. Lee, D.M. Lewis, P.J. Ashman, J. Appl. Phycol., **27**, 1881-1889 (2014)
19. I. Lee, J.I. Han, Bioresour. Technol., **186**, 246-251 (2015)
20. M.D. Kostic, N.M. Jokovic, O.S. Stamenkovic, K.M. Rajkovic, P.S. Milic, V.B. Veljkovic, Ind. Crops. Prod., **52**, 679- 686 (2014)
21. H. Topallar, U. Geçgel, U., Turk. J. Chem., **24**, 247 -253 (2000)

# Green synthesis of silver nanoparticle using rambutan (*Nephelium lappaceum L.*) peel extract and its antibacterial activity against *Salmonella parathypi A.*

*Puji* Lestari[1,*], *Indah* Pratiwi, and *Any* Juliani

Environmental Engineering Study Program, Universitas Islam Indonesia, Jln. Kaliurang KM 14.5, 55584 Yogyakarta, Indonesia

**Abstract.** Ag nanoparticles (AgNPs) have been synthesized via green method using rambutan (*Nephelium lappaceum L.*) peel (RP) extract. RP extract was prepared by washing the RP using tap water thoroughly and boiling it in distilled water at 70°C for 60 min. RP extract and $AgNO_3$ were used as the starting materials for the synthesis of AgNPs. RP extract was added to $10^{-3}$ M $AgNO_3$ solution with a ratio by volume of 1:10 (RP : $AgNO_3$), stirred at room temperature. The solution's color changes from reddish to dark brown, indicating the reduction of $Ag^+$ in the solution. The synthesized AgNPs were characterized using UV-Visible Spectrophotometer, FTIR Spectrophotometer, and Scanning Electron Microscopy-Energy-dispersive X-ray Spectroscopy (SEM-EDS) instruments. UV-Visible spectra show that the AgNPs have the maximum absorption band at 450 nm which is typical for AgNPs. The FTIR spectra revealed that the protein in RP extract acts as the capping agent for the synthesized AgNPs. The synthesized AgNPs were tested for their antibacterial activity against *Salmonella parathypi A*. The antibacterial test shows that 50 μL of AgNPs resulted in the inhibition zone of 4 mm against the aforementioned microorganism.

## 1 Introduction

Nanoparticles are known as particles with the size ranging from 1 to 100 nm [1]. Recently, nanoparticles have obtained great interest due to their unique properties. The physical and chemical properties of metallic nanoparticles are different with bulk metals. Metallic nanoparticles have lower melting points, higher specific surface areas, specific optical properties, mechanical strengths and specific magnetizations [2]. Nanoparticles have been widely applied in many areas such as medicine, chemistry, pharmacy, industry, environment, biology, etc. One of the uses of nanoparticles is as antibacterial agents. Among all metal nanoparticles, silver nanoparticles (AgNPs) show the highest antibacterial activity. Despite the fact that the toxicity of AgNPs is still not clear, AgNPs have been applied as anti bacterial agents in the health industry, food storage, textile coatings and a number of environmental applications [3].

The synthesis of AgNPs can be done using physical or chemical approaches. The physical approach involves some methods such as evaporation/condensation and laser ablation. The chemical approach uses strong reductants such as boro hydrate, citrate, ascorbate and elemental hydrogen. The reductants will reduce silver ions ($Ag^+$) to elemental form ($Ag^0$) which will eventually form the colloidal Ag particles [4]. Among these two approaches, chemical approach is more preferable. However, the chemical approach uses chemicals which are toxic and hazardous to the environment. Therefore, the development of the green process for the synthesis of AgNPs has been an emerging field of research interest. The green synthesis of AgNPs uses many types of plant extracts, bacteria, and fungi to eliminate the toxic chemicals needed for reducing the silver ions. The use of plant extract is most popular due to its simplicity and low cost [5]. The synthesis of AgNPs using plant extract is a rapid, eco-friendly, non-toxic and providing only a single step technique for the synthesis process [6].

Some plants' extracts have been reported to be successfully used for the synthesis of AgNPs. Other plants' extract which have been reported to be used for the synthesis of AgNPs are tea extract [7], inflorescence of *Cocous nucifera* [8], peel of *Citrus sinensis* [9], peel of *Musa paradisiacal* [10], leaves of *Carica papaya* [11], fruit of *Vitis vinifera* [12] and many other.

In this work, a native Indonesian plant, rambutan (*Nepphelium lappaceum*) peel extract was used to synthesize AgNPs. The synthesized AgNPs will then be studied for their anti bacterial activity against *Salmonella parathypi A*.

* Corresponding author: puji.lestari@uii.ac.id

# 2 Experimental Methods

## 2.1 Chemicals

Rambutan peel (RP) extract and $AgNO_3$ were used as starting materials for the synthesis of Ag nanoparticles (AgNPs) in this work. RP was washed using tap water and cut into pieces. RP was boiled in distilled water at 70°C for 1 h to obtain the RP extract. The RP extract was then filtered using Whatmann No.42. The extract was stored at 4°C for further use. The storage time for RP extract was 10 days at maximum. The chemicals used in this work were E.Merck production and were directly used without further purification. *Salmonella parathypi A.* strain was supplied by *Balai Laboratorium Kesehatan*, Yogyakarta. The distilled water was used as the solvent in this work.

## 2.2 Green Synthesis of Ag Nanoparticles (AgNPs)

The aqueous $AgNO_3$ solution with a concentration of $10^{-3}$ M was prepared in an Erlenmeyer flask. The RP extract with the ratio by volume of 1:10 (RP : AgNO3) was added to the solution to reduce the $Ag^+$ ions. The mixture was stirred at room temperature and the color change was observed. The AgNPs were considered to be formed when the color of the mixture changes into dark brown. The synthesized AgNPs were then characterized using UV-Visible Spectrophotometry, FTIR spectrophotometry, and Scanning Electron Microscopy-Energy-dispersive X-ray Spectroscopy (SEM-EDS).

## 2.3 Antibacterial Activity of AgNPs Against Salmonella parathypi A.

The synthesized AgNPs were tested for their antibacterial activity against *Salmonella parathypi A.* The antibacterial activity of AgNPs was studied using agar well diffusion method. The antibacterial activity of RP extract and chloramphenicol were used as comparisons. Agar media was prepared in a petri disk and added by the microorganism culture. The agar was divided into 4 parts and a well with a known diameter was prepared on each part of the agar. Well A was left blank, well B was filled with 50 μL RP extract, well C was filled with 50 μL AgNPs and well D was filled with 50 μL of 40 ppm Chloramphenicol and let to diffuse for some time. The petri disk was covered by paper and then incubated for 20 h. The inhibition zone for each substance was observed and measured.

# 3 Results and Discussion

## 3.1 Characterization of AgNPs

### *3.1.1 UV-Visible Spectrophotometry*

In the synthesis of AgNPs using RP extract, the color of the extract changed from reddish to dark brown due to the reduction of $Ag^+$ in the solution. Fig 1 shows the UV-Visible spectra of RP extract, $AgNO_3$ solution, and the synthesized AgNPs. The maximum absorption band of AgNPs in this work is 450 nm. The region of 350-450 nm is typical for the maximum absorption band of AgNPs [13]. The broad and asymmetrical peak of the UV-Visible absorption band of the synthesized AgNPs indicates AgNPs with a broad interval of size and non-uniform shape. This work focuses on the potential of RP extract for the green synthesis of AgNPs and its antibacterial activity against Salmonella parathypi A. Therefore, the size and shape of AgNPs were not controlled in this study.

### *3.1.2 Scanning Electron Microscopy – Energy-dispersive X-Ray Spectroscopy (SEM – EDS)*

The formation of AgNPs confirmed by UV-Visible spectra was further confirmed by analysis using SEM-EDS instrument. The result of SEM analysis and EDS spectra are presented in Fig 2. The SEM and EDS analyses confirm that the RP extract has been successfully used for synthesizing AgNPs. It is seen that the synthesized AgNPs have non-uniform shape and a wide range of particle size.

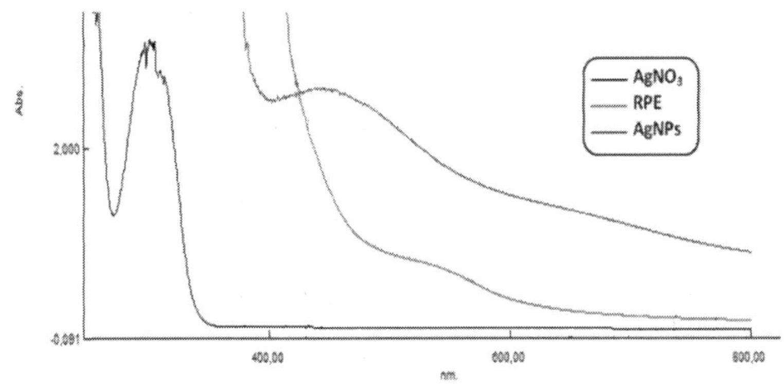


**Fig. 1.** UV Visible spectra of AgNPs synthesized using RP extract

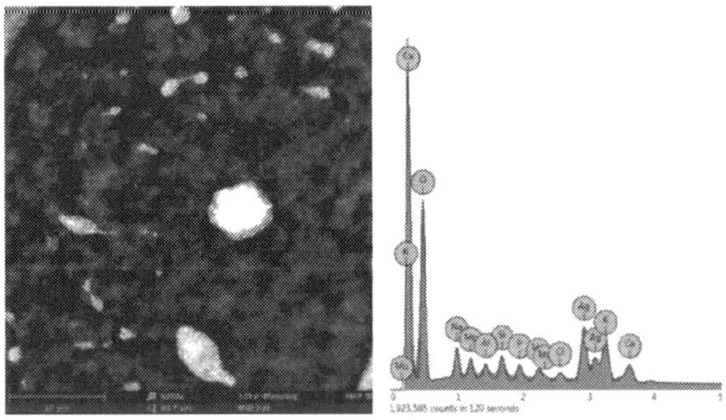

(a) (b)

**Fig. 2.** (a) SEM analysis result of AgNPs; (b) EDS spectra of AgNPs

### *3.1.3 FTIR Spectrophotometry*

FTIR analysis was conducted in order to know the functional groups of biomolecules in RP extract which are responsible for the synthesis and stabilization of AgNPs. The FTIR spectra of AgNPs synthesized using RP extract is displayed in Fig 3. The FTIR spectra of AgNPs shows some major peaks at 3339, 1574, 1322, 1039 and 820 $cm^{-1}$. The peak at 3339 $cm^{-1}$ is attributed to O-H stretching of phenolic and alcoholic –OH groups. The absorption band at 1574 $cm^{-1}$ corresponds to –COOH stretching vibration of flavonoids/phenolic groups. The peak at 1322 $cm^{-1}$ is due to the stretching of C-O bond in aromatic groups. the absorption band at 1039 $cm^{-1}$ is probably due to the C-N stretching of aromatic and aliphatic amines. It is known that the main biological active compounds of RP are anthocyanins, ellagitannins, ellagic acid, corilagin, geraniin, syringic acid and p-coumaric acid [14]. The presence of protein functional groups on the surface of AgNPs indicates that the proteins act as the capping agent for the synthesized AgNPs.

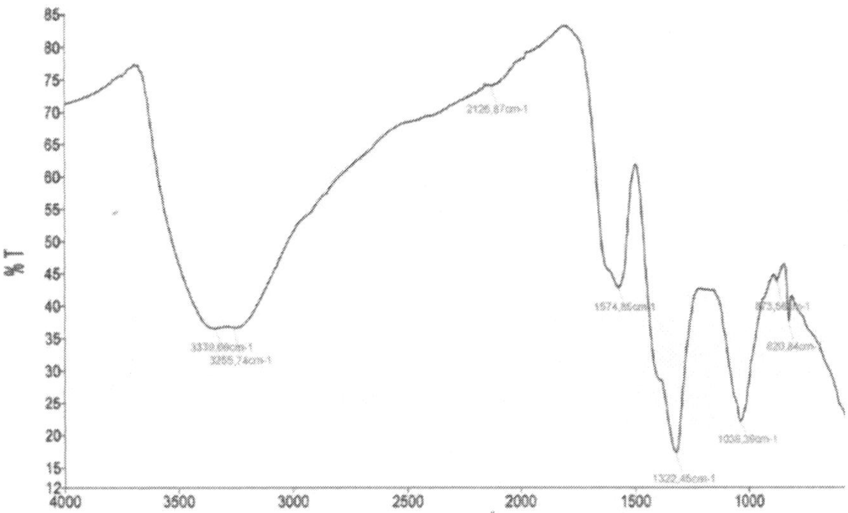


**Fig. 3.** FTIR spectra of the AgNPs synthesized using RP extrac

## 3.2 Antibacterial Activity of AgNPs Against *Salmonella parathypi A.*

Fig. 4 shows the result of the antibacterial test for each antibacterial agent. The inhibition zone of AgNPs, RP extract, and chloramphenicol against *Salmonella parathypi A.* is presented in Table 1. Among the three tested anti bacterial agent, Chloramphenicol shows the highest antibacterial activity against *Salmonella parathypi A.* The synthesized AgNPs exhibits a relatively much higher antibacterial activity against *Salmonella parathypi A.* compared to the RP extract.

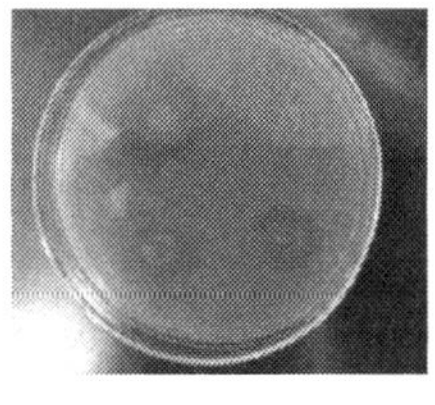

**Fig 4.** The antibacterial test of AgNPs, RP extract, and Chloramphenicol against *Salmonella parathypi A.*

**Table 1.** Inhibition zone of each antibacterial agent

| Antibacterial agent | Inhibition zone (mm) against *Salmonella parathypi A.* after 20 h incubation |
|---|---|
| RP extract | 1 |
| AgNPs | 4 |
| Chloramphenicol (40 ppm) | 6 |
| No addition of antibacterial agent | - |

## 4 Conclusions

The rambutan (*Nephelium lappaceum L.*) peel (RP) extract has been used to synthesize AgNPs using aqueous media at room temperature. The analysis of UV-Visible spectra confirms the formation of AgNPs. The SEM analysis shows that the synthesized AgNPs have non-uniform shape and wide distribution of particle size. The synthesized AgNPs have antibacterial activity against *Salmonella parathypi A.* with the inhibition zone of 4 mm, while the inhibition zone of Chloramphenicol against *Salmonella parathypi A.* used as a comparison is 6 mm.

## References

1. D. Williams, Biomaterials **29**, 1737-1988 (2008).
2. S. Horikoshi and N. Serpone, *Introduction to Nanoparticles* In S. Horikoshi and N. Serpone, *Microwaves in Nanoparticle Synthesis: Fundamentals and Application* (Wiley-VCH Verlag GmbH & Co. KGaA, 2013).
3. K. M. El-Nour, A. Eftaiha, A. Al-Warthan, and R. A. Ammar, Arab. J. Chem **3**, 135-140 (2010).
4. A. Tao, P. Sinsermsuksakul and P. Yang, Angew. Chem. Int. Ed **45**, 4597-4601 (2006).
5. A. K. Jha, K. Prasad and A. R. Kulkarni, Colloids Surf. B Biointerfaces **2**, 219-223 (2009).
6. S. Ahmed, M. Ahmad, B. L. Swami and S. Ikram, J. Adv. Res **7**, 17-28 (2016).
7. A. Nabikhan, K. Kandasamy, A. Raj and N. M. Alikunhi, Colloids Surf. B **79**, 488-493 (2010).
8. R. Mariselvam, A. J. Ranjitsingh, A. Usha Raja Nanthini, K. Kalirajan, C. Padmalatha and P. Mosae Selvakumar, Spectrochim. Acta Mol. Biomol. Spectrosc **129**, 537-541 (2014).
9. M. Dubey, S. Bhadauria and B. Kuswah, Dig. J. Nanomater. Biostruct **4**, 537-543 (2009).
10. A. Bankar, B. Joshi, A. R. Kumar and S. Zinjarde, Colloids Surf. B Biointerfaces **368**, 58-63 (2010).
11. D. Jain, H. K. Daima, S. Kachhwaha and S. Kothar, Dig. J. Nanomater. Biostruct **4**, 557-563 (2009).
12. G. Gnanajobitha, K. Paulkumar, M. Vanaja, S. Rajeshkumar, C. Malarkodi, G. Annadurai and C. Kannan, J. Nanostruct. Chem **3**, 1-6 (2013).
13. O. Velgosová, A. Mražíková and R. Marcinčáková, Mater. Lett **180**, 336-339 (2016).
14. T. Lim, *Nephelium lappaceum. In Lim, T., Edible Medicinal and Non-Medicinal Plants* (Dordrecht: Springer, 2013).

# Hydrothermal carbonization of biomass waste under low temperature condition

*Herlian Eriska* Putra[1,*], *Enri* Damanhuri[1], *Kania* Dewi[1] and *Ari Darmawan* Pasek[2]

[1]Faculty of Civil and Environmental Engineering, Institut Teknologi Bandung, Bandung 40132, Indonesia
[2]Faculty of Mechanical and Aerospace Engineering, Institut Teknologi Bandung, Bandung 40132 Indonesia

**Abstract.** In this paper, the use of banana peel for energy purposes was investigated. Banana peel is a lignocellulosic waste since it is the most widely produced and consumed fruit in Indonesia. Among the others, hydrothermal carbonization (HTC) was chosen as alternative themochemical process, suitable for high moisture biomass. Through a 1 L stirred reactor, hydrothermal treatments were performed under low temperature condition (190, 210 and 230 °C), residence times (30 and 60 min), and biomass to water ratio (1:3, 1:5, and 1:10). Three of product were collected from the process with primary material balance. Solid phase (hydrochar) was evaluated in terms of calorific value, proximate and ultimate analysis. The results suggested that the hydrothermal carbonization of banana peel gave high heating value (HHV) of 20.09 MJ/kg for its char after dried naturally.

## 1 Introduction

Biomass is one of the most abundant sources of renewable energy. Agricultural and forest residues, and even industrial biomass waste are cheap and non-edible feedstock. They can be an important part of more sustainable future energy system. One of promising technology that can be used for energy purposes is hydrothermal carbonization (HTC).

HTC is a thermochemical process which imitates the natural coalification of biomass. Compare to the natural process which requires some hundred (peat) to some million (black coal) years, HTC only needs less than a half day for the transformation of biomass into a material quite similar to brown coal. Its technique uses liquid sub-critical water as a reaction medium for conversion of wet biomass. According to several researchers [1-3]. HTC comprises a series of hydrolysis, condensation, decarboxylation and dehydration reactions. It utilizes subcritical water to heat feedstock at temperature typically ranging from 180 to 350 °C [4], even though the effect of carbonization can be achieved under temperature below 300 °C [5]. Not only process temperature, the type of feedstock is also important parameter to be considered in terms of sustainablity. The application of HTC process to several typologies of biomass has been discussed by many researcher [6-9].

In Indonesia, bananas are the most widely consumed fruit. In 2013, based on *Angka Tetap (ATAP)*, the production of bananas reached 6.28 million tons. More than 50% of Asian banana production is produced by Indonesia [10]. Most residual of consumed banana is still found as wet ligno-cellulosic residue, which must be properly disposed of. HTC could be an effective technology to handle wet feedstock. Considering that HTC is performed at lower temperature. This process has been investigated in this paper.

In this paper, the results of HTC of banana peel performed in a lab scale batch reactor were presented. Several reaction conditions such as reaction temperature, feed to water ratio, and residence times were determined. Three of product were collected from the process with primary material balance. Solid phase (hydrochar) was evaluated in terms of calorific value, proximate and ultimate analysis.

## 2 Materials and Methods

### 2.1. Experimental Setup

Banana peels (*Musa spp*) were purchased locally. Approximate size of it, in all dimensions, is <1 mm. The peels were air-dried prior to use in the carbonization experiments. Banana peels were used to represent cellulose and hemicellulose component. It comes from household residue. Banana is very popular fruit in Asia.

HTC of banana peels was carried out using 1 L batch reactor. Reactor, equipped with electrical heater and agitator, was made from stainless steel SS-304 as shown in insert in Figure 1. To analyze the effect of operating condition on feedstock, the experiments were performed at the different holding time period (30 and 60 minutes), feed to water ratio 1:3, 1:5, 1:10, and reaction temperatures (190, 210, and 230 °C) as

* Corresponding author: herl007@lipi.go.id

subcritical temperature for hydrolysis starts at 180 °C [11]. Nitrogen gas was passed through the reactor for 5-10 min to ensure oxygen free in the system. Figure 1 showed scheme of HTC experimental system.

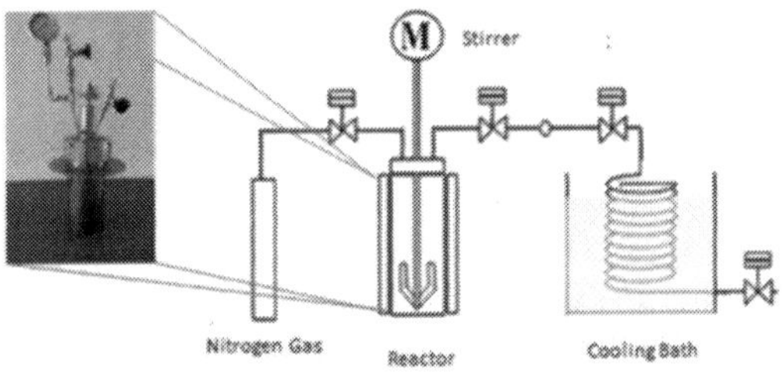


**Figure 1.** Experimental setup

50 gram of sample was loaded into reactor in every single run. Samples were mixed with deionized water with same amount to ensure complete soaking. Reactor was heated up in around 50 min and maintained at desired temperature for 30 and 60 minutes. Once the inside temperature of reactor dropped to room temperature, the pressure valve was released under fume hood to discharge gaseous products. Solid and liquid product were separated using filter paper (2 μm). Both gas and liquid were collected without further evaluation. Solid fuel (hydrochar) were dried naturally more than 24 hours before analysis.

### 2.2 Analysis

Proximate analysis (ash, volatile matter, and fixed carbon) was measured using Leco TGA-601. Ultimate analysis (carbon (C), hydrogen (H), oxygen (O), nitrogen (N), sulphur (S)) was determined using Elementar Vario Macro and for high heating value (HHV) were measured using bom calorimeter. All of analysis was under supervised by Research and Development Center of Mineral and Coal Technology (Tekmira). Mass yield (Eq.1), energy densification ratio (eq.2), and energy yield (eq.4) are three parameters that were calculated in the study and are expressed as:

$$mass\ yield\ = \frac{mass\ of\ dried\ hydrochar}{mass\ of\ dried\ feedstock} \times 100 \quad (1)$$

$$energy\ densification\ ratio\ = \frac{HHV\ of\ hydrochar}{HHV\ of\ feedstock} \times 100 \quad (2)$$

$$energy\ yield\ = mass\ yiel \times energy\ densification\ ratio \quad (3)$$

## 3 Results and Discussion

### 3.1 Effect of Operating Condition

Table 1 and Figure 2 showed that the experimental design that was completed in this study. Two kind of different operating condition; feed to water ratio and reaction temperature, during HTC process and their effect on calorific properties of each paper, banana peel, and sawdust. The results were presented in Table 2 and Figure 2, showed that both reaction temperature and holding time period, have influenced the properties of hydrochar mass yield, energy yield and high heating value.

**Table 1.** Experimental design of feedstock and hydrochar

| Temp. (°C) | Feed to Water Ratio | Residence time | Energ y yield | Mass Yield |
|---|---|---|---|---|
| 190 | 1/3 | 30 min | 67.62 | 64.21 |
| | | 60 min | 68.29 | 64.11 |
| | 1/5 | 30 min | 67.14 | 63.54 |
| | | 60 min | 67.47 | 63.28 |
| | 1/10 | 30 min | 66.42 | 62.87 |
| | | 60 min | 67.12 | 62.76 |
| 210 | 1/3 | 30 min | 70.56 | 63.25 |
| | | 60 min | 70.25 | 62.04 |
| | 1/5 | 30 min | 69.42 | 61.43 |
| | | 60 min | 70.96 | 60.95 |
| | 1/10 | 30 min | 69.08 | 60.06 |
| | | 60 min | 69.99 | 59.55 |
| 230 | 1/3 | 30 min | 68.02 | 59.26 |
| | | 60 min | 70.47 | 59.21 |
| | 1/5 | 30 min | 68.30 | 59.12 |
| | | 60 min | 68.81 | 58.86 |
| | 1/10 | 30 min | 67.46 | 58.46 |
| | | 60 min | 69.47 | 57.85 |

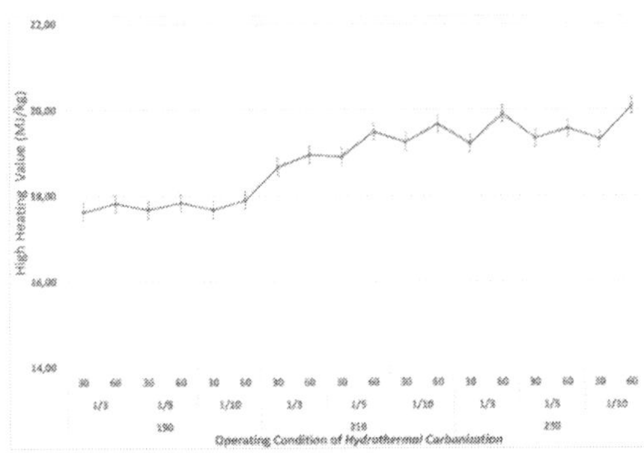


**Figure 2.** HHV from HTC biomass waste

The results suggest that even mass yield decreased, both energy yield and heat heating value increased along temperature rises. Banana peel contains cellulose and hemicellulose, it were easier to decompose than lignin feedstock during hydrothermal carbonization. It also explains the reason of the calorific value and carbon content increase of the most of feedstock after the hydrothermal treatment whose major component is cellulose. More than 90% hemicellulose starts to decompose in the temperature range 180-220 °C. Temperature of biomass hydrolysis starts at 180 °C

[11]. Lignin behaviour is different, less than 10% of lignin decomposed at below 250 °C [11,12].

From Figure 2 also shows the effect of reaction time and temperature on the mass yield of different feedstock during HTC. As expected, the mass yield less reduces with an increase in the holding time period and temperature. Banana peel showed very low in mass yield. Banana peel is one of biomass with high cellulose and hemicellulose content. Because of low thermal stability, banana peel tend to loss its mass easily [13].

The removal of hemicellulose via degradation and depolymerisation reactions during the thermal pre-treatment of a biomass results into the loss of structure [1]. The destroyed structure has improved friability. The example structures of the raw and hydrochar samples from banana peel have been observed under Scanning electron microscope (SEM) was shown in the Figure 3. The images show the variation in the structure a hydrochar with an increase in the process reaction temperature. For the raw feedstock, a clearly well-defined structure can be observed. However, in case of the hydrochar samples the unwrapping and rupture of polymer bundles and opening of pores is clearly noticeable. Almost complete loss of the structure can be observed for the samples pre-treated at 230°C.

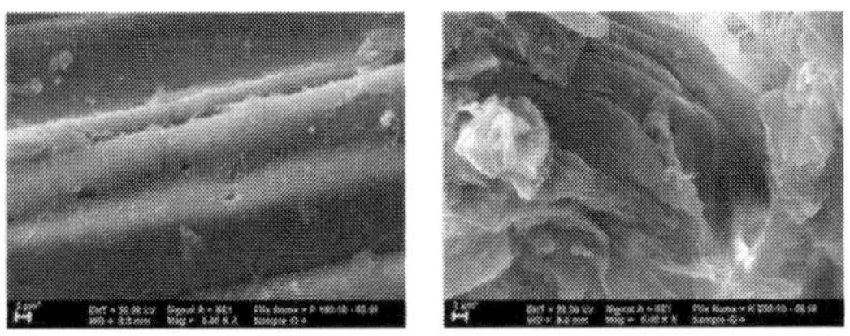

**Figure 3.** SEM of banana peel at (a) Raw, (b) HTC temperature 230 °C

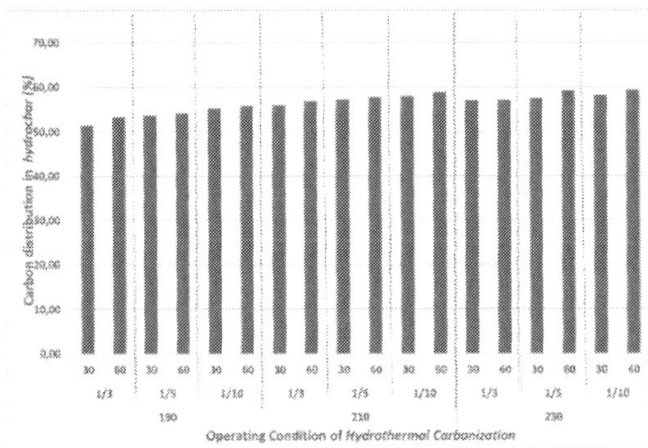


**Figure 4.** Carbon distribution in hydrochar

Mass balance analyses indicate that carbonization of the feedstock results in a significant fraction of carbon retained within the char (Figure 4). Carbonization of offie banana peel results in the smallest fraction of carbon remaining in the solid-phase. Hydrothermal process breaks the physical and chemical structure of material into smaller molecules. Biomass feedstock usually have high volatile matter and oxygen content. With the increase of the hydrothermal reaction temperature, the volatile matter and oxygen content decreased while the fixed carbon and high heating value slightly increased that influenced by hydrolysis reaction. The results suggested that the hydrothermal treatment of biomass waste to solid fuel gave high heating value (HHV) with value of 20.09 MJ/kg for fruit peel respectively after product dried naturally.

## 3.2. Proximate and Ultimate Analysis of Hydrochar

The proximate analysis is used to determine the quality of coal and other solid fuels. The analysis comprises the measurement of fixed carbon, volatile matter, and ash content. The carbon content of the produced hydrochar ranges from 42-73%. According to many literature [14-17], carbon contents of hydrochar from the carbonization process were vary depend on other compounds. So, it is very hard to compare between carbon-contents measured in this study with reported in other literature. There are so many factors, such as the different of temperature, pressure, reaction time, reactor design, feed/water ratio, water stream, and catalyst.

Fixed carbon is the combustible residue after volatile matter burned. Organic samples have high volatile matter content and low fixed carbon, but high moisture content. Increasing of volatile matter and fixed carbon contents after hydrothermal process will lead to increase calorific value. Figure 5 showed that the fixed carbon of each feedstock was increasing during the HTC process. The increasing of fixed carbon can influence the increasing of HHV.

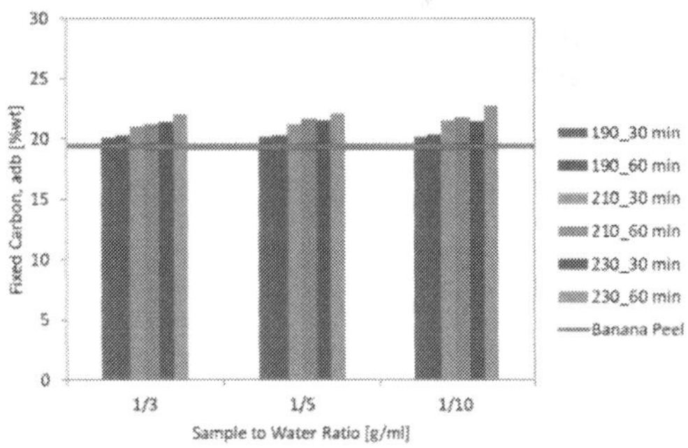


**Figure 5.** Fixed carbon of HTC of banana peel

Figure 6 showed that decreasing of volatile matter content in sample after process might lead to increase fixed carbon content. Increasing of volatile matter content in sample after process probably leads to increase calorific value. This is because volatile matter is combustible element and produce high flame. It can be seen that volatile matter content in feedstock decreases after hydrothermal process. This is because some volatile matters were probably evaporated and released into the air. In addition, it can be seen that using less process water will result lower volatile matter content in sample.

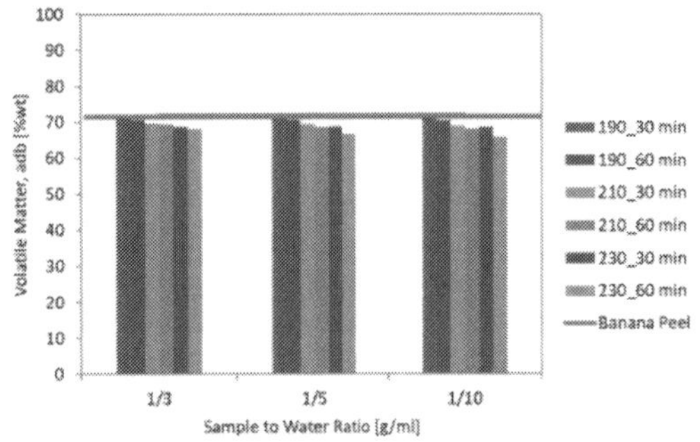


**Figure 6.** Volatile matter of HTC of banana peel

Figure 7 showed that ash content increases after process for feedstock. This because lignin content from banana peel is less than other wood feedstock. Lignin tends to decomposed at temperature more than 250 °C. Ash is the inorganic residue remaining after the sample is completely burned. Ash characteristics play an important role in system design in order to minimize slagging, fouling, erosion, and corrosion.

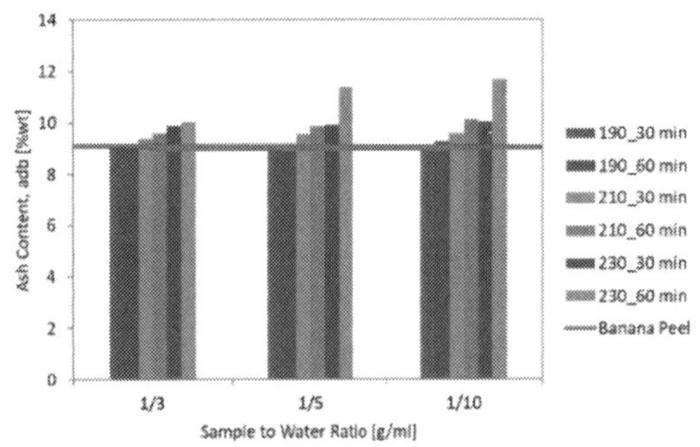


**Figure 7.** Ash content of HTC of banana peel

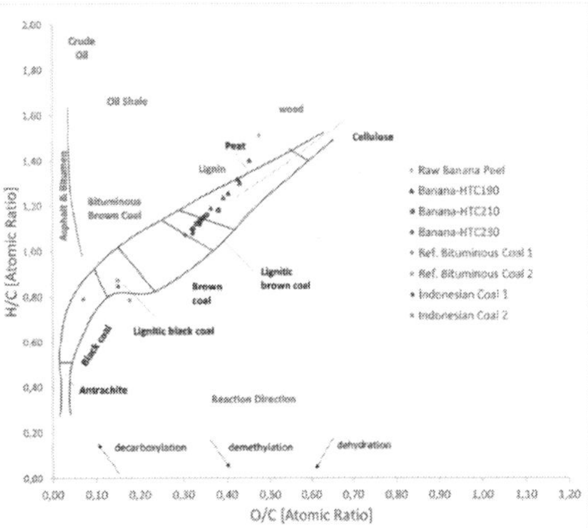


**Figure 8.** Coalification diagram

The hydrothermal carbonization or the coalification process that happen can be described (Figure 8) by the van Krevelen diagram or coalification diagram [18]. During carbonization as a result of demethylation, decarboxylation, and dehydration reaction, there is a reduction of hydrogen and oxygen content in the feedstock. It can be described by the molar ratios of H/C and O/C. In the coalification diagram, those ratios move from upper right to lower left as the carbonization advance. The process goes from cellulose and wood, peat and lignite to the various forms of brown coal and finally to coal and anthracite. From the preliminary test of HTC, hydrochar from banana peel fall in the region of lignite brown coal.

## 4 Conclusion

The results, both in terms of hydrochar yield and hydrochar chemical and thermal properties, showed that HTC represents a suitable way to energetically valorise the banana peeel. The resulted showed that increasing reaction temperature and holding time period during hydrothermal carbonization tend to increasing carbon content. Due to there was decreasing polarity and increasing temperature is associated with molecule breaking. In this perspective, high temperatures tend to increase the heating values of the hydrochar, although the yields resulted lowered. Thus, further experiments are required to assess the optimal values of the process parameters. In particular, a detailed reaction kinetics study could be useful to optimize the process and thus to obtain its maximum performance. Moreover, several parameters such as pure substances usage, variation of ratio feed to water, and catalyst is required in view of an actual assessment of the HTC potentiality for large-scale applications. Characteristic of liquid and gas product should be determined to gain more accurately about process and other carbonization products.

## References

1. A. Funke, F. Ziegler. Biofuels Bioproducts & Biorefining-Biofpr, **4**, (2010)
2. M. Sevilla, A.B. Fuertes. *Carbon*, **47**, (2009)
3. M.M. Titirici, A. Thomas, S-H. Yu, O.J. Müller, M. Antonietti. Chemistry of Materials, **19**, (2007)
4. I.H. Hwang, H. Aoyama, T. Matsuto, T. Nakagishi, T. Matsuo. Waste Management, **32**, (2012)
5. X. Lu, B. Jordan, N.D. Berge. Waste Management, **32**, (2012)
6. M. Pala, I.C. Kantarli, H.B. Buyukisik, J. Yanik. Bioresource technology, **161**, (2014)
7. H. Xiao, Y. Guo, X. Liang, C. Qi. Journal of Solid State Chemistry, **183**, (2010)
8. S. Kannan, Y. Gariepy, G.S.V Raghavan. Waste Management, **65**, (2017)
9. Oluwasola O.D. Afolabi, M. Sohail, C.L.P. Thomas. Energy, **134**, (2017)

10. Pusat Data dan Sistem Informasi Pertanian Kementerian Pertanian, (2014), accessed on 18 December 2015.
11. O. Bobleter. Progress in Polymer Science, **19**, (1994)
12. A.T. Yuliansyah, T. Hirajima, S. Kumagai, K. Sasaki, K. Waste and biomass valorization, **1**, (2010)
13. G. Garrote, H. Dominguez, J. Parajo. European Journal of Wood and Wood Products, **57**, (1999)
14. Kardono. Proceeding of International Symposium on EcoTopia Science 2007, ISETS07, (2007)
15. X. Lu, P. Pellechia, J.R.V. Flora, N. Berge. Bioresource Technology, **138**, (2013)
16. X. Lu, J.R.V. Flora, N. Berge. Bioresource Technology, **154**, (2014)
17. X. Lu, N.D. Berge. Bioresource Technology, **166**, (2014)
18. H.-G. Ramke, D. Blöhse, H.-J. Lehmann, J. Fettig. Twelfth International Waste Management and Landfill Symposium, (2009)

# Effect of $Na_2SO_3$ concentration to drug loading and drug release of ascorbic acid in chitosan edible film as drug delivery system membrane

*Lilis* Kistriyani*, *Muhammad Farid* Yulianto, and *Rifqi* Prasetya Y.

Department of Chemical Engineering, Islamic University of Indonesia, Yogyakarta, Indonesia

**Abstract.** Chitosan is a type of carbohydrate compounds produced from waste marine products, in particular the class of shrimp, crabs and clams. Chitosan is often process into edible films and utilized for food packaging also has potential as a membrane for drug delivery system. Drug loading and drug release can be controlled by improve the characteristics of the membrane by adding crosslinker. The purpose of this research is to study the effect of addition of crosslinker to the rate of loading and release of ascorbic acid in the chitosan edible film. $\mathbf{Na_2S_2O_3}$ was used as crosslinker. Two grams of chitosan was dissolved into 100 ml of distilled water. Acetic acid and **plasticizer** were added in the solution then heated at 50°C. $Na_2S_2O_3$ solution with mass various of $Na_2S_2O_3$ dissolved, 0.15; 0.3; and 0.5 grams were added about 30 mL to make edible film. The analysis include of drug loading, drug release and tensile strength. The result showed that the loading of edible film with crosslinker 0.15 g; 0.3 g; and 0.5 g respectively were 60.98 ppm; 52.53 ppm; and 40.88 ppm, meanwhile for the release with crosslinker 0.15 g; 0.3 g; and 0.5 g respectively were 3.78 ppm; 5.72 ppm; and 5.97 ppm.
**Keywords :** $Na_2SO_3$, drug loading, drug release, chitosan, edible film

## 1 Introduction

Drug Delivery System is a system that can deliver drugs as early as possible, provide pharmacological effects for as long as possible, and deliver the drug directly to the target. Therefore, the preparation of the drug is designed in such way that it has the characteristic that we want. There are several advantages of new drug delivery systems versus conventional preparations such as they can reduce the use of drug , optimize total amount of drugs needed to obtain the desired therapeutic response, and minimize undesirable effects (Indrawati, 2009).

(Skurtys et al., 2009) defines edible film as a thin film that can be consumed and used as a coating or barrier between food and the environment. Edible films are classified into three categories based on the component properties of hydrocolloids (proteins, polysaccharides), fat (fatty acids, acylglycerols), and composites (mixtures of hydrocolloids and fats). The main mechanism of film formation in the polysaccharides is the separation of the polymer segments and the reshaping of the polymer chain into the layer or gel matrix usually achieved by solvent evaporation thereby creating hydrophilic hydrogen bonds as well as electrolyte and ionic crosslinking (Butler et al., 1996).

Chitosan can be chemically modified to manipulate the properties needed in its application. Chitosan has some weakness they are less stable, not elastic, and hard. According to drug delivery applications, the desired properties are their solubility and release of controlled drugs. One of the chitosan modifications is the addition of cross-linker. The cross-linker will be bind to the chitosan amine group to form a network that will allow the drug to be trapped inside and the release of the drug can be controlled (Anitha, 2012).

## 2 Materials and Methods

### 2.1 Materials

The materials which used in this research were chitosan, glycerol, $Na_2S_2O_3$, aquadest, ascorbic acid, and buffer solution.

### 2.2 Methods

#### *2.2.1 Film Preparation*

The process of making edible film from chitosan begins with 2 gram of chitosan dissolved in 100 mL 1% acetic acid solution. Then, 1.2 gram of glycerol (as plasticizer) were added and heated at 60°C while stirred using magnetic stirrer for 60 minutes. After 60 minutes, $Na_2S_2O_3$ solution (as crosslinker) were added with mass variation of 0.15, 0.3, and 0.5 gram in 30 mL of aquadest while heated. The heating was continued for 30 minutes and then poured into the petridish plate for film casting and entered into the oven at 60°C for ± 24 hours. The dried film was inserted into the desiccator for ± 1 day to ensure the film was completely dry and ready to be released from the mold.

#### *2.2.2 Drug Loading Analysis*

Ascorbic acid was entered into edible film by diffusion. Edible film dipped in ascorbic acid solution with concentration 80 ppm for 2 hours. Edible film that already contained drug was dried in ambient temperature. The drug loading test was performed by spectrofotometer uv-vis method using 2 cuvet pieces. The ascorbic acid solution of residual film-immersion was incorporated into the cuvet and analyzed using a uv-vis spectrophotometer with a wavelength (λ) of ascorbic acid at 256.6 nm.

#### *2.2.3 Drug Release Analysis*

Edible film from drug loading analysis containing ascorbic acid with various concentrations was fed into 20 mL of buffer solution with pH 7.4 for 2 hours. Concentration of ascorbic acid detached from the film to buffer solution was measured with a uv-vis spectrophotometer with a wavelength (λ) of ascorbic acid at 265 nm.

## 3 Result and discussion

### 3.1 Drug Loading

The aim of crosslinker addition in chitosan was to get the properties that we need. The cross-linker would be bind to the amine group present in the chitosan and form a network, so that the drug could be loaded and the release of the drug could be controlled (Anitha, 2012). The profile of ascorbic acid loading in various concentration of crosslinker were shown in figure 1.

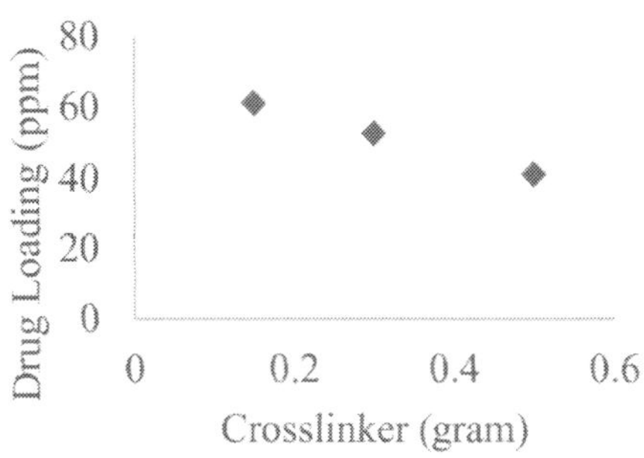


**Figure 1.** Drug Loading vs Crosslinker

From the results of the data obtained, the more cross-linker added to the edible film caused the ascorbic acid loading was reduced becaused of the formation of crosslinks or network minimized the pores that existed in the edible film so that ascorbic acid was entangled less slightly. According to the existing theory, the more crosslinkers were added, more crosslinks were formed so the smaller the pores caused the film was difficult to absorb the liquid.

### 3.2 Drug Release

Drug release was done by inserting a drug-filled film into the assumed buffer solution according to the liquid condition in the body. The concentration of the released drug into the liquid was observed until the concentration of the drug was constant (the drug concentration in the film was depleted) at a given time. The profile of ascorbic acid release in various concentration of crosslinker were shown in figure 2.

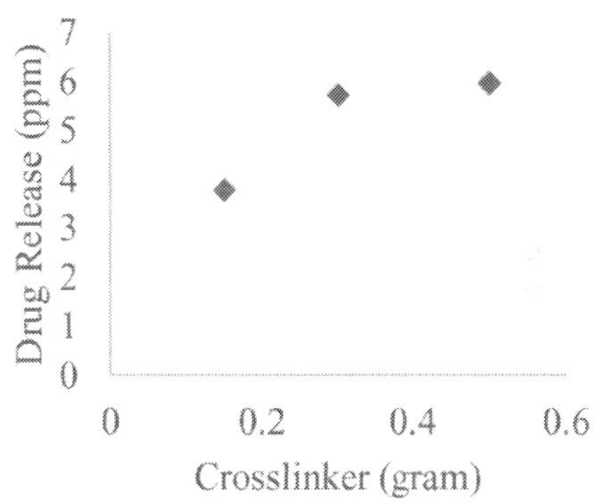


**Figure 2.** Drug Release Profile in Various Concentration of Crosslinker

Chitosan which dissolved in a solution with an acidic pH, converted an amine group ($-NH_2$) becomes ionized positively ($-NH_3^+$). This positive ionized group was further capable of forming ionic interactions with negatively charged drugs (Bhumkar and Pokharkar, 2006). Overall, the formed system tended to leave free ammonium groups that would reject each other, thus weakening the complex nanoparticles that had been formed. Therefore, it was necessary to add a crosslinker for stabilizing the remaining positive charge.

The more the addition of crosslinkers caused fewer ascorbic acid released of the film because of the small pores. But the results showed more ascorbic acid released. This happened because the non-bonding crosslinker molecule would interfere with the bonds that have formed in the edible film so that it could be weaken the strength of the bond that caused the ascorbic acid separated from the edible film easily.

By using ascorbic acid in the form of small grains, it could be assumed that the solute diffusion in solids to the surface was very rapid, so the extraction rate was determined by the speed of the solute transfer from the liquid to the surface of solid.

The solute mass transfer process from the liquid to the surface of solid were shown in the figure 3 :

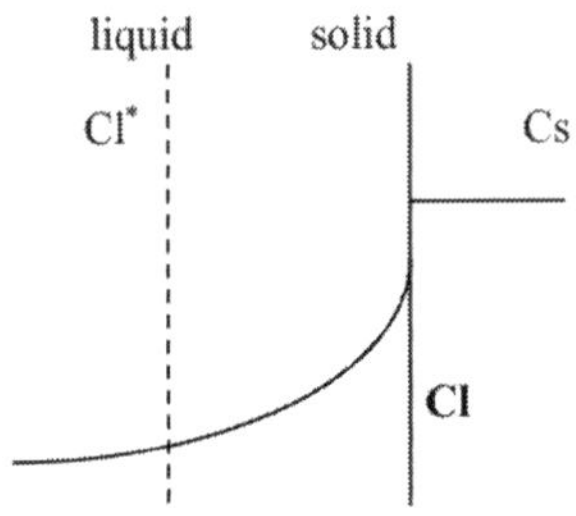


**Figure 3.** Mass Transfer between liquid to solids phase

The mass transfer rate of ascorbic acid from the liquid into the solid followed the equation:

$$Na = kc\ (Cl^{*}-Cl) \tag{1}$$

With:
Cl= Ascorbic acid concentration in solution, gram of ascorbic acid / mL of the sample solution
Cl*= The concentration of ascorbic acid in a solution that was equal to the concentration of ascorbic acid on the surface of the solid, gram of ascorbic acid / mL of the sample solution
kc= Mass transfer coefficient, cm / min
Na = The mass flux of ascorbic acid, gram of ascorbic acid / $cm^2$.min

Since the surface area was difficult to evaluate, we used a factor which showed the mass transfer surface area, so the equation becomes:

$$Na = kca\ (Cl^{*}-Cl) \tag{2}$$

kca= volumetric mass transfer coefficient, 1/ hour (Welty, 2008)

The equilibrium relationship between ascorbic acid concentration in solid and solution was approached to follow Henry's Law,

$$Cl^{*} = H\ Cs \tag{3}$$

H = Henry constant, gram ascorbic acid / gram sample solution
Cs = Ascorbic acid concentration in solid, gram ascorbic acid/gram film

The solubility of ascorbic acid into the solvent was approximated by the mathematical model as followed:
The equations below take the assumption:
- Diffusion through the membrane edge was ignored
- The volume in the system was constant
- Isothermal Process
- Drug concentration in the membranes were uniform
- Mass transfer rates from the surface to the membrane were considered fast
- The carrier medium (membrane) did not had significant swelling or erosion during drug release

The mathematical model proposed to represent release process of ascorbic acid with buffer solvents were:

$$kca = \frac{[Cl'.M]}{2[Cl'.V.H+Cl'.M+H.Cs.M]} \tag{4}$$

Where,

kca = Volumetric mass transfer coefficient, 1/ hour
Cl' = Ascorbic acid concentration in solution, gram of ascorbic acid / mL of the sample solution
M = Weight of solid, gram
V = Volume of buffer solution, mL
H = Henry constant, gram ascorbic acid/gram sample solution
Cs = Ascorbic acid concentration in solid, gram of ascorbic acid/gram solid

Based on the equation (4), the mass transfer coefficient of ascorbic acid (kCa) in various concentrations of crosslinker could be calculated and gave the results in Table 1 :

**Table 1.** Values of H and kca in Drug Release Process

| Crosslinker (gram) | H (gram ascorbic acid / gram sample solution) | Kca (1/hour) |
|---|---|---|
| 0,15 | 0,0959 | 0,1965 |
| 0,3 | 0,1113 | 0,2474 |
| 0,5 | 0,1430 | 0,2526 |

## 4 Conclusion

Crosslinker addition in edible film decreased drug loading but increased drug release. The effect of crosslinker concentration to the ascorbic acid mass transfer in drug release were increased mass transfer coefficient with a range of 0,1965 – 0,2526 $hour^{-1}$ and Henry's law constants with a range of 0,0959 – 0,1430. Based on the edible film result, the recomended treatment is the addition 0.5 gram of crosslinker.

Our greatest gratitude to the department of chemical engineering, faculty of industrial technology, Universitas Islam Indonesia for financial support and provide laboratory facilities to do this research.

## References

1. Anitha, A., Rejinold, N.S., Bungardner, J.D., Nair, S.V., & Jayakumar, R. (2012). Approaches for Fungtional Modification or Cross-linking of Chitosan. *Chitosan-based System for Biopharmaceuticals, Targetting and Polymer Therapeutic*.1. Department of Biomedical Engineering, University of Memphis, TN, USA.
2. Butler, B.L., Vernago, P.J., Testin, R.F., Bunn, J.M., & Wiles, J.L. (1996). Mechanical and Barier Properties of Edible Chitosan Films as affected by Composition and Storage. *J. Food Science*. 61(5), 953-955.

3. Bhumkar D.R. & Pokharkar, V.B. (2006). Studies on effect of pH on Cross-linking of Chitosan with Sodium Tripolyphosphate : A Technical Note, *AAPS PharmSciTech*, 7(2), E1-E6.
4. Indrawati, T. (2009). *Sistem Penghantaran Obat Peroral dengan Pelepasan Terkontrol Langsung ke Target*, Seminar Nasional 10 Desember 2009. Jakarta.
5. Skurtys O., Acevedo C., Pedreschi F., Enrione J., Osorio F., & Aguilera J.M (2009). *Food Hydrocolloid Edible films and Coatings*. Department of Food Science and Technology, Universidad de Santiago de Chile. Chile.
6. Welty, J.R., Wilson, R.E., & Wicks, C.E. (2008). *Fundamentals of Momentum, Heat and Mass Transfer Operation*. 5$^{th}$ ed. New York, USA : John Wiley and Sons Inc.

# Characteristic of ascorbic acid in crosslinked chitosan edible film as drug delivery system membrane

*Lilis* Kistriyani*, *Wahyu* Sulistyawan, and *Sapta Hadi* Kesuma

Department of Chemical Engineering, Universitas Islam Indonesia, Yogyakarta, Indonesia

**Abstract.** Chitosan is a polysaccharide compound in the form of a linear polysaccharide consisting of N-acetyl glucosamine (GlcNAc) and D-glucosamine (GlcN) monomer, which is a derivative of deacetylization of chitin polymer. Chitin is one of common type of polysaccharide on earth after the excess cellulose from inveterbrata skeletons. Chitosan has anti-microbial properties. Based on this properties, chitosan is potentially used to be an edible film as drug delivery system membrane. Edible film was made by dissolving chitosan in 100 mL acetic acid 1%, then the plasticizer and crosslinker was added while heated at 60° C. It was molded and dried in oven at 50°C for 48 hours. Drug loading in the edible film could be controlled by remodeling membrane characteristics in the presence of crosslinker additions. The purpose of this study was to estimate the mass transfer coefficient (kCa) of drug loading in various concentrations of ascorbic acid in the edible film. The characteristics of ascorbic acid in chitosan edible film could be seen from the number of drugs that could be loaded through the uv-vis spectrophotometric analysis. The higher concentration of ascorbic acid was added, the drug would be loaded more into edible film.

## 1 Introduction

In fact, the consumption of drugs is still done in a conventional way (oral). Conventional consumption of drugs has many disadvantages, such as taking a long time to reach the object and the excessive utilization of doses that can threaten drug users. This matter make a negative impact to the health of the user's body. Therefore, we need an innovation of drug consumption with utilization of edible film. Edible film is one type of membrane that can be used as drug delivery system medium. Drug Delivery System (DDS) is a term closely related to delivery of pharmaceutical compounds (drugs) in humans or animals [6]. The Drug Delivery System is usually made in a patch-shaped preparation that have characteristic like thin, elastic, and transparent for patient comfort and aesthetic side. This system can help drug users to get the same benefits in different ways. The drugs are entered into the edible film and applied in accordance with the functions of these drugs. One example of drugs that can be used is ascorbic acid which is anti-atherogenic [3].

Edible film can be produced from chitosan. Chitosan is a chemical compound derived from the chitin biological material, an organic compound that is abundant in nature after cellulose. Chitin is commonly obtained from invertebrate animal skeletons. The main source are shrimp shells, lobsters, crabs, and other shellfish in the sea. Chitosan has stronger anti-microbial properties for fungi than chitin [4]. Because of this character, chitosan produce a good edible film. Chitosan become one of the leading chemical and raw materials for industry especially for the pharmaceutical and health industries. This is possible because chitin and chitosan compounds have properties such as high coagulation and chemical reagent. Chitin and chitosan can be used in various industrial applications like additives in the fields of pharmaceutical, cosmetics, food, etc.

The main mechanism of film formation in the polysaccharides is the separation of the polymer segments and the reshape of the polymer chain into the layer or gel matrix. It is usually achieved by solvent evaporation thereby creating hydrophilic hydrogen bonds as well as electrolyte and ionic crosslinking [1]. The objectives of this research were to study the optimum concentration of ascorbic acid to drug loading, swelling precentage and estimated the mass transfer coefficient in ascorbic acid diffusion.

## 2 Materials and Method

### 2.1 Materials

The materials which used in this research were chitosan, glycerol, sodium thiosulfate, aquadest, ascorbic acid, and buffer solution.

* Corresponding author: lilis.kistriyani@uii.ac.id

## 2.2 Film Preparation [7]

The process of making edible film from chitosan begins with 2 gram of chitosan dissolved in 100 mL 1% acetic acid solution. Then, 1.2 gram of glycerol (as plasticizer) were added and heated at 60°C while stirred using magnetic stirrer for 60 minutes. After 60 minutes, 0.08 gram sodium thiosulfate solution (as crosslinker) in 30 mL of aquadest were added periodically for 30 minutes. The heating was continued for 30 minutes and then poured into the petridish plate for film casting and entered into the oven at 60°C for ± 24 hours. The dried film was inserted into the desiccator for ± 1 day to ensure the film was completely dry and ready to be released from the mold.

## 2.3 Mechanical Analysis of Edible Film [5]

### *2.3.1 Swelling test*

The 4x4 cm film was immersed in a beaker glass that had been filled with a buffer solution at pH 7.4. Every 15 minutes the film was lifted and weighed until it was constant.

### *2.3.2 Tensile strenght test and elongation*

Tensile strength and elongation testing used ASTM 882-91 procedure was measured using universal testing instrument (UTI) with load range specification 0-5000 N, extension range 0-1000 mm, crosshead speed 100.00 mm / min.

## 2.4 Drug loading Analysis [2]

Ascorbic acid was entered into edible film by diffusion. Edible film dipped in ascorbic acid solution with various concentrations (40 ppm, 60 ppm, 80 ppm, and 100 ppm) for 2 hours. Edible film that already contained drug was dried in ambient temperature. The drug loading test was performed by spectrophotometer uv-vis method using 2 cuvet pieces. The ascorbic acid solution of residual film immersion was incorporated into the cuvet and analyzed using a uv-vis spectrophotometer with a wavelength (λ) of ascorbic acid at 256.6 nm.

# 3 Result and Discussion

## 3.1 Swelling analysis

The addition of crosslinker in edible film would change the characteristic of edible film that caused the pores size of chitosan film were decreases. Decreasing the edible pores of the film would made an impact on the reduced absorbency of edible film. Swelling test results on edible film could be seen from the Fig. 3.1.

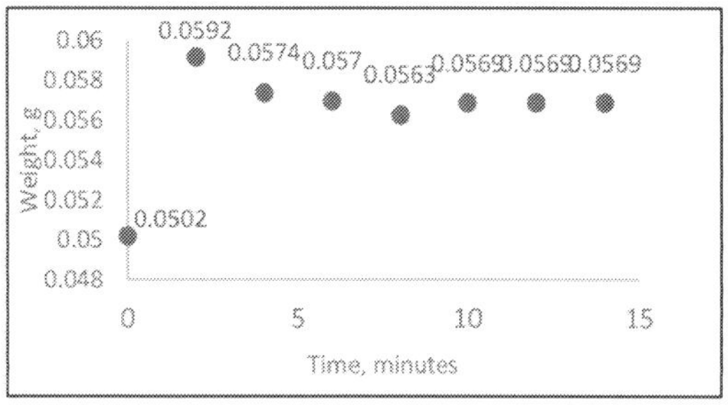


**Fig.** 3.1 Profile of Swelling in Various Times

Fig. 3.1 showed the results of swelling analysis from the edible film. The aims of this analysis were to determine the maximum capability of edible film to absorb liquid until the equilibrium occurs. In the first analysis, there was a significant increase in the weight of the film initially, due to the absorption of the liquid into edible films through the pores of the film. In the subsequent analysis, there was a decrease and increase until the liquid equilibrium stage was absorbed in the edible film that shown in the three last point. The absorption of liquid into edible film was about 13.3%.

## 3.2 Tensile strenght and elongation analysis

A strong tensile strength test was performed to determine the maximum pressure of material that could withstand its form when it was withdrawn before it was damaged or broken. While elongation was the extent of a material that occured when it was given the power to be damaged. The results of tensile strength and elongation analysis of sequential films were 980 cN and 17.8%.

## 3.3 Drug loading analysis

The ascorbic acid in liquid was carried out into edible film by diffusion. Narrowing of the pores on edible film was formed by a sodium thiosulfate crosslinking of edible film that could retain ascorbic acid.

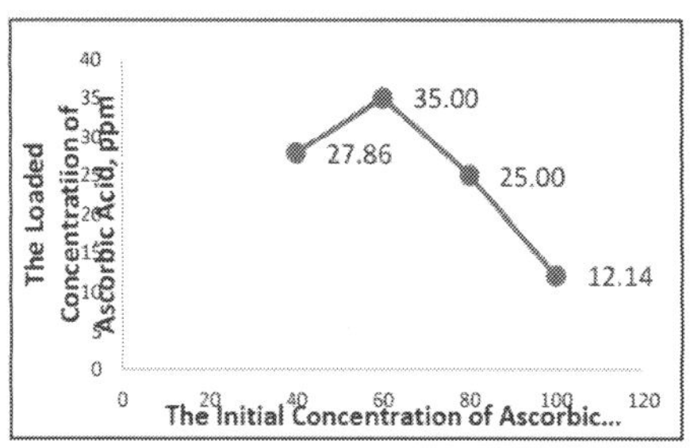


**Fig.** 3.2 Drug Loading Profile of Ascorbic

Figure 2 showed the results of ascorbic acid loading into edible films with initial concentrations of 40 ppm, 60 ppm, 80 ppm, and 100 ppm respectively were 27.85 ppm, 35 ppm, 25 ppm, and

12.14 ppm. The amount of ascorbic acid which could be loaded into edible film had same trendline with the mass transfer coefficients (kCa) of ascorbic acid to edible film. The ascorbic acid mass transfer coefficients (kCa) in various concentrations were estimated with the following mass balance:

The ascorbic acid mass balance in the element volume of the membrane:

[Mass velocity input] + [Mass velocity output] = [Mass accumulation in the membrane]

$$k\ (C^{*} - C)\,.\,V - 0 = \frac{dC\,V}{d} \tag{1}$$

$$k\ (C^{*} - C)\,.\,V = V\,.\,\frac{dC\,V}{d} \tag{2}$$

With boundary conditions:

$$t = 0\ \rightarrow C =\ C_0$$

$$t = t\ \rightarrow C = C$$

$$\int_{CC}^{C} d = kC\ (C^{*} - C)t\ \ ; t = 2\ hours \tag{3}$$

$$(C - C_0) = 2\,.kC\ (C^{*} - C) \tag{4}$$

$$C = \frac{2\ .\ kC\ .\ C^{*} + C_C}{(1 + 2\ .\ kC)} \tag{5}$$

The ascorbic acid mass balance in the solids:

[ascorbic acid in initial solid] + [ascorbic acid in initial liquid] = [ascorbic acid in solid at t=certain] + [ascorbic acid in solute at t=certain]

$$0 + C_0\ .\ V = C\ .\ M + C\ .\ V \tag{6}$$

$$C = \frac{V\ .\ (C_0 - C)}{M} \tag{7}$$

$$C^{*} = H.C \tag{8}$$

$$C^{*} = \frac{H\ .\ C_0\ .\ V - H\ .\ C\ .\ V}{M} \tag{9}$$

Equation (9) was inserted to equation (5), the equation becomes:

$$\left[C = \frac{2.\ kC\ .\ H\ .\ C + C_0}{(1 + 2\ kC)}\right] \tag{10}$$

The final equation proposed to estimate mass transfer coefficient:

$$kC\ = \frac{\left[\frac{C_0}{C} - 1\right]}{2\left[1 - \frac{H\ .\ C}{C}\right]} \tag{11}$$

Based on the equation above, the mass transfer coefficient of ascorbic acid (kCa) in various concentrations could be calculate and gave results in Table 3.1:

Table 3.1 The mass transfer coefficient of ascorbic acid (kCa) in various concentrations

| Ascorbic Acid Concentrations (ppm) | kCa (l/hour) |
|---|---|
| 40 | 1.147 |
| 60 | 0.70 |
| 80 | 0.227 |
| 100 | 0.069 |

Based on Table 1, it could be seen that decreasing of mass transfer coefficient along with increasing of ascorbic acid concentration. As the ascorbic acid concentration was increases, less ascorbic acid could be loaded in the edible film.

## 4 Conclusion

With the addition of crosslinked can be effected the characteristics of edible film. mechanical analytic of edible film showed swelling 13.3%; tensile strength 980cN; and elongation of break 17.8%. The mass transfer rate (kCa) is inversely proportional to yhe ascorbic acid concentration being used. The optimum value of drug loading mass transfer rate is shown on the result of ascorbic acid analysis with concentration of 40 ppm for 1.147 liter/hour.

Our greatest gratitude to the Department of Chemical Engineering, Faculty of Industrial Technology, Universitas Islam Indonesia for financial support and provide laboratory facilities to do this research.

## References

1. Butler, B.L., Vernago, P.J., Testin, R.F., Bunn, J.M., & Wiles, J.L.(1996). Mechanical and Barier Properties of Edible Chitosan Films as affected by Composition and Storage. *J. Food Science*. 61(5), 953-955.
2. Mishra, R. K., Datt, M., & Banthia, A. K. (2008). Synthesis and Characterization of Pectin/PVP Hydrogel Membranes for Drug Delivery System. *AAPS Pharmceutics Science and Technology*, 9(2), 395 – 403.
3. Naidu K.A. (2003). Vitamin C In Human Health And Disease Is Still A Mystery, An Overview. *Noutrition J*. 2, 7.
4. Mahatmanti, F. W., Sugiyo, W., & Sunarto, W. (2010). Sintesis Kitosan dan Pemanfaatannya sebagai Anti Mikrobia Ikan Segar, *Jurnal Sains dan Teknologi*, 8(2), 101-111.
5. Purwati, Ani. (2010). Analisis Kuat Tarik dan Elongasi Plastik Kitosan Terplastisasi Sorbitol. *Jurnal Teknologi*, 3(2), 99-106.
6. Upadhy, G., Verma, S., Parvez, N., & Sharma, P.K. (2014). Recent Trends in Transdermal Drug Delivery System - A Review. *Advance in Biological Research*, 3, 131–138.
7. Wirawan, S.K., Prasetya, A., & Ernie. (2012). Pengaruh Plasticizer pada Karakteristik Edible Film Dari Pektin. *Reaktor*. 14(1), 61-67.

# Esterification of glycerol with acetic acid over Lewatit catalyst

*Lucky* Setyaningsih*, *Fahmi* Siddiq, and *Aqila* Pramezy

Department of Chemical Engineering, Islamic University gof Indonesia, Yogyakarta, INDONESIA

**Abstract.** Tri Acetyl Glycerol or triacetin is one of the glycerol derivative products which can be used as additives in liquid fuels (octane boosters) to reduce knocking on the machine. The use of triacetin as octane booster is considered very promising because it comes from renewable and environmentally friendly raw materials. The resulting glycerol-based additives are **suitable** for gasoline, biodiesel and diesel. In this study glycerol esterification with acetic acid over Lewatit catalyst was investigated. The effects of various parameters, such as reaction temperature, weight of catalyst and molar ratio of glycerol to acetic acid were studied and the optimum condition for the triacetin synthesis are 100°C, 1: 7 mole of ratio and 3% weight of catalys

## 1 Introduction

Biodiesel is one of the renewable energy sources that can be an alternative to the use of petroleum. Biodiesel is an alternative fuel from renewable natural resources that have a positive impact, especially the emissions of gases produced more environmentally friendly. Biodiesel production will produce glycerol as a by-product of about 10% (by weight) of biodiesel production [1]. The increasing production of biodiesel will be followed by an increase in the amount of glycerol produced as a by-product, therefore an effort is needed to utilize glycerol as a raw material for the manufacture of other chemicals.

Tri Acetyl Glycerol or triacetin is one of the glycerol derivative products which can be used in various industries, e.g. emulsifiers, plasticizers, additives in the coating and paint industries, and can also be used as additives in liquid fuels (octane boosters) to reduce knocking on the machine [2]. The commonly used Octane booster is tetraethyl lead (TEL, $Pb(C_2H_5)_4$) but its use will form a thin layer of lead in an atmosphere which is harmful for living things. The use of triacetin as octane booster is considered very promising because it comes from renewable and environmentally friendly raw materials. The resulting glycerol-based additives are suitable for gasoline, biodiesel and diesel.

In this study triacetin will be produced from esterification process of glycerol and acetic acid. To increase the reaction rate and yield of product, in the synthesis process of glycerol derivative products into fuel additive materials used catalyst. The choice of catalyst type for a reaction is based on several considerations, namely: porous structure to produce large surface area as active site provider, thermal stability to temperature change and has high acidity.

The reaction of triacetin formation is an esterification reaction with glycerol and acetic acid reactants, the esterification reaction of glycerol is a slow reaction. In previous studies, triacetin formation of glycerol and acetic acid without catalyst takes more than 4 hours. Meanwhile, by using aminosulfonic acid catalyst it was obtained a yield of 90% [3]. Esterification reaction with a phosphotungstic acid catalyst, reactant ratio of 3.8%, a temperature of 135-155 °C and a reaction time of 7 hours yielded 84.6% [4]. The effect of the use of various solid catalysts for the preparation of triacetin was carried out by Gonsalves. The catalysts used were Amberlist-15, K 10, Niobic acid, HZMS-5 and HUSY. From the various catalysts the success of triacetin was achieved only when using Amberlist-15 and K 10 catalysts with selectivity of only 13% and 5%. While the use of other catalysts only get monoacetin and diacetin although its conversion can reach more than 90% [5]. Gelosa made triacetin with packed bed reactor with 36 grams Amberlyst catalyst stuffing material which run on column with internal diameter of 1.5 cm long and 44 cm long. When used the comparison of glycerol reactants and acetic acid 2: 9, with flow rate of 0.3 $cm^3$/min, 50% acetic acid conversion was obtained [6]. The selectivity of diacetylglycerol (DAG) and triacetylglycerol (TAG) mixtures after reacting for 4 hours can reach more than 80%. The reaction is carried out in an autoclave reactor at a temperature of 100-150°C and a pressure of about 4 atm [7].

In this research, triacetin synthesis process will be done through esterification reaction with glycerol and acetic acid using lewatit catalyst. The use of lewatit catalysts is intended to facilitate the separation of the reaction product [8]. In addition, the utilization of lewatit catalyst which is an ion exchange resin is expected to provide alternative materials supporting the development of glycerol derivative products.

## 2 Methodology

Prior to the use the catalyst is activated by using a HCl 1 N solution of 200% of the catalyst volume. The catalyst mixture is stirred using a magnetic stirrer for 30 minutes and rinsed with distilled water to neutral pH. Glycerol with a certain volume and the lewatit catalyst is heated in a three-neck flask, while acetic acid is heated in a separate container. After the second temperature of the reactant reaches the reaction temperature, acetic acid is introduced in a three-neck flask containing glycerol and a catalyst. During the reaction, stirring at a rate of 175 rpm and temperature is kept constant according to the reaction temperature. The research was conducted by varying the mole ratio of reactants (glycerol: acetic acid) 1: 6, 1: 7 and 1: 8, reaction temperature 90°C, 100°C and 110°C, 3% catalyst weight and 175 rpm stirring rate. Samples were taken every 15 minutes and the reaction was run for 90 minutes.

Data obtained from the study were free glycerol levels at each time interval of 15 min obtained in each process condition. This data is then used to evaluate the conversion value as a function of time. Glycerol levels can be calculated by the following formula:

$$glycerol\ (\%_{weight}) = \frac{2{,}302\ (B-S)(N)(\rho_a)(450)(170)}{50\ (w_1)(w_2)} \quad (1)$$

Where B is volume of $Na_2S_2O_3$ titration for blank solution (mL), S is volume of $Na_2S_2O_3$ titration for sample solution (mL), N is normality of $Na_2S_2O_3$ solution, $\rho_a$ is distilled density (g/mL), $w_1$ = initial sample weight (g), $w_2$ = diluted sample weight (g).

## 3 Result and Discussion

### 3.1 Effect of Temperature

The data of glycerol conversion versus time at various temperatures can be seen in Table 1 and Figure 1.

**Table 1.** Glycerol conversion at various temperatures

| Time (minute) | Conversion of Glycerol (%) | | |
|---|---|---|---|
| | T = 90°C | T = 100°C | T = 110°C |
| 0 | 0 | 0 | 0 |
| 15 | 50.55 | 50.40 | 60.92 |
| 30 | 54.76 | 56.80 | 62.43 |
| 45 | 55.34 | 58.46 | 63.31 |
| 60 | 57.00 | 59.33 | 64.82 |
| 75 | 57.91 | 63.45 | 66.03 |
| 90 | 58.56 | 66.91 | 66.38 |

Based on Table 1 and Figure 1, it appears that there is a spike in conversion shortly after the catalyst is added to the reactor. The same is also reported on the esterification of ethanol and acetic acid by using a monoplus s-100 lewatit catalyst [2]. The esterification of glycerol with myristic acid using a sulphated zirconia catalyst, when the catalyst is added, there is a spontaneous reaction that causes the acid conversion to rise. As the acid conversion increases, the water concentration formed also increases with the same trend. Water has the ability to deactivate acidic solid catalysts due to its high affinity to the acidic active site of the catalyst [9]. Thus it can be understood that after a high rate of conversion spikes, the reaction rate decreases as the activity of the catalyst decreases. The same is true of the use of other solid acid catalysts such as Amberlyst-15 [10].

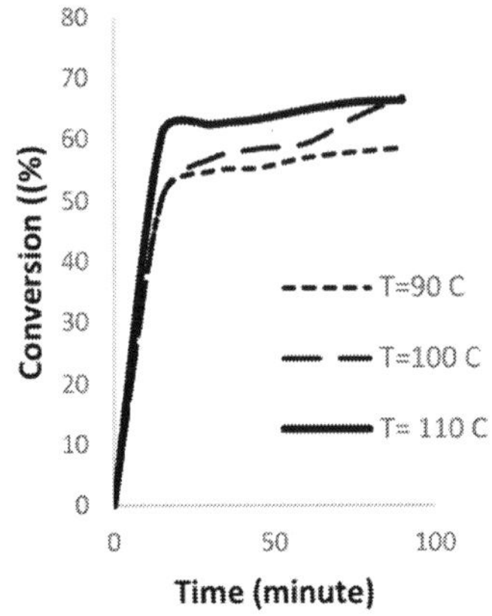


**Fig. 1.** Correlation between Glycerol conversion versus time at various temperatures

The observed reaction rate is the rate of reaction after the spontaneous reaction occurs, so the resulting graph is relatively sloping. The observed trend of Figure 1 is that the higher the temperature and the longer the reaction time, the conversion tends to increase. This is because with the rise of reaction temperature, the energy of reagent molecules increases in overcoming the activation energy. The reactant molecules will move faster so that the frequency of collisions between molecules will increase and the reaction rate becomes faster. The increase in reaction rate is indicated by the higher reactant conversion. This corresponds to the Arrhenius equation as follows:

$$k = A.e^{\frac{-E}{RT}} \quad (2)$$

Equation 2 shows that the temperature is directly proportional to the rate constant of the reaction, where as the reaction temperature increases the reaction rate constant increases so that the reaction rate was faster and the conversion was higher. The same phenomenon also demonstrated in experiment using an Indion 225 Na catalyst [2], a natural and physically activated Zeolite catalyst [11] and Amberlyst-15 as a catalyst [6].

At the end of reaction time, the highest conversion was obtained at temperature 100°C with conversion equal to 66.91%, while the lowest conversion was obtained at 80°C with conversion of 58.59%. According to the theory, the maximum conversion should be achieved at a temperature of 110°C, but in this study it did not occur. This is due to the temperature of 110°C of Lewatit catalyst which is used to change the color from yellow to dark brown indicating deactivation of the catalyst. The deactivation is caused by high experimental temperature near its maximum temperature, so the catalyst cannot retain under such conditions.

From the research it was obtained a fairly high conversion. This can be compared with other research. In esterification of glycerol carried out by Zhu using AgPW catalysts achieved 96.8% conversion [12] and 100%

conversion can be achieved in research conducted by Gao using Graphene Oxide catalyst at 120°C [13]. This is possible because Graphene Oxide and AgPW catalysts are more resistant to high temperatures for longer periods than Lewatit catalysts, so that the conversion becomes higher and the activated natural zeolite catalyst has a better ability to direct esterification reactions.

### 3.2 Effect of Molar Ratio of Reactant

From the experiment of various temperature variations obtained the optimum temperature is 100°C. Armed with the optimum temperature, the experiment of variation of reactant mole ratio was carried out at 100°C. The molar ratio of glycerol reactants to acetic acid was 1: 6, 1: 7 and 1: 8. Meanwhile, other variables were kept constant which was 3% by weight catalyst and 100 ° C reaction temperature. The data of glycerol to time conversion on various reactant variations can be seen in Table 2 and Figure 2.

**Table 2.** Glycerol conversion at various molar ratio of reactant

| Time (minute) | Conversion of Glycerol (%) | | |
|---|---|---|---|
| | **1:6** | **1:7** | **1:8** |
| 0 | 0 | 0 | 0 |
| 15 | 50.55 | 56.62 | 60.12 |
| 30 | 56.80 | 60.79 | 65.60 |
| 45 | 58.46 | 66.44 | 67.79 |
| 60 | 59.33 | 66.70 | 68.73 |
| 75 | 63.45 | 70.01 | 73.42 |
| 90 | 66.91 | 73.36 | 73.16 |

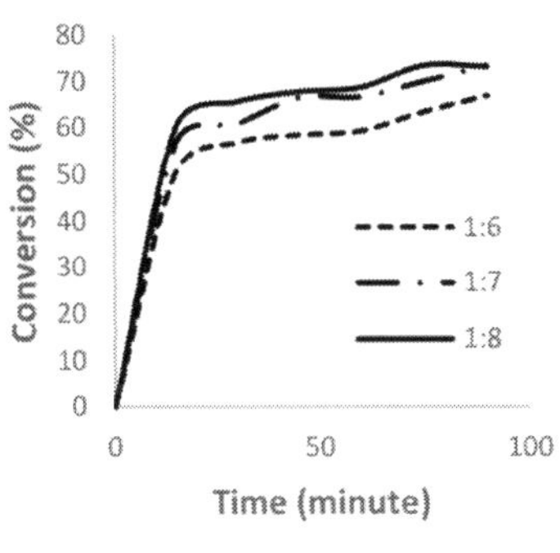


**Fig. 2.** Correlation between Glycerol conversion versus time at various molar ratio of reactant

The mechanism of the reaction of triacetin formation in an esterification reaction between glycerol and acetic acid is as follows:

$$\text{Glycerol} + 3\ \text{Acetic Acid} \leftrightarrow \text{Triacetin} + 3\ H_2O$$

Stoichiometrically, the esterification reaction to produce one mole of triacetin and three moles of water requires one mole of glycerol and three moles of acetic acid. However, because the esterification reaction is a reversible reaction so that acetic acid can be excessive to shift equilibrium toward the product.

From Table 2 and Figure 2 it can be seen that the higher the mole ratio of glycerol reactants than acetic acid, the higher the conversion value of acetic acid achieved. In a molar ratio of 1: 6 to 1: 7 reactants there was a considerable increase in conversion. However, in the molar ratio the 1: 7 to 1: 8 reactants produce a coincident graph which means that the increase in the mole ratio of the reactant is not very influential. From these results it can be concluded that the highest conversion is obtained at the mole ratio of 1: 7 reactant at 100°C temperature which was 73.36%, while the lowest conversion of 66.91% was obtained in the reagent mole ratio of 1: 6. This shows that the mole ratio of reactants gave a significant effect on the conversion of acetic acid. As the mole ratio of the larger reactant increases, it will affect the rate of the reaction that goes faster with the increase of conversion. According to Satriadi, if the concentrations of reactants get higher where the large concentrations of substances contain more number of particles, so the particles are composed more densely than the substances with low concentration. Particles that are closer together will more often collide so that the possibility of reaction is getting bigger [14].

The effect of the reactant mole ratio was very visible at the start of the reaction (from t = 0 to t = 15 minutes) as shown in Table 2 and Figure 2 above, where the resulting conversion is quite large with a very short time lag of 15 minutes. This is similar to the research conducted by Nuryoto where at the beginning of the reaction, the reaction goes very quickly and then the reaction goes slow [2]. This occurs because of the effect of water produced as a by-product, where in the presence of water the reaction will tend to shift to the left and water has a very strong attraction among the components present in the system that will greatly affect the rate of diffusion And the rate of reaction [6].

In a study conducted by Gao an increase in the mole ratio of glycerol/acetic acid from 1: 2 to 1: 6 leads to an increase in conversion, this is because the excess acetic acid will shift equilibrium toward the product, thereby increasing the formation of monoacetin and diacetin [13]. Increasing in the mole ratio of glycerol/acetic acid from 1: 6 to 1: 8 leads to a decrease in glycerol conversion, this is due to a sharp decrease in glycerol concentration to the total solution volume.

### 3.3 Effect of Weight Catalyst

The effect of percent weight of the catalyst will be studied on the basis of experiments carried out by varying the catalyst weight by 2%, 3% and 5% by weight of the catalyst to the glycerol reactant. Meanwhile other variables such as temperature and reactant mole ratios are maintained at 100 ° C and 1: 7. The conversion value obtained from each experiment will be compared to determine the effect of catalyst weight increase on conversion to glycerol. The relationship between conversion and time on various percent weight of the catalyst is presented in Table 3 and Figure 3.

From Table 3 and Figure 3 it is seen that the conversion value of glycerol to triacetin by esterification reaction increases with increasing catalyst concentration or percent weight of catalyst used. The observed trend of Fig. 3 is the weight of the catalyst increase from 2% to 3% of significant conversion increments. While the increase in catalyst weight from 3% to 5% increase in conversion is not too significant. although the resulting graph fluctuates but generally it can be concluded that there was no significant difference between catalyst weight 3% and 5%. Thus it can be concluded that the weight of the catalyst 3% is the weight of the optimum catalyst.

**Table 3.** Glycerol conversion at various weight of catalyst

| Time (minute) | Conversion of Glycerol (%) | | |
|---|---|---|---|
| | **2%** | **3%** | **5%** |
| 0 | 0 | 0 | 0 |
| 15 | 48.38 | 50.40 | 49.04 |
| 30 | 49.83 | 56.80 | 56.28 |
| 45 | 52.10 | 58.46 | 60.37 |
| 60 | 53.58 | 59.33 | 65.35 |
| 75 | 55.16 | 63.45 | 66.45 |
| 90 | 56.28 | 66.91 | 66.02 |

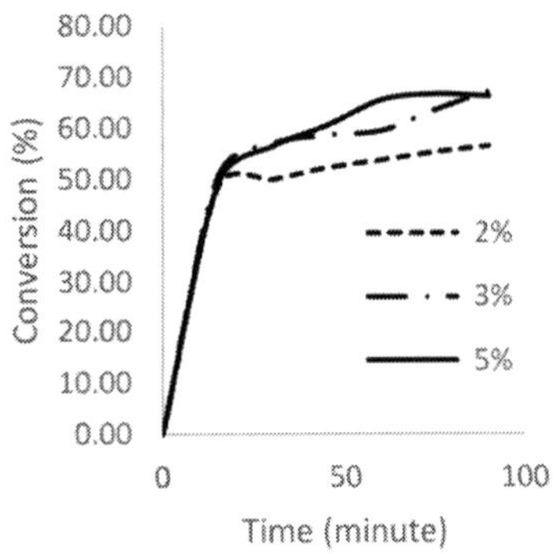


**Fig. 3.** Correlation between Glycerol conversion versus time at various weight of catalyst

At the end of the reaction time. the highest conversion was obtained at 3% catalyst weight with a conversion of 66.91%. while the lowest conversion was obtained at 2% percent catalyst weight with a conversion of 56.28%. Another phenomenon seen from Figure 3 is the increase in glycerol conversion as a function of time is quite gentle and volatile. This was because the number of active sites available on the catalyst was close to the maximum required by the reactants. The same phenomenon is also shown in the study conducted by Yadav where in percent weight of catalyst 5% it appears that the conversion did not increase compared with the use of 3% catalyst. This may be because at the weight of the catalyst 5% there is no perfect mixing due to the catalyst solid density in the solution being large enough to reduce the contact area between the catalyst and the reactants. In addition, the high concentration of solids in the solution allows the dead zone within the reactor.

## 4 Conclusion

1. The conversion value increase with increasing temperature, mole ratio of reactant and the weight of the catalyst.
2. The optimum condition for the triacetin synthesis are 100°C, 1: 7 mole of ratio and 3% weight of catalyst.

## References

1. G. Knothe, J. Van Gerpen, J. Krahl, *Encyclopedia of Chemical Technology*, **11**, 921 (2005)
2. Nuryoto, H. Sulistiyo, S.S. Rahayu, Sutijan, Jurnal Rekayasa Proses, *Kinetika Reaksi Esterifikasi Gliserol dengan Asam asetat menggunakan Katalisator Indion 225 Na* **5(2)**, 35-39 (2010)
3. Hou, Henan Huagon, *New Process for synthesis of Triacetin*, **15(6)**, 18-19 (1998)
4. M. Zang, X. Yuan, Hecheng Huaxue, *Synthesis of Glycerol Triacetate Catalized by Phosphotungstic Acid* **9 (5)**, 469-472 (2001)
5. G.F.C. Goncalves, B.P. Pinto, J.C. Silva, C.J.A. Mota, Catal. Today, *Acetylation of glycerol catalyzed by different solid acids* **133–135**, 673–677 (2008)
6. D. Gelosa, M. Ramaioli, G. Valente, M. Morbidelli, Ind. Eng. Chem. Res. *Chromatographic Reactors: Esterification of Glycerol with Acetic Acid Using Acidic Polymeric Resins* **42**, 6536-6544 (2003)
7. J.A. Melero, R.V. Grieken, G. Morales, M. Paniagua, Energ Fuel , *Acidic Mesoporous Silica for the Acetylation of Glycerol : Synthesis of Bioadditives to Petrol Fuel*, **21**, 1782-1791 (2007)
8. J.I. Choi, W.H. Hong, H.N. Cham, Int. J. Chem. Kinet, *Reaction Kinetics of Lactic Acid with Methanol Catalyzed by Acid Resins* **28**, 37-41 (1996)
9. Z. Lu, W. Ma, 1991, *Study on Esterification reaction (1), Esterification of Glycerin with Acetic Acid using Ion Exchage Resin and Drying Agent as Catalysts* **(2)**, 53-56 (1991)
10. A.A. Kiss, A.C. Dimian, Adv. Synth. Catal, *Solid Acid Catalyst for Biodiesel Production Towards Sustainable Energy* **348**, 75-81 (2006)
11. N. Sari, Z. Helwani, H. Rionaldo, Jom F Teknik, Esterifikasi Gliserol Dari Produk Samping Biodiesel Menjadi Triasetin Menggunakan Katalis Zeolit Alam **2(1)**, 1-7 (2015)
12. Zhu, Shanhui, X. Gao, D. Fang, Y. Zhu, Z. Hongyan, Y. Li, J Catal, *Design of a highly active silver-exchanged phosphotungstic acid catalyst for glycerol esterification with acetic acid* **306**, 155-163 (2013)
13. X. Gao, S. Zhu, Y. Li, (2015), Catal Commun, *Graphene oxide as a facile solid acid catalyst for the production of bioadditives from glycerol esterification* **62**, 48-51 (2015)
14. H. Satriadi, Jurnal Teknik, *Kinetika Reaksi Esterifikasi Gliserol Dan Asam Asetat Menjadi Triacetin Menggunakan Katalis Asam Sulfat* **36**, 75-80 (2015

# The potential of activated carbon derived from bio-char waste of bio-oil pyrolysis as adsorbent

*Ariany* Zulkania*, *Ghina* Hanum F., and *Amelia* Sri Rezki

Chemical Engineering Dept., Faculty of Industry Technology, University of Islam Indonesia, Jl. Kaliurang km 14, Yogyakarta, Indonesia

**Abstract.** Activated carbon from bio-char waste of bio oil pyrolysis of mixed sugarcane bagasse and Rambutan twigs was investigated. Bio-char as by-product of bio-oil pyrolysis has potential to be good adsorbed by activating process. Bio-chars waste was activated in fixed bed reactor inside furnace without presenting oxygen. Gas $N_2$ and $CO_2$ were employed to drive out oxygen from the reactor and as activator, respectively. One of the best activation treatments is achieved by performing activation in different temperature and time to produce standard activated carbon. The experiment was performed at different temperatures and activation time, i.e. 800, 850, and 900° C and 80 and 120 minutes, respectively, to determine the optimal operating condition. Activated carbon was characterized by analysis of moisture content, ash content pH, and methylene blue test. The results showed that optimum activation was at 850°C and 80 minute, where activated carbon produced indicated the best adsorption capacity. The ash content and pH had significant role in resulting good activated carbon.

## 1 Introduction

Activated carbon are commonly produced from carbonaceous material such as coal, wood, coconut shell, pine nut shell, rice straw, sugarcane bagasse, etc. Sugarcane bagasse is agricultural by-products that is available in large quantities. Based on Ikatan Ahli Gula Indonesia (IKAGI), about 9.64 million tones sugarcane baggase are produced. But only 60% of sugarcane bagasse is utilized, such as fuel, raw materials for paper, raw materials for brake canvas, mushroom industry, etc. Some studies utilizing sugarcane bagasse to generate activated carbon have been done [1 - 4]. Goncalves et.al. [4] who investigated activated carbon from pyrolysis of sugarcane bagasse and molasses and $CO^2$ activation found that it was possible to obtain activated carbon with surface up to 900 m2/g over 4.5 h activation time. Activation of carbon derived from sugarcane bagasse using microwave was explored by Foo et.al. [5]. It was revealed that activated carbon from sugarcane bagasse could be applied as adsorbent for semi-aerobic landfill leachate. Furthermore, the surface area obtain was up to 1620.69 m2/g and adsorption capacity for ammonical nitrogen and orthophosphate were 138.46 and 12.81 mg/g, respectively.

Wood waste is potential material for producing activated carbon. Most of wood waste especially twig are untapped. A report released by Winda Lismaya in 2014 [6] said that, in Indonesia, logging residues from 15 trees are 141.87 $m^3$ and around 13.8% of it is twig. Furthermore, both sugarcane bagasse and twig contain lignoselulose (lignin, selulose, and hemiselulose). Most of the agricultural lignocellulosic biomass is comprised of about 10–25% lignin, 20–30% hemicellulose, and 40–50% cellulose [7]. Therefore, lignoselulose has potential to generate renewable energy and activated carbon. Activated carbon usually produced by thermal or chemical activation. It is an amorphous solid with large surface area with range is given as 400 – 2000 $m^2/g$. Raw materials, activated method and processing condition influence characteristics of activated carbon [8 - 10].

Specific surface area (SSA) could be also predicted from Methyl blue test [11, 12]. El-Geundi et.al. [11] determined SSA of some clays and activated carbons using an equation. Then, the results were compared with the SSA using BET-analysis. There were deviations of the comparations depend on the material. However, this approach could be initial determining of specific surface area.

Activated carbon of the mixture of sugarcane bagasse and twigs has not been investigated. This study was focused on producing activated carbon from biochar as waste of bio-oil pyrolysis from sugarcane bagasse/twig of rambutan mixture. Bio-oil pyrolysis process was performed at temperature of 450°C and 15 minutes time retention. The effects of activation temperature and time retention on the biochar are investigated. Particularly, adsorption capacity and surface area are the aims of this study.

## 2 Material and Method

### 2.1 Material

* Corresponding author: ariany.zulkania@uii.ac.id

Biochar waste from bio-oil pyrolysis of sugarcane bagasse/twig of Rambutan mixture was sieved using the -20+30 Mesh sieve before being activated. Gas $N_2$ dan $CO_2$ are for activation process.

## 2.2 Apparatus

The experiment to produce activated carbon was performed using lab-scale fluidized bed reactor, according to layout shown on Figure 1.

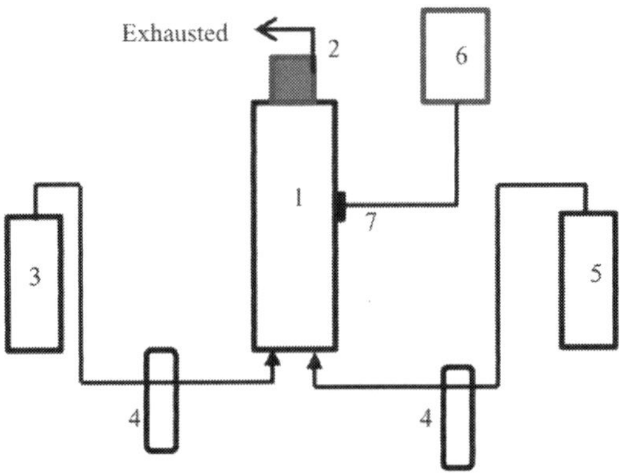


Note : 1. Furnance
2. Fluidized bed reactor
3. $CO_2$ gas tank
4. Flow meter
5. $N_2$ gas tank
6. Temperature control
7. Thermocouple

**Fig.1** Apparatus for activating bio-char

The reactor was composed a stainless steel tubular reactor (7.5 cm i.d. and 23.5 cm height) inserted in vertical tubular furnace. The thermocouple probe was inserted to a hole of furnace to measure temperature of the reactor.

## 2.3 Experimental Procedure

Ten grams of biochar poured into a reactor was purged under $N_2$ gas flow (0.5 cm³/min) until reaching activation temperature. Once reaching the temperature, the flow gas was switched to $CO_2$ (0.5cm³/min) and activation was happened over reaction/activating times. After activation, the sample was cooled to room temperature under $N_2$ gas flow. The sample was placed into desiccators for further characterization, except small part for water content analysis was put in a bowl. The following parameters were utilized in this study: activation temperatures of 800, 850 and 900°C and activation times of 80 and 120 min using heating rate of 10°C/min.

## 2.4 Activated carbon analysis

The analysis of activated carbon conducted in this study were water content testing, ash content, pH testing, and adsorption capacity using methylene blue test. All analysis was carried out in the research laboratories of Chemical Engineering Dept, University of Islamic Indonesia. The results of analysis were then compared to the standard quality of activated carbon properties. According to Indonesian Industrial Standard (SII No. 0258-88), the requirement for the quality of activated carbon is displayed on Table 1.

**Table 1**. Quality requirement of activated carbon (SII No. 0258-88) [13]

| Test type | Requirement (Solid) |
|---|---|
| Water content | Max. 15% |
| Ash content | Max. 10% |
| *Fixed* karbon (%) | Min. 65% |
| Adsorption capacity of *methylene blue* | Min. 120 mg/g |

### 2.4.1 Water content

Water content was determined to identify hygroscopic property of activated carbon. The calculation of moisture content was based on dry bases. Prior to be dried in oven for one hour at 105°C, one gram of activated carbon was weighed in a digital-scale. The sample was then, weighed after drying. The water content was calculated using Eq. (1)

$$\text{Water content} = \frac{a-b}{a} x100\% \quad (1)$$

Where:
a = initial activated carbon weight (g)
b = dried activated carbon weight (g)

### 2.4.2 Ash content

Ash content of activated carbon was determined by heating the sample in the reactor-furnace for two hours at temperature of 800°C. After heating process, the sample was located in the desiccator for cooling and weighed. The ash content was calculated using Eq.(2)

$$\text{Ash content} = \frac{ash\ weight}{sample\ weight} x100\% \quad (2)$$

### 2.4.3 pH

Determination of activated carbon's pH used Lutron PH-201 pHmeter. One gram of the sample was poured to water with ratio of 1:9. The mixture was then stirred for 10 minutes to homogenize. The pH of solution was measured using pH-meter.

### 2.4.5 Adsorption capacity of methyl blue

Adsorption capacity of methylene blue was determined by measuring adsorption isotherm. Here, the adsorption isotherms were determined by the bottle-point method [10], by contacting a constant mass of the adsorbent (0.1 g) with 25 mL of 100 ppm methylene blue solution. The mixture was stirred for 30 minutes to ensure reaching equilibrium. After that, the mixture was filtered to obtain sample solution. Concentration of methylene blue solution obtained at the end of all adsorption studies were measured by a UV–vis spectrophotometer with absorbance at 664.00-664.50 nm. Concentration of methylene blue that remained unabsorbed in the solution was determined and percentage of adsorption and amount of methylene blue adsorbed on per gram of activated carbon were calculated by the following Equations [14]

$$q = \frac{(C_o - C_e).V}{m} \quad (3)$$

$$Adsorption\% = \frac{(C_o - C_e)}{C_o} \quad (4)$$

Where:

q = amount of methylene blue adsorbed per gram of activated carbon ($mg.g^{-1}$)
Co =initial concentration of methylene blue solution ($g.mL^{-1}$)
Ce =concentration of methylene blue that remained unadsorbed in the solution ($g.mL^{-1}$)
V = volume of methylene blue solution (mL)
m = mass of activated carbon (g)

### *2.4.6 Analysis of Surface area*

The concentration of methylene blue remaining on the solution is able to predict surface area of the solid particles [11,12]. Surface area of activated carbon was predicted by using following formula

$$a = \frac{m_{mb}}{319.87 \; x \; m_s} A_v A_{mb} \quad (5)$$

Where:

a = specific surface area of activated carbon ($m^2g^{-1}$)
$m_{mb}$ = mass of adsorbed methylene blue (g)
$m_s$ = mass of activated carbon (g)
$A_v$ = Avogadro's number (6.02∗1023/mol)
$A_{mb}$ = area covered by one MB molecule (typically assumed to be 130 $Å^2$)

## 3 Result and Discussion

### 3.1. Yield of activated carbon

Figure 2 shows the yield of activated carbon resulted from varying two variables, those are temperature and activating time. As can be seen, the graphs indicate that increasing of temperature and activating time tend to reduce the yield. This phenomenon was comparable with the results from some researches [4,7,10,15].

Rising of activation time causes increasing burn-off grade and decrease the activated carbon yield. Obviously, increasing in burn-off represents inclining of weight losses due to activation process. Low burn-off is usually undesirable since they lead to low surface area which, in turn, may lead to a low adsorption capacity for target molecules [1].

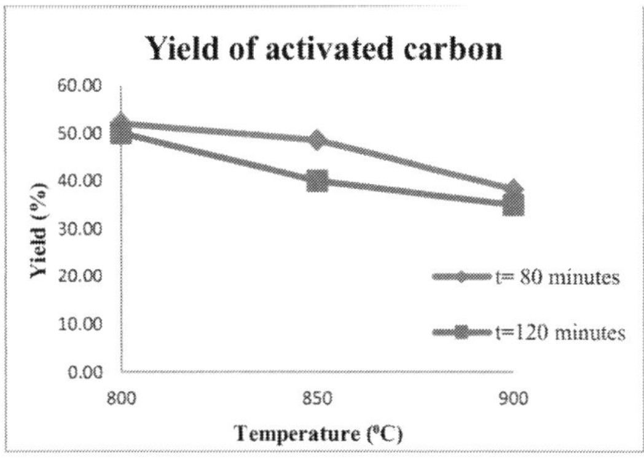


**Fig. 2.** Yield of activated carbon for temperature and activating time variables

It was evident from Figure 2 that activated carbon yielded from 120 minutes of activation time was lower than one of 80 minutes time. This revealed that prolong of activation time caused improving extent of activation and great reactions among carbon, steam ($H_2O$) and $CO_2$. Chen et.al. [15] summarized the main chemical reactions in the activation process as follow :

$C + H_2O \longrightarrow H_2 + CO$
$C + H_2O \longrightarrow H_2 + CO_2$
$C + H_2 \longrightarrow CH_4$
$C + CO_2 \longrightarrow CO$

Here, the volatile materials resulted from longer reaction time was higher, which, therefore it was released leading to reduce activated carbon yield.

Temperature increased would incline a more severe reaction as mention above which, in turn, raised the number of the volatile materials. As shown on Figure 2, activation in higher temperature obtained lower activated carbon yield since much more volatile matter was released.

### 3.2. Moisture content

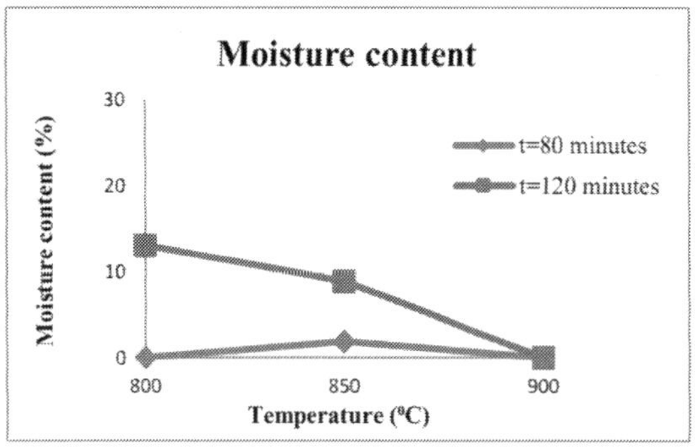


**Fig.3.** Moisture content at various temperature and activating time

Moisture content is amount of water bound to activated carbon under normal condition. As can be seen from Figure 3, it shows that moisture content of the activated carbon is in the range of 0 – 13.06 %. The moisture content of each activation time tended to decline with increasing the temperature.

Based on Indonesian Industrial Standard (SII No. 0258-88), moisture content of activated carbon allowed is not more than 15%. Less moisture content could increase the adsorption capacity of activated carbon.

For activated carbon whose was activation temperature of 800°C and 120 min, the moisture content was higher than before activation. This might be caused by activated carbon adsorbed water from the air, due to hygroscopic characteristic, during process of cooling, weighing and packing. Therefore, the handling of activated carbon during the experiment must be appropriate.

### 3.3. Ash content

Ash content indicates amount of residue contained minerals such as calium, natrium, magnesium, calsium, etc. spreads in pores of activated carbon [16]. Ash becomes more concentrate during activation and comprised of 1-20% depends on the type of raw material. As shown on Figure 4, ash content analyzed was in the range of 12.08-16.59%. It exhibits that ash content was still higher than standard activated carbon in Indonesia, which is 10 % (SII No. 0258-88).

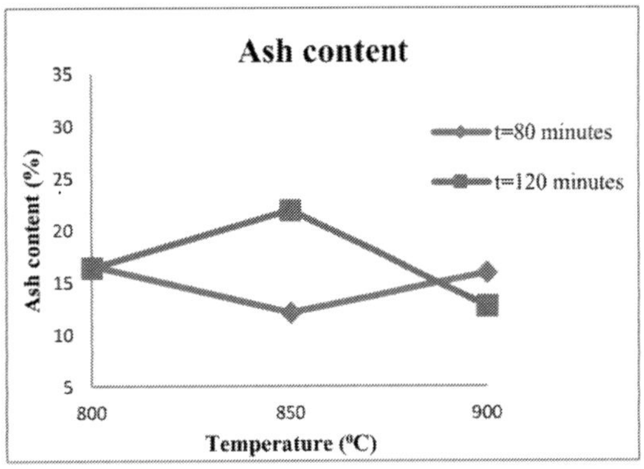


**Fig. 4.** Effect of activating temperature on ash content of activated carbons

As mentioned above, ash content is influenced by type of raw materials. Activated carbon based on soft by-product materials such as sugarcane bagasse, rice straw, rice husk, has higher ash content due to high minerals content [16]. The previous study obtained that ash content of activated carbon from sugarcane bagasse was various approximately from 9.26 – 20.26%. While, ash content of bagasse based product was up 42.74% [1]. Obviously, the presence of twigs (wood base) mixed with sugarcane bagasse in the activated carbon could have effect on the amount of the ash content. High ash content of activated carbon is undesirable since it reduces the mechanical strength of carbon leading to influence adsorption capacity.

### 3.4. pH

The result of pH measurement was shown on Figure 5. The pH values were in the range of 9.56-10.54 indicating that all activated carbon produced was alkaline. The pH values were comparable with some previous experiments [4,16,]. Their studies showed that activated carbon from sugarcane bagasse were alkaline, generally, pH values were above 8.0 due to inorganic compound in raw materials. To gain lower pH values, approximately 7-8, commercial activated carbon which had pH values 7.5-9 was washed with HCl 0.1M solution in order to reduce pH values from 9.4 to 8.5 [4].

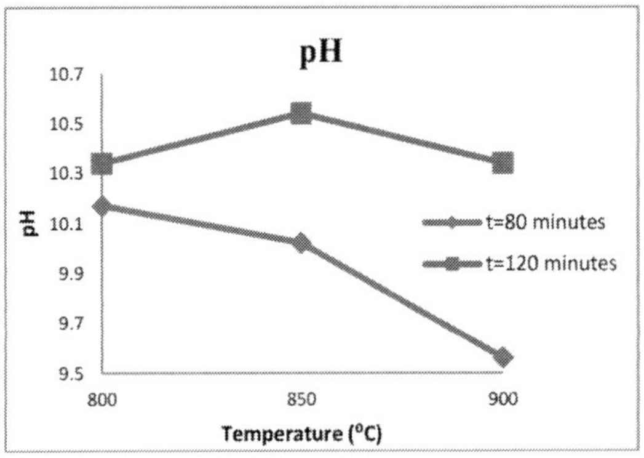


**Fig. 5.** Effect of activation time and temperature on pH of activated carbons

From Figure 5, it was shown that rising of activation time will increase pH value considerably. Increasing pH was probably caused reduction of $H_2O$ during activation process leading to increase OH- ion. As the time retention of activation increased, OH- ion constantly increased.

### 3.5 Methylene blue adsorption

The adsorptive properties of activated carbon were determined by methylene blue test. Methylene blue number of activated carbon is correlated with ability of the activated carbon to adsorb colour and high molecular weight substances. Figure 6 reflects values of methylene blue adsorbed on the activated carbon. The result of methylene blue adsorption calculated from Equation (3) was in the range of 241.875-242.500 mg/g. The highest adsorption showed at temperature of 850°C and 80 min. Besides, percentage of adsorption was also calculated using Equation (4) and the values are displayed on Figure 7. The graphs indicated that percentages of adsorption were quite high, those were in the range of 96.75 – 98.19%.

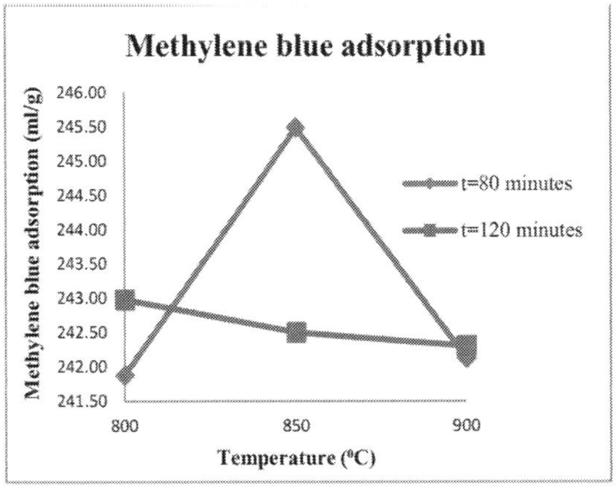


**Fig. 6.** Effect of activation temperature and time on methylene blue adsorption

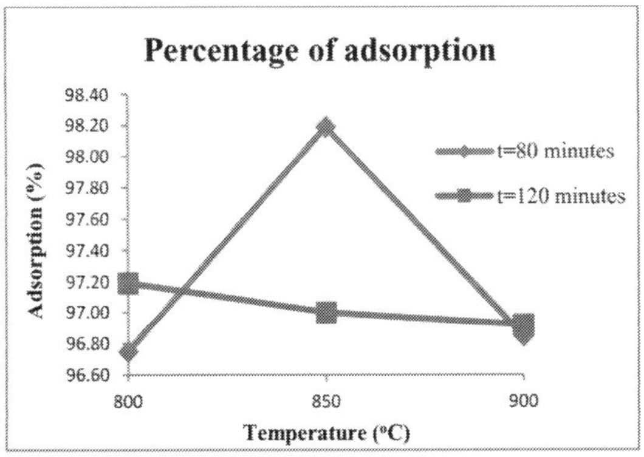


**Fig. 7.** Effect of activation temperature and time on percentage of adsorption

From Figure 4 reflects that the optimum adsorption of methylene blue was optimum for activated carbon which was activated at temperature of 850°C and activation time of 80 minutes. Here, reaction between $CO_2$ and biochar was constant leading to increase number of formed pores. Therefore, it induced to raise methylene blue adsorption significantly. Chen et.al. [15] from their study indicated that lower retention time and higher temperature (above 850°C) could transformed micropores into mesopores or even macropores, resulting in low value of methylene blue adsorption.

In outline, Figure 5 and 7 exhibited that high pH could increase methylene blue adsorption. This was proportionate with study done by Karacetin et.al. that used activated carbon from Hazelnut Husk [14] and Kumar [17] who used activated carbon from Foxnutshell. The increase in adsorption of MB with the high pH was due to the decrease in electrostatic repulsion between the positively charged methylene blue and the surface of the activated carbon. A study revealed that increasing of methylene blue adsorption would be significant in the optimum pH, around 8. However, the adsorption rate would be slower for pH above 8 [18]. Activated carbon produced in this study had pH values approximately 9-10, therefore averagely they would adsorb methylene blue quite highly. It was revealed from Figure 7 that all activated carbon had percentage of adsorption significant, above 90%.

Furthermore, ash content could have impact to adsorption properties. Ash could move freely to cover the existing pores effecting to obstruct adsorption process. It was reflected that methylene blue adsorbed on activated carbon resulted from 850 °C and 120 minutes activation (AC850/80) was lower than one resulted from 850oC and 80 minutes (AC850/80). On the other hand, the ash content of (AC850/80) was higher than (AC850/80) as reflected on Figure 4. Consequently, higher ash content trended to move freely covering existing pores leading to inhibit methylene blue adsorption.

As mentioned before, methylene blue adsorption was influenced by pH of activated carbon. (AC850/80) had pH closer to optimum pH (around 8). This could cause methylene blue adsorbed on (AC850/80) was higher compared to (AC850/120). Eventually, significant methylene blue adsorption on (AC 850/80) could be influenced by pH and ash content.

Meanwhile, activated carbons resulted from temperature 800°C and 900°C for two activation times exhibited almost the same adsorption capacity. It could also affected by ash content, pH and moisture content, where value of each parameter was close each other.

### 3.6 Surface area.

Predicted specific surface area (SSA) using methylene blue method was postulated by El-Geundi [11] and Yukselen [12]. The values showed a deviation compared to specific surface area value of BET analysis. For activated carbon which was alkaline, it exhibited that SSA of methylene blue method was always lower than SSA of BET analysis. The deviation showed up to 77%. While acidic material had higher SSA compared to SSA of BET analysis.

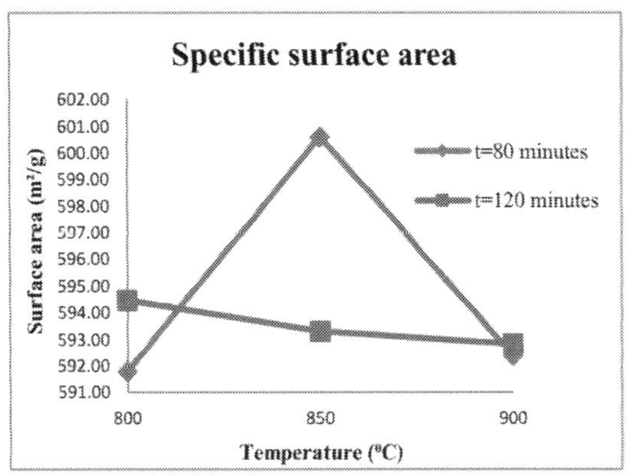


**Fig. 8.** Predicted specific surface area calculated from methylene blue adsorption

Mass of methylene blue adsorbed was included in the Equation (5) for calculating specific surface area of activated carbon. The surface area represented a variation trend similar to methylene blue adsorption.
The highest surface area showed at temperature 850°C and 80min.

## 4 Conclusion

Some properties of activated carbon have been analyzed. Moisture content, ash content, and pH could influence the adsorptive properties of activated carbon. However, the ash content and pH have significant role in methylene blue adsorption. At this experiment, the optimal activation conditions obtained were 850°C and 80 min, where under this condition, the activated carbon yield, moisture content, ash content, pH values, methylene blue adsorption, and surface area were 52%, 0%, 12.08%, 9.56, 245.475 mg/g, and 600.584 $m^2/g$, respectively. It revealed that biochar waste of bio-oil pyrolysis of sugarcane bagasse/twig of rambutan mixture had potential to be a good adsorbent.

The author acknowledge that the financial support provided by Chemical Engineering Dept., Faculty of Industrial Tecnology, University of Islamic Indonesia.

## References

1. K. Qureshi, I. Bhatti, R. Kazi, A. K. Ansari, Int'l. J. Che. Biomolec. Eng., **1:3**, 145-149 (2008)
2. B. Sutrisno, N.A. Rizka, A.S. Hidayat, A. Hidayat, IOP Conf. Series: Mat'l Sci. Eng. **105**, 12-27 (2016)
3. H.C. Tao, H.R. Zhang, J.B. Li, W.Y. Ding, Bio. Tech., **V.192**, 611-617 (2015)
4. G.da Cunha Goncalves, N.C. Pereira, M.T. Veit, Biomass and Bioenergy, **85**, 178-186 (2016)
5. K.Y. Foo, L.K. Lee, B.H. Hameed, Bio. Tech., **V. 134**, 166-172 (2013)
6. W. Lismaya, *Kayu Sisa Pohon Yang Ditebang Dan Tidak Ditebang Di Iuphhk-Ha Pt Inhutani Ii Unit Malinau Kalimantan Utara*, Forest Faculty, Institute Pertanian Bogor, Indonesia (2014)
7. Z. Anwar, M. Gulfraz, M. Irshad, J. Rad. Rsrch and Appl. Sci., **7**, 163-173 (2014)
8. A.S. Olawale, O.A. Ajayi, *Thermal Activation of Canarium Schweinfhurti Nutshell*, Ahmadu Bello University, Zaria, Nigeria (2009).
9. A. Shaabana, S.M. Se, N.M.M. Mitanb, M.F. Dimina, Proc. Eng., **68**, 365 – 371 ( 2013 )
10. E. David, J. Kopac, J. Anal. Appl. Pyr., **110**, 322-332 (2014)
11. M.S. El-Geundi, E.A. Ashour, R.M.A. Abobeah, N. Shehata, IJSETR, **V.3**, Issue 8 (2014)
12. Y. Yukselen, A. Kaya, Eng. Geo., **102**, 38-45 (2008)
13. Center for Documentation and Information Scientific, LIPI (1997)
14. G. Karacetin, S. Sivrikaya, M. Imamoˇglua, J. Anal. Appl. Pyr., **110**, 270-276 (2014)
15. D. Chen, X. Chen, J. Sun, Z. Zheng, K. Fu, Bio. Tech., **216**, 629-636 (2016)
16. M. Ahmedna, *LSU Agricultural Experiment Station Reports 456*, Louisiana State University Agricultural Center, USA. (2000)
17. A. Kumar, H.M. Jena, J. Clnr. Prod., **137**, 1246-1259 (2016)
18. A.F. Hassana, H. Elhadidy, J. Env. Che. Eng., 5, 955-963 (2017)

# Biodegradation of sulphide in biogas by biofilm on salak fruit seeds: The effect of intra-film-gradient

*Retno* A. S. Lestari[1,*], *Wahyudi* B. Sediawan[2], and *Sarto*[2]

[1]Chemical Engineering Department, 17 Agustus 1945 University, Semarang, Indonesia
[2]Chemical Engineering Department, Gadjah Mada University, Jl. Grafika 2 Yogyakarta 55281, Indonesia

**Abstract.** Elimination of $H_2S$ in biogas was performed by bio-filtration, in which $H_2S$ degrading bacteria immobilized on the packing material inside a column. Bacteria was taken from the local area, while salak fruit seed was used as packing material because it has a quite large carbon content, porous and has a high water absorption capability. To obtain a quantitative description of the process, which is useful in the design of large-scale bio-filters, two kinetics models were proposed. Model 1 assumes intra-film gradient does not exist and the gas phase is quasi-steady-state. Model 2 assumes intra-film gradient exists and the gas phase is quasi-steady-state. Comparison of the calculation results using the two models to the experimental data suggested the use of model 2. This conclusion is based on both the trend and the accuracy. The trend of model 1 is more conceivable. However, for rough calculation, model 1 could be used for the reason that it is simpler than model 2. The average absolute errors of model 1 and 2 are comparable, approximately 6 ppm. It can also be concluded that intra-film gradient significantly affects the overall rate processes, while the gas phase can be assumed to be quasi-steady-state.

## 1 Introduction

One of the potential biomass-based materials for alternative energy is biogas. It is commonly derived from the anaerobic decomposition of organic substances such as animal waste, household waste, agricultural waste, waste water [1]. Biogas is a gas produced by fermentation of organic material by microorganisms in anaerobic condition. This anaerobic process follows the stages of the hydrolysis, asidogenesis and methanogenesis [2]. According to Syed et al. [3], biogas was composed of methane ($CH_4$) 55-70%, carbon dioxide ($CO_2$) 30-45% and hydrogen sulphide ($H_2S$) 0-1.5%. Although the sulphide compounds contained in biogas are relatively small, they can be spoiling and even damage the equipment used on the utilization of biogas because it is corrosive [4].

Hydrogen sulphide removal process in biogas can be performed by physical and chemical processes such as absorption [5, 6] scrubbing [7], adsorption [8] and reactive absorption. Physical and chemical processes are costly and produce other waste.

Another way to remove sulphide compounds in biogas are biochemical conversion with bio-filtration method [9]. This process uses a bio-filter containing sulphide degrading bacteria immobilized on the surface of the packing material [10].

The use of salak fruit seeds for bio-filter media has never been done. Salak fruit seeds (SFS) easily absorb water, have a nearly uniform size, containing nutrition and are relatively hard. Furthermore their availability were abundant and cheap.

This research aims to eliminate sulphide compound contained in biogas using a bio-filter with SFS as packing material, and sulphide degrading bacteria taken from local area, as well as to get a kinetic model that could be used in designing a large scale system of sulphide elimination in biogas using bio-filter. Special focus in this study concerns with the effect of intra-film-gradient in the biofilm.

Studies on sulphide gas removal with bio-filters have been extensively conducted, but the use of salak fruit seeds as packing material have never be done. Studies on the removal of $H_2S$ in biogas using a bio-filter with other packing materials, mostly focused on semi quantitative descriptive aspects, namely efficiency of removal (RE) of $H_2S$ gas. Although there are researchers who studied quantitative descriptive, their approaches are still open for improvement.

This work is a quantitative descriptive study; the fundamental concepts of mass transfer and kinetics of biochemical reactions of sulphide compounds in the packed bed biogas are explored and laboratory experiments are also conducted to verify the kinetic model proposed.

Bio-filtration process is relatively simple. Gas containing $H_2S$ was flowed through a pile of packing material which has a surface coated by pollutant-degrading bacteria as a biofilm [11]. Hence, sulphide is absorbed by biofilm, and oxidized by microorganism to

* Corresponding author: retnotengaran@gmail.com

be elemental sulphur (if oxygen was limited) and $SO_4^{-2}$ (if oxygen is in excess) [12].

The packing material can be either organic or inorganic compounds. In selecting the support of bio-filter there are several criteria that must be fulfilled include nutrient content (source of C and N), moisture content, pH, porosity, ability to absorb contaminants, lightweight, and cheap [12, 13].

Biofilms are microorganisms that attached and stick together in a polymer matrix [14]. The attachment of microorganisms on the surface of the packing material is very important for the initiation of the development of biofilm [15].

Sulphide-oxidizing bacteria include: bacteria of the genus Thiobacillus (kemoautotrof/ litotrof), green and purple sulfur bacteria (photoautotrophs), and types of heterotrophic include Atrhrobacter, Bacillus, Mikrococcus, Mycobacterium and Pseudomonas [16].

In this research used biogas, the observation to investigated hydrogen sulphide concentration at some axial position in the column at the various time at 1 phase. At the other paper, the research used the liquid solution of hydrogen sulphide and with difference conditions and difference model. Mathematical model proposed in other paper assumed no intra-film-gradient. [17].

## 2 Mathematical models

Some previous researchers who have done modelling in the process of gas and liquid waste handling using bio-filters include Spigno et al. [18] in the elimination of phenol, Majumder et al. [19] in the process of removal of divalent copper dissolved in water, Jaber et al.[20] in modelling bio-filters for hydrogen sulphide contained in the air, Meena et al. [21] makes the mathematical modelling of the separation of a mixture of hydrophilic (methanol) and hydrophobic (αPinene), Santos et al. [22] used a mathematical model for the removal of hydrogen sulphide by oxidation in the bio-filter, Kim and Deshusses [23] calculate the mass transfer coefficient on the packing material used for bio-filter, either in the gas layer or the liquid layer. Furthermore Yang et al. [24] makes modelling of porosity variations of bio-filter equipment used in rotary drum, Ramirez et al. [25] studied the kinetics of microbial growth and biodegradation of methanol and toluene, De Visscher and Li [26] studied the disappearances modelling toluene contained in the air, Agarwal et al. [27] studied the mass balance for phenol degradation events. Modelling in this study was set-up by adopting and combining the basic ideas that emerged in previous studies.

The mathematical modelling of sulphide removal process with bio-filter columns proposed was based on the most fundamental concepts of mass transfer, phase equilibrium and kinetics of chemical reactions, and tried to avoid the use of special empirical equations. The solution of the mathematical equations formed is also based on common numerical methods, and avoiding special simplification. Two models were proposed with the objective to study the effects of intra-film-gradient to the rate of the overall process.

### 2.1 Model 1

Model 1 by Lestari et al. [28] assumes that biofilms are relatively thin, so there is no concentration gradient in the biofilm. Therefore the concentration of sulphide in the biofilm is uniform and the value is the same as the sulphide concentration at the surface of biofilm (Cs = Csi). Furthermore, the sulphide concentration in the biogas flowing in the column is considered to be quasi-steady-state.

In the model development, further following simplifications are applied:

1. The gas phase is plug-flow with no axial dispersion.
2. The biofilm is formed on the surface of the SFS only.
3. The thickness of the biofilm is relatively small compared to size of the SFS, so the geometry of the biofilm can be assumed as slab.
4. Biodegradation reaction occurs in the biofilm only, and follows Monod's type equation.
5. The biofilm thickness increases as time progresses.
6. The system is isothermal.

Sulphide material balance in the biofilm produces:

$$\frac{\partial C_S}{\partial t}=\frac{k_C}{\delta}(C_g-C_{gi})-r_A \qquad (1)$$

If the gas-bio-film equilibrium of sulphide follows Henry's type equation, then:

$$C_{gi}=H_S*C_S \qquad (2)$$

The rate of biodegradation follows Monod's equation:

$$r_A=\frac{\mu_{\max}C_s}{C_s+K_S} \qquad (3)$$

The thickness increase of the biofilm is caused by the the increase of the number of microbes, so:

$$\frac{d\delta}{dt}=\frac{Y_{\frac{X}{S}}}{\rho}.r_{A}.\delta \qquad (4)$$

The material balance of the sulphide in the gas, assuming quasi-steady-state, gives:

$$\frac{dC_g}{dz}=-\frac{k_C a_S S}{G}(C_g-C_{gi}) \qquad (5)$$

The models proposed were solved numerically by finite difference approximation as follows:

$$(C_S)_{j+1}=(C_S)_j+\frac{k_C\Delta t}{\delta}\left((C_G)_j-H_S(C_S)_j\right)-(r_A)_j \qquad (6)$$

$$\delta_{j+1}=\delta_j+\frac{AN}{\rho}r_A\delta_j\Delta t \qquad (7)$$

$$(C_G)_{i+1,j}=\frac{(C_G)_{i,j}+\gamma.H_S.\Delta z.(C_S)_{i+1,j}}{(1+\gamma\Delta z)} \qquad (8)$$

### 2.2 Model 2

Model 2 is the improvement of model 1, in which the biofilm is considered relatively thick, so the value of Cs in the biofilm is considered to be not uniform. Meanwhile the sulphide concentration in the biogas flowing in the column is considered to be quasi-steady-state.

The sulphide material balance in the volume element in the biofilm gives:

$$\frac{\partial^2 Cs}{\partial x^2} - \frac{1}{D_e} r_A = \frac{1}{D_e} \frac{\partial Cs}{\partial t} \tag{9}$$

The rate of biodegradation is the same as equation (3) in model 1. The gas-bio-film equilibrium of sulphide follows Henry's type equation as equation (2). The material balance of the sulphide in the gas, assuming quasi-steady-state, is also the same as equation (5).

The thickness increase of the biofilm is caused by the increase of the number of microbes and can be calculated as follows:

$$\int_{\delta_0}^{\delta} d\delta = \frac{Y_{\frac{x}{S}}}{\rho} \int_{X=0}^{X=\delta} r_{A.} dx \tag{10}$$

The models proposed were solved numerically by finite difference approximation.

$$(C_G)_{i+1,j} = \frac{(C_G)_{i,j} + \gamma . H_S . \Delta z . (C_S)_{i+1,j}}{(1+\gamma \Delta z)} \tag{11}$$

$$(C_S)_{i,j+1} = \frac{(C_S)_{i-1,j} + (M-2)(C_S)_{i,j} + (C_S)_{i+1,j}}{M} - \Delta t . (r_A)_{i,j} \tag{12}$$

$$M = \frac{(\Delta X)^2}{D_e(\Delta t)} \text{ ......................... } (i = 2, 3, 4, \ldots, N)$$

$$(C_S)_{1,j+1} = (C_S)_{1,j} + \frac{2k_C \Delta t}{\Delta x}\left((C_G)_j - H_S (C_S)_{1,j}\right) - \frac{2De\Delta t}{(\Delta x)^2}\left((C_S)_{2,j} - (C_S)_{1,j}\right) - \Delta t (r_A)_{1,J} \tag{13}$$

$$(C_S)_{N+1,j+1} = (C_S)_{N,j} - \frac{2De\Delta t}{(\Delta x)^2}\left((C_S)_{N+1,j} - (C_S)_{N,j}\right) - \Delta t (r_A)_{N+1,J} \tag{14}$$

If intra-film-gradient exists, the total sulphide degraded in the biofilm can be calculated by numerical integration as follows:

$$\delta_{j+1} = \delta_j + \frac{AN}{\rho} \frac{\Delta x}{3} \begin{pmatrix} (r_A)_{0,J} + 4(r_A)_{1,J} + 2(r_A)_{2,J} + \ldots \\ + 4(r_A)_{N,J} + (r_A)_{N+1,J} \end{pmatrix} \tag{15}$$

If the results of model 1 and 2 are not significantly different, it can be concluded that the effect of intra-film-gradient can be neglected. But for the opposite, the effect of intra-film-gradient is significant.

## 3 Research Methodology

### 3.1 Materials

Raw materials: salak fruit seeds for packing of bio-filter, sludge for source of sulphide degradation bacteria, biogas collected from Integrated Farming Unit, Agricultural Research and Development for Education (KP4) Facility, Gadjah Mada University.

Materials for nutrition: $KH_2PO_4$, $K_2HPO_4$, $MgSO_4.7H_2O$, $CaCl_2$, $FeCl_3.6H_2O$, $(NH4)_2SO4$, and yeast extract [29].

Materials for analysis: Zn-Acetic, $FeCl_3$ and N,N-Dimethyl-1,4-Phenylen Diamonium Diklorida solution, $MgCl_2.6H_2O$, $CH_3COONa.3H_20$, $CH_3COOH$, $BaCl_2$, and aquadest.

### 3.2 Research equipment

The main equipment used in this research is the bio-filter packed column. The column for the bio-filter reactor was made of acrylic resin which has an inside diameter of 8.0 cm and a height of 100 cm. The height of SFS packing was 80 cm. The column has 6 port holes at different axial positions for taking samples to analyse sulphide concentration [18].

### 3.3 Research Procedure

The research procedures include: characterization of salak fruit seeds (SFS), immobilized bacterial cultures on SFS, isolation of bacteria, test of isolate to sulphide oxidation, test of biofilm formation of isolate on SFS, explorative study of isolates immobilized on seeds to degrade sulphide, characterization and identification of isolates superior (all of experimental data no shown in this paper), test of sulphide in biogas degradation in the bio-filter and mathematical modelling as well as testing the mathematical models to experimental data of sulphide elimination in the bio-filter.

## 4 Result and Discussion

Before being used in the bio-filtration process, after immobilization bacteria on the surface of the SFS, the colonies of bacteria on SFS surface was 1,56x108 CFU/ml, after being used for 8 hours in bio-filter process, it has 2.43 x 109 CFU/ml. This shows that during bio-filtration process, the number of bacteria increased, caused by the growth and development of bacteria supported by nutrition from sulphide and nutrition in process bio-filtration [18].

### 4.1 Elimination of sulfide of biogas in the biofilter

The experiments of removal of sulphide in the biogas using a bio-filter containing the Bacillus cereus which forms biofilm on the surface of SFS operated at biogas flow rate of 30 L/h, 48 L/h, 66 L/h and 84 L/h. The axial position observed of the inlet biogas are 0 cm, 20 cm, 40 cm, 60 cm and 80 cm. The sulphide concentration in biogas was varied from 10 ppm to 180 ppm. The relationship between the sulphide concentration and RE versus axial position from gas inlet for biogas flow rate of 30 L/h are shown in Figure 1.

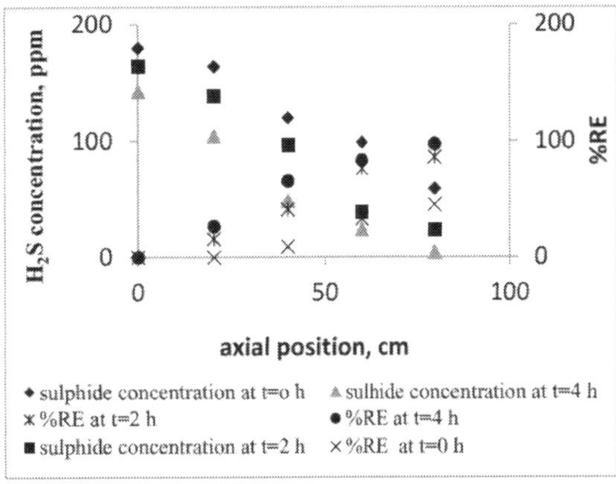


**Figure 1.** The relationship of the sulphide concentration and RE versus axial position at Q 30 L/h

From Figure 1, the influence of the axial position and operating time can be further discussed. The other figures of the relationship of the sulphide concentration and RE versus axial position at Q 48 L/h, 66 L/h and 84 L/h are presented in Figure 2, Figure 3, and Figure 4. From those figures the influence of the biogas flow rate in the bio-filter could also studied.

#### *4.1.1 Effect of axial position*

$H_2S$ removal efficiency will increase with the increasing distance from the bottom of the column. This happens as a larger distance from the bottom of the column will provide a longer contact time between the biogas and the biofilm, making the removal of $H_2S$ more effective. In this packing column (80 cm long and 8 cm inside diameter, thus L/D=10), the highest $H_2S$ removal efficiency achieved was 97.15%.

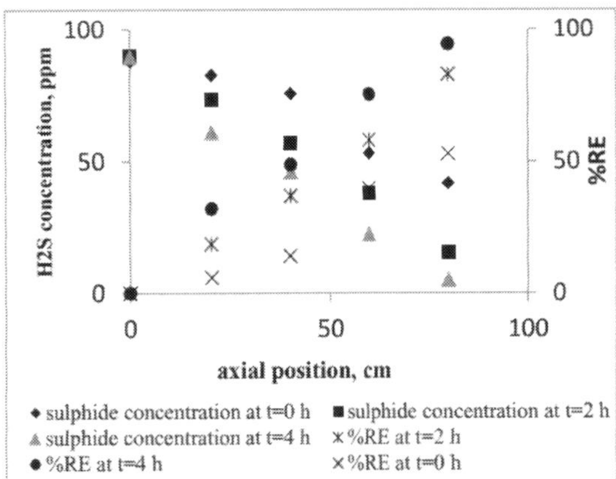


**Figure 2.** The relationship of the sulphide concentration and RE versus axial position at Q 48 L/h

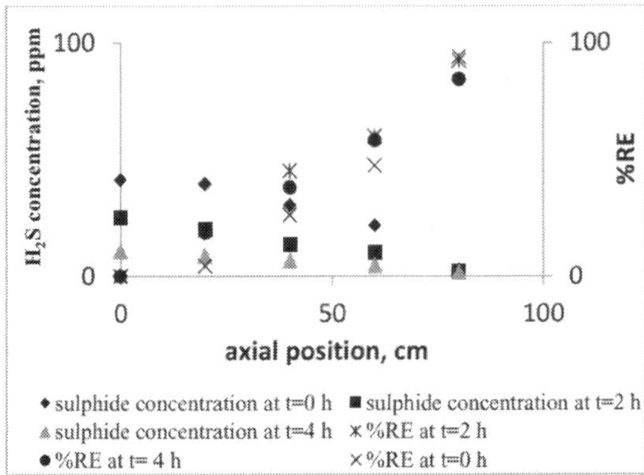


**Figure 3.** The relationship of the sulphide concentration and RE versus axial position at Q 66 L/h

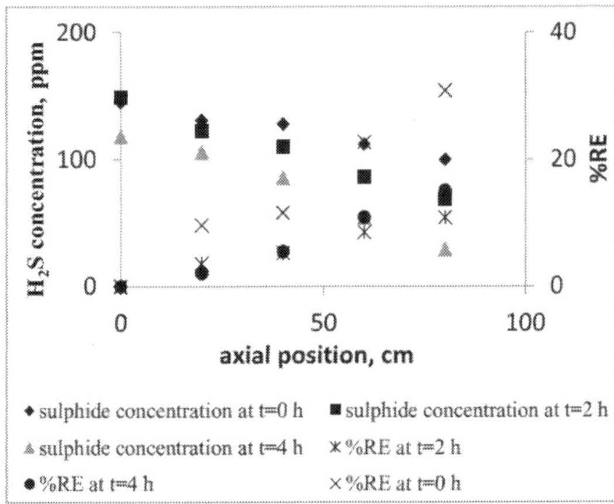


**Figure 4.** The relationship of the sulphide concentration and RE versus axial position at Q 84 L/h

#### *4.1.2 Effect of operation time*

The $H_2S$ removal efficiency increase with the increasing time of operation, for different initial concentration of $H_2S$ in biogas. This is conceivable since the biofilm grows on SFS surface throughout the column and a larger biofilm mass will consume more $H_2S$, resulting in the higher $H_2S$ removal. This microbial growth can be visually observed by the increase of the thickness of biofilm formed on the surface of the SFS.

#### *4.1.3 Effect of biogas flow rate*

The sulphide removal efficiency in a different position from the gas inlet indicates that the lower gas flow rate results in higher removal efficiency. This fact may be described as the following. As the gas flow rate increased, the contact time between $H_2S$ and microorganism on SFS is shorter, thus only a smaller part of the $H_2S$ could be degraded by the microorganism.

#### *4.1.4 The performance of SFS as a packing material in the bio-filter*

The results of this study concluded that SFS can be used as packing material in the bio-filter for elimination of sulphide in the gas phase. It turned out that performance of SFS was comparable with others packing materials studied in the literature.

### 4.2 Test of the mathematical modelling to experimental data of elimination of $H_2S$ in biogas

Simulations applying the two mathematical models proposed were performed using adjustable values of parameters involved until the results are close to the experimental data obtained. The comparison between the calculated results and the experimental data for the models are shown in Figure 5 and Figure 6. It can be observed that the calculation results using the two models are both close to the experimental data. It means that the two mathematical models proposed are adequate to quantitatively describe $H_2S$ removal from biogas using biofilm on packed bed of SFS.

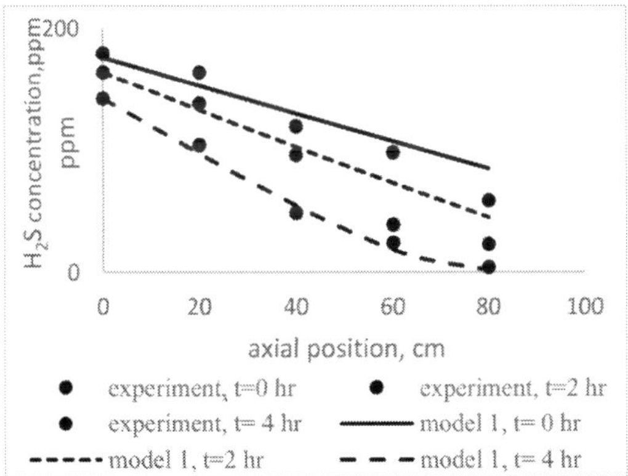


**Figure 5**. Comparison of modelling curve to data experiments of model 1

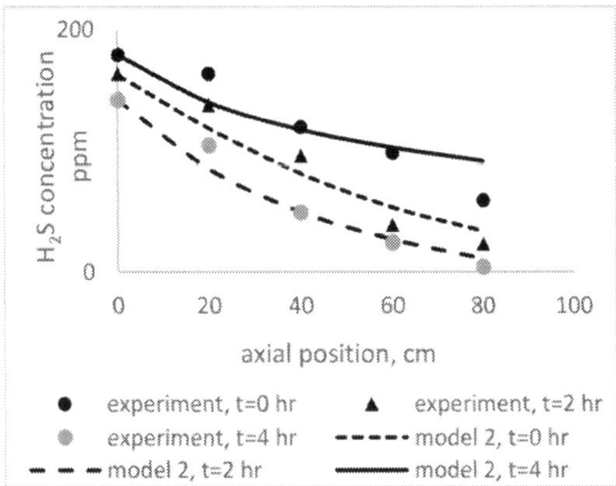


**Figure 6.** Comparison of modelling curve to data experiments of model 2

To be more comprehensively comparing the suitability of model 1 and model 2 to the experimental data, curves of model 1 and curves of models 2 were closely examined. The trends of the two curves are significantly different. Curves of model 1 are almost linear, while the curves of model 2 are non-linear. Logical interpretation of the process suggests that the curves of sulphide concentration versus time should be non-linear, because the sulphide concentration decreases due to biodegradation, resulting in the decrease of the rate of biodegradation. So, non- linear trend is more conceivable. As a consequence, model 2 is more conceivable than model 1. The average absolute errors of model 1 and model 2 are comparable, approximately 6 ppm. All in all, this comparison shows that even though both models have comparable accuracies, model 1 and model 2 are significantly different in the trends. Since the difference of model 1 and model 2 are on the consideration of intra-film gradient, it can be concluded that the effect of intra-film gradient needs to be considered.

Even though model 1 is inferior to model 2, but model 1 is still good enough to estimate the value of sulphide concentrations as well as the RE. The advantage of model 1 is the simplicity. It also needs simpler calculation. Time of calculation for model 2 is approximately four times to the one of model 1.

From model 1, it was obtained that: the maximum specific growth rate of bacteria ($\mu_{max}$) = 9.3E-7 $s^{-1}$, $K_S$ = 4.0E-7 g $cm^{-3}$, $H_S$ = 0.7, $Y_{X/S}$ = 60, and mass transfer coefficient $a_0$ = 0.038 and $a_1$ = 0.25. Meanwhile for model 2, the maximum specific growth rate of bacteria ($\mu_{max}$) = 1,8E-6 $s^{-1}$, half past saturation ($\alpha$) = 8,0E-7 g $cm^{-3}$, constant Henry ($H_S$) = 0.7, yield ($Y_{X/S}$) = 60, effective diffusivity (De) = 2E-4, and mass transfer coefficient $a_0$ = 0.0085 and $a_1$ = 0.2 for all flow rate.

## 5 Conclusion

This results of research can be concluded that the experiments of elimination of sulphide in biogas using bio-filter column shows that removal efficiency (RE) is affected by the axial position, operation time, biogas flow rate, and the concentration of $H_2S$ in the gas input.

Based on the observation of the trends, it can be concluded that of the two mathematical models proposed, the recommended one is model 2 (assuming there is a gradient of concentration in the biofilm, the flow of gas in the bio-filter quasi-steady-state, the rate of degradation reaction follows Monod's type equation).

The accuracies of model 1 and model 2 are actually comparable since the average absolute errors for both models are approximately 6 ppm.

Comparison to experimental data shows that model 2 works well. For model 2, it was obtained that maximum specific growth rate of bacteria ($\mu_{max}$) = 1.8E-6 $s^{-1}$, half past saturation ($\alpha$) = 8.0E-7 g / $cm^3$, constant Henry ($H_S$) = 0.7, yield ($Y_{X/S}$) = 60, effective diffusivity (De) = 2E-4 $cm^2$ $s^{-1}$, and mass transfer coefficient $a_0$ = 0.0085 and $a_1$ = 0.2 for all flow rates.

Tests of mathematical models against experimental data of the elimination of sulphide in the biogas using a bio-filter containing Bacillus cereus bacteria on surface of SFS also indicate that intra-film-gradient needs to be considered, and the gas phase can be assumed to be quasi-steady-state.

Even though not considering intra-film gradient, so that less accurate than model 2, the simpler model 1 is good enough for rough calculation.

The authors would like to acknowledge to Directorate of Research and Technology of Higher Educations of Indonesia, for financial support of the research through research grant of Hibah Bersaing 2015 and Hibah Disertasi Doktor 2016.

## References

1. H.R. Amini, D.R. Reinhart, Regional prediction of long-term landfill gas to energy potential. Waste Manage 31, 2020–2026 (2011).
2. Yadvika, T.R. Santosh, S. Sreekrishnan, Kohli, V. Rana, Enhancement of Biogas Production from Solid Substrates using Different Techniques - a Review. Bioresource Technology, 95, 1-10 (2004).
3. M. Syed, G. Soreanu, P. Falletta, M. Béland, Removal of hydrogen sulfide from gas streams using biological processes—a review, Can. Biosyst. Eng. 48 (2006) 2.1–2.14.
4. S. Pipatmanomai, S. Kaewluan, T. Vitidsant, Economic assessment of biogas to electricity generation system with H2S removal by activated carbon in small pig farm. Applied Energy, 86, 669-674 (2008).
5. Boumnijel, H.B. Amor, H. Chekir, N. Hajji, Hydrogen sulphide removal from the effluents of a phosphoric acid production unit by absorption into chlorinated seawater under alkaline conditions, Comptes Rendus Chimie, xxx, 1-8(2016).
6. M. Taheri, A. Mohebbi, H. Hashemipour, A.M. Rashidi, Simultaneous absorption of carbon dioxide ($CO_2$) and hydrogen sulfide ($H_2S$) from $CO_2$-$H_2S$-$CH_4$ gas mixture using amine-based nanofluids in a wetted wall column, J. Nat. Gas Sci. Eng. 28 (2016) 410–417
7. C.C. Lien J.L. Lin, C.H. Ting, Water Scrubbing for Removal of Hydrogen Sulfide ($H_2S$) Inbiogas from Hog Farms, Journal of Agricultural Chemistry and Environment 3, 1-6 (2014).
8. M.R.A. Al Mamun, S. Torii, Removal of Hydrogen Sulfide ($H_2S$) from Biogas Using Zero-Valent Iron, Journal of Clean Energy Technologies, 3, 6, 428-432, (2015).
9. C. Rattanapan, P. Boonsawang, D. Kantachote, Removal of $H_2S$ in downflow GAC biofiltration using sulfide oxidizing bacteria from concentrated latex waste water, Bioresource Technology, 100 125-130 (2009).
10. Elias, A. Barona, A. Arreguy, J. Rios, I. Aranguiz, J. Penas, Evaluation of a packing material for the biodegradation of $H_2S$ and product analysis, Process Biochemistry 37, 813–820 (2002)..
11. M.A. Deshusses, G. Hamer, and I.J. Dunn, Behavior of biofilters for waste air biotreatment. 1. Dynamic model development. Environ. Sci.Technol. 29. 1408–1458 (1995).
12. J.S. Devinny, M.A. Deshusses, and T.S. Webster, Biofiltration for air pollution control. Lewis Publishers. Boca Raton. 5–16 (1999).
13. M. Hirai, M. Ohtake, and M. Shoda, Removal Kinetic of Hydrogen Sulfide, Methanotiol and Dimethyl Sulfide by Pear Biofilter, Journal of Fermentation and Bioengineering (1990)
14. S. Craig, Microbiology: breaking down biofilm. Curr Biol;19:R132–4 (2002).
15. F.P. Chávez, F. Gordillo, C.A. Jerez, Adaptive responses and cellular behaviour of biphenyl degrading bacteria toward polychlorinated biphenyls. Biotechnol Adv 2006;24: 309–20 (2006).
16. P. Edmonds, Microbiology an environmental Perspective, Collier Macmillan Publisher. London-New York (1978).
17. R.A.S. Lestari, W.B. Sediawan and Sarto, Sulphide Removal from Liquid using Biofilm on Packed Bed of Salak Fruit Seeds, 2$^{nd}$ International Conference on Engineering and Technology for Sustainable Development, Yogyakarta, 13-14 September (2017).
18. G. Spigno, M. Zilli, C. Nicolella, Mathematical modelling and simulation of phenol degradation in biofilters, Biochemical Engineering Journal 19 (2004).
19. S. Majumder, G. Gangadhar, S. Raghuvanshi, S Gupta, Biofilter column for removal of divalent copper from aqueous solutions: Performance evaluation and kinetic modeling Journal of Water Process Engineering 6 136–143 (2015).
20. M.B. Jaber, A. Couverta, A. Amranea, F. Rouxel, P. Cloireca, L.E. Dumontd, Biofiltration of H2S in air—Experimental comparisons of originalpacking materials and modeling, Biochemical Engineering Journal 112, 153–160 (2016).
21. V. Meena, L. Rajendran, S. Kumar, P.G.J. Rani, Mathematical modeling of gas phase and biofilm phase biofilter performance, egyptian journal of basic and appliedsciences 3, 94–105 (2016).
22. J.M. Santos, E.S. Lopes, N.C.R. Junior, L. Melo de Sa, N.J. Horand, Mathematical modelling of hydrogen sulphide emission andremoval in aerobic biofilters comprising chemical oxidation, water research 43, 3355 – 3364 (2009).
23. S. Kim, and M.A. Deshusses, Determination of mass transfer coefficients for packing materials used in biofilters and biotrickling filters for air pollution control, Chemical Engineering Science 63, 841 – 855 (2008).
24. C. Yang, M.T. Suidan, X. Zhub, B.J. Kimc, G. Zeng, Effect of gas empty bed contact time on performances of various types of rotating drum biofilters for removal of VOCs, water research 42,3641-3650 (2008).
25. A.A. Ramirez, S. Bénard, A.G. Fendler, J.P. Jones, M. Heitz, Kinetics of microbial growth and biodegradation of methanol and toluene in biofilters

and an analysis of the energetic indicators, Journal of Biotechnology 138, 88–95 (2008).

26. Visscher, A.D., Li, G.Q., 2008, Toluene removal biofilter modeling: Optimization and case Study, process safety and environment protection 86, 277–282.
27. G.K. Agarwal, A.K. Ghoshal, Packed bed dynamics during microbial treatment of wastewater: Modelling and simulation, Bioresource Technology 99, 3765–3773 (2008).
28. R.A.S. Lestari, W.B. Sediawan, S. Syamsiah, and Sarto, Hydrogen sulfide removal from biogas using a salak fruit seeds packed bed reactor with sulfur oxidizing bacteria, Environmental Chemical Engineering Journal, 4, 2370–2377 (2016).
29. R.A.S. Lestari, W.B. Sediawan, S. Syamsiah, and Sarto, The Exploratory Study on Sulfide Removal Ability of Sulfur-Oxidizing Bacteria in Biofilm on Salak Fruit Seeds, International Proceeding of Chemical, Biological & Environmental Engineering, Chemical Engineering and Applications V, Taipei, Taiwan, 16-19 (2014).

# Sulphide removal from liquid using biofilm on packed bed of salak fruit seeds

*Retno* A. S. Lestari[1,*], *Wahyudi* B. Sediawan[2], and *Sarto*[2]

[1]Chemical Engineering Department, 17 Agustus 1945 University, Semarang, Indonesia
[2]Chemical Engineering Department, Gadjah Mada University, Jl. Grafika 2 Yogyakarta 55281, Indonesia

**Abstract.** This study focused on the removal of sulphide from liquid solution using biofilm on packed bed of salak fruit seeds. Bio-filter operation of 444 hours consists of 6 phases of operation. Each phase lasted for approximately 72 hours to 82 hours and run at various inlet concentration and flow rate. The highest removal efficiency is 92.01%, at the end of phase 7 at the inlet concentration of 60 ppm and the flow rate of 30 mL min-1. Mathematic model of sulphide removal was proposed to describe the operation of bio-filter. The model proposed can be applied to describe the removal of sulphide liquid using bio-filter in packed bed. The simulation results the value of the parameters in process. The value of the rate maximum specific growth is 4.15E-8 $s^{-1}$ , Saturation constant is 9.1E-8 g $cm^{-3}$, mass transfer coefficient of liquid is 0.5 cm $s^{-1}$, Henry's constant is 0.007, and mass of microorganisms growth to mass of sulphide consumed is 30. The value of the rate maximum specific growth in early process is 0.00000004 $s^{-1}$.

## 1 Introduction

Sulphide compounds are causing problems for the environment. For example, under anaerobic wastewater conditions, sulphides combine with hydrogen to produce hydrogen sulphide which in environment can be oxidized to sulphuric acid. Then the sulphuric acid formed reacts readily with the alkaline components of the concrete structures such as sewer pipes and manholes [1].

Several processes available to remove sulphide, such as physical, chemical, and biological processes. The physical and chemical processes such as absorption, combustion, and scrubbing have been used to remove sulphide from industrial waste water streams. But those physical and chemical methods need high energy requirements, high chemical disposal, and high costs [2]. A prospective biological process for sulphide removal is bio-filtration [3]. This paper report a study on the use of bio-filtration for sulphide removal from liquid. In this method microorganisms living in a support matrix were applied to degrade the pollutant. It is expected that this process is inexpensive. Furthermore this method is very attractive for low concentration gas streams due to the relatively moderate operating cost and minimum by-product generation [4].

The support matrix for the microorganism or packing material is an important factor in the design of a bio-filtration unit. The support matrix for the microorganism or packing material is an important factor in the design of a bio-filtration unit. The material need high porosity, high water retention capacity, and also nutrient presence and availability [4].

Organic materials have both the presence of nutrients and high water holding capacity than in organic packing materials. Most bio-filter media commonly applied are peat and compost with inert bulking agents such as activated carbon, wood chips or beads [5]. Composts are used frequently as they over dense and have more varied microbial population as well as good water holding capacity and nutrients. However, composts have low porosity, so the pressure drop in the bio-filter will be high. Soils are prone to short circuiting and clogging [4]. Activated carbon is good, but very expensive [6].

In this study, the removal of sulphide in liquid by bio-filter on packed bed of salak fruit seeds (SFS) in a column was investigated. These materials offer some advantages such as low cost, porous, and high water retention capacity. To quantitatively describe the process, a mathematical model considering liquid to bio-film mass transfer and bio-oxidation in the bio-film was developed and tested.

There were several mathematical model to describe the phenomena in bio-filter. Spigno and Nicolella [7] presented the model consist of mass balance equations and accounting for reaction, mass transfer and axial dispersion to describe the steady-state degradation of phenol in a bio-filter. Meanwhile Jaber et al. [8] were derived the $\alpha$lump parameter which enables the performance of the bio-filter as a whole to be characterized whatever its composition (mixture or layers of different packing materials) and whatever the EBRT (empty bed residence time).

Removal of hydrogen sulphide on wastewater by chemical reaction was studied by Santos et al. [9] which compares four different empirical expressions. Biodegradation and oxidation were found to be the two

* Corresponding author: retnotengaran@gmail.com

main removal mechanisms in bio-filter. Even though biodegradation was the main mechanism responsible for the $H_2S$ removal from the liquid phase, however oxidation being the most important one.

Agarwal and Goshal [10] worked on phenol removal in wastewater, claimed that their model has been able to predict the dynamics of the bio-filtration process. The system and condition of operations were input of substrates concentration, mass transfer coefficients in liquid phase, size reduction, Henry"s constants, flow rate of inlet, bed void fraction, growth and half saturation constants [10]. Inlet substrate concentration, inlet velocity, growth and half saturation constants and liquid phase mass transfer coefficients significantly control the operational dynamics.

The kinetics of microbial growth and the biodegradation of methanol and toluene in bio-filter and bio-trickling filters, packed with inert materials, has been studied and analysed by Ramirez et al. [11]. The specific growth rate was found to be a function of the methanol and toluene concentrations in the biofilm. In the bio-filter (BF) used for treating methanol, was found to be affected by the nitrogen concentration present in the nutrient solution, and the kind of packing material employed. A Michaelis-Menten model type provided a good fit for the elimination capacity (EC) of the Bio-trickling filter treating methanol, while a Haldane model type provided a good fit to the EC of the BF treating methanol and toluene.

Micro-kinetic models were expressed as functions of operating parameters such as the concentration of volatile organic compounds (VOCs) in the gas phase. Bio-filters treating methanol have shown that was also a function of the concentration present in the nutrient solution and the kind of packing material employed. The models of Michaelis-Menten and Haldane have been adapted to estimate the elimination capacities of the bio-filters, both with and without percolation.

The Michaelis-Menten model type was fitted to the EC of bio-trickling filters treating methanol. The Haldane model type has been fitted to the EC of bio-filters treating methanol and toluene, because these bio-filters presented inhibition by VOC concentration. The carbon dioxide production rate presented a linear correlation with the content of volatile solids in the biofilm, and with the difference of temperature between the packed bed and the environment. The carbon dioxide production rate was also identified as being an indicator of the energy generated during the VOC biodegradation.

In this research the sulphide to be removed was from liquid solution. The samples for sulphide observation were taken at the outlet of the column at various time and the process was done at 6 phases at different sulphide inlet concentration and debit (total operation time was 6 x 72 hours). Mathematical model proposed assumed no intra-film-gradient of sulphide concentration. A study on sulphide removal from gas phase at different conditions applying different mathematical model is presented by Lestari et al. [12].

## 2. Theoritical Development

### 2.1. Theoritical Development

Including figures and tables. SFS can be schematically shown in figure 1. The biofilm is attached on packing material and the liquid contain sulphide flows from the top of the column through outside the biofilm.

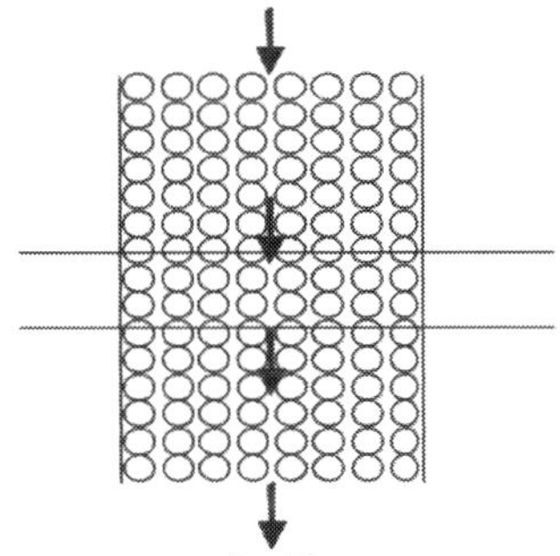


**Figure 1**. Schematic diagram of packed bed system

The detailed diagram of the mass transfer around the biofilm is taken from Lestari et.al. [12] and shown in figure 2. The sulphide move from the bulk of the liquid to the surface of the biofilm through the liquid film. Then the sulphide diffuses to the inner part of the biofilm. While diffusing part of the sulphide is consumed by the microorganism in the biofilm. So, the fundamental processes involved are the transfer of sulphide from liquid stream to the surface of the bio-film, sulphide diffusion in the bio-film, and bacterial action in the bio-film to degrade the sulphide compounds.

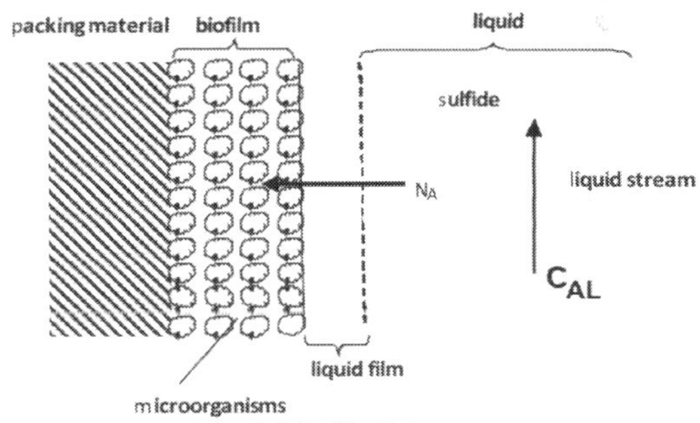


**Fig. 2**. Bio-film Scheme

The following kinetics model proposed is essentially similar to the model by Lestari et al. [13], in which the assumptions taken are as follows: 1. Sulphide concentration in the bio-filter for a given height is considered to be uniform because bio-film thickness ($\delta$) is relatively small, so the diffusion is relatively fast. 2

The microorganism concentration in the bio-film at certain axial position in the bed is also assumed to be uniform. Based on these assumptions, the concentration profile in the biofilm system can be schematically simplified as in Figure 3.

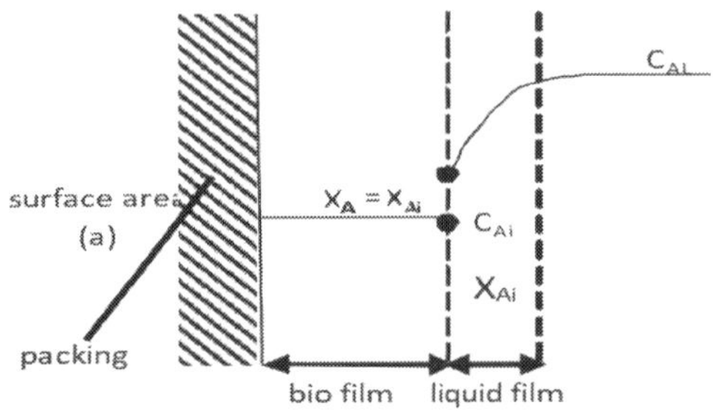


**Fig. 3.** Profile of concentration sulphide in the model

The mass balance of sulphide in liquid stream in the bed of thickness of $\Delta z$ as in Lestari et al. [14] based on figure 1 is as follows:

$$F_V \frac{\partial C_{AL}}{\partial z} + k_C(C_{AL} - C_{Ai})a_s S = -S\varepsilon \frac{\partial C_{AL}}{\partial t} \quad (1)$$

Similar assumption, pseudo steady-state as in Lestari et al. [13] is also taken. So in the liquid phase = 0, and equation (1) becomes:

$$\frac{F_V}{\varepsilon S}\frac{dC_{AL}}{dz} + \frac{k_C a_S}{\varepsilon}(C_{AL} - C_{Ai}) = 0 \quad (2)$$

$$\frac{dC_{AL}}{dz} = -\frac{k_C a_S S}{F_V}(C_{AL} - C_{Ai}) \quad (3)$$

The mass balance of sulphide in the bio-film at certain z position (figure 3) as in Lestari et al. [13] is as follows:

$$\frac{d\,X_A}{dt} = \frac{k_C}{\delta}(C_{AL} - C_{Ai}) - r_A \quad (4)$$

Because $C_{Ai}$ is in equilibrium with $X_{Ai}$, and the equilibrium is assumed to follow the model of distribution coefficient, the correlation between $C_{Ai}$ and $X_{Ai}$ is as the following:

$$C_{Ai} = K_A * X_{Ai} \quad (5)$$

Since the bio-film thickness is $\delta$, the rate of increase of the mass of microorganisms is [13]:

$$\frac{dm}{dt} = Y_{X/S}.r_A.a.\delta \quad (6)$$

where Y is contaminant yield coefficient (the ratio of mass of microorganism growth to mass of sulphide consumed). Since the mass of microorganism m = $\rho$ a $\delta$, where $\rho$ is mass of microorganism per volume of bio-film, equation (6) becomes:

$$\frac{d\delta}{dt} = \frac{Y_{X/S}}{\rho}.r_A.\delta \quad (7)$$

All in all, the mathematical model representing the sulphide removal from liquid using biofilm on packed bed of SFS is the set of equation of 2, 3, 4, 5 and 7. The model developed was solved by the finite difference approximation method. Experimental data will be applied to verify the accuracy of the mathematical model.

# 3 Methodology

## 3.1 Materials and Methods

### *3.1.1 Materials*

The material preparation was conducted according to Lestari et al. [15]. Salak fruit seeds were harvested from Sleman, Yogyakarta, Indonesia. The salak fruit seeds were washed and then dried in an oven at 55° C to eliminate all volatile impurities and the water. The microbes were isolate 12 which results of isolation from the sludge of the municipal waste water treatment plant in Srandakan, Bantul, Yogyakarta, Indonesia. For the growth of microbes, the sulphur oxidizing bacteria (SOB) medium of 0.4 g/l $NH_4Cl$, 0.2 g/l $MgCl .6H_2O$, 0.2 g/l $KH_2PO_4$, 0.2 g/l $K_2HPO_4$, 0.2 g/l yeast extract and 8 g/l $Na_2S_2O_3.5H_2O$ was added [15]. Medium and equipment were autoclaved and sterilized at 121°C, 1.2 atmospheric for 15 min.

### *3.1.2 Methods*

One ose of isolate was used as starter on 5 ml of medium SOB in 250 ml erlenmeyer flask, then was shaker at 180 rpm for overnight. The starter was added by 50 ml of medium SOB and was shaker at 180 rpm for overnight again, then added by 450 ml medium SOB and was shaker as before procedure. The method of bio-filter experiment was conducted according to Lestari et al. [15]. Isolate 12 was then immobilized on SFS. Approximately 2.5 kg of dried SFS and 3000 ml of SOB medium was mixed in a flask and was autoclaved. The mixture was then sterilized at 121°C and 1.2 atmospheric for 15 min. The sterilized mixture was then mixed with 300 ml of isolate 12, and was kept at 30° C at atmospheric pressure, for 6 days to let the biofilm grow on SFS surface. The SFS covered by biofilm was then used as packing material in the sulphide removal column. Prior to the continuous experiments, the SOB medium was added to the bio-filter to feed the microorganism.

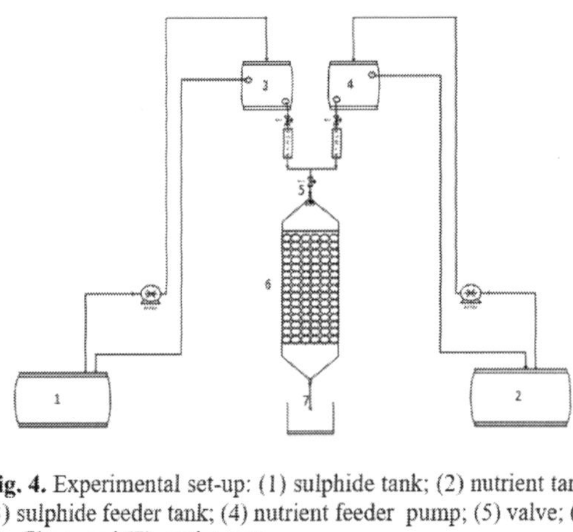


**Fig. 4.** Experimental set-up: (1) sulphide tank; (2) nutrient tank; (3) sulphide feeder tank; (4) nutrient feeder pump; (5) valve; (6) bio-filter, and (7) outlet port

**Table 1**. The operating conditions of the Bio-filter in each Phase

| Phase | Duration (hours) | Inlet Concentration (ppm) | Fow rate (mL/min) |
|---|---|---|---|
| I | 72 | 15 | 50 |
| II | 72 | 30 | 50 |
| III | 72 | 45 | 50 |
| IV | 74 | 60 | 50 |
| V | 82 | 60 | 40 |
| VI | 72 | 60 | 30 |

## 4 Results and Discussions

### 4.1 Performances of Bio-filter at Sequential Operations

The effects of operating conditions on outlet concentration and removal efficiency are shown in Fig 5. The time shown in Figure 5 is the operating time only, the switching time between phases is not counted. In phase 1 (acclimatization period), inlet concentration of $H_2S$ solution was 15 ppm and the flow rate was 50 mL min-1. It was observed that removal efficiency as 8.24 % at the end of 24th hour. It increased with time and reached 87.94% at the end of phase 1. The increase of sulphide removal by increasing time suggests that there is microbial growth during this period.

In phase 2, the inlet concentration of $H_2S$ solution was increased from 15 ppm to 30 ppm while the flow rate was kept to be constant of about 50 mL min-1. It was observed that by increasing inlet concentration, the removal efficiency decreased from 87.94% to 66.98%. This phenomena is conceivable since at the higher sulphide concentration the amount of sulphur to be removed was higher, so less portion of sulphide can be removed by the microbes. But by increasing time the microbes grew, so more sulphide can be removed. As a result, the removal efficiency increased by time and reached 95.63% at the end of phase 2. During phase 3, the inlet concentration was 45 ppm while the flow rate was 50 mL min-1. Similar phenomena were observed as in phase 2. The initial removal efficiency was 54.78% while the one at the end of this phase was 83.62%. During phase 4, the inlet concentration was increased to 60 ppm, but the flow rate was kept to be the same as in phase 3, which was 50 mL min-1. Again similar phenomena were observed in which the initial removal efficiency 53.23% while at the end of the phase, the removal efficiency was 70.06%. The effect of the increase of inlet concentration to the decrease of RE in early of each phases were similar to the study by Majumder et al. [16] that describes the Cu(II) removal using bio-filter. When the inlet concentration of Cu(II) was increased from 20 mg $L^{-1}$ to 27.5 mg $L^{-1}$, the RE was decreased from 97.5% to 64.7%.

In phase 5, the inlet concentration was kept to be the same as in phase 4, which was 60 ppm, while the flow rate was decreased to 40 mL $min^{-1}$. The removal efficiency at the initial phase was 63.58%, while the one at the end of phase was 87.46%. These data show that the decreased of removal efficiency at the time of period change from phase 4 to phase 5 is smaller than the ones before. This is logical since the flow rate at phase 5 is lower than the ones before, so the residence time in the column is longer. As a results the sulphide removal is more effective. In phase 6, the inlet concentration was 60 ppm while the flow rate was decreased to 30 mL min-1. The removal efficiency at the initial phase was 73.22%, while the one at the end of phase was 92.10%.

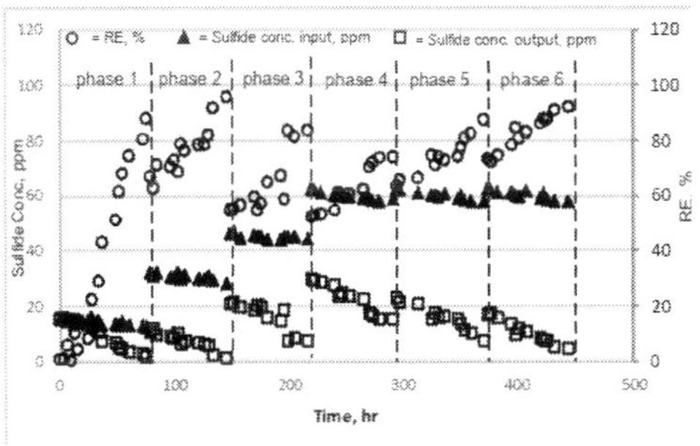


**Fig 5.** Performance of bio-filter at various flow rates and inlet concentrations

It was observed also that there is a drop in sulphide removal during every switching of operation. This may be caused by the shut-down of the bio-filter for approximately 3 hours, in which part of the microbes died so the removal capability becomes lower. It was also observed that drop of RE in the change of flow rate are relatively less than the one in change of inlet concentration. This phenomenon suggests that the disturbance caused by change of flow rate is less significant than the one caused by change of concentration. However as time progresses, the RE gradually increased as seem in figure 5. Figure 5 also shows that the lower the flow rate, the higher the RE would be. This result is similar to the research of Jaber et al. [8] which showed that decreasing the loading rate from 39.6 g $m^{-3}$ $h^{-1}$ to 9.9 g $m^{-3}$ $h^{-1}$ could increase RE from 45% to 100%.

### 4.2 Simulation Results

The comparison of simulation results and experimental data are shown in figure 6. The values of the parameters involved were obtained by curve fitting. During the entire operations the values of the parameters are constant, except for the first phase (acclimatization). The differences are due to the lower activity of the microbes during acclimatization.

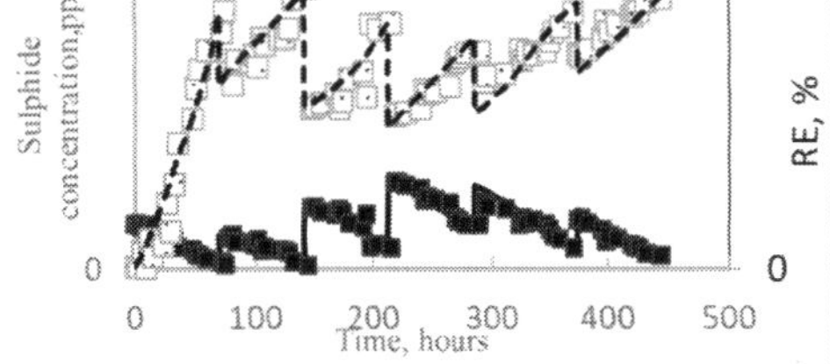


**Figure 6**. Concentrations and removal efficiency versus time from experiment and modelling

The thickness of the biofilm needs to be adjusted every switching of operations because of the death of part of the microbes during the shut-down period. Figure 6 shows that the model proposed can quantitatively describe the performances of the bio-filter.

Curve fitting resulted the appropriate values of the parameters for all phases as follows: $\mu_{max}$ = 0.0000000415 $s^{-1}$, α= 0.000000091 g $cm^3$, $k_C$ = 0.5 cm $s^{-1}$, $K_A$ = 0.007, and $Y_{X/S}$ = 30, except in early process, in which the value of $\mu_{max}$ is 0.00000004 $s^{-1}$, smaller than value of $\mu_{max}$ in the other phases. It is expected that the proposed model can be applied to quantitatively predict the performance of the sulphide removal at various condition (inlet sulphide concentration, flow rate, and removal efficiency).

Bio-filter column using the same packing material and microbes can also be applied to remove sulphide from gas phase [12]. The gas studied was biogas with sulphide content of 10 ppm to 79 ppm. The values of the parameters involved were found to be $\mu_{max}$ = 0.0000007 $s^{-1}$, $K_S$ = 0.0000039 g cm3, $k_G$ = 0.0086 cm $s^{-1}$, $H_S$ = 0.9 and $Y_{X/S}$ = 10. The comparison shows that the values of rate of maximum specific grow and half saturation constant for liquid phase are smaller than the ones of gas phase. However the values of mass transfer coefficient, Henry's constant and yield of microorganisms growth to mass of sulphide consumed for liquid phase were higher than the ones of the gas phase. Those suggest that the biofilm in gas environment is more active than the one in liquid environment (higher $\mu_{max}$). The mass transfer coefficient from the liquid to the biofilm is higher than the one of the gas. This is in accordance with mass transfer theory. The higher $K_A$ in the gas phase implies that sulphide in the gas is less absorbed in the biofilm than the sulphide in liquid phase.

### 4.3 Rate of degradation

The correlation between sulphide concentration in the biofilm and the rate of degradation modelled by Monod's equation using parameter obtained in this study are shown in figure 7. It is observed that the correlation is non linear. Hence, simplification of Monod equation to linear form is not recommended. Different conclusion was obtained by Ottengraf and Van den Oever [15] which reported that the correlation can be simplified to zero-order.

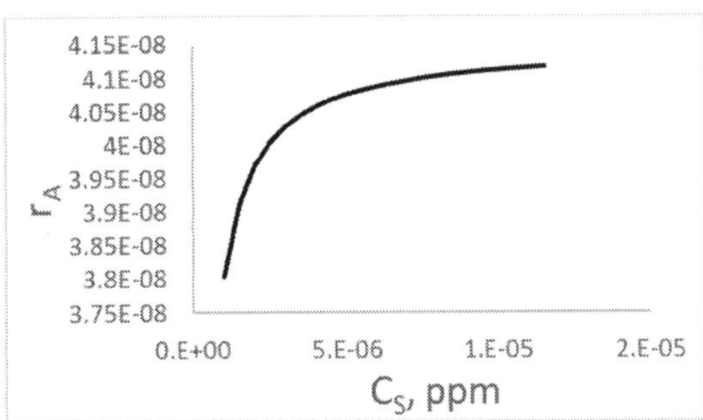


**Figure 7.** Effect of sulphide concentration in the biofilm to the rate of sulphide degradation by kinetic modelling

## 5 Conclusion

During every switching of operation, the sulphide removal efficiency drops. The drop of removal efficiency in the change of flow rate are relatively less than the one in change of inlet concentration. The highest removal efficiency is 92%, at the end of phase 6 at the inlet concentration of 60 ppm and the flow rate of 30 mL $min^{-1}$.

Monod's equation is recommended to applied for elimination of sulphide by biofilm on packed bed of salak fruit seeds. The model can quantitatively describe the performances of the bio-filter. Curve fitting resulted the appropriate values of the parameters for all phases as follows: μmax = 0.0000000415 $s^{-1}$, α= 0.000000091 g $cm^3$, $k_C$ = 0.5 cm $s^{-1}$, $K_A$ = 0.007, and $Y_{X/S}$ = 30, except in early process, in which the value of $\mu_{max}$ is 0.00000004 $s^{-1}$.

The value of $\mu_{max}$ at the first phase (acclimatization) is lower than the ones of the other phases because during the acclimatization, the performance of the microbes has not achieved the normal one. The values of $\mu_{max}$ and half saturation constant for liquid phase sulphide removal was found to be relatively lower than the ones of gas phase. Meanwhile, the values of mass transfer coefficient, Henry's constant and yield of microorganisms growth to mass of sulphide consumed for liquid phase were higher than the ones of the gas phase.

The authors would like to acknowledge to Directorate of Research and Technology of Higher Educations of Indonesia, for financial support of the research through research grant of Hibah Bersaing 2015.

## References

1. A.H. Nielsen, J. Vollertsen, H.S. Jensen, H.I. Madsen, H.J. Thorkild, Aerobic and Anaerobic Transformation of Sulfide in a Sewer System–Field Study and Model Simulation. Water Environment Foundation, 3653-3670 (2006).
2. M.T. Namini, S.M. Heydarian, B. Bonakdarpour, Removal of $H_2S$ from Synthetic Waste Gas Streams using a Biotrickling Filter, Iranian Chemical Engineering Journal, 5 (3) 40-51 (2007).
3. C. Rattanapan, P. Boonsawang, D. Kantachote, Removal of $H_2S$ in downflow GAC biofiltration using sulfide oxidizing bacteria from concentrated latex waste water, Bioresource Technology, 100 125-130 (2009).
4. M.C. Delhomenie, and M. Heitz, Bio-filtration of Air: A Review. Crit. Rev. Biotecchnol., 25 (1-2), 53-72 (2005).
5. M.E. Fischer, Biogas Purification: $H_2S$ Removal using Bio-filtration. M. A. Sc. thesis, Grad. Dept. of Chemical Engineering, University of Waterloo, Ontario, Canada (2010),
6. Y.C. Chung, C., Huang, C.P. Tseng, J.R. Pan, Microbial Oxidation of Hydrogen Sulphide with Bio-filter. J. Environ. Sci. Health Part A-Environ Sci. Eng. Toxic Hazard. Subst. Control, 31 (6) 1263-1278 (1996).

7. G. Spigno, M.Z.C. Nicolella, Mathematical Modelling and Simulation of Phenol Degradation in Biofilters, Biochemical Engineering Journal 19, 267–275 (2004).
8. M.B. Jaber, A. Couverta, A. Amrane, F.R.P.L Cloirec, E. Dumont, Biofiltration of $H_2S$ in Air-Experimental Comparisons of Original Packing Materials and Modeling, Biochemical Engineering Journal, 112, 153–160 (2016).
9. J.M. Santos, E.S. Lopes, N.C.R. Junior, L.M. Sa, N.J. Horan, Mathematical Modelling of Hydrogen Sulphide Emission and Removal in Aerobic Biofilters Comprising Chemical Oxidation, Water Research, 43, 3355-3364 (2009).
10. G.K. Agarwal, A.K. Ghoshal, Packed bed dynamics during microbial treatment of wastewater: Modelling and simulation, Bioresource Technology 99, 3765–3773 (2008).
11. A.A. Ramirez, S. Bénard, A.G. Fendler, J.P. Jones, M. Heitz, Kinetics of microbial growth and biodegradation of methanol and toluene in biofilters and an analysis of the energetic indicators, Journal of Biotechnology 138, 88–95 (2008).
12. R.A.S. Lestari, W.B. Sediawan, and Sarto, Biodegradation of Sulphide in Biogas by Biofilm on Salak Fruit Seeds: The Effect of Intra-film-gradient, 2nd International Conference on Engineering and Technology for Sustainable Development, Yogyakarta, 13-14 September (2017).
13. R.A.S. Lestari, W.B. Sediawan, S. Syamsiah, and Sarto, Hydrogen sulfide removal from biogas using a salak fruit seeds packed bed reactor with sulfur oxidizing bacteria, Environmental Chemical Engineering Journal, 4, 2370–2377 (2016).
14. R.A.S. Lestari, W.B. Sediawan, S. Syamsiah, Sarto, Kinetics of Sulfide Removal in Biofilter Employng Sulfur-Oxidizing Bacteria on Salak Fruit Seeds, 27th Symposium of Malaysian Chemical Engineers (SOMChE 2014) in conjunction with the 21st Regional Symposium on Chemical Engineering (RSCE 2014), 29-30 October 2014, Subang Jaya, Malaysia.
15. R.A.S. Lestari, W.B. Sediawan, S. Syamsiah, and Sarto, The Exploratory Study on Sulfide Removal Ability of Sulfur-Oxidizing Bacteria in Biofilm on Salak Fruit Seeds, International Proceeding of Chemical, Biological & Environmental Engineering, Chemical Engineering and Applications V, Taipei, Taiwan, 16-19 (2014).
16. S. Majumder, G. Gangadhar, S. Raghuvanshi, S. Gupta, Biofilter Column for Removal of Divalent Copper from Aqueous Solutions: Performance evaluation and kinetic modeling, Journal of Water Process Engineering, 6, 136–143 (2016).
17. S.P.P. Ottengraf, A.H.C. Van Den Oever, Kinetics of Organic Compound Removal from Waste Gases with a Biological Filter, Biotechnol. Bioeng. 25, 3089–3102 (1983).

# Preparation of porous carbon as ethylene adsorbent by pyrolysis of extraction waste *Mangosteen* rinds

*Nur Indah Fajar* Mukti [1,*], *Imam* Prasetyo [2], *Aswati* Mindaryani [2], *and Shofwatunnida'* Septarini [1]

[1]Department of Chemical engineering, Faculty of Industrial Technology, Universitas Islam Indonesia, Indonesia
[2]Department of Chemical engineering, Faculty of Engineering, Universitas Gadjah Mada, Indonesia

**Abstract.** *Mangosteen* rind is an important source of natural antioxidants. Due to the growing interest in extracting this anti cancer substances from the *mangosteen* rind, the amount of this lignocellulosic residu has been generated significantly as byproduct. In this research, extraction-waste *mangosteen* rind (EMP) was used as alternative precusor for production of carbon-based adsorbent for ethylene removal. Steam was used as activating agent and the effect of carbonization time and temperature on the development of pore structure were examined. Pyrolysis process was carried out by heating the *mangosteen* rinds powder (180 µm - 355 µm) from ambient temperature up to carbonization temperature of 848 K and kept for 3 hours then followed by heating up to 1123 K and kept for 15 minutes under flowing $N_2$ and steam. This process was repeated for several pyrolysis temperature (1053 K, 1073 K, 1083 K and 1103 K) and carbonization time (0 hours, 1 hour, 2 hours, and 3,5 hours). The carbon obtained was characterized in terms of its pore structure and ethylene uptake capacity. The results show that porous carbon obtained from pyrolysis of extraction-waste *mangosteen* rind can be characterized as mesoporous carbon. The highest surface area of 1080 $m^2/g$ was obtained from pyrolysis of extraction-waste *mangosteen* rinds with carbonization time of 3.5 hours and pyrolysis temperature of 1123 K. Furthermore, the mesopore portion and the specific surface area increased with the increasing carbonization time. From the ethylene uptake experiment, it was noted that the ethylene adsorption capacity of EMPC is 40.12 $cm^3/g$.

## 1 Introduction

Porous carbon with high surface area has been produced from a variety of carbonaceous materials. In practice, coal and lignocellulosic materials are often used as raw materials. The production of porous carbon from lignocellulosic material has been widely studied. There are the production of porous carbon from coconut shell, *mangosteen* rind, corn cob, bagasse, durian rind [2,3,8]. From the several previous studies, the *mangosteen* rinds has the potential as a raw material for the preparation of porous carbon with high surface area so that it can be used as an adsorbent with the large capacity [6,8].

*Mangosteen* fruit (Garcinia mangostana L.) is a fruit immensely produced in Indonesia. Every part of the fruits containts an antioxidants, especially on the rind. Antioxidant compounds in the form of xanthones in *mangosteen* rind can be obtained through extraction process. Extraction process will extract a compound to obtain xanthones and will leave a residue in the form of *mangosteen* rind which the antioxidant compounds has been extracted. Because of the increasing demand of *mangosteen* rind extraction recently, it will also increase the *mangosteen* rind residue leftover from the extraction process. Because of the potential of the *mangosteen* rind residual waste, the research will conduct the utilization of waste as a new potential that can be used as raw material for producing porous carbon to obtain economic value. Porous carbon produced will be used as an ethylene adsorbent.

The pore structure of the porous carbon was highly depended on a raw materials and production methods. Activation process was a very important step in the preparation of porous carbon which aims to increase the porosity of the carbon produced. The increasing porosity of the carbon, the specific surface area of the carbon will increase. The activation methods were used among others by using physical activation through a partial gasification with $CO_2$ and steam and chemical activation using chemicals such as $K_2CO_3$, KOH, $ZnCl_2$. Increased porosity with partial gasification consists of only one stage in which the pyrolysis beginning with carbonization process and proceed with partial gasification carried out in a single step [7]. While pyrolysis with the addition of chemicals was carried out in two steps. The first steps was the addition of chemicals and the next step was pyrolyzed the materials from ambient temperature to the desired temperature [1].

One of the methods in prolonging postharvest life of fruits is carried out by controlling ethylene concentration in the stored environments. It is necessary therefore to develop lower cost and more effective ethylene adsorption materials. This study was aimed to explore the feasibility of preparing ethylene adsorbent by pyrolysis of

* Corresponding author: nurindah_fm@uii.ac.id

extraction waste *mangosteen* rinds under different carbonization conditions. The carbon obtained was characterized in terms of its pore structure and ethylene uptake capacity.

## 2 Materials and methods

### 2.1 Material

Manggosteen rinds powder (180 μm – 355 μm) was obtained from Bina Agro Mandiri, Bantul, D.I. Yogyakarta. Nitrogen with a purity of 99.95% was used as an inert gas in the carbonization process, and to determine the $N_2$ adsorption-desorption at 77 K was used $N_2$ with ultra high purity (99.9995) as adsorbate. Ethylene with a purity of 99.9% was obtained from PT. Aneka Gas Industri as adsorbate.

### 2.2 Experimental

#### *2.2.1 Raw materials characterization*

*Mangosteen* rinds were characterized by lignocellulose content analysis using Chesson Method (Datta, 1981). Proximate analysis carried out by GB/T12496. Thermal Gravimetric Analysis (TGA) using Diamond TG/DTA Analyzer, PerkinElmer Instruments was carried out by heating a sample of ± 7 mg from 303 K up to 1123 K with a heating rate of 5 K/min under flowing nitrogen of 20 mL/min.

#### *2.2.2 Preparation of porous carbon*

Porous carbon for ethylene adsorbent in this study was prepared by pyrolysis of extraction waste *mangosteen* rinds (EMP). The carbon obtained from the former precursor is noted as EMPC. Furthermore EMPCT1 (1053 K), EMPCT2 (1073 K), EMPCT3 (1083 K), EMPCT4 (1103 K) and EMPCT5 (1123 K) were adjusted for variations in the temperature of pyrolysis and EMPCH1 (0 hour), EMPCH2 (1 hour), EMPCH3 (2 hours), EMPCH4 (3 hours) and EMPCH5 (3.5 hours) were adjusted for carbonization time. Precusors were pyrolized in furnace under flowing N2 from ambient temperature up to carbonization temperature of 848 K and kept for 3 hours, then followed by heating up to 1123 K and kept for 15 minutes under flowing N2 and steam. This process was repeated for several pyrolysis temperature (1053 K, 1073 K, 1083 K and 1103 K) and carbonization time (0 hours, 1 hour, 2 hours, and 3.5 hours).

#### *2.2.3 Porous carbon characterization*

$N_2$ adsorption-desorption obtained using NOVA 2000 adsorption analyzer (Quantachrome) at 77 K. Spesific surface area (S) calculated from isotherms data on the range of relative pressure (P/Po) between 0.05 to 0.30. Total pore volume(V) estimated from the amount of adsorbate adsorbed at P/Po of 0.99 and a mesopore area (Smes), mesopore volume (Vmes) determined using - plot method. Pore size distribution studied using Horvath-Kawazoe method (HK) and Barret-Joyner-Haleda method(BJH).

#### *2.2.4 Ethylene Adsorption*

Ethylene Adsorption capacity test were carried out using the volumetric method at temperature of 303 K. The ultra high vacuum adsorption measurement apartus rig was constructed by using Swagelok® VCR valve and fitting. The schematic diagram of adsorption system can be seen in Fig. 1. All carbon samples were first degassed at 423 K for 6 hours until the static pressure in the system was at least 1x10-2 Torr. The data obtained is presented as an adsorption isotherm curve. Ethylene adsorption analysis was began with ensuring all the valve in the closed state (V1, V2 and V3). To fill the loading cell, valve V1 was opened slowly to achieve the desired pressure and recorded as P1, then closed again. Furthermore, to make the process of adsorption, valve no. V2 was opened and observed a pressure decrease in the pressure gauge until the pressure in equilibrium obtained and recorded as P2. Adsorption equilibrium was indicated by stable pressure in the adsorption system. To proceed to the next equilibrium point, valve V2 was closed again and the steps were repeated until the maximum pressure that was able to operate (± 1.2 atm).

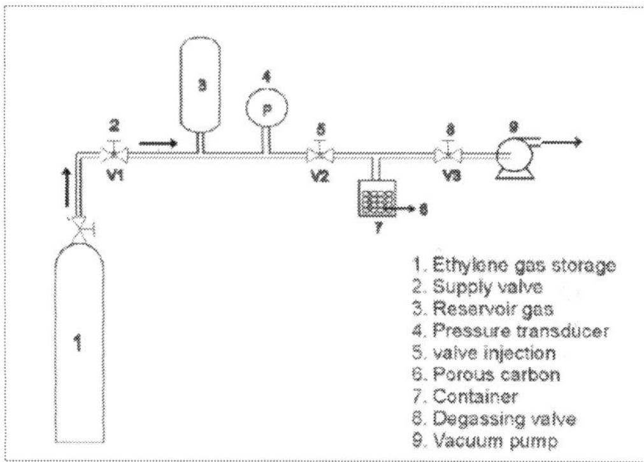


**Fig. 1.** Schematic Diagram of The Adsorpstion System by Static Volumetric

## 3 Results and Discussion

### 3.1 Raw materials characterization

#### *3.1.1 Proximate Analysis and Lignocellulose properties*

The results of lignocellulose analysis and proximate analysis are shown in Table 1. From the results show the composition of lignocellulosic of extraction waste *mangosteen* rind is mostly lignin of 48.52% and the proximate analysis shows that the maximum amount of carbon that can be obtained from *mangosteen* rind is 32.29%.

**Table 1.** Proximate Analysis and Lignocellulose Analysis

| Proximate Analysis | |
|---|---|
| Component | Composition, % |
| Water | 3.58 |
| Ash | 1.85 |
| Volatile matter | 62.28 |
| Fixed carbon | 32.29 |
| Lignocellulose Analysis | |
| Component | Composition, % |
| Cellulose | 26.22 |
| Hemicellulose | 15.39 |
| Lignin | 48.52 |

### *3.1.2 Thermal Gravimetric Analysis (TGA)*

TGA was carried out by heating a sample from 303 K up to 1123 K with a heating rate of 5 K/min under flowing nitrogen. The results of TGA are shown in Fig. 2 . TGA analysis result shows the carbonization process of *mangosteen* rind was divided into several steps. The first step at temperatures <150°C, the weight loss due to evaporation of free and bound water. The second step at temperature of 200–375°C, the weight loss due to the decomposition of cellulose and hemicellulose into condensable gases (methanol, tar and acetic acid) and uncondensable gases ($CO_2$, $CH_4$, etc). Furthermore, at a temperature of 375-575°C, the weight loss due to the decomposition of lignin. At temperatures > 575°C, the weight loss was not significantly . The temperature of 575°C (848 K) is taken as temperature carbonization.

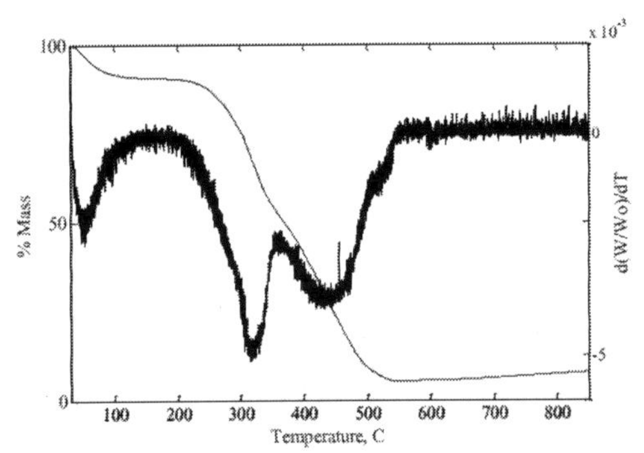


**Fig. 2.** TGA Curves of *Mangosteen* Rinds

## 3.2 Porous carbon characterization

### *3.2.1 Effect of pyrolysis temperature*

Fig. 3(a) shows burn off number from porous carbon at various pyrolysis temperatures. Burn off can be used for quality criteria of carbon. If thermal degradation generates porous on material, when the burn off is high, it will increase the number of porous and internal surface area. Fig. 3(a) shows burn off number increasing by increase pyrolysis temperature.

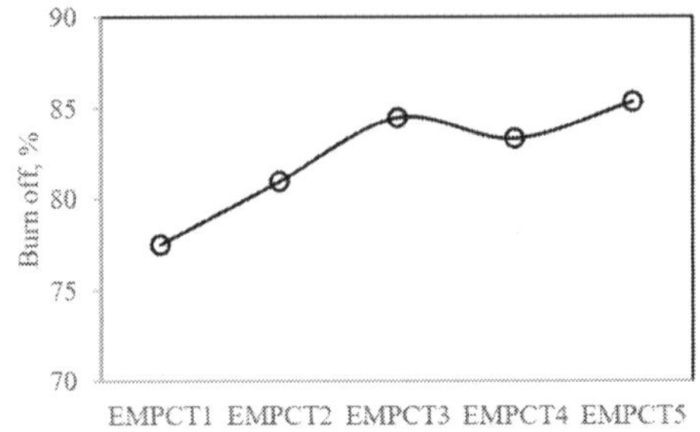


(a)

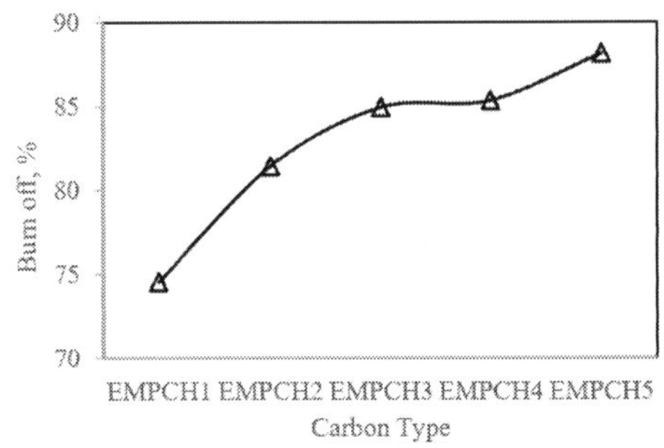


(b)

**Fig 3.** Burn off (a) EMPC at various T and t = 3 hours; (b) EMPC at various t and T= 1123 K

Fig. 3(b) shows number of burn off increasing when time of carbonization increased. Table 2 shows the pore structure of the carbon EMPC at various pyrolysis temperature. The Table 2 shows that the highest surface area is 1066.77 $m^2/g$ obtained from carbon obtained by pyrolysis of extraction waste *mangosteen* rinds up to 1123 K and carbonization time for 3 hours. Table 2 also shows that the specific surface area increased with the increasing pyrolysis temperature. Because of the formation of pore structure by $N_2$ gas or superheated $H_2O$ will be growing at high temperature. Table 2 also provides information that the resulting pore volume also increased with higher pyrolysis temperature. The similar thing is seen for mesoporous volume. Wherein the resulting mesoporous volume increases with the increasing pyrolysis temperature.

Pore size distribution for carbon EMPC at various temperatures are shown in Fig. 4 (a). In Fig. 4(a) can be seen that for all carbon EMPCT1 until EMPCT5 have tendency of similar pore size distribution with the peak between 2-100 nm and have not any significant change. However, it can be seen that the pore size distribution is widened with increasing pyrolysis temperature. This is caused by the higher the pyrolysis temperature made micropore size widening thus increasing the amount of mesopore size. Therefore, it can be concluded that the portion of mesopore size also increase with the increasing temperature.

**Table 2.** Characterization of EMPC at Various T and t = 3 h

| Sample | S, $m^2/g$ | $S_{mes}$, $m^2/g$ | % $S_{mes}$ | V, $cm^3/g$ | $V_{mes}$, $cm^3/g$ | % $V_{mes}$ | $D_{avg}$, nm |
|---|---|---|---|---|---|---|---|
| EMPCT1 | 657.61 | 565.96 | 86.06 | 1.115 | 0.835 | 74.89 | 6.78 |
| EMPCT2 | 839.31 | 527.32 | 62.83 | 1.051 | 0.749 | 71.27 | 6.62 |
| EMPCT3 | 979.55 | 543.35 | 55.47 | 1.172 | 0.969 | 82.68 | 4.78 |
| EMPCT4 | 1032.71 | 562.97 | 54.51 | 1.197 | 0.978 | 81.70 | 4.64 |
| EMPCT5 | 1066.77 | 603.61 | 56.58 | 1.276 | 1.069 | 83.78 | 4.79 |

**Table 3.** Characterization of EMPC at Various t and T=1123K

| Sample | S, $m^2/g$ | $S_{mes}$, $m^2/g$ | % $S_{mes}$ | V, $cm^3/g$ | $V_{mes}$, $cm^3/g$ | % $V_{mes}$ | $D_{avg}$, nm |
|---|---|---|---|---|---|---|---|
| EMPCH1 | 697.89 | 527.32 | 75.56 | 1.051 | 0.749 | 71.27 | 6.03 |
| EMPCH2 | 951.12 | 556.70 | 58.53 | 1.201 | 1.030 | 85.76 | 5.05 |
| EMPCH3 | 1018.57 | 662.88 | 65.08 | 1.230 | 1.069 | 86.91 | 5.32 |
| EMPCH4 | 1066.77 | 603.61 | 56.58 | 1.276 | 1.069 | 83.78 | 4.79 |
| EMPCH5 | 1079.94 | 762.06 | 70.57 | 1.528 | 1.389 | 90.90 | 5.66 |

### *3.2.2 Effect of carbonization time*

Carbonization process of natural materials is affected by the decomposition process of the constituent material composition such as cellulose, hemicellulose and lignin. Therefore, if the cellulose, hemicellulose and lignin of the material decomposed perfectly, the quality of the carbon produced will be better characterized by the increasing on specific area of the carbon. To enhance the decomposition process of cellulose, hemicellulose and lignin from EMPC, the process on hold at the final temperature decomposition of lignin (848 K). Carbonization time variation was performed to determine the effect of the produced carbon character.

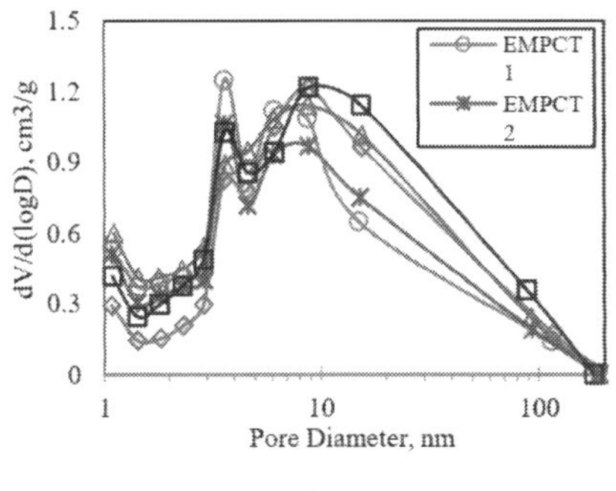


(a)

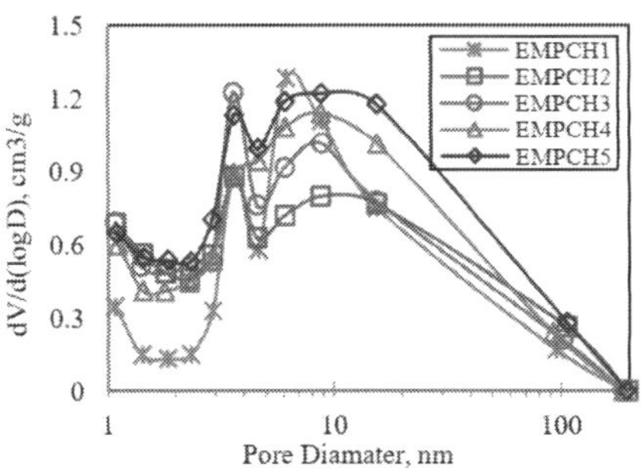


(b)

**Fig 4.** Pore Size Distribution BJH Method (a) EMPC at Various T and t = 3 Hours; (b) EMPC at Various t and T=1123 K

Table 3 shows the pore structure of the carbon EMPC at various carbonization time. The Table 3 provide information that the largest surface area is 1079.94 $m^2$/gram produced from carbon derived from extraction waste *mangosteen* rinds with carbonization time for 3.5 hours and a final pyrolysis temperature of 1123 K. From these data, it is known that the specific surface area of the carbon produced will increase with the longer time of carbonization. It is due to the cellulose, hemicellulose and lignin from extraction waste *mangosteen* rinds will decompose perfectly so that the porosity of the carbon produced will also increase. Table 3 also informs that the longer carbonization time, the pore volume generated will also increase. Similarly, for mesoporous volume, the mesoporous volume will increase with the longer carbonization time. From these results, it can be stated that the structure of mesoporous carbon increase with the longer carbonization time.

**Fig 5.** SEM Analysis for EMPC

Pore size distribution of carbon EMPC at various time can be seen in Fig. 4 (b). Fig. 4 (b) shows that all carbon EMPCH1 until EMPCH5 have the same pore size distribution with peaks between 2-100 nm. Pore size distribution was also widened with increasing carbonization time. This is caused by the higher the pyrolysis temperature made the micropore size widening thus it is increasing the amount of mesopore size. It can be concluded that the portion of mesopore size increase with the longer of carbonization time.

* Corresponding author: nurindah_fm@uii.ac.id

The morphology of carbon analyzed by Scanning Electron Microscope Analysis (SEM). Fig. 5 shows the result of SEM analysis for EMPC.

### 3.3 Ethylene Adsorption

Calculation of the ethylene adsorption capacity carried out by volumetric methods. The results of adsorption experiments evaluated using adsorption isotherm models. Some adsorption isotherm models used are (Do, 1998):

Langmuir equation: $$C_\mu = C_{\mu s}\left(\frac{bP}{1+bP}\right) \quad (1)$$

Freundlich equation: $$C_\mu = KP^{1/n} \quad (2)$$

Unilan equation: $$C_\mu = \frac{C_{\mu s}}{2s}\ln\left(\frac{1+be^{s}P}{1+be^{-s}P}\right) \quad (3)$$

Sips equation: $$C_\mu = C_{\mu s}\frac{(bP)^{1/n}}{1+(bP)^{1/n}} \quad (4)$$

Toth equation: $$C_\mu = C_{\mu s}\frac{bP}{\left[1+(bP)^{t}\right]^{1/t}} \quad (5)$$

**Table 4.** Parameter Value of Ethylene Adsorption Isotherm on EMPC at T = 303 K

| Adsorption isotherm parameter | Carbon Type EMPC |
|---|---|
| **Langmuir** | |
| $C_{\mu s}$ | 3.1808 |
| b | 1.1998 |
| **Freundlich** | |
| K | 1.6196 |
| n | 1.8361 |
| **Toth** | |
| $C_{\mu s}$ | 14.4335 |
| b | 0.5834 |
| t | 0.3805 |
| **Sips** | |
| $C_{\mu s}$ | 8.7359 |
| b | 0.1479 |
| n | 1.3929 |
| **Unilan** | |
| $C_{\mu s}$ | 3.1809 |
| b | 1.1998 |
| s | 0.0143 |

The calculation result of the parameter values on various equilibrium equation can be seen in Table 4. The result of adsorption ethylene capacity for KSME carbon can be find on Fig 6. Fig 6 shows the ethylene adsorption capacity of EMPC is 1,792 mmol/g or 40.12 $cm^3/g$.

## 4 Conclusions

The feasibility of preparing carbon-based adsorbent for ethylene removal from extraction waste *mangosteen* rind were examined. In this study, porous carbon obtained from pyrolysis of extraction-waste *mangosteen* rind (EMPC) can be characterized as mesoporous carbon. Carbon-based ethylene adsorbent with highest surface area of 1080 $m^2/g$ was obtained from pyrolysis of extraction waste *mangosteen* rinds with carbonization time of 3.5 hours and pyrolysis temperature of 1123 K. Mesopore portion and the specific surface area increased with the increasing pyrolysis temperature. In addition, the mesopore portion and the specific surface area increased with the increasing carbonization time. The ethylene adsorption capacity of EMPC is 40.12 $cm^3/g$.

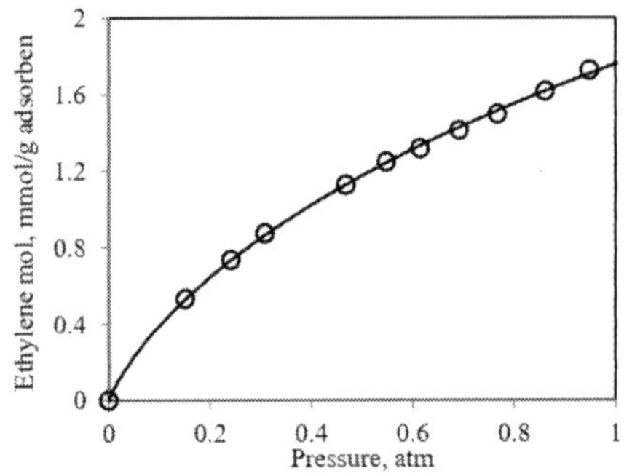


**Fig 6.** Ethylene Adsorption Isotherm on EMPC at T=303 K

## References

1. A. Ahmadpour, D.D. Do, The Preparation of Activated Carbon from Macadamia Nutshell by Chemical Activation. Carbon. **35**, 12, 1723-1732 (1997).
2. A. Okhovat, A. Ahmadpour, F. Ahmadpour, Z.K. Yadegar, Pore Size Distribution Analysis of Coal-Based Activated Carbons: Investigating the Effects of Activating Agent and Chemical Ratio. International Scholary Research Network, **2012**, 1-10 (2012).
3. A.S. Bujang, Utilization of Durian Biomass of Biorenewable Aplications. Graduate Theses and Disertations, Iowa State University (2014).
4. D. D. Do, Adsorption analysis : Equilibria and Kinetics. Imperial College, London (1998).
5. E. P. Barret, L. G. Joyner, P. P. Halenda, The Determination of Pore Volume and Area Distributions in Porous Substances: I. Computations from Nitrogen Isotherms, **73**, 373-380 (1951).
6. M.A. Ahmad, R. Alrozi, Optimization of Preparation Conditions for *Mangosteen* Peels Based Activated Carbon for The Removal of Remazol Brilliant Blue R Using Response Surface Methodology, Chem.Eng.Journal, **165**, 883-890 (2010).
7. S.C. Kim, I.K. Hong, Pore Development of the Activated Carbon Prepared by Steam Activation Process. Journal of Industrial and Engineering Chemistry, **4**, 3, 177-184 (1998).
8. Y. Chen, B.Huang, M. Huang, B. Cai, On the Preparation and Characterization of Activated Carbon from *Mangosteen* Shell, Journal of the Taiwan Institute of Chemical Engineers, **42**, 837-842 (2011).

# A kinetics study of acetic acid on cobalt leaching of spent LIBs: Shrinking Core Model

*Hendrik* Setiawan, *Himawan* T.B.M. Petrus, and *Indra* Perdana*

Center of Advanced Material and Mineral Processing, Department of Chemical Engineering, Universitas Gadjah Mada, Jl. Grafika No.2, Yogyakarta 55281, INDONESIA

**Abstract.** Lithium-ion batteries (LIBs) are secondary rechargeable power sources which increasing production also leads to large amount of waste. In order to environmentally friendly reduce the waste, this work aimed to use acetic acid as a substitute leaching agent to leach Co metals which constitutes about 72.39% wt of the battery cathode. The leaching process was done in a three-necked-flask where calcined LIB cathode powder was mixed with acetic acid solution. The variables of the leaching process under investigation were solution pH, concentration of $H_2O_2$ in the solution, S/L ratio, temperature and reaction time. Experimental results showed that only temperature significantly influenced the leaching rate of Co. Since the process was exothermic, the maximum recovery decreased as temperature increased. Conventional shrinking core model that considers diffusion and irreversible surface reaction resistances was found not sufficient to predict the kinetics of the Co leaching with acetic acid. A more representative kinetics model that considers a reversible reaction of Co complex formation needs to be further developed.

## 1 Introduction

Lithium-ion batteries (LIBs) are secondary rechargeable power sources dominantly used to power portable gadgets. This kind of batteries are mostly used in laptops, toys, handsets and many others. Nowadays, the usage of LIBs are being shifted to power more complicated and power-consuming electronics such are electric car and medical equipment [1]. With its high demand, the increasing waste of LIBs presents dire harm to environment in the form of its containing heavy metals and electrolyte. On the other side, these harmful materials can be recycled into valuable materials [2].

There are many methods in the recycling of LIBs, which are pyro-metallurgy, bio-metallurgy and hydrometallurgy. Hydrometallurgy method is the most used method now because of the high concentration of recycled metal, low energy cost and low emission [3]. The most important step in the hydrometallurgy itself is the leaching process. Leaching is a process in which a desired metal is extracted into a body of solution with the use of a certain solvent such as acid. The leaching process is usually assisted by the use of reducing agent which reduces metal ions to a more soluble state. Solute metal will then be purified and precipitated using electrochemistry, multi-step precipitation or solvent extraction [4].

Though the researches in perfecting the leaching process of LIBs have been widely conducted, the main problem present is the leaching agent. Commonly used leaching agents are inorganic acids such as HCl, $H_2SO_4$ and $HNO_3$ which were although efficient, yield side products such as $Cl_2$, $SO_3$ and $NO_x$ gases that can harm the environment [5]. In order to introduce a more environmentally friendly process, scientists have developed alternative process using organic acids such as citric acid [6] and oxalic acid [7]. The experimental results using organic acids showed promising recovery but the cost of recycle is still higher than that of inorganic acids.

In order to lower the cost from organic acid use, the present work proposed the use of acetic acid as a cheaper substitute. Acetic acid itself has been produced commercially for thousands of years and the synthesis process has been refined over and over in order to lower the price [8]. The potential of acetic acid as leaching agent has already been investigated. In the leaching of heavy metal contaminant from waste, acetic acid showed considerable recovery in a broader range of pH than other inorganic and organic acids. [9]. However, for LIBs there is still no research that specifically uses acetic acid.

In order to optimize the condition of the leaching process for scaling up, there is a need to study the kinetics and construct a mathematical model that can sufficiently predict those condition. Leaching process itself is a heterogeneous solid-liquid reaction which basic model has been described Yagi and Kunii as shrinking core model [10]. This model although accurate for some conditions need honing for various leaching process with different material and condition. The shrinking core model has been widely used in the scope of leaching research, for example in the leaching of gold ore with varying concentration and temperature [11], activation energy study of manganese ore leaching [12] and uranium carbonate reactor leaching [13]. In the present work, this

* Corresponding author: iperdana@ugm.ac.id

model will be used for acetic acid leaching of LIBs and then evaluated for its accuracy.

## 2 Mathematical Model

The leaching of Co from Li-Ion battery cathode powder proceeds through reaction between Co compounds with acetic acid that is hypothesized following mechanism described in eq. (1) – (3) [14].

$$LiCoO_{2(s)} \rightarrow Li^{+}_{(aq)} + Co^{3+}_{(aq)} + 2O^{2-}_{(aq)} \quad (1)$$

$$Co^{3+}_{(aq)} + 2H_2O_{2(aq)} + e^{-} \rightarrow Co^{2+}_{(aq)} + 2H_2O_{(l)} + O_{2(g)} \quad (2)$$

$$Co^{2+}_{(aq)} + 2CH_3COO^{-}_{(aq)} \rightarrow [Co(CH_3COO)_2]_{(aq)} \quad (3)$$

Solid-liquid reaction is following shrinking core model where the reaction initially occurred at the outer skin of the particle. Reaction zone then shifted deeper in the solid and left completely reacted material and inert solid ("ash") behind. This process would continue overtime until the shrinking unreacted core is depleted. Basically, the process consists of five steps:

a) Diffusion of acid reactant through film surrounding the particle to outer surface of particle.
b) Diffusion of acid reactant through ash layer to the surface of unreacted core.
c) Surface reaction between acid reactant and solid reactant on unreacted core in the reaction zone.
d) Diffusion of product through ash layer to the surface of particle.
e) Diffusion of product through film surrounding the particle to the solution.

For irreversible reaction, step d and step e usually do not contribute to the resistance of reaction and thus would not be the rate-controlling step. Therefore, the proposed model would only need to emulate step a to step c in the model as follows:

- Diffusion of acid reactant (A) through film surrounding the particle

$$-\frac{dN_A}{dt} = k_L 4\pi R^2 (C_{AL} - C_{ALs}) \quad (4)$$

- Diffusion of reactant through ash layer

$$-\frac{dN_A}{dt} = \frac{4\pi D_e}{\left(\frac{1}{r_c} - \frac{1}{R}\right)}(C_{ALS} - C_{AS}) \quad (5)$$

- Surface reaction

$$-\frac{dN_A}{dt} = k_r 4\pi r_c^2 C_{AS} \quad (6)$$

From the equation above, the rate of leaching of B (Co) could be derived as follow:

$$-\frac{dN_B}{dt} = \frac{C_{AL}/2}{\frac{1}{k_L 4\pi R^2} + \frac{1/r_c - 1/R}{4\pi D_e} + \frac{1}{k_r 4\pi r_c^2}} \quad (7)$$

where $-\frac{dN_B}{dt}$ =Total rate of decrease of Co (B) (mole s$^{-1}$)

$C_{AL}$ = Concentration of acetic acid (A) in liquid body (mole A/m$^3$ of liquid)

$C_{ALS}$ = Concentration of acetic acid (A) in ash layer (mole A/m$^3$ of liquid)

$C_{AS}$ = Concentration of acetic acid (A) on unreacted core surface (mole A/m$^3$ of liquid)

$k_L$ = Mass transfer coefficient through liquid film (m s$^{-1}$)

$D_e$ = Effective diffusion coefficient (m2 s$^{-1}$)

$k_r$ = Reaction rate coefficient (s$^{-1}$)

$R$ = Outer radius of particle (m)

$r_C$ = Radius of unreacted core (m)

In this process of leaching of LIBs using acetic acid, the effect of diffusion of acid reactant and products through film surrounding the particle (step a) were negligible. The usual particle size of cathode powder from LIBs are about 0.5-2.0 micron [15], this combined with adequate agitation would result in the resistance caused by film to become negligible.

The pre-treatment of the cathode of LIBs have already eliminate any inert materials that has the potential of forming an ash "layer" surrounding the particle. Thus the resistance that is caused by diffusion through ash layer is negligible. The radius of unreacted core can be expressed in term of conversion (XB) as follows:

$$r_c = R(1 - X_B)^{1/3} \quad (8)$$

Thus, the equation in favor of conversion ($X_B$) is:

$$\frac{dX_B}{dt} = \frac{C_{AL}}{C_{B0}} 2\pi R^2 k_r (1 - X_B)^{2/3} \quad (9)$$

Where $\frac{dX_B}{dt}$ is the total conversion rate of Co (B) and $X_B$ is the conversion of Co (B).

## 3 Material and method

### 3.1. Materials and reagents

Spent laptop LIBs were collected from local vendors and treated with saline solution for 2 hours to remove the residual charge. They were then dismantled from its plastic and steel shell using plier. Anode and the cathode of the batteries were then separated after the batteries coil was cut using knife. Metal-rich material from the cathode was scratched from the aluminium foil, powdered using spatula and calcined at 700°C for 4 hours at a heating rate of 5°C/ minute to remove the residual carbon. The powder was then analyzed for its metal content.

### 3.2. Metal Leaching

The leaching process used 0.5 gram of calcined cathode powder in 50 ml of 2.8 M acetic acid solution 2% v/v

$H_2O_2$ (pH=2). Experiments were done in a flask equipped with magnetic stirrer that runs at 400 rpm. Samples were taken after 5, 10, 15, 30, 60, and 120 minutes of experiment. The pH of acid solution, temperature, $H_2O_2$ concentration and S/L ratio of the leaching process were varied.

### 3.2. Elemental analysis

For analysis, 0.25 ml of the leachates were taken and diluted to 25 ml using distilled water. The resulting solution were then filtered using 0.22 µm membrane filter to separate possible suspended solid. The filtrate was then analyzed using ICP-AES (Optima 8300, Perkin Elmer, USA).

## 4 Result and discussion

### 4.1. Characterization of Cathode of LIB

Characterization of metal content in the cathode of LIBs was done using XRF analysis. The result of the analysis is shown in Table 1. It is shown that Co is dominating the cathode of LIB's elements used in the present work. This result is in line with characterization done by Xu et.al. [16] whose analysis showed that Co content was about 50 percent. This result also shows that the Co recovery is vital for LIB recycling process.

**Table 1**. Analysis Result of LIB's Cathode Content

| Metal | Content, ppm | Content, % wt |
|---|---|---|
| Co | 70.1920 | 72.39 |
| Li | 9.3957 | 8.59 |
| Mn | 6.9090 | 6.32 |
| Al | 5.2350 | 4.79 |
| Ni | 8.6630 | 7.92 |

### 4.2. Leaching Result

Experiment results showed that the rate of leaching is only affected by temperature. Acidity of solution (pH), concentration of $H_2O_2$ and S/L ratio of reactant mainly affect the resulting maximum recovery of Co. The result of Co recovery at varied variables is shown in Fig. 1.

Fig. 1 shows that with increasing temperature, the rate of which Co was recovered increases from 30 minutes at 50°C, 15 minutes at 70°C to 5 minutes at 90°C. The maximum recovery of Co was shown to be decreasing with increasing temperature from 37.5% at 50°C to 35.2 at 70°C and 33.8% at 90°C. This phenomenon shows that the leaching reaction of Co is an exothermic reaction.

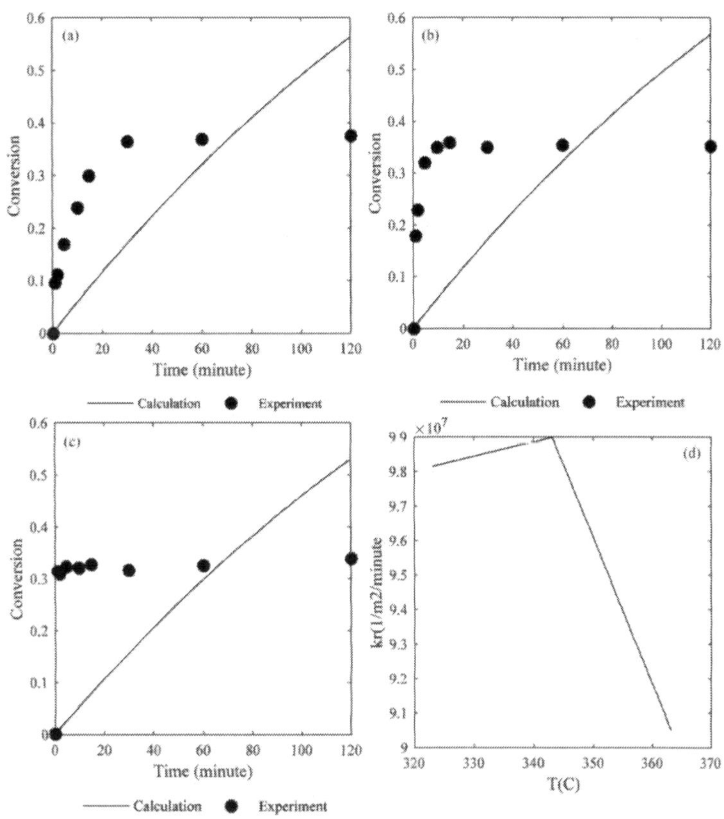


**Fig. 1.** Recovery Rate of Co at Various Factor of: (a) pH; (b) Concentration of $H_2O_2$; (c) S/L ratio and (d) Temperature

### 4.2. Calculation Result

The experimental results were used to evaluate the shrinking core model proposed in the present work as written in eq. (9). Calculation results were compared with experimental data by means of minimizing the sum of squares of errors between the data and results from calculation. As temperature was the only variable that affected the rate of recovery, the calculation was done for every temperature that was varied. The calculation results were shown in Fig. 2. Meanwhile, parameters of the proposed model are shown in Table 2.

**Table 2.** Calculation Result of Leaching Rate Parameter

| Temp. (°C) | A | $E_A$ (J/mole) | $k_f$ (1/m$^2$.minute) |
|---|---|---|---|
| 50 | | | 9,8151x10$^7$ |
| 70 | 4,7026x10$^7$ | -2,0257x10$^3$ | 9,8984x10$^7$ |
| 90 | | | 9,0510x10$^7$ |

Calculation result shows that the shrinking core model proposed in the present work failed to approach the experimental results. The calculation results are significantly far-off from the experimental data. As shown in Fig. 2(a), 2(b) and 2(c), with increasing temperature, the degree of inaccuracy seems to increase.

As depicted in Fig. 2, experimental data shows that there is a maximum limit of recovery at each temperature. This indicates that the leaching process might reach an equilibrium condition. However, this phenomenon would not be able to simulate with the use of the conventional shrinking core model. As shown in Fig. 2, the calculation results shows that the calculated recovery continuously increases with time and the value seems able to reach unity at prolong time. The leaching process of Co in this work seems to involve a reversible reaction that would limit the maximum recovery which value is depending on process condition such as temperature. A conventional shrinking model which considers diffusion resistance and irreversible surface reaction would be unable to predict the kinetics of leaching process studied in the present work.

Further proof that the leaching process of Co is possibly an equilibrium reaction is given by van Niekerk [17] and Donaldson [14] who stated that the reaction of Co and acetic acid is not a simple ionic salt formation but rather a complexation reaction which has an equilibrium limit. This results conclude that in order to more accurately model the kinetics of leaching process of Co there is a need to construct a model based on reversible reaction of Co as rate-controlling step. The conclusion that conventional shrinking core model need to be modified for different type of ore was also achieved in the previous research such as for the leaching of gold [11], manganese [12] and nickel [18], [19].

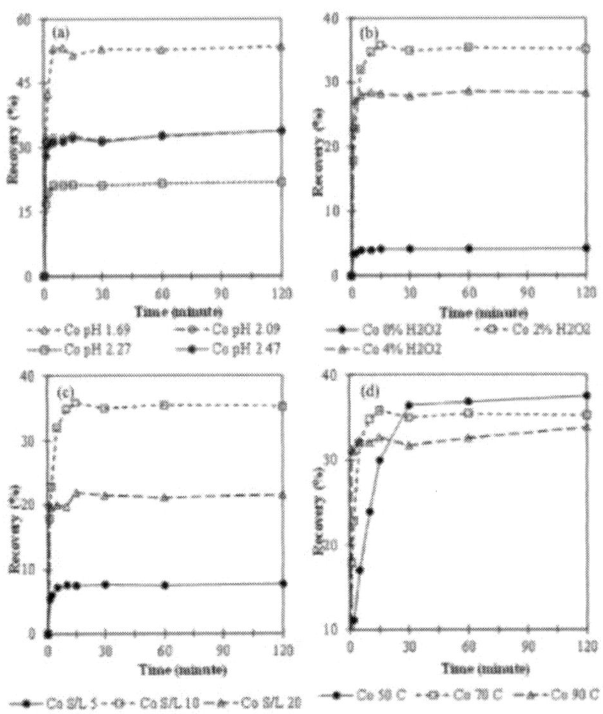


**Fig. 2.** Shrinking Core Model Calculation Result of Leaching Rate of Co for: (a) 50°C; (b) 70°C; (c) 90°C and (d) kr vs T
(b) 70°C; (c) 90°C and (d) kr vs T

## 5 Conclusion

The experimental result showed that the leaching process of Co with acetic acid is possibly exothermic as the maximum recovery of Co decreases with increasing temperature. To maximize the recovery of Co further research with lower temperature is needed but keeping in mind that lowering the temperature will also results in a lower reaction rate.

Conventional shrinking core model is unable to sufficiently approach the experimental data of the Co leaching which calculation result showed an increasing error as temperature increased. This is due to the model was developed based on the assumption of an irreversible surface reaction. In order to more accurately simulate Co leaching with acetic acid, there is a need to develop a kinetics model based on the possible equilibrium reaction that occurs in the complexation reaction of Co.

This research was funded by Sakura Science Project under Japan Science and Technology Agency and was supported by Kyushu University. This research was also supported by Universitas Gadjah Mada through LPDP project 2014-2015.

## References

1. A. M., Bernardes, D. C. R., Espinosa, J. A. S. Tenório, J. of Power Sources **130**, 291–298 (2004)
2. S. G. Zhu, W. Z. He, G. M. Li, X. Zhou, X. J. Zhang, J. W. Huang, Trans. Nonferrous Met. Soc. China (English Ed.) **22**, 2274–2281 (2012).
3. M. K. Jha, A. Kumari, A. K. Jha, V. Kumar, J. Hait, B. D. Pandey, Waste Manag. **33**, 1890–1897 (2013).
4. B. Swain, J. Jeong, J. Lee, and G. Lee, Sep. Purif. Technol. **63**, 360–369 (2008).
5. S. Saeki, J. Lee, Q. Zhang, F. Saito, Int. J. Miner. Process. **74**, 373–378 (2004).
6. L. Li, J. Ge, F. Wu, R. Chen, S. Chen, B. Wu, J. Hazard. Mater. **176**, 288–293 (2010).
7. L. Sun and K. Qiu, Waste Manag. **32**, 1575–1582 (2012).
8. H. Cheung, R. S. Tanke, and G. P. Torrence, *Ullmann's Encyclopedia of Industrial Chemistry* (McGraw-Hill, New York,2005).
9. Y. Yan, J. Gao, J. Wu, B. Li, An Interdiscip. Response to Mine Water Challenges **10**, 406–410 (2014).
10. O. Levenspiel. *Chemical Reaction Engineering* (John Wiley&Sons, New Jersey,1999).
11. S. C. Bouffard and D. G. Dixon, Hydrometallurgy **86**, 63–71 (2007).
12. F. Beolchini, M. Petrangeli Papini, L. Toro, M. Trifoni, F. Vegliò, Miner. Eng. **14**, 175–184 (2001).
13. F. K. Crundwell, N. Du Preez, and J. M. Lloyd, Hydrometallurgy **133**, 44–50 (2013).
14. J. D. Donaldson and D. Beyersmann, *Ullmann's Encyclopedia of Industrial Chemistry* (McGraw-Hill, New York,2005).
15. N. Yabuuchi, K. Kubota, Y. Aoki, and S. Komaba, J. Phys. Chem. **120**, 875–885 (2016).
16. J. Xu, H. R. Thomas, R. W. Francis, K. R. Lum, J. Wang, B. Liang, J. Power Sources **177**, 512–527 (2008).
17. J. N. van Niekerk and F. R. L. Schoening, Acta Cryst. **6**, 609–612 (1953).
18. K. C. Wanta, I. Perdana, H. T. B. M. Petrus, IOP Conf. Ser. Mater. Sci. Eng. **162**, 12018 (2016).
19. W. Astuti, T. Hirajima, K. Sasaki, and N. Okibe, Miner. Metall. Process. **32**, 176–185 (2015).

# The effect of temperature and time of extraction on the quality of Semi Refined Carrageenan (SRC)

*Heri* Heriyanto*, *Indar* Kustiningsih, and *Denni Kartika* Sari

Departemen of Chemical Engineering University of Sultan Ageng Tirtayasa, Indonesian, Jln Jenderal Sudirman Km 3,Cilegon, Indonesia

**Abstract.** Euchema cottonii is a good source of kappa-carrageenan and can be found cultivated in the Indonesia coastal areas in which one of them is in Banten Province. Carrageenans have many applications and are utilized in human food and pet-food industry. Carrageenans are also utilized in non-food industry such as pharmaceuticals, cosmetics, printing and textile formulations. Hence, the present study features on the cooking process cooking time and cooking temperature. The effects of these parameters on carrageenan quality such as gel viscosity and gel strength were studied. The process of extraction of carrageenan was conducted with variations temperature: 60, 70, and 80 °C and the variation of time: 1, 2, and 3 hours. Alkaline substance used was KOH with 8% concentration and the ratio of solvent to dry seaweed 8:1. From the present investigation, it was observed that SRC extraction reached the best condition at temperature 70 °C for 2 hours with the value of yield 30.20%, 5.90% moisture content, 18.34% ash content, sulfate content of 6.94%, viscosity of 190 cP, and the gel strength 714.45 g / $cm^2$. The treatment of temperature and extraction time significantly affected the quality of the SRC yield parameter, viscosity and gel strength.
Keywords: *Eucheuma cottonii, Semi Refined Carrageenan (SRC), Time, Temperature*

## 1 Introduction

Carrageenan is the sulfated product form of D-galactan obtained from different species of marine red alge (Zhu et.al, 2017). Then this type are taxonomically called Kappaphycus alvarezii. Seaweeds represent a natural resource of numerous valuable compounds such as polyphenols, vitamins, peptides and polysaccharides (Yousuf et.al, 2017). Some types of Euchema have an important role in the world of international trade as a producer of carrageenan extracts. Carrageenan levels within each species Euchema ranges from 54-73% depending on the type and location of the growth. The Southeast Asian countries have produced 96.5% of the total production of k-carrageenan extracted from red seaweed; of which 55% is contributed by Philippines, followed by Indonesia (38%) and Malaysia (2.5%) (Awang, et al, 2012)

Carrageenan is an alternating copolymer of α(1-3)-D-galactose and β(1-4)-3,6- anhydro-D-galactose and is classified as kappa (κ), iota (ι), and lambda (λ) based on the number of sulfate ester groups and presence of 3,6-anhydro-D-galacto. Among the polysaccharides, carrageenans have attracted intense attention due to its wide range of applications in food industry, such as thickening, gelling, stabilising agents and binder for the manufacture of different kinds of foods. Carrageenan consists of hydrocolloid compounds comprising ester potassium, sodium, magnesium and calcium sulfate. Carrageenan is a large molecules consisting of over 1,000 galactose residues, therefore carrageenan has a lot of variety (Balqis et al, 2017). In medicine Red seaweeds have also yielded a number of sulfated polysaccharides with potent anticoagulant activities (Jiao et al, 2011). Eucheuma cottonii or kappahycus alvarezi is good source of kappa carrageenan and it mainly harvested in Indonesia. The main component of Eucheuma cottonii is kappa carrageenan (Arif, 2014; Moses, 2015).

Banten had potential value for development of the Kappa Carrageenan to increase incomes value in banten province, especially for people who is living in the coast area as a center seaweed grass production. It can be seen from Kim (2015) area effectiv for seaweed growth in banten coast is about 907 ha. In effort to improve the economic value of seaweed made into the Semi Refined Carrageenan (SRC), that required a process to produce the SRC that meet the quality standards of food (food grade). Seaweed extraction process may affect the quality of the obtaining SRC. To get a quality that meets the standards of food SRC, research will be conduct on the temperature and time of extraction to know what is effect diffrent temprature to the quality of semi refned carrageenan. However, there are limited efforts to develop the optimization of process parameters in alkaline treatment. Hence, the present study emphasized on the production of SRC by alkaline treatment where the process parameters such as cooking temperature and cooking time would be investigated. The rheological properties of k-carrageenan (K. alvarezii species); gel strength and viscosity were studied to determine the best condition for the alkaline treatment.

* Corresponding author: heri.heriyanto@untirta.ac.id

## 2 Experimental

Raw seaweed was prepared using sun drying by fresh seaweed collected from farm. The first phase started with cleaning the seaweed which will be extracted to eliminate the salt, sand and undesirable seaweed. Carrageenan extraction process carried out at varied temperatures of 60, 70 and 80 oC. The extraction time was also varied during 1, 2, and 3 hours by using water as a solvent. Alkaline substance was used KOH with a concentration of 8%, which mixed with water as a solvent in the extraction process with the ratio of solvent and dried seaweed 8: 1. Results of the extraction process was filtered and neutralized to pH 7-8 with demin water. Then carrageenan neutralized, cutted and dried in the sun for 2-3 days before being made into carrageenan flour with disk mill.

### 2.1 Materials and Method

Seaweed Eucheuma cottonii was purchased from lontar, KOH 8% (w /v) (Sigma Alderich), Water, pH paper aluminum foil

### 2.2 Research Variables

Fixed variables in this research were seaweed mass 100 gr, Volume KOH: 8% (w / v), Comparison of the mass of seaweed and water: 1: 8 (w / v). Changed Variables in this research were Extraction temperature: 60oC, 70oC and 80 oC and Extraction time: 1 hour, 2 hours, and 3 hours

### 2.3 Characterisation

#### *2.3.1 Ash content*

Carrageenan approximately 2 grams was put in a porcelain cup (B) which was known the dry weight, then being ashed in a furnace at a temperature of 550 °C until free from ash. Further, the sample was being cooled in a desiccator and weighed as the final weight (A).

#### *2.3.2 Gel strength*

Sample was prepared by dissolving 1.5 g of SRC powder to 100 mL of distilled water with continuous magnetic stirring at 90°C around 20-30 min. The sample was allowed to stabilize in a water bath to eliminate bubbles forms at 80-90°C for 15 min. The viscous solution was poured into three 50 mL beakers and left for 20-30 min before sealing them. Samples were stored for 24 h and maintained at ambient temperature prior to analysis.

#### *2.3.3 Viscosity*

Sample of 7.5 gr was put into a 600 ml beaker glass and then stirred using a magnetic stirrer. Then weigh as much as 492.5 g distilled water in a 600 ml glass beaker, put the sample into a baker glass that already filled with distilled water. The solution was agitated for 10-15 minutes at room temperature, and heat to a temperature of 80 ° C and maintained for 10-15 minutes on a hot plate. Then remove and leave until the temperature ± 75oC, measuring viscosity using spindle 2 at 30 rpm spindle rpm and wait until stabilized (5-6 rounds).

#### *2.3.4 Calculation Sulfate Levels*

The sampel as much as 1 gr was put into erlenmeyer flask by adding 0.2 N HCl 50 ml and reflux for 1 hour. Then add a 25 ml solution of 10% H2O2, reflux forwarded to 5 hours. Then add 10 ml of 10% Barium Chloride. Then heated in water bath for 2 hours. Then filtered using Whatman paper number 42 nd washed with boiling distilled water until it's free of chloride, then stir in a Whatman cup along Barium Sulfate precipitate in furnace temperature 1000oC until white. After that, weigh the cup and calculate the results.

#### *2.3.5 Spectophotometer FT-IR*

Before testing the sample with FT-IR spectophotometer, prepared solid samples (KBr pellet manufacture) preceded by weighing 1-2 mg samples which free of water contained (preferably in freeze dry), then add 100-200 mg KBr, wherein the ratio of KBr = 1: 10 subsequently crushed to powder. The sample mixture press in a vacuum at a pressed of 7-8 ton for 10-15 minutes to produce a transparent disk. Then input into Spectrophotometer FT-IR instrument.

## 3 Results and Discussion

### 3.1 Effect of Temperature and Time of Extraction of the Value of the Yield

The yield was one of important parameter in assessing the effectiveness of process in making SRC. Yield calculation was performed to determine the percentage of SRC produced from dried seaweed which used based on the time and temperature of cooking. The Carrageenan yield testing results were shown in Figure 1

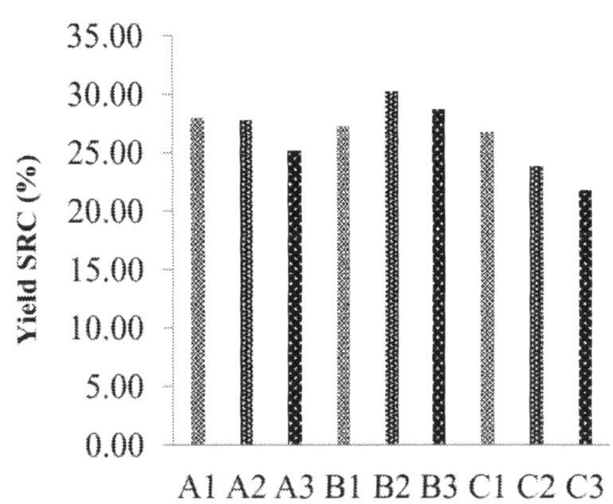


**Fig.1.** The test results SRC flour yield (%) Vs

Description: The difference in the letters A, B and C showed the old extraction, A = 1 hour, B = 2 and C = 3 hours. Figures 1, 2 and 3 show the extraction temperature treatment, 1 = 60 ° C, 2 = 3 = 70 ° C and 80 ° C.

Fig.1 shows the yield of SRC's research ranges from 21.731% - 30.1979%. The lowest yield was at the variation of 3 hours and temperature 80 $^0$C (C3), where as the highest yield was in the variation time of 2 hours and a temperature of 70 $^0$C (B2). The increase of temperatures will rise the melting point temperature of carrageenan, but will affect the carrageenan to come out from the talus and lowering the yield value. So the higher the temperature generated lower yield. Based on physical observations, the filtering process obtained more viscous alkaline solution at high temperature condition. This indicated a carrageenan which did not participate in the screening process filtered out and discarded. Time parameters as shown in Figure 1 has no effect on the temperature of 60 ° C, but at a temperature of 70 ° C and 80 ° C time parameters affect the time of 2 hours into the optimum time to get a high yield that had the same result with Webber (2012) which increase temperature and extraction time affected in yield, the best result in our study were at 70 $^0$ C not at 80 $^0$C that because At higher temperatures, biopolymer degradation may occur that the yield of carrageenan. The yield value increased because the longer extraction time will prolong sulfate binding process by alkaline materials.

### 3.2 Effect of Temperature and Time of Extraction of the Value of Viscosity

Viscosity was one of the physical properties of carrageenan which quite important. Tests were conducted to determine the viscosity of the carrageenan viscosity as a solution at a specific temperature and concentration. Here were the results of testing the viscosity analysis.

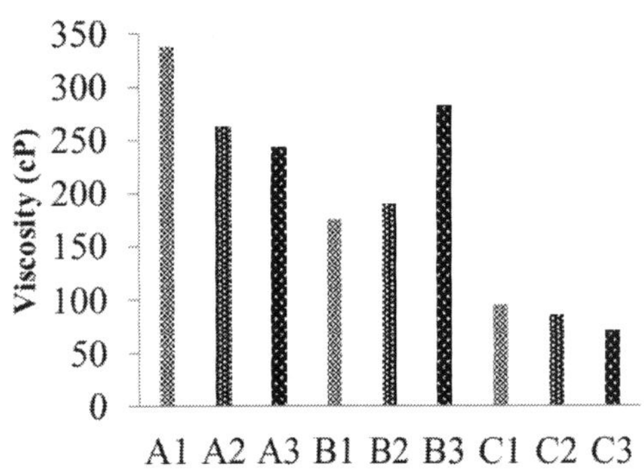


**Fig.2.** Viscosity (Cp)

Figure 2. showed that the viscosity of the analysis results still met FCC standards The highest values found in A1 which was treated with extraction time in 1 hour and 60 °C , the value was 337.5 cP, and the lowest result in treatment C3 with the extraction time 3 hours at temperature 80 °C which got value 70 cP. This confirmed that the time and temperature affected the viscosity values SRC. The longer the extraction time will be longer the $K^+$ ions in KOH bind sulfate groups resulting lower load the polymer chains so the electrostatic forces between the sulfate groups was reduced. According Velde and Ruiter (2002) due to the viscosity of the carrageenan caused by repulsive force in sulfate groups. It which had negative charge, which occur along whit polymer chain, causing the polymer chains were rigid and taut. Besides that their hydrophilic caused the molecules surrounded by the water not moving. The presence of water was not moving, as rising temperatures will cause the loss of the water and cause a decrease in viscosity. It has the simmiliar results that extraction done by Webber (2012) Carrageenan crude extract solution viscosity increased when the extraction temperature reached 60 °C, and at higher temperatures this parameter decreased.

### 3.3 Effect of Temperature and Time of Extraction of the Gel Strength Values

Gel strength was the main physical properties of carrageenan, because the main function of flour SRC turned liquids into solids. Normah and nazarifah (2003) stated that one of the important physical properties of carrageenan was the power to form a gel called gel strength a kappa carrageenan gel was characterised as being strong, brittle and subject to syneresis. It was not freeze/thaw stable (Koliandris, 2009). The results of the study can be seen in Figure 3.

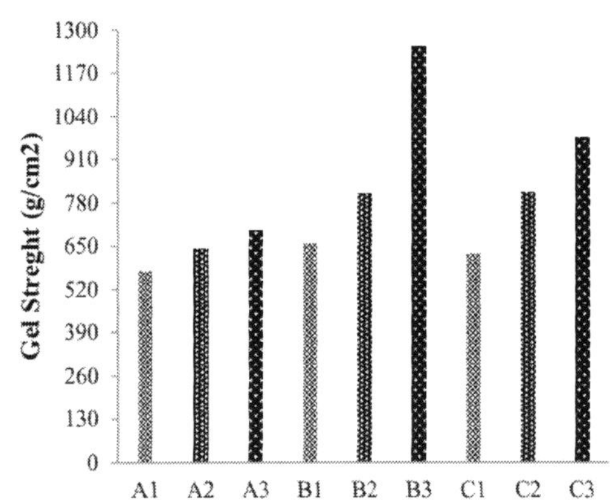


**Figure 3.** Gel Streght (g/cm$^2$)

Making SRC with C3 treatment at temperature of 80 °C for 2 hours obtained the highest gel strength value 1248.45 g/cm2, while the lowest value was 574.45 g / cm$^2$ with treatment A1 at 60 ° C for 1 hour. Gel strength values with a variety of treatments were still in accordance with the standards established FCC i.e ≥ 500 g/cm$^2$.

Figure 3 explained that the higher temperature and duration of extraction will cause a rise in the value of the gel strength. Pathik (2006) Reported that Kappaphycus alvarezii, when heated in an alkaline solution of potassium hydroxide by increase the time of extraction will increase gel strength. The hydroxide part of the reagent penetrated the seaweed, reduced the amount of sulfate in the carrageenan and increasing the 3,6-AG, thereby improving the gel strength of the carrageenan. Polysaccharide gel was a 3-dimensional structure which was formed from polymer solution. The process of gel formation occurs because of the bonds between the polymer chains formed a three-dimensional structure which contained a solvent in its gaps. K+ cations in the KOH can serve as a binder so the 3-dimensional structure of the polymer will maintain its shape when subjected to pressure (Peter, 1996; Velde and Ruiter, 2002). The rise in temperature and duration of the extraction will cause $K^+$ ions will be optimized to form a three-dimensional structure of the polymer.

### 3.4 Effect of Temperature and Duration of Extraction of the Value Sulfate Content

Sulfate content was one of the physical properties that were important to noted in the manufacturing of carrageenan. Kappa carrageenan corresponded to the less sulfated ideal polymer and leads to strong and elastic gels showing thermal hysteresis. In contrast to that, l-carrageenan corresponds to the more sulfated ideal structure which cannot form helices in solution and consequently did not exhibit gelling properties. The content of sulfate caused a repel force between negatively charged sulfate groups, so that the polymer chains got stiff and taut, causing an increase in viscosity (Manuhara, 2016). So that known why Sulfate content was directly proportional to viscosity values, where the higher levels of sulfates generate higher viscosity values obtained but inversely proportional to the value of the gel strength (Hilliou, 2006). The results of the study were shown in Figure 4.

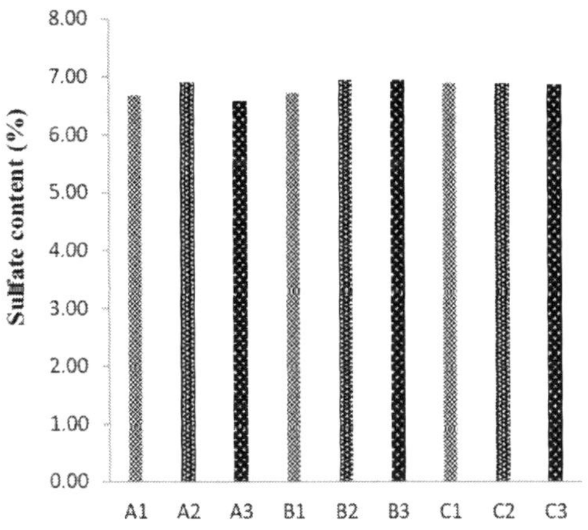


**Fig.4.** The Test Results SRC Sulfate Content (%)

Figure 4 showed that the value of the highest sulfate content of 6.94% with 2 hours extraction treatment at temperature 80 °C and the lowest value of sulfate was when treatment temperature 80 ° C for 1 hour, it's about 6.58%. Carrageenan can be distinguished by levels of sulfates into two fractions, namely kappa carrageenan containing sulfate less than 28% and iota carrageenan if it contained more than 30% (Diharmi et al. 2017). Based on FCC standards, which met maximum quality of SRC was 40% sulfate content. According to Moses (2015) the KOH solution given during processing would reacted with sulfate groups in the carrageenan form K2SO4 salt and sulfuric acid. Sulfate content decreased with increasing concentration of KOH solution. So the magnitude of the temperature and the length of the extraction process did not affect the value of sulfate content.

### 3.5 Effect of Temperature and Duration of Extraction of the Moisture Content

One of the quality requirements of carrageenan was the value of carrageenan moisture content that generated a value of water content that has required both in trade and standards (FCC, 1981) which was semi-refined carrageenan product for a maximum of 12%.

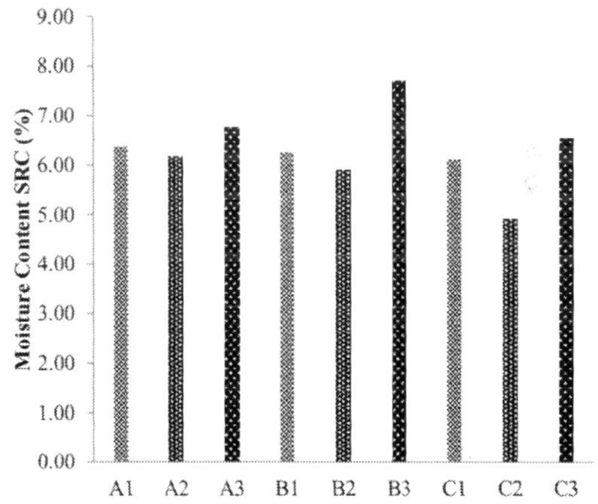


**Fig.5.** The Results Of The Analysis Of Flour SRC Moisture Content (%)

The moisture content quality of SRC flour produced in this study shown in Figure 4.5. FCC standards set with maximum moisture content 12%. The moisture content of all carrageenan samples met the FAO standard. The lowest value was 4.9097% in the samples C2 and the highest was 7.7038% in the sample B3. The size of the moisture content depended on the drying process. SRC drying process was done in the form of small pieces dried in a container with exposure to sunlight. The drying process was considered complete when the pieces SRC were already hard and easily broken (fragile). Drying process was not done using oven because it was feared product color scorched (brown). When it was used in the application will create a cloudy effect on product color. Drying factors by sunlight depended on the intensity of sunlight, humidity levels and wind speed.

### 3.6 Effect of Temperature and Duration of Extraction of the Ash Content Value

Standard specification of the required maximum ash content was 35% (FCC, 1991). Basma et al. (2008) suggested that the ash content was strongly influenced by carrageenan processing techniques and the level of clean liness in the washing process of grass that has been extracted using alkali salt sea. The washing process was less clean can lead to excess K+ ions on carrageenan, which would contribute to increased levels of ash. According to Ahmad (2005), potassium was a mineral element in the form of unburnt inorganic substances (ash).

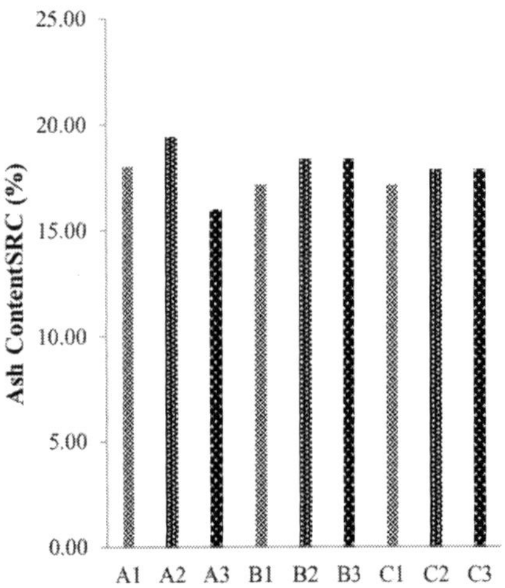


**Fig.6.** The Test Results SRC Ash Content (%)

SRC ash content in flour product ranged from 15.9764 to 19.3919% are shown in Figure 6, it showed all carrageenan powder produced in this study meet the standards set by the FCC to the maximum limit ash content 35%. The ash content product indicated the washing process of SRC after the extraction process had been carried out properly. The washing process was done by rinse SRC with water for 24 L or 6 times from the volume of alkali which being used. Washing was done until a minimum pH reached number 8 signifies that the pH was close to neutral so it could be assumed there was no excess of K + ions in the SRC.

### 3.7 Characteristics of SRC Product Produced

FTIR analysis was used to determine the presence of functional groups of molecules presented in a sample, where in the similarity of functional groups that existed between the standard and the sample state sample being analyzed was identical to the standard. Here was an FTIR spectrum of standard and samples that were generated, where both the spectrum comparable between groups function.

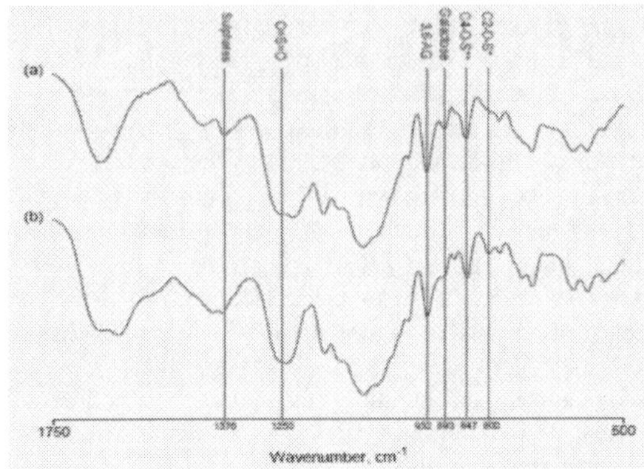


**Fig.7.** Standard FTIR Spectra Of Carrageenan (Mehta, 2008)

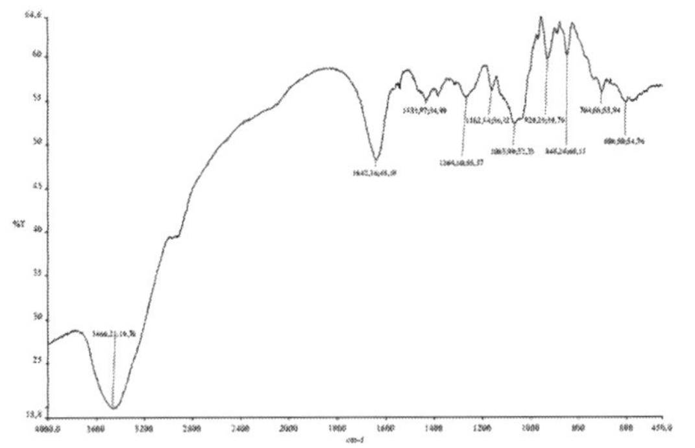

**Fig.8.** The FTIR Spectrum SRC Product Produced

In the spectra 1500-500 cm-1, FTIR spectroscopy showed a very strong beam absorption in the region 1162-1264 cm-1 (for S = O bond in the ester sulfate) and 848-1065 cm-1 region (considered glycosidic bonds) on all kinds of carrageenan. From carrageenan product spectrum which had been obtained, known that using FTIR identification. The product met the specifications for a standard carrageenan functional groups contained in the resulting sample spectrum was identical to the standard spectrum of carrageenan. In the infrared spectrum of the compound carrageenan isolation results contained in Figure 4.8 looks a glycoside bond, 3,6-anhidro-D-galactose, D-galactose-4-sulfate and 3,6-anhidro-D-galactose-2-sulfate ,

Below was the comparison of the characteristics of the SRC product produced to standards set by the FCC based on the optimum yield.

**Table 1.** Comparison of Product Characteristics SRC FCC

| Specification | FCC | Results |
|---|---|---|
| Sulfate (%) | Maks 40 | 6.94 |
| Viscosity (cP) | Min 5 | 190.00 |
| Ash Content (%) | Maks 35 | 18.36 |
| Moisture Content (%) | Maks 12 | 5.90 |
| Gel Strengt (g/cm$^2$) | > 500 | 806.20 |

## 5 Conclusion

Alkali treatment process conditions such as cooking temperature, cooking time were shown as important parameters for the production of semi refined carrageenan. The optimization of process parameters of alkaline treatmentwas developed in this study based on the characterization of carrageenan; gel strength,mousture content, sulfate content and viscosity. Semi Refined Carrageenan (SRC) of seaweed E.cottonii derived from Lontar (Banten) have been obtained with optimum extraction conditions at temperature 70 °C for 2 hours with yield of 30.20%, 5.90% moisture content, 18.34% ash content, sulfate content of 6.94%, 190 cP viscosity and gel strength of 714.45 g / cm2.The temperature and time of extraction influenced on the quality parameters of SRC produced at the yield, viscosity and gel strength.

This research was supported in part by IDB Islamic Development Bank

## References

1. Ahmad Rafiq and Rizwana Jabeen. Pak. J. Bot., 37(**4**): 913-920 (2005)
2. Arif Nur Nabillah Bt Mohd. Chemical Engineering. Universiti Malaysia Pahang. Thesis (2014)
3. Bono Awang, Anisuzzaman 1 S.M., Ong Wan Ding. Engineering Sciences **26**, 3–9 (2014)
4. Diharmi Andarini, Fardiaz Dedi, Nuri Andarwulan and Endang S. Heruwati, Phycological Research. 1-6., (2017)
5. Guiseley KB, Stanley NF, Whitehouse PA, 1980, Davids RL (editor), Hand Book of Water Soluble Gums and Resins, New York, Toronto, London: Mc Graw Hill Book Company
6. Hilliou a L. Larotonda , F.D.S., P. Abreu , A.M. Ramos , A.M. Sereno , M.P. Gonc¸Alves, Biomolecular Engineering **23**:201–208 (2006)
7. Jiao Guangling , Guangli Yu , Junzeng Zhang and H. Stephen Ewart, Mar. Drugs, 196-223, (2011)
8. Kim Se-Kwon. 2015. Seafood Science: Advances in Chemistry, Technology and Applications. pp 372. Korea.CRC Press
9. Koliandris, Andreas, Relationship between texture of gels and flavour release. University of Nottingham. Thesis, (2009)
10. Manuhara Godras Jati, Praseptiangga Danar, Rachmad Adi Riyanto. Aquatic Procedia **7** :106 – 111, (2016)
11. Mehta, Indian Journal of chemical technology vol. **15**, (2008)
12. Moses, J., Anandhakumar, R. and M. Shanmugam, African Jurnal of Biotechnology 14(**18**) :1584-1589, (2015)
13. Normah O. and Nazarifah, I.. J. Trop. Agric. and Fd. Sc. 31(**2**): 207–213, (2003)
14. Velde van de and Ruiter G.A. de., Carrageenan.Germany: Wiley-VCH, (2002)
15. Webber Vanessa, Matos De Carvalho Sabrina, Paulo José Ogliari1, Leila Hayashi1, Pedro Luiz Manique Barreto. Ciênc. Tecnol. Aliment., Campinas, 32(**4**): 812-818, (2012)
16. Youssouf, Latufa., Lallemand, Laura., Giraud, Pierre., Soule, Faiza., Bhaw-Luximon, Archana., Meilhac, Olivier., D'Hellencourt, Christian ´ Lef ebvre., Jhurry, Dhanjay., and Couprie, Joel. Carbohydrate Polymers, 1:**18**, 2017
17. Zhu, Mingjin., Ge, Liming., Lyu, Yongbo., Zi,Yaxin., Li, Xinying., Li, Defu., and Mu, Changdao. Carbohydrate Polymers. 1:**24** (2017)

# Pretreated of banana pseudo-stem as raw material for enzymatic hydrolysis and bioethanol production

Kusmiyati[1,*] and Ryzka Pratiwi Sukmaningtyas[2]

[1]Renewable Energy Research Centre, Department of Chemical Engineering, Universitas Muhammadiyah Surakarta, Indonesia
[2]Department of Chemical Engineering, Universitas Muhammadiyah Surakarta, Indonesia

**Abstract.** Development of alternative energy is needed to solve the energy problem, including bioethanol. Banana pseudo-stem is a lignocellulose material that can used to produce bioethanol. Banana pseudo-stem has 28.83% cellulose and 19.39% lignin. The amount of lignin will reduce by pretreatment process. Variations of pretreatment methods by autoclaving of banana-pseudo stem in a steam, 0.5N, 1N, 1.5N, 2N NaOH solutions for 90 minutes were employed. Then the preteated samples were further enzymatic hydrolysed for 24, 48, 72 hours. The fermentation method of simultaneous saccharification and fermentation (SSF) was applied using cellulase enzyme and yeast of *Saccharomyces cerevisiae* for 120 hours. The variation of the pretreatment process by increasing of NaOH concentration solutions led to decreased the lignin content while increased in cellulose content. The lowest lignin content was 11.44% and the highest cellulose was 51.66%. The highest sugar content was 29.8 g/L (at pretreatment 2N NaOH solution, 72 hours hydrolysis). The highest bioethanol amount (4.32 g/L) was produced from pretreated banana stem using 2N NaOH solution.
***Keywords***: bioethanol, banana pseudo-stem, alkali pretreatment, enzymatic hydrolysis, SSF, *Saccharomyces cerevisiae*

## 1 Introduction

One of the problems that appear in Indonesia is decreasing crude oil as a main source of energy. Development of renewable energy is needed to solve the problem of energy limitation, one way is a bioethanol development.

Bioethanol is made from raw materials such as fibers, molasses, fruits, and other materials that containing reduction sugar which can be fermented [1]. Second generation bioethanol is produced from lignocellulose material. The process of conversion lignocellulose material to bioethanol generally consists of pretreatment, hydrolysis, and fermentation [2].

Alkaline pretreatments are more effective to dissolve lignin than acid and hydrothermal process [3], due to decrease of cellulose crystallinity and increasing access for enzymes [4]. The most alkaline used are NaOH, KOH, $Ca(OH)_2$, and ammonia [5].

Banana pseudo-stem is one of lignocelllulosic material that could be fermented to ethanol. According to [6], banana pseudo-stem 44%.cellulose, and 8.1% lignin on a dry basis.

*Saccharomyces cerevisiae* is generally used in ethanol production industries. *S. cerevisiae* has tolerance to substrate concentration, high ethanol content, low pH, and low oxygen levels [1].

The aims of this study were to investigate the influence of pretreatment variation (autoclaving pretreatments of banana stem in steam, 0.5N, 1N, 1.5N, 2N NaOH solutions) on the cellulose, hemicellulose, and lignin contents of the pretreated materials. Subsequently, the effect of the pretreatments on glucose from hydrolysis process and bioethanol contents from fermentation were also evaluated.

## 2 Materials and Methods

### 2.1. Pretreatment

Dry banana pseudo-stem was cut to a size 5 cm, milled and sieved up to 60 mesh then dried at 140 °C to reduce the remaining moisture content. Alkaline solution (99% NaOH, Merck) was added to 25 gr of dry banana pseudo-stem placed in a 250 mL beaker glass to form a slurry. The concentrations of NaOH were 0.5 N, 1N, 1.5 N, and 2N. Pretreatment of the slurries was carried out for 90 minutes in the autoclave at 121 °C.

### 2.2 *Saccharomyces cerevisiae* preparations

*S. cerevisiae* was cultivated on agar slant containing 10 g/L glucose, 5 g/L peptone, 3 g/L yeast extract, 20 g/L agar at 28 °C [7]. The inoculum of *S. cerevisiae* was cultivated on culture medium containing 50 g/L glucose, 1 g/L $(NH4)_2HPO4$, 0.5 g/L $KH_2PO_4$, 0.25 g/L $MgSO_4.7H_2O$, 10 g/L peptone, 10 g/L yeast extract, at 30 °C for 24 hours [8].

* Corresponding author: kusmiyati@ums.ac.id

### 2.3 Enzymatic Hydrolysis

5 grams of pretreated banana pseudo-stem was added with citrate buffer solution at pH 5 (solid:solution ratio 1:10) then added with one gram of cellulase (SQzyme CS; 20,000 u/g). The hydrolysis process was carried out in rotary shaker water bath at 50°C, 150 rpm for 24, 48, and 72 hours.

### 2.4 Ethanol Fermentation

Ethanol fermentation was carried out by simultaneous saccharification and fermentation (SSF) process. The treated banana pseudo-stem was added with 100 mL of citrate buffer at solid concentration 10% (w/v), then sterilized at 121°C for 15 minutes. Cellulase enzim at a dosage 20 FPU/g $_{dry\ solid}$ and 10% (v/v) yeast *S. cerevisiae* was added to the samples of the pretreated banana-stem solution. The ratio of cellulase and yeast 1:5 (w/v). SSF process was performed at 37.5°C for 120 hours.

### 2.5 Distillation

After the fermentation process is complete, the fermentation product is then distilled to separate the ethanol from impurities.

### 2.6 Analysis

Analysis of lignin content was done by Klason method [9]. 1 gr dry sample (A) was extracted with alcohol benzene (ratio 1:2). The sample was transferred to a 50 mL beaker glass and 15 mL of 72% sulfuric acid was added at 20°C with stirring for 2-3 minutes and then idle for 2 hours. Then it was added with 300 mL of distilled water in a 100 mL erlenmeyer flask.The sample then was added to distilled water up to 575 mL (3% sulfuric acid concentration). It was boiled and leave for 4 hours at low heating. Sample was washed until the lignin acid-free was obtained then dried at 105°C, cool in desicator and weigh up to constant weight (B).

$$\text{lignin content} = \frac{B}{A} \times 100\% \quad (1)$$

Analysis of cellulose and hemicellulose content was done by Chesson method [10]. 1 gr dry sample (A) refluxed with 150 mL of distilled water at 100°C for 2 hours. The result is filtered, the residue is washed until neutral and dried (B). Dried residue refluxed with 150 mL of 0.5M sulfuric acid at 100°C for 2 hours. The result is filtered and dried (C). Dried residue treated with 10 mL of 72% sulfuric acid at room temperature for 4 hours, then diluted to 0.5 M sulfuric acid and refluxed at 100°C for 2 hours. The result is filtered, the residue washed and dried (D).

$$\text{hemicellulose content} = \frac{B - C}{A} \times 100\% \quad (2)$$

$$\text{cellulose content} = \frac{C - D}{A} \times 100\% \quad (3)$$

Analysis of glucose content was done by photometric method. Sample of 0.01 mL added with 1 mL of glucose reagent then incubated at photometer (BA-88A, Mindray) for 1 minute at 37°C.

Analysis of ethanol content was done by Gas Chromatography (GC) method. On the GC tool (Clarus 680, Perkin Elmer), the injector temperature is set to 130°C, the detector temperature 150°C, initial temperature 80°C, initial time 1 minute at heating rate 5°C/minute, final time 5 minutes.

## 3 Results and Discussion

### 3.1 Effect Different NaOH Concentration of Pretreatments on Lignin, Cellulose, and Hemicellulose Contents

Pretreatment process was carried out to decrease lignin content and increase cellulose content on banana stem. Physical and chemical processes were carried out by using grinding and autoclaving the banana stem using a steam or NaOH solution.

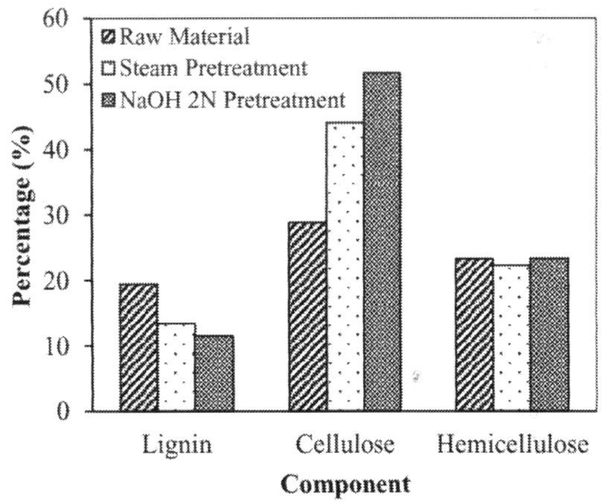


**Fig. 1.** Effect of pretreatments on lignin, cellulose, and hemicellulose contents. Pretreatment conditions at banana pseudo-stem: liquid ratio (1:4) autoclaved for 90 minutes.

Fig. 1 shows the optimal pretreatment was obtained using autoclaving banana stem in 2N NaOH solution that resulted lignin content 11.44%, cellulose 51.66%, and hemicellulose 23.29%. Lower cellulose was observed for steam pretreatment sample that yielded 13.37% lignin and 44.10% cellulose. While, for untreatment banana stem resulted the lowest cellulose that yielded 19.39% lignin and 28.83% cellulose. Physical pretreatment causes some hemicellulose degrades and lignin decomposes, while chemically is degrading the hemicellulose and dissolving lignin [11].

In the study by [12], pretreatment using alkali can increase cellulose content up to 8-10%. Moreover, the increase of NaOH concentration can give a significant influence to lignin degradation [13].

### 3.2 Effect of Different Pretreatments and Hydrolysis Time to Glucose Content

The hydrolysis process is aimed to convert cellulose into glucose which is then converted to bioethanol. The enzyme hydrolysis was carried out using 2% (w/w) cellulase enzyme with 5 gr of pretreated banana stem at 50 °C. Different pretreatment and time of hydrolysis on glucose content were investigated.

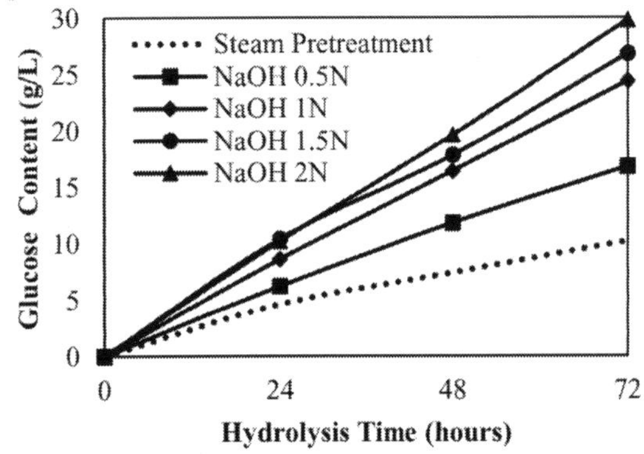


**Fig. 2.** Effect of different pretreatments and hydrolysis time on glucose content. Enzymatic hydrolysis at solid: liquid ratio 1:10 (w/v), 50 °C, 150 rpm, pH 5.

Fig. 2 shows the optimum glucose content obtained at banana stem pretreated with autoclaving in 2N NaOH solution for 72 hours hydrolysis that resulted 29.8 g/L glucose. Different pretreatments show that the glucose increased with increasing time. However the highest glucose were obtained at 72 h hydrolysis time that resulted 10.2 g/L for stem pretreatment, 16.8 g/L for NaOH 0.5 N pretreatment, 24.4 g/L for NaOH 1N pretreatment, 26.8 g/L for NaOH 1.5N pretreatment. The higher NaOH solution increased glucose from hydrolysis was similar with studied reported by [14].

### 3.3 Effect of Different Pretreatments on Bioethanol Concentration

The fermentation process was performed by Simultaneous Saccharification and Fermentation (SSF) method on 10 gram banana pseudo-stem by adding 100 mL of citrate buffer, 2% (w/w) cellulase enzyme, and 1:1 (w/v) *S. cerevisiae* at 37.5 °C for 120 hours.

The highest bioethanol was observed on banana stem pretreated using NaOH 2N solution that yielded 4.32 g/L ethanol. Ethanol content increased with increasing NaOH concentration in the pretreatment of banana stem. The increasing of ethanol due to cellulose content of banana increased with increasing NaOH concentration. The cellulose was converted to reduction sugar at hydrolysis process followed by fermentation of the sugar to bioethanol. Alkali pretreatment can increase ethanol concentration was also reported by a study by [15].

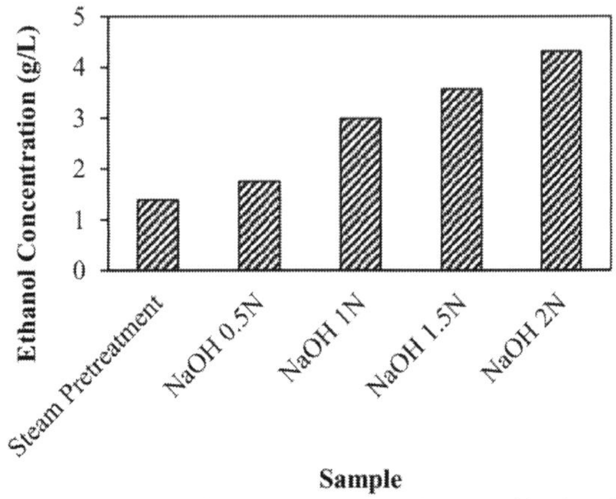


**Fig. 3.** Effect of NaOH concentration on bioethanol concentration. SSF conditions at 1:10 (w/v) raw material and citrate buffer ratio, 2% (w/w) cellulase enzyme, 1:1 (w/v) *S. cerevisiae*, 37.5°C, pH 5, 120 hours.

## 4 Conclusions

Different alkaline concentration of pretreatment can affect to the lignin, cellulose, and hemicellulose contained in banana pseudo-stem. It can also affect the ethanol content that is obtained from fermentation. Hydrolysis time also affected to glucose content from hydrolysis. The glucose content increased with increasing of hydrolysis time. The best result from pretreatment process were 11.44% lignin, 51.66% cellulose, and 23.29% hemicellulose from sample that pretreated by autoclaving banana stem in 2N NaOH . The highest result from hydrolysis process was 29.8 g/L glucose at 72 hours hydrolysis using sample pretreated by autoclaving using 2N NaOH that resulted 4.32 g/L ethanol.

The authors would like to thank the Directorate of Research and Community Service, Ministry of Education, Technology and Higher Education Indonesia (Kemenristekdikti) for funding this Research (PB Kompetensi 2017).

## References

1 E. Akponah, O. O. Akpomie, and M. Ubogu, "Bio-ethanol Production from Cassava Effluent using Zymomonas mobilis and Saccharomyces cerevisiae Isolated from Rafia Palm (Elaesis guineesi) SAP," *Eur. J. Exp. Biol.*, **3**, 247–253 (2013).

2 E. C. Bensah and M. Mensah, "Chemical Pretreatment Methods for the Production of Cellulosic Ethanol: Technologies and Innovations," *Int. J. Chem. Eng.*, **2013**, 1–21 (2013).

3 P. Alvira, E. Tomás-Pejó, M. Ballesteros, and M. J. Negro, "Bioresource Technology Pretreatment technologies for an efficient bioethanol production process based on enzymatic hydrolysis : A review," *Bioresour. Technol.*, **101**, 4851–4861 (2010).

4 R. M. Trevorah and M. Z. Othman, "Alkali Pretreatment and Enzymatic Hydrolysis of Australian Timber Mill Sawdust for Biofuel Production," *J.*

*Renew. Energy*, **2015**, 1–9 (2015).

5 S. H. Mood *et al.*, "Lignocellulosic biomass to bioethanol, a comprehensive review with a focus on pretreatment," *Renew. Sustain. Energy Revoews*, **27**, 77–93 (2013).

6 E. L. Souza, G. F. Liebl, C. Marangoni, N. Sellin, M. S. Montagnoli, and O. Souza, "Bioethanol from Fresh and Dried Banana Plant Pseudostem," *Chem. Eng. Trans.*, **38**, 271–276 (2014).

7 L. Peng and Y. Chen, "Conversion of Paper Sludge to Ethanol by Separate Hydrolysis and Fermentation (SHF) using *Saccharomyces cerevisiae*," *Biomass and Bioenergy*, **35**, 1600–1606 (2011).

8 F. A. Gonçalves, H. A. Ruiz, E. S. dos Santos, J. A. Teixeira, and G. R. de Macedo, "Bioethanol Production by *Saccharomyces cerevisiae, Pichia stipitis and Zymomonas mobilis* from Delignified Coconut Fibre Mature and Lignin Extraction According to Biorefinery Concept," *Renew. Energy*, **94**, 353–365 (2016).

9 M. Bunzel, A. Schüßler, and G. T. Saha, "Chemical characterization of Klason lignin preparations from plant-based foods," *J. Agric. Food Chem.*, **59**, 12506–12513 (2011).

10 R. Datta, "Acidogenic fermentation of lignocellulose-acid yield and conversion of components," *Biotechnol. Bioeng.*, **23**, 2167–2170 (1981).

11 K. Eisenhuber, K. Krennhuber, V. Steinmüller, and A. Jäger, "Comparison of Different Pre-Treatment Methods for Separating Hemicellulose from Straw during Lignocellulose Bioethanol Production," *Energy Procedia*, **40**, 172–181 (2013).

12 M. Idrees, A. Adnan, F. Malik, and F. A. Qureshi, "Enzymatic Saccharification and Lactic Acid Production from Banana Pseudo-Stem Through Optimized Pretreatment at Lowest Catalyst Concentration," *EXCLI J.*, **12**, 269–281 (2013).

13 N. I. Iberahim, J. M. Jahim, S. Harun, M. T. M. Nor, and O. Hassan, "Sodium Hydroxide Pretreatment and Enzymatic Hydrolysis of Oil Palm Mesocarp Fiber," *Int. J. Chem. Eng. Appl.*, **4**, 101–105 (2013).

14 D. Dahnum, S. O. Tasum, E. Triwahyuni, M. Nurdin, and H. Abimanyu, "Comparison of SHF and SSF Processes using Enzyme and Dry Yeast for Optimization of Bioethanol Production from Empty Fruit Bunch," *Energy Procedia*, **68**, 107–116 (2015).

15 S. Thakur, B. Shrivastava, S. Ingale, R. C. Kuhad, and A. Gupte, "Degradation and selective ligninolysis of wheat straw and banana stem for an efficient bioethanol production using fungal and chemical pretreatment," *3 Biotech*, **3**, 365–372 (2013).

# Pyrolysis of palm empty fruit bunch: Yields and analysis of bio-oil

*Bachrun* Sutrisno and *Arif* Hidayat*

Chemical Engineering Department, Universitas Islam Indonesia, Indonesia

**Abstract.** The palm oil industry is currently growing rapidly and generating large amounts of biomass waste that is not utilized properly. Palm empty fruit bunch (PEFB), by product of palm oil industry is considered as a promising alternative and renewable energy source that can be converted to a liquid product by pyrolysis process. In this work, pyrolysis of PEFB was studied to produce bio-oil. Pyrolysis experiments were carried out in a bench scale tubular furnace reactor. The effects of pyrolysis temperatures (400–600 °C) at heating rate of 10 °C/min to optimize the pyrolysis process for maximum liquid yield were investigated. The characteristics of bio-oil were analyzed using FTIR and GC–MS. The results showed that the maximum bio-oil yield was 44.5 wt. % of the product at 450 °C. The bio-oil products were mainly composed of acids, aldehydes, ketones, alcohols, phenols, and oligomers. The chemical characterization showed that the bio-oil obtained from PEFB may be potentially valuable as a fuel and chemical feedstock.

## 1 Introduction

The problems associated with the fulfillment of energy are the search for feedstock that do not compete with food needs. Biomass, such as agricultural and forest residues, low grade plants, and municipal solid wastes, have a great potential to replace the fossil fuels. Biomass can generally be defined as any hydrocarbon material which mainly consists of carbon, hydrogen, oxygen and nitrogen. Sulfur is also present in less proportion. Some biomass types also carry significant proportions of inorganic species. Utilization of biomass as energy feedstock has attracted attention in recent years. Biomass can be converted to energy by thermochemical processes, such as pyrolysis, gasification and combustion [1].

Pyrolysis is a thermal decomposition process of organic matter that occurs in the absence of oxygen or air. Pyrolysis can directly produce biofuel from biomass which can be readily stored or transported. This process is conducted at moderate temperature, between 400-800 °C. Product of pyrolysis can be divided into three phase: solid, liquid and gas. The solid product or char is useful as a renewable fuel or other applications such as metallurgical and leisure industries, soil amender and the production of activated carbon and electrodes. The gaseous products mainly consist of hydrogen, carbon dioxide, carbon monoxide, ethane, propane and methane which have a low to medium heating value. The gaseous products contain sufficient energy to supply the energy requirements of a pyrolysis process. Bio-oil is the liquid product of pyrolysis process. Bio-oil is a high density oxygenated liquid that can be used as a substitute for liquid fossil fuels in some applications. It can be burned in diesel engines, turbines or boilers.

Pyrolysis can be classified on the basis of heating rate i.e. slow, fast and flash pyrolysis. Slow pyrolysis is characterized by lengthy solids and gas residence times, and slow heating rates. During slow pyrolysis, tar and char are released as main products. In fast pyrolysis process, feedstock is rapidly heated to temperatures of 650 to 1000 °C depending on the desired amount of bio-oil or gas products. Flash pyrolysis occurs at rapid heating rates and moderate temperatures between 400 and 600 °C. However, vapor residence time of this process is less than 2 s. Flash pyrolysis produces fewer amounts of gas and tar when compared to slow pyrolysis.

Pyrolysis has been applied to various biomass such as coffee grounds [2], waste tea [3], sugarcane bagasse [4], corn cob, wheat straw, rice straw and rice husk [5], coconut shell [6], oil palm trunk [7], palm kernel shell [8], saw dust [9], aquatic biomass [10] and rubber seed shell and kernel [11]. Palm empty fruit bunch (PEFB) from an oil palm plantation is one of the potential biomass to produce biofuels due to its abundant supply and easy to handle. PEFB is a solid residue of fresh fruit bunch which produced after oil extraction at palm oil mills. On an average, every ton of palm oil produced from a fresh fruit bunch approximately 1 t of PEFBs is generated. Based on Indonesia's palm oil production of 31.5 million tons in 2014, the total PEFB generated from the palm oil industry is around 31.5 million tons. This huge amount of PEFB, if not handled properly, would evoke environmental problems. Composting and incineration are traditional methods which commonly applied to solve these environmental issues. PEFB is chemically composed of cellulose (up to 65.0%), hemicellulose (up to 33.52%), lignin (up to 30.45%) and only slightly of other organics. Cellulose, hemicellulose and lignin are lignocellulose

* Corresponding author: arif.hidayat@uii.ac.id

components. Lignocellulose components have potential to be utilized as energy source via pyrolysis process. The product distribution of the bio-oil depends on the lignocellulosic composition of the biomass [12]. The decomposition of hemicellulose occurred at 220-315 °C. Cellulose decomposed in the temperature range of 314-400 °C. The decomposition of lignin took place in a wide range of temperature from 160 to 900 °C, generating highest solid residue (40%).

The objectives of this works are: (i) investigate the effect of pyrolysis temperature on product yields, and (ii) characterize liquid product obtained under optimum pyrolysis conditions. The experiments have been carried out at temperatures of 400, 450, and 500 °C in nitrogen atmosphere. The bio-oil products have been characterized using Fourier Transform-Infrared Spectroscopy (FT-IR) and Gas Chromatography–Mass Spectrometry (GC–MS).

## 2 Experimental

### 2.1. Materials

Palm empty fruit bunches (PEFB) was obtained from Oil Palm plantation in Kalimantan Tengah province, Indonesia The pretreatment process, the material were processed by cutting to small size and then dried at 105 °C for 24 hours. The dried PEFB was grinded to and sieved to 0.5 – 1.0 m and then stored. The moisture content of the feedstock was also measured to ensure that it was less than 10 % on a dry feed basis.

### 2.2 Pyrolysis Experiments

The pyrolysis experiments were performed in a stainless steel fixed bed reactor under atmospheric pressure using a batch operation. The reactor was placed into an electrically heated tubular furnace. The reactor was controlled by a thermocouple. In a pyrolysis experiment, a quantity of 10 g of PEFB was loaded in the reactor. Then, the reactor temperature was increased at a heating rate of 10 °C min/K up to 400 °C and held at this temperature for 1 h. During heating, the nitrogen ($N_2$) gas was flowed into the reactor at the rate of 200 mL/min from the bottom. The gases formed from the pyrolysis flows out along with $N_2$ gas from the top of the reactor. The gas was passed through two condensers. The condenser was cooled with the flow of tap water. The liquid products from the condenser and second condenser were collected in conical flasks fitted to the condensers. The bio-oil obtained under experimental conditions that gave maximum oil yield was used for characterization. The chemical composition of bio-oil was analyzed using gas chromatography-mass spectrometry (GC-MS). The functional groups of bio-oil were characterized by using Fourier transform infrared spectroscopy (FTIR) instrument. The effect of temperature was investigated by carrying out the pyrolysis experiments over the temperature range of 400 to 500 °C.

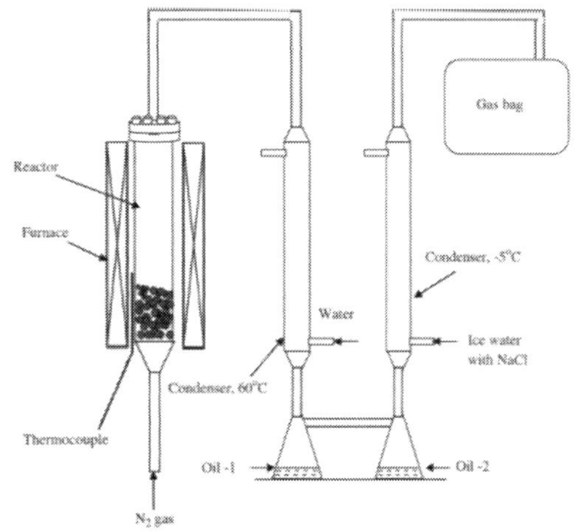


**Fig. 1.** Pyrolysis Experiments Apparatus.

Where $X_F$ is the FFA conversion, $AV_1$ is the initial acid value of the mixture and $AV_2$ is the acid value of mixture after reaction.

## 3 Results and Discussion

### 3.1. The effects of pyrolysis temperatures on the product yields

Figure 2 shows the bio-oil yields from the pyrolysis of the PEFB with heating rates of 10 and 30 °C/min in relation to final pyrolysis temperatures 400, 450, and 500 °C under nitrogen flow. For the heating rate of 10 °C/min, the bio-oil yield increased from 23.5 to 25.1% as the pyrolysis temperature increase from 400 to 450 °C. The highest bio-oil yield of 28.3 wt. % was obtained at the final pyrolysis temperature of 500 °C. At the heating rates of 30 °C/min, the bio-oil yields were 23.40% and 26.1%, at 400 and 450 °C. The bio-oil yield increased to 28.9% at the final pyrolysis temperature of 500 °C. Increasing the pyrolysis temperature and heating rate increased the bio-oil yield. It was known that the heavy-molecular-weight compounds of pyrolysis vapors become active at high temperature. At high temperatures secondary tar destruction may cause a decrease in the oil yield or increase in the gas yield. The gas product yield increased with pyrolysis temperature.

As shown in Figure 2, as the pyrolysis temperature increased from 400 to 450 °C, the char yields decreased from 33% to 29.5% and from 34% to 30.5% at heating rates of 10 °C/min and 30 °C/min, respectively. At the final pyrolysis temperature of 500 °C, the char yields decrease to 26.5% and 28% when heating rate were 10 and 30 °C, respectively. Higher char yields were obtained for lower heating rates over the entire temperature range. These higher yields may be related to the rapid heating rate, which promotes the fast depolymerisation of the solid material to primary volatiles, while at the lower heating rate, dehydration to more stable anhydrocellulose is limited and very slow. The char yields always decreased with increasing final pyrolysis temperature and heating rate, because the decomposition of the wood or secondary decomposition of the char leads to higher conversion with increasing final temperatures.

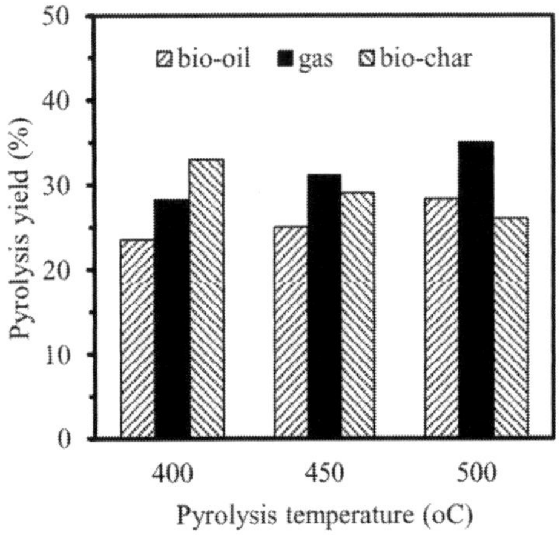


**Fig. 2.** Effect of final pyrolysis temperature on product yields at a heating rate of 10 °C/min.

The gas product yields consistently increased with increasing pyrolysis temperature. When the pyrolysis temperature was increased from 400 to 450 °C, the gas yields increased from 28.5% to 31.5% and from 29% to 31.5% at heating rates of 10 °C/min and 30 °C/min, respectively. At the pyrolysis temperature of 500 °C, the gas yields were 34% at 10 °C/min and 35.5% at 30 °C/min. The increase in the gaseous product is thought to be predominantly due to the secondary cracking of the pyrolysis vapours at higher temperatures. However, the secondary decomposition of the char at higher temperatures may also generate non-condensable gaseous products. The increase in gas products is thought to occur predominantly due to secondary cracking of the pyrolysis vapours at higher temperatures. However, secondary decompositions of the char at higher temperatures may also give non-condensable gas products.

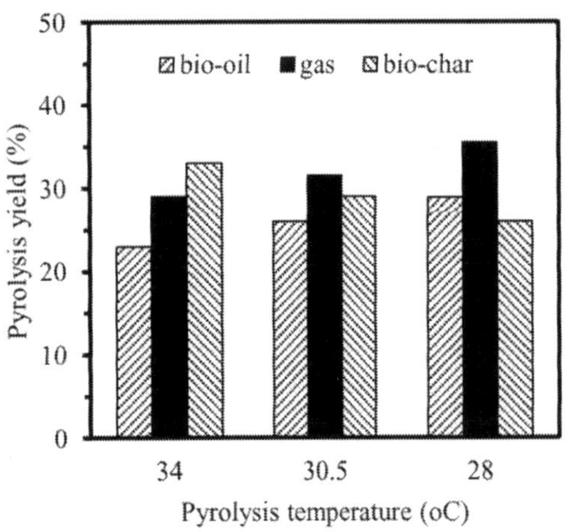


**Fig. 3.** Effect of pyrolysis temperature on product yields at a heating rate of 30 °C/min.

### 3.2 Chemical Composition

FT-IR spectra identify functional groups in bio-oils. FTIR spectra are shown in Figure 4. The spectra data showed that alcohol, ketone, ester and carboxylic acid groups are also major oxygen functions present in the bio-oil. The presence of O–H vibrations between 3050 and 3600 $cm^{-1}$ together with the presence of C–O stretching vibrations between 1650 and 1850 $cm^{-1}$ indicates the presence of carboxylic acids and their derivatives. The presence of the C–O stretching vibrations with absorbance between 1650 and 1850 $cm^{-1}$ may also indicate the presence of the presence of ketone, quinone, and aldehyde groups. The presence of aromatic compounds is confirmed by the aromatic C=C stretching vibrations between 1460 and 1455 $cm^{-1}$. The existence of a C–O stretch within 1075-1020 $cm^{-1}$ indicates the presence of primary, secondary and tertiary alcohols, phenols, and/or esters. A presence of alcohols and esters in bio-oil are confirmed by the peak appearing between 950 and 1300 $cm^{-1}$. Bio-oil is a complex mixture, containing mostly aliphatic and aromatic compounds. The presence of hydrocarbon groups indicates that bio-oil has the potential to be used as a fuel or a source of chemicals.

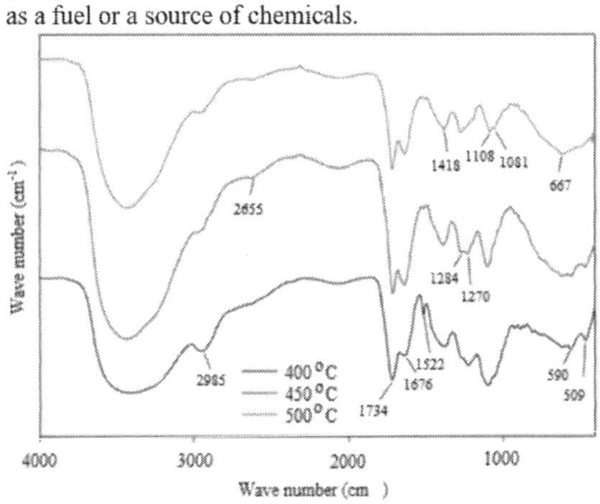


**Fig. 4.** FT-IR spectra analysis of bio-oil.

GC-MS analysis was carried out in order to get the nature and type of organic compounds in the bio-oil. In general bio-oils are mainly composed of aromatic, aliphatics, aldehydes, and phenolic compounds. Phenols and derivatives are primarily form by decomposition of lignin unit of biomass and the esters, aldehydes, alcohols and ketons are formed by degradation of cellulose and hemicellulose fraction of biomass. Identified compounds can be divided into four different classes of monoaromatics, aliphatics, oxygenated compounds and polyaromatic compounds. Monoaromatic compounds include toluene, furans, phenols, benzene and their derivatives. Aliphatics are mainly consisting of alkanes and alkenes and their derivatives. Oxygenated compounds composed of aldehydes, ketones, carboxylic acid and esters. Table 1 lists the possible compounds identified by the MS search libraries of the bio-oil. The main compounds identified from pyrolysis of PEFB phenolic derivatives and esters. The GC–MS analyzed result shows

 https://doi.org/10.1051/matecconf/201815401036

that 3-methyl- phenol, 3-ethyl-phenol, 2,6-dimethoxy-phenol, 4-Allyl-2,6-dimethoxyphenol, and phenol were the main phenolic products in the bio-oils. Table 1 also illustrates that there are several esters compounds such as 1-Pentadecene, n-Dodecane, Hexadecanoic acid, Palmitic acid, 9-Octadecenoic acid methyl ester.

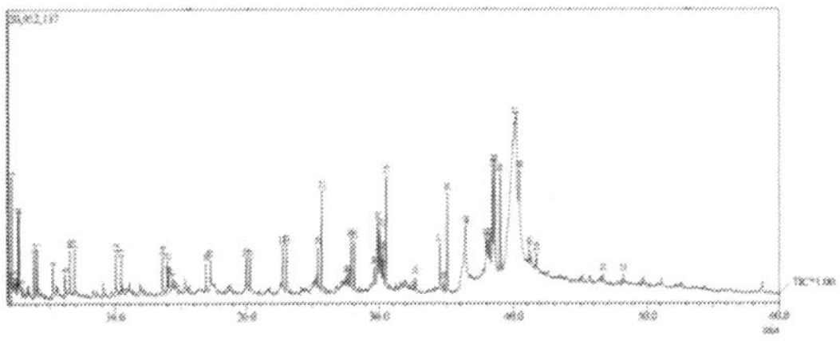

**Fig. 4.** GC/MS chromatogram of bio-oil from PEFB pyrolysis.

**Table 1.** Rate constant at different temperatures.

| Retention Time | Compounds identified in bio-oil | Area,% |
|---|---|---|
| 9,708 | 1,2,3-Trimethylbenzene | 1,34 |
| 10,097 | Phenol | 28,99 |
| 12,554 | o-Cresol | 5,44 |
| 12,953 | Guaiacol | 4,93 |
| 13,428 | 3-methyl- phenol | 5,69 |
| 16,396 | 3-ethyl-phenol | 5,08 |
| 19,026 | 1,4-dimethoxy-2-methyl-benzene | 2,19 |
| 20,978 | 2,6-dimethoxy-phenol | 5,72 |
| 29,826 | 4-Allyl-2,6-dimethoxyphenol | 2,94 |
| 25,412 | 1-Pentadecene | 1,24 |
| 25,682 | n-Dodecane | 3,66 |
| 35,032 | Hexadecanoic acid | 4,58 |
| 36,336 | Palmitic acid | 9,8 |
| 38,392 | 9-Octadecenoic acid methyl ester | 1,89 |
| 38,950 | Octadecanoic acid methyl ester | 0,85 |

## 4 Conclusions

In this study, pyrolysis experiments of PEFB were conducted in a fix-bed reactor. The effects of the pyrolysis temperature on the bio-oil yield were investigated for the pyrolysis experiment of PEFB. The optimal yield of bio-oil could be achieved at the pyrolysis temperature of 500 °C. Higher $N_2$ flows rates were more favorable for the production of the bio-oil. GC–MS and FT-IR spectra showed presence of phenolic, ketones and nitrogen-containing group in the bio-oils obtained from the PEFB pyrolysis.

The adsorption experiments indicated that modified coal fly ash was effective in removing of Remazol Blue. The percentage removal of dyes increased while the modified fly ash dosage increased. The percentage removal of dyes increased with decreased initial

## References

1. B. Sutrisno and A. Hidayat, Upgrading of bio-oil from the pyrolysis of biomass over the rice husk ash catalysts, IOP Conference Series: Materials Science and Engineering, **162** (1), 012014 (2016)
2. N. Chen, J. Ren, Z. Ye, Q. Xu, J. Liu, and S. Sun. Bioresource Technology, **221**, 534-540 (2016)
3. R. Soysa, Y.S. Choi, S.J. Kim, and S.K. Choi. International Journal of Hydrogen Energy, **41** (37), 16436-16443 (2016)
4. G.C. Goncalves, N.C. Pereira, and M.T. Veit. Biomass and Bioenergy, **85**, 178-186 (2016)
5. B. Biswas, N. Pandey, Y. Bisht, R. Singh, and T. Bhaskar. Bioresource Technology, **237**, 57-63 (2017)
6. I. Ali, H. Bahaitham, and R. Naebulharam. Bioresource Technology, **235**, 1-11 (2017)
7. H.A. Oramahi, Wahdina, F. Diba, Nurhaida, and T. Yoshimura. Procedia Environmental Sciences, **28**, 769-777 (2015)
8. S.J. Oh, G.G. Choi, and J.S. Kim. Energy, **113**, 108-115 (2016)
9. J.X. Wang, J.P. Cao, X.Y. Zhao, T.L. Liu, and X.Y.Wei. Journal of Analytical and Applied Pyrolysis, **125**, 279-288 (2017)
10. B. Biswas, R. Singh, B.B. Krishna, J. Kumar, and T. Bhaskar. Bioresource Technology, **242**, 139-145 (2017)
11. S.N.A.M. Hassan, M.A.M. Ishak, K. Ismail, S.N. Ali, and M.F. Yusop. Energy Procedia, **52**, 610-617 (2014)
12. H. Yang, R. Yan, H. Chen, D.H. Lee, and C. Zheng, Fuel, **86**, 1781-1788 (2007)

# Modified coal fly ash as low cost adsorbent for removal reactive dyes from batik industry

*Agus* Taufiq, *Pratikno* Hidayat, *Arif* Hidayat*

Chemical Engineering Department, Universitas Islam Indonesia, Indonesia

**Abstract.** The removal of reactive dyes on modified coal fly ash has been investigated during a series of batch adsorption experiments. Physical characteristics of modified coal fly ash was characterized by Brunauer Emmett Teller (BET) surface area analysis, X-ray powder diffraction (XRD), Fourier transform infrared spectrophotometer (FT-IR), and scanning electron microscope (SEM). The effects of operational parameters such as initial dye concentration (50–200 mg/L), solution pH (4–10) and adsorbent dosage (50–200 mg/L) were studied. The adsorption experiments indicated that modified coal fly ash was effective in removing of Remazol Blue. The percentage removal of dyes increased while the modified fly ash dosage increased. The percentage removal of dyes increased with decreased initial concentration of the dye and also increased with amount of adsorbent used. The optimum of removal of dyes was found to be 94% at initial dye concentration 50 g/mL, modified fly ash dosage 250 g/mL, and pH of 2.0.

## 1 Introduction

A large number of dyes are produced annually and applied in various industries, including textiles, cosmetics, paper, leather, pharmaceuticals and food. Synthetic dyes and pigments are extensively used for dyeing and printing in industry. The textile industry accounts for two-thirds of the total dye market, consuming mostly reactive azo dyes due to the high demand for cotton fabrics with brilliant colors. In process of dyeing and printing colored products on textile industry, wastewater contaminated with dyes is generated. Discharge of hazardous wastewater without further treatment can seriously damage the environment. The colored discharged effluents inhibit penetration of sunlight and oxygen which are crucial requirements of aquatic life. Biodegradation of dyes is not an easy process due to their toxic and complex aromatic structure.

The dyes used in the textile industry are classified into three classes: (i) anionic dyes (direct, acid, and reactive), (ii) cationic, and (iii) nonionic. Basic and reactive dyes are widely used in the textile industry because of its favorable characteristics with bright colors, easy soluble in water, simple application techniques, and low energy consumption. Some dyes are toxic and carcinogenic and require separation and further treatment of textile wastes prior to discharge into environment.

Removals of dyes from water and wastewaters have been investigated intensively in last decades. The methods of dyes removal include biological, physical (membrane filtration, adsorption, coagulation, flocculation, precipitation, reverse osmosis, ion exchange, etc.) and chemical (oxidation, ozonation, etc.) processes. However, most of these methods exhibit major drawbacks, such as a high operational cost and low efficiency. Adsorption is a suitable physical process for the removal of low concentration of dyes and pigments. On adsorption method, the dyes and pigments would accumulate on the surface of solids at the interface between two phases (liquid–solid interface or gas–solid interface). The substance that accumulates at the interface is called adsorbate and the solid on which adsorption occurs is adsorbent. The system is easily regenerated and desorption may take place with modification of physical and chemical conditions. Porous structure of the adsorbent, chemical nature of the surface and pH of the aqueous solution act as major parameters in adsorption and desorption processes. Adsorption can be classified into two types: chemical sorption and physical sorption. Chemical adsorption or chemisorption is illustrated by the formation of strong chemical associations between molecules or ions of adsorbate to adsorbent surface. Chemisorption is generally due to the exchange of electrons and thus chemisorption is irreversible. Physical adsorption or physisorption is characterised by weak van der Waals intraparticle bonds between adsorbate and adsorbent and thus reversible in most cases.

Coal fly ash is a byproduct of coal combustion generated in power plant. Coal fly ash is generated at temperature of 1200-1700 °C from the various constituents of the coal. Each year nearly 42 million tons of coal is used in the Indonesia to generate electricity at the power plant. Currently, its applications are limited to as cement and brick and as a filling in road works. Research is therefore needed to develop new alternative environmental friendly applications. The high percentage of silica and alumina in fly ash makes it a good candidate for utilization as an inexpensive adsorbent for dyes removal.

* Corresponding author: arif.hidayat@uii.ac.id

Numerous approaches have been found in the literature regarding the adsorption of chemical dyes on various adsorbents such as rice husk [1], peanut shell [2], coconut shell [2], coffee residues [4], eggshells [5], clay [6] etc. In the last few decades, fly ash, by-product from coal combustion in thermal power plants, has been used as an adsorbent in removal of dyes from water and wastewaters. Due to environmental considerations, coal fly ash are considered to be an adsorbent because they are cheap, available at large quantities and easily accusable sources. Saakshy et al. (2016) have studied the adsorption of direct black dye onto fly ash. It has been found that fly ash is a potential low cost adsorbent to treat effluent of handmade paper industry. Jarusiripot C. (2014) have investigated the removal of reactive dye in aqueous solution on coal bottom ash. The coal bottom ash was pretreated with 1 N HCl, 1 N $HNO_3$, and 30% $H_2O_2$ before the adsorption tests. Visa and Chelaru (2015) investigated the adsorption behavior of modified fly ash in wastewater treatment for removal of dyes by adsorption. Ferrero (2015) studied the removal of acid and basic dyes from aqueous solution using coal fly ash. In another work, the removal of disperse blue and disperse orange dyes by coal fly ash from aqueous solutions was studied by Kisku et al. (2015).

Although several investigations have been reported on removal of textile dyes using fly ash, more information is still needed for better understanding of the adsorption behavior of dyes on fly ash. The purpose of this study was to investigate the adsorption of reactive dyes on modified coal fly ash, which is a low cost adsorbent, very abundant and inexpensive material. The effects of initial dye concentration, solution pH, contact time, and adsorbent dosage on the adsorption of reactive dyes on modified coal fly ash were studied.

## 2 Methodology

### 2.1 Material

The main raw material, coal fly ash samples is collected from electrostatic precipitators of Suralaya power plants, West Java. The samples contains both amorphous (mainly $SiO_2$, $Al_2O_3$) and crystalline components (mainly quartz and mullite). Prior to an experiment, the the raw coal fly ash samples is dried at 300 °C for 48 h in order to remove any traces of moisture or other contaminants. Then, the size of coal fly ash is reduced to a powder by crushing it in the ball mill and screening to get uniform size distribution. The coal fly ash is treated with concentrated hydrochloric acid (acid/ash= 5 ml/g) at a temperature of 70 °C for 1 h. The acid treatment removed iron to a certain extent, thereby increasing the activity, thermal stability and acidity of the zeolite, all aiming for better catalytic applications. And then, the sample is then neutralized by adding $H_2O$ until the pH of the sample is 7. Then, the samples were characterized using Brunauer Emmett Teller (BET) surface area analysis, X-ray powder diffraction (XRD), Fourier transform infrared spectrophotometer (FT-IR), and scanning electron microscope (SEM). The dyes used in the adsorption experiments was Remazol Blue R which obtained from a batik industry.

### 2.2 Adsorption study

Batch experiments were carried out to measure the adsorption characteristics of Remazol Blue on modified coal fly ash. Prior to their use, the adsorbents were oven-dried at 378 K for 4 h to eliminate traces of moisture. A certain amount of the modified coal fly ash sample was added to 50 ml of dye solution of varying concentration (50, 100, 150, 200 mg/L) in a 250 ml conical flask. The mixture was agitated in a temperature controlled shaker at a constant speed of 250 rpm for 24 h at a constant temperature of 298 K. After stirring, the mixtures were separated by filtering. The residual concentration of dye solution was determined using a calibration curve prepared at the corresponding maximum wavelength using a UV–visible spectrometer. The effect of pH was studied for Remazol Blue dye of 50 mg/L concentration at pH 4, 6, 8 and 10. The pH was maintained with the addition of 1N NaOH mg/L and 1N HCl. The effect of adsorbent dosage was also studied at 4, 5, 10 and 20 g/L at optimized pH and initial dye concentration.

The percentage removal and amount adsorbed were calculated by the following equation:

$$\%\ removal\ of\ dye = \frac{C_0 - C_e}{C_0} x100 \tag{1}$$

Where: $C_0$ is the initial concentration of dye and $Ce$ is the concentration of dye at equilibrium

## 3 Results and Discussion

### 3.1 Material Characterization

The textural characteristics of fly ash were studied by using the $N_2$ adsorption method. The well-known Brunnauer-Emmett-Teller (BET) equation was used for the calculation of specific surface area, SBET, pore volume, V, and the average pore diameter, R. The results for raw (RFA) and modified fly ash (MFA) are given in Table 1. It reveals that the acid treatment enhances the surface area and pore volume of the RFA, while decreases its average pore diameter. This may be explained by the increasing micropore content of the MFA after the impurities and organics were removed by acid and water.

**Table 1.** Surface area and pore size of raw and treated fly ash

| Material | Raw FA | Modified FA |
|---|---|---|
| $S_{BET}$ (m$^2$/g) | 4,975 | 45,716 |
| $V_{total}$ (cm$^3$/g) | 0,0137 | 0,1572 |
| Pore diameter (nm) | 13,66 | 11,99 |

The SEM photographs of RFA and MFA are shown in Figure 2. It can be seen that the surface morphology of RFA is slightly changed by the acid treatment. The SEM

 https://doi.org/10.1051/matecconf/201815401037

image clearly shows that the particles are mainly composed of irregular and porous particles.

(a)

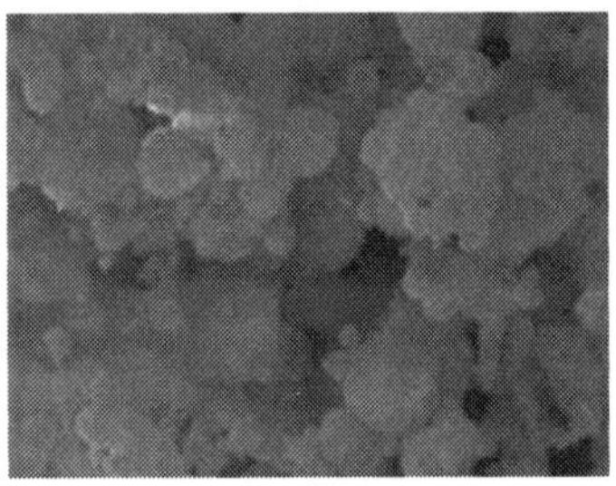

(b)

**Fig. 2.** SEM paragraphs of (a) RFA and (b) MFA.

The XRD of RFA and MFA are shown in Figure 2. It is seen that there are not significant difference for all XRD profiles. But acid-treated samples present higher intensity of diffraction peaks. The major phases for all samples are quartz and mullite. Some minor phases, hematite and magnetite, are also existed. Those results suggest that acidtreatment will not induce bulk phase changes.

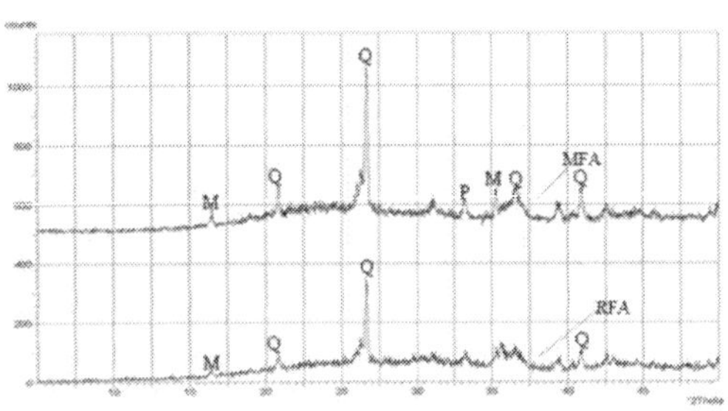


**Fig. 3.** The XRD patterns of (a) RFA and (b) MFA.

An FTIR analysis was performed in the range of 450–4000 $cm^{-1}$ in order to explore the surface characteristics of the samples. Figure 3 shows the FTIR spectrum of RFA and MFA. The band at 3200–3500 $cm^{-1}$ is due to stretch vibration (O–H). Moreover, the absorption bands at 1620–1650 $cm^{-1}$ were caused by deformational vibrations of adsorbed water molecules. Spectrum of the treated fly ash possesses three absorption bands, at 1084; 793, and 466 $cm^{-1}$. The 1084 and 466 $cm^{-1}$ bands are due to the asymmetric stretching and the bending vibration modes of M-O bonds, respectively (where M = Si or Al). The 793 $cm^{-1}$ band is assigned to the symmetric stretching vibration mode of O-M-O groups.

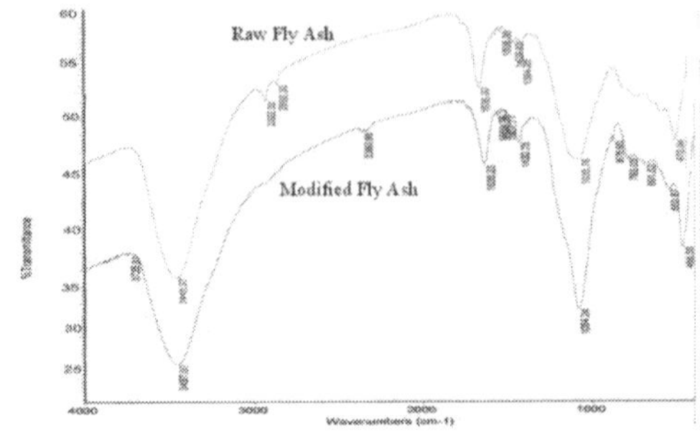


**Fig. 4.** FTIR spectra of raw and modified fly ash.

It has been seen that yield of the process increases with increase in molar ratio. With further increase in molar ratio the conversion efficiency more or less remains the same. However it can be concluded that the molar ratio increased from 10:1 to 12:1 there is no significant change observed with a higher molar ratio in the reduction of FFA. The maximum conversion efficiency is achieved very close to the molar ratio of 12:1.

### 3.2 Effect of initial dye concentration

The experimental results of removal of dyes on modified fly ash at various concentrations (50, 100, 150, and 200 mg/L) are shown in Figure 4. During removal of dyes, initially the dye molecules reach the boundary layer; then they have to diffuse into the adsorbent surface; and finally, diffuse into the porous structure of the adsorbent. It was observed that the percentage removal of dyes reached up to 94% at lower concentration. When the initial dye concentration was increased, the percentage removal of dyes increased. The percentage removal of dyes decreased with increasing the initial dye concentration because of nearly complete coverage of the binding sites at high dye concentration. The effect of initial dye concentration on removal of dyes could be attributed to the increase in the driving force of the concentration gradient with an increase in the initial dye concentration.

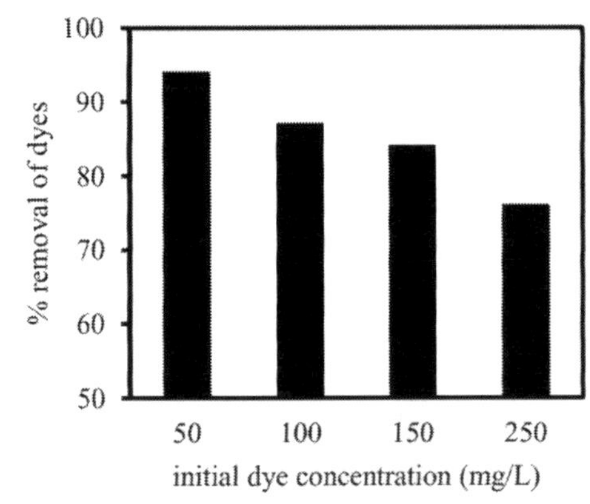


**Fig. 5.** Effect of initial dye concentration on removal of dyes.

### 3.3 Effect of modified fly ash dosage

The removal of dyes on modified fly ash dosage was studied by varying the modified fly ash dosage (50–200 mg/L) at pH 6.0. The removal of dyes at various dosages of modified fly ash is shown in Figure 5. The percentage removal of dyes increased while the modified fly ash dosage increased. The percentage removal of dyes increased from 78, 81, 89, to 94 for the modified fly ash dosage 50, 100, 150, and 200 mg/L, respectively. This can be attributed to increased adsorbent surface area and availability of more adsorption sites.

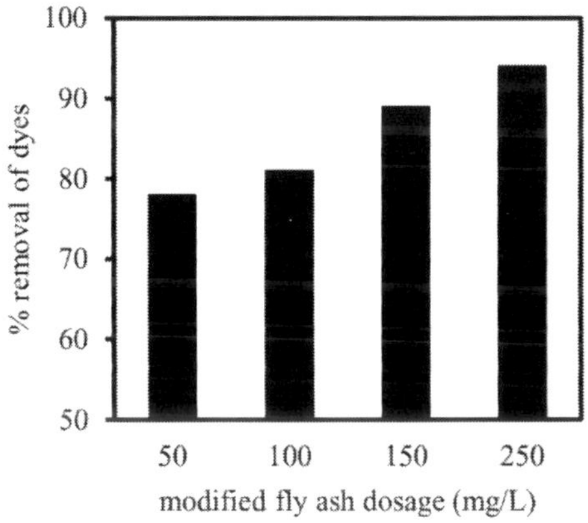


**Fig. 6.** Effect of modified fly ash dosage (pH 6, initial dye concentration 50 mg/l, temperature 30 °C, stirring speed 250 rpm).

### 3.4 Effect of solution pH

One of the most important factors affecting the capacity of adsorbent in removal of dyes is solution pH. The percentage removal of dyes is dependent on the solution pH, since variation in pH leads to the variation in the degree of ionisation of the adsorptive molecule and the surface properties of adsorbent (Nandi, 2009). The study was carried out with modified fly ash fly ash of dosage 5 g/L on reactive dye of 50 mg/L concentration maintained at pH from 4 to 10. The removal of dyes by fly ash at various pH is shown in Figure 6. The removal of dyes increased when the final pH was increased from 2.5 to 8.5. After these pH ranges, the removal of dyes slightly decreased. In general, at low solution pH, the percentage of removal of dyes will decrease for cationic dye adsorption, while for anionic dyes the percentage of dyes removal will increase. In contrast, high solution pH is preferable for cationic dye adsorption but shows a lower efficiency for anionic dye adsorption. At high solution pH, the positive charge at the solution interface will decrease while the adsorbent surface appears negatively charged. As a result, the cationic dye adsorption will show an increase and the anionic dye adsorption will decrease. At low pH solution, the positive charge on the solution interface will increase and the adsorbent surface will appear positively charged, which results in decrease in cationic dye adsorption and an increase in anionic dye adsorption.

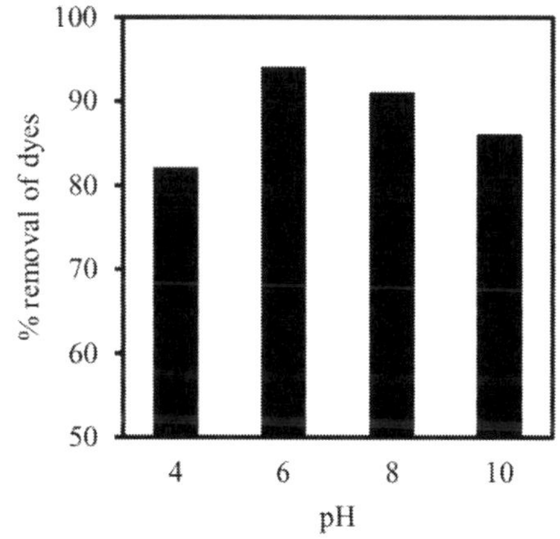


**Fig. 7.** Effect of pH (initial dye concentration 50 mg/L, fly ash dosage 50 mg/L, temperature 30 °C, stirring speed 250 rpm).

## 4 Conclusions

The adsorption experiments indicated that modified coal fly ash was effective in removing of Remazol Blue. The percentage removal of dyes increased while the modified fly ash dosage increased. The percentage removal of dyes increased with decreased initial concentration of the dye and also increased with amount of adsorbent used. The optimum of removal of dyes was found to be 94% at initial dye concentration 50 g/mL, modified fly ash dosage 250 g/mL, and pH of 2.0.

The adsorption experiments indicated that modified coal fly ash was effective in removing of Remazol Blue. The percentage removal of dyes increased while the modified fly ash dosage increased. The percentage removal of dyes increased with decreased initial

## References

1. L. Leng, X. Yuan, G. Zeng, J. Shao, X. Chen, Z. Wu, H. Wang, X. Peng, Fuel, **155**, 77-85 (2015)
2. J. Georgin, G.L. Dotto, M.A. Mazutti, E.L. Foletto, Journal of Environmental Chemical Engineering, **4** (1), 266-275 (2016)
3. A.M. Aljeboree, A.N. Alshirifi, A.F. Alkaim, Arabian Journal of Chemistry, **10**, Supplement 2, s3381-s3393 (2017)
4. G.Z. Kyzas, N.K. Lazaridis, A.Ch. Mitropoulos,. Chemical Engineering Journal, **189-190**, 148-159 (2012)
5. M.A. Abdel-Khalek, M.K. Abdel Rahman, A.A. Francis, Journal of Environmental Chemical Engineering, **5** (1), 319-327 (2017)
6. S.C.R. Santos, R.A.R. Boaventura, Journal of Environmental Chemical Engineering, **4** (2), 1473-1483 (2016)
7. Saakshy, K. Singh, A.B. Gupta, A.K. Sharma Journal of Cleaner Production, **112** (1), 1227-1240 (2016)

8. C. Jarusiripot, Procedia Chemistry, 9, 121-130 (2014)
9. Visa M., Chelaru A. M. Applied Surface Science, **303**, 14-22 (2014)
10. Ferrero F., Clean Technologies and Environmental Policy, **17** (7), 1907-1915 (2015)
11. G.C. Kisku, Markandeya, S.P. Shukla, D.Sen Singh, R.C. Murthy, **74** (2), 1125-1135 (2015)
12. B. Nandi, A. Goswami, M. Purkait, Appl Clay Sci, **42** (3–4), 583-590 (2009)

# Fractality evaluation for pulmonary crackle sound using the Degree of Self-Similarity

*Achmad* Rizal[1,2,*], *Risanuri* Hidayat[2], and *Hanung Adi* Nugroho[1]

[1] Universitas Gadjah Mada, Department of Electrical Engineering and Information Technology, 55281, Yogyakarta, Indonesia
[2] Telkom University, School of Electrical Engineering, 40287, Bandung, Indonesia

**Abstract.** Lung sound is a complex signal produced by the respiratory process. The complex signal has several properties including a chaotic behavior, fractality or self-similarity property. One of lung sounds that arise from abnormalities occurred in the respiratory tract is pulmonary crackle sound. In this study, we tested the degree of self-similarity of pulmonary crackle sound and examined whether the degree of similarity can be used as a feature to differentiate the pulmonary lung crackle sound with normal lung sound. The results showed the sufficient strength of the self-similarity nature of the pulmonary crackle sound. Meanwhile, a test using K-mean clustering produced an accuracy of 87.5% to differentiate between the pulmonary crackle sound and normal lung sound. It can be stated then that it is deemed important to take another feature to obtain higher accuracy. The high self-similarity degree indicates that a pulmonary crackle sound has fractals properties.

## 1 Introduction

Pulmonary crackle is an adventitious lung sound occurred in the respiratory sounds indicating several abnormalities in the airways. It might be caused by some diseases including lung fibrosis, pneumonia, and chronic bronchitis [1]. Lung sound could be heard using a stethoscope by doctors in diagnosing certain abnormalities in the respiratory system. It is commonly diagnosed in agreement with the expertise and experience of the physician. Along with the development of digital signal processing technology, many methods have been developed to analyze lung sounds automatically.

Various methods have been used to detect the pulmonary crackle sound automatically. In a study reported by Mendes and co-author [2], some Teager features such as energy, information entropy, and local Higuchi fractal dimension were calculated on the non-stationary part of the output of wavelet packed stationary transform-non-stationary transform filter (WPST-NST). In another study, some tests were conducted on the effects of the use of the window, wavelet, and machine learning for the pulmonary sound detection in the time-frequency domain and time-scale domain [3]. The results indicated that Support Vector Machine (SVM) as a classifier could produce the highest accuracy compared to multilayer perceptron and K-NN [3]. Rizal and co-workers used multi-order Tsallis entropy (TE) as a feature extraction method for pulmonary crackle [4]. The reported results showed that TE with the order of 2, 3, and 4 could produce accuracy up to 95.35%. Another study as reported took only three sub-bands (D3, D4, and D5) to analyze the crackle sound using a wavelet analysis [5]. The suggested mother wavelets were Daubechies 7 and Symlet 7.

Lung sound is a complex biological signal. One characteristic that often appears on a complex signal is fractal properties. Most of the fractals have a self-similarity feature. The self-similarity of the signal can be tested using several methods such as R/S method, variance method, the absolute moment method, and variance ratio of residuals [6]. In this study, we measured the degree of self-similarity (H) using the variance method to observe the characteristics of pulmonary crackle. Self-similarity is one of the characteristics of the fractal signal. Using the degree of self-similarity (H), we could find the fractality nature of pulmonary crackle sound. The H value would be used to determine whether the fractal method could be used for a lung sound analysis in future studies. It also tested whether the value of H could be utilized as a feature to distinguish between crackle sound and normal lung sound.

## 2 Materials and Methods

The flowchart of the process in this paper is displayed in Fig.1. First, normalization process was applied in the lung sound signal. Then, we calculated the degree of self-similarity (H) of the lung sound signal. We used maximum block size m = 100 in this paper. We analyzed the H value of normal and crackle lung sound using ANOVA then tested using K-mean clustering.

* Corresponding author: rizal.s3te14@mail.ugm.ac.id

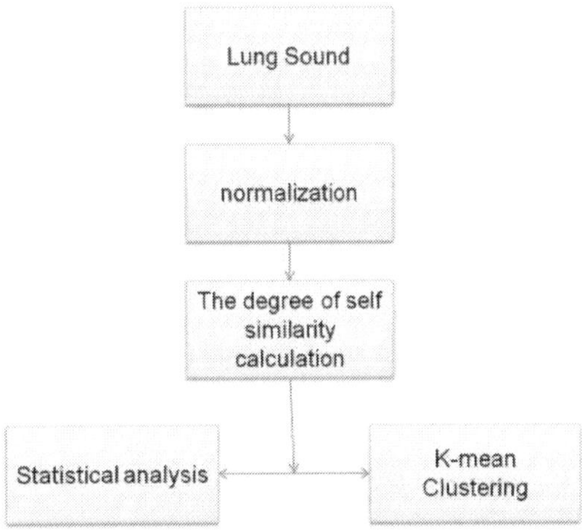


**Fig. 1.** Flowchart of the process in the research

K-mean clustering was used to test whether the H value could be used as feature for lung sound classification. The detail of each process will be described in next subsection.

### 2.1 Lung Sound Data

We used the same lung sound data set were collected from the internet as did in a previous paper [4]. The data consisted of 20 crackle sound data and 20 normal bronchial sounds. Crackle sounds were taken from cystic fibrosis and interstitial pulmonary fibrosis patient. Meanwhile, normal data was recorded from 26-year-old man [7]. Data had a single respiratory cycle with a sampling frequency of 8000 Hz. To overcome the differences in the recording process, we did a normalization process as shown in equation (1).

$$y(n) = x(n) - \frac{1}{N}\sum_{i=0}^{N-1} x(n) \quad (1)$$

where x (n) is the input signal, and y (n) is the result of normalization. This process was intended to eliminate the DC component in the signal. Amplitude normalization process was then performed as in equation (2).

$$y(n) = \frac{x(n)}{max|x|} \quad (2)$$

where |x| is a maximum absolute value of signal x. By the normalization process, the input signal would have a range from -1 to +1 and a zero mean.

### 2.2 The Degree of Self-Similarity

The degree of self-similarity (H) was proposed by Kalden and Ibrahim to measure self-similarity on GPRS signal [6]. By using H value, the characteristic of the traffic on the GPRS network could be predicted. If a sequence of a signal $x(i)$ is stated $y^m(k)$ as the aggregated process. $y^m(k)$ is formed from a non-overlapping block with the size m as in (3).

$$y^m(k) = \frac{1}{m}\sum_{l=0}^{m-1} x(km - l) \quad (3)$$

If the signal x (i) is stated as self-similar, the variance of the aggregated process will decay based upon the increase of block size m. The variance of $y^m$ is expressed as in (4).

$$var(y^m) \approx m^{-\beta} \quad (4)$$

where 0 < β < 1, hence the degree of self-similarity (H) can be expressed as in (5)

$$H = 1 - \beta/2 \quad (5)$$

The high value of H shows a strong self-similarity nature [8]. The H value can be an indication of the fractal properties of the signal.

## 3 Results and Discussion

Typical pulmonary crackle sound and normal bronchial are shown in Fig. 1 indicating that the duration of the crackle sound tends to be short and to have the largest frequency below 500 Hz. On the other hand, the normal bronchial sound has a clear pause between inspiration and expiration phase. Normal bronchial sound tends to continue.

Fig. 2 shows the variance decay as a function of block size m, as we used m = 1 to 100. Normal lung sound has slower decay than pulmonary sound so that the H value of normal lung sound is higher than pulmonary crackle as explained in (4) and (5). The boxplot of H value for pulmonary crackle sound and normal lung sound is shown in Fig. 3. Some H value of normal lung sound has a lower value than other so make it overlapped with H value of pulmonary crackle. In another hand, some H value of pulmonary crackle has a higher value to make it have the same value with normal lung sound.

To test the H value separation between normal bronchial sounds crackle, we used Analysis of Variance (ANOVA). Using single-factor ANOVA test, the F-value of 31.93 with a p-value < 0.001 were obtained. This indicated that the value of H in normal bronchial sounds and crackle sound statistically significant differences.

Further, we tested the possibility of using the value of H as a feature for lung sound classification using K-mean clustering. K-mean clustering groups the data into K clusters. The centroid of each cluster was calculated from the average of data grouped in it. Centroid and the distance calculation process were repeated until no data moved to another cluster. We used accuracy for performance evaluation parameter. We defined accuracy as:

$$accuracy(\%) = \frac{number\ of\ data\ grouped\ in\ correct\ cluster}{\sum data} \times 100\% \quad (6)$$

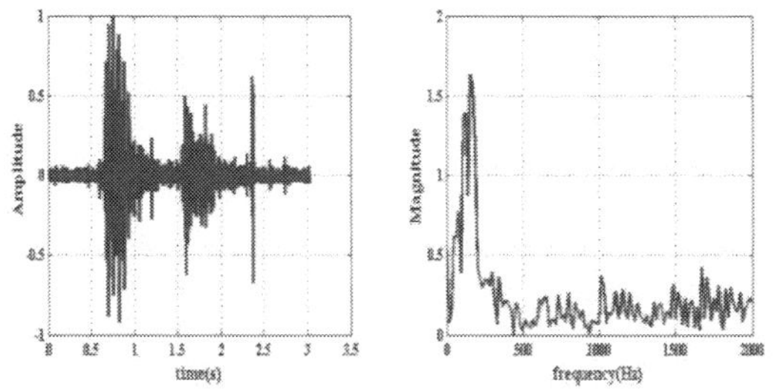


(a)

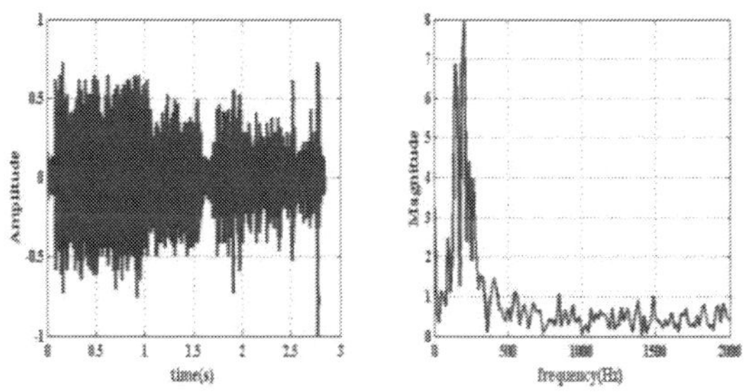


(b)

**Fig. 1.** (a) Crackle sound and the spectral (b) Normal bronchial sound and the spectral

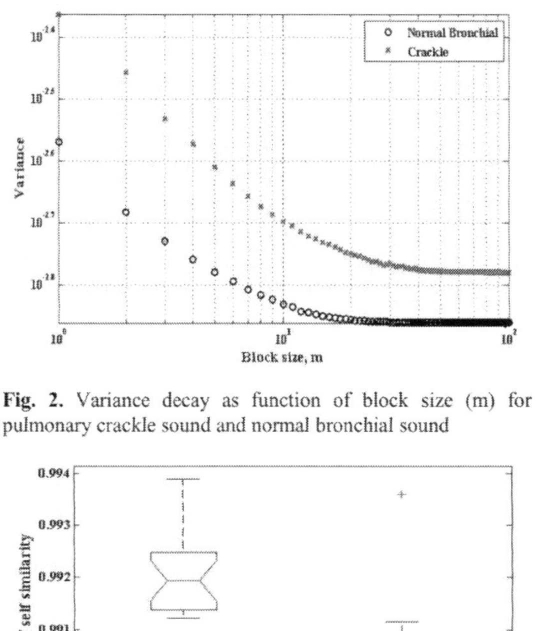


**Fig. 2.** Variance decay as function of block size (m) for pulmonary crackle sound and normal bronchial sound

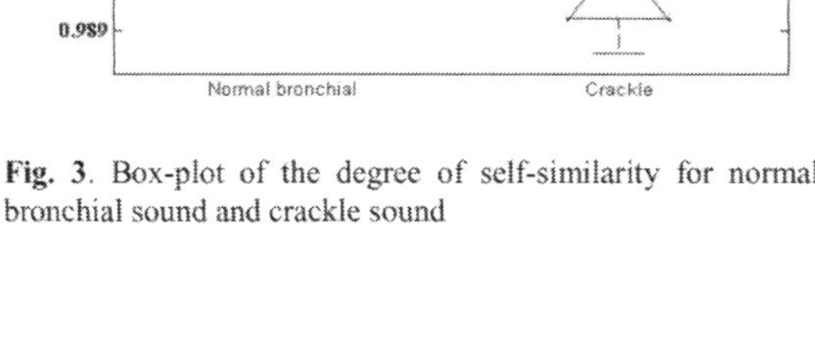


**Fig. 3**. Box-plot of the degree of self-similarity for normal bronchial sound and crackle sound

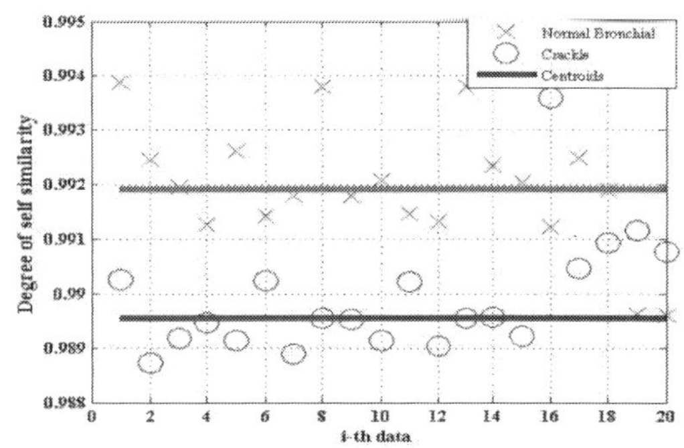


(a)

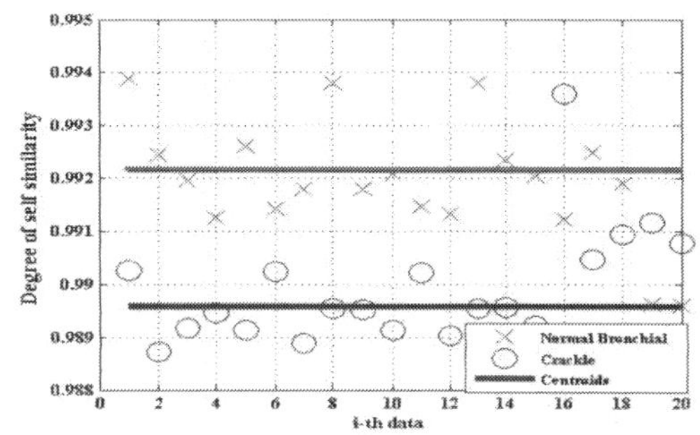


(b)

**Fig. 4.** (a) K-mean clustering result using city-block distance (b) K-mean clustering results using Euclidean distance

We used K-mean clustering with two distance measurement techniques as presented in Fig. 4. As seen in this Figure, some crackle sound data had a value closer to the normal bronchial centroid and vice versa. By using city-block distance, six data went to the wrong cluster, so the accuracy became 85%. On the other hand, by using Euclidean distance, five data were grouped into a wrong cluster in which the accuracy became 87.5%. The results showed that although the H value for pulmonary crackle was statistically different from the normal bronchial sound, it still requires another feature for enhancing the classification accuracy.

Some evaluations of lung sound characteristics had been done in the earlier study. The chaotic dynamics of respiration sound was reported in [9]. Lyapunov spectra, correlation dimension, and Kaplan-Yorke dimension indicated that respiratory sound had a chaotic behavior. Meanwhile, Gnitecki and Mousavi tested fractality of lung sounds using the Katz fractal dimension (KFD), Sevcik fractal dimension (SFD), and the variance fractal dimension (VFD)[10]. The results indicated that the measurement KFD and SFD could lead to defining respiratory sound as a fractal. With more lung sounds properties revealed, it would be easier for researchers to conduct further research. From these results, in future, it will be explored a fractal dimension as a feature for the classification of the various lung sound classes.

## 4. Conclusion

This paper presented the degree of self-similarity measurement for pulmonary crackle sound. The results showed that pulmonary crackle sound was very strong indicated as a self-similarity signal. Pulmonary crackle sound also had a different degree of self-similarity compare with a normal bronchial sound. Using K-mean clustering, H value achieved a classification accuracy of 87.5% for normal and crackle sound data. This result indicated a need for other features to enhance the accuracy. Based on this result we can state that fractal dimension can be used for the extraction of crackle sound features for the next study

## References

1. A. Bohadana, G. Izbicki, S. S. Kraman. N. Engl. J. Med., **370**, 8 (2014)
2. L. Mendes, P. Carvalho, C. A. Teixeira, R. P. Paiva, J. Henriques. *2014 36th Annual International Conference of the IEEE Engineering in Medicine and Biology Society* (2014)
3. G. Serbes, C. O. Sakar, Y. P. Kahya, N. Aydin. Digit. Signal Process. A Rev. J. **23**, 3(2013)
4. A. Rizal, R. Hidayat, H. A. Nugroho. *The 1st 2016 International Conference on Biomedical Engineering (iBioMed)* (2016)
5. V. I. Quandt, E. R. Pacola, S. F. Pichorim, H. R. Gamba. Res. Biomed. Eng. **31**, 2 (2015)
6. R. Kalden, S. Ibrahim. Lect. Notes Comput. Sci. **3015** (2004)
7. "The R.A.L.E Repository." [Online]. Available: http://www.rale.ca/Repository.htm. [Accessed: 22-Jul-2015].
8. D. K. Kumar, S. P. Arjunan, B. Aliahmad, *Fractal: Application in Biological Signaling and Image Processing* (CRC Press Book, New York, 2016)
9. C. Ahlstrom, A. Johansson, P. Hult, P. Ask, Chaos, Solitons Fractals, **29**, 5 (2006)
10. J. Gnitecki, Z. Moussavi. Chaos, Solitons Fractals, **26**, 4 (2005)

# Quantum dot solar cell studies on the influence of Cadmium Selenide(CdSe)QDs and the Zinc Sulfide(ZnS)QDs in the photoanode

*Mohammad Azren* Saad[1,*], *Noor Syafiqah* Samsi[1], *Oskar Hasdinor* Hassan[3], *Muhd Zu Azhan* Yahya[2,4], *Mohamad Fariz* Mohamad Taib[1,2] *Ab Malik Marwan* Ali[1], and *Rosnah* Zakaria[1,2,*]

[1]Faculty of Applied Sciences, Universiti Teknologi MARA, 40450 Shah Alam, Selangor, Malaysia
[2]Institute of Science, Universiti Teknologi MARA, 40450 Shah Alam, Selangor, Malaysia
[3]Faculty of Art and Design, Universiti Teknologi MARA, 40450 Shah Alam, Selangor, Malaysia
[4]Faculty of Defence Science & Technology, Universiti Pertahanan Nasional Malaysia, 57000

**Abstract.** The mixture between the difference semiconductor quantum dot sensitizer which is cadmium selenide(CdSe) and zinc sulfide (ZnS) into the Dye-synthesis solar cell (DSSCs) can affect the value of resistance and capacity photoanode in the system.In this experiment, each sample consists difference weight percent of Zinc sulfide and the constant weight percent of CdSe. Docter blade technique is used to stick and spread evenly the mixture CdSe/ZnS QD on the surface of the thin film. To prove the assembled of CdSe/ZnS on the thin film were observed using Scanning Electron Microscopy (SEM). The resistance and capacity of the photoanode were characterized by using impedance spectroscopy(EIS). The smallest resistance is 37.1kΩ produce by CdSe/ZnS(20 wt%) and the largest resistance 825KΩ produce by CdSe/ZnS(50 wt%) while the highest capacity is 12 μF in the CdSe/ZnS(40 wt%) and the lowest capacity is CdSe/ZnS(20 wt%) which is 538 nF. The most suitable composition to be used as photoanode is CdSe/ZnS(40 wt%) because it has high capacity and low resistance.

## 1 Introduction

Nowadays the world is facing the energy crisis, attracted attention of many researchers to study an alternative sources. Solar energy is an abundant energy and one of the most suitable renewables that could replace the hydrocarbon. DSSCs consist four main part which is an anode, cathode, sensitizing dye, and electrolyte. Anode electrode basically contains Indium tin oxide(ITO)glass, covers by titanium dioxide ($TiO_2$) on the conductive surface of the ITO glass[1].

The use of semiconductor as quantum dots as photoanode in Dye-sensitized solar cells(DSSCs) devices may reduce pollution and low cost-effective in converting solar energy to electric energy[2,3]. The nanoparticle of quantum dot can gives high photostability, wider absorbance spectrum and resistance to photobleaching and make it suitable to be a synthesizer in a photoanode of DSSCs[4]. The function of photoanode is to produce the current by emitting the electron from their valence shell into the conducting band[5].

CdSe/ZnS is the one of the QDs, has a very small band gap that can absorb a very wide absorbance spectrum to excited electron producing the hole on the surface on the sample[3].This process is called the recombination process. CdSe QDs alone can be toxic toward the biological environment. However, by doping ZnS with CdSe it can lower down the toxicity in CdSe[6].

$TiO_2$ with injection of dye is a medium to absorb light and excite electrons at below than visible light at 3.2 eV[7]. The converted energy will be used by the electron to release from the valence band into the conducting band. The flow of the electron will produce current in the circuit. On the other hand, the amount of light absorption is still low and this is the reason why combination of semiconductor quantum dot QDs into photoanode were introduced[8].

Resistance is the factor that slow down the flow of electron in the sample, if the sample have the low value of resistance it will produce a high current. Capacity is the ability of sample to store energy , more energy can be store in the sample that have high capacity[10].

## 2 Methodology

The conductivity of CdSe/ZnS QDs will be determined through the methodology of preparation of the material. Optimum wt% of the combination can give more number of hole in the structure which can increase the conductivity of the photoanode[9, 5].

### 2.1 Preparation of CdSe/ZnS solution

A control mass of ruthenium-based dye (N719) at 0.045 g was added into 9ml of ethanol and stirred using magnetic

*Corresponding author: azrensaad@yahoo.com.my

stirrer for 24 hours The mass of CdSe/ZnS was calculated to make it as 10 - 50 wt% using the Eq 1. Detail mass was tabulated in Table1. These samples were prepared in a dark environment.

$$W_{cdse/zns}=((wt\%)(wn719))/(1-(wt\%)) \quad \text{Eq. 1}$$

**Table 1.** Composition of the sample prepared with different percentage of CdSe/ZnS QD.

| Sample | CdSe/Zns QDs composition(wt%) | Mass of N719dye powder | Mass of CdSe/ ZnS QDs powder (g) |
|---|---|---|---|
| A | 10 | 0.045 | 0.005 |
| B | 20 | 0.045 | 0.011 |
| C | 30 | 0.045 | 0.019 |
| D | 40 | 0.045 | 0.030 |
| E | 50 | 0.045 | 0.045 |

## 2.2 preparation of photoanode

The ITO glass was clean using an ultrasonic bath. The $TiO_2$ paste was coated onto conductive ITO glass using the doctor blade technique. This ITO's coated glass then left to dry for 1 day in the room temperature.
The dry coated glass with $TiO_2$ was immersed in the mixture solution of N719/CdSe-ZnS in dark environment for 24 hours. Then, the sample was clean using ethanol to remove the residue and dried in the oven at 60 °C.

## 2.3 Characterization ITO/TiO2/N719/CdSe photoanode

### *2.3.1 Scanning electron microscope (SEM)*

Scanning electron microscope (SEM) used to analyze the appearance of ZnS QDs and CdSe QDs as glowing materials on the surface of the $TiO_2$ glass. This photoanode was analyzed using Energy Dispersive X-ray (SEM/EDX) spectroscopy to see the existence of CdSe/ZnS QDs on the surface of the photoanode [10].

### *2.3.2 Electrochemical Impedance Spectroscopy*

The Electrochemical Impedance Spectroscopy(EIS) method using Potentiostat Galvanostat will produce Cole-Cole plot to each sample.This plot shows the information about the capacity and resistance in the sample[9].The electrical impedance spectroscopy, EIS is an instrument to measure the resistance and capacity analysis of co-sensitized [11]. The QDs (ITO/TiO2/N719-CdSe/ZnS) were used as a working electrode ($W_e$), a platinum electrode is act as counter electrode ($C_e$) and Ag/AgCl/3MKCl is be the reference electrode. The model was used iodide/iodine as electrolyte of the cell. The characterization was used the frequency range of 0.009 - 100000 Hz for 50 readings. The Cole-Cole plot were analyzed using equivalent circuit by Nova software.

# 3 Results and Discussions

Figure1 shows the scanning electron microscope (SEM) at the resolution of 128x128 pixels, using the Energy Dispersive X-ray (SEM/EDX)spectroscopy, proved that the assembly of CdSe/ZnS QDs onto $TiO_2$. This finding is important to improve the efficiency of DQs ZnS/CdSe Photoanode in the QDSSCs.

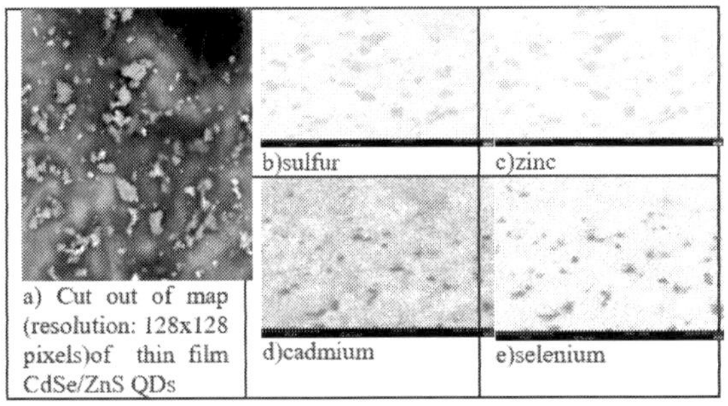


**Fig. 1**: The image of QDs (CdSe/ZnS) on the surface of Photoanode

The Figure 1b, 1c, 1d, 1e shows the different color which is represented the different type of material. Figure 1b shows the location of blue color represented the location of zinc and green spot in the figure 1c represent the location of sulfur and CdSe trace in figure 1d and 1e.Figure 1d show the presence of Cd(cadmium) and figure 1e show the presence of Se (selenide).The pattern dot color in each figure proves the element are interconnected each other, for example, figure1b and 1c have the same pattern this meant that this material is bonding each other and form ZnS.

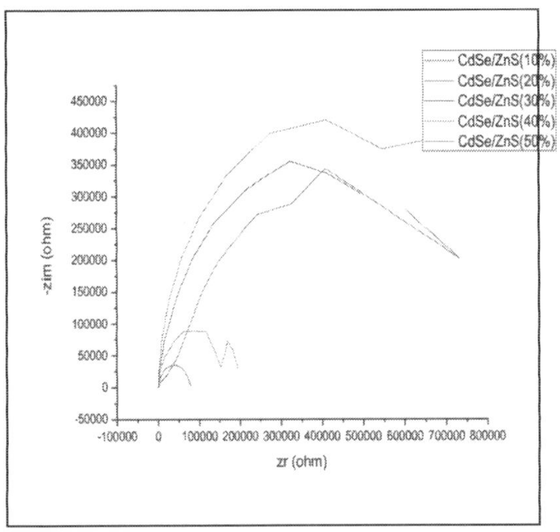


**Fig. 2**: The EIS graph of CdSe/ZnS QDs with different weight percent (Wt%)

The semicircle shown in graph 1 shows the difference value of resistance between the sample. The resistance value of each sample can be analyzed from the graph by

 https://doi.org/10.1051/matecconf/201815401039

measuring the diameter of the semicircle. The value of real impendence(Zr) represents the value of the resistance in the sample. The imaginary impendence shows the capacitance of the sample. This capacitance importance to analyze the ability of the sample to store the energy of electric charge. This graph proves that all sample produce the parallel resistance and capacity to each other and have single circuits, this single circuit represents the reaction of current flow in the photoanode and electrolyte. The resistance was calculated automatically by EIS device, by using the fitting and simulation setting on the device, it can create the circuit that is associated with our sample. This study used R(RC) meant single circuits.

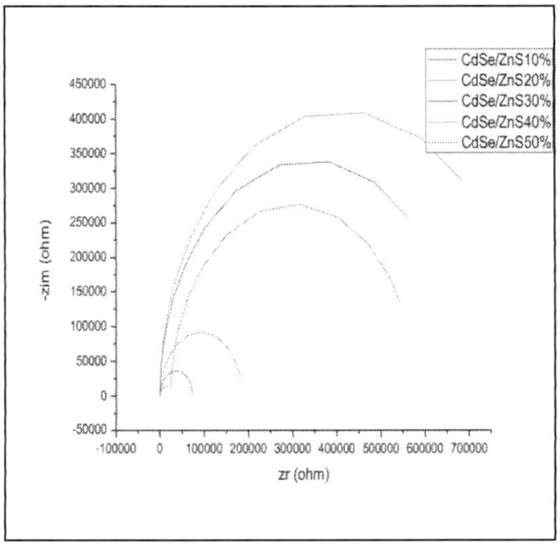


**Fig. 3**: The fitting EIS graph of CdSe/ZnS QDs with different weight percent (Wt%)

The fitting EIS graph for Figure 3 was drawn using the EIS fitting simulation. This simulation helps to find the magnitude of resistance in the sample. In Figure3 each of the samples shows the difference semicircle size, from observation it clearly found that the largest semicircle is produced by the combination between CdSe/ZnS (50 wt%). The curve of CdSe/ZnS(10 wt%), CdSe/ZnS(20 wt%) is bigger compared to the CdSe/ZnS(30 wt%) , CdSe/ZnS(40 wt%), from this result it show that 50%,10%and 20 wt% of ZnS in the photoanode produce the higher resistance compared to the 30 wt% and 40 wt% ZnS. The higher resistance in the material will slow down the movement of the electron, as the result, this will lower down the current produced by the system.

Resistance is the factor that can reduce the flow of electron as the result it will lower down the value of current. The factor that increases the value resistance is the structure of the material. In Dye-synthesis solar cell the structure that is needed to increase the flow of electron is the hole. The hole is the place for an electron to jump from one atom to another until it reaches anode terminal. The value of resistance in table 2 clearly show that the highest resistance is CdSe/ZnS(50 wt%).The value of resistance in the CdSe/ZnS(50 wt%) is 825kΩ.Photo anode with 10% ZnS produces a second larger value of resistance compared to the first one. The lowest resistance is 20 wt%ZnS with 37.1kΩ From the graph in figure 3, 20 wt%ZnS produce two semicircles 37.1kΩis the value of resistance of the small semicircles.

**Table 2.** Value of resistance and capacity in CdSe/ZnS QDs with different weight percent (Wt%)

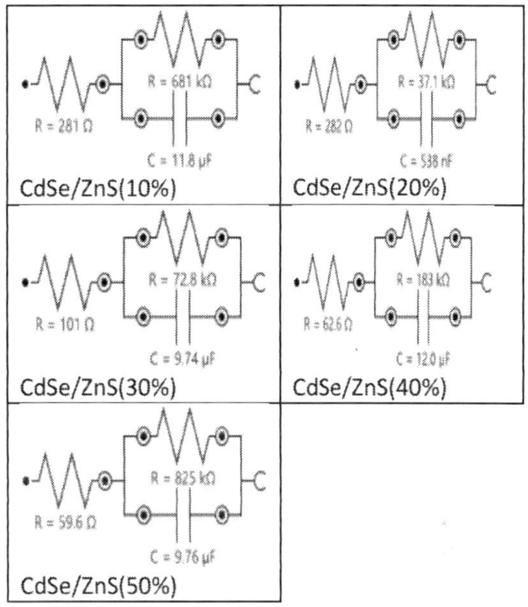


*Note: Adapted from potentiostat galvanostat (Auto Lab), using Nova software single circuit*

Capacity is the ability of a system to store an electric charge. Table 2 show the highest capacity is CdSe/ZnS 40 wt% with the 12 µF. The second large capacity produce is CdSe/ZnS(10 wt% ) and the value of capacitance is 11.8 µF. The capacity in CdSe/ZnS(20 wt%) is 538 nF, this is the lowest capacity produce compared to another sample.

## 4. Conclusions

The ZnS QDs combine with the CdSe QDs dissolve in the N719 effect the value of resistance and capacity. Electrochemical Impedance Spectroscopy (EIS) of varying the wt% of ZnS, found that the 40 wt% of ZnS is the most suitable to be added to the dye of photoanode because it produces the low resistant of 183kΩand high capacitance which is 12 µF compared to another sample. Furthermore, The SEM of ZnS / CdSe/N719 showed the existence of CdSe and Zinc sulfide and interconnected each other on the surface of ITO glass.

This work was partially supported by the Ministry of Higher Education (MOHE) Malaysia and financial support under the project FRGS/1/2015/SG06/UITM/02/5 grant and acknowledge the Universiti Teknologi MARA (UiTM) for the facilities provided.

## References

1. M. a M.Al-Alwani, A. B.Mohamad, N. a.Ludin, A. A. H.Kadhum, & K.Sopian,. Dye-sensitised solar

cells: Development, structure, operation principles, electron kinetics, characterisation, synthesis materials and natural photosensitisers. *Renewable and Sustainable Energy Reviews*, *65*, 183–213, (2016).

2. I.renewables eras-Saizarbitoria, E.Cilleruelo, & I.Zamanillo,Public acceptance of renewables and the media: An analysis of the Spanish PV solar experience. *Renewable and Sustainable Energy Reviews*, *15*(9), 4685–4696, (2011).
3. I. N.Obotowo, I. B.Obot, & U. J. Ekpe,. Organic sensitizers for dye-sensitized solar cell (DSSC): Properties from computation, progress and future perspectives. *Journal of Molecular Structure*, (2016)
4. K.Kim, , J. E EunS, Y.Chang, , J. Kim , C. Im, & M Lee, Electrochimica Acta ZnS-Passivated CdSe / CdS Co-sensitized Mesoporous Zn 2 SnO 4 Based Solar Cells. *Electrochimica Acta*, *121*, 223–232,(2014).
5. P. V. Kamat, (2008). Quantum Dot Solar Cells . Semiconductor Nanocrystals as Light Harvesters , 18737–18753.
6. S.Porcel, V.López-carrillo, C García-yebra, & M.Antonio,. Supporting Information ©, 1–13 (2008),
7. Z. Li, Y.Xie, H.Xu, T.Wang, Z.Xu, & H.Zhang,. Journal of Photochemistry and Photobiology A : Chemistry Expanding the photoresponse range of TiO 2 nanotube arrays by CdS / CdSe / ZnS quantum dots co-modification. "Journal of Photochemistry & Photobiology, A: Chemistry, (2011).
8. D.R.Ako, Peiris, U.S., P.Ekanayake, , A. L.Tan, , D. J.Young, Z.Zheng, , & V.Chellappan,. DSSCs with ZnO@TiO2 core-shell photoanodes showing improved Voc: Modification of energy gradients and potential barriers with Cd and Mg ion dopants. *Solar Energy Materials and Solar Cells*, *157*, 18–27, (2016).
9. E.Bayatloo, & E.Saievar-iranizad,dye-sensitized solar cells by carbon nanospheres in photoanode.
10. D.Nikolopoulos, I.Valais, C.Michail, A.Bakas, C.Fountzoula, D.Cantzos, G. Panayiotakis,. Radioluminescence properties of the CdSe / ZnS Quantum Dot nanocrystals with analysis of long-memory trends. *Radiation Measurements*, *92*, 19–31, (2016).

# Studies of the absorbance peak on the N719 dye influence by combination between Cadmium Selenide (CdSe)QDs and Zinc Sulfide(ZnS)QDs

*Mohammad Azren* Saad[1,*], *Noor Syafiqah* Samsi[1], *Oskar Hasdinor* Hassan[3], *Muhd Zu Azhan* Yahya[2,4], *Mohamad Fariz* Mohamad Taib[1,2] *Ab Malik Marwan* Ali[1], and *Rosnah* Zakaria[1,2,*]

[1]Faculty of Applied Sciences, Universiti Teknologi MARA, 40450 Shah Alam, Selangor, Malaysia
[2]Institute of Science, Universiti Teknologi MARA, 40450 Shah Alam, Selangor, Malaysia
[3]Faculty of Art and Design, Universiti Teknologi MARA, 40450 Shah Alam, Selangor, Malaysia
[4]Faculty of Defence Science & Technology, Universiti Pertahanan Nasional Malaysia, 57000

**Abstract.** The absorption rate of the photoanode can be influenced by the combination between the difference semiconductor quantum dot sensitizer. Six samples were prepared with difference weight percent (wt%) of ZnS from 0% to 50% and constant wt% of CdSe which then will be called as semiconductor QDs were immersed in 0.5mM of N719 dye. The purity of ZnS powder and CdSe powder was determined using x-ray diffraction (XRD).The ultraviolet-visible spectrophotometry (Uv-Vis) use to investigate the absorption spectrum and absorbance peak of this sample. 50 wt% of ZnS is the best composition to increase the absorbance peak of the photoanode. The Cyclic voltammetry (CV) of varying wt% of ZnS, found that the 40 wt% of ZnS is suitable combination for a DSSC's photoanode and produced the higher current.

## 1 Introduction

Nowadays people are looking for the artificial sources to replace the traditional source. Quantum –dot-sensitized solar cells (QDSSCs) is the most suitable artificial source to replace the traditional sources[1].The benefit using the QDSCs as the artificial source is low cost and the efficiencies boosted up to the traditional Shockley and Queisser limit of 30%[3]. The only difference between the Dye-synthesis solar cell (DSSCs) compare to QDSCs is the replacement of organometallic or organic dye with the narrow band gap semiconductor material QD and the combination between QD and dye is called QDSCs. The most important part in the QDSCs is the photoanode, this part consists of sensitizers which are a semiconductor QD. The semiconductor QDs will be combined with the $TiO_2$ and coated onto glass sustrate. The semiconductor quantum dot such as InAs, CdSe, CdS, PbSe, and InP have a narrow band gap that is suitable to be used as the sensitized in the QDSCs [3]. The narrow bandgap will couse of easily electron transport when exposes to visible light and even though the light carries a low energy the emitted electron will travel through the TiO2 nanoparticle until it reaches anode, the electron produce current and energy to a device [4]. This electron will arrive at the cathode and move into the electrolyte to be sent back into the QDs in the photoanode [6]. QDs afford extremely optical properties and admirable electrical properties as compared to traditional organic dyes [3]. Furthermore, the various properties of semiconductor QDs make them attractive to take a part in the QDSSCs as a sensitizer. The high absorption coefficients make it suitable for the generation of multiple electrons carrier under high-energy excitation [5].

The high potential of semiconductor QDs sensitizer for light harvesting in the visible –light region, drawn much attention to be explored by the researcher [11]. Semiconductor QDs which is CdSe/ZnS has higher conduction band compared to the $TiO_2$ layer, this can increase the rate of electron injection of photon excited electrons into the photoanode. The band gap of ZnS is 3.6eV, this will limit the absorption range below a wavelength of 344 nm and the band gap for the CdSe is 1.7 eV [3], therefore the wavelength absorption that can be absorbed is below the 720nm. Because of the low conduction band edge of CdSe compare to ZnS and $TiO_2$, the efficiency of injection electron could be less.

Light carries the source energy in the form of a photon, this energy will be permeated by a particle in the material [9]. If the particle absorbs the energy of light bigger than their band gap, this particle will emit electron out from their shell [12]. These studies were investigated the variation of absorption spectrum by the photoanode with the difference the weight percent of ZnS/CdSe.

## 2 Methodology

### 2.1 Preparation of CdSe/ZnS solution

The two main material are CdSe and ZnS purchased from Sigma Aldrich. Various wt% of the ZnS, were mixed into

* Corresponding author: azrensaad@yahoo.com.my

CdSe QDs. 0.045 g ruthenium-based dye (N719) were dissolve in 9mL of ethanol to produce 0.5mM of liquid N719

$$concentration=(number\ of\ mole)/(volume) \quad (1)$$

$$molar\ mass=(mass)/(number\ of\ mole) \quad (2)$$

Six samples of QDss were prepared with difference weight percent (wt%) of ZnS from 0% to 50% and constant 10% of CdSe. The solution of 0.5mM of N719 was added. The samples were stirred for 24 hours using magnetic stirrer to get a homogenous solution. The mass of QDs ZnS/CdSe powder for each sample were calculated using equation below:

$$Wcdse/zns=(x(wn719))/(1-x)) \quad (3)$$

Where $x$ is the percentage of ZnS/CdSe

**Table 1.** Percentage of CdSe/ZnS QDs Prepared.

| Sample | CdSe/Zns QDs composition (wt%) | Mass of N719dye powder | Mass of CdSe/ ZnS QDs powder (g) |
|---|---|---|---|
| A | 10 | 0.045 | 0.005 |
| B | 20 | 0.045 | 0.011 |
| C | 30 | 0.045 | 0.019 |
| D | 40 | 0.045 | 0.030 |
| E | 50 | 0.045 | 0.045 |

### 2.2 preparation of photoanode

Three steps of cleaning ITO glass using ultrasonic bath where the first step is using the distilled water then acetone and followed by ethanol. Then, the ITO glass was dried at 60°C for 21 hours. $TiO_2$ nanoparticle was paste onto ITO glass using the doctor blade technique. The thickness and the area of the $TIO_2$ on the surface of ITO glass will be control by two parallel adhesive tapes 1cm apart and the area of the $TiO_2$ film is 1.8 x 1.0 cm before heated in the oven for 24 hours at 450 °C. The dried $TiO_2$ film was then immersed in varied wt% of QDs for 24 hours in dark environment. Lastly, ethanol was used to remove and wash dye residue on the surface before dried again in the oven at 60°C for 24 hours

### 2.3 Characterization of N719/CdSe-ZnS solution

#### *2.3.1 X-ray diffraction technique (XRD)*

XRD was used to study the crystalline structure and phase components of QDs powder. The structure spacing was determined by the high of intensity peak of the spectra. The higher number of count on the same angle prove that the existing components of the CdSe and ZnS.

#### *2.3.2 Uv-vis spectrometry*

UV-Vis spectrometry uses to detect the band gap of each sample in the form of a solution. 0.0005 mM CdSe/ ZnS was analyzed at a range of 200 - 900nm and the blank solution is ethanol.

$$E=h(c/\lambda) \quad (4)$$

### 2.4 Characterization ITO/$TiO_2$/N719/CdSe-ZnS photoanode

#### *2.4.1 Cyclic voltammetry*

The conductivity of co-sensitized of photoanode were be analyzed using the Cyclic voltammetry device. The sample of difference wt% of QDs (ITO/$TiO_2$/N719-CdSe/ZnS) photoanode were used as a working electrode ($W_e$), while the counter electrode ($C_e$) is platinum and Ag/AgCI/3MKCI was act as the reference electrode. The applied voltage on the sample was from -1.0 V to 1.0 V and every sample was run for three cycles to get the average data [11]

## 3 Results and Discussions

Figure 1 shows the proof of the existing component CdSe and ZnS in the sample powder. From the result in Figure1a the experiment stick pattern of ZnS and CdSe is similar to the reference stick pattern from Institute of Chemical Dependency Studies (ICSD) by xu,y-N ching (1993).The h,k,l peak of CdSe which is(100) is the first clearly produce in the experimental result and another peak such as 002,101,102 also clearly be observed between this two pattern. In Table 2 the first three peak produce at 23.881,25.381,27.101 which is 100,002 and 101 show the strongest proof that this powder is CdSe.

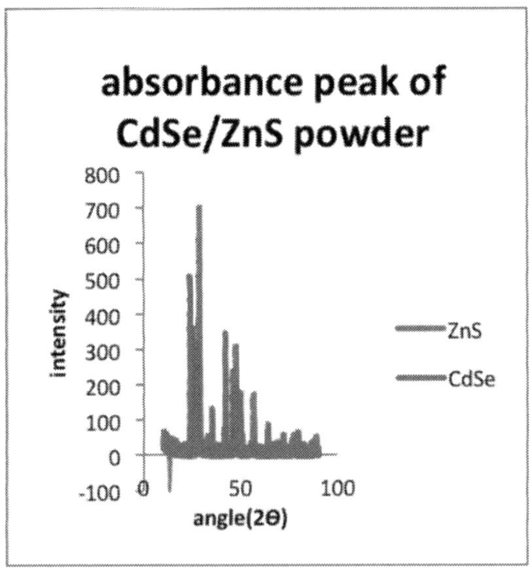


**Fig. 1(a).** Experimental absorbance peak CdSe/ZnS QDs

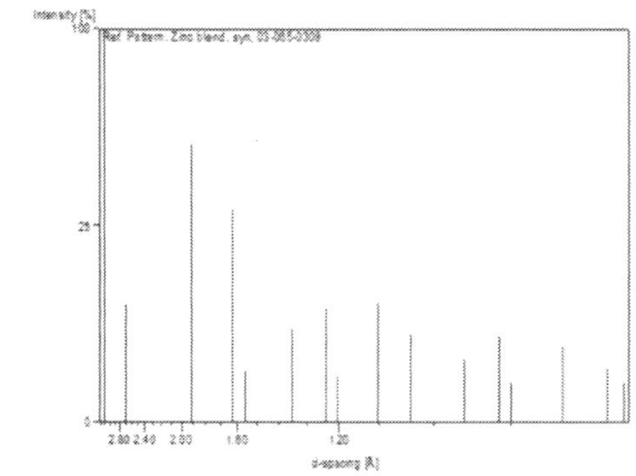

**Fig. 1(b)**: Absorbance Θpeak of CdSe QDs

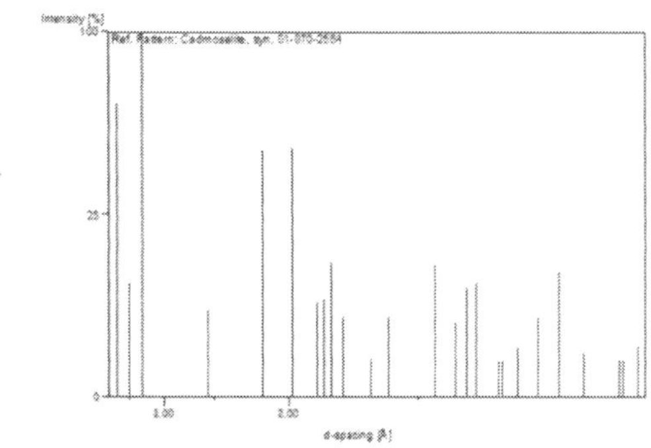

**Fig. 1(c)**: Absorbance Θpeak of ZnS QDs

**Table 2**. Comparison reference (R) result and experimental (Ep) result of CdSe powder

| No | h | k | l | (R) 2Θ | (R) d(A) | (Ep) 2Θ | (Ep) d(A) |
|---|---|---|---|---|---|---|---|
| 1 | 1 | 0 | 0 | 23.882 | 3.723 | 23.881 | 3.726 |
| 2 | 0 | 0 | 2 | 25.373 | 3.507 | 25.381 | 3.509 |
| 3 | 1 | 0 | 1 | 27.093 | 3.288 | 27.101 | 3.290 |
| 4 | 1 | 0 | 2 | 35.123 | 2.552 | 35.129 | 2.554 |

**Table 3.** Comparison reference (R) result and experimental(Ep) result of ZnS powder

| No | h | k | l | (R) 2Θ | (R) d(A) | (EP) 2Θ | (EP) d(A) |
|---|---|---|---|---|---|---|---|
| 1 | 1 | 1 | 1 | 28.609 | 3.117 | 28.589 | 3.122 |
| 2 | 2 | 0 | 0 | 33.153 | 2.700 | 33.157 | 2.702 |
| 3 | 2 | 2 | 0 | 47.591 | 1.909 | 47.581 | 1.911 |
| 4 | 3 | 1 | 1 | 56.473 | 1.628 | 56.441 | 1.630 |

In table3 the first three peak is at the angle $28.589^0$,$33.157^0$ and $47.581^0$ which is the h,k,l peak is 111,200 and 220, this peak is similar to the reference result from NIST by Dubrovin in 1983.

The highest number of count (intensity) by the CdSe samples is 505 at the angle 23.881, this means that the 505 particles reflects the x-ray source to that detector at this angle and from calculation the spacing particle is 3.72651(A), this is the spacing between particles d(A) in the CdSe powder that has been analyzed using XRD method compared to the reference result found that the spacing between the particle is 3.72304 (A) and their percentage difference is about 0.046%.The very small percentage difference between reference d(A) and sample d(A) prove that the powder consists the component of CdSe. For the ZnS powder, the same method is used by comparing the difference between the spacing particle d(A) in the sample with the reference result. The highest peak produce in the ZnS sample is 711 count at $28.569^0$ and the spacing between particle is about 3.12264(A), compared to the reference result the highest number count is at $28.609^0$ with the spacing between particles is 3.11769(A). The difference particle spacing between the reference value and ZnS sample is 0.0079%, the small difference percentage prove that this powder is ZnS. Table 2 and Table 3 shows the reference (R) and experiment (Ep) peak with comparison of the spacing between first four peaks.

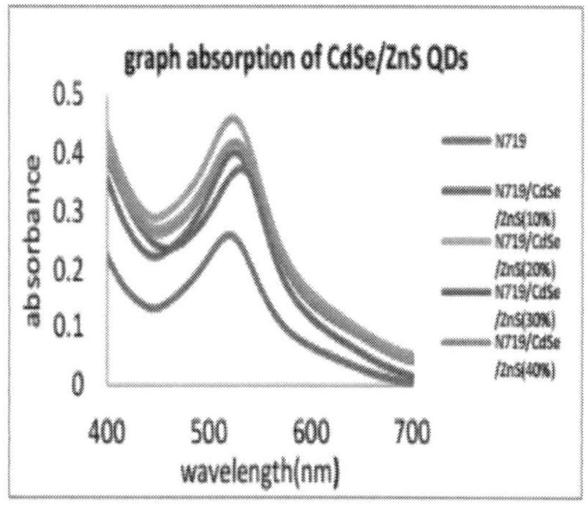


**Figure 2.** Graph absorption of CdSe/ZnS QDs

**Table 4:** The value absorbance of CdSe/ZnS QDs with difference weight present (wt%)

| Maximum wavelength (nm) | N719 | N719/CdSe/ZnS (10%) | N719/ CdSe/ ZnS (20%) | N719/ CdSe /ZnS (30%) | N719/ CdSe /ZnS (40%) | N719/ CdSe /ZnS (50%) |
|---|---|---|---|---|---|---|
| 525 | 0.2541 | 0.4015 | 0.4173 | 0.3687 | 0.4186 | 0.4584 |

Figure 2 shows the spectra absorption of six samples with various wt%. The highest peak of spectra is 525 nm of wavelength. This shows that the electron in the sample needs the energy of light with 525 nm to excited one electron out of the particle. Equation 4 was used to calculate the band gap of the CdSe/ZnS resulted 2.36 eV.

However, control variable, N719 dye showed the absorbance peak at 0.2541 as tabulated in Table 4. As for 10 wt% CdSe/ZnS QDs, the absorbance peak shifted from 0.2541 to 0.4015. This result shows that the combination of N719 and CdSe/ZnS QDs can increase the absorption of light. The 20% ZnS sample have the higher absorbance peak compared to the 10% ZnS sample. The difference between this two peak is 2% may due to formation hole in the 20% ZnS sample is higher compared to the 10 wt%

ZnS. The absorbance peak of 20 wt% ZnS is higher compared to 30 wt% ZnS samples. The absorbance peak of 40 wt% ZnS sample higher than 20 wt% ZnS sample. The absorbance peak of 50 wt% ZnS is higher compared to 40 wt% ZnS sample, this shows that the most suitable material to increase the absorption process is 50wt% of ZnS.

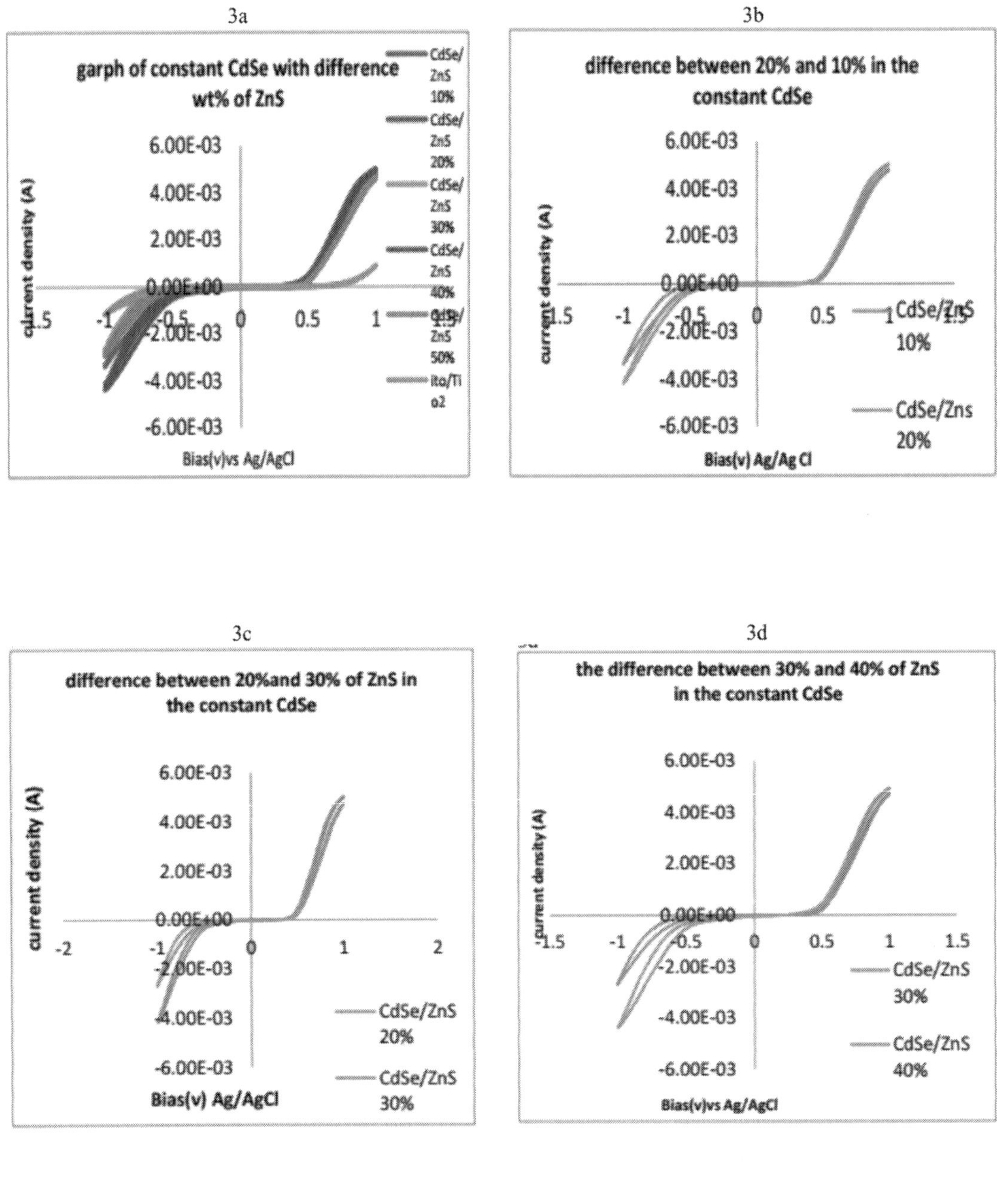

 https://doi.org/10.1051/matecconf/201815401040

3e

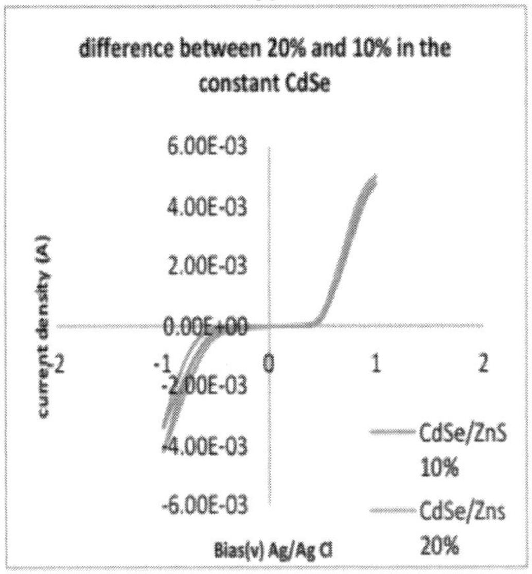


3f

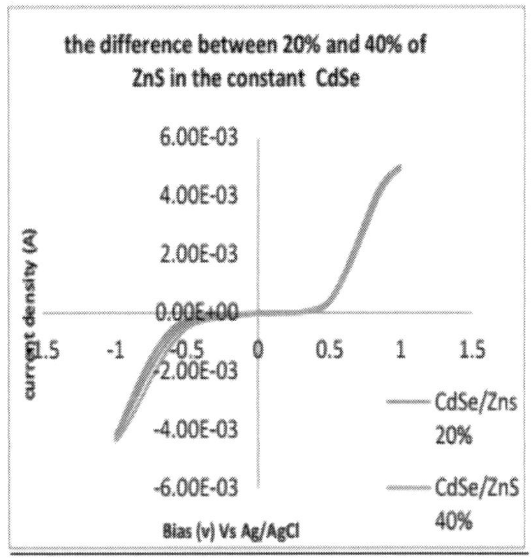


**Fig. 3.** Cyclic voltammetry of difference wt% of QDs

The I-V behavior of different sample with difference wt% of ZnS QD showed in Figure 3. Figure 3a shows the comparison of current produced between sample photoanode with difference wt% QD and sample without QDs. ITO/$TiO_2$ shows -0.00118 A while the combination of ITO/$TiO_2$/CdSe-ZnS(10%) is -0.00338 A, gives the difference of 0.00220 A. The current produced by the movement of an electron in the circuit, the more electrons flow the more current was produced [4]. However, the current from $TiO_2$ photoanode in Figure 3a is low may be due to the structure of the material which have a medium recombination process [6]. This recombination process reduces the acceleration of the electron to move from one particle to another in the structure of the $TiO_2$ thin film. Hence, the QDs photoanode consists of CdSe/ ZnS QDs provide a lot of hole at the valence shell of the particle, this valence shell is known as the hole act as the platform for the electron to jump from one particle to other until it reaches the conducting side of ITO glass [4]. The combination of ITO/$TiO_2$ with CdSe/ZnS QDs could increase the total number of the hole. As the result, the electron could flow easily.

Figure 3b shows the tabulated data of 10 wt% and 20wt% of ZnS. The current produce in CdSe/ZnS (20 wt%) is higher compared to the CdSe/ZnS(10 wt%). The 20 wt% ZnS produces -0.00418 A of current while the 10 wt% of ZnS is -0.00337952 A the difference between this two sample is -0.000801 A.

**Table 5**. The maximum current produced in each sample

| Sample | Applied voltage (v) | Current produce (A) | The difference current (A) between Tio2 and QDs(wt%) |
|---|---|---|---|
| ITO/TIO2 | -1.00098 | -0.00118 | |
| ITO/TIO2/ N719/CdSe-ZnS(10%) | -1.00098 | -0.00338 | -0.00220 |
| ITO/TIO2/ N719/CdSe-ZnS(20%) | -1.00098 | -0.00418 | -0.00300 |
| ITO/TIO2/ N719/CdSe-ZnS(30%) | -1.00098 | -0.0027 | -0.00152 |
| ITO/TIO2/ N719/CdSe-ZnS(40%) | -1.00098 | -0.00437 | -0.00319 |
| ITO/TIO2/ N719/CdSe-ZnS(50%) | -1.00098 | -0.00292 | -0.00174 |

The process of reduction in the system involved the donation of electrons from the electrolyte (Ag/AgCl 1M KCl) to the photoanode (CdSe/ZnS, QDs). The unstable particle which has the hole on their valence particle produce the easy way for an electron to pass through the barrier [6]. Table 5, shows the maximum current value produced by each sample. The highest value of current produced is CdSe/ZnS(40 wt%) which is -0.00437 A. The lowest current produce for the QDs photoanode is CdSe/ZnS(10%). The current increase -0.0008 A from 10 wt% to 20 wt% and -0.00167 from 30 – 40 wt% but the current reduce-0.00148 A from 20 wt% to 30 wt% and -0.00145 A from 40 wt% to 50 wt% of CdSe/ZnS.

## 4 Conclusions

CdSe/ZnS QDs immersed in N719 could increase the movement of electrons in the photoanode materials. 50 wt% of ZnS is the best composition to increase the absorbance peak of the photoanode. The Cyclic voltammetry (CV) of varying wt% of ZnS, found that the 40 wt% of ZnS at 0.00437 A is suitable combination to make a photoanode and could produce the highest current.

Authors would like to thank Ministry of Higher Education (MOHE) Malaysia for funding this research under the FRGS/1/2015/SG06/UITM grant and Universiti Teknologi MARA (UiTM) for the facilities provided

## References

1 D. Riehm,. Improving Dye-Sensitized Solar Cell Efficiency by Modification of Electrode Surface Charge.
2 Z. Li, Y.Xie, H.Xu, T.Wang, Z.Xu, & H.Zhang,. Journal of Photochemistry and Photobiology A : Chemistry Expanding the photoresponse range of TiO 2 nanotube arrays by CdS / CdSe / ZnS quantum dots co-modification. "Journal of Photochemistry & Photobiology, A: Chemistry, (2011).
3 K.Kim, J.Eun, E.Su, Y.Chang, J.Kim, C.Im, & M.Lee,. Electrochimica Acta ZnS-Passivated CdSe / CdS Co-sensitized Mesoporous Zn 2 SnO 4 Based Solar Cells. (2014).
4 P. V.Kamat,. Quantum Dot Solar Cells . Semiconductor Nanocrystals as Light(2008) .
5 A.Undergraduate,. Optical Properties of Quantum Dots : An Undergraduate Physics Laboratory,(2007).
6 B. I.Photovoltaics,. 2 Dye sensitized solar cells (DSSC), (n.d.).
7 B. O.Dabbousi, F. V Mikulec, J. R.Heine, H.Mattoussi, R.Ober, K. F.Jensen, & M. G.Bawendi,. ( CdSe ) ZnS Core - Shell Quantum Dots : Synthesis and Characterization of a Size Series of Highly Luminescent Nanocrystallites(1997).
8 I.Mora-Seró, V.Likodimos, S.Giménez, E.Martínez-Ferrero, J.Albero, E.Palomares, J.Bisquert, Fast Regeneration of CdSe Quantum Dots by Ru Dye in Sensitized TiO2 Electrodes. *The Journal of Physical Chemistry C*, 6755–6761, (2010).
9 M. A. M.Al-Alwani, A. B.Mohamad, N. A.Ludin, A. A. H.Kadhum, & K.Sopian,. Dye-sensitised solar cells: Development, structure, operation principles, electron kinetics, characterisation, synthesis materials and natural photosensitisers. *Renewable and Sustainable Energy* 183–213,(2016).
10 M. L.Landry, T. E.Morrell, T. K.Karagounis, C. H.Hsia, & C. Y Wang,. Simple syntheses of CdSe quantum dots. *Journal of Chemical Education*, 274–279. (2014).
11 E. M.Boatman, G. C.Lisensky, & K. J.Nordell, A Safer , Easier , Faster Synthesis for CdSe Quantum Dot Nanocrystals, *82*(11), 1697–1699, (2005).
12 E.Bayatloo, & E.Saievar-iranizad,Dye-sensitized solar cells by carbon nanospheres in photoanode.

# Classification of acute myeloid leukemia subtypes M1, M2 and M3 using active contour without edge segmentation and momentum backpropagation artificial neural network

*Agus* Harjoko[1], *Tri* Ratnaningsih[1], *Esti* Suryani[2], Wiharto[2], *Sarngadi* Palgunadi[2], and *Nurcahya Pradana Taufik* Prakisya[1*]

[1]Universitas Gadjah Mada, Yogyakarta, 55281, Indonesia
[2]Universitas Sebelas Maret, Surakarta, 57126, Indonesia

**Abstract.** Acute Myeloid Leukemia (AML) is a type of cancer which attacks white blood cells from myeloid. AML has eight subtypes, namely: M0, M1, M2, M3, M4, M5, M6, and M7. AML subtypes M1, M2 and M3 are affected by the same type of cells, *myeloblast*, making it needs more detailed analysis to distinguish. To overcome these obstacles, this research is applying digital image processing with Active Contour Without Edge (ACWE) and Momentum Backpropagation artificial neural network for AML subtypes M1, M2 and M3 classification based on the type of the cell. Six features required as training parameters from every cell obtained by using feature extraction. The features are: cell area, perimeter, circularity, nucleus ratio, mean and standard deviation. The results show that ACWE can be used for segmenting white blood cells with 83.789% success percentage of 876 total cell objects. The whole AML slides had been identified according to the cell types predicted number through training with momentum backpropagation. Five times testing calibration with the best parameter generated averages value of 84.754% precision, 75.887% sensitivity, 95.090% specificity and 93.569% accuracy.

## 1 Introduction

Leukemia is a cancer that attacks the white blood cells (WBC) formed in bone marrow. People with leukemia have a very large number of abnormal WBC. The excessive amount of WBC will result in decreasing number of healthy blood cells [13]. Leukemia is divided into four types, namely: Acute Myeloid Leukemia (AML), Acute Lymphocytic Leukemia (ALL), Chronic Myeloid Leukemia (CML) and Chronic Lymphocytic Leukemia (CLL). The AML type is subdivided into eight subtypes: M0, M1, M2, M3, M4, M5, M6, M7 [3][4][8]. There are three cell types whose percentages indicate AML subtypes M1, M2 and M3, namely: *myeloblast*, *promyelocyte*, and *monoblast* [6].

Leukemia can be detected by counting the total number of blood cells then performing a comparison of the number of white blood cells and red blood cells through a microscope by morphology by haematologists. However, the procedure of calculating blood cells with this microscope is quite time consuming and tiring [9]. The microscopic image inspection of bone marrow blood also has variative results and does not have clear standards due to the process relies heavily on the expertise of haematologists [20]. The other techniques used to calculate blood cell counts such as immunophenotyping, flowcytometer and molecular probing are also still used as a standard diagnostic of leukemia hence these techniques are considered relatively costly [17][22]. To help to overcome these problems, we proposed a combination of digital image processing and artificial neural network [18, 23].

This research combined active contour without edge segmentation (ACWE) process for WBC segmentation in blood microscopic images and artificial neural network with backpropagation momentum algorithm based on morphology and colour features of three types of cells mentioned above to identify AML subtypes: M1, M2 and M3.

## 2 Proposed Method

The research began from image acquisition of blood preparations. The image data were enhanced, segmented and then extracted to get the features value as input data in training and testing stage. The results from cell type classification were used to identify what AML class contained in each preparation. Research steps can be shown in Figure 1.

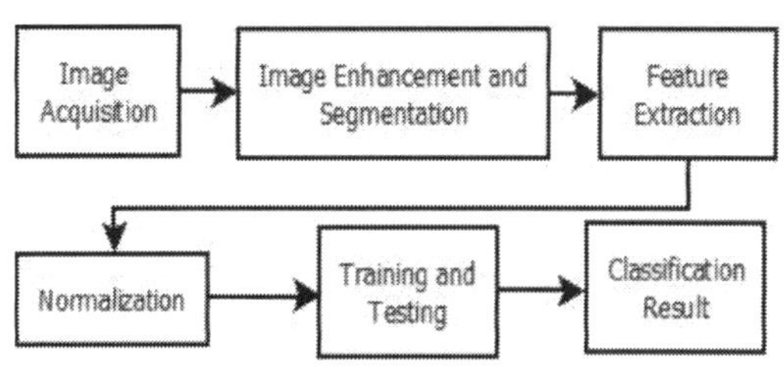


**Fig 1.** Research steps

* Corresponding author: nurcahya.pradana@gmail.com

## 2.1 Image Acquisition

AML M1, M2 and M3 preparations were obtained from Dr. Sardjito Hospital Yogyakarta through ethical clearance procedure. Image data were taken using a 21 megapixels camera attached to Olympus microscope's ocular lens with 1000 times magnification. The shooting was done 33 times for each preparation with different slide position.

## 2.2 Image Enhancement and Segmentation

Image enhancement for reducing image noise was conducted by applying median filter to the original image [7, 19]. Image segmentation consists of two important parts: WBC body segmentation and nucleus segmentation.

The concept of body segmentation is utilizing ACWE as the primary method of segmenting WBC candidates. ACWE minimizes the difference in energy value from background and foreground. The stopping term does not depend on the gradient of the image but related to a particular segmentation of the image [1, 5]. Having obtained the image of white blood cells intact, overlapping cells were separated by using Watershed Distance Transform [19, 24]. The result of the WBC body segmentation was a binary mask which was then pixel-multiplied with the corresponding pixel in original image to get the WBC object in RGB channel. The segmentation flow chart of the WBC body is shown in Figure 2.

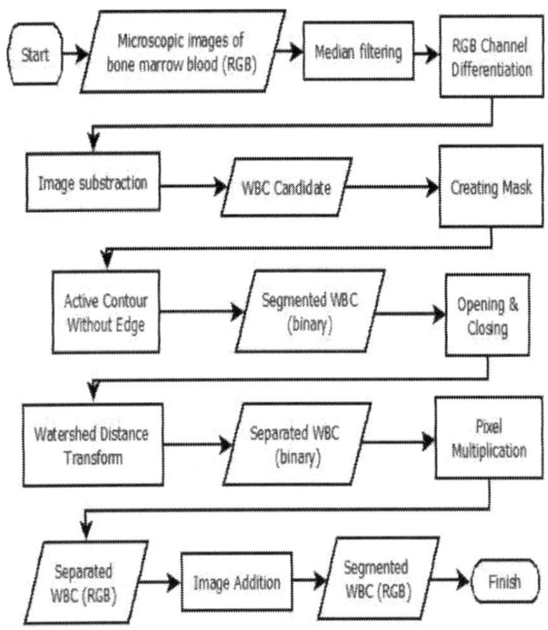


**Fig 2.** WBC body segmentation flowchart

Nucleus needed to be segmented to calculate nucleus ratio. Nucleus image should be converted to HIS channel because the nucleus had a high saturation making it easier to do thresholding on the saturation channel [2]. The segmentation flowchart of nucleus is shown in Figure 3.

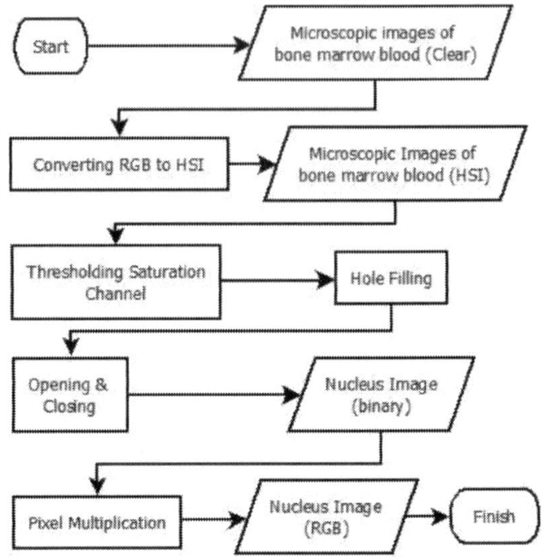


**Fig 3.** Nucleus segmentation flowchart

## 2.3 Feature Extraction

The feature extraction produced six numerical data which represent image characteristics: cell area, cell perimeter, cell circularity, nucleus ratio, mean and standard deviation. These six features were then used as artificial neural network inputs.

- Cell Area: The sum of white pixel from binary segmented cell [18].

$$A = \frac{1}{n.m}\sum_{x=0}^{m-1}\sum_{y=0}^{n-1}[\, i(x,y) |\, i(x,y) = 1, f(x,y)] \tag{1}$$

  The variable $i(x, y)$ is the pixel value in the image $f(x, y)$. Variable $A$ defines the area. The variable $n$ is the number of the length of image row whereas $m$ is the length of image column.

- Cell Perimeter: The sum of white pixel from the eroded binary segmented cell by 1-pixel strel [18].

$$p = A - (A \ominus B) \tag{2}$$

  Variable $B$ is a strel of image morphology. $A$ is the area of the object and the $p$ is the edge of the object. All variables are in units of number of pixels.

- Circularity: The roundness of cell within range 0 - 1 [18].

$$C = \frac{4\pi A}{p^2} \tag{3}$$

  $A$ is the area and $p$ is the perimeter of the area. The more rounded an object the closer the value of 1.

- Nucleus Ration: Nucleus area and cytoplasm area ratio within range 0 - 1 [18].

$$R = \frac{A(nucleus)}{A(cell)} \tag{4}$$

Variable *A(nucleus)* defines the area of the nucleus object while *A(cell)* defines the area of the body of a WBC. *R* is the ratio of the nucleus area to the WBC.

- Mean: Pixel colour intensity of red, green and blue channel [16].

$$\bar{x} = \frac{1}{n.m}\sum_{x=0}^{m-1}\sum_{y=0}^{n-1} i(x,y) \quad (5)$$

The variable *i(x, y)* is the pixel value of the cell. The variable *n* is the number of the length of image row whereas *m* is the length of the image column. The variable $\bar{x}$ defines the average pixel intensity value of the cell.

- Standard Deviation: The root of the squared difference in pixel colour intensity and pixel mean [16].

$$\sigma = \sqrt{\frac{1}{n.m}\sum_{x=0}^{m-1}\sum_{y=0}^{n-1}(i(x,y)-\bar{x})^2} \quad (6)$$

The variable *i(x, y)* is the pixel value of the cell. The variable *n* is the number of the length of image row whereas *m* is the length of the image column. The variable $\bar{x}$ defines the average pixel intensity value of the cell and σis the standard deviation of the cell.

### 2.4 Normalization

Due to varying values of the features, normalization needed to be done first before entering the training process with the following formula:

$$normalization = \frac{data[x]-\min(data)}{\max(data)-\min(data)} \quad (7)$$

The value of x is the value of the x-index, min (data) is the smallest value of the data set on each feature, while max (data) is the largest value of the data set on each feature.

### 2.5 Training and Testing

AML M1, M2 and M3 classification training stage used backpropagation momentum with one hidden layer as learning algorithm. The architectural design of this training can be explained in Figure 4.

Six normalized features were used as input neurons. The output layer consisted of two neurons called y1 and y2 in binary value. Cell type results were determined by adjusting the class to the output neuron pattern as shown in Table 1.

Learning rate (α) used were 0.1, 0.5, and 0.9. For the number of hidden layer (z) neurons used were 6, 8 and 10. Maximum epoch (ε) were 1000, 5000, and 10000. Momentum (μ) used were 0.1, 0.5, and 0.9. The tolerance limit for error (ξ) was 0.000001 thus the testing was done 81 times according to the combination of parameters.

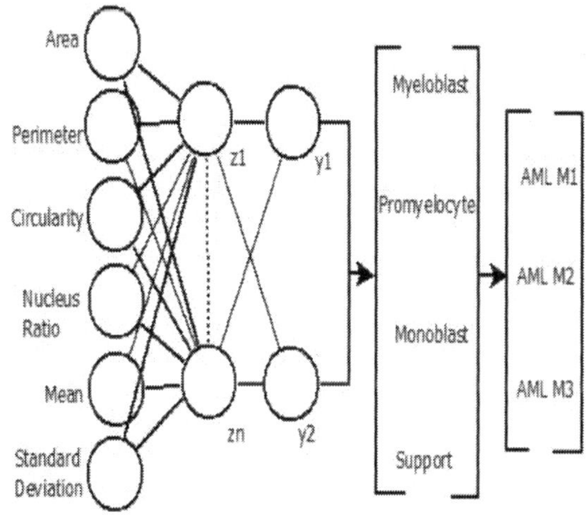


**Fig 4.** Momentum backpropagation architecture

**Table 1.** Output class based on output neuron pattern

| No. | Cell Type | y1 | y2 |
|---|---|---|---|
| 1 | *Myeloblast* | 0 | 0 |
| 2 | *Promyelocyte* | 0 | 1 |
| 3 | *Monoblast* | 1 | 0 |
| 4 | Support | 1 | 1 |

### 2.6 Classification Result and Validation

Validation for classification results used 3-fold cross-validation. Out of 734 total data, 250 data were used as fold-1 test data. 248 data were used in fold-2 and 236 data were used as fold-3 test data. Each result was then summed and inserted into a confusion matrix based on the prediction of each cell type. Those values were later used to calculate four predictive analysis values: precision, sensitivity, specificity and accuracy. The combination of parameters having the highest total predictive analysis was then used to recalibrate five times.

The identification of AML subtypes on each preparation was done by accumulating the number of predicted results of each cell and grouping them according to the origin of preparations. AML subtype later could be determined based on the percentage criteria of each cell type as outlined in Table 2. Preparation which was not meeting cell count criteria would be classified as "support" type cell.

**Table 2.** AML subtype class based on cell type

| Subtype | *Myeloblast* | *Promyelocyte* | *Monoblast* | Support |
|---|---|---|---|---|
| AML M1 | **>89%** | <10% | <10% | <10% |
| AML M2 | **30-89%** | **>10%** | **≤20%** | <10% |
| AML M3 | **≤30%** | **>20%** | <10% | <10% |
| NON-M1M2M3 | Does not meet cell count criteria | | | |

## 3 Research Result

### 3.1 Segmentation Result

Segmentation was done on each slide preparation containing dozens of blood cells. The number of each cell of each image on each preparation needed to be calculated to ensure the quantity of each cell type of a preparation. The samples of the segmentation results on AML M1, M2 and M3 preparations are shown in Table 3.

**Table 3.** Segmented image of AML M1, M2 and M3

| Subtype | AML M1 | AML M2 | AML M3 |
|---|---|---|---|
| Original | 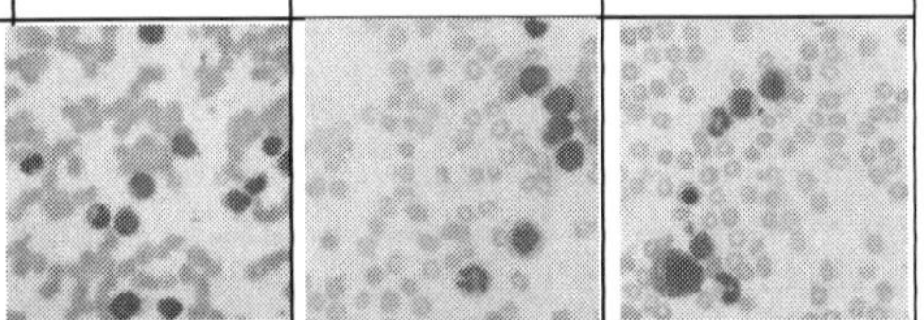 | | |
| Segmented | 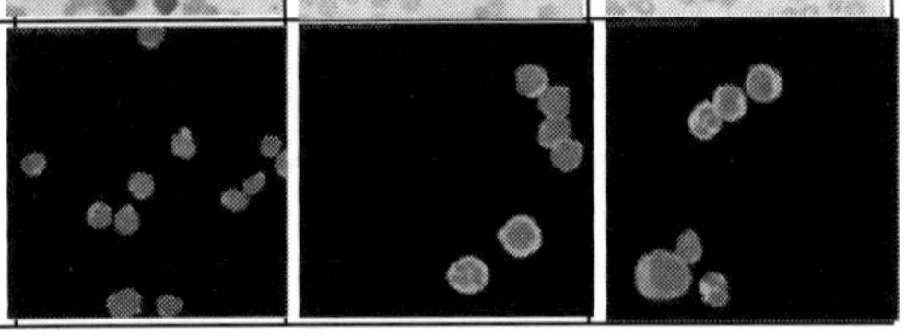 | | |

There were some cells which experience errors when segmented. The incorrectly segmented cells were not included as an artificial neural network training input data. Examples of incorrectly segmented cells and its reasons can be shown in the Table 4.

**Table 4.** Segmented image of AML M1, M2 and M3

| Reason | Non-WBC *(erythroid cell)* | Cell body is cut off by the image's edge | Cell body is cut off by WDT | Cytoplasm area beyond the cell body |
|---|---|---|---|---|
| Original | 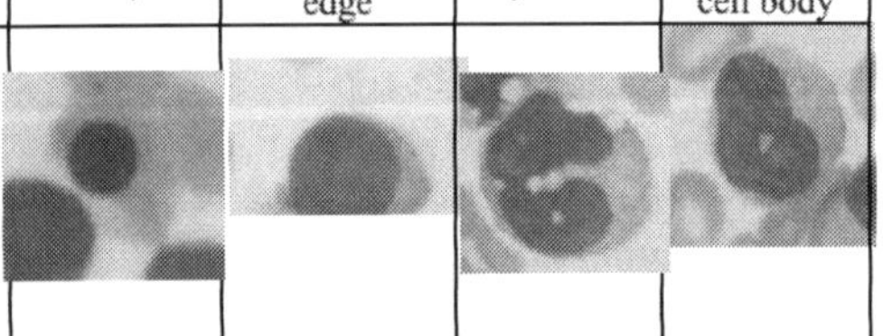 | | | |
| Segmented | 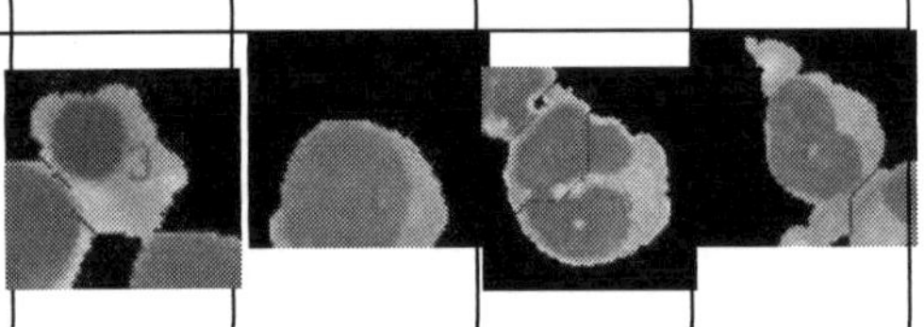 | | | |

The result of segmentation shows that from the total of 99 white blood cell images, 876 data of white blood cell object consisting of 734 segmented data is correct and the remaining 142 segmented data is wrong thus the percentage of cells segmented correctly amounted to 83.789%. Detailed comparison of the number of cell objects resulting from correctly and incorrectly segmented cell for each subtype of AML can be shown in Table 5.

**Table 5.** The number of objects resulted in the segmentation of each preparation

| Subtype | Correct | Incorrect | Total |
|---|---|---|---|
| AML M1 | 224 | 30 | **254** |
| AML M2 | 225 | 72 | **297** |
| AML M3 | 285 | 40 | **325** |
| **Total** | **734** | **142** | **876** |

### 3.2 Feature Extraction Result

The feature set of the cells was entered as artificial neural network input data. The average results of WBC feature extraction can be seen in Table 6. We can see that the values obtained from each feature are in accordance with the characteristics of the real cell type. For example, *myeloblast* has a higher average circularity while *monoblast* has a larger size of area than the other cell types.

**Table 6.** Average result of WBC feature extraction

| **Feature** | ***Myeloblast*** | ***Promyelocyte*** | ***Monoblast*** | **Support** |
|---|---|---|---|---|
| Area (pixels) | 7377,0530 | 13581,201 | 14923,724 | 9714,427 |
| Perimeter (pixels) | 322,482 | 454,659 | 470,482 | 378,412 |
| Circularity (0-1) | 0,8629 | 0,8153 | 0,8359 | 0,8245 |
| Nucleus ratio (0-1) | 0,8314 | 0,6016 | 0,6093 | 0,5238 |
| Mean (intensity) | 135,360 | 144,8222 | 149,6375 | 147,882 |
| Standard deviation (intensity) | 20,0017 | 24,9116 | 24,2194 | 31,0159 |

### 3.3 Training and Testing Result

Based on the results of training and testing conducted on each combination of momentum backpropagation parameters, it resulted the best parameters at $\alpha$=0.1, z=8 and $\mu$=0.1 in 10000 epochs. This pattern was then recalibrated five times. Details of the test result information of five times calibration can be presented in Table 7.

**Table 7.** Recalibration average result at $\alpha$=0.1, z=8 and $\mu$=0.1

| **No.** | **Presition (%)** | **Sensitivity (%)** | **Specificity (%)** | **Accuration (%)** |
|---|---|---|---|---|
| 1. | 83,306 | 73,052 | 94,777 | 93,120 |
| 2. | 84,005 | 77,671 | 95,295 | 93,665 |
| 3. | 82,852 | 73,621 | 95,092 | 93,597 |
| 4. | 87,484 | 73,434 | 94,990 | 93,460 |
| 5. | 86,125 | 81,658 | 95,247 | 94,005 |
| Avg. | **84,754** | **75,887** | **95,080** | **93,569** |

The identification result of AML subtype in each preparation is shown in Table 8. Preparation 1 was identified as a subtype of AML M1 because the counted cell of *myeloblast* exceeded 89%. Preparation 2 was identified as a subtype of AML M2 with 72% *myeloblast*, 13.333% *promyelocyte* and 2.667% *monoblast*. Preparation 3 was identified as AML M3 in which the percentage of *myeloblast* was 8.070% and 34.737% for *promyelocyte*.

**Table 8.** Percentage of cell count in each preparation

| Prep | *Myeloblast* | *Promyleocyte* | *Monoblast* | Support |
|---|---|---|---|---|
| #1 | 89,735% | 1,786% | 0% | 8,482% |
| #2 | 72% | 13,333% | 2,667% | 12% |
| #3 | 8,07% | 34,737% | 0% | 57,193% |

The three results of the inference were correct according to the type of preparation used. It means the classification of AML subtypes M1, M2 and M3 with backpropagation momentum can be said to be successful.

## 4 Conclusion

The ACWE algorithm could be used to segment WBC objects from image background with success rate 83.789%. The percentage of correctly segmented object data versus overall data from each preparations of AML M1, M2 and M3 were 88.189%, 75.757% and 87,692%, respectively. The feature extraction was performed on 734 cell objects from the total number of 876 segmented cell objects.

The best pattern of momentum backpropagation for AML subtype classification was 0,1 learning rate, eight neuron hidden layer and 0,1 momentum. Predictive analysis values obtained from five times calibration test were as follows: 84.754% average precision, 75.887% average sensitivity 95.090% average specificity and 93.569% average accuracy.

This research was supported by Grant PNBP 2016 Universitas Sebelas Maret, Faculty of Medicine Universitas Gadjah Mada and Dr. Sardjito Hospital, Yogyakarta, Indonesia.

## References

1. Abidin, S.R., Salamah, U., Nugroho, A.S., Segmentation of Malaria Parasite Candidates from Thick Blood Smear Microphotographs Image Using Active Contour Without Edge. 1st International Conference on Biomedical Engineering (2016).
2. Agaian, S., Madhukar, M., Chronopoulos, A.T., Automated Screening System for AcuteMyelogenous Leukemia Detection in Blood Microscopic Images. IEEE System Journal, **Vol. 8, No. 3.** (2014).
3. American Cancer Society, Cancer Facts & Figures 2014. Atlanta. (2014).
4. Bell, A. dan Sallah, S., *The Morphology of Human Blood Cells. Seventh Edition*. Abbott, A Promise for Live. (2005).
5. Chan, T. F., dan Vese, L. A., Active Contour Without Edges. IEEE Transactions on Image Processing, 10(2), 266-277. (2001).
6. Dacie, Lewis, Practical Haematology: Eleventh Edition. Elsevier Churcill Livingstone. ISBN-13: 9780702034084. (2011).
7. Gonzales, R.C., Woods, R.E. *Digital Image Processing: Second Edition*. New Jersey : Pearson Prentice Hall. Upper Saddle River, 07458. (2002).
8. Hamid, A.G., Classification of Acute Leukemia, Acute Leukemia: The Scientist's Perspective and Challenge. Prof. Mariastefania Antica (Ed.), ISBN: 978-953-307-553-2. (2011).
9. Kasmin F., Prabuwono, A.S., Abdullah A., Detection of Leukemia in Human Blood Sample Based on Microscopic Images: a Study. Journal of Theoretical and Applied Information Technology. **Vol.46 Num.2.** ISSN: 1992-8645. (2012).
10. Kjeldsberg, C.R., *Practical Diagnosis of Hematologic Disorders: Second Edition.* Chicago Illinois: American Society of Clinical Pathologists (ASCP Press). ISBN : 0891894012. (1995).
11. Kusumadewi, S., *Membangun Jaringan Syaraf Tiruan Menggunakan MATLAB & EXCEL LINK.* Yogyakarta: Graha Ilmu. (2004).
12. Losen Adnyana, I. W. Dan Suega, K., Perubahan Golongan Darah Pada Penderita Leukemia Mieloblastik Akut. J Peny Dalam, **Vol. 12 No.1** (2011).
13. Leukemia & Lymphoma Society. *The AML Guide Information for Patients and Caregivers Acute Myeloid Leukemia.* 1311 Mamaroneck Avenue, Suite 310, White Plains, NY 10605. (2012).
14. Madhukar, M., Agaian, S., and Chronopoulos, A.T., New Decision Support Tool for Acute Lymphoblastic Leukemia Classification. Journal of SPIE-IS&T/ **Vol. 8295 8295181**. (2012).
15. Mishra, A. K., Fieguth, P. W., & Clausi, D. A. Decoupled Active Contour (DAC) for Boundary Detection. IEEE Transactions on Pattern Analysis and Machine Intelligence, 33(2), 310-324. (2011).
16. Nasir, A.A.1, Mashor, M.Y.1, and Hassan, R.2. Classification of Acute Leukaemia Cells using Multilayer Perceptron and Simplified Fuzzy ARTMAP Neural Networks. 1Electronic and Biomedical Intelligent Systems Research Group, Universiti Malaysia Perlis, Malaysia. 2Department of Haematology, Universiti Sains Malaysia, Malaysia. (2013).
17. Patil T.G., Prof. Mr. V. B. Raskar, Automated Leukemia Detection By Using Contour Signature Method. E & TC Department JSPM's, Imperial College of Engineering & Research Wagholi, Pune. The International Arab Journal of Information Technology, **Vol. 10, No. 4**, (2015).
18. Pradana, T.P.N., Suryani, E., Wiharto, Pemanfaatan Seed Region Growing Segmentation dan Momentum Backpropagation Neural Network Untuk Klasifikasi

Jenis Sel Darah Putih. Seminar Nasional Teknoin 2013 FTI UII. (2013).

19. Prasetyo, Eko, *Pengolahan Citra Digital dan Aplikasinya Menggunakan Matlab*. Yogyakarta: Penerbit Andi. ISBN:978-979-270-0. (2011).
20. Putzu, L., Ruberto C.D., White Blood Cells Identification and Counting from Microscopic Blood Image. World Academy of Science, Engineering and Technology. International Journal of Medical, Health, Biomedical, Bioengineering and Pharmaceutical Engineering **Vol.7, No.1**. (2013).
21. Suryani, E., Wiharto, and Polvonov, N., Identifikasi Penyakit Acute Lymphoblastic Leukemia (ALL) Menggunakan Fuzzy Rule Based System Berdasarkan Morfologi Sel Darah Putih. Prosiding Seminar Nasional Teknologi Informasi dan Komunikasi Terapan (SEMANTIK), Udinus, Semarang, ISBN :979-26-0266-6. (2013).
22. Suryani, E., Salamah, U., Wiharto., Polvonov, N., Identification and Counting White Blood Study of Leukemia. International journal of Computer Science & Network Solutions. **Vol. 2 No. 6**. ISSN: 2345-3397. (2014a).
23. Suryani, E., Salamah, U., Wiharto., Wijaya., A.A., Identifikasi Penyakit Acute Myeloid Leukemia (AML) Menggunakan 'Fuzzy Rule Based System' Berdasarkan Morfologi Sel Darah Putih Studi Kasus : AML M2 dan AML M4. Seminar Nasional Semantik UDINUS Semarang. (2014b).
24. Suryani, E., Wiharto, Palgunadi, S., Pradana T.P.N., Classification of Acute Myelogenous Leukemia (AML M2 and AML M3) using Momentum Backpropagation from Watershed Distance Transform Segmented Images. International Conference on Computing and Applied Informatics. IOP Conf. Series: Journal of Physics: Conf. Series 801 (2017) 012044. doi:10.1088/1742-6596/801/1/012044. (2016).

# Fingerprint image reconstruction for swipe sensor using Predictive Overlap Method

*Ahmad Zafrullah* Mardiansyah[1,*], *Agus* Bejo[1], and *Risanuri* Hidayat[1]

[1]Department of Electrical Engineering and Information Technology, Universitas Gadjah Mada, Yogyakarta, Indonesia.

**Abstract.** Swipe sensor is one of many biometric authentication sensor types that widely applied to embedded devices. The sensor produces an overlap on every pixel block of the image, so the picture requires a reconstruction process before heading to the feature extraction process. Conventional reconstruction methods require extensive computation, causing difficult to apply to embedded devices that have limited computing process. In this paper, image reconstruction is proposed using predictive overlap method, which determines the image block shift from the previous set of change data. The experiments were performed using 36 images generated by a swipe sensor with 128 x 8 pixels size of the area, where each image has an overlap in each block. The results reveal computation can increase up to 86.44% compared with conventional methods, with accuracy decreasing to 0.008% in average.

## 1 Introduction

The fingerprint is still a biometric authentication standard that is still growing, especially in the industrial environment [1]. Today's development of technology likely tends to mobile computing. However, many mobile devices on the market have limited ability, especially regarding data processing. In general, fingerprint authentication requires a relatively large computation specification as a minimum. Consequently, fingerprint authentication only applied to devices with standard process capabilities only.

There are many implementations of fingerprint authentication on mobile devices due to the limitations of computing capabilities owned by mobile devices, especially on embedded devices. Efficiency in the computational process of fingerprint authentication system is still developed by many researchers [1-5]. The proposed method and algorithm were helping the development of implementation a fingerprint into embedded devices.

Unlike conventional fingerprint sensors (full sensor) [6], embedded devices have relatively small in sizes [7, 8]. This affect difficulty in implementation process by manufacturers. To overcome these challenges, fingerprint sensor has developed a swipe model that has a smaller size [1]. The swipe type sensor performs a fingerprint scan of every n-pixel (block) of the entire surface in vertically [9]. The results of the reconstruction produced in the research [6] show an overlapping ratio of 84% match.

Image mosaicking is comparing two images whether it has a correlation or not, and the correlation indicates a shift from one block image to another or vice versa [10]. Although the system successfully deployed, the computational likely leads to be heavy. In the real environment, there will be a delay during the authentication process. There is a time range during finger registration and matching results.

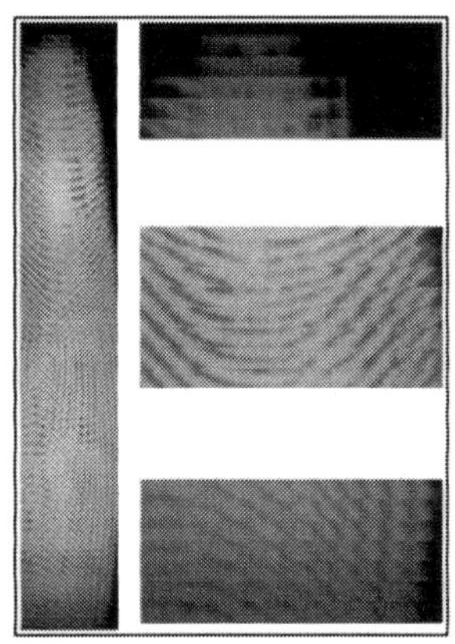

**Fig. 1.** Raw image from swipe sensor.

Regarding physical dimensions, swipe sensor is more advantageous because it has small in size that is enough to be implemented in embedded devices. However, on the other hand, swipe sensor needs one more process called reconstruction. This process has to reconstruct raw image data from the sensor to become a complete fingerprint image to perform extraction and matching features [11].

The reconstruction process requires a costly computational load. To solve the problem, many researchers have proposed methods and algorithms to reduce computational burdens. The method used in [3] reads pixel of the image by simultaneously to decrease the number of iterations. The relationship between phases of each block was introduced in [12] which was used to

* Corresponding author: zaf.mti15@mail.ugm.ac.id

indicate the pixel block correspondence, and the results reveal validated match regardless of computation.

Another research with the same approach was performed by [13] using the Sum of Absolute Difference (*SAD*) and Fast Normalized Cross Correlation (*FastNCC*) algorithms. The method shows an accuracy increase of about 1% using the *FastNCC* algorithm, but in implementation, *FastNCC* needs more computational load than SAD because of the number of iterations. In research [14] hybrid image mosaicking has been developed using Mean Absolute Error (*MEA*) to find the overlap transition value between two images. Mosaicking performed with a combination of pixels and sub-pixels.

From researchers above, there are two critical parameters relating to the performance of fingerprint authentication system. Increased accuracy tends to lead to increased in computational loads and vice versa. The combination of both parameters is capable of producing a better authentication system, but it is still difficult to apply. Nevertheless, research [15] proposed two levels of security with a combination of fingerprint and PIN instruments. Matching on fingerprints is not done based on the features found on the fingerprint, but can be solved using several techniques such as pressure on the sensor surface and the pixel image spreading rate (histogram).

Based on research in computation and accuracy produced by the conventional algorithm, in this paper proposed reconstruction algorithm using predictive overlap method which is used to reduce computation cost. This method applies shift value of previously calculated results to use as current shift value between two blocks; therefore the program does not need to calculate all change values from all blocks by visiting all pixel data.

The paper will be divided into several sections as follows: Section II describes the comparison of conventional reconstruction techniques with the proposed predictive overlap method. The result of experiments and discussions outlined in section IV. The conclusions described in section V.

## 2 Reconstruction Algorithm

### 2.1 Conventional Method

The digital image is composed of a collection of pixels in the form of a two-dimensional form, which is *x* and *y*. A block of the pixel is considered to be overlapping when its neighboring block has a pixel intensity composition that is mostly the same. Differences in pixel intensity in the block occurs due to differences in the speed of fingerprint shift when the process of acquiring images.

Reconstruct the swipe fingerprint image is to change the position of each block according to their real shift. A value of block's change can be resolved by comparing current block with the previous one; this can be done by calculating the difference between the total intensity of pixels in a block using *SAD*.

*SAD* needs to visit every pixel in the block while performing calculations, resulting in a computational load of $O(n)$. Matching is done by adjusting the magnitude of the shift ($d_y$) between two blocks to the maximum block height. The change with the lowest *SAD* value is the best overlap shift. Next, perform a reconstruction based on the generated in each iteration of the pixel block in the image.

### 2.2 Proposed Method

In practice, the $d_y$ value tends to change insignificantly from the previous $d_y$ value. The research in this paper utilizes the $d_y$ data set of past calculations to determine the probability of the next $d_y$ value.

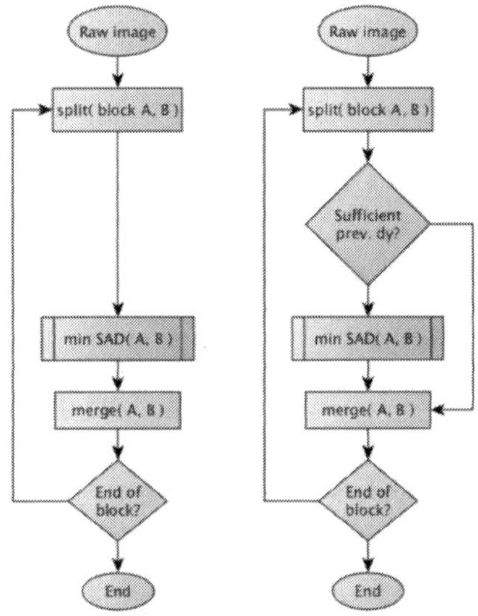


**Fig. 2.** Conventional (left) and proposed (right) method flow.

From the flowchart in figure 2 above, min SDA process using a large computing cost, the iteration process is done at the pixel level.

$$SAD = \sum_{i=1}^{n} |A_i - B_i|$$

where *A* and *B* are the two block that will be compared by finding the difference between pixel level, is the accumulation of the difference between block *A* and *B*.

The proposed approach adds a checking process to the previous $d_y$ values. In this paper, the reference value is two sets of former $d_y$ (*range*), then the value will be used as $d_y$ in the next predictive overlap ($d_{y-pred}$) by making it as a variable (*count*). The value of $d_{y-pred}$ obtained by calculating the middle value between the two former $d_y$. By minimizing min *SDA* process, it indirectly reduces the computational load required in reconstruction overall process.

$$d_{y-pred} = \frac{d_{yi-1} + d_{yi-2}}{2}$$

Where $d_{yi-1}$ is the first previous $d_y$ value, and $d_{yi-2}$ is the second last $d_y$ value, $d_{y-pred}$ is the predicted value obtained by calculating the average of both former $d_y$.

## 3 Results and Discussion

The experiments are done by reconstructing the set of fingerprint images captured using sensor swipe. Reconstruction is divided into two parts, first for the original and second after applying the predictive overlap method and adjusting the *count* variables.

There are 36 raw image data used in this experiments, images obtained from swipe sensor with six different fingers and six different shooting for each corresponding finger. In the original version, raw images reconstructed without using predictive overlap method.

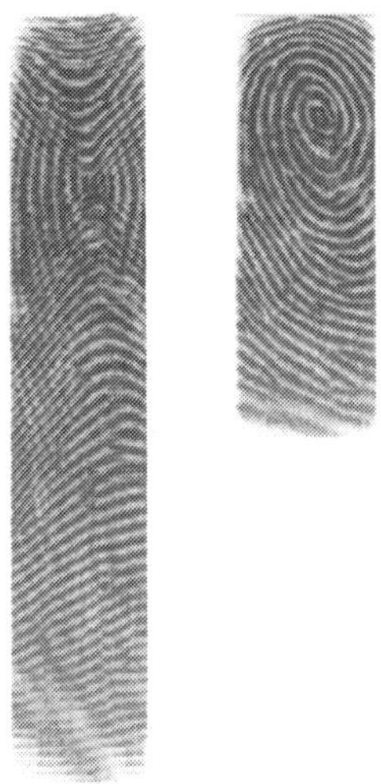

**Fig. 3.** Raw image from swipe sensor (left), original reconstructed image (right).

Reconstruction performed on all raw image data using predictive overlap method. Figure 4 shows the percentage of comparison between the computational time required by conventional reconstruction algorithm with reconstruction using predictive overlap approach. In the rebuilding process, the *count* variable indicates how often $d_{y-pred}$ used.

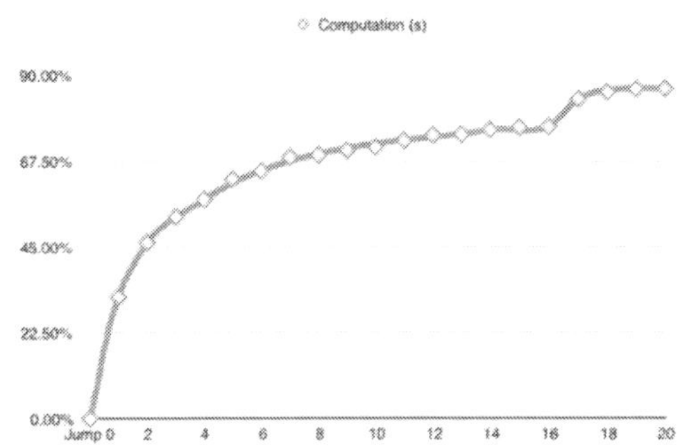


**Fig. 4.** Comparison computational time between conventional and predictive overlap.

By applying predictive overlap, the computational load tends to decrease in every turn of the *count* variables. When the *count* value equals 20, this method can reach 86.44% increase in computation time.

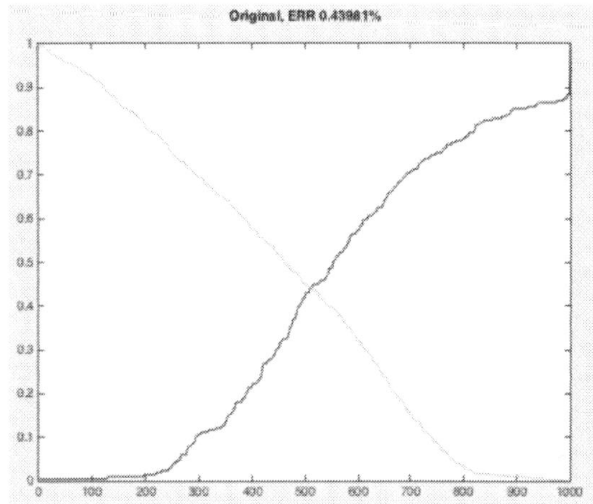

**Fig. 5.** EER for original reconstruction.

Next experiment is measuring the accuracy of the reconstructed image by matching the same finger (genuine) and the different finger (impostor). In figure 5, the original reconstructed image yields accuracy with *EER* 0.4398%.

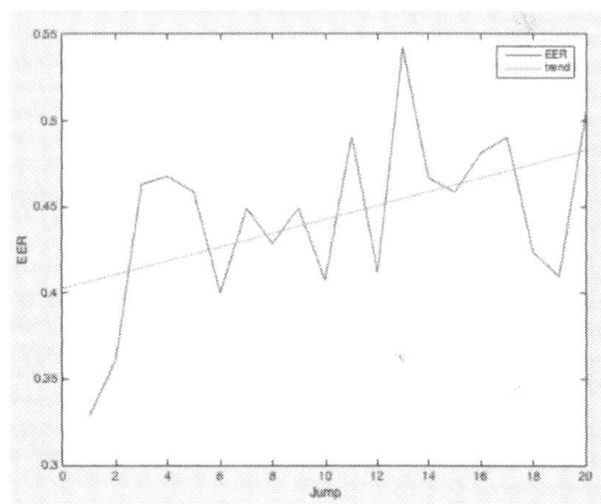


**Fig. 6.** Accuracy level by applying predictive overlap method.

On the other hand, with increasing in computational time, the accuracy of reconstructed images tends to decrease. As seen in figure 6 above, *EER* is fluctuating but shows an increasing tendency. The average *EER* obtained from all *count* values is 0.4446%, in other words, the *EER* rate increases by 0.0048%.

Variations in accuracy were caused by the difference of the extracted minutiae (feature) from the pattern on the reconstructed image. Figure 7 (left) is the result of conventional algorithm reconstruction, the number of minutiae features found is 27. Figure 7 (center) and (right) are the result of reconstruction with predictive overlap method, the number of minutiae features extracted respectively are 24 and 26. The reason is an incompatibility in placing block pixels while performing reconstruction, which is used to estimate the possibility of the next shift value.

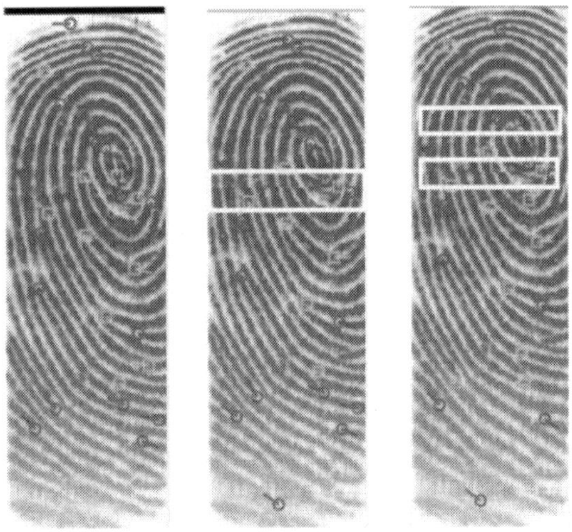

**Fig. 7.** Extracted minutiae for original image (left), predictive overlap with count = 10 (center), count = 20 (right).

Conducted experiments reveal that increasing in computation time, even by using the same $d_y$ for 20 times, while the changes in accuracy during the *count* tend to be stable. Thus this method is suitable to apply to devices that have limitations in the computing process.

## 4 Conclusion

In this paper, experiments of the predictive overlap method have successfully reduced the computational load for swipe fingerprint images. Increased computing time has no significant impact on matching accuracy. The iterations performed in the predictive overlap method based on the previous set of $d_y$, called $d_{y-pred}$. From the test results, obtained computation time can increase up to 86.44%, with decreasing in accuracy only 0.0048%.

## References

1. Kanich, O. and M. Drahanský, Currently Used Swipe Fingerprint Sensors. International Journal of Bio-Science and Bio-Technology, 2016. 8(1): p. 381-386.
2. DABBAH, M.A., W.L. WOO, and S.S. DLAY, Computationally Efficient Fingerprint Algorithm for Automatic Recognition. 2005.
3. Hamine, M., Y. Audet, and J.-P. David, a Real Time Image Reconstruction Algorithm For An Integrated Fingerprint Sensor. 2007.
4. Li, G., C. Busch, and B. Yang, A novel approach used for measuring fingerprint orientation of arch fingerprint. 2014.
5. Kasban, H., Fingerprints verification based on their spectrum. Neurocomputing, 2016. 171: p. 910-920.
6. Lee, J.-W., et al., A 600-dpi Capacitive Fingerprint Sensor Chip and Image-Synthesis Technique. 1999.
7. Xia, X. and L. O'Gorman, Innovations in ngerprint capture devices. 2003.
8. Kamaraju, M. and P.A. Kumar, DSP based Embedded Fingerprint Recognition System. 2013.
9. Clausen, S., A Single-Line AC Capacitive Fingerprint Swipe Sensor.
10. Habegger, A., et al., A Subpixel-Based Fingerprint Reconstruction Algorithm. 2012.
11. Galy, N., B.T. Charlot, and B. Courtois, A Full Fingerprint Verification System for a Single-Line Sweep Sensor. 2007.
12. Jie, L., H. Xuxiao, and W. Puyu, Sequence Fingerprint Image Mosaic Method and Its Experimental Study Based on Phase Correlation and Gray Correlation. 2009.
13. TAN, S.-Y., et al., A Sweeping Fingerprint Verification System using the Template Matching Method. 2010.
14. Zhang, Y.-l., J. Yang, and H.-t. Wu, a Hybrid Swipe Fingerprint Mosaicing Scheme. 2005.
15. Jacomet, M., J. Goette, and A. Eicher, On Using Fingerprint-Sensors for PIN-Pad Entry. 2008.

# Analyzing the potential and the load evaluation on Ubrug hydro power plant of Sukabumi, West Java

*Didik* Notosudjono[1,*], *Evita* Wismiana[1], *Fitrah* Alamsyah[1], *Bagus Dwi* Ramadhon[2]

[1] Pakuan University, Jl. Pakuan PB No.452 Bogor, Indonesia
[2] Brawijaya University, Jl. MT Haryono 66, Malang, Indonesia

**Abstract.** Built in 1923, the Ubrug Hydro Power Plant in Sukabumi was set up with an installed power of 5,94 MW on 1st unit and 2nd unit and 6,48 MW on the 3rd unit resulting in sum total of 18,36 MW. However, at present, the actual total power only generates 15,045 MW, a smaller number than its initial capacity, due to the lack of water discharge. Based on the measurement data, the load of the generator has experienced a decrease of power in normal field operation with a peak load of 9,5 MW and the lowest of 6 MW. On daily operations, only 2 generators are being operated, leaving the 3rd generator unoperated. This is due to the aging hydroelectric power plant on the location and the transition of forest functions in the upstream area, that serves as water absorption, to settlements of residents, and industrial estates. The other and the most salient cause is the waste carried by the Cicatih River coming along with the flow of water in the conduit.

## 1 Introduction

Hydropower energy is one type of a renewable energy resource which is produced by the motion of water through hydropower device and aims to generate electricity. When the water flows due to the force of gravity, its potential energy converts into kinetic energy. This kinetic energy of the flowing water subsequently turns blades or vanes in the hydro turbines, resulting in the energy being transformed into mechanical energy. The turbine turns the generator rotor which converts this mechanical energy into electrical energy. [3] Many other components may be used in a hydropower system to produce energy which presumably is the earliest type of renewable energy technique [1].

According to the power generation capacity, the Hydro power is classified into several types, namely large hydro power, medium hydro power, and small hydro power as explained in the following figures.

- Large hydro power: >100 MW
- Medium hydro power: 30-100 MW
- Small hydro power: 1-30 MW

Some hydro plants designated for a very small scale purpose are also available and classified into Mini hydro power engaging a capacity between 100 kW to 1MW and Micro hydro up to 100 kW. These are generally used to meet the need of a small community or rural industry where grid is not available [2]

The Ubrug Hydroelectric Power Plant built in 1923 is equipped with installed power of 18.36 MW and three generator units that are driven by three water turbines.

The result of the study of Hydro Power Ubrug, Sukabumi would help the Government review the already 90-year-old hydro power policy in Indonesia. In this case, the ouput of Hydro power and its potential have substantially decreased, as happens to the Hydro power in Ubrug, Sukabumi, which potentially generates output of 15,045 MW for three Generators, whereas in fact only two Generators work.

## 2 Methodology

The methods used in this analysis consist of:

a. Literature Study, where the authors study and collect data during the accomplishment of the Performance Studies of Sukabumi's Ubrug power plant as a case study, whereas the theoretical data is drawn from the referential books in the library, the relevant documents, and the scientific web site in the internet as well as other various sources.
b. Data Measurement, in which the Water Discharge (Q), height (h) and Potential Hydroelectric Ubrug are directly calculated and the Generator Load is measured by utilizing the existing measuring equipment available in the hydroelectric power in Ubrug.
c. Data Analysis, by which the research focuses on how to analyze the system, based on the calculation of the relevant equations using the data which is obtained from the hydro potential and the load performance analysis of Ubrug Power Plant.

The results of the preliminary study have revealed that .decreased, without eliminating the fact that prior to 1980, it was the Base Load which supplied energy for systems in Java Island. And, to date, due to the transition of forest function into housing and industries, the hydro power in Ubrug Sukabumi has functionally switched to a system

* Corresponding author: dnotosudjono@gmail.com

aimed at overcoming peak load with a maximum of 2 generators in operation.

## 3 Basic Theory

### 3.1 Generator

The AC generator or alternating current (also known as synchronous generator/alternator) is the main source of all electrical energy utilization. The machine is the largest energy converter in the world. Generators are machines that use magnets to convert mechanical energy into electrical energy. The principle of a generator is embodied when a conductor is driven in a magnetic field and, in turn, generates an electric motion force. The generator is driven by several mechanical machines such as steam, water turbine, gasoline engine and electric motor [4].

### 3.2 Hydro Power Plant

The generation of hydro power is a form of power transformation from hydropower with a certain height and water discharge into electricity, using a water turbine and generator [5].

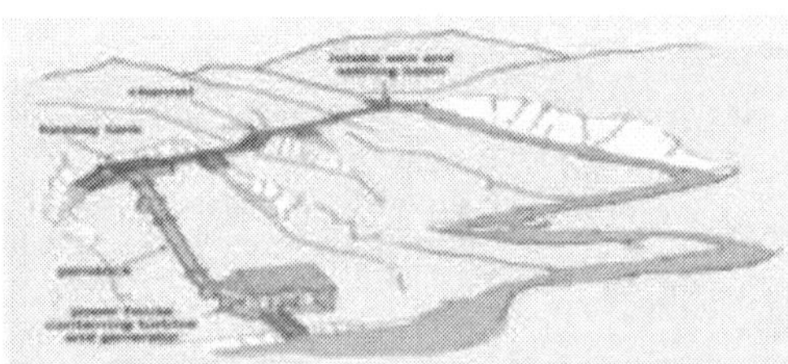

**Fig 1.** Run off River Hydro Power work principle [12]

#### *3.2.1 The calculation of output power generation from hydro turbine*

The potential in this regards refers to a description of the magnitude of the capacity of power plants that may be developed in a particular location plan. In accordance with the nature and the process or the mechanism of the generation of electrical energy derived from hydropower, there are 2 (two) main components that form the basis of the occurrence of the electricity generation, namely water discharge and water altitude (height) [6].

The output power generated from the hydro turbine (Pgen), can be estimated by the following expression [1]:

$$P_{gen} = \rho \cdot g \cdot H_n \cdot Q \cdot \eta_t \cdot \eta_{gen} \text{ [W]} \quad (1)$$

where $\rho = 1000$ [kg/m$^3$] as water density, $g = 9.81$ [m/s$^2$] as gravity acceleration, $\eta_t$ [%] as turbine efficiency, $\eta_{gen}$ [%] as generator efficiency, q as Water Discharge (m$^3$/s), h as Fall height (m).

#### *3.2.2 Water Discharge*

Discharge refers to the amount of water flowing through a particular cross-section of a river per unit of time. Water discharge is determined by several factors including rainfall, geological conditions, flora, temperature, and upstream. The discharge always changes from season to season and from day to day. Below is the calculation of water discharge:

$$Q = V \times A \quad (2)$$

Where:
Q = water discharge (m$^3$/second)
V = water flow velocity (m/second)
A = river cross section area (m$^2$)

The water discharge (Q) is the result of multiplication between the cross-sectional area (A) of the channel/stream and the water flow velocity (V) which can be identified in the equation:

$$V = S/T \quad (3)$$

Where:
V = Float flow velocity (m/s)
S = Distance from point A to point B (m)
T = Float reach time (s)

#### *3.2.3 Penstock*

Press pipes used to drain water from a head tank or directly from a water-take building to a water turbine are called Penstock. Channel pipes are the common name for the base or tunnel used for placing a penstock, anchor block and saddle, which will withstand the pipe [5].

To determine the cross-sectional area of the penstock, the equation below is used [8]:

$$A_n = \pi \times \left(\frac{D}{2}\right)^2 \quad (4)$$

Where:
$A_n$ = Penstock cross-sectional area (m$^2$)
D = Pipe Diameter (m)

a. To determine the velocity of water flow inside the penstock, the equation below is used:

$$V = \left(\frac{Q}{\frac{1}{4}\pi D^2}\right) \quad (5)$$

Where:
V = Flowing speed (m/det)
Q = Flow discharge (m$^3$/det)
D = Penstock diameter (m)

b. Losses on the penstock's duct due to the filter (trashrack) can be calculated using the following equation [9]:

$$hs = kt \left(\frac{t_k}{b_k}\right)^{\frac{4}{3}} Sin\ \alpha \frac{v^2}{2g} \quad (6)$$

Where:
$h_s$ = loss of height compression filter (m)
$k_t$ = Coefficient of energy loss due to lattice shape (0,5 for square-shaped, 0,05 For rounded shape)
$t_k$ = lattice thickness (m)
$b_k$ = lattice range (m)

$\alpha$ = mounting angle
v = flow speed (m/s)
g = gravitational acceleration (9.8 m/s$^2$)

c. Losses on the penstock's duct due to the inclusion can be calculated using the following equation:

$$hp = K \times V^2 / 2g \quad (7)$$

Where:
$h_p$ = loss of height compression due to inclusion (m)
K = velocity coefficient (0,95

d. Losses on the penstock's duct due to the water curving can be calculated using the following equation:

$$hp = Kb \times V^2 / 2g \quad (8)$$

Whereas:
$h_b$= Loss of height compression due to curving (m)
$K_b$= Curve coefficient (0,67 with a curve of 50°)

### 3.3 Prime Mover (Turbine)

Hydraulic turbines, closely related to the generator, is of the main function to convert water energy into electric power. Water flows through the turbine, energizing the mover (runner) of the turbine and making it rotate. The funnel of the driver is directly related to the generator, provided that the considerable mechanical power is channeled into the generator. Thus, turbines occupy key positions in the field of hydroelectric engineering (hydroelectric) and form a large part of all generation [10].

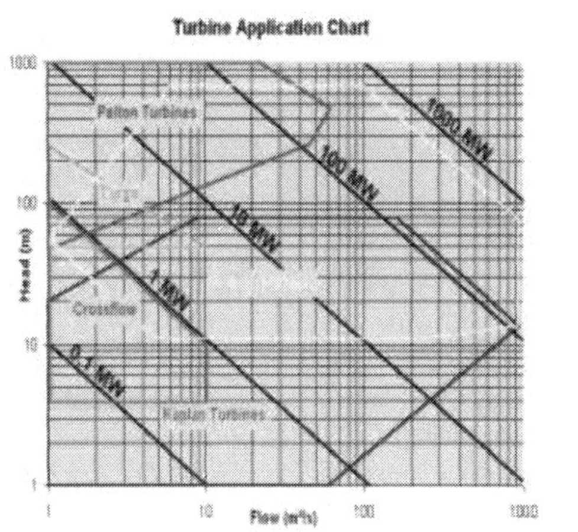


**Fig. 2.** Various turbine applications (https://yokealjauza.wordpress.com/2014/02/28/water turbine)

## 4 Analysis and Discussion

### 4.1 Water Flow Velocity

The measurement of water flow velocity is performed 3 times at each point, with a distance of 10 meters to obtain the velocity of the channel V (v) flow from the measurement result calculated by means of equation (2.13). For example, in the conduit at point A, 3 experiments are performed. The experiment 1 is with a distance of 10 meters and a time (T) of 14.56 second, then:

$$V_1 = \frac{10}{14{,}56} = 0.686 \text{ m/s}$$

So the water velocity time at point A in experiment 1 is 0.686 m/s. Similarly, we can calculate the water velocity at point A on experiment 2 and experiment 3 as follows:

$$V_2 = \frac{10}{14{,}72} = 0.678 \text{ m/s}$$

$$V_3 = \frac{10}{14{,}66} = 0.682 \text{ m/s.}$$

So the velocities of water at point A for the second and third experiment are 0.678 m/s and 0.682 m/s respectively. With the same method, the flow speeds at point B and point C are calculated and presented in the following Table 4.1.

**Tabel.1.** The Calculation Result of Water flow Velocity at an open duct

| Experiment | Range (m) | Flow Velocity (m/s) | | |
|---|---|---|---|---|
| | | **PointA** | **Point B** | **Point C** |
| **1** | 10 | 0,686 | 0,683 | 0,680 |
| **2** | 10 | 0,678 | 0,666 | 0,672 |
| **3** | 10 | 0,682 | 0,674 | 0,691 |

$$V_{rA} = \frac{0{,}686 + 0{,}678 + 0{,}682}{3}$$
$$= 0{,}682 \text{ m/s}$$

$$V_{rB} = \frac{0{,}683 + 0{,}666 + 0{,}674}{3}$$
$$= 0{,}674 \text{ m/s}$$

$$V_{rC} = \frac{0{,}680 + 0{,}672 + 0{,}691}{3}$$
$$= 0.681 \text{ m/s}$$

The total average velocity on the conducting duct is determined with:

$$V_{rtotal} = \frac{0{,}682 + 0{,}674 + 0{,}681}{3}$$
$$= 0{,}678 \text{ m/s.}$$

In conclusion, the average water flow speed on conducting duct is 0,678m/s.

### 4.2 Sectional Water Flow Analysis

The cross-sectional area of the conductor water channel with a base width (b) of 2 m, the water height (h) of 2.2 m and the slope (m) of 1.5 m is calculated by means of equation 2.5 as follows:

$$A = ( b + m \,.\, h ) h$$
$$= ( 2 + 1{,}5 \,.\, 2{,}2 ) 2{,}2$$
$$= 11.66 \text{ m}^2.$$

From the result, the area of water flow on the conducting duct is 11,66 m$^2$, with a water depth of 2,2 meter.

### 4.3 Water Discharge Analysis

After obtaining the cross-sectional area of the water channel (A) of 11.66 m2 and the average velocity of the conducting water channel (V) of 0.678 m/s, the discharge channel water flow (Q) can be calculated by the following equation (2.14):

$$Q = V \times A\ (m^3/s) = 0{,}678 \times 11{,}66 = 7{,}905\ m^3/s.$$

From the result, the water flow discharge happens to be 7,905 $m^3$/s. Based on the data obtained with current discharge of 10.25 $m^3$/s, such shrinkage of water discharge is visible which is due to the frequent swith functions of river upstream.

### 4.4 Effective waterfall height

The cross-sectional area for the penstock with a diameter of 1,7 m, for both unit 1 and unit 2 as observed from the technical data can be calculated using the following equation (2.17):

$$A_n = \pi \times \left(\frac{D}{2}\right)^2 = 3{,}14 \times \left(\frac{1{,}7}{2}\right)^2 = 2{,}268\ m^2.$$

The cross-sectional area of the penstock with a diameter of 1,7 m is 2,268 $m^2$.

The water velocity inside the penstock with a water discharge of (Q) 7,905 $m^3$/s and a diameter of 1,7 m can be calculated using equation (2.18) as follows:

$$V = \left(\frac{Q}{\frac{1}{4}\pi D^2}\right) = \left(\frac{7{,}905}{\frac{1}{4}x3{,}14x1{,}7^2}\right) = \left(\frac{7{,}905}{2{,}268}\right) = 3{,}485\ m/s.$$

So, the waterflow velocity obtained is 3,485 m/s.

The *losses* (energy loss) inside the *penstock duct* which is caused by filtering (*trashrack*) that has the energy loss coefficient of 0.5 due to the lattice shape, lattice thickness of 0.01 m, lattice spacing of 0.028 m, and the 45° mounting angle can be calculated by equation (2.19) provided:

$$ht = kt \left(\frac{t_k}{b_k}\right)^{4/3} Sin\ \alpha \frac{V^2}{2g}$$

$$= 0{,}5 \left(\frac{0{,}01}{0{,}028}\right)^{4/3} Sin\ 45 \frac{3{,}485^2}{2 \times 9{,}8}$$

$$= 0{,}5 \times 0{,}253 \times 0{,}7 \times 0{,}619 = 0{,}0548\ m.$$

So the *losses* (energy loss) inside the *penstock* duct due to filtering (*trashrack*) is 0,0548 m.

The *Losses* (loss of height compression) inside the *penstock duct* due to water inclusion with a velocity coefficient of 0,95 can be calculated with the following equation (2.20):

$$h_p = K \times \frac{V^2}{2g} = 0{,}95 \times \frac{3{,}485^2}{2 x 9{,}8} = 0{,}95 \times 0{,}619 = 0{,}5880\ m.$$

So the *losses* (energy loss) inside the penstock duct due to inclusion is 0,588 m.

The *Losses* (loss of height compression) on the *penstock* duct due to the curving with a curve coefficient of 0,67 can be calcuted with the equation (2.21) as follows:

$$h_b = K_b \times \frac{V^2}{2g} = 0{,}67 \times \frac{3{,}485^2}{2 x 9{,}8} = 0{,}67 \times 0{,}619 = 0{,}4147\ m.$$

So the *losses* (energy loss) inside the penstock duct due to curving on the pipe is 0,4147 m.

The total height compression loss can be calculated as:

$$h_{total} = 0{,}0548 + 0{,}5880 + 0{,}4147 = 1{,}0575\ m.$$

The effective available fall height with total height 69 m is:

$$H_{eff} = 69 - 1{,}0575 = 67{,}9425\ m.$$

While for unit 3 that has a size of 1,8 m as observed in the penstock technical data, the total height compression loss can be calculated as follows:

$$h_{total} = 0{,}0435 + 0{,}4674 + 0{,}3296 = 0{,}8405\ m.$$

So the effective available fall height with a total height of 69,4 m would be:

$$H_{eff} = 69{,}4 - 0{,}8405 = 68{,}5595\ m.$$

### 4.5. Power Potential

By inserting the efficiency value of generator turbine of 0,95, the power potential on Unit 1 would be:

$P_t$ = 9,8 m/$s^2$ x 7,905 $m^3$/s x 67,9425 m x 0,95 x 1000 kg/$m^3$

= 5.000.265,656 kg $m^2/s^3$

= 5.000.265,656 Joule/s

= 5.000.265,656 Watt

= 5.000,265 kW

= 5,000 MW

So the amount of power potential generated by unit 1 of turbine generator is 5,000 MW. With the same calculation method, the power potential on Unit 2 and Unit 3 can be obtained and therefore, the total power potential that can be generated at Ubrug's Hydropower plant would be:

$$P_{total} = \text{unit 1} + \text{unit 2} + \text{unit 3} = 5{,}000 + 5{,}000 + 5{,}045 = 15{,}045\ MW.$$

### 4.6 Installed Power Analysis

Theoretically, the installed power output generated by Ubrug Hydropower Plant can be calculated using equation (2.7) by referring to the nameplate or the technical data available on each generator and turbine at the Ubrug hydropower.

By calculating the data obtained from the nameplates or turbine technical data, the installed power of Unit 1 would be:

$P_{\#1} = \sqrt{3} \times 6.300 \text{ V} \times 605 \text{ A} \times 0{,}9$
$= 5.941 \text{ kW}$
$= 5{,}94 \text{ MW}.$

Thus the installed power output is 5,94 MW. With the same method, the installed power on unit 2 and unit 3 can be obtained, then

$P_{total} = \text{unit 1} + \text{unit 2} + \text{unit 3}$
$= 5{,}94 + 5{,}94 + 6{,}48$
$= 18{,}36 \text{ MW}.$

As a result, the total power output installed on Ubrug Hydropower Plant is 18,36 MW

## 4.7 Generator Load Characteristics

From the previous analysis, it is known that the water discharge of the duct amounting to 7,905 m3/s is only able to generate power potential of 15.045 MW whereas based on the preliminary data, the Ubrug Hydropower Plant is capable of generating power at the amount of 18.36 MW. Therefore the water discharge visibly shrinks.

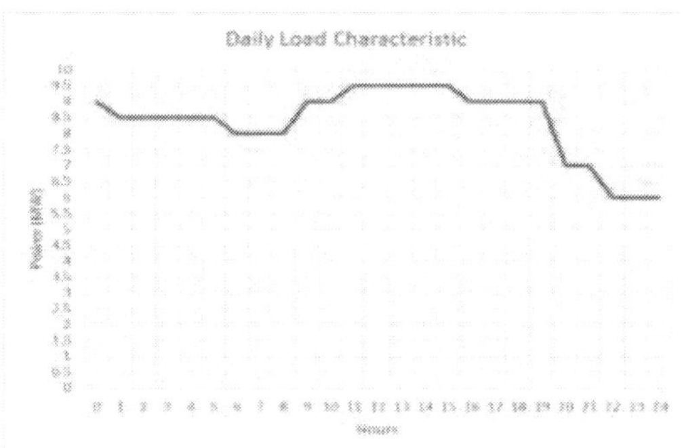


**Fig.3.** Daily Load Characteristic

According to data that was taken on November 3rd 2016, the peak load occurring between 11.00 - 15.00 WIB (Indonesian Western Time) was 9,5 MW. This was caused by the increase of water discharge in those hours, with 2 units of generators in operation. The lowest load of 6 MW occurred at 22.00 - 24.00 WIB. The following graph (figure 6.1) may help to describe the daily load characteristic.

Based on the observation in the field, the Generator in Ubrug hydro power evidently never fully operates that it only does with a maximum of 2 generators due to lack of water.

## 4.8 Performance Study on Ubrug Hydropower Plant

In the analysis made to calculate the performance of Ubrug Hydropower, a comparison is made between the potentially generated power, the measured power on turbine performance, and the installed power on the generating unit. The performance on each unit is mostly affected by the aging machines and the lack of water discharge.

The following contains the load comparison of each generating unit.

### *4.8.1 Performance comparison on Unit 1*

The following is the daily load comparison of Unit 1.

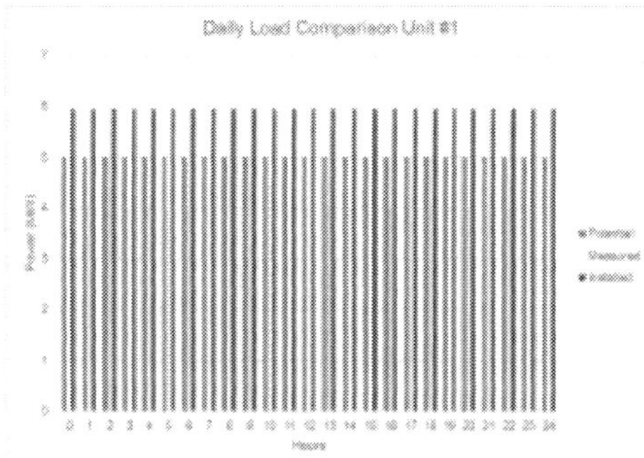


**Fig.4.** Daily load on Unit 1

Based on the graph in figure 7.1, on December 19th 2016, the observation focused only on Unit 1 daily load while Unit 2 and Unit 3 were unoperated. The amount of power at 00.00 WIB was 4 MW and it lasted until 06.00 WIB. At 07.00 WIB, the power increased to 4,5 MW due to increase of water level. The power state of 4,5 MW lasted until 10.00 WIB. At 11.00 WIB, the power dropped back to 4 MW, because the water level decreased, and lasted in such a state until 00.00 WIB.

### *4.8.2 Performance Comparison on Unit 2*

The following is the daily load comparison of Unit 2.

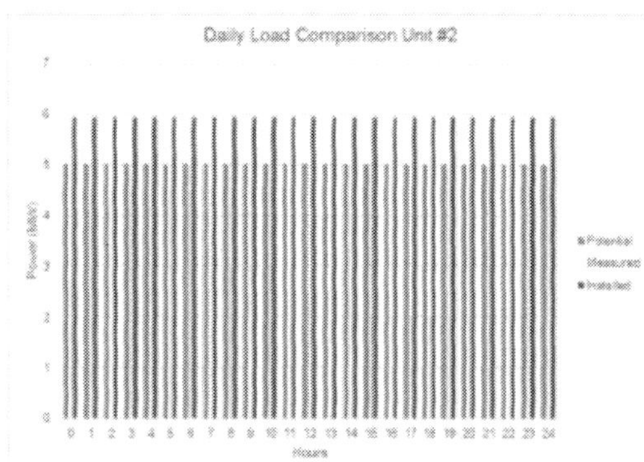


**Fig.5.** Daily load on Unit 2

Based on the graph in figure 7.2 which is derived from the data taken on November 23, 2016, the amount of power at 00.00 WIB was 2,5 MW because the water level was experiencing a decrease. At 07.00 WIB, the power load increased to 4,5 MW following the increase of water level. This amount lasted until 07.00 WIB. At 08.00 WIB the amount of power decreased to 3 MW due to the decrease of water level.

### *4.8.3 Performance Comparison on Unit 3*

The following is the daily load comparison of Unit 3.

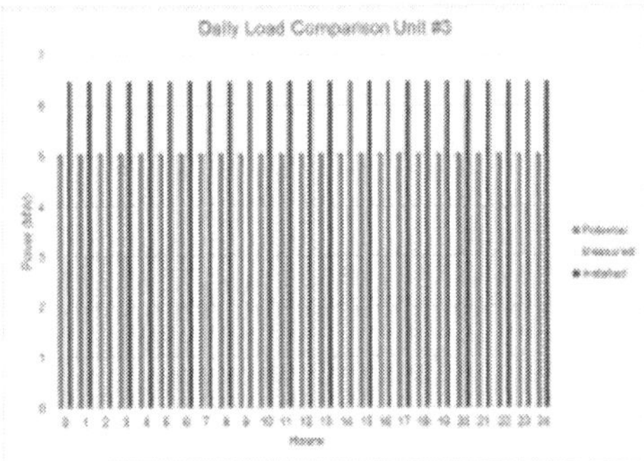


**Fig.6.** Daily Load on Unit 3

The graph in figure 7.3 above indicates the daily load comparison of Unit 3 which was taken on November 23rd 2016. At 00.00 WIB, the output power was 3 MW and lasted until 06.00 WIB. At 07.00 WIB, the power decreased to 2,5 MW due to decrease of water level and lasted until 10.00 WIB. At 11.00 WIB, the power increased to 3 MW and to 4 MW at the following hours until 00.00 WIB.

## 5 Conclusion

Based on the study result of Ubrug Hydroelectric power plant in Sukabumi, it can be concluded that:

1. The capacity of unit 1 and unit 2 is 5,94 MW for each, and that of unit 3 is 6,48 MW resulting in total power of 18,36 MW. However only 2 units are separately operated due to the limited amount of water.
2. The calculation of the potential power has revealed that the power which can individually be generated by Unit 1 and Unit 2 is 5 MW whereas it is 5,04 by Unit 3 resulting in total power of 15,045 MW. This figure is smaller than 18,36 MW installed power, due to lack of water discharge. Based on the analysis, the amount of water discharge in Cicatih River is only powerful to move 2 turbines in Ubrug Hydropower Plant. This is due to the frequent switching functions of the Forest in the Hydro power area into housing and Industries. Thus, the Government should issue a Policy that designates the catchment area in the forest around Ubrug Hydro Power, Sukabumi, to be a protected forest. This policy is very important to maintain the availability of water debit for Hydro power and restore the forest function so that the output power will potentially be improved.
3. From the performance analysis, the measured power is evidently lesser than the potential power and installed power. There is also data that indicates equal measured power to that of the potential power. This is due to the aging machine, the forest area in the upstream which is mostly converted into residential areas, and industrial areas, and most importantly, the wastes which are carried by the Cicatih River.

## References

1. Shpetim Lajqi, Naser Lajqi, Beqir Hamidi, Design and Construction of Mini Hydropower Plant with Propeller Turbine.International Journal, contemporary,Energy, vol 2, No1. pp.1 and 4. 2016 ,
2. Roshni Bhoi, Dr. S.M. Ali, Potential of Hydro Power Plant in India and its Impact on Environment, International journal Engineering Trends and Technology (IJETT) , volume 10 number 3, p.114, April 2014
3. Celso Penche Layman's guidebook on how to develop a small hydro site", Published by the European Small Hydropower Association (ESHA), Second edition, Belgium, June, 1998
4. Petruzella, Frank D. Elektronik Industri. Andi. Yogyakarta.p.311, 2002
5. Arismunandar, Artono. Dr dan Kuwahara Susumu. Dr. T,Electrical Engineering,volume 1, Water Power Engineering PT Pradnya Paramita. Jakarta.p.1 and p.44, 2004
6. Sebayang, Sedya. Ir. Evalution of Mini Hidro Power Bandung. p.3. 2002
7. Marsudi, Djiteng. Ir. Electric Power Generation. Erlangga. Jakarta.p.88 2005
8. Almahali, Jalaludin. Evaluation Hidro power in Ubrug Sukabumi Pakuan University Bogor.p.88, 2009
9. Setyo Putro, Yogi Suryo. Studi Planning of Micro Hydro Power in Sungai Atei desa Tumbang Atei , Distric Sanamang Province central of Kalimantan Universitas Brawijaya. 2016
10. Dandekar.M.M. dan Sharma. K.N. Electrical Power Generation. University of Indonesia, pp.391. 1991
11. https://yokealjauza.wordpress.com/2014/02/28/wate r turbine
12. http://sites.uoit.ca/sustainabilitytoday/blog-posts/blog-posts/2014/09/Capturing-the-power-of-water.php

# Design of a new PID controller based on Arduino Uno R3 with application to household refrigerator

*Eddy* Erham[1,*], Markus[1] , *Ary* Surjanto[1] ,and *Jaka* Rukmana[2]

[1]Department of Refrigeration and Air Conditioning, State Polytechnic of Bandung (POLBAN), Indonesia.
[2]Alumnus of Department of Refrigeration and Air Conditioning, State Polytechnic of Bandung, Indonesia.

**Abstract.** A refrigeration system to maintain a desired compartment temperature usually uses a thermostat as an on-off controller. In fact, the thermostat has some disadvantages. The main problem of system which is related to the thermostat is the biggest energy consumption in household appliances. In this paper, to solve the problem was designed a new PID controller based on an Arduino Uno R3 with application to a household refrigerator. In this case, the Arduino Uno was uploaded with PID controller algorithm. Then, in implementation to determine controller parameter values was defined new criteria. After that, to obtain the best refrigeration system performance was also proposed new performance criteria based on experimental data. The experimental results showed that the proposed control system was able to maintain the desired temperature with steady-state error of about 0.044°C. In addition, in steady state the control system for the refrigerator was able to the energy saving of about 30% and it almost did not depend on cooling load quantity.

## 1 Inroduction

A household refrigerator can maintain a desired compartment temperature since it employs a thermostat which serves as an on-off controller. In fact, the thermostat has some disadvantages. First, the present of control differential, difference between cut-in or cut-out temperature and setpoint temperature, causes response error that is set consciously according to the working principle of the on-off controller. Thus, the error is embedded in the thermostat. Second, if the differensial is set at a too small value, then the refrigerator will often turn on and off. Consequently, this action can breakdown the compressor[7,17]. Besides that, the action also requires high starting current equals to 4 to 10 times bigger than running current[21].Third , the energy consumption of refrigeration system is the biggest in household appliances[11,13,15].

To respond these disadvantages, the researchers developed a new on-off controller. However, this way could not reduce significantly both response error [20] and the energy consumption[12]. Then, PID (Proportional Integral Derivative) controller concept was used to improve the on-off controller[1, 2, 16, 22]. Then, it was followed adaptive control concept[8,9,23]. However, in implementation they had constraints, both of them were determination of control parameters, and control components that were needed to support a control system[5]. For example, to modify from the on-off control system to the PID one it needs to consider adding of safety components to avoid an overheat compressor. Since, the PID one is always working without ever turning off as in the on-off one.

To reduce the problem and also to give a better optional controller, then in this paper was designed a new PID control system based on Arduino Uno R3 with application to the refrigerator. In this case, Arduino Uno R3 was uploded with a new PID controller algorithm. Control parameters are proportional gain $K_p$, integral gain $K_i$, and derivative gain $K_d$ that were determined using a method of trial and error so that the control system response satisfies the proposed criteria. Moreover, it was investigated effect of various cooling loads to the refrigeration system performance. In this case, it was found settling time ($t_s$), steady-state error ($E_{ss}$), and also consumption of electrical power and energy. Then, the data was displayed on a LCD 20x4 and also on a Laptop in which was installed both softwares of Excel dan PLX-DAQ.

## 2 Design of control system

The design of control system consists of two parts, namely design of hardware and program. The design of hardware includes determination of controller, sensor, and actuator which will be selected. A controller is a device to maintain a quantity be the same as setpoint. It was a mechanical-electrical device or a electronic circuit, for instance a thermostat. However, currently the present of computer have supported control development. Thus, the working principle of old controller can be adopted into a computer program form so that the computer can be used as a

*Corresponding author: e_erham@yahoo.com

controller. This is known as control system based on computer. In this case, the design of program is of computer program for PID controller. To do this, it is initialized by designing a flow chart of the controller.

After all design stages were accomplished, then was continued determination of PID control parameters, namely $K_p$, $K_i$, and $K_d$. To determine the control parameters was used a method of trial and error . The proposed criteria are to find $K_p$, $K_i$, and $K_d$ value by substituting various sets of values of the control parameters into the algorithm of PID controller so that was obtained a set of value of $K_p$, $K_i$, and $K_d$ that results in $E_{ss}$ be within the limited range from 0 to 0.5. The method is the simplest, so it can be performed easily and quickly. However, the result is really effective and accurate.

## 2.1 Design of hardware

In this paper, the explaination focuses on designing of main components that is related to control system. Whereas, the other components are shown directly their connection paths to the Arduino which can be traced through a wiring diagram of the entire hardware design.

### *2.1.1 Selection of microcontroller.*

To design a PID controller was selected a microcontroller which is called the Arduino shown in Fig.1. To this purpose, an algorithm of PID controller was uploaded into the Arduino.

**Fig. 1.** Microcontroller board of Arduino UNO R3

### *2.1.2 Selection of actuator*

As an actuator was selected Variable Speed Drive (VSD) Telemechanique Altivar 11 or also called Inverter that has capacity of 1 HP (1 to 3 Phase) 220V shown in Fig.2. It can vary both its voltage output and frequency which will be sent to a compressor motor. It means that the compressor is operated at any certain frequency so that motor rotation per minute is also at any certain value. Thus, the compressor can respond the changes of frequency that are equivalent with the changes of cooling load value in the refrigerator compartment. If cooling load value is smaller, then the energy consumption is also smaller, and verse.

**Fig. 2.** Variable speed drive

### *2.1.3 Connnection of temperature (sensor)*

To measure compartment temperature was selected a sensor DS18B20. Connection of sensor DS18B20 with Arduino UNO R3 is shown in Fig. 3.

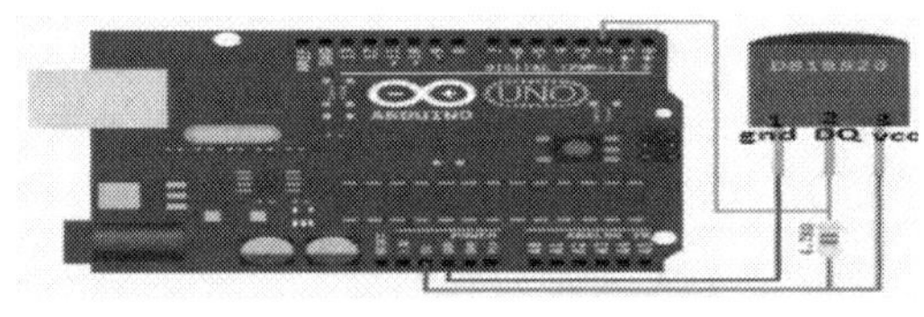

**Fig. 3.** Connection of temperature sensor to Arduino

### *2.1.4 Display*

Monitoring of data that includes temperature, $K_p$, $K_i$, $K_d$, electrical voltage and current was conducted through two displays, namely LCD and monitor of laptop. The usage of LCD display (20x4) with I2C is to minimalize pin usage of the Arduino. And the connection of LCD display 20x4 with I2C to the Arduino UNO R3 is shown in Fig.4. Whereas, a data display that uses a Laptop in which was installed a PLX-DAQ Software and Ms. Excel can be seen in Fig.5.

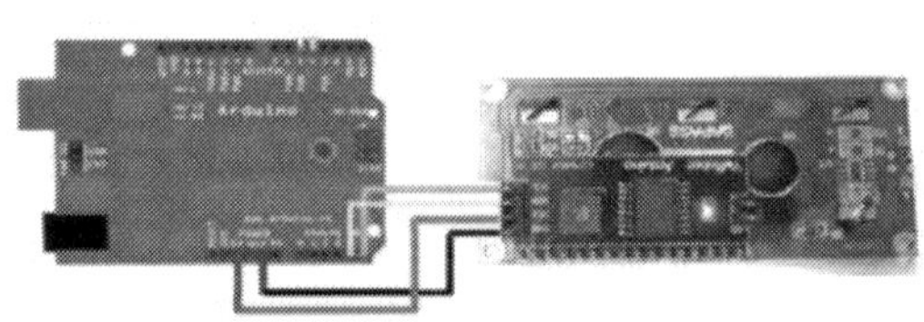

**Fig.4.** Connection of LCD display to Arduino

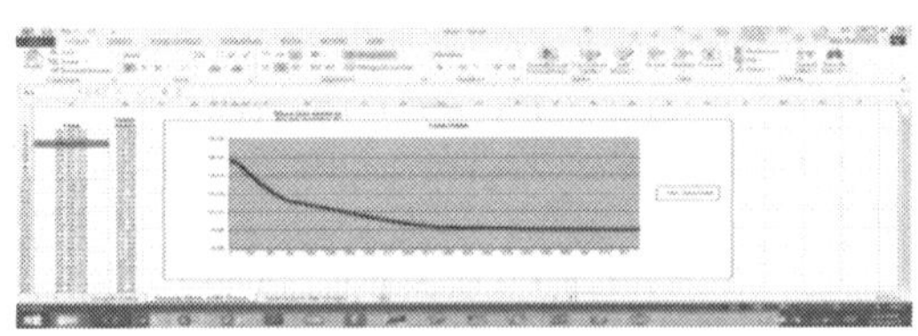

**Fig. 5.** Display of Laptop Monitor

### *2.1.5 Design of the entire hardware*

Finally, Fig. 6. shows how to all hardware components were integrated to become an unit in a wiring diagram form.In the diagram all components are displayed and shown their connections except Laptop which its connection has been provided a particular slot and a USB cable. The Arduino Uno as controller is connected directly to a temperature sensor DS18B20, Variable Speed Drive (Inverter) and LCD. However, to drive a compressor of the refrigerator, it does not have a connection directly and must be through an Inverter .

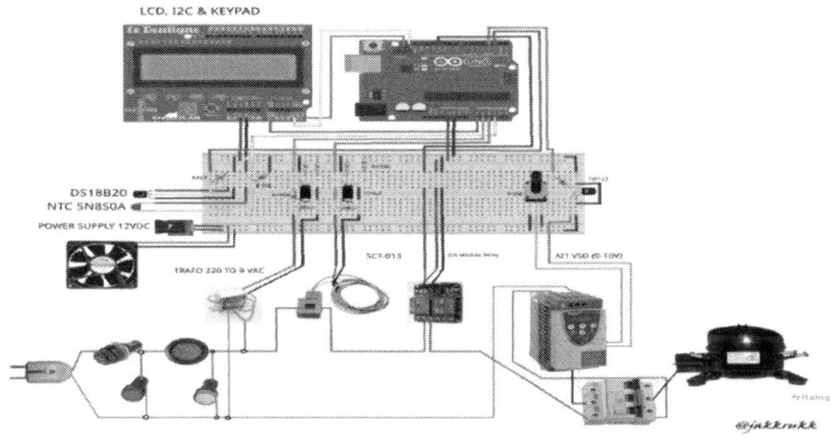


**Fig. 6.** Wiring diagram of the entire hardware design

## 2.2 Design of program

In this case, the program design does not only design a program of PID controller[3], but also it includes a data monitoring program. The controller design was conducted based on the flow chart shown in Fig. 7. This figure shows generally an algorithm of PID controller including input setpoint temperature (Ts) and read sensor output Tc. Then, calculate error, if error < 20 is true, this is caused maximum error be 20 which is $T_{set}$= $8^{0}$C subtract maximum compartent temperature Tc = $28^{0}$C , then calculate output of controller.If output > 255 is true, then output = 255 and If output < 0 is true, then output = 0, this is to limit a VSD (Inverter) maximum input of 10 and 0, respectively. So that cooling process can keep a setpoint value. Note that the designed algorithm is different from the existing algorithms including the algorithm in reference 2. Since, the method which is used is different from the existing ones, for example to calculate the output of controller or control signal was developed a new formulae in order to obtain more accurate results. Whereas, the reference 2 only provided results of simulation, not implementation or expremental and to do these was used the existing program, namely Mathlab and simulink

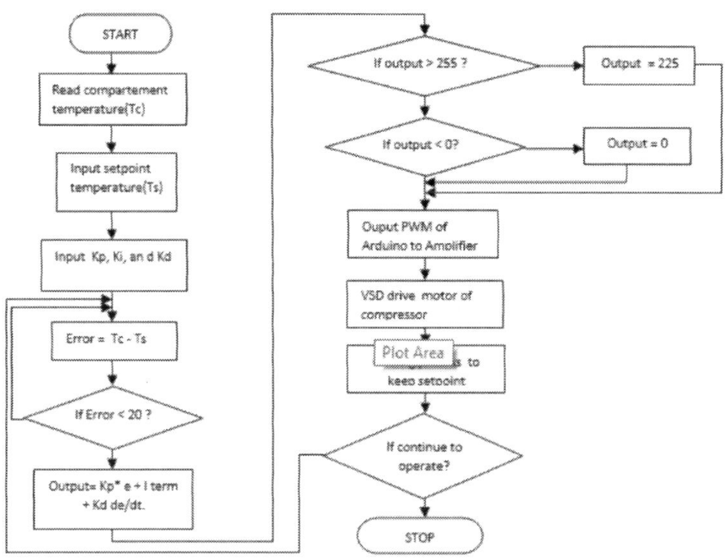


**Fig. 7.** Flow chart of PID controller

## 3 Results and discussion

The test of performance was conducted by using three types of cooling loads. In this case, it only includes determination of settling time, steady-state error ( $E_{ss}$) and energy consumption.The proposed performance criteria are a maximum steady-state error of 0.5°C and a maximum settling time value of 2 hours. Besides, it was also investigated effect of cooling loads on the electrical power and energy consumption. To this test was selected a setpoint temperature of 8°C so that the response reach chilling time quickly and does not take too long time.

### 3.1 Determination of $K_p$, $K_i$, and $K_d$

Before system performance could be determinated , first it was need to find appropriate parameters of $K_p$, $K_i$, and $K_d$ of PID controller.For this purpose was used a method of trial and error. $K_p$, $K_i$, and $K_d$ were selected by trial so that be obtained the best time response and it satisfies a particular criteria. The proposed criteria of parameter determination are the smallest steady-state error and the maximum error of 0.5°C as well as the smallest settling time ($t_s$) value that can be obtained by a response curve. Fig.8 shows temperature curves versus time for five parameter sets of $K_p$, $K_i$, and $K_d$ of PID controller. By comparing these curves was obtained the curve which satisfies the criteria, namely for $K_p$= 200, $K_i$ = 20, and $K_d$ = 0.01. It can be seen that its settling time ($t_s$) value was the smallest,while its Ess was almost the same as others.

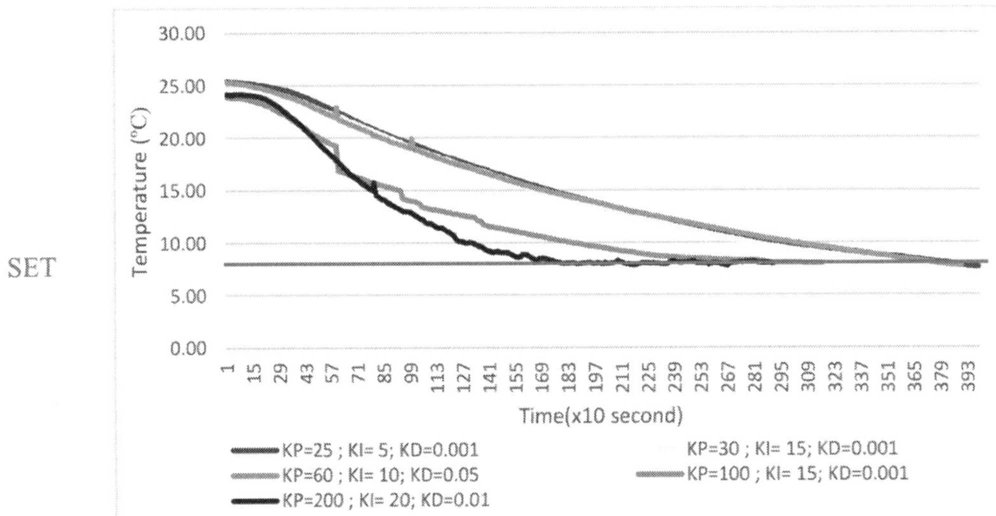


**Fig. 8.** Step response curves for some Kp, Ki, and Kd values

### 3.2 Analysis of performance

It was investigated the effect of three types of cooling loads to system performance. The cooling loads were formed from each a liter of water whose different temperature of 24°C, 50°C, and 75°C, respectively. In this case, the test of performance includes determination of settling time($t_s$), steady-state error ($E_{ss}$), and energy consumption.

Settling time value was determinated from exprimental data displayed in Fig. 9 For this purpose , first, it was need to find error 5% from setpoint 8°C, namely (5/100)x8 = 0.4°C, then compartement temperature = 8.4°C or to find error 2% from the setpoint, namely (2/100)x8 = 0.16°C, then the temperature=8.16°C. Then, settling time for each cooling load could be determinated when the temperature of 8.4°C ( criteria 5%) and then by using curves of cooling loads in Fig. 9 $t_s$ could be found, as an example for cooling load 1 was obtained $t_s$ = 29°C and similarly ts could be determined when the temperature of 8.16 °C ( criteria 2%). To make easier analysis the important data of Fig. 9. and Fig. 10 is summarized in Table 1. Then, the rest of settling time values for the three different cooling loads can be seen in Table 1. In this case, if cooling load quantity increases then ts value also increases. This makes sense.

Determination of $E_{ss}$ value was performed when refrigerator had worked for three hours assumed had reached steady state. Table 1 shows range of $E_{ss}$ from 0°C to 0.044°C. Thus, it satisfies criteria that Ess has to be smaller than 0.5° C. If it is compared with a thermostat error from 1.5°C to 1.95°C[7,17], then, a new PID controller is more accurate.

*Corresponding author: e_erham@yahoo.com

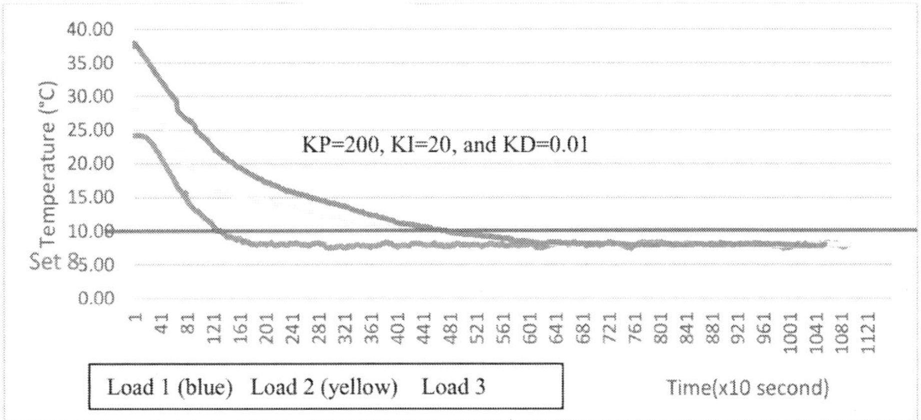


**Fig. 9.** Curves of compartment temperature for various cooling loads

**Table 1.** Experimental data for the three different cooling loads with $T_{set}$= 8$^0$C

| Cooling Load | For criteria of steady state error | | | | For other data after running 3 hours | | | |
|---|---|---|---|---|---|---|---|---|
| | 5% | | 2% | | | | | |
| | T (°C) | $t_s$ (min) | T (°C) | $t_s$ (min) | T(°C) | $E_{ss}$ (°C) | Power (Watt) | Energy (KWH) |
| **1(blue)** | 8.4 | 29 | 8.16 | 30 | 7.956 | 0.044 | 79.66 | 0.26946 |
| **2(yellow)** | 8.4 | 108 | 8.16 | 111 | 7.981 | 0.019 | 80.94 | 0.30505 |
| **3(green)** | 8.4 | 100 | 8.16 | 104 | 7.991 | 0.019 | 81.41 | 0.31273 |

Besides, it was also observed whether the consumption of electrical power and energy to be saving or not. Figure 10 shows the decreasing tendency of electrical power consumption for the three cooling loads from about 130 Watt at the part of beginning of curve became lower in range from 80 to 110 Watt. Moreover, from table 1 can be seen in steady state the power consumption for the three cooling loads turned to about 80 Watt. It meant that the consumption decrease of approximately 50 Watt or 30%.

Thus. in steady state the designed control system for household refrigerator could reduce the power consumption of about 30% and it almost did not depend on cooling load quantity. In the other words, the system performs energy saving of approximately 30%, since, the energy consumption is proportional to power one, then the energy saving of system is also proportional to the the power one. As extra information, Table 1 also shows the consumption of electrical energy in steady state for the three cooling loads was almost the same about 0.30 KWH.

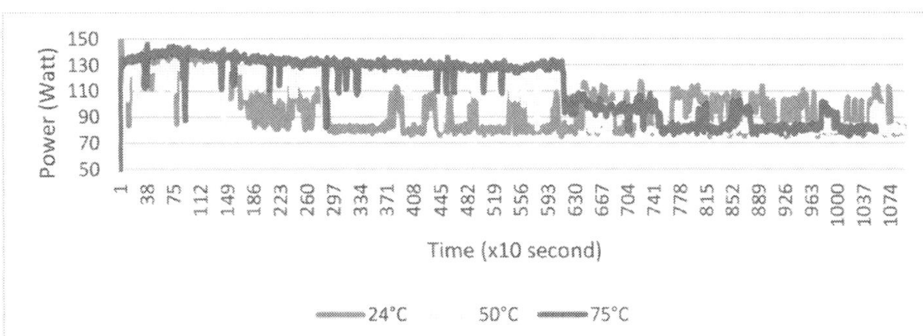


**Fig. 10.** Electrical power curves versus time for various cooling loads

## 4. Conclusion

The current research investigates the effect of cooling load on the designed control system response, the electrical power and the energy consumption for a refrigeration system, especially a household refrigerator. It can be considered as a promissing enough solution for the sustainable energy issues. The results showed that for three types of cooling loads were obtained range of steady-state error from 0°C to 0.044 °C. Thus, the control system has response more accurate than a conventional system with error from 1.5°C to 1.95°C. Moreover, in steady state the control system could reduce the power consumption of about 30% and it almost did not depend on cooling load quantity inside the refrigerator. In the other words, the system performs energy saving of approximately 30%.

## 5 References

1. E. Aridhi, M. Abbes, A. Mami, *Temperature control in a cavity of refrigeration using PI controller and predictive control.* Int. Conf. on Sci. and Tech. of Auto. Cont. & comp. Eng, : 123-128 (2014)
2. E. Aridhi, M. Abbes, A. Mami, S., Maarouf, R. Mhiri, *Cooling performance and energy saving of a refrigeration cavity supported by an outside cold air flow controlled by a PID controller.* Int. Conf. on Sci. and Tech. of Aut cont. & comp. Eng. : 1-7 (2015)
3. K.J. Astrom, T. Hagglund, *PID Controllers:Theory, Design, and Tuning.* NC:Instrument Society of America(1995)
4. M. S. Bhaat, *Domestic Refrigerators: Field Studies and Energy Efficiency Improvement.* J. of Sci. & Ind. Research, Vol. **60**, July 2001, pp 591-600 (2001).
5. W. R. Chang, D.Y. Liu, S.G. Chen, N. Y. Wu, *The Components and Control Methods for implementation of Inverter-Controlled Refrigerators/Freezers.* Int.Ref. and A/C Conf. Paper 696(2004)
6. A.E Coşgun, H. Demirel, *On-Off Controller Design for A Portable Refrigerator.* Proc. of 10th Int. Conf.ITELMS: 1- 4. (2015)
7. R. J. Dossat, *Principle of Refrigeration-SI Version, Second Edition.* John Wiley & Sons: Canada (1981).
8. M.D. Felice, L.Piroddi, A. Leva, A. Boer, *Adaptive Temperature Control of a Household Refrigerator.* American Contr. Conf. : 889 – 894. ( 2009)
9. E. P. B. Filho, F. E. M. Garcia, O.S.H. Mendoza, *Application of Adaptive Control in a Refrigeration System to Improve Performance.* J. of the Braz. Soc. of Mech. Sci. & Eng. *April- June 2011*, Vol. **XXXIII**, No. 2 : 176 – 182. (2011)
10. J. Fraden, *Handbook of Modern Sensors Physics, Design, and Applications, Third Edition.* Springer & Air Press: New York (2004)
11. Md. Hasanuzzaman, R. Saidur, H.H. Masjuki, *Investigation of Energy Consumption and Energy Savings of Refrigerator-Freezer During Open and Closed Door Condition.* J. of Appl. Sci., **8**(10): 1822-1831 (2008)
12. P. H. Lin, C. F. Tsai, *An Energy-saving Thermostat Controller for Electric Vehicle Air-Conditioning System.* Int. Elect. Vehical Symp. and Exh.: 1-5. (2015)
13. K. Mensah, J. M. Choi, *Energy Consumption and Stability Investigation of Constant Temperature and Humidity Test Chamber.* Int. J. of Air-Cond and Ref., Vol. **25**, No. 01(2017)
14. N. L. Djongyang, R.Tchinda, D. Njomo, *Thermal comfort: A review paper. Renewable and Sustainable Energy Reviews,* **14**: 2626–2640 (2010)
15. T. Olofsson, A. Meier, R. Lamberts, *Rating the Energy Performance Buildings.*The Int. J. of Low Energy and Sustainable Buildings, Vol. **3**. (2004)
16. Y. V. Parkale, *Comparison of ANN Controller and PID Controller for Industrial Water Bath Temperature Control System using MATLAB Environment.* Int. J. of Comp. Appl. (0975 – 8887) Volume **53**– No.2 : 1-6 (2012)
17. E. G. Pita, *Air Conditioning Principles and Systems, An Energy Approach, Fourth Edition.* Prentice Hall. (2002)
18. J. Park, S. Mackay, *Practical Data Acquisition for Instrumentation and Control Systems.* Britanian : Elveseir. (2003)
19. E. Premeaux, *Arduino Project to Save the World.* Technology in Action. (2011)
20. A. Ryniecki, J. Wawrzyniak, A. A. Pilarska, *Basics of Process Control: the on-off Control System. ResearchGate*: 25-29 (2015)
21. R. E. Smith, *Electricity for Refrigeration, Heating and Air Conditioning, 6 th Edition.* Delmar Learning, a division of Thomson Learning, Inc. (2003)
22. S. Srivastava, V.S. Pandit, *A PI/PID controller for time Delay systems with desired closed loop time response and guaranteed gain and phase margins.* J. of Proc. Cont. **37**: 70–77 (2015)
23. M. Umberger, I. Humar, *Energy savings of refrigerators in shopping centers with adaptive control and real-time energy management systems.* Przegląd Elektrotechniczny (Electrical Review), ISSN 0033-2097: 327 – 331 (2012)
24. F. Wang, H. Yoshida, B. Li, N. Umemiya, S. Hashimoto, T. Matsuda, H. Shinbayashi, *Evaluation And Optimization Of Air- Conditioner Energy Saving Control Considering Indoor Thermal Comfort.* Eleventh Int. Ibpsa Conf.Glasgow, Scotland : 88-95(2009)

# Early detection of LPG gas leakage based Wireless Sensor Networking

*Medilla* Kusriyanto[1], Firdaus[1], *Andik* Yulianto[2], and *Syakban* Kurniawan[1]

[1]Department of Electrical Engineering, Faculty of Industrial Technology, Islamic University of Indonesia
[2]Department of Environmental Engineering, Faculty of Civil Engineering and Planning, Islamic University of Indonesia

**Abstract.** LPG gas is the main fuel used in households. These days LPG gas leak triggered an accident that resulted in huge losses, especially if the accident occurred within the dense housing. LPG gas leakage monitoring and early detection of a gas leak and a leak warning is needed to prevent more fatal accident. Wireless sensor networking is a technology that can be used to build a centralized monitoring system by using a microcontroller as the interface and control system. This research developed system device monitoring and early detection of gas leaks by utilizing gas sensors MQ-4 and AVR microcontroller family as control devices. The system is also equipped with XBee PRO S2B nirkable devices as the interface to the wireless networking system that is used to transmit sensor data from the detection point to the monitoring center equipped with a PC and software integrated Visual Basic. Gas leak alert is sent with the message formats through social networking Gtalk.

## 1 Introduction.

The use of LPG gas in Indonesia is increasing. This situation is due to the implementation of government conversion subsidy from kerosene to LPG. With the increasing number of LPG gas consumer, then the case of leakage and explosion of LPG gas cylinders are also more and more. Based on data from the Indonesian Consumers Foundation ( YLKI ) , as many as 70% of cases are caused gas explosion gas stove accessories that do not meet standards and the remaining 30% due to consumer behavior . Tools used to detect this already developed in several previous studies exist that use sms or by using the buzzer. This system has flaws in remote communication. One of which should be available a cellular device to send sms and cannot be monitored on line and in real time.

In designing these tools, to detect the presence or absence of gas and LPG used a gas sensor MQ-4, as a controller and data processor using a microcontroller. Results of data processed is displayed on the LCD. In addition, data is also transmitted using wireless. At the receiver the data will be stored in a database and can be monitored via computer. The system is also capable of sending information to the Mobile users in the event of leakage of LPG gas. Tool is also equipped with LED indicator lights and a buzzer which indicates that the gas is detected by the sensor implies more.

## 2 Early Detection of LPG Gas Leakage

LPG is a mixture of various elements derived from natural gas. By increasing pressure and decreasing temperature, the gas turns into a liquid. LPG components are predominantly propane ($C_3H_8$) and butane ($C_4H_{10}$). LPG also contains other light hydrocarbons in small amounts, such as ethane ($C_2H_6$) and pentane ($C_5H_{12}$). Wireless Sensor Network (WSN) is a wireless network consisting of multiple sensors (sensor nodes) are placed in different places. The sensors will detect the object and send data wirelessly to the gateway and then to the server.

Fraiwan et.all (2011) in their research proposed a wireless safety device for gas leakage detection. The device is intended for use in household safety where appliances and heaters that use natural gas and liquid petroleum gas (LPG) may be a source of risk. The system also can be used for other applications in the industry or plants that depend on LPG and natural gas in their operations. The system design consists of two main modules: the detection and transmission module, and the receiving module. The detection and transmitting module detects the change of gas concentration using a special sensing circuit built for this purpose. This module checks if a change in concentration of gas(es) has exceeded a certain pre-determined threshold. If the sensor detects a change in gas concentration, it activates and audiovisual alarm and sends a signal to the receiver module. The receiver module acts as a mobile alarm device to allow the mobility within the house premises. The system was tested using LPG and the alarm was activated as a result of change in concentration.

[5] in their research use a prototype of a Wireless Sensor Network (WSN) to monitor and locate gas leaks of a complex indoor environment. Specifically, a mobile node is moving inside a building to monitor any leakage of carbon dioxide (CO2), supporting and displaying the level and the location of the leakage. Throughout the demonstration, the technological advantages of cognitive networking along with multihop routing are explored.

[1] in their work a safe Coal Mine Monitoring system which replaces the traditional coal mine monitoring systems which tend to be wired network systems. This play an important role in coal mine safe production. With continuous enlarging of exploiting areas and extension of depth in coal mine, many laneways become monitoring blind areas, where are lots of hidden dangers. Moreover, it is inconvenient to lay cables which are expensive and consume time.

In order to solve the problems, we designed a coal mine safety monitoring system based on wireless sensor network, which can improve the level of monitoring production safety and reduce accident in the coal mine. Zigbee technology provides a direction for scientists who commit to solve the safety monitoring problems of coal mine. The purpose of this study is to propose a solution suitable to mine wireless communication, safety monitoring, give a proof to the further study.

## 2.1 Material of Early Detection of LPG Gas Leakage

### *2.1.1 Gas sensors MQ-4/MQ-6*

MQ-4 gas sensor is a gas sensor has high sensitivity to methane, propane, and butane. Material of MQ-4 gas sensor is SnO2. Conductivity will rise in proportion to the increase in the concentration of the detected gas (Hanwei Electronics). Equivalent circuit of gas sensor MQ 4 is shown in Fig 1.

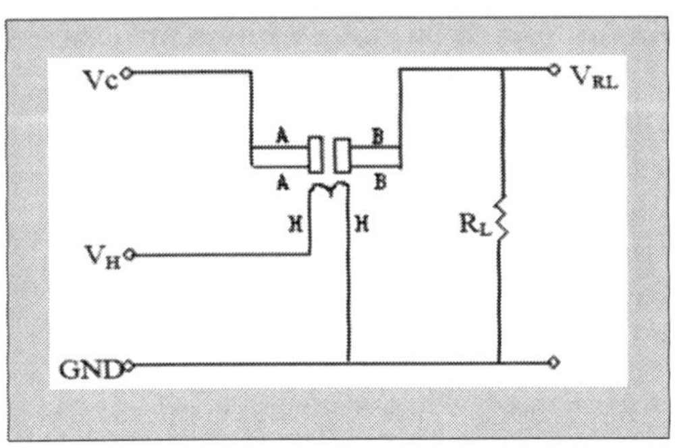


**Fig 1.** Equivalent circuit of gas sensor MQ-4

This sensor requires a supply voltage (VC) at 5 volts DC and voltage sensors are used for heating (VH). the output of the sensor MQ-4 is the change in the value of the sensor resistance (RS) contained between points A and B. The value of the sensor resistance MQ-4 will be converted into the output voltage by adding a series resistor RL. RS is resistant value can be obtained by using equation 1.

$$Rs = \frac{Vc}{VRL-1} \times RL \qquad (1)$$

LPG gas concentration can be obtained from ratio the RS and R0 with R0 is the resistance of reference which can be calculated by looking at the characteristics of the sensor are shown in figure 2.

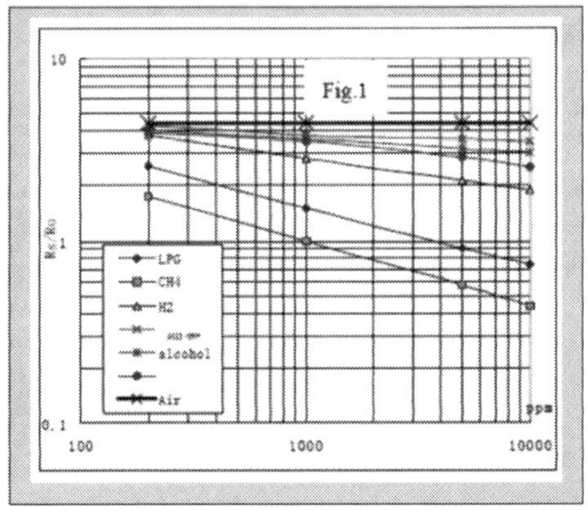


**Fig 2.** Characteristic of sensor MQ-4

### *2.1.2 AVR microcontroller family*

ATMega8 AVR microcontroller CMOS 8-bit AVR RISC architecture which has 8K bytes of In-System Programmable Flash. Microcontroller with low power consumption is capable of executing instructions at maximum speed 16MIPS at 16MHz frequency. ATMega8 has a 28 pin, each pin has its different functions both as a port and other functions.

### *2.1.3 XBee PRO S2B*

XBee PRO S2B is a RF module made by Digi International to meet the ZigBee standard, a technology with low data rate (Low Data Rate), and low cost (Low Cost). XBee PRO S2B module is easy to use for transceiver radio waves at a frequency of 2.4 GHz ISM band. The output power and frequency channel can be easily programmed through the X-CTU interface.

Addressing is used to distinguish between the XBee PRO S2B with other XBee PRO S2B and prevent duplication of data packets. Each module is XBee PRO S2B has a source address to prevent non - duplicate messages are not considered a duplicate message. XBee PRO S2B has two basic forms of addressing, namely Broadcast and Unicast. Broadcast message is a message that will be received by all modules that have a PAN ID (Personal Area Network) is the same.

### *2.1.4 Google Talk Application*

Google Talk (GTalk) is a web-based application service from Google that created as a means to communicate via instant messaging and VoIP.

Google Talk is provided free of charge and can be downloaded and accessed through Google. Google Talk service can only be used if already have an account on Google. Existing instant messaging between the Google Talk servers and its clients uses an open protocol, and XMPP which allows users to other XMPP clients to communicate with Google Talk users [2]. Meanwhile, Google Talk VoIP based on the Jingle protocol. The technology used in the Google server network cannot be known by the public. Google Talk software is available for BlackBerry, iPhone, and T-Mobile G1.

## 3 Systems and Methods of Research

Wireless sensor network applications for the early detection of LPG gas leakage includes hardware design and software design. The design of the system is divided into two parts, namely the design of the sender (transmitter) and receiver design part (receiver). Block diagram of the system design shown on fig 3.

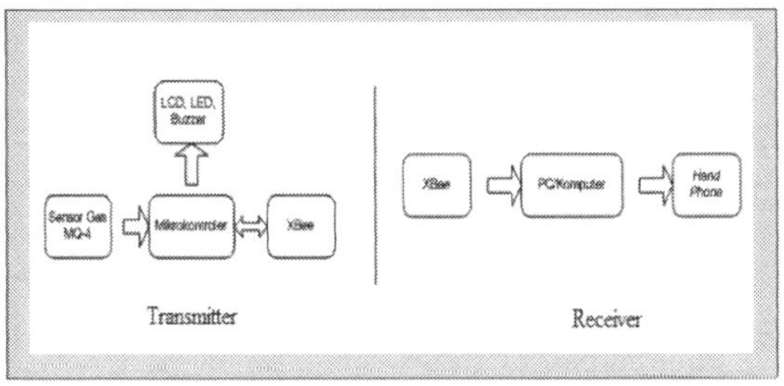


**Fig 3.** Block Diagram Design System

From the block diagram of Figure 3 can be explained how the system as a whole. Gas sensor will detect the presence of LPG gas in a room. The voltage value will then be converted into digital data by ADC (Analog to Digital Converter) on the microcontroller. Then the obtained data will be processed by the microcontroller and the results will be displayed on the LCD. The data is displayed on the LCD is a concentration of LPG gas.

LPG gas monitoring data is also transmitted wirelessly from the transmitter to the receiver. The devices used for the delivery of data is 2.4 GHz XBee PRO S2B. Once the data is received by the receiver and then the data is stored in a database on the server PC, the user interface of the software used in visual basic 2008. If the system detects a gas leak has occurred LPG then the information will also be sent to mobile home owners or users using GTalk application. So, when the LPG gas leak in a room, the system will provide early warning buzzer or through the LCD, PC servers, and mobile users.

## 4 Results and Discussion

### 4.1 Testing to Determine the Concentration of a Gas Sensor LPG

Testing done by varying the concentration of LPG gas to be detected by the sensor MQ 4. Values obtained from this test sensor resistance ($R_S$) which will be used to determine the level of leakage in LPG. To calibrate the sensor, then find the value of R0 as the reference ratio between RS and R0 as a determinant of the value of PPM. From several testing RS values obtained as shown in Table 1.

**Table 1.** Rs value for gas sensor MQ 4

| Voltage ($V_{rl}$) | Rs (Ω) |
|---|---|
| 2.35 | 74074.07 |
| 2.63 | 61349.69 |
| 2.77 | 56497.18 |
| 3.01 | 49751.24 |
| 3.27 | 44052.86 |

### 4.2 Testing Serial Communication Wirelessly with XBee PRO S2B

The test is performed to test the wireless communication between the transmitter and receiver. This test uses the XBee PRO S2B for each sender and receiver, the minimum system atmega 8 at the sender, as a voltage converter IC MAX232 RS232, USB to serial cable to connect the receiver to a laptop, as well as the X-CTU program. The results of testing the communication system is shown in Fig 4.

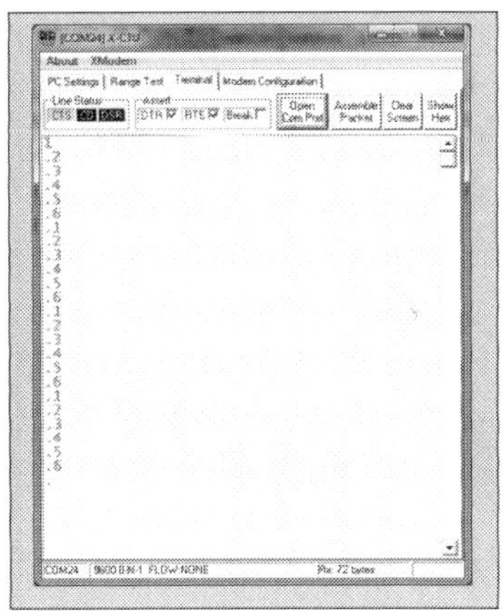

**Fig 4.** The results of testing the communication system

### 4.3 Testing the monitoring system with visual basic

Testing is done by observation data and graphs presented on the monitoring system using visual basic. The test results are shown in Fig 5.

### 4.4 Testing Software for Sending Messages to Gtalk Application

Software testing is performed to determine whether the software has been made appropriate design, which can send a message to the GTalk app. Software is created using the Java programming language. As a GUI (Graphical User Interface) using NetBeans IDE 7.3.1. test results are shown in Fig 6.

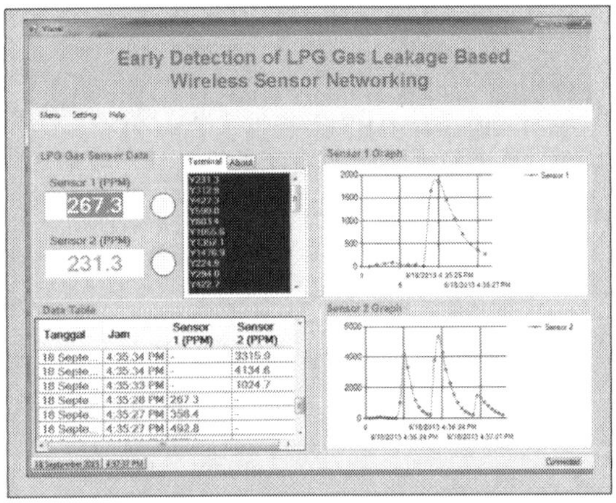


**Fig 5.** Monitoring system using visual basic

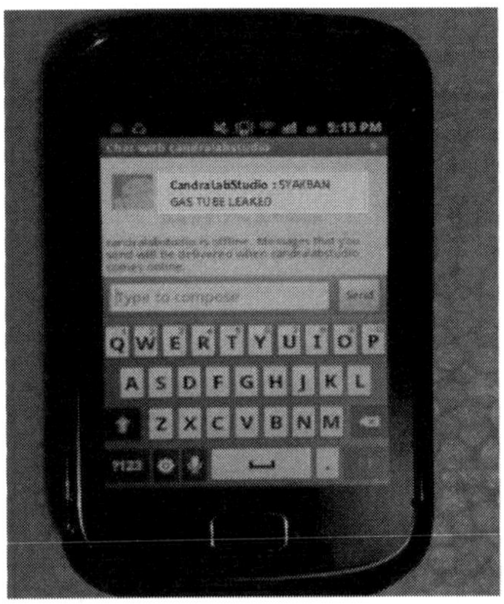

**Fig 5**. Message gas leak on Gtalk

## 5 Conclusion

Designing early detection of LPG gas leakage based wireless sensor network has been implemented properly in accordance with the system design. The output of the gas sensor MQ-4 amendments linear and proportional to the change in the input side, evidenced by the line equation $y = 1.000x + 0.004$ with a 3.46% error. The system interface can display and store data with good gas sensors and displays graphs for monitoring is done. Warning message about LPG cylinder leakage through GTalk application can be run in accordance with desired.

## References

1. Boddu, R, Balanagu, P & Babu, N Zigbee Based Mine Safety Monitoring System with GSM. International Journal of Computer & Communication Technology., (2012)
2. Chavan, P, Kadam, S., Zigbee Based Wireless A Remote Carbon Dioxide (CO2) Monitoring System. International Advanced Research Journal in Science, Engineering and Technology, (2017)
3. Fraiwan, L, Lweesy, K & Salma, A., *A wireless home safety gas leakage detection system*, 1st Middle East Conference on Biomedical Engineering (MECBME), IEEE, (2011)
4. Priyanto, R., *Langsung Bisa Visual Basic.Net 2008*. Yogyakarta: C.V Andi Offset. (2009).
5. Spachos, P, Song, L & Hatzinakos, D., *Gas leak detection and localization system through Wireless Sensor Networks.* Consumer Communications and Networking Conference (CCNC), IEEE, (2014)
6. Ustad, V, Mali, A & Kibile, S.,Zigbee Based Wireless Air Pollution Monitoring System Using Low Cost and Energy Efficient Sensors. International Journal of Engineering Trends and Technology, (2014)

# Artifacts removal from breath sound recordings in pediatric population

*Yusuf* A Amrulloh[1,*], *Jawahir* A K Haq[2]

[1]Electrical Engineering Department, Universitas Islam Indonesia, Yogyakarta, Indonesia
[2]B2TKS-BPPT, Kawasan Puspiptek Gedung 220, Tangerang Selatan, Indonesia

**Abstract.** Breath sound recordings from pediatric subjects pose more processing complications. Children, especially the younger ones, are not able to follow instructions to stay calm during recording. This makes their recordings not only contain stationary artifacts but also non-stationary artifacts such as movement of subjects and their heartbeats. Further, the breath sounds from pediatric subjects also have lower magnitude compared to adults. In this work, we proposed to address those problems by developing a method to remove the artifacts from breath sound recordings. We implemented a combination of a Butterworth band pass filter and a discrete wavelet filter. We tested three types of wavelets (Coiflet, Symlet and Daubechies). Ten level decompositions and a set of hard thresholds were implemented in our work. Our results show that our developed method was capable of removing the artifacts significantly while maintaining the signal of interest. The highest signal to noise ratio improvement (10.65dB) was achieved by 32 orders Symlet.

## 1 Introduction

Breath sound analysis is extremely important to support the initial diagnosis of respiratory tract infections [1]. It is used during physical examination at the first time patients admitted to hospitals or community health centres. In pediatric pneumonia, physicians use stethoscope to listen the breath sounds and to identify abnormal sounds such as crackle [2, 3], wheeze [4] or stridor [5].

Currently, electronics stethoscope is get used widely by physicians. In contrast with conventional stethoscope, the device is not only able to listen the breath sounds but also to record them for further analysis. Breath sound analysis of respiratory diseases in pediatric population is far different from adults due to the pathology and physiology. Dangerous diseases such as pneumonia mostly infect younger children less than five years [5]. Children in this age ranges are not able to follow instructions during physical examination. They tend to move and to cry. Further, their respiratory system is dissimilar with adults [6]. They have smaller lungs and airways that make the breath sounds magnitude weaker. The interference from their heart sound is higher because the number of heartbeat in younger children is much larger. All those artifacts impose the breath sounds and reduce their quality.

In this paper, we propose a method to remove the artifacts from breath sounds recorded in children with pneumonia. The novelties of this work are the subjects as well as the signal used in this study. Rather than involving adult subjects who are able to follow instruction to minimize artifacts during recording [7], we used actual breath sound recorded from from children with its natural artifacts. In facts, it is difficult to get a reference of artifact signals such as heartbeats during realtime recording. Therefore, a method for removing artifacts in the absence of reference signals is required.

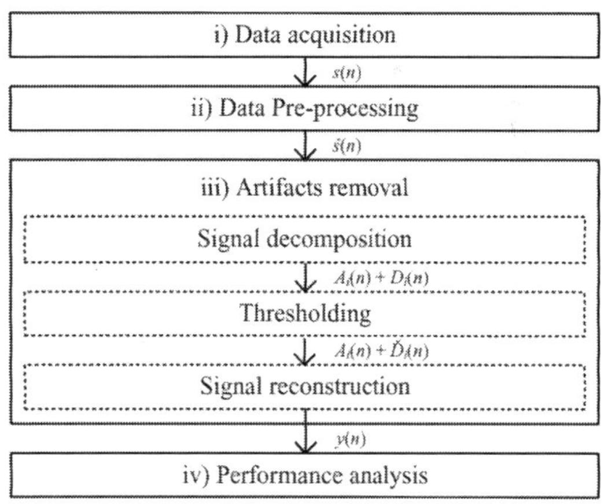


**Fig. 1**. Block diagram of the proposed method. It comprised of four main processes: i) data acquisition, ii) pre-processing, iii) artifacts removal and iv) performance analysis of the proposed method.

The work in presented in this paper consisted of four processes as shown in Fig 1. It comprised of i) data acquisition, ii) Pre-processing using band pass filter, iii) artifact removal using discrete wavelet transform filter, and iv) performance analysis on different types of

* Corresponding author: yusuf.amrulloh@uii.ac.id

wavelet. Details of the processes are described in the following sub sections.

## 2 Methods

### 2.1. Data Acquisition

We recorded the breath sounds from pediatric subjects diagnosed as pneumonia in the pediatric ward of PDHI Islamic Hospital Yogyakarta Indonesia. To record the breath sounds, we used a Thinklab One Digital Stethoscope (Thinklabs USA). We connected the electronic stethoscope to an Android mobile phone and a head set via Thinklink connector. TapeMachine Lite Recorder application was used to record the breath sounds. The connection of the hardware used for recording is illustrated in Fig 2.

The recording was started after informed consent had been completed. We apply the electronic stethoscope on the chest wall of the subjects to record the breath sounds. The sampling rate was set at 44.1 kHz, 16 bit to maintain the recording quality. The recorded breath sounds were saved in wav format files in mobile phone's memory. Once the recording had completed, we transferred the files to a computer for further processing.

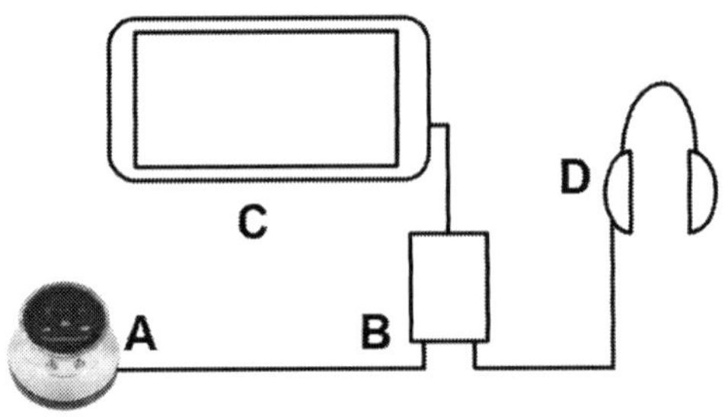


**Fig. 2.** Hardware used in this study. It comprised of A) Thinklab One Digital Stethoscope, B) Think Link connector, C) Android mobile phone and D) Headset for

### 2.2 Pre-processing

The pre-processing was carried to reduce the stationary artifacts. Let $s(n)$ denotes the breath sound recordings. After we had obtained the breath sounds data, then we pre-processed them to reduce the ambient noise and electronic noise having spectrum below 50 Hz and higher than 2000 Hz. We implemented a ten-order Butterworth band pass filter having cut off frequency of 50 Hz and 2000 Hz. The cut off frequencies were designed based on the noise profile of the recordings. The output signal from the pre-processing is denoted as $\hat{s}(n)$.

### 2.3 Artifacts removal

In this process, the output signal from the pre-processing $\hat{s}(n)$ was processed to remove the non-stationary artifacts. We developed a wavelet filter based-system for this purpose. Our system comprised of three stages:

#### *2.3.1 Decomposing sound signal using a discrete wavelet transform filter*

In this stage we decompose signal $\hat{s}(n)$ using discrete wavelet transform filter (DWT). Let $i$ ($i$ = 1, 2, ... $N$) denotes the decomposition level. Passing $\hat{s}(n)$ through $N$ level DWT generates approximation coefficient $A_i$ and details coefficient $D_i$ in each level as given in Eq 1 and Eq 2 [8].

$$A_i[n] = \sum_{k=-\infty}^{\infty} \hat{s}[n]g[2n-k] \quad (1)$$

$$D_i[n] = \sum_{k=-\infty}^{\infty} \hat{s}[n]h[2n-k] \quad (2)$$

where ($A_i$) and ($D_i$) are the approximation and details of the low and high frequency components, respectively. The coefficients were generated from the convolution of input signal $\hat{s}(n)$ with impulse response of high pass filter (HPF) $g$ and low pass filter (LPF) $h$ in each level of decomposition $i$. The term $2n$ means that the output of HPF and LPF were down sampled by 2. This means sound signal in each level will have half bandwidth from the previous level.

In this work, we set the maximum decomposition level $N$ = 10 to achieve the minimum frequency resolution below to 21 Hz. We tested three types of wavelets including Coiflet wavelet (*coifl5*), Symlet wavelet (*sym5*, *sym8*, *sym20*, *sym32*), and Daubechies wavelet (*db5*, *db8*, *db20*, *db32*). After decomposing the signal, we passed the signal to thresholding process described in the following sub-section.

#### *2.3.2 Comparing the decomposed signals with a set of thresholds*

To eliminate the artifacts, in this process we compared the details coefficients in each level $D_i$ with a set of thresholds. Coefficients with absolute values higher than the defined threshold $\tau$ were set to zero while coefficients below the threshold were kept. Threshold function is given in Eq 3.

$$\breve{D}_i = \begin{cases} 0, & |D_i| \geq \tau \\ D_i, & |D_i| < \tau \end{cases} \quad (3)$$

where $\breve{D}_i$ is the output of thresholding in i decomposition level. In this process, artifacts were assumed as the decomposed signal having absolute value higher than the threshold values. Therefore, the artifacts magnitudes were reduced to zero.

#### *2.3.3 Reconstructing signals using inverse discrete wavelet transform*

In this stage, we reconstructed the entire decomposed signal ($A_i$ and $\breve{D}_i$) from the thresholding process. Approximation coefficients $A_i$ and details coefficients $\breve{D}_i$

were up sampled by two and then processed through high pass filter and low pass filter, respectively. The filters were used to remove the aliased signal from previous process. Output signal from this process was the summation of the output signal from HPF and LPF denoted as $y(n)$, the breath sounds that free from the artifacts.

### 2.3.4 Performance evaluation if the methods

To evaluate the performance of our developed method, we compute the signal to noise ratio (SNR in decibel/dB) of the input signal $s(n)$ and output signal $y(n)$. The SNR is given in Eq. 4.

$$SNR_j = 10\log 10\left(\sum_{n=0}^{m} \frac{|\xi(n)|^2}{|\psi(n)|^2}\right) \quad (4)$$

where $\xi$ is the breath sound events, $\psi$is the artifacts and noise while $m$ is the length of data used for computation. The term $j$ represents the SNR for input signal ($j = 1$) and SNR for output signal from wavelet filter ($j = 2$). Then we computed the difference between input and output SNR ($\Delta$SNR) as given in Eq 5.

$$\Delta SNR = SNR_{j=2} - SNR_{j=1} \quad (5)$$

The best performance of the wavelet filter was indicated by the highest $\Delta$SNR.

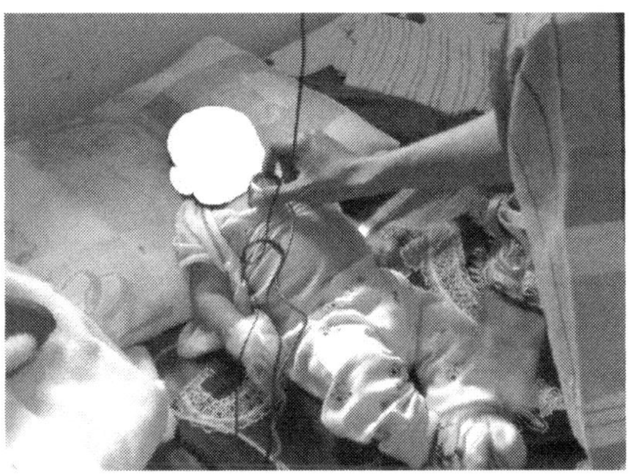

**Fig. 3.** Illustration of an auscultation process in a pediatric subject.

## 3 Results and discussion

### 3.1 Data acquisition

We recorded breath sounds from eleven (9 males and 2 female) pediatric subjects diagnosed as pneumonia. The age range was from two months to fifteen months with average of eight months. In the physical examination, there were symptoms such as fever (5 subjects), cold (5 subjects), and breathing difficulty (8 subjects). Cough was found in all of subjects. The auscultation process is illustrated in Fig 3.

Of the eleven subjects, we only able to use nine subjects (subject no 3 to 11) due to clipping problems in the two other subjects (subject no 1 and 2). Our working data consisted of 7 male and 2 female subjects. The average of recording duration was 118 second (minimum duration was 65 seconds and maximum duration 166 seconds). The duration of recording depended on the condition of the subjects.

In Fig 4 we illustrate the example of artifacts from rubbing sounds and heart sounds. It can be seen that artifacts contain frequency components lower than 50 Hz to 550 Hz wherein the spectrum of breath sound also exist. The artifacts may have duration from 0.05 seconds up to 0.2 seconds. This shows that artifacts in the recordings may overlap with breath sounds in both time domain as well as frequency domain.

The typical breathing sound obtained from the auscultation from pediatric subjects shown in Fig 5(A). It can be seen that stationary and non-stationary artifacts imposed the breath sound signal. The spectrum of the signal is shown in Fig 6(A). We processed the signal through pre-processing and artifact removal using discrete wavelet transform filter.

### 3.2 Pre-processing

Our frequency of interest was from 50 Hz to 2 kHz. Therefore, we developed ten order Butterworth band pass filter having low and high cut off frequency of $f_l$ = 50 Hz and $f_h$ = 2kHz, respectively. This filter was designed to reduce the stationary artifacts. The response of the developed filter is as follows. The attenuation in $f_l$ is 18,81dB and 7.62 dB in $f_h$. The attenuation in the first octave in $f_l$ and $f_h$ is 79,3dB and 63.4 dB, respectively.

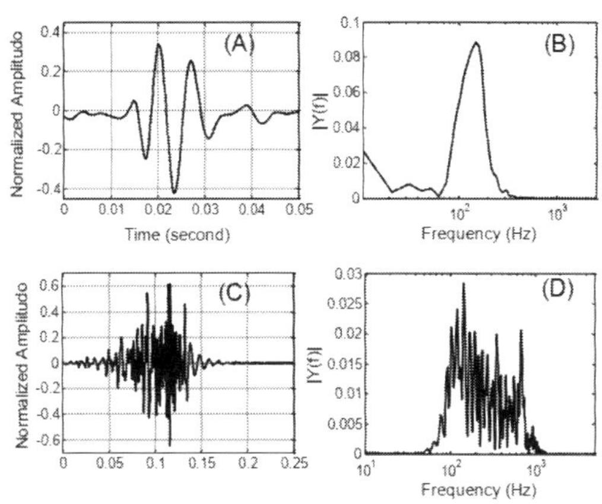


**Fig. 4.** Illustration of artifacts from heart sound (A) and rubbing sound (B). The spectrum of heart sound and rubbing sounds are illustrated in (C) and (D) respectively.

We illustrated the output of the pre-processing in Fig 5(B). It can be seen that the filter was able to reduce the low frequency noise. The spectrum of the signal after filtering is shown in Fig 6(B).

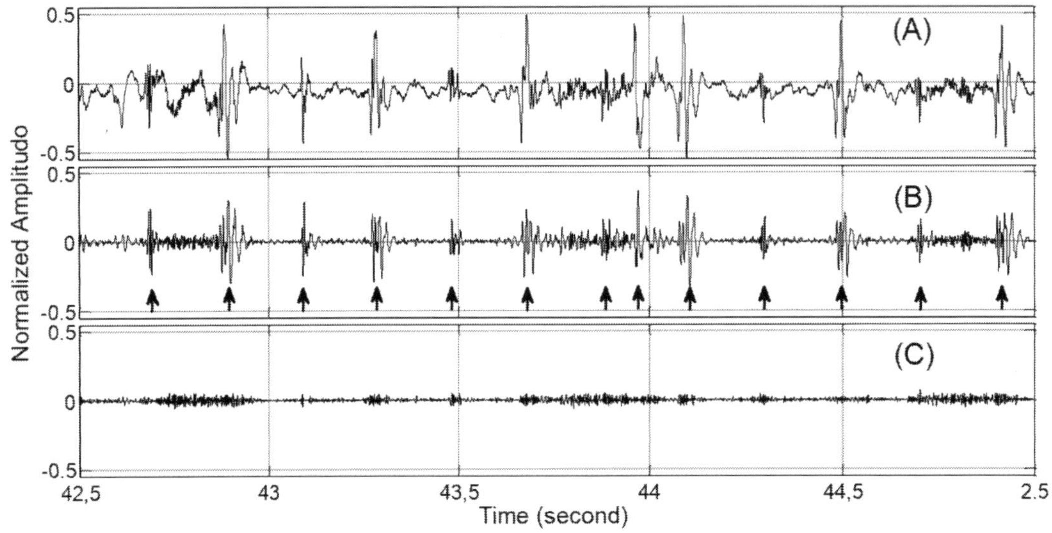


**Fig. 5**. (A) Breath sound signal from electronic stethoscope (B) Output signal from pre-processing and (C) Output signal from artifacts removal using discrete wavelet transform filter (*coift5*). Arrow signs show the artifacts in the signal.

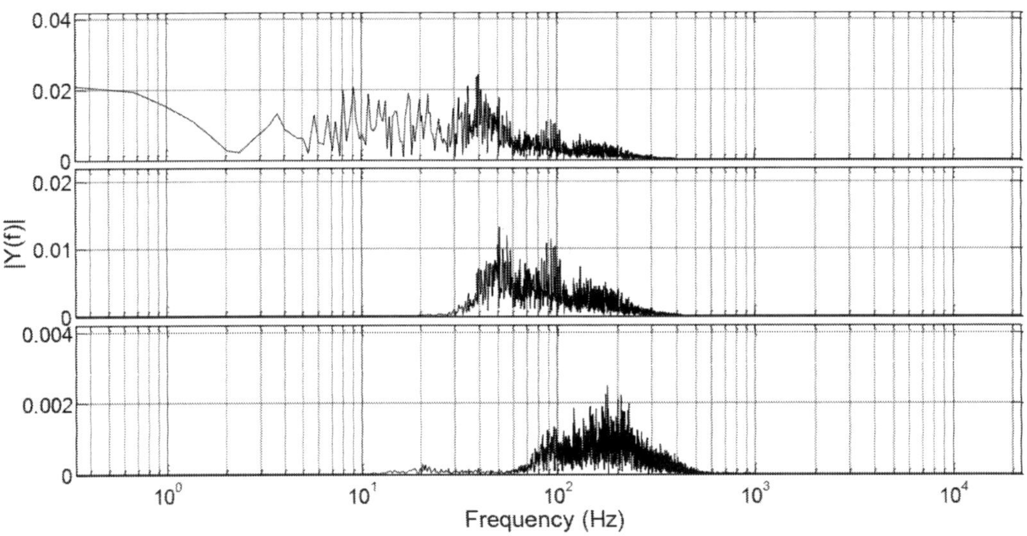


**Fig. 6**. The spectrum of (A) Breath sound signal from electronic stethoscope (B) Output signal from pre-processing and (C) Output signal from artifacts removal using discrete wavelet transform filter (*coift5*).

### 3.3 Artifacts removal

In order to reduce the non-stationary artifacts, we implemented a combination of decomposition, thresholding and reconstruction using discrete wavelet transform filter. As the spectrum of the artifacts spread lower than 50 Hz, we designed DWT with ten level decompositions. The artifact components mostly located in the Details coefficients $D_i$. Therefore, we attenuated magnitude of the artifacts by comparing with the thresholds given in Table 1. We followed Eq 3 in this process.

For decomposition level 1 to 5 we used threshold value of 10 as artifact signal was minimal in this spectrum. In spectrum where breath sound and artifacts were overlap, we implemented a moderate threshold (0.12 – 0.18). Meanwhile, we used lower threshold (0.016 – 0.023) in decomposition level of 9 and 10 to reduce the artifacts. We used the same threshold values for testing Coiflet, Symlet and Daubachies wavelet. The illustration of signal after discrete wavelet transform filter (*coift5*) is shown in Fig 5(C) whiles its spectrum in Fig 6(C). It can be seen that the artifacts were reduced significantly.

### 3.4 Performance of wavelet filter

In this work, we use three types of wavelet including Coiflet, Symlet and Daubachies. We tested 5, 8, 20 and 32 orders of wavelet filter. However, for Coiflet we only tested the fifth order as it is its maximum order. Figure 7 illustrates the average of ΔSNR from the three types wavelet applied in the nine subjects. The figure shows that 32 orders Symlet (*sym32*) achieved the highest SNR improvement (10.65 dB). The coift5 and sym20 showed relatively similar ΔSNR around 10.3 dB. The lowest performance showed in 5 orders symlet (*sym5*) with average of ΔSNR 9,61 dB. Figure 7 also shows that filters using symmetrical wavelet such as Symlet and Coiflet show better SNR improvement compared to asymmetrical wavelet (Daubechies).

**Table 1**. Threshold values used in ten decomposition levels of wavelet processing. The term $i$ is the decomposition level, $f_A$ is the approximation spectrum, $f_D$ is the spectrum of details and $\tau$ is the threshold values.

| p | $f_A$ (Hz) | $f_D$(Hz) | $\tau$ |
|---|---|---|---|
| 1 | 0-11025 | 11025-22050 | 10 |
| 2 | 0-5512 | 5512-11025 | 10 |
| 3 | 0-2756 | 2756-5512 | 10 |
| 4 | 0-1378 | 1378- 2756 | 10 |
| 5 | 0-689 | 689-1378 | 10 |
| 6 | 0-344 | 344-689 | 0,12 |
| 7 | 0-172 | 172-344 | 0,14 |
| 8 | 0-86 | 86-172 | 0,18 |
| 9 | 0- 43 | 43-86 | 0,023 |
| 10 | 0- 21 | 21-43 | 0,016 |

From the figure, it can be seen that higher order wavelet filters showing better SNR improvement. This occurs as higher order wavelet filters have higher vanishing moment in their wavelet function [9]. Theoretically, if a wavelet filter has more vanishing moment then its scaling function can represent signals that are more complex accurately. This means that it can decompose the artifacts accurately even though the signal was non-stationer. In the same order, the 5 orders Coiflet outperform both Symlet and Daubechies. This occurs as *Coif5* has more vanishing moment (10) compared to db5 and sym5 (5).

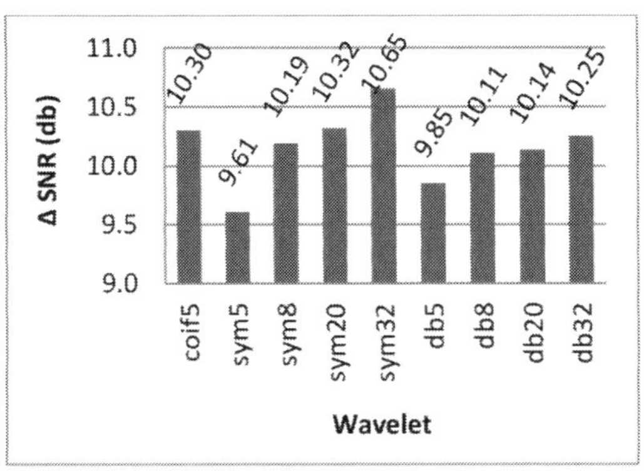


**Fig. 7**. Signal to noise ratio improvement (ΔSNR) before and after discrete wavelet transform filter.

Our results show that our developed method was able to remove the unwanted artifacts found in recordings from pediatric subjects. Unlike study in [7], our work does not used breath sound signals from adults. The combination of an electronic stethoscope and a mobile phone for acquiring breath sound is suitable for rural areas. The hardware can be deployed easily. The digital data is also easily saved and transferred for further analysis by other physicians. This method can be used to improve the diagnosis of pneumonia in high prevalence areas such as Nusa Tenggara and Papua [10].

## 4 Conclusion

In this work we address processing complications in breath sound recordings from pediatrics subjects. We employed Thinklab One Digital Stethoscope and a Mobile phone to record the breath sounds. We involved pediatric subject diagnosed as pneumonia. In these subjects, the breath sounds have lower magnitude but in contrast, it has many artifacts due to movements or sounds from other organ such as heart. In real situation, it is difficult to get a reference artifact signal for filtering process to omit the artifacts. Therefore, we proposed to develop a combination of band pass filter with discrete wavelet filter. The results show that our proposed method was able to improve the signal to noise ratio significantly up to 10.655 dB.

## References

1. L. J. Hadjileontiadis, *Lung Sounds: An Advanced Signal Processing Perspective*. Connecticut: Morgan & Claypool, 2009.
2. Y. Norisue, Y. Tokuda, M. Koizumi, T. Kishaba, and S. Miyagi, "*Phasic characteristics of inspiratory crackles of bacterial and atypical pneumonia*," Postgraduate Medical Journal, vol. 84, pp. 432-436, 2008.
3. A. Vyshedskiy, R. M. Alhashem, R. Paciej, M. Ebril, I. Rudman, and J. J. Fredberg, "*Mechanism of Inspiratory and Expiratory Crackles*," Chest, vol. 135, pp. 156-164, 2009.
4. B. Mathews, S. Shah, R. H. Cleveland, E. Y. Lee, R. G. Bachur, and M. I. Neuman, "*Clinical Predictors of Pneumonia Among Children With Wheezing*," Pediatrics, vol. 124, pp. e29-e36, 2009.
5. World Health Organization, "*IMCI: Integrated Management of Childhood Illnesses*," in Management of the child with a serious infection or severe malnutrition: Guidelines for care at the first-referral level in developing countries, ed: World Health Organization and UNICEF, 2000.
6. A. Chang, "*Cough: are children really different to adults?*" Cough, vol. 1, p. 7, 2005.
7. L. J. Hadjileontiadis and S. M. Panas, "*A wavelet-based reduction of heart sound noise from lung sounds*," International Journal of Medical Informatics, vol. 52, pp. 183-190, 1998.

8. O. Rioul and M. Vetterli, "*Wavelets and signal processing*," IEEE signal processing magazine, vol. 8, pp. 14-38, 1991.
9. H. Cui and G. Song, "*Study of the Wavelet Basis Selections*," in Computational Intelligence and Security: International Conference, CIS 2006. Guangzhou, China, November 3-6, 2006. Revised Selected Papers, Y. Wang, Y.-m. Cheung, and H. Liu, Eds., ed Berlin, Heidelberg: Springer Berlin Heidelberg, pp. 1009-1017, 2007.
10. Litbang-Kemenkes, "*Riset Kesehatan Dasar* " vol. 1, B. P. Pengembangan, Ed., ed. Jakarta: Kementrian Kesehatan Republik Indonesia, pp. 67-68, 2013.

# Improving public library service quality in local city of Indonesia

*Prima* Fithri*, *Afri* Adnan, and *Verra* Syahmer

Department of Industrial Engineering, Faculty of Engineering, University of Andalas, Padang

**Abstract.** The purpose of this study is to investigate the gap between perceptions and expectations of users on the services of West Sumatera Province Library as a public sector and determine the priority level of the requirements for the library in order to improve the quality of services based on the dimensions of the service quality. This was done using Servqual and Quality Function Deployment (QFD) method. It was found that, from overall attributes, only one attribute fulfills users' expectations (it has a positive gap), tangibles element for the attribute of cool air circulation. The ranking of the gap, for all dimensions, is the dimension of responsiveness, assurance, empathy, tangibles and reliability. It indicates that users's expectations were not being fulfilled yet. The main benefit of this research is to understand user's perceptions and expectations towards the services provided and to obtain the improvement priority to be taken by the West Sumatera Province Library.

## 1 Introduction

For many years libraries were major players in the information industry, dominating the field for centuries (Snoj and Petermanec, 2001). According to Line (Jordan, 1998, p.31) library responsibility are limited to collecting and processing prosedures and they engaged almost exclusively with themselves. They did not affected by the pressure from market forces and their profit did not depend on the successful selling of their services. Successful marketing depends on customer satisfaction. Nowadays, many researchers and practitioners concern about customer satisfaction in library (Hiller, 2001:606; Nitecki and Franklin, 1999:485; Allen, Ward, Wary and Lopez, 2003:138; Martensen and Granholdt, 2003:140; Calvert, 1998:296; Harwood and Bydder, 1998:161; Shi, Holahan, and Jurkat, 2004:122). Although user satisfaction is not an unfamiliar topic in library science, especially in university libraries (Thapisa, 1999; Luca, 2006; Kiran 2006; Mehran, 2008; Snoj and Pertamanec, 2001), there are not many research in the area of public library user satisfaction.

Padang, a local city of West Sumatera in Indonesia has one public library. Students, academicians, practitioners, and citizens use this library to get some information that they need. According to Deputy Development Resource Library at the National Library of Indonesia Republic, Supriyono, before the earthquake hit in 2009 West Sumatera province Library is one of four best libraries in Indonesia. Nowadays it's just a collapsed library, many collections and infrastructure were damaged or destroyed. The effort to rebuild West Sumatera province library reputation cannot completely depend on the others. West Sumatera province library need to promote quality services to produce continuous improvement on its performance. Performance improvements can be implemented through service factors enhancement which can be used to help deliver satisfaction and value added as well as bring a good image for the library.

If viewed from customer perspective, it is recognized that different backgrounds in terms of economic, educational, society etc can form the perceptions and expectations to library services in West Sumatera province. This study try to investigate the gap between perceptions and expectations of users on the services of West Sumatera Province Library. This study also determine the priority level of the requirements for the library in order to improve the quality of services based on the service quality dimensions. Therefore, perception and expectation aspects of consumers should be captured by the library workers in order to show good service as an effort to provide quality services that suit with consumer needs.

In recent years the service has grown dramatically. This resulted in various service industries that exists today, such as employment services, hospitals and schools offered by the government, airline services, entertainment companies and banks that offered by private non-profit organizations (Kotler and Armstrong, 2008).

According to Goetsch and Davis, 1994 (Tjiptono and Diana, 2003), quality is defined as an effort made to meet or exceed customer expectations that include products, services, people, processes and environments with changing conditions over time.

## 2 Research Methodology

In this research, data are obtained from questionnaires and interviews. The questionnaire addressed to users of the circulation and reference of West Sumatera Province library. After designing the questionnaire, a pilot test

* Corresponding author: author@e-mail.org

conducted to analyze whether the questionnaire is valid and reliable. The distribution of the questionnaire conducted in accordance with the number of samples that have been calculated before. Interviews are used to obtain preliminary data in this study.

Data from questionnaire are used as input for QFD and Servqual, to calculate the value gap and the importance of each service attribute of West Sumatera public Library.

a. Data collection
   Questionnaire are giving directly to respondents and researcher may give instructions or assistance directly to the respondent if the respondent had difficulty in filling out the questionnaire.
b. The number of questionnaire
   97 questionnaire are used for this study. 92 questionnaire filled out by respondents and 5 questionnaire filled by the Library employees. 5 questionnaire filled by library employees used as pilot test. Data processing consists of three stages:
   1. Validity and reliability test.
   2. The Calculation of Servqual
      Servqual used to calculate gap between the variables examined and any attributes that are experiencing a gap. In this study it has been established that the gap between user perceptions with user expectations as Gap 5. The calculation of Gap 5 follow this formula:
      Gap 5 = average user perceptions – average user expectation
   3. The Calculation of QFD
      QFD can help the company meet customer expectations to improve the quality of service provided. QFD is part of product planning which translate customer requirements into engineering characteristics. HQ phase I create using this following steps:
      a. Calculate the level of customer interest
      b. Determine the technical characteristics
      c. Determine the relation between consumer needs and the characteristics of engineering using relationship matrix
      d. Calculating the priority of engineering characteristics
      e. Determining correlation technique

From the analysis, the gap between the expectations and perceptions of the services quality can be determined and to improve what needs to be done to meet the user expectations. After the analysis, the conclusion can be drawn from this study. Suggestions could be put forward to improve the services provided.

## 3 Finding and Result

The finding and results are presented in two parts: Importance Performance Analysis and HOQ/House of Quality. However, pre-analysis was performed to determine the validity and reliability of the data gathered using the questionnaires from the population of 97 individuals. This research showed that the data and items collected for measuring customer satisfaction toward the product quality and service quality for perceived condition, and desired condition were valid and reliable. This data was then being processed using IPA and HOQ method.

### 3.1 Importance Performance Analysis

The steps used in applying the IPA method are as follows:

1. Determine the level of compatibility between the interest rate and the level of satisfaction by comparing scores with the score of interest.
   Processing the data using Importance Performance Analysis (IPA) consists of calculating the level of suitability and defines the results obtained in the Cartesian diagram. The first step is to measure the level of suitability. Level of suitability is the ratio between the perceived condition score with the score level desired condition. The levels of suitability are divided into three level; 0-65% for unsatisfied; 66%-100% for less satisfied; and more than 100% for very satisfied. 16 questions are used to calculate the level of suitability between perceived and desired conditions for product quality; 5 questions are unsatisfied and 11 questions are less satisfied. Furthermore, 14 questions are used for service quality; 2 questions are unsatisfied, 6 questions are less satisfied, and 6 questions are satisfied. The average level of suitability between perceived and desired conditions for service quality is 78% and for service quality is 76%.
2. Calculate the average for each attribute that is perceived by the respondents.
   For each questions, the average values of customers' interest and satisfaction were computed. Those values are then plotted in a Cartesian diagram. The horizontal axis of the diagram is the average of satisfaction level and its vertical axis is the average of interest level. The diagram is then divided into 4 quadrants. Analyses are focused on quadrant I and III which represent the highest level of interest and the lowest level of satisfaction. The scores were plotted into four quadrants and each quadrant covers related sub indicators as described in Figure 1 for product quality and Figure 2 for service quality. Sub indicators highlighted in quadrants I and III should be the greatest priority for improvement. The sub indicators in quadrants I and III for product quality are building quality, location factors, facilities and infrastructure and for service quality the sub indicators are reliability, assurance, and responsiveness.

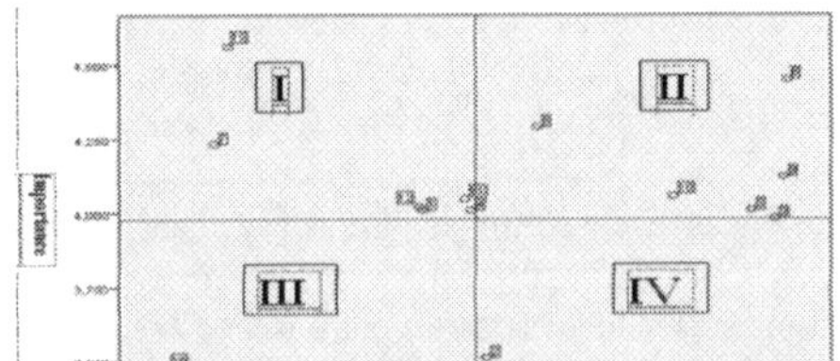


**Fig.1.** Cartesian Diagram of Product Quality

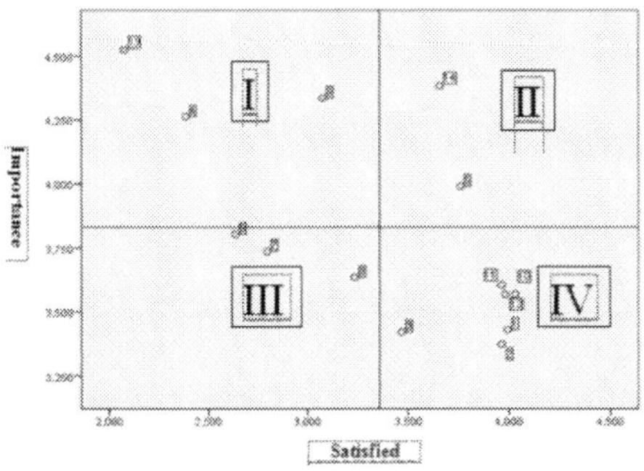


**Fig 2.** Cartesian Diagram for Service Quality

### 3.2 House of Quality

The steps used in applying the HOQ method are as follows:

1. Voice of Customer

   Sub indicators obtained in quadrant I and III of Cartesian diagram for product and service quality are applied to build the House of Quality. The sub indicators will be transformed into the voice of customers.
2. Technical Characteristics

   The needs of customers have to be responded technically by the home developer to increase the satisfaction level of customers. The technical characteristics are defined based on the interview conducted with the architects working in the company. The next step is to correlate the voices of customer and the technical characteristics in the form of a correlation matrix.
3. Technical Characteristics Priority

   Sometimes it is almost difficult to accommodate all the technical characteristics at the same time and therefore should be ranked according its priority. The processes to ordering the priority are calculated by multiplying the interest rate of costumers with the value of the matrix of relationship (0, 1, 3 and 9). The result is shown in Table 1.

Table 1 Priority of Technical Characteristics Score

| **No** | **Technical Characteristic** | **Voice of Costumer (VOC)** | **Priority Percentage** |
| --- | --- | --- | --- |
| 1 | Provide free renovations for a month after the handover of house key | Warrants within a specified period to prospective residents | 21.94 |
| 2 | Carry out routine maintenance and drainage cleaning | Locations of the houses were not flooded in the rainy weather<br>The availability of rain water disposal facilities / drainage | 20.08 |
| 3 | Casting and improving road quality on a regular basis | The ease of movement around the houses and are not damaged | 19.49 |
| 4 | Intense coordination with sub-contractor that handles the houses construction | The buildings were not damaged during the earthquake<br>The roof of the houses do not easily leak | 19.44 |
| 5 | Provide forms for critiques and suggestions | The ease of communication between the company and the consumers<br>Employee understand the consumer needs | 12.00 |
| 6 | The addition of working hours especially on Saturday | Employees attend office on holidays to help answering customers' Questions<br>Responsive in handling all consumers complaints<br>The ease of Communication between the company and the consumers | 4.98 |
| 7 | Provide vacant land for the establishment of the public facilities | Availability of Worship Facilities<br>Availability of open space / green | 2.07 |

| | |
|---|---|
| | outdoor |
| | Availability of sports facilities and open field |
| | Availability of trade and commerce facilities |
| | The accuracy of the developer promises related to residential facility |

4. Technical characteristics correlation

   The correlations among the technical characteristics are obtained based on an interview conducted with the marketing manager of the company. Based on the average levels of perceived condition, correlation matrix, and the relationship among the technical characteristics, a house of quality (HOQ) is constructed. The HOQ also concisely presents the priority level of the technical characteristics.

## 4 Conclusions

The research showed that the services quality provided by West Sumatera public library still not meet customer expectation. This can be seen from a negative value on each quality dimensions from gap calculation process. Furthermore, this study also produced priority of improvement actions that can be used by the library to improve their quality of service. The priority is proper construction of the library building and library standards, the addition of library infrastructure and collection, implement more effective oversight, the addition of service infrastructure, Information Management Training, adjustment and placement of personnel or officers who have competence in the implementation of service delivery and Technical Training Management Library.

Although this study has provided some interesting findings, there are still a number of limitations in this research. First, this study did not involve all kinds of services contained in the library. Second, the research done at the library occupying a temporary location so that the library has not been able to provide maximum service. Due to these limitations, it is recommended that further research is conducted, for example, conduct research for all types of services contained in the Library of West Sumatra Province, perform calculations gap 1, 2, 3 and 4 using methods Servqual and continuing the calculation of QFD to stage 2, 3 and 4.

## References

1. Allen, M., Ward S.M., Wary T. and Debus-Lopez K.E. (2003) *Patron-focused Services in Three US Libraries: Collaborative Interlibrary Loa.* Collection Development and Acquisitions, Interlending & Document Supply, 31(2), 138-141
2. Calvert, P.J. (1998) 'A Different Time, a Different Country: an Instrument for Measuring Service Quality in Singapore's Polytechnic Libraries'. *The Journal of Academic Librarianship*, 24 (4), 296-299.
3. Harwood, N. and Bydder J. (1998) 'Student Expectations of, and Satisfaction with, the University Library'. *The Journal of Academic Librarianship*, 24(2), 161-171.
4. Hiller, S. (2001) 'Assessing User Needs, Satisfaction, and Library Performance at the University of Washington Libraries'. *Library Trends*, 49(4), 605-625.
5. Jordan, P. (1998) *The Academic Library and Its Uers.* Gower, Aldershot.
6. Kiran Kaur Pauziaah Mohamad Sossamma George, (2006),"Quality management service at the University of Malaya Library", Library Management, Vol. 27 Iss 4/5 pp. 249 – 256
7. Luca Petruzzellis Angela Maria D'Uggento Salvatore Romanazzi, (2006),"Student satisfaction and quality of service in Italian universities", Managing Service Quality: An International Journal, Vol. 16 Iss 4 pp. 349 -364
8. Martensen, A. and Granholdt L. (2003) 'Improving Library Users' perceived Quality, Satisfaction and Loyalty: an Integrated Measurement and Management System'. *The Journal of Academic Librarianship.* 29(3), 140-147.
9. Mehran Nejati Mostafa Nejati, (2008),"Service quality at University of Tehran Central Library", Library Management, Vol. 29 Iss 6/7 pp. 571 – 582
10. Nitecki, D. and Franklin B. (1999) 'New Measures for Research Libraries'. *Journal of Academic Librarianship.* 25(6), 484-487.
11. Snoj, B. and Petermanec, Z. (2001) 'Let users judge the quality of faculty library services', New Library World, Vol. 102, No. 9, pp.314 – 324
12. Shi, X., Holahan, P.J, and Jurkat, P. (2004) 'Satisfaction Formation Processes in Library Users: Understanding Multisource Effects'. *The Journal of Academic Librarianship.* 30(2), 122-131.
13. Thapisa, A.P.N. and Gamini, V. (1999), 'Perceptions of quality service at the university of Botswana library: what Nova says', Library Management, Vol. 20 No. 7, pp. 373-83.

# Human resource allocation for humanitarian organizations: a systemic perspective

*Bertha Maya* Sopha*, *Anna Maria Sri* Asih

Industrial Engineering Programme, Department of Mechanical and Industrial Engineering, Universitas Gadjah Mada (UGM), Yogyakarta, INDONESIA

**Abstract.** Human resource allocation appears to be one of the important factors toward effective humanitarian relief operations. Particularly in developing countries, the role of volunteers which is mostly managed by humanitarian organizations has become prominent. Due to the limited human resources, the humanitarian organizations are challenged to allocate their resources to both provide assistance for disaster victims and build organization capacity effectively. The present research aims at identifying effective human resources allocation policy in humanitarian organizations. A formal simulation model was developed using system dynamics approach. Two scenarios, i.e., constant relief demand and empirical relief demand, were developed and tested. Experiments were conducted to examine various allocation policies for both scenarios. Results indicate there is allocation trade-off when it comes to allocate human resources to relief operations and capacity building. Results highlight that allocating resources to capacity building is necessary to sustain relief operations in the long-term.

## 1 Introduction

Many literatures such as [1] have indicated that disaster will increase both in number and impact in developed and developing countries. The rise of disasters requires for an increase of need for disaster relief operations. Therefore, the role of humanitarian organizations in providing humanitarian aids such as assessing needs, moving displaced people, caring the wounded, etc., becomes important. Based on empirical evidence, Humanitarian Organizations (HOs) have generally limited resources. But on the other hand, HOs face ever-increasing needs. Consequently, it challenges HO to improve operational efficiency of both relief operations and capacity building.

HOs have two main activities; one related to capacity building the organizations such as hiring, personnel training, structuring organization, administrative work, and one related to relief operations as their core business activities (assessing needs, designing and implementing transportation of first aid kit, food, equipment, and rescue personnel, transferring and helping disaster victims to the healthcare centres safely and rapidly, moving and helping, providing survival kits, etc.). Allocating the resources to humanitarian relief operations while building sufficient capacity is a challenging task [2, 3].

Unfortunately, although HOs have played an important role is disaster relief, only few literatures (e.g., [3, 4]) have studied the HOs. With respect to human resources, many literature has focused on a more macro-level such as capacity building of local people [5], and resource coordination among government [6]. Literature on the operation of HOs is however still limited. Reference [6] has indicated that there is still lack of process documentation, modelling and optimization efforts at humanitarian organizations. Based on the best knowledge of the authors, there are three researches which are specifically addressing the HO operations, i.e., [4] who investigated emergency preparedness strategies such as prepositioning and lead time reduction strategy by HOs, [7] who examined three types of fleet management system, i.e., centralized, decentralized, and hybrid, and [8] who explored the allocation strategies and trade-off between providing relief and building capacity in HOs.

Complementing the previous studies of HOs, the present study aims at exploring the allocation policy for human resources based on local humanitarian organizations in Yogyakarta, Indonesia. It is expected that the study would provide insights for local humanitarian organizations for a better management for volunteers. System dynamics is applied to build formal simulation model that quantifies the effects of various allocation policies.

It is interesting to note that the aforementioned HO literatures have applied system dynamics as an analytical tool. Following [9], system dynamics is advantageous dealing with dynamic complexity because system dynamics can evaluate both short and long term consequences of the decision, assess the consequences of interactions among variables, and represent dynamic complexity of a system of interest - predicting the effect of changes to the system over time. Further, Reference [7] has demonstrated the appropriateness of system dynamics to be applied in humanitarian operations to examine fleet management system. Briefly, system dynamics is a

* Corresponding author: bertha_sopha@ugm.ac.id

promising tool as to be used in modeling humanitarian operations which are characterized by multiple feedback loops, uncertainty, and time lags, to understand the consequences of the system change within a long-time horizon.

The rest of the paper is organized as follows. The brief description of system dynamics approach is presented in the next section. The conceptual model using system dynamics approach is presented in Section Three. Section Four describes a simulation model, comprising Stock-and Flow Diagram (SFD), model parameters, model testing, and scenario development in the same section. Section Five presents results and discussion of the scenario development. Finally, the paper is concluded by highlighting the main findings of the study and managerial implications in Section Six.

## 2 System Dynamics

The present study has applied the methodology of system dynamics (please refer to [9] for the details). This section outlines the method to develop system dynamics model which includes model building, model parameterization, and model testing.

The first step is model building. The model is developed using system dynamics approach. System dynamics, typically a top-down approach, is a method using system thinking [10] to enhance learning in a complex system where feedback interactions among system's component occur. It is grounded in the theory of nonlinear dynamics and feedback control developed in mathematics, physics, and engineering [9]. System dynamics is distinguishable from other methods when it comes to the ability to address feedback-loops which determine the dynamic behavior of the system, and to allow changing underlying structure to produce different pattern of behavior. A fundamental principle of system dynamics is based on the assumption that the structure of a system gives rise to its behavior. This is suitable approach when one variable relates with other variables in a complex way. System dynamics is therefore selected for a modeling and simulation tool in this study due to its capability to incorporate causal-links and feedback-loop among influential variables.

Systems dynamics have been widely used for examining several policy intervention models in various contexts such as transportation, population dynamics, product diffusion, inventory, human resources, etc. However, studies related to resource allocation in humanitarian organizations are still very few.

With respect to modeling process, the authors followed [9] who defined the process of system dynamics model consisting of problem articulation, formulating dynamic hypothesis, formulating a simulation model, testing a model, exploring and evaluating policy design. Based on the problem specified in the previous section, a dynamic hypothesis was formulated and a conceptual model was constructed. The conceptual model (presented as Causal-Loop Diagram - CLD) of this present study was built based on observations and interviews with relevant stakeholders of humanitarian organizations. Once the conceptual model has been constructed, the model was validated by respondents of the sampled humanitarian organizations. The respondents of the study are those who have responsibility for human resource management in humanitarian organization. After qualitative validation of the conceptual model, the model was then implemented as a simulation model in the form of Stock-Flow Diagram.

The second step is model parameterization. The simulation model was then parameterized by empirical data which was collected through sets of semi-structured interviews with respondents from the sampled humanitarian organization in 2015. The questionnaire was specifically designed to provide input parameters for the model. Section 4 specifies the model parameters in detail.

The third step is model testing. To build a model confidence and to assess the degree of representation of the model toward the real system, several steps of model testing from conceptual model to simulation results were conducted through different methods (see Section 4.3 for the details). Scenarios of various human resource allocation policy are then developed and examined with respect to their effectiveness to save affected people.

## 3 Conceptual Model

This section outlines a conceptual model which was presented as causal-loop diagram (CLD) following system dynamics approach [9]. CLD is used as a tool for articulating understanding of the system dynamic and interconnection between parts of the system. The causal loop diagram was drawn empirically through a set of interviews with managers of local HOs in Yogyakarta, Indonesia, in which Mount Merapi, the most active volcanoes in the world – more than 70 eruptions since 1548 [11], resides. The most recent volcano disaster occurred in 2010 which resulted in 400,000 Internally Displaced People (IDP). In addition, earthquake is another disaster that may occur in Yogyakarta. The most recent earthquake occurred in May 27, 2006 which caused 6000 victims. It is therefore HOs in Yogyakarta has played a vital role disaster response.

The HOs generally experience resource limitation particularly human resources because the nature of operation of the HOs in which disasters occur incidentally so that high number of workforces at all time is very costly. As the number of human resources in HOs are generally few, consequently, when disaster occurs, the HOs have to recruit new resources. There are two main tasks of human resources of the HOs, i.e., planning a rescue program, and recruiting new workforce. With respect to recruitment process, the HO staff should conduct interviews toward the workforce candidates in order to assess their capabilities. Both tasks are actually the backbone of the successful rescues. Planning rescue program affects the productivity in the field. Without planning and coordination of the rescue workforces, the rescue process will be ineffective and inefficient. In same way as the planning task, the recruitment of new resources has also impacted on the productivity because incapable human resource may not be able to help the victims. The number of recruited resources is so much depending on the number of victims/affected people by disaster. The more victims, the more volunteers/resources to be recruited. Once the victims have been saved, the need for volunteers becomes less and less.

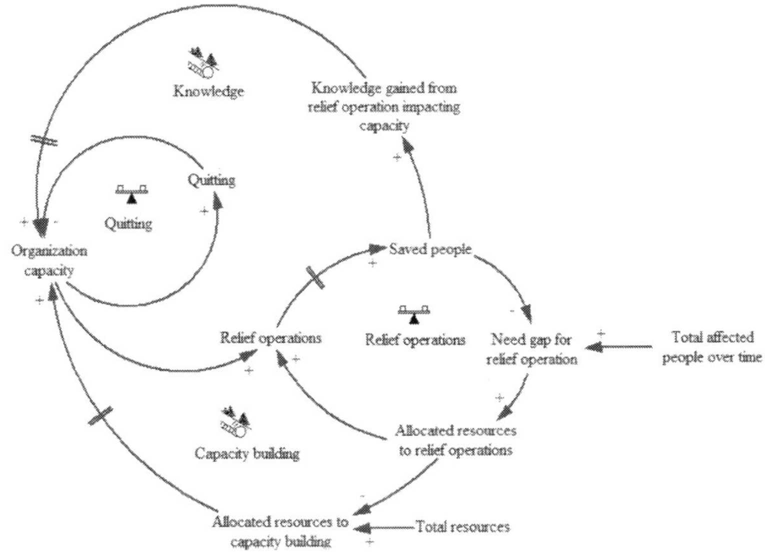


**Fig. 1.** Causal-Loop Diagram (CLD) of Human Resource Allocation during Disaster Response

Based on the aforementioned facts, the causal-loop diagram articulating the operation of the HOs is presented in Figure 1. The model comprises of two balancing feedback-loops (i.e., relief operation loop and quitting loop) and two reinforcing feedback-loops (i.e., capacity building loop and knowledge loop) detailed as follows.

Relief operations and recruitment are actually two basic operations of the HOs. Relief operation loop, which can be categorize as balancing loop, is triggered by the discrepancy between the total number of affected people and the number of saved people. The more human resources allocated to the rescuing operations leads to higher productivity of operation, which then leads to more affected people being rescued. In same way as relief operations, capacity building loop is depending on the discrepancy between the total number of affected people and the number of saved affected people. The discrepancy influences the human resource to be allocated for recruitment process. As the total available resource is limited, the more allocation to relief operations is corresponding to the lesser allocation for capacity building which then results in less organization capacity. Once the resource for capacity building is allocated, delay occurs to represent training of new resources before the resources can actually contribute to the organizational capacity. The organization capacity then influences the relief operations and consequently less saved people and high gap for the relief operations (Reinforcing loop of capacity building). The organization capacity is also affected by the number of quitting workforce (Balancing loop of quitting). Moreover, the more saved people by the HOs, the more knowledge gained from field which then impact the organization capacity to tackle the relief operations (Reinforcing loop of knowledge).

Based on the empirical conceptual model, it is hypothesized that if HOs allocates more on relief operations than capacity building, the saved people would be higher only in the short-term but fail to sustain in the long-term because the available capacity may not catch-up with the increased affected people. On the contrary, if more resources are allocated to capacity building, the number of saved people is lesser in early period but it would later on has more capacity for relief operations.

## 4 Simulation Model

This section are divided into four parts, i.e., description of Stock and Flow Diagram (SFD), parameters of the model, model testing, and scenario development related to human resource allocation policy.

### 4.1 Stock and Flow Diagram

Based on the developed causal-loop diagram of human resource allocation during disaster response (Figure 1), Stock and Flow Diagram (SFD) as a formalized simulation model is developed and shown in Figure 2. Stock and Flow Diagram (SFD) basically consists of sets of simultaneous differential equations. The model was implemented in VensimPLE (by Ventana Systems Inc.).

### 4.2 Model Parameters

The Stock and Flow Diagram model is then parameterized using empirical data. The data for parameterization purpose was acquired from secondary data. However, data which was not available in literatures and specifically characterizes Mount Merapi disaster in Yogyakarta was acquired from semi-structured interviews with the sampled humanitarian organizations. Table 1 describes model parameters and its associated definition and values used in the simulation model

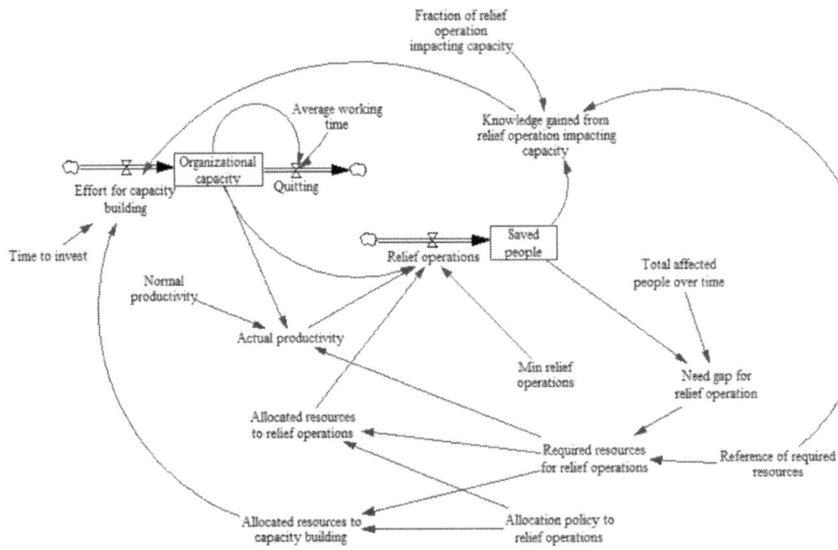


**Fig. 2.** Stock and Flow Diagram (SFD) of Relief Operations and Capacity Building in Humanitarian Organization

### 4.3 Model Testing

To develop the confidence of the model, a set of tests have been conducted. Three techniques were adopted in this study, i.e., structure verification test, dimensional consistency test, and face validation. Structure verification test examines whether the model is consistent with the knowledge of the real system and relevant to the purpose. This test was conducted through the series of interview with HOs during the process of building the conceptual model. Based on the interviews, the developed causal-loop model has represented actual humanitarian operation during Mount Merapi disaster response.

Dimensional consistency test was conducted using the feature in the VensimPLE which shows that the model is consistent with respect to unit, whereas face validation was conducted by comparing the general patterns produced by the simulation model with the historical pattern with respect to organizational capacity of the sampled HOs. The simulation model has successfully generated the similar pattern of that of the real system (most HOs currently allocates their resources more on relief operations so that the organizational capacity elicited by interviewees is similar to that of pattern Figure 4(b) for relief operations allocation of 80% of total resources).

### 4.4 Scenario Development

Two scenarios were developed to examine the impact of different pattern of affected people over time. First scenario (base model) uses the constant number of affected people at 1000 people over time. Second scenario uses empirical pattern of affected demand adopted from [12]. For each scenario, various allocation policies to relief operations – fraction of relief operations of 0, 0.2, 0.4, 0.6, 0.8, and 1 are simulated to examine the effect of the allocation policy on the dynamics of both organizational capacity and saved people.

**Table 1.** Model Parameters

| Parameters | Definition | Value | Unit | Sources |
|---|---|---|---|---|
| Total affected people | The number of affected people (victims) over time | 1000 | people/day | **Base model:** 1000 people (assumption) Empirical model: [12] |
| Reference of required resources | The ratio of standard required resources and victims | 0.00375 | dimensionless | Government Regulation No. 7 Year 2008 |
| Minimum relief operations | Minimum required relief operation if HO want to operate in the field | 10 | people/day | Interview with sampled HOs |
| Normal productivity | The capability of resources to help victims everyday | 5 | people/(people .day) | Interview with sampled HOs |
| Time to invest | The average time to hire and train new resources | 5 | days | Interview with sampled HOs |

**Table 2.** Model Parameters (continued)

| Parameters | Definition | Value | Unit | Sources |
|---|---|---|---|---|
| Average working time | The average time to work in HOs | 30 | days | Interview with sampled HOs |
| Fraction of relief operation impacting capacity | Fraction of relief operation impacting capacity due to knowledge gained from the field | 0.01 | Dimension-less | Interview with sampled HOs |

## 5 Results and Discussion

This section presents results of the two developed scenarios, i.e., base model (assuming the constant affected people of 1000 over time) and empirical model (based on historical pattern of the affected people).

Each scenario is simulated for various resource allocation policies for relief operation (0 – no resource allocated for relief operations, 0.2, 0.4, 0.6, 0.8, and 1 – all resource allocated for relief operations). Given the pattern of affected people (a), each scenario is examined its performances with respect to organizational capacity (b), saved people (c), and need gap (d). Figure 3 (a, b, c, d) and Figure 4 (a, b, c, d) present results of base model and empirical model respectively.

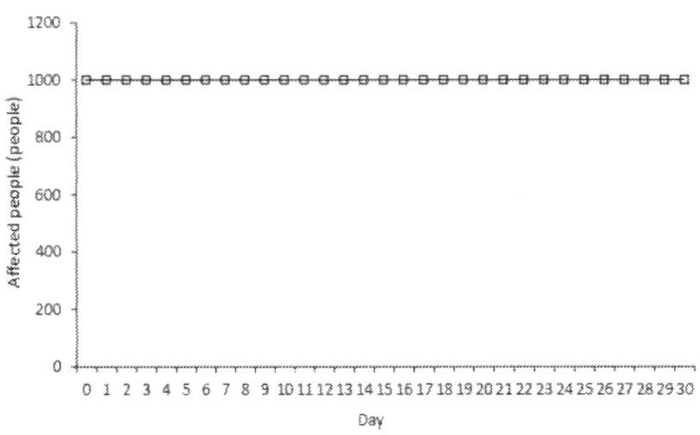

(a). Base Model – Affected People

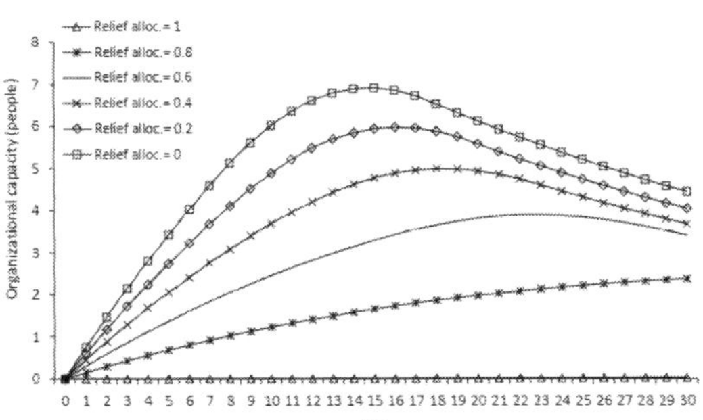

(b). Base Model – Organizational Capacity

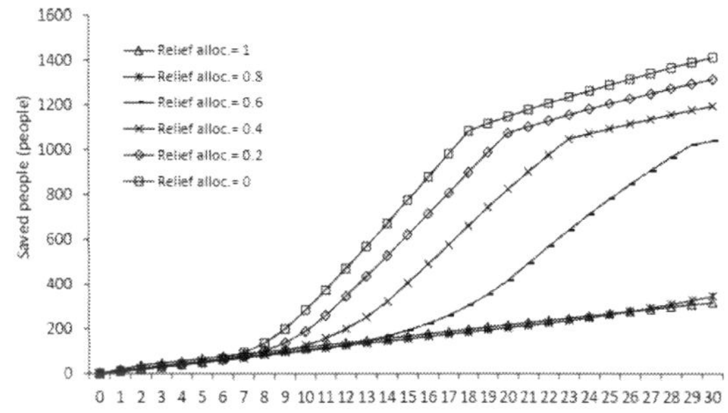

(c). Base Model – Saved People

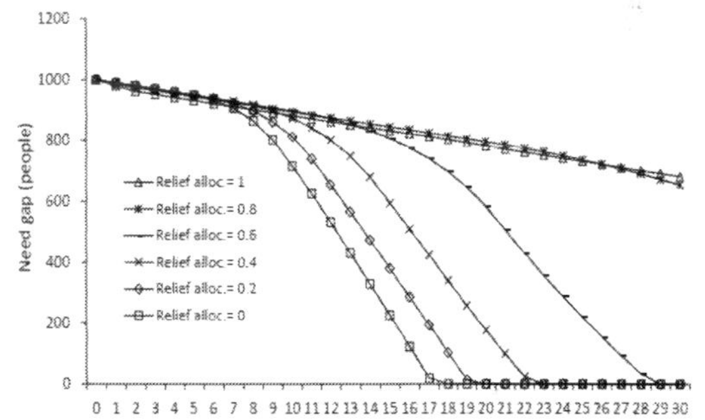

(d). Base Model – Need Gap

**Fig. 3.** Simulation Results for Base Model (Constant Relief Demand) for Various Human Resource Allocation Policies (a-d)

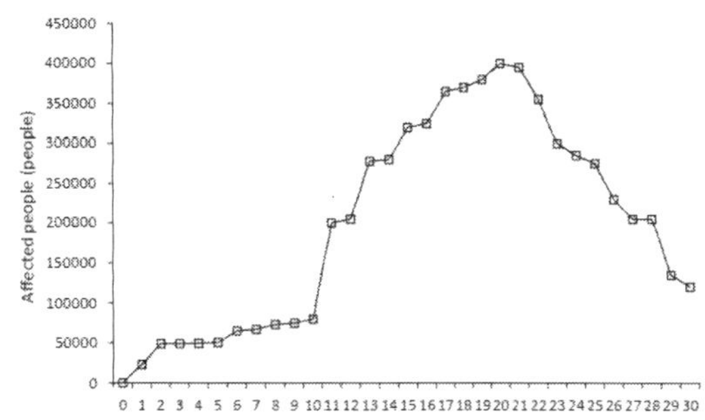

(a). Empirical Model – Affected people

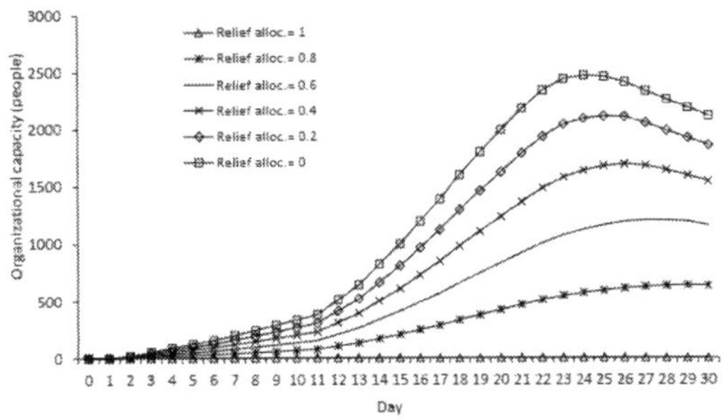


(b). Empirical Model – Organizational Capacity

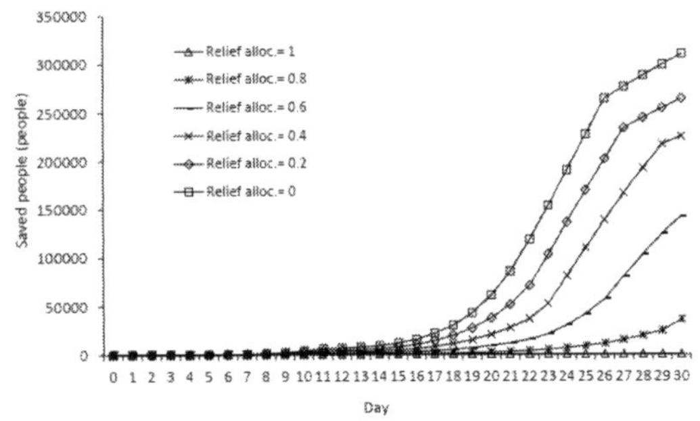


(c). Empirical Model – Saved People

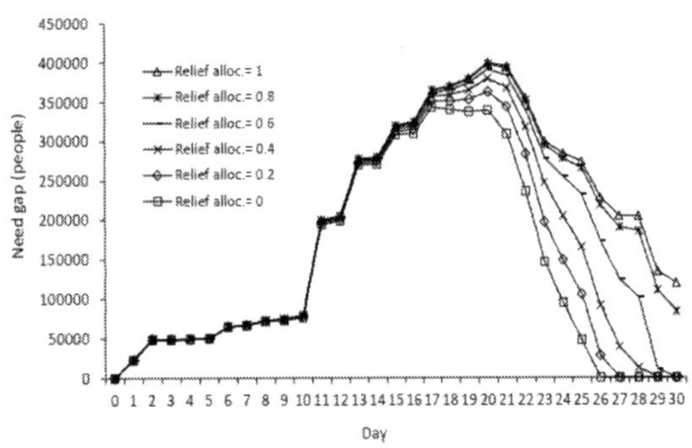


(d). Empirical Model – Need Gap

**Fig. 4.** Simulation Results for Empirical Model (Historical Pattern of Relief Demand) for Various Human Resource Allocation Policies (a-d)

Results indicate that allocating resource to relief operations without allocating the resources to capacity building causes slow and less people receiving assistance in the long run. As no resource is allocated to capacity building, there is no significant resource addition for relief operations. The increased organization capacity is only due to knowledge gained from the field. However, the organizational capacity increase is not significant so that, in the long run, the existing resources has less productivity due to continually high need gap. Consequently, it results in less people receiving help.

Because the effect of various allocation policy is not observable in the short-term, HOs tend to allocate more resources to relief operations due pressure of high surge of relief demand. However, based on the results, allocating resources to capacity building is necessary to sustain the effort of relief operations for the long-term. The results are supported by [8] who investigated that capacity building strategy is enduring strategy on HO performance, whereas relief strategy is a better-before-worse HO performance.

Results demonstrate that investing all resources to capacity building appears to be the best policy to save more people and meet relief demand in the long run. Although building capacity is incapable to save people in early days, the organizational capacity positively supports relief operations after some days in a way that the organizational capacity provide resources and increases productivity. Consequently, it can save more people quickly in later days. By saving more people rapidly, knowledge gained due to experience in saving people gives positive impact to more capability of the HOs in relief operations.

Although fully allocation to capacity building appears to be best policy, in reality, HOs cannot allocate resources fully on the capacity building because there is minimum requirement for relief operation if HOs want to get finance from donors and stay in the field. There is a trade-off of allocation policy to distribute more on either relief operations or capacity building. For both base model and empirical model, results indicate that at least 20% of human resources need to be allocated for capacity building in order to sustain relief operations in the long run.

It is also worth to mention that the time required to fulfil relief demand is very much depending on the pattern of relief demand. It seems that constant relief demand requires less time to fulfil the demand than that of non-linear relief demand.

## 6 Conclusions

The present study has developed a system dynamics simulation model for examining the allocation policy for relief operations and capacity building in humanitarian organizations. The conceptual model was built based on interviews with two HOs in Yogyakarta and parameterized with empirical data based on Mount Merapi disaster in 2010. Results indicate that, during disaster response, HOs need to allocate resource to capacity building, in addition to relief operations, in order to sustain the relief operations in the long run. It is suggested that at least 20% of the resources should be allocated to capacity building.

Although the model has been useful to demonstrate the allocation trade-off, some limitations still need to be highlighted. First, although the model is generic, the parameter values are only relevant for the typical disaster in the study (volcano eruption), different disaster and location of disaster may have different parameter values. Second, the conceptual model is elicited based on two HOs, more HOs are still needed to get representative model specification.

It is suggested that future research should thus focus on refining the conceptual model as well as testing the model to gain more confidence on the model. Last but not least, experiments could be conducted to explore the effect of productivity, knowledge gained from field, etc., which may be useful to increase efficiency in relief operations.

The authors gratefully acknowledge financial support from Higher Education Directorate (DIKTI), Indonesia, under research grant of Penelitian Unggulan Perguruan Tinggi (PUPT) – Universitas Gadjah Mada 2017. The authors also thank to Ryan Destryan for administering the interviews.

## References

1. A. Thomas, and L. Kopczak, *From logistics to supply chain management: the path forward in humanitarian sector*. Fritz Institute (2005)
2. D. Kaatrud, R. Samii, and L. M. Van Wassenhove, UN Joint Logistics Center: a coordinated response to common humanitarian logistics concernes, *Forced Migr. Rev.* **18**, 11-14 (2003)
3. L. Gustavsson, Humanitarian logistics: context and challenges, *Forced Migr. Rev.* **18**, 6-8 (2003)
4. C. Ni, R. de Souza, Q. Lu, and M. Goh, M, Emergency preparedness of humanitarian organizations: A system dynamics approach, In M. Klumpp, S. de Leeuw, A. Regattieri, and R. de Souza, *Humanitarian Logistics and Sustainability*. Lecture Notes in Logistics, Springer International Publishing, 113-127 (2015)
5. D. A., McEntire, Issues in disaster relief: progress, perpetual problems and prospective solutions, *Disaster Prev. Manag.* **8**, 351-366 (1999)
6. G. Heaslip, A. M. Sharif, and A. Althonayan, Employing a systems-based perspective to the identification of inter-relationships within humanitarian logistics, *Int. J. Prod. Econ.* **139**, 377-392 (2012)
7. M. Besiou, O. Stapleton, and L. N. Van Wassenhove, L. N, *Exploring the known and the unknown: Future possibilities of system dynamics for humanitarian operations*, Working Paper 2010/74/TOM/ISIC, INSEAD Social Innovation Center, The Business School for the World (2010)
8. P. Goncalves, Balancing provision of relief and recovery with capacity building in humanitarian operations, *Oper. Manag. Res.* **4**, 39-50 (2011)
9. J. D. Sterman. *Business Dynamics: System Thinking and Modeling for a Complex World*, McGraw-Hill Company, New York (2000)
10. P. M. Senge, *The Fifth Disciplines*. Doubleday; New York (1990)
11. B. Voight, E. Constantine, S. Siswowidjoyo, and R. Torley, Historical eruptions of Merapi Volcano, Central Java, Indonesia, 1768–1998, *J. Volcanol. Geotherm. Res.* **100**, 69–138 (2000)
12. E. T. W. Mei, F. Lavigne, A. Picquout, E. de Belizal, D. Brunstein, D. Grancher, J. Sartohadi, N. Cholik, and C. Vidal, Lessons learned from the 2010 evacuations at Merapi volcano, *J. Volcanol. Geotherm. Res.* **261**, 348-365 (2013)

MATEC Web of Conferences **154**, 01049 (2018) https://doi.org/10.1051/matecconf/201815401049
*ICET4SD 2017*

# Multi period pricing for managing local fruit supply chain

*Erwin* Widodo*, *Rahaditya Dimas* Prihadianto, and *Dody* Hartanto

Institut Teknologi Sepuluh Nopember, Industrial Engineering Department, Sukolilo, Surabaya, Indonesia

**Abstract.** Recently fresh fruit sector is grown not only due to increasing of demand that spirited by healthy lifestyle but also requirement of quality food should be eaten daily. Its complexity make many research considered fruit in certain supply chain, called as Fruit Supply Chain (FSC). In FSC, customers tend to purchase products with a longer remaining lifetime and avoid the ones which give aging signal. Customer willingness to pay decreases once the product start to be deteriorated, which may cause slower demand for aging fruits. Consequently, retailers should enable discounted price for aging fruits products to retain or improve demand rate. Main purpose of this research is explaining how to maximize supply chain profit by applying dynamic pricing. A set of mathematical model is optimized on this research. It addresses dynamic pricing for FSC players to achieve better profitability. The result proves that dynamic pricing is urgent to be done which is signaled by increment of FSC Profit. FSC players can separate selling period into three periods, which are forward buying period, normal price period, and markdown price period as implementation of this scheme.

## 1 Introduction

According to Soto-silva et al. [1], recently, fresh fruit sector is grown not only due to increasing of demand that spirited by healthy lifestyle but also requirement of quality food should be eaten daily. Based on FAO [2], fresh fruit is defined as portion of a plant housing seeds commonly eaten as dessert. Based on that definition, fresh fruit is considered as fruit which can be consumed raw. Repeatedly, Soto-silva et al. [1] argued that many regulation of food safety industrialization make fresh fruit should be produced by adopting good agricultural practices, automatization of sorting, selecting and packaging including sometimes some minor minimal processing tasks. Besides regulation, other factors such as globalization, competitor and customized products, make fresh fruits sector tends to be specialized and integrated vertically to become more competitive and dynamics. In this way, organizing fresh fruit under various related circumstances later called as Fruit Supply Chain (FSC).

Managing FSC is not as elementary as thought. Production fluctuation is one of challenge in fruit supply chain management. The fluctuation of agri-product, including fruits, its demand can be affected by several uncertain factors, such as weather, temperature, and customer preferences [3]. Fruit quality, like color, outlook appearance, and also best-before date label, become consideration aspects for customer to buy. Customer willingness to pay decreases once the product life is approached which may cause slower demand for aging fruits. Consequently, retailers should enable discounted price for aging fruits products to retain or improve demand rate. This research propose dymaic pricing model that accomoodate discounted price, normal price, and forward buying price as well. This model is extensive approach from preliminary research done by Maiti & Giri [4] that presents two period pricing on short life cycle product and research from Fauza && Lee [5] that presents revenue-based with EPQ/EOQ approach for perishable products. Authors previous work [6] also considered pricing in a specific structure or supply chain namely dual channel supply chain. In contorast to those aforementioned works, this research accommodates three period pricing on three echelon for fruit product handling. Common question will come up for proposed method whether dynamic pricing can help fruit supply chain players achieve better supply chain profit.

## 2 Problem Modelling

Conceptual model of problem is graphically presented below. There are three echelon of fruit supply chain players, which are supplier, wholesaler, and retailer.

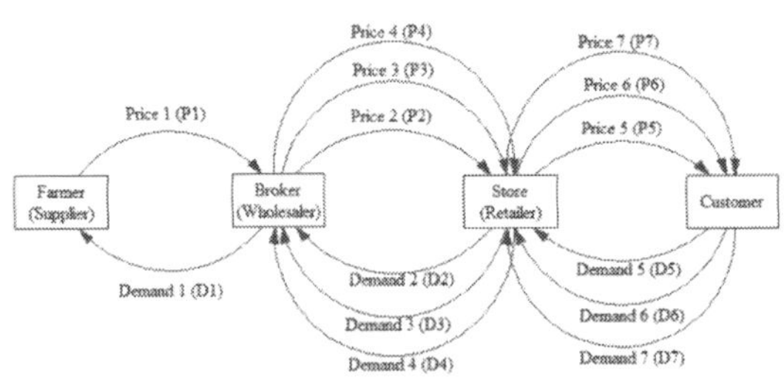


**Fig. 1.** Problem Mode

Based on fig. 1. Presented above, it is illustrated that there are different prices applied on wholesaler-retailer stage and retailer-customer stage. Price one (P1) is buyout price from wholesaler to retailer due to seasonal

* Corresponding author: erwin.widodo@ie.its.ac.id

properties of fruit product. Price four (P2) and price seven (P5) are forward buying from each stage, while price three (P3) and price six (P6) are normal price. Last but not least price four (P4) and price seven (P7) are discounted or markdown price. There is pricing rule applied here follows this pattern $P_1 \leq P_2 \leq P_3 \geq P_4 \leq P_5 \leq P_6 \geq P_7$ which is normal price is the highest price compare to forward and markdown prices.

## 3 Mathematical Formulation

Below, mathematical formulation for each demand will be elaborated. There is mathematical formulation represented demand one in tier one, which are farmer-wholesaler relation.

$$d_1 = \alpha - \beta * p_1 + \gamma * (c - p_1) \quad (1)$$

Equation (1) explained that demand for the farmer is generated price one ($p_1$) and product cost (c), where $\alpha$ represents base demand for whole supply chain. Price one ($p_1$) is price from farmer to wholesaler, which price is singular due to nature of harvesting period. In harvesting period, wholesaler mostly come to farmer's plant to buy goods in bulk amount.

While, demand formulation for tier two in fruit supply chain, which are wholesaler-retailer relationship. There are three formula represented each periodical demand gathered by price

$$d_2 = \alpha - \beta * p_2 + \gamma * (p_1 - p_2) \quad (2)$$

$$d_3 = \alpha - \beta * p_3 + \gamma * (p_2 - p_3) \quad (3)$$

$$d_4 = \alpha - \beta * p_4 + \gamma * (p_3 - p_4) \quad (4)$$

From equation (2) to (4), known that there are four different prices involved on tier two. In demand two, demand is influenced by difference of price one, which is buying price of product to farmer and price two, which is first launched price to retailer called forward buying price. This price is higher or equal to price one, to generate profit for wholesaler.

Demand three explained that difference of price involved are price two and three. Price three indicates normal price after wholesaler release forward buying previously. While in demand four, price three and four are involved in the demand equation. Price four will be lower than price three due to beginning of deterioration rate. Hence, wholesaler release markdown price to trigger more demand for selling remaining unsold products.

While, demand formula for tier three in fruit supply chain, which are retailer-customer relationship. There are three formula represented each periodical demand gathered by price

$$d_5 = \alpha - \beta * p_5 + \gamma * (p_2 - p_5) \quad (5)$$

$$d_6 = \alpha - \beta * p_6 + \gamma * (p_5 - p_6) \quad (6)$$

$$d_7 = \alpha - \beta * p_7 + \gamma * (p_6 - p_7) \quad (7)$$

Retailer to customer equations will be more or less the same with wholesaler to retailer equations. From equation (5) to (7), known that there are four different prices involved on tier three. In demand five, demand is influenced by difference of price two, which is first launched price to retailer called forward buying price and price five that is first launched price to customer also called forward buying price. Condition explained above represent condition of demand five that occurred on forward buying period. Price five is higher or equal to price two to generate profit for retailer.

Demand six explained that difference of price involved are price five and six. It indicates normal price after releasing forward buying previously. While in demand seven, price six and seven are involved in the demand equation. This equation indicates markdown price after releasing normal price period previously. Price seven will be lower than price six and less than or equal to price five due to beginning of deterioration rate. Hence, retailer release markdown price to trigger more demand for selling remaining unsold products.

Following after demand mathematical formulation, supply chain profit mathematical formulation will be elaborated. There are three profit function on, which are farmer profit, wholesaler profit, and fruit supply chain profit.

$$\pi_f = (P_1 - C) * D_1 \quad (8)$$

$$\pi_w = (P_2 - P_1) * D_2 + (P_3 - P_1) * D_3 + (P_4 - P_1) * D_4 \quad (9)$$

$$\pi_r = (P_5 - P_2) * D_5 + (P_6 - P_3) * D_6 + (P_7 - P_4) * D_7 \quad (10)$$

$$\pi_{sc} = (P_1 - C) * D_1 + (P_2 - C) * D_2 + (P_3 - C) * D_3 + (P_4 - C) * D_4 + (P_5 - C) * D_5 + (P_6 - C) * D_6 + (P_7 - C) * D_7 \quad (11)$$

Equation (8) to (11) presented above are profit equation for each echelon and whole fruit supply chain profit. Equation (8) explained that profit of farmer is difference of unit cost and price from farmer to wholesaler ($P_1$) multiply demand from wholesaler to farmer ($D_1$). While in equation (9), profit of wholesaler is defined by sum of difference of forward buying, normal price, markdown price and buying price of product multiply by demand two to four. Following equation, equation (10) represents retailer profit. Profit of retailer is defined by sum of difference of forward buying, normal, and markdown prices of wholesaler and retailer multiply by demand five to seven. Last but not least, profit supply chain is defined by sum of difference of each price on each tier and cost unit multiply by its demand, that formulated on equation (11).

Besides equation declared above, constraint that worked on this problem need to be formulated. First constraint is nonlinear constraint. Nonlinear constraint is declared below

$$p_1 * p_2 * p_3 * p_4 * p_5 * p_6 * p_7 \geq 10{,}000{,}000{,}000{,}000{,}000{,}000{,}000{,}000{,}000 \quad (12)$$

$$p_1^2 + p_2^2 + p_3^2 + p_4^2 + p_5^2 + p_6^2 + p_7^2 = 1{,}600{,}000{,}000 \quad (13)$$

Non-linear equation above is making sure optimization will go well. Equation (12) represents condition that multiply of each price will be exceed 10,000 billion rupiahs. This condition will trigger the optimization to yield different prices by result. Hence, there may be price value Rp 4,000 but later to be compensated by price value Rp 15,000 in cases. Therefore, equation (13) complement previous equation by equalities condition. This equation will give boundary to price combination, that will make sure that sum of square of each prices will not exceed 1,6 million.

Next, constraint is linear constraint. There are two sets of linear constraint design for this optimization process. First linear constraint set is dedicated to wholesaler echelon. The linear constraint is presented as follows:

$$p_1 - p_2 \leq 0 \tag{14}$$

$$p_2 - p_3 \leq 0 \tag{15}$$

$$-p_3 + p_4 \leq 0 \tag{16}$$

Equation (14) to (16) represents dynamic pricing rules. Equation (14) indicates condition that price one, which is buying price from farmer will be equal or lower than price two, which is forward buying price. Equation (150 indicates condition that price two, which is forward buying price will be lower or equal to price three, which is normal price. Equation (16) indicates condition that price three, which is normal price will be higher or equal to price four, which is markdown price.

Second linear constraint set is dedicated to retailer echelon. The linear constraint is presented as follows:

$$p_2 - p_5 \leq 0 \tag{17}$$

$$p_3 - p_6 \leq 0 \tag{18}$$

$$p_4 - p_7 \leq 0 \tag{19}$$

Equation (17) to (19) presented above are represent dynamic pricing happened on retailer echelon. On equation (17), price five on retailer is forward buying price for customer that must be higher than or equal to price two, which is forward buying from wholesaler. Consequently, equation (18) and (19) indicate normal price and markdown price on retailer must be higher than or equal to normal price and markdown price on wholesaler stage.

$$p_5 - p_6 \leq 0 \tag{20}$$

$$-p_6 + p_7 \leq 0 \tag{21}$$

$$-p_5 + p_7 \leq 0 \tag{22}$$

Equation (20) to (22) represents dynamic pricing rules. Equation (20) indicates condition that price five, which is forward buying price will be lower than or equal to price six, which is normal price. Equation (21) indicates condition that price six, which is normal price will be higher than or equal to price seven, which is markdown price. Equation (22) indicates condition that price five, which is forward buying price will be higher or equal to price seven, which is markdown price.

Following constraint is also linear constraint, but it express equality aspect of the problem. The equality constraint is declared below

$$p_1 + p_2 + p_3 + p_4 + p_5 + p_6 + p_7 = 100{,}000 \tag{23}$$

Equality constraint above represented condition that sum of each prices is equal to 100,000. This constraint triggers optimization process to yield unique price and sum of each prices do not excess that value.

Last step, all the equation is scripted into Matlab R2010 in order to optimize the problem. It is assisted by usage of FMINCON function to gather price should be made by each echelon in order to maximize the supply chain profit.

## 4 Optimization Process

Optimization Process begin with conducting numerical testing on three optimization scenarios. First scenario is using single prices on each echelon, which are IDR 10,000 as price one (P1), IDR 12,100 , IDR 12,500 , IDR 12,200 subsequently as price two (P2) to price four (P4), and IDR 15,000 , IDR 15,900 , IDR 15,200 subsequently as price five (P5) to price seven (P7). While, second and third scenario are dynamic prices which is automatically yield by optimization process. Results from Matlab are presented on figure two to four below.

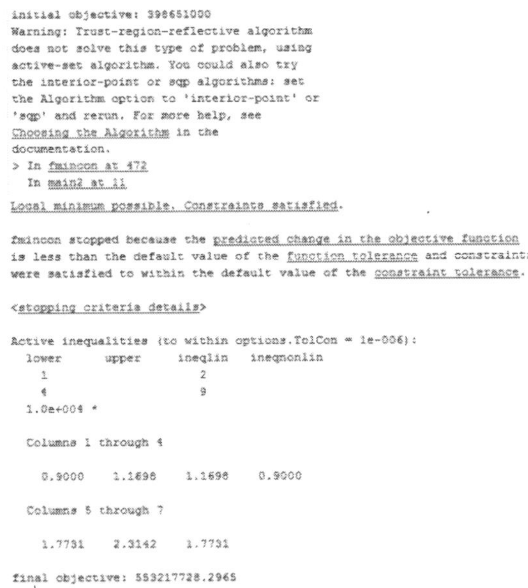

```
initial objective: 396651000
Warning: Trust-region-reflective algorithm
does not solve this type of problem, using
active-set algorithm. You could also try
the interior-point or sqp algorithms: set
the Algorithm option to 'interior-point' or
'sqp' and rerun. For more help, see
Choosing the Algorithm in the
documentation.
> In fmincon at 472
  In main2 at 11

Local minimum possible. Constraints satisfied.

fmincon stopped because the predicted change in the objective function
is less than the default value of the function tolerance and constraints
were satisfied to within the default value of the constraint tolerance.

<stopping criteria details>

Active inequalities (to within options.TolCon = 1e-006):
  lower      upper     ineqlin   ineqnonlin
    1                     2
    4                     9
  1.0e+004 *

  Columns 1 through 4

    0.9000    1.1698    1.1698    0.9000

  Columns 5 through 7

    1.7731    2.3142    1.7731

final objective: 553217728.2965
```

**Fig.2.** First Optimization Result

Based on optimization result presented above, known that optimization can answer question mentioned on introduction whether dynamic pricing could assist better profit fruit supply chain. Shown that final objective which is using dynamic pricing scheme yields higher profit compare to initial objective is using single price scheme, where initial objective value is 396 million rupiahs while using dynamic pricing is 553 million rupiahs.

## 5 Sensitivity Analysis

There are two sensitivity done on this research. First sensitivity analysis on pricing section. While, second sensitivity analysis is on parameters involved in optimization. First sensitivity is conducting sensitivity on optimization process, which adds There is pricing rule applied on this sensitivity, which is common pricing rule mainly used by company from tourism area. Forward buying price, regularly on weekdays, is given lower price which is around 80% from weekend price, which is normal price. This pricing strategy is mainly applied by Bakery Company, where expiration does exist on bakery products. Pricing rule mathematical formulation and optimization results presented consecutively as follows

$$p_2 - 0.8 * p_3 = 0 \quad (24)$$

$$-0.5 * p_3 + p_4 = 0 \quad (25)$$

```
Local minimum found that satisfies the constraints.

Optimization completed because the objective function is non-decreasing in
feasible directions, to within the default value of the function tolerance,
and constraints were satisfied to within the default value of the constraint tolerance.

<stopping criteria details>

Active inequalities (to within options.TolCon = 1e-006):
  lower      upper     ineqlin   inegnonlin
    1                     9
    4
  1.0e+004 *

  Columns 1 through 4

    0.8000    1.2800    1.6000    0.8000

  Columns 5 through 7

    1.5934    2.3332    1.5934

final objective: 554081522.6789
```

**Fig.3**. Second Optimization Result

Based on figure three presented above, final objective value is higher than preliminary optimization. Meaning that, this modified optimization is verified to seek better supply chain profit. Moreover, result gathered from this optimization is not possibilities of local optimum due to the objective function is non-decreasing on feasible directions. Wholesaler could apply dynamic pricing with scheme explained before, which was 80 – 50 from Price three. While, retailer could apply dynamic pricing, where price five and price seven is on the same position in Rp 15,934 with Rp 23,332 as normal price. Last but not least, farmer should reduce the initial price from Rp 10,000 to Rp 8,000. Therefore, it could become an aberration for farmer in spite of better profit for whole supply chain.

Second modified optimization try to solve first modified optimization and also attempt to achieve better supply chain profit. Following modification also using the pricing rule, but it is not scripted as equalization. Therefore, pricing scheme is scripted on non-equalization equation.

$$p_2 - 0.8 * p_3 \leq 0 \quad (26)$$

$$-0.5 * p_3 + p_4 \leq 0 \quad (27)$$

$$p_5 - 0.8 * p_6 \leq 0 \quad (28)$$

$$-0.5 * p_6 + p_7 \leq 0 \quad (29)$$

This modification is triggering the optimization to seek value of price which less than or equal to 0.8 – 0.5 of price three and price six, where those two prices are normal price. Hence, decision can be made by echelons by reviewing dynamic pricing and impact to whole supply chain profit. Optimization result is presented as follows

```
Local minimum found that satisfies the constraints.

Optimization completed because the objective function is non-decreasing in
feasible directions, to within the default value of the function tolerance,
and constraints were satisfied to within the default value of the constraint tolerance.

<stopping criteria details>

Active inequalities (to within options.TolCon = 1e-006):
  lower      upper     ineqlin   inegnonlin
               6          3
                          4
                          8
                          9
  1.0e+004 *

    1.0649    1.2000    1.9567    0.9784    1.2000    2.4000    1.2000

final objective: 559190067.1447
```

**Fig.4**. Second Optimization Result

Based on figure four presented above, final objective of second modified optimization value is higher than first optimization. Meaning that, this second modified optimization assist supply chain profit to better stage. Moreover, result gathered from this optimization is also local optimum that is fulfill all constraints due to the objective function is non-decreasing on feasible directions. By this optimization result, farmer is given bit higher price than initial price, in Rp 10,649 position. Wholesaler could apply dynamic pricing but not with exact 80 – 50 from normal price, the dynamic pricing scheme will be Rp 12,00 for forward buying, Rp 19,567 for normal price then decrease to Rp 9,784, which bit lower than price one, which is purchasing price from farmer. While, retailer place the same price for forward buying and markdown price on Rp 12,000 position, with Rp 24,000 on normal price.

Following sensitivity analysis is conducting sensitivity for parameters involved on mathematical model. Based on formulation used on this model, there are three parameters which are α β and γ. Each parameters are increased and decreased by 20% to simulate their effect on fruit supply chain players profit. Below recap table of sensitivity analysis is presented.

**Table 1.** Parameters Sensitivity Recap

| Parameter | Parameter Changes | Profit Changes (%) | | |
|---|---|---|---|---|
| | | Farmer | Wholesaler | Retailer |
| $\alpha = 50$ | -20% | -0.18 | -0.13 | -0.10 |
| | Threshold | 0 | 0 | 0 |
| | +20% | **0.18** | 0.13 | 0.10 |
| $\beta = 0.5$ | -20% | 19.16 | **19.78** | **19.65** |
| | Threshold | 0 | 0 | 0 |
| | +20% | -19.16 | -19.78 | -19.65 |
| $\gamma = 0.05$ | -20% | **1.02** | **0.35** | **0.45** |
| | Threshold | 0 | 0 | 0 |
| | +20% | -1.02 | -0.35 | -0.45 |

Based on table one presented above, each parameter has different effect to FSC players' profitability. Higher alpha will beneficial to farmer represented by higher profit changes to compare to other players. While, beta parameter that represents latest price review sensitivity constant, gives more profitability tendency to wholesaler. Nevertheless, beta can also useful to retailer due to higher profit changes when lower beta is set. Last but not least, lower gamma will give higher profitability tendency to farmer. While, changes of beta gives lower impact for wholesaler and retailer compare to farmer. This condition can be approached by anomaly effect of gamma to demand that will be comprehensively elaborated on section six.

## 6 Discussion & Findings

Mostly, Retailer Company on Indonesia offer banana on single price, for certain package of banana (no scaling process), for all selling period. Fruit like banana, has lifetime period before it ends as deteriorated product. Hence, retailer should be smart to push selling of deteriorative product such as fruit to prevent it is unsold, which can be overcame by applying forward buying. In example, based on figure four, retailer can offer Rp 12,000 per kg of banana when banana still on raw condition. Raw condition can be determined by color of banana skin and gentleness of banana fruit body. This price push demand into earlier period, therefore it can be used to prevent risk of unsold product. Then, retailer could begin mark up the price when there is indication of banana soon to be on well-consumed period. Again, based on figure four retailer should offer Rp 24,000 per kg of banana. This price still way too low from banana price on convenience store that observed by author. It indicates that retailer can play price since market acceptance level price still higher than threshold price. Last but not least, when fruits is indicating over well-consumed period, that can be determined by changes of banana fruit skin and gentleness of its fruit body as well, retailer could give lower price that commonly half price than normal price or just cost of goods sold price. This step also prevent risk of unsold product like forward buying price.

Besides prices, profit value is also need to be discussed. Based on optimization process known that supply chain profit growing better. Thing need to be discussed is prices involved on each profit. On first modified optimization model, FSC profit can be reach value Rp 554 million rupiahs, with pricing rules 80-50% on wholesaler stage, and decreasing farmer price 20% lower than initial price. While second modified optimization model could lead to better FSC profit in value Rp 559 million rupiahs. On this scheme, there is no 80%-50% pricing rules, but markdown price of wholesaler bit lower than buying price from farmer. Several consideration need to be elaborated here that is whatever the FSC profit considered by its player the decision still depend on the market. Meaning that, FSC players could take first modified optimization model, where farmer regain the price from Rp 8,000 to Rp 10,000 as long as wholesaler still willing to buy it and wholesaler market (retailer) tend to purchase markup price from wholesaler. Consequently, FSC players could take second modified optimization where it yields better FSC profit. Wholesaler also could regain their markdown price as long as retailer still willing to buy it and retailer market (customer) tend to purchase it. There are several thing can be concluded from here, which are

1. Dynamic pricing condition and its profit can be achieved by collaboration within FSC players, which its market need to be observed due to their obligation to adjust optimization done by FSC players
2. Dynamic pricing is worth to try due to this scheme can yield better whole supply chain profit. Besides, dynamic pricing is closer to the reality, where customer tend to purchase fruit in good condition. Hence, retailer could offer markdown price for fruit which start to deteriorate to prevent unsold product risks. Moreover, retailer also need to push forward buying with also offer price lower than normal price to prevent abundant amount of fruit start to deteriorate in markdown price selling period.

Second discussion is about parameters sensitivity analysis result. Alpha ($\alpha$) has more positive impact to farmer profit reflected by higher profit changes compare to wholesaler and retailer. It can be elaborated that farmer uses a single price to sell product to wholesaler. Hence, higher alpha, which is higher base demand placed by wholesaler to farmer, will raise the profit for farmer. Nevertheless, in real condition there are several consideration, which are nature of product and also market condition. Fruit product has nature to become deteriorative product in the end. Moreover, market condition can be so fluctuated. In previous chapter dynamic pricing assist to demand pushing, in the other side macro-economic condition that cannot be controlled. Commonly, fruit is bought in 50 kg amount from farmer on harvesting period which was set as initial base demand that used on the optimization. Based on consideration explained above, FSC players should consider certain number as a base demand to achieve targeted FSC profit.

Next parameter is beta. Based on table one from previous section presents that lower beta will give higher profit changes to wholesaler and retailer that come in second than farmer. Beta is represented latest price sensitivity, which is more related to wholesaler and retailer due to dynamic pricing scheme application. On dynamic pricing, there are several prices which can be lower or higher to one another. Declaring latest price has an impact to customer perception whether latest price is affordable or not. Beta parameter try to capture customer sensitivity to latest price. From previous paper, which is Maiti & Giri [1], valued default beta in 0.5. It insisted that customer sensitivity to latest price has great portion to effect demand for FSC players. Less than 0.5 value, demand will be higher and so does profit. In the other hand, the higher 0.5 value of beta, which is customer very sensitive to latest price declared or released by FSC players.

Condition explained above reflects real market condition. Higher latest price, may decrease customer willingness to pay. While, on lower latest price, it may increase customer willingness to pay. Nevertheless, FSC players applies higher price under condition of good condition fruit, which is period that has been waiting for by customer. While lower price is applied on before and after-good condition of fruit. This mechanism is compensated each other. Customer is given few options to buy before or after-good condition of fruit with lower price or good condition of fruit with higher price. Hence, FSC players should take an option to offer small gap between normal price and under normal price, remind that purpose of lowering the price to push demand to forward buying and prevent outdated fruits.

Last but not least is gamma. Gamma has unique impact on optimization. Based on Figure 6.2, gamma has positive impact to yield demand to FSC players. Nevertheless, table one also shows that lower gamma can give positive impact to higher profit changes to FSC players, especially farmer.

$$d_n = \alpha - \beta * p_n + \gamma * (c - p_n) \quad (30)$$

In farmer stage, this redundancy happened due to different of price offered to wholesaler. Gamma represents price correction from earlier price to latest price operate on its system. In farmer, due to application of single price, selling price to wholesaler directly minus by cost of product. Since selling price higher than cost of product, this price correction will increase the profitability where its price correction amplified by gamma.

In wholesaler and retailer stage, this redundancy happened due to application of dynamic pricing scheme. In example, on retailer stage different price of forward buying price to normal price, will be noted as negative reviewed by customer due to higher latest price than previous price. In the other hand, changes of normal price to markdown price will be noted as positive reviewed by customer due to decreasing price from previous prices. Gamma will amplify both of negative and positive review. In this dynamic pricing scheme, there is four negative review happened, two on each wholesaler and retailer, but only two positive review happened, one on each wholesaler and retailer. It can be concluded that gamma is following dynamic pricing scheme, which can be amplified both of negative or positive review of prices from customer.

One consideration need to be discussed, whether beta and gamma that captured as most sensitive parameter. From recent optimization, higher constant of beta make it as most sensitives compare to lower constant of gamma. Nevertheless, further market research need to be done to select which of those two parameters is the most influential for fruit supply chain. It can be indicated that on FSC, reviewing of price changes due to nature of product, can be considered as more sensitive compare to reviewing of latest price. Hence, higher constant can be given to gamma than beta.

## 7 Conclusion & Future Research Suggestion

From chapter five and six, several ideas can be underlined to be served as conclusion. Several conclusion are presented as follows

1. A set of mathematical model has been proposed to address dynamic pricing for FSC players to achieve better profitability. Dynamic pricing is urgent to be done due to nature of fruit product, where deteriorative period does exist. Hence, in order to avoid unsold product due to became deteriorated, dynamic pricing could become solution to push demand in several selling period, which are forward buying, normal price, and markdown price period compare to single price for all selling period, that is already proved by optimization.
2. Optimization parameters has their own effect on FSC profit. Alpha which is base demand could trigger higher demand for FSC players. Nevertheless, amount of base demand need to be considered due to nature of product and market condition. While, beta give negative impact on FSC profit, where it represent customer sensitivity to latest price, which is also considered as most sensitive parameter on optimization process. In addition, gamma could give both positive and negative impact depend on dynamic pricing scheme, which is could be considered as candidate to become most sensitive parameter on FSC.
3. FSC players could gather better profit supply chain, which affected by dynamic of demand due to dynamic pricing. Demand function reflect that customer preference to buy is more affected to latest price ($\beta$) than difference of latest price compare to previous price ($\gamma$). Hence, FSC players should prepare marketing technique for releasing latest price to their customer.

There are several suggestion that can be considered for future research, that is judgement or rules that apply for beta and gamma in order to capture price resulted from optimization process. Hence, constant of beta and gamma will follow price set by FSC players that represent customer or market sensitivity. Moreover, more consideration on mathematical formulation and model, in example inventory consideration and dynamic of base demand to enrich optimization result that could assist FSC players better. Last but not least, this research can be conducted by Graduate Grant of ITS Surabaya support.

## 8 References

1. Soto-silva, W.E. et al., 2015. Operational research models applied to the fresh fruit supply chain. *European Journal of Operational Research*, 0, pp.1–11.
2. FAO. 2014. *Crops Statistics*. Available at : http://www.fao.org/faostat/en/#data/QC (Accesed : 14 August 2017).
3. Chen, W., Li, J. & Jin, X., 2016. The replenishment policy of agri-products with stochastic demand in

integrated agricultural supply chains. *Expert Systems with Applications*, 48, pp.55–66.

4. Maiti, T. & Giri, B.C., 2016. Two-period pricing and decision strategies in a two-echelon supply chain under price-dependent demand. , 0, pp.1–20.
5. Fauza, G. & Lee, S., 2015. A Vendor-Buyer Inventory Model for Food Products Based On Shelf-Life Pricing. , 8(2), pp.67–73.
6. Widodo, E., Takahashi, K., Morikawa, K., Pujawan, I.N. and Santosa, B. 'Managing sales return in dual sales channel: an analysis of primary versus secondary resale strategies', *Int. J. Industrial and Systems Engineering*, Vol. 15, No. 2, pp. 119-151

# Predicting the probability of Mount Merapi eruption using Bayesian Event Tree_Eruption Forecasting

*Dyah Ika* Rinawati*, *Diana Puspita* Sari, *Naniek Utami* Handayani, and *Bramasta Raga* Siwi

Industrial Engineering Department, Faculty of Engineering, Diponegoro University, Prof. Soedarto, SH Street, Tembalang, Semarang, Indonesia

**Abstract.** Mount Merapi is one of the active volcanoes in Indonesia that had varied eruption periods from two to eight years. Due to the density of the population living around the slopes of Mount Merapi, its eruptions caused high number of victims. In order to avoid high number of victims, the disaster management should be improved. Disaster management consist of four phases i.e. mitigation, preparedness, response and reconstruction. In disaster mitigation phase, prediction of the Merapi unrest probability is needed. This paper focus on how to predict the probability of Merapi unrest based on volcano-logical information by using Bayesian Event Tree. Bayesian Event Tree (BET) is a probabilistic model that merges all kinds of volcano-logical information to obtain probability of any relevant volcanic event. The result showed that the probability of Merapi unrest is 0,822. In the next eruption, it has predicted that the volcanic explosivity index (VEI) 2 was biggest chance with the probability of 0,549. It showed that the eruption will take place in the main crater of Merapi with the probability of 0,938.

## 1 Introduction

Disaster refers to a ''disruption that physically affects a system as a whole and threatens its priorities and goals'' Based on the causes, it is possible to distinguish between a natural and man-made disaster. With respect to predictability and speed of occurrence, it is possible to distinguish between a sudden-onset and a slow-onset disaster 1]. Disaster that can be classified in sudden onset are earthquake, tsunami, eruption of volcano, mass movement, flood and storm. While slow onset disaster are epidemic, dryness and extreme temperature.

It is widely known that Indonesia is a country that is passed by the pathway of active volcanoes that is called Ring of Fire. In consequence, Indonesia has 400 volcanoes and 129 of them are active. Therefore, Indonesia is the country with the most active volcanoes where most of them located in Sumatera and Java. Mount Merapi is one of the active volcanoes in Indonesia. The period of the eruption cycle is varied. Merapi's danger level is very high, due to the density of the population living around the slopes of Merapi Mount. The victim died in the eruption of Merapi at 2010 reached 353 peoples. In avoiding the high number of the victims, the disaster management need to be improved. The disaster management consist of four phases, i.e. mitigation, preparedness, response and reconstruction [2]. The mitigation phase refers to rules and mechanisms that reduce social vulnerability. The preparation phase refers to various operations that occur during the period before a disaster strikes. This phase prepare the strategies that allow the implementation of a successful response. The response phase refers to the various operations that are instantly implemented after a disaster strikes. The reconstruction phase refers to operations that conducted in the effect of a disaster.

The mitigation phase is a series efforts to reduce disaster risk, either through physical development as well as awareness and capacity building is facing the threat of disaster. Disaster mitigation is an activity that acts as disaster risk reduction, or efforts made to reduce the victims of the disaster, both fatalities and property. In a volcanic eruption disaster mitigation, risk estimates are required in decision making, especially in land use and management in emergency. According to [3], one of the basic materials science to achieve both of these is the use of reliable and quantitative eruption forecasting. According to [4] eruption forecasting refers to assess the occurrence probability of a given eruptive event, whereas volcanic hazards are normally associated with the analysis of superficial and evident phenomena that usually accompany eruptions (e.g., lava, pyroclastic flows, tephra/ ash fall, lahars, etc.). Nevertheless, several hazards of volcanic origin may occur in non-eruptive phases during unrest episodes. Among others, remarkable examples are gas emissions, phreatic explosions, ground deformation, and seismic swarms. This study aim is predict the Merapi Mount unrest by using Bayesian Event Tree_Eruption Forecasting (BET_EF) in order to help the mitigation phase of disaster management for Merapi eruption.

* Corresponding author: dyah.ika@gmail.com

# 2 Research Method

## 2.1 Bayesian Event Tree Model

Eruption forecasting is complicated to do because there are many data and relatively poor knowledge in the eruption process. In contrast with this reality, the extreme risk caused by many volcanoes eruption push to solve this problem. [3]. According to [5], Bayesian statistic is good for forecast the volcano eruption forecasting. Bayesian Event Tree (BET) is a quantitative model for volcanic hazard assessment. BET also offers a short-term and long-term forecast, therefore it can be used in many aspects such as: (1) to compare various types of risk, (2) to analyze the cost / benefit of the risk of mitigation, (3) to show the right land use planning and residential location, and (4) suggesting direct risk reduction like evacuation for people living in a dangerous area [6]. BET represents a development of the method proposed by [7] Marzocchi. BET is probabilistic model that merges all kinds of volcano-logical information – from theoretical/empirical models, geological and historical data, and monitoring observations – to obtain probability of any relevant volcanic event. Such probabilities represent a homogeneous and quantitative synthesis of the present knowledge about the volcano [4]. In general, BET does not rule out any possibility, but it shapes the probability distribution of the event considered around the most likely outcome accounting for all the information reported above.

This model is based on the event tree concept, which is a graphic representation of every event. Every branch is referring to every step to gain a more specific event and the end of the branch is representation of future event. This Event tree shows all the relevant results that might happen in every volcanoes activity. The point of every branch is named nodes. Here is a node in event tree scheme [3]:

Node 1: shows if there is or there is not a volcano activity in time interval ($t0$, $t0 + \tau$), where $t0$ is the current time, and $\tau$ is a time window that

Node 2: shows if a volcano activity is caused by magma or something else (for example, hydrothermal, tectonic, etc.)

Node 3: shows that magma will reach the surface (there will be an eruption) or not, in time interval ($t0$, $t0 + \tau$)

Node 4: shows the location of eruption

Node 5: shows the size of the eruption (for example, Volcanic Eruption Index)

"Node *k*" indicate one of the possible events of the *k*th of the event tree. A probability function was attributed at each nodes. For each nodes, there are probability of the conditional events that have been specify, $[\theta_1^{(unrest)}]$, $[\theta_2^{(magma)}]$, $[\theta_3^{(erup)}]$, $[\theta_4^{(loc)}]$, $[\theta_5^{(size)}]$, where the square brackets stand for a generic "probability density function

(PDF)". Because the first three nodes have only two possible states that are mutually exclusive and exhaustive (unrest and no unrest, magma and no magma, eruption and no eruption), so for simplicity, it has been set, $[\theta_1^{(unrest)}] = [\theta_1]$, $[\theta_2^{(magma)}] = [\theta_2]$, $[\theta_3^{(erup)}] = [\theta_3]$.

BET combines all PDFs at each nodes to get the probability of each events. The PDF of the probability of a *k* type eruption occur in the time interval ($t_0$, $t_0+\tau$] at the *j*th vent location, i.e., $\Phi_1$, is

$$[\Phi_1] = [\theta_1]\,[\theta_2]\,[\theta_3]\,[\theta_4^{(j)}]\,[\theta_5^{(k)}] \quad (1)$$

The probability of having eruption at any location, i.e., $\Phi_2$, is

$$[\Phi_2] = \sum_{j=1}^{M} [\theta_1]\,[\theta_2]\,[\theta_3]\,[\theta_4^{(j)}]\,[\theta_5^{(k)}] \quad (2)$$

## 2.2 Parameter of Mount Merapi Activity

Volcano status or volcano warning scheme in Indonesia are classified into four phases, i.e. "Normal", "Alert", "Standby" and "Caution" [8]. The status was determined by several parameters have been observed. Normal status is a condition where volcanic activity is normal and does not show any increase in activity based on visual observations, or instrumental research results. Alert status is a condition that show increase in seismic activity and other volcanic events. This light changes in activity caused by magma, tectonic and hydrothermal activity. While Standby status happen if there are intensive increase of seismic activities that indicates that the volcano moving toward eruption. Caution Status indicates that the volcano is immediate or erupting or there is a critical state of disaster.

Table 1 show the parameter and value limit that being observed to determine the status of Mount Merapi according to Center for Research and Technological Development Geology Disaster [8].

**Table 1** Parameter of status determination of Mount Merapi

| Status | Parameter | Value Limit |
|---|---|---|
| Normal | Swarm | None |
| | Deformation rate | 0-0,3 mm day-1 |
| | CO2 | 5-10 % mol |
| | SO2 | 0-1 % mol |
| | H2S | 0-1 % mol |
| | HCl | 0-0,3 % mol |
| | NH3 | 0-1,5 % mol |
| | H2O | 80–100 % mol |
| | Volcanic Earthquake | None |
| | Multi-Phases Earthquake | None |
| | Low-Frequence Earthquake | None |
| | Earthquake Avalanches | None |
| | Pyroclastic flows | None |
| | RSAM | None |
| Alert | Swarm | Exist |
| | Deformation rate | > 0,3 mm/ day |

| Status | Parameter | Value Limit |
|---|---|---|
| Standby | $CO_2$ | > 10 % mol |
| | $SO_2$ | > 1 % mol |
| | $H_2S$ | > 1 % mol |
| | HCl | > 0,3 % mol |
| | $NH_3$ | > 1,5 % mol |
| | $H_2O$ | < 80 % mol |
| | Volcanic Earthquake | > 1 event/ day |
| | Multi-Phases Earthquake | > 5 event/ day |
| Caution | Volcanic Earthquake | > 25 event/ day |
| | Multi-Phases Earthquake | > 25 event/ day |
| | Low-Frequence Earthquake | Exist |
| | Earthquake Avalanches | Exist |
| | Pyroclastic flows | Exist |
| | RSAM | Exist |

### 2.3 Historical data of Mount Merapi Eruption

Based on observation of Mount Merapi from 1961 to 2010, Merapi has recorded 146 times in unrest condition. For 146 times, has recorded 55 periods caused magmatic activity. In an eruption, Mount Merapi erupted as long as 2 until 3 months. From 1780, Merapi erupted 57 times with average rest period between eruption 3-5 years and average scale of eruption in VEI (Volcanic Explosivity Index) is VEI 2. Table 2 show historical data of Merapi eruptions.

**Table 2.** Historical data of Merapi eruptions

| Year of Eruption | Rest Period (Year) | VEI |
|---|---|---|
| 1786 | 18 | 1 |
| 1791 | 5 | 2 |
| 1797 | 6 | 1 |
| 1807 | 10 | 1 |
| 1810 | 3 | 1 |
| 1812 | 2 | 1 |
| 1822 | 10 | 4 |
| 1832 | 10 | 3 |
| 1837 | 5 | 2 |
| 1840 | 3 | 1 |
| 1846 | 6 | 3 |
| 1849 | 3 | 3 |
| 1861 | 12 | 2 |
| 1862 | 1 | 2 |
| 1865 | 3 | 2 |
| 1872 | 7 | 4 |
| 1878 | 6 | 2 |
| 1883 | 5 | 1 |
| 1885 | 2 | 1 |
| 1888 | 3 | 2 |
| 1889 | 1 | 1 |
| 1891 | 2 | 2 |
| 1897 | 6 | 2 |
| 1902 | 5 | 2 |
| 1905 | 3 | 2 |
| 1907 | 2 | 1 |
| 1909 | 2 | 1 |
| 1915 | 6 | 1 |
| 1920 | 5 | 2 |
| 1924 | 4 | 1 |
| 1925 | 1 | 1 |
| 1930 | 5 | 3 |
| 1933 | 3 | 2 |
| 1935 | 2 | 2 |
| 1939 | 4 | 2 |
| 1942 | 3 | 2 |
| 1948 | 6 | 2 |
| 1949 | 1 | 2 |
| 1953 | 4 | 1 |
| 1954 | 1 | 1 |
| 1956 | 2 | 2 |
| 1957 | 1 | 1 |
| 1961 | 4 | 3 |
| 1967 | 6 | 2 |
| 1969 | 2 | 2 |
| 1972 | 3 | 2 |
| 1976 | 4 | 2 |
| 1980 | 4 | 2 |
| 1986 | 6 | 2 |
| 1994 | 8 | 2 |
| 2001 | 7 | 2 |
| 2010 | 9 | 4 |

## 3 Data Processing

Data processing have done by using BET_EF/ Bayesian Event Tree for Eruption Forecasting software that being developed by [3]. Each steps of data input were performed sequentially according to the nodes.

### 3.1 Data Input for Node 1

Starting from 1961 to 2010, Mount Merapi is actively monitored for 146 months from a total of 600 months of observation. Parameters monitored at node 1 are early warning parameters at alert level, i.e. swarm (small earthquake sequence) and also deformation rate from lava dome formation. Figure 1. shows the input process on Node 1.

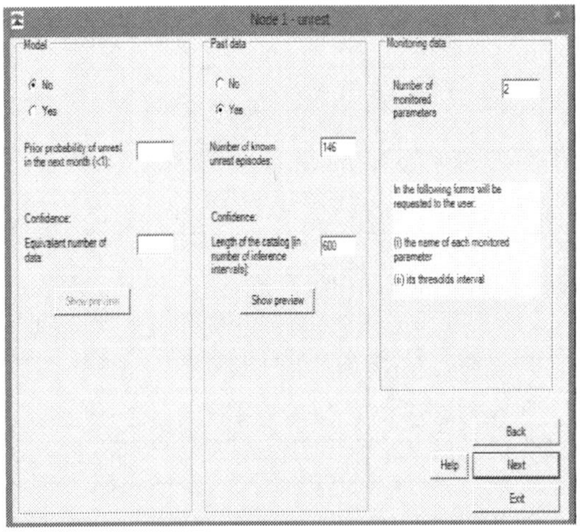


**Fig. 1** Data Input for Node 1

### 3.2 Data Input for Node 2

Mont Merapi is observed as active caused by magmatic activity of 55 months from a total of 146 months of active states. The parameters monitored at node 2 are early warning parameters at the standby level, i.e. volcanic gas content (CO2, SO2, H2S, HCl, NH3), water content (H2O), volcanic earthquakes, and multi-phases earthquakes. Figure 2. Describes data input for node 2.

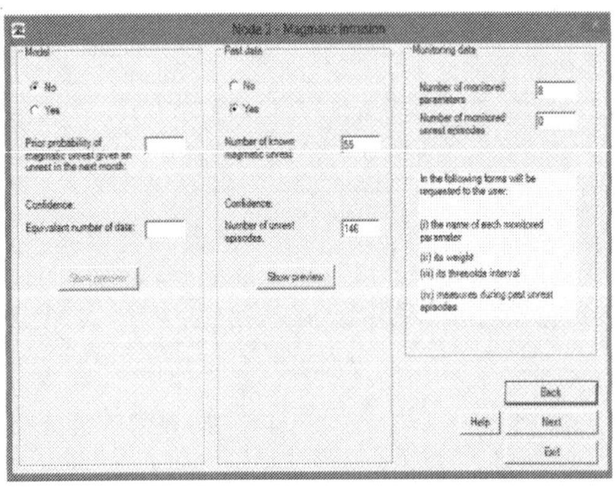


**Fig. 2.** Data Input for Node 2

### 3.3 Data Input for Node 3

The required data in this stage is how many observed eruptions in a matter of months from the total of Mount Merapi is actively monitored due to the magmatic activity and the parameters monitored at the early warning level. Counted Mount Merapi monitored eruption as much as 46 months from a total of 55 months of active Merapi observed as magmatic activity. Parameters monitored in Node 3 are early warning parameters at standby level, i.e. volcanic earthquakes, multi-phases earthquakes, low-frequency earthquakes, falling earthquakes, hot clouds, and RSAM (Realtime Seismic Amplitude Measurement). In Figure 3. the data input for Node 3 is given.

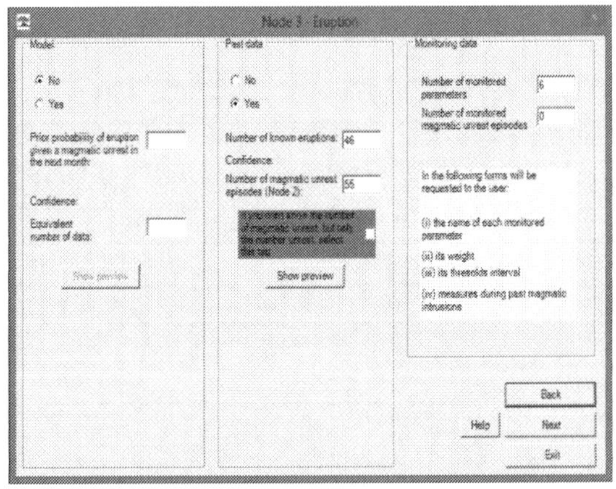


**Fig. 3** Data Input for Node 3

### 3.4 Data Input for Node 4

At this node the required data are the central coordinates of Mount Merapi, the radius of the eruption range, the direction of the next eruption, and the historical data of the location of the eruption. Based on the Center for Volcanology and Geological Hazard Mitigation, Mount Merapi lies at 7032'30 "LS and 110026'30" BT. If converted to Universal Transverse Mercator (UTM) unit Mount Merapi lies 9166432.32 m at latitude and 438908.54 m at longitude. The closest range radius is 5 km while the most distant radius is 17 km. The eruption data from 1961-2010 shows that the eruption always appears in the central crater which means it is located at location 1.

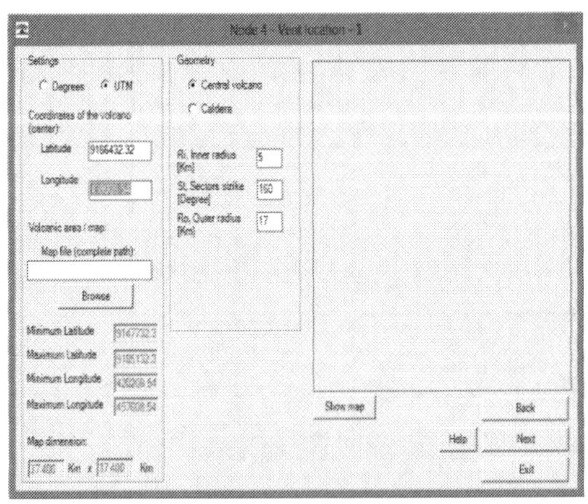


**Fig. 4.** Data Input for Node 4

### 3.5 Data Input for Node 5

The required data on this node is the historical data of eruption size (VEI) of Mount Merapi. Figure 5. shows the input of Node 5 for VEI 1. Input for Node 5 is then performed for each VEIs according to the eruption frequency at each VEIs.

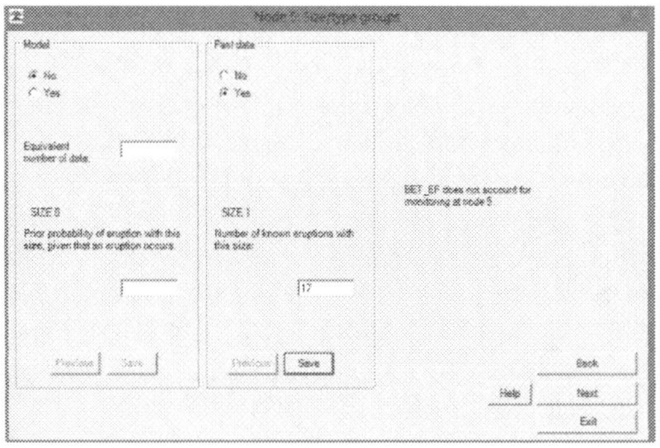


**Fig. 5**. Data Input for Node 5

## 4 Result and Discussion

The result is shown in Figure 6, Figure 7 and Figure 8. Figure 6 shows the probability of Merapi eruption in the future is 0,82. After getting probability of Merapi eruption, we can get the probability of the location of the next eruption as shown in Figure 7. There are 5 locations that probably being center of eruption.

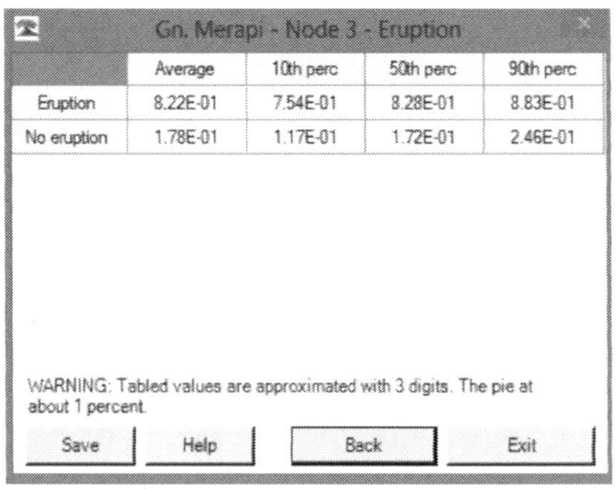
Gn. Merapi - Node 3 - Eruption

| | Average | 10th perc | 50th perc | 90th perc |
|---|---|---|---|---|
| Eruption | 8.22E-01 | 7.54E-01 | 8.28E-01 | 8.83E-01 |
| No eruption | 1.78E-01 | 1.17E-01 | 1.72E-01 | 2.46E-01 |

WARNING: Tabled values are approximated with 3 digits. The pie at about 1 percent.

Save Help Back Exit

**Fig. 6.** The probability of Mount Merapi eruption

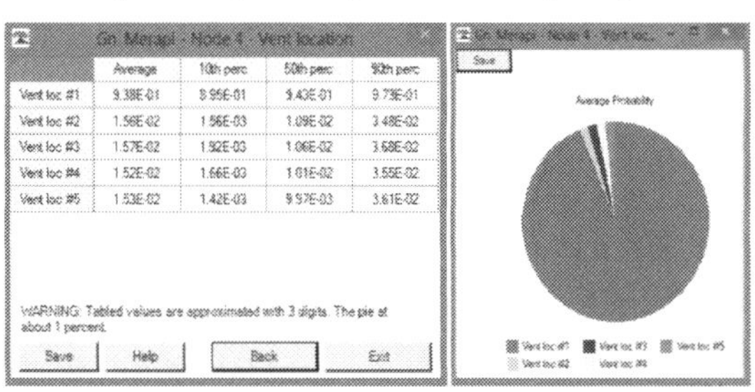

| | Average | 10th perc | 50th perc | 90th perc |
|---|---|---|---|---|
| Vent loc #1 | 9.38E-01 | 8.95E-01 | 9.43E-01 | 9.73E-01 |
| Vent loc #2 | 1.56E-02 | 1.56E-03 | 1.09E-02 | 3.48E-02 |
| Vent loc #3 | 1.57E-02 | 1.92E-03 | 1.06E-02 | 3.68E-02 |
| Vent loc #4 | 1.52E-02 | 1.66E-03 | 1.01E-02 | 3.55E-02 |
| Vent loc #5 | 1.53E-02 | 1.42E-03 | 9.57E-03 | 3.61E-02 |

WARNING: Tabled values are approximated with 3 digits. The pie at about 1 percent.

**Fig 7.** The probability of Mount Merapi Eruption Location

The purpose of predicting eruption's size is to know the future size of the eruption so we can determine which area that will be affected by the eruption of Merapi. As has been stated in the regulation of Ministry of Human Resources No. 15 year 2011 about Mitigation Guidelines of Volcano Disaster, Ground Movement, and Tsunami, disaster-prone zone is divided into three zone: Disaster-prone Zone I, Disaster-prone Zone II, and Disaster-prone Zone 3. Disaster-prone Zone I is 17 Km from the central of eruption. Disaster-prone Zone II is 10 Km from the center of eruption. While Disaster-prone Zone III is 5 Km from the center of eruption. The size of the eruption determines which areas affected by Mount Merapi's eruption and the safe zone from the eruption. The 2010 eruption was a high level eruption with VEI of 4, so the government of Sleman did not impose the Disaster-prone zone regulation, hence Disaster-prone Zone I became a dangerous zone too. Therefore, the government of Sleman set the safe radius as 30 Km seeing how big the eruption was in 2010. The size of the future eruption can affect which areas will be affected by it, so the related parties can make a decision regarding safe zone. Software BET_EF measures the eruption size in future VEI units. Figure 5 shows the output software that shows the probability of every future VEI. In Figure 8, it can be seen that the biggest probability is VEI 2 and the smallest probability is VEI 4.

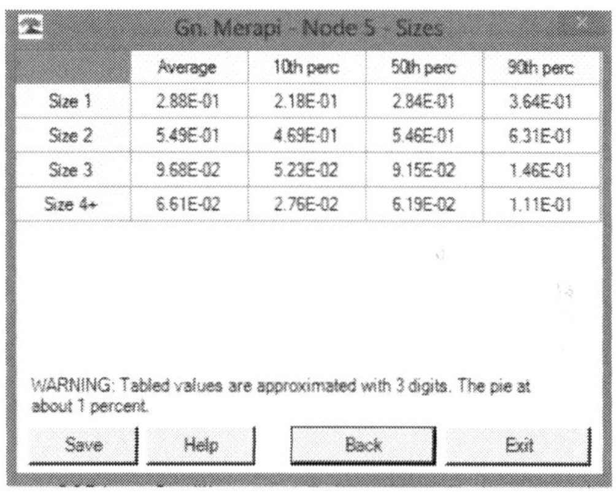
Gn. Merapi - Node 5 - Sizes

| | Average | 10th perc | 50th perc | 90th perc |
|---|---|---|---|---|
| Size 1 | 2.88E-01 | 2.18E-01 | 2.84E-01 | 3.64E-01 |
| Size 2 | 5.49E-01 | 4.69E-01 | 5.46E-01 | 6.31E-01 |
| Size 3 | 9.68E-02 | 5.23E-02 | 9.15E-02 | 1.46E-01 |
| Size 4+ | 6.61E-02 | 2.76E-02 | 6.19E-02 | 1.11E-01 |

WARNING: Tabled values are approximated with 3 digits. The pie at about 1 percent.

Save Help Back Exit

**Fig. 3** The probability of eruption size

Based on the data processing using software BET_EF, it shows a probability with beta distribution for every possible size of future eruption. VEI 1 has probability of 0,288. VEI 2 has probability of 0,549. VEI 3 has probability of 0,097. VEI 4 has probability of 0,066. It can be seen from the output software BET_EF that VEI 2 probability is the highest, so it can be concluded that the level of the next eruption will be the same as VEI 2 level. This is supported by historical data and the characteristics of Merapi Mount that big eruption will occur once every 30-40 years and after that happens, the magma chamber will be emptied and needs time to make another big eruption.

If the next eruption is the same level as VEI 2, it is estimated that Disaster-prone Zone II and some parts of Disaster-prone Zone I will be affected. By seeing the Hazard Map, the areas that will be affected by the eruption are Ngemplak, Tempel, Turi, Pakem, and Cangkringan districts. From these 5 districts, there are 21 villages that are estimated to be affected by the eruption. The choosing of affected areas is also based on the morphology of Merapi Mount, which the eruption is estimated to be going to these 5 districts.

## Conclusion

The result of the data processing shows that Merapi will erupt again in the future with the probability of 0,822. The eruption will be take place in the main crater of Merapi with the probability of 0,938. The next eruption will have the power of VEI 2 with the probability of 0,549.

In accordance with the data processing using software BET_EF that predicts the location of eruption (main crater) with the power VEI 2, the areas that will be affected by the eruption are Disaster-prone Zone I and Disaster-prone Zone II with the radius of 17 Km from the center of the eruption and directed 1600 from the north, which is Sleman. Ngemplak and Tempel are in Disaster-prone Zone I, while Turi, Pakem, and Cangkringan are in Disaster-prone Zone II. So, it can predict the five districts that will be affected by the Merapi eruption.

## Future Work

After getting prediction of Mount Merapi eruption involve the location and size of the next disaster we will design preparedness in humanitarian logistic based on eruption forecasting.

## References

1. A. Cozzolino, *Humanitarian Logistics: Cross-sector Cooperation in Disaster Relief Management, Springer*, Heidelberg (2012)
2. G. O'Brien, P. O'Keefe, Z. Gadema, J. Swords, "Approaching disaster management through social learning", *Disaster Prevention and Management*, **19** 498-508 (2010)
3. W. Marzocchi, L. Sandri, J. Selva, "BET_EF: a probabilistic tool for long- and short-term eruption forecasting", *Bull Volcano*, **80**, 623-632 (2008)
4. A. Garcia-Aristizabal, J. Selva, E. Fujita, "Integration of stochastic models for long-term eruption forecasting into a Bayesian event tree scheme: a basis method to estimate the probability of volcanic unrest", *Bulletin of Volcanology*, **75**, 689 (2013)
5. A. Gelman, J.B. Carlin, H.S. Stern, D. B. Rubin, Bayesian data analysis. CRC, Boca Raton, FL (1995)
6. E. M. Fournier d'Albe, Objectives of volcanic monitoring and prediction. *Journal of the Geological Society London*, **136**, 321–326 (1979)
7. W. Marzocchi, L. Sandri , P. Gasparini, C. Newhall , E. Boschi, Quantifying probabilities of volcanic events: the example of volcanic hazard at Mt. Vesuvius. J Geophys Res 109 (2004)
8. "____",Balai Penyelidikan dan Pengembangan Teknologi Kebencanaan Geologi, 2014

# Factor identification of higher education choice to enhance brand awareness of state university

*Almira* Puan Rachmadhani[1,*], *Naniek* Utami Handayani[1], *Mochammad* Agung Wibowo[2], *Ratna* Purwaningsih[1] and *Hery* Suliantoro[1]

[1]Department of Industrial Engineering, Faculty of Engineering, Diponegoro University, Indonesia
[2]Department of Civil Engineering, Faculty of Engineering, Diponegoro University, Indonesia

**Abstract.** The commercialization of higher education through the concept of *Perguruan Tinggi Negeri Berbadan Hukum* (PTN-BH) has led the higher competition among State University and Private University. It becomes a threat and challenge faced by State University in recruiting new students. A clearer understanding of why and how students choose universities is needed to develop the brand awareness enhancement strategies of State University. This paper aims to identify the factors that influence student's decision of university choice, without regarding the individual aspects. The Deplhi technique is used at the stage of information gathering, in which four experts from higher education were being interviewed. The combination of HEdPERF dimensions and brand awareness indicators were used to develop the variables. The pilot study done through the sample of 1st-year undergraduate students enrolled in the Top 8 State Universities in Indonesia for the validation test using SPSS 16.0. This findings result that there are six factors in choosing higher education, including academics, facilities, campus life, reputation, industry linkage and access. The test and evaluations result that from 28 items proposed, there are 23 items that can be considered as the student's decision in their choice of State University.

## 1 Introduction

Higher education in Indonesia has grown rapidly, which was only demanded by 200 students in the post-World War II, now reaching 4,3 million students that spread over 83 (2.7%) State Universities and 2987 (97.3%) Private Universities. Although the number of Private University in Indonesia is much higher than the State University, the enthusiasts of State University continues to increase every year. In 2015, the number of prospective students that interest in State University increased 5 to 10 times.

Nowadays, the trend of business is starting to change. People are customers who will drive the company, so that the business must be obtained with the provision of understanding the needs and desires of their customers. Higher education is being driven towards commercial competition as the impact of economic forces caused by the development of global education markets and the reduction of government funds that forces them to seek other financial resources [1]. In Indonesia, based on UU No. 12 Tahun 2012, the govenrment has organized a new concept of higher education that called as *Perguruan Tinggi Negeri Berbadan Hukum* (PTN-BH), in which State University has more autonomy at self-regulate, or given an independency at managing three aspects of business, including finance, infrastructure and employment.

The commercialization of higher education has led the higher competition among State and Private University. In the last decade, Private University has proven their ability in providing an educational services that conform to the needs and desires of their customers, both in the aspects of finance and infrastructure. This might become a threat and challenges that must faced by State University. They must be able to keep the loyalty of their customers up and raising awareness of their existence. State University had to increase their brand awareness, the recognition and recall of a brand and its differentiation from other brands in the field [2] that commonly measured by 4 indicators, including recall, recognition, purchase and consumption. A clearer understanding of why and how students choose higher education has to be considered as the basis in developing marketing strategies for State University [3], to make the State University become top of mind in Indonesian society.

Various studies on the impact of brand on higher education have been conducted. One of them prove that the reputation of the university becomes the main indicator which determines the uniqueness of higher education [4]. With the adoption of the concept of branding, reputation becomes more important to develop the identity of higher education institution to improve their reputation in the competitive global environment.

The purpose of this study is to identify and validate the factors that influence the student's decision about their choice of university. However, understanding the process of student decision making in choosing university is complex. Hence, this study restricted to the internal

* Corresponding author: apuanr@gmail.com

factors that could maintained by the university, without regarding the individual aspects.

## 2 Literature review

There are several groups of customers in higher education, such as students, employers, academic staff, government and families, with a diversity of requirements [1]. General view defines that the primary participant in educational services is the student [5]. Students form their perceptions of brand image, identity and meaning before enrolling at university, that continued during their study and even after they were graduated [6]. Also, their ability to recognize the brand of university are varies depending on its brand communication. So that, the existence of university depend on the quality of their students that will equally impacted to brand awareness.

Education is related to knowledge enhancement (teaching resources), co-curricular activities (campus life), and the industrial opportunities [7]. There are three stages in student's decision making of selecting higher education: (i) predisposition, (ii) search and (iii) choice decision [8]. Another researcher defined that higher education need to interact and impress their customer in the different way to build an organization's brand at the three touchpoints of customer experience: (i) pre-admission, (ii) during the course and (iii) post-passing, that represented by every action, tactic or strategy taken to reach their customers [7]. Meanwhile, this research focus on the choice decision of higher education that related to pre-admission touchpoints.

Prestige of an university is sometimes more important that its actual quality, because it represents their perceived excellence as the guidance to the decisions of prospective students to enroll the university [9]. There are six factors affecting customers satisfaction in higher education, that called HEdPERF, namely non-academic aspects, academic aspects, reputation, access, programmes issues and understanding [1]. On the other hand, there are 7 factors of reputation in higher education, including teaching staff, course materials, facilities, administrative support, campus life, industry linkage and reputation [10].

Hence, there are six constructs developed in this research related to the existing literature that can be described as follows:

- Academics, related to the importance of offering wide ranging and reputable of academic programmes and teaching staff.
- Facilities, related to the soft and hard infrastructure that will supports students on their learning process.
- Campus life, related to the co-curricular and extra-curricular activities that the campus provides to develop students' softskill besides the academic activities.
- Reputation, related to the professional image projection of higher education that impacted by the media influence, track record of alumnus and achievement of university at both national and international scope.
- Industry linkage, related to the cooperation of higher education with various national and international industries that will affect the career prospects of their students.
- Access, related to the approachability, ease of contact, availability and convenience given by the higher education.

## 3 Research methods

This research conducted using both qualitative and quantitative measures to identify and validate the factors that influence the student's decision about their choice of higher education. In particular, this study attempts to qualitatively identify the factors of student's decision in choosing higher education, and incorporate the quantitatively validation to a sample of population. A detailed description of the stages described below.

### 3.1. Information gathering

This stage involves an in-depth literature searching to identify factors that influence students in choosing their university. A total 8 carefully selected students from different universities were invited to participate in the focus groups. A set of questions was created by the researcher and participants were asked to describe various aspects related to the choice of university.

### 3.2. Development of research variables

The results of focus group interviews then generated into the draft of questionnaire, consist of 6 constructs and 28 items related to different factors in the higher education decision making. The Delphi technique used as an initial validation and the items were measured on a 2-point scale, agree and disagree, by four experts (academicians and practitioners). They were asked to give an approval and comments concerning the questionnaire.

### 3.3. Pilot testing and validation

The final questionnaire were presented in functional and dysfunctional question and measured on a five-point Likert scale that varied from 1 = strongly disagree to 5 = strongly agree. Data were collected by personal approach from the 1st-year undergraduate students over Top 8 State Universities in Indonesia. A total 50 students have been surveyed, 68% were female in the age of 18-20 years old.

## 4 Results and discussion

The factors were selected based on literature regarding the customer needs and branding of higher education. The results presented in Table 1.

### 4.1. Research variables

Academics are identified as the teaching and learning process. Reputation of the teaching staff (A1) is the educational backgrounds and achievement of the lecturers in various research and community services. Diversity of the majors (A2) are the varied and specalized majors provided at the university. Accreditation of the majors

(A3) are the is the formal recognition from accreditation entity against the majors competency and quality. Uniqueness of the majors (A4) are defined as the favorite and desirable majors from year to year.

**Table 1.** Research variables.

| Constructs | Items | Code | Source |
|---|---|---|---|
| Academics | Reputation of the teaching staff | A1 | [3], [7], [10] |
| | Diversity of the majors | A2 | |
| | Accreditation of the majors | A3 | |
| | Uniqueness of the majors | A4 | |
| Facilities | Academic facilities | F1 | [12] |
| | Health centre | F2 | |
| | Sports facilities | F3 | |
| | Lifestyle facilitators | F4 | |
| | Students dormitory | F5 | |
| Campus life | Students union | C1 | [10], [12] |
| | Campus events | C2 | |
| | Co-curricular activities | C3 | |
| | Scholarship programmes | C4 | |
| | Internship programmes | C5 | |
| Reputation | Reputation within the community | R1 | [7], [10] |
| | Accreditation of the university | R2 | |
| | National rankings of the university | R3 | |
| | International rankings of the university | R4 | |
| | Prestige of the university | R5 | |
| | Information about university in mass media | R6 | |
| | Alumni recommendation | R7 | |
| | Research quality | R8 | |
| Industry linkage | National industry partners | I1 | [10] |
| | International industry partners | I2 | |
| | Career prospects | I3 | |
| Access | Location of the university | S1 | [7], [12] |
| | Fees of the program | S2 | |
| | Transportation links | S3 | |

Facilities are related to the infrastructure provided by the university. Academic facilities (F1) are the facilities that supports the learning process, such as classrooms, libraries and laboratories. Health centre (F2) is the medical clinics provided for students. Sports facilities (F3) are the facilities that supports sports activities, such as field, sports equipment and swimming pool. Lifestyle facilitators (F4) are defined as the facilities that supports the basic needs of students, such as ATM centre and print shop. While, students dormitory (F5) is the house provided for the 1-st year or underprivileged students.

Campus life is considered as another activities provided besides the academic activities. Students union (C1) is the college organizations that accomodated the students' talents, interests and their potency undertaken in extra-curricular activities. Campus events (C2) is defined as the top events held by university as their icon and promotion to the public society. Co-curricular activities (C3) or known as *Unit Kegiatan Mahasiswa* (UKM) is identified as student organizations at sports and arts. Scholarship programmes (C4) are the number of financial aid given to students as the appreciation for achievement at their study. Also, internship programmes (C5) are the chance of students on experiencing as the professional worker at various companies for the short-term period.

Reputation is described the higher educational image. Reputation of university (R1) is identified as the image of university that has been formed in Indonesian society. Accreditation of university (R2) is the formal recognition from accreditation entity against the universities competency and quality. National ranking (R3) and international ranking (R4) are the achievement of university assessed in both national and international scope. Prestige of the university (R5) has the similar meaning with R1. Information in mass media (R6) is identified as the existence of university on television, newspaper, magazines or another online mass media. While, alumni recommendation (R7) is the ability of alumni to compete in builing their career path and research quality (R8) is the number of indexed research that conducted by the university.

Industry linkage is decribes the career opportunity of students. National industry partners (I1) and international industry partners (I2) are the number of stakeholders from industries or companies owned by the university. Career prospects (I3) are the amount of opportunity for students to get jobs after they finish their study.

Access is related to the approachability of university. Location of the university (S1) is defined as how strategic the university located geographically. Fees of the program

(S2) is the the cost that have to paid by students to finish their study at the university. Transportation links (S3) is defined as the ease of getting public transportation to go to the university.

**Table 2.** Approval score of constructs.

| Constructs | Approval score |
|---|---|
| Academics | 1.00 |
| Facilities | 1.00 |
| Campus life | 1.00 |
| Reputation | 1.00 |
| Industry linkage | 1.00 |
| Access | 1.00 |

**Table 3.** Approval score of items.

| Items | Approval score | Items | Approval score |
|---|---|---|---|
| A1 | 1.00 | R1 | 1.00 |
| A2 | 1.00 | R2 | 1.00 |
| A3 | 1.00 | R3 | 1.00 |
| A4 | 1.00 | R4 | 0.25 |
| F1 | 1.00 | R5 | 0.25 |
| F2 | 1.00 | R6 | 1.00 |
| F3 | 1.00 | R7 | 1.00 |
| F4 | 1.00 | R8 | 0.00 |
| F5 | 1.00 | I1 | 1.00 |
| C1 | 1.00 | I2 | 0.25 |
| C2 | 1.00 | I3 | 1.00 |
| C3 | 1.00 | S1 | 1.00 |
| C4 | 1.00 | S2 | 1.00 |
| C5 | 0.25 | S3 | 1.00 |

## 4.2. Content validity

Content validity shows how representative and comprehensive the variables were. Table 2 and 3 presents the approval scores for 28 items being proposed. Variables whose approval scores were lower than 0.5 will be dropped from further analysis [11]. It can be inferred that there are five items dropped with approval scores 0.25 (C5), 0.25 (R4), 0.25 (R5), 0.00 (R8) and 0.25 (I2).

The experts argued that students won't regard that variables at the time they choosing university. The internship programmes (C5) and research quality (R8) were dropped because that variables were the part of during course touchpoints, it means that the students will realize it by the time they studied there. International ranking of university (R4) was dropped because it only would be regarded if the students want to continue their

**Table 4.** Validity and reliability statistics.

| Cronbach's α | Number of items | Items | Total Correlation |
|---|---|---|---|
| 0.758 | 4 | A1 | 0.553 |
| | | A2 | 0.534 |
| | | A3 | 0.604 |
| | | A4 | 0.557 |
| 0.874 | 5 | F1 | 0.698 |
| | | F2 | 0.869 |
| | | F3 | 0.782 |
| | | F4 | 0.646 |
| | | F5 | 0.572 |
| 0.835 | 4 | C1 | 0.662 |
| | | C2 | 0.680 |
| | | C3 | 0.711 |
| | | C4 | 0.613 |
| 0.916 | 5 | R1 | 0.840 |
| | | R2 | 0.835 |
| | | R3 | 0.769 |
| | | R6 | 0.691 |
| | | R7 | 0.860 |
| 0.693 | 2 | I1 | 0.549 |
| | | I3 | 0.549 |
| 0.808 | 3 | S1 | 0.747 |
| | | S2 | 0.577 |
| | | S3 | 0.664 |

study at an International University, while prestige of university (R5) was dropped because it has the same meaning with variable R1. Also, the international industry partners (I2) was dropped regarding the high competition among national industry in Indonesia.

### 4.3. Construct validity and reliability

Validity of the items depends on the sample size [13]. For a sample size of 50, a corrected item-total correlation more than 0.279 is considered valid. While the reliability figured from the Cronbach's αwith the lowest acceptable limit is 0.6 [14]. Table 4 shows the results of validity and reliability test using SPSS 16.0.

Total correlation for all items were greater than 0.5, indicating good discriminant validity to describe the real situation of this research.

The two items of industry linkage have a moderate reliability with Cronbach's α 0.693. The 21 items of academics, facilities, campus life, reputation and access have a high reliability with Cronbach's α 0.758, 0.874, 0.835, 0.916 and 0.808. It shows that the respondent's answers were reliable to uncover the research objective.

## 5 Conclusion

This study identifies the factors influencing the student's decision to continue their study at higher education. It shown that there are 6 factors that affecting the choice of university, including academics, facilities, campus life, reputation, industry linkage and access. From 28 items proposed, there are five items dropped from further analysis and resulting 23 significant items that can be considered as important aspects of choosing universities.

This findings shows that university experience is now concerned by students besides the skills and ability of their graduates. It provides a guidance for State University in developing their marketing strategies.

This research was partially funded by Diponegoro University in "Penelitian Dasar Department of Industrial Engineering" Program.

## References

1. A. Firdaus, Int. J. Cons. Stud. **30**, 6, 569-581 (2006)
2. R. East, Prentice Hall. (1997)
3. S. Munisamy, N. I. M. Jaafar, S. Nagaraj, Asia-Pac. Educ. Res. **23**, 3, 451-462 (2014)
4. J. H. Brown, T. Melewar, B. Nguyen, E. Wilson, J. Bus. Res. **69**, 8, 3019-3022 (2016)
5. L. Galloway, TQM Mag. **10**, 1, 20-26 (1998)
6. C. Dennis, S. Papagiannidis, E. Alamanos, M. Bourlakis, J. Bus. Res. **69**, 8, 3049-3057 (2016)
7. M. Khanna, I. Jacob, N. Yadav, J. Mark. High. Educ. **24**, 1, 122-143 (2014)
8. D. Hossler, K.S. Gallagher, Coll. Univ. **62**, 3, 207-221 (1987)
9. T. Gatfield, M. Barker, P. Graham, Corp. Comm. Int. J. **4**, 2, 73-79 (1999)
10. C. Plewa, J. Ho, J. Conduit, I. O. Karpen, J. Bus. Res. **69**, 8, 3087-3095 (2016)
11. K. Williamson, Chandos Publishing. (2002)
12. T. Woodall, A. Hiller, S. Resnick, Stud. High. Educ. **39**, 1, 48-67 (2014)
13. A. Field, SAGE Publications. (2009)
14. J. F. Hair, R. E. Anderson, R. L. Tatham, W. C. Black, Prentice Hall. (1998)

https://doi.org/10.1051/matecconf/201815401052

# A case study of supply chain simulation for determining the best stock allocation

Puryani, *Yuli Dwi* Astanti*, and *Gity Nur* Fitriani

Department of Industrial Engineering, Pembangunan Nasional "Veteran" University, Yogyakarta, Indonesia

**Abstract.** You *Dare* is a Small and Medium Enterprise (SME) engaged in fashion, namely the sale of shoes. The problems that occurred in *Dare* are their limit on the amount of the shoes supply from suppliers and the inventory allocation policies are unplanned, so the profit is not maximized. This study simulates inventory allocation policy that *Dare* applies to see the impact on profits of *Dare* then proceed with the design of new system scenarios. The new scenario aims to implement policies more appropriately to increase the profits of *Dare*. Simulations are performed with ProModel software.The results of this study indicate that the policy of only using intuition cannot maximize the profits. An appropriate policy to increase profits of *Dare* is to consider the demand and cost allocation. Based on simulation results, *Dare* will only have an average profit of IDR 52.187.000, whereas if *Dare* uses the proposed policy, the average profit will increase to IDR 55.948.000 within two years.

## 1 Introduction

Inventory inaccuracy is a main issue in businesses dealing with goods supply chain. The allocation policy of inventory is a significant concern for the smooth flow of goods. Less inventory will have an impact on loss of sales, while excess inventory will be detrimental due to costs incurred. An appropriate allocation policy analysis is required for each retail supply in the supply chain. Inventory problems not only occur in large companies, but also in small and medium industries. One of them is *Dare*. The aim of this paper is to examine the best stock allocation from supplier to their retail with different cooperation system in *Dare*.

*Dare* is one of the fashion industries that sell products such as snikers shoes. *Dare* is currently unable to meet the demand for shoes in real time. This is shown by the inventory shortages that occurred in retail of *Dare* during 2015. Based on the historical data, *Dare* in 2015 discovered that the average of total inventory shortage reaches 133 requests per month. The inability of *Dare* to meet the market demand is caused by two factors, firstly the limit on the amount of supply from suppliers, i.e. a maximum of 500 shoes per month, and secondly the inventory allocation policy from *Dare* factory to the retail shop is not well planned. The allocation policy is not well planned because it does not consider the system of cooperation established with retail. In the supply chain system, one of *Dare's* suppliers is located in Bandung. The supplier will send the shoes by an amount that *Dare* determines, but it should not be more than 500 shoes per month. After that, *Dare* will sell the shoes through the online store and through cooperation with nine distributions that are willing to become a retail of *Dare*. These distributions are located in Java and Bali.

There are two systems of cooperation between retail and *Dare*. The buying system is bifurcated. Firstly, the direct buying system. *Dare* would provide special pricing on retail and the retail of *Dare* is willing to buy shoes for resale. Shipping costs will be paid by the retailer. Secondly, the consignment system. In this system, the retailer does not buy shoes from *Dare*, but *Dare* leaves shoes at the retail. *Dare* had to pay the shipping costs and the cost of the consignment. Consignment fee is the remuneration given to the retail which *Dare* deposited on each shoe. There are six retails using consignment cooperation system and there are only three retails using direct buying system. Each retailer has a stochastic demand. The data in the data request in 2015 showed that retail consignment had a higher level of demand.

The allocation policy to the retail inventory implemented by *Dare* currently only relies on the intuition of the owners without considering both systems. Under the policy, *Dare* will lose whenever there is a remaining inventory at retail consignment arrangement and when there is a shortage of inventory in retail using direct buying system. This is because *Dare* already spent more allocation to retail by the consignment system. If the loss is not prevented, it will have an impact on *Dare's* inability to survive in the competitive business of fashion. Thus the main objective of conducting this research is to get the inventory allocation policy to increase the revenue of *Dare* orders and not to experience losses.

The approach will be used to find a product allocation policy of *Dare* which is a simulation method. Simulation is a clone of the system using a computer model to evaluate and improve system performance [1]. The

* Corresponding author: yuli.upnyk@yahoo.com

simulation method is chosen because it can see the behavior of the real system as a whole, have a high flexibility, and provide representative results.

In supply chain context, simulation can undoubtedly play an important role, above all for its main property to provide what-if analysis and to evaluate quantitatively bene fi ts and issues deriving from operating in a co-operative environment rather than playing a pure transaction role with the upstream / downstream tiers [2]. Siprelle et al. [3] performs a simulation to find the best strategy in terms of supply allocation at the distribution center to avoid cross-shipment of product. Another research is conduct by Arreola et al. [4]. This research consider a two-stage decision problem, in which an online retailer first makes optimal decisions on his profit margin and free-shipping threshold, and then determines his inventory level. Arreola et al. [4] use Arena to simulate the operation of the online retailer, and then use OptQuest to find the optimal solutions.

The software that is employed in this paper is ProModel. ProModel is a simulation and animation tool designed to quickly yet accurately model manufacturing systems of all types, particularly supply chain systems [5]. In this study, the model will be designed by adjusting the existing policy system at *Dare*. The results of the development of the model will be used to create inventory allocation policy scenarios that can increase revenue of *Dare*.

## 2 *Dare* System

*Dare* sells its shoes through its online store and by collaborating with nine distros (distribution outlets) who are willing to be retail. For sales through online stores, all activities are carried out at the office of *Dare*. *Dare* provides wages to 10% of the shoes selling price for its employees who concurrently administer the online store. Shipping costs will be borne by the buyer. The retail demand of any nature is uncertain or stochastic. The system of cooperation is the direct buying system and the consignment system.

1. Direct buying system. The Purchasing system is bifurcated into a cooperation established between *Dare* and retailing in which the retailer is willing to pay the shipping costs and buy shoes from *Dare* for resale. In addition, *Dare* is also willing to give 20% discount of the shoes selling price. This co-operation system is more advantageous for *Dare* compared with consignment system because *Dare* will not lose any money if these shoes are not sold. However, there are only three retails willing to use this cooperation system. Retail usually comes from a small town and has not had loyal customers.
2. Consignment system. The Consignment system is a system of mutually beneficial cooperation between *Dare* and retailing in which the retailer acts as a place to leave shoes to *Dare*. To entrust the shoes to retail, *Dare* charges a consignment of 15% -20% of the selling price of these shoes. The gain derived by *Dare* is able to sell its products on a distro that has loyal consumers. Consumers at the retail consignment are usually consumers of the community, such as the skateboard community, band, BMX bikes, etc. Thus *Dare* shoes can be more quickly recognized and can draw a definite interest. The drawback of this system is *Dare* consignment must pay a fee for every shoe that is deposited and has to pay shipping costs. Thus, if the deposited shoes are unsold, then *Dare* will lose the shipping costs and the cost of the consignment.

At the end of every month *Dare* will order the shoes to the supplier. The number of orders is determined by *Dare*, but the supplier limits the number of shoes that are ordered. It should not exceed 500 pairs per month. The shoes are ordered on the 30$^{th}$ every month and will come one month later on the 1$^{st}$ of the following month. So, if *Dare* orders on January 30$^{th}$, then shoes will come on March 1$^{st}$. Every beginning of the month, *Dare* receives the supply of shoes. The shoes are then repackaged using a cardboard shoe from *Dare* and carried out allotment allocation. The time required for packing and rationing the allocation of shoes is one day. Once packaged, the shoes are then allocated to every *Dare* retail. *Dare* has nine retails and one online store. *Dare* Retail is located in various regions of Java Island and one retail in Bali Island, while the online shop is operated in *Dare* office. The list of retail and its location can be seen in Table 1. The amount of allocation for each retail is determined based on the policies and intuition of the *Dare* owner. After receiving the allocation of shoes from *Dare*, every retail will make sales until the end of the month. Every 30$^{th}$, the retail will send a sales report containing the number of shoes that have been sold, the remaining shoes, and the number of unmet requests. This report of each retail will be used by *Dare* to place orders to suppliers and determine the amount of inventory allocation for each retail in the next month.

**Table 1**. The list of retail and the location

| No | Retail Name | Locations | *Lead Time (day)* | System |
|---|---|---|---|---|
| 1. | Electrohell | Jogja | 1 | Consig-ment |
| 2. | Koffin | Jogja | 1 | |
| 3. | Rumbble | Jogja | 1 | |
| 4. | Koffin Jgj | Semarang | 3 | |
| 5. | Dominion | Surabaya | 3 | |
| 6. | Rumbble Bl | Bali | 3 | |
| 7. | Blind | Tuban | 3 | Direct Buying |
| 8. | Maxx | Jember | 3 | |
| 9. | Pride | Banyuwangi | 3 | |

*Dare* has nine retails and one online store where each retail has a varying demand level. There are some data used in this study, both primary data and secondary data. Primary data in this study are drawn to complete the secondary data through interviews with the owners of *Dare*. Secondary data are taken from historical data of the company during the period from January 2015 to April 2016. The secondary data are as follows; consumer demand of each retail, product allocation of *Dare*, the shortage that occurred in each retail, supply of shoes from suppliers, the retail characteristic, and the data of shipping and consignment.

## 3 Research Method

Simulation is an imitation of a dynamic system using computer models to evaluate and improve system performance [1]. Simulation is one of the most widely used operations research and management methods. Simulation is used as a mean of numerical analysis of the models to see which inputs that affect the measurement of output on the system performance. Simulation is not an optimization tool that gives a decision but it is the result of a decision support tool, thus the interpretation of the results depends on the purpose of the decision maker. ProModel is one of the windows-based simulation softwares that can be used to model and analyze a system of various types and sizes. ProModel provides a good combination of usage, flexibility, and ability to model a real system. ProModel is equipped with animation feature, so that the model looks more realistic. ProModel is closely related to resource utilization, optimization of production capacity, productivity, production scheduling, and inventory levels. Development of the real system model simulation is constructed with the following steps :

### 3.1 Fitting Distribution

Fitting distribution is applied to determine the pattern of the data. Determination of distribution pattern for consumer demand data to each retail is done by Stat :: Fit in ProModel. By using Stat :: Fit, there will be some theoretical distribution corresponding to the data in the real system. The used theoretical distribution is a discrete distribution.

### 3.2 Calculate the percentage of shoe supply arrival from supplier

To determine the product supply data from supplier to *Dare*, this research will employ user distribution menu that is also found in ProModel. User Distribution was chosen because the number of shoe supplies are known, i.e. among 400, 450, or 500 shoes. For this purpose, the user distribution is used to determine the percentage of supply values emerging during the simulation process. The percentage is determined based on the company historical data.

### 3.3 Calculating the proportion of allocation of products of the real system

In the real system, the number of products allocation from *Dare* to each retail is determined based on the owner intuition. On the construction of a real system model simulation, the allocation amount is entered in the form of proportions. To determine the proportion, the test is carried out by summing the number of allocations of each retail for 16 months.

### 3.4 Development of real system simulation

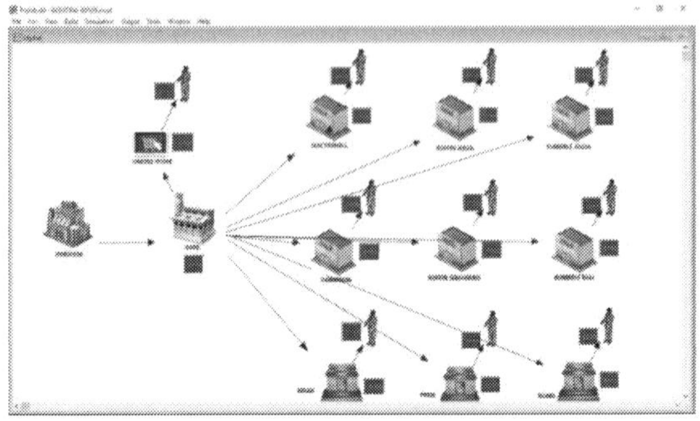

**Fig. 1.** The illustration of real simulation system

Development of real models is constructed using ProModel. Data inputs that used in the construction of the model are distribution data that are requested, the data distribution of the arrival of the supply of shoes, and the proportion of allocation. The construction of the real system simulation, at least needs some components in ProModel, those are Location, Entities, Variable, Arrivals and Processing. Location is where the entity of the system will be processed. Entities are components that are processed in the system. Variables are used to record the movement of entities. Arrivals is a coding process that defines the entity arrival time in the system. Processing is a coding process used to describe commands. The real system simulation using ProModel is illustrated in Figure 1.

Figure 1 is depiction of shoes that come from supplier to *Dare*. After that, shoes will wait for one day to be allocated to the retails. The proportion of allocations is adjusted to the current *Dare* policy. Retail will receive shoes at the beginning of the month, then at the end of the month retail will receive the orders from consumers. Retail will send shoes supplies to consumers. All delivery processes are recorded by variables in the simulation. Variables are indicated with blue boxes. From these variables can be seen whether the retail experiences a shortage of inventory or excesses inventory.

### 3.5 Verify the simulation model

Verification is a process to ensure whether the model is correct. The verification process identifies the errors in the simulation logic and the input data. The first step of the verification process is checking the code of the model. The model codes check is used to check whether coding entered into the system simulation is in conformity with the logic in the conceptual model. Secondly, checking output. An output check is performed to see if the simulation results are reasonable. Examination is done by comparing the result of simulation running with original data. This check only ensures that the simulation results make sense. The last is checking the animation. Animation checks are performed to determine whether the entity flow is in conformity with the conceptual model.

### 3.6 Validation of the model

Validation is implemented to determine whether the simulation model that has been made already represents the real system. Validation is comparing simulation result with the real system output. Law and Kelton [6] describe validation as illustrated in Figure 2.

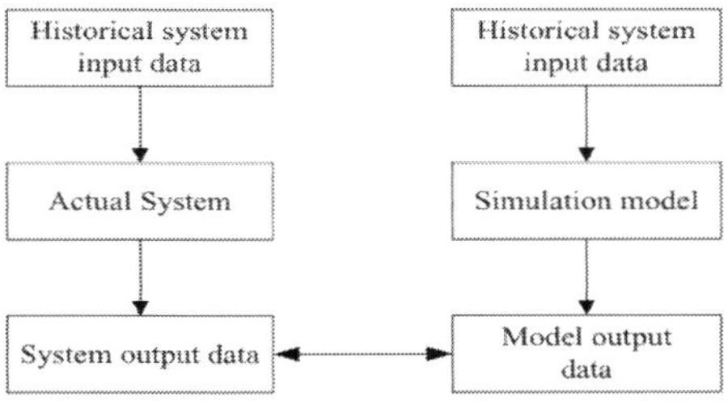


**Fig. 2.** Validation of Simulation [6]

The entity to be tested is the number of demand and the amount of inventory shortage. The number of demand represents the data used as input, while the amount of inventory shortage represents the simulation output. Prior to the validation, there are data normality test and homogeneity test. Normaly distributed data validation is done by using an independent t-test, while another distributed data validation is performed by the Mann-Whitney test. The confident level that is used is 95%. Validation test results can be seen in Table 2.

**Table 2.** Validity test of demand and shortage data

| No | Retail | Statistic Test Method | | | | Result |
|---|---|---|---|---|---|---|
| | | Demand | P-value | Shortage | P-value | |
| 1 | Online | *Ind. T test* | 0,928 | *Ind. T-test* | 0,279 | Valid |
| 2 | Electrohell | *Ind. T test* | 0,546 | *MW Test* | 0,056 | Valid |
| 3 | Kofiin Jgj | *Ind. T test* | 0,989 | *Ind. T-test* | 0,254 | Valid |
| 4 | Rumbble | *Ind. T test* | 0,255 | *MW Test* | 0,090 | Valid |
| 5 | Koffin Smg | *Ind. T test* | 0,391 | *MW Test* | 0,509 | Valid |
| 6 | Dominion | *Ind. T test* | 0,980 | *Ind. T-test* | 0,253 | Valid |
| 7 | Rumbble Bl | *Ind. T test* | 0,071 | *Ind. T-test* | 0,820 | Valid |
| 8 | Maxx | *Ind. T test* | 0,236 | *Ind. T-test* | 0,911 | Valid |
| 9 | Pride | *Ind. T test* | 0,385 | *MW Test* | 0,776 | Valid |
| 10 | Blind | *Ind. T test* | 0,913 | *Ind. T-test* | 0,826 | Valid |

### 3.7 Simulation result

The simulation runs with 24 times replication. The simulation result shows the average of the shoes allocated to each retail, the number of demand, and the number of shortages that occur in each retail. The result of the real system simulation can be seen in Figure 3. Figure 3 shows there are three boxes, i.e A, B, and C. These three boxes are the result of temporary records (variables) that have been made during the simulation. Box A shows the number of shoe allocations to each retail. Box B shows the number of consumer demand to each retail. Box C denotes the end condition of the demand. If the value in box C is positive, there is a shortage of inventory. If the value in box C is negative, then the remaining inventory is at retail accounts (absolute value).

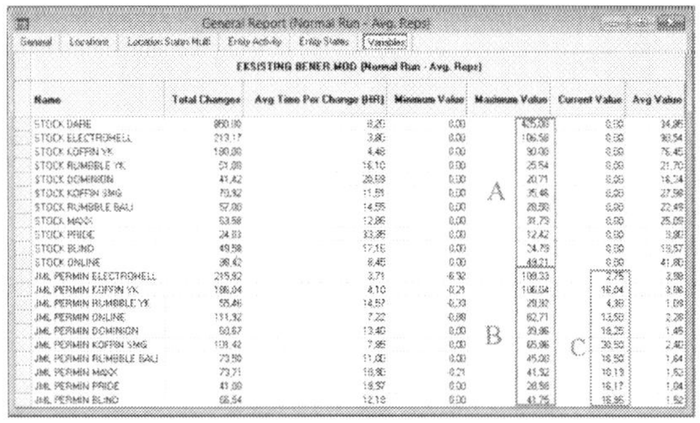

**Fig. 3.** Real system simulation output

In Figure 3, it can be seen that the average supply of shoes to *Dare* is 425 shoes per month. *Dare* policy calculation results can be seen in Table 3.

**Table 3.** Real system simulation result (Average)

| No | Retail | Allocation | Demand | *shortage* | Sold |
|---|---|---|---|---|---|
| 1 | Online | 49 | 63 | 14 | 49 |
| 2 | Electrohell | 107 | 109 | 2 | 107 |
| 3 | Koffin jgj | 90 | 106 | 16 | 90 |
| 4 | Rumbble | 26 | 30 | 4 | 26 |
| 5 | Koffin | 35 | 66 | 31 | 35 |
| 6 | Dominion | 21 | 40 | 19 | 21 |
| 7 | Rumbble Bl | 29 | 45 | 16 | 29 |
| 8 | Maxx | 32 | 42 | 10 | 32 |
| 9 | Blind | 25 | 42 | 17 | 25 |
| 10 | Pride | 12 | 29 | 17 | 12 |

From the ProModel simulation output results, the calculation is carried out as follows:

a. Numbers of sold shoes : if the allocation average < demand average, the average sold is equal to the allocation average. If not, then the sold average is equal to the demand average.
b. Average of cost allocation : (allocation average x shipping costs) + (allocation average + cost cooperation).
c. Revenue average: (selling average x IDR 280.000) - cost allocation average.
d. The average profit : The average amount of revenue - Cost of Procurement
e. Procurement costs : average number of supply shoes x IDR 120,000, -

## 4 Development of Scenarios

Meeting the optimal policy of the stock allocation simulation can help *Dare* improve the profit. There are 4 (four) scenarios generated in the simulation model.

**Scenario 1** (Applying the old policy, i.e the number of supply is 500 shoes per month). On the real system, *Dare* experiences a lot of lack in each retail inventory. This happens because during 2015 the supply of shoes from suppliers ranged only around 400 to 450 shoes per month. Thus, incoming demand can not be fulfilled. In 2016, *Dare* got more quotas from suppliers of 500 pairs of shoes per month. This first scenario will simulate conditions from *Dare* in 2016.

**Scenario 2** (Prioritizing retail in high demand). In this second scenario, the policy will be amended to prioritize the allocation of retail that has a high demand. The demand average used in the preparation of this scenario refers to the results of running the real system simulation. The demand average is then sorted from largest to smallest. Once it is sorted, it is also proceeded with the determination of the allocation proportion to each retail.

**Scenario 3** (Prioritizing low retail cost allocation). This scenario will prioritize retail that has low cost allocation. This scenario is created with the aim to raise the income of *Dare*. Compiling this scenario requires the calculation of the cost allocation for each shoe retail. After obtaining the retailers are prioritized to receive an allocation, the next is to determine what proportion of the allocation is given to each retail. Calculations are based on the proportion of the consumer demand average to each retail.

**Scenario 4** (Weighting). This scenario will weight to see the level of demand and cost allocation. Given these two aspects are equally important, the proportion of demand interest rate compared to the cost allocation is the same, namely 50:50. With these proportions, weights can be calculated for each retail. Retail discount level of high demand and low cost allocation will have greater weight. Inventory allocation will be based on weight of the gain. Retail has the highest weight, then the demand will be prioritized to be met first. The simulation of four scenarios above can be seen in Table 4.

**Table 4**. The Results of the scenarios simulation

| Scenario | Description | Cost Allocation (IDR) | Profits (IDR) |
|---|---|---|---|
| Existing | Simulating real condition | 20.137.500,- | 48.022.500,- |
| 1 | Policy of Dare for 2016 | 23.613.000,- | 52.187.000,- |
| 2 | High demand retail | 23.175.000,- | 54.865.000,- |
| 3 | Low cost retail | 22.509.000,- | 54.131.000,- |
| 4 | Average demand and costs of cooperation | 22.652.000,- | 55.948.000,- |

## 5 Conclusions

Based on data processing and analysis that have been done in this study, it can be seen that policies introduced by *Dare* are currently not effective to increase profits of Dare. From the simulation scenarios, all of the policies applied in scenario 1 to 4 can increase profits for *Dare*. The best scenario is scenario 4 because it mostly increases the profits for *Dare*. There are eight retails that get shoe inventory allocation, they are Online, Koffin Jogja, Electrohell, Koffin Semarang, Bali Rummble, Pride, Maxx, and Blind.

## 6 Further Research

Further research should be simulated by considering the addition of supplier to see if it can increase profits for *Dare*. Simulations should be performed to calculate how many shoes exactly to book *Dare* for suppliers, It can be done by changing the proportion of development scenarios corresponding to weighting another viewpoint. The consumer behavior needs to be considered in each retail when there is the closure of retail. For example, if a retail Rumbble Jogja closed, whether it will affect the distribution of demand in the Electrohell, Koffin Jogja, and Online Retails.

## References

1. C. Harrel, B. K Ghosh, R. Bowden, *Simulation Using ProModel.* (2004).
2. S. Terzi, S. Cavalieri, Sim. in The SC Context: a Survey. *Computers In Industry, Vol. 53*, 3-16. (2004).
3. J. A. Siprelle, J. D. Parsons, J. R. Clark, Benefits of Using a SC Sim. Tool to Study Inv. All. *Winter Simulation Conference*. 238-245 (2003).
4. B. R Arreola, M. Leng, M. Parlar. OL ret. prom. pric, free-ship. thres., and inv. dec. *European Journal of Operational Research, Volume 230*, 272-283 (2013).
5. D. Benson. Sim. Mod. & Opt. Using Prom. *Winter Simulation Conference*, pp. 587-593 (1997)
6. 6. M. A. Law, W.D Kelton. *Sim. Mod. An.* Singapore: McGraw-Hill Book Co (2000)

# Urban logistics profile – Yogyakarta city, Indonesia

*Anna* Maria Sri Asih, *Bertha* Maya Sopha, and *Siti* Mardiyah

Industrial Engineering Programme, Department of Mechanical and Industrial Engineering

**Abstract.** In recent years, more rural areas are becoming urban areas accompanied by escalation of logistics activities. Unlike passenger transport, the planning, policy and control of freight transport in developing countries have not been integrated into a reliable and efficient logistics system. Therefore, generating city logistics profile is necessary in order to support the planning of urban logistics system. This study aims to establish a logistics profile of Yogyakarta city, Indonesia, by dividing urban zones into several homogeneous groups, judging from several aspects, including city area features, product characteristics and agents/delivery profile. Logistics profile variables were calculated based on administrative boundaries, resulting in 45 areas to be investigated. Profiles were matched in groups of homogeneous stores (A), large commercial stores (C) and residential areas with local trade (D),with one overlapping profile, i.e. profile A and profile D, in some locations in the middle of the city.

## 1 Introduction

In recent years, there has been a shift of population concentration in rural areas to the city, especially in developing countries. This phenomenon has caused escalation of logistics activities to meet the needs of industrial sector as well as the household sector. This may allow for negative impacts if those activities are not handled properly, i.e. congestion, noise, decreased air quality due to vehicle emissions (Quak, 2008). In contrast to passenger transport, generally the planning, policy and control of freight transport have not been yet integrated in a reliable and efficient logistics system. Therefore, in order to support the planning of urban logistics system, logistics profile maps is necessary to be developed.

Logistics profile is a concept introduced by Macario *et al.* (2008). This concept is to divide urban zones in several homogeneous groups, judged from several criteria including city area features, product characteristics, and agents / deliveries profiles. These three criteria are then called logistic profile elements. This concept was used to establish logistical profiles in Lisbon (Alho & de Abreu e Silva, 2014). The result of this research is the formation of four profiles and from these results, four types of logistics regulations can be made. The enactment of logistics regulations may have different effects in regions with different characteristics (Anderson *et al.*, 2010). By using a logistics profile, logistics problems can be anticipated by adjusting logistics regulations based on the profiles developed. One example of its application is on TURBLOG (Transferability of urban logistics) project (2011). In this project, five logistic profiles were created which were then matched to the conditions of several different cities. The cities in the TURBLOG project are Paris (France), Utrecht (Netherlands), Belo Horizonte (Brazil), Santiago de Chile (Chile), Tokyo (Japan), Beijing (China), New York (USA), and Mumbai (India).

Results of TURBLOG project showed that an area in Mumbai, India, was categorized in profiles requiring special conditions on the distribution process, e.g. for perishable or frozen foods and solid distribution path conditions at certain hours. Mumbah Dabbawalas Operation System implemented an effective logistics model under these conditions, i.e. door-to-door delivery using public transport and non-pollutant vehicles or bicycles. In addition to Mumbai, in Belo Horizonte, Brazil, the Business-to-Consumer concept was used to deliver goods from producers directly to consumers. Other areas with different profiles, for example areas with many retail stores such as the Monoprix's supermarket in Paris, the policy of using the Monoprix rail train was effectively used. Other policies under different profile conditions may also be applied such as vehicle restrictions, time window applicability, and logistics routes as applied in Utrecht for the majority of logistics profiles adjacent to residential areas.

Logistics profile created in Alho and de Abreu e Silva (2014) used combination of logistic profile elements similar to those used in the TURBLOG project but with some adjustments to the condition of the research object i.e. city of Lisbon, Portugal. An example of adjustment is that the restriction variable in the city area features elements is ignored because this variable did not have too much effect on the logistics profile. In addition to restriction variables, logistics accessibility variables are

*Corresponding author: amsriasih@ugm.ac.id

also adjusted due to unavailability of data on the level of congestion so that data is replaced with available loading/unloading bay data in Lisbon city. In addition to the adjustment of several variables, another thing that distinguished this research (Alho and de Abreu e Silva, 2014) with the TURBLOG project is the use of quantitative methods to determine the value of the frequency of delivery variable, which is one of the agents/deliveries profile elements. Unlike other logistic profile variables, this variable requires another approach for the calculation process, which can be done by computing the freight trip generation or the number of trips/trips resulting from the logistics activities. The calculation of freight trip generation can be carried out in different ways, both qualitative and quantitative and using different data according to city conditions and availability of data (Alho and de Abreu e Silva, 2014).

The combination of elements in the logistics profile can also be tailored to the purpose of making it more flexible in combination with other urban logistics cases. This paper aims to use this concept to built logistics profile in Yogyakarta city, Indonesia. In the Special Region of Yogyakarta, the development of the trade sector in 2009-2014 is generally experiencing positive developments. Trading business in DIY is dominated by micro and small entrepreneurs and contributed by availability of trading facilities such as modern stores and traditional markets. The number of modern stores until 2014 amounted to 15% more than traditional markets. Increasing the number of modern stores becomes a rising concern in managing logistics problems, e.g. congestions and negative effect on environment and social life. Thus, logistics policy is required in order to anticipate them. This policy includes many considerations based on characteristics owned by each area in the city, which can be seen if logistics profile map is created.

**Figure 1.** Yogyakarta city's administrative boundaries

## 2 Methodology

Yogyakarta city, as a studied case, has total area of 32.5 $km^2$ (1.02% of the total area of Yogyakarta Special Province). Administratively, the city of Yogyakarta consists of 45 sub-districts. This city is divided into fishnet based on those sub-districts boundary (see Figure 1). Information used as the object of this study is data of stores registered in the city of Yogyakarta, with the total number of 4190 stores. Figure 2 is the result of the entire store map, shown alongside with the boundaries.

This paper combine logistic profile concept from Macario *et al.* (2008) and TURBLOG concept. Table 1 shows the elements of logistic profile from Macario *et al.* (2008), adjusted to the conditions of Yogyakarta city. Implementation of this concept to logistics practices, i.e. how is the logistics profile for specific area or land uses, was combined with logistics profile developed by TURBLOG (2011), which is summarized as follows:

- *Groups of homogeneous stores, which sellthe same type of product or services*
- *Hotels, restaurants*
- *City Business District*
- *Large shopping centers*
- *Residential areas*

**Figure 2.** Distribution of stores' location within Yogyakarta city, represented by black dots

In the process of developing Yogyakarta's logistics profile, B profile is not used because of many constraints in collecting data of this profile. Adjusted logistics profile (TURBLOG, 2011) to Yogyakarta city is shown in Table 2

**Table 1**. Elements of Logistics Profile (Macario *et al.* 2008) – adjusted for Yogyakarta city

| Name of variable | | Type of variable | Classification | | |
|---|---|---|---|---|---|
| ***City area features*** | | | | | |
| 1.1 | Commercial density | continuous | Low<br>< 20% | Medium<br>30% - 70% | High<br>> 70% |
| | | | Commercial face to residencies service / industries | | |
| 1.2 | Homogeneity | continuous | Low<br>Several types of services and products | Medium<br>Mix of residential areas with offices and commercial stores | High<br>Cluster of one type of services or similar products |
| ***Product characteristics*** | | | | | |
| 2.1 | Easiness of Handling | ordinal/categorical | difficult | reasonable | Easy |
| 2.1.1 | Size | ordinal/categorical | Large (wheelbarrow, crane) | Medium ( >1 person to carry one unit) | Small ( > 1 unit per person to carry) |
| 2.1.2 | Weights | ordinal/categorical | Heavy (wheelbarrow, crane) | Medium (>1 person to carry one unit) | Light ( > 1 unit per person to carry) |
| 2.2 | Special Conditions | ordinal/categorical | Special needs (e.g. valuable products and frozen products) | Might have special needs (e.g. open packages, if food handled ambient temperature and chilled) | No special needs |
| ***Agents/deliveries profile*** | | | | | |
| 3.1 | Frequency of deliveries | continuous | Low ( < once a week) | Medium (several days per week) | High (daily) |

**Table 2.** Logistics Profile (TURBLOG, 2011) – adjusted for Yogyakarta city

| Elements/variables | Profile | | | | |
|---|---|---|---|---|---|
| | A | B | C | D | E |
| Commercial density | High | - | High | High | Low/medium |
| Homogeinity | High | - | Low | Low | Low/medium |
| Easiness of handling | - | - | Easy | - | - |
| Special condition | - | Special needs | No special needs | - | - |
| Frequency of deliveries | - | High | High | Medium/high | Low/medium |

In order to get quantitative interpretation of logistics profile in Table 2, variables defined should be measured and calculated. Values obtained from above calculations were used to evaluate the fishnets that meet the values. Detail of calculations refers to Alho and de Abreu e Silva (2015) and it is explained briefly as follows:

1. Commercial density was calculated by dividing the number of stores in each fishnet by the area of the related fishnet.
2. Homogeneity calculation was based on entropy concept. The lower the number obtained is associated with increased homogeneity.
3. Easiness of handling was calculated based on two indicators, i.e. weight and volume as representation of size (see Table 1)
4. Special conditions has three attributes, i.e. fragile, perishable and refrigerated (see Table 1)
5. Frequency of deliveries was calculated using two indicators, i.e. sum of establishment deliveries and average of establishment of deliveries.

## 3 Results and Discussion

Results summary of continuous variables and ordinal/categorical variables are shown in Table 3 and Table 4. Values of continuous variables are presented by *Min, Max, Average* and *Standard Deviation,* whereas ordinal/categorical variables are presented by proportion of each variable in each industry category (Table 5).

Quantitative interpretation of logistics profile obtained from calculations above is shown in Table 6. This table was then used to determine which fishnet that meet the specification for each profile mentioned in Section 2 (see Table 2). Maps of the results are shown in Figure 3. In Table 2, *high is defined as a* half standard deviation above average value and *low* is a half standard deviation below average value

**Table 3.** Summary of continuous variables

| Name of variable | Min | Max | Average | Standard Deviation |
|---|---|---|---|---|
| Commercial density | 6 | 542 | 155 | 123 |
| Sum of establishments of weekly deliveries | 6 | 1400 | 384 | 309 |
| Average of establishments of weekly deliveries | 1 | 12 | 4 | 2 |
| Homogeneity | 0.24 | 0.92 | 0.78 | 0.12 |

**Table 4.** Summary of ordinal/categorical variables

| Name of variable | distribution | | | | | | | | | |
|---|---|---|---|---|---|---|---|---|---|---|
| | 1 | 2 | 3 | 4 | 5 | 6 | 7 | 8 | 9 | 10 |
| Industry category | 24% | 17% | 6% | 1% | 5% | 6% | 2% | 10% | 11% | 18% |
| Number of special item attributes | 48% | 52% | 0% | - | - | - | - | - | - | - |
| Item's weight | 9% | 76% | 15% | - | - | - | - | - | - | - |
| Item's volume | 9% | 76% | 15% | - | - | - | - | - | - | - |

**Table 5.** List of industry categories

| No | Industry category |
|---|---|
| 1 | Food/beverage |
| 2 | Textile/clothing/shoes |
| 3 | Paper/printing |
| 4 | Petro-chemical |
| 5 | Non-metallic |
| 6 | Primary metal |
| 7 | Machinery |
| 8 | Electronics |
| 9 | Transportation equipment |
| 10 | others |

**Table 6.** Summary of quantitative interpretation of logistics profiles

| Variable name | Profile | | | | |
|---|---|---|---|---|---|
| | A | B | C | D | E |
| Commercial density | ≥ 217 | - | ≥ 217 | ≥ 217 | ≤ 155 |
| Homogeneity | ≤ 0.72 | - | ≥ 0.84 | ≥ 0.84 | ≥ 0.78 |
| Easiness of handling (weight) | - | - | 1 | - | - |
| Easiness of handling (volume) | - | - | 1 | - | - |
| Special conditions | - | - | 1 | - | - |
| Frequency of deliveries (sum of establishment deliveries) | - | - | ≥ 539 | ≥ 384 | ≤ 384 |
| Frequency of deliveries (sum of establishment deliveries) | - | - | ≥ 5 | ≥ 4 | ≤ 4 |

Figure 3 shows that profile A laid in three areas, one at the north, one area in the middle of the city and one in the southeast part of the city. *Bausasran* sub-district, which is located at the north is dominated by shops selling food and basic necessities (56%). Other area, *Sosromenduran* in the middle of the city, has 50% stores selling clothing and textiles with the rest consists of various types of stores, including flower shops alongside a street inside this sub-district. *Prenggan*, which is located at the southeast is dominated by numerous craft shops, especially from silver. Silver handicraft industry in this area is dominated by small enterprises.

Profile C or business center / CBD (courier, small deliveries, B2C), does not have members, or in other words there is no district in the city that matched the characteristics of the profile elements of logistics in accordance with the profile of C. This is due to the high requirements to be able to fit profile C, such as high frequency of delivery, while it is only medium/high for profile D.

Profile D, which is large commercial (retail, shopping centers, distribution warehouse), is found mostly in the middle of the city as there are lots of big stores alongside the streets. In addition of the other three profiles, there are three sub-district, i.e. *Sosromenduran, Suryatmajan and Ngupasan* (red color), which represent overlapping of profile A and D. These areas are actually corresponds to *Malioboro* street and its surroundings, which is well known as major shopping street in Yogyakarta with some large shopping malls in it. Therefore, delivery of goods is very frequent in this area. The last profile, E, is residential areas with local trade which is spread in several locations in the north, west, south and middle of the city. These areas are characterized with medium population area. The findings shows that the result of logistics profile is indeed influenced by combination of variables forming the logistic profile and conditions of the city. Hence, different regional area with different characteristics will have combination of different logistics profile elements and thus different map of logistics profile.

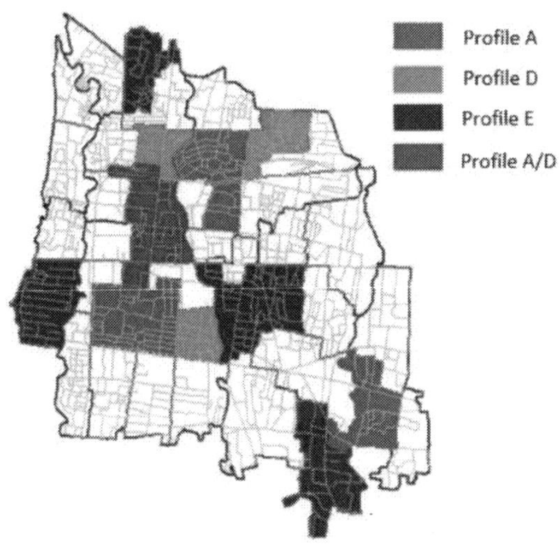


**Figure 3.** Logistics profile map of Yogyakarta city

## 4 Conclusion

Logistics profile concept has been adapted to the conditions of Yogyakarta city, producing three profiles matched, i.e. profile A (groups of homogeneous stores), profile D (large commercial stores), and E (residential areas with local trade) with one additional overlapping profile (profile A and profile D). Profile B was not formed as many constraints in data collection and unavailability of data on urgency and planned delivery. On the other hand, profile C was not formed because none of the areas can match the requirements of this profile. The profiles obtained is expected to be used in the next stage in formulating policies related to the logistics management in this city.

The authors wish to thank Higher Education Directorate (DIKTI) Indonesia for their financial supports under the research grant of Master Pengembangan dan Percepatan Pertumbuhan Ekonomi Indonesia (MP3EI) 2017.

## References

1. Alho, A.R. & de Abreu e Silva, J. Utilizing urban form characteristics in urban logistics analysis: a case study in Lisbon, Portugal. *Journal of Transport Geography, 42,* 57-71. (2014).
2. Alho, A.R. & de Abreu e Silva, J. Modeling retail establishment's freight trip generation: a comparison of methodologies to predict total weekly deliveries. *Transportation.* doi:10.1007/s11116-015-9670-6. (2015).
3. Anderson, S., Allen, J., Browne, M. Urban logistics – how can it meet policy makers sustainability objectives. *Journal of Transport Geography. 13*(1). 71-81. (2005).
4. Macario, R., Galelo, A., Martins, P.M. Business models in urban logistics. *Ingenieria Desarrollo.* 77-96. (2008).
5. Quak, H.J. Sustainability of urban freight transport retail distribution and local regulations in cities (Ph.D. thesis, Erasmus University Rotterdam, the Netherlands). (2008).
6. TURBLOG. Transferability of urban logistics concepts and practices from a worldwide perspective. Deliverable 2: Business Concepts and Models for Urban Logistics. TIS.pt, Portugal. (2011).

# Implementation of supply chain management in supplier performance assessment using Analytical Hierarchy Process (AHP) Objective Matrix (OMAX) and Traffic Light System

*Irwan* Sukendar, Nurwidiana, *Diah Nurul* Hidayati

Industrial Engineering Department, Universitas Islam Sultan Agung, Kaligawe Street km 4 Semarang, Indonesia.

**Abstract.** This study focuses on researching the implementation of supply chain management in supplier performance measurement using Analytical Hierarchy Process (AHP) Objective Matrix (OMAX) and Traffic Light System. The method used consisted of several stages: Identification stage of criteria variable and subcriteria variable, designing hierarchy AHP structure, calculating the weight of each criterion and subcriteria, supplier performance appraisal using OMAX method, calculating the total value of each supplier perfomance and identifying the performance of each supplier Using color indicator from Traffic Light System. Based on the results of the research, it can be seen that 9 out of 10 suppliers that are considered have a good performance while 1 supplier has a performance under the target.

## 1 Introduction

A manufacturing company in Indonesia that produces Al-Quran has a pretty good performance in product marketing. The company holds a market share of about 30% of all domestic Quran publishers. Even the company has been certified ISO 9001: 2008 in terms of quality management. As one of the companies that have been certified ISO 9001: 2008 then the performance evaluation of supplier activities are regularly implemented. Implementation of supplier performance evaluation is done based on several criteria such as quality, price, payment and delivery where each criteria already has the standard set by the company.

However, supplier performance evaluation is still fairly simple and has not used the weighting of the criteria. In addition, there is also one criterion that has two assessments to consider but the criteria are not broken down as subcriteria but are made into one. Assessment criteria need to be weighted because the importance of each criterion is different. Whereas if a criterion has more than one appraisal it is better if each assessment is separated into several subcriteria

Based on the above explanation it is seen that the evaluation of suppliers made has some shortcomings. The first deficiency is that there is no weighting on the criteria and subcriteria, it is not considered to separate criteria that have more than one assessment to be broken down into subcriteria, as well as identify the weaknesses and advantages of each qualitative supplier.

The problem is part of the supply chain theme. Much research on the subject has grown tremendously. Some of them serve as literature in this research
[1] analyze the supplier selection process, [2] conduct supplier evaluation with Vendor Performance Indicator and Analytical Hierarchy Process (AHP) method, meanwhile [3] designing the Supplier Rating and Selection System by Using AHP Method and Traffic Light System. [4] Write a measurement of supply chain performance by reviewing the literature. Meanwhile,[5] Write the Total Quality Management (TQM) measurement in the supply chain section with research objects of health institutions. [6] Writing the implementation of supply chain management on aviation manufacturing, [7] Researching the collaboration between supply chain networks, while [8] Writing about the optimal delivery time to minimize tardiness and shipping costs. [9] Write about drivers that affect supply chain complexity, while [10] Wrote a key success factor determination to implement reverse logistics using fuzzy-topsis.

On the basis of these literatures, one of the structured methods that can be used to measure the criteria and subcriteria weighting based on the level of importance between the criteria and the level of importance between subcriteria is by using the Analytical Hierarchy Process (AHP).

Furthermore, to see how the overall performance of each supplier and also to know the advantages of the shortage of each supplier method that can be applied properly is the method of Objective Martix (OMAX) and Traffic Light System. This is because the OMAX method not only analyzes the overall supplier's performance but also analyzes the performance of each criterion.

On the basis of the problem solving hypothesis, this study aims to evaluate supplier performance in this study by using AHP, OMAX and Traffic Light System methods.

* Corresponding author: irwan@unissula.ac.id

## 2 Methods

There are three methods used in this research. The first method is analytical hierarchy process (AHP), the second is Objective Matrix (OMAX) and the third is Traffic Light System.

Analytical hierarchy process (AHP) method will be used to find out how the relationship between criteria and subcriteria based on hierarchy structure and to know the importance of each criterion and subcriteria. The AHP method will be used to see what criteria and subcriteria will be used to perform performance evaluations and also to find out how much weight each of the criteria and subcriteria

Objective Matrix (OMAX) and Traffic Light System methods will be used to determine the performance of each supplier and also to know the advantages and disadvantages of each supplier.

The OMAX method is used to know the performance value of each supplier while the traffic light system is used as an indicator of the value of each performance value. If the performance value is at level 8-10 then the supplier's performance gets the green indicator color indicating that the supplier's performance is the same or exceed the target, if the performance value is at level 4-7 then the supplier's performance gets the yellow indicator color indicates that the performance has not been reached even though the value Is approaching the target. Whereas if the performance value is at the 0-3 level then the supplier's performance gets the red indicator color indicating that the performance is really below the set target and needs immediate improvement.

## 3 Results

### 3.1 Identify criteria and subcriteria

The first step of the research is to identify the criteria variable in supplier performance measurement. Based on the results of brainstorming with the company then it can be identified six criteria variables. The six criteria are: quality, price, delivery, payment, flexibility and responsiveness

The next step, each criterion identified subcriteria. Subcriteria can be identified: on Quality criteria, four sub criteria are identified. In the price criteria identified two subcriteria, on the delivery criteria identified two sub criteria, on the criteria of payer identified two sub criteria, on the criteria of flexibility identified two criteria and on the criteria of responsiveness identified four sub criteria. All six criteria and 16 criteria can be seen in Table 1.

### 3.2 Design of AHP Hierarchy Structure and weighting

The criteria and subcriteria tables should be described in the Hierarchy AHP Structure form. In addition to the criteria and sub criteria, all suppliers totaling 10 are also included in the hierarchy structure.

Subsequently all criteria and sub criteria need to be weighted. This weight is derived from the judgments made by the most authoritative official in the company. In this case is the head of purchasing department. The assessment is normalized so that it becomes a weight.

Tabel 1. Criteria and Subcriteria

| No | Criteria | SubCriteria |
|---|---|---|
| 1 | Quality | Quality consistency (SK1) |
| | | The existence of warranty (SK2) |
| | | Inclusion of product specification sheet (SK3) |
| | | Consistency of packaging form (SK4) |
| 2 | Price | Competitive product prices (SK5) |
| | | Discounting (SK6) |
| 3 | delivery | Timeliness of delivery (SK7) |
| | | Accuracy of quantity of product shipped (SK8) |
| 4 | payment | Grace period of payment (SK9) |
| | | Way of payment (SK10) |
| 5 | Flexibility | Eligibility to meet delivery time shift (SK11) |
| | | The ability to meet changes in the number of products ordered (SK12) |
| 6 | Responsiveness | Speed confirmation of order availability (13) |
| | | Speed responding to change of delivery time (14) |
| | | Speed responds to changes in order quantity (SK15) |
| | | Speed of responding to product complaints (SK16) |

### 3.3. Performance Supplier Rating and Traffic Light System

Performance appraisal of suppliers is done by OMAX method. Temporal Traffic Light System method is used to give the indicator color on the performance of the supplier. The performance value of each supplier and the indicator color of each supplier can be seen below:

Tabel 2. Performance and Traffic Light System

| No | Supplier | *Performance* | *Level* |
|---|---|---|---|
| 1 | CV.Multiwarna | 9.250 | 9 |
| 2 | PT.Hana Grafika | 9.168 | 9 |
| 3 | PT.Pura Barutama | 8.969 | 8 |
| 4 | PT.Pabrik Kertas Noree | 8.969 | 8 |
| 5 | CV.Dirgantara | 8.947 | 8 |
| 6 | PT.Manggala Indo Nusa | 8.908 | 8 |
| 7 | PT.Maritama | 8.674 | 8 |
| 8 | CV. Berkat Lamandau | 8.531 | 8 |
| 9 | PT.Hasta Yuga Halim | 8.531 | 8 |
| 10 | PT.Puri Nusa Eka Persada | 7.474 | 7 |

Based on the performance table, it shows that from 10 suppliers, two suppliers have excellent performance, seven suppliers have good performance, and one supplier has enough performance.

Traffic Light System method shows that Nine suppliers are given green indicator color which means having good performance and very good. So it is recommended to be maintained. Meanwhile, one supplier is colored yellow indicator which means to have to improve its performance again.

## 4 Conclusion

There are 6 criteria and 16 subcriteria used to assess supplier performance. Based on the assessment of the 6 criteria and the 16 subcriteria, it can be concluded that the total performance score for 9 suppliers is at level 8 and also level 9 which means that the performance on the nine suppliers is equal or exceeds the target while there is one supplier that has total value Performance at level 7 which means under the performance of the supplier is under the target but sudha approaching, so it needs to be done follow-up in the form of two-way communication so that the performance on the supplier can be in accordance with the target

## References

1 Verma, R., and Pullman, M.E.: 'An analysis of the supplier selection process', Omega, 1998, 26, (6), pp. 739-750

2 Andika, D., Anggraeni, S.K., and Sirajuddin, S.: 'Usulan Pemilihan Supplier Bahan Baku Tetap Menggunakan Vendor Performance Indicator dan Analytical Hierarchy Process (AHP)', Jurnal Teknik Industri Untirta, 2013, 1, (2)

3 Iriani, Y.: 'PERANCANGAN SISITEM PENILAIAN DAN SELEKSI SUPPLIER DENGAN MENGGUNAKAN METODE AHP DAN TRAFFICT LIGHT SYSTEM', 2009

4 Arzu Akyuz, G., and Erman Erkan, T.: 'Supply chain performance measurement: a literature review', International Journal of Production Research, 2010, 48, (17), pp. 5137-5155

5 Turkyilmaz, A., Bulak, M.E., and Zaim, S.: 'Assessment of TQM Practices as a Part of Supply Chain Management in Healthcare Institutions', International Journal of Supply Chain Management, 2015, 4, (4)

6 Axtman, S.J., and Wilck, J.: 'A Review of Aviation Manufacturing and Supply Chain Processes', International Journal of Supply Chain Management, 2015, 4, (4)

7 Kampstra, R., Ashayeri, J., and Gattorna, J.: 'Realities of supply chain collaboration', The International Journal of Logistics Management, 2006, 17, (3), pp. 312-330

8 Dumitrescu, S., Steiner, G., and Zhang, R.: 'Optimal delivery time quotation in supply chains to minimize tardiness and delivery costs', Journal of Scheduling, 2015, 18, (1), pp. 3-13

9 Serdarasan, S.: 'A review of supply chain complexity drivers', Computers & Industrial Engineering, 2013, 66, (3), pp. 533-540

10 Agrawal, S., Singh, R.K., and Murtaza, Q.: 'Prioritizing critical success factors for reverse logistics implementation using fuzzy-TOPSIS methodology', Journal of Industrial Engineering International, 2016, 12, (1), pp. 15-27

# Sustainable-value stream mapping to evaluate sustainability performance: case study in an Indonesian furniture company

*Sri* Hartini[1], *Udisubakti* Ciptomulyono[2], *Maria* Anityasari[2], *Sriyanto*[1], *Darminto* Pudjotomo[1]

[1]Department of Industrial Engineering, Diponegoro University, Indonesia
[2]Department of Industrial Engineering, Institut Teknologi Sepuluh Nopember (ITS), Indonesia

**Abstract.** Lean manufacturing tools do not consider environmental and societal benefits. The conventional value stream mapping (VSM) methodology examines the economics of a manufacturing line, most of which are in regards to time (cycle time, lead time, change-out time, etc.). Incorporating the capability to capture environmental and societal performance visually through VSMs will increase its usefulness as a tool that can be used to assess manufacturing operations from a sustainability perspective. A number of studies have addressed the extension of VSM to incorporate additional criteria. A vast majority of these efforts have focused on adding energy-related metrics to VSMs, while several other studies refer to 'sustainable' VSM by including environmental performance in conventional VSMs. This research has developed a method for VSM integrated with environment metric and social metric for ensuring sustainable manufacture. The proposed technique is capable of visualizing and evaluating manufacturing process performance from sustainability view point. The capability of proposed technique has been tested by an application study on furniture company. The study provides insights to practitioners to visualize process performance in economic, environment and social metric.

## 1 Introduction

The concept of Lean Manufacturing (LM) was pioneered by a Japanese automotive company, Toyota, during 1950's which was famously known as Toyota Production System (TPS). The primary goal of TPS was to reduce the cost and to improve productivity by eliminating wastes or non-value added activities [1-2]. Manufacturing firms operating in the rapidly changing and highly competitive market of the past two decades have embraced the principles of lean thinking. In doing so, they are reorganized into cells and value streams to improve the quality, flexibility, and customer response time of their manufacturing processes [3]. Studies at the operation level concluded that LM was very powerful for increased operational performance on quality [4-5], reduced inventory [6], productivity [4]; [7], and reduced cost [8 - 9]. Even LM can also be used as a strategy to improve business performance in terms of profitability [4], sales, customer satisfaction and business performance [10]. But, traditional lean manufacturing tools do not take into account for environmental and societal benefits [11].

During the recent days, the modern manufacturing systems are expected to be lean and sustainable. Lean concepts ensure waste elimination and cost reduction. Sustainable concepts focus on the development of environmentally friendlier products and processes considering the economical and societal constraints as well. There is a potential to extend lean tools for ensuring sustainable benefits.

Few studies have been reported on the context of applying lean tools for developing greener processes and strategies for environmentally benign manufacturing. There is positive impact of integration lean and sustainable manufacturing on three bottom line performance [12]. Value stream mapping (VSM) is a fundamental lean technique to identify waste and value improvement opportunities. VSM is a diagnostic technique that originated from lean manufacturing principles. Its purpose is to identify value-adding and non-value-adding activities in the value stream so that wasteful activities can be eliminated, and production can be aligned with demand [13].

It has been emphasized that traditional VSM framework does not account for environmental and societal performance. A number of studies have addressed the extension of VSM to incorporate additional criteria. A vast majority of these efforts have focused on adding environment and energy-related metrics to VSM [14 - 22]. Hence, Faulkner and Badurdeen [23] and Brown et al [24] suggested sustainable VSM (Sus-VSM) to evaluate triple bottom line (TBL) performance of manufacturing and applied to an industry study. The Sus-VSM added environment and social metrics to VSM. However, Faulkner and Badurdeen [23] only focus to identify a general set of metrics that will have broad application across many industries. Further customization may be needed to assess industry specific aspects during application on a case-by-case basis by first identifying key

* Corresponding author: srihartini@undip.ac.id

performance drivers for that industry and then selecting relevant metrics to evaluate those aspects.

Goals of this research is to design an analysis technique based on lean and green to achieve sustainable manufacturing, map of wastage and waste from furniture factory (current state mapping) and suggestion strategy to improve production sustainability. This research is very relevant to the problems of Indonesian manufacturing company. Indonesian manufacturing company still has a low level of efficiency, especially small and medium enterprises. A Furniture factory in Semarang has efficiency rate about 50% [25]. In this case the traditional value stream mapping can determine the source of waste, non-value-added activity, i.e. high setup, inappropriate processes, inventory and defects. The development of jig management information systems can reduce setup time up to 65% [25]. Redesigning the working environment with 5S can reduce the change over time about 41.6% and can increase the machine capacity [26].

The VSM, however, could not identify the waste (solid, wastewater or emission) in the furniture factory. even though, many forms of non-product output occurs in the furniture industry that may not be realized by the management, such as wood waste from defective product, spraying when doing finishing, excessive use of office paper, glue, water, paint and others. The production of furniture is among the ones that drives global economy : intense environmental degradation and high levels of waste in operational activities and high-volume consumption of raw forest materials [27].

This study developed a sustainable-VSM to evaluate sustainability performance in furniture company after selecting relevant metrics with furniture company. The selecting relevant metrics. The selecting relevant metrics is done based on lean and sustainable manufacturing thinking. It is necessary for furniture company to understand the environmental impacts of their products at each phase of the production processes. Integrating sustainable concept and VSM allows improvement process to focus on specific environmental improvement actions. It also provides benefits in terms of monitoring the environmental impacts during the lean improvement initiatives. To validate the capability of sustainable-VSM, case study was conducted in a furniture company.

## 2 Literature review: Value Stream Mapping to enhance sustainability

Simons and Mason [14] proposed a method called Sustainable Value Stream Mapping (SVSM) as a means to enhancing sustainability in product manufacture by analysing emissions of the greenhouse gas, CO2, in addition to value-adding time, throughout the value stream. The research used SVSM in a study carried out for the UK Department for Transport in order to model CO2 emissions from farm gate to retail outlet for alternative distribution scenarios for three supply chains: lettuce, apples and cherries. Norton [13] was extended the SVSM to include additional environmental performance indicators (EPIs) in order to evaluate the occurrence of physical wastes and wasted resources in the manufacture of a specific product or product family, and to attribute those wastes to discrete activities. This extended form of SVSM provides an applied and systematic approach to the improvement of operational efficiency and environmental performance, especially for small companies with limited resources. US Environmental Protection Agency [15] discussed about the toolkit framed by the US EPA's to track the environmental wastes using lean techniques. This tool kit helps in identifying and eliminating environmental wastes by tracking and visualizing environmental metrics namely material and water usage along with conventional VSM metrics. This toolkit fails to include energy consumption metric in VSM. Subsequently, US EPA created another toolkit which adds energy consumption metric with the existing environmental metrics in VSM which helps in tracking energy flow and consumption. However, both the tools fail to include societal metrics and emphasize more on inclusion of environmental and energy consumption metrics in conventional VSM. Torres and Gati [17] developed a managerial tool to align the economical and environmental aspects in production process by mapping strategy. This study was applied in an alcohol and sugar manufacturing industry, but only for water consumption. The tool is called Environmental VSM (EVSM). After the tool development, future development alternatives are discussed. Paju et al. [28] developed a methodology named sustainable manufacturing mapping (SMM) by combining discrete event simulation (DES) and life-cycle analysis (LCA) along with traditional VSM. They also identified sustainable manufacturing indicators and modeled current maps to assess the sustainable performance of manufacturing organizations. It has also been identified that no attempt was made in constructing the desired future state for the current manufacturing system from sustainability viewpoint. Kuriger and Chen [18] modeled an assessment tool integrating environmental and energy metrics with VSM to evaluate the sustainability of manufacturing firms. This tool captures only the energy consumption for processes and fails to capture energy consumed for transportation.

Faulkner and Badurdeen et al. [23] developed a methodology for Sus-VSM to capture economic, environmental, and societal sustainability of manufacturing firms. They also identified suitable metrics and visual symbols to develop Sus-VSM and conducted a study at satellite television dishes manufacturing firm. Brown et al. [24] extended the conventional VSM and incorporated the metrics for sustainability and created a Sus-VSM. The applicability of the model was demonstrated using three case studies and the results are summarized for each case. Folinas et al [22] explored the application of the VSM tool so as to determine the waste that have environmental impact in a specific agrifood supply chain; the production of the canned peach. The main objective of their study is to propose a four-step approach for measuring the environmental performance of supply chains in the food sector based on lean thinking techniques so as to identify sources of waste in the selected supply chain.

Hartini et al. [29] collected some literature on extended VSM to enhance sustainable manufacturing in

2002 - 2016. It was found that most literature were case study articles, followed by theoretical papers and simulation modeling papers. While, based on three bottom line indicator, that most metrics are environmental, followed by social and then economic metrics. The most of the environment metrics were energy, material and water consumption. Based on the application in the industrial sector, the extended VSMs were applied in the automotive industry, followed by agricultural food and electronic industry. While the sustainable-VSM for the furniture industry has never been applied.

Sustainable-VSM previously developed [23] has a general set of metrics that will have broad application across many industries. However, further customization may be needed to assess specific aspects of furniture company. This study developed sustainable-VSM for furniture company with the specific metrics and relevant with production process of furniture company. And, the study provides insights to practitioners of furniture company to visualize process performance from traditional and environmental.

## 3 Design of Sustainable-VSM for Furniture Industry

### 3.1. Design of Metric for Sus-VSM

Value stream mapping is a system analysis tool designed to find potential waste in the production process. This research develop VSM so that is can analyze the impact of production process towards sustainability.

Design of the metrics in sustainable-VSM for furniture company based on the results of questionnaires from top level of furniture companies (10 companies). Questionnaire based on lean metrics [29] and sustainable manufacturing metrics [30] was prepared using the Likert scale. The selection of metrics is using the level of consensus (cut off value = 0.5) and weighted average (cut off value 4.2). Methodology of metric development for sustainable – VSM for furniture company is shown in the figure 1. Design of metrics in sustainable – VSM for furniture company is shown in Table 1. Selected metrics for sustainable performance of furniture based on lean manufacturing are time, work in process, cost, defect (economic aspect), material waste, material consumption energy consumption (environment aspect), environment health and safety, physical load index, sickness rate, employee training, hazard material (social aspect). Selected metrics for sustainable performance of furniture based on sustainable manufacturing are revenue, cost (economic aspect), recycling of waste, energy efficiency, material waste and consumption (environment aspect), employee satisfaction, accidents rate (frequency and number), employee health evaluation, local supplier (social aspect). Design of metrics for furniture company based lean and sustainable manufacturing are time, work in process, cost, product defect (economic aspect), energy consumption, material consumption (environment aspect), health rate of employee, safety rate of employee, satisfaction rate of employee, training of employee (social aspect).

Due to revenue is assumed to be represented by cost, so revenue is not selected. Waste treatment is not incorporated in VSM because reuse of wood waste is commonly done in furniture companies. While the chemical waste from finishing process will be done by third party.

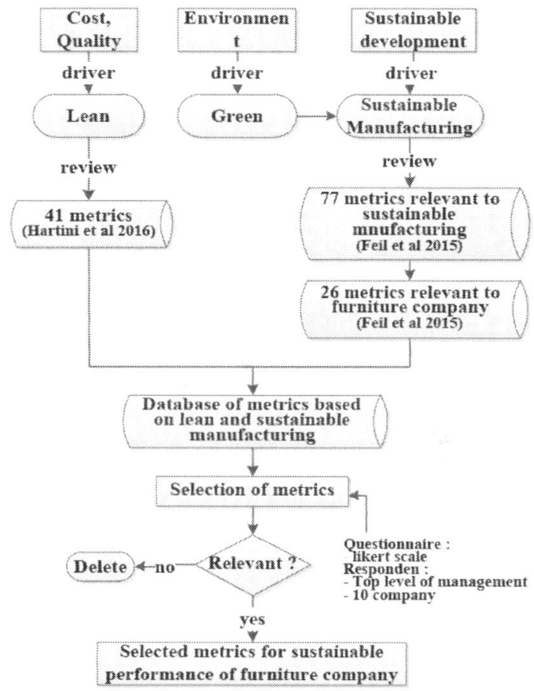


**Figure 1** Methodology of metric selection for Sus-VSM in furniture company

### 3.2. Design of Sustainable-Value Stream Mapping

From Table 1, it can be shown that Sustainable-VSM as analysis tool is able to measure economic metric, environment metric, and social metric. With the measured values from the three aspect of sustainability, the sustainability analysis can be done. The proposed sustainable-VSM has multi layer analysis line, including time analysis line for analysis about value-added ratio, material analysis line for analysis about material efficiency, energy analysis line for analysis about energy efficiency, defect analysis line for analysis about quality efficiency, cost analysis line for analysis about cost efficiency and health-safety analysis line for analysis about labour efficiency. Material waste is not include in analysis line because material waste has been incorporated in material analysis line.

**Table 1**. Design of metrics in sustainable – VSM for furniture company

| **Initial metrics from lean and sustainable manufacturing** | **Selected metrics** |
|---|---|
| Metrics in value stream mapping for ensure sustainable manufacturing [29] :<br>**Economic** : time , work in process, cost, defect<br>**Environment** : CO2 emission, energy consumption, hazard waste, material waste , Packaging waste, waste metal container, water waste, food waste, material consumption, water consumption<br>**Social** : Biodiversity, number of man hour, environment healty safety, absence day, physical load index, noise, local community, sickness rate, employee training, hazard material | **Economic** : time, work in process, cost, product defect<br>**Environment** : energy consumption, material waste and consumption<br>**Social** : environment health & safety, physical load index, sickness rate, Employee training, hazard material |
| Design of metrics in sustainable-value stream mapping for furniture company based on lean and sustainable manufacturing | **Economic** : time, work in process, cost, product defect<br>**Environment** : energy consumption, material waste and material consumption, recycling of waste<br>**Social** : health rate of employee, safety rate of employee, satisfaction rate of employee, training of employee |
| Metrics for measuring sustainability in furniture industries [30]:<br>**Environment** : Generation of dangerous waste, waste disposal, effluent treatment, recycling of waste, atmospheric emissions, recycling of products, reuse of products, renewable energy consumption, efficiency of energy consumed, consuming renewable materials, adherence to environmental standards, water consumption<br>**Social** : employee satisfaction, employee training and development, serious and fatal accidents (frequency and number), employee health evaluation, child labor, complaints of dust, business ethics<br>**Economic** : revenue, operating profit, net profit, tax payments, operational costs, comparison of the wages with the value set by the market, local suppliers | **Environment** : recycling of waste, energy efficiency, material consumption<br>**Economic** : revenue, operational costs<br>**Social** : employee satisfaction, acccidents rate (frequency and number), employee health evaluation, local supplier |

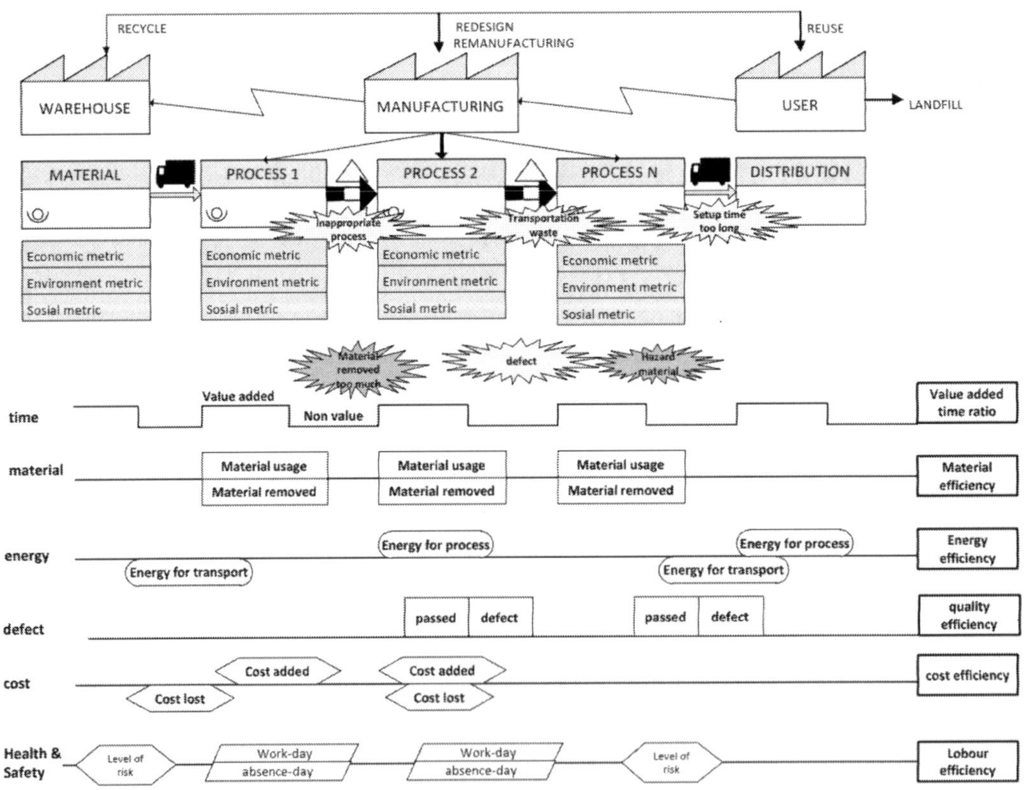


**Figure 2.** Design of sustainable-value stream mapping that has involved three pillars of sustainable 2

https://doi.org/10.1051/matecconf/201815401055

If the efficiencies of each aspect can be improved then system sustainability can also be improved. Design of sustainable-value stream mapping that has involved three pillars of sustainable manufacturing is shown in Figure 2.

## 4. Application in A Furniture Company

### 4.1. Data Collection

Case study was done at PT X, furniture company in Semarang, Indonesia. The reason to choose this company is that it has relatively high standard product with several product variations. The object are Totem Coat and Hat Stand, because those products were produced in the survey period.

Generally, the production process in PT X are raw material preparation, construction, assembly, finishing, quality control and packaging. The first process is raw material preparation, which is turning wood logs into boards to be processed. Construction process is forming the boards according to the product design. Assembly is the process of combining the various components. Finishing is the last process to finalize the product, e.g. smoothening of the wood surface. After the finishing is done, quality control is done before packaging to ensure that products are in accord to the specifications.

Measurements of production time was designed with 20 observations in each process. Material consumption was calculated using approximated unit of weight by multiplying the wood volume and the density of tamarind wood (0,45 gr/cm3 or 450 Kg/M3) (www.fao.org/docrep/w4095e/w4095e0c.htm).
Production of wood waste was calculated by subtracting material input with its output. Energy consumption was approximated by calculating electrical energy needed for production process of Totem Coat and Hat Stand at each machine in kwh. Result of data processing in Fig 3.

### 4.2 Discussion

From the result of sustainable-VSM it can be seen that potential waste from economic and social aspect almost spread in almost all production processes while potential of environmental impact occurs at sawmill and construction process. Waste from economic aspect is marked with yellow spot, environmental aspect with green, and social with blue.

Types of waste in economic aspect includes too much setup and processing time, too many defects and work in process, and machine failure. These can be reduced by improving the work system. Too much unneeded time such as waiting, setup, and transportation time results in large processing time. Old machines and breakdowns are also a problem in production and result in idle time. Careless production process during construction results in out of specification final products. Defects will result in waste of material, workforce, and energy. This will increase the production cost thus reducing profit.
Types of waste in environmental aspect include the high amount of wood waste, especially in sawmill process. Wood consumption efficiency is 145 kg/266kh (54,5%). The high amount of energy consumption during sawmill process is the most (60%), and followed by oven process (10%). Reduction in environmental impact can be done by turning wood waste into alternative product that can increase the value. Energy consumption during oven process can be reduced by trying to find another alternative of oven with less energy consumption. Benchmark from other companies suggested, that oven could be operated by burning wood waste that cannot be used anymore. However, burning wood can also produce carbon dioxide emission and has negative impact to the environment. Comparison between using electricity and burning wood waste can be done by looking at the production cost and the emission level.

Social aspect in this research is more focused on the work posture which is not ergonomic. The high amount unergonomic wood lifting and machining may cause muscle strain. This can be reduced by improving the work environment, especially by using aid tools during production process and material handling.

If the potential problems that result in economic, environment, and social aspect can be improved, then sustainability level can also improve.

This research has succeeded in designing analysis tool based on lean and green manufacturing and performing case study in a furniture company. The result is that the developed tools can analyze problems in economic, environment, and social aspects as the basic of improvement.

Up to now, the improvement process still undergone. Once completed, the improvement will provide input for further analysis.

## 5 Conclusion

Improvement design of analysis tool based on lean and sustainable manufacturing has been done and validated by performing case study in a furniture company. Analysis tool is able to identify potential problem in economic, environment, and social aspects which are the pillars of sustainable manufacturing. If these problems can be reduced the sustainability level will improve. Problems in economic aspect include the high amount of process and setup time, amount of defect, breakdown machine and work in process in almost all processes. While in the environmental aspect is the high amount of wood waste during sawmill process and energy during oven process. While in the social aspect is unergonomic the work posture and potential of accident rate still high in sawmill process.

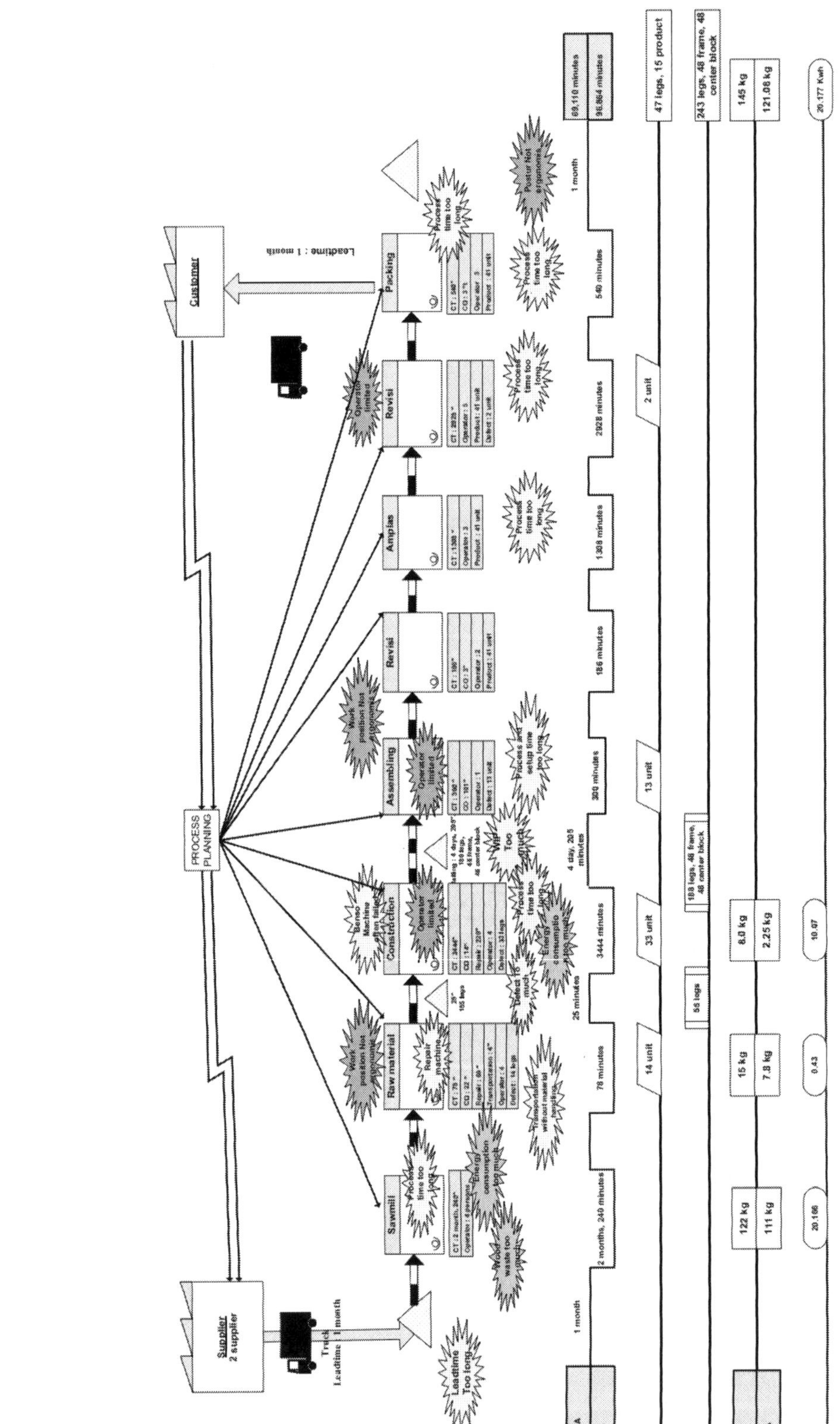


**Figure 3**. Sustainable – value stream mapping of furniture company

This study successfully developed an analysis tools based on lean and green for evaluate of sustainability performance: sustainable - VSM. If this proposed tools was compared to other sustainable-VSM findings [22-24], this tool involves the relevant metrics with the characteristics of the company under study. The proposed analysis tool involves the level of defect and level of risk that have not been included by other researchers.

However, due to limitations of data access, cost metrics cannot be measured even though they are highly relevant. Suggestions for further research is to do simulation of various strategies that can reduce problems in economic aspect. By doing simulation, a strategy with the highest impact on the improvement of sustainability can be found.

## References


1. J. P. Womack, D. T. Jones, and D. Roos, Bus. Horiz., 35, 81–82 (1992).
2. N. Jasti and R. Kodali, Int. J. Oper. Prod. Manag., 34,1080–1122 (2014).
3. R. R. Fullerton, F. A. Kennedy, and S. K. Widener, J. Oper. Manag., 32, 414–428 (2014).
4. R. R. Fullerton and W. F. Wempe, Int. J. Oper. Prod. Manag., 29, 214–240 (2009).
5. R. Shah and P. T. Ward, J. Oper. Manag., vol. 21,129–149 (2003).
6. A. Chong, H. Cheah, W. P. Wong, and Q. Deng, *Proceedings of the 2012 International Conference on Industrial Engineering and Operations Management Istanbul, Turkey,* 2012.
7. C. Singh, B., Garg, S.K., Sharma, S.K. and Grewal, Int. J. Lean Six Sigma, 1, 57–168 (2010).
8. R. . Cua, K.O., McKone, K.E. and Schroeder, J. Oper. Manag., vol. 19, 675–694(2001).
9. M. Hallgren and J. Olhager, *Int. J. Oper. Prod. Manag.*, 29, 976–999 (2009).
10. G. Nawanir, L. K. Teong, and S. N. Othman, J. of Manufacturing Tech. Manag., 24, 7(2013).
11. Vinodh, R. B. P. Ruben, and P. Asokan, Clean Technol. Environ. Policy, 18,1(2016).
12. S. Hartini and U. Ciptomulyono, *Procedia Manufacturing* 4 (Iess, 2015)
13. A. Norton and A. Fearne, Sustainable Value Stream Mapping: A Practical Aid to Sustainable Production. 2007.
14. D. Simons and R. Mason, Ecr, 84–91, 2003.
15. US EPA, *http://www.epa.gov/lean/environment/toolkits/energy/index.htm*, 2007. .
16. US EPA, *The U.S. Environmental Protection Agency (EPA)*, 2007. .
17. A. S. Torres and A. M. Gati, *PICMET '09 - 2009*
18. G. W. Kuriger and F. F. Chen, *Proceedings of the 2010 Industrial Engineering Research Conference*, 2010.
19. M. D. Dadashzadeh and T. J. Wharton, *Int. J. Manag. Inf. Syst.*, 16, 125–136 (2012).
20. H. Li, H. Cao, and X. Pan, Int. J. Comput. Integr. Manuf., 25, 12(2012).
21. D. Folinas, D. Aidonis, G. Malindretos, N. Voulgarakis, and D. Triantafillou, Int. J. Agric. Resour. Gov. Ecol. **10**, 129 (2014).
22. D. K. Folinas, D. Aidonis, and P. Karayannakidis, J. Agric. Informatics 6, 24–39 (2015).
23. W. Faulkner and F. Badurdeen, J. Clean. Prod., 85, pp. 8–18 (2014).
24. A. Brown, J. Amundson, and F.Badurdeen, J. Clean. Prod., 1–16 (2014).
25. S. Hartini and I. Rizkiya, *Seminar Nasional Teknik Industri dan Manajemen Produksi IV* 2009.
26. S. Hartini, Susatyo, and D. Subekti, *Apchi-Ergofuture International Joint Conference 2010*
27. J. Gabiati, M. F. Deimling, and R. Barichello, *International Administration Conference*, 2014.
28. M. Paju, J. Heilala, M. Hentula, A. Heikkila, and B. Johansson, *Proceeding of the 2010 Winter Simulation Conference*, 2010
29. S. Hartini, U. Ciptomulyono, and M. Anityasari, *Annual Conference on Industrial and System Engineering (ACISE)*, 2016.
30. A. A. Feil, D. M. De Quevedo, and D. Schreiber, Sustain. Prod. Consum.,3, 34–44 (2015).

# Filling machine preventive maintenance using age replacement method in PT Lucas Djaja

*Fifi Herni* Mustofa*, *Ria Ferdian* Utomo, and *Kusmaningrum* Soemadi

Industrial Engineering Department, Institut Teknologi Nasional, PHH. Mustafa No. 23 Bandung 40124, Indonesia

**Abstract.** PT Lucas Djaja is a company engaged in the pharmaceutical industry which produce sterile drugs and non-sterile. Filling machine has a high failure rate and expensive corrective maintenance cost. PT Lucas Djaja has a policy to perform engine maintenance by way of corrective maintenance. The study focused on the critical components, namely bearing R2, bearing 625 and bearing 626. When the replacement of the failure done by the company is currently using the formula mean time to failure with the result of bearing R2 at point 165 days, bearing 625 at a point 205 days, and bearing 626 at a point 182 days. Solutions generated by using age replacement method with minimization of total maintenance cost given on the bearing R2 at a point 60 days, bearing 625 at the point of 80 days and bearing 626 at a point 40 days.

## 1 Introduction

In the drug production every pharmaceutical industry must be able to find the way in producing good medicine to guarantee and produce qualified product. Qualified products are supported by machines that can increase productivity and minimize production costs. PT Lucas Djaja is a company engaged in the pharmaceutical industry that produces sterile and non-sterile drugs. Non-sterile drugs consist of syrup, tablets, capsules and effervescent. The sterile drug consists of an ampoule injection and a vial injection. Injection drug is a sterile drug in the solution form that must be injected by tearing skin or mucous membranes.

PT Lucas Djaja has a policy to perform maintenance of the machine by means of preventive maintenance. But the maintenance process is not done regularly, so that breakdown machine can hamper production activities. Inhibition of production activities can lead to downtime, therefore by applying a good maintenance policy can improve productivity and machine performance. Machines used for the production process in manufacturing injectable drugs such as distillation machines, washing machines, mixing machines, filling machines, and autoclave machines. Among all production process machines, the machines that have an interval of failure is filling machine. PT Lucas Djaja has two injection filling machines that have the same specification but different capacity.

Filling machine is a machine that support the production process at PT Lucas Djaja, the machine has a high failure rate and expensive maintenance cost. Failure due to irregular maintenance will cause problems such as production schedules and costs incurred. At present, the company does not yet have a model to determine the schedule for preventive maintenance. Company need a maintenance schedule for prevent failure that can reduce the cost of the maintenance. Cumulative failure cost is more expensive than the cumulative preventive cost. The schedule for preventive maintenance can prevent the occurrence of failure that causes inhibition of the production process due to unplaned maintenance.

## 2 Preventive Maintenance Model

In analyzing those issues, the age replacement method is used. The age replacement method is a component replacement scheduling method based on the optimal component life [3]. This is done to minimize costs incurred by the company due to the engine failure. Preventive maintenance is done at the optimal time before the failures occurs. The determination of optimal preventive maintenance times depends on the objectives, ie minimizing costs or maximizing availability [4]. Here is a model of care policy:

1. Block Replacement, as a constant interval, because the preventive maintenance is done at definite intervals and corrective maintenance whenever required.
2. Age Replacement, the preventive maintenance time depends on the component age. If failure occurs, the component time reset to zero.

In age replacement model, the replacement policy is to perform preventive maintenance when the component reaches a certain $t_p$ age, as well as corrective maintenance if necessary. The purpose of this model is to determine the optimal interval of component preventive maintenance by minimizing the predicted total cost per time unit. The formula to find out the expected total maintenance cost per time unit can be seen in this equation [3];

* Corresponding author: fifi.mustofa@gmail.com

$$C(t_p) = \frac{Cp \times R(tp) + (Cf \times [1 - R(tp)]}{(tp+Tp) x R(tp) + (M(tp) + Tf[1-R(tp)])} \quad (1)$$

Where:

$C(t_p)$ = expectation of total maintenance cost per unit time ($t_p$).
$C_f$ = the cost of corrective maintenance to preventive maintenance.
$C_p$ = preventive maintenance cost.
$f(t_p)$ = probability density function.
$T_p$ = time of preventive maintenance performance.
$T_f$ = time moments of failure performance.
$R(t_p)$ = equipment and machine reliability function at $t_p$.
$M(t_p)$= meantime error of time moments of failure performance at $t_p$.

## 3 Research Method

The steps taken in the research method are formulate the problem, literature study, determination of resolution method, data collection and processing, analysis and conclusions.

### 3.1 Formulation of the Problem

The phase of problem identification is the first step done on this research. In the injection production there are distillation machine, washing machine, oven machine, mixer machine, filling machine and autoclave machine. PT Lucas Djaja has a policy to perform maintenance of the machine by means of preventive maintenance. But the maintenance process is not done regularly.

Based on the data of damages obtained from the company and the selection of critical machinery, filling machine is a machine that has a major impact on the production of injection. The filling machine is the most commonly damaged machine with high cost and filling machine has many machine compiler components.

No regular schedule can result in the quickly broken components, and the machine can not be operated. This can result in huge losses to the company due to the faltered of production process. Company need a maintenance schedule for prevent components failure that can reduce the cost of the maintenance.

### 3.2 Literature Study

The study of literature is the supporting theories to find resolution method. These theories are necessary in assessing the problems that occur and as a reference to solve and analyze. Theories used such as the theory of model preventive and corrective maintenance, the distribution of statistical opportunities used to test the existing failure rate [2], the distribution parameters used, the calculation of the distribution functions, the calculation of maintenance costs, and decision making for component replacement [3].

### 3.3 Determination of Resolution Method

The solution to solve the problem is to use the age replacement method. The age replacement method is a component replacement scheduling method based on the optimal component life [3]. This is done to minimize costs incurred by the company due to maintenance. The policy solution in the scheduling of preventive maintenance has a parameter that is the working day of the machine as a reference time interval for doing preventive maintenance.

### 3.4 Data Collection and Analyze

Data collections contain data of producing process of injection drugs, the failure data of each machine, critical machine, and selected machine characteristics. Component failure data consists of; maintenance time data and component failure, data of preventive maintenance cost, and component malfunction.

After the data collected and classified, it analyzes by following steps:

a. Critical Component Determination
   The critical component is the component that has the biggest maintenance cost which be effected on all component maintenance cost, and it is selected using a pareto chart.
b. Calculation of Failure Interval [2]
   In the data of failure component, there is information about failure date occurred; from the date can be calculated the intervals between the failure in days.
c. Calculation Pearson Product Moment Test (Index Of Fit)
   The critical failure interval data is tested using 4 (four) distribution patterns, ie normal, lognormal, exponential and weibull distributions. By doing the test, it can be known that the tendency of critical component failure data follows a certain distribution pattern.
d. Calculation of Selected Failure Distribution Parameters
   The failure distribution parameter is calculated based on selected distribution pattern, while the selected distribution parameters calculation aims to determine the rate of failure component and support for the calculation of the distribution function.
e. Calculation of Component Preventive Maintenance Interval [3]

In age replacement model, the replacement time interval depends on day which result the smallest value of expected total maintenance cost.

### 3.4 Analysis and Conclusions

After the data processing phase it will be done analysis of the data processing results. Includes failure data analysis, selected distribution patterns, time interval analysis between failure and cost comparison analysis of preventive replacement intervals optimally under current conditions.

The conclusion contains results from research at PT Lucas Djaja. Based on these conclusions, the suggestion is in the form of improvement.

## 4 Numerical Problem

### 4.1 Data Collecting

Data collections contain data of the failure data of each machine, critical machine, selected machine characteristics, and component critical machine are failure data that consists; maintenance time and failure of the components and the cost of preventive maintenance and corrective maintenance.

#### *4.1.1 Critical Machine and Component Selection*

The critical machine selected is the filling machine based on the highest failure frequency. Selection of the critical machine uses Pareto chart.

The filling machine failures is the result of components failure. Good component conditions can have a good effect on the machine. But if the component condition is badly it will result in defective product or the machine will not be used for production process. Filling machine has five of components that often suffered damage are bearing R2, bearing 625, bearing 626, bearing 607 and roller.

#### *4.1.2 Preventive Maintenance and Corrective Maintenance Duration*

Preventive and corrective maintenance activities are the failure component replacement, including time open cover and shut down the machine. The duration time is based on interviews with the maintenance technicians who directly handle the machine breakdown. Duration of preventive and corrective maintenance component of filling machine can be seen on Table 1.

**Table 1**. Duration Preventive and Corrective Maintenance

| Component | Preventive maintenance duration (hour) | Corrective maintenance duration (hour) |
|---|---|---|
| Bearing R 2 | 9 | 11 |
| Bearing 625 | 12 | 14 |
| Bearing 626 | 12 | 14 |
| Bearing 608 | 12 | 14 |
| Roller | 12 | 14 |

#### *4.1.3 Preventive Maintenance Cost*

Cost incurred by the company, including:

1. Component Purchase Cost
   The purchase price component of filling machine that failure can be seen in Table 2.

**Table 2**. Component Price

| Component | Component Number (Unit/replace) | Component Price (Rp) |
|---|---|---|
| Bearing R 2 | 12 | 15.000 |
| Bearing 625 | 12 | 30.000 |
| Bearing 626 | 12 | 30.000 |
| Bearing 608 | 4 | 25.000 |
| Roller | 6 | 20.000 |

2. Maintenance Technician Cost
   Preventive and corrective maintenance is done by a permanent technician. Technicians perform maintenance for 8 hours work/day, if maintenance is done beyond working hours it will be done overtime. The overtime fee is Rp 15.000/hour.
3. Unemployed Operators Cost
   The cost of idle operators is the cost incurred when the machine is undergoing preventive or corrective maintenance, while the operator is still paid every month. There are six operators. The operator fee is Rp 15.000/hour.
4. Lost Profits
   If the machine is broken suddenly, the production process will be halt. It causes the production target not will be reached and the company will lose their profit. The profit lost is Rp 423.998/hour.
5. Downtime and Failure Frequency
   Downtime of the machine is required to know how long the machine will breakdown. Downtime occurs due to expected technician waiting time and failure time. Expected technician waiting time is 5,67 hours. This downtime will influence the idle operator and the lost profits on the failure impact cost.

Failure frequency is the number of failures that occurred over a span of three years from 2014 to 2016. Downtime dan failure frequency can be seen on Table 3.

**Table 3**. Downtime dan Failure Frequency

| Component | Average Downtime (hour) | Failure Frequency |
|---|---|---|
| Bearing R 2 | 17,67 | 7 |
| Bearing 625 | 19,67 | 4 |
| Bearing 626 | 19,67 | 3 |
| Bearing 608 | 19,67 | 3 |
| Roller | 19,67 | 2 |

### 4.2 Data Processing

Data processing contains the determination of critical components, determination of failure distribution, parameter determination and calculation of component preventive maintenance interval.

### *4.2.1 Determination of Critical Component*

Critical components are selected based on the highest maintenance cost and affect to the other components if it's get failure, which can cause the company loses their profit. The calculation of the selection of critical components in the filling machine can be seen in Table 4.

**Table 4.** Calculation of *Filling* Machine Critical Component

| Component | Failure Impact Cost (Rp) | Replace-ment Cost (Rp) | Total Cost (Rp) | Cumm % |
|---|---|---|---|---|
| Bearing R 2 | 63.996.413 | 1.260.000 | 65.256.413 | 37% |
| Bearing 625 | 36.569.379 | 1.440.000 | 38.009.379 | 58% |
| Bearing 626 | 27.427.034 | 1.080.000 | 28.507.034 | 74% |
| Bearing 608 | 27.427.034 | 300.000 | 27.727.034 | 90% |
| Roller | 18.284.689 | 240.000 | 18.524.689 | 100% |
| | | Total | 178.024.549 | |

Example Calculation of Component Bearing R 2:
- Component Purchase Cost
  = Number of components x Component Price
  = 12 unit x Rp 15.000
  = Rp180.000/unit bearing R2

- Replacement Cost = Purchase Cost x Frequency of Failure
  = Rp180.000 x 7
  = Rp1.260.000

- Failure Impact Cost = (Unemployed Operator Cost + Maintenance Technician Cost + Lost Profit) x frequency of failure
  = (6 people x Rp 15.000 x 17,67 hours) + (4 hours x Rp15.000) + (Rp 423.998 x 17,67 hours) x 7
  = Rp 63.996.413

- Total Bearing Cost R 2 = Replacement Cost + Failure Impact Cost
  = Rp1.260.000 + Rp 63.996.413
  = Rp 65.256.413

- Percentage = (Total Component Bearing Cost R2/Total Cost) x 100%
  = Rp 65.256.413 / Rp178.024.549 = 37%

The critical components selected are the most influential components of bearing R2, bearing 625 and bearing 626. Components bearings 608 and roller are not selected because the cost percentage is not more than 20% and the total cost does not have a big effect.

### 4.2.2*Determination of Failure Distribution*

The component corrective interval is obtained from the day moments of failure performance and the machine can be used until the component is failure again. Below, for example time interval of failure component bearing R2 can be seen in Table 5.

**Table 5.** Interval of Failure Bearing R2

| No | Date of Failure | Failure Interval (days) |
|---|---|---|
| 1 | 04-Feb-14 | |
| 2 | 21-Jul-14 | 167 |
| 3 | 10-Nov-14 | 112 |
| 4 | 16-Mar-15 | 126 |
| 5 | 10-Nov-15 | 239 |
| 6 | 03-Feb-16 | 85 |
| 7 | 23-Agu-16 | 202 |

The selected distribution pattern is the largest index of fit value, the recapitulation of selected failure distribution pattern can be seen in Table 6.

**Table 6.** Recapitulation of Failure Distribution Pattern

| No | Components | Selected Failure Distribution Pattern |
|---|---|---|
| 1 | Bearing R 2 | Lognormal |
| 2 | Bearing 625 | Weibull |
| 3 | Bearing 626 | Lognormal |

### *4.2.3 Parameter Failure Distribution Determination*

The calculation of distribution parameters based on the selected failure distribution for each component. Example for bearing R2 can be seen on Table 7.

**Table 7.** Lognormal Distribution Parameters

| No(i) | $T_i$ | $X_i$ | M | $(X_i-\mu)^2$ | σ |
|---|---|---|---|---|---|
| 1 | 85 | 4,443 | 4,984 | 0,292 | 0,354 |
| 2 | 112 | 4,718 | | 0,070 | |
| 3 | 126 | 4,836 | | 0,022 | |
| 4 | 167 | 5,118 | | 0,018 | |
| 5 | 202 | 5,308 | | 0,106 | |
| 6 | 239 | 5,476 | | 0,243 | |
| Total | 931 | 29,900 | | 0,751 | |

Calculation Example:

$Xi = \ln ti = \ln 85 = 4{,}442651$

$$\mu = \overline{Xi} = \frac{\sum_{i=1}^{n} \ln ti}{n} = \frac{29{,}9001}{4} = 4{,}984$$

$$\sigma = \sqrt{\frac{\sum_{i=1}^{n} (Xi-\mu)^2}{n}} = \sqrt{\frac{0{,}75099}{6}} = 0{,}354$$

$$\text{MTTF} = exp\left(\mu + \frac{\sigma^2}{2}\right) = exp\left(29{,}90016 + \frac{0{,}3537^2}{2}\right) = 165{,}43 \text{ days}$$

### *4.2.4 Calculation of Component Preventive Maintenance Interval*

Before calculation of preventive maintenance interval using equation (1), recapitulation of preventive and corrective maintenance costs of bearing R2, bearing 625, and bearing 626 can be seen on Table 8.

**Table 8.** Preventive and Corrective Maintenance Cost

| Component | Preventive Cost (Rp) | Corrective Cost (Rp) |
|---|---|---|
| Bearing R 2 | 1.095.000 | 9.322.338 |
| Bearing 625 | 1.395.000 | 10.560.334 |
| Bearing 626 | 1.395.000 | 10.560.334 |

Calculation of component replacement intervals based on calculation of distribution functions selected at a particular time unit and the determination of optimal component replacement intervals selected based on the expectation of total replacement cost per unit time (C (tp)) the smallest [3].

Interval calculation component Bearing R2 based on lognormal distribution, 60 days is the optimum point because it has the smallest value of C (tp) which is worth Rp 5.059/day can be seen on Table 9.

**Table 9.** Preventive Maintenance Interval Bearing R2

| tp (day) | $F_{(tp)}$ | $R_{(tp)}$ | $(tp+Tp).R_{(tp)}$ + $(M_{(tp)}+Tf).(1-R_{(tp)})$ | $C_{(tp)}$ (Rp/day) |
|---|---|---|---|---|
| 10 | 0,0000 | 1,0000 | 176,545 | 6.202 |
| 20 | 0,0000 | 1,0000 | 186,545 | 5.870 |
| 30 | 0,0000 | 1,0000 | 196,545 | 5.571 |
| 40 | 0,0001 | 0,9999 | 206,540 | 5.307 |
| 50 | 0,0012 | 0,9988 | 216,484 | 5.105 |
| 60 | 0,0060 | 0,9940 | 226,189 | 5.059 |
| 70 | 0,0189 | 0,9811 | 235,231 | 5.315 |
| 80 | 0,0446 | 0,9554 | 242,997 | 6.015 |
| 90 | 0,0858 | 0,9142 | 248,856 | 7.237 |
| 100 | 0,1425 | 0,8575 | 252,350 | 8.984 |

The calculation of component Bearing 625 the interval based on the weilbull distribution, 80 days is the optimal point because it has the smallest value of *C (tp)* which is worth Rp 29.118/day. The calculation of component preventive maintenance intervals can be seen in Table 10.

**Table 10.** Preventive Maintenance Interval Bearing 625

| tp (day) | $F_{(tp)}$ | $R_{(tp)}$ | $(tp+Tp).R_{(tp)}$ + $(M_{(tp)}+Tf).(1-R_{(tp)})$ | $C_{(tp)}$ (Rp/day) |
|---|---|---|---|---|
| 40 | 0,0010 | 0,9820 | 40,658 | 38.389 |
| 50 | 0,0013 | 0,9701 | 49,891 | 33.456 |
| 60 | 0,0017 | 0,9550 | 58,693 | 30.791 |
| 70 | 0,0020 | 0,9367 | 66,967 | 29.495 |
| 80 | 0,0023 | 0,9151 | 74,619 | 29.118 |
| 90 | 0,0026 | 0,8905 | 81,560 | 29.411 |
| 100 | 0,0029 | 0,8629 | 87,714 | 30.232 |
| 110 | 0,0032 | 0,8325 | 93,017 | 31.499 |
| 120 | 0,0034 | 0,7997 | 97,417 | 33.164 |
| 130 | 0,0036 | 0,7647 | 100,878 | 35.206 |
| 140 | 0,0038 | 0,7279 | 103,381 | 37.620 |
| 150 | 0,0039 | 0,6895 | 104,923 | 40.416 |

The calculation of component Bearing 626 the interval based on the lognormal distribution, 40 days is the optimal point because it has the smallest value of *C (tp)* which is worth Rp 6.564/day. The calculation of component preventive maintenance intervals can be seen in Table 11.

**Table 11.** Preventive Maintenance Interval Bearing 626

| tp (day) | $F_{(tp)}$ | $R_{(tp)}$ | $(tp+Tp).R_{(tp)}$ + $(M_{(tp)}+Tf).(1-R_{(tp)})$ | $C_{(tp)}$ (Rp/day) |
|---|---|---|---|---|
| 10 | 0,0000 | 1,0000 | 193,325 | 7.216 |
| 20 | 0,0002 | 0,9998 | 203,321 | 6.870 |
| 30 | 0,0019 | 0,9981 | 213,268 | 6.623 |
| 40 | 0,0075 | 0,9925 | 223,027 | 6.564 |
| 50 | 0,0191 | 0,9809 | 232,379 | 6.755 |
| 60 | 0,0374 | 0,9626 | 241,093 | 7.209 |
| 70 | 0,0625 | 0,9375 | 248,973 | 7.904 |
| 80 | 0,0934 | 0,9066 | 255,886 | 8.798 |
| 90 | 0,1290 | 0,8710 | 261,762 | 9.847 |
| 100 | 0,1680 | 0,8320 | 266,588 | 11.009 |
| 110 | 0,2092 | 0,7908 | 270,394 | 12.250 |

## 5 Analysis

PT Lucas Djaja has 6 machines used during injection production process. The six machines include distillation machines, washing machines, oven machines, mixing machines, filling machines, and autoclave machines. The criteria for this critical machine selection diagram pareto is seen from the frequency of error occurring in the past. According to pareto chart, the most frequently error happened in filling machine, the cause of this filling machine error due to bearing composing components.

These criteria are seen from the amount of component maintenance cost and the consequent of letting the component failure. Components selection are selected by showing the biggest cost incurred by the company if the component is error. The error components can cause hampering of production process and the lost of company's profit. The critical components selected on filling machine are bearing R2, bearing 625 and bearing 626.

Determination of time interval for critical components maintenance is done by using age replacement method with minimization total maintenance cost. For the determination of preventive maintenance interval is determined by calculating *C (tp)* which is repeatedly calculated to obtain the smallest C(tp). The method used to determine the optimal preventive maintenance interval of filling machine for component bearing R2, bearing 625, and bearing 626 is age replacement method developed by Jardine (1973). The purpose of this method is to know the optimal preventive maintenance interval with minimal maintenance cost. So, the interval with the smallest *C (tp)* is the optimal component maintenance time. This method is suitable to use in research to remember the lifetime of the components affected by its usage.

## 6 Conclusions

Based on the finding and discussion of the data above, it can be concluded as below:

1. On the corrective maintenance schedule done by the company, calculated using the formula of mean time to failure, it obtained the result of maintenance of filling machine component bearing R2 at 165 days, bearing 625 at 205 days, bearing 626 at 182 days.

2. On the calculation result using age replacement method, the calculation interval of preventive maintenance for filling machine component bearing R2 at point 60 days. bearing 625 at 80 days. bearing 626 at point 40 days.
3. On the result of the failures cost, and if the cost is done only the preventive maintenance has the difference on bearing R2 Rp 8.227.338, bearing 625 Rp 9.165.334, and bearing 626 Rp 9.165.334. The cost has a big difference because at the cost of corrective maintenance there are advantages lost and the downtime is too long.

## References

1. A. Corder, S,. *Teknik Manajemen Pemeliharaan,* (Erlangga, Jakata, 1988)
2. C. Ebeling, *An Introduction to Reliability and Maintainability Engineering* (McGraw-Hill Companies, Singapore, 1997)
3. A. Jardine, K.S., *Maintenance, Replacement, and Reliability* (Pitman Publishing Corporation, Canada, 1973)
4. J. Usher, S., Kamal, A.H., Syed, W.H., IIE Transactions, **30**, 1121-1128 (1998)
5. O'Connor, P.D.T., *Practical Reliability Engineering* (John Willey & Sons, New York, 1991)
6. R. Walpole, E. & Myers, R.H. *Probability and Statistics for Engineers and Scientists*, (Macmillan Publishing Co, New York, 5th ed., 1993)
7. S. Wignjosoebroto, *Pengantar Teknik & Manajemen Industri* (Guna Widya, 2006)

# Interpretive structural modeling for technology-related barriers to energy efficiency improvement in Indonesia steel industry

*Apriani Soepardi*[1,2,*], *Pratikto*[2], *Purnomo Budi Santoso*[3], and *Ishardita Pambudi Tama*[3]

[1]Industrial Engineering Department, University of National Development "Veteran" Yogyakarta 55584, Indonesia
[2]Mechanical Engineering Department, Brawijaya University, Malang 65145, Indonesia
[3]Industrial Engineering Department, Brawijaya University, Malang 65145,

**Abstract.** This paper investigates the existence of the barriers in Indonesia steel industry, namely: existing technologies were still considered efficient, new technologies not available, new technology not adequate or inappropriate at the site, the energy efficient technology not compatible with other elements of the plant, the uncertainty about future technology standards, lack of technical skills, technology installation period was too long, and poor information quality regarding access to new energy efficient technology. Five practitioners from the national steel industry were retrieved to identify and analyze critical elements and the causal interrelationships among the technology-related barriers. Finally, using the approach of interpretive structural modelling, the barriers were clustered in accordance with their driving power and dependence power. Lack of reliable information access about new technologies is the only barriers at the bottom level of the hierarchy, implying higher driving power. It is also observed that the availability of new technologies, existing technology still efficient, and the uncertainty about future technology standardization have a weak dependent on the other constructs but strongly driver. The four remaining constructs are identified as dependent obstacles. It means that these barriers are the automatic followers of other impediments. The result of this model will assist managers to understand the relative importance and the interdependencies among the barriers. It can guide them to resolve these barriers.

## 1 Introduction

The iron and steel industry is one of the highly energy-intensive industrial sectors. The energy consumption in Indonesia's steel industry depletes about 20-35 % of total production costs. As a consequence, energy use efficiency is the important measure to reduce energy intensity that eventually results declining production costs. Although some national industries have focused on rectifying energy efficiency by implementation of energy efficient technologies, but the result had not significantly decreased the energy use rate [1]. It can denote that there are still a number of difficulties which impedes implementation of improved energy efficiency. These difficulties show the existence of barriers to energy use efficiency, in particular related to technology.

Nevertheless, the implementation of improved energy efficiency measures becomes an imperative for manufacturing industries, a number of barriers still exist which prevents industry involvement in reducing energy use [2]. To obtain and understand the suitable policies and the know-how of energy-efficient technologies and practices adopted in the firm may be possible with thorough understanding and reforming studies of these hindrances [3]. Therefore, it is still needed to find new and more effective methods to evaluate the importance of barriers to the firm's decision-making process for adopting energy saving technologies.

Furthermore, there is a trend to explore how each hindrance interconnected. In other words, it is still significantly needed to develop a barrier model involving the relationship between barrier. For these reasons above, this research has conducted to identify and analyze the interaction of the technology-related barriers to energy efficiency improvement from energy use in the Indonesia steel industry using interpretive structural modeling (ISM) approach.

## 2 Technology-related Barriers

According to [4] said that "*A barrier is a postulated mechanism that inhibits in technologies that are both energy efficient and (apparently) economically efficient.*"

Whereas in accordance with [5], the technical risk is related to technological barriers in energy efficiency implementation lead to production failure and product quality losses.

Limited in skilled technical workforce can impede the improved energy efficiency because staffs only focus on daily production target [6].

The same situation have been mentioned by [7] that lack of technical skill and expertise or skilled man power are barriers to energy efficiency improvement. Beside

*Corresponding author: apriani.soepardi@gmail.com

these obstacles, [7] reveals that lack of qualified information access to new technologies may bring out a barrier.

## 3 Methodology

The focus of this research is to understand the mutual relationship of technology-related barriers to energy efficiency using ISM approach and identify these barriers which influence the other barriers and are influenced by others. Interpretive structural modeling (ISM) is a well-established methodology for identifying and summarizing relationships among specific variables, which define a problem [8].

Formerly, the technology-related barriers which relevant to energy use issue were identified based on literature review. Five practitioners from five steel industries were consulted to identify the relationships among related elements. The practitioners' opinions were taken separately using a questionnaire to avoid possible influence among their answers. Further, the questionnaire results were combined and analyzed by three academics to achieve a final matrix.

### 3.1 Structural Self-Interaction Matrix (SSIM)

The eight barriers identified previously are arranged in rows and columns for developing the barrier matrix. A contextual relationship is symbolized with four symbols that denote the direction of the relationship between the barriers (*i* and *j*) [9] :

V: Barrier *i* will influence/cause Barrier *j*;
A: Barrier *j* will be influenced/caused by Barrier *i*;
X: Barriers *i* and *j* will influence each other; and
O: Barriers *i* and *j* are unrelated.

On the basis of pairwise relationships between barrier issues, the SSIM is developed, as shown in Table 1.

**Table 1.** Structural Self-Interaction Matrix

| Technology-related barriers | 8 | 7 | 6 | 5 | 4 | 3 | 2 | 1 |
|---|---|---|---|---|---|---|---|---|
| 1. Availability of new energy efficient technologies | O | X | O | O | O | V | V | |
| 2. New technologies not adequate or compatible at the site | A | A | A | O | O | A | | |
| 3. Lack of know-how in integration of new and existing technology | A | A | O | X | X | | | |
| 4. The installation period of new technologies was too long | O | O | O | A | | | | |
| 5. Lack of skilled technical staff for new technology implementation | O | A | A | | | | | |
| 6. Lack of reliable information access to new technologies | O | V | | | | | | |
| 7. Existing technology still efficient | O | | | | | | | |
| 8. The uncertainty about future technology standardization | | | | | | | | |

### 3.2 Reachability Matrix (RM)

An initial reachability matrix is then developed in the next step. This matrix (Table 2) is constructed from the SSIM by converting into binary numbers, "1" and "0". In this regard, rules for transformation are given.

1. If the (*i*, *j*) entry in the SSIM is V, then substitute in the (*i*, *j*) entry in the reachability metrics becomes 1 and the (*j*, *i*) entry as 0.
2. If the (*i*, *j*) entry in the SSIM is A, then substitute in the (*i*, *j*) entry in the reachability metrics becomes 0 and (*j*, *i*) entry as 1.
3. If the (*i*, *j*) entry in the SSIM is X then substitute in the (*i*, *j*) entry in the reachability metrics becomes 1 and (*j*, *i*) entry as 1.
4. If the (*i*, *j*) entry in the SSIM is O then substitute in the (*i*, *j*) entry in the reachability metrics becomes 0 and (*j*, *i*) entry as 0.

**Table 2**. Initial Reachability Matrix

| Technology-related barriers | 8 | 7 | 6 | 5 | 4 | 3 | 2 | 1 |
|---|---|---|---|---|---|---|---|---|
| 1.Availability of new energy efficient technologies | 0 | 1 | 0 | 0 | 0 | 1 | 1 | 1 |
| 2.New technologies not adequate or compatible at the site | 0 | 0 | 0 | 0 | 0 | 0 | 1 | 0 |
| 3.Lack of know-how in integration of new and existing technology | 0 | 0 | 0 | 1 | 1 | 1 | 1 | 0 |
| 4.The installation period of new technologies was too long | 0 | 0 | 0 | 0 | 1 | 1 | 0 | 0 |
| 5.Lack of skilled technical staff for new technology implementation | 0 | 0 | 0 | 1 | 1 | 1 | 0 | 0 |
| 6.Lack of reliable information access to new technologies | 0 | 1 | 1 | 1 | 0 | 0 | 1 | 0 |
| 7.Existing technology still efficient | 0 | 1 | 0 | 1 | 0 | 1 | 1 | 1 |
| 8.The uncertainty about future technology standardization | 1 | 0 | 0 | 0 | 0 | 1 | 1 | 0 |

The initial reachability matrix is examined for incorporating the transitivity concept and making modifications (if any). This concept states that if a barrier *i* is related to *j* and if the barrier *j* is related to a third barrier *k*, then *i* is necessarily related to *k*. Thus, a final reachability matrix is achieved, as shown in Table 3.

**Table 3**. Final Reachability Matrix

| Technology-related barriers | 8 | 7 | 6 | 5 | 4 | 3 | 2 | 1 |
|---|---|---|---|---|---|---|---|---|
| 1 Availability of new energy efficient technologies | 0 | 1 | 0 | **1** | **1** | 1 | 1 | 1 |
| 1.New technologies not adequate or compatible at the site | 0 | 0 | 0 | 0 | 0 | 0 | 1 | 0 |
| 2.Lack of know-how in integration of new and existing technology | 0 | 0 | 0 | 1 | 1 | 1 | 1 | 0 |
| 3.The installation period of new technologies was too long | 0 | 0 | 0 | **1** | 1 | 1 | **1** | 0 |
| 4.Lack of skilled technical staff for new technology implementation | 0 | 0 | 0 | 1 | 1 | 1 | **1** | 0 |
| 5.Lack of reliable information access to new technologies | 0 | 1 | 1 | 1 | **1** | **1** | 1 | **1** |
| 6.Existing technology still efficient | 0 | 1 | 0 | 1 | **1** | 1 | 1 | 1 |

| | | | | | | | | |
|---|---|---|---|---|---|---|---|---|
| 7.The uncertainty about future technology standardization | 1 | 0 | 0 | 1 | 1 | 1 | 1 | 0 |

### 3.3 Level partitions and Canonical Matrix

The reachability and antecedent set for each barrier is obtained from the final reachability matrix. The reachability set consists of the barrier itself and the other barriers, which it may reach. Whereas, the antecedent set consists of the barrier elements itself and the other barrier elements, which may reach it. The intersection for the reachability and antecedent sets is derived for all the barriers. The barriers are considered as a top-level barrier in the ISM hierarchy for which these sets are the same. To obtain the next level, the top-level barrier is separated out from the other remaining barriers. The whole process of partitioning is continued until the levels of all barriers are determined (Table 4-7).

**Table 4**. Partition of RM Interaction – Iteration 1

| Barriers # | Reachability set | Antecendent set | Intersection set | Level |
|---|---|---|---|---|
| 1 | 1,2,3,4,5,7 | 1,6,7 | 1,7 | |
| 2 | 2 | 1,2,3,4,5,6,7,8 | 2 | I |
| 3 | 2,3,4,5 | 1,2,3,4,5,6,7 | 3,4,5 | |
| 4 | 2,3,4,5 | 1,2,3,4,5,6,7 | 3,4,5 | |
| 5 | 2,3,4,5 | 1,2,3,4,5,6,7 | 3,4,5 | |
| 6 | 1,2,3,4,5,6,7 | 6 | 6 | |
| 7 | 1,2,3,4,5,7 | 1,6,7 | 1,7 | |
| 8 | 2,3,4,5,8 | 8 | 8 | |

**Table 5**. Partition of RM Interaction – Iteration 2

| Barriers # | Reachability set | Antecendent set | Intersection set | Level |
|---|---|---|---|---|
| 1 | 1,3,4,5,7 | 1,6,7 | 1,7 | |
| 3 | 3,4,5 | 1,3,4,5,6,7,8 | 3,4,5 | II |
| 4 | 3,4,5 | 1,3,4,5,6,7,8 | 3,4,5 | II |
| 5 | 3,4,5 | 1,3,4,5,6,7,8 | 3,4,5 | II |
| 6 | 1,3,4,5,6,7 | 6 | 6 | |
| 7 | 1,3,4,5,7 | 1,6,7 | 1,7 | |
| 8 | 3,4,5,8 | 8 | 8 | |

**Table 6**. Partition of RM Interaction – Iteration 3

| Barriers # | Reachability set | Antecendent set | Intersection set | Level |
|---|---|---|---|---|
| 1 | 1,7 | 1,6,7 | 1,7 | III |
| 6 | 1,6,7 | 6 | 6 | |
| 7 | 1,7 | 1,6,7 | 1,7 | III |
| 8 | 8 | 8 | 8 | III |

**Table 7**. Partition of RM Interaction – Iteration 4

| Barriers # | Reachability set | Antecendent set | Intersection set | Level |
|---|---|---|---|---|
| 6 | 6 | 6 | 6 | IV |

The final reachability matrix is converted into a canonical format by clustering barriers in the same level across the rows and columns of the matrix, as shown in Table 8. In this table, the driver power is the total number of first in the raw and the dependence power is the total number of first in the column

**Table 8**. Canonical matrix

| Barrier # | 6 | 8 | 7 | 1 | 5 | 4 | 3 | 2 | Driver power | Ranks |
|---|---|---|---|---|---|---|---|---|---|---|
| 2 | 0 | 0 | 0 | 0 | 0 | 0 | 0 | 1 | 1 | 5 |
| 3 | 0 | 0 | 0 | 0 | 1 | 1 | 1 | 1 | 4 | 4 |
| 4 | 0 | 0 | 0 | 0 | 1 | 1 | 1 | 1 | 4 | 4 |
| 5 | 0 | 0 | 0 | 0 | 1 | 1 | 1 | 1 | 4 | 4 |
| 1 | 0 | 0 | 1 | 1 | 1 | 1 | 1 | 1 | 6 | 2 |
| 7 | 0 | 0 | 1 | 1 | 1 | 1 | 1 | 1 | 6 | 2 |
| 8 | 0 | 1 | 0 | 0 | 1 | 1 | 1 | 1 | 5 | 3 |
| 6 | 1 | 0 | 1 | 1 | 1 | 1 | 1 | 1 | 7 | 1 |
| Dependence | 8 | 7 | 7 | 7 | 3 | 3 | 1 | 1 | | |
| Ranks | 1 | 2 | 2 | 2 | 3 | 3 | 4 | 4 | | |

### 3.4 ISM-based model formation

You are free to use colour illustrations for the online version of the proceedings but any print version will be printed in black and white unless special arrangements have been made with the conference organiser. Please check with the conference organiser whether or not this is the case. If the print version will be black and white only, you should check your figure captions carefully and remove any reference to colour in the illustration and text. In addition, some colour figures will degrade or suffer loss of information when converted to black and white, and this should be taken into account when preparing them.

## 4 Result and discussion

In MICMAC (Matrice d'Impacts Croises-Multiplication Applique' an Classment or Cross-Impact Matrix-Multiplication Applied to a Classification) analysis, the barriers will be classified into four groups, namely autonomous, dependent, linkage and independent [10]. The driver-dependence power diagram is built.

It can be seen that there are no autonomous variables (see Figure 2). The absence of these barriers in the present research shows that all the considered barriers influence the energy efficiency improvement which related to energy efficient technology adoption. New technologies not adequate or compatible at the site (barrier 2), lack of know-how in integration of new and existing technology (barrier 3), the installation period of new technology was too long (barrier 4) and lack of skilled technical staff for new technology implementation (barrier 5) are in the category of dependent obstacles. These barriers have little driving power, but strong dependence. These barriers show the unfavorable outcome to the management. The handling of these barriers depends on overcoming other obstacles formerly.

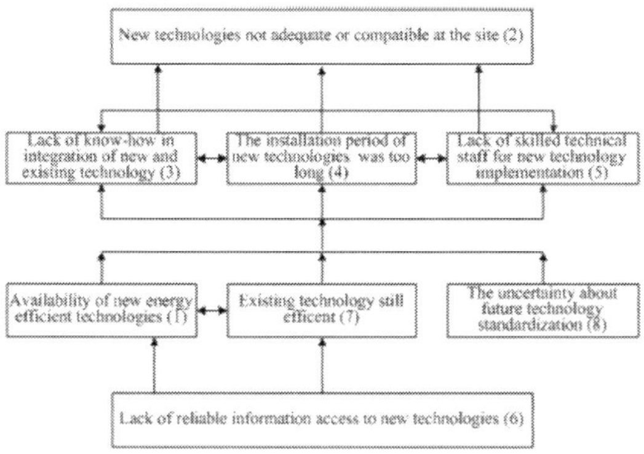


**Fig. 1**. Final of the Relationships among Technology-related Barriers

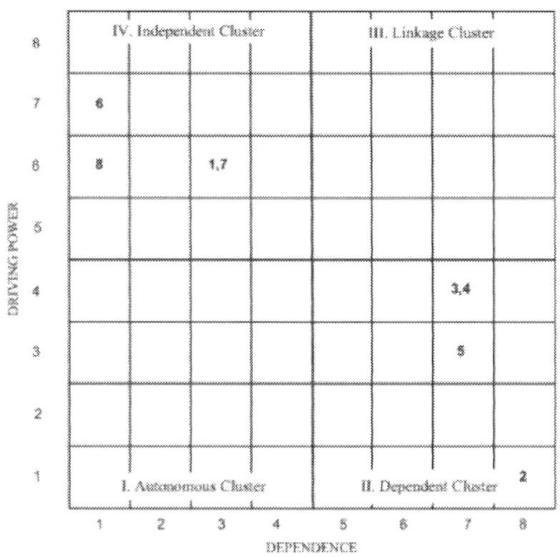


**Fig. 2.** Driving Power and Dependence Diagram

There is no barrier found in linkage cluster that has a strong driving power as well as strong dependence. It can be inferred that all the barriers of this study are stable. The driver-dependence power diagram indicates that barriers such as availability of new energy efficient technologies (barrier 1), lack of reliable information access to new technologies (barrier 6), existing technology still efficient (barrier 7), and the uncertainty about future technology standardization (barrier 8) are in the fourth cluster. This category includes independent barriers that have weak dependence, but strong driving power. They may be treated as the root cause of barriers or the major barriers.

## 5 Conclusion

Four issues have been identified as the driver or key variables and four as the dependent variable. No issue is seen as a linkage and autonomous barrier. Lack of reliable information access to new technologies is at the bottom level of the structural model which implies as the most important level of the hierarchy. Managers should focus on resolving this barrier that has high driving power. Thus, the management has the capability to tackle other barriers.

ISM model will help managers to understand the relationship among the barriers. The MICMAC analysis allows meaningful insights about the relative importance and the interdependencies among the barriers. This can help the top management to handle these barriers. The ISM model has not been statistically tested and validated. Thus, in the future research, it is needed to use another approach to test the validity and reliability of the model.

### References

1. H. Yurismono, J. Santosa, E. Hilmawan, Hariyanto, E. Djubaedah, T. Hartadi, S. Palaloi, N.R. Iskandar, Yasmin, *Planning of Energy Efficiency and Elasticity*. Agency for The Assessment and Application of Technology, Banten, Indonesia (in Indonesian) (2013)
2 J. Schleich, E. Gruber, Beyond case studies: Barriers to energy efficiency in commerce and the services sector. *Energy Economics*, **30**, 449-464 (2008)
3 P. Rohdin, P.Thollander, Barriers to and driving forces for energy efficiency in the non-energy intensive manufacturing industry in Sweden. *Energy*, **31**(12),1836-1844 (2006)
4 S. Sorrell, J. Schleich, S. Scott, E. O'Malley, F. Trace, U. Boede, K. Ostertag, P. Radgen, *Reducing Barriers to Energy Efficiency in Public and Private Organizations*. Brighton, UK (2000).
5 T. Fleiter, J. Schleich, P. Ravivanpong, Adoption of energy-efficiency measures in SMEs—An empirical analysis based on energy audit data from Germany. *Energy Policy*. **51**, 863-875 (2012).
6 H. Shi, S.Z. Peng, Y. Liu, P. Zhong, Barriers to the implementation of cleaner production in Chinese SMEs: government, industry and expert stakeholders' perspectives. *J.of Cleaner Production*, **16**(7), 842-852 (2008).
7 N. Cooray, Cleaner production assessment in small and medium industries of Sri Lanka, *Global Competitiveness through Cleaner Production: Second Asia Pacific Cleaner Production Roundtable* (1999).
8 A.P. Sage, *Interpretive Structural Modeling: Methodology for Large-scale System*. McGraw-Hill, New York (1977)
9 V. Ravi, R. Shankar, Analysis of interactions among the barriers of reverse logistics. *Technological Forecasting & Social Change*, **72**(8), 1011-1029 (2005)
10 T. Raj, R. Shankar, M. Suhaib, An ISM approach for modelling the enablers of flexible manufacturing system: the case for India. *Int'l J. of Production Research*, **46**(24), 6883-6912 (2008)

# Determining key performance indicators for warehouse performance measurement – a case study in construction materials warehouse

*Elisa* Kusrini*, *Fadrizal* Novendri, and *Vembri Noor* Helia

Department of Industrial Engineering, Universitas Islam Indonesia, Jl.Kaliurang km 14.5, Yogyakarta, Indonesia

**Abstract.** Warehouse performance measurement is needed to improve performance of logistics system. In order to improve warehouse performance, it is necessary to identify Key Performance Indicator (KPI). Different warehouses have different KPI, therefore this research aims to identify the most important KPI of warehouse so that warehouse manager can determine corrective actions in their warehouse efficiently and effectively. In this research, 25 KPI of warehouse are identified in five warehouse activities based on Frazelle model. The most important KPI are then determined in each warehouse activity using Analytical Hierarchy Process (AHP). Warehouse performance are measured and final score is determined using SNORM. Improvement steps are proposed base on benchmarking among warehouses. Warehouse performance measurement is conducted in 5 construction's material warehouses located in Yogyakarta, Indonesia. From this study, it is found that most important KPI for receiving is productivity (receive per man-hour), KPI for put away is cycle time (put away cycle time), KPI for storage is utilization (% location and cube occupied), KPI for order picking is cycle time (order picking cycle time) and KPI for shipping is productivity (order prepared for shipment per man-hour). Improving warehouse performance could be done by comparing warehouse performance with the best performance among peer group.

## 1 Introduction

The increasingly competition requires company to improve their logistics performance. One of the important component in logistic performance is warehouse performance. Good warehouse performance will improve quality performance, delivery time, customer satisfaction and reduce cost in logistics system [1]. Warehouses are dedicated space for storage and handling of materials. In supply driven chain, warehouses hold stocks to feed internal department such as manufacturing and production. In demand-driven chain, warehouses store goods to fulfill the needs of external customers [2].

Due to the increasing complexity of logistic network, the warehouse performance analysis has become an important issue [3]. Warehouse performance evaluation has been studied in different ways by many researchers. Each focuses on different studies, in term of objectives, indicators, warehouse systems, measuring instruments etc. Regarding all this diversity, there is not a consensus of a group of measures used to assess warehouse performance [4]. Different types of warehouse require different indicators/metric. Therefore, this research aims to identify the most importance key performance indicators for warehouse in order to help warehouse manager to improve warehouse performance effectively. This paper studies warehouse performance measurement on five different construction material warehouse located in Yogyakarta, Indonesia.

## 2 Literature Review

### 2.1 Warehouse Activities

In general, warehouse activity consists of receiving, put away, storage, picking and shipping [5] Receiving is operation that involve the assignment of trucks to dock and the scheduling and execution of unloading Activities [6]. Put away is activity of placing a product or material that has been purchased in the warehouse. This activity including material handling activities verifying the location of the product material and the placement of the product [5]. Storage is movement of material from unloading area to its designated place [7]. Order Picking is order preparation. It is regarded as the main and labor-intensive activity of warehouses [8, 9]. Shipping is activity that involves scheduling and assignment of trucks to docks the orders, packing after picking and the loading of trucks [6]. Warehouse activity based on Frazelle model [5] is depicted in Figure 1.

*Corresponding author: elisa_kusrini@uii.ac.id

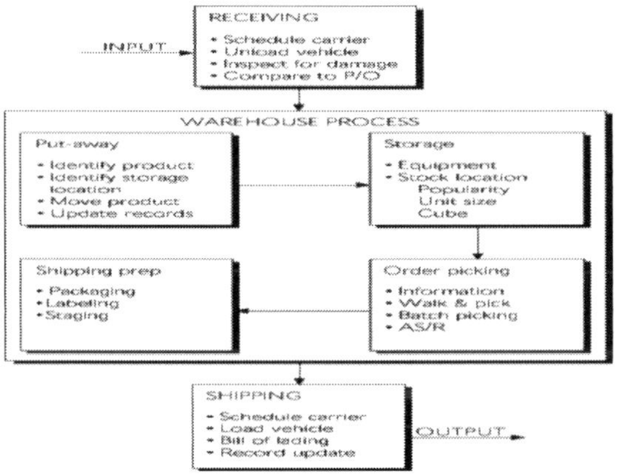


**Fig.1.** Warehouse activity [5]

## 2.2 Warehouse Performance Measurement

A warehouse performance measurement is a method to measure activity performance, program or service which is provided by a warehouse. Performance measurement system as the sets of metric used to quantify both the efficiency and effectiveness of action [10]. Performance measurement can be divided in 4 categories: input, output, efficiency and effectiveness [11]. In general, efficiency and effectiveness is the most widely utilized as a measure of performance [11, 12].

There are many methods for classifying warehouse performance measures. Performance warehouse based on traditional logistics performance measures include 'hard' and' soft' metrics. The first one treats quantitative measures such as order cycle time, fill rates and costs; the second deals with qualitative measures like the manager's perceptions of customer satisfaction and loyalty [13]. [14] distinguishes over direct and indirect indicators, where hard indices as indirect indicators. Different classifications are given by [15] that classify the performance of the warehouse into 130 indicators. In line with traditional logistic performance and direct performance, Frezelle model classify performance measured into 25 KPIs throughout the warehouse activity, i.e. financial, productivity, utilization, quality and cycle time.

Until recently, there is no agreement among researcher which KPI suitable for warehouse. Therefore, this study aims to identify the KPI on warehouse, especially for construction material's warehouse. In this research, we will measure hard or direct indicator using Frazelle model because it is in line with measure proposed by previous researcher and considered relevant for construction material warehouse. According to Frazelle model, warehouse key performance indicator (KPI) is shown in Figure 2.

## 3 Warehouse Performance Measurement in Construction Materials Warehouse

In this study, performance measurement is conducted in five warehouses that store construction materials such as brick, paving blocks, piles and concrete electric poles. This warehouse is owned by small medium's construction company in Yogyakarta, Indonesia. It is done in 3 steps as follows.

**Table 4-1: Warehouse key performance indicators (Frazelle, 2002, 56)**

| | Financial | Productivity | Utilization | Quality | Cycle Time |
|---|---|---|---|---|---|
| **Receiving** | Receiving cost per line | Receipts per man-hour | % Dock door utilization | %Receipts processed accurately | Receipt processing time per receipts |
| **Putaway** | Putaway cost per line | Putaways per man-hour | % Utilization of putaway labor and equipment | % Perfect putaways | Putaways cycle time (per putaway) |
| **Storage** | Storage space cost per item | Inventory per square foot | % Locations and cube occupied | % Locations without inventory discrepancies | Inventory days on hand |
| **Order picking** | Picking cost per order line | Order lines picked per man-hour | % Utilization of picking labor and equipment | % Perfect picking lines | Order picking cycle time (per order) |
| **Shipping** | Shipping cost per customer order | Orders prepared for shipment per man-hour | % Utilization of shipping docks | % Perfect shipments | Warehouse order cycle time |

**Fig.2.**Warehouse Key Performance Indicators [5]

Step 1. Identifying importance weight of KPI. A questionnaires of pair wise comparisons among indicators (KPIs) is prepared and a total of five senior manager of construction materials warehouse were asked to give their preferences on the questionnaires. The scale of 1 to 9 (1 = equally, 3 = moderate, 5 = strong, 7 = very strong, 9 = extreme) is utilized to rate the preferences of the expert. Consistency Ratio (CR) is used to check the consistency of the comparison. CR less than 0.1 means that comparison is accepted. The results of questionnaires is analyzed using Analytical Hierarchy Process (AHP) model. Analytical Hierarchy Process was first introduced by Thomas L. Saaty in 1971 and became the most widely method for Multiple Criteria Decision Making (MCDM). It is a decision-making approach that is designed to find solutions to complex problems with multiple criteria into multiple domains. It is also known as an essential tool for practitioners and academics to conduct research in decision making and measure management theory. Analytic Hierarchy Process as a flexible problem-solving methods [16]. The result of importance weight of each KPIs can be seen in Table 1.

**Table 1.** The importance weight of each warehouse KPIs

| *Activity* | *Finan-cial* | *Produc-tivity* | *Utili-zation* | *Quality* | *Cycle time* |
|---|---|---|---|---|---|
| Receiving | 0,19 | **0,41** | 0,08 | 0,16 | 0,17 |
| Put Away | 0,12 | 0,23 | 0,19 | 0,14 | **0,32** |
| Storage | 0,15 | 0,17 | **0,34** | 0,20 | 0,13 |
| Order picking | 0,16 | 0,18 | 0,15 | 0,20 | **0,31** |
| Shipping | 0,14 | **0,30** | 0,16 | 0,24 | 0,15 |

Step 2. Determining the most important KPI for warehouse. It is carried out by selecting the highest weight among KPI on each warehouse activity. KPI with the highest weight regarded as the most important KPI. KPI for receiving is productivity (Receipt per man-hour), KPI for put away is cycle time (put away cycle time), KPI for storage is utilization (% location and cube occupied), KPI for order picking is cycle time (order picking cycle time) and KPI for shipping is productivity (order prepared for shipment per man-hour).

Step 3. Measuring warehouse performance using selected KPIs. Warehouse performance is measured using selected KPIs and then scored them using standardized method, i.e SNORM based on benchmark data among warehouses. The process of normalization (SNORM) De Boer [17] employed scale of 0 to 100 to get the value of each metric with the following formula:

Larger is better

$$Snorm=((Si-SMin)/(Smax-SMin))x100 \quad (1)$$

Lower is better

$$Snorm=((SMax-Si)/(SMax-Smin))x100 \quad (2)$$

where,

Si = average value
S max = the value of achieving the best performance indicators
S min = the value of achieving the worst performance indicators

The performance of each warehouse presented in Table 2. Final score of warehouse performance is average of each score of warehouse activity as presented in Table 3.

Base on the result, improvement of warehouse performance could be executed through benchmarking based on the final score of each warehouse. Benchmarking is a measure for improving performance by learning from best practices and understanding the processes by which they are achieved [18]. There are 7 types of benchmarking, one of them is called process benchmarking. Process benchmarking is used when the focus is on improving specific critical processes and operations by learning from best practice organizations that perform similar work or deliver similar services [19]. Therefore, in this study, process benchmarking is employed by comparing each warehouse performance activity/operation with the best activity performance among peer group.

Referring to the warehouse score as presented in Table 2, warehouse 1 has the best performance on the activity of put away, storage and order picking. While warehouse 3 has the best performance in receiving activity. Warehouse 5 is the best warehouse on shipping activity. This result can be used by the warehouse manager to improve its performance by referring to the best warehouse performance. The highest performance on each activity could be a target setting for other warehouses in order to improve their performance. The advantage of process benchmark in this study is simple and easy to do. However, it is only useful for the short term improvement, not for long term. A more advanced benchmarking method is using mathematical model such as Data Envelopment analysis (DEA) which is a tool for the evaluation of the performance of an activity within a unit of entities or organizations [20]. DEA is a mathematical model by calculating the efficiency level and can be used for benchmarking with clear and objective methodologies [21]. This can be an opportunity for further research.

**Table 2.** Warehouse performance

| KPIs | WH 1 | WH 2 | WH 3 | WH 4 | WH 5 |
|---|---|---|---|---|---|
| Receipt per man-hour (unit/hour) | 5,56 | 187,5 | 945 | 187,5 | 333 |
| Put away cycle time (unit/hour) | 1,52 | 3 | 2,5 | 2 | 10 |
| % location and cube occupied (%) | 75 | 60 | 65 | 43,33 | 46,2 |
| Order picking cycle time (unit/order) | 0,5 | 3 | 1,5 | 1 | 28 |
| Order prepared for shipment per man-hour (unit/hour) | 6,44 | 9,3 | 187,5 | 162,5 | 226 |
| Receipt per man-hour (unit/hour) | 5,56 | 187,5 | 945 | 187,5 | 333 |
| Put away cycle time (unit/hour) | 1,52 | 3 | 2,5 | 2 | 10 |
| % location and cube occupied (%) | 75 | 60 | 65 | 43,33 | 46,2 |
| Order picking cycle time (unit/order) | 0,5 | 3 | 1,5 | 1 | 28 |
| Order prepared for shipment per man-hour (unit/hour) | 6,44 | 9,3 | 187,5 | 162,5 | 226 |

**Table 3.** Performance score using SNORM

| KPI | WH 1 | WH 2 | WH 3 | WH 4 | WH 5 |
|---|---|---|---|---|---|
| Receipt per man-hour (unit/hour) | 0 | 19.04 | 100 | 19.04 | 34.08.00 |
| Put away cycle time (unit/hour) | 100 | 64.06.00 | 88.04.00 | 94.03.00 | 0 |
| % location and cube occupied (%) | 100 | 52.06.00 | 68.04.00 | 0 | 09.01 |
| Order picking cycle time (unit/order) | 100 | 89.01.00 | 96.04.00 | 98.02.00 | 0 |
| Order prepared for shipment per man-hour (unit/hour) | 0 | 01.03 | 82.05.00 | 71.01.00 | 100 |
| Average Score | 60 | 45.04.00 | 87.01.00 | 56.06.00 | 28.08.00 |

## 4 Conclusion

Warehouse activity improvement can be done based on the most important KPI in each activity. This is to facilitate warehouse managers in determining strategies and actions to improve warehouse performance. From this research, it is found that the most important KPI for receiving is productivity (receive per man-hour), KPI for put away is cycle time (put away cycle time), KPI for storage is utilization (% location and cube occupied), KPI for order picking is cycle time (order picking cycle time) and KPI for shipping is productivity (order prepared for shipment per man-hour).Based on the KPI's score, warehouse performance improvement can be done through process benchmarking. Although this research is focused on measuring performance in construction materials warehouse, the result and methodology can be employed for other type of warehouses. Further research is to improve warehouse performance through benchmarking using mathematical methods, such as Data Envelopment Analysis.

## References

1. A. D. Marco and G. Mangano, Relationship between logistic service and maintenance costs of warehouses. Facilities Emerald Group Publishing Limited, Vol. 29 No. 9/10, pp. 411-421 (2011)
2. S. Emmet, Excellence in Warehouse Management. *John Wiley and Sons Ltd*, West Sussex (2005)
3. Y. Wu, and M.Dong, Combining multi-class queueing networks and inventory models for performance analysis of multiproduct manufacturing logistics chains. *The International Journal of Advanced Manufacturing Technology*, 37 (5–6): 564–575 (2007)
4. J.S. Keebler, and R.E. Plank, Logistics performance measurement in the supply chain: a benchmark. *Benchmarking: An International Journal* 16 (6):785–798 (2009)
5. E. H. Frazelle, World-Class Warehousing and Material Handling. McGraw-Hill Book Company(2002)
6. J.X. Gu, M. Goetschalckx, and L.F. McGinnis, Research on warehouse operation: a comprehensive review. *European Journal of Operational Research* 177 (1): 1–21 (2007)
7. A. Johnson, and L. Mcginnis, Performance measurement in the warehousing industry, *IIE Transactions* (2011) **43**, 220–230 (2011)
8. M. Dotoli, M.P. Fanti, G. Iacobellis, G. Stecco, and W. Ukovich, Performance analysis and management of an automated distribution center. *35th Annual Conference of IEEE Industrial Electronics*, November 4371–4376 (2009)
9. E.H. Grosse, C.H.G. Carlo, and K. Giersch. An experimental investigation of learning effects in order picking systems. *Journal of Manufacturing Technology Management* Vol. 24 No. 6, pp. 850-872(2013)
10. A.Neely, M.Gregory, and K. Platts, Performance measurement system design: a literature review and research agenda, *International Journal of Operation and Production Management*, Vol 25, (2005) 1228 – 1263 (2005)
11. D.N. Ammons, Overcoming the inadequacies of performance measurement in local government: the case of libraries and leisure services. *Public Administration Review*, 55(1) 37-47( 1995)
12. E. Kusrini, Subagyo, and N.A. Masruroh, Good criteria for supply chain performance measurement. *International Journal of Engineering Business Management*, **6**:9. doi: 10.5772/58435 (2014)
13. G. Chow, T.D. Heaver, and L.E. Henriksson, Logistics Performance: Definition and Measurement. *International Journal of Physical Distribution & Logistics Management* 24 (1): 17–28 (1994)
14. F.H. Staudt, G. Alpan, M. Mascol, C.M.T.Rodriguez, Warehouse performance measurement: a literature review. *International Journal of Production Research*, Vol. 53, No. 18, 5524–5544 (2015)
15. E. Krauth, H. Moonen, V. Popova, and M. Schut, Performance indicators in logistics service provision and warehouse management – a literature review and framework, on-line at http://www.cs.vu.nl/~schut/pubs/mcs-Krauth2005a.pdf (2005)
16. T.L. Saaty, The analytic hierarchy and analytic network measurenment process: application to decisions under risk. *European Journal of Pure and Applied Mathematics* **1**(1) 122-96 (2008)
17. J.H. Trienekens, and H.H. Hvolby, Performance measurement and improvement in supply chain, *Proceeding on the third CINET Conference, CI 2000 from Improvement to Inovation*, Aalborg (2000)
18. R.E. Mittelstaedt, Robert E, Jr, Benchmarking: How To Learn from Best-In-Class Practices, *National Productivity Review (1986-1998)*, 11, 3; ABI/INFORM Collection, pg. 301 (1992)
19. M. Isoraite, Theoretical Aspects of Benchmarking Theory, vieðoji politika ir administravimas, ISSN 1648-2603 (2004)
20. A. Charnes, W.Cooper, and E. Rhodes, Measuring the efficiency of decision making units. *European Journal of Operational Research*, Vol. 2 No. 6, 429-444 (1978)
21. Coelli, T., A Guide to DEAP version 2.1. A Data Envelopment Analysis (Computer) Program, *Departement of Economic University of New England* (1996)

# Effect of internal R&D activities to the accumulation of organizational technical knowledge with the mediation role of absorptive capacity to establish the innovation capability of electronic industries

*Bambang* Purwanggono* and *Yohana Aeria* Damyana

*Department of Industrial Engineering, Faculty of Engineering, Diponegoro University*
*Jl. Prof. Sudharto, Tembalang, Semarang Indonesia 50239 Telp/Fax. (+6224) 7460052*

**Abstract.** Innovation is a strategy for the electronics industry to create a sustainable competitive advantage, in the midst of a rapidly changing environment with all its complexity. Seven AT program as an effort for PT. Hartono Istana Teknologi (Polytron) into enterprise knowledge, will accelerate the innovation process, combined with good organizational technical knowledge management. Organizational technical knowledge will be instrumental in innovation capabilities properly if there is an internal R & D activities that support and absorptive capacity as a mediator. This study reviewed the organizational technical knowledge influence to innovation capability, the influence of R & D activities to organizational technical knowledge, as well as the role of absorptive capacity as a mediator. The study was conducted by distributing questionnaires to 130 employees of PT. Hartono Istana Teknologi. Data processing was conducted using SEM. The results showed that the absorptive capacity mediate the relationship between R & D activities and organizational technical knowledge by 51%, and organizational technical knowledge affect innovation capabilities by 64%.

## 1 Introduction

Currently business competition and Indonesian economy are increasing, coupled with the policy of ASEAN *Economic Community* (AEC) which was implemented in 2015. In realizing AEC, all ASEAN countries should be liberalizing trade. This requires industries to become more competitive and responsive to market change and consumer trends. Indonesia, as the ASEAN member, continues to increase its value-added strategy, especially for export products in order to compete with other ASEAN member countries. Among the 10 Indonesian export commodity, electronic products is one commodity that considered to have potentials to be improved. (Ministry of Commerce of the Republic of Indonesia, 2010)

Innovation becomes an important part in the progress of electronic industry of PT. Hartono Istana Teknologi with brand name Polytron. This is demonstrated by the inclusion the role of innovation in the second point of 7AT *corporate identity*. The second point of 7AT *corporate identity* defined as superior in producing new products that are innovative, unique, and consiederd as customers' favourite.' .

However, 7AT *corporate identity* has not been yet well implemented. Based on interviews, employee management feels a lesser role in providing knowledge-based innovation advice to the firm. In fact, to become a more competitive industry, PT. Hartono Istana Teknologi required to be able to optimize existing resources, including knowledge resources. For this reason the organization's ability to optimize existing knowledge resources are extremely important to accelerate the innovation capabilities (Subramaniam and Youndt, 2005). Knowledge management then deemed capable of managing knowledge that circulating in each and between departments.

Various studies state that the performance of innovation has a positive effect on knowledge management, especially the *organizational technical knowledge's* firm accumulation. *Organizational technical knowledge* defined as a credible information that has potential value for an organization (Hult, 2003 in Wang et al., 2009). This positive correlation is described by Cohen and Levinthal (1990) who argued that the ability to evaluate and utilize *technical organizational knowledge* is a function of the level of innovation. At the most basic level, *organizational technical knowledge* to include basic skills, or even usage the same language, must also include the latest scientific knowledge or technological developments in the firm.

According to Pavitt, Tornatzky and Fleischer in Attewell (1992), *organizational technical knowledge* accumulated through *output* of knowledge from the activities *of research and development centers* or hereinafter referred as internal R & D activities. Murovec (2008) explains further that the *output of* internal R & D's firm activities cannot be directly used by the firm without absorptive capacity of the firm itself, so that the function of absorptive capacity is to mediate the relationship between internal R & D activities and *organizational technical knowledge*. The firm with higher absorptive capacity will be more sensitive to the flow of knowledge from internal R & D *activities*. The absorptive capacity of the firm determines which knowledge will be applied in the firm, while the absorptive capacity itself depends on the extent to how the flow of knowledge of internal R & D activities of the firm. Internal R & D activities and absorptive capacity need to be examined as variables to determine the effect *of organizational technical knowledge* accumulation on innovation capabilities.

In this study a problem that can be defined is how the innovation capabilities of PT. Hartono Istana Teknologi influenced by *organizational technical knowledge* accumulation, taking into account the internal R & D activities as a source of knowledge in the firm mediated by absorptive capacity. The purpose of this study was to investigate the influence between variables research on enterprise innovation capabilities using *Structural Equation Modeling (SEM)*, the result will be basic of recommendations for PT. Hartono Istana Teknologi improvements to increase the innovation capability.

## 2 Research Methodology (*Organizational Technical Knowledge)*

*Organizational technical knowledge*, which is part of the *tacit* knowledge in the context of knowledge management, is considered important in the development of the firm's knowledge. This knowledge affects the perception and intensity of employees in the use of technology in the firm because of higher individual *organizational technical knowledge* will allow employees to work more with sophisticated equipment.

Attewell (1992) claims that *organizational technical knowledge* considered essential in the success of the firm innovation achievement. Learn and communicate existing *organizational technical knowledge* needed for successful innovation better because of the amount and clarity of information within the firm will continue to accumulate along with increasing experience of the firm.

Bedard, et al. (2003) adds that the individual *organizational technical knowledge* that optimally accumulated will make people to have the perception that the technology applied by the firm is quite easy to apply. The firm will tend to delay the adoption of complex technologies to companies that have already obtained sufficient *organizational technical knowledge* as a basis to implement and operate the technology successfully.

### 2.1 Internal R&D Activities

According to Pavitt, Tornatzky and Fleischer in Attewell (1992) basic *organizational technical knowledge* innovation can come from internal and external sources. Internal source is based on the activity of *research and development centers*. Attewell (1992) explains, the success of an innovation depends on the development of new *organizational technical knowledge* in the innovation process, one of which *organizational technical knowledge* derived from R & D activities within the firm.

Internal R & D activities are sources of internal knowledge that are important for a firm and have shown to have a significant influence on innovation, because if investment in internal R & D activities grew, the firm will be more motivated to innovate (Rustika, 2011).

Ellitan (2006) adds that companies which innovate, improve internal sources of innovation usage through R & D, and investing in technology will be able to achieve better performance than the firm with low level implementation of the innovation strategy or do not have the innovation strategy at all. Companies that decide to be innovative need actively understand and complement R & D activities' program and procure external agreements and technology acquisitions.

### 2.2 Absorptive Capacity

The concept of absorptive capacity first introduced by Cohen and Levinthal (1990), they define the absorptive capacity as the extent of the firm's ability to identify new *organizational technical knowledge*, assimilate, and apply *organizational technical knowledge* to achieve organizational goals.

Cassiman (2002) and Murovec (2008) adds that the organization/firm trying to improve gains in *organizational technical knowledge* through investments in absorptive capacity. Absorptive capacity is one of the important concepts in organizational research in the recent years. *Organizational technical knowledge* that can be practiced in the firm is dependent on the firm's absorptive capacity. The firm with higher absorptive capacity will be more sensitive to the presence of *organizational technical knowledge*. In other words, the firm's ability to identify and recognize *organizational technical knowledge* also depends on the absorptive capacity of the firm.

### 2.3 Innovation Capability

Innovation is an effort to maintain the existence of the organization in the competitive environment. Innovation is expected to respond to the complexity of the environment and the dynamic changes in the environment, especially in a tight competition and create resources for competitive advantage. This can be achieved through the introduction of new technologies, new applications in products and services, development of new markets, and the introduction of new forms of organization. According to Santoso in Iksan (2010), the main objective of innovation is increasing the sources of power, money, and means including organizational structures and procedures. Innovation as a strategic component in many companies to respond to the speed of technological change that cannot be predicted, and motivated by a dependence on high technology and firm experienced a rapid response in the form of products and procedures for the delivery of the product. Innovation is done with the aim of lowering the level of costs, improve efficiency, delivering good quality at fair prices, the greater possibility of obtaining profit and growth.

According to Lawson and Samson (2001), organizational innovation capability is the ability to integrate the capabilities and resources of their companies to successfully stimulate the development of innovation in the firm. Includes activities to define the purpose and objectives of innovation, selecting strategic arena, develop *strategic map*, and allocate resources.

### 2.4 SEM *(Structural Equation Modeling)* Methods

Based on the conceptual model and hypotheses made, this study uses *Structural Equation Modeling (SEM)*. SEM can explain and illustrate the relationship patterns between variables observed with latent variables through the indicators. Unlike regression in general can only explain the causal relationship between the variables observed only.

This study focuses on firm's behavior or perception on innovation capabilities. Behavior is a variable that cannot be observed directly, or so-called latent variables. The use of latent variables in a multiple regression cause measurement errors that affect the estimation of the parameters of the *biased-Unbiased* angle and the size of the *variance*. By SEM, this measurement issue is addressed through the existing equations in the measurement model. The parameters of the equation is the load factor of the latent variables of indicators or variables observed. So, SEM models in addition to providing information about the causal relationship between the simultaneous the variables, also provide information about the load factor and errors of measurement.

Viewing excellence SEM and conceptual models that have been described, it is suitable to test the model innovation capabilities using SEM method, because there is latent variables such as internal R & D activities, the absorptive capacity, *organizational technical knowledge*, and innovation capabilities. In a model of SEM, a latent variable can serve as exogenous or endogenous variables. A dependent variable can be the independent variable to another variable.

## 3 Hypotheses (*Organizational Technical Knowledge* Accumulation)

Innovation is a complex process in which new knowledge is applied for commercial purposes. The new knowledge generated through a cumulative process in which the knowledge that already exists will be changed, modified, and improved. The role of knowledge flows to the success of the innovation process has long been recognized. The success of an innovation depends on the development of new knowledge in the innovation process, one of which is the knowledge that comes from the firm's R & D activity. Firm's R & D activity is also a source of *organizational technical knowledge* development (Attewell, 1992).

In a product development, internal R & D activities must be able to dig information up whether they will develop products that other companies have done (Huston, 2006). Growing knowledge from information exploring activity will be useful for increasing the firm's profits and competitiveness against the firm's competitors. The absorptive capacity is the extent to which the firm's ability to identify new external knowledge, assimilate, and apply knowledge to achieve organizational goals. Absorptive capacity of the firm determines which *organizational technical knowledge* that will be applied in the firm, while the absorptive capacity itself depends on the extent of knowledge of internal R & D activities of the firm. Knowledge that can be implemented in firm depending on the absorptive capacity of the firm itself (Murovec, 2008). Firms with higher absorptive capacity will be more sensitive to the knowledge from internal R & *D*. In addition, the firm's ability to identify and recognize the type of *organizational technical knowledge* also depends on the absorptive capacity of the firm itself. If the Internal R & D Activities have complex knowledge, the firm will more easily acquire knowledge. Absorptive capacity become intercede so that the flow of knowledge from Internal R & D Activities can be processed with optimal become accumulated *organizational technical knowledge*. Because of its nature as an intermediary between Internal R & D Activities and *organizational technical knowledge accumulation*, the absorptive capacity is referred as *a mediator* (Hatani, 2013).

**H1: internal R & D activities have a positive influence to the accumulation *of organizational technical knowledge***

**H2: internal R & D activities have a positive influence to the absorptive capacity**

**H3: The absorptive capacity has a positive effect to the accumulation *of organizational technical knowledge***

**H4: The relationship between internal R & D activities to the accumulation *of organizational technical knowledge* influenced indirectly (mediation) by the absorptive capacity**

### 3.1 Innovation Capability

*Organizational technical knowledge* is credible information that has potential value for an organization (Hult, 2003 in Wang et al., 2009). This positive correlation is described by Cohen and Levinthal (1990), who argued that the ability to evaluate and utilize *organizational technical knowledge* is a function of the level of innovation. At the most basic level, *organizational technical knowledge* including basic skills, or even use the same language, can also include the latest scientific knowledge or technological developments in the firm. Companies tend to delay the adoption of complex technologies, untill the firm acquire sufficient *organizational technical knowledge* to implement and operate the technology successfully. *Organizational technical knowledge* considered essential in the success of the firm achievement of innovation. Learning and communicating *organizational technical knowledge* are needed for successful innovation, because the bigger number and clarity of the information in the firm would be a greater advantage than the previous corporate's experiences (Attewell, 1992).

**H5: Accumulation *of organizational technical knowledge* has a positive effect on innovation capabilities**

Based on the description of the previous hypotheses, the conceptual model can be made as follows:

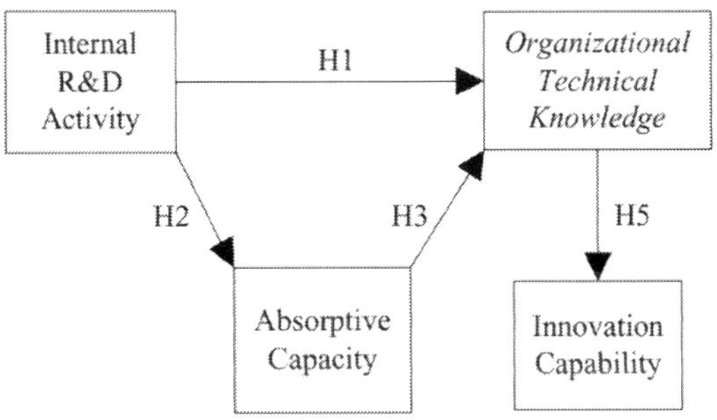


**Figure 1** Model Research

## 4. Results and Discussion

After we distributed questionnaires, then data processing is done with the help of *software* SmartPLS version 2.0.

### 4.1 Convergent Validity

*Convergent validity* is defined as a measure of whether these indicators can measure a construct precisely. An indicator is said have good validity and reliability if the *loading* factor value is larger than 0.7, but the *loading* factor 0.5 to 0.6 can still be maintained for a development stage model. Based on the results of *the outer loading,* all the indicators already meet the *convergent validity* because it has a *loading factor* values greater than 0.5.

### 4.2 Discriminant Validity

Reflective indicators also need to be tested using *discriminant validity. Discriminant validity* used to test unidimensionality of each construct. This test is done by looking *cross loading* value shown in Table 1.

### 4.3 Composite Reliability

Reliability test is done by looking at the value of *composite reliability* in the indicator block that measures the construct. The results will show the *composite reliability* satisfactory if the value above 0.7. In the results test, all the variables already demonstrated the value *of composite reliability*> 0.7 to be considered reliable.

### 4.4 Inner model

*Inner model* is a model that specifies the relationship between the constructs or also referred as *structural models. Inner model* formed after *the outer models* have been significant overall. In general, *the inner models* are evaluated by looking at the magnitude of the coefficient lines and also to see the *R-square* for the construct. In addition, the significant value of T-statistics obtained from *bootstrapping* procedure used to test the hypothesis that built.

**Table 1**. *Cross Loading*

| | KAPIN | KAPSER | OTK | R&D |
|---|---|---|---|---|
| I1 | 0.6297 | 0.2153 | 0.2573 | 0.3117 |
| I2 | 0.6954 | 0.3567 | 0.299 | 0.257 |
| I3 | 0.6996 | 0.4204 | 0.376 | 0.4453 |
| I4 | 0.7042 | 0.3966 | 0.3393 | 0.4368 |
| I5 | 0.6843 | 0.365 | 0.4843 | 0.4202 |
| I6 | 0.7307 | 0.4495 | 0.6504 | 0.4 |
| O1 | 0.4975 | 0.6043 | 0.8096 | 0.5091 |
| O2 | 0.4091 | 0.2011 | 0.5169 | 0.2526 |
| O3 | 0.2989 | 0.251 | 0.5633 | 0.3032 |
| O5 | 0.4927 | 0.6315 | 0.7972 | 0.5527 |
| O6 | 0.4591 | 0.5368 | 0.8275 | 0.3737 |
| O7 | 0.4266 | 0.4096 | 0.59 | 0.3606 |
| O8 | 0.4935 | 0.5342 | 0.8579 | 0.4461 |
| O9 | 0.6051 | 0.561 | 0.8138 | 0.4736 |
| R1 | 0.2619 | 0.2983 | 0.2019 | 0.6282 |
| R2 | 0.3328 | 0.4628 | 0.3459 | 0.6978 |
| R3 | 0.2257 | 0.2964 | 0.2328 | 0.6589 |
| R4 | 0.4363 | 0.4381 | 0.4945 | 0.7628 |
| R5 | 0.4077 | 0.4127 | 0.4594 | 0.7241 |
| R6 | 0.4703 | 0.3967 | 0.4735 | 0.6685 |
| R7 | 0.4199 | 0.4456 | 0.3577 | 0.6088 |
| R8 | 0.2364 | 0.3323 | 0.2658 | 0.5368 |
| R9 | 0.3717 | 0.569 | 0.391 | 0.5893 |
| S1 | 0.3948 | 0.5909 | 0.4306 | 0.4426 |
| S2 | 0.3818 | 0.6216 | 0.3998 | 0.3957 |
| S3 | 0.3825 | 0.5751 | 0.3461 | 0.2202 |
| S4 | 0.1857 | 0.5159 | 0.2468 | 0.3737 |
| S5 | 0.2495 | 0.699 | 0.4041 | 0.4045 |
| S6 | 0.2717 | 0.5484 | 0.4222 | 0.3685 |
| S7 | 0.3602 | 0.6197 | 0.497 | 0.3521 |
| S8 | 0.358 | 0.6907 | 0.4606 | 0.513 |
| S9 | 0.467 | 0.7126 | 0.4734 | 0.4485 |

**Table 2**. *Path Coefficient*

| | Original Sample | Sample Mean | StDev | Standard Error | T Statistics |
|---|---|---|---|---|---|
| Kapser -> Otk | 0.508 | 0.528 | 0.138 | 0.138 | 3.676 |
| Otk -> Kapin | 0.637 | 0.682 | 0.070 | 0.070 | 9.077 |
| R&D -> Kapser | 0.640 | 0.665 | 0.079 | 0.079 | 8.025 |
| R&D -> Otk | 0.247 | 0.228 | 0.148 | 0.148 | 1.668 |

Table 2 is the result of T-statistical significance values and coefficient parameters of the structural model to test the hypothesis that was built in the study and to determine the influence coefficient between variables in the study.

*Original sample estimate* value for relationship between internal R & D activities with *organizational technical knowledge* accumulation is 0.2479 (≈0.25), which means that the internal R & D activities have a positive effect to the accumulation *of organizational technical knowledge* by 25%. While the value of the T-statistic is 1.6688, the value does not meet the criteria of the T-statistic> 1.96. Thus **H1 is rejected.** This is because knowledge that generated by R & D activities can not be directly used by employees of other departments. For example, employees of the HR department and GA have a different understanding of the limits of the R & D department in the *'self design product'* knowledge. Whereas the HR department and GA have the same potential to provide innovative ideas. Therefore basic knowledge of existing products and that will be developed as well should be a better understanding of the ability to be shared by all employees of PT. Hartono Istana Teknologi.

*Original sample estimate* value for relationship between internal R & D activities to the absorptive capacity is 0.6402 (≈ 0.64), which means that the Internal R & D activities positively influence the absorptive capacity by 64%. While the value of the T-statistic of 8.0258, the value meets the criteria of the T-statistic> 1.96. Thus **H2 is accepted.** The absorptive capacity is the ability of employees to identify, assimilate, and apply new knowledge derived from internal R & D activities. This ability to determine how much information, which comes from internal R & D activities, which will be useful as an *organizational technical knowledge.* In terms of the influence of internal R & D activities of the absorptive capacity by 64%, PT. Hartono Istana Teknologi consider the absorptive capacity is important as a basis for using knowledge of the results of R & D activity.

*Original sample estimate* value for relationship between the absorptive capacity *to organizational technical knowledge* accumulation is 0.5089 (≈0.51), which means that the absorptive capacity have a positive effect to the accumulation *of organizational technical knowledge* by 51%. While the value of the T-statistic of 3.6765, the value meets the criteria of the T-statistic> 1.96. Thus **H3 is accepted.**

### 4.5 Mediation Model Testing

According Urbayatun (2012), a variable called mediating variables *(intervening)* if the variables influencing the relationship between the independent variables and the dependent variable. Mediation hypothesis testing can be done with a procedure developed by Sobel in 1982 and known as the Sobel test *(Sobel Test).* Mediating variables test using this method is done by testing the strength of the independent variable indirect effect (internal R & D activities) on the dependent variable *(organizational technical knowledge)* through the mediating variables *(absorptive capacity).* Based on the results obtained Sobel test T-statistics value 3.342 which means **H4 is accepted.** Through the absorptive capacity, the relationship between Internal R & D activity and the accumulation of *organizational technical knowledge* becomes stronger. It can be seen from the value of the T-statistical relationship between R & D activities Internal and accumulation of *organizational technical knowledge.* The value of the T-statistic direct relationships at 1.6688, while involving variable absorptive capacity as a mediating variable then the value of the T-statistic of 3.3426. Increasing the value of T-statistics are much different, this suggests a role absorptive capacity variable proved to strengthen the relationship between internal R & D activities and *organizational technical knowledge* accumulation as mediating variables.

*The original sample estimate* value of relationship between the accumulation of *organizational technical knowledge* with innovation capabilities is 0.6372 (≈ 0.64), which means that the accumulation *of organizational technical knowledge* positively influence innovation capabilities by 64%. While the value of the T-statistic of 9.0771, the value meets the criteria of the T-statistic> 1.96. Thus **H5 is accepted.** The existence of an innovative organizational culture that makes employees motivated to improve absorptive capability of knowledge for the accumulation *of organizational technical knowledge.*

## 5. Conclusions and Recommendations

Based on the results of data processing and analysis it can be seen that the result of internal R & D activities cannot be directly used as an accumulated *organizational technical knowledge* of the firm. To function optimally, the knowledge which is the result of internal R & D activities must be identified and understood before it can be implemented into accumulated *organizational technical knowledge.* The process of identifying, understanding, and applying this knowledge is much dependent on the existence of the absorptive capacity. Maximum accumulation of *organizational technical knowledge* mediated by absorptive capacity within the firm will boost the firm's innovation capabilities.

Recommendations that can be given to PT. Hartono Istana Teknologi in order to improve the performance of services are based on the relationship of accumulation *of organizational technical knowledge* and innovation capabilities that have the strongest relationship by 64%, then it should be an effort to improve accumulation *of organizational technical knowledge* to get the maximum increase of innovation capabilities. Efforts to improve accumulation *of organizational technical knowledge* cannot be separated from the influence of internal R & D activities and the absorptive capacity as a mediator, then the firm must do two things: (1) the company should be able to build a knowledge enterprise communications system which serves to establish cooperation between the various internal departments. This system starts with forming a functional unit in the firm that manages the firm's knowledge and is responsible for the exchange of knowledge between departments. Furthermore, making internal corporate portals, such as Telkom internal portal

(POINT), which serves as a vehicle for the grouping of knowledge, knowledge storage, and *sharing knowledge,* which in turn will improve the *value* of companies with innovations that have been created by combining, synthesizing and organizing new knowledge. (2) The company should be able to enhance innovation capabilities in the firm, which is to encourage and facilitate employees to produce, combine, and apply new knowledge in order to increase employee competence. This capability can be built with holding training in the use of new technologies as one program among others.

The authors suggest that further studies should be considering other variables that affect the relationship between accumulation of *organizational technical knowledge* and innovation capability, such as involving external accumulated *technical organizational knowledge* factors such *knowledge spillover* variable (an abundance of knowledge from outside the firm).

Regardless of the limitations, this study is expected to provide benefits as an essential consideration to the firm when implementing the accumulation *of organizational technical knowledge* to improve innovation capabilities. Being expected from this research is that it can provide guidance for managers to choose the appropriate knowledge management strategy in accordance with the firm's goals. The results of this study are expected to also contribute to academics and practitioners, especially in developing knowledge management literature and innovation management. The results of this study should lead to further research that are much more perfect and provide greater benefits.

## References

1. Ministry of Commerce of the Republic of Indonesia. Menuju ASEAN Economic Community 2015 (2010). Accessed on August 2nd, 2014, from the website ditjenkpi.kemendag.go.id/.../Buku%20Menuju%20ASEAN%20ECONOMI...
2. M. Subramaniam and M.A. Youndt, The Influence of Intellectual Capital on The Types of Innovative Capabilities. Academy of Management Journal, **48**, 450 (2005)
3. C.L. Wang, et al. Knowledge Management Orientation, Market Orientation, and Firm Performance: An Integration and Empirical Examination. Journal of Strategic Marketing, **17**, 147 (2009)
4. W. Cohen and D. Levinthal, Absorptive Capacity: A New Perspective on Learning and Innovation. Administrative Science Quarterly, **35**, 128 (1990)
5. P. Attewell, Technology Diffusion and Organizational Learning: The Case of Business Computing. Organization Science, **3**, 1 (1992)
6. N. Murovec and I. Prodan, The Influence of Organizational Absorptive Capacity on Product and Process Innovation. Organizacija, **41**, 43 (2008)
7. J.C. Bedard et al. The Roles of Task and Technical Knowledge in Acceptance of Information Technology Among Preparers and Reviewers of Audit Workpapers (2003) Accessed on May 24th, 2014, from the website https://aaahq.org/audit/midyear/02midyear/papers/TAM-Auditing%20Midyear%2012-5.pdf
8. N. Rustika. Analisis Pengaruh Penerapan Akuntansi Manajemen Lingkungan Dan Strategi Terhadap Inovasi Perusahaan (2011). Accessed on May 24th, 2014, from the *website* eprints.undip.ac.id/30899/1/Skripsi002.pdf
9. L. Ellitan. Strategi Inovasi dan Kinerja Perusahaan Manufaktur di Indonesia Pendekatan Model Simultan dan Model Sekuensial. Jurnal Manajemen, **6**, 1 (2006)
10. B. Cassiman and R. Veugelers, R&D Cooperation and Spillovers: Some Empirical Evidence from Belgium. The American Economic Review, **92**, 1169 (2002)
11. F. Iksan. *Dasar – Dasar Kependidikan*. (Rineka Cipta, Jakarta, 2010)
12. B. Lawson and D. Samson. Developing Innovation Capability in Organisations: a Dynamic Capabilities Approach. International Journal of Innovation Management, **5**, 377 (2001)
13. L. Huston and N. Sakkab. Connect and Develop Inside P&G New Model for Innovation. Harvard Business Review, **R0603C** (2006). Accessed on May 14th, 2014, from the Harvard Business School Publishing Corporation
14. L. Hatani, D. Zain, B. Wirjodirjo and Djumahir. The Role of Competitiveness as Mediator for the Relation between Supply Chain Flexibility and Firm Performance. Journal of Management Research, **5**, 269 (2013)
15. S. Urbayatun and W. Widhiarso. Variabel Mediator dan Moderator dalam Penelitian Psikologi Kesehatan Masyarakat. Jurnal Psikologi, **39**, 180 (2012)

MATEC Web of Conferences **154**, 01060 (2018) https://doi.org/10.1051/matecconf/201815401060
*ICET4SD 2017*

# The effect of Quran murottal's audio on short term memory

*Perdana* Suteja Putra, *Retno* Gumilar, *Sasmita* Rahma Kusuma*, *Hari* Purnomo, and *Chancard* Basumerda

Islamic University of Indonesia , Jalan Kaliurang Km. 14,5 Sleman, Yogyakarta, Indonesia

**Abstract.** In all activities, people need memory as information storage. A good memory will make it easier for people to remember so they can do their job well. One of the things that affect memory is the sound of music. Music is commonly used as a stress treatment or as one of the stimuli of human brain memory. Not only classical music can be one of the stimuli of brain memory, but listening to the Quran can also be one of the brain memory stimulants. Several studies have proven that by listening to the Qur'an, the ability of the brain can be increased significantly. In this study used the variables pre and posted Qur'an listening to 50 respondents. Respondents consisted of 26 men and 24 women aged 19-22 years old. All respondents will get two treatments. The first treatment is without listening to murottal and the second treatment is done after listening to the Quran. All respondents will do memory test using the program running memory in software design tools. The results showed that the average ability of short-term memory without listening to murottal 11.8 and the average ability of short-term memory by listening to murottal is 12.62. Based on these results it is known that by listening to the Quran provides a significant increase in the ability of the brain.

## 1 Introduction

In carrying out daily work, human beings need memory to store information. Good memory performance will make it easier for people to remember so that work can be completed faster and minimize errors [1]. The performance of memory is not only be influenced by body condition and human health, but also influenced by work environment factor. Therefore, the environment and working atmosphere must be designed in such a way as to support better memory performance.

Based on the duration and level of use, memory is divided into two, namely long-term memory and short-term memory. Long term memory is a memory that is often used and stored up to a long time even up to years. While short-term memory only shortly after the information obtained, without any repetition of the information. The information will be lost in maximum time after 30 seconds (Xu, 2002). In addition, the amount of information that can be stored is limited. The amount of information that can be stored by short term memory is 7 + 2 units [2]. Based on the above explanation, short term memory needs to be considered its performance because of its easy to lose in a short time.

So to optimize the work performance, need to note the state of the physical work environment. One of the physical work environments that can affect performance is sound. Various sources of sound can be a stimulus for someone in work, one of which is music. Music is sound, a collection of musical instruments, and a combination of sound and musical instruments that produce harmony, melody, and rhythm. Music is widely used for the treatment of stress or as one of the brain memory stimulants. Not only classical music can be one of the brain's memories stimulants. Several studies have shown that reading the verses of the Qur'an can also provide a good stimulus for the brain. In addition, there is a stigma in the community that by reading or listening to the verses of the Quran can be a drug stress and improve the ability of the brain.

A study was conducted on four respondents using two different treatments. After a memory test using a free recall test, respondents in the experimental group were recited by the Qur'an for 15 minutes. While the respondents in the control group received a placebo treatment of reading the story of the animal world. Then all the respondents performed a memory test again. The results obtained by the experimental group experienced significant improvement in memory capacity after treatment. While the control group there was no significant difference between before and after treatment [3]. Research from Abdurrochman, Perdana, and Andhika shows when listening to the Qur'an occurs a significant rise in brainwaves before and after the Al-Quran [4]. Therefore, researchers conducted research on the effect of murattal Quran on short-term memory performance.

## 2 Literature Review

### 2.1 Short-Term Memory

Short-term memory has a small capacity, but a very large role in the memory process, which is where we process the stimuli that come from our environment. This small information storage capability corresponds to a limited processing capacity. Short-term memory serves as a transitory storage that can store very limited information and transform and use that information to produce a response to a stimulus [8].

Sensory memory records information or stimuli that enter through one or a combination of the five senses, i.e., visually through the eyes, ear hearing, smell through the nose, taste through the tongue, and touch the skin. If the information or stimuli are not noticed will be immediately forgotten, but when noted then the information is transferred to the short-term memory system. Short-term memory systems store information or stimuli for about 30 seconds, and only about seven chunks of information (chunks) can be stored and maintained in short-term memory systems at a time. Once in the short-term memory system, the information can be transferred again by repetition process to the long-term memory system for storage, or it may be lost/forgotten because it is replaced by additional displacement [5].

The temporal dynamics of the embedded component model are being mapped out in the behavioral literature. For example, memory items that are no longer relevant for behavior can be removed (within 1–2 sec) from the focus of attention, thereby reducing the load on the system's limited capacity and consequently reducing reaction time to memory probes of the behaviorally relevant items still in the focus [6]. Information removed from the focus remains in a state of heightened availability for several seconds, as shown by the finding that lures from a recently encoded memory list are harder to reject than lures not recently encountered [6]. This information can be refocused if needed again otherwise, it is prone to forgetting by decay or by interference [7].

### 2.2 Holy Qur'an

The Qur'an is a holy book of Muslims. According to the Islamic ummah is the revelation of the God revealed to the prophet Muhammad 14 centuries ago by the mediation of the angel Gabriel. The Qur'an consists of 114 chapters called letters. Each letter consists of various verses with different amounts. Muslims say that in the Qur'an contains important information that can provide answers and solutions to the problems facing humans [9].

Hawari (1997) [10] mentions that many Qur'anic verses contain the guidance of how humans in life in this world free from anxiety, tension, and depression. Research conducted by Chalfant and Heller in Hawari (1997) showed 40% of people who experience mental anxiety are brought to religious experts for treatment. Also in the Qur'an also found verses about the peace Research from a doctor in Florida named Qadhi found that when reading the verses of the Quran, it will feel the serenity [3].

## 3 Research Method

### 3.1 Subject and Object Experiment

a. Subject Experiment

There were 50 university students (26 male and 24 female) in the range of 19 to 22 years old was appointed to be the subject of this research. Assuming each participant has the same understanding of the ability of the experiment, in this case, is the short term memory.

b. Object Experiment

Short term memory testing using software design tools. One of the components contained in this software is a program running memory task. Running memory task program is a program used for short-term memory testing. In this program a sequence of numbers will appear for a few seconds, then disappears. After that will be shown a dialog box to enter a sequence of numbers that appear earlier.

### 3.2 Research Procedure

a. Preparation

In this step, researchers prepared needed apparatus to do the short term memory test performance influenced by murottal. Apparatus were used in this research are notebook Toshiba 11", handsfree, design tools software and experimental room.

b. Experimental Design

Respondents will perform two experimental conditions. In the first condition, participants do the tests given without listening to murrotal. After working on the first experiment, participants were given a chance to rest for a few minutes. Then participants will experiment as before along with listening to murrotal juz amma. The correct numerical answer data is used as a measurement parameter. This is done 15 times using the Design Tools software to test the respondents.

c. Data analysis

The method used in this study is the experimental method, because according to its importance the experimental method is used to find out the cause [11] and to know there is no effect of listening murrotal to the ability of short-term memory. Data processing in this research using paired sample t-test with a confidence level 95%.

Hypothesis :

Ho = There are no difference in the average short term memory between pre and post
listening murottal.

H1 = There are an difference average of short term memory between pre and post
listening murottal.

## 4 Result and Discussion

The results of this study used to determine the effect of the murattal Quran on the ability of short-term memory. Data obtained from the testing of short-term memory performance using Design Tools software.

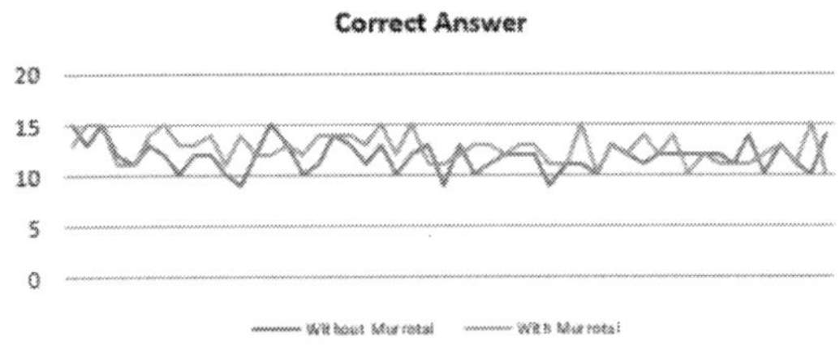


**Figure 1.** Result of the Experiment

From 15 experiments, the graph shows that there is a difference in experimental results when were not listening to murattals and when were listening to murattals. Respondents who listened to murattals during the experiment conducted, having good concentration. So it can remember and reply the numbers displayed correctly. While respondents who did not listen to murattals during the experiment, had low concentrations. This resulted in only part of the correct answer.

Data were analyzed using statistical analysis paired sample t-test. During the test, found that there was a difference between the two treatments in the experiments performed. Based on the results of paired sample t-test shows there are differences caused by the use of murattal. The test was performed using IBM SPSS software version 22, with the following results.

**Table 1.** Result of the paired samples test

| | | Paired Differences | | | T | df | Sig. (2-tailed) |
|---|---|---|---|---|---|---|---|
| | | Mean | Std. Deviation | Std. Error Mean | | | |
| Pair 1 | With Murottal – Without Murottal | 0.82 | 1.945 | 0.275 | 2.981 | 49 | 0.004 |

From the result of Paired Sample T-test, the decision were taken based on the significant value. The data shows that the Sig Value < 0.05 is 0.004 so that Ho is rejected. The result showed that there is a difference in the average value of short-term memory between pre and post murattal listening. It proved after the respondents listening to the murottal, and the concentration will be increase. This is supported by [12], which has conducted a study to find out whether music can enhance the ability of people with Alzheimer's to recognize the nurse face according to the name of the nurse. This study used two treatments, there are with and without the music. The results showed that four out of seven subjects get better results when using music.

Another alternative besides music therapy according to [13] is a religious therapy. Religious therapy can be an accelerated healing, as evidenced by various experts has been done by Ahmad Al Qadi, managing director of Islamic Medicine Institute for Education and Research in Florida, USA. The results of these studies showed positive results that listen to the Qur'an has a significant effect in reducing a nervous tension [13]. Holy verses of the Qur'an have a therapeutic effect in reading and listening [14]. Murattal and music therapy works on the brain, which driven by external simultaneous, then the brain will produce a chemical called neuropeptide. These molecules will be hooking into their receptors in the body and will provide feedback in the form of pleasure or comfort [15]. Based on some opinions above, it can be concluded that listening to Holy Qur'an can give comforts to people who could be a stimulant to increasing people concentration level.

The authors thank to all parties for sponsoring and facilitating this research, and also for all respondents in this research.

## References

1. Xu Y. *Limitation of Object-Based Feature Encoding in Visual Short Term Memory*. http://www.sciencedirect.com, [online, accesed 19 March 2010]. (2002).
2. Miller G., A. *The Magic Number Seven plus-minus Two: Some Limits on Our Capacity to Process Information*. Psychological Review, 63, pp 81-97. (1956).
3. Julianto V., & Etsem, M., B. *The Effect of Reciting Holy Qur'an toward Short-term Memory Ability Analysed trought the Changing Brain Wave*. Jurnal Psikologi Fakultas Psikologi Universitas Gajah Mada. (2011).
4. Abdurrochman A., P. *Muratal Al Quran : Alternatif Terapi Suara Baru*. Prosiding Seminar Sains dan Teknologi-II. (2008).
5. Solso R.,L. *Cognitive Psychology*. (4th ed). Boston: Allyn and Bacon, Inc. (1995).
6. Oberauer, K. *Removing irrelevant information from working memory: A cognitive aging study with the modified Sternberg task*. Journal of Experimental Psychology: Learning, Memory, & Cognition, 27, 948–957. (2001).
7. Oberauer, K. *Control of the contents of working memory—A comparison of two paradigms and two age groups*. Journal of Experimental Psychology: Learning, Memory, & Cognition, 31, 714–728. (2005).
8. Bhinnety M., Sugiyanto dan M. Pudjono. *Pengaruh Intensitas Kebisingan Terhadap Memori Jangka Pendek*. Jurnal Psikologi, XXI, 1, Juni, 28-38. (1994).
9. Firmanto A. *Implementasi Principal Component Analysis Dan Backpropagation Neural Network Dalam Pengklasifikasian Terjemahan Ayat-Ayat Ilmu Pengetahuan Dalam Alquran*. Tugas Akhir. Telkom University. (2011).
10. Hawari D. *Al-Qur'an Ilmu Kedokteran Jiwa dan Kesehatan Jiwa*. Yogyakarta: PT. Dana Bhakti Yasa. (1997).
11. Akmal R. *Pengaruh Pengulangan Terknik Permainan Olahraga Bola Basket Terhadap Peningkatan Kemampuan Motorik Kasar (Gross Motor) Anak Tunagrahita Ringan Usia Mental 5 - 9 Tahun*. Universitas Pendidikan Indonesia. (2013).

https://doi.org/10.1051/matecconf/201815401060

12. Carruth E. *The Effects Of Singing And The Spaced Retrieval Technique On Improving Face-Name Recognition In Nursing Home Residents With Memory Loss*. Journal of Music Therapy, 34, 165-186. (1997).
13. Remolda P. *Pengaruh Al-Quran pada Manusia dalam Perspektif Fisiologi dan Psikologi*. http://www.theedc.com. (2009).
14. Oken B., S. *Complementary Therapies in Neurology: An Evidence-Based Approach. USA:The Parthenon Publishing Group*. (2004).
15. O'riordan R., N., L. *Seni Penyembuhan Alami:Rahasia Penyembuhan Melalui Energi* Ilahi diterjemahkan oleh Sulaiman Al-Kumaiyi dari judul asli The Art of Sufi Healing. Bekasi: Gugus Press. (2002).

# Measuring the usability of e-banking in Special Region of Yogyakarta

*Anindyya Agripina* Hadyanawati[1,*], *Ayudyah Eka* Apsari[2], and *Hari* Purnomo[3]

[1,3] Industrial Engineering Departement, Universitas Islam Indonesia
[2]Industrial Engineerin Departement, Universitas Islam Batik Surakarta

**Abstract.** Online shopping is become an option because customers do not need to go to the store. Internet banking is a bank facility to help customers in common. This research is aim to know usability aspect of e-commerce banking. Data collection was done by distributing questionnaires in Yogyakarta Special Region. The number of respondents is 140 with the range of age 17-22 years. Anova is used as method research. The results showed that in each group show significantly different. Post hoc analysis explained that learnability and efficiency variable has significantly different, with value p<0,05. The highest average value is in the learnability variable and the lowest is the memorability variable.

## 1. Introduction

Nowadays internet has grown in the last two decades and many used by education institution, business and finance institution. Education institution have been using internet to study. Social networking and online shop are used for business by the people (Laohapensang, 2009). Potential e-commerce is up to 77% which is used for search the information product and online shopping. While online shopping users reach 8,7 million people with the prediction of transaction value is US$ 4,89 billion in 2016 (Kominfo, 2015). Shopping using internet is become an option because customers just only sit in front of the computer and make transaction via e-banking (Khodijah, 2013). Customer can also request the design with payment option using online (Ollie, 2008). According to Kotler (2002), the benefit of online service for the distributor are lower cost and adapt in market condition

Increasing number of e-commerce every year it easier for the people to search items through the internet. Buying item can using transaction via e-banking. Electronic banking is facility given by the bank for customer to get information. (Widiastuti, 2010). Application information technology in banking is used to processing the information and increase the service for customers. (Widiastuti, 2010) explained that quality system is significantly affect to customer's trust in increasing internet banking frequency usage.

Related research of internet banking is already studied. (Szopiński, 2016) stated the effect of e-banking. The use of e-banking is influenced by features, compatibility and service to consumers (Boateng, et al., 2016). And Ahmad & Al-Zu'bi, (2011) research about e-banking features. (Oruç and Tatar, 2017) said that communication and convenience significantly influence e-banking.

Based on review related research about e-banking, then it is important to study about e-commerce banking usage. The purpose this research is to know about usability e-commerce banking. In addition to the result of this research can be used as a suggestion to improve the quality e-commerce banking. Improving the quality of e-commerce banking can affect the customers satisfaction. The level of satisfaction will be linear to the level of loyalty so that customers become loyal (Kassim dan Abdullah, 2010; Ahmad & Al-Zu'bi, 2011)

## 2 Method

### 2.1 Object and subject of research

Object research is banking e-commerce in Yogyakarta, 140 banking e-commerce customers that using facilities e-banking are the subject of this research. Respondents with criteria (1) male and female (2) banking e-commerce usage (3) age between 18-50 years old.

### 2.2 Method of collecting data

Collecting data method is use observation and questionnaires. Observation technique in this research was done by direct observation of banking e-commerce and how to using e-banking. Questionnaire was distributed by google form as long as 2 months. Collecting data technique is using accidental sampling. Respondents selected according to specified criteria

## 3 Questionnaire Variables

The research variables in the questionnaire consist of five attributes:

* anindyagripina@gmail.com

1. Learnability, with measurement indicator is internet banking is easy to learn and ease of accessing e-banking anywhere and anytime
2. Efficiency, with measurement indicator is website fast responds and completeness of the information provided
3. Memorability, with measurement indicator is website design layout and placement of menus available on the website easy to remember
4. Errors, with measurement indicator is server down frequency transaction failure
5. Satisfaction, with measurement indicator are satisfied with the systems of the services provided by the bank related

# 4 Results and Discussion

## 4.1 Characteristic of respondents

Respondents who fill questionnaires as many as 140 respondents with age between 17-22 years. Respondents with age 19 years are 42.5% percentage, followed with respondents with age of 18 years old are 22.5%, 27.5% at age 20 years, 5% 21 years old and 2.5% 22 years old. Respondents male is 50% and female 50%

Differential tests result using anova for each variable are sown in Table 1 below

**Table 1.** Differential test result

| ANOVA | | | | | |
|---|---|---|---|---|---|
| | Sum of Squares | df | Mean Square | F | Sig. |
| Between Groups | 8.584 | 4 | 2.146 | 2.666 | .034 |
| Within Groups | 156.987 | 195 | .805 | | |
| Total | 165.571 | 199 | | | |

Based on the differential test result on Table 1 indicated that there is significantly differences between groups with P value is 0,034 (p>0,05). It explained that each variable has different value, the importance of the respondent to variable is different. The average value of each variable is shown in Figure 1.

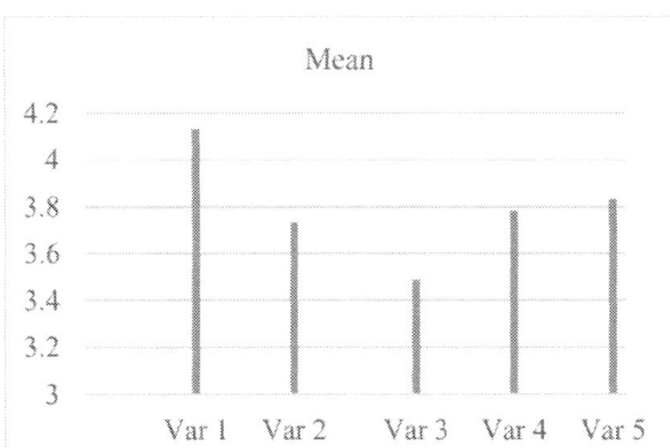


**Figure 1.** The mean of variables

Note:
Var1 : Learnability
Var2 : Efficiency
Var3 : Memorability
Var4 : Errors
Var5 : Satisfaction

Based on figure 1, the highest value is learnability with an average 4.13. It can be explained that banking e commerce is very easy to use. The highest value is variable efficiency, variable errors and variable satisfaction has a little differential result with p value is 3.73; 3.78; 3.83. This result is followed by Sangar and Rastari (2015) that explained Learnability, Satisfaction, and Efficiency is first grade factors. The lowest value is variable memorability with an average 3.49 explained that e-commerce is less memorable. All variables have value higher than 3 in scale 5. This explained that learnability, efficiency, memorability, errors and satisfaction variable have good value in banking e-commerce application. Rastari and Sangar (2015) explained that usability is factor that need to improve for increase quality and efficiency. Independent pair t-test among variables can be seen in table 2.

**Table 2.** Independent pair t-test among variables

| Multiple Comparisons | | | | | | |
|---|---|---|---|---|---|---|
| | | Mean | | | 95% Confidence Interval | |
| (I) Factor | (J) Factor | Difference (I-J) | Std. Error | Sig. | Lower Bound | Upper Bound |
| 1 | 2 | .40075* | .20063 | .047 | .0051 | .7964 |
| | 3 | .64600* | .20063 | .002 | .2503 | 1.0417 |
| | 4 | .35025 | .20063 | .082 | -.0454 | .7459 |
| | 5 | .30025 | .20063 | .136 | -.0954 | .6959 |
| 2 | 3 | .24525 | .20063 | .223 | -.1504 | .6409 |
| | 4 | -.05050 | .20063 | .802 | -.4462 | .3452 |
| | 5 | -.10050 | .20063 | .617 | -.4962 | .2952 |
| 3 | 4 | -.29575 | .20063 | .142 | -.6914 | .0999 |
| | 5 | -.34575 | .20063 | .086 | -.7414 | .0499 |
| 4 | 5 | -.05000 | .20063 | .803 | -.4457 | .3457 |

*. The mean difference is significant at the 0.05 level.

The result of independent pair t-test of variable 1 (learnability) and variable 2 (efficiency) is significantly different with p value 0,047 (p<0,05) with mean difference is 0,4. Variable 1 (learnability) and variable 3 (memorability) are significantly different with p value 0,02 (p<0,05) with mean different 0,64. Another variable has p value 0,05 (p>0,05)

# 5. Conclusion

Respondent importance level in each variable has different value. Differential test result for groups there is significantly different with p value is 0,034. Meanwhile, independent pair t-test shown that variable 1 (learnability) and variable 2 (efficiency) is significantly different with p value is 0,047 and mean difference 0,4. Variable 1 (learnability) and variable 3 (memorability) is significantly different with p value 0,02 and mean difference 0.64. meanwhile, the other variables are not significantly different with p value higher that 0,05. Mean value of learnability has the highest value 4,13 and memorability has the lowest variable 3,49. Variable 2, 4 and 5 has the mean value 3,73; 3,78; 3,83, all variables has value higher that 3 of scale 5. Then it's known that banking e-commerce is easy to use.

## References


1. Laohanpengsang, O., "Factor influencing internet shopping behavior: a survey of consumers in Thailand." journal of fashion marketing and management vol.13 No.4,2009 pp.501-513 (2009)
2. Kementrian Komunikasi dan Informasi Republik Indonesia. Potensi Besar untuk Pemasukan Negara. Edisi XXI. Kominfo.go.id (2015)
3. Khodijah F.F., Analisis Pengaruh Kualitas Layanan dan Kualitas Produk terhadap Loyalitas Pelanggan pada Onlineshop menggunakan Structural Equation Modeling. Skripsi. Universitas Diponegoro. Semarang (2013)
4. F. De Lillo, F. Cecconi, G. Lacorata, A. Vulpiani, EPL, **84** (2008)
5. Widiastuti B. Studi tentang Intensitas Penggunaan Electronic Banking oleh Nasabah PT.Bank Central Asia (BCA) Kantor Cabang Utama Semarang. Tesis. Universitas Diponegoro. Semarang. (2010)
6. Szopiński T., S. Factors affecting the adoption of online banking in Poland. Journal of Business Research, Vol. 69 (11), p 4763–4768 (2016)
7. Boateng, H., Adam, D., R. Okoe ,A., F., Anning-Dorson, T. Assessing the determinants of internet banking adoption intentions: A social cognitive theory perspective. Computers in Human Behavior. Vol. 65, p. 468–478 (2016)
8. Ahmad, A.E.M.K. dan Al-Zu'bi, H.A., E-banking functionality and outcomes of customer satisfaction: an empirical investigation. International Journal of Marketing Studies, 3(1), pp.50-65 (2011)
9. Oruç, O., E., and Tatar, C. An investigation of factors that affect internet banking usage based on structural equation modeling. Computers in Human Behavior. Vol. 66, p 232–235 (2017)
10. Kassim, N. and Abdullah, N., A. The effect of perceived pervice quality dimensions on customer satisfaction, trust, and loyalty in e-commerce settings: A cross cultural analysis. Asia Pacific Journal of Marketing and Logistics, Vol. 22 Iss: 3, pp.351 – 371 (2010)
11. Sangar, A., B. and Rastari, S. A Model for Increasing Usability of Mobile Banking Apps on Smart Phones. Indian Journal of Science and Technology, Vol 8(30) (2015)
12. Rastari, S and Sangar, A., B. Investigating and comparing the usability models in the banking applications, Engineering Research Journal Vol 3, Issue 6, pp. 77-85 (2015)

# Product-service system method to measure sustainability level of traditional smoked fish processing industries

*Ratna* Purwaningsih[1*], *Anggaina Elfandora* Cahyantari[1], *Zulfaida* Ariyani[2], *Aries* Susanty[1], *Ary* Arvianto[1], and *Haryo* Santoso[1]

[1]Diponegoro University, Industrial Engineering Department, Jl Prof Soedharto SH, Tembalang, Semarang, 50275, Indonesia
[2]Diponegoro University, Vocational School, Jl Prof Soedharto SH, Tembalang, Semarang, 50275, Indonesia

**Abstract.** Small Medium Enterprise's (SME) of traditional fish processing at Semarang, Central Java, Indonesia still focus their business on gain more profits. Sustainability aspect has not received enough attention yet. This study aims to review the sustainability level of SME smoked fish Semarang using product service system (PSS) method. PSS consists of three dimensions (1) Environment, (2) Socio-cultural and (3) Economic. Each dimension consists of 6 criteria's. PSS not only assess the level of sustainability but also formulated the recommendation to increase the industries sustainability level. Sustainability assessment and recommendations formulation is guided by a check-list form. Then, the portfolio diagram used to select these recommendations according to its feasibility to be implemented and its importance for the industries. The result of sustainability assessment for traditional fish processing is 0.44, categorized as medium level. The recommendations for the environmental dimension are (1) use of liquid smoke on fish processing and (2) use of wastewater treatment with anaerobic ponds Recommendation for the socio-cultural dimension is use personal protective tool to reduce worker risk on safety and health. Recommendation for the economic dimension is used social media for product marketing and increasing the economic value of fish lung wastes. Recommendations are then illustrated in a diagram in the form of radar sustainability.

## 1 Introduction

Sustainable development has been defined in many ways, but the most frequently quoted definition is from Brundtland Report: "Sustainable development is the development that meets the needs of the present without compromising the ability of future generations to meet their own needs [1]. Product-service system (PSS) is a method to measure industry sustainability level. There are three dimensions of sustainability that will be the basis for assessing and formulating recommendations to increase sustainability level. These three dimensions are the environmental, socio-cultural and economic dimensions in which each dimension contains six criteria [2]. Application of PSS in the industry was believed to be a source of innovation [3], improving consumer satisfaction, product life-cycle management and information flow in the manufacturing cycle [4], also increase company productivity levels [5].

Assessment of the sustainability level of a region or a business unit has been developed in various fields. Sustainability indicator on mining and mineral sectors was developed by MMSD (Mining, Minerals and Sustainable Development). In the forestry sector, the impact assessment of the forest wood supply chain is measured using ToSIA / A Tool for Sustainability Impact Assessment [6]. In the chemical sector the characterization factors for human eco-toxicity and toxicity were measured by the Uniform System for the Evaluation of Substance or USES-LCA) method [7]. In industry has been developed a measurement method and indicators of Sustainable Development for Industry (ISDI). Sustainability is assessed through environmental impact, environmental efficiency, and voluntary action. ISDI follows the lifecycle approach and considers the complete life cycle of materials and energy utilized [8].

There has been a significant decrease which reaches 48% in production volume of SME smoked fish at Semarang. Although, the raw material for the production of smoke fish was increased by 14%, and fish consumption of Semarang community increased at 15% [9]. Besides, there are environmental and health problems of workers and the surrounding community due to the smoke from the fish processing. The smoke that arises during the production process contains chemicals that potential to cause a decline in lung function and threaten the health of workers. The surrounding community is also disturbed by the pollution. Worker health problems and their impact on surrounding communities are included in the problems of the socio-cultural dimension [10]. The purpose of this study is to measure the sustainability level, formulate recommendations and assess the recommendations.

Since 1970, marketing and innovation management have been trying to find ways to connect the functions of consumer desires into new product and service design systems. PSS is one method that connects various stakeholders in the design and assessment, as well as system development both production system and in product and service development. The global green

* Corresponding author: ratna.tiundip@gmail.com

economic transition calls for the transformation of business strategy from traditional business-oriented model to service-oriented business model. One of the new business models that fit this transformation is the product -service system [11]. Geum and Park [12] created a road map of PSS in the form of a systematic methodology for PSS to be easily applied.

Application of Product Service System in System Design is on the design of supply chain. Wang and Fu Jia [13] applied PSS to develop a supply chain that considers multi-stakeholders. Wangphanich [14] examines the impact of applying PSS to the supply chain of a washing machine product to a dynamic system simulation and found that PSS can reduce the environmental impact of the product to consumers. Durugbo [15] combines PSS with Collaborative Networks on a product delivery system. Application of PSS in logistic system design done by Xing and Heon Lee [16] they used PSS to design logistic system and combine it with life cycle model approach.

## 2 Research Method

PSS method is carried out in a series of stages developed by UNEP and DELFT University of Technology. PSS consists of 4 stages (1) Identify system sustainability dimension, (2) Formulate and select recommendation, (3) Conduct assessment of detail of recommendation, and (4) Conduct feasibility test on a recommendation to choose the best one. The first stage is to identify the dimensions of system sustainability. There are 3 activities in this stage, drawing system maps, making a SWOT of sustainability and assessing system sustainability.

The second stage is to formulate and select recommendations. At this stage, the questionnaire was distributed using the PSS worksheet. The results of the questionnaire are then used to make detailed recommendations. The recommendation categorize on 4 class; H = high, M = Medium, L = Low and N = No, or have no relation. The recommendation will be detailed are the H categorize.

The third stage is to choose the recommendations to be applied. At this stage, the recommendations selected in stage 2 are then detailed. We assess the recommendation by giving the 'mark' ++ for much better, + for better, = for nothing change and – mean worse than existing condition. This mark then became the-y axis value of sustainability improvement possibility in portfolio diagram. The fourth stage is to test the feasibility of recommendations by making portfolio diagrams and PSS radar. Radar serves to compare the future conditions if the recommendation applied to the current system. Then, we need to create a portfolio diagram use a feasibility recommendation questionnaire. Respondents were asked to assess detailed recommendations with judgment based considerations, (a) Duration of recommendations (long term or short term), and (b) Easy-ness of implementation/ implementation (difficult to implement, or easy to implement). This feasibility value became the value of the x-axis on portfolio diagram. This portfolio diagram has the y-axis for sustainability improvement possibility and the x-axis for feasibility/ implementation. Put all recommendations in the matrix according to its value on these two dimensions.

## 3 Results and Discussion

The sustainability assessment was given to 21 traditional smoked fish processing at Bandarhardjo Semarang. The system map and sustainability SWOT were done by interviewing the owner of traditional fish processing industries and related agencies. The result of system map given in figure 1 and sustainability SWOT in detail given in table 1. This SWOT analysis will be useful in identifying recommendation for sustainability improvement.

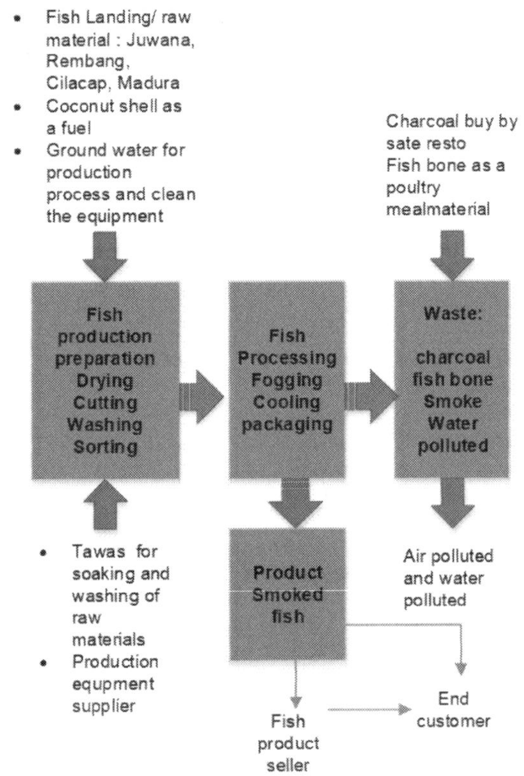


**Fig. 1.** System map of traditional smoked fish processing industries

Figure 1 shows the parties involved in the smoked fish production. Related parties include fresh fish suppliers, coconut shell suppliers, processed fish traders, waste collectors, and consumers. The system begins when the industry owner purchase of fish as raw materials in the fish market. The amount or the availability of raw material was following the natural conditions. The fuel (the coconut shell) purchased from coconut shell collectors who came to the traditional fish processing center. The raw materials obtained on one day then processed entirely on the same day. Smoked fish are immediately distributed on the same day to traders in the market, or to consumers who come directly to the industries location. The wastes are partially sold, and some are still poorly managed.

**Table 1**. Result of SWOT sustainability analysis

| SW | State the existing condition | |
|---|---|---|
| | **Strength** | **Weakness** |
| Environment | Several types of waste have been re-utilized<br>There is waste has a high selling value | The production process is traditional.<br>The result of polluted waste.<br>Poor water quality |
| Social Cultural | The industry in Bandarharjo has been established for almost half a century | Health and safety risks<br>Air pollution disturb the community<br>Not hygienist process<br>Low-level education |
| Economic | economic support from the government | Products (potentially containing bacteria and carcinogenic substances) |
| | **State of future condition** | |
| **OT** | **Opportunity** | **Threat** |
| Environment | Liquid curing techniques to reduce air pollution and carcinogenic | Long-term environmental degradation. |
| Social Cultural | Future generations who are educated can change to better production process | The health of workers is threatened.<br>Poor quality of the production environment |
| Economic | The culinary tourism<br>Further waste treatment | Competition that is not addressed properly will threaten the existence of industries |

Next step is the categorization of sustainability criteria. This step is conducted to find out what sustainability criteria need to be developed by giving the categorization as No, Low, Medium or High categories. Categorization of criteria conducted after the spreading of questionnaires to 21 fish processing industries. Then, the average calculation of all criteria was scored [17]. Results of calculation found that the level of sustainability of smoked fish industries of Semarang City is 0.44. It indicates that the sustainability of the industries center is in the medium class. The sustainability assessment results were given in table 2.

The third step is the formulation the recommendations, obtained from the results of the questionnaires referring to the Recommendation Development Guide. Questionnaires were distributed to 23 academicians, 2 industries practitioners, and 3 industry and trade ministry officer of Semarang district. The recommendation coding is given in table 3.

**Table 2.** Result of sustainability assessment

| | Criteria's | Score | Categorize |
|---|---|---|---|
| Environment | System Age Optimization | 0.36 | *Medium* |
| | Reduction of Transportation / | 0.09 | *Low* |
| | Reduction of Resources | 0.11 | *Low* |
| | Minimization / Waste Management | 1 | *High* |
| | Conservation | 0 | *No* |
| | Toxicity | 0.5 | *Medium* |
| Socio Cultural | Social Responsibility | 0 | *No* |
| | Occupational Health and | 1 | *High* |
| | Product quality | 0.28 | *Low* |
| | Employment | 0.38 | *Medium* |
| | Stakeholders relationship | 0.5 | *Medium* |
| Economic | Market position and competitiveness | 0.9 | *High* |
| | Profitability | 1 | *High* |
| | Value-added for consumers | 0.33 | *Low* |
| | Long-Term Risk | 0.7 | *High* |
| | Partnership / Cooperation | 0.4 | *Medium* |
| | Macroeconomic Effect | 0 | *No* |
| **Mean** | | **0.44** | ***Medium*** |

**Table 3**. Alternative recommendation

| Criteria | Alternative Recommendation |
|---|---|
| A. Minimization / Waste Management | A.1. Using liquid smoke<br>A.2. Processing liquid waste with the concept of anaerobic pond |
| B. Occupational Health and Safety | B.1. Using personal protective equipment (PPE)<br>B.2. Using a smoking cabinet |
| C. Market Position and Competitiveness | C.1. Using airtight packaging (vacuum packaging)<br>C.2. Expand the marketing area<br>C.3. Clean water by filtration<br>C.4. Promoting through social media |
| D. Profitability | D.1. Increasing the economic value of waste from fresh fish<br>D.2. Promotion of culinary tourism<br>D.3. Cooperate with parties involved in the culinary world |
| E. Long-Term Risk | E.1. Maintain sanitation and hygiene, and improve bad habits<br>E.2. Government reward to fish processors implementing clean, healthy, and safe production process<br>E.3.Industrialization smoked fish in Bandarharjo |

Alternative recommendations that have been compared with the existing condition of fish processing then conducted to a feasibility testing to find out how difficult to be applied. The results of the feasibility test are then visualized into a portfolio diagram given in Figure 2.

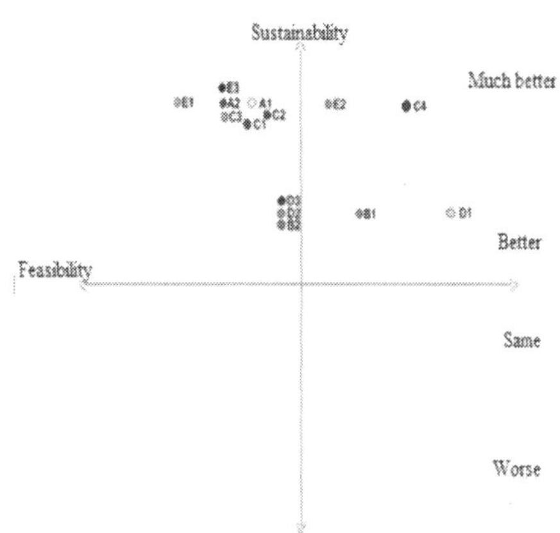


**Fig. 2**. Portfolio Diagram to map the recommendation

Figure 2 shows the recommendation at the top right means is the alternative best that can improve the sustainability of the fish processing industries and have the highest possibility to be applied. Some recommendations for fish processing industries are:

- C4 (promoting through social media) and
- D1 (increasing the economic value of product)
- B1 (Using personal protective equipment) and
- E2 (Government reward to fish processors implementing clean, healthy, and safe production process).

The fourth step is to describe the condition and recommendation on a radar diagram for each aspect. The radar for environmental aspect given in Figure 3, the radar for socio-cultural aspect dimension on Figure 4, and the radar for economic aspect on Figure 5.

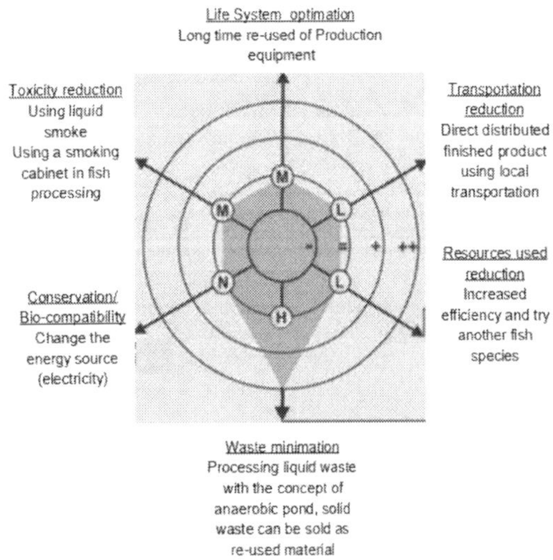


**Fig. 3**. Radar diagram for environmental dimension

Recommendations for the minimization/waste management criteria suggest the industry to use liquid smoke and installed an equipment of liquid waste treatment with the concept of anaerobic ponds. Liquid smoke has advantages such as easy to apply, smoke concentration can be arranged according to consumer tastes, the product has a uniform appearance and environmentally friendly. Another important point is that liquid smoke not only plays a role in shaping the sensory characteristics but also in terms of food security assurance [18].

Figure 4 is a radar diagram for socio-cultural dimension. Criteria on the high category are Health and Safety, and the recommendation suggests the worker of the industry to using a personal protection equipment. Workers health complaint are on eyes pain and respiratory organ which caused by air pollution. The industries also suggested to using a liquid smoke and smoked cabinet while processing the fish to minimize pollution.

Criteria on a high category on economic dimension are market position and competitiveness, Profitability, and long-term risk. Recommendations for the economic dimension are marketing through social media, and cooperation with supplier and local transportation service provider to distribute the finished product.

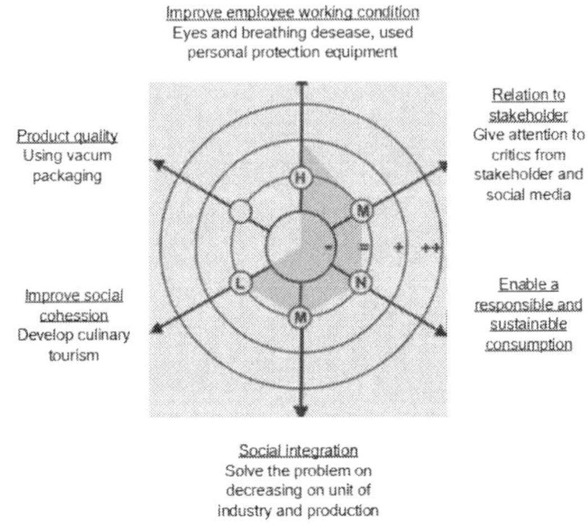


**Fig. 4**. Radar diagram for socio cultural dimension

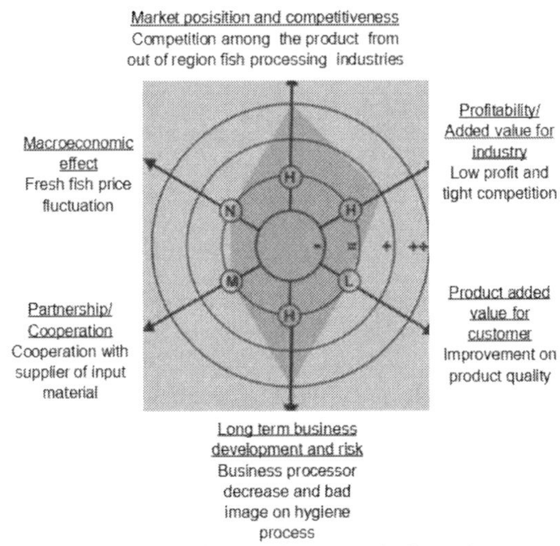


**Fig. 5**. Radar diagram for Economic dimension

Recommendations to increase profitability is Utilizing fish raw material waste. The smoked fish processors sell back Manyung fish (*Arius Thalassinus*) lung without further processing, at a price of Rp. 100,000 / kilogram. If it is sold in the form of crackers the price can reach Rp. 250,000 / kilogram.

Recommendations to face long-term risk are creating a reward for an industry which already applied a hygienist production process. The bad image of the production process can cause a threat to sustainability. The fish processors should be able to maintain sanitation and hygiene in the production process and achieve the standard of Good Manufacturing Practices (GMP). Reward in the form of money can be used as an initial motivation.

## 4 Conclusions

Measurement of sustainability level of traditional smoked fish processing at Semarang gives value 0.44 from 0 - 1 scale, included in the medium category. Sustainability criteria that need to be improved are Minimization of Waste, Health and Work Safety, Market Position and Competitiveness, Profitability, and Long-term Risk.

The most feasible recommendations to apply are the use of personal protected equipment for a worker, marketing product through social media, increasing the economic value of waste, and government rewarding for industries implement sustainability in its production process. Waste minimization can be applied by use of liquid smoke and installed liquid waste treatment equipment anaerobic ponds.

## References

1. WCED Our Common Future (The Brundlandt Report). Oxford University Press (1987).
2. M. Crul, J. C. Diehl. *Design for Sustainability: A Step by Step Approach*. UNEP (2009)
3. T. Mc Aloone, C. and M. Andreasen. Proceedings of *the 8th International Design Conference*, Dubrovnik, Croatia (2004)
4. K. Haapala, R., K. L. Brown, and J. W. Sutherland. Transactions of NAMRI/SME, **36**: 333-340 (2008)
5. E. Moreau, B., B. Tether, Proceeding of *DRUID Conference*, Copenhagen Business School (2011)
6. T. Palosuo, T. Suominen, W. Werhahn-Mess, J. Gonzalo, G., M. Lindner, Ecological Modelling, **221** pp.2215-2225 (2010)
7. D. Jager,.T., T. Vermeire,.G., W. Slooff, H. Roelfzema, Chemosphere, **29**, Issue 2, pp. 319-335 (1994)
8. A. Azapagic, Journal of clean production. **12**. 6, pp. 639-662 (2004).
9. Indonesian statistician Bureau. BPS Kota Semarang. (2015).
10. U. Tischner, C. Vezzoli. Module C: Product-Service Systems–Tools and Cases. UNEP. Paris. (2009)
11. Jing Hao. Proceeding of *the International Conference on Management Science and Engineering (ICMSE)* Sept. 2012, pp 790 - 795 (2012).
12. Geum, Y., Park, Y. (2010). Proceeding of *Industrial Engineering and Engineering Management* (IEEM) IEEE International Conference, pp 410 – 414 (2010)
13. Wang, Li Ping. , Fu Jia Proceeding of *Emergency Management and Management Sciences* (ICEMMS), 2010, IEEE International Conference, Beijing, pp 126 – 129 (2010)
14. Wangphanich, P Proceeding of *Management Science and Industrial Engineering* (MSIE) International Conference, pp 554 – 559 (2011)
15. Durugbo, C *Concurrent Enterprising* (ICE) 17th International Conference, pp 1-8 (2011)
16. Xing, L.Y., Heon Lee, S *Service Operations and Logistics and Informatics* (SOLI) IEEE International Conference, pp 23 – 28 (2010)
17. R. Purwaningsih, N. Susanto, N. Apsari. Proceeding of *International Conference on Electric vehicular Technology and Industrial, Mechanical, electrical, Chemical Engineering (ICEVT & IMECE)* C(5) 1 – 6 (2015).
18. Swastawati, F. Jurnal Dinamika Ekonomi Pembangunan, **1**(1): 18 – 24 (2011)

# Incorporating Hofstede' National Culture in Human Factor Analysis and Classification System (HFACS): Cases of Indonesian Aviation Safety

*Gradiyan Budi* Pratama, *Ari* Widyanti*, and *Iftikar Zahedi* Sutalaksana

Department of Industrial Engineering, Bandung Institute of Technology (ITB), Bandung, Indonesia

**Abstract.** National culture plays an important role in the application of ergonomics and safety. This research examined role of national culture in accident analysis of Indonesian aviation using framework of Human Factors Analysis and Classification System (HFACS). 53 Indonesian aviation accidents during year of 2001-2012 were analyzed using the HFACS framework by authors and were validated to 14 air-transport experts in Indonesia. National culture is viewed with Hofstede' lens of national culture. Result shows that high collectivistic, low uncertainty avoidance, high power distance, and masculinity dimension which are characteristics of Indonesian culture, play an important role in Indonesian aviation accident and should be incorporated within HFACS. Result is discussed in relation with HFACS and Indonesian aviation accident analysis.

## 1 Introduction

In global aviation accident, Shappell & Wiegmann (2005) stated that human factors contribute to 80-90% of those accident. Moreover, it has been known that aviation-accident rates differ across countries and regions. In general, developing countries in Asia and Africa have higher accident rates than Western American or Western Europe. The reason why developing countries have higher aviation accident rate than developed countries relates with mismatch between aviation system and the user. It should be noted that majority of the aviation system has been developed from a Western (North American/Western European) perspective [10]. The most possible cause of aviation accident in developing countries in relation with the mismatch is the different national culture.

National culture, defined as collective mental programming of the mind that distinguishes the members of one group or category of people from others (in this case, the group is a nation, [3]) plays important role in many ergonomics application. For example, in mental workload measures, national culture influence the sensitivity of mental workload measure between Eastern and Western culture, both for subjective measures (see [9] for further review) as well as objective measures (see [18] for further review).

Hofstede [3-5] proposed six dimensions of national culture: power distance (PD), uncertainty avoidance (UAI), individualism (IDV), masculinity (MAS), Long-term orientation (LTO), and indulgence vs. retraint (IVR). PD is the extent to which member of society accept equality among them in society. High PD culture is characterized by inequality between ordinate and subordinate. UAI is the extent to which the members of society perceive a threat in uncertain or unfamiliar situations. High UAI culture is characterized by acceptance of possibility of such situations by no strict behavioral codes, laws and rules, and approval of deviant opinion. IDV is the extent to which society reinforce individual or collective achievement and interpersonal relationship. High IDV culture is characterized by high respect of individual rights and individual tends to form a larger number of moderately distant relationship. MAS is the extent to which the society distinguish male and female in achievement, control, and power. High MAS culture is characterized by male domination. LTO related to the choice of focus for people's efforts: the future or the present and past. High LTO is characterized by perseverance, thrift, ordering relationships by status, and having a sense of shame. IVR related to the gratification versus control of basic human desires related to enjoying life. High indulgence is characterized by allowing relatively free gratification of basic and natural human desires related to enjoying life and having fun.

There are many research have been conducted to explore the relationship between national culture and aviation safety, i.e., different cultures hypothesized to be differ in promoting aviation safety. According to Li and Harris (2007), the short-term orientation culture (i.e., the US) seems to be superior for promoting aviation safety than the long-term orientation culture (i.e. Taiwan) [12]. Jing et al. (2001) also offers indirect proof of the relationship between world aircraft accident rates and national cultural background, in which authoritarianism explain more than half the variance in accident rates [8].

* Corresponding author: widyanti@mail.ti.itb.ac.id

They showed that the 'normal' activities in the cockpit designed in the West are highly vulnerable to an authoritarian mentality

The crucial role of national culture in aviation safety raises a need of accident analysis tool which incorporating national culture in the analysis of aviation accident. One tool that has been applied in investigating aviation accident is Human Factors Analysis and Classification System (HFACS) [14-16]. HFACS is based on Reasons' model of human error (Reason, 1990) as can be seen in Figure 1. According to Reason, active failures are related with the performance of operators in complex system, whereas latent failures are characterized as inadequacies or miss-specifications that lie within a system for a long time and only triggered when combined with other factors. The first level of HFACS is "unsafe acts of operators" and can be classified into two categories that are errors and violation. The second level is "precondition for unsafe acts" related with the latent failures within the causal sequence of events. The third level is "unsafe supervision" which traces the causal chain of events producing unsafe acts up to the level of the frontline supervisor. The fourth level is "organizational influences" which describes upper level management' fallible decisions that directly influence both supervisory and front-line operators' practices.

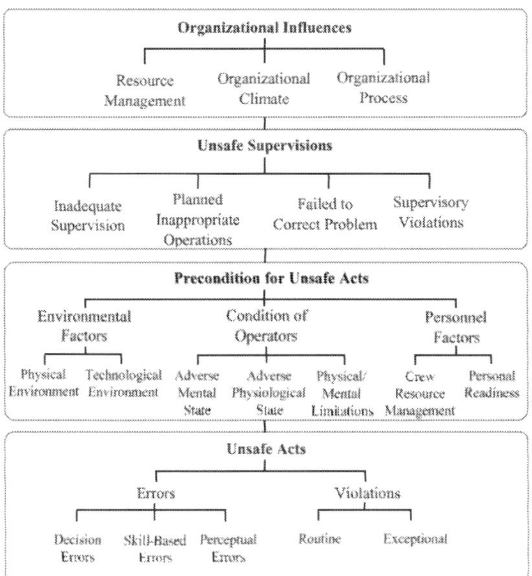


**Fig. 1.** Human Factors Analysis and Classification System

Indonesia, a fast growing developing countries in Asia, has high aviation accident among other countries, even worst, Indonesian airlines rank lowest in world for safety, after Brazil and China (www.aviation-safety.net). Although more modern planes and better navigation equipment have helped cut crash rates, the number of fatal crashes hasn't changed very much. Indonesian National Transportation Safety Committee (Komite Nasional Keselamatan Transportasi/KNKT in Indonesian) stated that human factors contribute to 52% of Indonesian aviation accident. Preliminary interview with KNKT reveal many factors related to the role of human factors in Indonesian aviation accident, including Indonesian national culture. Regarding national culture, Hofstede' study [3] as well as Johnson & Widyanti' study [9] found that Indonesia has high PD, low UAI, low IDV, high MAS, low LTO, and high IVR.

The purpose of this study is to observe role of culture in Indonesian aviation safety and to incorporate national culture in HFACS. Accident reports have been observed and experts or accident investigators were asked to apply HFACS in these accident reports with possible cultural role in the accident investigation.

## 2 Method

### 2.1 Material

Report of 53 accidents or near-miss accidents in Indonesian aviation were obtained from Indonesian National Committee of Transportation Safety (*Komite Nasional Keselamatan Transportasi*/KNKT in Indonesian) between 2001 and 2012. The report of KNKT consists of chronology of the accident as well as general report of the cause(s) of the accident.

### 2.2 Participants

There are 14 participants in this study. They are two lecturers of engineering aviation department Bandung Institute of Technology (ITB), one director of flight carrier, one senior pilot, one senior flight engineer, one operational risk and hazard manager, one chief pilot and senior manager incident management, one senior manager corporate quality assurance, IOSA compliance manager, senior manager safety information management, one commercial director of flight carrier, one senior flight attendant and senior manager safety information management. Participation was voluntarily.

### 2.3 Procedure

Accident analysis report from KNKT is used as a base in identification and classification of accident based on HFACS taxonomy. In addition, in-depth interviews with participants were conducted to address specific events in which participants were ever involved or well-informed. The participants were asked to make a comment on aviation safety in Indonesia in general, and then relate concrete situation in one or two specific accidents. There are also some participants described multiple independent accident during an interview session. Interviews were audiotaped and then transcribed. The interviews for each participants is about 30-60 minutes.

From the accident report from KNKT and transcript, result of interviews were identified and classified based on HFACS taxonomy. The initial classification was conducted by the second author and discussed with first author until consensus was reached. An additional reliability check was conducted by confirmed the classification to all participants. Discussion and consensus method was applied in HFACS taxonomy for 53 accident reports in Indonesian aviation.

Based on classification of HFACS, Hofstede' indicator [3] was used as a guide for the identification process of cultural factors in each layer of HFACS. Hofstede describes 10 indicators of PDI, 7 indicators of IDV, 10 indicators of MAS, 17 indicators of UAI, 12 indicators of LTO, 10 indicators of IVR. For example, indicators of small power distance are: (1) use of power should be legitimate and is subject to criteria of good and evil, (2) parents treat children as equal, (3) older people are neither respected nor feared, (4) student-centered education, (5) hierarchy means inequity of roles, established for convenience, (6) subordinate expect to be consulted, (7) pluralist governments based on majority vote and change peacefully, (8) corruption rare and scandals end political careers, (9) income distribution in society rather even, (10) religions stressing equality of believers. (see Hofstede, 2011 for further review).

## 3 Results and Discussions

For the 53 accident reports, application of HFACS based on analysis of KNKT and interview of participants, following by classification of accident cause conducted by first author and second author, confirmed by all participants, can be seen in Table 1.

**Table 1.** Result of HFACS for 53 Indonesian aviation accidents

| Sub Level 1 | Sub Level 2 | Number | % |
|---|---|---|---|
| Errors | Decision error | 22 | 9.09 |
| | Skill-based error | 38 | 15.70 |
| | Perceptual errors | 10 | 4.13 |
| Violations | Routine | 4 | 1.65 |
| | Exceptional | 2 | 0.83 |
| Environment factors | Physical | 16 | 6.61 |
| | Technological | 14 | 5.79 |
| Condition of operators | Adverse mental states | 21 | 8.68 |
| | Adverse physiological states | 0 | 0 |
| | Physical/mental limitations | 8 | 3.31 |
| Personnel factors | Crew resource management | 23 | 9.50 |
| | Personal readiness | 0 | 0 |
| Inadequate supervision | | 18 | 7.44 |
| Planned inappropriate operations | | 8 | 3.31 |
| Failed to correct problem | | 7 | 2.89 |
| Supervisory violations | | 5 | 2.07 |
| Resource management | | 13 | 5.37 |
| Organizational climate | | 3 | 1.24 |
| Organizational process | | 30 | 12.40 |

From table 1, it can be seen that skill-based error own first rank of accident cause in HFACS analysis, followed by organizational process, crew resource management, and decision errors. Using Hofstede' indicators, cultural factors that contribute to HFACS' layer have been identified as can be seen in Table 2. It can be seen that national culture that plays an important role in Indonesian aviation safety is PD, IDV, MAS, and UAI. Result of this study supports previous study. As stated by Hofstede [4] and Helmreich & Merrit [2], some aspects of national culture that have been identified as crucial in aviation include IDV, PDI, UAI.

There are 3 Indonesian aviation accidents happened in 2007, 2011 and 2012 which correlate with high PD and decision error. In these three accidents, senior pilot ignores alert from instrument as well as warning from co-pilot. Participants agree that it is very common in a cockpit that senior pilot have over confidence with his capability and feel superior to co-pilot. Ignoring co-pilot is usually happen. This is in line with common facts of high PD in Indonesia. According to Helmreich (1999) high PD is characterized by unwillingness to question the decision or action of leaders, which leads to accident [1].

**Table 2.** Hofstede' cultural factors identified in HFACS for 53 Indonesian aviation accidents.

| HFACS Taxonomy | Hofstede' Culture Dimensions | | | | | | | | | | | |
|---|---|---|---|---|---|---|---|---|---|---|---|---|
| | PD | | IDV | | MAS | | UAI | | LTO | | IVR | |
| | H | L | H | L | H | L | H | L | H | L | H | L |
| Rsc. Mgt | | | | | | | | 8 | | | | |
| Org.Clim. | | | | | | 1 | | 2 | | | | |
| Org. Process | | | | | | 1 | | 16 | | | | |
| Inadqt. Sprvision | | | | | | | | 5 | | | | |
| Planned Inapp. Opt | | | | | | | | 5 | | | | |
| Fail to Correct | 1 | | | | | 1 | | 5 | | | | |
| Supervisory Violations | | | | | | 1 | | 3 | | | | |
| Physical Enviro. | | | | | | | | 1 | | | | |
| Techno. Enviro. | | | | | | | | | | | | |
| Adverse Mental Sts. | 1 | | | | | | | 3 | | | | |
| Adverse Physio Sts. | | | | | | | | | | | | |
| Physical/ Mental Lim. | | | | | | | | | | | | |
| CRM | 3 | | | | | 4 | | 2 | | | | |
| Personal Readiness | | | | | | | | | | | | |
| Skill-based Errors | | | | | | | | | | | | |
| Decision Errors | 3 | | | | | 1 | | 13 | | | | |
| Perceptual Errors | | | | | | | | | | | | |
| Exceptional Violations | | | | | | | | 4 | | | | |
| Routine Violations | | | | | | | | 4 | | | | |

UAI plays an important role in the Indonesian aviation safety since 13 accidents have been influenced by this cultural dimension. Such an accident happened because pilot take a risk by flying in bad weather because experiences teach him that flying during bad weather is safe. Another pilot force to land during unstable condition of aircraft. The tendency to take a risk is related with low UAI index of Indonesian culture which give tolerance to risk and use common sense in some situation. UAI was also identified in organizational process for 13 aviation accidents. According to Helmreich (1999), low UAI may be more prone to disregard SOPs [1].

Indonesian low of IDV also influence 1 aviation accident. There is a tendency that a decision must be discussed first and consensus must be reached upon the decision, so that decision cannot be taken immediately, which in turn affect safety. In term of IDV, Helmreich (2000) stated that IDV is contributing to aviation safety, in particular IDV influence effective communication in aviation safety (Hayward, 1999; Metscher et al., 2009). In addition, Li et al. (2007) stated that collectivistic culture or low IDV relates with bad problem solving particularly with collective decision making, so that individualistic culture have lower accident cause than collectivistic culture [12]. Collectivism which is often associated with high power distance is characterized by unwillingness to question the decision or action of leaders [1]. In addition, Soeters and Boer (2000) suggested that aviation safety was enhanced in individualistic culture [17]. They note that in collectivistic culture, the pre-disposition towards regulation led to an increased likelihood of accident.

The fact that Indonesian high MAS also influence aviation safety enrich analysis of global aviation safety. Relation between unsafe act-decision error and MAS can be seen from accident in 2008. In this accident, pilot use his intuition to estimate fuel that must be carry on during the flight. This fact is in line with one indicator of Hofstede' masculinity that is a habit to use intuition during activity. Result of this study that LTO and IVR do not influence HFACS must be noted as well since in other non-Western culture (i.e., Chinese culture), LTO influence aviation safety [7]. The difference may due to different nature of Indonesia and China. Whereas China has high LTO, Indonesia has no clear pattern of LTO [3]. Similar reason for the absence of IVR in HFACS.

This study has several limitation worth noting. First, the accidents reports which have been analyzed restricted to year of 2001-2012 based on KNKT data. Further study with more complete and new data collection is needed to draw more concrete conclusions. Second, further study should be conducted in order to identify in which level the cultural factors are included within HFACS.

Overall, the evidence from this research supports the observation that national cultures have an impact on aviation safety. Further, the compability of national culture and aviation system is a must to obtain aviation safety. As stated by Jing and Batteau (2015) in cases of Chinese culture, crucial insights into national culture and how it manifests itself during flight operations is needed as well as highlighting ways in which Western technology and national culture clash within the cockpit [7].

## 4 Conclusion

Results of this study show that in cases of Indonesian aviation safety, culture plays an important role in aviation accidents investigation. For Indonesian aviation safety, this study gives valuable contribution in considering cultural factors during investigation process of aviation accident, in particular giving a new paradigm for KNKT. Implication of the study is that not to ignore cross-cultural issues in aviation safety. This study also give valuable insight of cultural role in global aviation safety. Countries having similar culture with Indonesia can learn from cases of Indonesia aviation safety. HFACS that is created based on Western culture is argued to be cultural bias. The challenge for aviation safety is not to ignore the cultural issue rather to manage the potential risk that may present.

## References

1. Helmreich, R.L., Proceed. IATA Hum. Fctrs. Sem. 39-43 (1999)
2. Helmreich, R.L., Merrit, A.C., *Culture at work in aviation and medicine: National, organizational, and professional influence* (Aldershot, UK: Ashgate, 1998)
3. Hofstede, G. & Hoftstede G.J., *Cultures and Organizations – Software of the mind.* (New York: McGraw-Hill USA 2010)
4. Hofstede, G., *National Cultures and Corporate Cultures. In L.A. Samovar & R.E. Porter (Eds.), Communication Between Cultures* (Belmont, CA : Wadsworth 1984)
5. Hofstede, G., *Cultural Constraints in Management Theories* (The Academy of Management Executive 1993)
6. Hofstede, G., Psychology & Culture, **2** (2011)
7. Jing, H., Batteau, A., *The Dragon in the Cockpit: How Western Aviation Concepts Conflict with Chinese Value Systems* (Ashgate Publishing, Ltd. 2015)
8. Jing, H.S., Lu, C.J., Peng, S.J., Hum. Fctrs. & Aero. Safe., **1**, 341-359 (2001)
9. Johnson A, Widyanti A., Ergonomics, **54**, 509-518 (2010)
10. Klein, H.A., *Cultural differences in cognition: Barriers in multinational collaborations* (Lawrence Erlbaum, MArwah. 2005)
11. KNKT, *Laporan Penyelidikan Kecelakaan Pesawat Udara.* (Kementrian Perhubungan Republik Indonesia. 2013)
12. Li, W.C., Harris D., Chen A., Aviatn., Spc., Enviro., Med., **78**, Sct. 1 (2007)
13. Reason, J., *Human Error* (Cambridge University, New York 1990)
14. Shappell, S.A. & Wiegmann, D.A., Hum. Fctrs. & Aero. Safe., **1**, 59-86 (2001)
15. Shappell, S.A. & Wiegmann, D.A., *A human error analysis of general aviation controlled flight into*

*terrain accidents occurring between 1990-1998* (Washington DC: Federal Aviation Administration, Office of Aviation Medicine 2003)

16. Shappell, S.A. et. al., *Human Error and Commercial Aviation Accidents : An Analysis Using the Human Factors Analysis and Classification System* (2007)
17. Soeters, J.L. and Boer, P.C., Int. J. Aviat. Psych., **10**, 111-133 (2000)
18. Widyanti A, De Waard D, Johnson A, Mulder B., Ergonomics, **56**, 182-194 (2013)

MATEC Web of Conferences **154**, 01064 (2018) https://doi.org/10.1051/matecconf/201815401064
*ICET4SD 2017*

# The effectiveness of single minute exchange of dies for lean changeover process in printing industry

*Sri* Indrawati*, *Mentari Endah* Pratiwi, Sunaryo, and *Abdullah* 'Azzam

Industrial Engineering, Universitas Islam Indonesia, Yogyakarta, Indonesia

**Abstract.** The changeover time is a factor that greatly affects the lean production implementation in industry with make to order system. Large product variations and unpredictable quantity of orders will trigger some kind of production wastes if changeover time is done in a longer time. One industry with make to order system is printing industry. In general, to produce several types of products such as books takes quite a long time because of long production changeover process. The general problem faced is the delay in book's production completion. Based on this problem, changeover time reduction is needed to overcome the delay of book's production completion using single minute exchanges of dies (SMED) method. The SMED method is the method that separates the changeover activity into two, i.e. internal setup and external setup. The research shows that changeover time for printing workstation is 18 minutes 29 seconds, which consists of internal setup activities 14 minutes 37 seconds and external setup 4 minutes 33 seconds. By converting 45% of the internal setup activity into an external setup, then the setup time can be reduced. The initial setup activities performed when machine is stop, now can be done when the machine is running. In addition, a changeover process improvement also done using 5S method in workstation tools area so the internal setup time is reduced 46% becomes 7 minutes 59 seconds. Under these conditions, the printing industry can increase production by 2%.

## 1 Introduction

The growth of industrial competition and customer demand are influencing every industry with a make-to-order system to improve production performance in fulfilling customer demand. Fast order completion time and short delivery time are key to today's industry competition [1]. If an industry does not increase the speed of production, then the industry cannot compete with other industries. Because customers will prefer industries that fulfil orders quickly.

To increase the speed of customer order fulfilment, the industry must analyse several factors that affect its productivity. Factors influencing the productivity of newspapers printing industry are changeover time, information flow, machine condition, quality, production rate and workers [2]. Changeover time and process time greatly affect the cycle time of producing a product. To improve the speed, every industry should be able to minimize the changeover and processing time, so that customer satisfaction will be achieved [3]. Changeover time is a preparation time for processing new lots or new products [4]. Changeover time can be reduced by using the single minute exchange of dies (SMED) method [5].

SMED implementation of critical components replacement in Extrusion machine can reduce replacement time by 19 minutes and 25 minutes for caster machine [6]. SMED also successfully applied with metal forming objects in the automotive industry to improve work processes so that production time can be reduced [7]. SMED implementation allows for changeover time reduction, through reorganization of the industry's internal resources without requiring significant investment [8].

By changeover time reduction, an industry can save production costs and increase sales by 2% [9]. The SMED method can be integrated with other methods such as Theory of Inventive Problem Solving (TRIZ). The integration of these two methods can reduce the changeover time from 240 minutes to 32 minutes in the intel technology production process [10].

The SMED method can be used to reduce non value added activity. A total of 75 internal set up and 2 idle activities can be identified through SMED implementation in injection mold machine [11]. SMED is implement in the automotive industry with the result of a 30% reduction in production cost and 97 seconds in production time [12]. The SMED method is one of lean tools that used to reduce production waste in current value stream map [13]. Using SMED method, activity that does not add value to the product can be reduced [14]. The reduction of some non-value added and non-necessary non value added activities can improve production efficiency [15].

One industry with make to order system is printing industry. This industry is mainly producing books, but there are also other products, i.e. calendars, invitation cards and notes. In general, the production process is done

* Corresponding author: sriindrawati@uii.ac.id

according to the number of orders by customers. To fulfil the customer's order, it is required to complete orders based on customer specification. The problem faced is the delay in production completion that lead to delivery lateness. It is caused by a different changeover time of each workstation. For cutting ± 5 minutes and printing ± 18 minutes. Since a long changeover time appear in production, it will cut the available time for production. That will shorten production time, so production quantity will be less. In fact, to complete one order, sometimes it takes more than a day. With a long changeover time, it is necessary to make improvement so the production process can be completed on time by reducing changeover time. Changeover time consist of internal setup and external setup. Reducing changeover time can be done using SMED method.

## 2 Basic Theory

Single minutes exchange of dies (SMED) is one of improvement methods for lean manufacturing implementation. In lean manufacturing, all production wastes are being eliminated through a systematic approach [15]. Changeover time is one form of waste in the lean concept that must be eliminated because it is not provide an added value to customers and result in inefficient process. The SMED method can be used to reduce changeover time. This method can be implemented for all types of industries and all types of machines [16]. Single minute does not mean that setup time takes only one minute, but takes under 10 minutes or single digit minute [17].

The setup time is defined as a certain time to prepare production process and system required in production [18]. In setup time there are organizational time such as machine stoppage and maintenance, equipment setup, machine set up, changeover and startup, adjustment and production trials. Changeover time is the time to prepare production process with new type of product or lot size as seen in Figure 1[4].

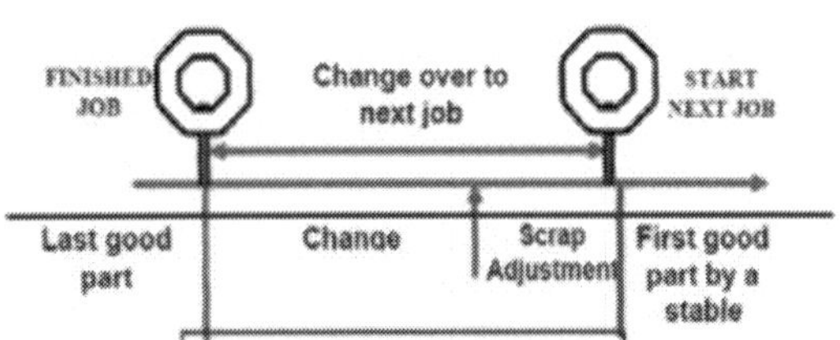


**Fig 1.** Changeover process [19] Changeover time should be reduced so the production process will be more efficient.

There are four major stages in SMED method, i.e. [20]:

a. The documentation process
The process of recording all setup activities is done in this stage. The setup activities consist of removal, installation of new equipment on machine and others. The documentation process can be done by recording all activities and time during the set up process using a stopwatch and recorded on a worksheet.
b. Identify internal and external activities
Determine all setup activities and identify the internal and external setup. Internal setup is an activity that should be done when machine stopped. External setup is an activity that can be done when the machine still producing parts.
c. Convert internal setup to external
The main goal of reducing setup time is to reduce internal setup time. This process is done by analyzing each internal setup activity and determine the possibilities to convert it into external setup activity.
d. Reduced internal activity
This process is done by simplifying movement, reducing movement and elimination of movement. 50% of setup reduction usually determined as SMED implementation target for many industry [21].

## 3 Research Framework

This research is conducted at printing and publisher AB located in Yogyakarta, Indonesia. This printing and publishing industry is producing books, brochures, calendars, tabloids and invitation cards. Research begins by collecting data of production time, setup time, production cost, product types and production process. The data then being verified with statistics, i.e. sufficiency and uniformity data test. Then it is subsequently used to calculate cycle time, normal time and standard time to analyze the production process time. The SMED analysis is then performed by identifying setup activities and separating the internal and external setup activities. Identification of several factors affecting the production setup activities is done to find out some potential internal setup activities that can be eliminated. Simplification of movement and lean improvement of changeover process is done based on the results of previous process evaluation. Further, a comparative analysis of initial and final production quantity is done to determine the effectiveness of SMED implementation in lean changeover process.

## 4 Discussion

There are four workstations used in the book production process, i.e. printing, cutting, drafting and bending. The standard time of the book production process on each workstation is measured using time study method. Based on statistical testing, the data already sufficient and uniform. The largest standard time is on printing workstation, 83,88 sec./ unit with production capability of 300 units/ day. For production process on cutting and drafting workstation have no significant standard time difference that is equal to 74,2 and 75,47 sec./ unit with production capability equal to 339 and 334 units/ day. While bending time is the smallest that is equal to 47,18 sec./ unit with a largest production capability of 534 units/ day.

Potential production bottleneck is printing workstation with largest standard time. So production activities in printing workstation needs to be analyzed by SMED method. Observations on changeover process in printing workstation is done to classify internal and external setup elements. The internal setup element is an activity that can only be done when the machine stops, while external elements can be performed while machine is working. The results of changeover process documentation and its classification indicates that there are 41 types of book production changeover activities consisting of 75,6% internal setup 838,2 seconds or 14 minutes 37 seconds and 24,4% external setup 259,36 seconds or 4 minutes 33 seconds.

Further, some activities are changed from internal setup to external setup. Setup is done when the machine is off, can be done when the machine is working. When the machine produces previous product, the operator can prepare for the next product, so changeover time becomes shorter. Because there are no specific reference for converting internal activity to external activity, then the conversion is done based on industry condition. To reduce delays, the operator should do some setup activity to produce the next product while the machine still proceed the previous product. There are also several activities performed in parallel, so two jobs can be done simultaneously such as when the operator runs the paper, the also take the paper plate. There are 45% of internal activity transformed into external activity. The remaining 17 internal activities still exist with 460,11 seconds or 8 minutes 7 seconds.

Based on potential cause analysis of high intenal setup time, there are several factors that influence, i.e. the number of equipment is less so that operators have to wait each other to use a certain equipment. In addition there is no certain place used for placing the production tools so that operators need to find a tool because the tool does not have a permanent place. Therefore, it is necessary to improve toolbox using the principles of 5S (Seiri, Seiton, Seiso, Seiketsu, Shitsuke), so the time wasted for activity searching and tidying tools can be eliminated. The lean changeover process based on SMED and 5S is developed as seen in Appendix 1. The original internal setup time is 838,2 seconds or 14 minutes 37 seconds. By improving changeover process, the new internal setup time is 455,55 seconds or 7 minutes 59 seconds.

The effectiveness of changeover process improvement that has been done using the SMED method is measured using production quantity parameter. The initial production capability in printing workstation is 300 units/ day whereas after improving the changeover process becomes 306 units/ day. There is an increase in book production and revenues of 2% per day.

## Conclussion

Based on the research that has been conducted, the production activity of printing workstation becomes the biggest potential bottleneck in printing industry. Proven with the largest standard time among other workstation activities. The changeover time for the printing workstation is 1097,56 seconds or 18 minutes 29 seconds, which consists of 838,2 seconds or 14 minutes 37 seconds for internal setup activities and an external setup of 4,32 minutes. After improving the changeover process using SMED method and 5S principle, the internal setup time is reduced 46% becomes 455,55 seconds or 7 minutes 59 seconds. It can increase production quantity by 2% per day.

Researchers thank to the support of Industrial Engineering Department, Faculty of Industrial Technology, Universitas Islam Indonesia.

## References

1. Shtub, A. & Cohen, Y., *Introduction to Industrial Engineering 2nd Edition*, CRC Press, USA, (2016)
2. Ifra Research, Optimising Productivity in Newspaper, Production Lines, *Ifra Special Report* 3.33: 01-23, (2001)
3. Esa, M.M., Rahman, N.A.A., Jamaludin, M., Reducing High Setup Time in Assembly Line: A Case Study of Automotive Manufacturing Company in Malaysia, *Procedia Social and Behavioral Science*, Vol. **211**: 215-220, (2015)
4. Bicheno, J. & Holweg, M., The Lean Toolbox: *The Essential Guide to Lean Transformation*, Production and Inventory Control, System and Industrial Books, UK, (2009)
5. Mcintosh, R.I., Culley, S.J., Mileham, A.R. Owen, G.W., *Improving Changeover Performance: A Strategy for Becoming a Lean, Responsive Manufacturer*, Butterworth Heinemann, UK, (2001)
6. Nurtrianto, R., *Analisis Dan Perancangan Sistem Informasi Penjadwalan Preventive Maintenance Pada PT Starmas Inti Aluminium Industry*, Binus University, Jakarta, (2011)
7. Refrizal & Sudarmadji, H., Aplikasi Metoda SMED Untuk Perbaikan Waktu Proses Ganti Model (Changeover Time) Dan Waktu Penyetelan (Setup Time), *Jurnal Me Trik Polban*, Vol. **.5**, No. 02, (2011)
8. Arvianto, A. & Arista, R., Usulan Perbaikan Operation Point Sheet Pada Mesin Feeder Aida 1100 Dengan Menggunakan Metode SMED, *J@Ti Undip*, Vol **VI**, No 2: 125-136, (2011)
9. António, C.M & Gil, C.S.P., Single Minute Exchange of Die A Case Study Implementation, *Journal of Technology Management Innovation*, Vol. 6, Issue 1: 130-146, (2011)
10. Kartik, S.K. & Muhamad, Z.M.S, Integration of SMED and TRIZ in Improving Productivity at Semiconductor Industry, *Jurnal Mekanikal*. no **33**: 40-55, (2011)
11. Novitasari, P.P., Arif, R., Ceria, F.M.T., Perbaikan Waktu Setup Dengan Pendekatan Single Minute Exchange of Dies (SMED) Untuk Mengurangi Unnecessary Motion Waste, *Jurnal Rekayasa dan Manajemen Sistem Industri*, Vol. **1**. No. 1: 1-10. (2013)
12. Rahul, R. J. & Naik, G.R., Application Of Smed Methodology- A Case Study In Small Scale Industry, *International Journal of Scientific and Research Publications*, Vol. **2**, Issue 8: 2250-3153, (2012)

13. Saputra, F. & Widyadana, I.G.A., Penurunan Waste Proses Produksi Produk RJK Di Departemen DEW PT. X Dengan Value Stream Mapping, *Jurnal Tirta*, Vol 1. No.2 :233-240, (2013)
14. Abdul, W.N., Surachman., Nasir, W.S., & Rudi, S., Implementasi Konsep Lean Manufacturing Untuk Meminimalkan Waktu Keterlambatan Penyelesaian Produk "A" Sebagai Value Pelanggan Studi Kasus PT TSW (Tuban Steel Work), *Jurnal Rekayasa Mesin*, Vol.**4**: 147-156, (2013)
15. Indrawati, S. & Ridwansyah, M., Manufacturing Performance Improvement Through Lean Six Sigma Method: An Iron Ores Industry Case Application, *Procedia Manufacturing*, Vol **4:** 528-534, (2015)
16. Shingo, S. A., *Revolution in Manufacturing: The SMED System*. Productivity Press, Cambridge, (1983)
17. Dave, Y. & Sohani, N., Single Minute Exchange of Dies, *Journal of Lean Thinking*, Vol. **3**, Issue 2: 27-37, (2012)
18. Rother, M. & Shook, J., *Learning to See: Value Stream Mapping to Add Value and Eliminate MUDA*, Lean Enterprise Institute Inc., USA, (2009)
19. Ulutas, B., An Application of SMED Methodology, *World Academy of Science, Engineering and Technology*, Vol **79**: 100-103, (2011)
20. Raikar, N.A., Reduction in Setup Time by SMED Methodology: A Case Study, *International Journal of Latest Trends in Engineering and Technology*, Vol.**5**, Issue 4: 56-60, (2015)
21. Kusar, J., Berlec, T., Zefran, F., Starbek, M., Reduction of Machine Setup Time, *Journal of Mechanical Engineering*, Vol. **56,** Issue 12: 833-845, (2010)

**Appendix 1.** Lean Changeover Process for Printing Industry

| No | Activity | Time (s) | Internal | External |
|---|---|---|---|---|
| 1 | Arranging the plate | 3,46 | | √ |
| 2 | Take a ruler and plate | 13,78 | | √ |
| 3 | Measure the plate | 59,52 | | √ |
| 4 | Preparing the paper | 116,09 | | √ |
| 5 | Loading paper | 61,28 | | √ |
| 6 | Set the paper position | 85,1 | √ | |
| 7 | Set the place of printed paper | 33,76 | √ | |
| 8 | Running paper | 52,02 | | √ |
| 9 | Take the paper | 13,78 | | √ |
| 10 | Take the plate | 32,96 | | √ |
| 11 | Take a screwdriver | 8,68 | √ | |
| 12 | Rotate the bolt | 73,15 | √ | |
| 13 | Returns the screwdriver | 2,66 | √ | |
| 14 | Placing the plate | 55,69 | √ | |
| 15 | Take a screwdriver | 6,54 | √ | |
| 16 | Rotate the bolt | 15,64 | √ | |
| 17 | Returns the screwdriver | 2,17 | √ | |
| 18 | Setting up ink | 25,59 | | √ |
| 19 | Cleaning plate and roller | 37,9 | √ | |
| 20 | Insert ink | 29,08 | √ | |
| 21 | Adjust the ink volume | 15,59 | √ | |
| 22 | Rotate the machine to flatten the ink | 46,26 | | √ |
| 23 | Try printing | 18,14 | | √ |
| 24 | Checking print results | 30,5 | | √ |
| 25 | Set the paper size | 31,23 | √ | |
| 26 | Giving lubricant | 35,53 | | √ |
| 27 | Set the ink | 5,6 | | √ |
| 28 | Try printing | 20,33 | | √ |
| 29 | Checking print results | 20,74 | | √ |
| 30 | Ensure the paper position | 43,67 | √ | |
| 31 | Try printing | 4,55 | | √ |
| 32 | Checking print results | 31,82 | | √ |
| 33 | Set the paper | 14,69 | √ | |
| 34 | Try printing | 9,58 | | √ |
| 35 | Checking print results | 25,42 | | √ |
| | Total | 1082,5 | 455,55 | 626,95 |

* Corresponding author: sriindrawati@uii.ac.id

# Work facility design for VCO operator using integrated quality function deployment and Analytic Hierarchy Process

*Sri* Indrawati*, *Abdullah* 'Azzam, and *Wini* Wiliyanti

Industrial Engineering, Universitas Islam Indonesia, Yogyakarta, Indonesia

**Abstract.** The Musculoskeletal disorder (MSDs) is skeletal muscle disorders that lead to damage in nerves, muscles, joints and cartilages. The damage is caused by muscle tension and degeneration. From the preliminary research in virgin coconut oil (VCO) processing industry, Central Java, Indonesia, there are potential risk of MSDs for workers. The majority of operators complaining on the right thigh (score 37) and left thigh (score 35). In addition, 50% of operators is work with moderate risk (rapid entire body assessment scores 4-7) and 50% others at a high risk level (rapid entire body assessment scores 8-10). Therefore, improvement is needed to prevent MSDs through work facility design using an integrated quality function deployment (QFD) method. In QFD, there is a concept screening phase to select some product design concept. To determine the priority, the QFD method can be integrated with analytic hierarchy process (AHP) method. The research shows that attributes of work facility needed by user are ergonomic, portable, durable and aesthetic. Concept design B have the highest priority score 0.32 and selected as the final work facility design. The technical specification are made from mahogany wood with an attractive design, red colour combination and developed based on user's anthropometry data.

## 1 Introduction

The coconut oil processing industry is one small and medium size enterprise in Indonesia that have a good business development with abundant raw materials availability. The area of coconut plantations in Indonesia is the largest in the world of 3,88 million ha [1]. The coconut plantation owned by civilians of 96,6% and the remaining is state sectors (0,7%) and private sectors (2,7%). The average coconut fruit production is 15,9 billion pieces per year with a total of 3,75 million tons of water, 0,75 million tons of shell charcoal, 1,8 million tons of coir fibre and 3,3 million tons of coconut husk dust [2].

Based on preliminary research in virgin coconut oil (VCO) processing industry, Central Java, Indonesia, the work posture is at risk of causing musculoskeletal disorders as seen in Figure 1. The potential of musculoskeletal disorders is caused by over bending posture and sitting with static position in repetitive movement within eight hours.

**Figure 1.** Working Postures of VCO Operators

Improving the work facility design in VCO processing industry is a way to reduce the risk of musculoskeletal disorders [3]. The work facility required by VCO processing workstation operators consist of temporary chairs and coconut storage box. There are several methods that can be used to design the work facility. Quality function deployment (QFD) is one of the many methods used by manufacturing companies for products based on consumer needs [4].

Many researchers have studied and documented the results of the QFD application. This method has been successfully used in developing food trolley products in India [5]. Based on that research, there are several consumer needs developed into the product, i.e. reusability, modular design, ergonomic and aesthetics. The method of QFD is also an integrated approach to modeling some product design criterias that are prioritized in work chair products for office environment [6]. QFD also sucessfully used for sailing boat material selection in India [7].

QFD can be used to design loser lift with a good quality [8]. The new product of mineral water that being improved using QFD have a significant potential product sucess [9]. This method is used to develop a concept design to fulfil the customer needs for multifunction scisor product [10]. QFD can also sucessfuly applied in a challenging environment of social housing and other service sectors through some model modifications [11]. The QFD is a powerful tool for deploying consumer needs to product design in steel manufacturing industry [12].

A new product strategy for direct open moulds product already develop to produce an innovative product through integration of QFD and theory of creative problem solving

* Corresponding author: sriindrawati@uii.ac.id

(TRIZ) [13]. In QFD, there is concept screening phase to select the design concept based on user needs and then developed into a prototype [14]. In that phase, an assessment of each design concept is done to create a product development priority. Analytic hierarchy process (AHP) is one method that can be integrated with QFD to prioritize the product design concept. This method has been widely implemented in various industries both manufacturing and services. Using AHP, the importance of each user's needs for the product can be clearly identified [15].

# 2 Basic Theory

## 2.1 Quality Function Deployment (QFD)

QFD is a product development method with a systematic process framework and focuses on consumer needs (voice of customer), improving the process to achieve maximum effectiveness [16]. There are three main phase for implementing QFD [17]:

a. Voice of customer identification that consists of determining attributes that consumers are interested in (in the form of qualitative data) and measuring the importance of each attributes.
b. Technical specifications determination that consist of several stages, i.e. identifying and measuring the importance of each technical specification to satisfy every customers needs attributes that obtained by the following equation (Annuraj *et al*, 2013):

$$B_j = \sum_{i=1}^{n} K_{ij} H_i \quad (1)$$

where, J is row vector for technical specification weight (j = 1 , ... , m), Kij is weight indicated by the relationship matrix (i = 1 , ... , n, j = 1 , ... , m), Hi is column vector of consumer needs weight ( i = 1 , ... , n), m is number of technical specification and n is number of consumer spesification. Then the other stages are development direction and target value, develop the concept design and concept screening.
c. Analysis and interpretation to analyze the overall product development process that already done.

## 2.2 Analytic Hierarchy Process (AHP)

AHP is one of decision making process tool for a simple or complex problem. A decision priority is chosen based on several criterias decomposed from a problem [18]. So AHP helps to solve complex problems by compiling a hierarchy of criteria, subjectively assessed by expert or decision maker and then draw a consideration for developing weights or priorities for problem solution. There are several phase in AHP [19]:

a. Identify the problem.
b. Develop a problem hierarchical structure that consist of criteria and alternative solutions.
c. Perform a pairwise comparison matrix with numbers that represent the relative importance of an attribute to other attributes for prioritizing attributes and alternative problem solutions. Priority decisions can be calculated from priority weight of criteria (PW criteria) and priority weight of alternative solution (PW alternative) using this formulation:

$$PK = \sum PW\,Criteria \times PW\,Alternative \quad (2)$$

d. Compute eigenvalues and test the consistency. If the concistency value is more than 100%, then the input data assessment should be corrected. However, if the consistency ratio is less than or equal to 0,1, then the data can be used to support decicion making process [20].

# 3 Research Framework

Research on work facility design for VCO processing operators consists of three main stages, i.e. identification of user needs, determination of technical specifications, house of quality development, concept generation and concept screening. The research was conducted in VCO processing industry in Central Java, Indonesia. Identification of user needs is conducted through a survey to all operators at VCO processing workstation. This survey is used to determine the user needs and its priority. Only valid and reliable data that used in the work facility design and development. The importance level of user requirement attributes is obtained using AHP method. Furthermore, the technical specifications are determined to fulfill the user needs. After developing a list of technical specifications, a correlation analysis between technical specifications and user needs is done to prioritize technical specifications due to resource constraints to fulfil user's needs. Then concept generation is developed to bring the idea of work facility development. In this phase, concept design of work facility is determined in detail. In concept screening, all concept design alternatives performance is evaluated based on user needs using AHP method. The best concept design alternative then being developed into a VCO work facility prototype.

# 4 Discussion

There are four types of user needs as VCO work facility atttributes that must be developed. Ergonomic becomes highest priority requirement of 0,33. Security and convenience in use are some needs required by users. That needs serves to reduce the potential risk of musculoskeletal disorders. The type of wood used must have a strong characteristic. While the dimensions of VCO work facility should base on the user's body anthropometry. The second attribute is portable with a priority score of 0,27 because the operators need a work facility that is easy to move. To realize that needs, the work facility should have a light weight so it can be moved by one person. Durable and aesthetic are the last factors that have the same priority score of 0,2. User needs a durable work facility that easy to maintain and not easily damaged during use. Types of wood and a robust design can be used to fulfill that user needs. In addition, the physical appearance of VCO work facility required by the user are interesting and simple design. That characteristics can be influenced by several factors, i.e. design of chairs and temporary coconut shelter, type of wood and finishing process.

The technical specification that developed based on user needs contain some of variables as shown in Table 1. There are four user needs for VCO work facility design consisting of chair and coconut shelter box. Seven technical specification of VCO work facility design are determined to fulfil that needs.

**Table 1.** List of Technical Specification Metrics

| No. Metrics | No. Needs | Technical Specification | Technical Weight | Unit | Direction of Goodness | Value |
|---|---|---|---|---|---|---|
| 1 | 3,4 | Shape of a chair | 1,2 | Subj | ○ | Standard |
| 2 | 1 | Size of product bearing | 1,65 | Cm | ○ | 42x32 |
| 3 | 3,4 | Wood type | 1,2 | List | ↑ | Standard |
| 4 | 1,2 | Chair dimensions | 2,46 | Cm | ↑ | 57x32x80 |
| 5 | 1 | Box dimensions | 1,65 | Cm | ↑ | 60x32x100 |
| 6 | 2 | Total mass of products | 1,35 | Kg | ↓ | <10 |
| 7 | 4 | Color | 1 | Subj | ↑ | Standard |

Furthermore, the priority of technical specifications is determined as a basis of work facility development process. The technical specification with the highest priority score is the chair dimension at 2,46. While the smallest priority score is color.

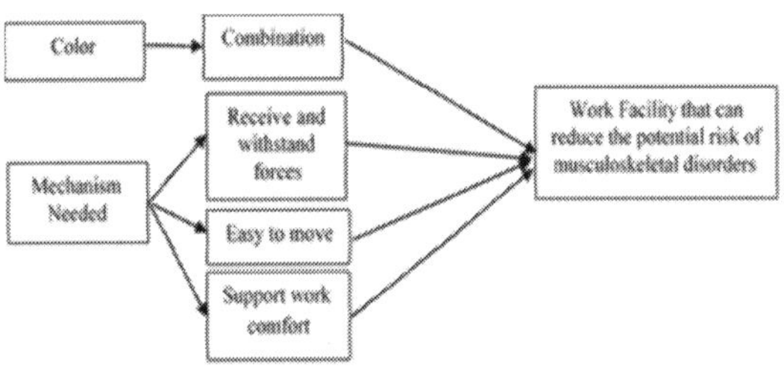


**Figure 2.** Problem decomposition diagram of VCO work facility design

The priority score indicates that the chair dimension has an important role in VCO work facility design to meet the user needs. Chair dimension have a strong positive correlation to the two technical specification of VCO work facility design, i.e. the temporary coconut storage box dimension and the total mass of product. The temporary coconut storage box dimension should be adjusted to chair dimension so that users feel comfortable during work. From the focus list of VCO work facility development, the product design problem can be decomposed as shown in Figure 2.

After identifying the problems of VCO work facility design, some feasible concept design alternatives are generated in a concept generation as shown in Table 2. There are four design concept alternatives that being generated based on problem decomposition diagram, i.e. design concept A, B, C and D. Each design concept alternatives are developed based on specification combination of wood type, color, box position and seat bearing.

**Table 2.** Concept Design of VCO Work Facility

| No. | Type of Concept | Specification |
|---|---|---|
| 1 | A | Teak dutch wood, bright colors, box position in front, with seat bearings |
| 2 | B | Mahogany wood, bright colors, box position in front, with seat bearings |
| 3 | C | Teak dutch wood, dark color, box position in side, no seat bearings |
| 4 | D | Mahogany wood, dark color, box position in side, no seat bearings |

Further a design concept screening is done to select the best concept design using AHP method. Based on the calculation of consistency ratio, all user answer to the work facility attributes that being asked have a value below 0,1. So the data can be used for the next phase. The concept design B have a highest work facility performance. It is shown from the highest priority score of 0,32 compared with other concept design alternatives. It is indicated that concept design B becomes the first priority to be produced. The specification of concept design B is using mahogany wood with virtual prototype as shown in Figure 3.

**Figure 3.** VCO work facility design

## Conclusion

Based on the result of research that have been done for VCO work facility design by integrating QFD and AHP method, it can be concluded that the work facility attributes that needed by user are ergonomic, portable, durable and aesthetic. The alternative concept design that choosed based on the highest priority of development is the concept design B. The priority score of concept design B is 0,32. The final VCO work facility design is made with mahogany wood. This product also have an attractive design with a combination of red color and have dimension that matches to the user's body anthropometry.

Researchers thank to the support of Industrial Engineering Department, Faculty of Industrial Technology and Board of Planning Universitas Islam Indonesia.

## References

1. Ministry of Industry, *Roadmap Industri Pengolahan Kelapa*, Directorate of Agricultural and Chemistry Department, Indonesia, (2009)
2. Board of Agricultural Research and Development, *Prospek dan Arah Pengembangan Agribisnis Kelapa*, Departemen Pertanian. Indonesia, (2009)

3. Rostykus, W., Winnie, I., Mallon, J., Musculoskeletal Disorders: Five Common Management Approach, *Professional Safety*, Vol. **December 2013**: 35-42, (2013)
4. Kaebernick, H., Farmer, L.E, Mozar, S., *Concurrent Product and Process Design*, The University of New South Wales, USA, (1996)
5. Anuraj, N.P., Srinivasa, V. Ravichandran, Design of a Food Trolley for Hospital India, *SASTECH Journal*, Vol. **12**: 90-98, (2013)
6. Mohanty, P.P. & Mahaptra, S.S., Design Of Office Chair: A Quality Function Deployment Approach, *Journal of Advanced Materials Manufacturing and Characterization*, Vol **3**.: 520-523, (2013)
7. Prasad, K. & Chakraborty, S. A., Quality Function Deployment-Based Model for Materials Selection, *Journal of Materials & Designs*, Vol. **49**: 525-535, (2013)
8. Widodo, Y. & Ikatrinasari, Z.F., Implementasi Quality Function Deployment Untuk Meningkatkan Kualitas Produk Lift, *Jurnal Ilmiah Teknik Industri*, Vo.**2.,** No.3: 195-203, (2014)
9. Moldovan, L., QFD Employment for a New Product Design of a Mineral Water Company, *Procedia Technology*, Vol. **12**: 462-468, (2014)
10. Mustakim. Integrasi Quality Function Deployment dan KANO Guna Perencanaan dan Pengembangan Produk Gunting Multifungsi. *Widya Teknika*, Vol. **23**, No.1: 21-28, (2015)
11. Kassela, K., Papalexi, M., Bamford, D., Applying Quality Function Deployment to Social Housing, *The TQM Journal*, Vol. **29**. Issue.3 : 422-437, (2017)
12. Moody, A.K., Lal, R.R., Pandey, V., Analysis of Customer Oriented Product Development with Quality Function Deployment, *International Journal of Mechanical Engineering and Technology*, Vol. **8**, Issue 5: 270-279, (2017)
13. Caligiana, G., Liverani, A., Francia, D., Frizziero, L., Donnici, G., Integrating QFD and TRIZ for Innovative Design, *Journal of Advanced Mechanical Design, Systems and Manufacturing*, Vol. **11**., No. 2: 1-15, (2017)
14. Ikmar, S.B.S. & Firdaus, M.B.I., Implementation of Quality Function Deployment (QFD) in Design Anti-Theft Motorcycle Device, *Journal of Mechanical Manufacturing*, Vol. **1**: 132-146, (2014)
15. Battistoni, E., Colladon, A.F., Scarabotti, L., Schiraldi, M.M., Analytical Hierarchy Process for New Product Development, *International Journal of Engineering Business Management*, Vol.**5**, No.42: 1-8, (2013)
16. Walden, J., Performance Excellence: A QFD Approach, *International Journal of Quality & Reliability Management*, Vol. **20**, Issue 1: 123-133, (2003)
17. Cohen, L., *Quality Function Deployment: How to Make QFD Work for You*, Addison-Wesley, Michigan, (1995)
18. Saaty, T.L., *The Analytic Hierarchy Process*, University of Pittsburgh, Pittsburgh, (1988)
19. Mulyono, S., *Teori Pengambilan Keputusan*, Lembaga Penerbit Fakultas Ekonomi Universitas Indonesia, Jakarta, (1996)
20. Davoodi, A., On Inconsistency of a Pairwise Comparison Matrix, *International Journal of Industrial Mathematics*, Vol. **4:** 343-350, (2009)

# Development of ergonomic website for engineering education

*Arnes* Faradilla[1], and *Muhammad Ridwan Andi* Purnomo[2]

[1]Industrial Engineering of Universitas Trisakti, Jakarta, Indonesia
[2]Industrial Engineering of Universitas Islam Indonesia, Yogyakarta, Indonesia

**Abstract**. Engineering education contains two aspects; theory and practice. Practice activity supports the students to implement the theory obtained in the class. Therefore, it will be better if the theory has material visualization in order to support and make easier the student and lecturer in teaching activity. Analisis Perancangan Kerja dan Ergonomi (APK & E) is one of the laboratories in Industrial Engineering, Universitas Islam Indonesia. APK & E laboratory which uses website as a tool to communicate to the students. Currently, the existing website still show standard display by conventional way. Based on the questionnaire, the student difficult to understand the material especially for Micromotion Study since this material requires the video to show the student in assembly activity. Augmented Reality (AR) combines the real and virtual objects that interactive in the real time and registered in 3D space. AR also combines the real and virtual in displaying information in the real time, in a way that enhances the individual abilities to operate the system in the real world. The system which combines between AR and Micromotion Study was come up. The required software to develop the online AR system is FLARToolKit 2.5.1. In this research, marker detection accuracy measurement of the used toolkit has been conducted and it showed 85% of accuracy level.

## 1 Introduction

Engineering education contains two aspects; theory and practice. Nowadays, theory is not enough anymore if there is no practice of material inside. Therefore, the theory is needed to visualize the material in order to support the student and lecturer in teaching activity. The APK & E is one of the laboratories in Industrial Engineering which concerns in analysis of work design and ergonomics. APK & E Laboratory which uses website as a tool to communicate to the students. Currently, the APK & E lab work contains 6 subjects; Anthropometrics, Micromotion study, Biomechanics, Physical Work Environment, Physiology and Work Sampling. The APK & E Laboratory provides lab work subject in the website and updated every week continuously. In fact, the website just show standard display and conventional. The provided modules are still standard and not attractive. Based on the preliminary questionnaire, the weaknesses of APK & E website are color combination as much as 21.6%, website's display as much as 18.9%, picture as much as 13.5%, font's type as much as 13.5%, animation as much as 13.5%, assembly visualization on the web as much as 13.5% and the other weaknesses as much as 5.4%.

There is one subject of lab work, which is Micromotion study, which requires a video to show an example of assembly activity conducted in the lab work to increase the student understanding. Currently, the students just download the module without understand the content leads to inefficient study time in the class. The assistant must explain all over again the subject because the students have not understood the material.

AR combines the real and virtual that interactive in the real time and registered in 3D space. AR combines the real and virtual in displaying information in the real time, in a way that enhances the individual abilities to operate the system in the real world. Based on the problem, the system to combine between AR and Micromotion Study module was come up. In this research, FLARToolKit will be utilized to develop the online AR system.

## 2 Literature Review

Reinhart & Patron (2003) have discussed the integrating Augmented Reality into assembly domain. AR is a new technology of human-machine interaction based on the computers that show the information of a real world. The information is shown on display. AR can increase the worker understanding in virtual object based on the real condition. The functions of AR in assembly domain are guiding the assembly worker and as the visual assembly instruction. Siltanen (2012) stated that marker-based systems have many advantages in such cases, as we justify later in this work. In consequence, the author's main contribution is in marker-based applications. Often, the ultimate goal is a mobile solution, even though the demonstration may run on a PC environment. Hence, the focus is on methods that require little processing capacity and little memory. Ong (2011) present the augmented assembly technology based on the 3D bare-hand interaction. Augmented reality was applied into augmented assembly. AR is required to display the information and interaction between the environment and the users. This paper presents the 3D bare hand interaction in an augmented assembly environment to manipulate and assembly the virtual components. Dangelmaier et al (2004) describe the Virtual and Augmented Reality Support for Discrete Manufacturing System Simulation. The purpose of this research is because the dynamics of the innovation is increase and the product life cycle becomes shorter. Therefore, they use the digital factory/virtual production method to reduce the time and cost. The result of this research is a general application of the VR and AR-technologies and to realize the simulation by developing the appropriate link and the integration mechanism.

Tateno et al (2005) explain about the nested for Augmented Reality. They require camera calibration in developing the marker. After design the marker, then develop the AR system to demonstrate the ability of the nested marker. It requires the ARToolKit framework as software. The purposes of this research are visualize the marker design and the experiment evaluation using artificially rendered marker images.

Alkhamisi and Monowar (2013) explain the prospect of AR is used to facilitate the learning whereas it enables students to access location, specific information provided through various sources.

Domingues et al (2010) describe the virtual laboratories in (bio) chemical engineering education. This paper describes the design and the implementation of two virtual labs for biochemical engineering education intended for students at the BSc degree. The effectiveness of the implemented system was evaluated through direct experimentation and survey (through questionnaires) with students taking the chemical technology lab course. The survey results show the value of 93 % students consider the virtual labs of great utility.

Khapre & Chandramohan (2011) present personalized web selection. They describe the display selection on the web appropriate with people's preferences. In this paper, they present an extensive study of different selection techniques towards advanced personalization of the web in service selection.

## 3 Research Methodology

In this research, there are several steps in developing the online AR system and measuring the successful test. The first step of this research is identifying the problem of APK & E Laboratory. It was obtained from questionnaire according to student understanding problem. The second step is creating the video objects to be visualized in the online AR system. The well known 3D graphic design software called *3D Studio Max 9* is utilized. The reason of using the 3D Studio Max software is because the software can also be used to create an animation of the assembly process of T-shape electrical plug based on the built 3D objects. The third step is creating the proper unique marker for the online AR system. The software to be used to design a proper marker for the online AR system is *Adobe Photoshop*. The next step is developing the online AR system using *FLARToolKit 2.5.1*. After that, testing the online AR system. First step in testing the online AR system is testing a marker which can display the related video in the virtual environment. The second step is rotaring the marker in Y axis and Z axis. The online AR system to be deployed directly to the internet.

## 4 AR System Development

Technically, the aim of developing an online AR system is to visualize the video objects in a web-based virtual environment using unique markers. Basically, when the user accesses the web page, the AR system will be downloaded automatically to the user computer as *activeX* object. The *activeX* object will access webcam installed in the user computer to get real time picture of the user. The user needs to show a printed marker to the webcam in order to see the virtual object in the virtual environment. The detail step of developing the AR system will be explained as follows:

1. Creating the Video Objects
   Based on the questionnaire, the selected object of assembly activity is T-shape electrical plug. The video objects to be visualized are the video on T-shape electrical plug assembly process, measurement activity, initial layout of assembly work station and suggested layout of assembly work station.
2. Creating the Marker
   The function of marker is to trigger related video appearance in virtual environment. The software to be used to design a marker is *Adobe Photoshop*. The aim of using *Adobe Photoshop* is because the software can be used to create the proper marker and also to determine the pixel size in each marker. The first step in creating the marker is designing a rectangle marker with 2 cm width black frame. The second step is creating an image with 4 cm in the middle of marker. In this research, there are 4 designed marker with picture in the middle represents the video object to be displayed in the virtual environment.
3. Development of Marker Detection Module
   The steps in development of marker detection module are finding potential marker, marker pattern identification and matching found marker with database of marker. The detail step in the development of marker detection module will be explained as follows:
   a. Finding the Potential Marker
      The first step in finding the potential marker is doing the grayscaling of the original image. There is a quite contrast object; a black bordered on white paper. Thus, most probably a good direction is to search for black image frame surrounded by white areas and do their analysis. The second step is to try doing thresholding for finding the black image frame. One of method to be used to threshold the black image frame is Otsu method.
   b. Marker Pattern Identification

The first step of marker pattern identification is the image marker divided into cells. The marker image is divided by dark gray lines of cells in the same width and height. The second step is counting the number of white pixels in each cell. If the number is greater than cell's area then assumed that the cell is filled by white color which corresponds to "1". And if the number is less than half of the cell's area then the black cell will correspond to "0".

c. Matching Found Marker with Data Base of Marker
The finding a matching marker's database needs to do compares of extracted marker data with every object in the database - compare the possible rotations of extracted object data with the database. The required FLARToolKit is saved in *".dat"* file.

4. Pseudocode
The language to be used to explain the process of developing program based on natural language is pseudocode. The pseudocode of this system is shown in Figure 2.

```
startprogram()

function startprogram()
{
import pattern file
set video to play
if (grabAndDisplayCamera() === pattern file)
        {
        papervision(video to play)
        }
        else
        {
        return false
        }
}

function grabAndDisplayCamera()
{
set width of cam picture
set height of cam picture
get camera source(camera index)
add camera signal to cam picture
add cam picture to scene
return bitmap per frame
}

function papervision(videoOBJ)
{
set width of videoOBJ
set height of videoOBJ
set position of videoOBJ
create new 3d scene
render videoOBJ
create papervision object
}
```

**Figure 2**. Pseudocode

Based on the pseudocode above, the first step is starting to initialize the program with importing pattern file and set the video to be played. After that, set the dimension of bitmap file to be captured from camera and get source the camera to grab the scene per frame and add to display. The output from this step will be an image file to be compared with pattern file. If it is similar (Boolean value return true), then it will go to create paper vision object. The first step was to set the dimension of the video that mentioned in previous, and then set the position video object. The last step was created the 3d scene and renders the object.

5. Testing the AR System
Basically, the video's position can be controlled using marker. The aim of controlling the video objects using marker is in order to show the video objects in various angles. The screenshot of video in normal position is shown in Figure 3.

**Figure 3.** Video in Normal Position

When the marker turned 45° with Y axis as center, thus video will rotate 45°. The screenshot of rotated video in AR system is shown in Figure 4 and Figure 5.

**Figure 4**. Rotated Video in Y axis as far as +45°

**Figure 5.** Rotated Video in Y axis as far as - 45°

When the marker turned 45° with Z axis as center, thus video will rotate 45°. The screenshot of rotated video in AR system is shown in Figure 6 and Figure 7.

**Figure 6.** Rotated Video in Z axis as far as + 45°

**Figure 7.** Rotated Video in Z axis as far as - 45°

## 5 Discussion

The next step is measuring the marker detection accuracy of the used toolkit. The marker detection accuracy is needed since marker has function to trigger the related video in the virtual environment. The detail step of measuring the marker detection accuracy of the used toolkit will be explained as follows:

1. Noising the Marker
   In this step, the required data is collecting the marker which is able to show the video in the online AR system. Noise is a disturbance of brightness in the image. Marker to be given the noise in various sizes. The aim of giving the noise in the marker is to measure the ability of marker contained noise in detecting the object. The formula to measure the noise in the marker is shown in Equation 1.

   $$A = W \times H \quad (1)$$

   Where:
   A = Area of marker/noise ( pixel$^2$ )
   W = Width of marker/noise ( pixel )
   H = Height of marker/noise ( pixel )

   There are 5 treatments in noising the marker. The 5 treatments as follows:
   a. In the first treatment, there is no noise in marker. The size of whole marker is 800 pixels for width and 1000 pixels for height. The result of calculation is 2,500,000 pixels$^2$
   b. In the second treatment, the noise's size is 100 pixels for width and 190 pixels for height. Based on the result, camera still read the marker and is able to visualize the video. The result of the second treatment is 19,000 pixels$^2$
   c. In the third treatment, the noise's size is 200 pixels for width and 340 pixels for height. Based on the result, camera still read the marker and is able to visualize the video. The result of the third treatment is 68,000 pixels$^2$
   d. In the fourth treatment, the noise's size is 340 pixels for width and 400 pixels for height. Based on the result, camera still read the marker and is able to visualize the video but in weak detection. The result of the fourth treatment is 136,000 pixels$^2$
   e. In the fifth treatment, the noise's size is 500 pixels for width and 420 pixels for height. Based on the result, camera still read the marker and is able to visualize the video but in the weak detection. The result of the fifth treatment is 210,000 pixels$^2$
   f. In the sixth treatment, the noise's size is 750 pixels for width and 500 pixels for height. Based on the result, camera cannot read the marker and visualize the video. The result of the sixth treatment is 385,000 pixels$^2$
2. Accuracy Level of Marker
   The accuracy level of marker is measuring the capability level of marker in detecting the video and also to measure the noise level in order to create a new marker. It can be determined by testing the noise of

each marker. The accuracy level of marker can be determined by calculating the percentage of each treatment. The formula to calculate the accuracy level of marker is shown in Equation 2.

$$ND = \frac{AN}{AM} x100\% \quad (2)$$

Where :
ND = Noise Detection (Percentage)
AN = Area of Noise (Pixel)
AM = Area of Marker (Pixel)
Based on the result above, the accuracy level of the used toolkit has 85 %. It means that webcam still able to visualize the video which has marker contained noise less than 15 %. Thus, the marker can be varied if marker value higher than 15 %.

## 6 Conclusion and Suggestion

The result of this research can be concluded as follows:

**a. Conclusion**

1. The developed AR system is able to visualize Micromotion Study module.
2. The accuracy level of marker is 85%. It means that system can read the marker with noise less than 15 %.

**b. Suggestion**

1. For the next research can use dynamic threshold to decrease the noise level because it can reduce the noise level.
2. Develop further research for another APK & E subjects using AR technology in order to make understand better of the students.

## References

1. G. Reinhart, C. Patron. Integrating augmented reality in the assembly domain – fundamentals, benefits and applications. *Proceeding of TU Munchen, Garching.* Germany. (2003).
2. S. Siltanen. Theory and applications of maker-based augmented reality. *VTT Science 3*. (2012).
3. W. Ong. Augmented assembly technologies based on 3D bare-hand interaction. *National University of Singapore* 11757. Singapore. (2011).
4. Dangelmaier et al. Virtual and augmented reality support for discrete manufacturing system simulation. *Computers in Industry* 56*: 371–383*. (2004).
5. K. Tateno. I. Kitahara, Y. Ohta. A nested marker of augmented reality. *Proceedings of the 2nd International Workshop on Augmented Reality*. Japan. (2005).
6. A.O. Alkhamisi, M.M. Monowar. Rise of Augmented Reality : Current and Future Application Areas. *International Journal of Internet and Distributed Systems*. 25-34. (2013).
7. Domingues et al. Virtual laboratories in (bio)chemical engineering education. *IBB–Institute for Biotechnology and Bioengineering, Centre of Biological Engineering, Universidade do Minho, Campus de Gualtar*. Portugal. (2010).
8. Khapre, Chandramohan. Personalized web service selection. *Department of Computer Science, Pondicherry University, Pondicherry*. India. (2011)

# Evaluation of SME (Small Medium Enterprise) production system with discrete system simulation method

*Reno* Dias Anggara Purba*, M Iqbal Sabit, and *Joko* Sulistio

Department of Industrial Engineering, Faculty of Industrial Technology, Indonesia

**Abstract.** Like manufacturing companies that have flowshop production lines, SME facing the amount of flow time and makespan. Unfulfilled production targets, buildup on some machines that result in other machines being idle, increasing waiting times across multiple machines, and poor performance of workers are a series of problems facing SME. This article uses a discrete system simulation method to analyze and evaluate SME production lines to improve performance. Simulation is an appropriate tool used when experiments are needed in order to find the best response from system components. From the results of modeling and simulation done found the root of the problem is due to accumulation that occurs in one machine and the lack of utility of the operator in producing bags. So do the experimental design with 3 scenarios on the system that have been modeled and obtained some solutions that can be offered to solve the problem. Furthermore, after the alternative selection is obtained the best scenario based on the alternative selection test is the scenario3 by adding 2 new machines and 1 new operator on the production line obtained a significant output increase compared to other scenarios of approximately 30%. While the best scenario based on the minimum cost is the scenario2 by adding 2 new operators obtained output that is not much different than the scenario3. However, it has not been able to solve a series of problems facing SME. Therefore, the results of this study to solve a series of problems faced by SME is to add facilities in the production line of 2 new machines and 1 operator (scenario3).

## 1 Introduction

Flowshop scheduling is a continuous movement of units through a series of work stations arranged by product. The arrangement of a flowshop production process can be applied appropriately to products with a stable design and produced in large quantities (product volumes), so special purpose investments that can be quickly returned. Flow Shop, a production system that arranges machines based on the order of processing (routing) products, so often referred to as the product layout (product layout) (Baker, 1974). In general, scheduling is a process in production planning and control that is used to plan production and allocation of resources at a given time with respect to the capacity of existing resources.

Scheduling in outline based on the sequence of production process is divided into 2 kinds (Masruroh, 2008). First, Job Shop is the flow pattern of N job through M machine with any flow pattern and job sorting process for irregular product trajectory. Scheduling on the production process of job shop one of its characteristics is the form of job shop layout is usually classified from equipment that has similar functionality in an area (Badworth, 1987). Second, Flow Shop is the process of determining the sequence of jobs that have the same product trajectory. Basically there are several kinds of flow shop patterns, among others: (a) Pure Flow Shop, Flow shop which has the same production line for all tasks. The machine is structured according to the existing process flow where a job is required to undergo a one-time process for each condition. (B) General Flow Shop has a different flow pattern, where the General Flow Shop flow pattern has characteristics: The flow of the process always shows the direction to the right, the sequence of operations may not be sequential (Farouq, 2013).

Problems or constraints that SME often face are production problems. SME are often unable to meet market demand because they can not meet production targets. This research uses a SME engaged in handicrafts in Yogyakkarta. Based on the interviews, it is known that SME have a minimum market demand of 55 units per day. However, the production output per day is less than 40 units. Then, the manufacturing process time for one product. Moreover, less optimal use of existing machines that result in utility of workers decreased.

This is because 2 of 19 machines owned damaged. While the number of workers owned 17 people. Thus each operator is assigned on 1 machine. In addition, the operator also duty to deliver flowitem from one process to another process. Such conditions lead to increased waiting time for the operator to deliver the flowitem from one of the machines. Thus, the impact is the occurrence of blocked or bottleneck on some machines. Therefore, the utility of the operator becomes reduced in processing the flowitem. This study aims to improve

* Corresponding author: 15522226@students.uii.ac.id

work performance and optimization of production factors from the company by analyzing the behavior of production systems using simulation methods and to evaluate in order to achieve the purpose of this research.

Simulation is a technique to mimic the operations or processes that occur in a system with the help of computer devices and is based on certain assumptions so that the system can be studied scientifically. In the computer simulation used to study the system numerically, then doing data collection to make statistical estimation to obtain the original characteristics of the system. Simulation is the right tool to use especially if required to conduct experiments in order to find the best comments from system components. This is because it is very expensive and takes a long time if the experiment is tried in real terms. By doing a simulation study then in a short time can be determined the right decision and with a cost that is not too large because everything is done with the computer. The simulation approach begins with the development of a real system model. The model should be able to show how the various components in the system interact so that it really describes the behavior of the system. Once the model is created then the model is transformed into a computer program allowing it to be simulated (Law & Kelton, 1991).

## 2 Method

This research uses qualitative description method. The data used are primary data obtained from observation and interview. The first stage in this study is to observe in a predetermined location. Then do the data collection needed for modeling. The data required is the data processing time of each machine and the flow of production process. Next, done data processing and modeling using software flexsim 6. After the model is made, then the simulation. Simulation output data is used to perform validation to ensure the credibility of the model, if the output data of the simulation results are not in accordance with the output data in the real system then do the data collection again. While if appropriate then can be done analysis on the model. Furthermore, an experimental design was created to solve the problem at hand. Next, make the best alternative selection by doing anova test followed by bonferroni test to see the significant difference of each scenario. The final stage of this research is to analyze the cost of each scenario so that some suitable solutions are obtained.

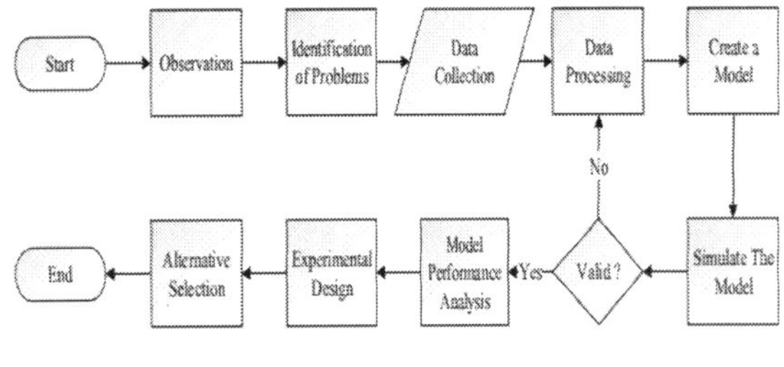


**Fig. 1.** Scheme of research methodology

### 2.1 Object of Research

The production process in SME Zanoproduction has 7 stages of the process: (1) pattern making and cutting, (2) tailoring the front, (3) rear and side of the bag, (4) fitting bag handles, (5) zipper installation, 6) inspection, and (7) packing. The sequence of production processes for all jobs is the same through the sequence from start to finish with the same groove and 1 machine has only 1 function (flowshop).

## 3 Result

Once the model is built, then enhanced its credibility by validating. The following historical data is obtained from obervation for 30 days, while for the simulation result data obtained from the simulation run of 30 replication. This validation process uses a chi-square test to test the results of the simulation model having a match with the observed real system.

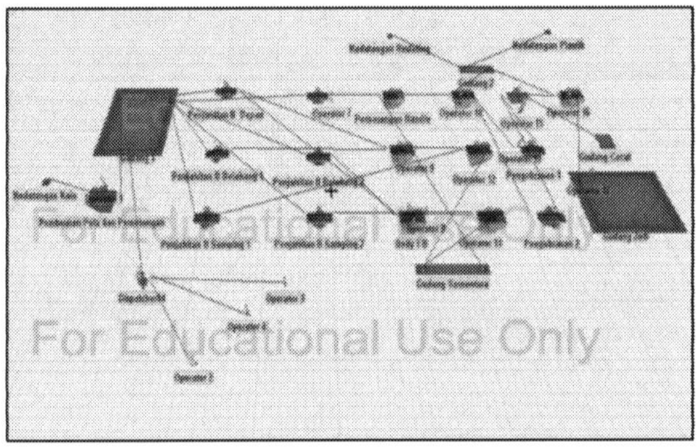


**Fig. 2.** Model Built in Flexsim

**Table 1.** Historical Output Data and Simulation

| Historical | | | | Simulation | | | |
|---|---|---|---|---|---|---|---|
| **Data to-** | | Data to- | | Data to- | | Data to- | |
| **1** | 30 | **16** | 37 | **1** | 36 | **16** | 31 |
| **2** | 38 | **17** | 37 | **2** | 39 | **17** | 32 |
| **3** | 37 | **18** | 35 | **3** | 37 | **18** | 40 |
| **4** | 38 | **19** | 30 | **4** | 37 | **19** | 37 |
| **5** | 37 | **20** | 38 | **5** | 31 | **20** | 30 |
| **6** | 35 | **21** | 35 | **6** | 30 | **21** | 37 |
| **7** | 36 | **22** | 36 | **7** | 39 | **22** | 37 |
| **8** | 37 | **23** | 35 | **8** | 33 | **23** | 41 |
| **9** | 39 | **24** | 39 | **9** | 37 | **24** | 30 |
| **10** | 38 | **25** | 38 | **10** | 39 | **25** | 35 |
| **11** | 37 | **26** | 30 | **11** | 36 | **26** | 34 |
| **12** | 38 | **27** | 39 | **12** | 30 | **27** | 37 |
| **13** | 37 | **28** | 37 | **13** | 34 | **28** | 38 |
| **14** | 39 | **29** | 30 | **14** | 35 | **29** | 31 |
| **15** | 36 | **30** | 38 | **15** | 32 | **30** | 39 |

H0: $\mu 1 = \mu 2$, The average output of the real system is the same as the average output of the Simulation model.

H1: $\mu 1 \neq \mu 2$, The average output of a real system is not the same as the average output of the Simulation model.

The model has been built, then validated the output by applying the two-equality test, the two-variance equation test, and the chi-square. After the calculation of statistical tests then found that the value of Z arithmetic obtained for 0.455830058. According to the test criteria Since -Z 0.025 < Z arithmetic < Z 0.025, ie: - 1.96 < Z count < 1.96. Then H0 is accepted which means the average output of the real system corresponds to the average output of the simulation model. After the calculation of statistical tests looking for F arithmetic, then found that the value of F arithmetic of 1.7358. According to the test criteria because 0.975 < F arithmetic < F 0.025 tab ie: 0.475965 < 1.7358 < 2.100996. Then H0 is accepted which means the average output of the real system is equal to the average output of the simulation model.

From the test results obtained chi square calculated value is 37.74, whereas with $\alpha$: 0.05 obtained chi-square table is 42.56. Thus, the observed frequency of the simulation model is consistent with the theoretical frequencies of the actual system. After testing, the model is said to be valid and has met the test. So the model can be said to have sufficient credibility, as well as built models can represent the real system well. So the results of this model can be used to improve the performance and problem solving faced.

## 4 Discussion

**Table 2.** Simulation Report

| Model Stop Time : 28800 (Second) | | | |
|---|---|---|---|
| **Object** | **Class** | **Idle** | **Blocked** |
| Hemming 1 | Processor | 78.03% | 0.94% |
| Hemming 2 | Processor | 74.21% | 0.59% |
| Sewing B Side 1 | Processor | 7.25% | 67.36% |
| Sewing B Backward 1 | Processor | 0.33% | 76.00% |
| Sewing B Backward 2 | Processor | 0.25% | 80.05% |
| **Object** | **Class** | **Idle** | **Utilize** |
| Operator 15 | Operator | 92.20% | 3.44% |
| Operator 16 | Operator | 92.81% | 4.37% |

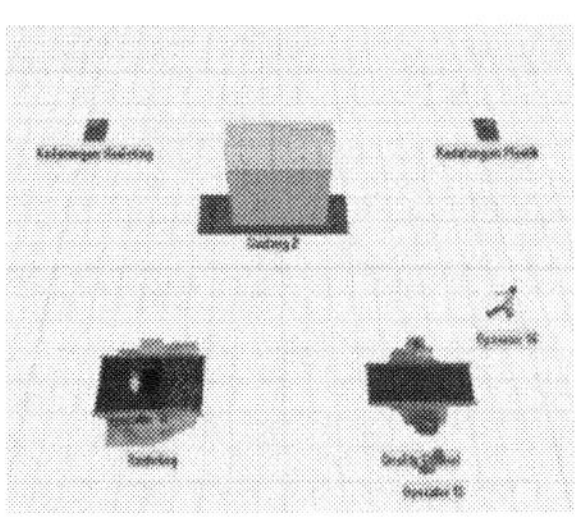

**Fig. 3.** Problem state

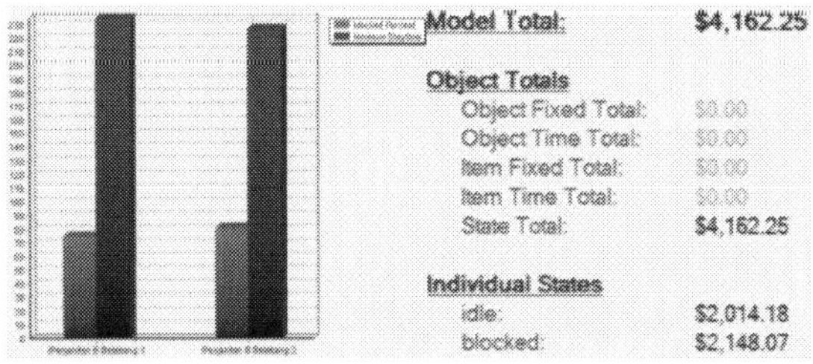


**Fig. 4.** Blocked, staytime object comparison, financial report

From Figure 3 shows the visual observations after the simulation ended with a simulation time of 28800 seconds. Simulation results show that there are still semi-finished goods or flowitem on some machines and in semi-finished warehouse. In addition, there are some operators who are not in the workplace, one of which Operator16. It appears blocked due to the accumulation that occurs on some machines that cause the production process on the machine after it is hampered and there idle. This is due to the flowitem being retained in the initial process in the Tailoring machine. So a little flowitem that goes into the final process in the Hemming machine, Quality Control, and Packing. In addition, idle occurs in some operators who work in the final stages of production due to the delay flowitem to be produced and cause the addition of waiting time. As a result of buildup on some machines, causing blocked occurrence in many engines. Blocked machines are: Sewing B Side 1, Sewing B Backward 1, and Sewing B Backward 2. In addition, increased waiting time and idle causes in some machines and operators. Machines that experience an increase in idle are: Hemming1 and Hemming2 engines. While the operators who experience an increase in idle are: Operator 15 and Operator 16 which served on Quality Control and Packing machine. In addition, the greater the idle value of the operator causes the utilize value of the operator in processing the increased flowitem. So automatically makespan and flowtime in the bag production process is also enlarged thus affecting the quantity of the final output produced. Then from figure 4 can be seen the comparison between the staytime with the blocked that occurs on the machine Sewing Backward B 1 and the Hemming 1. Where there is increased waiting time and blocked in both machines. Furthermore, from figure 5 the amount of blocked and idle values on the Sewing Backward B 1 and Hemming machine is converted to a cost of approximately $ 4000/day. The fee is only for 2 machines as a sample. Because the cost of conversion results is large, SME experience losses. So it is necessary to make improvements on the SME production line.

### 4.1 Design of Experiment

In this research, the experiments are 3 scenarios. Scenario 1 by adding 2 new machines to complete blocked on 3 machine that is: (1) Sewing B Side 1 (2) Sewing B Backward 1 (3) Sewing Backward B 2. Then to lower idle on 2 machine that is: (1) Hemming 1 (2) Hemming 2 and lower idle of 2 operators are: (1)

Operator 15 (2) Operator 16. Scenario 2 by adding 2 new operators to improve utilize of 2 operators: (1) Operator 15 (2) Operator 16 Scenario 3 by adding 2 machines and 1 new operator to solve a series of problems facing SME.

**Gudang Jadi**

| | Mean (90% Confidence) | Sample Standard Deviation | Min | Max |
|---|---|---|---|---|
| **Scenario 1** | 42.96 < 44.83 < 46.7 | 6.02 | 32 | 55 |
| **Scenario 2** | 43.17 < 45.07 < 46.96 | 6.12 | 34 | 57 |
| **Scenario 3** | 42.52 < 44.57 < 46.61 | 6.6 | 33 | 59 |

**Fig. 5.** The improvement of output's number

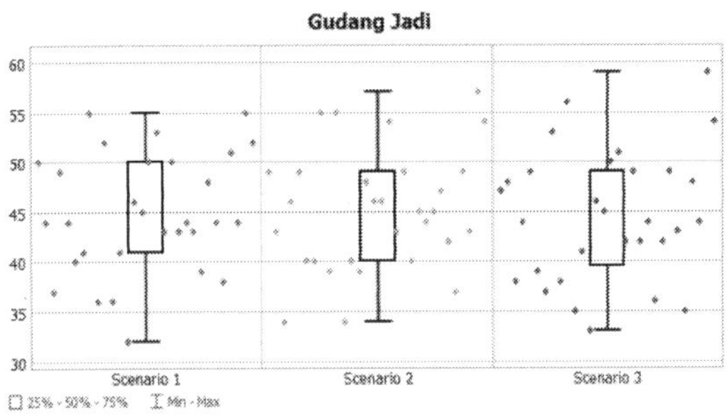


**Fig. 6.** The improvement of output's number

### 4.2 Alternative Selection

In the test results anova value F count > F table, ie 3.382155 > 3.101296 then H0 rejected and Ha accepted, which means there is difference in output output scenario1, scenario2 and scenario3. This distinction is a requirement in the Bonfferoni test that will be a further step in alternative selection. Furthermore, the selection of the best alternative is selected by testing the bonffernoni with the following hypothesis:

Ho : There is no significant mean difference in the output that is influenced between the skenario

Ha : There is a significant mean difference in the output that is influenced between the skenario

Testing Criteria:

Ho rejected, if P(T<=t) two-tail < $\alpha$/n

Ho accepted, jika P(T<=t) two-tail > $\alpha$/n

(1) In the bonferroni test it is found that the value of P (T <= t) one-tail (0.464635525) > alfa/n (0.016666667) then H0 is accepted which means there is no significant mean difference between scenario output1 and scenario2.

(2) In the bonferroni test it is found that the value of P (T <= t) one-tail (0.042712284) > alpha/n (0.016666667) then H0 is accepted which means there is no significant mean difference between scenario2 and scenario3.

(3) In the bonferroni test it is found that the value of P (T <= t) one-tail (0.008349887) < alfa/n (0.016666667) then H0 is rejected and Ha is rejected which means there is a significant mean difference between scenario1 and scenario 3.

### 4.3 Alternative Cost Analisys

**Table 3.** Simulation Report

| No | Facility | Quantity | Cost |
|---|---|---|---|
| 1 | Sewing Machine | 1 | Rp. 2.000.000 |
| 2 | Operator | 1 | Rp. 1.000.000 |

From the result of alternative selection test above can be seen that there is no significant average difference between the output of scenario1 with the output of scenario2. This means both alternatives are just as good or nothing better. The same can be seen by comparing the mean and variance of the two alternatives that have almost the same value. Then because scenario2 is no better than scenario1 then scenario 2 can not be selected if it refers to result of alternative selection test. Then from the result of alternative selection test above can be seen that there is no significant average difference between the output of scenario2 with the output of scenario3. This means scenarios and scenarios3 are equally good or there is nothing better between them. However, when viewed from the mean and variance between the two alternatives having a mean difference of 3 and the variance difference of 15, this indicates that there is a considerable difference in the difference to choose the best alternative. Then from the alternative selection test results above can be seen there is a significant average difference between the output scenario1 with the output scenario3. This means the two alternatives are not as good or scenario3 is better than scenario1. The same can be seen by comparing the mean and variance of the two alternatives having the difference for the mean of 4 and the variance of 19.

Then because scenario3 is better than scenario1 then scenario3 can be chosen as the best alternative when referring to alternative selection test result. Scenario 3 is done by adding 1 operator, adding 1 machine Body 1 B and 1 machine Body 2 B. This will surely increase the cost incurred by company to hire 1 operator, and to buy 2 new machine, then scenario1 that is done addition of 1 machine Body 1 B and 1 machine Body 2 B, while scenario2 is done by adding 2 operators. This of course will increase the cost incurred by the company to pay 2. If it is assumed the cost to pay 1 operator of Rp. 1,000,000 and the cost to buy 1 new machine is Rp. 2,000,000 then the total cost incurred for scenario3 is Rp. 5,000,000, then the total cost incurred for scenario1 is Rp. 4,000,000, and the total cost incurred for scenario2 is Rp. 2,000,000. When compared in terms of cost, the scenario2 can be selected as the best alternative because it costs a minimum of about half of the scenario3 and has an average output that is not much different. However, it has not been able to solve a series of problems facing SME. Therefore, to solve a series of problems facing SME it is advisable to choose scenarios3 even with slightly larger costs compared to scenarios2.

Previous research conducted by (Ramírez-Granados, Hernández, & Lyons, 2014). In Their journal entitled "A Discrete-event Simulation Model for Supporting the First-tier Supplier Decision-Making in a UK's

Automotive Industry" has a problem that the company can not meet production targets. By using discrete system simulation modeling and simulation. From 5 scenarios that have been made then got the best scenario is by adding 3 new operators in different production lines obtained output that can meet the company's production target and greater than other scenario. Various similarities with this research that has the same problem that the company can not meet production targets. The method used the same, namely by using a discrete system simulation. Then both do the verification and validation of the model built. Then the difference is the software used to build different models. In the previous study did not use alternative selection test to determine the best scenario by considering the significant difference of output between scenarios. Furthermore, do not use the cost analysis of the scenario that has been created to analyze the scenario that has the least cost with maximum output. Therefore, this study is better than previous research.

## 5 Conclusion

The performance of SME production system is less than optimal, this can be seen from the many flowitem that still accumulate in the system when the simulation ended. This is due to the time factor of the production process and the capacity it processes. In addition, the reduced utility of operators in processing flowitem causes SME unable to meet production targets. Then the best scenario based on alternative selection test result is scenario3 that is adding 2 new machine and 1 new operator in production line obtained significant increase of output compared to other scenario about 30%. While the best scenario based on the minimum cost is the scenario2 that is adding 2 new operators obtained output that is not much different than the scenario3. However, it has not been able to solve a series of problems facing SME. Therefore, the result of this research to solve a series of problems faced by SME is by adding facilities in the production line ie 2 new machines and 1 new operator (scenario3).

## References

1. Badworth, D. *Integrated Production Control System.* New York: John Wiley & Sons, Inc. . (1987)
2. Baker, K. *Introduction to Sequencing and Scheduling.* Canada: John Wiley & Sons, Inc. (1974).
3. Farouq, E. *Simulasi Aturan Johnson Untuk Penjadwalan Produksi Flowshop di Perusahaan Furniture.* Malang: Jurusan Ilmu Komputer Program Studi Teknologi Informasi dan Ilmu Komputer Universitas Brawijaya. (2013).
4. Law, A. M., & Kelton, W. D. (1991). *Simulation Modeling & Analysis, second edition.* McGraw-Hill: International. (1991).
5. Masruroh, N. *Analisa Penjadwalan Produksi dengan Menggunakan Metode Campbell Dudeck Smith, Palmer dan Dannenbring di PT. Loka Refraktoris Surabaya.* Surabaya: Jurusan Teknik Industri Fakultas Teknik Industri Universitas Veteran UPN Surabaya. (2008).
6. Ramírez-Granados, M., Hernández, J., & Lyons, A. *A Discrete-event Simulation Model for Supporting the First-tier Supplier Decision-Making in a UK's Automotive Industry.* Journal of Applied Research and Technology. (2014).

# Evaluation on performance and SME's productivity with discrete system simulation approach

*M. Iqbal* Sabit*, *Reno Dias Anggara* Purba, and *Joko* Sulistio

Department of Industrial Engineering, Faculty of Industrial Technology, Indonesia

**Abstract.** Inventory becomes one of important aspects to identify whether a company has fulfilled the consumers' demands or not. Shortage of stock, finished goods particularly could emerge the failure of accomplishing the target of consumers' demand that could also lead to loss of sales. Therefore, it needs good inventory control. This paper is objected improve SME's productivity by using the approach of discrete system simulation of aluminum industry that rapidly developed in Yogyakarta. Simulation is one of the appropriate tools for experiments that purposed to identify best response from system components. Based from Observation from line production, there are semi-finished products were not finished. Other than that the firm has no clue to solve the problem. Design of experiment and scenario are developed in a model that was built to obtain solution in enhancing the performance and productivity. As the result of initial performance that is not optimal, rescheduling on transporter schedule and the replenishment of machines and operators on the lathing process could be performed as improvements to enhance the productivity. This experiment can be used as the solution in improving the product's output as 64.36%.

## 1 Introduction

Recently, the numbers of micro, small and medium enterprises' actors in Indonesia have increased massively among other countries since 2014. The numbers keep growing in 2015, 2016 up to 2017. There are 57,8 million and growing of MSME's actors in the next few years as stated by Ministry of Cooperatives and SMEs recorded in 2014 (Ministry of Cooperatives and SMEs, 2014).

From the perspective of economics sector, MSMEs that have the biggest business unit, consecutively are agriculture sector, livestock, forestry and fishery (48,85%), trading, hotel and restaurant (28,83); transportation and communication (6, 88%), processing industry (6,41%), services (4,52%; financial, renting and company services (2,37%); construction (1,57%); mining and excavation (0,53%); electricity, gas and clean water (0,03%). One of the MSMEs that could possibly developing rapidly is aluminum processing industry (Bank of Indonesia, 2015).

Special Region of Yogyakarta (DIY) has huge potential in aluminum. DIY is the biggest pan's producer. Around 15.000 pans are produced by 10 companies in DIY (Okezone Finance, 2016). One of the SMEs companies that engages in aluminum processing is SP Aluminum.

SP Aluminum is the company of casting, smelting and household appliances' manufacturing that made from aluminum, such as: frying pans, saucepan, cormorant and other pans. Later, the research will be only focused on the manufacturing of frying pan. Production system is determined as make to order repetitive, which means that the company will produce the products based on consumers' orders, but the company also will produce numbers of frying pans as stocks.

According to Nasution (2008), production planning has specific purpose to provide optimum decision based on available resources in company to fulfill products' demand. Hence, it considers as action to balance company's demands and supplies. Stock is considered as the important factor to decide whether the company has fulfilled the consumers' demands or not. Shortage of stock particularly finished goods could emerge the failure in accomplishing the target of consumers' demand that could also lead to loss of sales. On the contrary, the excessive stocks could cause additional cost. Therefore, it needs good inventory management as the control process in regulating the stocks. Besides, from observation, there are stacks of half-finished goods in every production department that indicates the poor performance of company's production system.

Based on existing above problem, therefore a simulation method is used to solve problems in production floor, through this simulation method, a model will be constructed to represent real system on the production system at SP Aluminum.

## 2 Method

This research applies qualitative descriptive method. The data are primary data that was obtained from observation and interview result. The data are used to form system behavior to simulate model from real system. The subject of the experiment is *discrete event simulation*. Discrete

* Corresponding author: 15522331@students.uii.ac.id

event simulation (DES) is a flexible modeling method characterized by the ability to represent complex behavior within, and interactions between individuals, populations, and their environments (Pidd, 2004). According to Banks (2000), simulation is the imitation of real system operation in the unit time, where simulation involves history from system to draw conclusion out of real system characteristics. Simulation is used before an existing system is altered or a new system built, to reduce the chances of failure to meet specifications, to eliminate unforeseen bottlenecks, to prevent under or over-utilization of resources, and to optimize system performance (Maria, 1997).

The first thing to be conducted is process time data collecting for every production process. These data are employed to find out the distribution of process time data to establish model, so it can represent the real system. Based on above data, then a model is constructed by using flexim 6.0 software as the simulation assisting tools towards constructed model. Finally, it can be resumed the final output as result of simulation that will be the validation reference for simulation model's result that has the fitness with observed real system. The manner in which a model is implemented determines the range and level of outputs that can be used in the validation and final analyses (Karnon, 2012), two-sided test of equality and two variances test of equality will be the methods to be applied. Besides, from simulation, the performance of SMEs could be obtained and employed as data to be further analyzed as a basis of determination on new policies in solving existing problems. The alternative of selection is applied for better policy out of other offered policies by implementing ANOVA test as condition to conduct the next test, which is bonferoni test.

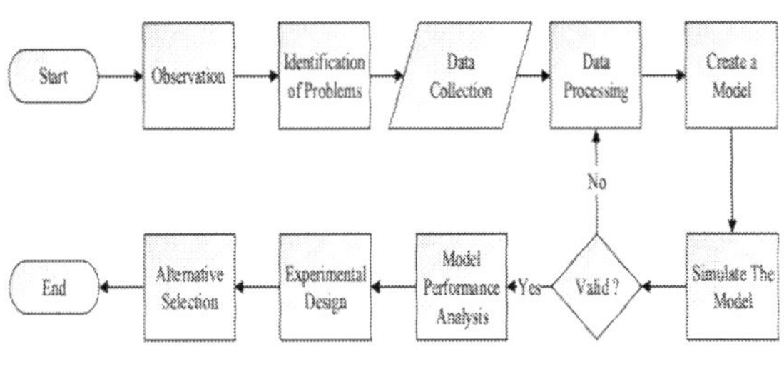


**Fig. 1.** Scheme of research methodology

## 2.1 Object of Research

Researcher only focuses on production process of thin frying pan manufacturing with dimension of 12, 14, 15, 16, 18, 20 and 24 with type of plain and polished frying pans. The production process in SP Aluminum covers 7 stages, which are material smelting (aluminum bar), stamping, quality control 1, finishing 1 (grinders and files), finishing 2 (lathing and polishing), quality control 2 and final stage of finishing 3 (numbering and labeling). The sequence of production process for all jobs are the same from starting stage to the final one.

# 3 Result

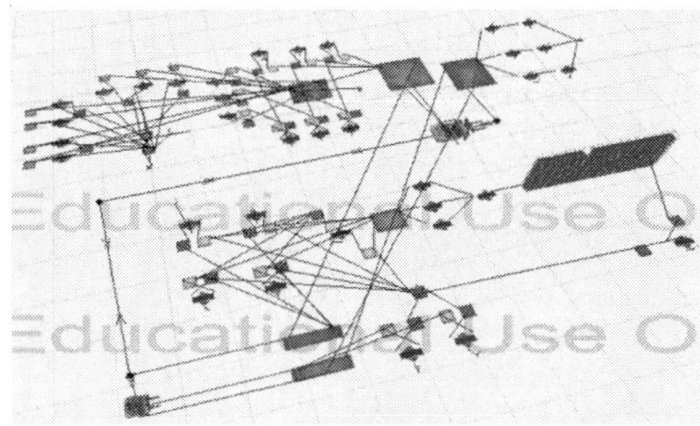

**Fig. 2.** Model Built in Flexsim

After the model is constructed, then the credibility will be enhanced by performing the validation. Following historical data were obtained from all seven types of frying pans in 30 days of production. Validation process will apply two-sided of equality test and two variance of equality test, to assess whether the resulted simulation model will have the fitness with observed real system.

**Table 1.** Historical Output Data and Simulation

| **Historical** | | | | **Simulation** | | | |
|---|---|---|---|---|---|---|---|
| **Data of-** | | **Data of-** | | **Data of-** | | **Data of-** | |
| **1** | 500 | **16** | 500 | **1** | 505 | **16** | 500 |
| **2** | 490 | **17** | 500 | **2** | 505 | **17** | 504 |
| **3** | 480 | **18** | 480 | **3** | 500 | **18** | 500 |
| **4** | 510 | **19** | 530 | **4** | 504 | **19** | 550 |
| **5** | 500 | **20** | 500 | **5** | 504 | **20** | 502 |
| **6** | 500 | **21** | 490 | **6** | 504 | **21** | 500 |
| **7** | 510 | **22** | 540 | **7** | 500 | **22** | 550 |
| **8** | 490 | **23** | 550 | **8** | 505 | **23** | 550 |
| **9** | 450 | **24** | 540 | **9** | 454 | **24** | 527 |
| **10** | 500 | **25** | 510 | **10** | 504 | **25** | 500 |
| **11** | 460 | **26** | 500 | **11** | 455 | **26** | 502 |
| **12** | 500 | **27** | 450 | **12** | 504 | **27** | 455 |
| **13** | 510 | **28** | 460 | **13** | 500 | **28** | 454 |
| **14** | 490 | **29** | 510 | **14** | 500 | **29** | 503 |
| **15** | 520 | **30** | 520 | **15** | 505 | **30** | 505 |

First validation is conducted by testing the similarity of two means data of simulation result with real system data. After the calculation of statistics test is conducted, it can be resumed that Z count is -0.325455019. According to the calculation criteria stated that since $-Z\ (0.025) < Z_{count} < Z\ (0.025)$, which is: $-\ 1.96 < Z_{count} < 1.96$. Hence, $H_0$ is accepted, which means that the output mean of real system is similar with output mean of simulation model.

Second validation is conducted by testing the similarity of two variance data of simulation result with real system. After F count is calculated, the result of F count is found as 1.038072, according to the criteria of testing, since $F_{0.975} < F_{count} < F_{0.025}$ which is: 0.475965 < 1.038072 < 2.100996. Hence, $H_0$ is considered as accepted, which means that the output mean of the real system is similar with the output mean of simulation model.

On the both similarity tests, model is considered as valid. Therefore, the result of model could be assumed as representative to represent real model. Later, it can be applied to improve the performance of SMEs.

## 4 Discussion

**Table 2.** Simulation Report

| Model Stop Time: 28800 | | |
|---|---|---|
| **Object** | **Class** | **State_Content** |
| **Queue Lathe 1** | Queue | 31 |
| **Queue Lathe 2** | Queue | 65 |
| **Queue Lathe 3** | Queue | 394 |
| **Queue Lathe 4** | Queue | 237 |

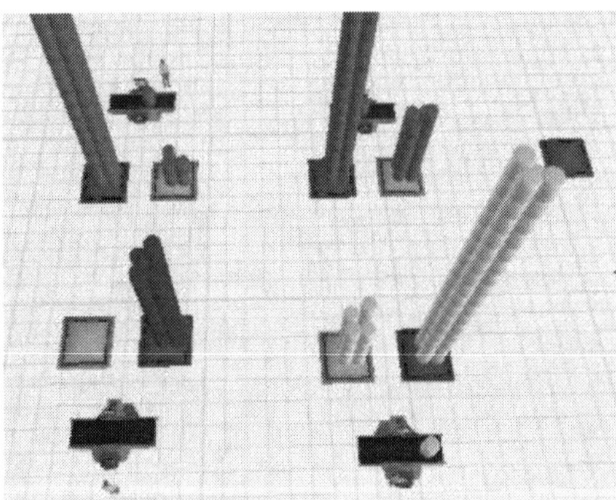

**Fig. 3.** Problem state

Based from Observation from line production, there are semi-finished products were not finished. Other than that the firm has no clue to solve the problem. Therefore using discrete system simulation could find cause of the problem. Based on the summary report, in the lathe process department, there was frying pans' shelter where they will be forwarded to the lathe machine for being processed. Nevertheless, based on the state content, there were numbers of flow items that left behind when the simulation ended. This was caused by limited time of process and the capacity to process lots of flow items, on the other hand, flow items that were carried by transporters from products' shelter of finishing 1 to the shelter of finishing 2 considered to have wide schedule, at 09.00, at 12.00, at 14.00 and at 15.30, those are the only times to accommodate the drop of the flow items. Therefore, process in finishing 2 will also experience postponement for loading and processing so it will also be effected the results of final production output.

### 4.1 Design of Experiment

Improvement is made on the first scenario to enhance final input of finished products, where rescheduling on transporters is applied with the additional of extra schedule, which is at 10.00. While on the second scenario, the additional machines for lathing process department are placed subjected to reduce state content with the combination of first scenario.

**PFM1**

| | Mean (90% Confidence) | Sample Standard Deviation | Min | Max |
|---|---|---|---|---|
| **Scenario 1** | 753.79 < 766.9 < 780.01 | 42.27 | 700 | 850 |
| **Scenario 2** | 803.93 < 814.4 < 824.87 | 33.76 | 762 | 868 |
| **Scenario 3** | 826.34 < 835.8 < 845.26 | 30.5 | 803 | 904 |

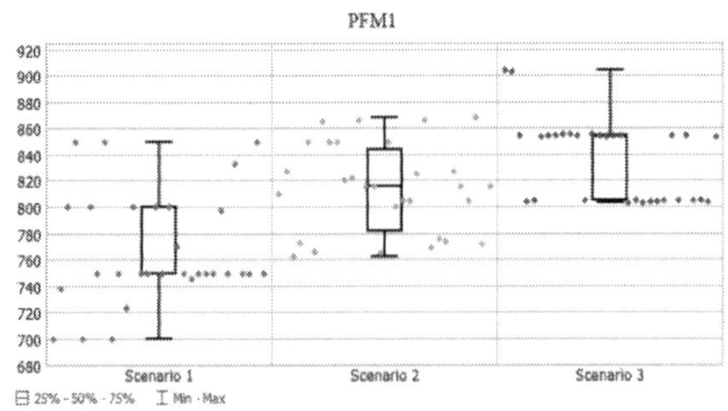


**Fig. 4.** The improvement of output's number

### 4.2 Alternative Selection

Based on the design of experiment's result, scenario 1 and scenario 2 are suggested to be able to improve final output of production process. In determining policies that will be decided, alternative selection based on comparison of scenarios' final output will be performed. Below is the replication table of simulation results from both scenarios.

**Table 3.** Design of Experiment's Result

| Initial Model | | | | Design of Experiment 1 | | | | Design of Experiment 2 | | | |
|---|---|---|---|---|---|---|---|---|---|---|---|
| **Data of-** | | **Data of-** | | **Data of-** | | **Data of-** | | **Data of-** | | **Data of-** | |
| **1** | 505 | **16** | 500 | **1** | 753 | **16** | 754 | **1** | 900 | **16** | 900 |
| **2** | 505 | **17** | 504 | **2** | 754 | **17** | 800 | **2** | 900 | **17** | 900 |
| **3** | 500 | **18** | 500 | **3** | 805 | **18** | 804 | **3** | 900 | **18** | 900 |
| **4** | 504 | **19** | 550 | **4** | 805 | **19** | 754 | **4** | 900 | **19** | 900 |
| **5** | 504 | **20** | 502 | **5** | 754 | **20** | 804 | **5** | 900 | **20** | 900 |
| **6** | 504 | **21** | 500 | **6** | 755 | **21** | 755 | **6** | 900 | **21** | 900 |
| **7** | 500 | **22** | 550 | **7** | 755 | **22** | 753 | **7** | 900 | **22** | 900 |
| **8** | 505 | **23** | 550 | **8** | 803 | **23** | 804 | **8** | 904 | **23** | 900 |
| **9** | 454 | **24** | 527 | **9** | 753 | **24** | 755 | **9** | 900 | **24** | 900 |
| **10** | 504 | **25** | 500 | **10** | 755 | **25** | 755 | **10** | 900 | **25** | 900 |
| **11** | 455 | **26** | 502 | **11** | 754 | **26** | 755 | **11** | 900 | **26** | 900 |
| **12** | 504 | **27** | 455 | **12** | 753 | **27** | 754 | **12** | 900 | **27** | 855 |
| **13** | 500 | **28** | 454 | **13** | 753 | **28** | 755 | **13** | 900 | **28** | 900 |
| **14** | 500 | **29** | 503 | **14** | 755 | **29** | 754 | **14** | 900 | **29** | 900 |
| **15** | 505 | **30** | 505 | **15** | 804 | **30** | 755 | **15** | 900 | **30** | 904 |

 https://doi.org/10.1051/matecconf/201815401068

To choose the best alternative, anova test is firsly performed on output of each scenario. Based on data of table 3, anova results are listed, as follows:

- $H_0$: There is no difference on output mean among three designs of experiment.
- Ha: There is difference on output mean among three designs of experiment.

**Table 4.** Anova Test

SUMMARY

| *Groups* | *Count* | *Sum* | *Mean* | *Variance* |
|---|---|---|---|---|
| Model Awal | 30 | 15051 | 501.7 | 574.5621 |
| D.E 1 | 30 | 23022 | 767.4 | 494.731 |
| D.E 2 | 30 | 26963 | 898.7667 | 69.35747 |

ANOVA

| *Source of Variation* | *SS* | *Df* | *MS* | *F* | *P-value* | *F crit* |
|---|---|---|---|---|---|---|
| Between Groups | 2455156 | 2 | 1227578 | 3234.297 | 2.22056E-82 | 3.101296 |
| Within Groups | 33020.87 | 87 | 379.5502 | | | |
| Total | 2488177 | 89 | | | | |

From table 4 above, it suggested that if f count > f table, hence based on testing criteria, $H_0$ is rejected, then there is the difference on output mean among three designs of experiment. The difference becomes the condition for Bonfferoni test as advance step in selecting the alternatives. Then, the best alternative will be chosen by performing bonffernoni test under following hypothesis:

Ho : if there is output mean that effected among design of experiment

Ha : if there is no difference on output mean that affected among experiments

Testing criteria:

Ho is accepted, if P(T<=t) two-tail < $\alpha/n$

Ho is rejected, if P(T<=t) two-tail > $\alpha/n$

1 Initial model and Design of Experiment 1
   Based on calculation, it is obtained that P(T<=t) one-tail (1.7953E-46) < alfa/n (0.016666667), then $H_0$ is accepted, which means that there is significant difference on mean between initial model output with the design of experiment 1
2 Initial model and Design of Experiment 2
   Based on calculation, it is obtained that P(T<=t) one-tail (1.72063E-62) < alfa/n (0.016666667), then $H_0$ is accepted, which means that there is significant difference on mean between initial model output with the design of experiment 2
3 Design of Experiment 1 & Design of Experiment 2
   Based on calculation, it is obtained that P(T<=t) one-tail (3.01E-37) < alfa/n (0.016666667), then $H_0$ is accepted, which means that there is significant difference on mean between output of design of experiment 1 with the design of experiment 2

Based on statistics calculation by comparing all three final outputs from three alternatives, it can be stated that there is difference on final output's mean that resumed by three alternatives. It also can be listed the mean on each alternative for initial model, design of experiment 1 and design of experiment 2 consecutively: 499,6, 767,4 and 898,76. From those means, it can be concluded that design of experiment 2 has the biggest mean if compared with others. Hence, it can be suggested that the company could employ design of experiment 2 as alternative to increase its final output significantly.

## 5 Conclusion

Initial performance of company is considered as not optimal, which was showed by numbers of flow items that was stacked in lathing department by the time the simulation was ended. It was caused by the process allocation time and capacity, other than that, transporter's scheduling is the main factor of state content on the processed flow items. Based on the result of that initial performance, rescheduling on transporter schedule and the replenishment of machines and operators on the lathing process are suggested as improvements to enhance the productivity. This experiment can be used as the solution in improving the product's output as 64,36%.

## References

1. Bank Indonesia. *Profil Bisnis Usaha Mikro, Kecil Dan Menengah (UMKM).* Retrieved June 18, 2017, from www.bi.go.id, (2015)
2. Banks, J., Carson, J., Nelson, B., & Nicol, D. *Discrete event system simulation 3rd edition.* Englewood Cliffs, (2000)
3. Karnon, J. *Modeling using Discrete Event Simulation: A Report of the ISPOR-SMDM*, Value In Health, (2012)
4. Kementrian Negara Koperasi dan Usaha Kecil dan Menengah, *Peran Usaha Mikro, Kecil dan Menengah dalam Pembangunan Ekonomi Nasional.* Retrieved June 18, 2017, from www.depkop.go.id, (2014)
5. Maria, A. *Introduction to Modeling and Simulation.* Proceedings of the Winter Simulation Conference, (1997)
6. Nasution, A., & Prasetyawan, Y. *Perencanaan dan Pengendalian Produksi Edisi Pertama.* Yogyakarta: Graha Ilmu, (2008)
7. Okezone Finance Retrieved June 10, 2017, from http://economy.okezone.com/read/2016/06/14/320/1414532/yogya-jadi-produsen-terbesar-wajan-di-indonesia, (2016)
8. Pidd, M. *Computer Simulation in Management Science* (5th ed). John Wiley & Sons, (2004)

# Determining the vehicles routes by considering its different types and capacity (heterogeneous) to minimize the total distribution cost

Nurhidayat and *Annie* Purwani*

Industrial Engineering University of Ahmad Dahlan, Jl. Dr. SoepomoJanturan Yogyakarta Indonesia

**Abstract.** Packaged sugar is one of the products in sugar cane manufacture, PT Madubaru Yogyakarta. Currently, the company deals with a high distribution cost because there is no plan to determine the vehicle route, vehicle type, and capacity for distributing the product. In this research, the optimum route of the distributing vehicles is developed. The company has three different types and capacity of vehicles: L300, HD, and PS. A problem in determining the distribution route here is called as Vehicle Routing Problem (VRP). The basic form of classic VRP says that all vehicles owned by a company have the same capacity (homogenous), meanwhile not all companies have the vehicles with same capacity. The heterogeneous variant is used to minimize the fixed cost of vehicles and distribution variant cost using Sequential Insertion Algorithm. This research has three purposes; are minimizing the number of vehicles used (NV), Total time of Completion tour (TCT) and Distribution Total Cost (TCD). The results based on a test calculation of a shipment date (August 19, 2016) of the company show three alternative solutions to distribute the packaged sugar to 12 consumers. The third alternative solution is the optimal solution and chosen as the decision result of the packaged sugar shipment. Based on the calculation results, it is needed 2 vehicles type HD with capacity 7,000 kg and type PS with capacity 3,500 kg, with total time of completion tour (TCT) is 828.49 minutes or 13.81 hours, and distribution total cost is IDR. 959,011.

## 1 Introduction

*PT. Madubaru* Yogyakarta is a sugar cane manufacture which has packaged sugar as the final product. *PT. Madubaru* does not have yet optimal configuration of distribution line in distributing the products to retails. Lack of delivery products planning along with the selection of vehicles routes, vehicles capacity, to the vehicles schedules affects improperly configured distribution channels which leads to the high transportation cost. The selection of unorganized vehicle routes leads to a longer vehicle route with longer distance. Therefore, the product distribution requires effective and efficient distribution strategy and planning.

Researches on the distribution of packaged sugar at PT. Madubaru had previously conducted by Ikhsan (2013) which discusses about optimizing the products distribution using *saving matrix* method, considers that all vehicles have same capacity (homogeneous), with optimizing indicator is the shortest distance. The next study is from Azizah (2015) who compared the saving matrix and general assignment method by considering the same parameters as Ikhsan used. The saving matrix and general assignment methods are part of the heuristic algorithm used to minimize the distance and cost by considering the constraints. In fact, PT. Madubaru has different types and capacity of vehicles so that the method is less effective to optimize the distribution cost of the company.

The problem of determining the distribution route is often known as the Vehicle Routing Problem which refers to a problem where there are several routes must be passed by a number of vehicles that depart from a depot to some destinations that have been determined and ends at the same depot (Hasanah, 2013). To decrease the transportation costs and also improve the services to the customer, it is necessary to find the best path or route, which can minimize the distance and time. Problems that aim to create an optimal route, for a group of vehicles, in order to serve a number of consumers, referred to as Vehicle Routing Problems (Hermansyah, 2011). The basic form of VRP is to assume that all vehicles owned have the same capacity (homogeneous) (Arvianto, 2014). In fact, not all companies have vehicles in the same capacity, either big or small companies. Therefore, the old VRP settlement is difficult to implement to solve this problem. The new VRP variant of Heterogenous Vehicle is used to solve the problems of companies that have vehicles with different capacities in the delivery of their products.

One of the methods that can be applied to solve VRP and its variation is the heuristic method. The heuristic

*Corresponding author: annie.purwani@ie.uad.ac.id

method is a technique for solving problems with more emphasis on simple computing performance. The heuristic algorithm used in this research is sequential insertion algorithm. Arvianto (2014) explains that the sequential insertion algorithm that has been developed in his model can build a viable solution by repeatedly trying to include customers who have not entered yet to any route should be inserted and the selection of location where the customer is inserted. This algorithm has advantages in the selection of customers, named by considering the customer's position on the available insertion, so the best results can be obtained. This heuristic algorithm is expected to overcome the problems of previous research in the settlement of new variants of the heterogeneous vehicle.

## 2 Research Methods

This study aims to facilitate the decision-making process in determining vehicle routes by considering the type and capacity of different vehicles. The stages of route determination, selection, and scheduling of vehicles consist of preliminary studies and literature, defining problems and objectives, collecting data, processing data, and analyzing the route determination, vehicle selection, and vehicle scheduling

## 3 System Characterization

This study adopts the model of the route and scheduling of vehicles from Arvianto(2014). Products that become the objects of research in PT. Madubaru are packaged sugar consisting of 1 kg and 0.5 kg. The calculation of the algorithm is done on one of the company's shipment date, August 19, 2016, with 12 customers. Delivery has a planning horizon according to working hours of 8.5 hours by ignoring the rest time. The vehicles used are as follows:

a. Vehicle type L300 capacity 1,500 kg of 1 unit,
b. Vehicle type PS capacity of 3500 kg as much as 2 units,
c. Vehicle type HD capacity of 7,000 kg as much as 1 unit

The speed of the vehicle used is 50 km / hour. The loading speed (LT) per minute is 5 sacks and unloading (UT) per minute is 3 sacks. 1 plastic bag containing 25 kg and loading time (WLT) for each different type of vehicle

## 4 Vehicle Routing Model

The model to be developed in the distribution of packaged sugar has several restrictions as follows:

### 4.1 Heterogeneity

The vehicles used in this research have different types and capacities.

### 4.2 Multiple Trips

One vehicle can serve one or more customers or routes within a single product delivery tour.

### 4.3 Split Delivery

Delivery of products to consumers can be sent directly in one delivery (entirely delivery) or several times the delivery process (partial delivery) based on the condition and the capacity of the remaining vehicles at the time of product delivery

### 4.4 Periodic

Delivery is done by considering the planning horizons that have been determined. If the condition of the cargo on the vehicle has been empty in sending the goods but planning horizon is still there, then the vehicle will return to the warehouse to take the goods and send back to the customer until the planning horizon runs out

### 4.5 Stochastic

The number of subscribers is random or unsystematic. Each customer has the possibility to not be visited every day.

## 5 Mathematical Formulations

According to the distributed characteristics above, a mathematical model will be used as an interpreter of the algorithm to solve the problem. This model adopts from the algorithm which is conducted in the study (Arvianto, 2014):

| | |
|---|---|
| $i$ | = Location index; I = 0 is the warehouse, i = 1,2,3 ..., N is the store/retail. |
| $t$ | = Tour index; $t = 1,2,3\ldots, NT$ |
| $r$ | = Route index; $r = 1,2,3\ldots,NR_{[t]}$ |
| $k$ | = Position index, k = 1,2,3..., $NL_{[t,r]}$ |
| $z$ | = Vechile type index; $z = 1,2\ldots, Z$ |
| $N$ | = Number of stores / retail sets $i$ |
| $q_{Ltz,r,k}$ | = The amount of sugar demand at locations with $r$ routes, $t$ tours and $z$ vehicles (kg units) |
| $LT$ | = Loading speed (units: number of products per unit time) |
| $UT$ | = Unloading speed (units: number of products per unit time) |
| $V_z$ | = Speed of vehicle $z$ (unit: distance per unit time) |
| $\tau_{[Ltz,z,r,k],[Lt,z,r,k+1]}$ | =Travel time of location $k$ to location k + 1, on route $r$, tour $t$ vehicle $z$ (time unit) |
| $Q_z$ | =Capacity on vehicle $z$ |
| $H$ | = Horison planning (time unit) |
| $NV$ | = Total number of vehicles (unit units) |
| $NT$z | = Number of tour vehicles $z$ |

| | |
|---|---|
| $NR_{[t,z]}$ | = Number of routes in tour $t$ by vehicle $z$ |
| $NL_{[t,z,r]}$ | = Number of locations on route $r$ in tour $t$ by vehicle $z$ |
| $L_{t,z,r,k}$ | = Location at position $k$, route $r$ in tour $t$ vehicle $z$ |
| $b_{Lt,z,r}$ | = The amount of charge sent on route $r$, the tour of vehicle $z$ |
| $Wp$ | = Travel time |
| $WLT$ | = Loading time (time unit) |
| $WUT$ | = Unloading time (time unit) |
| $CT_{[t,z]}$ | = Tour completion time $t$ by vehicle $z$ |
| $TCT$ | = Total completion tour time (unit time) |
| $TCD$ | = Total cost distribution |
| $C_{SK}$ | =The fixed cost of the vehicle |
| $C_{BB}$ | = Fuel costs per km |
| $C_G$ | =Driver's salary visits |

## 6 Function Goals

This research has three objective functions of minimizing as follows:

### 6.1 To minimize the number of vehicles (NV)

In the context of this research problem, by considering the existence of multiple trips, the purpose function is to minimize the number of vehicles (*NV*). It is because minimizing the number of vessels is the main target model of multiple trips aspects in VRP.

### 6.2 To minimize total completion time (TCT)

This function is required to see the best combination of tours and routes which is generated from the algorithm by considering the work balance represented by the *TCT* variable

### 6.3 To minimize cost distribution (TCD)

Minimizing total distribution costs (*TCD*) is a measure of performance as a result of the goal of minimizing *NV* and *TCT*.

$$\text{Min TCD} = \sum_{r=1}^{NR}\sum_{i=0}^{N}\sum_{j=1}^{N+1} C_{SKZ}N_{TZ} + C_G N_R + \sum_{r=1}^{NR}\sum_{Z=1}^{Z} C_{BB}\, V_Z W_P \quad (1)$$

#### *6.3.1 Warehouse and Store*

The travel time taken by each vehicle from the starting point to each location is expressed by $\tau_{[Lt,z,r,k],[Lt,z,r,k+1]}$. Each customer has a different request that is stated with the $q_{Ltz}$ notation. For the notation $L_{t,z,r,1}$ describes the location of the warehouse where $i = 0$ and will end up in the same location, meaning the journey starts from the warehouse and ends in the warehouse.
The location of the warehouse is described as follows:

$$L_{t,z,r,1} = L_{t,z,r,NL[t,r]} = 0;\quad t=1,2,\ldots NT;\quad z = 1,2,\ldots,Z;\quad r=1,2,\ldots, NR_{[t]} \quad (2)$$

The store/retail location is defined as:

$$L_{t,z,r,k} = i; t = 1,2, \ldots, NR_{[t]}; z = 1,2, \ldots , Z; k = 2, NR_{[t,r]} - 1; i = 1,2 \ldots N \quad (3)$$

#### *6.3.2 Route and Tour*

Route $r$ is a trip made by one vehicle from a single depot (warehouse) and ends at the same location. The route represents the order of visits of a vehicle to customers, starting from the warehouse and back to the warehouse. A route consists of one depot and at least one customer. The tour is the set of routes that must be passed by vehicle $z$.

#### *6.3.3 Number of Vehicle*

A number of vehicles denoted by *NV*, for VRP issues with multiple trip characteristics of one vehicle doing one tour. In accordance to the characteristic of the time window, one vehicle may only operate within work hours on a work day determined by the company. This statement is written with the following equation:

Number of vehicle tour type $z$ = number of vehicles

$$NT_Z = NV \quad (4)$$

The product capacity in vehicle $z$ is expressed by $Qz$ and vehicle speed is $v$. The number of routes in the tour is expressed by the tour $t$ ($t$ = 1, …,$NT$) is $NR_{[t]}$. This type of z-vehicle serves multiple routes in a single trip (multiple trips) with a heteroneous fleet so the vehicle begins its journey to route $r$ in the tour $t$ which is the final route location of the previous $r$-1 route in the same $t$ tour.

$$L_{t,z,r,1} = L_{t,z,r-1,NL[t,z,r-1]} = 0; t = 1,2, \ldots NT; z = 1,2, \ldots, Z; r = 1,2, \ldots, NR_{[tz]} \quad (5)$$

A tour consists of a set of routes with $NR_{[t,z]}$ shows the number of routes which is included in a single tour by vehicle $z$. A vehicle route describes the order of vehicle visits to a retail/store which goes and ends to the depot. The split delivery constraint ensures that each subscriber will get at least one delivery so that the demand on location or customer can be delivered with more than one delivery.

$$L_{t,z,r,k} \geq 1; t = 1,2, \ldots NT; z = 1,2, \ldots, r = 1,2, \ldots, NR_{[tz]}; k = 2, \ldots NL_{[t,z,r]} - 1 \quad (6)$$

The load on each route should be smaller or equal to the capacity of the vehicle. In the case of product delivery, the amount of cargo of each vehicle $z$ which serves a route $r$ in tour $t$ shall not exceed the capacity of the vehicle $z$.

$$b_{L,t,z,r,} \leq Q_z, \quad \forall t = 1,2, \ldots NT; z = 1,2, \ldots, Z; r = 1,2, \ldots NR_{tz}; \quad (7)$$

The amount of cargo carried by the vehicle in a tour *t* by vehicle *z (t,z)*, route *r* is less or equal to the number of customer requests on route *r* from tour *(t,z)*.

$$b_{L,t,z,r,} \leq \sum_{k=2}^{NL_{[t,z,r]-1}} q_{Lt,z,r,k}\ ; t = 1,2, \dots NT\ ; z = 1,2, \dots, Z; r = 1,2, \dots NR_{[tz]} \quad (8)$$

Constraint (9) ensures that each subscriber will receive a full demand shipment.

$$\sum_{r=1}^{NT} y_{Lt,z,r,k} = 1\ ; t = 1,2, \dots NT\ ; = 1,2, \dots, z\ ; r = 1,2, \dots NR_{[t,z]}; \quad (9)$$

### 6.3.3 Tour Completion Time

The completion time of a tour which is expressed by $CT_{[t,z]}$ includes the travel time, and the loading / unloading time available on the tour. The completion time of each tour should not exceed the planning horizon. This tour time is the sum of trip time, loading and unloading time.

$$CT_{[t]} = (WDT + WUT) \sum_{r=1}^{NR_{[t,z]}} \sum_{i=1}^{n} b_{L_{t,z,r}} + \sum_{r=1}^{NR_{[t,z]}} \sum_{k=1}^{NL_{[t,z,r]-1}} \tau_{[L[t,z,r,k]],[L[t,z,r,k+1]]} \quad (10)$$

### 6.3.4 Travel Time

Travel time $\tau_{[Ltz,z,r,k],[Lt,z,r,k+1]}$, is the total time needed by the vehicle z from the store / retail warehouse up to the warehouse location again. The speed of the ship and the distance between the first location to the second location determine the time of travel.

$$WP = \tau_{[L[t,z,r,k]],[L[t,z,r,k+1]]} \quad (11)$$

### 6.3.5 Loading Time

WLT loading time is the time to load the cargo onto the vehicle z in the warehouse. The loading time is generated from the amount of packaged sugar which are transported. It means the more cargo to load, the more time needed.

$$WLT = \frac{b_{[L[t,z,r]]}}{LT} \quad (12)$$

### 6.3.6 Unloading Time

The unloading time of WUT is the unloading time from the vehicle at the store / retail location. The unloading time of WUT is also obtained from the amount of packed sugar taken out from the vehicle. It means the more cargo to unload, the more time needed.

$$WUT = \frac{b_{[L[t,z,r]]}}{LT} \quad (13)$$

### 6.3.7 Horizon Planning

This planning horizon limits the total tour of completion time. The length of the planning horizon is denoted by H. completion time for the z-t ride tour should not exceed the planning horizon.

$$CT_{[t,z]} \leq H \quad (14)$$

### 6.3.8 Total Completion Time

Total Completion Time (TCT) is the sum of all the completion time of all the t tours needed as a solution.

$$TCT = \sum_{t=1}^{NT} CT_{[t,z]} \quad (15)$$

The algorithm used to solve the problem in this research is Sequential Insertion Algorithm. Arvianto (2014) explains that this algorithm builds a viable solution by repeatedly trying to include customers who have not entered any route into the temporary portion of the current route. The advantage of the Sequential Insertion Algorithm is trying to generate the smallest possible number of vehicles (tour) by utilizing the capacity of the vehicle as much as possible.

Suprayogi (2005) in his research explains that this algorithm begins by choosing a customer to occupy a position as a seed customer on the first route and tour. In the first customer selection, there are several criteria: earliest deadline, earliest ready time, shortest time window, longest travel time, and random. The criteria for selecting the initial customers (seed customers) included in the next route (second, third, and so on) on a tour that is using completion time of the tour.

## 7 Discussion and Conclusion

The algorithm described previously will be used in the completion of the company's distribution. The selection of stores / retails that deserve to be inserted to the arc insertion is the retails / stores that have the smallest completion time of tour (CT). Calculation will be done for all vehicle types. The optimal decision-making process is done by comparing 3 alternatives for each vehicle type and options of each alternative, which has options and vehicles type combination on the established tour and route.

Comparative analysis results related to the objective functions which are minimizing the number of vehicles used (NV), the total time of completion of the Tour (TCT), and the total cost of distribution (TCD) are presented in the Table 1.

The result of minimizing vehicle is on the third alternative solution with 2 types of vehicle used; those are type HD (7,000 kg) and type PS (3,500kg). Using these vehicles to serve all customers requests of packaged sugar, it needs 2 times of tour with 2 vehicles. This indicates that the more amount of demand, the small capacity vehicles will have many tours and routes which

https://doi.org/10.1051/matecconf/201815401069

lead to the longer total tour completion time and higher distribution cost as well. The shortest total tour completion time is the third alternative solution which results 828.49 minutes.

**Table 1** Comparison Results Based on Objective Functions

| Information | *NV* (units) | *TCT* (minute) | *TCD* (Rp) |
|---|---|---|---|
| Exsisting | 4 | 856,70 | Rp 1.810.826 |
| Solution 1 | 2 | 879,97 | Rp 1.703.960 |
| Solution 2 | 3 | 1053,01 | Rp 1.370.303 |
| Solution 3 | 2 | 828,49 | Rp 959.011 |

This indicates that TCT is influenced by the duration of tour completion time; the more routes formed in one tour, the longer completion time of the tour will be generated. The duration of the tour completion time is influenced by the vehicle used; the greater of the capacity use, the longer the time will take. The more vehicles to use, the shorter the time will be to serve all customer needs. The impact of the total number of vehicles used is seen on the total cost of packaged sugar distribution. Alternative solution 3 has the least total cost of distribution (TDT) which is IDR 959,011. This shows that if the tours run are fewer, equal to the number of vehicles used (NV), so the total cost of distribution (TDT) will be smaller.

## References

1. Arvianto, A., Setiawan, A.H., Saptadi, S. (2014). *Model Vehicle Routing Problem dengan Karakteristik Rute Majemuk, Multiple Time Windows, Multiple Products dan Heterogeneous Fleet untuk Depot Tunggal*, Jurnal Teknik Industri, Vol. 16, No. 2, Desember 2014, 85-96.
2. Azizah, U.N., Oesman, T.I. (2015). *Optimalisasi Biaya Distribusi Produk PT. Madubaru Dengan Pendekatan Metode Saving Matrix dan Generalized Assignment*, Jurnal REKAVASI, Vol. 3, No. 2, Desember 2015, 104-107.
3. Hasanah, Matondang, N., Ishak, A. (2013). *Penentuan Rute Distribusi Barang Yang Optimal Dengan Menggunakan Algoritma Heuristik Pada PT. XYZ*, e-Jurnal Teknik Industri FT USU Vol 3, No. 3, Oktober 2013 pp. 48-51.
4. Hermansyah, B. (2011). *Penyelesaian Vehicle Routing Problem (VRP) Menggunakan Algoritma Genetika*, Skripsi Fakultas Sains Dan Teknologi Universitas Islam Negeri Sultan Syarif Kasim, Riau Pekanbaru.
5. Ikhsan, A.N., Oesman, T.I., Yusuf, M. (2013). *Optimalisasi Distribusi Produk Menggunakan Daerah Penghubung Dan Metode Saving Matrix*, Jurnal REKAVASI Vol.1 No.1, 1-10, Desember, 2013.
6. Munawir, H., Narima, A. (2012). *Penentuan Rute Pendistribsian Kertas Karton Model Studi Kasus: PT. Papertech Indonesia Unit Ii Magelang*, Simposium Nasional RAPI XI FT UMS – 2012.
7. Suprayogi dan Imawati, D. (2005). *Algoritma Sequential Insertion dengan Forward dan Backward Pass untuk memecahkan Vehicle Routing Problem dengan Multiple Trips dan Time Windows*, Jurnal Teknik dan Manajemen Industri, Vol. 25 No. 1, April 2005

# Non-cooperative and cooperative solutions of government subsidy on public transportation

*Hennie* Husniah[1,*], *Udjiana S.* Pasaribu[2], *Yuza Z.* Beta[2] , and *Asep K.* Supriatna[3]

[1]Engineering Department, Langlangbuana University, Bandung, Indonesia
[2]Mathematics Department, Institut Teknologi Bandung, Bandung, Indonesia
[3]Mathematics Department, Universitas Padjadjaran, Bandung, Indonesia

**Abstract.** The paper deals with two models of government subsidy given to a public transport operator: (i) the subsidy for buying bus from an appointed public transport manufacturer, and (ii) the subsidy for reimbursing reduced ticket price for passengers. The models are developed to determine the maximum profit for both the public transport operator and the manufacturer. Since we consider two parties – the public transport operator and the manufacturer of the bus, then we use game theoretical approach by considering non-cooperative and cooperative solutions. Furthermore, since the bus is repairable we consider virtual age to model the preventive maintenance and we consider minimal repair to model the corrective maintenance. We analyse both type of subsidy models and give some numerical examples which show the effects of different subsidies to the profit of operator and manufacturer. The result of the numerical examples indicates that reducing ticket price would give a higher profit both to the operator and the manufacturer.

## 1 Introduction

The exesive use of natural resources is becoming prevalent nowadays. It is reported that due to the over-using and over-consuming of this natural resources of the planet, future sustainability of the vital resources is questioned. The Living Planet reported that the human consumption of natural resources is far above the capability of the Earth to replenish the resources. It is more than 30% higher than the Earth can replenish each year. This undoubtly leads to deforestation, degraded soils, polluted air and water, and the collapse many commercial fisheries and the extinction other species (The Guardian, Oct 2008). The world mostly agrees that something needs to be done to resolve this problem. The UNDP agenda of 12th SDG's goal points out that in "[a]chieving economic growth and sustainable development requires that we urgently reduce our ecological footprint by changing the way we produce and consume goods and resources" (UNDP, 2017).

Some large parts of energy consumption have gone to the transportation sector. Unwanted consumption in transportation is due to severe traffic congestion. In many urban areas, severe traffic congestion has been blamed to cause air pollution. It is also regarded to hinder economic activity to some extent and wasting unnecessary energy consumption. According to Indonesia's transportation ministry report, vehicles contribute 70% as the source of air pollutant in Jakarta. In 2009 alone there were 9.9 million vehicles in Jakarta, which made the air pollutant in the city was getting worse. In order to reduce the traffic congestion, the Indonesian government has arranged a subsidy policy in public transportation sector. In general there should be an optimal subsidy scheme to reduce emissions in urban cities. Some methods for designing optimal subsidy scheme in complex urban areas are readily available (Qin and Zhang, 2015).

A subsidy is often defined as a form of financial aid or support extended to an economic sector aiming of promoting economic and social policy [1]. The economic sector could be an institution, bussiness, or individual; and the form of the subsidies could be direct, such as cash grants and interest-free loans, and could be indirect, such as tax breaks, insurance, low-interest loans, accelerated depreciation, rent rebates, and reimbursement [2, 3]. A subsidy is given by a government to an economic sector or even to a public service institution whenever there is a foreseen benefits generated by the economic sector or the public service institution, directly or indirectly. The benefits might be in the form of economic benefit, health benefit, environmental benefit, or any other recognized benefits. The American Public Transportation Association explored the health impacts of public transportation and found six benefits compared to the use of individual vehicles, i.e. it can make the users are more active physically, it is safer than individual vehicles, it reduces stress, it keeps air cleaner, it saves money of the user, and it provides access to essential needs later in life (http://transloc.com/6-health-benefits-of-public-transportation/).

Considering the many benefits of public transportation, in 2016 the Indonesia's Transportation Ministry gave subsidy to one of public transport operator, i.e. Damri, by supplying buses in 11 big cities, that is

* Corresponding author: hennie.husniah@gmail.com

Medan, Padang, Palembang, Bandar Lampung, Bandung, Yogyakarta, Surakarta, Surabaya, Denpasar, Mataram, and Makassar (liputan6.com, 2016). This has been done to reduce the use of private or individual vehicles as the transportation mode in the cities. The reduction of the use of private or individual vehicles is expected to reduce the severe level of traffic congestion.

An early literature in public transport research has shown that the lack of efficient pricing scheme has been blamed as one of the responsible factors causes the severe traffic congestion in urban areas (Jackson, 1975). Hence, other attempt to increase the number of passengers of the public transport is by making an efficient pricing scheme, e.g. by reducing the ticket of the transportation. The rationale is that a lower public transport ticket price will discourage the use of private and individual vehicles, and hence favouring the use of public transport (Parry and Small, 2009).

Levinson and King (2013) pointed out that subsidy should be considered two ways - capital subsidy and operating subsidy. They argued that these two are related, although different enough so that the two should be considered separately. Providing assets (such as giving new buses) is considered as the former form of subsidy while giving reimbursement of reduced ticket price to the operator is regarded as the later form of subsidy. Inline to this argument, the present paper deals with government subsidy model given to the Damri in the aim to increase the use of buses as the main mode of transportation. There are two subsidy models that will be studied, i.e. subsidy given to the Damri for buying buses and subsidy given to the Damri for reimbursing of reduced ticket price.

Considering a leader-follower relationship between the Damri and the manufacturer of the buses, we consider non-cooperative solution and cooperative solution in order to maximize profit. In this scheme, the Damri acts as the leader which has the first opportunity in devising policy to maximize profit. The Manufacturer, as a follower, maximizes profit based on the Damri's policy that has been chosen. This paper is organized as follows. Section 2 gives model formulation and section 3 gives solution and analyzes the model. Section 4 gives numerical example to see which model is better and finally conclude in section 5.

## 2 Mathematical Model

The following notations will be used in the model formulation.

$q$ : number of passengers per bus per year
$Q$ : total number of passengers per year
$n$ : bus demand
$PM$ : Preventive Maintenance
$CM$ : Corrective Maintenance
$K$ : bus operating year
$N$ : number of PM per bus
$\delta$ : degree of repair
$\lambda_0(t)$ : failure rate without PM
$\lambda(t)$ : failure rate after PM
$p$ : ticket price per passenger
$Cm$ : expected total cost of CM
$Cf$ : cost per CM
$Cp$ : expected total cost of PM
$Cr$ : manufacture production cost
$\tau$ : PM's period
$u$ : subsidy amount per year
$w$ : bus price
$\Psi_d$ : operator's profit
$\Psi_m$ : manufacture's profit
$\Psi$ : total profit

To formulate the model we make the assumption that wholesale the bus purchase price is determined by the manufacturer and the ticket price is determined by the Damri. It is also assumed that every failure item of the bus only need minimal repair so that time between failures is negligible. In the following section we define operator revenue model, operator expenses model, the PM and CM maintenance models, and other related concepts needed in the subsequent sections.

### 2.1 Operator Revenue Model

In *ceteris paribus* condition the demand's law said that if the product price increase then the demand will decrease and if the product price decrease then the demand will increase. In this case, if the number of passengers per year is $q$ and ticket price is $p$, assuming a linear relationship will have a demand function

$$q(p) = \gamma_0 + \gamma_1 p \tag{1}$$

with $\gamma_1 < 0$ and $\gamma_0 > 0$.

The number of buses per year ($n$) can be obtained by dividing $q$ with bus capacity $m$ so that $n = \frac{q(p)}{m}$ ; $n, m > 0$. If the Damri operates in $K$ years and the number of passengers per year constant, then the total number of passengers is

$$Q(p,K) = q(p).K \;\; ; K > 0 \tag{2}$$

The Damri's revenue, $R_d$, is obtained from total passengers multiplied by the ticket price

$$R_d(p,K) = Q(p,K).p \tag{3}$$

In this paper, we will use two government subsidy models. First, subsidy for buying bus from manufacturer and the second one is subsidy for reducing ticket price. For simplicity, we use the index 1 for the first model and the index 2 for the second model.

For the first model, the subsidy doesn't influence demand function so $q(p_1) = \gamma_0 + \gamma_1 p_1$. For the second model, the subsidy amount $u$ influences the demand function so $q(p_2, u) = \gamma_0 + \gamma_1 p_2 + \gamma_2 u$ where

$u, \gamma_2 > 0$. Thus, based on equation (3) we have revenue function $R_d$:

$$R_d = \begin{cases} (\gamma_0 + \gamma_1 p_1) p_1 K & ; \\ (\gamma_0 + \gamma_1 p_2 + \gamma_2 u) p_2 K & ; \end{cases} \tag{4}$$

The first row is for the 1st model and the second row is for the 2nd model. The function of the number of buses for the 1st model and the 2nd model is respectively given by the first and the second row of the following formula:

$$n = \begin{cases} \dfrac{(\gamma_0 + \gamma_1 p_1)}{m}, \\ \dfrac{(\gamma_0 + \gamma_1 p_2 + \gamma_2 u)}{m}. \end{cases} \tag{5}$$

## 2.2 Operator Expenses Model

The main Damri's expenses are the costs of preventive maintenance (PM), the cost of corrective maintenance (CM), and the cost of purchasing the buses. The maintenance is purported to reduce failure intensity. In this paper, we will use the two parameter Weibull failure intensity function $\lambda_0(t)$ with the scale parameter $\alpha = (2/\theta)^{1/2}$ and the shape parameter $\beta = 2$. Thus, we have a linear failure function $\lambda_0(t) = \theta t;\ \theta \geq 0$. The following sub-sections derive the cost functions for the PM and CM respectively.

### *2.2.1 Preventive Maintenance Cost Model*

Let the public transport operator undertakes $N$ times PM in $K$-years period, then the time interval between two PMs, i.e. $\tau$ is formulated by $\tau = K/(N+1)$ years. According to Kijima-Type 1 model, PM turns the age of bus $t$ into a virtual age $v(t) < t$. Assumed that every PM has the same degree of repair $0 \leq \delta \leq 1$, where $\delta = 1$ means a minimal repair and $\delta = 0$ means a perfect repair (Hamidi *et al.*, 2016). By refering to Fig. 1 in (Hamidi *et al.*, 2016), a PM reduces the failure intensity function to $\lambda(t) < \lambda_0(t)$ and the normal age by $(1-\delta)\tau$. As a result the bus virtual age for $i\tau \leq t < (i+1)\tau$ is $v(t) = i\delta\tau + t - i\tau$. So, the failure intensity function of the bus will become

$$\lambda(t) = \lambda_0(v(t)) = \theta(i\delta\tau + t - i\tau).$$

If every cost PM is $C_p^* = a + (1-\delta)\tau b$ then total cost for $N$ times of PM is given by

$$C_p = Na + N(1-\delta)\tau b = Na + \frac{NbK(1-\delta)}{(N+1)}. \tag{6}$$

### *2.2.2 Corrective Maintenance Cost Model*

While operating, bus may have failure at a random time. When failures occur, bus need to be repaired. Every failure is assumed minimally repaired so that the failure intensity just the same as that just before the failure happened. Without any PM, failure follows a Non-Homogenous Poisson Process (NHPP) with failure intensity $\lambda_0(t)$ (Jaturonnatee, *et al.*, 2006; Chang & Lo, 2011). After PM, failure process in the interval $[i\tau, (i+1)\tau)$, for $i = 0, 1, 2, \ldots N$ still follows the NHPP with the intensity function $\lambda(t) = \lambda_0(v(t))$ (Kim *et al.*, 2004). The expected total number of failure is

$$\sum_{i=0}^{N} \int_{i\tau}^{(1+i)\tau} \lambda_0(v(t))\,dt = \sum_{i=0}^{N} \int_{i\tau}^{(1+i)\tau} \theta(i\delta\tau + t - i\tau)\,dt = \frac{K^2\theta(N\delta+1)}{2(N+1)}.$$

If the cost for a PM is $C_f$, then the expected total amount of PM for $K$-years operating time is

$$C_m = \frac{C_f K^2 \theta(N\delta+1)}{2(N+1)}. \tag{7}$$

### *2.2.3 Bus Purchase Price*

Another expense for the public transport operator (Damri) is the bus purchasing price $w$. For the first subsidy model the Damri has received government subsidy for purchasing the buses, so that the Damri needs only to pay $w_1 < w$. However, for the second model the Damri needs to pay $w_2 = w$.

### *2.2.4 Operator Profit function*

The Profit function is the difference between the revenue (4) and the expenses for PM (6), CM (7), and bus purchasing price. Damri's profit function for first subsidy model is given by

$$\begin{aligned} \Psi_{d_1}(p_1, w_1, K, N, \delta) &= R_{d_1}(p_1, K) - (w_1 + C_p + C_m) n_1 \\ &= (\gamma_0 + \gamma_1 p_1) p_1 K \\ &\quad - (w_1 + C_p + C_m) \frac{(\gamma_0 + \gamma_1 p_1)}{m}, \end{aligned} \tag{8}$$

and for the second model is

$$\Psi_{d_2}(p_2,w_2,K,u,N,\delta)=R_{d_2}(p_2,K)-(w_2+C_p+C_m)n_2$$
$$=(\gamma_0+\gamma_1p_2+\gamma_2u)Kp_2$$
$$-(w_2+C_p+C_m)\frac{(\gamma_0+\gamma_1p_2+\gamma_2u)}{m}. \quad (9)$$

*2.2.5 Manufacturefr Profit function*

If the production cost for every is $c_m$ then manufacture's profit function for the first model is

$$\Psi_{m_1}(p_1,u,w_1,K)=uK+(w_1-C_r)n$$
$$=uK+(w_1-C_r)\frac{(\gamma_0+\gamma_1p_1)}{m}, \quad (10)$$

and for second model is

$$\Psi_{m_2}(p_2,w_2,u)=(w_2-C_r)n$$
$$=(w_2-C_r)\frac{(\gamma_0+\gamma_1p_2+\gamma_2u)}{m}. \quad (11)$$

## 3 The Optimal Solution

### 3.1 Non Cooperative Solution

In the non-cooperative solution, manufacturer will act as a leader and makes a decision policy first. Damri will then act as a follower and make profit policy based on manufacturer's policy. In the first subsidy model, we determine ticket price $p_1$ that maximizes profit function (8) by differentiating $\frac{\partial\Psi_{d_1}}{\partial p_1}=0$ and $\frac{\partial^2\Psi_{d_1}}{\partial p_1^2}<0$, which yields in

$$p_1=\frac{1}{2mK}(w_1+C_p+C_m)-\frac{\gamma_0}{2\gamma_1} \quad (12)$$

We substitute (12) into (9) to obtain

$$\Psi_{m_1}(p_1,u,w_1,K)=uK+\frac{\gamma_1}{2m^2K}(w_1-C_r)\left(w_1+C_p+C_m+\frac{\gamma_0mK}{\gamma_1}\right). \quad (13)$$

To get manufacturer's maximum profit, we determine $w_1$ so that $\frac{\partial\Psi_{w_1}}{\partial p_1}=0$ and $\frac{\partial^2\Psi_{w_1}}{\partial p_1^2}<0$, which yields in

$$w_1=\frac{1}{2}\left(C_r-C_p-C_m-\frac{\gamma_0mK}{\gamma_1}\right). \quad (14)$$

Next we substitute (14) to (13) to obtain manufacturer maximum profit $\Psi_{m_1}(\max)=uK-\frac{\gamma_1}{8m^2K}A^2$; $\gamma_1<0$ with $A=\left(C_r+C_p+C_m+\frac{\gamma_0mK}{\gamma_1}\right)$. The profit of Damri can be obtain by substituting (14) to (12) to have

$$p_1=\frac{1}{4mK}\left(C_r+C_p+C_m-\frac{3\gamma_0mK}{\gamma_1}\right). \quad (15)$$

Substitute (15) to (8) to obtain $\Psi_{d_1}(\max)=-\frac{\gamma_1}{16m^2K}A^2$. Analogously, we can find the maximum profit for Damri and manufacturer for the second model. The Damri's maximum profit is given by $\Psi_{d_2}(\max)=-\frac{\gamma_1}{16m^2K}B^2$ with $B=\left(C_r+C_p+C_m+\frac{\gamma_0mK}{\gamma_1}+\frac{\gamma_2umK}{\gamma_1}\right)$ and the manufacturer's maximum profit is $\Psi_{m_2}(\max)=-\frac{\gamma_1}{8m^2K}B^2$.

### 3.2 Cooperative Solution

In cooperative solution, we consider the sum of Damri and manufacturer profit function. For the first subsidy model we have

$$\Psi_1=\Psi_{d_1}+\Psi_{m_1}=R_{d_1}(p_1,K)-(C_p+C_m+C_r)n_1+uK. \quad (16)$$

The maximum total profit will be obtained by differentiating this sum with the respect to $p_1$, that is $\frac{\partial\Psi_1}{\partial p_1}=0$ and $\frac{\partial^2\Psi_1}{\partial p_1^2}<0$, which yields in

$$p_1=\frac{1}{2mK}(C_p+C_m+C_r)-\frac{\gamma_0}{2\gamma_1}. \quad (17)$$

By substituting (17) into (16), the maximum total profit of Damri and manufacturer is given by

$$\Psi_1=uK-\frac{\gamma_1}{4m^2K}A^2. \quad (18)$$

Analogously, cooperative solution for the second subsidy model is given by.

$$\Psi_2 = -\frac{\gamma_1}{4m^2K}B^2. \quad (19)$$

It is straight forward to have the following propositions for these maximum profits.

**Proposition 1:** *In both subsidy model, cooperative solution gives total profit of Damri and manufacturer better than non-cooperative solution* $\Psi_1 - \left(\Psi_{d_1} + \Psi_{m_1}\right) > 0$ *and* $\Psi_2 - \left(\Psi_{d_2} + \Psi_{m_2}\right) > 0$.

**Proposition 2:** *The degree of repair and the number of PM that make optimum profit are* $\delta = 0$ *and* $N = \sqrt{\frac{K\left(C_f K\theta - 2b\right)}{2a}} - 1$ *with* $N$ *is integer. The value* $\delta = 0$ *means Damri does a perfect repair PM.*

To illustrate the propositions we give the following numerical example.

## 4 Numerical Example

For example, we have data number of passengers per year $q$, ticket price $p$, and subsidy amount per year $u$ as in Table1.

**Table 1.** Data number of passengers per year $q$, ticket price $p$, and subsidy amount per year $u$.

| $q$ | $p$ (in IDR) | $u$ (in $10^6$ IDR ) |
|---|---|---|
| 1250000 | 3500 | 600 |
| 1200000 | 3750 | 550 |
| 1190000 | 4000 | 500 |
| 1150000 | 4300 | 450 |
| 1100000 | 4500 | 400 |
| 1000000 | 4800 | 350 |
| 920000 | 5000 | 300 |
| 860000 | 5500 | 250 |
| 820000 | 5700 | 200 |
| 700000 | 6000 | 150 |

Using *R* application program, we have a linear model for the number of passengers per year $q(p,u) = \gamma_0 + \gamma_1 p + \gamma_2 u$, where $\gamma_0 = 1{,}975{,}000$, $\gamma_1 = -208$ and $\gamma_2 = 0.00006$. If the bus capacity for one year is $m = 54{,}000$ passengers then we have number of buses is $n = \frac{q(p,u)}{54{,}000}$. Let failure intensity function of bus is Weibull with $\alpha = \sqrt{0{,}5}$ and $\beta = 2$ so the failure intensity function become $\lambda_0(t) = 4t$. Cost for every CM is $C_f = IDR\ 700{,}000$ and for every PM is $C_p^* = a + (1-\delta)\tau b$ with $a = IDR\ 300{,}000$ and $b = IDR\ 200{,}000$. According to proposition 2, we have $\delta = 0$ and $N = \sqrt{\frac{K\left(C_f K\theta - 2b\right)}{2a}} - 1$ to maximize the profit. The following figures show the result for 3 and 5 years bus operation with D1/2 indicates operator's profit for subsidy model 1/2 and M1/2 indicates manufacturer's profit for subsidy model 1/2.

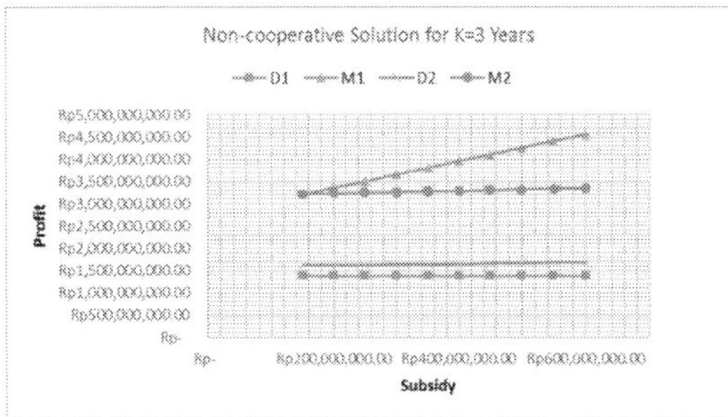


**Fig. 1.** Non-cooperative solution for 3 years operation.

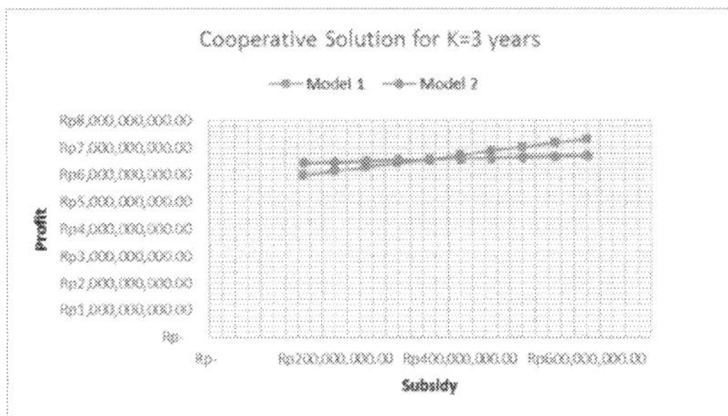


**Fig. 2.** Cooperative solution for 3 years operation.

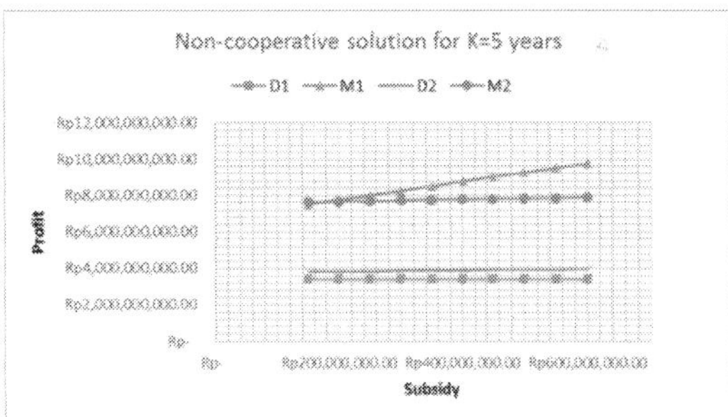


**Fig. 3.** Non-cooperative solution for 5 years operation.

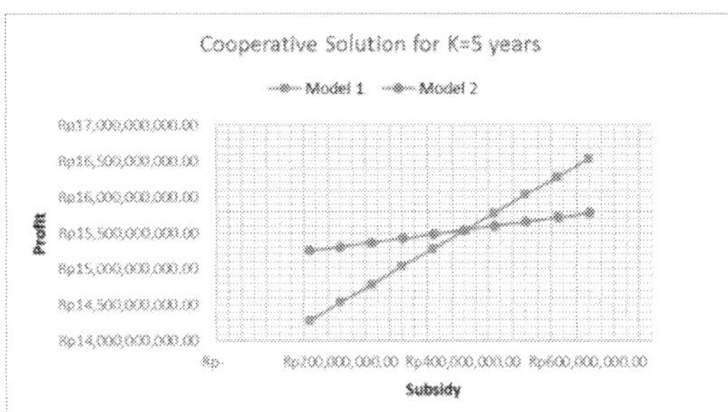


**Fig. 4.** Cooperative solution for 5 years operation.

From the examples above, clearly we can see that profit from cooperative solution is higher than non-cooperative. In cooperative solution, when subsidy is getting bigger, the first model is better to use than the second one.

## 5 Conclusion

We have studied two mathematical models of government subsidy to public transportation. In this case we applied the theory to the Indonesian bus public transport agency, i.e. the DAMRI. The subsidy is purported to increase people's interest in using bus for their main transportation. This is done as government attempt to reduce the severity of traffic congestion, which has a damaging effect to the environment. We analysed two different kinds of subsidy: (i) the subsidy in purchasing bus from an appointed public transport manufacturer, and (ii) the subsidy for reimbursing reduced ticket price for passengers. Numerical examples show that:

a) PM reduces the number of failure which makes the buses operating in a longer time of service,
b) Cooperative solution gives a higher profit for both the public transport operator and the manufacturer,
c) From manufacturer point of view, the government subsidy in purchasing bus is better,
d) From public transport operator, the government the subsidy for reimbursing reduced ticket price for passengers is better.

H.H. acknowledges support from Kemenristekdikti for the PUPT research grant contract number SP DIPA/042.06.1.401516/2017 and A.K.S. acknowledges support from Universitas Padjadjaran, due to the funding to parts of this work through the scheme of Academic Leadership Grant (ALG) with contract number 855/UN6.3.1/PL/2017.

## References

1. Hamidi, M., Liao, H., and Szidarovszky, F. (2016). Non-cooperative and cooperative game theoretic models for usage-based lease contracts. *European Journal of Operational Research*, 4(064).
2. Iskandar, B. P., Husniah, H., and Pasaribu, U. S. (2014). Maintenance service contract for equipment sold with two dimensional warranties. *Quality Technology and Quantitative Management*, 11(3).
3. Jaturonnatee, J., Murthy, D., and Boondiskulchok, R. (2006). Optional preventive maintenance of leased equipment with corrective minimal repairs. *European Journal of Operational Research*, 174(1), 201-215.
4. Murthy, D. N. P. and Jack, N. (2016). Game theoretic modelling of service agent warranty fraud. *Journal of the Operational Research Society*, 68(11).
5. Ross, M. S. (1996). *Stochastic Processes second edition*. Canada: John Wiley and Sons.
6. Xu, Ying, Zhang, Xuemei and Hong, Yu. (2016). The game model between government subsidies act and green supply chain. *International Journal of u- and e- Service, Science and Technology*, 9(11), 121-130.
7. https://www.theguardian.com/environment/2008/oct / 29/climatechange-endangeredhabitats
8. http://www.undp.org/content/undp/en/home/sustain able-development-goals/goal-12-responsible-consumption-and-production.html
9. Myers, N.; Kent, J. (2001). *Perverse subsidies: how tax dollars can undercut the environment and the economy*. Washington, DC: Island Press. ISBN 1-55963-835-4.
10. Collins Dictionary of Economics". Retrieved 2013-09-05.
11. https://en.wikipedia.org/wiki/Subsidy
12. R. Jacksons (1975). Optimal subsidies for public transit, *Journal of Transport Economics and Policy*, 9(1).
13. Feifei Qin, F and Zhang, Xiaoning (2015). Designing an Optimal Subsidy Scheme to Reduce Emissions for a Competitive Urban Transport Market. Sustainability 2015, 7, 11933-11948; doi:10.3390/su70911933.
14. Ian W.H. Parry Kenneth A. Small (2009). Should Urban Transit Subsidies Be Reduced? *American Economic Review* 99(3), 2009.
15. Here's How The Government Should Be Subsidizing Public Transit
16. David Levinson and David King Apr. 22, 2013, 12:11 PM (Bussines insider) http://www.businessinsider.com/the-right-way-to-subsidize-public-transit-2013-4/?IR=T

# Mixed integer linear programming model for dynamic supplier selection problem considering discounts

*Purnawan* Adi Wicaksono[1]*, I *Nyoman* Pujawan[2], *Erwin* Widodo[2], *Sutrisno*[3], and *Laila* Izzatunnisa[1]

[1]Department of Industrial Engineering Diponegoro University, Semarang 50275 Indonesia
[2]Department of Industrial Engineering, Sepuluh Nopember Institute of Technology, Surabaya 60111 Indonesia
[3]Department of Mathematics, Diponegoro University, Semarang 50275 Indonesia

**Abstract.** Supplier selection is one of the most important elements in supply chain management. This function involves evaluation of many factors such as, material costs, transportation costs, quality, delays, supplier capacity, storage capacity and others. Each of these factors varies with time, therefore, supplier identified for one period is not necessarily be same for the next period to supply the same product. So, mixed integer linear programming (MILP) was developed to overcome the dynamic supplier selection problem (DSSP). In this paper, a mixed integer linear programming model is built to solve the lot-sizing problem with multiple suppliers, multiple periods, multiple products and quantity discounts. The buyer has to make a decision for some products which will be supplied by some suppliers for some periods cosidering by discount. To validate the MILP model with randomly generated data. The model is solved by Lingo 16.

## 1 Introduction

In a highly competitive industry today, firms must realize the importance of supplier selection that meet the required quality and time. Firms are now facing increased consumer demand, shorter product life cycles, and sharper price cuts. This condition causes firms to cut costs and improve supply chain. Supply chain improvements become critical in rising the competitiveness of the firm. It can be done through supplier selection process (Sagar, 2012).

Fluctuative and dynamic demand make it difficult for firms to determine the order quantity and and when the order must be made to fulfill the demand and minimize the inventory costs. Suppliers are also expected to be dynamic as well. This can happen because a firm usually has a demand regarding supplier capacity, quality level, lead time, unit cost part, and time-varied fixed transportation cost. (Ware et al., 2014). Supplier selection problems are usually classified in terms of multi-echelon, multi-product, multi-supplier and the time period's nature (Lin & Lei, 2009). Firm's requests always fluctuate so that time value requirements are considered dynamic, not with short or long-term agreements. Therefore, this paper will discuss Dynamic Supplier Selection Problem (DSSP). In DSSP which a set of suppliers is selected supplier for each period. (Ware et al., 2014).

Most of the models that have been developed for supplier selection problems ignore transportation elements that significantly impact the total procurement costs. (Wicaksono et al., 2016). However, there are already some researchers who consider transportation costs in the selection of single product suppliers such as (Burke et al., 2007; Liao & Rittscher, 2007; Aguezzoul, A., & Ladet, 2007; Choudhary & Shankar, 2011; Choudhary & Shankar, 2013). Some researchers are choosing suppliers with multi-period, multi-product and multi-supplier (Rezaei & Davoodi, 2011; Ware et al., 2014; Wicaksono et al., 2016). In another case, the supplier provides discounts for each purchase in large quantities to increase sales of a product goods / services. With discounts, buyers may be interested to buy a product. In practice, ordering in large amount with far distance can reduce transportation costs per unit (Shinn et al., 1996). Therefore, this study will develop a model that integrates procurement decisions with transportation costs and discounts.

To minimize procurement expenditures, purchasing and transportation costs need to be considered. Suppliers offer total cost discount and transportation costs based on trucking rates. The goal is to select several suppliers in fulfilling the demand of the product with a minimum total cost. In this paper, the researcher will make a model to solve supplier selection problem with multi-period, multi-product and multi-supplier condition based on the development of the previous model (Ware et al., 2014) with discount consideration. The method used is Mixed Integer Linear Programming (MILP). Random data used for model validation. This model then solved using LINGO 16 software.

Furthermore, in part 2 of this paper, a brief literature review of models or techniques related to the dynamic supplier selection will be presented. Section 3 will present the model and its mathematical formulation. Section 4 will present a case study to test the proposed model. Finally, section 5 will provide conclusions and

*Corresponding author: purnawan@undip.ac.id

suggestions for further research.

## 2 Literature Review

In supplier selection, there are two categories: quantitative model and qualitative model. Quantitative model includes linear programming, MILP, MINLP, dynamic programming, and multi-objective programming (Masella & Rangone, 2000; Demirtas & Üstün, 2008). Masella & Rangone (2000) propose four supplier selections that are time and content based (logistics or strategic). Cakravastia (2004) propose a multi-objective model for supplier selection and consider the impact of decisions to be made.

Treviño-Garza et al., (2015) presented MILP for multi-product and multi-period to solve supplier selection problem. Ware et al., (2014) developed MINLP to address the issue of dynamic supplier selection (DSSP). Transportation costs have a large impact in procurement decisions, such as dividing orders on some suppliers will beresulting less quantity of goods and greater transportation costs (Aguezzoul & Ladet, 2007). Wicaksono et al., (2016) studied DSSP considering truckload shipping using MILP.

Considering discount in supplier selection is important because discount can affect the decision of supplier selection (Chaudhry et al., 1993). Howefer, limited paper especially in DSSP consider discount in there model. There are two types of discounts: quantity discount and volume discount. In quantity discounts, product sales do not affect the price and discounts of other products. This discount strategy can be either incremental or all-unit. The linear programming model considers various prices including quantity discounts and inflated prices according to the number of orders (Aissaoui et al., 2007).

The supplier selection process for DSSP involves higher complexity in terms of selecting suppliers for all periods to produce the minimum total cost with the shortest lead time and meet the desired quality level for each product set by the company. In addition, to the best author knowledge there is not DSSP paper considering discount. Therefore, this paper will present a MILP formulation for DSSP by considering transportation costs and discount if multi-period, multi-product and multi-supplier conditions.

## 3 Model Development

The problem can be explained briefly as follows: in a procurement management, there were supplier selections with consideration of the transportation costs and discounts provided by the supplier and there was various demand for various products during T period. The objective is to select one or more suppliers, determine which period the procurement will take place and the lot size to be purchased during that period.

### 3.1 Model Assumption

The proposed model was limited by several assumptions such as fluctuative and constant demand. Shortages are permitted and backlog are assumed when the total supply available in a given period is insufficient to meet demand in that period. The production capacity of suppliers and the storage capacity of buyers in any period were limited. The delivery trucks each had the same capacity. The transportation costs of each supplier could vary depending on the distance of supplier's location from the manufacturer.

### 3.2 Model Parameters and Decision Variables

Indices

T Set of time periods; 1,2,...,t
R Set of all-unit price break levels; 1,2,…,r
S Set of suppliers; 1,2,...,s
P Set of products; 1,2,...,p

Parameters

$UP_{trsp}$ : Unit price of product p supplied by supplier s at price break level r in period t
$b_{rsp}$ : quantity at which all-unit price break r occur at supplier s of product p
$TC_{ts}$ : FTL cost from supplier s to the buyer in period t
$NC_{ts}$ : Contract cost of new supplier s in period t
$SOC_{tp}$ : Shortage cost per unit of product p in period t
$C$ : FTL capacity
$D_{tp}$ : Demand of product p for time period t
$SC_{tsp}$ : Supplier capacity of product p in period t
$l_{trsp}$ : Percentage of product delivered late by supplier s at price break level r in period t
$d_{trsp}$ : Percentage of rejected product delivered by supplier s at price break level r in period t
$P^{l}_{tsp}$ : Penalty cost for late delivery product p by supplier s in period t
$P^{d}_{tsp}$ : Penalty cost for defected product p by supplier s in period t
$O_{ts}$ : Cost of ordering for a purchase quantity to supplier s in period t
$H_{tp}$ : Holding cost of product p per unit time in period t
$MS_{tp}$ : Buyer's storage capacity for product p in period t
$\emptyset$ : Buyer's service level requirement in period t so (1-$\emptyset$) is the proportion of buyer's demand that are not met by supplier in period t
$M$ : Big number

Decision Variables

$X_{trsp}$ : Number of product p supplied by supplier s at price break level r in time period t
$S_{ts}$ : Frequency of truck delivered product from supplier s in period t
$Y_{trsp}$ : Binary variable used in separating price levels r for product p in a transaction between buyer and supplier s in period t

$Z_{ts}$ : Binary variable (1 if an order is placed to supplier s in period t and hence order cost is charged, 0 otherwise)

$V_{tsp}$ : Binary variable indicating whether supplier s is chosen or not in period t

$W_{ts}$ : Binary variable for choosing new supplier (1 if a new contract is established with supplier s in period t, 0 otherwise)

$i^{+}_{tp}$ : Inventory of product p in period t

$i^{-}_{tp}$ : Shortage of product p in period t

### 3.3 Mathematical Formulation

With the above parameters and decision variables, the mixed integer non linear program model is expressed as follows:

Minimize $Z = Z_1 + Z_2 + Z_3 + Z_4 + Z_5 + Z_6 + Z_7 + Z_8$ (1)

$$Z_1 = \sum_{t=1}^{T}\sum_{r=1}^{R}\sum_{s=1}^{S}\sum_{p=1}^{P} X_{trsp} * UP_{trsp} \quad (2)$$

$$Z_2 = \sum_{t=1}^{T}\sum_{s=1}^{S} TC_{ts} * S_{ts} \quad (3)$$

$$Z_3 = \sum_{t=1}^{T}\sum_{s=1}^{S} O_{ts} * Z_{ts} \quad (4)$$

$$Z_4 = \sum_{t=1}^{T}\sum_{s=1}^{S} NC_{ts} * W_{ts} \quad (5)$$

$$Z_5 = \sum_{t=1}^{T}\sum_{p=1}^{P} H_{tp} * i^{+}_{tp} \quad (6)$$

$$Z_6 = \sum_{t=1}^{T}\sum_{p=1}^{P} SOC_{tp} * i^{-}_{tp} \quad (7)$$

$$Z_7 = \sum_{t=1}^{T}\sum_{r=1}^{R}\sum_{s=1}^{S}\sum_{p=1}^{P} P^{d}_{tsp} * d_{trsp} * X_{trsp} \quad (8)$$

$$Z_8 = \sum_{t=1}^{T}\sum_{r=1}^{R}\sum_{s=1}^{S}\sum_{p=1}^{P} P^{l}_{tsp} * l_{trsp} * X_{trsp} \quad (9)$$

Subject to:

$$i^{+}_{(t-1)p} + \sum_{r=1}^{R}\sum_{s=1}^{S} X_{trsp} + \sum_{r=1}^{R}\sum_{s=1}^{S} l_{trsp}X_{(t-1)rsp} - \sum_{r=1}^{R}\sum_{s=1}^{S} l_{trsp}X_{trsp} - \sum_{r=1}^{R}\sum_{s=1}^{S} d_{trsp}X_{trsp}$$
$$\geq D_{tp} + i^{-}_{(t-1)p} + i^{+}_{tp} - i^{-}_{tp} \qquad \forall t \in T, \forall p \in P \quad (10)$$

$$\sum_{t=1}^{T}\sum_{r=1}^{R}\sum_{s=1}^{S} X_{trsp} \geq \sum_{t=1}^{T} D_{tp} \qquad \forall p \in P \quad (11)$$

$$b_{(r-1)sp}Y_{trsp} \leq X_{trsp} \leq b_{rsp}Y_{trsp}$$
$$\forall t \in T, \forall r \in R, \forall s \in S, \forall p \in P \quad (12)$$

$$\sum_{r=1}^{R} Y_{trsp} = V_{tsp} \quad \forall t \in T, \forall r \in R, \forall s \in S, \forall p \in P \quad (13)$$

$$\left\lceil \frac{\sum_{p=1}^{P} X_{trsp}}{C} \right\rceil \leq S_{ts} \quad \forall t \in T, \forall r \in R, \forall s \in S \quad (14)$$

$$X_{trsp} \leq SC_{tsp} \quad \forall t \in T, \forall r \in R, \forall s \in S, \forall p \in P \quad (15)$$

$$\sum_{p=1}^{P} X_{trsp} \leq M * Z_{ts} \quad \forall t \in T, \forall r \in R, \forall s \in S \quad (16)$$

$$i^{+}_{tp} \leq MS_{tp} \qquad \forall t \in T, \forall p \in P \quad (17)$$

$$\sum_{t=1}^{T} Z_{ts} \leq M * W_{ts} \qquad \forall s \in S \quad (18)$$

$$\sum_{t}^{T} i^{-}_{tp} \leq (1-\emptyset)\sum_{t}^{T} D_{tp} \quad \forall p \in P \quad (19)$$

$$X_{trsp}, i^{+}_{tp}, i^{-}_{tp} \geq 0 \quad (20)$$

$$Y_{trsp}, Z_{ts}, V_{tsp}, W_{ts} \in \{0,1\} \quad (21)$$

## 4 Numerical Example

The effectiveness of the proposed MILP model was shown through numerical examples. Buyers wanted to buy products for two periods. Firm's demand for each part changed over times due to market fluctuations and customer demand patterns. In the illustration example there are two periods, two price breaks, four suppliers, and three products (2T-2R-4S-3P). Under the same scenario, a re-calculation was performed but with a constant demand of 1000. Here is an example illustration:

**Table 1.** Randomly generated data for the first period (2T-2R-4S-3P problem)

| $T_1$ | $P_1$ | $P_2$ | $P_3$ |
|---|---|---|---|
| $S_1$ | 600, 0.08, 0.01, 4, 2* | 1000, 0.05, 0.021, 3, 0 | 2000, 0.1, 0.01, 5, 3 |
| $S_2$ | 1000, 0.12, 0.031, 4, 1 | 700, 0.13, 0.035, 3, 1 | 1500, 0.06, 0.02, 5, 0 |
| $S_3$ | 800, 0.07, 0.025, 4, 4 | 800, 0.14, 0.033, 3, 1 | 2000, 0.08, 0.03, 5, 2 |
| $S_4$ | 800, 0.07, 0.025, 4, 3 | 800, 0.1, 0.033, 3, 2 | 2500, 0.12, 0.02, 5, 0 |
| D | 1000 | 1600 | 2200 |
| MS | 3200 | 4000 | 3000 |
| H | 2 | 1 | 1 |
| SOC | 4 | 2 | 3 |

* $SC_{tsp}$, $d_{trsp}$, $l_{trsp}$, $P^{d}_{tsp}$, $P^{l}_{tsp}$

**Table 2.** Randomly generated data for the second period (2T-2R-4S-3P problem)

| $T_1$ | $P_1$ | $P_2$ | $P_3$ |
|---|---|---|---|
| $S_1$ | 1200, 0.14, 0.015, 5, 2* | 1800, 0.11, 0.015, 7, 0 | 2200, 0.13, 0.02, 6, 3 |
| $S_2$ | 2000, 0.01, 0.01, 5, 1 | 1600, 0.12, 0.01, 7, 1 | 1500, 0.15, 0.02, 6, 0 |
| $S_3$ | 1400, 0.08, 0.005, 5, 4 | 1200, 0.08, 0.012, 7, 1 | 1500, 0.08, 0.005, 6, 2 |
| $S_4$ | 1500, 0.03, 0.015, 5, 3 | 2000, 0.12, 0.01, 7, 2 | 1500, 0.11, 0.01, 6, 0 |
| D | 2500 | 3200 | 4500 |
| MS | 4500 | 3900 | 5200 |
| H | 1 | 2 | 1 |
| SOC | 2 | 3 | 2 |

* $SC_{tsp}$, $d_{trsp}$, $l_{trsp}$, $P^d_{tsp}$, $P^l_{tsp}$

**Table 3.** The values of UP, TC, O, NC, D, MS, H and SOC for the first period (2T-2R-4S-3P problem)

| | UP | | | | | | TC | O | NC |
|---|---|---|---|---|---|---|---|---|---|
| | $R_1$ | | | $R_2$ | | | | | |
| $T_1$ | $P_1$ | $P_2$ | $P_3$ | $P_1$ | $P_2$ | $P_3$ | | | |
| $S_1$ | 30 | 15 | 30 | 27 | 13 | 28 | 600 | 280 | 520 |
| $S_2$ | 35 | 14 | 22 | 33 | 13 | 20 | 750 | 360 | 680 |
| $S_3$ | 32 | 20 | 25 | 29 | 18 | 23 | 650 | 315 | 570 |
| $S_4$ | 35 | 22 | 30 | 32 | 19 | 27 | 650 | 310 | 575 |

**Table 4.** The values of UP, TC, O, NC, D, MS, H and SOC for the second period (2T-2R-4S-3P problem)

| | UP | | | | | | TC | O | NC |
|---|---|---|---|---|---|---|---|---|---|
| | $R_1$ | | | $R_2$ | | | | | |
| $T_1$ | $P_1$ | $P_2$ | $P_3$ | $P_1$ | $P_2$ | $P_3$ | | | |
| $S_1$ | 40 | 25 | 40 | 37 | 22 | 39 | 1000 | 490 | 920 |
| $S_2$ | 45 | 24 | 32 | 40 | 23 | 34 | 1200 | 590 | 1125 |
| $S_3$ | 42 | 30 | 35 | 41 | 25 | 34 | 1500 | 730 | 1420 |
| $S_4$ | 45 | 32 | 40 | 43 | 30 | 36 | 1200 | 580 | 1125 |

**Table 5.** Final result of supplier assignment and quantity supplied for the respective product (2T-2R-4S-3P problem).

| | | $R_1$ | | | | $R_2$ | | | |
|---|---|---|---|---|---|---|---|---|---|
| | | $S_1$ | $S_2$ | $S_3$ | $S_4$ | $S_1$ | $S_2$ | $S_3$ | $S_4$ |
| $T_1$ | $P_1$ | 600 | 800 | 800 | 0 | 0 | 0 | 0 | 800 |
| | $P_2$ | 1000 | 700 | 700 | 616 | 0 | 0 | 0 | 0 |
| | $P_3$ | 0 | 0 | 0 | 284 | 2000 | 1500 | 2000 | 0 |
| $T_2$ | $P_1$ | 340 | 160 | 0 | 0 | 0 | 0 | 0 | 340 |
| | $P_2$ | 0 | 0 | 0 | 0 | 1000 | 900 | 0 | 0 |
| | $P_3$ | 216 | 700 | 0 | 0 | 0 | 0 | 0 | 216 |

**Table 6.** Final result of supplier assignment and quantity supplied for the respective product (2T-2R-4S-3P problem) with a constant demand

| | | $R_1$ | | | | $R_2$ | | | |
|---|---|---|---|---|---|---|---|---|---|
| | | $S_1$ | $S_2$ | $S_3$ | $S_4$ | $S_1$ | $S_2$ | $S_3$ | $S_4$ |
| $T_1$ | $P_1$ | 600 | 602 | 800 | 0 | 0 | 0 | 0 | 0 |
| | $P_2$ | 0 | 698 | 319 | 0 | 1000 | 0 | 0 | 0 |
| | $P_3$ | 0 | 0 | 0 | 0 | 0 | 1000 | 1000 | 0 |
| $T_2$ | $P_1$ | 0 | 0 | 0 | 0 | 0 | 0 | 0 | 0 |
| | $P_2$ | 0 | 0 | 0 | 0 | 0 | 0 | 0 | 0 |
| | $P_3$ | 0 | 0 | 0 | 0 | 0 | 0 | 0 | 0 |

The results for the above scenarios are listed in Table 5. From Table 5 we can see that in period one all suppliers supplied the product. Product one was supplied by $S_1$, $S_2$ and $S_3$ at the first price break level and $S_4$ at the second price break level. Product two was supplied by S1, $S_2$, $S_3$ and $S_4$ at the first price break level. And product three was supplied by $S_4$ at first price break and $S_1$, $S_2$ and $S_3$ at second price break level. Period two was only supplied by $S_1$, $S_2$, and $S_4$ with the allocation of products shown on the table. The value of the objective function for that model was 472476,5.

In the second calculation result shown by Table 6, it is shown that only $S_1$, $S_2$ and $S_3$ supply product 1, 2 and 3. It is seen that there are more number of units allocated. In the second period, order is not necessary as it tends to store inventory in the first period.

## 5 Conclusion

In this paper, the supplier selection problem was solved by MILP mathematical model. Numerical illustration with scenario 2T-2R-4S-3P and a second experiment was used to demonstrate the model. The second experiment used the same value for all parameters except the demand parameter using a constant value of 1000. Optimization solved for less than 10 seconds using LINGO 16. When the demand was fluctuating, all suppliers contributed in supplying products, and when it was constant there were only a few suppliers who shipped products and prefered to store inventory in the first period. This model can be applied in real conditions. The resulting random data can be replaced with real time data. However, the number of periods, products, and suppliers used were limited because Lingo software was unable to solve high complexity problems. Therefore, further research can develop this model using heuristic methods to solve more complex problems.

## References

1. Aguezzoul, A., & Ladet, P. (2007). A nonlinear multiobjective approach for the supplier selection , integrating transportation policies. https://doi.org/10.1108/17465660710763434
2. Aissaoui, N., Haouari, M., & Hassini, E. (2007). Supplier selection and order lot sizing modeling: A review. *Computers and Operations Research*, *34*(12), 3516–3540. https://doi.org/10.1016/j.cor.2006.01.016
3. Burke, G. J., Carrillo, J. E., & Vakharia, A. J. (2007). Single versus multiple supplier sourcing strategies, *182*, 95–112. https://doi.org/10.1016/j.ejor.2006.07.007
4. Cakravastia, A., & Takahashi, K. (2004). Integrated model for supplier selection and negotiation in a make-to-order environment. *International Journal of Production Research*, *42*(21), 4457–4474. https://doi.org/10.1080/00207540410001727622
5. Cárdenas-Barrón, L. E., González-Velarde, J. L., & Treviño-Garza, G. (2015). A new approach to solve the multi-product multi-period inventory lot sizing with supplier selection problem. *Computers & Operations Research*, *64*, 225–232. https://doi.org/10.1016/j.cor.2015.06.008
6. Chaudhry, S. S., Forst, F. G., & Zydiak, J. L. (1993). Vendor selection with price breaks. *European Journal of Operational Research*, *70*(1), 52–66. https://doi.org/10.1016/0377-2217(93)90232-C
7. Choudhary, D., & Shankar, R. (2011). Modeling and analysis of single item multi-period procurement lot-sizing problem considering rejections and late deliveries. *Computers and Industrial Engineering*, *61*(4), 1318–1323. https://doi.org/10.1016/j.cie.2011.08.005
8. Choudhary, D., & Shankar, R. (2013). Joint decision of procurement lot-size, supplier selection, and carrier selection. *Journal of Purchasing and Supply Management*, *19*(1), 16–26. https://doi.org/10.1016/j.pursup.2012.08.002
9. Demirtas, E. A., & Üstün, Ö. (2008). An integrated multiobjective decision making process for supplier selection and order allocation. *Omega*, *36*(1), 76–90. https://doi.org/10.1016/j.omega.2005.11.003
10. Liao, Z., & Rittscher, J. (2007). Integration of supplier selection , procurement lot sizing and carrier selection under dynamic demand conditions, *107*, 502–510. https://doi.org/10.1016/j.ijpe.2006.10.003
11. Lin, J. R., & Lei, H. C. (2009). Distribution systems design with two-level routing considerations. *Annals of Operations Research*, *172*(1), 329–347. https://doi.org/10.1007/s10479-009-0628-y
12. Masella, C., & Rangone, A. (2000). A contingent approach to the design of vendor selection systems for different types of co-operative customer/ supplier relationships. *International Journal of Operations & Production Management*, *20*(1), 70–84.
13. Rezaei, J., & Davoodi, M. (2011). Multi-objective models for lot-sizing with supplier selection. *International Journal of Production Economics*, *130*(1), 77–86. https://doi.org/10.1016/j.ijpe.2010.11.017
14. Sagar, M., & Singh, D. (2012). Supplier Selection Criteria: Study of Automobile Sector in India. *International Journal of Engineering Research and Development*, *4*(4), 34–39. Retrieved from http://scholar.google.com/scholar?hl=en&btnG=Search&q=intitle:Supplier+Selection+Criteria+:+Study+of+Automobile+Sector+in+India#0
15. Shinn, S. W., Hwang, H., & Park, S. S. (1996). Joint price and lot size determination under conditions of permissible delay in payments and quantity discounts for freight cost. *European Journal of Operational Research*, *91*(3), 528–542. https://doi.org/10.1016/0377-2217(94)00357-2
16. Ware, N. R., Singh, S. P., & Banwet, D. K. (2014). Expert Systems with Applications A mixed-integer non-linear program to model dynamic supplier selection problem. *Expert Systems With Applications*, *41*(2), 671–678. https://doi.org/10.1016/j.eswa.2013.07.092
17. Wicaksono. (2016). Supplier Selection Model Considering Truckload Shipping.

# Innovative design of ironing board based on Kansei Engineering and usability test

*Nadia* Zhafira, *Perdana Suteja* Putra, *Fety Ilma* Rahmillah, and *Amarria Dila* Sari*

Department of Industrial Engineering, Islamic University of Indonesia, Yogyakarta, Indonesia

**Abstract.** Laundry business is popular service business in Indonesia that mostly involve muscle activity for lifting and ironing activity. It causes the worker get muscular fatigue and injury especially for ironing activity that is included in the static and monotonous activity. moreover, existed ironing board design in the market is not easy to be adjusted to standing or sitting position of workers. Hence, this paper aimed to design the innovative and adjustable ironing board to minimize the musculoskeletal complaint. The preliminary study is conducted by NBM questionnaires found that fatigue experienced mostly in the back, right upper arm, and calves. The product designed by Kansei engineering to satisfy consumer's feeling and assessed by USE questionnaire to fulfil consumer's usability. Orthogonal array and conjoint analysis were conducted to incorporate all attribute potentials in developing a new single concept design. Statistical analysis was also done to test the hypothesis. 80 respondents have participated in this study. The result of this study is an innovative design of ironing board which is valid to meet consumers' needs such as comfort, secure, attractive, sturdy, durable and multifunction and practical at 5% of the significant level.

## 1 Introduction

One of success standard in developed country is its prosperous [1]. The prosperous is identified based on its society including the role of workers in existed industry. moreover, female worker commonly supports the economic growth of family in order to reach the prosperous condition. In fact, recently the number of female workers is increased in two cases which are necessity and choice [2]. In addition, female workers are mostly worked for the informal sector such as laundry industry [3]. As found that laundry industry has been growing in many towns in Indonesia.

The laundry industry developed in many towns where there are many students, workers, and households using the laundry services to facilitate everyday activities [4]. Especially in Jogjakarta, the industrial development is also due to tourism and education that can increase the economic development and employment equity. Laundry workers have responded in weighing and sorting dirty clothes, washing, drying, ironing, and customer services that limited to certain conditions [5]. Workers have to do the job in limited time due to the deadline of customer pick up and face different customer characteristics. This condition leads to musculoskeletal disorders such as while lifting the bucket of laundry and drying the clothes by ironing. Ironing is the most exhausting and riskiest activity among other work processes and workers feel a pain of the waist and arm after did it [6,7]. Moreover, most of them work eight to fifteen hours per day [7].

The musculoskeletal disorders are mostly caused by static and monotony work, a station which is unfit with the workers' anthropometry, a long duration, and heat posed because of the ironing tool [8]. Moreover, study by [9] shows that this complaint is dominantly by 57% for low back pain. This condition is at high risk and need improvement based on posture assessment [6]. On the other measurement, it stated that complaint in ironing activity is felt by hand, knee, elbow, arm, and shoulder [4]. Hence, one of the methods used to minimize the users' complaint is redesigning the equipment used by workers such as ironing board. This innovative design has to meet users' expectation in addition to a favourable impression or Kansei.

Kansei is a Japanese word which has multi-faceted expression and it is the impression somebody gets from a certain artefact, environment or situation using all the senses of sight, hearing, feeling, smell, taste as well as recognition [10,11]. Kansei Engineering (KE) is one method to take into account human feeling or emotion to translate it into design elements [12]. Thus, KE is a proper method in developing products which provide a way to realize human feelings into products.

Some previous researchers applied KE to develop consumers' product design such as sunglasses selection [13], design of baby bags based on the principle of Kansei Engineering in combination with fuzzy linguistic method [14], and design of maternity pillow [18]. However, the usage of KE and its combination with usability test are rarely used. Therefore, this study aimed to design innovative ironing board in order to minimize the musculoskeletal disorders complaint based on users' feeling and usability.

* Corresponding author: amarria@uii.ac.id

## 2 Research Methodology

### 2.1 Nordic Body Map

In order to strengthen the statement that ironing activity leads to musculoskeletal disorders for the users because of its posture, supporting method is needed. Nordic Body Map questionnaire is chosen as the supporting method in order to determine the parts of muscle experiencing symptoms ranging from discomfort to very discomfort [15]. Nordic Body Map questionnaire uses the image of the human body which is divided into 9 main body parts that are the neck, shoulders, upper back, elbows, lower back, hips, knees, and ankles. From 9 body parts are then broken down into 28 parts of the body map questionnaire which distributed into 30 respondents who did standing position as their ironing activity position. In NBM questionnaire, the respondents should fill the checklist boxes which represent the grade of pain that they felt in each part of the body. The table 1 below shows the description of the grade of pain in NBM.

**Table 1**. Grade of Pain Description

| Symbol | Description |
|---|---|
| A | No Pain |
| B | Moderately Pain |
| C | Painful |
| D | Very Painful |

### 2.2 Kansei Engineering

In this study, 50 respondents of laundry workers and housewives in Sleman, Yogyakarta that have been experienced laundry activity for more than 1 year are being interviewed. In addition, the object of this study is user complaints, feelings and requirements, body dimensions, and experience.

There are three main steps in this study to design the ironing board which conducted by the laundry workers. Those main steps are Kansei Engineering which consists of Kansei word collection, Semantic differential scale questionnaire, and sufficiency, validity, and reliability data test. After the valid user's feelings are collected, those aspects are being converted into product requirements. The next step in this study is product visualization with 3D design based on product requirements which has been obtained. Then, usability testing was conducted at the end of this study in order to test the usability aspects which consists of usefulness, ease of use, ease of learning, and user satisfaction of the ironing board design.

### 2.3 Research Procedures

The design process of this study has followed several steps as in Figure 1. The Nordic Body Map questionnaires is distributed in order to identify the muscle complain for ironing board activity. Then, Kansei words was collected to subjects aimed to know the customer's need of product. Furthermore, the Kansei words was proceed based on Kansei Engineering process type I which is semantic differential and statistical analysis.

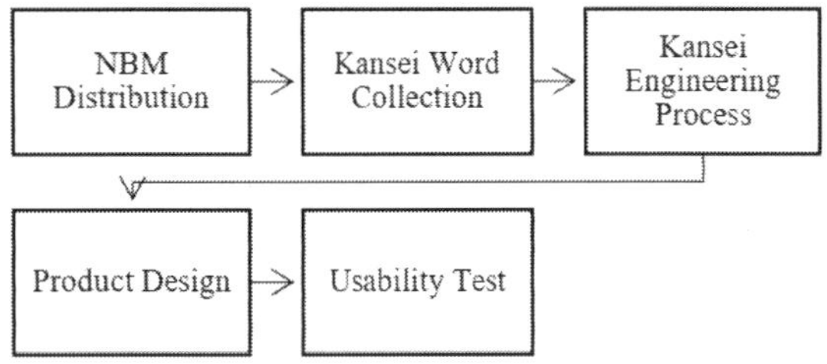


**Fig. 1**. Research procedures

Based on the result of analysis, the product was designed by researchers representatively. Moreover, in order to assess the usability of design, the usability test was designed based on the USE questionnaire.

## 3 Results and Discussion

### 3.1 Nordic Body Map

Most of the complaints that arise in the ironing activity are because the ironing activity tended to be static and monotony, and they felt fatigue in their back which can lead to musculoskeletal disorders. Nordic Body Map questionnaire is chosen as the supporting method in order to strengthen the background statement that ironing activity leads to musculoskeletal disorders for the users because of its posture. From 28 body parts based on the questionnaire that was already given to 30 respondents, most of the respondents feel pain in their back and right upper arm and feel moderate pain on their calves when the ironing activity taking too much time. The result of the NBM questionnaire is shown in Table 2.

**Table 2**. Nordic body map questionanire result

| No | Location | Grade of Complaints | | | | | | | |
|---|---|---|---|---|---|---|---|---|---|
| | | A | | B | | C | | D | |
| | | Total | % | Total | % | Total | % | Total | % |
| 5 | Back | 1 | 3% | 6 | 20% | 23 | 77% | 0 | 0% |
| 6 | Right upper | 0 | 0% | 7 | 23% | 23 | 77% | 0 | 0% |
| 22 | Left calf | 4 | 13% | 18 | 60% | 8 | 27% | 0 | 0% |
| 23 | Right calf | 4 | 13% | 18 | 60% | 8 | 27% | 0 | 0% |

### 3.1. Kansei Engineering

In designing the ironing board, Kansei Engineering are used as the initial step in order to obtain the user requirements based on their feelings. In this study, Kansei Engineering type-1 is conducted in order to obtain the Kansei words. The Kansei word collected using open questionnaires that followed by semantic differential questionnaires. Based on the questionnaire spread, 7 Kansei words presented by consumers' voices were found.

The words were tested by using fault tolerance of 5% as mentioned in Table 3.

**Table 3**. Kansei Words

| Kansei Words | Description | Validity Score | Reliability Score |
|---|---|---|---|
| Comfort | Provide comfortable design | 0.82430 | 0.869 |
| Secure | Has strong and safe part | 0.82431 | 0.869 |
| Practical | Easy to be brought and simple adjustment | 0.31182 | 0.95 |
| Attractive | Attract attention by design and colour | 0.38491 | 0.94 |
| Sturdy | Strong and solid design | 0.82431 | 0.869 |
| Durable | Good quality of material | 0.31182 | 0.95 |
| Multifunction | Has more than one setting to be used | 0.82440 | 0.869 |

It shown on the result above that validity and reliability result gain more than r table value (0.284). it means that seven Kansei words are reliable and valid to be inputted as parameter design of ironing board. Hence, this product made was based on the psychological feelings of consumers which are comfort, secure, practical, attractive, sturdy, durable and has multifunction.

After the user's feelings are obtained, it can be converted into product requirements which can be continued to product design visualization. In determining the proper body dimension for the ironing board design, anthropometric percentile is used. Percentiles provide a basis for estimating the proportion of a population accommodated by a design. Thus, percentile can be applied to assess size limitations imposed by a design [16]. There are four dimensions which involved in this study; sum of thigh thickness (TP) and popliteal height (TPO) to determine the minimum height of the sitting ironing table, elbow height standing (TSB) to determine the maximum height of the standing ironing table, and arm's reach (JT) to determine the minimum ironing table length.

Figure 2 shows the side view of the ironing board design. The 5th percentile of elbow height standing (TSB) which gives value of 94.46 cm added by 4.5 cm as the allowance which gives 98.96 cm is chosen as the height of the standing ironing table. The 5th percentile of both thigh thickness (TP) and popliteal height (TPO) which gives value of 46.59 cm added by 4.5 cm as the allowance which gives 51.09 cm is chosen as height of sitting ironing table.

While figure 3 shows the top view of the ironing board design. The 95th percentile of arm's reach (JT) which gives value of 83.19 cm added by 4.5 cm as the allowance which gives 87.69 cm is chosen as the length of the sitting ironing table.

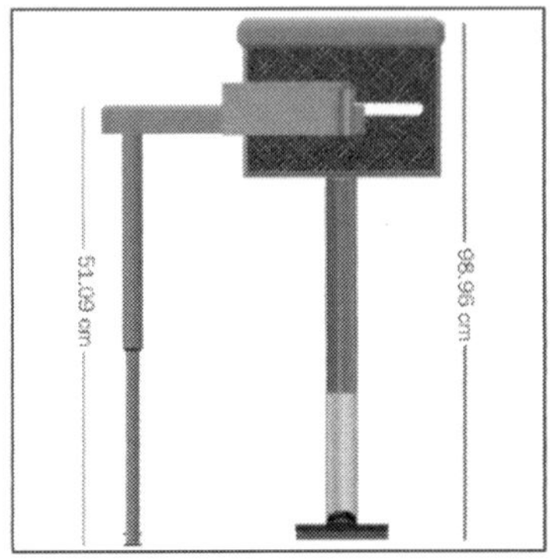


**Fig. 2**. Side view of ironing board design

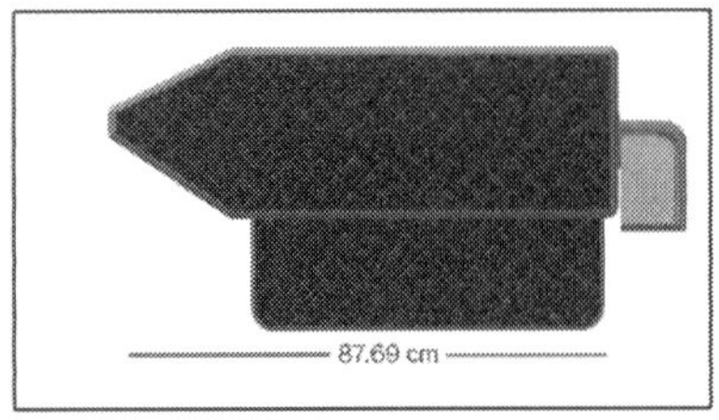


**Fig. 3**. Top view of ironing board design

Figure 4 below is the overall view of the ironing board design while the sitting position table is opened.

**Fig. 4**. Ironing board design isometric view

### 3.3. Usability Testing

In order to test the usability aspects in the design, usability testing is conducted. In usability testing, there are there are 3 independent variables (Usefulness, Ease of Use (X2), Ease of Learning) and 1 dependent variable (Satisfaction) which is assessed by USE questionnaires through the 50 respondents. After the data is collected, validity and reliability test should be performed first before analyze the usability of ironing board design. Based on the reliability test with 5% confidence interval, it shown that all of the data are valid and reliable and can

https://doi.org/10.1051/matecconf/201815401072

be continued. The table below shows the result of usability test.

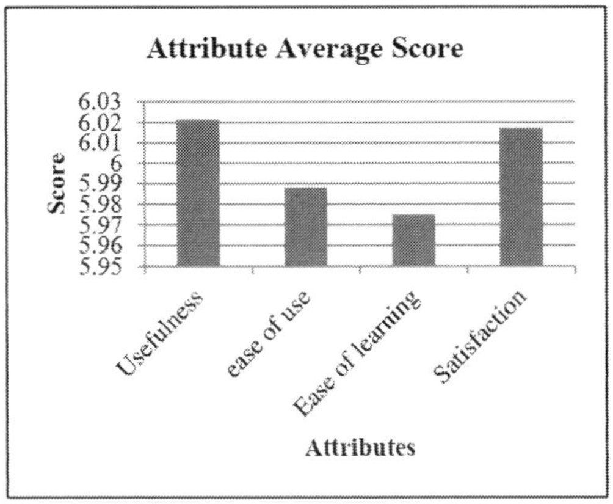


**Fig. 5.** Usability test result

Based on the result above, it can be found that Usefulness and Ease of Use variable have significant effect on the variable Satisfaction, while Ease of Learning did not affect satisfaction variables. It can be caused due to the product being tested is a product with a new innovation that is still unfamiliar to the user. However, looking at the average value of the Ease of Learning variable which is 5,975 out of a scale of 1 to 6. This shows that the variable ease of learning can be tolerated because of the high level and the support level of satisfaction in the use of the product [17].

## 4 Conclusion

Most of the women workers in laundry business sector felt low back pain while ironing in a long period, it can lead to musculoskeletal disorders due to the static position in ironing activity. Innovative design of ironing table board is designed based on customer's feelings by using Kansei Engineering method. The Kansei words used as the specification of the design are comfort, secure, practical, attractive, sturdy, durable and multifunction. The recommendation product dimensions are 98.96 cm as the height of the standing ironing table, 51.09 cm as the height of sitting ironing table, and 87.69 cm as the length of the ironing table. USE questionnaire is used in order to test the usability aspects in the design with Usefulness, Ease of use, Ease of learning, and Satisfaction as the attributes.

## References

1. P.M. Dewi, *JEKT*, 119-124 (2012)
2. H. Ware, *Australian National University* (1991)
3. N. Artini, Handayani, *Piramida*, 9-15 (2009)
4. A.D. Sari, M.R. Suryoputro, M.D. Pramaningtyas, P.S. Putra, S.B. Maulidyawati, *IEEE Int. Conf. on Know. Eng. and Appl.*, 1-6 (2016)
5. OHSAH, *Occ. Health Safety Agency for Healthcare in BC* (2003)
6. G.C. Angkoso, *Unpublished Thesis, Faculty of Medicine and Public Health Islamic State University Syarif Hidayatullah* (2012)
7. I.W. Ekawati, *Proc. Int. Sem. Workshop on Pub. Health Act.* (2015)
8. I. Dinata, N. Adiputra, I. Adiatmika. The Indonesian J. of Erg. 30-40 (2015)
9. Setyawati. *Unpublished Thesis, Universitas Pembangunan Nasional Veteran* (2003)
10. M. Nagamachi, App. Erg. 289–294 (2002)
11. M. Nagamachi, M. Tachikawa, N. Imanishi, T. Ishizawa, S. Yano, *Linköping Elect. Conf. Proc.* 234-254 (2008)
12. M. Nagamachi, *CRC Press* (2011)
13. N.K. Chuan, A. Sivaji, M.M. Shahimin, N. Saad., *Procedia—Soc. and Beh. Sci.***97**, 707–714 (2013)
14. H. Soewardi, M.I. Nasution, *IOP Conf. Series: Mat. Sci. Eng.* 165-171 (2016)
15. Corlett. *Static Muscle Loading and the Evaluation of Posture, Taylor & Francis* (1992)
16. F. Tayyari, J.L. Smith, *Occupational Ergonomics; Principles and applications, Kluwer Academic Publishers* (1997)
17. V. Sahfitri, M. Ulfa, *Proceeding SNaPP2014 Sci., Tech. and Health 373* (2014)
18. F.I. Rahmillah, R.N. Kartika, *IOP Conf. Series: Mat. Sci. Eng.* **215** (2017)

# Assessing university's sustainability programs from the perspective of university students: a gap analysis

*M. Mujiya* Ulkhaq*, *Wismar R.* Wijayanti, *Rahmaningtyas* Wiganingrum, *Wini R.* Dewi, and *Fahmi* Ardi

Department of Industrial Engineering, Faculty of Engineering, Diponegoro University, Semarang - Indonesia

**Abstract.** The role of universities in promoting sustainability has outstretched over the past decades as a result of abundant declarations related to the need for sustainability in higher education. As consequences, several universities have integrated sustainability into their curricula, research, programs, projects, partnerships, and assessments. However, despite the need for sustainable development in today's highly competitive global network, many university members especially students are still unaware or ignorant of sustainability practices. This research tried to analyse the gap between what the university has done in campaigning for sustainability and what the students have perceived. A case study is conducted in Diponegoro University, Indonesia. Twelve indicators of four variables, i.e., community outreach; sustainability commitment and monitoring; waste and energy; as well as land use and planning are used to accomplish this research. Three hundred and fortyfive students spreading from all faculties participated in the survey. The result shows that an average gap of –0.38 is revealed; indicating that the respondents did not consider a significant impact on the sustainability programs. In addition, more than 50% of the respondents were not familiar with the sustainability programs. It is suggested that the university has to put more attention in fostering sustainability to its biggest stakeholders.

## 1 Introduction

Universities recently become a main player in fostering sustainability through knowledge dissemination. The role has grown progressively across countries since the successful accomplishment of Stockholm Declaration on the Human Environment in 1972 [1] which has been acknowledged as the first reference about sustainability in higher education. Since then, there are numerous conferences, and declarations about the role of higher education in promoting sustainability. Some of them are Belgrade Charter in 1975, Rio de Janeiro's Earth Summit in 1992, Kyoto Declaration in 1993, The Lüneburg Declaration in 2001, Lucerne Declaration in 2007, and Tokyo Declaration in 2009 (see [2] for the detail). In consequence, the concept of sustainability has been integrated into the universities' curricula, researches, operations, partnerships, as well as their assessment and reporting [3 - 6].

These initiatives should be based on the universities'' visions along with their managements' concept of sustainability. The rationale is that the concept of sustainability presents diverging interpretations according to the actors that are inferred [7]. Some have a master plan, environmental plan, guidelines, or statement about being sustainable campus [6]; some by signing of national or international declarations [8]; and others create individual institutional policies or framework to achieve campus sustainability, such as ISO 14001 [9 - 11], eco-management and audit scheme (EMAS) [12], green building initiative [13], environmental impact assessment [14], graphical assessment of sustainability in universities (GASU) [15, 16], sustainability tracking, assessment and rating system (STARS) [17, 18], and Alshuwaikat and Abubakar's framework [2, 19, 20].

The sustainability goals, in essence, will not be achieved without cooperation and participation from all stakeholders, i.e., the students, the faculties and staff, funding agencies, as well as the communities [21, 22]. The students, which can be considered as the biggest stakeholders, play a critical role in a bottom-up approach to promote awareness across the institution and urge interactions among various stakeholders [8, 23]. However, despite the tangible and intangible benefits gained from the sustainability practices, many university members, especially the students are still unaware of it [24].

The objective of this research is to analyse the gap between what are the sustainability programs conducted by the universities and what the students have perceived about the programs. Twelve indicators from four variables for the key factors of a sustainable university from the perspective of students by [24] are used, i.e., community outreach, sustainability commitment and monitoring, waste and energy, and land use and planning. A case study to exhibit the applicability of the methods was conducted in Diponegoro University,

* Corresponding author: ulkhaq@live.undip.ac.id

which is located in Semarang, Central Java Province, Indonesia.

**Table 1** Variables and indicators of assessing universities' sustainability programs

| Variable | Indicators | |
|---|---|---|
| Community outreach | S1 | The university engages in community outreach programs that benefit the local environment |
| | S2 | The university has created green community centres to benefit the local environment |
| | S3 | The university has created partnerships with government, non-governmental organizations, and industry working toward sustainability |
| Sustainability commitment & monitoring | S4 | The university has created a written statement of their commitment to sustainability |
| | S5 | Each department within the sustainable university has created its own written statement of their commitment to sustainability |
| | S6 | Sustainability audits are performed on the surrounding community |
| | S7 | Regular sustainability audits are performed on campus |
| Waste and energy | S8 | The university reuses campus waste |
| | S9 | The university uses renewable and safe energy sources |
| | S10 | The university emphasizes sustainability through support services |
| Land use and planning | S11 | Campus building planning is a top priority in university sustainability issues |
| | S12 | Campus landuse is a top priority in university sustainability issues |

## 2 Research methods

Since there are limited researches addressing the issue of sustainability from the perspective of students, we employed the study by [24] to accomplish the objective of this research. There are four variables for the key factors of a sustainable university, i.e., community outreach, sustainability commitment and monitoring, waste and energy, as well as land use and planning. Each variable has their own indicators to assess the universities' sustainability programs.

Community outreach is the activities that engaged by the university outside campus. It consists of three indicators, such as: (i) the university engages in community outreach programs that benefit the local environment, (ii) the university has created green community centres to benefit the local environment, and (iii) the university has created partnerships with government, non-governmental organizations, and industry working toward sustainability. The second variable is sustainability commitment and monitoring. The university as long as each department supposed to create a written statement of its commitment to sustainability, as well as monitoring and audit are performed regularly. It consists of three indicators as well, i.e., (i) the university has created a written statement of their commitment to sustainability; (ii) each department within the sustainable university has created its own written statement of their commitment to sustainability, (iii) sustainability audits are performed on the surrounding community, and (iv) regular sustainability audits are performed on campus. The third one is waste and energy variable. It consists of three indicators, i.e., (i) the university reuses campus waste, (ii) the university uses renewable and safe energy sources, and (iii) the university emphasizes sustainability through support services. The last is land use and planning, which refers to the priority to both build sustainable campus building and use campus land for the sustainability practice. It consists of two indicators, i.e., (i) campus building planning is a top priority in university sustainability issues and (ii) campus landuse is a top priority in university sustainability issues. The total of twelve indicators for each variable are shown in Table 1.

To calculate the gap, there are required two group of respondents. The first group is the university faculties or staff that are associated with the sustainability programs while the second group is the university students. The gap for each indicator is calculated as follows:

$$Gap_i = SP_{Si} - SP_{Fi}, \tag{1}$$

where $SP_{Si}$ is the total score from the perspective of the university students of the sustainability programs for indicator $i$ and $SP_{Fi}$ is the total score of the sustainability programs for indicator $i$ from the perspective of the university faculties or staff. The $SP_{Si}$ and $SP_{Fi}$ can be calculated as follows:

$$SP_{Si} = \frac{\sum_{i=1}^{n_S} P_{ij}}{n_S}, \tag{2}$$

$$SP_{Fi} = \frac{\sum_{i=1}^{n_F} P_{ij}}{n_F}, \tag{3}$$

where $P_{ij}$ is the score obtained from individual $j$ with respect to the performance of sustainability programs on indicator $i$, $n_S$ is the number of the university students to be sampled, and $n_F$ refers to the number of the first group, i.e., the university faculties or staff to be questioned.

## 3 Case study: result and discussion

To exhibit the applicability of the methods, a case study was conducted in the Diponegoro University, Indonesia. As has been previously mentioned, there are two kind of groups to be investigated. The first consists of university faculties or staff that are associated with the university's sustainability programs. It can be the professors or the lecturers that are aware of the programs; or the university staff that monitor or maintain the performance of the programs. The second group is the university students that supposed to perceive the university's sustainability programs that have been designed and operated by the university.

**Table 2** Respondents' profile of the second group

| Variable | | Percentage |
|---|---|---|
| Gender | Male | 29 |
| | Female | 71 |
| Age | < 20 years | 37 |
| | 20 – 23 years | 62 |
| | > 23 years | 1 |
| Current level of the study | Diploma | 3 |
| | Bachelor/under-graduate | 95 |
| | Post-graduate | 2 |
| Faculty/ School | Law | 4 |
| | Economics and Business | 18 |
| | Engineering | 28 |
| | Medicine | 7 |
| | Animal and Agricultural Science | 4 |
| | Humanities | 4 |
| | Social Science and Political Science | 4 |
| | Science and Mathematics | 10 |
| | Public Health | 9 |
| | Fisheries and Marine Science | 5 |
| | Psychology | 4 |
| | Vocational School | 3 |
| Period of the study | < 1 year | 23 |
| | 1 – 2 years | 14 |
| | 2 – 3 years | 42 |
| | > 3 years | 21 |
| Have attended lectures about environment or engaged in social activities? | Yes | 78 |
| | No | 22 |

The potential participants of the second group were first approached and asked if they agreed to participate in the survey. Three hundred and forty five students spreading from all faculties participated in this survey, indicates the diversity for the purpose of the research. Profile of the respondents of the second group is shown in Table 2. The respondents are then questioned about the university's sustainability programs for each indicator. The score "0" is given when the programs do not comply at all with the statement (or the indicator); the score "1" is given when the statement is more or less fulfilled, but where there is still room for improvement; and finally when the situation in which the statement appears to be completely fulfilled the score or "2" is applied. The respondents also are able to answer "do not know" if he or she has no idea about the statement.

The previous procedure was also applied for the first group of the respondents. There are only three respondents participated in the survey, i.e., two respondents are the university staff that monitor the university's environment programs and one is the lecturer of the Department of Environmental Engineering.

**Table 3** Case study result

| Indicators | $SP_{Si}$ | $SP_{Fi}$ | $Gap_i$ |
|---|---|---|---|
| S1 | 1.114 | 1.667 | –0.552 |
| S2 | 1.037 | 1.000 | 0.037 |
| S3 | 1.070 | 2.000 | –0.930 |
| S4 | 1.174 | 1.333 | –0.159 |
| S5 | 0.982 | 1.667 | –0.684 |
| S6 | 0.854 | 1.667 | –0.813 |
| S7 | 0.682 | 1.667 | –0.985 |
| S8 | 0.566 | 1.000 | –0.434 |
| S9 | 0.664 | 1.000 | –0.336 |
| S10 | 0.991 | 1.000 | –0.009 |
| S11 | 1.000 | 1.000 | 0 |
| S12 | 1.300 | 1.000 | 0.300 |
| Average | 0.953 | 1.333 | –0.380 |

The gap for each indicator is then calculated using (1) by calculating (2) and (3) at first. The result is shown in Table 3. Note that we exclude the "do not know" answers coming from the respondents of the first group. Surprisingly, there are 140 "do not know" answers or more than 50% of the total respondents, indicates that the students are not familiar with the university's sustainability programs. We strongly recommend that the university has to put more attention in promoting its sustainability programs for its biggest stakeholders.

### 3.1 Community outreach

In the community outreach variable, S3 (the university has created partnerships with government, non-governmental organizations, and industry working toward sustainability) becomes the indicator that has the highest gap, i.e., –0.930. The students consider that the results gained from partnership created by Diponegoro University cannot be felt properly. However, the university staff claim that the indicator has been fulfilled and realized well. For example, there are several communities that are engaged into the environmental issues. Some of them are "Atmosphere", "Environmental Study Group", "Oxygen 16", "Kesemat", and "Diponegoro Social Care." They have several programs such as *green competition*, *youth dialog*, and *youth in action* by "Atmosphere"; *tarakura home method composting* by "Environmental Study Group"; *campus cleaning* and *composting* by "Oxygen 16"; *training for creating biopores*; *mangrove conservation* in Semarang coastal area by "Kesemat"; and *garbage bank* and *plastic and paper campaign* by "Diponegoro Social Care" [2].

The University also has established partnership with the French government in mangrove forest conservation. Another partnership that has been created is with the Indonesian government, such as the Ministry of Environment and the Ministry of Forestry. The example of partnership with the first party is a preservation program of Rawa Pening, a swamp located near Semarang City, and example partnership with the later party is trees planting program.

However, due to the existence of the gap, the university is suggested to give a promotion or socialization about the partnership with external parties, such as government, non-governmental organizations, as well as industries. In addition, the university is strongly recommended to financially support the local communities and local environment related to sustainability.

It is fortunate that in this variable, there exist one indicator that has positive gap, i.e., S2 (the university has created green community centres to benefit the local environment). It means that the students felt that the green community centres established surrounding Diponegoro University affect the local environment.

### 3.2 Sustainability commitment and monitoring

Indicators from the smallest to the largest gap in sustainability commitment and monitoring variable are S4, S5, S6, and S7, i.e., S4 with –0.159; S5 with –0.684; S6 with –0.813; and S7 with –0.985. Note that all indicators have negative gaps, indicates that sustainability programs related to university's commitment and monitoring are not felt properly by the students. Therefore, it is strongly recommended that the university does socialization about its commitment to sustainability to the students, even though the university has stated it clearly. Moreover, regular audits and monitoring related to sustainability practices have to be performed transparently.

### 3.3 Waste and energy

Waste or garbage separation as a part of waste management has been completely done by the university. The waste is separated into two categories: organic and inorganic which is thrown away into different trash bin placed in various corners of the university. While organic waste is fully composted, the inorganic waste could be recycled into several recycled products, such as lumber, bottle, rope, garbage cans, scrapers, etc. However, only 25% to 50% of wastes that have been recycled. This performance has to be improved significantly in order to gain the number of 90% to be recycled.

Diponegoro University also has sewage treatment facility that is operated by the Department of Environmental Engineering and the Department of Chemical Engineering. In term of energy utilization, Diponegoro University has replaced conventional energy appliances with energy efficient appliances by 20% to 40%.

Despite of the sustainability programs abovementioned by the university staff, the students are not familiar with those programs; hence the gap does exist. All indicator have negative gaps: S8 (the university reuses campus waste) is –0.434; S9 (the university uses renewable and safe energy sources) is –0.336; S10 (the university emphasizes sustainability through support services) is –0.009. Therefore, it is recommended to give a socialization to the students and endorse students so that they can enthusiastically participate into those kind of activities.

### 3.4 Land use and planning

It is arguable that some of the buildings in the university have been designed with the concept of energy efficient, such as the building in the Department of Environmental Engineering and in the Department of Industrial Engineering. The rooms are designed so that one big room is partitioned into several small chambers. Only one central air conditioner and one lighting are used for more those partitioned rooms. It could minimize the energy consumption.

The priority to build a building is not based on a sustainability needs. The university faculties stated that the university will build a campus depend on the necessity, such as the number of existing students. The university has a master plan to integrate some classes from different departments into one shared lecturer class. In contrast, the students consider that each department has to have each own campus.

In term of land use, Diponegoro University has a deer park that is built to preserve the deer. In front of deer park, there is an open space for preserving endangered plants. Forest campus and agrotechnopark are new programs which are still in the development stage.

## 4 Conclusion

This study tried to analyse the gap between what the students of the university perceive about sustainability programs and what the performance of the programs. This is motivated by the fact that the students, as the biggest stakeholders of the university are mostly unaware or ignorant of the sustainability programs managed by their university. This is an irony since the sustainability in this today's global competitive network plays an important role.

Four variables to measure university's sustainability practices which consist of total twelve indicators by [24] are used in this research. The university faculties or staff and the students are required to score what they perceive about the performance of the university's sustainability programs. The gap is calculated by subtracting the scores obtained from the university faculties or staff into the scores obtained from the students.

A case study has been conducted in the Diponegoro University, Indonesia, to apply the proposed methods. Three university faculties and three hundred and forty five students participated in the survey. Based on the case study result, from twelve indicators, only one indicator that has positive gap (S12), one has zero gap (S11), and the rest have negative gaps; while the average gap is –0.380 (see Table 3 for the detail). The negative score shows that the sustainability programs are not well known by the students. Moreover, more than 50% of the students are not aware of the university's sustainability programs. It indicates that the university has to put more attention in promoting sustainability programs to its students.

In addition, we strongly recommend that the university has to do socialization to the students about the sustainability programs so that the students can excitedly participated into the programs.

The authors thank to the university faculties, staff, and students that have supported in this research.

## References

1. United Nations, *Report of the United Nations Conference on the Human Environment* (UN, 1973)
2. M.M. Ulkhaq, P.I. Prayogo, M. Firmansyah, D. Agustina, Int. J. Infor. Educ. Tech **6**, 8 (2016)
3. M. Brinkhurst, P. Rose, G. Maurice, J.D. Ackerman, Int. J. Sust. Higher Educ. **12**, 4 (2011)
4. W. Calder, R.M. Clugston, Plan. Higher Educ. **31**, 3 (2003)
5. A.D. Cortese, Plan. Higher Educ. **31**, 3 (2003)
6. L. Velazquez, N. Munguia, A. Platt, J. Taddei, J. Cleaner Prod. **14**, 9-11 (2006)
7. N. Lourdel, N. Gondran, V. Laforest, C. Brodhag, Int. J. Sust. Higher Educ. **6**, 3 (2005)
8. T.S.A. Wright, Higher Educ. Policy, **15**, 2 (2002)
9. R.M. Fisher, Int. J. Sust. Higher Educ. **4**, 2 (2003)
10. T.J. Price, Int. J. Sust. Higher Educ. **6**, 2 (2005)
11. N.A. Setyorini, M.M. Ulkhaq, D.R. Rasyida, P.R. Setiowati, R. Trianto, Int. J. Adv. Agric. Env. Engg. **3**, 2 (2016)
12. B. Delakowitz, A. Hoffmann, Int. J. Sust. Higher Educ. **1**, 1 (2000)
13. K.A. Owens, A. Halfacre-Hitchcock, Int. J. Sust. Higher Educ.**7**, 2 (2006)
14. T.B. Ramosa, T. Cecílio, J.J. de Melo, J. Cleaner Prod. **16**, 5 (2008)
15. R. Lozano, J. of Cleaner Prod. **14**, 9 (2006)
16. R. Lozano, J. Sust. Higher Educ. **12**, 1 (2011)
17. M. Urbanski, W.L. Filho, Env. Dev. Sust. **17**, 2 (2015)
18. A. Wigmore, M. Ruiz, Ramon Llull J. App. Ethics **1**, 1 (2010)
19. H. M. Alshuwaikat, I. Abubakar, J. Cleaner Prod. **16**, 16 (2008)
20. R. de Castro, C.J.C. Jabbour, J. Cleaner Prod. **61** (2013)
21. H. van Weenen, Int. J. Sust. Higher Educ. **1**, 1 (2000)
22. T. Wright, Int. J. Sust. Higher Educ. **11**, 1 (2010)
23. P. Christensen, M. Thrane, T. Herreborg Jørgensen, M. Lehmann, Int. J. Sust. Higher Educ. **10**, 1 (2009)
24. M. Nejati, M. Nejati, J. Cleaner Prod. **48** (2013)

# The effect of clothing fit and material of women's Islamic sportswear on physiological and subjective responses during exercise in warm and humid environment

*Astrid Wahyu Adventri* Wibowo*, *Titis* Wijayanto*, *Watri* Widyastuti, and *Muhammad Kusumawan* Herliansyah

Gadjah Mada University, Departement of Mechanical and Industrial Engineering, Jl. Grafika 2, Yogyakarta, Indonesia

**Abstract.** The purpose of this study was to investigate the effects of clothing fit and material of Islamic sportswear for female on physiological responses and body heat balance during exercise in warm and humid environment. Twelve healthy female students (20.3±0.4 years) exercised wearing four types of women's Islamic sportswear comprised of two level of clothing fit: loose-fit and tight-fit, and two types of material for sportswear: cotton and polyester on four separate occasions, and in random order. They performed a 30-min treadmill exercise at an intensity of 70% $HR_{max}$ and then rested on a chair for 20 min for recovery in a chamber set at an ambient temperature of 34°C and relative humidity of 80%. The results showed that clothing fit did not significantly affect physiological and subjective responses, but clothing material did; sportswear made of cotton resulted in a higher increase of tympanic temperature during exercise and recovery compared to that made of polyester ($P<0.05$). In addition, sportswear made of cotton have lower conductive and evaporative heat loss than sportswear made of polyester ($P<0.05$). Clothing fit only had significant effect on conductive heat loss; that is tight-fit sportswear showed greater conductive heat loss than loose-fit one ($P <0.05$). Regarding subjective responses, participants reported lower thermal comfort, greater thermal sensation, and greater skin wetness sensation when performing exercise wearing tight-fit sportswear made of polyester.

## 1 Introduction

In 2012, sales of sportswear in global markets exceeded $ 244 billion [1]. The sale included 46 countries around the world who estimated that sales of clothing and sports shoes would grow more than 7.5% in 2012 and continued to increase until 2017 with additional sales of $ 55 billion. This increasing demand should be balanced with ergonomic sportswear design in an attempt to optimize an athlete's performance.

Sportswear is generally designed with a certain level of clothing fit to give freedom of movement. The level of clothing fit plays an important role in providing freedom of movement and comfort [2]. The loose-fit sportswear generally has a larger shirt size thus it widens the distance between the skin and the shirt that facilitating the occurrence of air circulation and heat release through convection. Therefore, wearing loose-fit clothing may decrease physiological strain during exercise through convective heat lose [3]. However, loose-fit sportswear can block sweat evaporation as a result of an increase of thermal insulation [4]

In addition to the clothing fit, the clothing material of a sportswear also plays important role. Sportswear made of fabric that has large fibber pores will tend to easily release heat through the pores [5]. Cotton fabric has a characteristic to absorb sweat, wool has the characteristic to regulate body temperature in all conditions (hot or cold) but not suitable for sportswear because this type of cloth is not light. Silk has a very compact characteristic but it is expensive. polyester (spandex) has a flexible characteristic that follows body shape so that it can move freely, and so on [2]. Among these fabrics, cotton and polyester are commonly used as clothing material for sportswear. In comparison with cotton, exercise wearing the polyester produced a greater sweating efficiency and less clothing regain. However, thermo-physiological and subjective sensations during the intermittent high-intensity exercise were similar for both fabrics. In addition, the thermal and rating of shivering/sweating sensation were lower after exercise in the warm and humid environment in subjects wearing polyester than when wearing cotton fabric [6]. Based on this explanation, the selection of material types highly contributes to the comfort of its wearer during exercise.

There has been a less number of study on encapsulated women's clothing, especially for Islamic sportswear. Al Ajmi et al. [7] may be the first who measured the value of thermal insulation of clothing (Clo) of some of the commonly used clothing in the Arabian Gulf region. However, they did not include the investigation of thermo-physiological effect of the clothing. A study by

* Corresponding author: twijaya@ugm.ac.id; astrid.w.adventri@gmail.com

Davis et al. [8] investigated the effects of women Islamic sportswear that encapsulated subjects whole body except hands and face on physiological responses. They reported that subjects felt thermally discomfort when wearing the Islamic sportswear, but they did not find any significant different in physiological responses and dehydration level in comparison with traditional soccer outfit. Another study by Wijayanto et al. [3] compared tight-fit and loose-fit Islamic sportswear made of polyester worn during exercise. They found that wearing tight-fit sportswear resulted in greater thermal strain than wearing the loose-fit one [3].

Thus far, there were no study investigating the clothing material for sportswear, especially for sportswear that encapsulate whole body except face and hands. From the aforementioned, this study investigated the effect of clothing fit and material used in encapsulated women's sportswear on physiological and subjective response during exercise in hot and humid environment.

## 2 Methods

### 2.1 Participants

Twelve healthy females with a mean (±SD) age of 20.3 years (±0.4), body mass of 52.7 kg (±0.95), height of 1.59 m (±1.9), and body surface are of 1.53 m$^2$ (±0.02) participated in this study. They were not pregnant or not in their menstruation period during the experiment period. Prior to participation, the details and risks associated with the experiment protocol were explained. Written informed consent was obtained in accordance with the local institutional review board policy for use of human subjects.

### 2.2 Clothing Ensemble

Four types of Islamic sportswear for female comprised of two level of clothing fit: loose-fit and tight-fit, and two types of material for sportswear: cotton and polyester were selected for the main experiment. The garments covered and encapsulate the entire participant's body, except face and hands.

The tight-fit garment was designed with the clothing tightly fit to the participant's body (air gap between the skin and garment was less than 10 mm) and head cover prepared to cover and compressed the head and the neck. The loose-fit garment was designed with the gap between the skin and garment was between 14 and 21 mm and head cover prepared to cover head, neck, and chest with compression on the head only, so as not to outline the shape of the body. These sportswears were made of two types of material: one made of 93% polyester and 7% elastin (PES), the other of 100% cotton fabric (CT). The clothing ensembles in this study are shown in Figure 1.

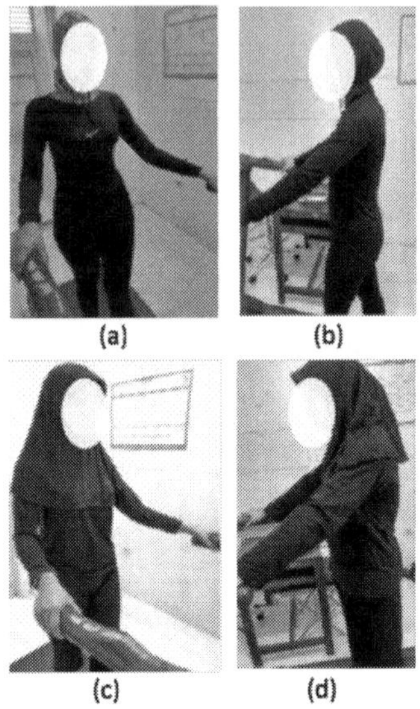


**Fig. 1.** **(a)** Clothing ensemble for tight-fit polyester, **(b)** tight-fit cotton, **(c)** loose-fit polyester, **(d)** loose-fit cotton.

### 2.3 Measurements

Tympanic temperature ($T_{ty}$) was monitored every second using an infrared tympanic temperature sensor (CE Thermo, Nipro Corporation, JAPAN). The $T_{ty}$ probe was gently introduced into the ear canal, so that the probe correctly positioned towards the tympanic membrane. Then the outer ear was cover throughout the whole exposure.

Body heat balance was estimated from heat gain and heat loss at the end of exercise. Heat gain was calculated as external work subtracted from metabolic heat production (M-W), whereas heat loss was estimated from conductive heat loss ($K_{sk}$), sensible heat loss (C+R), and evaporative heat loss ($E_{sk}$). Body heat storage (S) was estimated from the product of the body mass (BM), human body specific heat ($c_{body}$) and the rate of increase in mean body temperature ($\Delta T_{body} \cdot \Delta t^{-1}$) from the baseline to the end of exercise ($S = BM \cdot c_{body} \cdot \Delta T_{body} \cdot \Delta t^{-1}$).

For subjective thermal measurements, each participant gave a verbal evaluation of their thermal condition using nine points rating thermal sensation (TSV, 4: very hot, 3: hot, 2: warm, 1: slightly warm, 0: neither, -1: slightly cool, -2: cool, -3: cold, and -4: very cold), seven points of thermal comfort (TCV, 3: very comfortable, 2: comfortable, 1: slightly comfortable, 0: neither, -1: slightly uncomfortable, -2: uncomfortable, and -3: very uncomfortable), Rating of Perceived Exertion (RPE) based on Borg's scale, and skin wetness sensation (SW) [9] every 10-min.

### 2.4 Data analysis

Data are presented in mean ± standard deviation (SD). Prior to the main statistical analysis, data were tested for normal distribution using the Kolmogorov-Smirnov test, and all scale data were normally distributed. Data from time-course measurements were evaluated using a repeated measure analysis of variance (RM ANOVA). Multiple paired comparison with Bonferroni correction was used to determine to determine significant difference between conditions. The nonparametric Wilcoxon signed-rank test was performed to compare the changes in subjective ratings of perceptions (TSV, TCV, RPE, and SW). Significant level was established at $P<0.05$.

# 3 Results

## 3.1 Tympanic Temperature

Figure 2 depicts the changes in $T_{ty}$ ($\Delta T_{ty}$) at rest, during exercise and recovery when the participants wore tight-fit polyester, loose-fit polyester, tight-fit cotton, and loose-fit cotton. $\Delta T_{ty}$ was significantly higher in comparison between tight-fit cotton compared to the and tight-fit polyester condition ($P<0.05$) and was also significantly higher in comparison between loose-fit polyester compared to the tight polyester condition at the end of exercise($P<0.05$). During the recovery period, $\Delta T_{ty}$ was significantly higher for the tight cotton compared to the tight polyester condition, significantly higher for the loose polyester compared to the tight polyester condition, and also significantly higher for the loose cotton compared to the tight polyester condition ($P<0.05$).

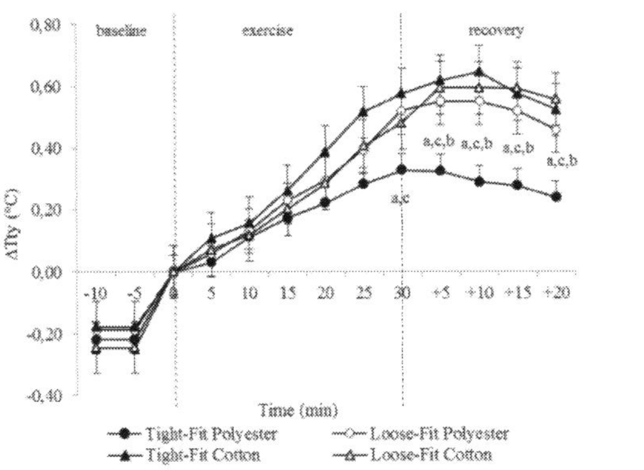


**Fig. 2.** Change of tympanic temperature ($\Delta T_{ty}$) during baseline, exercise, and recovery. [a] shows significant between tight-fit polyester and tight-fit cotton at $P<0.05$; [b] shows significant between tight-fit polyester and loose-fit cotton at $P<0.05$; [c] shows significant between tight-fit polyester and loose-fit polyester at $P<0.05$.

## 3.2 Thermal Balance

Table 1 shows body heat balance at the end of exercise phase. $E_{sk}$ was significantly higher when wearing polyester, either tight-fit or loose-fit, than when wearing tight-fit cotton during exercise ($P<0.05$). $K_{sk}$ was significantly higher when wearing tight-fit garment than when wearing loose-fit garment during exercise ($P<0.05$). Sensible heat loss (C+R) was significantly higher when wearing tight-fit cotton than when wearing loose-fit cotton ($P<0.05$). There were no significant differences in M-W and S among conditions at the end of exercise.

**Table 1.** Body heat balance at the end of exercise (mean±SD)

| | Tight-fit | | Loose-fit | |
|---|---|---|---|---|
| | **Polyester** | **Cotton** | **Polyester** | **Cotton** |
| M-W ($Wm^{-2}$) | 150.6±18.1 | 127.7±22.1 | 154.7±20.8 | 145.6±24.4 |
| $E_{sk}$ ($Wm^{-2}$) | $60.5\pm1.9^{c,d}$ | $19.6\pm0.4^{d}$ | $59.6\pm1.7^{c,d}$ | 18.3±0.5 |
| $K_{sk}$ ($Wm^{-2}$) | $36.9\pm4.6^{b,c,d}$ | $19.9\pm3.8^{d}$ | $2.3\pm0.4^{c}$ | 1.6±0.2 |
| C+R ($Wm^{-2}$) | 15.7±1.6 | $18.3\pm1.7^{d}$ | 14.5±1.2 | 14.3±1.3 |
| S ($Wm^{-2}$) | 123.2±15.8 | 152.8±17.5 | 140.5±15.7 | 147.9±14.1 |

a = significantly different to tight-fit polyester at $P<0.05$
b = significantly different to loose-fit polyester at $P<0.05$
c = significantly different to tight-fit cotton at $P<0.05$
d = significantly different to loose-fit cotton at $P<0.05$

## 3.3. Subjective Responses

Table 2 depicts the subjective responses during exercise. Participants reported greater thermal discomfort (TCV) during exercise in tight-fit polyester than in loose-fit polyester ($P<0.05$). Thermal sensations (TSV) were significantly warmer when wearing tight-fit polyester than when wearing loose-fit polyester or tight-fit cotton garment during exercise ($P<0.05$).

Perceived exertion during exercise was significantly harder when wearing tight-fit polyester than when wearing tight-fit cotton ($P<0.05$) and was also harder when wearing tight-fit cotton than when wearing loose-fit cotton ($P<0.05$). Finally, skin wetness sensation (SW) was greater in tight-fit polyester than when wearing tight-fit cotton garments during recovery ($P<0.05$). Skin wetness was sensed to be significantly worse when wearing tight polyester garment than when wearing loose polyester or loose cotton garment during exercise ($P<0.05$).

**Table 2.** Subjective responses (mean±SD)

| | Tight-fit | | Loose-fit | |
|---|---|---|---|---|
| | **Polyester** | **Cotton** | **Polyester** | **Cotton** |
| TCV | $-2\pm0.3^{b}$ | -1±0.4 | -1±0.4 | -1±0.4 |
| TSV | $3\pm0.3^{c,b}$ | 2±0.3 | 2±0.3 | 3±0.2 |
| RPE | $12\pm0.8^{c,d}$ | 11±0.7 | $12\pm0.7^{d}$ | 11±0.8 |
| SW | $76.6\pm6.7^{b,d}$ | 72.4±7.5 | 63.4±7.9 | 69.3±7.5 |

a = significantly different to tight-fit polyester at $P<0.05$
b = significantly different to loose-fit polyester at $P<0.05$
c = significantly different to tight-fit cotton at $P<0.05$
d = significantly different to loose-fit cotton at $P<0.05$

# 4 Discussion

Wearing Islamic Sportswear made of cotton resulted in a higher increase of tympanic temperature during exercise than wearing the other made of polyester. This may be due to water-transfer properties of polyester. The water permeability of polyester (528.1 mm/s) is higher than cotton (327.1 mm/s). Thus, polyester absorbs less sweat but spread and evaporate sweat faster than cotton [6]. Evaporative heat loss when using polyester fabric is higher than when using cotton fabric. It is possible that higher evaporative heat loss when wearing polyester fabric released heat from within the body more rapidly. As a result, a smaller increase of tympanic temperature was observed when wearing polyester fabric during exercise in comparison with other of cotton.

In this study, thermal sensation, thermal comfort, and skin wetness were significantly higher when wearing tight-fit garment, in comparison with loose-fit garment. These findings were in agreement to [3]; wearing tight-fit garment were sensed to be hotter and produced more thermal discomfort compared to wearing the loose-fit one. Wearing tight-fit clothing was reported to be uncomfortable because direct contact between the fabric and the skin. There were some evidences that thermal sensation and clothing sensation are associated with skin wetness [10, 11]. The more widely moisture is distributed over the skin, the stronger the discomfort sensation [12, 13]. In the present study, the participants reported higher thermal sensation, greater thermal discomfort, and more wetness sensation when wearing the tight-fit than the loose-fit clothing. A study suggested that increase contact of the skin with the clothing in tight-fit clothing increase dampness and lead to negative impact of thermal comfort [12]. In addition, more sensation of skin wetness would affect the weight of the garment, impair heat dissipation from the skin, and post-exercise impairment evaporative cooling [14, 15]. Compared with exercise wearing the cotton, exercise wearing the polyester fabric produced a greater sweating efficiency and less clothing regain (i.e., less sweat retention), but subjective sensations during the intermittent high-intensity exercise were similar for both fabrics [6].

## 5 Conclusion

In conclusion, our findings suggest tight-fit Islamic sportswear for women that made from polyester had no detrimental effects on physiological responses and had less body heat storage during performing moderate exercise in hot environment. However, it should be considered that wearing tight-fit sportswear that encapsulates whole body except hands and face may produce greater thermal discomfort during exercise. Further studies investigating the relationship between physiological indices and subjective responses during exercise wearing this type of sportswear is necessary to gain more insight in designing Islamic sportswear for women.

## References

1. R. Shishoo, Introduction to textiles in sport, *Textile for sportswear* (Woodhead Publishing, Sweden, 2015)
2. J. McCann, Environmentally conscious fabric selection in sportswear design, *Textile for sportswear* (Woodhead Publishing, United Kingdom, 2015)
3. T. Wijayanto, M.F. Fathna, Y. Tochihara, Physiological responses during exercise wearing women's islamic sportwear in warm humid environment, *The fifth international conference on human –environment system* (2016)
4. Y.S. Chen, J. Fan, X. Qian, W. Zhang, Effect on garment fit on thermal insulation and evaporative resistance, *Textile research journal,* **74**, 742 (2004)
5. R.M. Rossi, Cold weather sports clothing, *Textile for sportswear* (Woodhead Publishing, Switzerland, 2015)
6. M. Brazaitis, S. Kamandulis, A. Skurvydas, G. Havenith, The effect of two kinds of t-shirt on physiological and psychological thermal responses during exercise and recovery, *Applied ergiomic,* **42,** 46 (2010)
7. F.F. Al-ajmi, D.L. Loveday, K.H. Bedwell, G. Havenith, Thermal insulation and clothing area factors of typical Arabian gulf clothing ensembles for males and females: measurements using thermal manikins, *Applied ergonomics*, **39**, 407 (2008)
8. J.K. Davis, P.A. Bishop, Y. Zhang, J.M. Green, C. Casaru, K.D. Orrick, M.C. Smith, M.T. Richardson, R.E. Schumacker, Fluid balance, thermal stress, Nd post exercise response in women's islamic athletic clothing, *European journal of applied physiology,* **112**, 725 (2012)
9. J.Y. Lee, K. Nakao, Y. Tochihara, Validity of perceived skin wettedness mapping to evaluate heat strain, *Eur j appl physiol,* **111,** 2581 (2011)
10. D.D. Pascoe, T.A. Bellingar, B.S. McCluskey, Clothing and exercise: II. Influence of clothing, during exercise/work in environment extremes, *Sports med*, **18**, 94 (1994)
11. Y. Aoyagi, T.M. McLellan, R.J. Shephard, Effects of endurance training and heat acclimation on psychological strain in exercise men wearing protective clothing, *Ergonomics,* **41**, 328 (1998)
12. A.R. Gwosdow, J.C. Berglund, J.A.J. Stolwijck, Skin friction and fabric sensation in neutral and warm environment, *Textile res. j,* **56**, 754 (1986)
13. D.M. Scheurell, S.M. Spivak, N.R.S. Hollies, Dynamic surface wetness of fabric in relation to clothing comfort, *Textile res. j,* **55**, 392 (1985)
14. T.P. Gavis, Clothing and thermoregulation during exercise, *Sports med*, **33**, 941 (2003)
15. X.Q. Dai, R. Imamura, G.L. Liu, F.P. Zhou, Effect of moisture transport on microclimate under t-shirt, *Eur j. appl physiol,* **104**, 337 (2008)

https://doi.org/10.1051/matecconf/201815401075

# Quantitative analytical hierarchy process to marketing store location selection

*Harwati** and *Intan* Utami

Industrial Engineering Department, Universitas Islam Indonesia, 55501, Yogyakarta

**Abstract.** The selection of Store to market the product is belong to Multi Criteria Decision Making problem. The criteria used have conflict of interest with each other to produce an optimal location. This research uses four important criteria to select new location of marketing store appropriate with the references: distance to location, competition level with competitor, number of potential customer, and location rent cost. Quantitative data is used to determine the optimum location with AHP method. Quantitative data are preferred to avoid inconsistency when using expert opinion. The AHP result optimum location among three alternatives places.

## 1 Introduction

Marketing is one of the keys to success to realize the goals of the company, especially in increasing the value of sales followed by increased profits. A strong marketing capability of a firm provides communication and promotion of a product in the commercialization [9]. One effective marketing strategy is to expand the marketing area. Location is often regarded as trivial, but the location of the business largely determines business strategic planning and operational profit [6]. Location selection should also refer to the development of an area. Both positive and negative impacts can affect performance and business development. The location selection is one of critical decision of distribution network and it is belong to a multi-criteria decision making (MCDM) problem [4]. As one of the problems of MCDM, the location consists of various criteria that are considered in choosing the best alternative from several alternatives. There are several criteria commonly used as a consideration of location selection. It is said that the decision making of location selection is influenced by many factors, such as the impact of resources, population, transportation, education, the level of development of industry, consumption levels, government policies and other factors affected the region, the impact of the economic activity and production processes [2]. It is also said that from his literature review it is indicated that there are four aspect to select best location: Economic Impacts i.e. less cost, functional impacts i.e. higher productivity, public acceptance i.e. better community relations, and quality of life i.e. better living conditions [8]. Even, a researcher was suggested parking facilities and located at a street corner to be criteria in location selection [7]. Conflict between criteria has become one of the features in MCDM problems. There are many methods successfully formulated to solve this MCDM problem. ELEKTRE, SAW, TOPSIS, AHP, ANP, BWM are some method that usually used to solve the MCDM problems. Fuzzy SAW was used to select facility location [10]. TOPSIS combined with AHP is used to select join distribution center location [11]. Determination of emergency logistic locations can be also done by MCDM method which is in this case uses AHP [5].

Expert opinion is usually used to compare the degree of importance among criteria. Opinions are subjective so it is possible to inconsistent [1]. This paper focus on selection of distribution location using quantitative data. Quantitative data is used partially to compare alternatives in each criteria. The use of this technique is expected to minimize the bias that occurs when using expert judgment.

## 2 Case Study

This study aims to determine the optimal location among three alternatives of locations for marketing traditional canned food PT X. PT X is a famous traditional canned food factory in Yogyakarta. The problem firstly identified when the sales rate did not reach the target. The company has targeted sales of 6000-8000 cans per month or 180-240 mil IDR but the amount sold was only 3000-4000 or worth with only 90-120 mil IDR per month. PT X wanted to open another branch of the store to increase product sales. There are three alternative places of marketing that match with the purpose of making the product as souvenirs: airport, train station and bus station. The location of each alternatives places comparing with the factory is presented in the figure 1.

Based on previous literature, the criteria which are commonly used on store selection decision for marketing stores have been studied by various number of research. There are many criteria which are importantly influence the selection location decision. Distance, from factory or manufacturer to the store, has been determined as one of factors to select location (et al). Distance will affect the

* Corresponding author: harwati@uii.ac.id

manufacturer's access to the selected store. To get optimum decision in location selection, many researchers also consider competition level and potential customer. Competition level can be represented by number of competitor's store in the location. Store selection is also influenced by the sellers desire to reach as many consumers as possible (Akalin, et al 2013). That is why the number of potential customer is also considered to be criterion in store location. Store selection should also consider the costs include the costs to buy or rent, renovation costs, and store set up costs (Irwing, 1987). So, in this case study there are four criteria and three alternatives to find the optimum location. That system can be presented is the hierarchy structure as shown in figure 2.

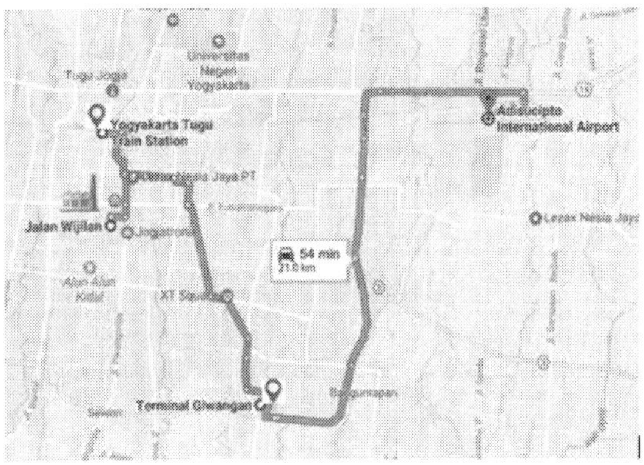


**Fig. 1.** Map of case study location

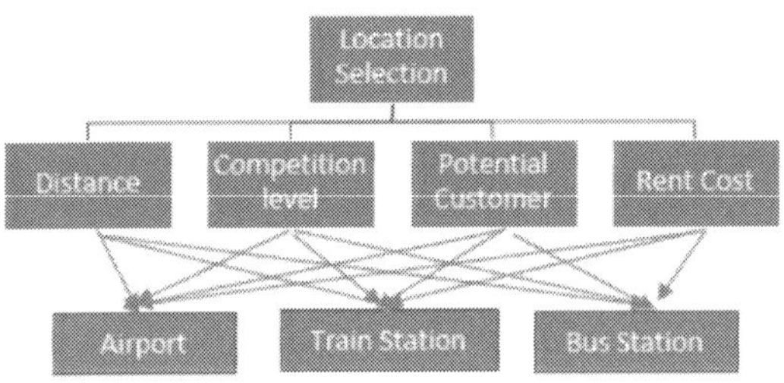


**Fig. 2.** Hierarchy structure of location selection case

## 2.1 Data Collection

The first step in the calculation is to define the definitions and perspectives for each of the criteria. Description of each criteria related to the case study is shown in the Table 1.

To collect quantitative data for each criteria we used primary and secondary survey. Distance between factories to each location is calculated from the fastest route in Google Map. This calculation ignores the jam and other disruption that may happen on the way. The result is shown in the Table 2.

**Table 1.** Criteria description

| Criteria | Description |
|---|---|
| Distance | Distance between Factory to each location |
| Competition Level | Number of competitor store in location |
| Potential Customer | Number of passenger |
| Cost | Rent cost |

**Table 2.** Duration from factory to location

| From Factory to | Duration |
|---|---|
| Airport | 29 minutes |
| Train Station | 7 minutes |
| Bus Station | 18 minutes |

Number of competitors are counted from the number of stores that sell similar products in each location. The result is shown in the Table 3.

**Table 3.** Number of Competitors

| Location | Competitors |
|---|---|
| Airport | 3 stores |
| Train Station | 5 stores |
| Bus Station | 13 stores |

Potential customer is counted from number of passenger arrival in a month. The result is shown in the Table 4.

**Table 4.** Number of passengers

| Location | Passengers |
|---|---|
| Airport | 633.638 |
| Train Station | 246.111 |
| Bus Station | 228.432 |

Assuming there is no new location development, all alternative locations will be leased from the transport facility management, then this following data for rental price per $m^2$ per month for each location:

**Table 5.** Rental cost

| Location | Cost |
|---|---|
| Airport | 200.000 |
| Train Station | 62.500 |
| Bus Station | 49.500 |

## 2.2 Pairwise Comparison

Based on quantitative data above, then pairwise comparison is compiled to show the importance of an alternative to others. The following are presented pairwise comparison data for each alternative:

**Table 6.** Duration pairwise comparison

| Alternatives | Airport | Train Station | Bus Station |
|---|---|---|---|
| Airport | 1 | 29/18 | 7/18 |
| Train Station | 18/29 | 1 | 7/29 |
| Bus Station | 18/7 | 29/7 | 1 |

**Table 7.** Duration pairwise comparison

| Alternatives | Airport | Train Station | Bus Station |
|---|---|---|---|
| Airport | 1 | 3/13 | 5/13 |
| Train Station | 13/3 | 1 | 5/3 |
| Bus Station | 13/5 | 3/5 | 1 |

**Table 8.** Duration pairwise comparison

| Alternatives | Airport | Train Station | Bus Station |
|---|---|---|---|
| Airport | 1 | 246.111/633.838 | 228.342/633.838 |
| Train Station | 633.838/246.111 | 1 | 228.342/246.111 |
| Bus Station | 633.838/228.342 | 246.111/228.342 | 1 |

**Table 9.** Duration pairwise comparison

| Alternatives | Airport | Train Station | Bus Station |
|---|---|---|---|
| Airport | 1 | 200.000/62.500 | 200.000/49.500 |
| Train Station | 62.500/200.000 | 1 | 62.500/49.500 |
| Bus Station | 45.900/200.000 | 62.500/49.500 | 1 |

## 2.3 Consistency Testing

Pairwise comparison for criteria level is done by asking judgement from owner to compare criteria each other depend on their interest. Before calculating the weight of each alternative, consistency testing is done on each criteria comparison. This step is necessary to be done to see the consistency of quantitative data obtained. The result of the consistency testing is shown as the Table 10.

From the Table 10, it is known that the most importance criteria according to owner judgement is distance with percentage 33% followed by competition level 23%; potential customer 24% and rent cost 19%. Since the consistency ratio is more than 0.1 (in this case the consistency ratio is 0.18) it is recommended to query the comparison between these criteria. The rest of consistency testing result for each alternatives in every criteria is shown in the Table 11.

**Table 10.** Matrix comparison for main criteria, degree of importance, and consistency ratio

| **Criteria** | Distance | Competition Level | Potential Customer | Cost | Degree of Importance |
|---|---|---|---|---|---|
| Distance | 1 | 2 | 1 | 2 | 0,330 |
| Competition Level | 1/2 | 1 | 1/2 | 1 | 0,233 |
| Potential Customer | 1/2 | 1/2 | 1/2 | 1/2 | 0,241 |
| Cost | 1 | 1 | 1 | 1 | 0,196 |
| CR | | | | | 0.18 |

**Table 11.** Criteria description

| **Alternatives** | **Degree of Importance** | | | |
|---|---|---|---|---|
| | Distance | Competition Level | Potential Customer | Cost |
| Airport | 0.149 | 0.126 | 0.157 | 0.117 |
| Train Station | 0.606 | 0.546 | 0.405 | 0.385 |
| Bus Station | 0.245 | 0.328 | 0.437 | 0.498 |
| CR | 0 | 0 | 0 | 0 |

# 3 Result and Discussion

The calculation of weighted alternatives is done by summing the value of each alternative based on each criterion, so that the AHP result is as follows

**Table 12.** AHP result

| Alternatives | Weight | Ranking |
|---|---|---|
| Airport | 0.20 | 3 |
| Train Station | 0.35 | 2 |
| Bus Station | 0.44 | 1 |

Based on the previous calculation we get the weight of each alternative according to the location access that Train Station has the highest weight of 0.606, then Bus Station has a weight of 0.245 and the last is Airport with a weight of 0.149. The easier access to a location, the lower the cost of transportation. Otherwise, the more difficult access to a location, the higher the cost of transportation. Based on the level of competition, train station is also the best alternative (weight 0.546) due to low number of sellers with the same type of product. The lower competition number in the same location, the more people who probably buy in the area, and the more competitors of the same product in that location, the less people will buy That means, the lower the level of competition, the higher the chances of the company in selling its products. Based on level of crowd, it is resulted that bus station is the best alternative. The more crowded a location then the more chances of people who do shopping at that location. Otherwise, the more lonely a location will be the lower the chances of people shopping at that location. This will affect the company's sales volume. The calculation of the weight of each location based rental price obtained that the Bus Station with weight 0.495 is suitable to be the location of business. The more costs the company incurs, the higher the sacrifices the company makes for that location.

With the calculation of AHP method, the weight of each alternative in terms of the overall criteria is as follows: (1) Train Station with percentage 49.8% (2) Airport with percentage 38.5% and (3) Bus Station with percentage 11.7%. This indicates that the alternative location 3 or Bus Station is the best location to be the location of the new marketing branch when viewed from the criteria of location access, the level of competition, potential customer, and the rental cost.

Overall, the best alternative to build new location of product marketing facility is in train station. In determining this location, many criteria should be used as decision-making factors. But very few criteria that can be obtained from quantitative data. This study only focuses on quantitative data, therefore qualitative data such as questionnaires or public opinion are not included in this study. In addition, the authors have not considered or separated the criteria into two types, namely positive criteria and negative criteria. The problem structure in this study is in still single hierarchy and not considered in the relationship or interconnection among sub criteria. The main advantages of problem solving using a hierarchical structure are easier to do and more quickly get a decision

## References

1. Arto Haaraa, Ron Store, Pekka Leskinen, 2017, Analyzing Uncertainties And Estimating Priorities Of Landscape Sensitivity Based On Expert Opinions, *Landscape and Urban Planning* 163, 56–66
2. H. S. Lei and Y. S. Chen, "The right tree for the right bird: Location choice decision of Taiwanese firms" FDI in China and Vietnam," International Business Review, vol. 20, pp. 338-352, 2011
3. Irwing, B. (1986). *Basic retailing*. Irwin, Inc. Illinois, USA
4. J. Raymond and J. Mataloni, "The structure of location choice for new U.S. manufacturing investments in Asia-Pacific," Journal of World Business, vol. 46, no. 2, pp. 154-165, 2004
5. Liu Hong, Zhang Xiaohua, 2011, Study on location selection of multi-objective emergency logistics center based on AHP, *Procedia Engineering*, Volume 15, Pages 2128-2132
6. Liu, Weng-Kun, 2015, Using FDM and DEMATEL Approaches to Evaluate the Location Selection of Investment, *International Journal of Information and Education Technology*; Vol. 5, No. 10, p: 732-739
7. Mehmet Akalin, Gulden Turhan, Azize Sahin. (2013). The Application of AHP Approach for Evaluating Location Selection Elements for Retail Store: A Case of Clothing Store, *International Journal of Research in Business and Social Science*, IJRBS Vol.2 No.4
8. Mahalik, Debendra Kumar, (2012), Selection of a Plant Site: A Multi Criteria Decision Making Using AHP and GRA, *Journal of Supply Chain Management Systems*; No 1 Vol 4 , p: 23-29.
9. Nanwoon Kim, Sohyoun Shin, Sungwok Min, 2016, Strategic Marketing Capability: Mobilizing Technological Resources For New Product Advantage, *Journal of Business Research*, Volume 69, Issue 12, Pages 5644-5652
10. Shuo-Yan Chou. Yao-HuiChang, Chun-YingShen,(2008) A Fuzzy Simple Additive Weighting System Under Group Decision-Making For Facility Location Selection With Objective/Subjective Attributes, *European Journal of Operational Research*, Volume 189, Issue 1, Pages 132-145
11. Yandong He, Xu Wang, Yun Lin , Fuli Zhou , Lin Zhou, (2017), Sustainable Decision Making For Joint Distribution Center Location Choice, *Transportation Research* Part D 55 202–216.

# Production waste analysis using value stream mapping and waste assessment model in a handwritten batik industry

*Putri Citra* Marifa*, *Feny Yuliana* Andriani, *Sri* Indrawati, *Anggita Noviyanti* Parmasari, *Hardiyanti* Budiman, and *Atika* Kamilia

Industrial Engineering, Universitas Islam Indonesia, Yogyakarta, Indonesia

**Abstract.** Batik is one of Indonesian cultural heritage that confirmed by United Nations of Educational, Scientific, and Cultural Organization (UNESCO) on October 2009. This legal confirmation improves the number of batik industry from many regions based its local unique characteristic. The increasing number of batik SMEs in Indonesia requires a strategy that can create competitive advantage. This strategy can be done by reducing production waste. One of Indonesian batik SMEs is SME Batik CM located in Yogyakarta. There are several problems that occur in the industry, i.e. length of the production process, spots on Batik and excessive raw materials inventory. Based on that problems, this research is done by applying lean manufacturing concept using value stream mapping (VSM) method to evaluate production wastes. Based on the result of the research, there are seven types of production waste: overproduction (9,62%), inventory (17,3%), defect (23,08%), motion (9,62%), transportation (9,62% ), Over processing (9,62%) and waiting (21,15%). Process improvement is done to reduce the highest waste, defect, using quality filter mapping (QFM).

## 1 Introduction

The economic and trading development in recent years creates a tight business competition. All industries are forced to improve business competitive advantage [1]. Batik industry is one of industry that needs to improve the manufacturing and business process. Batik is one of Indonesian cultural heritage that confirmed by United Nations Educational, Scientific, and Cultural Organization (UNESCO) on October 2009. This legal confirmation improves the number of batik industry from many regions based its local unique characteristic. The number of batik industry in Indonesia is increasing from year to year as seen in Figure 1.

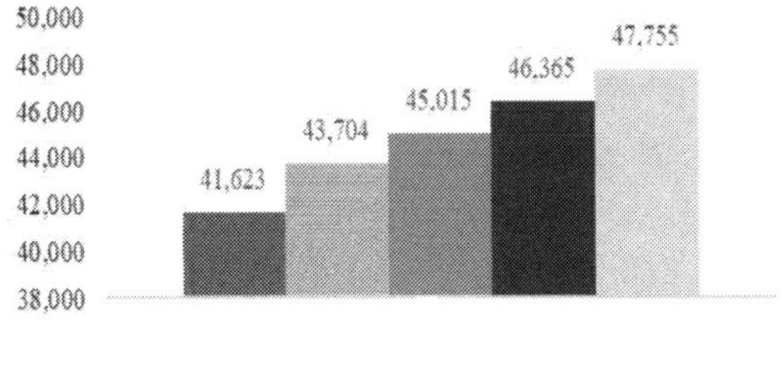


**Figure 1.** The development of batik industry in Indonesia [2]

One of batik industry in Indonesia is Batik CM that produces handwritten batik cloth with Javanese classic pattern. The process of making handwritten batik is different and takes longer time than stamped batik. It needs an accuracy so that Batik CM only produces 5 products per day. In the production process there are some steps such as pattern design, batik work, colouring, *ngeblok, ngerok, mbironi, nglorot*, giving starch, drying and packaging. There are several problems that occur in the industry, i.e. length of the production process, spots on Batik and excessive raw materials inventory. A production waste reduction is needed to improve the productivity of batik industry.

Production waste is any production activity that doesn't provide a value in processing input into output along value steam [3]. To reduce production wastes and cost, lean manufacturing can be implemented [4]. Production wastes are being eliminated through a systematic approach through lean manufacturing implementation [5]. Lean is a continual effort to reduce production wastes so that it can increase productivity and product quality [6].

Elimination of production waste can be effectively done using five criteria, i.e. process & equipment enhancement (PEE), production rules & discipline (PRD), Workforce Engagement (WFE), suppliers' integration (SUI) and customer focus (CUF) [7]. Most of research have reported that workforce development and total quality management (TQM) are commonly used on lean practices [8]. To achieve lean goals, a value stream mapping (VSM) is a powerful tool to evaluate the

* Corresponding author: 14522030@students.uii.ac.id

relationship of people, process, tools and information along the value sream [9]. VSM can help the industry to get better understanding of lean concepts and continuous improvement process [10].

## 2 Basic Theory

### 2.1 Value Stream Mapping (VSM)

VSM is a tools that can lead an industry to a better condition through continuous improvements, communication and process inefficiencies reduction [11]. Flows of information and materials of current production process is visually summarized and evaluated then envisioning a future state map with a better performance [12]. Process activity analysis is used in VSM to evaluate value added and non-value added activities. Value added activity is activity that add value to the product. Taiichi Ohno stated that non-value added activities should be eliminated [13].

### 2.2 Value Stream Analysis Tool (VALSAT)

Value Stream Analysis Tool (VALSAT) is used to analyse production waste and give recommendation to improve manufacturing performance. Based on Shigeo Shingo, there are seven wastes that can be identified in production [14]. The main function of VALSAT is to identify and evaluate production [15]. There are seven type of VALSAT tools, i.e. process activity mapping, supply chain response matrix, production variety funnel, quality filter mapping, demand amplification mapping and decision point analysis.

## 3 Research Method

This research is done through six main stages, i.e. production waste identification, production waste prioritization and production waste evaluation. Production waste identification is done by identifying all production activities. After knowing the activities in batik production process, then production waste analysis is done using VSM. Production waste prioritization is done through waste relationship matrix and waste assessment model. The purpose of this step is to know the biggest production waste that occurs. After knowing the biggest waste, then VALSAT is used to evaluate the waste in Batik CM Industry.

## 4 Discussion

### 4.1 Value Stream Mapping (VSM)

Based on current state map of VSM that describe production process in Batik CM, this industry value added activity is 9175 minutes and non value added activity 11547 minutes. There are product defect in handwritten batik production process. In batik work, there are 5% product defect. From ngeblok process, the product defect rate is at 3%, 15% in ngerok process, 10% in bironi process, 2% in second coloring process and 2% in nglorot process.

### 4.2 Waste Assessment Model

A waste relationship matrix is develop to evaluate the correlation between production waste. This matrix is used to analyze the influence of a production waste to others. There are seven production waste in batik processing, i.e. overproduction (O), inventory (I), product defect (D), unnecessary motion (M), transportation (T), overprocessing (P) and waiting time (W) as seen in Table 1. The production waste that becomes the most influential to others is product defect at 23,08 %. It means that product defect give a big impact in causing another waste. Moreover, the most affected production waste is overproduction. It is shown with a highest score from WRM column, the overproduction score is at 17,31%.

**Table 1.** Waste relationship matrix

| F/T | O | I | D | M | T | P | W | Score | % |
|---|---|---|---|---|---|---|---|---|---|
| O | 10 | | | | | | | 10 | 9,62 |
| I | 2 | 10 | 2 | 2 | 2 | | | 18 | 17,31 |
| D | 2 | 2 | 10 | 2 | 2 | 2 | 4 | 24 | 23,08 |
| M | | | | 10 | | | | 10 | 9,62 |
| T | | | | | 10 | | | 10 | 9,62 |
| P | | | | | | 10 | | 10 | 9,62 |
| W | 4 | 4 | 4 | | | | 10 | 22 | 21,15 |
| Score | 18 | 16 | 16 | 14 | 14 | 12 | 14 | 104 | 100,00 |
| % | 17,31 | 15,38 | 15,38 | 13,46 | 13,46 | 11,54 | 13,46 | 100,00 | |

### 4.3 Waste Assessment Model

A waste assessment questionnaire (WAQ) that consisted of 68 questions is used in this model to identify and allocate the production waste [16]. In WAQ, there are two types of question, i.e."from" which explained the type of waste that lead to other waste occurance and "to" which explained waste arising due to other waste. Each question consisted of three types of answer, i.e. "yes", "moderate", and "no" that have a point of 1, 0,5 and 0. Product defect have the highest score at 24,45% as seen in Table 2. Furthermore, the right of VALSAT is determined to analyze the prioritized production waste. Quality filter mapping (QFM) is VALSAT that can be used to analyze product defect.

**Table 2.** Waste assessment result

| | O | I | D | M | T | P | W |
|---|---|---|---|---|---|---|---|
| Score (Yj) | 0,46 | 0,47 | 0,53 | 0,63 | 0,63 | 0,74 | 0,47 |
| Pj Factor | 0,02 | 0,03 | 0,04 | 0,01 | 0,01 | 0,01 | 0,03 |
| Final result (Y final) | 0,01 | 0,01 | 0,02 | 0,01 | 0,01 | 0,01 | 0,01 |
| Final result (%) | 10,04 | 16,11 | 24,45 | 10,66 | 10,66 | 10,67 | 17,41 |
| Rank | 7 | 3 | 1 | 6 | 5 | 4 | 2 |

### 4.4 VALSAT: Quality Filter Mapping (QFM)

Quality filter mapping (QFM) is VALSAT that used to analyze product quality along the supply chain. The highest scrap deffect is in *ngerok* process at 15% level as seen in Figure 2. In this process, the wax in batik cloth is scrapped with a metal plate. This process is done manually, so it should be done carefully to minimize the defect.

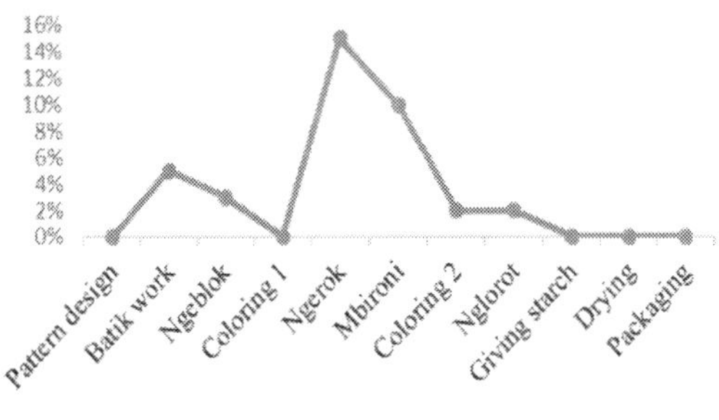


**Figure 2.** QFM of Batik CM

To identify the potential cause of product defect, a cause and effect analysis is conducted using fishbone diagram as seen in Figure 3. This diagram is a tool that can be used to identify possible causes of problems by listing all causes and effects [17]. The main cause of scrap defect in ngerok process is a low skill labor. To improve the product quality, a labor empowerment should be done by conducting training for the workers. This program is expected to be able to increase the labor skill in performing the job and reduce product defect.

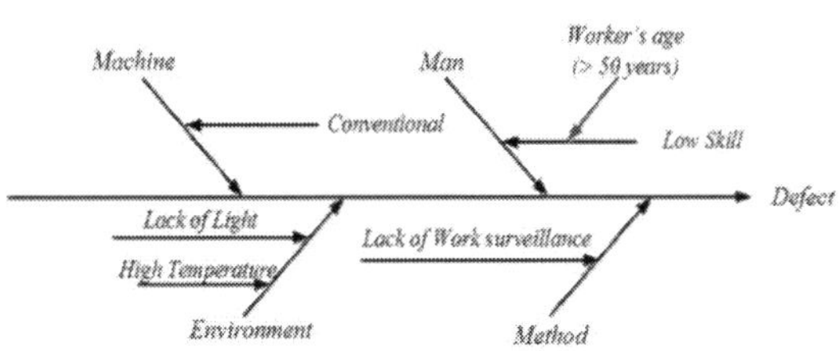


**Figure 3.** Fishbone diagram of defect in *Ngerok* process

## 5 Conclussion

Based on the research, the production process in Batik CM is dominated with a non value added activity at 50% from total lead time. The highest production waste occurs product defect at 23,08% from all production waste. This production waste becomes the main production waste that influence manufacturing performance of handwritten batik in Batik CM. The highest product defects rate is in ngerok process with 15% defect. Labor skill improvement by conducting a training should be done to minimize the product defects.

Researchers thank to the support of Industrial Engineering Department, Faculty of Industrial Technology, Universitas Islam Indonesia.

## References

1. Urban, W., Organizational Culture Assessment as a Means of Lean Management Maturity Diagnosis, *Journal of Management and Finance*, Vol. **13**, No. 4, 131-139, (2015)
2. Ministry of Industry, *Jumlah Unit Usaha Batik 2011-2015*, http://kemenperin.go.id/, Indonesia, (2015)
3. Murugunathan, V.R., Govindaraj, K, Sakhthimurugan, D., Process Planning Through Value Stream Mapping In Foundry, International *Journal of Innovative Research in Science, Engineering and Technology*, Vol.**3**, Issue 3, 1140-1143, (2014)
4. Motwani, J. A., Business Process Change Framework for Examining Lean Manufacturing: A Case Study, *Industrial Management & Data System*, 339-346, (2003)
5. Indrawati, S. & Ridwansyah, M., Manufacturing Performance Improvement Through Lean Six Sigma Method: An Iron Ores Industry Case Application, *Procedia Manufacturing*, Vol **4**, 528-534, (2015)
6. Marshall, R.E., Farahbakhsh, K., Systems Approaches to Integrated Solid Waste Management in Developing Countries, *Waste Management*, Vol. **33**, 988–1003, (2013)
7. Nordin, N., Osman, A.A., Adom, A.H., A Review on Lean Assessment Models and Performance Measures, *Journal of Advanced Review on Scientific Research*, Vol. **21**, No.1., 1-26 (2016)
8. Wahaba, A.N.A., Mukhtara, M., Sulaiman, R., A Conceptual Model of Lean Manufacturing Dimensions, *Procedia Technology*, Vol. **11**, 1292 – 1298, ( 2013 )
9. Manjunath, M., Shrivaprasad, H.C., Kumar, K.S.K, Puthran, D., Value Stream Mapping as a Tool for Lean Implementation: A Case Study, International *Journal of Innovative Research & Development*, Vol. **3**, Issue 5, 477-481, (2014)
10. Xia, W, Sun, J., Simulation Guided Value Stream Mapping and Lean Improvement: aCase Study of a Tubular Machining Facility, *Journal of Industrial Engineering and Management*, Vol. **6**(2), 456-476, (2013)
11. Roother, M., & Shoock, J., *Learning to See Value Stream Mapping to Create Value and Eliminate Muda*, Brookline, Massachusetts, USA, The Lean Enterprise Institute, (1998)
12. Voelkel, J., & Chapman, C., Value Stream Mapping, *Quality Progress*, Vol.**25**, No.5 , 65-68, (2003)
13. Tjahjadi, B., *Just-In-Time Purchasing, Just-In-Time Production System: Pengaruhnya Terhadap Kinerja Priduktifitas*, Yogyakarta: Majalah Ekonomi, (2001)
14. Hines, Peter, & Rich, *Going Lean*, Lean Enterprise Research Centre, Cardiff Business School, USA, (2000).

15. Hines, P., & Rich, N., The Seven Value Stream Mapping Tools, *International Journal of Operation and Production Management*, Vol. **17**, (1997)
16. Rawabdeh, I. A., *A Model for The Assessment of Waste in Job Shop Environments*, Jordan: University of Jordan, (2005)
17. Wahid, A. N., Surachman, Setyanto, N. W., & Soenoko, R., *Implementasi Konsep Lean Manufacturing Untuk Meminimalkan Waktu Keterlambatan Penyelesaian Produk A Sebagai Value Pelanggan*. Jurnal Rekayasa Mesin, Vol.**4**, No.2, 147-156, (2013)

# Development of decision support system for employee selection using Adaptive Neuro Fuzzy Inference System

Abdullah 'Azzam* and Sri Indrawati

Industrial Engineering, Universitas Islam Indonesia, Yogyakarta, Indonesia

**Abstract.** The number of children day care is increasing from year to year. Children day care is categorized as service industry that help parents in caring and educate children. This type of service industry plays a substitute for the family at certain hours, usually during work hours. The common problems in this industry is related to the employee performance. Most of employees have a less understanding about the whole job. Some employees only perform a routine task, i.e. feeding, cleaning and putting the child to sleep. The role in educating children is not performed as well as possible. Therefore, the employee selection is an important process to solve a children day care problem. An effective decision support system is required to optimize the employee selection process. Adaptive neuro fuzzy inference system (ANFIS) is used to develop the decision support system for employee selection process. The data used to build the system is the historical data of employee selection process in children day care. The data shows the characteristic of job applicant that qualified and not qualified. From that data, the system can perform a learning process and give the right decision. The system is able to provide the right decision with an error of 0,00016249. It means that the decision support system that developed using ANFIS can give the right recommendation for employee selection process.

## 1 Introduction

Employee plays an important role for a service industry, therefore an organization should hire a qualified people. An effective employee selection process is needed to achieve that condition. The Children day care is one of service industry that urgently need a qualified employee from a right selection process. The common problems that exist in a children day care is associated with employees with lack of ability in performing a full role of a day care. This problem can disrupt the performance stability of a children daycare as a whole. Therefore, an effective decision support system for employee selection process is required to solve a children day care problem. A decision support system based on some common variables, i.e. age, years of experience and qualification can improve the human decision making capabilities [1].

There are four popular methods used in personel selection, i.e. technique for order preference by similarity to ideal solution (TOPSIS), analytic hierarchy process (AHP), analytic network process (ANP) and expert system (ES) [2]. Human resources selection process is done using a fuzzy technique for order preference by similarity to ideal solution (FTOPSIS) to minimize errors in decision making process [3]. System dynamics is effectively used in human resource dynamic problems [4].

A decision support system for employee recruitment process for textile industry in Indonesia have been designed using fuzzy Mamdani [5]. There are four types of data that being used, i.e. psychological tests, physical tests, interview tests and health tests. The system can perform a decision with an error point of 0-5%. A fuzzy analytic hierarchy process is used to develop a decision making system for employee recruitment in an Indonesian telecommunication industry [6]. There are five important data in this system, i.e. age, gender, education, work experience and desired salary. A decision support system developed for employee selection using the fuzzy MADM (multiple attribute decision making) [7]. The system can provide rankings according to the quality of each applicant. A fuzzy technique of order preference by similarity to ideal solution (TOPSIS) is applied for personnel selection to reduce the risks of hiring unqualified employee [8].

A decision support system also can be done with adaptive neuro fuzzy inference system (ANFIS). A decision support system developed for flood forecasting in Korea using ANFIS [9]. This system can forecast and give an early warning from heavy storm in real time. A computer aided decision support system successfully created using ANFIS to detect glaucoma in monocular fundus images [10]. Based on that research, the proposed method can perform a better accuracy for glaucoma detection that neural network and support vector machine classifier (SVM). ANFIS have already successfully used to develop a decision support system in many medical fields to diagnose and monitor a various disease [11].

ANFIS also can be applied in other fields. ANFIS based system has been developed to select bank loan installments package in Saudi Arabia [12]. The research

* Corresponding author: abdullah.azzam@uii.ac.id

generated 10,000 random sample data with different possible. ANFIS also can create an effective and accurate decision support system to determine major of the prospective student in a vocational school [13]. There are three determinant variables, i.e. national exam scores, interesting to the majors and placement test score.

The use of ANFIS method on decision support system with employee selection process data as training data is expected to give the right decision in recruitment process. The employee selection process data consist of a good performance employee and unqualified applicants. ANFIS method can speed up the process of knowledge translation where if using fuzzy inferences it takes time in setting the function of membership while in artificial neural network can be used learning technique that can shorten the time. The instructional technique in this study uses supervised learning techniques where the technique requires previous available data.

## 2 Basic Theory

ANFIS is a method that perform as artificial neural network based on Takagi-Sugeno fuzzy inference system. This method is developed in 1990 where this method a strength that can be drawn from artificial neural networks concept and fuzzy logic principles because ANFIS combines these two things [14]. The inference system used is to apply IF-THEN rules that have the ability to understand non-linear function problems [15]. The ANFIS methods can be used more efficiently and optimally, if the best parameters can be used from the parameters obtained from the genetic algorithm [16].

## 3 Research Method

The research is done in a chidren day care located in Yogyakarta, Indonesia. There are three main stage that being used, i.e. data collection, development of decision support system with ANFIS method and performance analysis. Historical data of employee reqruitment process is used as initial data. There are two kind data, interview and observation process result. The data then being adjusted before it is used to develop a fuzzy system using MATLAB software. The next process is setting the data training parameters and train the initial data. After ANFIS generates a model, the model is tested for validation. The purpose of this process is know the accuracy in modeling the system. The ANFIS model is validated by comparing the output of the data that has been done from data training process with the actual data. The best model if preform with a small error. A graphical user interface then being develop to make an easier interaction between the user with the decision support system.

## 4 Result and Discussion

Decision support systems that developed for employee selection can displays or deliver the result after the user already input the data from applicants. The context diagram can be seen in Figure 1. The system can help the decision maker to decide an applicant is accepted or not based on two types of selection process, observation and interview as seen in Appendix 1. There are eight determinants variables that being used, i.e. innitiative score, friendly score, responsiveness score, interaction with parents score, cooperation score, interaction with kids score, commitment score and expected sallary.

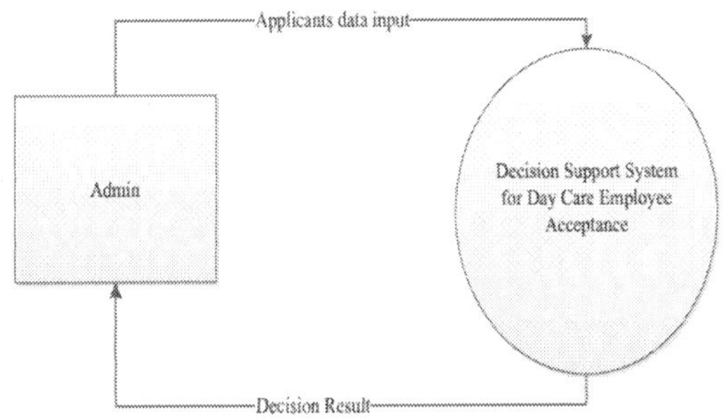


**Fig. 1.** Context diagram of Decision Support System for employee selection

The decision support system is built using Matlab and ANFIS method. The first step is to adjust the employee acceptance data in Appendix 1 so it can be process in decision support system. The adjustment is done with changing the status data into binary numbers, 1 is for accepted applicants and 0 for rejected applicants. Data adjustment also being done with summing the value at each type of employee selection process, i.e. observation and interview. The results of the initial data adjustment can be seen in Table 1.

**Table 1.** Data Adjustment

| No | Name | Observation | Interview | Status | No | Name | Observation | Interview | Status |
|---|---|---|---|---|---|---|---|---|---|
| 1 | Widyarti | 2,5 | 3,5 | 1 | 6 | Santi | 2,5 | 2 | 0 |
| 2 | Putri | 2,7 | 3,5 | 1 | 7 | Susan | 2,5 | 1,5 | 0 |
| 3 | April | 2,3 | 3,5 | 1 | 8 | Umi | 1,5 | 2 | 0 |
| 4 | Siska | 3,7 | 3 | 1 | 9 | Anggi | 1,3 | 1 | 0 |
| 5 | Ningsih | 3,3 | 3 | 1 | 10 | Lestari | 2,3 | 1 | 0 |

The next step in developing a decision support system for employee selection is to determine membership function. There are three types of membership functions that being used in this research, i.e. high, medium and low. In addition, it is also necessary to specify the type of membership function using *Gauss* type in order to produce a smoother output. A hybrid training method which is a combination of backpropagation and least square methods is used. This training process is use epoch 100. The result of the training process can be seen in Figure 2.

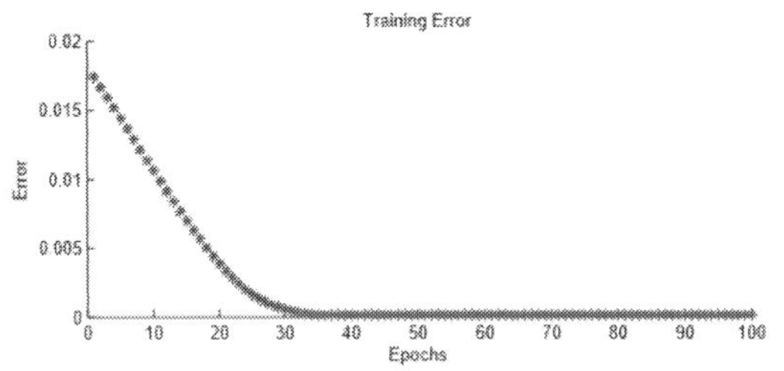


**Fig. 2**. Training Error

The training error is decreasing starting from epochs 0 to 100. The error rate resulting from the training process is 0,00016249 and remains at that value starting at epochs 31. The ANFIS performance that already trained using initial data from previous stage can be known by doing a test. The test can be done by plotting the initial data and ANFIS using the same input value. The results of the fuzzy inference system test (FIS) on the ANFIS editor can be seen in Figure 3. The results from the initial data plotting in the form of a blue circle, as a whole coincide with the ANFIS output that symbolized by the red star. This means that there almost no error when the user inputs the data in that range of values.

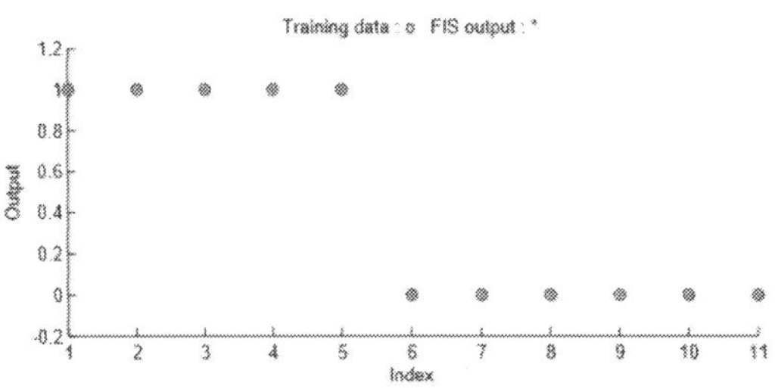


**Fig. 3**. FIS Test

The data training process is the way used in ANFIS to understand existing data patterns that being used to provide solutions for employee selection problems in children day care. The training process can be run by providing initial input data. The training process in this system uses 11 initial data and will generate the rules of the membership function that have been specified previously. There are nine rules that being generated as seen in Figure 4.

1. If (input1 is in1mf1) and (input2 is in2mf1) then (output is out1mf1) (1)
2. If (input1 is in1mf1) and (input2 is in2mf2) then (output is out1mf2) (1)
3. If (input1 is in1mf1) and (input2 is in2mf3) then (output is out1mf3) (1)
4. If (input1 is in1mf2) and (input2 is in2mf1) then (output is out1mf4) (1)
5. If (input1 is in1mf2) and (input2 is in2mf2) then (output is out1mf5) (1)
6. If (input1 is in1mf2) and (input2 is in2mf3) then (output is out1mf6) (1)
7. If (input1 is in1mf3) and (input2 is in2mf1) then (output is out1mf7) (1)
8. If (input1 is in1mf3) and (input2 is in2mf2) then (output is out1mf8) (1)
9. If (input1 is in1mf3) and (input2 is in2mf3) then (output is out1mf9) (1)

**Fig. 4**. The rules

Each new applicants data entered and in accordance with the restrictions or rules that exist then the system will provide the right solution or decision. The result of decision support system for employee selection in children daycare can be seen in Figure 5.

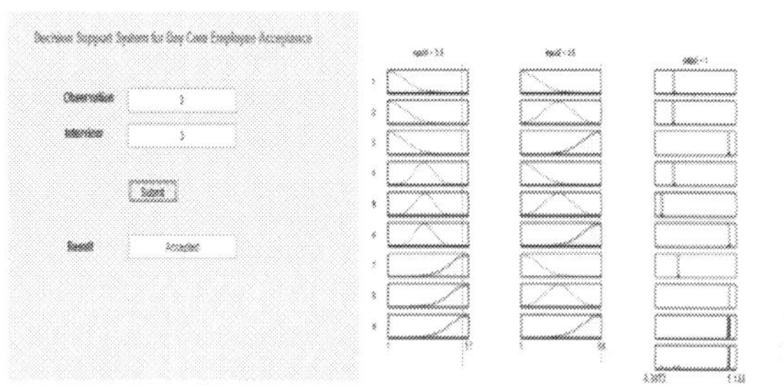

**Fig. 5**. Implementation

## 5 Conclusion

The decision support system for employee selection in children day care using ANFIS method is successfully being implemented. The system can provide a decision whether the applicants that follows the selection process is accepted or not with a small error value. The error rate resulting from the training process is 0,00016249. This error is more smaller then other method. The ANFIS performance will be more optimal if it is used a large numbers of data. Because system can learn with more data.

Researchers thank to the support of Industrial Engineering Department, Faculty of Industrial Technology, Universitas Islam Indonesia.

## Reference

1. B. Ngwenya, Application of Decision Support Systems and Its Impact on Human Resources Output: A Study of Selected Universities in Zimbabwe, *Journal of Computer Sciences and Applications*, Volume **1**, No. 3: 46-54, (2013)
2. A.R Afshari,, M. Nikolic, D. Cockalo, Applications Of Fuzzy Decision Making For Personnel Selection Problem - A Review, *Journal Of Engineering Management And Competitiveness*, Vol. **4**, No. 2, pp. 68-77, (2014)
3. K. M. Doka, F. Ahmad, S. N. W. Shamsuddin, W. S. W. Awang, N. Ghazali, Integrated Decision Support System for Human Resource Selection Using

* Corresponding author: abdullah.azzam@uii.ac.id

TOPSIS Based Models, *Applied Mathematical Sciences*, Vol. **9**, No. 129, pp. 6403 – 6414, http://dx.doi.org/10.12988/ams.2015.53288, (2015)

4. C.B. Richa, Application Of Decision Support System In Human Resource Management In Mfis- A System Dynamics Approach, *International Journal of Business and Economic Research*, Vol.**14**, No. 10, pp. 6849-6862, (2016)
5. A. T. Suseno, *Penerapan Metode Fuzzy Mamdani Untuk Sistem Pendukung Keputusan Penerimaan Karyawan PT. Primatexco Indonesia*, Program Studi Teknik Informatika Universitas Dian Nuswantoro, Semarang, (2009)
6. R. K. Ablhamid, B. Santoso, M. A. Muslim, Decision Making and Evaluation System for Employee Recruitment Using Fuzzy Analytic Hierarchy Process, *International Refereed Journal of Engineering and Science (IRJES)*, Volume **2**, Issue 7: 24-31, (2013)
7. D. A. Ramadhani, S. Astuti, *Sistem Pendukung Keputusan Penerimaan Pegawai dengan Metode Fuzzy MADM*, Techno.com, (2014)
8. M. Mammadova, Z. Jabrayilova, Application of Fuzzy Optimization Method in Decision-Making for Personnel Selection, *Intelligent Control and Automation*, Volume **5**: 190-204, (2014)
9. Jaeeun, Development of Decision Support System for Flood Forecasting and Warning in Urban Stream, *Journal of The Korean Society of Civil Engineers*, Vol. **28**, Issue 6B: 743-750, (2008)
10. S. Kavitha, K. Duraiswamy, An Efficient Decision Support System for Detection of Glaukoma in Fundus Images Using ANFIS, International *Journal of Advances in Engineering & Technology*., Vol. **2**, Issue 1: 227-240, (2012)
11. N. Walia, S. Kumar, H. Singh, A Survey on Applications of Adaptive Neuro Fuzzy Inference System, *International Journal of Hybrid Information Technology*, Vol.**8**. No.11: 343-350, (2015)
12. I. Albidewi, R. Alouhali, An Intelligent Decision Support System Using Adaptive Network-Based Fuzzy Inference System (ANFIS) For Choosing Suitable Bank Loan Installments, *IOSR Journal of Computer Engineering (IOSR-JCE)*, Volume 16, Issue 3: 130-137.
13. A. Pranolo, F. In'ammurrohman, Y. Hendriana, D. Octaviani, A Decision Support System using ANFIS to Determine the Major of Prospective Students in A Vocational School of Indonesia, *International Journal of Computer Trends and Technology (IJCTT)*. Volume **27**. Number 2: 100-105, (2015)
14. JSR. Jang, ANFIS: Adaptive-Network-Based Fuzzy Inference System, *IEEE Transactions on System, Man, and Cybernetics,* Volume **23**: 665- 685, (1993)
15. N. Nedjah, L. de Macedo Mourelle, Fuzzy Systems Engineering: Theory and Practice, *Studies in Fuzziness and Soft Computing*, Vol. **181**: 53–83.
16. P. Tahmasebi, A Hybrid Neural Networks-Fuzzy Logic-Genetic Algorithm For Grade Estimation, *Computers & Geosciences*. 42: 18–27, (2012)

# The mental workload analysis of safety workers in an Indonesian oil mining industry

*Sri* Indrawati, *Atyanti Dyah* Prabaswari*, and *Tasya* Pradipta

Industrial Engineering, Universitas Islam Indonesia, Yogyakarta, Indonesia

**Abstract.** The responsibilities of occupational health and safety workers are very hard to ensure other workers is safety. The responsibility make the workers of occupational health and safety has some affecting to their job. Some effect can cause over the mental workload. This research aims to determine the score of mental workload from three professions in occupational health and safety, i.e. safetyman contractor, safetyman field and safetyman officer. Six indicators in the NASA-TLX method, i.e. mental demand (MD), physical demand (PD), temporal demand (TD), performance (OP), effort (EF) and frustration level (FR) are used to determine the worker's mental workload. The result shows mental demand (MD) is the most dominant indicators affecting the mental workload between safetyman contractor, safetyman field and safety officer. The highest mental workload score among safety workers is on the safetyman field with WWL score at 62,38, because among the three types safety workers, the highest MD is on the safetyman field due to the responsibility.

## 1 Introduction

The rapid development of industrialization and technological innovation creates a tight competition between industries. In this case, each industry required to perform a better manufacturing performance. Human resources management plays a vital role for manufacturing excellence. Human resources are the main assets of the company in improving products quality [1]

[2] states that workers involvement in industry have an important role in the sustainability of industry's business. The right people in the right places at the right time with the right policies is the key of an industrial success [3]. Based on that research, 56% industry in Slovakia consider the human resources for increasing productivity. An effective workload planning by optimizing workers capacity should be done to improve productivity [4]. Employee dissatisfaction and excessive mental workload are two main factors that can reduce the employee productivity in 21st century [5].

The worker fatigue conditions can decrease productivity level of each worker [6]. An excessive workload leads to physical, mental and emotional fatigue, i.e. headaches, indigestion and irritability [7]. There are some factors that affecting worker productivity, i.e. age (25%), gender (15%), work tenure (10%) and 50% workload and fatigue [8]. Various factors have successfully increase a worker productivity, i.e. a workload based salary, interesting job, career path, policies and work characteristics [9]. There are four factors that influenced the worker productivity in a musical instrument production, i.e. job stress (23,5%), work motivation (25,55), nutritional status (25,5%) and workload (25,5%) [10].

Generally, workers get workload both physically and mentally. Mental workload have a more significant relations with the musculoskeletal problem (neck and upper back) [11]. Mental workload consisting of some factors that influence the worker's mental processing of information, decision making process and individual reactions [12]. Some factors, i.e. worker age, gender, martial status, type of duties and profession have a positive relationship with mental workload [13]. The NASA Task Load Index (NASA-TLX) is a method for subjective mental workload measurement that broadly used in both manufacturing and service industry [14]. The integration of NASA-TLX and the goals, operators, methods and selection rules (GOMS) model is successfuly used to evaluate mental workload in human computer interaction process [15]. NASA-TLX also used to evaluate mental workload in electrical power control and operations center in Portugal and Northteast Brazil [16]. The research give some action plan to reduce a human error operation, economic losses, physical damage and treats to worker life. NASA-TLX is used to validate the result of a novel electroencephalography (EEG) approach [17]. This research is done to estimate task mental workload in constructions project.

In an Indonesian oil mining industry, the need to evaluate mental workload is increasingly important. Workers safety and health both mentally and physically are managed by an occupational health and safety department. The workers in that department are work to assure other workers use personal protective equipment (PPE) for work safety. Based on observation in that area

* Corresponding author: atyanti.dyah@uii.ac.id

and interview some worker, sometimes this work is very hard and complicated due to other workers resistance, ignorance and defensiveness. A mental workload analysis is needed to improve worker productivity in occupational health and safety department. So, researcher want to examine mental workload of three jobs position in occupational health and safety department. The three jobs position are safety contractor, safetyman, and safety officer.

## 2 Basic Theory

### 2.1 Mental Workload

Mental workload is a working condition where there is information needs to be processed in the brain that requires a creativity [18]. Mental workload can be evaluated from the difference between the workload demands of a task and the maximum capacity of a workers mental load in a motivated condition. An excessive mental workload will lead to job stress such as fear, anxiety, guilt, anger sad, despair and boredom [19]. Stoner (1986). Different jobs for each worker will cause different levels of work stress [20]. There are several symptoms as the impact of excessive mental workload, i.e.: physical symptoms such as headache and abdominal pain, mental symptoms such as easy to forget, difficulty concentrating, anxious, irritable and desperate [21]. The last symptom is behavior such as smoking, drinking alcohol and avoiding others people.

### 2.2 NASA-TLX

Measurement method in subjective workload is a measurement of mental workload based on perceptions of respondent/workers of mental workload in subjective perspective, i.e. [22]:

a. National Aeronautics and Space Administration Task Load Index (NASATLX)
b. Subjective Workload Assessment Technique (SWAT)
c. Modified Cooper Harper Scaling
d. Multi-descriptor Scale
e. Rating Scale Mental Effort (RSME)

Some method to measurement mental workload from subjective perspective which more used and give good result is NASA-TLX and SWAT [23]. This method is measuring mental workload from the types of job, not measuring mental workload from the workers.

The NASA-TLX method is a method used to analyze the mental workload faced by workers who must perform various activities to finish the job. There are six main steps of NASA-TLX, i.e. (Hancock and Meshkati, 1988):

a. Weighting, in which section respondents are asked to choose one dominant to cause a mental workload.
b. Rating for six mental workload indicators.
c. Product scoring is done by multiplying the rating with factor weight for each descriptor. This is done for six mental workload indicators.
d. Weighted workload (WWL) that is derived by summing up the six product values.
e. Average of WWL that calculated by dividing WWL by total weight.
f. Interpretation scores of average WWL using Table 1.

**Table 1**. The interpretation score of workload

| Workload | Value |
|---|---|
| Low | 0-9 |
| Medium | 10-29 |
| Somewhat high | 30-49 |
| High | 50-79 |
| Very high | 80-100 |

## 3 Research Method

This research is done for workers that work in an occupational health and safety department of an Indonesian oil mining industry to evaluate the mental workload. Three types of profession are evaluated, i.e. safetyman contractor, safetyman field and safetyman officer. Six indicators of NASA-TLX method, i.e. mental demand (MD), physical demand (PD), temporal demand (TD), performance (OP), effort (EF) and frustration level (FR). MD is an indicator of how much mental and perceptual activity it takes to see, remember and search. Besides how kind of work is done (difficult, simple or complex). PD is related to the amount of physical activity required. TD is used to evaluate working time. The working time that felt slow, relax or fast and tired. OP is an indicator to analyze a person succeeds in doing job and how satisfied with the results. FR is related how insecure, hopeless, offended and disturbed feeling in doing a job. EF is an indicator how hard work is needed to reach the target.

The data are collected through direct observation, interview, and a survey using NASA-TLX questionnaire. A survey is done for each worker in occupational health and safety department with 30 respondents. NASA-TLX questionnaire is used for mental workload assessment. After data collection is done, data processing is performed to determine the value of mental workload of each operator using NASA-TLX method and statistical test. A mental workload analysis is developed for each NASA-TLX indicators. The NASA TLX steps used in this research are:

a. Weighting
Comparing for each factor, which more important to worker as seen in Table 2. Compare using every two factor. Example MD comparing to PD, MD comparing to TD, MD comparing to OP, MD Comparing to EF, MD comparing to FR. And if comparing MD to every factor is done, weighting can continued to other factor, example PD comparing to each other factor. So, the total number of factor which chosen is 15.
b. Rating
Worker giving the rating to six factor which influencing the mental workload. The rating has subjective poin of view because the perception ech worker.
c. Product score

Product = rating x weighting
The result of product can get from multiplication of weghting and rating.

d. Weighted Workload (WWL)
The result of WWL is sum of all product score.
e. Average of WWL
Score of WWL divided 15.

**Table 2.** Comparing factor

| | MD | PD | TD | OP | EF | FR |
|---|---|---|---|---|---|---|
| MD | | | | | | |
| PD | | | | | | |
| TD | | | | | | |
| OP | | | | | | |
| EF | | | | | | |
| FR | | | | | | |

# 4. Result and Discussion

## 4.1 Result

The research shows that the average recapitulation of weighted workload (WWL) for three types of profession in occupational health and safety department is categorized as high as seen in Table 3. The safetyman field have a highest mental workload at score 62,38. Followed by mental workload of safetyman contractor with WWL score is 56,23. The last is the safety officer with an average result at 54,5. Mental demand (MD) is the greatest indicators affecting the WWL score for all type of profession. It means that a great mental and perceptual activity needed to see, remember and searching are occurs.

MD values are equally high in value compared to other categories. This condition indicates a big responsibility for ensuring workers to work safely and try to reduce the number of accidents. The highest mental demand is on the safetyman field due to the responsibility for a large oil refinery area and an urgent condition of safetyman straight down the spaciousness to finish the problems related to damaged equipment, gas leaks and fires. The other task and responsibility is to intervene on unsafe conditions /actions. This tasks and responsibilities becomes harder due to some arrogant workers.

A high physical demand value (PD) also obtained on the safetyman field because this profession should travel around the field in all working hours. The fatigue possibility occurs because of other additional task, i.e. check the PPE compliance report, a report of good housekeeping (GHK) compliance and also for hazardous and toxic materials. Besides that the safetyman field also check permit to work (PTW) document that consist of a requirement to perform dangerous tasks with regard there are potential hazards and preventive.

The temporal demand (TD) score for safety officer is the highest at 197. It is caused by responsibilities to follow-up the safety field document result with intervention, evaluation and monitoring in a short time. The safety officer always make some analysis and evaluation as well as prepare data for improvement program.

**Table 3.** The recapitulation of WWL and interpretation score

| Catego-ry | Safetyman contractor | | Safetyman field | | Safetyman officer | |
|---|---|---|---|---|---|---|
| | Total | Aver age | Total | Aver age | Tot al | Avera ge |
| MD | 2080 | 208 | 2302 | 230,2 | 202 5 | 202,5 |
| PD | 1030 | 103 | 1156 | 115,6 | 985 | 98,5 |
| TD | 1600 | 160 | 1781 | 178,1 | 197 0 | 197 |
| OP | 1525 | 152, 5 | 1679, 5 | 167,9 | 895 | 89,5 |
| EF | 2040 | 204 | 2244 | 224.4 | 181 0 | 181 |
| FR | 160 | 16 | 195 | 19.5 | 490 | 49 |
| WWL Average | 56,23 | | 62,38 | | 54,5 | |
| Catego-ry | High | | High | | High | |

Based on the average results for performance (OP), the safety officer get a best score at 89,5. This condition is caused by a work support from safety field who always try to supervise the workers in oil refining areas. For effort indicator (EF), safety field obtain a high value at 224,4. The safetyman field always trying to solve existing problems and help the workers to reach target in oil refining areas. The frustration level (FR) for safety officer becomes the highest because a big responsibility related to any safety or accident problems that must be resolved.

Based on mental workload assessment result using NASA-TLX, an improvement program is proposed to reduce mental workload risk. The improvement program consist of some activity, i.e. 5 S program (*seiri, seiton, seiso, seiketsu* and *shitsuke*) for workplace improvement design, improvement of communication pattern to be more effective and focused on problem solving, occupational stress management and some informal activities. The occupational stress management for the department leaders should be done so that leaders get a better understanding in some causes and early symptons of work stress and how to to minimize the cause. Informal activities is done with the aim of improving relaxation and relationships among workers. Some informal activities that can be done include happy hours, afternoon tea, outbound activities, and family gathering. Workers have some positive responses from relaxation exercises that stress can be reduced, reduced mental burden, body and soul become fresher and fitter, good for health, and feel more calm [22].

# 5 Conclussion

The research shows that safetyman field have a highest mental workload with WWL score of 62,38 and including in high category. The most dominant indicators affecting the mental workload between safetyman contractor, safetyman field and safety officer is mental demand

(MD). Among the three types safety workers, the highest MD is on the safetyman field due to the responsibility for a large oil refinery area and an urgent condition of safetyman straight down the spaciousness to finish the problems related to damaged equipment, gas leaks and fires. The highest MD caused by an indicator of how much mental and perceptual activity it takes to see, remember and search. Safetyman field have the highest value of MD score 203,2. This is due to the great responsibility for the safety and security of workers.

Researchers thank to the support of Industrial Engineering Department, Faculty of Industrial Technology, Universitas Islam Indonesia.

## References

1 Setyawati, L.S., *Selintas tentang kelelahan kerja*, Amara Books, Yogyakarta, (2010)

2 Schroeder, H., The Importance of Human Resource Management in Strategic Sustainability: An Art and Science Perspective, *Journal of Environmental Sustainability*, Vol. 2. Issue 2: 75-82, (2012).

3 Koltnerova, K., Chlpekova, A., Samakova, J., The Importance of Human Resource Planning in Industrial Enterprises, *Research Papers Faculty of Materials Science and Technology Slovak University of Technology*, Vol. **20: 62-67**, (2012)

4 Berseneva, K.V., Natsubidze, A.S., Chaschin, V.V., Human Resource Planning in Implementation of Industrial Enterprises` Strategy, *World Applied Sciences Journal*, Vol. 29 **(3): 433-437**, (2014)

5 Aslam, H.D., Aslam, M., Ali, N., Habib., B.,. Importance of Human Resource Management in 21st Century: A Theoretical Perspective, *International Journal of Human Resource Studies*, Vol. 3, No. **3: 87-96**, (2013)

6 Ulfah, N., Nurcahyo, J., & Dwiandhono, I., Model Kuantitatif Manajemen Kelelahan dan Beban Kerja untuk Peningkatan Produktivitas Pekerja Penggilingan Padi, *Jurnal Kesehatan Masyarakat Nasional*, Vol.7. No.**10: 477-480**, (2013)

7 Manuaba, A., & Sriwatomo, W., *Ergonomi untuk keselamatan, kesehatan kerja dan produktivitas*, UNIBA PRESS, Surakarta, Indonesia, (2000)

8 Katsuro P.C., Gadzirayi, Taruwona M, & Suzanna M., Impact of occupationalhealth and safety on worker productivity: a case of Zimbabwe food industry, *African journal of Business Manajement*, Vol. 4, No.**13: 2644-2651**, (2010)

9 Adnyani, I.G., Membina semangat kerja untuk meningkatkan produktivitas kerja karyawan, *Jurnal Studi Ekonomi Universitas Udayana*, Vol. 13, No.**2: 202-209**, (2008).

10 Reniati R. A., Faktor-faktor yang mempengaruhi produktivitas kerja pada pekerja bagian pembuatan drum unit drum plant PT X 2006, *Tesis Universitas Diponegoro*, Universitas Diponegoro, (2006)

11 Larsman, P. & Hanse, J.J., Psychological and Physical Workload and the Development of Musculoskeletal Symptoms Among Female Elderly-Care Workers, *The Ergonomics Open Journal*, Vol. **1: 34-38**, (2008)

12 Sartang, G., Ashnagar, M., Habibi, E., Sadeghi, S., Evaluation of Rating Scale Mental Effort (RSME) effectiveness for mental workload assessment in nurses, *Journal of Occupational Health and Epidemology*, Vol.5. Issue **4: 211-217**, (2016)

13 Seker, A., Using Outputs of NASA-TLX for Building a Mental Workload Expert System, *Gazi University Journal of Science*, Vol. 27. No. **4: 1131-1142**, (2014)

14 Cao, A., Chintamani, K.K., Pandya, A.K., Ellis, D., NASA TLX: Software for assessing subjective mental workload, *Behavior Research Methods,* Vol 41. Issue **1: 113-117**, (2009)

15 Ramkumar, A., Stappers, P.J., Niessen, W.J., Adebahr, S., Schimek-Jasch, T., Nestle, U., Song, Y., Using GOMS and NASA-TLX to Evaluate Human-Computer Interaction Process in Interactive Segmentation, *International Journal of Human–Computer Interaction,* Vol. December 2016: **1-12**, (2016)

16 Melo, M.O., Silva, L.B., Rebelo, F.S, Ergonomics Aspects And Workload On The Operators In The Electric Power Control And Operation Centers: Multi-Case Studies In Portugal And Brazil, *Iberoamerican Journal of Industrial Engineering*, Volume 8 No. **16: 35-55**, (2016)

17 Chen, J., Taylor, J.E., Comu, S., Assessing Task Mental Workload in Construction Projects: A Novel Electroencephalography Approach, *Journal of Construction Engineering and Management*, Vol. 143. Issue **8: 4017053**, (2017)

18 Khroemer, K.H.E. & Grandjean, E., *Fitting the Task to Human*. 6th edition, CRC Press, USA, (1997)

19 Fraser, P., *Stres dan Kepuasan Kerja*, Erlangga, Jakarta, (1992).

20 Rauf, F., Analisis Beban Mental Menggunakan Metode National Aeronautics And Space Administration-Task Load Index (Nasa-Tlx) Di PPPPTK Bmti Di Departemen Mesin Bandung, *Tugas Akhir Universitas Komputer Indonesia,* Indonesia. (2012)

21 Hancock, P. A. & Meshkati, N., *Human Mental Workload*, Advances in psychology, Elsevier Science Publisher, Netherlands, (1988)

22 Widyanti, A., Johnson, A. & Waard, D.d. 2010. Pengukuran Beban Kerja Mental Dalam Searching Task Dengan Metode Rating Scale Mental Effort (RSME). JTI Universitas Diponegoro, 1(V).

23 Puspawardhani, E. H., Suryoputro, M. R., Sari, A. D., Kurnia, R. D., & Purnomo, H., Mental Workload Analysis Using NASA-TLX Method Between Various Level of Work in Plastic Injection Division of Manufacturing Company, *Advances in Safety Management and Human Factors,* Vol. July 27-31 2016: **311-319**, (2016)

# Fuel distribution process risk analysis in East Borneo

*Raizsa* Laksmita* and *Imam* Djati Widodo

Industrial Engineering, Universitas Islam Indonesia, Yogyakarta, 55584, Indonesia

**Abstract**. Fuel distribution is an important aspect of fulfilling the customer's need. It is risky because it can cause tardiness that can cause fuel scarcity. In the process of distribution, many risks are occurring. House of Risk is a method used for mitigating the risk. It identifies seven risk events and nine risk agents. Matrix occurrence and severity are used for eliminating the minor impact risk. House of Risk 1 is used for determining the Aggregate Risk Potential (ARP). Pareto diagram is applied to prioritize risk that must be mitigated by preventive actions based on ARP. It identifies 4 priority risks, namely A8 (Car trouble), A4 (Human Error), A3 (Error deposit via bank and underpayment), and A6 (traffic accident) which should be mitigated. House of Risk 2 makes for mapping between the preventive action and risk agent. It gets the Effectiveness to Difficulty Ratio (ETD) for mitigating action. Conducting safety talk routine once every three days with ETD 2088 is the primary preventive actions.

## 1 Introduction

Supply chain management has an important role in every business process in a company because it influences efficiency and effectivity of relation to the supplier, manufacturer, and the customer. "Generally, all activities related to material flow, information flow, and financial flow along the supply chain are the activities within the scope of SCM" [1].

The supply chain is a set of firms that pass materials forward. Normally, several independent firms are involved in manufacturing a product and placing it in the hands of the end user in a supply chain. Raw material and component producers, product assemblers, wholesalers, retailer merchants and transportation companies are all members of a supply chain [2].

Every company's business activity, including supply chain, always has the risk that can disrupt the success of the company and even endanger the safety of worker [3]. Therefore, risk management is needed. British Standard Institute's Guide to project management defines risk management as "processes whereby decisions are made to accept a known or assessed risk or the implementation of action to reduce the consequences or the probability of occurrence of an adverse event" [4]. In the context of supply chain risk management, events are characterized by their probability of occurrence and their related consequences within the supply chain[5].

Supply chain risk management is not only as the cost Saving but rather a value creation activity has given that it can lead to more sustainable supply chains[6, 7].

The fuel distribution is an important process of fuel industry to deliver the fuel to the customer in the right quantity and on time. It is risky activity because if tardiness of delivering happens, it can cause fuel scarcity that endangers other related business. In the process of distribution, some potential risks can happen especially in filling and delivery process. Identifying risk and preventive action are needed for mitigating and minimizing the risks. Some risk analysis researches have been done in some aspect of oil and gas industries such as in financial [8, 9] and supply chain modeling [10, 11].

Delivery flexibility measures adopted by the firms are largely internal to the firm. Delivery or shipment risk is in not only to the internal process and distribution flexibility related aspects of the firm but also to the poor logistics infrastructure that is external to the firm.[12]

House of risk is a renewable method of risk analysis. It is modification of the Failure Mode and Effects Analysis (FMEA) method for quantification of risk by quantification and the House of Quality (HOQ) model to prioritize which risk agents should be addressed first and to select the most effective action to reduce Potential risks posed by risk agent [13]. Implementation house of risk in supply chain have been done by some researchers. Kusmantini et al. implemented House of Risk in Furniture supply chain risk [14]. Data collection uses a triangular method that combined in depth interview technique with a number of successful exporters and distributing questioners to more than 80 SMEs in DIY. It identifies 23 agents and 43 incidence risks in the supply chain. It is also implemented in the supply chain of shoe company. By using two phases of house of risk, Kristanto [15] identifies 27 risk events and 52 risk agents. It gets 6 mitigation actions.

Thus, the advantages of prevention and mitigation in the distribution process are crucial because if the risk occurs it will cause enormous losses. It encourages researchers to conduct research on the mitigation and preventive actions to overcome the risk. House of risk will

* Corresponding author: 13522169@students.uii.ac.id

be implemented to identify and mitigate risks to minimize losses. This study is done in a national oil company in Balikpapan.

## 2 Research Method

This step is identifying and formulating problems for doing research that uses in identifying the problems that occurred in the process. Data are collected by observation and interview. A deep discussion is done to determine risk rating scale with the company. Firstly, it collects and processes the data. Collecting data consist of primary data and secondary data. Primary data consists of result of observation, result of interview, and a rating scale from the company. The secondary data consist of House of Risk Application. The rating scale of severity can be shown in the Table 1.

**Table 1.** Severity rank

| Rank | Subjective Estimate for Severity | Description |
|---|---|---|
| 1 | No impact | Insignificant |
| 2 | Minor impact | Single small impact |
| 3 | Medium impact | Cause short-period difficulties |
| 4 | Major impact | Causes long-period difficulties |
| 5 | Catastrophic impact | Directly affect highest rates |

Before interviewing an expert, it should need the occurrence rating scale, the occurrence rating scale shown in Table 2.

**Table 2.** Occurrence rating scale

| Rank | Subjective Estimate of Occurrence | Description | |
|---|---|---|---|
| 1 | Very unlikely | Very rare event (very rarely) | Occurs 1 times a year |
| 2 | Improbable | There is indirect evidence of event (rarely) | Occurs 2-3 times a year |
| 3 | Moderate | There is direct evidence of event (sometimes) | Occurs 3-4 times a year |
| 4 | Probable | There is strong direct evidence of the event (often) | Occurs 4-5 times a year |
| 5 | Very Probable | Event recurs frequently (Very Often) | Occurs > 5 times a year |

Data analysis is started by identifying activities based on SCOR-model. SCOR-model is the product of the Supply-Chain Council (SCC), an independent, not-for-profit, global corporation with membership open to all companies and organizations interested in applying and advancing the state-of-the-art in supply-chain management systems and practices. There are three process levels of SCOR, level-1 process is known as SCOR process and are defined for each element such as a producer or wholesaler in the supply chain [16].

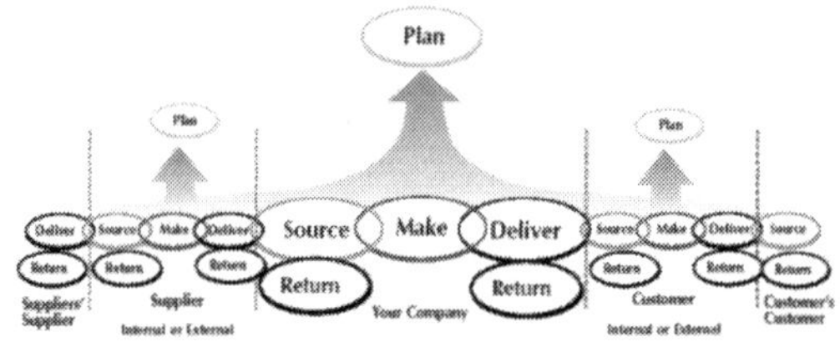


**Fig. 1.** SCOR Process (*sources: supply chain council.inc. copyright 2016*)

Occurrence and severity matrix is needed to eliminate the risk that has the minor impact. The matrix can be shown in Fig 2.

| Occurrence | | | | | | | |
|---|---|---|---|---|---|---|---|
| | 5 | Very Probable | Medium | High | High | Extreme | Extreme |
| | 4 | Probable | Low | Medium | High | High | Extreme |
| | 3 | Moderate | Low | Medium | Medium | High | High |
| | 2 | Improbable | Low | Low | Medium | Medium | High |
| | 1 | Very Unlikely | Very Low | Low | Low | Low | Medium |
| | | | No impact | Minor Impact | Medium Impact | Major Impact | Catastrophic impact |
| | | | 1 | 2 | 3 | 4 | 5 |
| | | | Severity | | | | |

**Fig. 2.** Occurrence and severity matrix

The secondary data is the application of House of Risk 1, House of risk phase one is trying to relate a set of requirements (what) and a set of responses (how) where each response could address one or more requirements. The degree of correlation is typically classified as none (and given an equivalent value of 0), low (one), moderate (three), and high (nine). Each requirement has a certain gap to fill and each response would require some types of resources and funds. In HOR 1 calculate the Aggregate Risk Potential (ARP). The formula of ARP is below:

$$ARPj = Oj \sum_i Si\ Rij \tag{1}$$

where:
Oj: Occurrence {1, 2, 3, 4, 5}
Si: Severity {1, 2, 3, 4, 5}
Rij: Relationship between risk agent and risk event {0, 1, 3, 9}

After making the HOR 1, Pareto analysis, based on the ARP, is run to identify risk priority that should be to mitigate. Then, The HOR 2 is used to determine which actions are to be done first, considering their differing effectiveness as well as resources involved and the degree of difficulties in performing. It needs to calculate the total effectiveness of each action. The formula to calculate the effectiveness is:

$$TEk = \sum_j ARPj\ Ek \qquad (\forall k) \tag{2}$$

where:
ARP : Aggregate Risk Priority
Ek : Relationship between risk agent and preventive action {0, 1, 3, 9}

It needs to calculate the total effectiveness to difficulty, the formula is:

$$ETDk = \frac{TEk}{Dk} \tag{3}$$

where:
ETDk : Effectiveness to difficulty ratio
TEk : Total effectiveness of action k
Dk : Degree of difficulty performing action k

It chooses the high ETD is the first preventive action that should be doing first.

## 3 Result and Discussion

Based on data analysis, some results are found:

### 3.1 Risk Event

Risk event is the risk that may be occurring. Risk event and the severity of the distribution process can be shown in Table 3.

**Table 3.** Risk event

| No | Major Processes | Sub-processes | Risk events | Code | Severity |
|---|---|---|---|---|---|
| 1 | Plan | Scheduling | the emptiness stock on fuel station | E1 | 5 |
| 2 | Deliver | Filling | There is oil spill and overflow | E2 | 3 |
| | | | fire and explosion | E3 | 4 |
| | | | The pulling clutch | E4 | 3 |
| | | Deliver process | Tardiness of deliver | E5 | 2 |
| | | | Not delivery fuel oil | E6 | 3 |
| 3 | Return | Handling return from customer | tardiness of cross ferry | E7 | 3 |

Determining the major process is from SCOR-method then divided into three plans, deliver, and return. Each Sub process has risk event. Risk event was given by code with E1 until E7. The severity has number 1 until 5; it means minor impact until catastrophic impact.

### 3.2 Risk Agent

Risk agent is the cause of the risk event. It can show the risk agent and occurrence in the distribution process in Table 4.

There are nine identified risk agents. The risk agent gives a code with A code (A1 to A9). Each risk agent has occurrence from 1 to 5. The value of occurrence means very unlikely until very probable. The occurrence rating scale is from the expert by interview.

**Table 4.** Risk agent

| No | Code | Risk Agent | Occurrence |
|---|---|---|---|
| 1 | A1 | The failure of delivering schedule | 3 |
| 2 | A2 | Fuel station not deliver stock report actually | 2 |
| 3 | A3 | Error deposit via bank and underpayment | 2 |
| 4 | A4 | Human Error | 2 |
| 5 | A5 | Not implement SOP (Standard Operation Process) | 1 |
| 6 | A6 | Traffic accident | 3 |
| 7 | A7 | Traffic jam | 5 |
| 8 | A8 | Car trouble | 3 |
| 9 | A9 | Negligence car tank crew | 1 |

### 3.3 Occurrence and Severity Code

The occurrence and severity code has aims to determine the matrix of occurrence and severity to gets the minor and major impact. The severity code and occurrence code match to get the new code of E and A for put on the matrix of occurrence and severity. It aims to make a new code to determine the risk in what position - minor or major impact. It gets some code of risk agent and risk events (E1A1 until E7A9). The code can be shown in Table 5.

### 3.4 Occurrence and Severity Matrix

The occurrence and severity matrix show the minor and major impact. It can be shown in the Fig. 3.

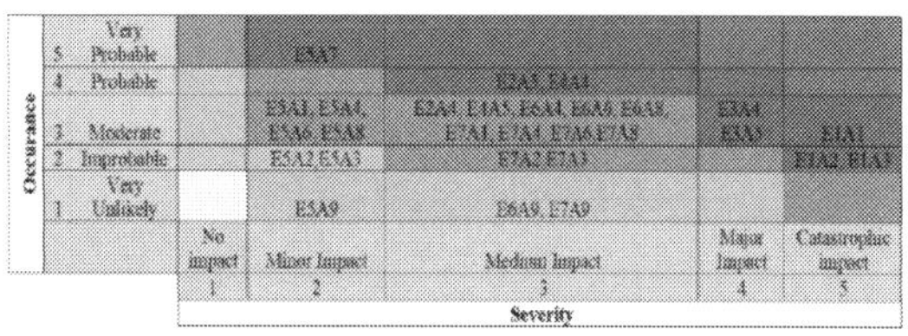


**Fig. 3.** Occurrence and severity matrix

From the matrix of occurrence and severity, it gets the major impact and minor impact of the risks. It is identified that E5A9 has a minor impact. It needs to eliminate the risk because it only focusses on the risk that has the major and medium impact and immediate mitigation is required.

**Table 5.** Risk event and risk agent

| Code | Severity | Code | Occurrence | Code E and A |
|---|---|---|---|---|
| E1 | 5 | A1 | 3 | E1A1 |
| | | A2 | 2 | E1A2 |
| | | A3 | 2 | E1A3 |
| E2 | 3 | A4 | 3 | E2A4 |
| | | A5 | 4 | E2A5 |
| E3 | 4 | A4 | 3 | E3A4 |
| | | A5 | 3 | E3A5 |
| E4 | 3 | A4 | 4 | E4A4 |
| | | A5 | 3 | E4A5 |
| E5 | 2 | A1 | 3 | E5A1 |
| | | A2 | 2 | E5A2 |
| | | A3 | 2 | E5A3 |
| | | A4 | 3 | E5A4 |
| | | A6 | 3 | E5A6 |
| | | A7 | 5 | E5A7 |
| | | A8 | 3 | E5A8 |
| | | A9 | 1 | E5A9 |
| E6 | 3 | A3 | 2 | E6A3 |
| | | A4 | 3 | E6A4 |
| | | A6 | 3 | E6A6 |
| | | A8 | 3 | E6A8 |
| | | A9 | 1 | E6A9 |
| E7 | 3 | A1 | 3 | E7A1 |
| | | A2 | 2 | E7A2 |
| | | A3 | 2 | E7A3 |
| | | A4 | 3 | E7A4 |
| | | A6 | 3 | E7A6 |
| | | A8 | 3 | E7A8 |
| | | A9 | 1 | E7A9 |

### 3.5 House of Risk 1

House of Risk is trying to relate a set of requirements (what) and a set of responses (how) where each response could address one or more requirements. It gives degree correlation such as low (one), moderate (three), and high (nine). Result House of Risk 1 can be shown in Table 6.

**Table 6.** House of risk 1

| No | Risk Event | Risk Agents | | | | | | | | |
|---|---|---|---|---|---|---|---|---|---|---|
| | | A1 | A2 | A3 | A4 | A5 | A6 | A7 | A8 | Si |
| 1 | E1 | 3 | 3 | 9 | | | | | | 5 |
| 2 | E2 | | | | 9 | 9 | | | | 3 |
| 3 | E3 | | | | 9 | 9 | | | 9 | 4 |
| 4 | E4 | | | | 9 | 3 | | | | 3 |
| 5 | E5 | 9 | 3 | 9 | 3 | | 9 | 9 | 9 | 2 |
| 6 | E6 | | | 9 | 3 | | 9 | | 3 | 3 |
| 7 | E7 | 3 | 1 | 1 | 9 | | 3 | 3 | 9 | 3 |
| | Oj | 3 | 2 | 2 | 2 | 1 | 3 | 5 | 3 | |
| | ARPj | 126 | 48 | 186 | 264 | 72 | 162 | 135 | 270 | |
| | Pj | 6 | 9 | 3 | 2 | 7 | 4 | 5 | 1 | |

ARP (Aggregate Risk Potential) is calculated, the result is shown in Table VI. It indicates the importance of each risk agent. It is the result of the likelihood of occurrence of the risk agent j and the aggregate impacts generated by the risk events caused by the risk agent j.

### 3.6 Pareto Calculation

Pareto Calculation needed to determine the number of risk agents with high-priority rank based on the ARPj, (from HOR 2). The calculation can be shown in Table 7.

**Table 7.** Pareto Calculation

| Risk Agent | Pj | ARPj | Cumulative ARP | % ARP | Cumulative ARP | |
|---|---|---|---|---|---|---|
| A8 | 1 | 270 | 270 | 21% | 21% | Priority |
| A4 | 2 | 264 | 534 | 21% | 42% | |
| A3 | 3 | 186 | 720 | 15% | 57% | |
| A6 | 4 | 162 | 882 | 13% | 70% | |
| A7 | 5 | 135 | 1017 | 11% | 81% | non-priority |
| A1 | 6 | 126 | 1143 | 10% | 90% | |
| A5 | 7 | 72 | 1215 | 6% | 96% | |
| A2 | 8 | 48 | 1263 | 4% | 100% | |

Based on Table VII. It gets the priority and non-priority risk agent from the Pareto calculation. It calculates the ARP percentage and cumulative ARP. The Cumulative ARP above 80% is not a priority and below 80% is a priority.

Diagram Pareto is used for the show between the ARP and cumulative ARP to know what the immediate mitigation is required. By using 80% cut off based on the value of ARP, four risk agents are identified to be "prioritized risk agent", (Table 7). They are A8 (Car trouble), A4 (Human Error), A3 (Error deposit via bank and underpayment) and A6 (traffic accident). The diagram Pareto can be shown in Fig.4.

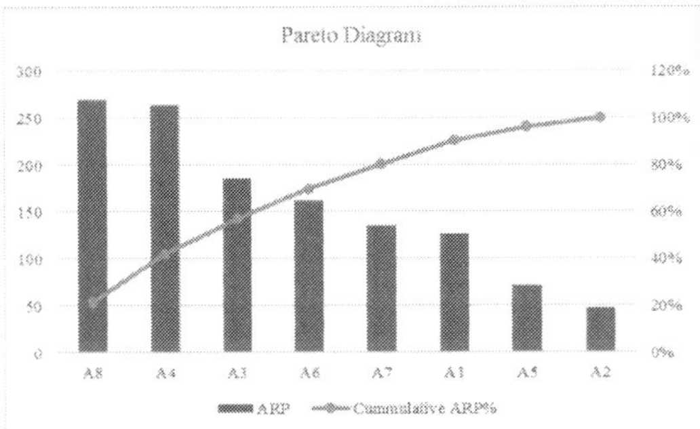


**Fig 4.** Pareto diagram

From the diagram, it can determine which the risk agent that should be the priority to mitigate. A8, A4, A3, and A6 become the priority that should be needed mitigation. The priority risk can be shown in Table 8.

**Table 8.** Priority Risk Agent

| Risk Agent | ARP | Cumulative | Risk Agent |
|---|---|---|---|
| A8 | 270 | 21% | Car trouble |
| A4 | 264 | 21% | Human Error |
| A3 | 186 | 15% | Error deposit via bank and underpayment |
| A6 | 162 | 13% | traffic accident |

After Pareto diagram, it has priority risk agent that has the ARP 270, 264, 186, and 162. It becomes the priority to mitigate. Based on risk agent identified, an intensive discussion has been done to identify all possible action to prevent the problem. Nine preventive actions are identified Table 9.

**Table 9.** Preventive Action

| No | Code | Preventive Actions |
|---|---|---|
| 1 | PA1 | SOP socialization |
| 2 | PA2 | Display SOP in the place of activity |
| 3 | PA3 | Conduct safety talk routine once every three days |
| 4 | PA4 | Conduct monitoring in the place of activity |
| 5 | PA5 | Conduct Standard Operating Procedure training |
| 6 | PA6 | Conduct check car tank routine |
| 7 | PA7 | Servicing regularly |
| 8 | PA8 | More communicative to bank and customer |
| 9 | PA9 | Reexamination of the data existing data |

To get effective and efficient action, some important preventive actions will be selected from all possible action. HOR 2 is applied; the result is in Table 10.

**Table 10.** House of Risk 2

| No | Risk Agent | Preventive Action | | | | | | | | | ARP |
|---|---|---|---|---|---|---|---|---|---|---|---|
| | | PA1 | PA2 | PA3 | PA4 | PA5 | PA6 | PA7 | PA8 | PA9 | |
| 1 | A8 | | | 9 | 3 | 9 | 9 | 9 | | | 270 |
| 2 | A4 | 9 | 9 | 9 | 9 | 9 | | | | 1 | 264 |
| 3 | A3 | | | | 1 | 1 | | | 9 | 9 | 186 |
| 4 | A6 | 9 | 9 | 9 | 1 | 9 | 9 | 9 | | | 162 |
| Total Effectiveness of Action (TEk) | | 3834 | 3834 | 6264 | 3534 | 6450 | 3888 | 3888 | 1674 | 1938 | |
| Degree of difficulty performing action (DK) | | 3 | 3 | 3 | 3 | 4 | 3 | 3 | 3 | 3 | |
| Effectiveness to Difficulty Ratio (ETD) | | 1278 | 1278 | 2088 | 1178 | 1612,5 | 1296 | 1296 | 558 | 646 | |
| Rank priority | | 5 | 6 | 1 | 7 | 2 | 3 | 4 | 9 | 8 | |

Based on nine preventive actions, a mapping analysis is done by house of risk 2. An expert judgment is applied to give a rating for each relation among risk agent and preventive actions. Based on the calculation of the total effectiveness, degree of the difficulty, effectiveness to difficulty ratio, and rank priority are gained. The priority preventive action to mitigate the selected risk agents can be shown in Table 11.

**Table 11.** Chosen preventive action

| No | Code | Preventive Actions | ETD |
|---|---|---|---|
| 1 | PA3 | Conduct safety talk routine once every three days | 2088 |
| 2 | PA5 | Conduct Standard Operating Procedure training | 1612,5 |
| 3 | PA6 | Conduct check car tank routine | 1296 |
| 4 | PA7 | Servicing regularly | 1296 |
| 5 | PA1 | SOP socialization | 1278 |
| 6 | PA2 | Display SOP in the place of activity | 1278 |
| 7 | PA4 | Conduct monitoring in the place of activity | 1178 |
| 8 | PA9 | Examination of the data existing data | 646 |
| 9 | PA8 | More communicative to bank and customer | 558 |

The priority preventive action is from Effectiveness to Difficulty (ETD). It gets the preventive action that should be the priority to mitigate the risk agent. The first priority preventive action is conducting safety talk routine once every three days and followed by conducting Standard Operating Procedure training.

## 4 Conclusion

Based on the result, the conclusion from this research is:
1. The distribution process of fuel gets 7 risks events and 9 risk agent identifications.
2. From the mapping of house risk 1 gets 5 risk agents that should be used to determine the mitigation. The risk agents are trouble car tank (A8) with ARP 270, human error (A4) with ARP 264, error deposit via bank and underpayment (A3) with ARP 186, and traffic accident (A6) with ARP 162.
3. Based on house of risk 2, two most significant preventive action are conducting safety talk routine once every three days (PA3) with ETD 2088 and conducting Standard Operating Procedure training (PA5) with ETD 1612,5

## References

1. L.D. Frendendall & J, E. Hill.) *Basics of Supply Chain Management*, CRC, (1999)
2. J. T. Mentzer, J. S. Keebler, N. W. Nix, C. D. Smith, & Z. G. Zacharia, *Journal Of Business Logistics*, Vol.**22**, No. 2, 1, *22*(2), 1–25, (2001)
3. D. Walters, *Supply Chain Risk Management*. London and Philadelphia Kogan Page Limite, (2006)
4. T.C.E Cheng, F.K.Yip, A.C.L Yeung, Supply risk management via guanxi in the Chinese business context: The buyer's perspective, *International Journal of Production Economics* **139**, 3–13, (2012)
5. Heckmann, T. Comes, S. Nicke,. A critical review on supply chain risk – Definition, measure. *Omega*, 121. (2014)
6. M. Giannakis, T. Papadopoulos,. Supply chain sustainability: A risk management approach. *Int. J. Production Economics*, 455-470. (2015)
7. G. Behzadi, M. J. O'Sullivan, T. L. Olsen, A. Zhang. Agribusiness Supply Chain Risk Management: A Review of Quantitative Decision Models. *Omega*, 1-22, (2017).
8. J. Kakeu, & M. Bouaddi Empirical evidence of news about future prospects in the risk-pricing of oil assets, *Energy Economics*, **64**, 458-468, (2017)
9. P. L Chu, C. Vanderghem, H. L. MacLean, B.A. Saville, Financial analysis and risk assessment of hydroprocessed renewable jet fuel production from camelina, carinata and used cooking oil, *Applied Energy*, 401-409, (2017)
10. X. Sun, C. Liu, X. Cheng, X. Li, Modeling systemic risk of crude oil imports: Case of China's global oil supply chain, *Energy*, **121**, 449-465, (2017)
11. M. H. Paydar, V. Babaveisi, A. S. Safaei, An engine oil closed-loop supply chain design considering collection risk, *Computers & Chemical Engineering*, **104**, 38-55, (2017)
12. R. Sreedevi & H. Saranga.. Uncertainty and supply chain risk: The moderating role of supply chain. *International Journal of Production Economics*, 332-342. (2017)
13. N. Pujawan and L. H. Geraldin.. House of risk: a model. *Business Process Management*, 953-967, (2009)
14. T. Kusmantini, A.D. Guritno, & H. Rustamaji, Mapping of Supply Chain Risk in Industrial Furniture Base on House of Risk Framework, **7(34)**, 104–115,.(2015)
15. B. Kristanto, B.. Aplikasi Model House of Risk ( Hor ) Untuk Mitigasi Risiko Proyek Pembangunan Jalan. *Jurnal Ilmiah Teknik Industri*, **13(2)**, 1–10, (2014)
16. M. Golparvar, & M. Seifbarghy, Application of SCOR Model in an Oil- producing Company, **4**, 59–69, (2009)

# Macroergonomic approaches as a solution to local wisdom-based tourist village development planning

*Ahmad* Padhil and *Hari* Purnomo

Faculty of Industrial Technology, Department of Industrial Engineering, Islamic University of Indonesia, Yogyakarta,

**Abstract.** Tourist villages are among the tourist attractions which are often visited by tourists. The development of these tourist destinations is undertaken by referring to the data of the Ministry of Tourism and Creative Economy. Currently, there are a total of 978 tourist villages in Indonesia. This number has increased sharply since 2009, in which there were only 144 tourist villages. Unfortunately, the existence of tourist villages might result in a shift in land conversion, tourism use, and clothes. Moreover, this might cause damage to cultural assets which can be reduced only by rigorous rules of the local wisdom. This research tries to improve the system adopted in the development of tourist villages using the approaches of Kansai Engineering and Macroergonomic Analysis and Design (MEAD). Based on the research findings using Kansai Engineering, there are 3 kansai words found to cause problems to the development of tourist villages, which were then repaired using the approach of the MEAD concept so that the proposed work system design is to integrate the existing components by formulating and creating visions & missions, work instructions, and Standard Operating Procedures (SOP).

## 1 Introduction

In Indonesia, the tourism sector is one of the third largest contributors of foreign exchange income (Ministry of Tourism, 2016). Foreign tourists visiting Indonesia increase as reported by Statistics Indonesia (BPS) which recorded the growth in the number of foreign tourists visiting Indonesia by 3.71 (CNN Indonesia, 2016). Among the tourist destinations which they visited were tourist villages. Currently, there are a total of 978 tourist villages in Indonesia. This number has increased sharply since 2009, in which there were only 144 tourist villages (Ministry of Tourism, 2016). The term *tourist village* itself refers to rural tourism where a small group of tourists lives in or near the traditional atmosphere (Inskeep, 1991).

The Province of the Special Region of Yogyakarta is one of the most successful areas in the development of tourist villages and Sleman Regency located in this area is one of the regencies that has 10 tourist villages and has always been a destination to host the international jamboree (*Pikiran Rakyat*, 2017). Unfortunately, the existence of tourist villages might result in a shift in land conversion, tourism use, and tradition which results in damage to cultural assets (Isnaini et al., 2015). Such a condition can be solved by reducing the rigorous rules of local wisdom (Hermawan, 2016). Among the methods that can be used to solve this problem is the concept of Macroergonomic Analysis and Design *(MEAD)*. This method accommodates the whole aspects in the evaluation of a system (Hendrick et al., 2001). Research on improving a working system using *MEAD* was conducted by Realyvásquez et al. (2015), Cahyaditha et al. (2013), Kazemi et al. (2017), and Purnomo et al. (2017). This method is also used to enhance work productivity (Robertson et al., 2017).

Based on the aforesaid problems, this research analyzes the *MEAD* approach that remains rarely applied in the development of tourist villages. It is expected that the findings of this research will provide an input for the government with regard to sustainable village development. In accordance with the mandate of the Law on tourism which stipulates that one of the objectives of tourism is to increase regional income in order to increase the welfare and prosperity of the people (UU, 1990).

## 2 Method

### 2.1 Data Collection

Data were collected by conducting interviews and distributing questionnaires. The interviews were conducted with Civil Servants of the Office of Tourism of the Province of the Special Region of Yogyakarta, and several village apparatuses and the management of tourist villages. Questionnaires were used to search for *kansai* words (Yuqing et al., 2013). The research sample consisted of 30 respondents (Nagamachi & Lukman, 2009) taken randomly using the following criteria: (a) male and female; (b) aged

20 to 50 years. As for the secondary data, they were obtained by studying literature and documentation on urban area planning.

### 2.2 Identification using Macroergonomic Analysis and Design *(MEAD)*

The steps of this research were then carried out based on the *MEAD* approach, namely the stages to obtain information so as to realize research objectives. The stages of the macroergonomic approach using the method of Macroergonomic Analysis and Design (Hendrick, 1997; Hendrick & Kleiner, 2001) among others are: (1) analyzing the environmental system and organizational subsystems; (2) defining the type of a production system and determining the level of performance expected; (3) defining operating units and work processes; (4) identifying variances; (5) creating a variance matrix; (6) creating a table of key variance control and the analysis of personnel's roles; (7) allocating functions and combining designs; (8) analyzing stakeholders' perceptions and responsibilities; (9) redesigning support and combining subsystems; and (10) improving the system.

## 3 Findings and Discussion

### 3.1 Establishment of the Problem Factor Tree

The problem factor tree was used to find out the cause of a problem. It was formulated based on interview and questionnaire results. Results of questionnaire validation using Cronbach's alpha by less than 0.7 (Eubank et al., 2016) discover the following words: (1) planning without community participation, 0.879; (2) implementation without community participation, 0.884; and (3) less strict rules, 0.873. The Problem Factor Tree developed in this research is presented in Figure 1.

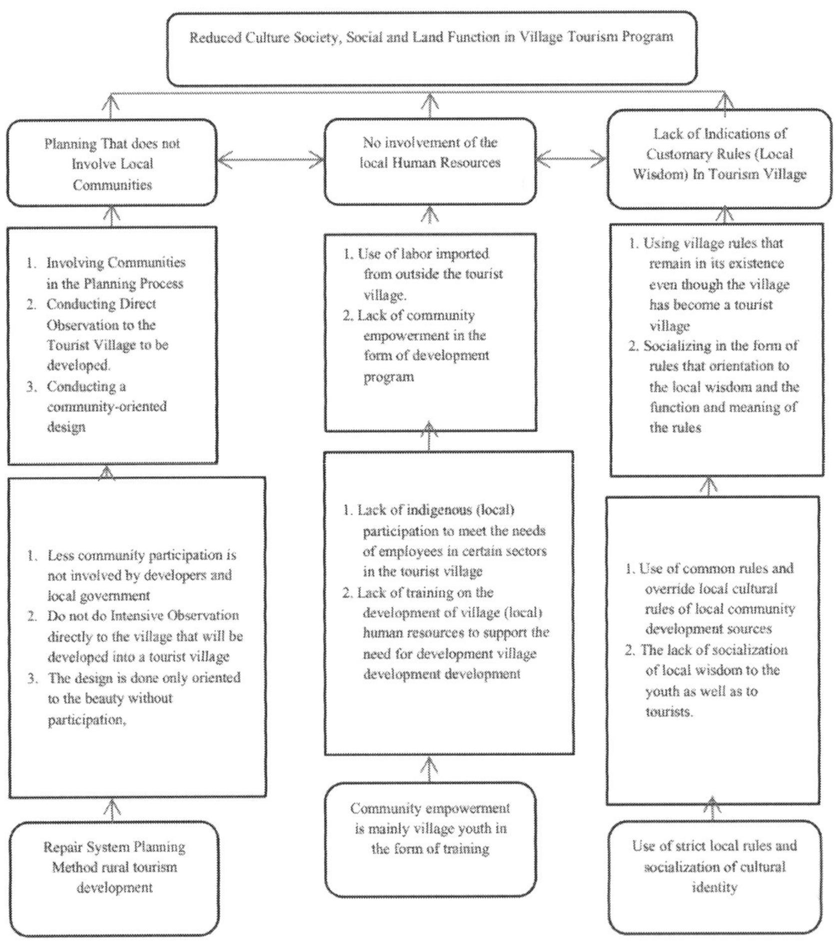


**Fig 1.** The problem factor tree

### 3.2 Macroergonomic Analysis and Design (MEAD) Identification

1. Analyzing the Environmental System and Organizational Subsystems (Step 1)
   During the development of tourist villages, the Provincial Government of the Special Region of Yogyakarta serves as a facilitator as well as a planner which is then in charge of holding an auction or making direct appointment in terms of the execution of village tourism development (Regional Regulation of the Special Region of Yogyakarta No. 3 of 2015). The regulation stipulates that the regional government shall work in conjunction with the developer to carry out the physical development of tourist villages.
2. Defining the Type of the System and Determining the Performance Level as well as Defining Work Processes (Steps 2 and 3)
   In the service-oriented production system, the Regional Government of the Special Region of Yogyakarta offers the tourism service concept where tourist satisfaction will serve as a vital parameter to measure the target performance (Ramseook et al., 2010). Work operating units are the work operating units of the tourism officewith the following divisions: (1) Destination Development Division tasked with developing a tourist village concept and development of new tourist village; (2) Capacity Improvement Division tasked with improving the tourist visit capacity in the Province of the Special Region of Yogyakarta; (3) Marketing Division tasked with marketing tourist destinations situated in the Province of the Special Region of Yogyakarta; and (4) Regional Technical Implementation Unit (UPTD), which is an Implementing Unit that performs the functions of implementation and supervision.
3. Identifying Variances (Step 4).
   Here are some of the variances or problems which occurred in the process of designing tourist villages as mapped in the problem factor tree. Table1.VarianceData.
4. Creating a Variance Matrix (Step 5).
   This stage aims to examine the relationship between the variances found in the planning process of village tourists. This matrix was made to determine the relationship level between variances and their data type.

**Table 1.** Variance data

| No | Process Stages | Variance | Possibility | Cause | Impact |
|---|---|---|---|---|---|
| 1 | Planning Does not Involve Local Communities | a. Planning does not match the demography of the village tourist areas | Only use data of similar project development experience | Previous projects may not necessarily have the same specifications | The existence of the design is not in accordance with the demographics of the community |
| | | b. Planning does not do direct observation To the tourist village | Rely solely on subjective data without direct appraisal to Target rural tourism project | No direct observation and data collection | Error in determining the allocation of both capital resources and materials to be used |
| | | c. Planning design is only developer oriented | The planning process is only profit-oriented | Pressing the amount of project cost | Less suitability of design results and wishes of the people in the tourist village |
| 2 | Not involved the human resources of the local community | Least Participation of local communities involved | The people around the tourist village do not have the skills | The level of community education around the village is still low | The existence of social turmoil that occurred between the developer and the community |
| 3 | Less stringent implementati on of the rules to local wisdom | Not in the socialization of local people's wisdom | The rules of Culture and local wisdom are perceived as a barrier to progress | The influence of western culture that is already deeply rooted in life | Reduced cultural rules of the original community will even disappear at a certain time |

**Table 2.** The variance data matrix

| Variance | Unit Operation | Variance data types | | |
|---|---|---|---|---|
| | | Significant Impacts | Has a lot to do with other variations | Has significant impact as a single variation |
| Planning does not match the demography of the village tourist areas | Destination Development Division | √ | √ | √ |
| Planning does not make observations To the tourist village | Capacity Improvement Division | √ | √ | X |
| Design design is only oriented to the developer | Destination Development Division | √ | X | √ |
| Least Participation of local communities involved | UPTD | √ | X | X |
| Not in the socialization of local people's wisdom | Marketing Division | √ | X | X |

Description:
√ = To indicate that a particular variable has an effect
X = To indicate that a particular variable has an effect

5. Creating a Table of Key Variance Control and the Analysis of Personnel's Roles (Step 6).
   This section identifies the control of variance which already existed in the Tourism Office of the Province of the Special Region of Yogyakarta and the roles of personnel in charge of the work unit where each variance occurred. Table 3. Key Variance Control and Personnel's Roles.

**Table 3.** Key variance control and personnel's roles.

| **Variance** | **Supevisior** | **Parties involved directly** | **Supporting activities that already exist** |
|---|---|---|---|
| Planning does not match the demography of the village tourist areas | Head of destination development section | Field survey officer and developer. | Mapping the potential of tourist areas |
| Planning does not make observations To the tourist village | Head of destination development section | Employee of tourism and developer sections | Development of tourist villages with infrastructure support |
| Design design is only oriented to the developer | Head of destination development section | Field survey officer and developer | Design using software facilities and technical personnel of the developer |
| Least Participation of local communities involved | Head of UPTD | UPTD employees as technical implement | The use of local labor at the bottom of a working structure |
| Not in the socialization of local people's wisdom | Head of marketing development section | Employee of promotion and marketing | Development of tourist villages with infrastructure support |

6. Allocating Functions and Combining Designs (Step 7).
   This stage aims to make some alternative improvements and allocate functions so as to get a better alternative. For information about the alternative generated, see Figure 2 which shows the possible alternative that can be implemented by the tourism office of the Province of

the Special Region of Yogyakarta. According to (Robertson et al., 1998), to allocate functions and combine designs, there are a number of alternative improvements that can be made with some adjustments to the existing conditions. According to (Mosard, 1982), to decide which alternative is better and can be applied to the scheme presented in Table 4, scoring can be undertaken by giving four assessment criteria. Table 5 presents the assessment criteria for scoring to assess the three alternatives. Table 5. Assessment criteria for scoring.

**Table 4.** The problem solving scheme

| Alternative | Implementation Technique | Solution | Goal |
|---|---|---|---|
| Alternative 1 | a. Conduct a live survey in planning<br>b. Provide Work Instructions and SOP | Improved work standards | Improvement of Village Tourism Development Implementation Process |
| Alternative 2 | a. Conduct FGDs with government, developers and communities<br>b. There is an Official Report of the Meeting | Maintain and improve the quality of tourist villages with community participation | |
| Alternatve 3 | a. Pemberdayaan Masyarakat<br>b. Pelatiahan Pemuda dan Masyarakat | Community empowerment | |

**Table 5.** Assessment criteria for scoring

| Criteria | | | |
|---|---|---|---|
| Access to the organization | Possible risks/obstacles to successful implementation | Benefits/effectiveness | The effect on the amount of spending |
| 1. Helping to improve kinship within the organization | 1. Employees are opposed to change | 1. Help improve the quality of life in the workplace | 1. Increase training costs |
| 2. Creating effectiveness in the organization | 2. The inability of employees to learn something new | 2. Decrease in Absence | 2. The cost of human resources |
| | 3. Lack of support from management | 3. Decrease Case of compensation | 3. Cost reduction performance |
| | 4. Training programs are lagging behind with current knowledge | 4. Improved Quality of service planning for community satisfaction | 4. Cost of ergonomic design work and training program. |
| | 5. Lack of active employee participation | 5. Increase work productivity | 5. Cost reduction design quality |
| | 6. The concept design is now incompatible with the new place design | 6. Decrease in Interest Participation | 6. Maintenance and replacement costs |
| | 7. Stress due to organizational change | 7. Decreased work stress | |
| | 8. Failed to use Work Instructions | 8. Improved Health and wellbeing | |
| | 9. Lack of feedback from management | 9. Decrease in the number of claims | |
| | | 10. Improve group calibration | |

 https://doi.org/10.1051/matecconf/201815401080

The scoring system for each criterion is described below:

1. Access to the organization = 1 to 2
2. Possible risks/obstacles to successful implementation = 1 to 9.
3. Benefits/effectiveness = 1 to 10.
4. The effect on the amount of spending = 1 to 6.

7. Analyzing Stakeholders' Perceptions and Responsibilities; (Step 8)
   This stage aims to give a score to each alternative improvement so as to get an alternative which is better and can be applied by the Tourism Office of the Province of the Special Region of Yogyakarta. The scoring of the three alternatives was undertaken by referring to Table 5 and results of this scoring are presented in Table 6.

**Table 6.** Table scoring alternatif

| | **Kategory** | | | | |
|---|---|---|---|---|---|
| Alternative | Organizational Reach (Table 5, Column 1) | Risks in Success (Table 5, Column 2) | Advantages / effectiveness (Table 5, Column 3) | Influence on costs (Table 5, Column 4) | Total Weights |
| Direct Survey of spaciousness, Preparation of SOP and WI | 2 | (-2) | 5 | (-2) | 3 |
| Bringing FGD Participation at the design stage with the Community | 2 | (-1) | 5 | (-2) | 4 |
| Community empowerment and training | 1 | (-2) | 8 | (-4) | 3 |

In Table 6, a negative sign (-) was put on the categories *possible risks/obstacles to successful implementation* and *the effect on the amount of spending* because they potentially produce a negative effect. According to Mosard (1982), the assessment scale used ranged from 0 to 10, where a score between 0 and 3 indicates low preference, a score between 4 and 7 indicates moderate preference, and a score between 8 and 10 indicates high preference.

8. Redesigning Support and Improving the System (Steps 9 & 10).
   Based on the scoring results in the previous stage, it is revealed that Alternative 2 has the highest score by 4, namely improvement by conducting FGD by involving the community in the planning and development of Work Instructions (WI) and meeting reports as requirements to decide whether physical work is allowed to be started or not as specified in the Standard Operating Procedures (SOP) and to formulate visions & missions.

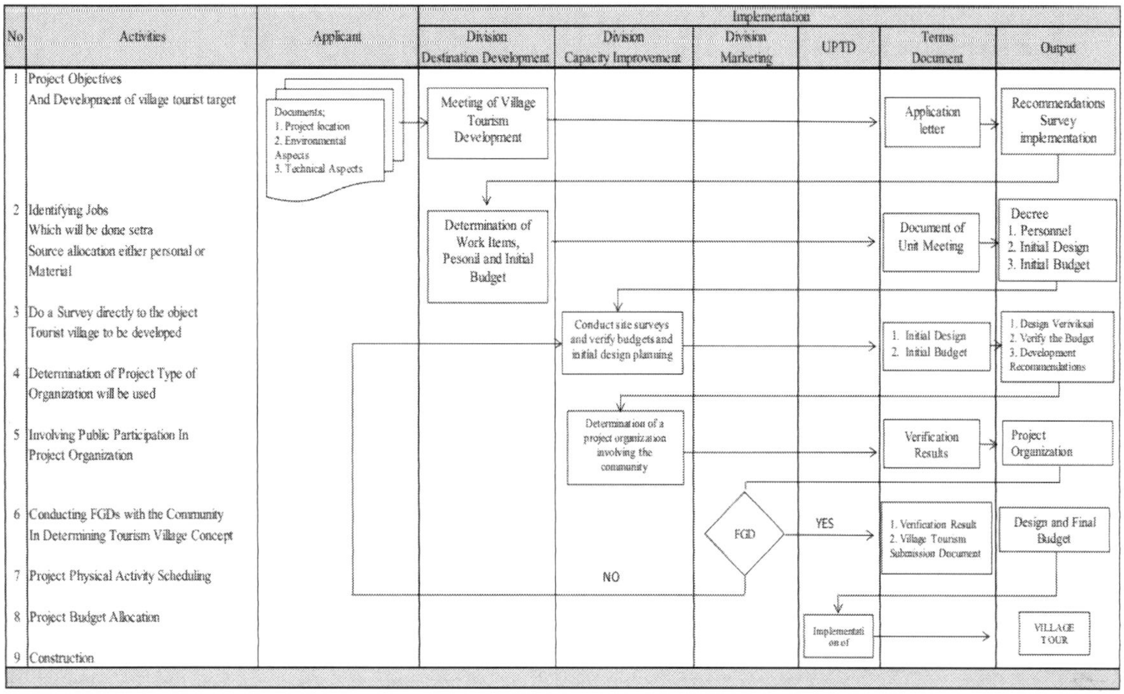


**Fig 2.** Below presents the scheme for tourist village development

## 4 Conclusions

Based on data processing and analysis results that have been conducted, the following conclusions are drawn:

1. Results of problem identification using the Macroergonomic Analysis and Design *(MEAD)* method suggest that the Tourism Office of the Province of the Special Region of Yogyakarta had not encouraged community participation properly.
2. The proposed improvement was obtained by selecting Alternative 2, i.e. improving the planning standards by conducting FGD with the government, the community, and the developer. This was done based on the four categories and their respective score, namely access to the organization (2), possible risks/obstacles to successful implementation (-1), benefits/effectiveness (5), and the effect on the amount of spending (-2).
3. The proposed work system design is to integrate the existing components by formulating and developing visions & missions, work instructions, and Standard Operating Procedures (SOP), which are expected to encourage the sustainability of local culture and community empowerment as well as development which suit the characters of a tourist village.

## Reference

1. Cahyaditha.A.A, Nazlinaz, Tambunan M.M ( 2013) *Perbaikan fasilitas penumpang kereta api pada stasiun x dengan pendekatan ergonomi makro*, e-Jurnal Teknik Industri FT USU Vol 3, No. 1, September 2013 pp. 22-29.
2. CNN Indonesia 2016, Pertumbuhan Kunjungan Wisatawan. [16 April 2017]. http:// www.cnnindonsia.com.
3. Eubank, B.H, Mohtadi, N. G Lavane, M. R, Willey, J.P & Ermy J.C.H. 2016, Further Validation and Reability Testing of the Rotator cuff Quality of life Index (RC-QOL) According to consensus-based standards for the selection of health measurenment instrument, Journal of Shoulder and elbow surgery (2016)
4. Hendrick. W, Kleinner.M.B, 2001, Macroergonomics: An Introduction to Work System Design, Human Factor and Ergonomics Society, Volume 2 ( 2001 )
5. Hermawan.H, 2016, Dampak pengembangan Desa wisata Ngelanggeran terhadap social budaya masyarakat lokal, Jurnal Ilmu pengetahuan dan teknologi computer, sekolah Internasional no 1-6.
6. Inskeep. E, 1991, Tourism Planning an Integrated and Sustainable Devolpment Approach. Hal 166. Wiley, Indonesia
7. Isnaini.W.N, Muktiali M. 2015, Pengaruh keberadaan desa wisata samiran terhadap perubahan lahan, ekonomi, sosial, dan lingkungan, Jurnal Teknik PWK Volume 4 Nomor 3.

8. Kazemi M, Safari S, Mooudi A.M, Mahaki. B. (2013), Macro-ergonomic risk assessment with the relative stress index method in textile industry, International Journal of Environmental Health Engineering, Vol. 2 Issue 6,| November-December 2013
9. Kemenpar, 2016, Rekap Wisata Nasional [15 April 2017]. http:// www. Kemenpar.go.id
10. Kemenpar, 2016, Tingkat Perumbuhan Desa Wisata [19 April 2017]. http:// www. Kemenpar.go.id
11. Mosard, 1982, Ergonomic desain for organizational effectivenees, Technology and engineering, CRC Press, 1988.
12. Nagamachi, M & Lokman A. M, 2009 Kansei Engineering: A beginners Persepctive, [21 April 2017], dari KDRU Publishing.
13. Perda Provinsi DIY, 2015, Tentang Kelembagaan Pemerintah Daerah Istimewa Yogyakarta, [26 April 2017]. http:// www.jogjaprov.go.id
14. Purnomo, H. Giyono, E. Apsari, A.E, the use of macro-ergonomic work system designs to reduce musculoskeletal disorders and injury risk in training, South African Journal of Industrial Engineering May 2017 Vol 28(1), pp 47-56
15. Ramseook. P, Bhiwaaje-Lukea D.S, Naidoo P., 2010 Service Quality in the public service International journal of management and marketing research volume 3 number 1
16. Realyvásqueza, A, Maldonado-Macíasa, A.A, Alcaraz. L.J, Fernándezb.B.JB, (2015) Effects of organizational macroergonomic compatibility elementsover manufacturing systems' performance, 6th International Conference on Applied Human Factors and Ergonomics (AHFE 2015) and the Affiliated Conferences, AHFE 2015.
17. Robertson M.M, Huang. H.Y, Lee .J, (2017) Improvements in musculoskeletal health and computing behaviors: Effects of a macroergonomics office workplace and training intervention, Applied Ergonomics, 62; 182 -196.
18. Undang- Undang Republik Indonesia No 9 Tahun 1990. [20 April 2017]. http;//www. Kemenpar.go.id.
19. Utami N.R, 2014, Usulan Perancangan Sistem Kerja, Universitas Muhammdiayah Surakarta, Fakultas Teknik, UMS, Surakarata
20. Yuqing, X., Kun C., Haibin, Q., & Zhaoyang, W. 2013. Study on application of kansei engineering in product from desaign. Switzerland: Applied mechanics and material Vol.274 (2013) pp 513-516.

# Workload assessment on foundry SME to enhance productivity using full time equivalent

*Amarria Dila* Sari, *Fajri* Hardiansa, and *Muhamad Ragil* Suryoputro

Departement of Industrial Engiineering, Islamic University of Indonesia, Indonesia

**Abstract**. Aluminium SME aims to increase the production amount by producing wok as much as 300 Units. The problem is workload analysis of operator on the wok production line in the wok foundry SME as well as the production cycle-making cycle time and analyze the workload received by the operator when producing 300 woks using the full time equivalent (FTE) method. This study aims to measure the workload of each division worker in the production process with a total of 13 workers observed. This study provides a work division recommendation based on the workload that has been carefully examined. This research involves percentage of workload effectiveness and the wages of workers. In lathe division have overload workload. While the printing division, melting inspection division, packaging and transportation division including normal workload category and the percentage of good work effectiveness. The result provides recommendations for the addition of 2 workers in each division that includes the category of overload of the lathe division with the number of initial workers as many as 13 workers to 15 workers. In the last stage perform a simulation by comparing the system of prefix work and proposal. The simulation results obtained with the initial work system to get an average of 223 woks / day. Meanwhile, for the proposed work system to get an average output of 291 woks.

## 1 Introduction

Determination of the amount of labour is sometimes done only based on estimates, however it would result in incompatibilities use of human resources in the company [1]. Every worker has the responsibility to complete a number of jobs within a certain time which is a workload for the workers [2]. Workload analysis aims to determine how the number of workers required to complete a job and how much load the appropriate delegated to one worker [3]. Workload is the number of job targets or target outcomes to be achieved within a certain time unit [4]. Workload is also defined as a management technique systematically performed to obtain information about the effectiveness and efficiency of the organization based on the work volume [3]. The determination of workload is based on the employee's ability to perform any standard tasks. The standards of employees average ability refers to the standard capabilities showing the size of the average energy given by an employee or group to obtain a unit of outcome [4]. This research will be used method of Full Time Equivalent supported by the stopwatch method to determine standard time of a job.

Full Time Equivalent (FTE) aims to change the hours of workload to the number of people needed to complete a particular job [5]. This FTE method will provide information on the allocation of employee resources to complete a job and time in each work activity that can be seen based on the results of the measurement of work time directly with the Stopwatch method. The Stopwatch method is a direct observed work-time measurement commonly applied to jobs that are generally short-lived and repetitive [6].

Related research to the workload has been carried out through various methods, for example the study of mental workload using NASA-Task Load Index (TLX) for industrial employees conducted by [7, 8]. Meanwhile, the study of physical workload for laundry worker using cardio vascular load method has been discussed by Sari [9]. Research regarding workload using workload indicators of staffing need (WISN) in Namibia [10], WISN in staffing nurse [11], also WISN Method for staffing in family health care [12] . Furthermore, research related workload and foundry has been conducted by Sharma & Singh [13], regarding work-related musculoskeletal disorders, job stressors and gender responses in foundry industry. The study concerning implementation of lean manufacturing principles in Foundries [14]. Standard time measurement method with a stopwatch can result in an accurate time with some activities if it is followed by a clear working order [15].

Based on the research workload that has been done on previous researcher it is necessary to conduct further research on workload measurement of workers in Foundry SME at wok production line at Bantul Yogyakarta (Figure 1). Aluminum SME is a home

* Corresponding author: amarria@uii.ac.id

manufacturing industry that produces kitchen utensils. The determination of workload is based on the standard of worker's ability to perform its duties [15]. Excessive workload can cause work atmosphere that is less comfortable for workers because it can trigger the emergence of work stress faster [16]. Therefore, must determine many workers based on production requirements so that every job done to get optimal results so that the results of improvement can be a reference or guidance to recruit the number of employees who must be accepted. The work load that has been obtained will then be converted to worker incentives. From the workload it can be known how much appropriate salary for the worker. As well as applying simulations to prefix work systems and recommendations to fit the actual work system [17].

## 2 Literature Review

The workload is the calculation and comparison between the needs of human resources or capacity required and the real capability of the human ([18, 19]). Besides that, according to [20] workload not only calculate the time spent for productive work but also includes the calculation about human factors, such as fatigue, personal needs, and looseness factor.

One method that can be used to measure the workload is a full time equivalent. Full time equivalent employees equal the number of employees on full-time schedules plus the number of employees on part-time schedules converted to a full-time basis. The number of full time equivalent employees in each industry is the product of the total number of employees and the ratio of average weekly hours per employee on full-time schedules. An industry's full-time equivalent employment will be less than the number of its employees on full- and part-time schedules, unless it has no part-time employees. The full time equivalent could be defined as the total amount of time needed by the worker to perform all the job in a certain time that analysed [18].

## 3 Method

### 3.1 Object and subject

Research was conducted at the Wok Aluminium SME with a variety job including melting, printing, thought, latheing, inspection, packaging, transpenim as shown in Figures 1 - 4. While in each bank two employees have been selected in accordance with some criteria including male, aged between 20 years to 35 years old, volunteer, with a minimum work experience is one year. Meanwhile, employees were selected based on the management, recommendation in consideration to the skills and effort.

### 3.2 Research Procedure

Research procedures were divided into the following phases:

a. **Preparation Phase**: The research preparation was an early stage prior to the study. determining the observation location; setting the standard time measurement form; preparing the tools used as a stopwatch and recording devices; and preparing the workforce to be observed.
b. **Determination of Allowance**: At this stage will be determined allowance consisting of leniency for personal needs, allowance to eliminate the sense of fatique, allowance for inevitable obstacles and work environment.

**Fig 1.** Melting division

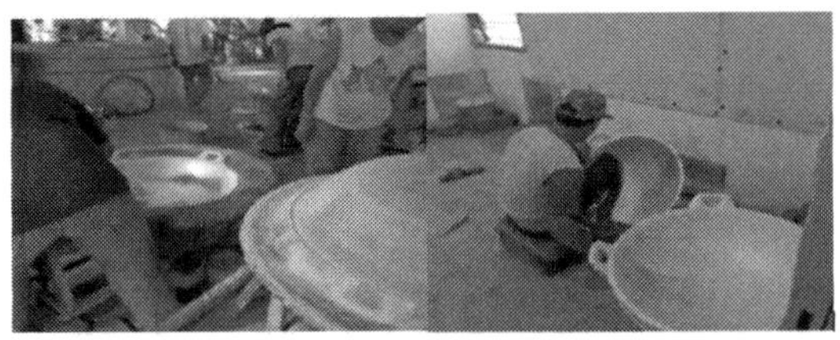

**Fig 2.** Printing and measurement divisions

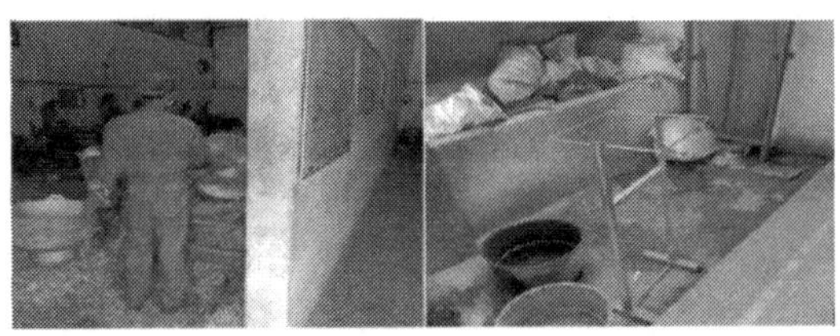

**Fig 3.** Lathe and inspection divisions

**Fig 4.** Packaging and transportation divisions

c. **Determining Effective Working Time**. At this stage will be determined effective working time is the number of days in the calendar minus holidays and leave. After the determination of the effective working days later determined the effective working hours per year: Total workday in hours multiplied by the percentage of the effectiveness of the work.

d. **Workload Calculation**. Full Time Equivalent aims to simplify work measurements by changing the hours of workload to the number of people needed to complete a particular job. The implication of FTE value is divided into 3 types: overload, normal, and underload. Based on the workload analysis guidance, the total FTE index value above 1.28 is considered overload, between 1 to 1.28 is considered normal whereas if the FTE index value is between 0 to 0.99 Considered underload or workload is still less than the maximum [21].
e. **Manpower Requirement Calculation**: At this stage, the optimal labor requirement is calculated for a certain position with the task approach per job assignment obtained from the calculation of the number of task completion times for one year divided by the number of working hours effective for one year. This approach is the result of FTE calculations to measure workload.
f. **Effectiveness Persetanse Workload Calculation and Actual Work Hours**. Every employee is also entitled to a break and time for worship. Working Hours are time to do work, can be done during the day and/or at night [1]. To determine the time can be done by calculating the time division work or individual work time of the worker. For employees with 5 working days per week, worker's obligation to work is 8 hours / day and 40 hours/week.
g. **System Modeling**: Simulation is an imitation of a facility or process of an operation, usually using a computer [22]. Software used is flexim software, Flexsim is a set of 3-D computer image processing technology, simulation techniques, artificial intelligence technology, and data handling techniques [17].

## 4 Result and Discussion

Respondents in this study were 13 male operators, aged between 24-34 years. At the level of education as many as eight workers have a high school education, while for 5 other workers is junior high. Work Experience owned by each operator is between 1 to 4 years. Based on Table 1, it can be seen that the amount of allowance for the wok production employee is 15%. Allowance is the adjustment done to the normal time to obtain the standard time for the purpose to recover the lost time due to personal needs, fatigue, and unavoidable delays [5].

### 4.1 Workload Measurement

The results of the calculation of adequacy and uniformity of data can be explained that the data collected is enough with the value of N '<N for all elements of the activity. Meanwhile, uniformity calculation data can also be concluded that all the data entered in the control limits, so that the data collected is already eligible to be counted workload. The calculation of workload with the full time equivalent method on the production of pan is shown in Table 2.

**Table 1.** Allowance

| Allowance Factor | Category | % |
|---|---|---|
| Energi Release | Low | 3 |
| Work Postur | Standing on two legs | 2 |
| Work Movement | Normal | 1 |
| Eye Fatigue | View is disconnected | 2 |
| Temperature | Medium | 2 |
| Environment | Circumstances exceptional | 2 |
| Atmosphere | Adequete | 2 |
| Personal needs | Male | 1 |
| **Total** | | **15** |

**Table 2.** Recapitulation workload per division

| Existing | | | | | | | | Recommendation | | | | | | | |
|---|---|---|---|---|---|---|---|---|---|---|---|---|---|---|---|
| No | Div | Work Element | No of Worker | FTE Ave | % Wload | Work Hour | Remarks | No | Div | Work Element | No of Worker | FTE Ave | % Wload | Work Hour | Remarks |
| 1 | Melt | 3 | 1 | 1.15 | 98 | 8 | Fit | 1 | Melt | 3 | 1 | 1.15 | 98 | 8 | Fit |
| 2 | Print | 6 | 3 | 1.04 | 88 | 7 | Fit | 2 | Print | 6 | 3 | 1.04 | 88 | 7 | Fit |
| 3 | Grind | 6 | 3 | 1.38 | 117 | 9.3 | Overload | 3 | Grind | 6 | 4 | 1.23 | 88 | 7 | Fit |
| 4 | Lathe | 9 | 3 | 1.34 | 114 | 9.1 | Overload | 4 | Lathe | 9 | 4 | 1.18 | 85 | 7 | Fit |
| 5 | Inspect | 5 | 1 | 1.05 | 94 | 7.5 | Fit | 5 | Inspect | 5 | 1 | 1.05 | 94 | 7.5 | Fit |
| 6 | Packaging | 7 | 1 | 1.11 | 94 | 7.5 | Fit | 6 | Packaging | 7 | 1 | 1.11 | 94 | 7.5 | Fit |
| 7 | Transp | 5 | 1 | 1.13 | 96 | 8 | Fit | 7 | Transp | 5 | 1 | 1.13 | 96 | 8 | Fit |
| Total | | | 13 | | | | | Total | | | 15 | | | | |

The full-time equivalent calculation for each division is only the smelting, printing, packaging, transport, and inspection division that has normal loads of 1.15, 1.04, 1.05, 1.11 and 1.13 since it is between 1.00 up to 1.28. Meanwhile, the grinding division is in the category of overload because the value of FTE is above 1.28. The calculation of the effectiveness of the work issued in each division is only the smelting division, printing, packaging, inspection, and transportation, which has a normal load with a value between 80% -100% i.e. 97.70%, 88.14%, 94.39%, 94.10%, 95.72%. Meanwhile, in the other two divisions in the division of thinking and division is in the category of overload because the value of workload is above 100%. Workloads charged to employees may occur under three conditions: normal workload (overwork), overload and underload [23]

Through good human resource planning, it can be seen how many human resources are available with the required human resources [24]. Therefore, it is of the problems found then get a recommendation addition of 1 worker in smelting division and division transportation whose value is in the normal category. The calculation of the effectiveness of work issued based on recommendations can be said to be normal workload of each division. This is because the effectiveness of the given workload is between 80% - 100%.

### 4.2 Simulation Result

The simulation technique is not an optimization process and does not produce an answer but only produces a set of outputs from the system under different conditions [17].

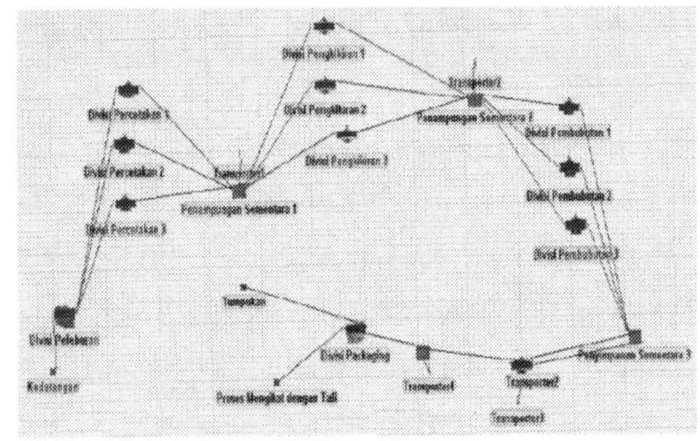

**Fig 5.** System modelling existing

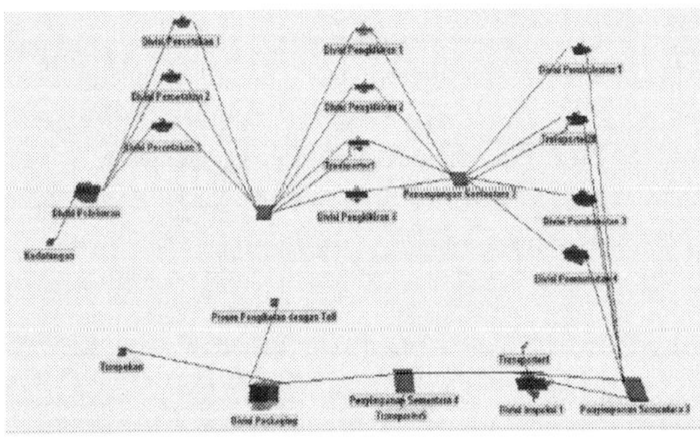

**Fig 6.** Proposed system modelling

Based on Figures 5 and 6, explaining the estimation simulation of the production process when performing production with workers and the available work system. In the simulation serves to estimate about the system under study and serves to estimate about the system created. In the simulation explains the stages of production of wok. The production process consists of 13 workers with 7 divisions with production capacity of 300 woks/day. In the printing division, the thinking division, the division of each worker is given a target of 74 woks/day. Meanwhile, for the division that employs 1 worker to process 300 woks/days. Figure 2.b describes the stages of the wok production of the recommendations. The production process consists of 15 workers with 7 divisions with production capacity of 300 woks/day. In the printing division, the grinding division, the division of each worker is given a target of 100 woks/day. Meanwhile, for the division that employs 1 worker to process 300 woks/days.

The pre-work system to get a big bottleneck that is in the first shelter of the printing division has a high pile that is an average of 48 units/day and at the temporary shelter division gets a stack averaging as many as 19 units. The total amount of bottleneck pile that occurred is 76 wok. From the simulation experiment, 25% of finished goods are not produced. Simulation is an application process building a model of a real system or proposed system, experimenting with the model to explain system behavior, studying system performance or to build new systems according to desired performance [25]. Simulation results from recommendation system get less bottleneck than prefix system. However, to overcome the bottleneck that occurs can be given treatment to the workers to be trained to work more quickly and consistently. In the first shelter of the printing, division shelter has a high pile that is an average of 1 unit and at the temporary shelter 2 that is the sheltered division of the lathe get a pile of wok average of 5 units/day. In some divisions ie division, inspection, inspection and transportation and weighing there is 1 stack average with the percentage of a stack of 3%.

## 5 Conclusion

Based on the research that has been done, it can be concluded that the workload capacity of the worker in the work system is the division of 1.15 with the effectiveness of 9, the printing division of 1.04 with the effectiveness of 88, the division of thinking 1.38 with the effectiveness of 117, the division 1.34 with the effectiveness 113, the inspection division of 1.05 with 94 effectiveness, the packaging division 1.10 with effectiveness 94, And the transpenim division 1.13 with the effectiveness of 95. Then the total number of workers to be employed in each division in the prefix system at the one-workers melting division, the printing press 3 workers, 4 workers' mind divisions, the division division adds 2 workers to 4 workers. For the inspection division of 1 worker. The last two divisions are the weighing and packaging transport divisions of each worker. The simulation results of many pre-loaded items with an average of 76 units of wok with average wok produced 223 units of wok/day. Meanwhile, when producing with a proposed work system, the workflow will be better with an average production of 291 pans/day with a total pan that is stuck for 9 woks.

## References

1. R. Ridha, A. Bakar, C. Nugraha. Usulan Kebutuhan Jumlah Tenaga Kerja di Bagian Water Based PT.X Berdasarkan Analisis Beban Kerja. *Reka Integra, Jurnal Online Industri Institut Nasional*, 1(2) 52-61 (2013)
2. R. M. Afia, M.E Ranu. Kontribusi Beban Kerja, Disiplin Kerja, Hubungan dengan Teman Sekerja Terhadap Produktivitas Kerja di PT. Viccon Modern Industry. Jurnal Administrasi Perkantoran (JPAP), 1 (3) (2013)
3. W. Adawiyah and A. Sukmawati. Analisis Beban Kerja Sumber Daya Manusia Dalam Aktivitas Produksi Komoditi Sayuran Selada DI CV Spirit Wira Utama. *Jurnal Manajemen Organisasi* (JMO) 4 ,128-143 (2016)
4. Menpan, "Pedoman Perhitungan Kebutuhan Pegawai Berdasarkan Beban Kerja dalam Rangka Penyusunan Formasi Pegawai Negeri Sipil (*Kep. Men. PAN Nomor: KEP/75/M. PAN/7/2004)*, Jakarta, Kementerian Pendayagunaan Aparatur Negara Republik Indonesia, (2014).
5. . Hurd, *FTE for Definition.Calculations,Examples And System Usage Information For HRMS And GL Budget Purpuses*, Colorado, (2004)

6. J. Hurd, *FTE for Definition.Calculations,Examples And System Usage Information For HRMS And GL Budget Purpuses*, Colorado, (2004)
7. A. Freivalds, B. Y. Niebels. *Methods, Standards, and Work Design*. Singapore: McGraw-Hill. (1999)
8. EH. Puspawardhani, MR. Suryoputro, AD. Sari, RD. Kurnia, and H. Purnomo. Mental workload analysis using NASA-TLX method between various level of work in plastic injection division of manufacturing company. *Advances in Intelligent Systems and Computing Advances in Safety Management and Human Factors*, Orlando (2016)
9. O. J. Sealetsa and A. Thatcher. Ergonomics issues among sewing machine operators in the textile manufacturing industry in Botswana. *Work*, 38, pp. 279–289 (2011)
10. A.D Sari, M.R. Suryoputro, M.D. Pramaningtyas, P.S Putra and S.B. Maulidyawati. Work Physiology Evaluation of Laundry Workers. *IOP Conference Series: Materials Science and Engineering* (2016)
11. P.A McQuid, R.L Kolehmainen-Aitken, and N. Forster. Applying the workload indicators of staffing need (WISN) method in Namibia: challenges and implications for human resources for health policy. *Human Resources for Health*, 11, 64 (2013)
12. D. Bonfim, A.M Laus, A.E Leal, F.M.T Fugulin, R.R Gaidzinski. Application of the Workload Indicators of Staffing Need method to predict nursing human resources at a Family Health Service. *Revista Latino-Americana de Enfermagem*, 24 (2016)
13. R. Sharma, R. Singh. Work-related musculoskeletal disorders, job stressors and gender responses in foundry industry. *International Journal of Occupational Safety and Ergonomics* (JOSE), 20, 363-373 (2014)
14. P. Tandon, A. Tiwari, S. Tamrakar. Implementation of Lean Manufacturing Principles in Foundries. *International Journal of Modern Engineering Research (IJMER)* IJMER, 4 (2). 46-50 (2014)
15. H. Purnomo. Workload Analysis For Determining The Number Of Employees At Bank Companies. GEMA Jurnal 201. 2068-2076(2015)
16. R. P. Wibawa, Sugiono et R. Y. Efranto. Analisis Beban Kerja Dengan Metode Workload Analisis Sebagai Pertimbangan Pemberian Insentif Pekerja di Bidang PPIP PT Barata Indonesia, Jurnal Universitas Brawijaya, pp. 672-683, (2014)
17. Zhu, X., Zhang, R., He, Z., Chu, F., & Li, J. Flexsim-Based Optimization for The Operation Process of Cold-Chain Logistics Distribution Centre. *5School of Mechanical, Electronic and Control Engineering*, 270-278. (2014)
18. OrientPoint. FTE Analysis and Models, Retrived from : http://www.orientpoint.com/FTE.htm (2014)
19. Wu, Hsin-Chieh, and Mao-Jiun J. Wang. "Relationship between maximum acceptable work time and physical workload." Ergonomics 45, no. 4 280-289 (2002)
20. R. M. Barnes, Motion and Time Study. Design and Measurement of Work. Wiley., 1980
21. Karo, G. Karo, and E.Adianto. "Pengukuran Produktivitas Karyawan Dengan Metode Full Time Equivalent (FTE) PT. Astra International Tbk Divisi Astra Motor Penempatan Jakarta Honda Center." JIEMS (Journal of Industrial *Engineering and Management Systems)* 7, no. 1 (2017)
22. Y. S. Leung, W. K. Wong et L. C. Mak, A simulation analysis of the impact of production lot size and its interaction with operator competence on manufacturing system performance,» *Simulation Modelling Practice and Theory 49*, pp. 203-214, (2014)
23. W. Novera, Analisis Beban Kerja Dan Kebutuhan Karyawan Bagian Administrasi Akademik Dan Kemahasiswaan Di Tata Usaha Institut Pertanian Bogor, (2010)
24. Hendrayanti, Endang. "Analisis Beban Kerja Sebagai Dasar Perencanaan Kebutuhan SDM." Jurnal LPPM: PARADIGMA 9, no. 01 (2011)
25. Y. G. Sandanayake , F. C. Oduoza et G. D. Proverbs, Systematic Modelling and Simulation Approach For JIT Performance,» *Robotics and Computer-Integrated Manufacturing 24*, pp. 735-743, (2008)

# Fuzzy subtractive clustering based prediction model for brand association analysis

*Imam Djati* Widodo*

Industrial Engineering Department, Universitas Islam Indonesia, Yogyakarta, 55584, Indonesia

**Abstract**. The brand is one of the crucial elements that determine the success of a product. Consumers in determining the choice of a product will always consider product attributes (such as features, shape, and color), however consumers are also considering the brand. Brand will guide someone to associate a product with specific attributes and qualities. This study was designed to identify the product attributes and predict brand performance with those attributes. A survey was run to obtain the attributes affecting the brand. Subtractive Fuzzy Clustering was used to classify and predict product brand association based aspects of the product under investigation. The result indicates that the five attributes namely shape, ease, image, quality and price can be used to classify and predict the brand. Training step gives best FSC model with radii (ra) = 0.1. It develops 70 clusters/rules with MSE (Training) is 9.7093e-016. By using 14 data testing, the model can predict brand very well (close to the target) with MSE is 0.6005 and its' accuracy rate is 71%.

## 1 Introduction

Marketing efforts include creating strategies, marketing tactics and operational activities has very significant role in product success. One of important aspect of marketing is branding management. Brand is important because it is not just a name for the product, but also an identity to differentiate it from the products resulting from other companies. Some studies showed that brand, one important aspect of marketing, has significantly contributed to the success.

Product that has positive brand association tends to get better probability to be purchased in the future [1] and recommended to other customer [2]. It can also increase customer's loyalty [3]. Brand is a symbol of the company name on a product that consistently delivers the features, benefits and services to its customers. Consumers in determining the choice of a product will always consider product attributes (such as features, shape, and color), however consumers are also pay attention on brand [4] [5]. Brand will guide someone to associate a product by selecting information elements in the product selection [6]. This was confirmed by Page and Herr which states that the design and brand have different influences on preferences and quality assessment [7]. Preference is determined by the design, while the quality is determined by both design and brand.

Brand equity will give special value to customers. It can help customers to interpret, process and store information related to the product. Brand equity is built by brand awareness, perceived quality, brand association and brand loyalty [8]. Brand association is claimed to the most important by some expert because it can drive to the better brand performance [9] to survive in this dynamic environment, brand association should involve institutional theory [10].

Brand association known can be used by consumers to identify other products from a company [11]. Many studies showed that positive brand association positively affects behavior. There are many different techniques to measure and develop brand association network [12] such as brand concept maps (BCM) [13]. However, it is still needed to develop techniques and methods to evaluate and predict the brand association.

Based on some facts and studies above, this research is intended to develop prediction method of the brand association based on customer perception of product attributes. Fuzzy clustering is implemented to divide the data space into fuzzy clusters, in which each representing a specific part of the system behavior. The fuzzy rules can be found for each cluster. By applying fuzzy systems, uncertainty of the case can be involved in linguistic labels to more adapted results [14, 15].

Fuzzy subtractive clustering (FSC) method is applied to get the prediction. The FSC method is used because it is fast method to estimating clusters of the data as basis to get estimation and it is more consistent than fuzzy C Mean method [16, 17].

FSC has been applied in some field of researches such as three phase induction control [18], Non Linear process modeling [19], Data mining [20] and web design [21].

Hand Phone (HP) or smart phone is chosen as the object of this study because it has a high growth rate. Based on research conducted by the Research on Asia

* Corresponding author: imamdjati@uii.ac.id

(ROA), the average rate of growth of HP in Indonesia from 2006 to 2010 are around 17% per year [22]. In fact, surveying HP Nielsen Company Indonesia stated that the average level of ownership HP increased by about 23% per year from 2006 to 2010 [23]. Sales growth is also expected to remain high for the next years. International Data Corporation (IDC) estimates that mobile phone sales will grow about 10%, equivalent to 48.8 million units in 2012 [24]. Besides, HP has a wide range and variety associated with features and price. Number of series HP-powered GSM in the market is more than 180 series. HP's low end product has the most variants. It is about 42% of the total series in the market.

## 2 Basic Concept

### 2.1 Brand

The brand is a name, term, sign, symbol, design or a combination of these things, intended to identify the goods or services and to differentiate them from competitors' products. According to Aaker "A brand is a name or symbol discriminatory (such as logo, seal or packaging) with a view to identify the goods or services of one seller or group specific seller thus distinguishing the goods and services produced by the competitor" [8].

Brand equity is an asset that can provide its own value in the eyes of its customers. Asset that it contains help customers interpret, process and store information related to the product and the brand. Besides giving value for the consumer, it also provides the brand equity value for the company. Aaker defines brand equity by "A set of assets (and liabilities) linked to a brand's name symbol that adds to the value provided by a product or service to a firm and for that firm' customers" [8]. Five categories of assets includes brand loyalty, image quality is captured, associations brand, brand awareness and the asset of other brands (other proprietary asset) such as patents, cap, distribution channels and others, all of which can provide value to the customer and the company. Five categories of assets underlying brand equity is illustrated in figure 1.

Romaniuk and Nenycz-Thiel underlined that the knowledge and experience of a consumer with a certain brand will influence the total amount of associations a consumer related to a brand [3]. Based on sport shoes study, Mühlbacher et al. claimed that cognitive brad research showed that brand association characteristics - such as favorability, number, uniqueness, and consensus - influence brand strength [25]. Moreover, the finding supported the theory of cognitive brand equity. Favorability, number, uniqueness, and consensus can be used as predictor of brand strength.

### 2.2 Fuzzy Subtractive Clustering

The idea of fuzzy clustering is to divide the data space into fuzzy clusters, each of which represents a particular part of the system behavior. After projecting the cluster to the input space, the antecedent part of the fuzzy rules can be found. In the fuzzy clustering algorithm, the membership function can be determined according to two possible methods [26].

In the first method, the cluster is projected orthogonal to the axis of the antecedent variables and membership functions attached to these projections. The second method uses multi-dimensional membership functions, the fuzzy cluster is projected to input space. Membership degree of data points are directly calculated in this cluster and is projected by distance from the center of the cluster is projected.

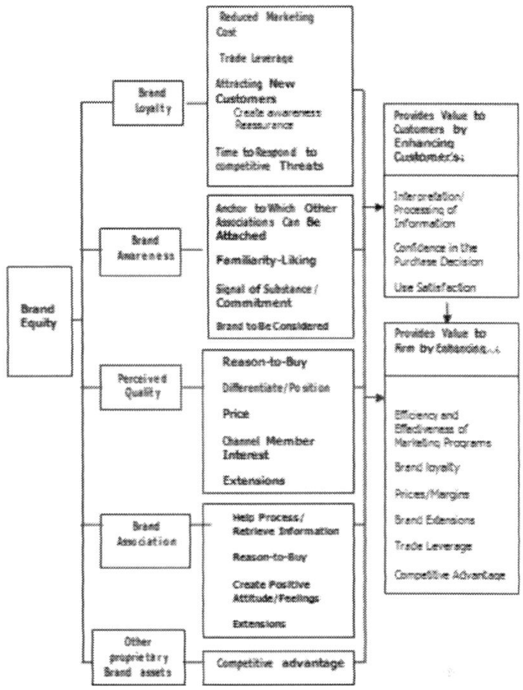


**Fig. 1**. Brand Equity Value [8]

Some fuzzy clustering methods have been developed in the literature. The most common method is fuzzy C-means clustering and subtractive clustering. Fuzzy C-Means (FCM) is a supervised filtering algorithm. It is developed based on a determined (known) cluster numbers o be performed. If the number of clusters to be formed is not known before, then an unsupervised algorithm should be performed. The most common unsupervised algorithm is Fuzzy Subtractive Clustering (FSC). FSC is based on the size of the density (potential) of data points in a space (variable).Another difference is related to the position of cluster central. In FCM method, the cluster central may not be in one of the clustered data. It will not happen in the subtractive clustering method. In the subtractive clustering method, a cluster center must be one of the clustered data, where the degree of membership in the cluster is equal to 1. The sum of all degrees of membership in FCM is always equal to 1, but not so with subtractive clustering. In the subtractive clustering method, the sum of all degrees of membership is not necessarily (even rarely) equal to 1.

FSC is a fuzzy clustering method that applied Sugeno fuzzy inference System (FIS) and it is unsupervised model. Figure 1 is the structure of the Sugeno FIS.

Subtractive Fuzzy Clustering (FSC), an extension of the mountain clustering method proposed by Yager and Filev is a method that is expected to resolve issues that arise in the previous grouping method in which the calculation is growing exponentially with the dimension of the problem, namely because the function must be evaluated at each grid point.

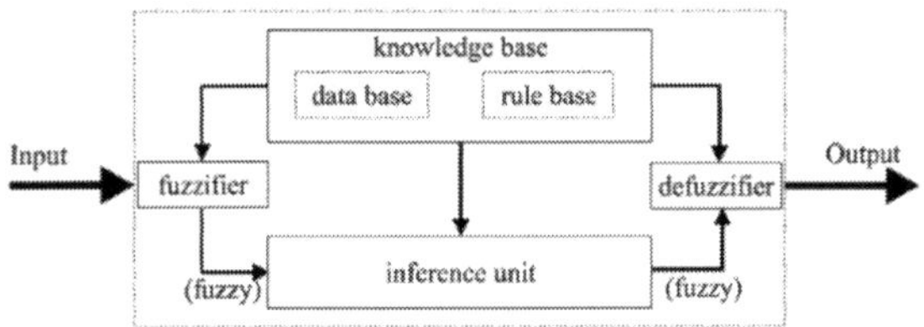


**Fig. 2.** Structure of FIS

FSC solves the problem by using a data point as a potential center of the cluster, not the grid points as in the mountain cluster. This means that the calculation is now comparable to the dimensions of the problem size; not dimension [27, 28]. Although the exact center of the cluster is not necessarily located at one data point, but in many cases it is a good approach. Each data point is the candidate for the center cluster, the size of the density of data points as shown in (2.1).

$$D_k = \sum_{j=1}^{N} \exp\left( - \frac{\left\| X_k - X_j \right\|}{(r/2)^2} \right), \qquad (2.1)$$

when ra is a positive constant that describes to a radius around a central point.

The first cluster center $x_{c1}$ be selected that have the greatest value of $D_{C1}$. Then measures the density corrected by equation 2.2.

$$D_k^{'} = D_k - D_{C1} * \exp\left( - \frac{\left\| X_k - X_{C1} \right\|}{(r_b/2)^2} \right) \qquad (2.2)$$

By Dk is a solid potential for the center cluster to k (ck). Rb is a positive constant that defines a radius around a central point which has a measurable reduction of solids. Rb value describes the range subtractive efficient and have a greater value than the value ra.

$$r_b = \eta\ r_a, \qquad (2.3)$$

with $\eta$ is a value greater than 1 is referred to as squash factor. After subtraction, any potential new cluster centers associated with upper acceptance limit and lower rejection limit and a distance criterion selects the second cluster centers.

In the implementation, there are 2 fractions as a comparator factor. They are Accept ratio and Reject ratio. Accept ratio and reject ratio is a fractional number that has value 0 to 1. Accept ratio is the lower limit where a data point becomes a candidate (candidate) cluster center is allowed to become the center of the cluster. While the reject ratio is the upper limit in which a data point to be a candidate (candidate) cluster center is not allowed to be the center of the cluster. On an iteration, if it has been found a data point with the highest potential (eg. Xk with potential Dk), then it will be continued by searching for the potential Ratios of the data point with the highest potential of a data point at the beginning of the iteration (eg. Xh with potential Dh). The result for Dk with Dh is then called Ratio (Ratio = Dk / Dh). There are 3 conditions that can occur in an iteration (figure 2):

- If Ratio> Accept ratio, then the data point is accepted as a new cluster center.
- If the Reject Ratio <Ratio ≤ Accept ratio then the new data point will be accepted as the new cluster center only if the data point is located at a considerable distance with the other cluster center (the sum of the Ratio and the nearest distance of the data point with a center other existing clusters ≥ 1). If the sum between the ratio and the longest distance of the data point with the other existing cluster center <1, then other than that point the data will not be accepted as the center of the cluster, it will no longer be considered to be the center of the new cluster (its potential is set equal to zero ).
- If the ratio ≤Reject ratio, then no more data points will be considered to be a cluster central candidate, iteration is stopped

## 3 Research Method

This study was designed to cluster the samples and develop FSC model to predict brand of products with regard to the nature of contingency products segment. Research design was explanatory or confirmatory research that aimed to find the relationship of research variable. A survey was conducted to determine the relationship between some HP independent variables and brand choice for low end segment. The independent variables were shape, quality, convenience, price and image.

This research consists of four steps namely data instrument development and instrument testing, data gathering, modeling, and model testing and prediction.

First, research instrument was developed based on some variables as found in the literature review. This study used consumer product attributes preferences questionnaire with Summated Rating Scale Method. A questionnaire was composed to get information from respondents.

Secondly, based on tryout data, instrument was tested for their reliability and validity. Thirty five respondents were involved in tryout. Thirdly, research data gathering was run based on cross-sectional research design type, in which one or more samples were taken from the population at certain of time.

The population of this research was the HP consumers aged between 17-55 years and use low/medium end HP. The sample was taken by purposive sampling method from Yogyakarta and Surakarta. Questions regarding the background of respondents asked in the questionnaire and if it did not meet the criteria will be sorted. Minimum number of samples was determined based on normal assumption (based Kolmogorov Smirnov test). This research used 97 samples. The next steps, FSC modeling was developed based on data gathered. Eighty-three data was used for model development. Finally, the model was used to predict the target. It was done by some other data (using 14 data). The result was tested to the fact.

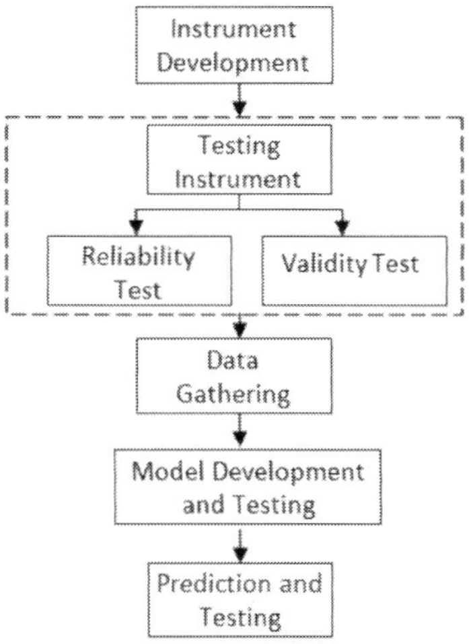


**Fig 3.** Research flowchart

## 4 Result and Discussion

There were three steps of analysis in this research; namely validity and reliability analysis, model development, and testing & prediction.

### 4.1 Validity and Reliability Analysis

Before it was applied, research questionnaire must be analyzed for their validity and reliability. Thirty seven respondents participated in try out step. All items were tested and the result was valid (r > 0.3) and reliable (alpha=0.8898).

### 4.2 Model Development and Evaluation

Data analysis was performed using Matlab 14. The program was done with the help function genfis2 and evalfis. Models were generated by using functions genfis. After doing several trials with different radii (r), the best model was found by considering the magnitude of the error rate and number clusters. Genfis2 was applied to generate a Sugeno-type Fuzzy Interface System (FIS) structure using subtractive clustering for adaptive neuro fuzzy interface system (ANFIS) training. Evalfis function was to perform fuzzy interface calculation.

The first step to develop a model is running training. Source code forthis model development (training) based on 5 independent variables is performed below:

```
In=P(:,2:6); Out=P(:,7);
fismat=genfis2(In,Out, r);
fuzout=evalfis(In,fismat);
trnrmse=norm(fuzout-Out)/sqrt(length(fuzout));
salah=fuzout-Out;
hasil1=[fuzout]
selisih1=[salah]
MSE1=sqrt(mse(salah))
plot(no(:,1),Out,'bo',no(:,1),fuzout,'r*');
pause;
```

The study applied Mean Square Error (MSE) to measure the error in this training stage. Best model selection is done by simulating the value of radii. The result can be seen in Table 1.

**Table 1.** The training evaluation result

| Radii (R) | Training | |
|---|---|---|
| | # rule | MSE |
| 0.9 | 2 | 0.5375 |
| 0.7 | 3 | 0.5230 |
| 0.5. | 10 | 0.2522 |
| 0.3 | 51 | 6.9609e-015 |
| 0.1 | 70 | 9.7093e-016 |
| 0.05 | 73 | 5.8138e-016 |
| 0.01 | 73 | 4.1362e-016 |
| 0.001 | 73 | 4.1362e-016 |

Based on the calculation subtractive clustering source code above, the results obtained were very close to the target (brand). In training step, with 84 data, the best model was developed. Fuzzy inference rules were generated by genfis function. Table 1 simulates model fuzzy inference rules in training steep. Number rules will increase with the decrease of radii. The smaller radii was used, the more clusters will be generated. The fewer clusters were generated, the number rules will be generated and the fewer MSE (training) found. However, rules number was constant for radii less than 0.1. Based on this condition, the study decides to choose model with ra = 0.1 for the next prediction. It gives 70 rules/clusters and its MSE is 9.709 e-16.

The prediction and its test source code is performed bellow.

```
CekIn=Y(:,1:5);
CekOut=Y(:,6);
Fuzcek=evalfis(CekIn,fismat);
plot(CekIn(:,5),CekOut,'bo',CekIn(:,5),Fuzcek,'r*');
pause;
err=Fuzcek-CekOut;
hasil=[Fuzcek]
selisih=[err]
MSE=sqrt(mse(Fuzcek-CekOut))
plot(err,'r*');hold on;plot(err);
pause;
```

By using 14 testing data, brand association prediction was done and evaluated based on both prediction MSE and accuracy. The best model (with ra = 0.1) gave very good result. The result showed that only four of the fourteen results that was significantly difference (more than 0.5) to the targets. It means the accuracy rate was 71% (Table 2).

**Table 2.** Brand association prediction accuracy (+/-0.5)

| No | Target | Prediction Result | Difference | Similar |
|---|---|---|---|---|
| 1 | 1 | 1.6666 | 0.6666 | No |
| 2 | 2 | 2.9143 | 0.9143 | No |
| 3 | 2 | 2.0000 | -0.0000 | Yes |
| 4 | 2 | 3.3070 | 1.3070 | No |
| 5 | 2 | 1.9375 | -0.0625 | Yes |
| 6 | 2 | 3.3501 | 1.3501 | No |
| 7 | 3 | 2.7500 | -0.2500 | Yes |
| 8 | 3 | 2.9732 | -0.0268 | Yes |
| 9 | 3 | 2.5877 | -0.4123 | Yes |
| 10 | 3 | 3.0000 | -0.0000 | Yes |
| 11 | 3 | 3.0000 | -0.0000 | Yes |
| 12 | 3 | 2.9906 | -0.0094 | Yes |
| 13 | 4 | 4.0000 | -0.0000 | Yes |
| 14 | 4 | 4.0000 | 0.0000 | Yes |

Figure 2 shows how close the prediction and the target visually for radii is 0.1. It looks that prediction is very close to the target. Numerically, it also gives smallest MSE (test) by 0.6005. It can be seen in Table 3.

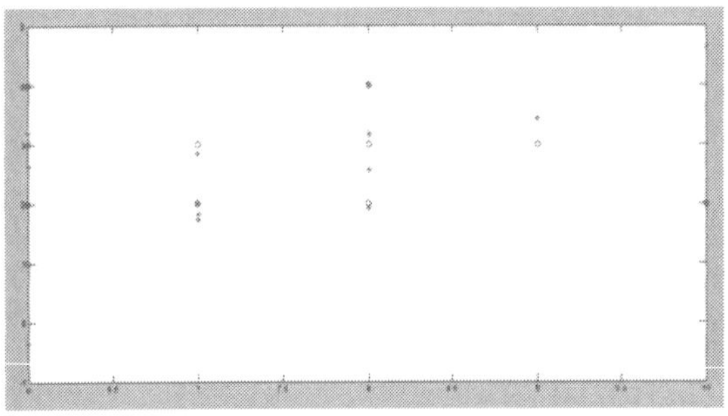

**Fig. 2.** Comparison between fuzzy subtractive clustering prediction results and target

The prediction performance (MSE Test) and accuracy is influenced much by the characteristic training and testing data.

If the data are similar to the training data, the performance of prediction will be very good. In contrary, if testing data have many outliers, data are higher than the highest training data or lower than the lowest training data, the model built by training data will not predict target of testing data well.

This research also has limitation in determining the best cluster used to get best prediction. It still uses simulation and by trial and error. Thus, further research in optimizing cluster number is needed. Some meta-heuristic such as genetic algorithm, simulated annealing can be applied.

**Table 3.** The test evaluation result

| Radii (R) | # rule | Testing | |
|---|---|---|---|
| | | MSE | Accuracy (%) |
| 0.9 | 2 | 0.8265 | 42% |
| 0.7 | 3 | 0.8430 | 42% |
| 0.5. | 10 | 2.7409 | 42% |
| 0.3 | 51 | 1.0881 | 57% |
| 0.1 | 70 | 0.6005 | 71% |
| 0.05 | 73 | 0.6467 | 71% |
| 0.01 | 73 | 0.5669 | 28% |
| 0.001 | 73 | 0.5825 | 28% |

## 5 Conclusion

1. The result indicates that the five attributes namely shape, ease, image, quality and price can be used to classify and predict the brand.
2. Training step gives best FSC model with radii ra = 0.1. It develops 70 clusters with MSE (Training) is 9.7093e-016.
3. By using 14 testing data, the model can predict brand very well (close to the target) with MSE is 0.6005 and its' accuracy rate is 71%.

## References

1. X. Bian, L. Moutinho, The role of brand image, product involvement, and knowledge in explaining consumer purchasing behavior of counterfeits: direct and indirect effects, *European Journal of Marketing*, **45**, 1919-216 (2013)
2. A. Belen, R. Vazquez, V. Iglesias, The effect of brand association on customer response, *Journal of Consumer Marketing,* **18**(5), 410-425 (2001)
3. Romniuk, Nanycz, , Behavioral brand loyalty and consumer brand association, *Journal of Business Research*, **66** (1), 67-72 (2013)
4. Randall, T Ulrich, Andreibstein, , Brand equity and vertical product line extent,*Marketing Science,* **17**(4) 361-365 (1998)
5. G. R. Foxall, V. K. James, The behavioralecology of brand choice: how and what do customersmaximize? *Psychology and Marketing*, **20**(9), 811-836, (2003)
6. R. Pieters, L. Warlop, Visual attention during brand choice: the impact of time pressure and task motivation, *International Journal of Research in Marketing*, **16**, 1-16 (1999)
7. C. Page, M. Herr,An investigation of the processes by which product design and brand strength interact to determine initial affect and quality judgments, *Journal Of Consumer Psychology*, **12**(2), 133–147 (2003)
8. A.D. Aaker, Measuring brand equity across products and markets, *California Management Review*, **38**, 3, 102-120(1996)
9. H. He and Y. Li, CSR and service brand: the mediating effect of brand identification and

moderating effect of service quality, *Journal of Business Ethics*, Vol. 100, 673-688, 2011

10. P. Virutamasena, K. Wongpreedee, W. Kumnungwutab, Strengthen brand association through se: institutional theory, *Procedia - Social and Behavioral Sciences,*195, 192 – 196 (2015)
11. D. W.Cravens, *Pemasaran Strategis*, Jilid 2, Edisi 4, Penerbit Erlangga (1998)
12. S.Gensler, F.Völckner, M. Egger, K. Fischbach, D. Schoder,. Listen to your customers: Insights into brand image using online consumer-generated product reviews. *International Journal of Electronic Commerce*, **20**(1), 112–141, (2016)
13. D. Bögera, P. Kottemanna, M. Meißnerb, R. Deckera, A mechanism for aggregating association network data: An application to brand concept maps, *Journal of Business Research*. 79, 90–106, (2017)
14. A, Y. Chamzini, M. Razani, S. Yakhchali, E.K. Zavadskas, Z. Turskis, Developing a fuzzy model based on subtractive clustering for road header performance prediction, *Automation in Construction*, 35, 111–120, (2013)
15. R. Pereira, A. Fagundes, R. Melício,V.M. Mendes, J. Figueiredo, J.C.Quadrado, Fuzzy subtractive clustering technique applied to demand response in a smart grid scope, *Procedia Technology*,17 ,478 – 486 (2015)
16. K.M.Bataineh,M. Najia, M. A. Saqera ,A.comparison study between various fuzzy clustering algorithms, *Jordan Journal of Mechanical and Industrial Engineering*, **5**(4), 335 – 343 (2011)
17. U. M. Raoa, Y.R. Soodb, K.Jarialb, Subtractive clustering fuzzy expert system for engineering applications, *Procedia Computer Science*, 48, 77 – 83 (2015)
18. J. Dhandapani, R. Chandralekha, Analysis of faults in three phase induction motor using neuro fuzzy logic and wavelet transformation, *International Journal of Control Theory and Applications*,**10**(34), 1-14 (2017)
19. A. M. Mohan, and M. J. Rani, Application of Soft Computing Techniques for Non-linear Process Modeling, *International Journal Of Innovative Research In Technology*, **2** (10) (2017)
20. M. Casta˜n´on-Puga, J. Flores-Parra, J. Castro, C. Gaxiola-Pacheco, and L. E. Palafox-Maestre, Building java intelligent applications data mining for java type-2 fuzzy inference systems, *Procedia Computer Science*, 51, 2719–2728 (2015)
21. D. S. Sisodia, S.Verma, O. P. Vyas,, A subtractive relational fuzzy c-medoids clustering approach to cluster web user sessions from web server logs, *International Journal of Applied Engineering Research*, **12**(7), 1142-1150 (2017)
22. Firdaus,Bisnis ICT di Indonesia, Siapa yang Menikmati?, *HarianBisnis Indonesia,* 17 Januari 2012
23. Nugroho, F, *PerkembanganPasarHandphone di Indonesia Dari Tahun 2005 Hingga 2010*, available: http://www.teknojurnal.com/2011/03/03/perkembangan-pasar-handphone-di-indonesia-dari-tahun-2005-hingga-2010/ (2012)
24. A. Wicaksono, *IDC perkirakanpenjualanponseltumbuh 10%*, available: http://industri.kontan.co.id/ news/2012-idc-perkirakan-penjualan-ponsel-tumbuh-10 (2012)
25. H. Mühlbacher, K. Raies, R. Grohs, O. Koll, Drivers of brand strength: Configural paths to strong cognitive brand equity. *Journal of Business Research,* 69, 2774–2780, (2016)
26. A. Priyono, P. Ridwan, A.J. Alias, R. Atiq, O. K. Rahmad, A. Hassan, and A. M. Ali, Generation of fuzzy rules with subtractive clustering, *Jurnal Teknologi*, 43(D), 143–153 (2005)
27. R.S.Sidhu, K. Sunil, P. S.Sandu,R. P. S.Bedi, K. Kaur, A subtractive clustering based approach for early prediction of fault proneness in software modules, *Engineering and Technology*, 67, 828-832, (2010)
28. M. T. Hayajneh, A. M. Hassan, F. Al-Wedyan, Monitoring defects of ceramic tiles using fuzzy subtractive clustering-based system identification method, *Soft Comput*, Mey, Springer, and Online Published (2009)

# Work postural analysis and musculoskeletal injury risk in critical working station at XYZ Ceramics Yogyakarta

*Amarria Dila* Sari*, *Annisa Rahmadani* Anwar, and *Muhammad Ragil* Suryoputro

Faculty of Industrial Technology, Industrial Engineering, Islamic University of Indonesia, Indonesia

**Abstract**. Companies that have manual work, muscle injury is one of the major health issues that often occur when working. XYZ Ceramics Yogyakarta is still doing work activities involving manual processes and there has been identified that worker posture has the potential to cause injury. From the Nordic Body Map questionnaire, complaint was identified in the back were 88%, the knees were 80%, the upper arm 75%, and the right *shoulder* by 70% from 15 workers. The Ovako Work Posture Analysis System (OWAS) method is used to evaluate and analyze work attitudes so that improved categories of work methods and recommendations could be obtained. Plan for *Identifiering av Belastningsfaktorer* (PLIBEL) was used to identify the factors that cause muscle injury or harmful effect. In addition, the Nordic Body Map was applied to determine the critical work station. The purpose of this research is to analyze work posture and description of risk level related to musculoskeletal injury at critical work station. The three highest percentage of risk of muscle injury identified at raw material processing station, based on calculation result of PLIBEL Checklist, were on elbow, forearm, and hand. Based on the assessment of working posture using OWAS, the highest action category is identified in the activity of moving the cart containing the soil to the automatic pest that is the action category of 4, which means that this activity is very dangerous for the musculoskeletal system.

## 1 Introduction

In the industry, working in non-ergonomics conditions is definitely not comfortable and easy to get fatigue, which ultimately decreases productivity. The impact of this non-ergonomics condition is a musculoskeletal complaint. Musculoskeletal complaint is a complaint on the part of the skeletal muscles felt by someone, varied from very mild to very painful [1]. Musculoskeletal system allows humans to stand up and move. In addition, the muscular and skeletal system also serves to protect the vital organs in the body. Muscular and skeletal system is closely related to limb, every injury or interference with this system would lead to disruption of the movement of a person whether temporarily or permanently.

In Yogyakarta, precisely in Kasongan located in Banguntapan, Kasihan, Bantul, 95% of its citizens living as a potter. There are 408 business units with the number of workers over 2,000 people and has entered the commercial export market in the industrial district Kasongan [2]. In the printing process in ceramics pottery observed, the posture of operator's body did not comply with the ergonomics rules. The position of the lower rotary tool makes workers should bend their body during the manufacturing process, lift, and put the pottery. From Nordic Body Map questionnaires provided for the workers in the XYZ Ceramics show complaints of 68% in most of the body felt by the operator at the work station processing for raw materials, complaints by 64% on the portion of the body for operator at the workstation of formation, complaints by 59% in some body parts of the operator's in burning work stations, complaints by 60% in most part of the body in finishing workstations, and complaints by 61% from packing workstation. Thus, indicating that the work activities in the XYZ Ceramics cause discomfort on the musculoskeletal.

Some study that support this research was varied, listed two major contributions for this research. The first was using Quick Exposure Checklist (QEC), NIOSH, and RULA that designing a proper seating for Transformer manufacturing industry in Germany on workers at coil winding machine workstation [3]. In addition, the result suggested that the analysis and implementation of ergonomic chairs could provide a major difference that can minimize the pain of the operator's body. Move on to the second research, which performed an assessment of operator posture in steel pipe production using OWAS, concluded that to develop a system of rotation of employees at workstations and ergonomic changes at the same time implementing preventive programs [4]. Both researches could emphasized on the need of using postural analysis to describe the occupational condition of a workstation and provide better solutions to improve the productivity. Moreover, this research also strengthened by

* Corresponding author: Amarria@uii.ac.id.

the use of PLIBEL (Plan for Identifering avbelastnings faktorer) to identify root cause of injury [5]. Therefore, this research was needed to be conducted in five stages. The determination of critical workstation was the first stage to identify the workstation that has the highest risk of musculoskeletal injury. There were 15 operators of XYZ Ceramics that fulfilling the questionnaire of Nordic body map. After the identification of the root causes of muscle injury that can cause harmful effects by using PLIBEL. Then, this research performed the assessment work posture or attitude using OWAS to evaluate and analyze the work attitude uncomfortable and result in injury musculoskeletal [1]. After the result of the identification of the risk factors of injury by using PLIBEL and posture assessment using OWAS were used to make suggestions or recommendations for work improvement. Moreover, these recommendations should be limited to create working conditions that are effective, convenient, safe, efficient, and also minimize the risk of injury to the operator who perform the work.

Objects in this study was the small businesses that addresses XYZ Ceramics in Kasongan Bantul, Yogyakarta. The samples taken which covered in the criteria for inclusion were 15 workers in the XYZ Ceramics Yogyakarta, which are respectively 3 work stations processing of raw materials, forming, firing, finishing and packing. Collecting data in this study were preliminary observations in the form of evaluation of the potential and symptoms of musculoskeletal complaints using Nordic Body Map questionnaires and PLIBEL checklist to identify the cause of the risk of injury.

## 2 Results and Discussions

This research was described on three main results based on the method used and continued with the future recommendations to ensure the improvement were made.

### 2.1 Musculoskeletal Complaint Analysis using Nordic Body Map Questionnaire

Nordic body map questionnaire describing the nuisance or complaints (pain, tenderness and stiffness) in the body during work performed. Using this questionnaire will impose the parts of muscle that have complaints with the various level of symptoms, ranging from discomfort (not ill) until severely ill (very ill) [6]. There were five work stations in the production at the XYZ Ceramics, namely raw material processing, forming, firing, finishing and packing. Raw material processing work station consisted of three operators, three operators at the work station formation, three operators on combustion, three operators on finishing, and three operators on the packing. The data resulted from the questionnaire from Nordic Body Map of an operator on raw material workstation can be seen in Table 1.

The weighting rate of Nordic Body Map questionnaire are categorized as follows:

1) Weights for no pain (discomfort/not ill) scale is 1,
2) The weights for the scale a bit sore (mildly ill) is 2,
3) Weights for pain (ill) scale is 3, and
4) Weights for scales of very sick (severely ill/very ill) was 4.

Furthermore, then calculate the score of each skeletal muscle complaints, the following is an example of how the calculation processed:

Sore / stiff in the upper neck = the total number of weights = total x weights

$= (0 \times 1) + (0 \times 2) + (1 \times 3) + (0 \times 4) = 3$

After the scores from each section of musculoskeletal has been obtained, then calculate the overall in total. To demonstrate clear the level of risk in each part of the body, it is provided marks with different colors according to the following conditions: the color blue is used to lower the risk level with a range of score 28 to 49, the color green is used to level the risk of being with a range of scores from 50 to 70, the yellow color is used for high-risk level with a range score of 71 up to 91, while red is used for the level of risk is very high with a range of scores 92 to the 112.

The results of the overall total score of each workstation operator is described on Table 2. The total score obtained indicate the level of risk of injury on the job every workstation observed. From the results of the questionnaire calculation of Nordic Body Map at each workstation that is part of the production can be seen that there are critical work stations at the workstation for processing of raw materials. Especially in the third operator processing of raw materials workstation that obtained results that indicate a high risk level.

**Table 1.** Nordic body map questionnaire for operator 1 in raw material workstation

| Complaint Type | Discomfort/ Not ill | Mildly ill | ill | Severely ill/ very ill | Score |
|---|---|---|---|---|---|
| Sick/ stiff in the upper neck | 0 | 0 | 1 | 0 | 3 |
| Hospitals / stiff in the lower neck | 0 | 1 | 0 | 0 | 2 |
| Pain in the left shoulder | 0 | 1 | 0 | 0 | 2 |
| Pain in right shoulder | 0 | 0 | 0 | 1 | 4 |
| Pain in the upper arm left | 1 | 0 | 0 | 0 | 1 |
| Pain in the back | 0 | 0 | 0 | 1 | 4 |
| Sick The upper arm right | 0 | 0 | 1 | 0 | 3 |
| Hospital at waist | 0 | 1 | 0 | 0 | 2 |
| Pain in the butt | 0 | 0 | 1 | 0 | 3 |
| Hospital dipantat | 1 | 0 | 0 | 0 | 1 |
| Pain in the left elbow | 1 | 0 | 0 | 0 | 1 |
| Pain in right elbow | 0 | 1 | 0 | 0 | 2 |
| pain in the left forearm | 0 | 0 | 1 | 0 | 3 |
| pain in the right forearm | 0 | 0 | 1 | 0 | 3 |
| pain in the left wrist | 0 | 1 | 0 | 0 | 2 |
| pain in the right wrist | 0 | 0 | 1 | 0 | 3 |
| pain in his left | 0 | 0 | 0 | 1 | 4 |
| ill The right hand | 0 | 0 | 1 | 0 | 3 |
| Pain in the left thigh | 0 | 0 | 1 | 0 | 3 |
| Pain in the right thigh | 0 | 0 | 0 | 1 | 4 |
| Pain in the left knee | 0 | 0 | 1 | 0 | 3 |
| Pain in the right knee | 0 | 0 | 1 | 0 | 3 |
| Pain in his left calf | 0 | 0 | 1 | 0 | 3 |
| Pain in right calf | 0 | 0 | 1 | 0 | 3 |
| Pain in the left ankle | 0 | 1 | 0 | 0 | 2 |
| Pain in the right ankle | 0 | 1 | 0 | 0 | 2 |
| Pain in the left leg | 0 | 0 | 1 | 0 | 3 |
| Hospital on the right foot | 0 | 0 | 1 | 0 | 3 |
| Total | 3 | 7 | 14 | 4 | 75 |

From the calculation of questionnaire, the Nordic body map in raw material processing workstation, the operator first obtained results score of 75, the operator 2 by 75, and the operator 3 for 76. All three scores indicate that level of risk of injury. musculoskeletal a high Additionally obtained complaints highest limb is on the right shoulder and back. Activities of workers when bent spine resulted in moves to the front side of the body, the muscles of the stomach and the front side of disc in the vertebratal in the lumbar suppressed. On the ligament back side of in vertebratal disk instead to stretch or sag [7]. This condition causes pain in the back. Besides the shoulder including a risky position if the position of the lift at the shoulders to form an angle of ≥ 450 from the vertical axis of the body, either laterally of the body and towards the front of the body [8].

**Table 2.** Summary of total score nordic body map

| Work Station | Operator | Score | Risk Level |
|---|---|---|---|
| Processing of Raw Materials | Operator 1 | 75 | High |
| | Operator 2 | 75 | High |
| | Operator 3 | 76 | |
| High-Formation | Operator 1 | 74 | High |
| | Operator 2 | 73 | high |
| | Operator 3 | 69 | Average |
| Combustion | Operator 1 | 63 | Average |
| | Operator 2 | 65 | Average |
| | Operator 3 | 70 | Average |
| *Finishing* | Operator 1 | 66 | Average |
| | Operator 2 | 69 | Average |
| | Operator 3 | 67 | Average |
| *Packing* | Operator 1 | 71 | high |
| | Operator 2 | 69 | Medium |
| | Operator 3 | 66 | Medium |

## 2.2 Injury Risk Factors Analysis Using PLIBEL Checklist

Data processing PLIBEL checklist obtained by calculating the percentage value obtained from any member of the body. This percentage value obtained by counting the number of "yes" in each column PLIBEL checklist that has been filled, then dividing by the total number of questions in the answer and multiplied by 100%. The results of data processing PLIBEL checklist for the operators in raw material processing work station are presented next. The results of data processing showed that the percentage value of the risk of injury to the operator 1 at the work station processing of raw materials to the neck, shoulders and upper back are 62%, elbow, forearm and hand by 73%, leg by 38%, knees and hips by 38%, and for the lower back by 62%. As for the environmental score of 50%. The results showed that the value of the percentage of the risk of injury to the operator 2 at the work station processing of raw materials to the neck, shoulders and upper back are 58%, elbow, forearm and hand by 73%, the feet of 38%, knees and hips by 38%, and for the lower back by 57%. As for the environmental score of 50%. Moreover, the results showed that the value of the percentage of the risk of injury to the operator 3 at the work station processing of raw materials to the neck, shoulders and upper back are 58%, elbow, forearm and hand by 64%, foot by 38 %, knees and hips by 38%, and for the lower back by 57%. For more details, can be seen in the chart below, Figure 1.

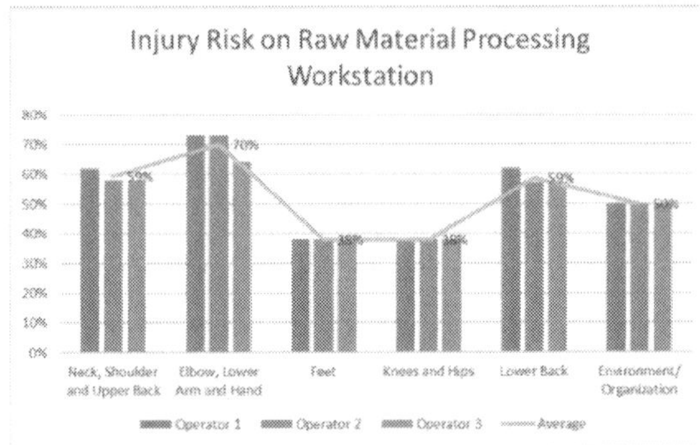


**Fig 1.** Injury risk on raw material processing workstation

The figure shows that the average on the body of the elbow, forearm and hand have a percentage of the highest risk of injury when compared with other body parts. Activities during the digging work resulted in the movement full extension where the elbow is driven repeatedly into top and bottom.

Several different result from other researches describing the highest score depends on the working environment, not only type of work. The calculation from PLIBEL checklist obtained at the installation workstation of granito shows that part of the neck, shoulders and upper back has the most severe musculoskeletal complaints compared to other body parts [9]. In other case, during research on the design of the critical workstations on the assembly at PT. Primarindo Asia Infrastructure Tbk., the data obtained from the calculation PLIBEL concluded that the percentage of the risk of injury at the buffing workstation, the highest score in the elbow, forearm and hand, that is equal to 72.73% [5].

Based on the analysis for the musculoskeletal risk factors, which was conducted by reviewing the 'yes' answer on the PLIBEL checklist data for every body parts, the risk factors were the inappropriate work facility design on the raw material processing workstation, including the height of the automatic pestle that exceed the height of operator's shoulder. In this case, the elbow, arm and hand should work a lot harder, prolong repetitive action, and physical workload (lifting heavy 22 kg weight load, transferring the dirt contented basket under the feet of operator into the automatic pounder). The object that has more than 12.5 kg in weight, raised above the shoulder, under the knees, or along the arms to more than 25 times per day could inducing the muscle fatigue, muscle damage, tendons and surrounding tissues [10]. Moreover, unnatural grip on the basket resulting uncomfortable hand position. Lifting object using only the power of fingers is not allowed, because of the limited capability of finger muscle, that could lead to finger injury [11].

## 2.3 Working Posture Analysis Assessment OWAS

The first step is assessing the posture of workers to the critical work stations starting from the initial stage to the final stage of the process. Based on the result from the activity category, the operator working posture 1 on the processing of raw materials are as follow. The activities are hoeing raw materials, move the raw materials into the basket and lift the basket containing ground pounder to the automatic. The following are the steps in determining the categories of worker attitudes for phase 1 activities hoeing raw materials:

1. Code OWAS for phase 1 at the work station processing of raw materials is 2141, because of the attitude of the back shows the code OWAS 2, the attitude of the arms shows the code OWAS 1, attitude feet (legs) shows the code OWAS 4, and the weight of the load shows the code OWAS 1 (Table 3).
2. Stage column, enter the code back to the line of code attitude BACK with shading in figure 2. Still one line of code BACK, that has been shading then to the right side for shading in figure 2 ARMS code.
3. Stage on the line, put the foot attitude code to the line of code LEGS by shading in figure 4. Still one column at code LEGS that has been in shading, then shading towards the bottom for the number 1 on the LOAD line code.
4. Pull shading on columns LOAD and rows ARMS, so both traction intersect. Box Meeting this cross is the category/level work attitude, which in 2141 had OWAS code category/level 3, which means in this attitude harmful to the system musculoskeletal (working attitude has resulted in a very significant influence of tension) and in need of repair as soon as possible.

**Table 2.** Category rating OWAS in Phase Activity 1

| BACK | ARMS | 1 | | | 2 | | | 3 | | | 4 | | | 5 | | | 6 | | | 7 | | | LEGS |
|---|---|---|---|---|---|---|---|---|---|---|---|---|---|---|---|---|---|---|---|---|---|---|---|
| | | 1 | 2 | 3 | 1 | 2 | 3 | 1 | 2 | 3 | 1 | 2 | 3 | 1 | 2 | 3 | 1 | 2 | 3 | 1 | 2 | 3 | LOAD |
| 1 | 1 | 1 | 1 | 1 | 1 | 1 | 1 | 1 | 1 | 1 | 2 | 2 | 2 | 2 | 2 | 2 | 1 | 1 | 1 | 1 | 1 | 1 | |
| | 2 | 1 | 1 | 1 | 1 | 1 | 1 | 1 | 1 | 1 | 2 | 2 | 2 | 2 | 2 | 2 | 1 | 1 | 1 | 1 | 1 | 1 | |
| | 3 | 1 | 1 | 1 | 1 | 1 | 1 | 1 | 1 | 1 | 2 | 2 | 3 | 2 | 2 | 3 | 1 | 1 | 1 | 1 | 1 | 2 | |
| 2 | 1 | 2 | 2 | 3 | 2 | 2 | 3 | 2 | 2 | 3 | [illegible] | 3 | 3 | 3 | 3 | 3 | 2 | 2 | 4 | 2 | 3 | 3 | |
| | 2 | 2 | 2 | 3 | 2 | 2 | 3 | 2 | 3 | 3 | 3 | 4 | 4 | 3 | 4 | 4 | 3 | 3 | 4 | 2 | 3 | 4 | |
| | 3 | 3 | 3 | 4 | 2 | 2 | 3 | 3 | 3 | 3 | 3 | 4 | 4 | 4 | 4 | 4 | 4 | 4 | 4 | 2 | 3 | 4 | |
| 3 | 1 | 1 | 1 | 1 | 1 | 1 | 1 | 1 | 1 | 4 | 3 | 3 | 3 | 4 | 4 | 4 | 1 | 1 | 1 | 1 | 1 | 1 | |
| | 2 | 2 | 2 | 3 | 1 | 1 | 1 | 1 | 1 | 4 | 4 | 4 | 4 | 4 | 4 | 4 | 3 | 3 | 3 | 1 | 1 | 1 | |
| | 3 | 2 | 2 | 3 | 1 | 1 | 1 | 2 | 3 | 4 | 4 | 4 | 4 | 4 | 4 | 4 | 4 | 4 | 4 | 1 | 1 | 1 | |
| 4 | 1 | 2 | 3 | 3 | 2 | 2 | 3 | 2 | 2 | 4 | 4 | 4 | 4 | 4 | 4 | 4 | 4 | 4 | 4 | 2 | 3 | 4 | |
| | 2 | 3 | 3 | 4 | 2 | 3 | 4 | 3 | 3 | 4 | 4 | 4 | 4 | 4 | 4 | 4 | 4 | 4 | 4 | 2 | 3 | 4 | |
| | 3 | 4 | 4 | 4 | 2 | 3 | 4 | 3 | 3 | 4 | 4 | 4 | 4 | 4 | 4 | 4 | 4 | 4 | 4 | 2 | 3 | 4 | |

Based on Table 3 the results of the activities of Phase 1 work attitude categorization, hoeing raw material value is 3, which means in this attitude harmful to the musculoskeletal system (work postures lead to a very significant influence of tension) and need immediate repair perhaps. Based on the summary of the three operators, it can be seen that the results of the categorization of the working posture at all three operators generate action categories of action Similarly. The three operators due to movement to work almost the same activity. The value of the highest category is on the activities of transferring the basket containing soil into the automatic pounder that is resulted on the action category 4, which shows that this attitude is very dangerous on the system. In addition, musculoskeletal work postures also lead to an obvious risk and in need of repair directly or current too.

**Fig 2.** Operator hoeing raw materials

Can be seen in Figure 2 that the working attitude at this third category is the attitude of on the foot stand, rests with both feet with your knees slightly bent. Similarly, research conducted by [12], work attitude in category 3 obtained has a foot position that rests on both legs bent. This foot position does not provide a stable body of work, so that the body easily slips workers. Working attitude in this category have a significant harmful effect on the musculoskeletal system, so that the necessary repairs had to be made as soon as possible.

**Fig 3.** The operator insert raw materials into a basket

Shown in Figure 3, the placement position of raw materials under the reach of the hand or foot is parallel to the worker, forced the workers to perform the process of moving raw materials by means of bending and standing leg stance with their knees bent. Moreover, working in a squatting position, bending and kneeling is also a posture awkward, the attitude or position of the body parts that deviate from the neutral position, significant deviation to the normal position will increase the workload of the muscle so that the amount of force needed is greater. This is because the result of the transfer of power from muscle to the skeletal system that are not efficient [10]. Therefore, this position needs improvement as soon as possible to avoid accidents and reduce musculoskeletal complaints seriously.

**Fig 4.** Operator entering raw material into automatic pounder

Based on Figure 4, the work attitude is very influential on the back. Heavy load and in the upper body (arms, shoulders, and upper back) resulted in a big moment in the spine caused by the distance the load to the body and the load is lifted exceed 10 kg. This condition, if performed repeatedly will cause disturbances in the back and waist in the form of musculoskeletal complaints [12]. Therefore, try to back and hip attitude cultivated in line when doing lifting activity. Because of these conditions can cause loading on the spine is relatively small, because the moment the body weight in the back does not happen.

### 2.4 Proposed Product

Based on the results of the above analysis, in order to minimize muscle injury in the XYZ Ceramics Yogyakarta operator is required to facilitate workers' tools while performing the job. In hence, the proposed products supplied in the form of a footrest to assist the operator when inserting the raw material into the automatic pounder and is expected to minimize the risk of muscle injury, especially on the neck, shoulders and arms. In designing this product proposal needed anthropometric data, size and determination percentile values are used to create a foothold in accordance with the rules of ergonomics. The dimensions of the body are taken into account are as follows: Standing shoulder height, Long and width of foot. Percentile used for the design of the footrests to be made is the fifth percentile that is 127.4 cm, then high pestle with a size of 138 cm automatically reduced by 127.4 cm so that the measures taken to footrest height is 10.6 cm. Meanwhile, the percentiles were used to draft a footrest to be made is the 95th percentile, which is 26.1 cm. Proposed footrest to be made used to assist or facilitate the operator when inserting the raw material into the automatic pounder for automatic grinding position is too high. The proposed design of the footrest can be seen in the figure below (Figure 5).

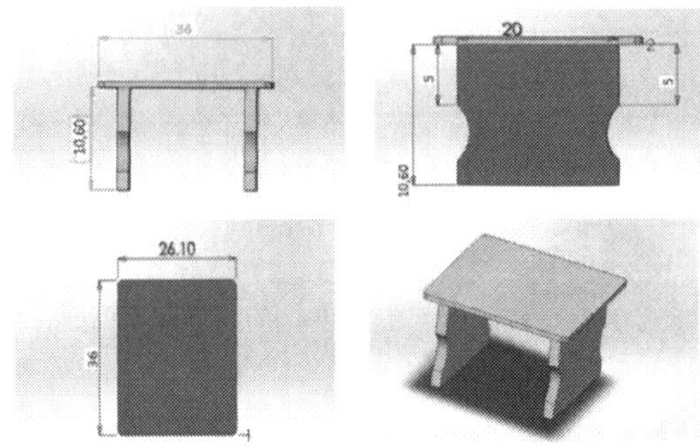


**Fig 5.** Proposed foot rest design

The measurement used for the design of the proposed design of the footrest obtained from percentile calculation for each section. Footrest height is 18.6 cm high, 5th percentile obtained from a shoulder stand, which is 127.4 cm, then high pestle with a size of 138 cm automatically reduced by 127.4 cm so that the measurement taken to footrest height is 10.6 cm. The length of 26.1 cm footrest is derived from the 95th percentile for the long dimension of the feet, while the width is 36 cm footrest length obtained from the position of the foot while standing parallel to the shoulders.

## 3 Conclusions

Based on the results of data processing and analysis has been done, the conclusions obtained are as follows:

a. a The factors that would cause musculoskeletal risk based on PLIBEL Checklist analysis are working facility design on processing of raw materials workstation, the physical load in the form of heavy lifting, and the not natural grips of the basket.
b. OWAS working posture assessment categories obtained, showed that the value of the highest action category is on the activities of transferring the basket containing ground into automatic pounder, which is on action category 4, that shows the very dangerous attitude on the musculoskeletal system and the need of direct improvement.
c. Recommendations given to improve the posture is a good working posture changes based on OWAS work attitude and add a footrest to assist the operator when inserting the raw material to automatic pounder

## References

1. E. Grandjean *Fitting the task to the Man*. Taylor & Francis. Philadelphia. (1993)

2. Disperindagkop. *Daftar Sentra Industri Kecil Kabupaten Bantul.* Bantul: Dinas Perindustrian Perdagangan dan Koperasi Kabupaten Bantul. (2006)
3. Wanave, B. Sandip, Bhadke, K Manish. *An Ergonomics Intervention in a Transformer Manufacturing Industry to Improve the Productivity.* IOSR Journal of Mechanical and Civil Engineering (IOSR-JMCE). PP 52-57. (2014)
4. A.M Lasota, K. Hankiewicz. *Working Postures of Operators in The Pipe Production Process*. A Case Study. *Logistyka*. (2014)
5. Y. Helianty, F.W Sachriadi, S.Caecilia, *Rancangan Stasiun kerja kritis pada Bagian Assembly di PT. Primarindo Asia Infrastructure, Tbk Berdasarkan Analisis Plibel Checklist.* Jurusan Teknik Industri, Institut Teknologi Nasional Bandung. (2010)
6. Tarwaka, Sudiajeng, L. Bakri, S.H.A., *Ergonomi Untuk Kesehatan dan Keselamatan Kerja dan Produtivitas*. UNIBA Press: Surakarta. (2004)
7. Triyono. *Analisis Sikap Kerja Pekerja Manual Material Handling UD. Tetap Semangat Dengan Metode OWAS.* Universitas Sebelas Maret, Surakarta. (2006)
8. Masliah. *Faktor Yang Berhubungan Dengan Keluhan Musculoskeletal Disorders (MSDs) Pada pekerja Manual Handling Di Pelabuhan Makassar.* Universitas Hasanuddin, Makassar. (2015)
9. R Barley, B. Aribowo. *Perancangan Perbaikan Stasiun Kerja Pemasangan Granito Menggunakan Analisis Metode PLIBEL Checklist di PT. Louserindo Megah Permai. Prosiding Semnastek.* Jakarta(2015).
10. Utomo. *Gambaran Tingkat Risiko Ergonomi dan Keluhan Gangguan Trauma Kumulatif pada Pekerja Pabrik Rahmat Tempe di Pancoran Jakarta Selatan.* Universitas Indonesia, Depok. (2012)
11. H.Laraswati, *Analisis Risiko Musculoskeletal Disorder (MSDs) Pada Pekerja Laundry (Studi Kasus Pada 12 Laundry Sektor Usaha Informasi di Kecamatan Beji Kota Depok).* Thesis. Universitas Indonesia. (2009)
12. R.D Astuti, B. Suhardi. *Analisis Postur Kerja Manual Material Handling Menggunakan Metode OWAS.* GEMA TEKNIK Majalah Ilmiah Teknik, ***10***(1), pp-67. (2007)

MATEC Web of Conferences **154**, 01084 (2018) https://doi.org/10.1051/matecconf/201815401084
*ICET4SD 2017*

# Operational risk analysis with Fuzzy FMEA (Failure Mode and Effect Analysis) approach (Case study: Optimus Creative Bandung)

*Taufiq* Immawan*, *Wahyudhi* Sutrisno, and *Annisa Kamilia* Rachman

Universitas Islam Indonesia, 55584 DI Yogyakarta, Indonesia

**Abstract.** Industrial development in Indonesia, manufacturing and services, are required to be able to manage the company very well. However, in practice, the company's activities are always faced with risks. In general, the risk can be defined as a situation faced by a person or a company in which there is a possibility that harm. The level of risk faced losses due to highly variable depending on the cause and effect influence. To be able to manage (risk management), it can use FMEA (Failure Mode and Effect Analysis). FMEA is a method of analyzing potential failure are applied in product development, system engineering and operational management and is one of a qualitative risk assessment. Using FMEA can also note the value of the RPN (Risk Priority Number) to determine improvement priorities at risk. But there are weaknesses in the use of FMEA, namely RPN calculation is only done by multiplying the severity, occurence and detection alone and irrespective of the degree of importance of each input, to the FMEA method is integrated using fuzzy logic. Fuzzy FMEA is aimed at obtaining the highest fuzzyRPN value which will be used as the focus of improvements to minimize the possibility of these risks occur back. The results were obtained 7 out of 18 types of risks that have a high priority for repairs. Risk troublesome computer (hank / die) while doing photo editing scored the highest RPN 540 (scale 1-1000) and also the highest FRPN 9 (scale 1-10). There is a difference in value between RPN and FRPN. FRPN value obtained from the fuzzification, generate value by taking into account the degree of interest of any given input.

## 1 Introduction

The risk is related to uncertainty, this happens because of lack or unavailability of adequate information about what is going to happen. Something uncertain might be beneficial or detrimental to the company. According to Wideman (1992), which raises uncertainty the potential to benefit known as the chance (opportunity), while uncertainties which give rise to these adverse effects is called the risk. For the creation of good corporate governance cannot be separated from the application of risk management so that risk management is a vital force for good corporate governance. Risk management is very important so that it becomes an urgent matter that should be applied, because it is through the implementation of a formal risk management, structured and integrated company will be able to adapt to the business environment.

The main objective of risk management is to keep operational activities that do not cause losses which exceed the ability of business entities to absorb such losses or endangering its survival. One element of risk that will inevitably arise in the business life activity at a business entity is operational risk, and is also the oldest and the inherent risk that comes before other risks (Muslich, 2007). Basically operational risk management is the understanding that the operational risks faced by the company can be identified, measured, controlled, mitigated and reported correctly through the implementation of the risk management framework. Optimus Creative Company is one of the services in the field of graphic design, photography, yearbook and event organizer. In the process of operation, the company is always confronted with a variety of business operational risks derived from changes in external and internal environment relating to business management. Some of the business operating risks among which the delay in payment by the consumer, the delay in the printing process, errors in the editing photos, errors in the printing results, delays in delivery, complaints from consumers and others. In order to carry out its activities in order to minimize the potential occurrence of such risks, the company is in need of policies and guidelines for risk management so that goals and objectives can be achieved. The method used to overcome this problem is by applying Failure Mode and Effects Analysis (FMEA) and the use of fuzzy logic.

Failure Mode and Effects Analysis (FMEA) is one method of failure analysis / potential failure are applied in product development, system engineering and operational management and is one risk measurement / risk measurement category qualitative risk assessment

* Corresponding author: immawan16@gmail.com

(Hanggraeni, 2010). Fuzzy logic is one method for analyzing systems that contain uncertainty (Kusumadewi S, 2002). The use of fuzzy logic in the FMEA to help determine the value of the Risk Priority Number (RPN) of failure occurring.

## 2 Literature Review

### 2.1 Risk Definition

Understanding the risks within the scope of the general public better known as something harmful from the presence of an incident. There are several definitions of risk, according to some experts, including:

- Risks in the Great Dictionary of Indonesian is a result of less favorable (adverse harm) of an acts.
- According Hanggraeni 2010 risk is an events that the potential to occur that might cause harm to a company.
- ISO / IEC Guide 73: 2002 Risk Management-Vocabulary-Guidelines for use in standards (2002) states that the risk is a combination of the probability of occurrence and consequences.

### 2.2 Operational Risk

Operational risk is the risk of most types of 'old' but least understood compared with other types of risks (eg market risk or interest rate). In this type of operational risk, there are inherent operational risk categories, namely risk arises because the company does business. For example, operational risk, namely, internal control systems are inadequate, work accidents, computer system failures, data corruption, failure to manage human resources (employees) and others. Basel II (which regulates international banking institutions) defines operational risk as the risk arising from the failure of internal processes, people, systems or from external events. There are two types of business operational risks that can be faced by the company (Hanggraeni, 2010), namely:

- The risk of non-entrepreneurial, that is a risk that was not caused by an entrepreneurial decision taken by the company. Companies usually use the services of insurance as a way to protect the company from this risk. Example: natural disasters, fires, and pollution.
- Entrepreneurial risk, namely the risk posed by the entrepreneurial decision taken by the company. If the company's business decisions and not precisely predict the decision, the company may experience significant losses. Example: the risk of build a new building, the risk of launching new products and the risk of merger.

### 2.3 Risk Management

Risk management is closely related to the company's sustainability. If the company carries out the management of risk, the company can avoid bankruptcy or even be able to generate increased profits. Risk management is a series of procedures and methodologies used to identify, measure, monitor and control risks arising from the business operations of a company (Hanggraeni, 2010). On the Project Risk Management PMBOK (2004), risk management is a systematic process of planning, identifying, analyzing, responding to and controlling risk.

Risk management is intended to ensure the sustainability of profitability and growth in line with the vision and mission of the company. In the control strategy and business risk management, corporate identification and risk maps (risk mapping), quantification and measurement of risk (risk measurement and assessment), risk management (risk treatment) as well as risk management policies (Hanggraeni, 2010). Risk Management process starting from identifying risk, analyzing risk, manage risk, risk management implementation and control of the application of risk management (ISO 31000-risk management system).

### 2.4 FMEA (Failure Mode and Effect Analysis)

Use of the method FMEA (Failure Mode and Effect Analysis) was first created in 1960 by the Aerospace Industry. In 1974, the Navy developed the so-called FMEA Procedure Mil-Std-1629 in which FMEA format is often used today. FMEA became popular after being used by the automotive industry to be implemented Integration into the product development process of their cars to overcome the problems of low product quality. FMEA until now still used primarily to prevent accidents and incidents at work (Mc Dermott, 2009).

**Table 1.** Severity rating scale

| Rating | Description | Criteria |
|---|---|---|
| 10 | Extremely dangerous | Failure could cause the death of a customer (patient, visitor, employee, staff member, business partner) and/or total system breakdown, without any prior warning |
| 9<br>8 | Very dangerous | Failure could cause a major or permanent injury and/or serious system disruption with interruption in service, with prior warning. |
| 7<br>6 | Dangerous | Failure could cause a minor to moderate injury with a high degree of customer dissatisfaction and/or major system problems requiring major repairs or significant re-work. |
| 5 | Moderate danger | Failure could cause a minor injury with some customer dissatisfaction and/or major system problems. |
| 4<br>3 | Low to moderate danger | Failure could cause a very minor or no injury but annoys customers and/or results in minor system problems that can be overcome with minor modifications to the system or process. |
| 2 | Slight danger | Failure could cause no injury and the customer is unaware of the problem; however, the potential for minor injury exists. There is little or no effect on the system. |

| Rating | Description | Criteria |
|---|---|---|
| 1 | No danger | Failure causes no injury and has no impact on the system. |

Source: Silva, M. M, 2014

FMEA is a structured procedure to identify and prevent as much as possible modes of failure. FMEA is used to identify the sources and root causes of quality problems. According to Wideman (1992), FMEA can be done by:

- Identify and evaluate the potential failure of a product and its effects.
- Registration process (document the process).
- Identify actions that could eliminate or reduce the chance of potential failure occurs.

In the use of FMEA methods have involved three things that assist in determining the risk to be able to know the value of the RPN (Mc Dermott, 2009), namely:

- Severity is the impact that arises when an error occurs. Table scale levels to the severity (damage) can be seen in Table 1.
- Occurrence is the likelihood or probability or frequency of occurrence of errors. Table scale levels for occurrence can be seen in Table 2.
- Detection is possible to detect a fault will occur or before the impact of these errors occur. Table scale levels for detection can be seen in Table 3.

**Table 2. Occurence rating scale**

| Rating | Description | Potential failure rate |
|---|---|---|
| 10 | Certain probability of occurence | Failure occurs at least once a day, or failure occurs almost every time. |
| 9 | Failure is almost inevitable | Failure occurs predictably, or failure occurs every 3–4 days |
| 8<br>7 | Very high probability of occurence | Failure occurs frequently, or failure occurs about once per week. |
| 6<br>5 | Moderately high probability of occurence | Failure occurs approximately once per month. |
| 4<br>3 | Moderate probability of occurence | Failure occurs occasionally, or failure occurs once every 3 months. |
| 2 | Low probability of occurence | Failure occurs rarely, or failure occurs about once per year. |
| 1 | Remote probability of occurence | Failure almost never occurs; no one remembers the last failure. |

Source: Silva, M. M, 2014

**Table 3.** Detection rating scale

| Rating | Detection | Definition |
|---|---|---|
| 1 | No chance of detection | There is no known mechanism for detecting the failure. |
| 2<br>3 | Very remote/unreliable chance of detection | The failure can be detected only with a thorough inspection, and this is not feasible or cannot be readily performed. |
| 4<br>5 | Remote chance of detection | The error can be detected with a manual inspection, but no process is in place, so that detection left to chance. |
| 6 | Moderate chance of detection | There is a process for double-checks or inspections, but it is not automated and/or is applied only to a sample and/or relies on vigilance. |
| 7<br>8 | High chance of detection | There is 100% inspection or review of the process, but it is not automated. |
| 9 | Very high chance of detection | There is 100% inspection of the process, and it is automated. |
| 10 | Almost certain chance of detection | There are automatic "shut-offs" or constraints that prevent failure. |

Source: Silva, M. M, 2014

## 2.5 Types of FMEA (Failure Mode and Effect Analysis)

There are two types of FMEA according to Mc Dermott, Mikulak & Beauregard (2009), namely:

1. Design / product FMEA.
   Design / product FMEA is an analytical technique based on the design of engineering / team that includes a potential failure mode causes of mechanical failures that arise in the process. The systematic approach is taken in parallel, formal and all documents related to the design engineer through several processes. This design FMEA is used to analyze the product before input into the production process. Design FMEA focus on the mode of failure caused by the design.
2. Process FMEA.
   Process FMEA is an analytical technique manufacturing or assembly process where in it contains a potential failure mode causes of mechanical failures that appeared in the production process. Each item of all the systems, sub-systems all the components should be evaluated. FMEA process is used to analyze the production and assembly process. FMEA of this type focused on the mode of failure caused by the production or assembly process.

## 2.6 Procedure of FMEA (Failure Mode and Effect Analysis)

There are ten steps in the procedure FMEA according to Mc Dermott, Mikulak & Beauregard (2009). Even this procedure applies to all types of FMEA, good product / design and process. The ten steps are:

1. Review the process or product.
2. Conduct brainstorming failure mode (a potential failure mode).
3. Register effects of the failure of each failure mode.
4. Establish severity weights for each effect of failure.
5. Establish the weight of occurrence for each failure mode.
6. Establish the weight of detection for each failure mode/effect failure

7. Calculate the value of the RPN (risk priority number) for each effect of failure. RPN is a measurement of the relative risk by multiplying the value of severity, occurence and detection. Within the scope of FMEA, this value can range from 1 to 1000. Mathematically it can be expressed as follows:
   RPN = (S) x (O) x (D)

In equation is the multiplication between the severity (S), occurence (O) and detection (D). RPN is determined before setting a corrective action recommendations and used to prioritize action.

1. Prioritize failure modes that have the highest RPN.
2. Take measures to eliminate or reduce the failure modes that have the highest risk.
3. Recalculating the value of the RPN after the failure mode have been reduced or eliminated.

### 2.7 Fuzzy Logic

Fuzzy logic is a proper way to map an input space into an output space. There are several reasons why people use fuzzy logic, among others (Kusumadewi, 2002):

1. The concept of fuzzy logic is easy to understand. Mathematical concepts underlying the fuzzy reasoning is very simple and easy to understand.
2. Fuzzy logic is very flexible.
3. Fuzzy logic can tolerate data that is not appropriate.
4. Fuzzy logic is able to model non-linear functions are very complex.
5. Fuzzy logic can develop and apply the experiences of experts directly without having to go through the training process.
6. Fuzzy logic can work with techniques conventional control.
7. Fuzzy logic is based on natural language.

### 2.8 Fuzzy FMEA (Failure Mode and Effect Analysis)

Fuzzy logic is one method to analyze systems that contain uncertainty (Kusumadewi, 2002). Research using Fuzzy Logic will get more accurate results than using conventional FMEA. According to Xu et al. (2002) and Yeh & Hsieh (2007), some weaknesses Conventional FMEA namely:

1. Statement of the FMEA often subjective and qualitative described in natural language.
2. Three levels of severity parameter (S), occurance (O) and detection (D) are assumed to have similar interests, it turns out in practice the weight of the interests of all three parameters are not the same.
3. The value of risk priority number (RPN) were produced by multiplying the level of S, O and D may imply a representation of risk.

To overcome the weakness - the weakness of the method is based on fuzzy logic is often used to manipulate the linguistic terms used directly in making a critical assessment. According to Iqbal et al. (2013), Fuzzy system is a knowledge-based system that is built on the expertise and experience in the form of fuzzy IF-THEN rules. Fuzzy inference method FMEA performed using Mamdani or often known by the method of Max - Min introduced by Ebrahim Mamdani in 1975. To get a fuzzy output requires four steps:

1. Arrange the Fuzzy membership function.
2. Create a rule-based fuzzy logic.
3. Perform Fuzzy inference process.

## 3 Case Study

Optimus Creative was founded in 2010, located in Sarimanah, Bandung, West Java. Optimus Creative is a creative company advertising agency, Graphic Design, Photography, Videography and Event Organizer. Services offered by this company is more focused on the annual book production services (yearbook) for each of the schools in the city of Bandung. Optimus Creative has opened various branches in Yogyakarta, Jakarta, Tasikmalaya, so today has been a lot of work together in making the school yearbook with more than 100 school.

### 3.1 FMEA Application

In making the yearbook through six processes, namely the presentation, meeting concept, photo session, photo editing, printing and delivery. The entire process is not immune from the risks experienced. To be able to identify the overall risk and knowing the weights for each risk, then the distribution of questionnaires to all employees who work at the company's operations. The results of the questionnaire were as follows:

**Table 4.** Recapitulation questionnaire

| No | Production Process | Risk Identification | Weighted Average |
|---|---|---|---|
| 1. | Presentation | 1. Presentation materials deleted/lost/not getting carried away | 4 |
| | | 2. Facilities on site presentations inadequate | 2 |
| | | 3. Team members did not attend the presentation | 2 |
| | | 4. Termination of unilateral presentation by the client | 5 |
| 2. | Meeting Concept | 5. Disagreements bout the concept of team and client | 2 |
| | | 6. The objection about the payment schedule | 4 |
| 3. | Photo Session | 7. Equipment for photo session is damaged | 5 |
| | | 8. Bad weather when the concept of photo out door | 2 |
| | | 9. Some students were absent/late when the photo session | 3 |
| | | 10. The late photographer team attendance | 2 |
| | | 11. Errors photo location information | 2 |
| | | 12. Unable to permit the shooting lovation (location not in a public area) | 3 |

| | | | |
|---|---|---|---|
| 4. | Photo Editing | 13. Computer problems (hank on/off) | 5 |
| | | 14. *File* data is lost | 5 |
| | | 15. Misinformation (different between the photo and description) | 3 |
| 5. | Printing | 16. The printer is trouble | 5 |
| | | 17. Result of the finishing not appropriate | 5 |
| 6. | Delivery | 18. Delay in delivery | 5 |

The results in Table 4 are then selected risks that have the highest average weight (score 5) for further processed using FMEA method. The main objective was to determine the method FMEA RPN value for each risk. RPN value is obtained by multiplying the value of severity, occurence and earlier detection that these values have been obtained through brainstorming and document review. RPN value can be seen in Table 5.

**Table 5.** Calculation results of RPN

| Potential Failure | Potential Effect(s) of Failure | SEV | Potential Cause(s) of Failure | OCC | Current Process Control | DET | RPN | Code |
|---|---|---|---|---|---|---|---|---|
| Termination of unilateral presentation by the client | Presentation materials conveyed not so big possibility of cooperation is not established. | 6 | Appearance and delivery of material less attractive presentation of the company's presentation team. | 2 | Re-discuss with the client to schedule a presentation. | 8 | 96 | a1 |
| Equipment for photo session is damaged | The scheduled completion of the photo session is too late, the catch is less good photos. | 5 | Treatment of equipment was not regularly | 4 | Replacement of the components of the equipment have been damaged when the photo | 4 | 80 | a2 |
| Computer problems (hank on/off) | Photo editing workmanship exceeds the specified deadline | 10 | Treatment of the computer is not on a regular basis, companies do not have a generator or a UPS (Uninterruptible Power Supply) to avoid the computer suddenly died due to power cuts. | 6 | Repairs done when knowing troubled computer | 9 | 540 | a3 |
| *File* data is lost | Photo editing workmanship exceeds the specified deadline and also can lead to the repetition of the previous process. | 10 | The process data storage is not good, data storage is only performed in a data storage medium | 2 | Storage of data stored on more than two data storage media | 6 | 120 | a4 |
| The printer is trouble | Book page printout was not optimal (there is a shadow, the color is not clear, etc.) and workmanship mold exceeds specified deadline. | 9 | Printing work overload and lack of oversight and periodic maintenance of the printing press. | 7 | The execution stops after finding trouble printing machine. | 6 | 378 | a5 |
| Result of the finishing not appropriate | Poor final quality books and pages in the book are easily separated from the book cover stitching. | 7 | Employees who work on finishing the book lacks focus and is not thorough. | 6 | Conduct an examination of the results of the book before the book entered the process of packaging and shipping. | 8 | 336 | a6 |
| Delay in delivery | Causing complaints or protests from the students and the school. | 9 | Work on editing and printing is delayed, stalled in the journey time of delivery, fault location information delivery. | 4 | Confirm beforehand with the school about delays in the delivery process | 9 | 324 | a7 |

### 3.2. Application of Fuzzy FMEA

In the processing of fuzzy FMEA using conventional FMEA result, Matlab 8 is used as a tool pengoalahan data. Fuzzy Logic Matlab toolbox is a software program that has been used in calculating the values of Fuzzy RPN. A model that was built in the techniques of Fuzzy FMEA has 3 inputs (severity, occurence and detection) and 1 output variable (FuzzyRPN) (Kumru, 2012).

There are 5 levels of membership function in the input variables, ie Almost None, Low, Medium, High and Very High (Kumru, 2012). Of the five membership functions are then producing 125 fuzzy rules, which are used as a factor in determining the value fuzzyRPN. As for the variable output used 10 level membership function in output variables, namely None, Very Low, Low, High Low, Medium Low, Medium, High Medium, Low High, High and Very High (Kumru, 2012)

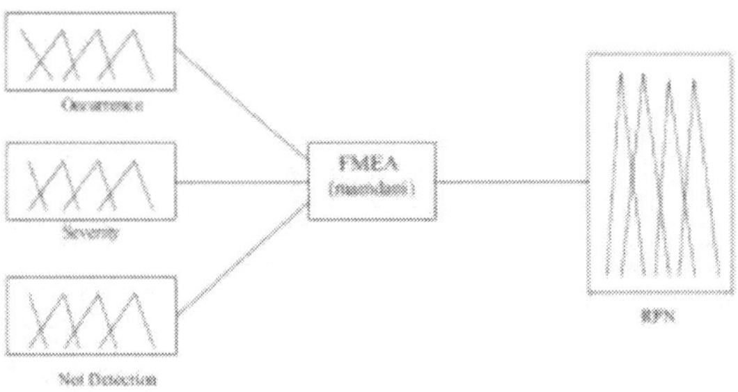


**Fig 1**. Fuzzy model

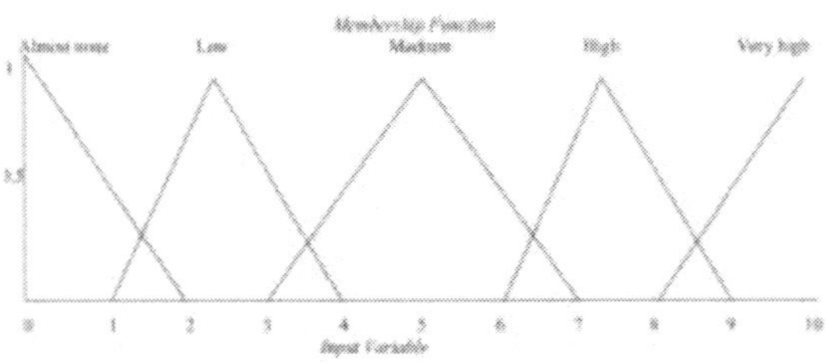

**Fig 2**. Input variable membership function

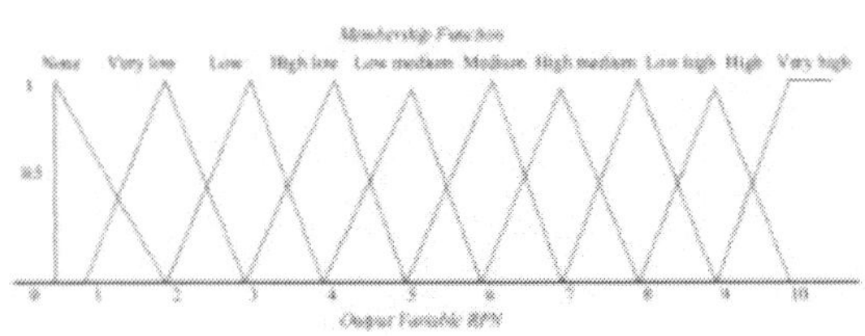

**Fig 3**. Output variable membership function

### 3.3 Deffuzyfication

Input from the process is the set Fuzzy defuzzyfication resulting from the process of composition and output is a value of One technique used in defuzzyfication is Center Of Gravity (Centroid). In Centroid, the output value is obtained based on the gravity of the decision-making process yield curve. In this study, defuzzification is used to find the value of output in the form of FRPN value of the input that has been entered. Inputs come from both the severity, occurence and detection that have been obtained from the results of FMEA risk identification using conventional methods (Table 5). Here are the results obtained from the FRPN defuzzification process using Matlab 8:

**Table 6.** Comparison of RPN and FRPN

| *Potential Failure* | S | O | D | RPN | Priority | FRPN | Priority |
|---|---|---|---|---|---|---|---|
| Computer problems (hank on/off) | 10 | 6 | 9 | 540 | 1 | 9 | 1 |
| The printer is trouble | 9 | 7 | 6 | 378 | 2 | 8 | 3 |
| Result of the finishing not appropriate | 7 | 6 | 8 | 336 | 3 | 7 | 4 |
| Delay in delivery | 9 | 4 | 9 | 324 | 4 | 9 | 2 |
| *File* data is lost | 10 | 2 | 6 | 120 | 5 | 6 | 7 |
| Termination of unilateral presentation by the client | 6 | 2 | 8 | 96 | 6 | 7 | 5 |
| Equipment for photo session is damaged | 5 | 4 | 4 | 80 | 7 | 7 | 6 |

Based on Table 6 there are the difference between the value and ranking among the NDP and FRPN. This is due to calculations using RPN simply done by multiplying the severity, occurence and detection alone and irrespective of the degree of importance of each input. While FRPN value obtained from the fuzzification, generate value by taking into account the degree of interest of any given input. In the process of defuzzification calculations have put the rules that prioritize the handling of the problem over to the cause of the risk. Whereas in the calculation of RPN, RPN value generated simply by multiplying the value of severity, occurence and detection only causing less accurate calculations with the NDP and the different results with calculations FRPN. Basically the method of fuzzy FMEA FMEA is more consistent than conventional methods.

### 3.4 Risk Handling

Here are some strategies for each risk than risk having the highest priority to low priority based on the calculation FRPN (Fuzzy Risk Priority Number):

1. Risk Priority 1: Computer problems (hang on / off) when editing pictures with a value FRPN 9
   - Risk Mitigation
     - Installation of double generator or UPS (Uninterruptible Power Supply) for all workstations
     - Regular checks of the computer used (at least 1x1 month)
     - Reset the computer's power cable installation on the layout of photo editing special work
2. Risk Priority 2: The delay in the delivery process with a value of RPN 9

- Risk Mitigation
  - Confirm the address with the recipient before sending
  - Avoid delivery time during peak hours on the go
  - Check the condition of the vehicle before transporting (avoid crashes vehicle)
3. Risk Priority 3: The print engine trouble during the process of printing the value FRPN 8
   - Risk Mitigation Strategies
     - The treatment is good against the printing press, especially after use
     - Regular checks periodically for printing machines (minimum 1x1 month)
     - Reset the printing schedule (avoid overload)
     - Do not use a printing machine that has been aged> 10 years
     - Setting up a special printing machine technician
4. Risk Priority 4: Results of faulty finishing (disagree) with the value FRPN 7
   - Risk Mitigation Strategies
     - Application written SOP for the process of finishing a book
     - Routine supervision of the workforce
     - Procurement training for old and new workers
     - Perform quality control before packaging
5. Risk Priority 5: Dismissal of unilateral presentation by the client (the school) with a value FRPN 7
   - Risk Mitigation Strategies
     - Prepare materials with interesting
     - Election of members of attractive, physically attractive to the team presentation
     - Using language that remained polite during the presentation
     - Procurement training (public speaking) routine for the team presentation
6. Risk Priority 6: Equipment damaged when the photo photo session with the value FRPN 7
   - Risk Mitigation Strategies
     - Treatment of photo equipment properly and regularly
     - Examination of photo equipment on a regular basis (at least 1x1 month)
     - Storage equipment was stored in a place that is not wet or humid
     - Avoid carry equipment that is not good (easy to fall)
7. Risk Priority 7: File photo editing data is lost with the value FRPN 6
   - Risk Mitigation Strategies
     - Preparation of a data base with the online system
     - Implementation of additional applications of the data autosaved
     - Storage of data is done in more than two data storage media (external hard drive, etc.)
     - Implementation of data recovery application if at any time the data is lost / deleted

### 3.5 Risk Control

Controlling risk is done by organizing regular meetings periodically to evaluate the handling of such risks during this time. The formation of a team or a special division that handles risk management can also be done as a way of controlling risk. During the meeting, it can be evaluated the risks that usually happens, if there are new risks arising, eating could be immediately documented to be analyzed great danger and action to handle, and if there is a risk that is no longer relevant can also be eliminated. In addition, through regular meetings could be used to discuss current risk management strategies and continue to improve the strategy

## 4 Conclusion

Of the many known risks identified are 7 risks that have the highest weight for processing by the FMEA. The use of FMEA simply multiplying severity, occurence and detection alone to obtain the value of the RPN regardless of the degree the interests of each input. To cover the shortfall which is owned by the FMEA method is then integrated using fuzzy logic, then the difference of the results obtained with the conventional FMEA FuzzyFMEA. From the results obtained fuzzyRPN knowable risks becoming a priority for rapid repairs. The company can make the handling and controlling the risks that these risks do not occur in the future. One way is by applying risk mitigation actions for each risk and set up a special division that handles risk management.

## References

1. Hanggraeni, D. *Pengelolaan Risiko Usaha*. Jakarta: Lembaga Penerbit Fakultas Ekonomi Universitas Indonesia (2010).
2. ISO/IEC Guide *Risk Management-Vocabulary-Guidelines for use in standards* (2002)
3. Iqbal, M., Muflikhah, L., & Setiawan, N. Y.. Penggunaan Fuzzy Failure Mode and Effect Analysis (Fuzzy FMEA) dalam Mengidentifikasi Resiko Kegagalan Proses Pemasangan dan Perbaikan AC. Jurnal Informatika/Ilmu Komputer , **1-6**. (2013)
4. Kumru, M. Fuzzy FMEA application to improve purchasing process in a public hospital. Turkey : Dept of Industrial Engineering, Dogus University (2012).
5. Kusumadewi, S. *Analisis dan Desain Sistem Fuzzy Menggunakan Toolbox Matlab*. Yogyakarta : Graha Ilmu (2002).
6. McDermott, R.E., Mikulak, J.E., Beauregard, M.R. *The Basics of FMEA (2nd edition)*. New York : Productivity Press (2009).
7. Muslich, M. *Manajemen Risiko Operasional*, Edisi Pertama. PT Bumi Aksara (2007).
8. Project Management Institute (PMI). *A guide to the project management body of knowledge (PMBOK®Guide)*. Pennsylvania: Project Management Institute (2004).
9. Silva, M,.M. A multi dimensional approach to information security risk management using FMEA and fuzzy theory. Brazil: International Journal of Information Management, Department of Production

Engineering, Universidade Federal de Pernambuco (2014).

10. Wideman, M.R. *Project and Program Risk Management: A Guide To Managing Project Risk Opportunities*. Amerika Serikat: Project Management Institute (1992).
11. Xu, K., Tang, L. C., Xie, M., Ho, S. L., & Zhu, M. L. Fuzzy Assessment of FMEA for Engine System. Reliability Engineering and System Safety, **17-29 (2002).**
12. Yeh, R. H., & Hsieh, M. H.. Fuzzy Assessment of FMEA for a Sewage Plant. Journal of the Chinese Institute of Industrial Engineers , **505-512** (2007)

# Analysis of manual material handling activity to increase work productivity (Case study: manufacturing company)

*Muhammad Ragil* Suryoputro*, *Khaerul* Wildani, and *Amarria Dila* Sari

Universitas Islam Indonesia, Jl. Kaliurang KM 14.5, Yogyakarta, Indonesia

**Abstract.** Manual material handling is one of work activities that have an effect on the physical aspect of workers in manufacturing industry, it is necessary to do the analysis of the risks from such activities. Analysis was performed on worker when performing manual lifting activity and when using two tools (Automatic Handlift and Manual Handlift). In addition to analyse in ergonomics aspect, time study analysis and productivity measurement were carried out to determine the effects of the equipment. Nordic Body Map (NBM) questionnaire on worker using Automatic Handlift showed declining level of musculoskeletal disorders by 22%. REBA method obtained score of 10 and was declined to 4 after using these tool. The results of MPL method showed declining of Fc value from 4756.37 N to 1346.56 N. The results of RWL method showed declining of LI value (Lifting Index) origin and destination from 1.84 and 1.18 to 1.12 and 0.89 respectively. As for worker using Manual Handlift, NBM questionnaire result shows declining level of musculoskeletal disorders by 57%. REBA method obtained scores of 8 and was reduced to 5. For MPL methode result, the Fc value is 4906.99 N and reduced to 2047.88 N. RWL method results showed declining of LI value (Lifting Index) origin and destination from 1.02 and 0.67 to 0.74 and 0.58. The results of time study analysis showed declining of Standard Time when use the two tools and make the increasing productivity of 9% by worker using Automatic Hadlift and 4% by worker using Manual Handlift.

## 1 Introduction

PT. XYZ is one of the manufacturing company engaged in the production of musical instruments named piano. In the manufacturing industry, human labor is still needed on the production floor to ensure the smoothness of the process by transferring materials (handling) from one machine to another machine or from one workstation to another. At PT. XYZ, manual handling material handling that often done by workers is the lifting activity.

MMH is the most common cause of fatigue, waist injury and spinal cord injury [1]. From some material handling manual activity, lifting activity proved to be one cause of the high level of injury [2],with the risks faced by workers when performing manual material handling activities, it needs an intervention or improvement of ergonomics aspect to minimize the risk of injury to workers. The improvements made by PT. XYZ is to provide lifting tools Handlift Manual and Automatic Handlift for workers. However, good ergonomics interventions those that can achieve two objectives: to reduce the risk and improve work productivity [3]. Therefore in this study will be discussed how far ergonomic improvements can reduce risks and increase work productivity.

## 2 Methodology

### 2.1 Sample

This study focused on workers in the woodworking department of PT. XYZ. Woodworking workers were chosen due to manual handling material handling were majority performed especially lifting activity. In addition, the lifting object is quite large because the woodworking department is one of the early stages of the piano production process. The workers observed were Double-sizer machine operator from Cutting Sizer section and Ban Saw machine operator from Machine Bridge section.

### 2.2 Research Scheme

The focus of this study is to compare the risk level and productivity of workers when performing manual lifting and lifting activities using lifting tools so that it can be seen how much effect the improvements made. Therefore, several methods such as Nordic Body Map, REBA, RWL, MPL, time study and productivity measurement are performed on both conditions (manual lifting and using lifting tools).

*Corresponding author: ragil.suryoputro@uii.ac.id

### 2.3 Nordic Body Map

Nordic Body Map is subjective identification questionnaire that is used to identify discomfort or complaints from body parts when performing work activities [4]. Nordic Body Map is based on the 28 points of the body to judge it based on the level of complaints are divided into 4 levels of complaints [5].

### 2.4 REBA

REBA (Rapid Entire Body Assessment) is a method developed to investigate the risks experienced by workers based on posture while performing work activities by considering the load, coupling (grip) and posture changes [6]. In work posture evaluation using REBA, the body is divided into two groups were observed, namely Group A and Group B. Group A consists of the neck, back and legs while Group B consists of the upper arm, forearm and wrist. The results of REBA is the level of risk experienced by workers based on work posture. The score and the risk level of REBA method can be seen on the table 2.3 below.

### 2.5 Maximum Permissible Limit

One method in biomechanics that can be used as an analytical tool for lifting acivity is Maximum permissible limit (MPL) [7]. MPL is used to determine the value of the Force Compression (Fc) at the L5/S1. MPL is a limit of the amount of compressive force on the segment L5/S1 standardized by NIOSH (National Instiute of Occupational Safety and Health) in 1981. According to NIOSH, maximum limit of compressive force on L5/S1 is 6400 N, while limit for normal force (Action Limit) is 3400 N on the L5/S1. Therefore the limit for lifting activity according to NIOSH based on the MPL, Navy and Air Force Compression (Fc) is as follows:

LC = Load Constant = 23 kg
HM = Horizontal multiplier = 25/H
FM = Frequency multiplier
CM = Coupling multiplier

Based on the principles set by NIOSH, if the LI value more than 1, it can be said liifting activity has a high risk of the occurrence of low back pain. So the good lifting activity is if it has LI value of 1 or less than 1 [8].

### 2.6 Time Study (Stopwatch)

Stopwatch method is a direct measurement of working time commonly applied for the jobs of short duration and repetitive [9]. In this study, stopwatch readout method used is Repetitive Timings. For this method, stopwatch read simultaneously and the numbers on the stopwatch is returned to zero after each work element is completed.

### 2.7 Productivity

Productivity is a value that indicates the ratio between output and input, increase productivity value can be seen when comparing the numbers getting bigger [10]. While other opinion explains that productivity is a comparison between the results achieved with the participation of labor per unit time [11].

## 3 Result

### 3.1 Nordic Body Map

The identification results of musculoskeletal complaints using Nordic Body Map questionnaire shows a decrease in the level of complaints in Double Sizer worker at 22% and Ban Saw worker by 57% as shown in Figure 1.

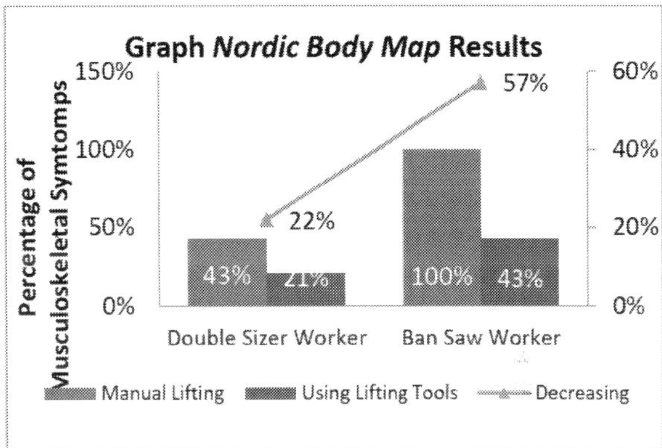


**Fig 1.** Graph nordic body map results

### 3.2 REBA

Work posture analysis using the REBA method gives the result that after the use of both lifting apparatus, the work posture score of Double Sizer worker and Ban Saw worker has decreased, which means the risk of injury also decreases. Decreasing score that occurs can be seen in Figure 2 below.

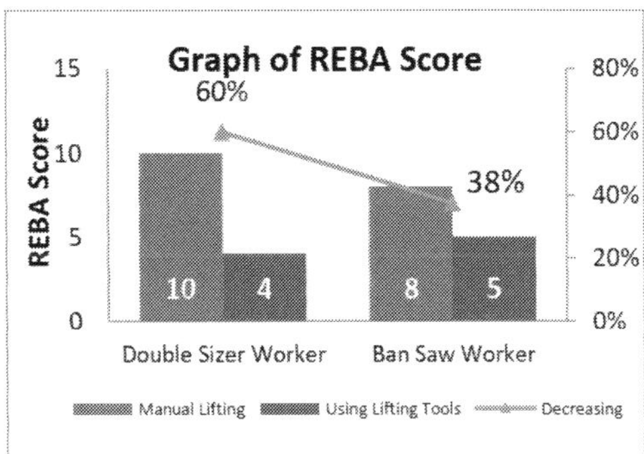


**Fig 2.** Graph of REBA score

### 3.4 Maximum Permissible Limit (MPL)

On the results of the MPL method, the value of Fc (Compression Force) of the two worker also declined after using the lifting tools as shown in Figure 3.

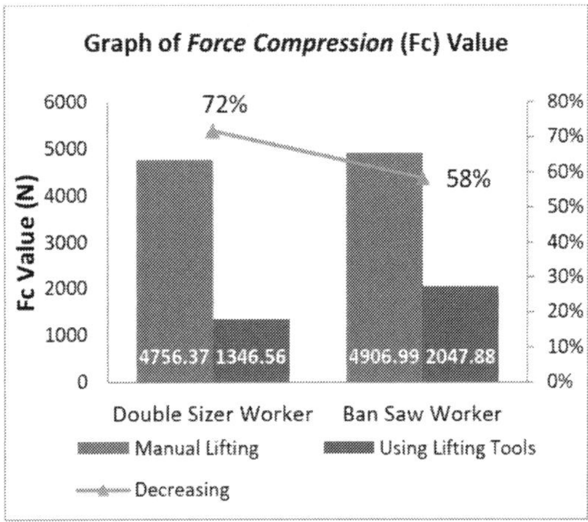


**Fig 3.** Graph of Force Compression (Fc) value

### 3.5 Recommended Weight Limit (RWL)

From the results of RWL method, the value of LI (Lifting Index) was obtained showed decline of two workers both origin and destination position as shown in Figure 4.

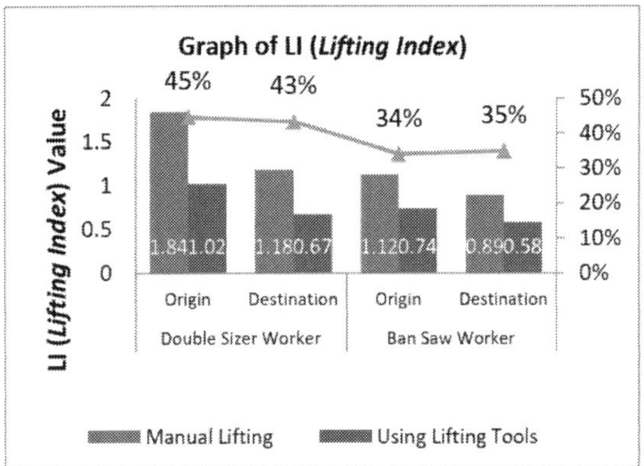


**Fig 4.** Graph of LI (Lifting Index)

### 3.6 Time Study (Stopwatch)

Time study analysis using the stopwatch method showed a decrease of standard time after the two workers use lifting tools as shown in Figure 5.

### 3.6 Productivity

And lastly from productivity measurement, both workers have increasing productivity after using the lifting tools as shown in Figure 6.

## 4 Analysis

Based on the results of NBM questionnaires that have been filled by the operator, there are subjective complaints experienced by Double Sizer worker by 43% of the total 28 points of the body. Then after using the Automatic Hand lift tool, there was a decrease in the complaint rate to 21%. The decrease in complaints rate also occurred in Ban Saw worker who previously experienced a complaint rate of 100% and then dropped to 43%.

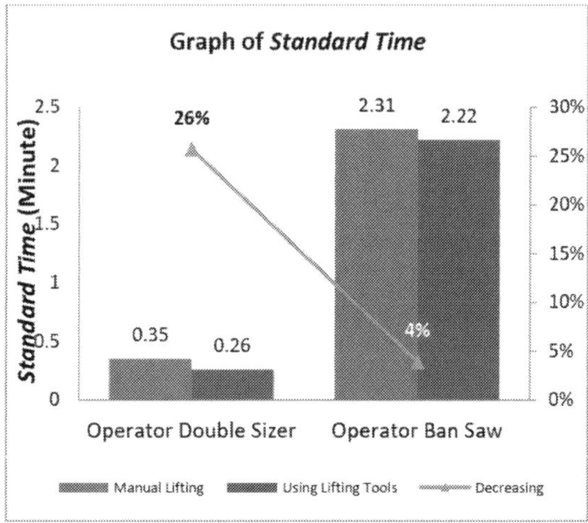


**Fig 5.** Graph of standard time

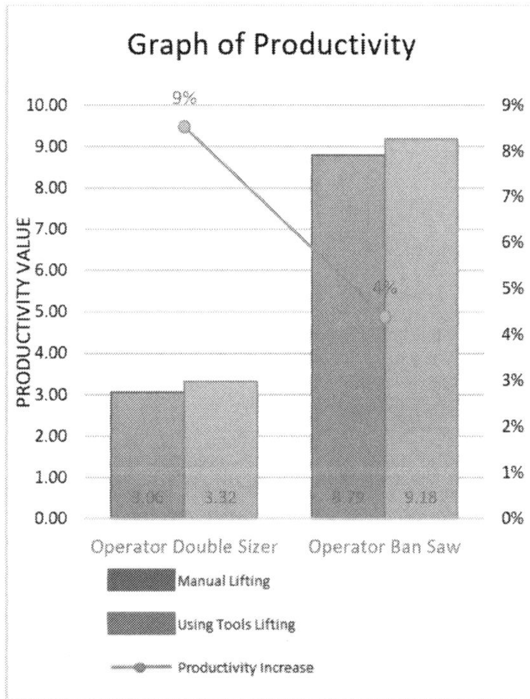


**Fig 6.** Graph of productivity

On the REBA score, REBA score of Double Sizer worker has decreased from 10 to 4, reducing the risk of injury. The amount of angles on the shoulders and elbows due to manual lifting movements is the cause of the high risk of injury [12]. In addition, the bending back position also causes the high risk of injury [13]. Decrising score of REBA also occur in Ban Saw worker after using Handlift Manual tool. Initial score of 8 down to 5 after using these tool. The high risk of injury of manual lifting activity is caused by an extension of the neck that moves extension with the lifting distance around the elbow so that the use of muscle strength up to 50% which causes muscle strain on the back of the neck [14, 15].

MPL results showed decreasing of Fc value from Double Sizer worker after using Automatic Handlift. Meanwhile, Ban Saw worker also experienced a decrease in Fc value which caused a decrease in risk level after using Manual Handlift. The high value of Fc at the manual lifting of both workers is caused by the distance of initial lifting position is lower, causing the increase of force and force moment in the segment L5/S1 [16],

From the results of RWL method, it can be seen that declining Lifting Index (LI) value of both workers after using these two tools. Decrease in LI value occurs in the position of Origin and Destination. The high value of LI on manual lifting of both operators is caused by the distance of initial lifting position is too low and make the larger angle of waist and back inclination which causing dangerous bending posture [17]. However, the LI value of Ban Saw worker is smaller than Double Sizer worker. This is due to one other factor that is the lifting frequency (FM). Where the Double Sizer worker has higher lifting frequency causes higher muscle tension and injury [18]. While the analysis of Time Study using stopwatch method showed that a decline in Standard Time of two workers after using these tools. The decline can be seen when worker lift the materials to the machine. Both of these tools help the worker to lift the pile of the materials in desired height so that the lifting activity can be done more quickly.

In accordance with the decline of Standard Time, then the productivity measurement of both workers increased after using the lifting tools. This shows that the use of two tools was a positive impact for the productivity of both workers..

## 5 Conclusion

Based on the results of several methods and analysis has been presented it can be concluded that the improvement of the ergonomic aspects by using Manual Handlift and Automatic Handlift give the effect of decreasing the risk level of the injury suffered by both workers. In addition, the use of these two tools also proved to have positive impact on the performance aspects of the operator with the Standard Time reduction and productivity improvement

## References

1. Canadian Center of Occupational Health and Safety, "www.ccohs.ca," 2016. [Online]. Available: https://www.ccohs.ca/oshanswers/ergonomics/mmh/mmhintro.html. [Diakses 31 Oktober 2016].
2. D. Stambolian, M. Eltoukhy dan S. Asfour, "Development and Validation of a Three dimensional Dynamic Diomechanical Lifting Model for Lower Back Evaluation for Careful Box Placement," *International Journal of Industrial Ergonomics,* vol. LIV, pp. 10-18, 2016
3. M. Resnick dan A. Zanotti, "Using Ergonomics To Target Productivity Improvements," *Computers & Industrial Engineering,* vol. XXXIII, no. 2, pp. 185-188, 1997
4. W. Sukania, L. Widodo dan D. Natalia, "Identifikasi Keluhan Biomekanik dan Kebutuhan Operator Proses Packing di PT X," *Jurnal Energi dan Manufaktur,* vol. VI, no. 1, pp. 1-94, 2013.
5. K. Tirtayasa, I. N. Adiputra dan I. G. Djestawana, "The Change of Working Posture in Manggur Decreases Cardiovascular Load and Musculoskeletal Complaints Among Balinese Craftmen. Jurnal Human Ergology," *Jurnal Human Ergology,* vol. 32, pp. 71-76, 2003.
6. S. Hignett dan L. McAtamney, "Technical Note Rapid Entire Body Assessment (REBA)," *Applied Ergonomics,* vol. XXXI, p. 201}205, 2000.
7. F. Tayyari dan J. L. Smith, Occupational Ergonomics : Principles and Applications, Chapman & Hall, 1997
8. T. R. Waters, V. Putz-Anderson dan A. Garg, Application Manual for The Revised NIOSH Lifting Equation, Ohio: U.S Department of Health and Human Services, 1994.
9. R. M. Barnes, Motion and Time Study : Design and Measurement of Work, 7th penyunt., Canada: John Wiley and Sons, Inc., 1980
10. T. Nakajima, N. Koji dan M. Toshiyuki, Total Factor Productivity Growth, Tokyo: Asian Productivity Organization, 2004.
11. J. Ravianto, Produktivitas dan Manusia Indonesia, Jakarta: Lembaga Sarana Informasi Usaha dan Produktivitas, 1985.
12. N. D. Nath, R. Akhvian dan A. H. Behzadan, "Ergonomic Analysis of Construction Worker's Body Postures Using Wearable Mobile Sensors," *Aplied Ergonomics,* vol. LXII, p. 107e117, 2017
13. N. J. Delleman, C. M. Haslegrave dan D. B. Chaffin, Working Postures and Movements : Tools for Evaluation and Engineering, Boca Raton, Florida: CRC Press, 2004
14. A. D. Nimbarte, "Risk of Neck Musculoskeletal Disorders Among Males and Females in Lifting Exertions," *International Journal of Industrial Ergonomics,* vol. XLIV, pp. 253-259, 2014
15. A. Peolsson, E. Marstein, T. McNamara dan D. Nolan, "Does posture of the cervical spine influence dorsal neck muscle, " *Manual Therapy,* vol. XXX, pp. 1-5, 2013.
16. S. A. Lavender, G. B. Andersson, O. D. Schipplein dan H. J. Fuentes, "The effects of initial lifting height, load magnitude, and lifting speed on the peak dynamic L5/S1 moments," *International Journal of Industrial Ergonomics,* vol. XXXI, p. 51–59, 2003.
17. Plamondon, C. Lariviere, A. Delisle, D. Denisa dan D. Gagnon, "Relative importance of expertise, lifting height and weight lifted on posture and lumbar external loading during a transfer task in manual

material handling," *Journal Ergonomics,* vol. LV, no. 1, p. 87–102, 2012.

18. R. A. Al-Ashaik, M. Z. Ramadan, K. S. Al-Saleh &T. M. Khalaf, "Effect of safety shoes type, lifting frequency, and ambient temperature on subject's MAWL and physiological responses," *International Journal of Industrial Ergonomics,* vol. L, pp. 43-51, 2015.
19. B. Klein, R. .. Jensen dan L. M. Sanderson, "Assesment of Workers' Compensation Claims for Back Strains/Sprains," *Journal of Occupational Medicine,* vol. XXVI, no. 6, pp. 443-448, 1984.

 https://doi.org/10.1051/matecconf/201815401086

# System design of blood supply chain management based on Supplier Customer Relationship Management (SCRM) approach

*Zuhdi* Allamsyah and *Agus* Mansur*

Industrial Engineering, Islamic University of Indonesia, Yogyakarta, Indonesia

**Abstract.** Blood supply chain management is very complex system. By considering blood's demand that characterized as stochastic, therefore information and communication among stakeholders are considered as important cycle in supply chain. This research is objected to design interactive communication model based on information technology with android as platform to create effective customer-supplier relationship to support Customer Relationship Management or better known as CRM. The methods that suggested in designing its information system are Quality Function Deployment (QFD) approach, Kano Model and Morphological Chart. From mentioned methods, attributes of design and specification priorities of purposed android system technology design are established. Attributes of design will be adjusted with customers' requirements derived from customer voice that will further be integrated with technical responds in QFD. Technical responds will provide priorities on each design's attribute of android information system and will be forwarded to the specifications determination by using morphological chart. Later, work flow of design system will be generated by using Data Flow Diagram (DFD). The result of the research will demonstrate the design of information system technology based on android that already adjusted with the requirement of donors as means of communication between suppliers and customers in blood supply chain.

## 1 Introduction

Blood is a vulnerable commodity yet a vital transport system component to every person. The number of minimum requirement of blood in Indonesia was counted around 5,1 million blood packs per year or equal to 2% of total Indonesian residences. Until recently, it only covers 4,1 million blood packs sourced from 2,7 million donations, in which 87,74% of them are derived from voluntary blood donation [1]. The fact there was the imbalance supply and demand in DIY province [2]. Generally, worldwide, there are hospitals that suffer from insufficiency of blood stock, hence they require proper management [3]. The source of blood unavailability is the stochastic products' demand. While, only 5% of population that consider as proper as blood's donors [4]. The scarcity of blood also emerged by the stochastic number of donors and/or the blood's stocks that are stored in the blood bank [5].

As the efforts in improving good service to customers, PMI DIY provides SMS gateway service to deliver information about blood's stock, reminder to conduct the donation and various events of blood donation. This service is expected to maximize the information distribution from PMI to customers. Nevertheless, this SMS gateway service still covers several shortages, for instance, it only could deliver information based on text that apparently will be one of the causes of information retardation. The availability of information system has effect on its user's performance in accessing the information [6]. Appropriate information system can possibly facilitate two-way communication for better information sharing. Information sharing and information quality on the implementation of supply chain management could result on customers' satisfaction [7]. Customers' satisfaction is considered as immense capital for loyalty [8]. Accessible and accurate information could maximize the CRM [9].

Based on described background, the researchers proposed information system design based on location-based service to improve services and to fulfill the requirements of stakeholders in blood management. The result of CRM system design is expected to enhance services as well as provide satisfaction to all donors in PMI.

## 2 Research Methods

Established system design is based on concepts of CRM as the integration of coordinated selling, marketing and service strategy [10]. Initial step is to find out customers' requirements toward established system. QFD and Kano model are employed as tools. QFD is defined as analysis tool to convince that development of products' features, characteristics and specification that covers manufacturing and required processes, has met the customers' necessities [11].

While, Kano model is described as theoretical model that connects requirements that supposed to be filled by a

* Corresponding author: agusmansur@uii.ac.id

product or service with customer's satisfaction by identifying three types of necessities that could influence it, which are: must be requirement, one dimensional requirement and attractive requirement [12]. Second step is to determine technical requirement derived from voice of customers by using morphological chart. The final step is designing system data flow by applying Data Flow Diagram (DFD). DFD is considered as a method that describes data process flow from a system to its end [13].

## 3 Result

### 3.1 Technical Requirements of the System

Obtained Voice of customer will be graded by using Kano Model that implements Blauth's formula, later, several attributes that have influences toward customers' satisfaction will be selected. Attributes that are categorized as indifference will no longer be processed, as they have low influence on customers' dissatisfaction [14]. The result of Kano model will be the input for Quality Function Deployment (QFD) processing to translate costumers' requirements to company's technical requirements that also described by morphological chart stages to determine system's specification in detail. Hence, the aspects of system requirements are illustrated as follows:

**Table 1**. Morphological chart

| No | Function | Means |
|---|---|---|
| 1 | Providing appropriate application's content | Appropriate information for donors' requirements (information about blood's stock, events, history of blood's donor, personal identification, etc) |
| 2 | Presenting information on blood's availability | Entire information about blood (type of blood, rhesus, blood's component) |
| 3 | Displaying information status for donor activity | Available |
| 4 | Displaying notification about information | Only notification related to information of donor's events, routine reminders, birthday greeting and friends' request |
| 5 | Integrated with blood's supplier institutions | Local BDRS and UTD |
| 6 | Displaying accurate information | All features |
| 7 | Requesting location checking while application is being operated | Available |
| 8 | Recommending the nearest location with blood's supplier institutions | Available |
| 9 | Recommended radius of distance with blood's supplier institutions | ≤ 10 km from user's location |
| 10 | Providing information of contact person and the nearest location of other Donors (other users) | Available |
| 11 | Radius of recommended distance with the Donors | ≤ 10 km from user's location |
| 12 | Providing information and recommendation on an event and its nearest location | Available |
| 13 | Recommended radius of the nearest blood donation event | ≤ 25 km from user's location |
| 14 | Accomplishing user's account to improve privacy, data security and accessibility while operating information system of blood's donor | Available, but optional so the users could access the application without user's account |
| 15 | Displaying profile picture on its information system's user account | Available |
| 16 | Providing the features for utilization information | Help Menu |

### 3.2 System Data Flow

#### *3.2.1 System Context Diagram*

In this diagram, scope of stakeholders' relationship in system is determined, which are: customers that consist of the donors and patients, Blood Transfusion Unit (UTD) PMI and hospitals, either Blood's bank hospital (BDRS) or not.

#### *3.2.2 Data Flow Diagram Level 1*

Blood donor's information system, consists of 9 main processes, which are: login, location service, management of customers, management on history of the donors, management of blood's stock, management of events, notifications, the donors' access, help. This information system also equipped with donors' schedule reminder system to ease the donors in performing the activity of blood donation. It also equipped with location based service to identify distance's estimation to institutions of blood suppliers, blood's donation events and the nearest location of the donor from user's location.

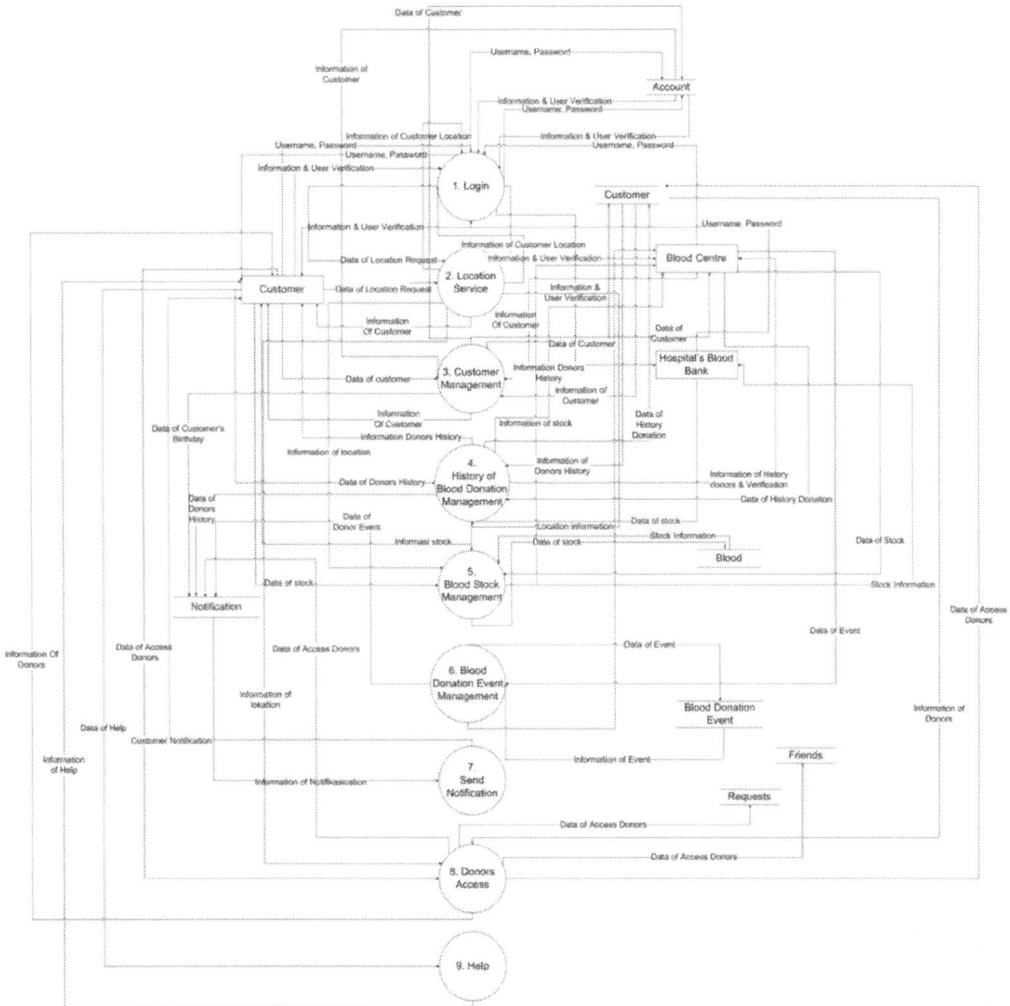


**Fig 1.** DFD Level 1

### 3.3 Relations among Tables on Diagram

Relation among tables for blood's donor information system is described on Figure 2.

### 3.4 Description of System Design

This information system ensures 5 features, which are: My profile that has specific function to identify customer's personal identification and history of blood's donation. Check stock is designated to check on recent blood's stock from available suppliers that completed with the nearest location recommendation from user's recent location. Donor's access is created to locate appropriate and the nearest donors for transfusion purpose. Event of blood donation is established to identify recent blood's donation activities based on location consideration and Help menu that functioned to assisst users in operating the information system.

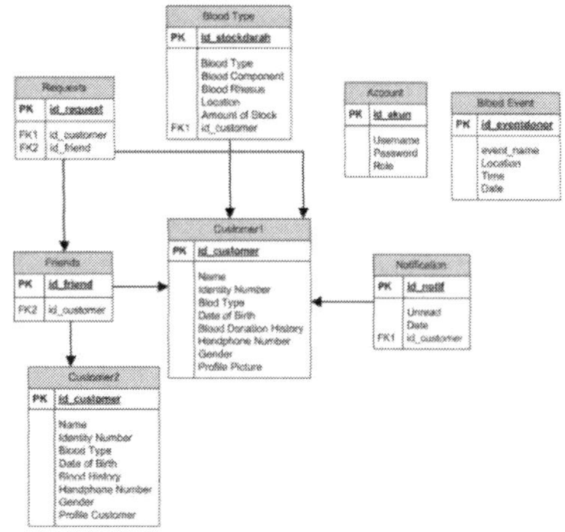


**Fig 2.** Entity relationship diagram

## 4 Conclusion

Based on the results of the test which has been performed, it can be concluded as follows:

Producing shared information system that appropriates with donors' requirements and existing means, by diplaying information that adjusted with donor's requirement (availability information on blood's stock, information on events of blood donation, information on blood's donor history, personal identification, etc). Displaying information on blood's stock (type of blood, rhesus and blood's components), performing information status on donation schedule, notification that relates to blood's donation events, routine's reminder, birthday's greeting and feedback on application. Integration with local BDRS and UTD, announcing accurate information on all features, conducting location scanning when application is operated, recommending the nearest location with supplier institutions ≤ 10 km from user's location, providing information on contact person and recommending the nearest location of the donor ≤ 10 km from user's location, providing information related to blood's donation events by radius of ≤ 25 km from user's location, establishing user's account to improve privacy, data security and accessibility. Yet, this will be optional to allow users to be able to access the application without user's account. Displaying profile picture to ease the identity recognition and providing utilization assistance feature with Help menu.

## References


1. Tribunnews, *Indonesia Kekurangan Satu Juta Kantong Darah, Ini yang DilakukanATSI dan Kemenkominfo*, (http://www.tribunnews.com/kesehatan/2016/02/15/indonesia-kekurangan-satu-juta-kantong-darah-ini-yang-dilakukan-atsi-dan-kemenkominfo, 2016)
2. D. Widianto, *Pasien RS Sardjito Dihimbau Minta Rujukan Darah ke PMI*, (accesed from www.krjogja.com, 2015)
3. S.H. Stanger, R. Wilding, N. Yates, & S. Cotton, What drives perishable inventory management performance? Lessons learnt from the UK blood supply chain. SCM: An. Inter. Jour **17/2**, 107-123 (2012)
4. S. Lin, & B. Lin,. *Accessing information sharing and information quality in supply chain management.* Decision Support Systems **42**, 1641–1656 (2006)
5. S.X. Li Na, & Z.M. Wei Yang, *Decision Making Model Based on QFD Method for Power Utility Service Improvement.* System Engineering Procedia **4**, (2011)
6. R. Kalakota, & M. Robinson, *E – Business 2.0 Roadmap For Success.* USA: Addison – Wesley (2001)
7. N. Kano, K. Seraku, F. Takahashi, S. Tsuji, Attractive Quality and Must-be Quality. The Jour of the Jpn Soci for Q C **14**, 39-48 (1984)
8. E. Bendoly & M. Swink, Moderating effects of information access on project management behavior, performance and perceptions. Jour of Op Mgmt **25**, 604-622 (2007)
9. M. Rifai, *Customer Relationship Management (CRM) dengan Blog.* Elex Media Komputindo (2009)
10. R. Lupiyoadi, *Intim dengan Konsumen.* Jakarta: Salemba Empat (2013)
11. Indrajani, *Perancangan Basis Data dalam Al lin 1.* Jakarta: PT. Elex Media Komputindo (2011)
12. P. Gupta, *Analysis of Customer Satisfaction of The Hotel Industry in India Using Kano Model & QFD.* Motilal Nehru National Institute of Technology (2012)
13. J. Belien & H. Force, Supply Chain Management of Blood Products: A Literature Review. Eur J of Op Rrch (2012)
14. L. Cohen, *Quality Function Deployment: How To Make QFD Work For You.* New York: Addison-Wesley Publishing Company (1995)

# Job safety and awareness analysis of safety implementation among electrical workers in airport service company

*Perdana Suteja* Putra*, *Amarria Dila* Sari, and *Amirah Shafa* Husna

Department of Industrial Engineering, Islamic University of Indonesia, Yogyakarta, Indonesia

**Abstract.** Electrical is a fundamental process in the company that has high risk and responsibility especially in public service company such as an airport. Hence, the company that operates activities in the airport has to identify and control the safety activities of workers. On the safety implementation, the lack of workers' awareness is fundamental aspects to the safety failure. Therefore, this study aimed to analyse the safety awareness and identify risk in the electrical workplace. Safety awareness questionnaires are distributed to ten workers in order to analyse their awareness. Job safety analysis method used to identify the risk in the electrical workplace. The preliminary study stated that workers were not aware of personal protective equipment usage so that the awareness and behavioural need to be analysed. The result is the hazard was found such as electrical shock and noise for various intensity in the workplace. While electrical workers were aware of safety implementation but less of safety behaviour. Furthermore, the recommendation can be implemented are the implementation of behaviour-based safety (BBS), 5S implementation and accident report list.

## 1 Introduction

Occupational safety and health is an important concern for all working individuals [1]. In lack condition of information and awareness of occupational safety in the workplace would like to tend accident in the workplace. Among the main contributing factors that influenced in occupational safety and health implementation to unintentional occupational injuries identified by World Health Organization [2] are poor worker-employer collaborative mechanism, poor training and lack of knowledge, and lack of incentive-based compensation system. Each of these contributing factors falls within the traditional purview of psychological or behavioural of workers among safety system. Therefore, the safe design and safe of use of instruments or tool are broad subjects that involve nearly all procedures in order to do the work [3].

These kinds of procedure were involved all of work such as in operating, machine, production, inspection, etc. Safety consideration must be applied in all of workplaces in order to minimize the assurance cost and worker's accident especially in special type of workplace such as mechanical, electrical, chemical workplace, etc. Over the last few years, there has been tremendous attention given to the arc flash hazard phenomenon. Even though this hazard needs attention to reduce the number of injuries occurring because of it, in some ways, it has become the predominate focus of electrical safety [4]. Electrical safety is more than just the arc flash hazard. Electrical safety needs to be a holistic approach that includes all aspects of electrical safety.

Moreover, even there are routine inspection that has already done by safety health and environment section, the electrical workers are not concern in the safety workplace and their procedure that based on the observation conducted. The electrical workers were not followed the regulation in personal protective Equipment (PPE) usage and avoiding the inspection. These acts of workers that ignoring the safety regulation especially in safety procedures can be caused by several factors such as lack of training or lack of knowledge so that the awareness among safety in the workplace and job procedures was not tend to their safety behaviour [5]. Therefore, it is important to identify appropriate procedures for applying occupational safety and health system in different local conditions and the key factors in effectively using these procedures [6].

Safety procedural behaviour of electrical workers must be considered in order to make safety job in the workplace. Therefore, company such as PT. XYZ as largest company in airport operation has responsible to conduct safety management system must focus in consideration of safety procedural and their awareness in dangerous workplace such as electrical workplace.

In this case, electrical workers have one of safety concerned in the safety system. Electrical workers were gained bad implementation in safety inspection especially in PPE usage on their procedural job and safety behavioural. Therefore, this research was aimed to analyse the procedural of job that has been done and identifying all of hazard in every step of procedural then it was in consideration of worker's awareness in the implementation of safety management system. The

* Corresponding author: perdanasuteja@gmail.com

procedural safety job analysis would like to provide hazard and the risk in every step to do the work that and the awareness level would consider commitment of safety management system, safety compliance, safety awareness, safety behaviour, work stress, safety teamwork and its influences. Therefore, by identifying the safety job and awareness analysis, the company can know the factors of safety awareness and create innovation on the safety management system in order to minimize the risk in steps of job then increase the worker's awareness in the implementation of safety system by analysing the most affected factors that influenced in awareness especially for electrical workers.

## 2 Research Methodology

### 2.1 Subject and Object

A total of ten electrical workers were employed for purpose of safety awareness. They all agreed to participate in this study, and filled out the questionnaires. Then, safety expert and all of electrical workers involved on the job safety analysis and accident reported data. The object of research are accident data, job safety activity and awareness of electrical workers.

### 2.2 Research Method

The research has been conducted by using job safety analysis in order to identify the hazard and risk in workplace. In addition, workers' awareness was assessed by questionnaires and direct interview. Moreover, direct interview is conducted to safety expert in company to support the data on job safety and awareness analysis.

### 2.3 Research Procedures

In this paper, the examination of awareness is following questions regarding safety awareness aspects.

1. Safety commitment of company
2. Safety compliance
3. Safety awareness of workers
4. Safety behaviour
5. Safety teamwork in electrical unit

The above aspects were analysed in deep interview and questionnaires that supported by expert analysis.
In addition, the flow and research procedures has been conducted following figure 1.

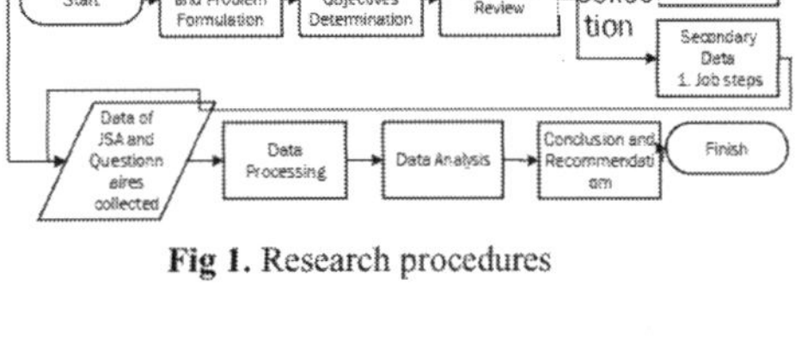


**Fig 1.** Research procedures

## 3 Result and Discussion

### 3.1 Job Safety Analysis

A Job Safety Analysis (JSA) is a qualitative analysis that must be done for job practice in order to identify the hazard and potential accidents that may occurred during the execution of the job [7]. Since the job in electrical unit is related directly to the machinery and electrical wave as potential risk may be occurred.

There are four main steps job as described on figure 2. Job Safety Analysis Lists. These steps were taken by procedural documents that has been provided. Since there is similarity of job procedurals so that the job safety analysis can be conducted as one analysis of job steps that depend on the direct field needed [8].

The unit has responsible to maintain all of electrical in the airport so that all of maintenance must be scheduled as soon as possible. There is no direct hazard identified so that the management must be accurate in creating maintenance schedule. Therefore, the condition of creating schedule must be in standard condition specially to minimize human error [6].

After scheduling was created, the next job procedural was in order to prepare all of tools, equipment and stuffs needed. Since majority population of workers was outsourcing workers so that workers has their own equipment. However, in the real activity of electrical maintenance, the tolerance was found between outsource workers to borrow the tools each other's. Unfixed properly location of tools that might be confused the engineers and waste their time to find it. In emergency condition, this problem would be turned into huge issues since the electrical problem must be maintained soon [9].

On the action activities, these steps can be focused as main activity that must be done by workers in order to finish, repair and maintain. The hazard depends on the electrical problems occurred in the airport. In these case of electrical problem in airport such as airfield lighting system, the hazard and risk potential might be electrical shocks. The electrical shock felt was also depends on the electrical problems found which is small, medium or large range of problems that has potential of explosion. Hence, all of personal protective equipment must be used in proper condition even in the short or long time hazards [10].

On the report process, there is no significantly hazard found. However, the proper standard for administrative of report must be considered such as standard temperature, dust and another physical work environments aspects that must be affected to the workers. Therefore, the condition of creating report must be in standard condition specially to minimize human error [6]. The result and discussion would be explained briefly based on sub-explanation.

Moreover, accidents can be avoided by understanding near misses and eliminating the root cause [11]. Therefore, it is important to create a list of accident occurred recently as described by Table 1.

**Table 1.** Accident list found

| No | Accidents | Frequency | Comment |
|---|---|---|---|
| 1 | Small Range of electrical shock | almost every work | Prevented by using PPE properly |
| 2 | Med Range of electrical shock | large maintenance | Prevented by using PPE properly |

## 3.2 Safety Awareness Analysis

Safety commitments of company has safety management system as authority of Safety and Health Environment Section. The management in safety can demonstrates how importance the usage of PPE. It has safety campaign program that has to be done in every month. It provides PPE needed for all workers in the airport as described on Figure 2.

**Fig 2.** PPE provision.

The training has been conducted in first work period. Therefore, the commitment of safety management to fulfil their responsibility has been fulfilled to conduct needed training in electrical unit. Routine inspection for every week so that all of incident, accident or hazard found can be minimized [12]. The management eliminate and cover hazard and communicate to the related section that has responsible to the hazard control.

Safety compliance shown that the unit is not actively discussing about safety. Hence, in case of accident and incident occurred has not been reported as list so that the safety management system also did not know the accident. It might be the habits and behavioural factors of workers that has impact to pretend the safety problem such as electrical shock has not been reported [5]. The cleanness is already arranged well. Moreover, the physical work environment was indicated in standard ratio for electrical workplace.

Safety awareness includes understanding of safety management is basic that begin by understanding the policy and rule of occupational safety and health [13]. The understanding as described on the Figure 3.

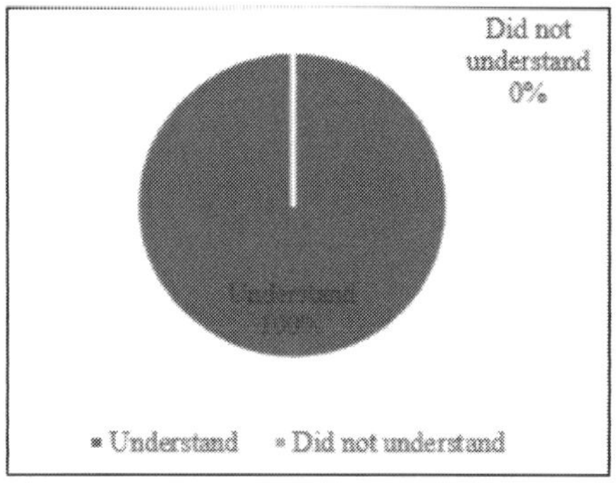


**Fig 3.** Safety understanding of workers

Then, in the activity of job they were not following the safety policy was not caused by the lack of understanding but it caused by behavioural and habits [5].

Safety behaviour measures the behaviour of workers in the workplace to implement their safety to finish their jobs. The priority made was the objectives to maintain problems rather than the safety principles. This case is main point that affected the safety worker's behaviour. In fact, the lack of PPE usage was caused by the confident level and the habits that they would be fine without any equipment. Therefore, the main cause in lack of safety behaviour is the habits of workers and confident level [5]. Then, there is no reward and punishment system on safety behavioural.

Safety teamwork shown that supervisor has given advices and control to prevent the hazard risk and accident. It is importance that must be often done by the supervisor because the supervising system in the safety management is one of best way to increase the safety awareness [14]. Then, workers have aware to give advice and comment to their team that does not use PPE and bend the safety policy. Moreover, all of machine procedures has been known by all of workers that maintained the machine. On the other hand, the reporting of accident occurred was not been controlled in all range of accident. In addition, the discussion to prevent the hazard almost done by safety management that applied by electrical units.

# 4 Conclusion

Based on the results and analysis that has been conducted the hazard found based on the job safety analysis are unfixed properly location of equipment that has risk to human error and confusion in wasting time to find it. Moreover, the electrical shock was hazard in small to medium range of shock. In addition, accidents in the last month occurred are small electrical shock in every activity of job and medium range of electrical shock in larger project maintenance. Moreover, the awareness level in the electrical unit has divided into several aspects. The safety commitment was done properly but needs to be improved, safety compliance was the unit did not actively discuss about safety yet, safety awareness is the workers understand and aware among safety implementation, safety behaviour in electrical has to be improved and safety teamwork was done properly and still can be improved. Furthermore, the recommendation can be implemented are the implementation of behaviour-based safety (BBS) which is effectively increase the safety awareness among workers [15]. In addition, 5S implementation and accident report list also has to be conducted to maintain the sustain safety [16, 17].

## References

1. J. Barling,M.R.Frone.Amrcn.Psy.Ass.(2004)
2. J. Takala. *Introductory report: Decent work- safe work*.(2002)
3. W.H.Olson.Phy.Eff.of.Elec.(2008)
4. D.Liggett.Ref.Elec.Sty. *IEEE Transactions on Industry Applications*, 1340-1345. (2006).
5. Azuddin,Hanum,A.Osman.Aw.o.Reg.a.Ind.Cod.Pra c.VaV.Occ.Sty.Hlt.Act94.*Mediterranean Journal of Social Science*, 269-273. (2015)
6. K.Kogi.Wk.Imp.Occ.Sty.Hlt.Mgt.Sys.*Industrial health*, 121-133.(2002)
7. M.Rausand.*System Reliability Theory*.(2005)
8. V.Jaiswal,Banodha,P.Patel.Ris.Ass.Main.Wk.Dsl.lo c.Wshp.*International Journal on Emerging Technologies*, 2249 – 3255. (2014)
9. D.Haiduven,DeMaio,D.Stevens.5yr.std.ndle.stk.inj. *Infect Control Hosp Edidemol*, 265 - 271.(1992)
10. V.Gopinath,K.Johansen.Ris.Ass.Pro.Cll.Asbl.*6th CIRP Conference on Assembly Technologies and Systems (CATS)*, 199 - 203.(2016)
11. Greenfield, *Normal Accident Theory*. (1998).
12. A.Bahari,Jainoo,H.Hassan.Sus.Occ.Sty.Hlt.Wk.Pl.*P enerbit Universiti Malaysia Perlis*.(2013)
13. H.SyniAn,Marta,Alice,TonyaLowery,Cristian,Eric,S haron. *American Journal of Industrial Medicine*.**38**, 71 - 81. (2000)
14. K.Agyekum,Ayarkwa,Amoah.Fr.Sty.Aw.Mgt.MSt.S tu.Hos.*International Conference on Infrastructure Development in Africa*, 426 - 438.(2016)
15. D.Cooper. *Behavioral Safety a Framework for Success*.(2009)
16. Y.Ennin,Obi,.5S.gd.hkpg.tchq.wh.usd.tl.sysc.app.pr od.*International Trade Center*.(2012)
17. T.Osada.5S.fv.kys.tot.qlt.env.*Asian Productivity Organization*.(1993)

# Quality improvements on creative industry by using Six Sigma: a study case

*Dwi Adi* Purnama, *Riadho Clara* Shinta, and *Vembri Noor* Helia

Department of Industrial Engineering, Islamic University of Indonesia, Jl. Kaliurang Km 14.5, Yogyakarta, Indonesia

**Abstract.** Indonesia has several creative economy sectors that rapidly developed and potentially can be improved to 7% annually. One of them is textile industry specifically bag industry. This industry's development could enhance the production of commodities and emerge other SMEs to initiate the similar field. Therefore, to raise its profit, customer's satisfaction must be elevated by minimizing defected products. In order to support it, an improvement on production process is a must. This research employs Six Sigma methods integration with Fuzzy Analytical Hierarchy Process (Fuzzy AHP)-Failure Mode and Effect Anlysis (FMEA). Six Sigma is designated to analyze the level of products' defect through Define, Measure, Analyze, Improve, and Control (DMAIC) stages and assess the factors that cause the defects by using Fuzzy AHP decision support system and FMEA in risk analysis on defect. The result of the research shows the shortage of defects that demonstrated by DPMO indicator and Sigma level. It is identified that the number of DPMO was being reduced from 20003.75 to 11185.73. While Six Sigma level demonstrated the enhancement from 3.61 to 3.86. Those suggest that the improvement solution provides has significant effect on the defects decreasing at the bag convection industry.

## 1 Introduction

In recent modern era, economic matter becomes one of the measurement standards of country's development. World's economic development that could support the worldwide' people life includes Indonesia as one. According to the Indonesian Creative Economy Board, there are several creative economy sectors developed rapidly and potentially to be further advanced. It is stated that there are three sectors, which are fashion, culinary and handcraft [1]. The existing creative industry could sharpen the creativity, skill and individual talent in creating proper welfare and job opportunities by producing and empower one's creation and creativity [2].

In Indonesia, creative industry plays important role in economic growth, by 7 % annual growth [3]. One of the creative industries that rapidly developed is textile industry or handcraft. It is considered popular since there are various of ethnic, race and culture that grow in Indonesia along with its existing natural resources. As illustrated by Central Bureau of Statistics (BPS), the growth of textile industry in Indonesia shows the annual elevation up to 142,72 in 2016 [4].

The development of textile industry, specifically bag convection could push the trade's commodities up and emerge other SMEs to work on the similar fields. As other SMEs actively involves in this sector, hence new innovations will be invented and quality will be improved. It can be fulfilled by providing qualified products and minimizing defect product, which can influence profit percentage. Therefore, profit can be achieved by increasing customers' satisfaction and minimizing defect products [5]. One of the SMEs that produced bag in Yogyakarta is Zano Production that produces innovated bags with make to stock and make to order system and employs 26 workers. Zano production is the case study of this research. In every production process, it still employs conventional process by applying simple technology, material and tools. That possibly triggered products' defect at each process.

This research is objected to minimize the numbers of defects due to existing failures, to enhance customers' satisfaction by using Fuzzy AHP-FMEA and Six Sigma method through phases of Define, Measure, Analyze, Improve and Control (DMAIC). Quality control plays important role to identify product's quality characteristics, by comparing specifications or requirements and taking appropriate actions for differences between its actual performance with the standard [6]. It will be conducted to fulfill customers' service [7]. Most of the Small and Middle Enterprises do not aware on the importance of quality control by using Six Sigma. Moreover, if the management could relate customers' Six Sigma with business strategy [8]. The difference between recent research with the previous ones is the implementation of Six Sigma method that designated to analyze the products' defect level and assess the root cause of the defect, which is integrated with Fuzzy AHP decision support system and FMEA in risk analysis. Therefore, from the assessment, the

* Corresponding author: vembri@uii.ac.id

improvement on products could be performed to reduce the defect that will directly enhance their quality.

## 2 Methodology

The research was taken place in SME of Zano Production located in Godean street, Yogyakarta. This research implements Fuzzy AHP-FMEA by using Six Sigma calculation to improve quality. It will be conducted in several stages as described in following figure 1.

## 3 Implementation of Six Sigma

There are five stages in this implementation, which are Define, Measure, Analyze, Improve, and Control that abbreviated with DMAIC. It will be used to measure the quality of products and services as well as to control their quality [9].

### 3.1 Stage of Define

Define is purposed to identify the production process and types of defects in industry [10] This stage is objected to identify defects on bag products based on interview conducted with production manager of Zano Production. There are 3 types of defects that frequently occurred in this company, which are: defects in cutting activity, sewing and embroidery.

### 3.2 Stage of Measure

Measure is conducted by using DPMO (*Deffect Per Million Opportunities*) to rate the recent company's performance, specifically in quality of management and to calculate sigma level from DPMO. The observation was conducted on 47 batches of bag production with value of DPMO and sigma level are presented on table 3.

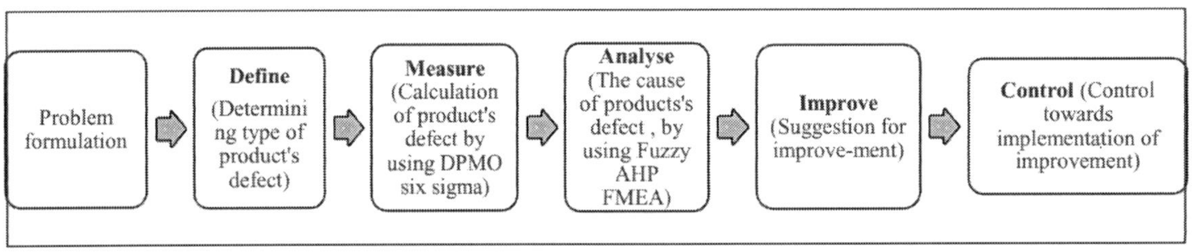


**Fig 1.** Research stage

**Table 1.** Recapitulation of products' defect data

| Batch | Type | Amount of bag | Defect Sewing | Defect Cutting | Defect Embroidery | Number of defect | DPO | DPMO | Sigma |
|---|---|---|---|---|---|---|---|---|---|
| 1 | Ladies Bag | 50 | 0 | 3 | 0 | 3 | 0,02 | 20000 | 3,55 |
| 2 | Telkomsel Bag | 60 | 3 | 1 | 1 | 5 | 0,027777 | 27777,77 | 3,41 |
| 3 | Timnas Black Canvas Bag | 125 | 2 | 1 | 0 | 3 | 0,008 | 8000 | 3,91 |
| 4 | Timnas Canvas Bag | 25 | 0 | 1 | 0 | 1 | 0,013333 | 13333,33 | 3,72 |
| 5 | Timnas Black Cordura Bag | 100 | 3 | 2 | 1 | 6 | 0,02 | 20000 | 3,55 |
| 6 | Baby Gray Back Pack Bag | 150 | 8 | 4 | 1 | 13 | 0,028888 | 28888,88 | 3,39 |
| 7 | Pemkot Demak Bag | 35 | 1 | 2 | 0 | 3 | 0,028571 | 28571,42 | 3,40 |
| 8 | Sibalec Bag | 500 | 3 | 1 | 1 | 5 | 0,003333 | 3333,333 | 4,21 |
| 9 | Trolley Bag | 200 | 10 | 10 | 2 | 22 | 0,036666 | 36666,66 | 3,29 |
| 10 | PMI Black Bag | 310 | 5 | 3 | 3 | 11 | 0,011827 | 11827,95 | 3,76 |
| 11 | Ladies Black Vinyl Bag | 31 | 0 | 2 | 0 | 2 | 0,021505 | 21505,37 | 3,52 |
| 12 | Ladies Maroon Vinyl Bag | 31 | 0 | 1 | 0 | 1 | 0,010752 | 10752,68 | 3,79 |
| 13 | Gray Gontor Bag | 50 | 0 | 2 | 0 | 2 | 0,013333 | 13333,33 | 3,72 |
| 14 | Black Gontor Bag | 50 | 0 | 3 | 1 | 4 | 0,026666 | 26666,66 | 3,43 |
| 15 | Brown Gontor Bag | 50 | 0 | 1 | 0 | 1 | 0,006666 | 6666,666 | 3,97 |
| 16 | Black UGM Tote Bag | 30 | 0 | 1 | 1 | 2 | 0,022222 | 22222,22 | 3,51 |
| 17 | Black AFI Bag | 235 | 4 | 3 | 1 | 8 | 0,011347 | 11347,51 | 3,78 |

| Batch | Type | Amount of bag | Defect | | | Number of defect | DPO | DPMO | Sigma |
|---|---|---|---|---|---|---|---|---|---|
| | | | Sewing | Cutting | Embroidery | | | | |
| 18 | Blue Supreme Bag | 200 | 3 | 2 | 2 | 7 | 0,011666 | 11666,66 | 3,77 |
| 19 | Green Trolley Bag | 100 | 2 | 4 | 2 | 8 | 0,026666 | 26666,66 | 3,43 |
| 20 | Distro Bag | 160 | 3 | 2 | 3 | 8 | 0,016666 | 16666,66 | 3,63 |
| 21 | Black Distro Sling Bag | 20 | 0 | 0 | 1 | 1 | 0,016666 | 16666,66 | 3,63 |
| 22 | Brown Distro Sling Bag | 20 | 0 | 0 | 2 | 2 | 0,033333 | 33333,33 | 3,33 |
| 23 | Read Waist Bag | 20 | 0 | 1 | 0 | 1 | 0,016666 | 16666,66 | 3,63 |
| 24 | Black Waist Bag | 20 | 0 | 1 | 0 | 1 | 0,016666 | 16666,66 | 3,63 |
| 25 | Green Distro Bag | 126 | 4 | 2 | 3 | 9 | 0,023809 | 23809,52 | 3,48 |
| 26 | Shopper Bag | 50 | 0 | 1 | 1 | 2 | 0,013333 | 13333,33 | 3,72 |
| 27 | PMI Bag | 30 | 1 | 0 | 0 | 1 | 0,01111 | 11111,11 | 3,79 |
| 28 | Juventus Back Pack | 50 | 0 | 1 | 1 | 2 | 0,013333 | 13333,33 | 3,72 |
| 29 | Juventus Sling Bag | 50 | 0 | 2 | 0 | 2 | 0,013333 | 13333,33 | 3,72 |
| 30 | Timnas Waist Bag | 50 | 0 | 3 | 0 | 3 | 0,02 | 20000 | 3,55 |
| 31 | PMI Back Pack | 160 | 1 | 1 | 1 | 3 | 0,00625 | 6250 | 3,99 |
| 32 | Mutu Bag | 150 | 1 | 2 | 1 | 4 | 0,008888 | 8888,888 | 3,87 |
| 33 | Workshop Bag | 1355 | 16 | 8 | 8 | 32 | 0,007872 | 7872,078 | 3,91 |
| 34 | SD Sapen Back Pack | 362 | 3 | 5 | 2 | 10 | 0,009208 | 9208,103 | 3,86 |
| 35 | PMI BPO Bag | 150 | 4 | 2 | 0 | 6 | 0,013333 | 13333,33 | 3,72 |
| 36 | SD Sapen Tote Bag | 350 | 2 | 2 | 0 | 4 | 0,003809 | 3809,523 | 4,17 |
| 37 | Parcel Tote Bag | 100 | 0 | 2 | 1 | 3 | 0,01 | 10000 | 3,83 |
| 38 | Hartono Trolley Bag | 200 | 8 | 4 | 2 | 14 | 0,023333 | 23333,33 | 3,49 |
| 39 | Pink Yummy Ladies bag | 50 | 1 | 2 | 2 | 5 | 0,033333 | 33333,33 | 3,33 |
| 40 | Navy Blue Yummy Ladies Bag | 50 | 1 | 3 | 0 | 4 | 0,026666 | 26666,66 | 3,43 |
| 41 | Green Yummy Ladies Bag | 50 | 0 | 2 | 0 | 2 | 0,013333 | 13333,33 | 3,72 |
| 42 | Black Yummy Ladies Bag | 50 | 2 | 6 | 3 | 11 | 0,073333 | 73333,33 | 2,95 |
| 43 | Brown Yummy Ladies Bag | 50 | 1 | 3 | 1 | 5 | 0,033333 | 33333,33 | 3,33 |
| 44 | Maroon Yummy Ladies Bag | 50 | 1 | 5 | 0 | 6 | 0,04 | 40000 | 3,25 |
| 45 | Light Blue Yummy Ladies Bag | 50 | 1 | 4 | 1 | 6 | 0,04 | 40000 | 3,25 |
| 46 | Blue White Combination Yummy Ladies | 50 | 0 | 3 | 1 | 4 | 0,026666 | 26666,66 | 3,43 |
| 47 | Zano Bag | 250 | 15 | 5 | 0 | 20 | 0,026666 | 26666,66 | 3,43 |
| | | **Average** | | | | | | **20003,75** | **3,62** |

To measure the level of six sigma, below calculation can be conducted:

$$DPO = \frac{Number\ of\ Defect}{(amounts\ of\ defect's\ possibility)x(numbers\ of\ examination)} \quad (1)$$

$$DPMO = DPO\ x\ 1.000.000 \quad (2)$$

After resuming the value of DPMO, sigma level can be calculated by using following formula.

$$Sigma\ Level\ = normsinv\left(\frac{1000000-DPMO}{1000000}\right) + 1,5 \quad (3)$$

On every batch, DPMO and Sigma level can be calculated to get overall DPMO value and sigma level from average of 47 production batches. Hence, from average calculation, DPMO valued as 20.003,75361 and sigma level of 3,615702857. DPMO of 20.003,75361 indicates that in one million chances of bag production, there are 20.003,75361 possibilities of bag for experiencing product's defect.

### 3.3 Stage of Analyze

Analyze stage is carried out to identify the cause of defect on bag production. It was performed by involving Failure Mode and Effect Analysis (FMEA) to identify and assess the risk that turns to be the potential cause of failure [11]. The result of FMEA is shown on table 2.

**Table 2.** Failure mode and effect analysis

| Function Item | *Potential Failure Mode* | *Potential Effect on Customer Because of Defect* | *SEV* | *Potential Causes* | *OCC* | *Current Process Control* | *DET* | *RPN* | Ranking |
|---|---|---|---|---|---|---|---|---|---|
| Level of Accuracy | Sewing | Bag becomes easy to break | 7 | Inappropriateness of sewing distance (F!) | 6 | Distance's adjustment for sewing process (F1) | 4 | 168 | 2 |
| | | | | Inappropriateness on the depth of sewing (F2) | 7 | No regulation on the depth of sewing process (F2) | 8 | 392 | 1 |
| | Cutting | The lack of accuracy on the dimension of the bag | 8 | Cutting inaccuracy based on pattern (F3) | 4 | Adjustment on machine's platting towards pattern (F3) | 3 | 96 | 3 |
| | | | | The workers are careless in employing cutting machine (F4) | 2 | The workers should focus in carrying out the cutting based on the pattern (F4) | 3 | 48 | 4 |
| Esthetic | Embroidery | Less of esthetic value of the bag | 3 | Machine's quality in creating embroidery (F5) | 4 | Machine's Set up (F5) | 3 | 36 | 5 |
| | | | | The workers are careless in processing the embroidery (F6) | 2 | Assisted by using pattern's molding (F6) | 2 | 12 | 6 |

Based on above table, it can be identified the occurred risk value and the effect based on the value of RPN. The value of RPN is derived from the result of multiplication between rating severity (S) probability (P), and detectability (D). On the analysis of FMEA, the highest value of RPN (Risk Priority Number) is recorded on the sewing defect due to inappropriateness towards the sewing depth with RPN of 392. But, FMEA only considers 3 types of RPN's assessments equally or similar, which are: severity, probability and detectability, but in real case, the criteria of the risks are weighted differently [12]. Hence, it takes risk analysis criteria weighting by using AHP method (Analytical Hierarchy Process).

Coupled comparison on Analytical Hierarchy Process (AHP) will be more accurate if the comparison scale specified by experts applies fuzzy AHP. It is assumed that AHP method includes high subjectivity. Fuzzy AHP effectively could reduce the human thoughts' unclarity by providing value's interval [13]. There is the weighting comparison's result for AHP and Fuzzy AHP provided by experts in table 3.

### 3.4 Calculation of RPN Value with the Weighting of Fuzzy AHP

Calculation on RPN (Risk Priority Number) applies the result of Fuzzy AHP's weighting by multiplying RPN's value with Fuzzy AHP's weighting. Therefore, recent RPN's value is verified [14].

$$RPN = (W_S \times S) + (W_O \times O) + (W_D \times D) \quad (4)$$

where: $W_S$, $W_O$, and $W_D$ are the relative weights of severity, occurrence, detectability factors while S, O, and D each represented the score of S, O, and D value.

The following is the result of multiplication between relative weights with the value of severity, occurrence, and detectability (table 4).

**Table 3.** The result of fuzzy AHP weighting comparison

| Criteria | Weight's Comparison of Initial AHP | Weight's Comparison of Fuzzy AHP |
|---|---|---|
| Severity | 0,696531334 | 0,845492751 |
| Occurrence | 0,231613959 | 0,269892123 |
| Detectability | 0,071854707 | 0,083141035 |

**Table 4.** The result of calculation of RPN on fuzzy AHP-FMEA

| No | Code | Severity | Occurrence | Detectability | Ws | Wo | WD | RPN | Ranking |
|---|---|---|---|---|---|---|---|---|---|
| 1 | F1 | 7 | 6 | 5 | 0,845492751 | 0,269892123 | 0,083141035 | 7,95350717 | 3 |
| 2 | F2 | 7 | 7 | 8 | 0,845492751 | 0,269892123 | 0,083141035 | 8,472822399 | 1 |
| 3 | F3 | 8 | 4 | 3 | 0,845492751 | 0,269892123 | 0,083141035 | 8,092933605 | 2 |
| 4 | F4 | 8 | 2 | 3 | 0,845492751 | 0,269892123 | 0,083141035 | 7,55314936 | 4 |
| 5 | F5 | 3 | 4 | 3 | 0,845492751 | 0,269892123 | 0,083141035 | 3,86546985 | 5 |
| 6 | F6 | 3 | 2 | 2 | 0,845492751 | 0,269892123 | 0,083141035 | 3,242544569 | 6 |

Based on RPN's calculation by considering the weighting using Fuzzy AHP, hence it can be resumed the highest ranking of RPN as 8,472822399 with process code of F2, the sewing defect that caused by sewing depth inappropriateness. It is similar with conventional RPN calculation. Yet, based on the ranking order, after fixing the depth of the sewing, the cutting inappropriateness from pattern should be the next concern. Therefore, it takes improvements for six activities on bag production system that could possibly cause the defect and considers the ranking system on designated risks.

### 3.5 Stage of Improve

After performing the observation towards the cause of defect, then next step is conducting the improvement by determining solution to minimize the defects on bags. Several options that could be taken to overcome defects on products are illustrated by table 5.

**Table 5.** Solution on defects of bags

| *Potential Failure Mode* | *Potential Causes* | *Improve* |
|---|---|---|
| Sewing | Inappropriateness distance of the sewing | By using needle from the sewing machine that already adjusted with the size of the bag. |
| | Inappropriateness on the depth of the sewing | Needle's adjustment with the thickness of material |
| Cutting | The cutting inappropriateness from pattern | It needs pattern's clamp when performing the cutting |
| | The workers are careless in employing cutting machine | It needs clearer pattern by emphasizing lines on pattern |
| Embroidery | Quality of embroidery's machines | The workers should conduct the initial set up and inspection on the machine's readiness before performing the embroidery process |
| | The workers are careless in making the embroidery | The use of molding to simplify the pattern's construction |

**Table 6.** The comparison of DPMO and sigma level before and after improvements

| Condition | DPMO | Level Sigma |
|---|---|---|
| Before Improvement | 20003.75 | 3,61 |
| After Improvement | 11185.73 | 3,86 |

### 3.6 The stage of Control

In this stage, it is conducted the actions of reducing the number of defect's bags by comparing the value of DPMO and Sigma level before and after improvements. Consequently, it is suggested to provide structured improvement actions. Below, is the table covers result's comparison between DPMO value and Sigma level before and after improvements.

## 4 Conclusion and Suggestion

From this research, it can be concluded that the defects occurred the most at Zano Production are the lack of accuracy at the cutting and the sewing as well as at the embroidery process. The value of DPMO before the improvement is 20003.75 while the sigma level is recorded as 3,61.

To reduce those defects, several improvements were conducted, hence the value of DPMO was decreased with difference of 8818,02 and sigma level of 0,25. FMEA table showed that the defects mostly caused by depth inappropriateness on the sewing process. It will decrease the bags' quality that will lead to easily-broken bag with RPN value of 392 while RPN with multiplication of Fuzzy AHP-FMEA counted as 8,472822399. Suggestion that can be delivered to company is by paying more attention to the process of controlling of the bag production that needs structured improvement actions.

## References

1. Munaf, Triawan, *KOMPAS.COM.* [Online] Available at: http://bisniskeuangan.kompas.com/read/2017/04/13/203000326/inilah.tiga.sektor.ekonomi.kreatif.yang.sedang.naik.daun [Accessed 28 Juni 2017] (2017)
2. Kemendag, *Studi Industri Kreatif Indonesia.* Jakarta: Kementerian Perdagangan Republik Indonesia (2007)
3. Menperin, *Kementrian Perindustrian Republik Indonesia.* [Online] Available at: http://www.kemenperin.go.id/artikel/12797/Menperin:-Industri-Kreatif-Tumbuh-7-Per-Tahun [Accessed 28 Juni 2017] (2016)
4. Central Bureau of Statistics (Badan Pusat Statistik), *Indeks Produksi Triwulanan Industri Mikro dan Kecil (2010=100) menurut 2-digit KBLI, 2011-2016.* [Online] Available at: https://www.bps.go.id/linkTableDinamis/view/id/974 [Accessed 28 Juni 2017] (2016)
5. BisnisUKM, *BisnisUKM.COM.* [Online] Available at http://bisnisukm.com/strategi-meningkatkan-kepuasan-pelanggan.html [Accessed 28 Juni 2017] (2017)
6. M. Djunaidi, A. Suryadarmawan V, Pengendalian dan Perbaikan Kualitas Produk Kawat Baja dengan Metode Aplikasi *Six Sigma* (DMAIC) dan Kaizen (5W+1H) pada Divisi *Wire Rod Mill* (Studi Kasus: PT. Krakatau Steel Tbk). Surakarta: Universitas Muhammadiyah Surakarta (2014)
7. J. Susetyo, Winarno, H.A. Wahyuni, Aplikasi *Six Sigma* DMAIC dan Kaizen Sebagai Metode Pengendalian dan Perbaikan Kualitas Produk. *Jurnal Teknologi*, **4** (1), Juni 2011, 61-53 (2011)
8. J. Antony, M. Kumar, C.N. Madu, Six Sigma in small and medium sized UK manufacturing enterprises : Some empirical observations. *Int. J. of Quality and Reliability Management.* **22**(8), 860-874 (2005)
9. A. Syukron, M. Kholil, *Six Sigma Quality For Business Improvement,* Jakarta: Graha Ilmu (2012)
10. A.Fauziah, A. Harsono, G.P. Liansari, Usulan Perbaikan Kualitas Menggunakan Metode Six Sigma Untuk Mengurangi Jumlah Cacat Produk Tahu Pada Perusahaan Pengrajin Tahu Boga Rasa, *J. Online Institut Teknologi Nasional* , **2**(4) (2014)
11. I. Vitho, E. Ginting, Anizar, Aplikasi Six Sigma Untuk Menganalisis Faktor-Faktor Penyebab Kecacatan Produk Crumb Rubber Sir 20 Pada PT. XYZ, *Jurnal Teknik Industri FT USU, **3*** (2013)
12. R.K. Aslani, A hybrid of fuzzy FMEA-AHP to determine factors affecting alternator failure causes, *Management Science Letters*, 1981-1984 (2014)
13. E.B. Beşikçi, T. Kececi, O. Arslan, O. Turan,. An Application of Fuzzy-AHP to Ship Operational Energy Eficiency Measures, *Ocean Engineering, 121, 392-402* (2016)
14. A. Basuki, Manajemen Resiko Kerusakan Di Unit Pengemasan Pt.Semen Indonesia, Tbk, Pabrik Tuban. *Prosiding Seminar Nasional Manajemen Teknologi XXII* (hal. 1-6). Surabaya: Program Studi MMT-ITS (2015)

# Risk analysis of warehouse operation in a power plant through a Modified FMEA

*Sri* Indrawati*, *Kharina Novia Karunia* Ningtyas, *Alfina Budi* Khoirani, and *Riadho Clara* Shinta

Industrial Engineering, Universitas Islam Indonesia, Yogyakarta, Indonesia

**Abstract.** Currently, electricity becomes basic needs for human's life sustainability. Most of activities require electricity. Some power plant are demanded to be able to fulfil above necessity by distributing electricity as it required within time. Therefore, to accommodate good performance, it needs assessment on risk analysis, specifically at the warehousing division. A risk analysis is needed for assuring a good performance warehouse. A Modified FMEA method is used to analyse the risk. This method id done by identifying sources and root causes of a problem based on the value of risk priority number (RPN). The research is conducted in an Indonesian power plant, located in West Java. There are 10 types of failure modes. The result shows that the failure mode priority is inventory discrepancies. There are no difference ranking on the most impacted failure to be prioritized using FMEA and modified FMEA method.

## 1 Introduction

Electricity is one of human basic needs to support every human activity. If there are no electricity supply then all activities undertaken by humans will be stopped or obstructed. Therefore, every power plant produces electricity with continuous production, as well as minimizing the risk of occurrence of bottleneck or other production constrain. The process can be run with a good system in order to increase productivity.

A good system is done to minimize the risk of an industry [1]. There are some industries that fail to run their business processes because triggered by bad risk management. Most industries maximize the potential of assets and resources to make a profit without identifying potential risks that may occur in each of the assets or resources managed [2].

Risk is a combination of the probability and severity of an event. The amount of risk is determined by several factors, such as exposure, location, usage, quantity and the vulnerability of the elements involved [3]. Risks related to business activities should be identified, measured, assessed, mitigated, and controlled by the company. Management aims to minimize the possibility of potential losses that may be experienced by the company [4].

One unit in a company that has a big enough risk is a warehouse. Warehouse has a big influence on the company because without the warehouse then the production will be difficult to control. There are several risks in an Indonesian power plant that may cause a failure of the warehousing process, such as errors in the item code to be picked up, the amount of the stock does not match the existing records and the goods become damaged/ expired. Therefore, risk management is required to eliminate or reduce the risks by adding the necessary precautions.

The Failure Mode and Effect Analysis (FMEA) is one of the most commonly used methods to assess risk. FMEA is a structured procedure for identifying and preventing as many failure modes as possible. FMEA is used to identify the sources and root causes of a problem. Determining the cause of potential failure on FMEA is based on the highest value of risk priority number (RPN) which is the result of multiplication between severity, occurrence, and detectability. However, conventional FMEA usage often results in the same RPN, but presents different risk representations. Calculations of conventional FMEA methods weigh the severity (S), occurrence (O), and detectability (D) equally or in proportion, but in real cases these criteria have different weights [5]. To overcome that problem, weighted criteria can be done using analytic hierarchy process (AHP) method.

There are three most significant risk factors, i.e. dust, noise and gas in the occupational disease of mining industry in the south west of Hubei Province by using FMEA and AHP [6]. A multi-criteria decision making (MCDM) method is introduced in fuzzy FMEA based on fuzzy set theory in FMEA [7]. AHP-PROMETHEE method is used to prioritize failure modes in FMEA. This integration can enhance the precision of FMEA results in the field of risk assessment [8].

An improved AHP model is used to asess the risk in occupational health and safety management [9]. The most influential failure modes is determined to optimize the designing process in offshore engineering using probability network evaluation technique (PNET) and FMEA [10].

* Corresponding author: sriindrawati@uii.ac.id

## 2 Basic Theory

### 2.1 Failure Mode and Effect Analysis (FMEA)

FMEA is a structured procedure for identifying and preventing as many failure modes as possible. FMEA is used to identify the sources and root causes of a quality problem. There are two types of FMEA, i.e. [11]:

a. FMEA Design
   The FMEA design is used after system/ product design already determined. The FMEA design is used to analyze the product design before the manufacturing process.
b. FMEA Process
   FMEA process is used to test the error or failure mode of each stage of manufacturing and assembly process of a product.

FMEA considers three main criteria, i.e.:

- Severity is a rating of the seriousness of the consequences of failure.
- Occurrence is possible causes of failure.
- Detectability is the quantitative of the controls or procedures that exist to regulate the function or that make the failure detectable.

Risk analysis of failure modes is done using RPN (Risk Priority Number) that can be calculated based on formulation (1), where S is the severity, O is for occurrence and D is for detection value. Any failure mode with high value of the RPN is prioritized to be corrected [12].

$$RPN = SxOxD \qquad (1)$$

### 2.2 Analytical Hierarchy Process (AHP)

AHP allows decision makers to use simple hierarchical forms to solve complex problems and to evaluate qualitative and quantitative data in systematic methodologies with attention to multi-criteria [13]. There are seven pillars in the AHP [14], i.e. ratio scale, pairwise ratio, conditions for sensitivity of Eigen vector, homogeneity and clustering, synthesis, retention and reversal sequence and group considerations.

## 3 Research Method

This research was conducted in an Indonesian power plant, located in West Java using five main stages. The first stage is the observation process to identify the problems found in the company especially in the warehouse. The second stage is identification of potential failures, potential causes and potential effects caused if the failure occurs in the warehouse operations process. The next stage is conducting a survey to determine the risk priority number in each failure mode and weighting FMEA criteria. FMEA modification is then performed on the FMEA criteria weighting stage, i.e. severity, occurrence and detectability using the AHP method. The value of RPN in modified FMEA can be calculated using formulation (2) [15]:

$$RPN = (WSxS) + (WOxO) + (WDxD) \qquad (2)$$

Where WS, WO and WD are the relative weight of the severity, occurrence, and detectability criteria of the calculation using the AHP method. Analysis is done by comparing the RPN value of FMEA and modified FMEA. Then a failure analysis is performed to reduce the risk of failure mode that has the highest RPN value.

## 4 Discussion

There are ten types of failure modes identified from the warehouse operational process in Indonesian Power Plant as seen in Appendix 1. F4 and F6 are failure modes with high severity value, which can give very high effect. Both modes of failure are very influential, if F4 occurs for critical components in a frequent time, so the failure will disrupt the system. For expired materials (F6) also have a very high value of Severity, this is because some materials require special care during storage period. If the item is expired or damaged, then it cannot be used for production. On the criteria of occurrence, there are three failure modes that cannot be avoided, i.e. F1, F9 and F10. The processes that performed on the receiving process are the inspection towards delivery permit letter, amount of loads, the appropriateness of order specification. Occasionally, the numbers of incoming loads with the ordered ones are different. Failure mode F1 is considered as repetition and human errors. Later, for failure mode F9, duration of delivery process, could disturb 25% of system work, especially when the materials are considered as critical materials that urgently used in a short time. If they are late, the selling value will be reduced per day. The probability of failure is quite high, which is repeated failures. Failure mode F10 (the failure on material code that will be taken/accepted) is caused by the human error. This occurrence usually happens when there is a failure on material code or quantity. This failure mode is considered as disturbing and effects 50% of work system with the high probability of repetition.

There are two failure modes with high detectable value, among others, F8 and F10. The failure mode F8 and F10 are considered as low for detection, since on F8 or the identification of dead stock can be resumed on stock opname activity that always conducted annually, hence the possibility of detection is extremely low. On the failure mode F10, which is, the failure on material code that will be taken/ received, the detection process will be low since on the material pickup and acceptance process, sometimes the officer in charge is absent and replaced by others. It will emerge the material placement impropriate with the material code.

The calculation of RPN value by using FMEA is conducted by employing the value of severity, occurrence and detectability. F10 is determined as failure mode with the highest RPN value of 168, as described by Figure 1.

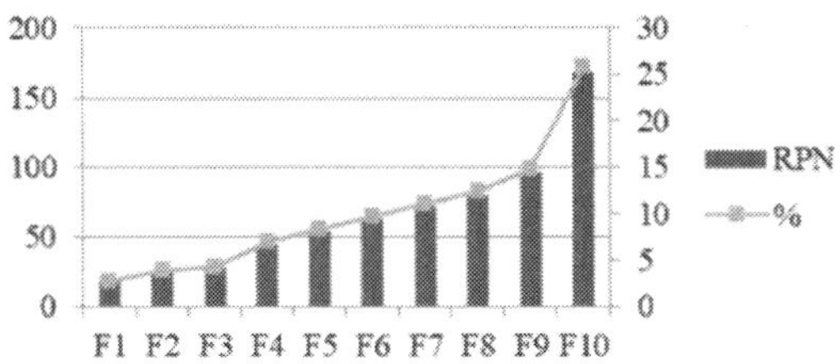


**Fig 1.** RPN values

This failure mode is triggered by human errors in performing material code or quantity inputting. It disturbs 50% system work with probability of repeated failures, hence the detection possibility is considered as low, probability checking will detect designated failure mode.

On the FMEA method, the weight of criteria is assumed as similar, but in real, similar RPN value will establish different effect. For instance: the value of severity, occurrence, and detectability are 6, 3 and 1 consecutively, while other failure modes are 9, 2 and 1. Both values have the same RPN value yet provide different effect. Hence, FMEA is modified with AHP to gain weight of severity, occurrence and detectability criteria. The weight of severity criteria is higher than two other criteria as illustrated by Figure 2.

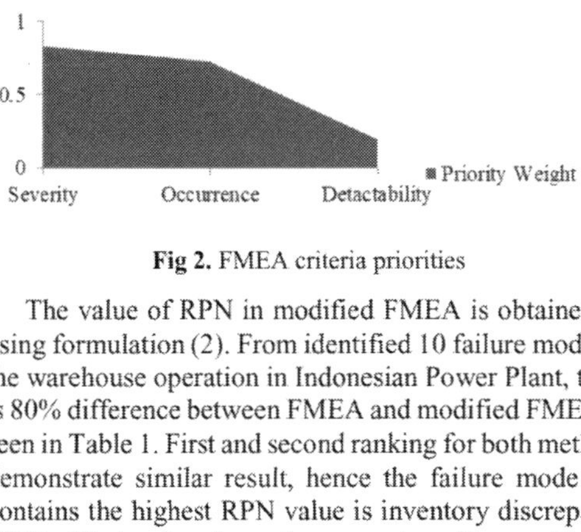


**Fig 2.** FMEA criteria priorities

The value of RPN in modified FMEA is obtained by using formulation (2). From identified 10 failure modes at the warehouse operation in Indonesian Power Plant, there is 80% difference between FMEA and modified FMEA as seen in Table 1. First and second ranking for both methods demonstrate similar result, hence the failure mode that contains the highest RPN value is inventory discrepancy (F10). F9 is placed on the second ranking, which is described as the delivery lead time.

**Table 1.** Comparison of FMEA and Modified FMEA result

| No. | Code of Failure Modes | Risk Priority | | No. | Code of Failure Modes | Risk Priority | |
|---|---|---|---|---|---|---|---|
| | | FMEA | Modified FMEA | | | FMEA | Modified FMEA |
| 1 | F1 | 8 | 3 | 6 | F6 | 3 | 8 |
| 2 | F2 | 10 | 9 | 7 | F7 | 7 | 6 |
| 3 | F3 | 9 | 10 | 8 | F8 | 6 | 4 |
| 4 | F4 | 4 | 5 | **9** | **F9** | **2** | **2** |
| 5 | F5 | 5 | 7 | **10** | **F10** | **1** | **1** |

## 5 Conclusion

From the research, it can be concluded that the most influenced failure mode at warehouse operation in Indonesian Power Plant to be prioritized is the failure on taken/received material code. The ranking determination of the failure is based on RPN value. There is no classification on ranking for the most influenced failure to be prioritized by FMEA and modified FMEA method. Further research is needed using fuzzy logic modification to minimize errors.

Researchers thank to the support of Industrial Engineering Department, Faculty of Industrial Technology, Universitas Islam Indonesia.

## References

1. Rosih, A.R., Choiri, M., Yuniarti, R., Analisis Risiko Operasional Pada Departemen Logistik Dengan Menggunakan Metode FMEA, *Jurnal Rekayasa dan Manajemen Sistem Industri*, Vol **3**., No. 3: 580-591, (2015)
2. Nitisusastro, M., *Asuransi dan Usaha Perasuransian di Indonesia*, Alfabeta. Bandung, (2013)
3. Ramli,S., *Sistem Manajemen Keselamatan & Kesehatan Kerja OHS AS 18001*, Dian Rakyat, Jakarta, (2010)
4. Osabutey, D., Adibo, O., Agbodohu, G., Kumi, W.P., Analysis of Risk Management Practices in the Oil and Gas Industry in Ghana: Case Study of Oil Refinery (Tor), *European Journal of Business and Management*, Vol.**5**, No.29: 139-150, (2013)
5. Aslani, R. K., A Hybrid of Fuzzy FMEA-AHP to Determine Factors Affecting Alternator Failure Causes, *Management Science Letters*: 1981-1984, (2014)
6. Allen, H.H., Hsu, C.W., Kuo, T.C., Wu., W.C., Risk Evaluation of Green Components to Hazardous Substance Using FMEA and FAHP, *Expert Systems with Applications*, Vol. **36**, Issue 3, Part 2: 7142-7147, (2009)
7. Chen Liu, H., Liu, L., Liu, N., Xiang Mao, L., Risk Evaluation in Failure Mode And Effects Analysis With Extended VIKOR Method Under Fuzzy Environment, *Expert Systems with Applications*, Vol.**9**, Issue 17: 12926-12934, (2012)
8. Maheswaran, K. & Loganathan, T., A Novel Approach for Prioritization of Failure modes in FMEA using MCDM, *International Journal of Engineering Research and Applications*, Vol. **3**, Issue 4: 733-739, (2013)
9. Jiangdong B., Johansson J., Zhang, J., An Occupational Disease Assessment of the Mining Industry's Occupational Health and Safety Management System Based on FMEA and an Improved AHP Model, *Sustainability*, Vol. **9**, No. 94: 1-10, (2017)
10. Kang, J., Sun, L., Sun, H., Wu, C., Risk Assessment of Floating Offshore Wind Turbine Based On Correlation-FMEA, *Ocean Engineering*, Vol. **129**: 382-388, (2017)
11. Mc Dermott, R.E., Mikulak, R.J., Beauregard, MR., *The Basics of FMEA 2nd Edition*, CRC Press, New York, (2008)

12. Namdari, M., Rafiee, S., Jafari, A., Using FMEA Method to Optimize Fuel Consumption in Tillage by Moldboard Plow, *International Journal of Applied Engineering Research*, Vol.**1**: 734-742, (2011)
13. Mustafa, A., Jia-Pei, F., Siaw-Peng, L., Hamid, H.A., The Evaluation of Airline Service Quality Using AHP, *International Conference on Tourism Development,* 382-390, (2005)
14. Saaty, T.L., *The Analytic Hierarchy Process*, McGraw-Hill, New York, (1980)
15. Basuki, A., Manajemen Resiko Kerusakan Di Unit Pengemasan PT Semen Indonesia, *Prosiding Seminar Nasional Manajemen Teknologi* **XXII**: A-28-1 -6, (2015)

# Production quality improvement analysis of grey cambric using Six Sigma Method

*Yosa Permata* Shafira and *Agus* Mansur

Department of Industrial Engineering, Faculty of Industrial Technology, Universitas Islam Indonesia, Jalan Kaliurang Km. 14,5 Sleman, Yogyakarta, Indonesia

**Abstract.** Quality has an important role in business. A good quality means it is able to meet the required standards. The concept of quality must be comprehensive, whether the quality of the product or its process. PT Textile Yogyakarta is still having a high rate of defect products. The Waste Assessment Model (WAM) was used to identify the wastes that frequently occur by scoring each type of waste. The results show the waste rank are defect 26.12%, inventory 20.36%, motion 15.26%, transportation 14.51%, over production 12.91%, waiting 7.17%, and process 3.67%. While using the Six Sigma from the 4 CTQ process, the result of sigma level is 3.3. It means that the sigma level of PT Textile Yogyakarta is still under the average sigma level in Japan which is 5 Sigma. As for the improvement part, FMEA AHP was used and the results show that the first rank is the uncleaned woof residue, the second rank is the uncut thread which has an RPN value of 8.04 that was caused when the operator did not clean the machine from the woof residue, and the third rank is the malfunctioned machine which has an RPN value of 7.32 that was caused by broken kleting.

## 1 Introduction

In 2010, textile industries were able to employ up to 11% of total industrial workforce or 1.34 million people in 2,853 companies throughout Indonesia and donate up to 8.9 percent of total domestic exports. Until 2014, the realization of textile industry investment in Indonesia reached Rp 9.53 trillion or it grew 9.4% from the previous year [1]. To sustain the growth, it needs a trigger to boost the competitiveness based on work productivity. Quality is everything that satisfies the customer and fits the requirements [2]. Product quality is the key to win the market competition, because it can earn the customer's trust to the product offered by the company, then in the end it will lead to customer loyalty.

Quality problems are often experienced by all kinds of companies, including PT. Tekstil Yogyakarta which frequently encounters defected cloth on the finished goods. To cope with that problem, the company should optimize the quality control to minimize the amount of defected goods and ensure the finished goods to meet the required standards of quality in the company. Quality control should not only focus on the finished goods, but also the whole process of production starting from the raw materials, work-in-process goods, and finished goods ready to be sold. Based on this background, the aim of this research is to analyze the quality control of the highest defect in PT Tekstil Yogyakarta from September 2016 to February 2017 using the Six Sigma DMAIC method.

## 2 Basic Theory

Alkatiri et al. [4] explains that by combining Six Sigma with seven tools such as fishbone diagram, pareto diagram and histogram is able to analyze and identify causes of defects product so it can be reduced [3]. As the result of analysis on textile and garment industry in PT SSP is known the cause of DPMO (Defect Per Million Opportunity) average value and sigma value before implementation of 6,523,27 and 3,98 sigma. DPMO average value and sigma value after implementation of 4,753.80 and 4,09 sigma in other words the average DPMO decreased by 27.13% and the sigma value increased by 0.11 sigma [4].

Further research on six sigma obtained the result of increasing the sigma value from the previous 3.05 to 3.80 sigma and the defective product defect by 29.87% [5]. The improvement is done by applying Failure Mode Effect and Analys is method and statistical process control (SPC). Failure Modes and Effects Analysis (FMEA) is able to identify the risk of failure that occurred during the production process in making sarong weaving at PT. Asaputex Jaya [6]. Integrates Lean Six Sigma with other statistical tools can extend effectiveness and sustain improvements. This study compares the performance before and after the implementation of the proposed solution for waste reduction by integrating with Analytic Hierarchy Process (AHP) the objective is to analyze and assess the effect of the highest waste causes to increase effectiveness [7].

* Corresponding author: agusmansur@uii.ac.id

## 3 Method

The phases of this research are referring to Gasperz to settle problems and enhance the process using the DMAIC (Define, Measure, Analyze, Improve, Control).

### 3.1 Define

It is the goal setting phase of the Six Sigma quality improvement activities. This phase determines the plans that have to be done to carry out the improvement of each stage in the key business process [8]. This phase will identify the waste by using the Waste Assessment Model (WAM) which can show the relations of each waste and how the rank of defect types occurred so they can be minimized [9]. After that, this phase should be continued by arranging the diagram of SIPOC (Suppliers, Inputs, Process, Outputs and Customers) which summarize the input and output from one process to another, in the form of a table. Then finally this phase should determine the Critical to Quality.

### 3.2 Measure

It is the second phase as an effort to improve the quality of Six Sigma that measure the control chart and performance baseline. Furthermore, the U chart is used as the control chart to find out whether the defected goods of the resulting product are within the required limits or not. Performance baseline measurement uses the measurement unit of sigma level to detect the level position of the company which indicated by PPM (Part per Million Opportunities) value. PPM is the number of defected goods in a million opportunities [10].

### 3.3 Analyze

This phase will analyze the process capability by using the cause and effect diagram or also known as fishbone. This diagram establishes ways to make better products and how to reach the best results.

### 3.4 Improve

This phase is used as an effort to improve production quality. FMEA gives best solution after the analyze phase, while AHP has a mathematic approach to find out the right solution n. One of the weaknesses of FMEA is the possibility of earning the same RPN value with different nt purposes. It causes the relative importance value of Severity, Occurrence and Detection should be considered by integrating the FMEA with the AHP method.

## 4 Result and Discussion

### 4.1 Define

#### *4.1.1 Waste Identification*

The results of Waste Assessment Model (WAM) show the rank of waste from the highest to lowest percentage respectively, i.e. defect 26.12%, inventory 20.36%, motion 15.26%, transportation 14.51%, overproduction 12.91%, waiting 7.17% and process 3.67%. By referring the Pareto principle 80/20 which means that 20% of the cause is responsible for 80% of the problem, the waste type that has to be the taken to analyze is Defect.

#### *4.1.2 SIPOC Diagram*

Identifying the key processes in Six Sigma method can be done by creating a process model Suppliers - Input - Process - Outputs - Customers or SIPOC diagram. Figure 1 shows the diagram SIPOC for the production process of grey cloth.

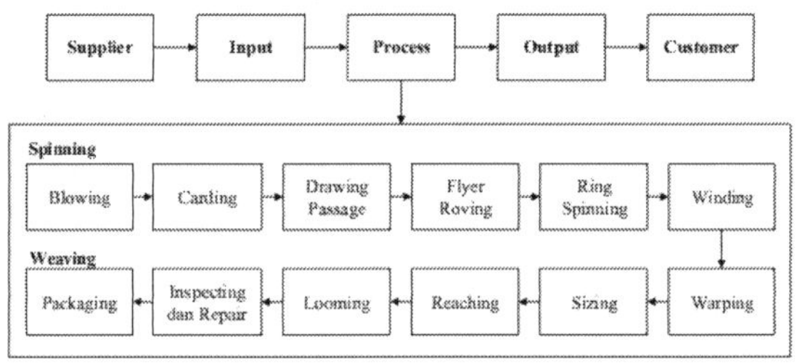


**Fig 1.** SIPOC diagram

#### *4.1.3 Critical to Quality*

Identification is done towards the quality characteristics that have potential in causing the defected goods 217. The quality characteristics of this research are the types of defected goods that affect the output.

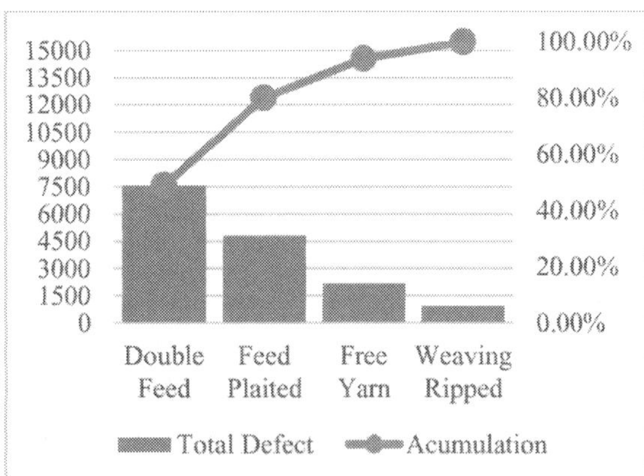


**Fig 2.** Diagram Pareto CTQ 217

## 4.2 Measure

### *4.2.1 U Chart*

Chart control has median value which is the average of the quality characteristics or value of the defects proportion which relates to controlled situation, and horizontal lines called the Upper Control Limit (UCL) and the Lower Control Limit (LCL) [11]. A process is called as controlled if the data is between the LCL and UCL. The reason why this research uses the U Chart is because the data which are defective in one product may have more than one type of defects with the sample value (n) used is not constant.

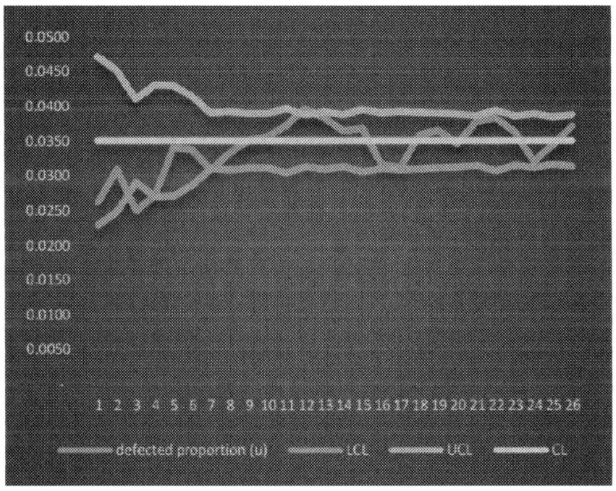


**Fig 3.** U chart

Based on the figure above, it is known that the data is not consistent, which there are some data above and under control. Data that has the cause of special variations should be removed. As the data in September where the number of machines used is below average, the data outside the control limit must be removed. Meanwhile, for data that has no special cause variations is maintained. Calculation during the September 2016 – February 2017 shows the average value or the center line is at 0.03522.

### *4.2.2 Baseline Measurement*

Performance baseline measurement at the output level was carried out directly on the final product to find out how far the final outputs of the process meet the specific needs of the customer, before they are handed over to the customer. The performance baseline outcome measurement used in Six Sigma is the level of PPM (Parts per Millions) and sigma level. Based on the calculation, the average value of the PPM is 34,412.00 with sigma value 3.32. This indicates that the possibility of 34,412.00 units of defected goods occurred per one millio n events. The sigma value is categorized in the sigma level of 3 and if it is analyzed using the COPQ (Cost of Poor Quality) table by Gaspersz (2005) it will cause 25% - 40% losses of sales.

## 4.3 Analyze

Based on CTQ that has been determined and the calculation that has been done, it can be identified the root cause of the problem by using a fishbone diagram. It will help to find out what actions to solve the causes of the problem. The figure below is an example of the fishbone diagram of each defect:

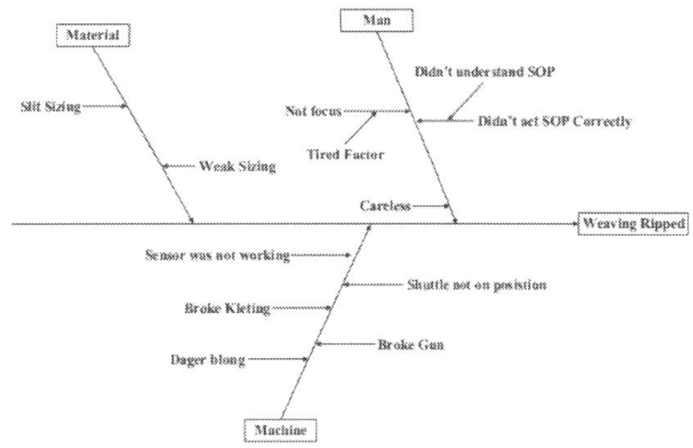


**Fig 4.** Fishbone of weaving ripped

## 4.4 Improve

It is generally a technique used for product and process design, by identifying the potential failures and prioritizing them, subsequently reducing the errors in the process. A successful FMEA activity helps a team to identify potential failure modes, based on past experience with similar products or processes [12]. The conventional calculation of FMEA method is weighing the Severity (S), Occurrence (O), and Detection (D) proportionally. One of its weaknesses is the possibility of earning the same RPN value with different purposes. Meanwhile, in reality, those criteria may have different weight [13]. Therefore, the relative importance value of Severity, Occurrence and Detection should be considered by integrating with AHP method.

**Tabel 1.** Consistency ratio

| Matrix Multiplication | Eugen Value | $\lambda$ maks | CI | IR | CR |
|---|---|---|---|---|---|
| 0.54 | 3.02 | 3.03 | 0.01 | 0.58 | 0.03 |
| 0.21 | 3.01 | | | | |
| 2.29 | 3.06 | | | | |
| 3.05 | 9.09 | | | | |

Conventional FMEA RPN:

$$Severity \times Occurrence \times Detection \quad (1)$$

FMEA AHP RPN:

$$(WS \times S) + (WO \times O) + (WD \times D) \quad (2)$$

**Tabel 2.** Conventional FMEA and FMEA AHP

| *Potential Failure Mode* | *CODE* | Conventional FMEA | | | | | FMEA AHP | | | | |
|---|---|---|---|---|---|---|---|---|---|---|---|
| | | *SEV* | *OCC* | *DET* | *RPN* | *RANK* | *SEV 0.18* | *OCC 0.87* | *DET 0.75* | *RPN* | *RANK* |
| Not focus | F1 | 5 | 3 | 3 | 45 | 17 | 5 | 3 | 3 | 3.36 | 17 |
| Careless | F2 | 5 | 3 | 3 | 45 | 17 | 5 | 3 | 3 | 3.36 | 17 |
| Dont understand SOP | F3 | 5 | 3 | 3 | 45 | 17 | 5 | 3 | 3 | 3.36 | 17 |
| Slit Sizing | F4 | 9 | 3 | 6 | 162 | 7 | 9 | 3 | 6 | 6.33 | 5 |
| Weak Sizing | F5 | 9 | 3 | 6 | 162 | 7 | 9 | 3 | 6 | 6.33 | 5 |
| Pallet Double | F6 | 6 | 6 | 6 | 216 | 5 | 6 | 6 | 6 | 6.00 | 7 |
| Cone dobel | F7 | 6 | 3 | 6 | 108 | 9 | 6 | 3 | 6 | 5.79 | 9 |
| The initial scrolls are woven | F8 | 6 | 6 | 6 | 216 | 5 | 6 | 6 | 6 | 6.00 | 7 |
| Broken Feed | F9 | 9 | 6 | 6 | 324 | 3 | 9 | 6 | 6 | 6.54 | 4 |
| The remaining yarn is not cut offf | F10 | 9 | 6 | 8 | 432 | 1 | 9 | 6 | 8 | 8.04 | 1 |
| Remaining Feed from Battery | F11 | 9 | 6 | 8 | 432 | 1 | 9 | 6 | 8 | 8.04 | 1 |
| Machine is not function | F12 | 5 | 3 | 6 | 90 | 10 | 5 | 3 | 6 | 5.61 | 10 |
| | F13 | 5 | 3 | 6 | 90 | 10 | 5 | 3 | 6 | 5.61 | 10 |
| | F14 | 5 | 3 | 6 | 90 | 10 | 5 | 3 | 6 | 5.61 | 10 |
| | F15 | 5 | 3 | 6 | 90 | 10 | 5 | 3 | 6 | 5.61 | 10 |
| | F16 | 5 | 3 | 6 | 90 | 10 | 5 | 3 | 6 | 5.61 | 10 |
| | F17 | 5 | 3 | 6 | 90 | 10 | 5 | 3 | 6 | 5.61 | 10 |
| | F18 | 5 | 3 | 6 | 90 | 10 | 5 | 3 | 6 | 5.61 | 10 |
| | F19 | 5 | 6 | 8 | 240 | 4 | 5 | 6 | 8 | 7.32 | 3 |
| | F20 | 5 | 3 | 3 | 45 | 17 | 5 | 3 | 3 | 3.36 | 17 |

The conventional FMEA and FMEA AHP have the different priorities in determining what causes of defect are priorities that must be solved first.

## 5 Conclusion and Recommendation

1) The results of the Waste Assessment Model show that the highest waste is defect 26.12% with 4 defect types that frequently occur, i.e. double feed, remaining feed, free yarn and weaving ripped.
2) Based on measurement obtained average sigma value are 3.32. The sigma value included in the sigma 3 and it will losses of 25-40% of sales.
3) Implementing 5S (Seiri, Seiton, Seiso, Seiketsu, Shitsuke) to eliminate or minimize the waste in the working environment.
4) Adding the fast reed warp protector to ensure that the shuttle is in place and if there is a change or it is out of position, the machine will stop automatically.
5) Adding photo cell censors that capable to detect the passing weft, therefore it will stop automatically when the feed is cut.
6) To minimize the human error, the company can use the Poka Yoke anti-fault device by making special control board completed with tools image for operator, therefore the production process can be done correctly and meet the required standards.

## References


1. BKPM, "Peluang Investasi," 26 Desember 2016. [Online]. Available: http://www.bkpm.go.id/id/peluang-investasi/peluang-berdasarkan-sektor/industri.
2. V. Gaspersz, (2003)
3. H. A. Alkatiri, H. Adianto and D. Novirani, *Reka Integra Itenas*, **3**, 148-159, (2015)
4. H. A. Alkatiri, H. Adianto and D. Novirani, **3** (2015)
5. S. K. Dewi, *Jurnal Teknik Industri UMM*, 43-50, (2012)
6. N. B. Puspitasari and A. Martanto, *J@TI Undip*, 93-98, (2014)
7. M. K. Hassan, *American Journal of Industrial Engineering*, **1**, 28-35, (2013)
8. V. Gaspersz, *Pedoman Implementasi Program Six Sigma Terintegrasi dengan ISO: 2000*, (2005)
9. I. A. Rawabdeh, *International Journal of Operations & Production Management*, **25**, 800-822, (2005)
10. N. R. R. C. Pande, The Six Sigma Way: Bagaimana GE, Motorola dan Perusahaan Terkenal Lainnya Mengasah Kinerja Mereka, (2002)
11. H. Purnomo, Pengantar Teknik Industri, **II**, (2004)
12. V. S. Kamble and T. Z. Quazi, (2014)
13. R. K. Aslani, H. R. Feili and H. Javanshir, *Management Science Letters*, 1981-1984, (2014)

# Warehousing performance improvement using Frazelle Model and per group benchmarking: A case study in retail warehouse in Yogyakarta and Central Java

*Elisa* Kusrini*, *Citra Indah* Asmarawati, *Galuh Masita* Sari, *Anisa* Nurjanah, *Alex* Kisanjani, *Sadiq Ardo* Wibowo, and *Indro* Prakoso

Master of Industrial Engineering, Faculty of Industrial Technology, Islamic University of Indonesia, Kaliurang Km 14,5 Yogyakarta, Indonesia

**Abstract.** Warehouse performance management has an important role in improving logistic's business activities. Good warehouse management could increase profit, time delivery, quality and customer service. This study is conducted to assess performance of retail warehouses in some supermarket located in Central Java and Yogyakarta. Performance improvement is proposed base on the warehouse measurement using Frazelle model (2002), that measure on five indicators, namely Financial, Productivity, Utility, Quality and Cycle time along five business process in warehousing, i.e. Receiving, Put Away, Storage, Order picking and shipping. In order to obtain more precise performance, the indicators are weighted using Analytic Hierarchy Analysis (AHP) method. Then, warehouse performance are measured and final score is determined using SNORM method. From this study, it is found the final score of each warehouse and opportunity to improve warehouse performance using peer group benchmarking

## 1 Introduction

Warehousing management becomes very important because it contributes to cost improvement in logistics system (Liviu et al., 2009). Good warehouse management contribute cost and the quality both in terms of service to customers and the quality of the resulting product (Ackah and Eric, 2016; Speh, 2009). One of the best practice in good management warehouse is performance measurement which can be used to improve warehouse to be more effective and efficient. Performance measurement is useful for manager in order to carry out some improvement action base on periodic review.

Base on literature study on warehouse performance, many models and indicators for measuring warehouse performance are existed, that differ in term of time, cost, quality and productivity dimensions (Staudt et al., 2015). Therefore, warehouse managers need to choose which model comply with their warehouse environment. One of the model for measuring warehouse performance that measures operational activity is Frazelle model (2002). This model use measurement metrics that commonly used in industries. The advantage of Frazelle model is comprehensive, since it does not only measures quality, cost, time and productivity dimensions, but also it measures in terms of warehouse activities (receiving, put away, storage, picking, shipping. Therefore this study decide to focus on warehouse performance measurement using Frazelle model (2002) that measure on five indicators, namely Financial, Productivity, Utility, Quality and Cycle time along with five business process in warehousing, i.e. : Receiving, Put Away, Storage, Order picking and shipping. A case study on warehouse performance measurement in retail warehouse is conducted. Based on the result, then improvement is proposed using benchmarking by comparing warehouse performance with the best performance among peer group.

## 2 Literature Review

Now days, there is growing interest in the analysis and measurement of performance, because performance analysis allows the development of decision support systems for the design, planning and management of enterprise. A performance measurement system can be defined as the set of metrics used to quantify both the efficiency and effectiveness of actions (Neely et al., 1994). Although warehouse performance can be measured in to hard metric (such as order cycle time, fill rates and costs) and soft metric (such as manager's perceptions of customer satisfaction and loyalty (Staudt et al, 2015), however, in general, efficiency and effectiveness is the most widely utilized as a measure of performance (Ammons, 1995; Kusrini et al., 2014).

Warehouse performance measurement has been explored in different ways by researchers. These works differ from one another with respect to the objectives (long- or short-term decisions), the way to measure these objectives (variety of performance indicators), the type of

*Corresponding author: elisakusrini@uii.ac.id

warehouse systems (distribution centre, cross-dock platforms, etc.), the focus area inside the warehouse (storage, picking, etc.) and the tools used for measurement (statistical tools, mathematical programming, etc.) (Staudt et al., 2015). Warehouse manager needs to choose which type of suitable measurement with their warehouse environment. This is not an easy job since all of the measurement are different (Keebler and Plank, 2009).

Warehouses could have different activities according to product specification, customer requirements and service levels offered. The complexity of the warehouse activities depends mainly on: (i) the number and variety of items to be handled; (ii) the amount of daily workload to be done; and (iii) the number, nature and variety of processes necessary to fulfill the needs and demands of the customers and suppliers (De Koster and Warffemius, 2005). According to Frazelle (2002), warehouse activities are divide into 5 activity, i.e.: receiving, put away, storage, order picking and shipping. Each activity can be measured by using 5 Key Performance Indicator (KPI), namely financial, productivity, utility, quality and cycle time. Therefore there are 25 KPIs for measuring warehouse as depicted in Figure 1

**Table 4-1: Warehouse key performance indicators (Frazelle, 2002, 56)**

| | Financial | Productivity | Utilization | Quality | Cycle Time |
|---|---|---|---|---|---|
| Receiving | Receiving cost per line | Receipts per man-hour | % Dock door utilization | %Receipts processed accurately | Receipt processing time per receipts |
| Putaway | Putaway cost per line | Putaways per man-hour | % Utilization of putaway labor and equipment | % Perfect putaways | Putaways cycle time (per putaway) |
| Storage | Storage space cost per item | Inventory per square foot | % Locations and cube occupied | % Locations without inventory discrepancies | Inventory days on hand |
| Order picking | Picking cost per order line | Order lines picked per man-hour | % Utilization of picking labor and equipment | % Perfect picking lines | Order picking cycle time (per order) |
| Shipping | Shipping cost per customer order | Orders prepared for shipment per man-hour | % Utilization of shipping docks | % Perfect shipments | Warehouse order cycle time |

**Fig 1**. Warehouse Key Performance Indicators (Frazelle, 2002)

## 3 Research Methods

### 3.1 Research Objects

Warehouse performance measurement is conducted in four retail/supermarket warehouses in Yogyakarta and Central Java. The retail/supermarket sell the needs of households, schools and offices. This warehouses has an average space ranging from 175m$^2$ to 200m2 and an average 3-5 labor dedicated to this warehouse. In general, business processes in this retail warehouse starts from the goods received and the process of checking goods, the transfer of goods into the warehouse, put the goods into storage racks. The goods in the warehouse is picked and delivered if there is an order from the store.

### 3.2 Research Stages

The stages in this research are as follow: (1) Determine warehouse criteria based on Frazelle model; (2) Identifying importance weight of KPIs (3). Conduct a field survey and measuring warehouse performance (4) Normalize data using SNORM (5) Perform peer group benchmarking.

## 4 Results and Discussion

### 4.1 Result

This research employ criteria base on frazelle model to find warehouse performance and identify opportunity for improvement. In order to identify the importance weight of indicators, Analytic Hierarchy Process (AHP) model is used. A questionnaires of pair wise comparisons among indicators (KPIs) is prepared and a total of four manager of retail warehouse were asked to give their preferences on the questionnaires. The scale of 1 to 9 (1 = equally, 3 = moderate, 5 = strong, 7 = very strong, 9 = extreme) is utilized to rate the preferences of the expert. Consistency Ratio (CR) is used to check the consistency of the comparison. CR less than 0.1 means that comparison is accepted. The results of questionnaires is analyzed using Analytic Hierarchy Process (AHP) model. Analytic Hierarchy Process was first introduced by Thomas L. Saaty in 1971 and became the most widely method for Multiple Criteria Decision Making (MCDM) (Saaty, 2008). The result of importance weight of each KPIs can be seen in Figure 2.

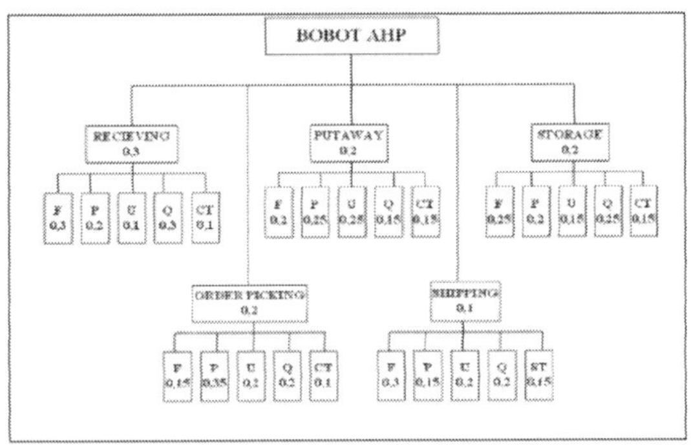


**Fig 2**. The important weight

The results of field surveys and measuring warehouse performance for four retail warehouse, i e: retail 1 (M1), retail 2 (M2), retail 3 (M3) and retail 4 (M4) as seen in Table 1.

Warehouse performance indicators then scored using standardized method, i.e. SNORM based on benchmark data among warehouses. The process of normalization (SNORM) De Boer (Trienekens and Hvolby, 2000) employed scale of 0 to 100 to get the value of each metric. The final score is obtain by multiplying normalized data with importance weight, as seen on Table 1.

**Table 1**. Warehouse Performance Indicators from observation

| Activity | M1 | M2 | M3 | M4 |
|---|---|---|---|---|
| 1. Receiving | | | | |
| - Financial (Rp/line receiving) | 39.550 | 31.250 | 15.052 | 39.625 |
| - Productivity (box/man-hour) | 106,67 | 216 | 108,4 | 100 |
| - Utilization (%) | 33,3 | 90 | 98 | 31,3 |
| - Quality (%) | 98,75 | 100 | 58 | 99 |
| - Cycle Time | 1,5 | 2,08 | 0,56 | 1,5 |
| 2. Put Away | | | | |
| - Financial (Rp/line putaway) | 46.151 | 62.500 | 38.461 | 79.250 |
| - Productivity (box/man-hour) | 11,4 | 12 | 18 | 16,67 |
| - Utilization (%) | 95,24 | 90 | 55,78 | 62,5 |
| - Quality (%) | 98,75 | 100 | 97 | 100 |
| - Cycle Time | 0,0875 | 0,0833 | 0,055 | 0,06 |
| 3. Storage | | | | |
| - Financial (Rp/line storage) | 13.186 | 16.667 | 8.333 | 23.775 |
| - Productivity (box/m$^2$) | 4,76 | 2,48 | 7,58 | 10 |
| - Utilization (%) | 70 | 90 | 95 | 50 |
| - Quality (%) | 90 | 99 | 87 | 99,2 |
| - Cycle Time | 12,5 | 35 | 20 | 14,2 |
| 4. Order picking | | | | |
| - Financial (Rp/line order picking) | 5.492 | 4.375 | 7.692 | 33.021 |
| - Productivity (box/man-hour) | 119 | 193 | 80 | 125 |
| - Utilization (%) | 75 | 50 | 90 | 50 |
| - Quality (%) | 100 | 100 | 94 | 100 |
| - Cycle Time | 0,0083 | 0,005 | 0,0125 | 0,008 |
| 5. Shipping | | | | |
| - Financial (Rp/line receiving) | 32.965 | 58.333 | 15.384 | 33.021 |
| - Productivity (box/man-hour) | 20 | 112 | 10 | 60 |
| - Utilization (%) | 50 | 90 | 90 | 26 |
| - Quality (%) | 100 | 100 | 98 | 100 |
| - Cycle Time | 0,05 | 0,008 | 0,1 | 0,017 |

The final score of the warehouse is then arranged into the spider diagram as it shows the lowest and the highest score on each indicators along five warehouse activities, as shown on Figure 3.

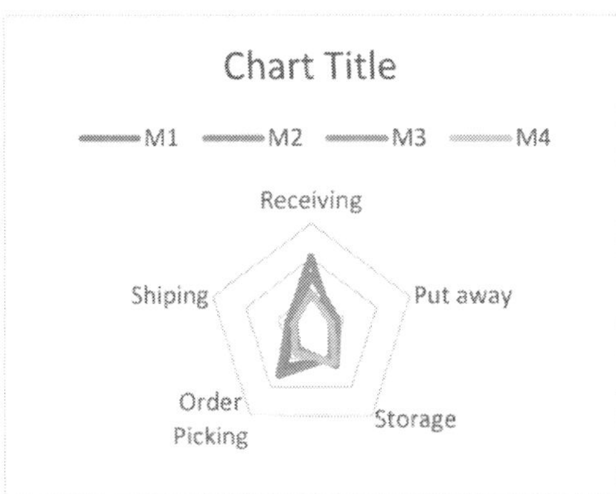


**Fig 3**. Spider diagram of retail warehouse performance

## 4.2. Discussion

Base on the performance score, it can be seen that M2 warehouse has the highest overall performance value of 59.92. M2 warehouse is superior for receiving, order picking and shipping. This is because in receiving, order picking and shipping process, M2 warehouse maximize its utility that increase its productivity and reduce cost. Especially in order picking, many authors assume that order picking is one of the most labor- and time-intensive internal logistics processes, and that in many cases it is responsible for more than 50 percent of the operating costs of a warehouse (Tompkins et al., 2003; Richards, 2011). Studies suggest that order picking is automatized only to a small degree in many companies, and that more than 80 percent of all orders are picked manually on average (De Koster et al., 2007). Meanwhile, M3 warehouse has the highest score of put away and storage values. This is because there is no stacking of goods in the staging area and using trolley effectively so speeding up the duration of the put away process. While good performance in storage due to high inventory turnover, in other words there is no buildup of goods in the warehouse in the long term in M3 Warehouse. This result can be used by the warehouse manager to improve its performance by referring to the best warehouse performance. The highest performance on each activity could be a target setting for other warehouses in order to improve their performance. Opportunity for improvement for each warehouse is described on Table 3.

## 5 Conclusions

This research focus on measuring warehouse performance using Frazelle model (2002). Frazelle model is selected because its metric commonly used in industries and comprehensive, because it not only measures quality, cost, time and productivity dimensions, but also measures in terms of warehouse activities (receiving, put away, storage, picking, shipping). To find performance score of warehouse, the performance indicators are weighted using Analytic Hierarchy Analysis (AHP) method. Warehouse performance are then measured and final score is determined using SNORM method. From this study, it is found the final score of each warehouse and opportunity to improve warehouse performance using peer group benchmarking

## References

1. Ackah, M. R. & Erick, E. G. (2016). Assessing Inventory management on Performance of the Production Sector in Ghana. *Dama International Journal of Researches*. Vol **1**, issue 7, Page 17-27.
2. Ammons, D. N. (1995). Overcoming the inadequacies of performance measurement in local government: the case of libraries and leisure services. *Public Administration Review*, **55**(1) 37-47.
3. De Koster, M.B.M. andWarffemius, P.M.J. (2005) American, Asian and third-party international warehouse operations in Europe: a comparison. *International Journal of Operations and Production Management*, **25**(8), 762–780.
4. De Koster, R., Le-Duc, T. and Roodbergen, K.J. (2007), "Design and control of warehouse order picking: a literature review", European Journal of Operational Research, Vol. **182** No. 2, pp. 481-501.
5. Frazelle, E., H. (2002). World-Class Warehousing and Material Handling. McGraw-Hill Book Company.
6. Keebler, J. S., and R. E. Plank. (2009). Logistics performance measurement in the supply chain: a benchmark. *Benchmarking: An International Journal* 16 (**6**):785–798.
7. Kusrini, E. Subagyo, and Masruroh, N.A (2014). Good criteria for supply chain performance measurement. *International Journal of Engineering Business Management*, **6**:9. doi: 10.5772/58435.
8. Liviu, L., Turdean, A. M. & Emil, C. (2009). *Warehouse Performance Measurement – A Case Study.*
9. Neely, A., Gregory,M. and Platts, K. (2005). Performance measurement system design: a literature review and research agenda, *International Journal of Operation and Production Management*, Vol **25**, (2005) 1228 – 1263.
10. Richards, G. (2011), Warehouse Management – A Complete Guide to Improving Efficiency and Minimizing Costs in the Modern Warehouse, Kogan Page, London.
11. Saaty TL.(2008). The analytic hierarchy and analytic network measurenment process: application to decisions under risk. *European Journal of Pure and Applied Mathematics* **1**(1) 122-96.
12. Speh, T. W. (2009). Understanding Warehouse Cost and Risk. *Ackherman Warehouse Forum*, Vol. **4**, No. 7.
13. Staudt, F.H., Alpan G., Mascolo M., Rodriguez, C.M,T. (2015). Warehouse performance measurement: a literature review. *International Journal of Production Research*, Vol. **53**, No. 18, 5524–5544.
14. Trienekens, J.H. and Hvolby, H.H. (2000). Performance measurement and improvement in supply chain, *Proceeding on the third CINET Conference, CI 2000 from Improvement to Inovation*, Aalborg

# Challenge and opportunity research in blood supply chain management: a literature review

*Agus* Mansur[1, 2, *], *Iwan* Vanany[1], and *Niniet Indah* Arvitrida[1]

[1]Industrial Engineering Department, Institut Teknologi Sepuluh Nopember, Surabaya, Indonesia
[2]Industrial Engineering Department, Universitas Islam Indonesia, Yogyakarta, Indonesia

**Abstract.** An interconnected series of the blood management is called blood supply chain management (BSCM). The stages of BSCM consisted of blood collecting, production, inventory, and distribution. The main challenges in BSCM are related to shortage, outdate, and supply chain cost which needed to minimize. Naturally, problems in BSCM are complex, it is not an easy task to find the solution. This complexity brought by several factors as follows: its inflicted risk, the uncertainty of supply and demand, blood nature as perishable commodity, demand uniqueness, and cost occurred. This research purposes was to review of various research related to BSCM and highlight opportunities to develop further research in blood supply chain (BSC). The result of this research is a suggestion on various possible future research to be explored in BSC, for example, developing an adaptive inventory model to support blood supply chain management that could be responsive toward demand fluctuation and developing collecting strategy to minimize shortage, outdate and incurred cost in supply chain level.

## 1 Introduction

Generally, blood management is a complex system. There are a lot of factors must be involved in the system. Some previous research considered health-related risk, supply availability, demand fluctuation, short blood lifetime, to the cost incurred during the process. An interconnected series of the blood management is called Blood Supply Chain Management (BSCM). The stages of BSCM consisted of blood collecting, production, inventory, and distribution [40]. The effectivity on the previous stage of BSCM will influence the future stage. Therefore, well-being coordination on each echelon and stakeholder in BSCM is vital to guarantee its success [16, 32, 53].

BSCM is one of important research areas because of its value for human life. However, blood management has some special characteristics. First, blood is a commodity that regulated by strict health regulation. It is also perishable and it has tendency to mismatch between the number of blood donors (supply) and recipients (demand) [48]. Increasing the effectiveness of blood supply chain management is expected to optimize the number of blood stock. Practically, the combination of using precise planning system on blood production and well-scheduled blood donors can support the balance creation between supplier and demand. Both strategies will lessen the shortage or outdated risk [27].

Balancing the blood supply and demand is not an easy job because of the stochastic phenomenon in the supply and demand [5]. Furthermore, the richness of the blood derivatives, starting from the kind, type, rhesus, to quantity, increases the management complexity [45]. According to Kendall [45], blood is a perishable product which has limited usage lifetime. Therefore, if blood has outdated or expired, it will turn to waste. Dillon et al., [16] stated that whole blood lifetime is 35 days maximum, while Blood Red Cell and Platelet lifetime, respectively, is 21 days and 5 days.

Researchers have tried to develop blood management improvement in every aspect, including procurement, production, storage/inventory and delivery. Actually, every solution offered is not comprehensive solution and sometimes it cannot applied directly. Therefore, there are still many problems which must be solved on the upstream nor the downstream of blood supply chain.

In this case, this research will conduct literature review related to various research in BSCM. Highlighting future research chances will lead to advancement in the BSCM study. Realized or not, BSCM studies also have given benefit by supporting people welfare.

This paper relates the review of published articles from the 1960s up to 2017, which the articles will be classified on several categories as following:

1. Blood Product Type
2. Performance Measurement
3. Hierarchy Level
4. Inventory Problem

Afterwards, this paper will focus about BSCM model in blood inventory, probabilities on its scenarios, and the importance of developing an adaptive model in blood supply chain.

* Corresponding author: agusmansur@uii.ac.id

## 2 Blood Product Type based BSCM Research Review

After series of laboratory tests, blood fractionation is performed. Transforming it into several types of blood product. In this research, only the dominant significant product will be discussed, which are Red Blood Cells (RBC), Blood Platelets, Plasma, Whole Blood, and Frozen Blood. Those products are the highly produced and directly transfused one. The reason why blood product type become one of this study review parts is that each blood product type has its own distinct type in function and characteristic. It leads researchers to conduct specific and focused study on one or more particular blood product type, in order to lessen the complexity of research.

**Table 1.** Research classification in BSCM based on product type

| Type | Author | Count |
|---|---|---|
| Red Blood Cell | 1,16,18,19,23,28,33,37,46,49 | 10 |
| Platelet | 8,10,15,20,22,27,31,43,57 | 9 |
| Plasma | 39,42,47,49 | 4 |
| Whole Blood | 11,34,39,49 | 4 |

## 3 Performance Measurement based BSCM Review

Research on BSCM was conducted by some researchers that focused one various performance measurement, mostly in minimization of outdate and shortage. Yet, the latest study has leaned on cost factor and other consideration. BSCM performance management that was run by researchers could be categorized as follows:

**Table 3.** Research classification in BSCM based on performance indicator

| Performance Indicator | Author | Count |
|---|---|---|
| Outdate | 1,9,14,16,18,25,32,37,51,52,54, 58 | 12 |
| Shortage | 1,4,14,15,16,18,19,21,23,25,32,37,48,51,54,58,59 | 17 |
| Transportation Cost | 11,14,27,30,44,51 | 6 |
| Service Level | 11,26,27,30,43,44 | 6 |
| Safety | 1,3,16,18,33,36 | 6 |
| Supply Chain Cost | 16 | 1 |

## 4 Coordination Hierarchy Level based BSCM Research Review

Researchers have been put out a various study on blood management in different hierarchy level. There is a consideration regarding urgency and depth of the study that was conducted. Hierarchy levels that commonly discussed in research are individual hospital, regional blood center level, and supply chain level.

**Table 2.** Research classification in BSCM based on coordination level

| Coordination | Author | Count |
|---|---|---|
| Individual Hospital | 12,17,25,41,47,55,56,60 | 8 |
| Regional Blood Center | 13,22,24,27,35,42,44,50 | 8 |
| Supply Chain Hierarchy | 6,14,20,34,49,58 | 6 |

## 5 Blood Inventory Models Review

Research on blood inventory problem has grown in number and variation. The nature of blood demand is uncertain. Thus creating challenges to overcome in forecasting the blood demand. In order to ease computation process, researchers would use certain distribution, like Poisson or normal distribution. Another challenge in managing the inventory could come from its perishable characteristic, where it also affected the number of wastage and cost incurred. Civelek et al. [10] tried to build protection model on platelet that was ordered by the hospital using simulation approach. As for Mole [38] had to use Markov chain approach to building relation model between a variable which served a purpose in minimizing shortage and minimizing outdate in the hospital.

**Table 5.** Research classification in BSCM based on Inventory Model

| Year | Author | Goal | Method |
|---|---|---|---|
| 1975 | 38 | Balancing blood stock in hospital to avoid wastage and shortage | Markov Decision Process (MDP) |
| 1979 | 11 | Developed managerial processes at each level of the hierarchy in a regional blood bank system | Mathematical |
| 1980 | 13 | Designing information systems for regional blood service | Information System Design |
| 1982 | 39 | Review perishable inventory | Mathematical |
| 2007 | 34 | Improvement procedure dan outcome BSC (ordering policies) | Discrete Event Simulation |
| 2008 | 37 | Build a blood inventory system that takes into consideration the demand rate | Simulation |
| 2009 | 15 | Determine order period and optimum order for platelet in the hospital to minimize the cost | Stochastic dynamic programming (SDP) model and simulation |

| | | | |
|---|---|---|---|
| 2013 | 18 | Minimize the expected system outdate rate under a predetermined maximally allowable shortage level | Simulation optimization (SO) |
| 2013 | 6 | Evaluate and design facilities and measure the impact on customer service | Simulation |
| 2013 | 52 | Improve blood supply chain management in order to use the scarce resource of blood more efficiently | Discrete Event Simulation |
| 2013 | 29 | Identify the optimal order levels that minimizes the total cost, shortage and wastage levels of blood products at a hospital. | Integer Programming |
| 2014 | 19 | Constructed inventory model based on optimization framework simulation to minimize outdate rate of platelet | Old inventory ratio (OIR) |
| 2015 | 10 | A simple inventory replenishment and allocation heuristic to minimize the expected total cost over an infinite time horizon | Markov decision process (MDP) |
| 2015 | 59 | Design a robust blood supply chain system to improve the blood supply chain efficiency | Dynamic simulation and taguchi |
| 2015 | 61 | Transhipment model between rescue and the affected blood bank by considering the age of blood so as to obtain a minimum expired date | Simulation |
| 2016 | 32 | Design centralized-based inventory model and collaboration of the hospital to maintain its inventory | Inventory model approach (S-1, S) |
| 2017 | 46 | Determine periodic-review fixed lifetime perishable inventory literature by considering multiple independent sources of supply. | Dynamic program. |
| 2017 | 16 | Determine periodic review and optimum order target | Two-steps stochastic programming approach |

## 6 Indonesia's Blood Management

In Indonesia, only few studies are dedicated to investigate the blood management, particularly operation management and supply chain management. [62] Investigated the elevated value of average shortage and expired product in regional Red Cross (PMI), approximately 29% and 59%. Discrete simulation method was used to construct the shortage and expired scenario out of several possible scenarios to decrease the shortage and expire items. They underlined several improvements that should be appointed in blood supply chain management. This research is only focused on blood production combination, which cannot determine safe blood stock that could be possibly handled by PMI.

[63] pointed the low quality of information will lead to the poor coordination among echelons in blood supply chain management. The researcher proposes traceability blood system to reinforce the coordination among echelons in blood supply chain, specifically in Indonesia. The model could show the potential benefits that will be extended from each stake holders (donor, PMI and hospital) in improving the performance in blood supply chain. The research does not provide the solution on safety stock and does not discuss the effect of behaviour alteration towards relationship pattern

A phenomenon in Indonesia's blood management that still has not been solved is its large gap related to awareness level of the donor. It is normal to find an area that has been well informed, on the other hand, also found another area that has less to none understanding regarding blood donor. This will cause the large discrepancy between blood supply from one area to another. This will cause a problem in the long run since blood demand will eventually exceed the supply. Therefore, good coordination system amongst nearby blood center will be needed

Blood demand across Indonesia could be classified into two; normal condition demand and special condition demand. The normal condition is a condition when there is no extraordinary situation that could cause a sudden spike in blood demand. A special condition is a certain condition that affects blood demand and directly spiked the blood demand, in example endemic disease outbreak such as dengue fever. It has been known that several cities in Indonesia have high potential as endemic disease area. In certain months, it is also common to have a sudden high number of a traffic accident that will obviously trigger blood demand to increase.

Based on this phenomenon, adaptive blood management model is needed. Since blood management has to be responsive to every system changes, rapidly and flexibly. Blood management is expected to be adaptive on demand fluctuation and blood supply pattern. One of the most important things to be done is strategizing blood procurement that could cater to the demand. This situation called for a study on optimum alternative blood procurement, mobile unit location determination, and arranging ideal capacity. The adaptive framework is shown in Figure 1.

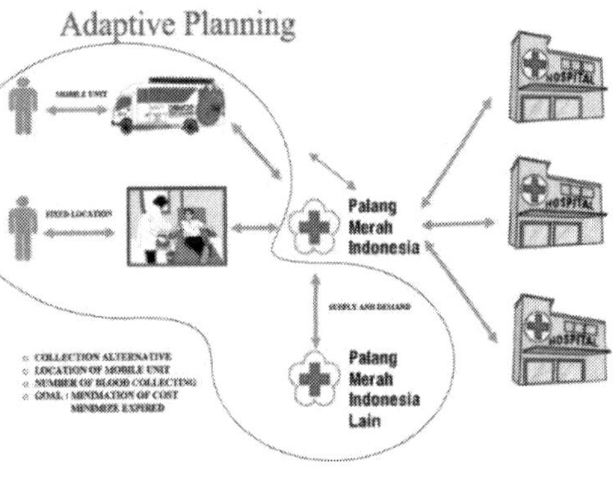


**Fig 1.** Adaptive planning scheme in BSCM

Adaptive planning will be affected by demand characteristic. Demand pattern changes will affect blood center's supply target level. Target level will varied, depends on demand behavior and condition, such as, whether the demand in normal, emergency or special condition.

Some research on blood inventory have been conducted by some researchers. Yet, those research have not discussed policy related flexibility level in inventory, as a writer has illustrated previously. This is an opportunity for future research to be explored more.

Even though, Dillon et al. [38] stated that he conducted optimum periodic review analysis on platelet. However, his research has not mentioned the anticipated demand changes phenomenon, so it was only focused on service level and outdate tolerance to determine inventory target level. The model above could not give a rapid response to determining optimum stock and replenishment lead time target. Opportunity to develop this research is highly possible to build inventory system by periodic review approach, considering its adaptability to supply and demand changes, to minimize the possibility of shortage and outdate.

## 7 Conclusion

Research in Blood Supply Chain can be considered as an active and dynamic study. Many studies, which appear every year, have a variety of themes, aspects and methodologies. This study presents the results of an article review in the field of the supply chain of blood from upstream to downstream with various products, differences in blood coordinate management system and performance measurement. This study also reveals research opportunities in the field of blood supply chain at the strategic level, tactical level to operational level. One of the research opportunities required is to build a reliable inventory management system, making it adaptive to environmental change

Opportunity for substantial research is still wide open, especially developing an adaptive model for blood supply chain management that could be responsive toward demand fluctuation and developing procurement strategy to minimize shortage, outdated and occurred cost in supply chain level.

## References

1. B. Abbasi, S.Z. Hosseinifard, A., *On the Issuing Policies for Perishable Items such as Red Blood Cells and Platelets in Blood Service,* Decision Sciences **45**, 995-1020 (2014)
2. E. Alfonso, X. Xie, V. Augusto, O. Garraud, *Modeling and simulation of blood collection systems,* Health Care Management Science **15**, 63-78 (2012)
3. E. Alfonso, X. Xie, V. Augusto, P. Garraud, *Modeling and simulation of blood collection systems: Improvement of human resources allocation for better cost-effectiveness and reduction of candidate donor abandonment,* Vox Sanguinis **104**, 225-233 (2013)
4. R.K. Bedi, K. Mittal, T. Sood, P. Kaur, G. Kaur, *Segregation of blood inventory: A key driver for optimum blood stock management in a resource - poor setting* (2016)
5. J. Beliën, H. Force, *Supply chain management of blood products: A literature review*, Eur J of Opr Research **217**, 1-16 (2012)
6. J. Blake, M. Hardy,*Using simulation to evaluate a blood supply network in the Canadian maritime provinces,* J of Enterprise Infor Manag **26**, 119-134 (2013)
7. J. Blake, M. Hardy,*Using simulation to evaluate a blood supply network in the Canadian maritime provinces* (2013)
8. J.T. Blake, M. Hardy, *A generic modeling framework to evaluate network blood management policies: The Canadian Blood Services experience,* Opr Research for Health Care **3**, 116-128 (2014)
9. J.F. Chapman, C. Hyam, R. Hick, *Blood inventory management*, Vox Sanguinis **2**, 143-145 (2004)
10. I. Civilex, I. Karaesmen, A. Scheller-Wolf, *Blood platelet inventory management with protection levels,* Eur J of Opr Research ***243**, 826-838 (2015)*
11. M.A. Cohen, W.P. Pierskalla, R.J. Sassetti, J. Consolo, *An overview of a hierarchy of planning models for Regional Blood Bank Management,* Transfusion **19**, 526-534 (1979)
12. A. Coleman, O.S. Akinsola, *Effective blood distribution in rural hospitals through ICT service-oriented architecture (SOA) framework: A case study in rural hospitals in South Africa,* In Studies on Ethno-Medicine **6**, 141-147 (2012)
13. P.D. Cumming, K.E. Kendall, C.C. Pegels, J.P. Seagle, *The impact of an information system on a regional blood service,* Information and Management **3**, 63-72 (1980)
14. D. Delen, M. Erraguntla, R.J. Mayer, C.N. Wu, *Better management of blood supply-chain with GIS-based analytics,* Annals of Opr Research **185**, 181-193 (2011)
15. N.V. Dijk, J.V.D. Wal, C. Smith, *Blood platelet production with breaks: optimization by SDP and simulation,* **121**, 464-473 (2009)

16. M. Dillon, F. Oliveira, B. Abbasi, *A two-stage stochastic programming model for inventory management in the blood supply chain A Two-Stage Stochastic Programming Model for Inventory Management in the Blood Supply Chain,* J of Production Economics **187**, 27-41 (2017)
17. G. Dobson, E.J. Pinker, O. Yildiz, *An EOQ model for perishable goods with age-dependent demand rate,* Eur J of Opr Research **0**, 1-5 (2016)
18. Q. Duan, T.W. Liao, *A new age-based replenishment policy for supply chain inventory optimization of highly perishable products.* Int J of Prod Economics **145,** 658-671 (2013)
19. Q. Duan, T.W. Liao, *Optimization of blood supply chain with shortened shelf lives and ABO compatibility*, In Int J of Prod Economics **153**, 113-129 (2014)
20. H. Ensafian, S. Yaghoubi, *A robust optimization model for integrated procurement, production, and distribution in platelet supply chain*, Transportation Research Part E **103**, 32-55 (2017)
21. B. Fahimnia, A. Jabbarzadeh, A. Ghavamifar, M. Bell, *Supply chain design for efficient and effective blood supply in disasters*, In Int J of Prod Economics 1-10 (2015)
22. O.S.S. Filho, M.A. Carvalho, W. Cezarino, R. Silva, G. Salviano, *Demand Forecasting for Blood Components Distribution of a Blood Supply Chain*, In IFAC Proceedings Volumes **46**, 565-571 (2013)
23. O.S.S. Filho, W. Cezarino, G.R. Salviano, *A decision-making tool for demand forecasting of blood components*, In IFAC Proceedings Volumes **14**, 1499-1504 (2012)
24. S.M. Fortsch, E.A. Khapalova, *Reducing uncertainty in demand for blood*, In Opr Research for Health Care **9**, 16-28 (2016)
25. B.A. Friedman, R.D. Abbot, G.W. Williams, *A blood ordering strategy for hospital blood banks derived from a computer simulation*, American J of Clinical Pathology **78**, 154-160 (2008)
26. M.J. Galloway, G. Jane, L. Sudlow, J. Trattles, J. Watson, *A tabletop exercise to assess a hospital emergency blood management contingency plan in a simulated acute blood shortage*, Transfusion Medicine **18**, 302-307 (2008)
27. P. Ghandforoush, T.K. Sen, *A DSS to manage the platelet production supply chain for regional blood centers*, In Decision Support Systems **50**, 32-42 (2010)
28. A. Grasas, A. Pereira, M. Bosch, P. Ortiz, L. Puig, *Feasibility of reducing the maximum shelf life of red blood cells stored in additive solution: a dynamic simulation study involving a large regional blood system*, 233-242 (2015)
29. S. Gunpinar, *Supply Chain Optimization of Blood Products*, In A Dissertation, Departement of Industrial and Management System Engineering, University of South Florida (2013)
30. V. Hemmelmayr, K.F. Doerner, R.F. Hartl, M.W.P. Savelsbergh, *Delivery strategies for blood products supplies*, OR Spectrum **31**, 707-725 (2009)
31. S.M. Hesse, C.R. Coullard, M.S. Daskin, A.P. Hurter, A case study in platelet inventory management, Proceedings of the 6th Ind Engineering Research Conf, 801-806 (1997)
32. Z. Hosseeinifard, B. Abbasi, The inventory centralization impacts on the sustainability of the blood supply chain, Elsevier, 1-7 (2016)
33. K. Katsaliaki, Cost-effective practices in the blood service sector, **86**, 276-287 (2015)
34. K. Katsaliaki, S.C. Brailsford, *Using simulation to improve the blood supply chain*, J of the Opr Research Society **58**, 219-227 (2007)
35. K.E. Kendall, *A Decentralized Information Blood Management and Control System for*, In the J of System and Software **1**, 299-306 (1980)
36. K.E. Kendall, S.M. Lee, *Formulating Blood Rotation Policies with Multiple Objectives*, Manag Science **26**, 1145-1157 (1980)
37. R. Kopach, B. Balcioglu, M. Carter, *Tutorial on constructing a red blood cell inventory management system with two demand rates*, In Eur J of Opr Research **185**, 1051-1059 (2008)
38. R.H. Mole, *Inventory Control in Hospital*, **3**, 461-473 (1975)
39. S. Nahmias, *Perishable inventory theory: a review*, Opr Research **30**, 680-708 (1982)
40. A.F. Osorio, S.C. Brailsford, H.K. Smith, *A structured review of quantitative models in the blood supply chain: a taxonomic framework for decision-making*, Int J of Prod Research **1**, 1-22 (2015)
41. G. Perera, C. Hyam, C. Taylor, J.F. Chapman, *Hospital Blood Inventory Practice: The factors affecting stock level and wastage,* Transfusion Medicine **19**, 99-104 (2009)
42. C. Pitocco, T.R. Sexton, *Alleviating blood shortages in a resource-constrained environment,* Transfusion **45**, 1118-1126 (2005)
43. G.P Prastacos, *Blood inventory management: an overview of theory and practice,* Manag Science (1984)
44. G.P. Prastacos, E. Brodheim, *Computer-based regional blood distribution,* Computers and Opr Research **6**, 69-77 (1979)
45. K. Puranam, D.C. Novak, M.T. Lucas, M. Fung, *Managing Blood Inventory with Multiple Independent Sources of Supply,* In Eur J of Opr Research (2016)
46. K. Puranam, D.C. Novak, M.T. Lucas, M. Fung, *Managing blood inventory with multiple independent sources of supply,* In Eur J of Opr Research **259**, 500-511 (2016)
47. J. Rautonen, J. MacPherson, *Redesigning supply chain management together with the hospitals,* Transfusion **47**, 197-200 (2007)

48. E. Reynolds, C. Wickenden, A. Oliver, *The impact of improved safety on maintaining a sufficient blood supply,* Transfusion Clinique et Biologique **8**, 235-239 (2001)
49. J.S. Rytila, K.M. Spens, *Using simulation to increase efficiency in blood supply chains,* Manag Research News **29**, 801-819 (2006)
50. G. Sahin, H. Sural, S. Meral, *Locational analysis for regionalization of Turkish Red Crescent blood services,* Comp and Opr Research **34**, 692-704 (2007)
51. W.M. Smid, R. Buining, W.D. Kort, *Blood supply management: experience and recommendations from the Netherlands,* 50-53 (2013)
52. K.M. Spens, *Using simulation to increase efficiency in blood supply chains* (2013)
53. S.H.W. Stanger, *Vendor managed inventory in the blood supply chain in Germany Evidence from multiple case studies, (2013)*
54. S.H.W. Stanger, R. Wilding, N. Yates, S. Cotton, *What drives perishable inventory management performance? Lessons learned from the UK blood supply chain,* (2012)
55. S.H.W. Stanger, N. Yates, R. Wilding, S. Cotton, *Blood Inventory Management: Hospital Best Practice,* In Transfusion Medicine Reviews **26**, 153-163 (2012)
56. R.W. Toner, L. Pizzi, B. Leas, S.K. Ballas, A. Quigley, N.I. Goldfarb, *Costs to hospitals of acquiring and processing blood in the US: A survey of hospital-based blood banks and transfusion services,* Applied Health Economics and Health Policy **9**, 29-37 (2011)
57. L.M. Williamson, D.V. Devine, *Challenges in the management of the blood supply,* In the Lancet **381**, 1866-1875 (2013)
58. E. Yuzgec, Y. Han, N. Nagarur, *A simulation model for blood supply chain systems,* In IIE Annual Conference and Expo, 1703-1711 (2013)
59. S.M. Zahraee, J.M. Rohani, A. Firouzi, A. Shahpanah, *Efficiency Improvement of Blood Supply Chain System Using Taguchi Method and Dynamic Simulation*, Procedia Manufacturing **2**,1-5 (2015)
60. E.D. Zepeda, G.N. Nyaga, G.J. Young, *Supply chain risk management and hospital inventory: Effects of system affiliation,* In J of Opr Manag **44**, 30-47 (2016)
61. K.M Wang, Z.J Ma *Age-based policy for blood transshipment during blood shortage,* Transportation Research Part E 80 ,166–183, (2015)
62. A.W. Suwardie, B.M.Sopha, M.K. Herliansyah, *A Simulation Model of Blood Supply Chain at Indonesian Regional Red-Cross,* Proceedings of The 2013 International Conference on Logistics and Maritime Systems (2013).
63. I. Vanany, A. Maryani, B. Amaliah, F. Rinaldy, F. Muhammad, *Blood traceability system for Indonesian blood supply chain,* Procedia Manufacturing 4, 535 – 542, ( 2015 )

# Productivity improvement in the production line with lean manufacturing approach: case study PT. XYZ

Halimatussa'diah*, *Ali* Parkhan, and *Muchamad* Sugarindra

Industrial Department, Universitas Islam Indonesia, Jl. Kaliurang, Km 14.5, Yogyakarta, Indonesia

**Abstract.** The increaseing productivity is one of the competitive strategies that can be applied in a company in order to survive in an intense competitive presure. PT. XYZ is a textile industry manufacturing golf gloves and caddy bag. Every day, the company has a production target of 600 pieces per line. However, the desired target is not achieved that it will affect the delay in delivery of products to customers. In this research, a case study on implementing value stream mapping and Kaizen as the lean manufacturing concept is reported. The purpose of this study is to map the current production line, analyse and design the future value stream mapping by eliminating waste occured. It is obtained a lead time reduction as much as 440.4 seconds through eliminating 17 non-value added activities. Then, the output can be increased up to 21% which is equal to 502 pieces.

## 1 Introduction

PT.XYZ is a textile company that manufactures golf glove as primary production. In conducting the production process, the company has a production target of 600 pcs each line with a total production target of 2400 pcs. However, the desired targets are often not able to be achieved which affects the delay in delivery of products to customers. Based on the observation, some causes identified are work stations cutting, sewing, and preparation before sewing (PSP) as a core process in the production of golf gloves. The root causes of waste consist of these following wait time situations: (1) waste of production lead-time due to delaying of raw materials from suppliers; (2) waste of defect product due to inaccuracy or incorrect execution by operator so the defect products have to be reworked; and (3) waste of iinappropriate processing and unnecessary motion caused by indiscipline and lack of skilled operators.

To enhance productivity, the company should be able to improve the production process by eliminating all non-value added activities int their plants using the application of lean manufacturing [1]. Value stream mapping is a set of methods to map activity in the application of lean manufacturing production [2]. It has also been applied to analyse the flow of production as well as eliminating or minimizing non-value adding activities [3]. While lean concepts with value stream mapping approach is proved to be effectively implemented on the production floor [4]. On the other hand, lean manufacturing can be used as a good strategy of process improvements and an alternative to increase the production output without a big amount of investment [5].

In order to overcome the waste occured in process above, this study aims to determine the value-added activities, non-value added activities, nnecessary but non-value added (NNVA) activityies, and to propose an improvement of each identified major waste using VSM tools .

## 2 Literature Review

### 2.1 Lean Manufacturing

Lean manufacturing is every activityy described in a timeline begun with customer demand pushing towards producer, which aims to reduce lead-time by eliminating waste or removing activities that contribute no value to the product produced from the arrival of raw materials from suppliers to the customers' final product through continuous improvement method [6]

### 2.2 Waste

Waste is any activity that consumes time, resources and space but do not contribute to satisfying the needs of cos [7]. There are 7 types of wastes occured in the manufacturing process [8]:

1. Overproduction
   Overproduction, where production exceeds customer demand, is considered the most harmful waste as it itself creates additional waste including excess inventory and transportation.
2. Defect
   Defects are products or services that are out of specification. They need extra effort to fix and lead directlyto increased costs and lost time, either through repairs or having start over.

* Corresponding author: halimatussadiah13522198@gmail.com

3. Unnecessary Inventory
Inventory, such as raw materials, work-in-progress or goods which are sitting idle, is considered waste as it has not yet contributed value to the end customer.
4. Inappropriate processing
Innappropriate processing or excess process is any activity that is not needed to produce a functioning product or service and can occur due to unnecessarily drawn-out processes.
5. Excessive transportation
Transportation occurs whenever products are transported from one location to another. This can increase the risk of damages, additional waiting time and costs.
6. Waiting
Waiting occures whenever items are not in transport or being processed. For example, employees wait for materials, equipment waits for maintenance, or work in process waits for the employee to return.
7. Unnecessary Motion
Unnecessary motion can occur due to insufficient layouts or searching for mislaid items. It can also increase the risk of damage to equipment or cause employee injuries.

### 2.3 Value Stream Mapping (VSM)

Value Stream Mapping (VSM) is a tool used to map the value stream and identify any waste that add value and do not add value [2]. VSM is used to increase the cycle time by showing how the actual operating process in every activity with time details. VSM can be used to analyse the process and performance improvement by identifying and reducing the time utilization on non-value-added activities [9].

## 3 Methodology

The method used is the approach of lean manufacturing and value stream mapping. Primary data is obtained by employing interviews, documentation using video recorder, and a stopwatch to create the Current State Value Mapping. For creating the Current State Value Mapping, data needed are the production process of the research object, the cycle time, the number of available time, and waste identification using 7 types of waste. While the secondary data is referred to the company production data. The next step is developing a Future State of Value Mapping using the current state value map which the major wastes have been reduced.

## 4 Data Analysis

### 4.1 Current State Value Mapping

The Current state map (CSM) charts the present flow of information and material as product goes through the manufactuing process. This is vital both to understand the need for change and to understand where opportunities lie. CSM is a condition that is used to identify the waste for the repair and improvement of a company [10]. To develop CSM, we conduct a direct observation and interviews with the head of production who understand the most of PT.XYZ production. There are several golf gloves with the similar processes. Yet, they have different level of complexity in the form of variations and fantasies so Lucky 39 is chosen as the object of this study. The data of operator and available time in producing golf gloves can be seen in Table 1.

**Table 1.** Data of operator and available time

| Proses | Operator | Available Time |
|---|---|---|
| Cutting | 12 | 25200 |
| PSP | 60 | 25200 |
| Sewing | 82 | 25200 |

Current state value mapping of golf gloves based on the data proceeded showed in Figure 1.

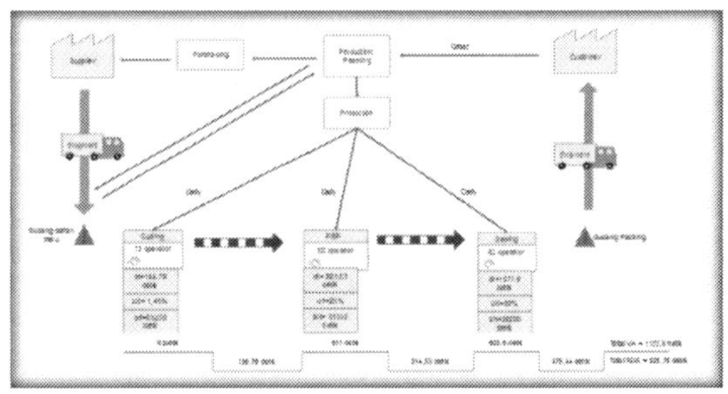

**Fig 1.** Current state mapping section sanding model A

### 4.2 Waste Identification

There are 5 wastes identified in the process of cutting, PSP, and sewing:

1. Defects: Throughout 2016, there were 369 pcs wrong cut product in the cutting process and 2228 pcs of defective products in the sewing process.
2. Unnecessary inventory: excess materials stored for reuse.
3. Inappropriate processing: It is caused by human error, the less scrupulous operators to inspect the products and raw materials as well as the presence of broken machines when the production processes proceed that disrupt the operators' works.
4. Waiting: delayed delivery of raw materials from suppliers, operators are not discipline, and production capacity is not appropriate.
5. Unnecessary motion: The undisciple operators who do a lot of activities outside of their main job such as a lot of chat among operators, take a drink frequently, and go to the bathroom for some time.

### 4.3 Future State Value Mapping

The future state value mapping or Future State Map (FSM) is a chart that suggests how to create a lean flow. It is a picture that will be used in the future and has been fixed from the current state value mapping [10]. The FSM is used to help make decisions which create future process improvements. The picture of the future state mapping is as follows:

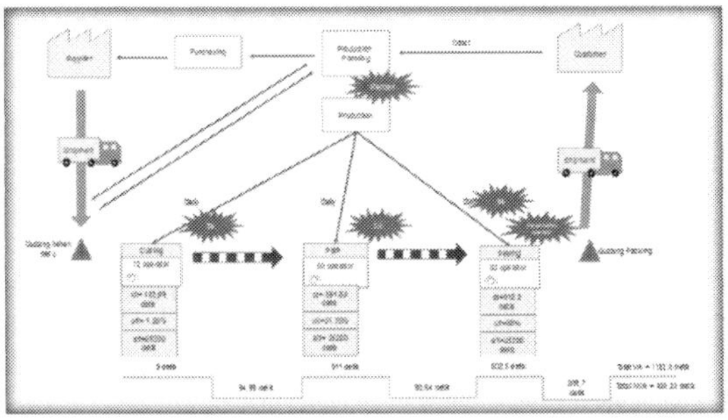

**Fig 2.** Current state mapping section sanding model A

## 5 Result and Discussion

According to the current state value mapping in Figure 1, it is known that lead time in producing a golf gloves is as much as 2048.21 seconds. The output produced by the company is 1902 pcs, which is lack of 498 pcs production (20.73%) of the company's target. The future state value in Figure 2 showed the lead time of 1607.81 seconds. The decreasing lead time of 440.4 seconds is due to the reduction of non-value added activities. To achieve the target of 2400 pcs in the cutting process, the number of operators required is 10 people, 57 operators in PSP, and 87 operators in the sewing process. Therefore, in order to overcome these problems, 2 operators of the cutting process and 3 operators of the PSP will be moved to the sewing process. With this operator, transferring, sewing training is highly required for new/backup operators. Here is a comparison between the current state and future state value mapping.

**Table 2.** The comparison between CSM and FSM

| Comparison | Current state | Future state | Improvements |
|---|---|---|---|
| Lead time | 2048,21 (s) | 1607,81 (s) | 440,4 (s) |
| Non value added | 925,76 (s) | 440,41 (s) | 485,25 (s) |
| Output | 1902,41 pcs | 2403,94 pcs | 501,53 pcs |
| Persentase | 79,27 % | 100,16 % | 20,89 % |

The comparison in Figure 2 indicates a 440.4 seconds lead time reduction in the critical areas of defects, unnecessary inventory, inappropriate processing, waiting, and unnecessary motion. The productivity increased 502 pcs or is up t0 21% so the company's target can be reached. These improvements are gained by removing non-value added activities including waste. In the cutting process, the discharged non-added value activities are: (1) calculating the amount of the pattern. Since the inspection section has already done it, so the operator does not need to perform this activity again. (2) Sharpening blade can be removed by purchasing a new blade. (3) Searching stationeries can be eliminated by applying the principles of 5S so the operator may spend less time. (4) Recording quantities can be eliminated using kanban or it will be recorded by administrative staff or section leader, not the operator. In the PSP process, there are several non-added activities consisted of sewing machine preparation, ribbon rolls separation, thread repairing, checking, repairing fantasy omo face, and ribbon rolls separation. They are part of Inappropriate processing and Uunnecessary motion. The sewing machine repairing can be removed by a regular maintenance. Whereas in the sewing process, the non-added value are checking, separating winding thread, and place the ribbon. These are Inappropriate processing and Uunnecessary motion. The activities of seeking machi size, taking or separating folding omo, and placing or separating can be removed by applying 5s. Labeling every machi size and providing information on any work in process (WIP) will guide the operator to look for or separate WIP. Some of the proposed improvements of the identified wastes are clearly explained in Table 3.

**Table 3.** The proposed improvements of identified wastes

| Causes | Impacts | Proposed improvement |
|---|---|---|
| Less scrupulously operator | Defect<br>Inappropriate processing<br>Waiting | - The leader must conduct a strict supervision whenever the operator loose focus particularly after break time and work is almost over<br>- The leader must conduct a regular and random checking |
| Less focus operator | Defect<br>Waiting<br>Inappropriate processing | - Adding break time<br>- Design a particular chair for pressing activity in cutting process |
| Undiscipline operator | Waiting<br>Inappropriate processing<br>Unnecessary motion | - Motivate and encourage them, and evaluate the working performance before production process started<br>- The leader must strictly warn the undiscipline operator<br>- Provide a reward and punishment for discipline and |

| Causes | Impacts | Proposed improvement |
|---|---|---|
| | | undiscipline operator |
| Lack of skilled operator | Waiting<br>Inappropriate processing<br>Unnecessary motion<br>Defect | - Conduct a continous training for every operator. It will be useful to leverage the skill as well as refresh the energy in order to improve the working performance<br>- Add a new requirement when recruiting a new operator such as experience in sewing process |
| Broken machine | Waiting<br>Inappropriate processing<br>Unnecessary motion<br>Defect | - Routine maintenance<br>- Purchase a new tool/machine |
| Delay delivery of raw materials | Waiting | Renew the suppliers' contracts so they will deliver the materials needed on schedule |
| Uncomfortable/unclean working environment | Defect<br>Inappropriate processing | - Add the cleaning service staff at the production floor<br>- Implement 5s methods consistently. |

## 6 Conclusion and Recommendation

### 6.1 Conclusion

1. The current state mapping value shows 28 value added activities, 55 non-value added activities, and 7 necessary but non-value added activities. While on the future state mapping value, 17 non-value added activities have been reduced.
2. It is obtained a 1607.8 seconds lead time using the future state value mapping. By reducing the non-value added activities, the FSM has indicated a decreasing lead time of 440.4 seconds. Thus, it will increase the output of 502 pcs as well as the increasing productivity can be achieved from 79.27% to more than 100%.
3. The major root causes of identified wastes are less scrupulously operator, less focus operator, undiscipline operator, lack of skilled operator, the broken machine, delay delivery of raw materials, and uncomfortable/unclean working environment.

### 6.2 Recommendation

Future research should look at simulation of the proposed improvements combining with the cost consideration, downtime machine, and other factors to eliminate waste that would contribute positively to the bottom line.

## References

1. Bhati, S. & Porwal, S., A Case Study on Improving Process and Eliminating Waste through Lean Manufacturing Techniques, 3 (12), pp.3–6 (2015)
2. Taylor, P., Gupta, S. & Jain, S.K., Literature review of lean manufacturing. International Journal of Management Science and Engineering Management, (November 2014), pp.37–41 (2013)
3. Ar, R. & Al-Ashraf, M., Production Flow Analysis through Value Stream Mapping: A Lean Manufacturing Process Case Study. , 41(Iris), pp.1727–1734. Available at: http://dx.doi.org/10.1016/j.proeng.2012.07.375 (2012)
4. Goriwondo, W.M., Mhlanga, S. & Marecha, A., Use of the Value Stream Mapping Tool for Waste Reduction in Manufacturing. Case Study for Bread. , pp.236–241 (2011)
5. Murugesan, M., Rajenthirakumar, D. & Chandrasekar, M., Manufacturing Process Improvement Using Lean Tools. Annals of faculty Engineering Hunedoara International Journal of Engineering, pp.151–154 (2016)
6. Ohno, T., 1995. Toyota Production System, Beyond Large-Scale Production. Terjemahan:
7. Liker, J.K. & Mcgraw-hill, T., The Toyota Way (2004)
8. Liker, J. & Meier, D., The Toyota Way Fieldbook (2006)
9. Capital, M., Introduction to Lean Manufacturing for Vietnam. , (June), pp.1–20 (2004)
10. Tilak, Minakshi, Aken, E. V., McDonald, T., & Kannan, R. Value Stream Mapping: A Review and Comparative Analysis of Recent Applications (2010)

# Active and passive fire protection system in academic building KH. Mas Mansur, Islamic University of Indonesia

*M. Ragil* Suryoputro*, *Fajar Aga* Buana, *Amarria Dila* Sari, and *Fety Ilma* Rahmillah

Department of Industrial Engineering, Islamic University of Indonesia, Jl. Kaliurang Km 14.5, Yogyakarta, Indonesia

**Abstract.** According to the theory of fire triangle, the existence of combustible materials, heat, and oxygen can cause fire disaster. KH. Mas Mansur building, Islamic University of Indonesia has a fire protection, but rarely to be checked regularly and the number of equipment is less standard as well as the lack of an evacuation route map to facilitate the evacuation process. Inside the building also does not provide safety signs such as the evacuation directions, exit, and warning in case of fire. Therefore, researchers analysed the infrastructure of prevention and control in the building KH. Mas Mansur. Researchers used the method of observation, interviews, and checklist to know the condition directly, and compare with the standard regulations. Results concordance rate of existing infrastructure is 67% fire extinguisher, hydrant box 56%, 71% alarms, sprinkler 0%, 40% detectors, emergency doors 71%, 50% emergency stairs, assembly point 0% and directions 0%. The current results were below the standard of at least 80%. As for recommendations, researchers create a new evacuation map then put the existing infrastructure according to standard regulations, and it had consulted with the specialist of Occupational Safety and Health in the field of fire.

## 1 Introduction

Fire is a flame both small and large in an undesired place, situation and time which detrimental and generally difficult to control. Fires in some areas increase annually, including in Sleman, Yogyakarta as presented in Fig 1. Figure 1 is derived from the Technical Implementation Unit (UPT) of the fire department in Sleman, Yogyakarta. It shows that almost every year incidence of fires in Sleman, Yogyakarta increases. There were 661 events of fire incidents during 2007 until 2015 with an average of 73 events per year, so there were six times fires in a month, and the highest incidence frequency was in 2015.

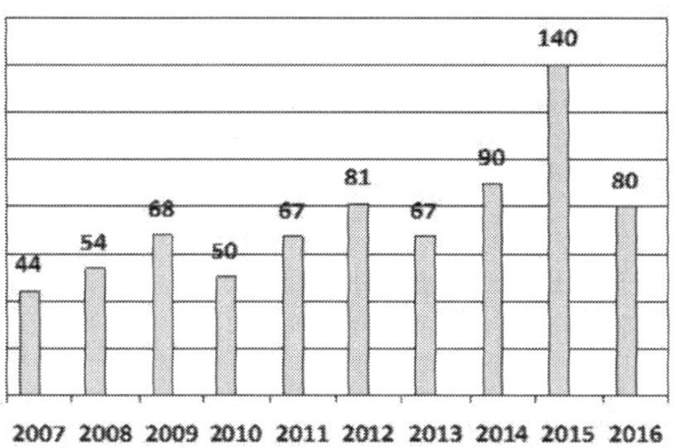


**Fig 1.** The incidence of fires in Sleman, Yogyakarta (Source: UPT Firefighters, Sleman, Yogyakarta)

Some scholars do research about safety management system of fire protection in universities [1, 2] and student accommodation [3] by using various methods such as observing the documents, interview, and simulation. While [4] and [5] study the factors caused the fire by using fault tree analysis; [6] compare the safety level private and state school in Iran; [7] develops fire risk assessment method for heritage building in Malaysia. The frequent fire accidents in some universities [2] show a lack of student understanding of fire safety, campus-resistant construction and distrust in security situations in the event of a fire. Some fire accidents also happened in some universities in Indonesia [8-10]. Therefore, it is necessary to do a prevention. However, none research has done toward fire protection system in KH. Mas Mansur building, Islamic University of Indonesia which was established in 2010. Some of the existing rooms in KH. Mas Mansur building has fire extinguishers (APAR) but rarely checked regularly, and the number of equipment was insufficient as well as less standard and lack of an evacuation route map to facilitate the evacuation process. In the building also does not have safety signs such as evacuation signs, exit, and warning in case of fire. Hence, the importance of the prevention of firefighting infrastructure needs to be improved so that every person who was in the building KH. Mas Mansur, the Islamic University of Indonesia feel safe.

This study emphasizes the method of observation, interviews, and a checklist to make a comparison between the current level of fire protection system of KH. Mas Mansur building and the regulatory standards as well as to make the evacuation map for fire prevention and suppression.

* Corresponding author: ragil.suryoputro@uii.ac.id

## 2 Theory

Traditionally the concept of Fire Triangle, which visually represents Fuel, Heat, and oxygen, is used to symbolize the necessary conditions for the creation of fire. Fire cannot be created if there is a missing part of the triangle. Once the fire has created, the fourth component then appears, that a complex sequence of Chemical Chain Reaction. Fire will not be created if one component of fire triangle (oxygen/fuel/heat) is lost. In addition, fire will not continue if one of the tetrahedron components (oxygen/fuel/heat/chemical chain reaction) is lost. According to [11], fire can be classified as follows: class A is while the fire of solid materials except for metal, and dry chemical can be used to extinguish the fire; class B is liquid or a gas fire; class C is the fire of electronic materials; while class D is the fire of metallic materials. Dry chemical, halon gas, and Carbon Dioxide ($CO_2$) can be used to extinguish the fire of class B, C, and D.

According to [12], the active fire protection system is a fire protection system that completely comprises either manual or automatic fire detection systems including fire extinguisher (APAR), hydrant box, fire alarm, sprinkler and fire detectors. While the passive fire protection including the emergency door, emergency stairs, exits and assembly point to prevent accidents or injuries during evacuation in the event of an emergency.

## 3 Methodology

The method in this study is using interviews, observation, and checklist. Review methods are as follow:

### 3.1 Interviews

Interviews were conducted from several sources; respondent 1 is a security guard in the building KH. Mas Mansur, respondent 2 is an employee in charge of infrastructure building, and respondent 3 is a fire expert from the Department of Labor and Social Affairs, Sleman, DIY.

### 3.2 Observation

The collection data based on observation carried out in building KH. Mas Mansur and its surrounding. It includes infrastructure conditions of active and passive fire protection including fire extinguisher (APAR), hydrant box, fire alarm, sprinkler, fire detectors, the emergency door, emergency stairs, exits and assembly point [12].

### 3.3 Checklist

A checklist is used in comparing the actual condition of the existing infrastructure and the standard fire regulations such as from [12]-[19]. The weight rating refers to [15] which is shown in Table 1. Then the result of this comparative analysis was used to create evacuation route mapping.

**Table 1**. Level of fire audit assessment

| Score | Conformance | Reliability |
|---|---|---|
| >80- 100 | As required | Good |
| 60-80 | Installed but small amount of installation unfulfilled the requirement | Enough |
| <60 | Not proper at all | Less |

Source: Research and Development Center of Settlement (2005)

## 4 Results and Discussion

### 4.1 Interview

Three people were interviewed in this study. Respondent 1 said that the infrastructure is good enough such as a fire extinguisher and hydrant box, but the only problem is regarding maintenance. The check is only done if the fire training will be held. There is also no data collection regarding the number of APAR. A fire training conducted by the university was held for security personnel while indoor and outdoor fire workshop sponsored by industrial engineering department which held a few months ago. The training provides more insight about the importance of fire prevention and how to countermeasure.

The second interview was conducted with respondent 2 as the person in charge in term of an infrastructure of the building KH. Mas Mansur. The building was built in 2004 and completed by fire prevention and control tools such as fire extinguisher, hydrant box, alarms and heat detectors. However, the number of existing infrastructure is not known because of the lack of data and technical information constraints. Hydrants ever used to clean volcanic ash when an eruption of volcanic mountain happened, and APAR was periodically replaced according to the validity period. However, the thermal detector was never being tested. The faculty tends to prioritize fire extinguisher on every laboratory than that in the office space and lecture room, due to the danger level. The laboratory has higher danger because of many chemical solutions, electronic tools, etc.

The third Interview was done with Mr S as the fire expert from Department of Labor and Social Affairs, Sleman, DIY. Before interviewing respondent 3, prior observation towards building KH. Mas Mansur had been done. Actually, the building already equipped with fire protection such as fire extinguisher, hydrants, alarms and heat detectors. However, when the observation conducted, the alarm condition was not working so it cannot be tested whether the detector work or not. Second, the hydrant boxes were incomplete such as no nozzle and a fire hose. It can be stated that the maintenance of fire hydrant and alarm has not run optimally. Then, there was no APAR at some points, even if it is available, it was not placed properly. So, it is necessary to optimize the function of available alarm and hydrant, place APAR properly, and give mark as well as socialize how to use it. Regarding the evacuation route, since many activities held in the building, when there is a fire can cause panic, potentially. Therefore, the evacuation path must exist.

https://doi.org/10.1051/matecconf/201815401094

### 4.2 Observation

Some active and passive fire protection in the building KH. Mas Mansur were found. The condition of the infrastructure suits with the result of interview toward three people. Figure 2 shows some of the fire protection tools.

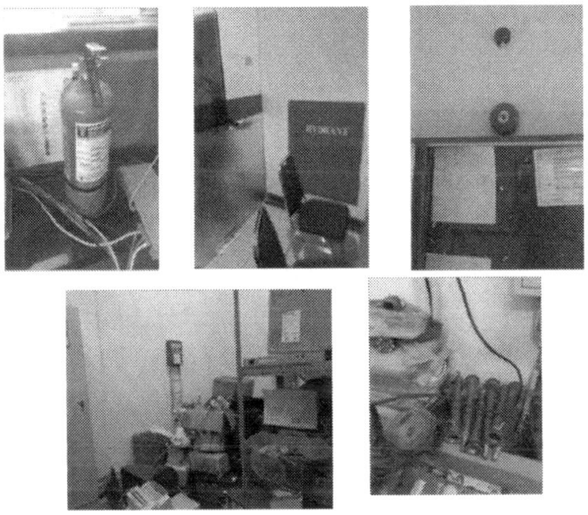

**Fig 2.** Active fire protection

**Fig 3.** Passive fire protection

### 4.3 Checklist

The suitability of the existing condition towards the standard of active fire protection as follows. 10 out of 15 requirements related to APAR [12] were fulfilled, so that the score is 67% means conformance level is enough. While 5 out of 9 requirements of hydrant according to [13]. It means the suitability with Indonesian National Standard is less. Regarding the alarm, there are two specifications needed which are the availability of the alarm and the signal of the alarm that is tested. Since only one fulfilled, the score is 50% or not in accordance with SNI. Whereas the existing condition of sprinkler at building KH. Mas Mansur does not fulfil 13 requirements of [17] which means not good. Regarding the heat detector, 2 out of 5 requirements are fulfilled which are the availability of detector in all rooms and the ease of reaching the detector. It means 40% and categorized as not good level. Therefore, the average suitability of active fire protection facilities in the building KH. Mas Mansur is 42.6% which means less good. The low fulfilled requirements because there are still many active protection facilities that do not meet the standard terms and conditions. It can be seen from each active protection that is less than 80 % (Fig. 4).

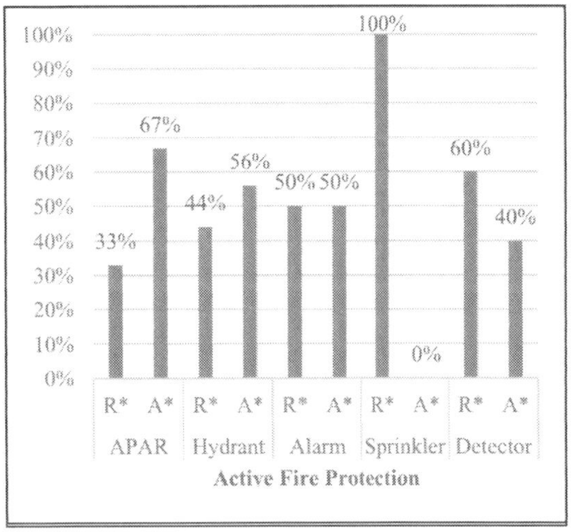


**Fig 4.** Comparison between R* (Required) and A* (Actual) Active Fire Protection

The passive fire protection consists of the emergency door, emergency stairs, exit direction, and assembly point. According to [16], there are seven requirements for the emergency door, but only five is fulfilled (71%). Emergency doors can utilize side hinges that are designed to swing from any position and completed by exit sign marked out in the corridor [12]. Emergency stairs should meet six specifications, but only three fulfilled (50%). It means not suitable with [12] due to no special identification, no floor level hints, and the use of space under the stairs to store goods, even not to store goods. The exit direction does not get any point (0%) due to no installed exit directions yet. This also happens to the fourth passive fire protection (0%) in which nothing assembly point installed. Whereas, according to [15], there should be three requirements that are met following [18] and [19]. Assembly point is critical so that people who save themselves know where the safe gathering place is. Only 8 out of 35 total requirements of passive fire protection fulfilled. The low percentage because many infrastructures do not meet the terms and conditions applied. It can be seen from each of the passive fire protection which less than 80% (Fig 5).

### 4.4 Evacuation Map

The evacuation route map is made based on the analysis of checklist result and derived from the old map in accordance with the fire regulation standards infrastructure. The map can be seen on the attachment (Fig. 6). For active protection devices such as fire extinguisher is placed every 15 meters [11], hydrant is available on each floor with the place close to the toilet, fire alarms are set near the stairs, and heat detectors are attached to the ceiling of each room [20]. While passive protection facilities such as emergency exits are located to the west, east, and hall where each is connected to the assembly point to facilitate the evacuation process [12,

18]. Emergency stairs are located on four sides, namely west, east, near mosque, and near chemical engineering department. Each emergency staircase is equipped with floor number marks, exit directions [12], fire-resistant exit (approximately 2 hours) and panic bar as the handle so it is easily opened from the outside of the stairs to prevent the smoke entry into the fire escape [21]. Assembly points are located in the west and east of car park, and in front of the building hall. By dividing the gathering point, it is hoped that people can be more easily and quickly in the evacuation process.

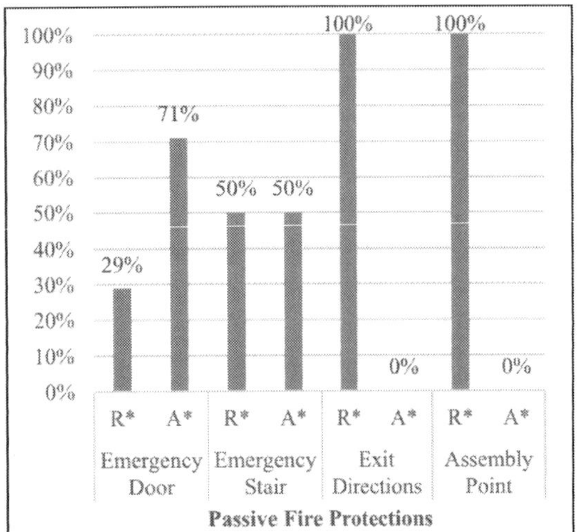


**Fig 5.** Comparison between R* (Required) and A* (Actual) passive fire protection

## 5 Conclusions and Suggestions

### 5.1 Conclusions

From the results of data processing and analysis that has been done, it can be concluded that:

a. According to the checklist results, most of the active and passive protection is still far from standard conformance rules. In total, only 38% of the protection met the standard in which 40% from active protection and 35% from passive protection. The active protection comprises of APAR 67%; hydrant boxes 56%; fire alarm 50%; sprinkler 0%; and fire detectors 40%. Whereas the passive protection consists of door emergency 71%; emergency staircase 50%; assembly point 0%; and directions 0%.
b. Based on the observation and interviews in the building KH. Mas Mansur, Islamic University of Indonesia, it needs an evacuation route map which is made based on the existing regulatory standards and consultation with experts.

### 5.2 Suggestions

Some suggestions for further actions are:

a. Adjust both active and passive infrastructure by regulatory standards until reach enough category, so that awareness of prevention and fire countermeasure can be created.
b. Apply the proposed evacuation route map so that everyone inside the building could escape easily in case of a fire disaster.

## References

1. S.K. Jia, X.-D. Liu, S.-J. Cai, J. Saf. Sci. Tech. **4** (2008)
2. D. Meng, H.-W. Yao, T.-Y. Cui, Y.-C. Sun, *Procedia Eng.* **135**, 25-28 (2016)
3. I. Ismail, M. Taib, M.A.O. Mydin, *Matec Web of Conf.* **10**, 06002, 1-7 (2014)
4. Y.N. Hu, *Procedia Eng.* **135**, 524-530 (2016)
5. X. Liu, H. Zhang, Q. Zhu, *Procedia Eng.* **45**, 643-648 (2012)
6. D. Ketabi, G.H. Halvani, *Int. J. Occ. Hyg.* **2**(1), 6-9 (2010)
7. M.N. Ibrahim, K. Abdul-Hamida, M.S. Ibrahima, A. Mohd-Dina, R.M. Yunus, M.R. Yahya, *Procedia Eng.* **20**, 317-324 (2011)
8. City News, Department of Fire and Rescue Management of Special Region Jakarta, Available at [http://www.jakartafire.net/news/detail/308/kampus-stie-perbanas-terbakar] (2006)
9. A. Sudarno, Liputan 6, Available at [http://news.liputan6.com/read/2903007/kebakaran-hanguskan-laboratorium-fateta-ipb-bogor] (2017)
10. Metro, Tribun Lampung, Available at [http://lampung.tribunnews.com/2017/11/05/kampus-stkip-kumala-kota-metro-ludes-terbakar-api] (2017)
11. Rule of the Ministry of Manpower and Transmigration, *SNI No. Per.04/Men/1980 about the Terms and Maintenance of Fire Extinguisher,* Article 2 (1980)
12. Rule of the Ministry of Public Works, *No.26/PRT/M/2008 about the Technical Requirements of Fire Protection System on Building and Environment* (2008)
13. Rule of the Ministry of Public Works, *SNI No.20/PRT/M/2008 about the Technical Guidelines of Fire Protection Management in Urban Areas* (2008)
14. Ministry of Public Works, *Technical Guidance of Fire Building Fire Inspection* (2005)
15. National Fire Protection Association, *NFPA 101, Life Safety Codes* (1995)
16. National Standardization Agency, *SNI 03-3985-2000 about Procedure of Planning, Installation, and Testing of Fire Detection and Alarm System for Fire Hazards Prevention in Building* (2000)
17. National Standardization Agency, *SNI 03-3989-2000 about Planning Procedures, Installation of*

 https://doi.org/10.1051/matecconf/201815401094

*Automotive Sprinkler System for Fire Prevention in Building* (2000)

18. National Standardization Agency, *SNI 03-1746-2000 about the Procedures for Planning and Installation of Out-of-Road Facilities for Rescue of Fire Hazards in Building* (2000)
19. National Standardization Agency, *SNI 03-6571-2001 about Fire Smoke Control System in Building* (2001)
20. National Standardization Agency, *SNI 03-6574-2000 about Emergency Detection and Alarm System* (2000)
21. National Standardization Agency, SNI 03-1735-2000 *about Planning Procedures for Building and Environmental Access* (2000)

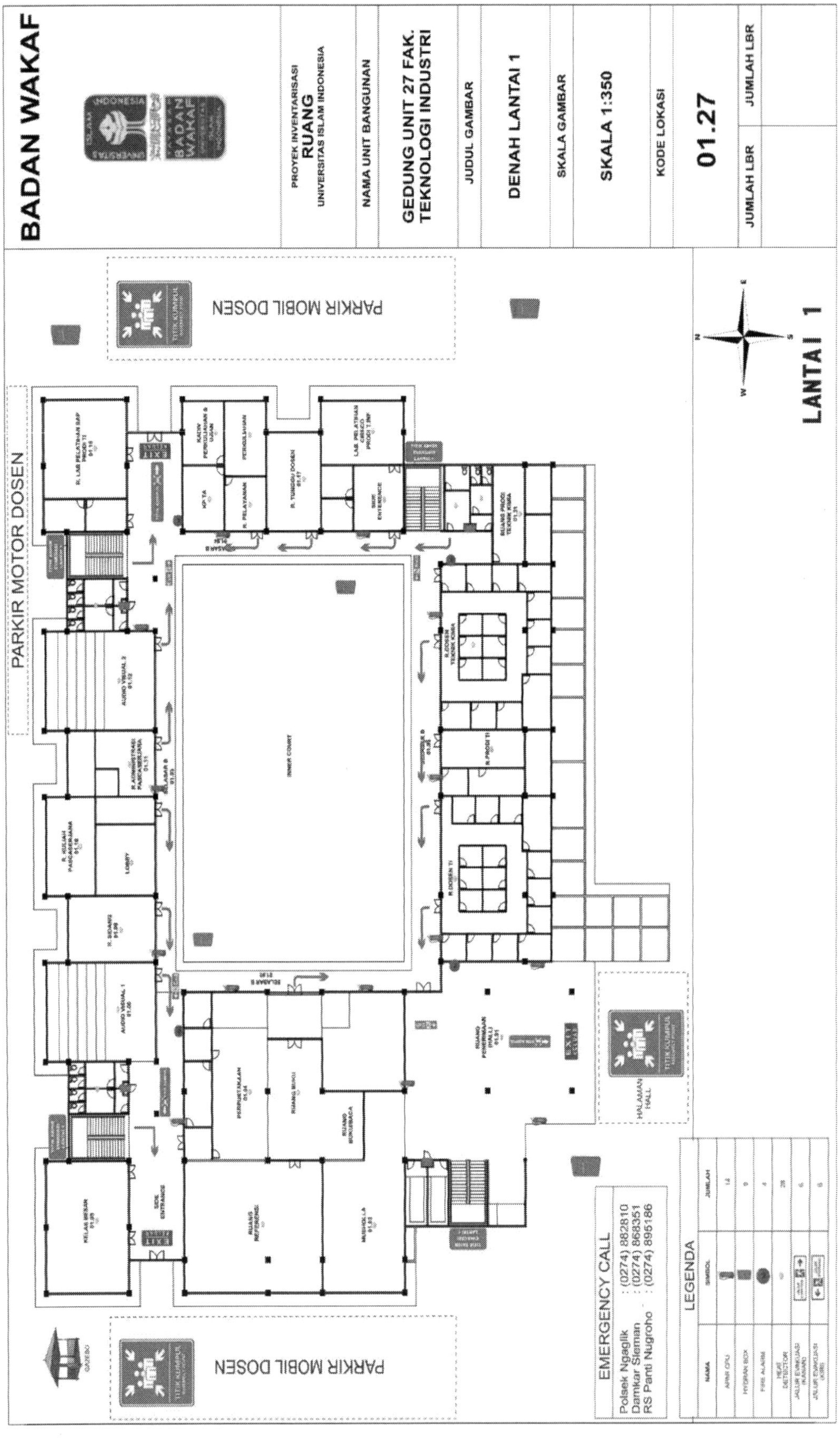


**Fig 6**. Evacuation map 1st Floor of KH. Mas Mansur Building

# Implementation of lean manufacturing to reduce waste in production line with value stream mapping approach and Kaizen in division sanding upright piano, case study in: PT. X

W. A. Santosa* and M. Sugarindra

Industrial Department, Universitas Islam Indonesia, Jl. Kaliurang, Km 14.5, Yogyakarta, Indonesia

**Abstract.** PT. XY produces musical instruments such as Upright Pianos and Grand Pianos. Due to a competitive competition, a good quality product is highly required as well as increasing the production scale. To achieve these objectives, company needs to reduce wastes occurred in its production lines, particularly in the division of sanding panel upright piano (UP) which produces type of PE B1 pianos. High cycle time and lead-time are caused by wastes in UP panel sanding division. Therefore, it is needed improvements to be applied here so that the production lines will be run more effectively and efficiently. This study aims to identify wastes using Value Stream Mapping (VSM) as a tool of lean manufacturing and to implement the improvements using Kaizen. It is found that the wastes are motion and waiting. Furthermore, the improvements (kaizen) are focused on reducing motion and waiting wastes. It is shown that cycle time decreased from 51.16 minutes to 41.90 minutes, work in process or inventory can be reduced to 24 pcs over 32 pcs, and the lead-time is 0.167 days of 0.222 days.

## 1 Introduction

The increasing competition among manufacturing industries has pushed every company to leverage its productivity in various ways. PT. XY is a manufacturing company that produces the type of pianos as musical instruments. The company is growing with a wide range of improvement projects in order to increase productivity. A piano production process is broken down into several units including the division of Sanding Panel Upright Piano (UP). According to the observation, there are wastes occurred such as less effective process, the layout is not set properly, and the cycle time to produce a cabinet panel piano upright piano into one unit X models is longer than the normal time. X model has a cycle time by 51.16 minutes which is caused the daily target of this division not reached. Thus, lead-time becomes 0.222 days. In order to increase its production, PT. XY needs to create a shorter cycle time and lead time so that daily goals can be achieved as well as create a production line more effectively and efficiently.

A balancing between effective and efficient can be created by reducing waste [1]. Waste is all activities that have non-value added [2]. To eliminate waste, companies can use lean concepts. Lean is a continuous effort to eliminate waste and improve value-added products in order to deliver value to customers [3]. Tools that can used to reduce waste are Value Stream Mapping (VSM) and Kaizen. VSM can assist companies in identifying non-value-added activities [4] and Kaizen is continuous improvements to improve productivity [5].

This study aims to help company understands the real condition of production lines to identify existing waste, design the improvements needed, and determine the changes occurred after the improvement applied in the division of sanding panel UP. Improvements carried out with a focus on improvements that can reduce cycle time and lead time in accordance with the company expected.

## 2 Literature Review

### 2.1 Lean Manufacturing

Lean manufacturing can be defined as a systemic approach and systematic to identify and eliminate waste or activities that are not value-added through the continuous improvement radically by flowing the product (material, work in process, output) and information using pull system of internal and external customers for the pursuit of excellence and perfection in the manufacturing industry [6].

### 2.2 Waste

Waste is anything other than the minimum amount of equipment, materials, parts, and labor (working time) which are essential to add value to the product [7]. There are seven kinds of waste defined according to [8]:

- Overproduction
  The waste of overproduction is considered as the most serious waste as it discourages a smooth flow of goods

* Corresponding author: wijiagungsantosa@icloud.com

or services and is likely to inhibit quality and productivity. Overproduction also tends to lead excessive lead and storage times.

- Defect
  The waste of defects are errors occured when working processes, product quality and bad delivery performance.
- Unnecessary Inventory
  The waste of unnecessary inventory is excess storage of goods, as well as the delay of a product or material information that lead to increased costs and decreased quality of service to the customer.
- Inappropriate processing
  The waste of inappropriate processing is production process which is not appropriate for any of the procedures, the use of equipment or machinery that is not in accordance with the capacity and ability to work.
- Excessive transportation
  The waste of excessive transportation includes any movement of people, information, product or material.
- Waiting
  The waste of waiting occurs when time is not being used effectively, worker inactivity, information, material or product, each spending time waiting in a long period.
- Motion Unnecessary
  The waste of motion unnecessary is whenever use of time does not add any value to the product nor process.

### 2.3 Value Stream Mapping

Value Stream Mapping (VSM) is a tool that is ideal as a first step in the process of change to get condition lean manufacturing or lean enterprises [9]. VSM is a tool to identify activities that value added and non-value added in the manufacturing industry, making it easier to find the root problems in the process [10, 11]. State that the VSM is a method of lean process flow that can be reached by a three-stage method [12]. 1). Describe a current state map. 2). Identify the root causes of the problems that hinder the process of improvement, repair process determines what can done in the process flow, and then described it in a future state map. 3). Determine the implementation plan improvements to the production process of the company. Making the VSM uses certain symbols to describe the process of waiting, storage, decision-making, queues and inspection [13].

### 2.4 Kaizen

Kaizen is a term in the Japanese language that can interpreted as continuous improvement [6]. Kaizen means continuous improvement involving everyone, both the top management, managers and employees and principal strategy kaizen is to realize that the management should strive to satisfy customers and meet customer needs [14].

## 3 Method

Data collected are standard time, Standard Process Flow, and Work in Process by using video recorder and a stopwatch. Then, the current state mapping is drawn to understand a clear picture related to the processing time. After the wastes have been identified, some improvements are designed to fix these problems using Kaizen. The Kaizen approved will implemented, as a further corrective action will be carried out a new mapping to the condition after the repair in order to know what changes occur.

## 4 Result and Discussion

The data used has been tested the adequacy and the uniformity of the data with the overall results of data fairly and uniformly. Then, collect the data of WIP or inventory that will be included in the mapping of the Current State Value Stream Mapping. The process flow map Current State Value Stream Mapping illustrated as follows:

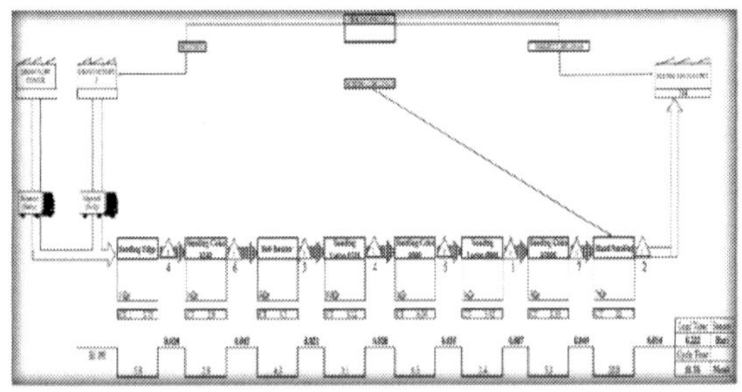

**Fig 1.** Current state mapping section sanding model A

Current state mapping is drawn from the previous process, which is part of Painting and will be sent to the buffing UP Panel. Before sending the material to buffing, material goes through the process of sanding edge (#240 - #1000) until hand sanding process.

The process of hand sanding is the longest cycle time of 22 minutes. It is because of a manual process without machines utilization, just *ategi* and *abrasives* involved to the process. For the longest cycle time process where the machine involved is sanding machine surfaces sanding #500 by 6:34 minutes. The machine sanding #500 is an automatic machine so that the processing time will depend on the machine setting. While, the fastest cycle time is 2.90 minutes which belongs to the process of sanding surface machine sanding# 320. Then, the total cycle time on production lines Sanding UP Panel is 51.16 minutes. Besides the cycle time, it is found that the sanding surface machine sanding #240, the process of sanding surface machine sanding #500, and the sanding surface machine sanding #1000 have bottleneck. This is caused by the unbalanced cycle time and the total lead-time occured in UP Panel Sanding is 0.222 days. The waste identified and proposals kaizen will be explained clearly in Table 1.

**Table 1.** Identification of waste and proposals Kaizen.

| No | Problem Finding | Propose Kaizen | Waste Category |
|---|---|---|---|
| 1 | Process sanding surface machine sanding #240 and machine #500 sanding capacity of setting only 2 pcs per process | Promvide Stopper in order to increase setting process sanding surface machine sanding# 240 capacity and Top Board, machine sanding# 500 to be 3 pcs per process for cabinet Fall Center. | Motion |
| 2 | Process sanding surface sanding machine #1000 capacity setting only 1 pcs per process | Provide Stopper so that setting the process sanding surface sanding machine #1000 capacity can be 2 pcs per process for cabinet Fall Center. | Motion |
| 3 | Adjusment button Foot Pad cylinder on the sanding machine #1000 too Far. | Changing the position or add a key regulator of the cylinder pad foot next to the operator. | Motion |
| 4 | Process mentored sanding is still using a hand sanding (ategi) so it takes a longer time. | The use orbital sanding to simplify and make the process mentored sanding faster. | Motion |
| 5 | Sanding machine #500 often does not work because solenoid often broke | Repair sanding machine #500 so that it will more stable and work properly | Waiting |
| 6 | Motor meja *sanding #500* sering terbakar, sehingga level naik turunnya meja tidak dapat dilakukan. | Perbaikan mesin *sanding #500* agar kondisinya stabil dan berfungsi dengan baik | *Waiting* |
| 7 | Karena mesin #500 adalah mesin *automatic* membuat proses *Sanding* mesin *#500* lama karena kecepatan maju mundur meja mesin yang lama | Mempercepat kecepatan maju mundur meja mesin *#500* | *Waiting* |

After Kaizen applied, a Future State Value Stream Mapping is built. The process drawn in Future State Value Stream Mapping is similar with the existing process in the Current State Value Stream Mapping. This is because there is no kaizen eliminating processes in sanding UP Panel. The Future State Value Stream Mapping is illustrated in Fig 2.

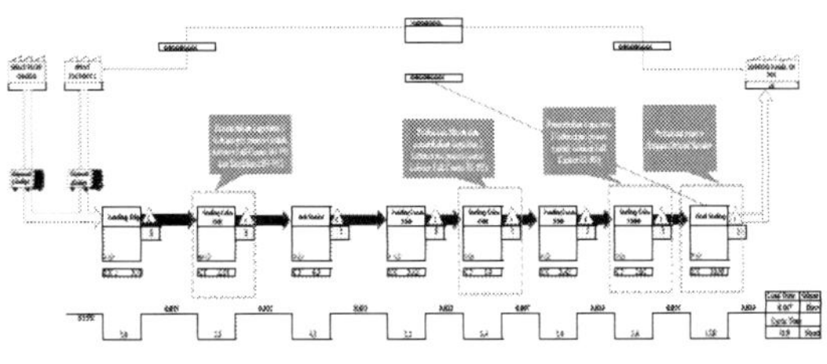

**Fig 2.** Future state value stream mapping section sanding model A.

# 5 Conclusion and Recommendation

## 5.1 Conclusion

Based on current state mapping, the hand sanding has the longest cycle time by 22 minutes. The longest cycle time of machine process is sanding machine surfaces sanding #500 up to 6:34 minutes. Meanwhile, the fastest cycle time is process sanding surface machine sanding #320 reaching 2.90 minutes. Thus, the total cycle time on production lines Sanding UP Panel is 51.16 minutes. There are a few processes having bottleneck: the process of sanding surface machine sanding #240, the process of sanding surface sanding machine #500, and the sanding surface sanding machine #1000.

The implementation of Kaizen has given a couple of changes in decreasing cycle time: (1) sanding machine #240 has become 0.57 minutes from 2.90 minutes; (2) machine sanding #500 is 3.40 minutes compared to 6.34 minutes; (3) machine sanding #1000 is eliminated more than 0.70 minutes or it is 2.61 minutes now; and (4) process of hand sanding is exactly 17 minutes, 5 minutes faster. Thus, the total of cycle time of division sanding panel up has been reduced for almost 9.3 minutes from the previous total of 51.16 minutes. Inventory reaches 24 pcs or it decreases 6 pcs. Finally, the previous 0.222-day lead time has been increased to 0.167 days.

## 5.2 Recommendation

According to these results, PT. XY should keep maintaining the continous improvements into the production line so it will always be lean and any wastes occured can be reduced. For further, an improvement or kaizen may be design for processes using machines of straight sanding such as sanding process of the surface sanding machine #320 and sanding process of the surface sanding machine #800. It is also interesting to conduct a simulations in order to know whether there is any impact if the sanding machine #500 is changed from the automatic into manual machine.

# References

1. D. M. Utama, S. K. Dewi, V. I. Mawarti, Identifikasi Waste Pada Proses Produksi Key Set Clarinet Dengan Pendekatan Lean Manufacturing, **15**, 1 (2016).
2. A. Maulana, L. Herlina, B. Kurniawan, Usulan Lean Manufacturing System untuk Mereduksi Waste Dan

Efisiensi Biaya Produksi Di PT. ABC Divisi Slab Steel Plant 1. **4**, 3 (2016).

3. V. Gaspersz, A. Fontana, *Lean Six Sigma for Manufacturing and Service Industries,* (2011).
4. T. Prayogo, T. Octavia, Identifikasi Waste dengan Menggunakan Value Stream Mapping di Gudang PT. XYZ, **1**, 2 (2013).
5. E. A. P. H. Putra, Z. F. Ikatrinasari, Penerapan Lean Manufacturing Melalui Metode Gemba Kaizen Dengan Pendekatan Siklus Pdca Untuk Peningkatan Produktivitas Di PT. XYZ, BEKASI, **3**, 2 (2012).
6. V. Gaspersz, *Lean Six Sigma for manufacturing and service industries*, (2007).
7. T. Narusawa, J. Shook, *Kaizen Express 2nd edition*, (2008).
8. P. Hines, D. Taylor, *Going Lean*, (2000).
9. Goriwondo, M. William, M. Samson, M. Alphonce, *Use of The Value Stream Mapping Tool for Waste Reduction in Manufacturing (Case Study for Bread Manufacturing in Zimbabwe),* (2011).
10. Mc. Williams, L. Douglas, E. G. Tetteh, *Value-Stream Mapping to Improve Productivity in Transmission Case Machining*, (2008).
11. M. Tilak, E. Van Aken, T. McDonald, K. Ravi, *Value Stream Mapping: A Review and Comparative Analysis of Recent Applications*, (2002).
12. A. Rahani, M. al-Ashraf, *Production flow analysis through value stream mapping: a lean manufacturing process study*, (2012).
13. B. El-Haik, R. Al-Omar, *Simulation Based Lean Six Sigma and Design For Six Sigma*, (2006).
14. M. Imai, *Kaizen Kunci Sukses Jepang Dalam Persaingan*, (2001).

# Research in halal logistics and halal supply chain: Issue and area development

Qurtubi[1,*] and *Elisa* Kusrini[2]

[1,2]Department of Industrial Engineering, Universitas Islam Indonesia, Yogyakarta, Indonesia
[1]Student of Post Graduate Program of Economics Faculty, Universitas Islam Indonesia, Yogyakarta, Indonesia

**Abstract.** Today the study of halal logistics and halal supply chain is growing and provides a great opportunity to do research both empirical and theoretical development. However, until now research in the field of logistic and supply chain is still limited. Therefore, this paper aims to explore opportunities and areas of development in this field by conducting literature review of previous research. Research opportunities in halal logistics and halal supply chain has been reviewed by many researchers and grouped into many categories, namely (1) Principles, Conceptual Model and Framework, (2) Innovation, Development and Challenges, (3) Critical Control Point, Critical Success Factors and Halal Control Activities, (4) Consumers Willingness to Pay, (5) Transportation and Information Technology, (6) Strategic approach. The issue in each category and research opportunities will be discussed in this paper.

## 1 Introduction

The development of the field of halal logistics and halal supply chain on all aspect of the halal business have provided great opportunities to do research. But as far as the author's knowledge, the paper discusses problems and research opportunities in the field of halal logistics and halal supply chain does not exist yet. While Jaafar et al. [5] only wrote the number of halal logistics and halal supply chain publications in the range of year 2004-2013. Therefore, this paper is expected to provide additional insight for researchers. Research opportunities in halal logistics and halal supply chain has been reviewed by many researchers. Principal component bodies (PCB) and the issues that related to halal logistics and halal supply chain consist of 6 issues. For detail of classification of each paper is showed on the Table 1.

## 2 Research Opportunities in Halal Logistics and Halal Supply Chain

The halal logistics and halal supply chain issue can be classified into 6 categories, which includes: (1) Principles, Conceptual Model andFramework, (2) Innovation, Development and Challenges, (3) Critical Control Point, Critical Success Factors and Halal Control Activities, (4) Drivers of Consumers Willingness to Pay, (5) Transportation and Information Technology, (6) Strategic Approach. The following sections will explain each of the category.

### 2.1 Principles, Conceptual Model and Framework

A variety topics that fall into this category, such as Empirical Research, Qualitative Research, Different Markets, Good Manufacturing Practice (cGMP), Assesment Tools and Area Storage. Various areas of research opportunities based on the literature review include: Consumer perception of the Muslim consumer; Consumer perception Muslim and non-Muslim countries; Perception toward halal logistics; Testing the relationship between halal logistics, marketing mix, and the seven element, halal logistics market by performing SWOT analysis; The relationship between factors affecting the integrity of halal supply chain; Halal purchasing, halal network development, halal trade and halal parks; the application of halal in management of cosmetics and pharmaceutical supply chains; The management of halal ports; The principles in organising halal supply chains for different market (for example, between Muslim and non-Muslim countries); Integrate Good Manufacturing Practice (cGMP) and halal compliance into a holistic framework; Measuring of resilience impact of the attributes of halal logistics.

### 2.2 Innovation, Development and Challenges

The needs to be innovative in initiating more logistics services that are based on halal concept are crucial in meeting the needs of the increasing demand by the customers especially the Muslims [20]. Some issues are widely discussed in Innovation, Development and Challenges such as Halal Training Program and Research on Other Halal Business. Research opportunities in this area includes: Human capital training and development

*Corresponding author: qurtubi@uii.ac.id

strategies and planning; Research on tourism, food product, pharmaceuticals, healthcare, cosmetics, and banking.

**Table 1.**Halal logistics and Halal Supply Chain Issue Classification

| Issue Classification | Authors |
|---|---|
| Principles, Conceptual Model and Framework | [1-15] |
| Innovation, Development and Challenges | [16-21] |
| Critical Control Point, Critical Success Factors and Halal Control Activities | [22-26] |
| Consumers Willingness to Pay | [27-28] |
| Transportation and Information Technology | [29-31] |
| Strategic Approach | [32-34] |

## 2.3 Critical Control Point, Critical Success Factors & Halal Control Activities

A variety topics that fall into this category, such as Critical Control Point, Extra Cost, Empirical Evidence and Share the new knowledge and Techniques. Some research opportunities in this category such as Financial aspect of Halalan Toyyiban; Quantify of CSFs for a more exact outcome by clarifying with CSFs are applicable; Assist the smaller organizations toward the realization of halal supply chain.

## 2.4 Consumer Willingness to Pay (WTP)

Understanding consumer willingness to pay can lead manufacturers to implement the concept of halal in its value chain [27]. Some issues are widely discussed in Consumer Willingness to Pay include: Consumer Behavior and People's Concerns Regarding Halal. Some research opportunities in this area includes: Identifying a more appropriate method of measuring the long term effects of consumer behavior; Apply the research model to other countries.

## 2.5 Transportation and Information Technology

The Halaltracer tracking system was developed using the combinations of Global Positioning System (GPS) and geofence algorithm. The Halaltracer tracking system is significantly contributing in terms of automatic detection on the spatial activity taken during the halal product's shipment process [31]. Some issue that appear in the halal logistics and halal supply chain related to transportation and information technology such as: Adaption of Halal Supply Chain, Sensor and Improvement. Some research opportunities in this category such as: Finding what factors could be a driver for halal manufacturers; Research with better contextual adoption theory; Barcode sensor, QR code or RFID can be integrated in the system; Mobile version for tablet either in Android or IOS version.

## 2.6 Strategic Approach

Good governance requires the examination of external factors such as industry regulation, market forces and competition and internal factors such as business strategy and goals and corporate culture [33]. Some issues are widely discussed in Strategic Approach such as Halal Certification, SWOT analysis and Sample Size. Some research opportunities in this area are Develop a framework of halal certification; Find the barriers and critical success factors; SWOT analysis on other halal market (pharmaceutical and healthcare, tourism, and food products), Research with a larger sample size.

Various categories of issues and research opportunities are given in Table 2.

**Table 2.**Research opportunities in halal logistics and halal supply chain.

| Categories | Issue | Development areas/Research opportunities |
|---|---|---|
| Principles, Conceptual Model and Framework | Empirical research; Qualitative research; Different markets; Good Manufacturin g Practice; Assessment tool; Area of storage. | Consumer perception of the Muslim consumer [2], Consumer perception Muslim and non-Muslim countries [4], Perception toward halal logistics [24], Testing the relationship between halal logistics, marketing mix and the seven elements [7], Halal logistics market by performing SWOT analysis [7], The relationship between factors affecting the integrity of halal food supply chain [8], Halal purchasing, halal network development, halal trade and halal parks [11] |
| Innovation, Development, and Challenges | Halal traning program; Research on other halal business. | Human capital training and development strategies and planning [19], Tourism, food product, pharmaceuticals, healthcare, cosmetics, bank [16]. |
| Critical Control Point, Critical | Critical Control Point; | |

| | | |
|---|---|---|
| Success Factors and Halal Control Activities | Extra cost; Empirical evidence; Share the new knowledge and techniques. | Financial aspectof halalan toyyiban supply chain [6], Further research could be embarked in these areas [22], Quantify the CSFs for a more exact outcome by clarifying which CSFs are applicable [25], Assist the smaller organizations towards the realization of halal supply chain [26]. |
| Consumers Willingness to Pay | Consumer behavior; People's concerns regarding halal. | Identifying a more appropriate method of measuring the long term effects of consumer behavior [27], Apply the research model to other countries [28]. |
| Transportation and Information Technology | Adoption of halal supply chain; Sensor and improvement. | Finding what factors could be a driver for halal manufacturers [30], Research with better contextual adoption theory [30], Barcode sensor, QR code or RFID can be integrated in the system [31], Mobile version for tablet either inAndroid or IOS version [31]. |
| Strategic Approach | Halal certification; SWOT analysis. | Develop a framework [33], On other halal markets such as pharmaceutical and healthcare, tourism, and food products [32], Finding the barriers and critical success factors [32]. |

## 3 Conclusion

Today, the increasing competitiveness in industry force many company to increase capability in all aspect along their business. One of the challenges is to provide halal product and services that create opportunity for researcher to explore this filed. From study literature , it can be conclude that research in halal logistic and halal supply chain can be classified into 6 categories, namely Principles, Conceptual Model and Framework, Innovation, Development and Challenges, Critical Control Point, Critical Success Factors and Halal Control Activities, Consumers Willingness to Pay, Transportation and Information Technology, Strategic approach. The development of each categories still in early phase, therefore it provide great opportunity to study halal logistic and halal supply chain by considering the maturity of traditional logistic and supply chain.

## References


1. E.N. Omar, H.S. Jaafar, *IEEE Symp on Bus, Eng and Appl (ISBEIA)* (2011)
2. M. Tieman, J. of Isl Mark **2**, 186-195 (2011)
3. M. Tieman, J.G.A.J. van der Vost, M.C. Ghozali, J. of Isl Mark **3**, 217-243 (2012)
4. M. Tieman, J. of Emerg Eco and Isl Res **1**, 1 (2013)
5. H.S. Jaafar, E.N. Omar, M.R. Osman, Faisol N. *Paper Presented at ICLT* (2013)
6. E.N. Omar, H.S. Jaafar, M.R. Osman, J. of Emerg Eco and Isl Res **1**, 3 (2013)
7. A.B.A. Hamid, M.S.A. Talib, N. Mohamad *Intellectual Discourse* **22**, 191-214 (2014)
8. M.H. Zulfakar, M.M. Anuar, M.S.A. Talib, Proc-Soc and Beh Sci **121**, 58-67 (2014)
9. E.N. Omar, H.S. Jaafar, M.R. Osman, J. of Emerg Eco and Isl Res **1**, 3 (2013)
10. O. Aigbogun, Z. Ghazali, R. Razali, Glob Bus and Man Res: An Inter J. **7**, 3 (2015)
11. M. Tieman, J. of Isl Mark **6**, 2-21 (2015)
12. H.S. Jaafar, N. Faisol, F.A. Rahman, A. Muhammad, *Sprin Sci+Bus Med* (2016)
13. S.A. Rahim, B. Mohammad, N.A.A. Rahman, *Sprin Sci+Bus Med* (2016)
14. A. Asiaei, A.A. Aziz, S. Zailani, *Sprin Sci+Bus Med* (2016)
15. S. Nasir, H.S. Jaafar, Sci. Int(Lahore) **28(2)**, 1769-1774 (2016)
16. M.S.A. Talib, L. Rubin, V.K. Zhengyi, J. of Emerg Eco and Isl Res **1**, 2 (2013)
17. N. Karia, M.H.A.H. Asaari, *Vision 2020: Sust Growt, Eco Dev, and Glob Comp* (2014)
18. M.H.A. Kadir, R.Z.R.M. Rasi, S.S. Omar, Z.I.A. Manap, *IOP Conf. Seri: Mat Sci and Eng* **160**, 012070 (2016)
19. H.I.C. Hasim, S.M.M. Shariff, *Proc Eco and Fin* **37**, 33-38 (2016)
20. H.S. Jaafar, I.R. Endut, N. Faisol & E.N. Omar, *IMPRA Paper* **34655**, Posted 12 (2011)
21. N. Ahmad, S.M. Shariff, Proc Eco and Fin **37**, 418-425 (2016)
22. E.N. Omar, H.S. Jaafar, M.R. Osman, Proc–Soc and Beh Sci **00**, 000-000 (2011)
23. R.C. Din, S. Daud, Proc-Soc and Beh Sci **121**, 96-103 (2014)
24. M. Tieman, M.C. Ghazali, Proc-Soc and Beh Sci **121**, 44–57 (2014)
25. M.S.A. Talib, A.B.A. Hamid, M.H. Zulfakar, J. of Isl Mark **6**, 44-71 (2015)

26. S.M. Shariff, H.S. Jaafar & A. Muhammad. *Sprin Sci+Bus Med* (2016)
27. R. Kamaruddin, H. Iberahim, A. Shabudin, Proc-Soc and Beh Sci **50**, 722–729 (2012)
28. E. Fathi, S. Zailani, M. Iranmanesh, K. Kanapathy, British Food J. **118**, 464-479 (2016)
29. H.A. Tarmizi, N.H. Kamarulzaman, I.A. Latiff, A.A. Rahman, UMK Procedia **1**, 42–49 (2014)
30. A.H. Ngah, Y. Zainuddin, R. Thurasamy R. Proc-Soc and Beh Sci **129**, 388–395 (2014)
31. M.A. Mohamad, S. Mansor, N. Ahmad, W.A.W. Adnan, I.M. Wali, J. of Theori and Appl Inf Tech **90**, 2 (2016)
32. M.S.A. Talib, A.B.A. Hamid, J. of Isl Mark **5**, 3 (2014)
33. N. Noordin, N.L.M. Noor, Z. Samicho, Proc-Soc and Beh Sci **121**, 79-95 (2014)
34. O. Aigbogun, Z. Ghazali, R. Razali, Int Rev of Man and Mark **6(S4)**, 209-216 (2016)

# House of risk approach for assessing supply chain risk management strategies: A case study in Crumb Rubber Company Ltd

*Taufiq* Immawan* and *Dea* Kusuma Putri

Industrial Engineering Department, Faculty of Industrial Technology, Universitas Islam Indonesia, Indonesia

**Abstract.** Risk is an uncertain and can have both negative and positive impacts. If the risks have a negative impact then a company will incur losses. CRUMB RUBBER COMPANY LTD is one of crumb rubber company in West Kalimantan. The length of the supply chain contained in CRUMB RUBBER COMPANY LTD and the high dependence on suppliers leads to vulnerability. So the purpose of this research is to identify the risk and determine the priority of source of risk along with the priority of handling it on CRUMB RUBBER COMPANY LTD supply chain with House of Risk approach. House of risk approach consists of two phases. Phase 1 is used to determine the dominant risk agent and phase 2 determines the effective action to deal with the dominant risk agent. From the research results, there are 19 risk events and 29 risk agents identified. The result of house of risk in phase 1 is known that 13 of 28 risk agents are dominant risk agent. Then the priority handling strategy in house of risk in phase 2, in this phase obtained 18 priority risk handling strategies.
**Keywords:** HOR, risk management, SCM

## 1 Introduction

The term SCM was first proposed [2]. Supply chain is a physical network, such as companies involved in supplying raw materials, producing goods, or sending, while SCM is a method, tool, or management approach. Supply chain management is very important to support the smoothness of business activities in the company because supply chain management pay attention to the scope from upstream to downstream.

Risks are uncertain and can have both negative and positive impacts. According [1] the risk is the amount of deviation between expected return and actual return. Meanwhile, supply chain risk is the uncertainty of an event that can lead to disruption of a smooth supply chain in the company.

CRUMB RUBBER COMPANY LTD is a company engaged in manufacturing. This company is one of the manufacturers of crumb rubber which supplies its products to domestic and foreign. The length of the supply chain at CRUMB RUBBER COMPANY LTD and the high dependence on suppliers lead to vulnerability to possible risks. The factors causing the risk of supply chain activity according to [3] include a very complex supply chain network, high dependence on suppliers, the existence of organizational interaction interactions within the supply chain, the short life cycle of a product. Thus, the risk that may occur can adversely affect the company. Therefore, the need for supply chain management to overcome the risks may occur so that companies do not lose and do not hinder the achievement of corporate goals.

Currently PT XYZ does not yet have a structured supply chain risk management in handling the risks. Supply Chain Risk Management or a risk management strategy that can be controlled by a company. The benefits of supply chain risk management are to identify and assess supply chain disruptions and address them.

Based on the background description, it is known that the importance of risk management in the supply chain, so that this study aims to identify the risks and determine the priority of the source of risk along with the priority of the handling of CRUMB RUBBER COMPANY LTD supply chain with the House of Risk approach.

## 2 Research Methods

According to [4] House of risk is a method focused on formulating preventive, reduction and handling of risk factors that potentially lead to more than one risk. House of risk or HOR consists of 2 phases, namely phase 1 and phase 2. In the first phase of House of risk, first mapping supply chain activity and risk identification at CRUMB RUBBER COMPANY LTD by using Supply Chain Operation Reference (SCOR) model. Then performed risk assessment using Failure Mode and Effect Analysis (FMEA) method. This phase is used to determine priority risk agents for precautionary measures. Table 1 is a HOR phase 1 template [4].

**Table 1.** HOR Phase 1

| Proses | Risk Event (Ei) | Risk Agent (Ai) | | | | |
|---|---|---|---|---|---|---|
| | | A1 | A2 | A3 | An | Severity of risk event i (Si) |
| Plan | E1 | R11 | R12 | | Rnn | S1 |
| | E2 | R21 | | | | S2 |
| Source | E3 | | | | | S3 |
| Make | | | | | | |
| Deliver | | | | | | |
| Return | En | | | | | Sn |
| Occurance of agent j | | O1 | O2 | O3 | On | |
| Aggregate risk potential j | | ARP1 | ARP2 | ARP3 | | |
| Priority rank of agent j | | P1 | P2 | P3 | Pn | |

Information:

| | |
|---|---|
| A1, A2, A3…An | = Risk Agent |
| E1, E2, E3….En | = Risk Event |
| O1, O2, O3,…On | = Occurrence value of the risk agent (Ai) |
| S1, S2, S3…Sn | = Severity value of risk event (Ei) |
| ARP1, ARP2, …ARPn | = Aggregate Risk Priority |
| P1, P2, P3…Pn | = Ranking of risk agents based on ARP values |

Then perform a 2nd phase HOR calculation which in this phase determines to the priority of risk management strategies of identification risk agents at high risk level. Where the output of HOR phase 1 will be used as input on this 2nd phase. Table 2 below is a HOR phase 2 template [4]:

**Table 2.** HOR Phase 2

| To be treated risk agent (Ai) | Preventive Action (PAk) | | | | Aggregate risk priority (ARP) |
|---|---|---|---|---|---|
| | PA1 | PA2 | PA3 | PAN | |
| A1 | E11 | E12 | E13 | PAn | ARP1 |
| A2 | E21 | | | | ARP2 |
| A3 | | | | | ARP3 |
| An | | | | Enn | ARPn |
| Total Effectiveness of Action (TEk) | TE1 | TE2 | TE3 | TEN | |
| Degree of Difficulty preforming action (Dk) | D1 | D2 | D3 | Dn | |
| Effectiveness to Difficulty Ratio (ETD) | ETD1 | ETD2 | ETD3 | ETDn | |
| Rank | R1 | R2 | R3 | Rn | |

Information:

- A1, A2, A3…An =Risk agent who was elected to do the handling of
- P1, P2, P3 …Pn =Preventive action to be performed
- E11, E12,… Enn =Correlation between preventive action and risk agent
- ARP1, ARP2,..ARPn =Aggregate Risk Priority from risk agent
- TE1, TE2, TE3…Ten =Total effectiveness of preventive action
- D1, D2, D3…Dn =Degree of difficulty in the implementation of preventive action
- ETD1, ETD2, ETD3…ETDn =Total effectiveness of divide by degree difficulty
- R1, R2, R3…Rn =Rank of preventive action based on sequence of grades ETD highest

## 3 Result

Based on observation and interview with expert, it can be identified the HOR 1 and HOR 2.

### 3.1 Risk Identification

Risk identification is using Failure Mode and Effect Analysis approach. According to [5] that the reason of FMEA usage is FMEA is a technique that can be used to analyze the potential cause of a disturbance, the probability of its emergence and how to prevent it. Giving the value of the variable is done by expert judgment method. Expert judgment is a collection of data provided by an expert on a technical problem [6]. The following is a list of identified risk:

**Table 3** Risk Event

| Process | Activity | Risk Event | Code |
|---|---|---|---|
| Plan | Production planning | Forecasting improper demand | E1 |
| | | Sudden production changes | E2 |
| | Inspection of raw material inventory | The actual amount of stocks recorded | E3 |
| Source | Procurement process | Price negotiations do not reach agreement | E4 |
| | | No offer from supplier | E5 |
| | Arrival of raw materials | Delay of raw material arrival | E6 |
| | | Non-conformity of raw materials ordered | E7 |
| | | Mismatch amount ordered | E8 |
| Make | Production process | Raw material damage | E9 |
| | | Defective product | E10 |
| | | Delayed production | E11 |
| | | Accident workers | E12 |
| | | Engine failure | E13 |
| | | Production is not on target | E14 |
| | Quality control | Not testing product quality during process | E15 |
| Delivery | Delivery of product to buyer | Delivery to buyer late | E16 |
| | | Product damage during shipment | E17 |
| Return | Return of raw material to supplier | Delay in complaint to suppliers | E18 |
| | Handling of products returned from buyers | Delay of product return from buyer | E19 |

**Table 3** Risk Agent

| Risk Agent | Code |
|---|---|
| A mistake in forecasting | A1 |
| Demand a sudden products | A2 |
| Lacking capacity supply raw materials in a warehouse | A3 |
| Record-keeping error in raw materials in and out | A4 |
| Less coordination between parties internal | A5 |
| The company did not take into account price increases that significant | A6 |
| Lack of communication with external sides | A7 |
| Accident occurs transportation | A8 |
| Supplier not fulfill a promise | A9 |
| The supplier in producing fluctuant | A10 |
| The storage of the raw material that inadequate | A11 |
| A mistake in selected raw materials | A12 |
| Contaminated with other objects | A13 |
| A mistake in setting an oven manual | A14 |
| The quality of raw materials low | A15 |
| Natural disasters | A16 |
| To limited human resource | A17 |
| Employees arrived late | A18 |
| Occurring bottle neck | A19 |
| The machine breaks | A20 |
| Employees disobedient in used a protective own according to standard operating procedures | A21 |
| Not held routine maintenance | A22 |
| Limited machine capacity | A23 |
| Schedule a ship erratic | A24 |
| Production late | A25 |
| Interruption along the way | A26 |
| Delays in dealing with the raw material come | A27 |
| Inspection acceptance of raw materials not scrupulous | A28 |

### 3.2 House of Risk Phase 1

Assessment of risk event and risk agent is done by field observation, interview to company side by forming team and questionnaire data for severity assessment from risk event and occurrence valuation from risk agent and correlation from both. Based on ARP values that have been obtained from the calculation of HOR phase 1, then further determine the dominant risk agents with pareto approach. According to [7] pareto diagrams are created using the cumulative percentage of each ARP from the risk agent. Figure 1 below shows the pareto risk agent diagram:

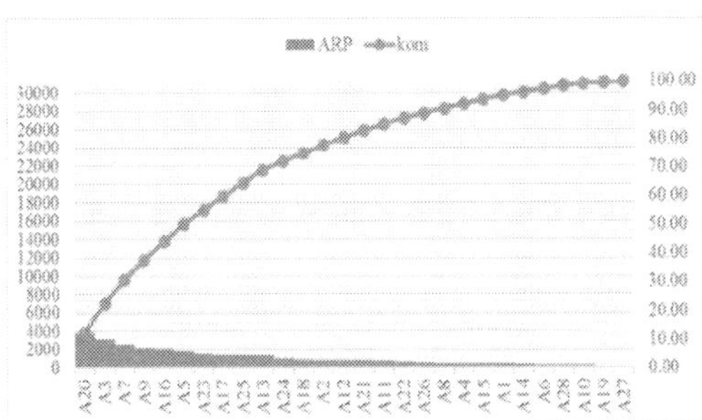


**Figure 1** Pareto diagram

From the results of pareto known there are 13 dominant risk agent, including the troubled engine, lack of capacity to supply raw materials in warehouses, lack of communication with external parties, supplier not fulfill the promise, natural disasters, lack of coordination between internal parties, limitations of engine capacity, limitations of human resource, late production, contaminated with other objects, ship's erratic schedule, the employee was late coming, and demand a sudden products After knowing the dominant risk agent then further create a risk map. The risk map is useful to see the risk condition before handling. Here is a risk maps before drafting strategy priority handling:

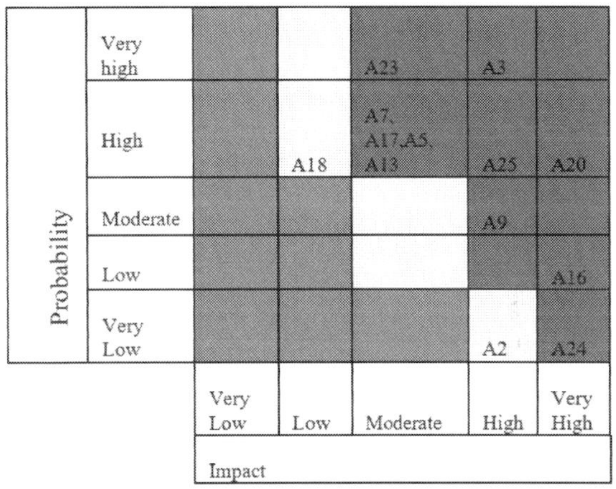


**Fig 2.** Risk maps before drafting strategy priority handling

### 3.3 House of Risk Phase 2

After completing phase 1 of HOR phase then subsequently entering phase HOR phase 2. From result of focus group discussion resulted 18 strategy preventive action. The following Table 4 of HOR phase 2

**Table 4.** Result of HOR Phase 2

| | Preventive Action (PAk) | | | | | | | | | | | | | | | | | | |
|---|---|---|---|---|---|---|---|---|---|---|---|---|---|---|---|---|---|---|---|
| Ai | PA1 | PA2 | PA3 | PA4 | PA5 | PA6 | PA7 | PA8 | PA9 | PA10 | PA11 | PA12 | PA13 | PA14 | PA15 | PA16 | PA17 | PA18 | ARP |
| Troubled engine | 9 | 9 | | | | | | | | | | | | | | | | | 3888 |
| Lack of capacity supply raw materials in a warehouse | | | 9 | | | | | | | | | | | | | | | | 3159 |
| Lack of communication with external parties | | | | 9 | 9 | | | | | | | | | | | | | | 2680 |
| Supplier not fulfill the promise | | | | | 9 | 9 | 9 | | | | | | | | | | | | 2148 |
| Natural disasters | | | 3 | | | | | 9 | 9 | | | | | | | | | | 2065 |
| Lack of coordination between internal parties | | | | | | | | | | 9 | 3 | | | | | | | | 3841 |
| Limitations of engine capacity | | 9 | | | | | | | | | | | | | | | | | 3539 |
| Limitations of human resource | | | | | | | | | | | | 9 | 9 | | | | | | 3470 |
| Late production | | | | | | | | | | 9 | | | | 9 | | | | | 3466 |
| Contaminated with other objects | | | | | | | | 9 | | | | | | | 9 | | | | 1487 |
| Ship's erratic schedule | | | | | | | 9 | | | | | | | | | 9 | | | 972 |
| The employee was late coming | | | | | | | | | | | | | | | | | 9 | 9 | 938 |
| Demand a sudden products | | | 9 | 1 | | | 3 | | 9 | | | | | | | | | | 866 |
| Total Effectiveness of Action ($TE_k$) | 34997 | 48843 | 42264 | 26364 | 47732 | 19652 | 30672 | 34866 | 26640 | 29745 | 5523 | 13730 | 13746 | 13176 | 13665 | 8748 | 8898 | 8890 | |
| Degree of Difficulty performing action ($D_k$) | 3 | 5 | 3 | 3 | 5 | 3 | 4 | 5 | 4 | 3 | 3 | 5 | 4 | 3 | 4 | 3 | 3 | 4 | |
| Effectiveness to Difficulty Ratio (ETD) | 11666.0 | 9768.6 | 14087.0 | 8888.0 | 14244.0 | 6444.0 | 7668.0 | 6213.6 | 6645.3 | 9915.0 | 1841.0 | 2646.0 | 3387.5 | 4392.0 | 3365.8 | 2916.0 | 2738.8 | 2847.5 | |
| Rank | 3 | 5 | 2 | 6 | 1 | 9 | 7 | 10 | 8 | 4 | 18 | 16 | 12 | 11 | 13 | 14 | 15 | 17 | |

It is expected that after conducted design priority handling this, agent risk no in the category of red area. So some changes good to overcome a source of risk potential to appear. Figure 3 under showed position risk agent after conducted design priority strategy handling.

| Probability | | Impact: Very Low | Low | Moderate | High | Very High |
|---|---|---|---|---|---|---|
| | Very high | | | | | |
| | High | | A3, A24 | | | |
| | Moderate | A2 | | | | |
| | Low | A9, A23, A25 | | A20, A7, A16, A13 | | |
| | Very Low | A5, A18 | | A17 | | |

**Fig 3.** Risk map after conducted design priority strategy handling

## 4 Conclusion

Based on the results of the test which has been done we can conclude that:

1. There are 19 risk event and 29 risk agent identified. From the house of risk phase 1 which has been pareto to come by 13 risk agent that are risk agent priority to handling.
2. After conducting calculation using the house of risk phase 2, obtained priority strategy handling risk of them building relationships closely and communication with supplier, do management safety stock raw materials, discipline implement routine maintenance, improve coordination among parts, buying new machinery, building relationships closely and communication with buyer, make and obey MOU with related parties, planning safety stock finished products, evaluate performance supplier, redesign a work environment where better, rescheduled production, the use of labor off, the use of transportation closed which protected from heat and rain, coordination with the transporters, guidance individual regarding the responsibility of the and discipline time, the addition of labor, the provision of rewards for civil servants discipline and sanctions to did not act discipline, and do briefing routine before activity.

## References

1. Hanafi, M. M. (2012). *Manajemen Risiko*. Yogyakarta: UPP STIM YKPN.
2. Oliver, R. K., & Webber, M. D. (1982). *Supply chain management: Logistics catches up with strategy*. London.
3. Punniyamoorthy, M., Thamaraiselvan, N., & Manikandan, L. (2013). Assessment of Supply Chain Risk: Scale Development and Validation. *Benchmarking: An Internation Journal*, 79-105.
4. Pujawan, P. I., & Geraldin, L. H. (2009). House Of Risk: A Model For Proactive Supply Chain Risk Management. *Jurnal Businnes Process Manegement*, 963-967.
5. Christopher, M., & Peck, H. (2004). Building the Resilient supply chain. *International Journal Logistics Management*.
6. Meyer & Booker. (1991). *Eliciting and Analyzing Expert Judgement: A Practical Guide*. London: Academia Press Limited.
7. Caesaron, D., & Tandianto. (2014). Penerapan Metode Six Sigma Dengan Pendekatan DMAIC Proses Handling Painted Body BMW X3 (Studi Kasus: PT. Tjahja Sakti Motor). *PASTI*, 248-256.

# Analysis of customer satisfaction in hospital by using Importance-Performance Analysis (IPA) and Customer Satisfaction Index (CSI)

*Vembri Noor* Helia*, *Cahya Putra* Abdurrahman, and *Fety Ilma* Rahmillah

Department of Industrial Engineering, Islamic University of Indonesia, Jl. Kaliurang Km 14.5, Yogyakarta, Indonesia

**Abstract.** As a major health referral centre, the hospital demanded to provide comprehensive services provided by a multi-disciplinary team according to the needs of patient. In the expansion of a growing number of hospitals in Yogyakarta especially Sleman area where retreat acquire customers even more stringent, it is necessary to increase the quality of services. There are five determinants of service quality namely: reliability, responsiveness, assurance, empathy, and tangible. Collecting data is done by observation and distributing questionnaires to 70 respondents, and Importance-Performance Analysis and Customer Satisfaction Index are used to determine satisfaction level of outpatients. As a results, six out of twenty attributes need improvement, namely: a) The condition of the hospital is clean, comfortable and tidy; b) Guidance and information boards poly is easy to see and read; c) The ease and accuracy of obtaining information for patients (referrals, lab results, etc.); d) The hospital staff has a quick response to the needs of patients; e) Attention physicians in managing patients and willingness to provide a particular time for consultation; and f) The prescribed medication is suitable and safe. It is known that the value of CSI by 76% means the service is not satisfied the patient.

## 1 Introduction

Quality is a dynamic condition associated with the products, services, people, processes, and environments that meet or exceed expectations. The quality of service is addressing the needs and desires of customers as well as the accuracy of delivery to offset the customers' expectations. Currently, people increasingly expect the quality of service that can satisfy and provide ease of the registration process or inspection.

According to [1], five attributes can be used to evaluate the quality of services. First, Reliability is the ability to perform the services promised appropriately. It can be seen from the reception of patients quickly and accurately, examination and treatment quickly and precisely, service schedules carried out correctly, and the service procedure is straightforward. Second, Responsiveness is the ability to help customers and provide faster services. It can be seen on the willingness of officials to be responsive to resolve patient complaints; officers provide clear information and easy to understand; as well as quick action when required. Assurance is about the knowledge, ability and courtesy service providers to generate trust and confidence, ability to assign officers to solve problems, personnel skills in working, polite and friendly service, as well as security and trust services. Empathy is the personal attention given to the customers by providing individual attention to each patient, attention to complaints of patients and families, and care for all patients regardless of social status. Tangible is physical appearance, equipment, personal cleanliness, neatness and comfort of the room, the arrangement of the exterior and interior, completeness, readiness and hygiene tool equipment used, neatness and cleanliness of the hospital staff appearance.

Many researches have been conducted analysis of service quality gap in hospital by using Importance-Performance Analysis (IPA) from the view of patients [8, 10], nurse [4], and physicians [9], while [6] applied SERVQUAL scale to evaluate patient's perception of nine Chinese cities. Scholar [5] proposes an integrative model of the health care consumer satisfaction based on established relationships among service quality, value, patient satisfaction and behavioural intention, and tests it in the context of the South Korean healthcare market. Results showed that service quality emerged as a more important determinant of patient satisfaction than value and both service quality and value have a significant direct impact on behavioural intention while perceived service quality influence value assessment.

Considering the importance of research on the level of satisfaction and attributes of what needs to be improved to meet patient satisfaction, this study aims to measure the quality of service that is performed by using Importance-Performance Analysis (IPA) and measuring customer satisfaction by using Customer Satisfaction Index (CSI). The results will show how the patients' satisfaction level towards services of the hospital as well as it can be utilized

* Corresponding author: vembri@uii.ac.id

as an evaluation in decision-making to improve service quality.

# 2 Theory

## 2.1 Importance-Performance Analysis (IPA)

It was first introduced by [3] with the aim to measure the relationship between the perception of consumers and improving the quality of services also known as the quadrant analysis. Importance-performance analysis has the function of displaying information relating to the factors which, according to customer service significantly affect their satisfaction and loyalty, and the factors that need to be improved service because the conditions do not currently satisfy the client. The importance-performance analysis combines measurement of the level of interest and the level of satisfaction in the two-dimensional graph that facilitates explanation of data and gains practical proposals. Graph interpretation of importance-performance analysis is elementary, where the graph divided into four quadrants based on the measurement results [3].

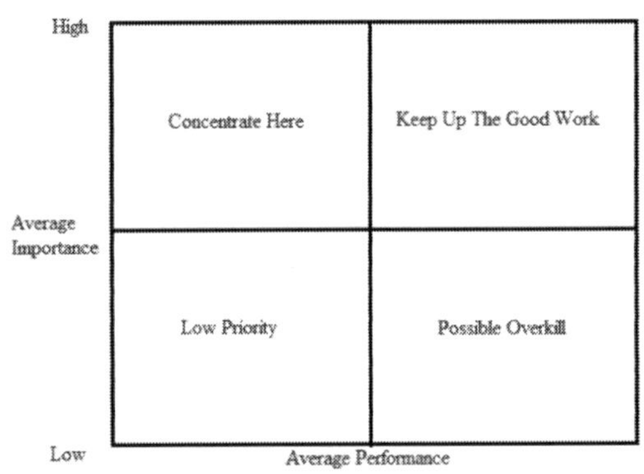


**Fig. 1.** Cartesius diagram [3]

The formula used as follows:

$TK_i = x_i / Y_i \, x100$ (1)

Where:

i : Research attributes

$TK_i$ : Suitability of respondents

$x_i$ : Score of $i^{th}$ performance level

$Y_i$ : Score of $i^{th}$ importance level

$Q = PE$ (2)

Where:

Q : Performance quality

P : Perceived service

E : Expected service

The meaning of the calculation using the formula above is as follows: positive gap means the service is said to be a surprise and very satisfying; zero means the perceived as expected; while the negative value means unsatisfactory.

## 2.2 Customer Satisfaction Index

Customer Satisfaction Index can be calculated by using steps as follow [7]:

1. Calculate weighting factors (WF),
2. Calculate Weighted Score (WS),
3. Calculate Weight Median Total (WMT), and
4. Calculate Customer Satisfaction Index.

CSI percentage value in this study was divided into seven criteria, Table 1 shows the percentage CSI criteria.

**Table 1.** Criteria of CSI

| No | Index (%) | Criteria of CSI |
|---|---|---|
| 1 | X≤64 % | *Very Poor* |
| 2 | 64%<X≤71% | *Poor* |
| 3 | 71%<X≤77% | *Cause for Concern* |
| 4 | 77%<X80% | *Borderline* |
| 5 | 80%<X≤84% | *Good* |
| 6 | 84%<X≤87% | *Very Good* |
| 7 | 87%<X | *Excellent* |

Source : *Customer Satisfaction Measurement, Satisfaction Index*

# 3 Methodology

This study conducted at the Hospital X, Sleman, Yogyakarta, Indonesia. The object of research in this study was patients as users of hospital services. Primary data was obtained from the results of questionnaires which are important services and the level of service performance and direct interview to patients. While secondary data acquired from journals. Accident sampling method especially simple random sampling technique was used in this study, so that each element of the population has an equal opportunity to be selected into the sample. To determine how large a sample as a representative of the population to be used, the researchers used the following formula:

$$n = \frac{Z^2.p(1-p)}{E^2}$$

$$f(p) = p - p^2$$

$$\frac{df(p)}{d(p)} = 1 - 2p$$

$$\frac{df(p)}{d(p)} \text{ maximum, if } \frac{df(p)}{d(p)} = 0$$

$$0 = 1 - 2p$$

$$-1 = -2p$$

$$p = 0{,}5$$

with confidence level 90% and degree of accuracy ($\alpha$) 10% = 0.1; $\alpha / 2 = 0.05$; $Z_{\alpha/2} = 1.645$

$$n = \frac{Z^2.p(1-p)}{E^2}$$

$$n = \frac{1{,}645.0{,}5(1-0{,}5)}{0{,}1^2}$$

$$n = 67{,}65 \approx 68 \text{ sample}$$

Therefore, 70 data from respondents was sufficient (n' (70) > n (68)). The questionnaire in this study is based on five dimensions of service quality as shown in Table 2.

**Table 2.** Dimensions and attributes questionnaire

| Dimension | Attribute |
|---|---|
| *Tangible* | The condition of the hospital is clean, comfortable and tidy |
| | Board instructions and information poly easily visible and readable |
| | Examination room complete, neat, clean, comfortable and hygienic |
| | Toilets are clean and comfortable |
| | Waiting room administration is clean and comfortable |
| *Reliability* | The hospital staffs are tidy and polite |
| | The process of administrative services patients is quick and easy |
| | Ease and accuracy of the information for patients (referrals, lab results, etc.) |
| | Officer pharmacy tells the rules of taking the medication in a clear and complete |
| *Responsiveness* | Medical personnel have the proper competence to diagnose and provide medical solutions needed |
| | The hospital staff quickly in responding to the needs of patients |
| | The hospital staff showed enthusiasm in working |
| *Assurance* | Alertness officers in managing patients |
| | No discriminacy to deliver services |
| | Confidentiality of patient data is well preserved |
| | Medical staff have the competence and the appropriate certificate to the areas cultivated |
| *Empathy* | Prescription drugs appropriate and safe |
| | Attention doctor in managing patients and willingness to give a specific time for the patient to consult |
| | Hospital staff patience in providing services |

In this study, the calculation of reliability test using SPSS 16.0 software can be seen by the value of Cronbach's alpha ranging from 0.824 to 0.920 > 0.6 which means the questionnaire is reliable. While the validity test was calculated from the value of $r_{calculation}$ on each of the questions with a view looking output table Correct Item - Total Correlation in which every question has greater value than $r_{table}$ (X > 0.2352) means valid.

# 4 Results and Discussion

## 4.1 Customer Satisfaction Index (CSI)

Value of Customer Satisfaction Index (CSI) is an overall customer satisfaction with the services provided by the hospital. Based on calculations that have been done, CSI values obtained for services provided by the hospital is 76%. This means that the attributes of services provided overall still not satisfied the customers and it is necessary to improve and maintain the service performance.

## 4.2 Importance Performance Analysis (IPA)

After distributing questionnaires and calculating the data, then Cartesian diagram is built by using SPSS 16.

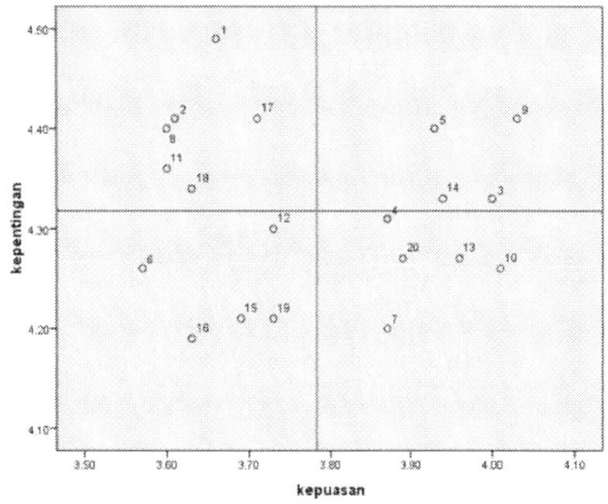


**Fig. 2.** Cartesian diagram

Quadrant 1. Concentrate Here (high importance & low performance). Factors that lies in this quadrant is regarded as a very important factor to consumers, but the conditions at this time have not been satisfactory, so that management is obliged to allocate qualified resources to improve the performance of these factors. These factors include:

**Table 3.** Attributes in quadrant I

| No. | Attribute | Gap Score | Priority |
|---|---|---|---|
| 1 | The condition of the hospital is clean, comfortable and tidy | -0,83 | 1 |
| 2 | Board instructions and information poly easily visible and readable | -0,80 | 2 |
| 8 | Ease and accuracy of the information for patients (referrals, lab results, etc.) | -0,80 | 3 |
| 11 | The hospital staff quickly in responding to the needs of patients | -0,76 | 4 |
| 18 | Attention doctor in managing patients and willingness to give a specific time for the patient to consult | -0,71 | 5 |
| 17 | Prescription drugs appropriate and safe | -0,70 | 6 |

From the results obtained in the first quadrant, then these attributes need to be improved. The improvements that can be made:

a. Monitoring and evaluating the staff to improve hospital cleanliness as well as checking of existing facilities on regularly.
b. Use color gradation for more clear information as well as visual comfort.
c. The hospital should provide the results and detailed explanation to the patient for lab results, referral, etc.
d. Training to improve the competence of officers in managing patients, so that patients will be more comfortable.
e. Doctors give special attention to each patient and give the motivation to help cure patients faster.

f. Standard Operational Procedure should be socialized as well as implemented effectively and efficiently.
g. Pharmacist gives clearer explanation towards drugs given to patients such as the quality of the drug, the dosage, and also the side effects.

Quadrant 2. Keep up the good work (high importance & high performance). The four attributes lie here called by additional factors in which management is obliged to ensure that the performance of the institution under its management can sustain the achievements.

**Table 2**. Attributes in Quadrant 2

| No. | Attributes |
|---|---|
| 3 | Examination room complete, neat, clean, comfortable and hygienic |
| 5 | Waiting room administration is clean and comfortable |
| 9 | Officer pharmacy tells the rules of taking the medication in a clear and complete |
| 14 | No discriminacy to deliver services |

Quadrant 3. Low priority (low importance & low performance). Factors that lies in this quadrant have a low level of satisfaction and once considered too important to the consumer so that the management does not need to prioritize or pay attention to these factors.

**Table 3**. Attributes in Quadrant 3

| No. | Attributes |
|---|---|
| 6 | The hospital staffs are tidy and polite |
| 12 | The hospital staff showed enthusiasm in working |
| 15 | Confidentiality of patient data is well preserved |
| 16 | Medical staff have the competence and the appropriate certificate to the areas cultivated |
| 19 | Hospital staff patience in providing services |

Quadrant 4. Possible overkill (low importance & high performance). Factors which is located in this quadrant are considered less important so that it is better to allocate resources to other higher priority which needs improvement.

**Table 4.** Attributes in Quadrant 4

| No. | Attributes |
|---|---|
| 4 | Toilets are clean and comfortable |
| 7 | The process of administrative services patients is quick and easy |
| 10 | Medical personnel have the proper competence to diagnose and provide medical solutions needed |
| 13 | Alertness officers in managing patients |
| 20 | Easy to contact management if there are complaints and its follow-up |

Based on the research that has been done, it can be seen that the attributes that need to be improved at each hospital is different. A study conducted by [2] on RUMKITDIK, there are seven attributes that a top priority that needs to be fixed whereas in this study there were six attributes that need to be fixed. However, the similar attributes that need to be improved, namely: a) the results of the examination, b) service to patients, c) the attention of doctors towards patient.

## 5 Conclusions and Suggestions

### 5.1 Conclusions

Based on the results of research and discussion that has been done, it can be summarized that the quality of existing services at the Hospital X is not satisfactory for the patient due to the presence of attributes that need to be improved and therefore required an increase in service which is essential to meet patient satisfaction. From this research study conducted by several other sources are similar attributes that need to be improved, namely: a) the results of the examination, b) service to patients, c) the patient's towards doctor's attention.

CSI value obtained in this research is 76% means the service is not satisfied the patient yet, so that it needs to be improved. According to IPA method, improvements to meet patients' satisfaction should be prioritized in several aspects. There are : a) The condition of the Hospital clean, comfortable and tidy; ) Guidance and information boards poly easy to see and read; c) The ease and accuracy of obtaining information for patients (referrals, lab results, etc.); d) The hospital staff have a quick response to the needs of the patient; e) Attention physicians in managing patients and willingness to provide a special time for patient consultation. f) The prescribed medication is suitable and safe.

### 5.2 Suggestions

Based on the above conclusions, the authors can provide recommendations as follows:

a. Monitoring and evaluating the staff to improve hospital cleanliness as well as checking of existing facilities on regularly.
b. Use color gradation for more clear information as well as visual comfort.
c. The hospital should provide the results and detailed explanation to the patient for lab results, referral, etc.
d. Training to improve the competence of officers in managing patients, so that patients will be more comfortable.
e. Doctors give special attention to each patient and give the motivation to help cure patients faster.
f. Standard Operational Procedure should be socialized as well as implemented effectively and efficiently.
g. Pharmacist gives clearer explanation towards drugs given to patients such as the quality of the drug, the dosage, and also the side effects.

## References

1. A. Parasuraman, V. A. Zeithaml, and L. L. Berry, "Servqual: A Multiple-Item Scale For Measuring Consumer Perc," *J. Retail.*, **vol. 64**, no. 1, p. 12 (1988).
2. Istiani Maulina, 2014. "*Analisis Tingkat Kepuasan Pasien Umum Rawat Inap Terhadap Kualitas Pelayanan RUMKITDIK Kramat Jati, Jakarta Timur* (Unpublished thesis). Bogor : Institut Pertanian Bogor. Indonesia. (2014).

3. J. C. Martilla, J.A.; James, "Importance-Performance Analysis," *J. Mark.*, **vol. 41**, no. 1, pp. 77–79, 1977.
4. J. F. Cohen, E. Coleman, and M. J. Kangethe, "An importance-performance analysis of hospital information system attributes: A nurses' perspective," *Int. J. Med. Inform.*, **vol. 86**, pp. 82–90 (2016).
5. K. S. Choi, W. H. Cho, S. Lee, H. Lee, and C. Kim, "The relationships among quality, value, satisfaction and behavioral intention in health care provider choice: A South Korean study," *J. Bus. Res.*, **vol. 57**, no. 8, pp. 913–921, (2004).
6. M. Li *et al.*, "Evaluating patients' perception of service quality at hospitals in nine Chinese cities by use of the ServQual scale," *Asian Pac. J. Trop. Biomed.*, **vol. 5**, no. 6, pp. 497–504 (2015).
7. N. Hill, J. Brierley, & R. MacDougall. How to measure customer satisfaction (2nd ed.). Burlington, VT, USA: Gower. (2003).
8. R. Mohebifar, H. Hasani, A. Barikani, and S. Rafiei, "Evaluating Service Quality from Patients' Perceptions: Application of Importance–performance Analysis Method," *Osong Public Heal. Res. Perspect.*, **vol. 7**, no. 4, pp. 233–238 (2016).
9. S. K. Dixit, "Integration of importance-performance analysis into the strategy of hospitals," *Int. J. Healthc. Manag.*, **vol. 10**, no. 3, pp. 178–183 (2017).
10. S. S. Andaleeb, "Service quality perceptions and patient satisfaction: a study of hospitals in a developing country," *Soc. Sci. Med.*, **vol. 52**, no. 9, pp. 1359–1370 (2001).

# An application of data mining techniques in designing catalogue for a laundry service

*Annisa Uswatun* Khasanah*, *Deliana Ardhitama* Erlangga, and *Ahmad Mustopa* Jamil

Department of Industrial Engineering, Universitas Islam Indonesia

**Abstract.** Catalogues are the media that companies use to promote their products or services. Since catalogue is one of marketing media, the first essential step before designing product catalogue is determining the market target. Besides, it is also important to put some information that appeal to the target market, such as discount or promos by analysing customer pattern preferences in using services or buying product. This study conduct two data mining technique. The first is clustering analysis to segment customer and the second one is association rule mining to discover an interesting pattern about the services that commonly used by the customer at the same service time. Thus, the results will be used as a recommendation to make an attractive marketing strategy to be put in the service catalogue promo for a laundry in Sleman Yogyakarta. The clustering result showed that the biggest customer segment is university student who come 3 until 5 times in a month on weekends, while the association rule result showed that clothes, shoes, and bed sheet have strong relationship. The catalogue design is presented in the end of the paper.

## 1 Introduction

Marketing is an important activity that must be done by a company in order to maintain the business continuity process. According to Stanton [1], the definition of marketing is an overall system of business activities aimed at planning, pricing, promoting and distributing goods or services that satisfy the needs of both existing and potential buyers. Marketing is a company activity which is directly related to the consumer and the main purpose of this activity is to maximize the company's profits. Marketing can be conducted with several ways and the common one that have been used by lot of companies is using catalogue.

Catalogues are the media that companies use to promote their products or services. Catalogues are used to describe in detail the products or services the company offers to consumers. It usually produces in a given period, in which each period issuing different offers. Offers are usually associated with new products or services, product or service promos, discounts, or special events. In designing a catalogue, there are several things that must be considered by a company. Beside catalogue give the consumer detail information about products or services, such as figure, price, detail specification, contact person, and so on, a catalogue should be also contained some information that appeal to consumers. To create an attractive product catalogue, company should know who the customer is, and what the customer wants. If the company can design a good catalogue, it can encourage the customer to look inside the catalogue, and then entice them to buy the products or use the services offered.

One important thing to be succeed in marketing is knowing the market target [2]. Since catalogue is one of marketing media, the first essential step before designing product catalogue is determining the market target. This step can be conducted through customer segmentation. Segmentation is greatly embedded in a marketing strategy because different customer groups will require a different marketing mix [3]. Customer or market segmentation is the division of the market into different customer groups where specific customer groups can be selected as market targets to be achieved with a specific marketing mix [4]. After knowing the consumer group who became the target market then the marketing strategy, in this case is the catalogue contain and design, can be adjusted.

In designing product or service catalogue, it is very important to put some information that appeal to the target market, such as discount or promos. Usually a company will see the pattern of consumer buying in giving promo or discount. By analysing the consumer buying pattern, or in the service industry is about the services that commonly used by the customer, the company can get valuable information to make the right marketing strategy. Examples of discounts or promos given to consumers are usually a discount on a good or service if it is bought or used in conjunction with another item or service.

The purpose of this study is to give suggestion related with the contain and the design of a service catalogue by conducting customer segmentation and analysing the services that commonly used by the customer at the same service time. This kind of analysis can be obtained using data mining techniques. Data mining is the process of automatically discovering useful information in large data repositories [5]. Data mining has been used widely in

* Corresponding author: annisa.uswatun@uii.ac.id

many different scopes of study and one of them is in marketing. Data mining has several popular techniques such as clustering analysis and association rule mining. There are many methods that can be used in clustering analysis, one of the most popular method is K-means algorithm. In this study clustering analysis, using K-means algorithm, is used to conduct the customer segmentation, and association rule is used to discover an interesting pattern about the services that commonly used by the customer at the same service time, then those results will be used as a recommendation to make an attractive marketing strategy to be put in the catalogue. This study was conducted in a laundry in Sleman, Yogyakarta. The rest of this paper is organized as follows, Section 2 present the literature review, Section 3 presents the research method related to this study, Section 4 shows the results and discussion. The concluding is finally made in Section 5

### 2.1 Clustering Analysis using K-means Algorithm

Clustering analysis is the process of identifying natural groupings or clusters within multidimensional data, based on some similarities, like Euclidean distance [6]. Its main purpose is to group samples with the same statistical characteristics together into same cluster in order to achieve higher similarities within same cluster, also there are more significant differences between different clusters [7]. There are many methods that usually used to deal with clustering problems, however K-means is the most widely used clustering method even though it was proposed over 50 years ago [8].

K-means was firstly proposed by MacQueen in 1967 [9,10]. K-means algorithm is also well known for its efficiency in clustering large data set [11]. K-means algorithm is easy to be implemented and computationally efficient [5,10,12]. The K-means algorithm starts with K cluster centroids, which are initially randomly selected or derived from some a priori information. Each point in data set is then assigned to the closest centroid, and each collection of points assigned to a centroid is a cluster. The centroid of each cluster is then updated based on the point assigned to each cluster. This process is repeated until no point change clusters, or equivalently, until the centroids remain the same [5].

Some scholars have already used K-means algorithm to conduct customer segmentation [13-15]. Hosseini et al. [13] used K-means to classify customer product loyalty under business to business concept in a company in Iran. It used K-means algorithm with K-optimum according to Davies–Bouldin Index. The results show that with this method the company was able to assess customer loyalty with a better marketing strategy compared to most other companies in Iran. Hong & Kim [14] integrated self-organizing map and K-means algorithm to segment customer based on customer's psychographic data for an online store to offer customized marketing. Wei et al. [15] also combined K-means algorithm and self-organizing map (SOM) to segment customer and develop marketing strategy in a hair salon in Taiwan. The results indicated four customer types including loyal customers, potential customers, new customers and lost customers. The marketing strategy for each customer types was also developed

### 2.2 Association Rule Mining

Association rule mining can be illustrated as follows, suppose that the set I=(i1,i2,…,ik) is the total number of items. The item set X and Y are part of the total set I. If Y happen in one transaction when X happens, as association rule can be expressed as [16]. In market basket analysis, association rule can be can be express as "A customer who buys product X1 and X2 will also buy product Y with probability c%" are found [17]. For example, in a restaurant egg and coffee may be ordered together primarily between 7 am and 11 am, and maybe turkey and pumpkins are sold at the same time in the week before Thanksgiving [18]. In service industry, association rule mining can be express as a customer who use a certain service will also use another service at the same service time.

The strength of an association rule can be measured in terms of its support and confidence. Tan et al. [5] explain that support determines how often a rule is applicable to a given data set. A low support rule is uninteresting from a business perspective and sometimes it will be eliminated. While, confidence determines how frequently items in Y appear in transaction that contain X, the higher the confidence the more likely it is for Y to present in transaction that contain X. The formal definitions of these metrics are

$$Support, s(X \Rightarrow Y) = \frac{\sigma(X \cup Y)}{N} \tag{1}$$

$$Confidence, c(X \Rightarrow Y) = \frac{\sigma(X \cup Y)}{\sigma(X)} \tag{2}$$

Some scholars have applied association rule mining. Abdulsalam et al [19] implement association rule mining which is also known as market basket analysis to reveal sales pattern in a supermarket. This study showed rules of purchasing one product would lead to the purchase of another product. This study concluded that apple and chocolate had strong correlation. This information helps the company to create marketing and advertising strategies that outshine the competitors. Verma et al. [20] apply association rule mining to find out the pattern incidents in a steel plant in India. Czibula et al. [21] used association rule to predict defect of a software system in order to conduct continuous improvement. Kumar, & Toshniwal [22] apply data mining techniques to analyse data related with road accident in India. Cluster analysis was used to cluster the accident data, then association rule mining was used to identify the situation when an accident occurred in each cluster. The results can be used to make some preventions for each different categories, thus it can minimize the number of accidents happened.

## 3 Research Method

The research was conducted in laundry "A" in Sleman Yogyakarta. The data consist of 100 data from customer of the laundry and collected using questioner. This research consists of three steps, the first step is clustering, the second step is association rule mining, and the third step is designing the content of service catalogue that contain an interesting information based on the result of clustering and association rule analysis. Clustering is used to segment the customer and association rule mining is used to know the several services of the laundry that commonly used by the customer at the same service time. The laundry that become the object of this research offer laundry service such as clothes, bed sheets, blankets, shoes, dolls, bag, curtain, prayer kit, and so on. The result from clustering and association rule than will be used to create suitable business strategy for a certain customer segment. The business strategy than will be put in the service catalogue of the laundry. Software SPSS was used to apply K-means clustering and Rapid Miner was used to apply association rule mining. The research step for this study is represented in Figure 1.

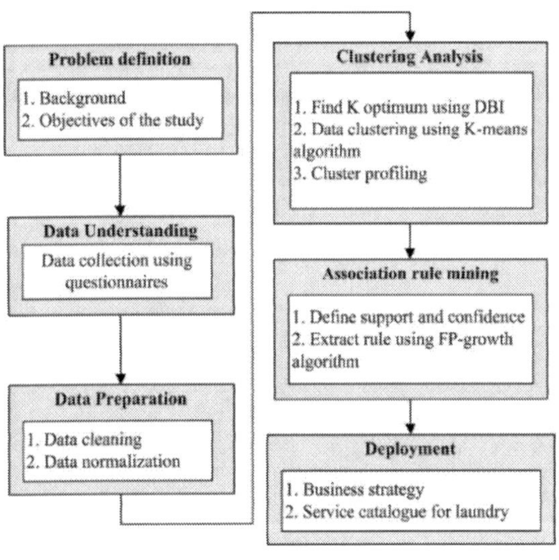


**Fig 1.** Research step

## 4 Result and Discussion

### 4.1 Clustering Analysis

In this step, laundry customer was segmented based on several attributes including occupation, arrival frequency, arrival day, arrival time, and laundry weight. Several number of cluster (K) were implemented and Davies Bouldin Index (DBI) was used to evaluated the performance. Lower DBI index indicates better result. Table 1 shows the DBI value for each K.

Based on those result, K=3 has the smallest value of DBI, it means that the best cluster number for this data is 3. Thus for the next step, the data is divided into 3 clusters. The cluster centroid and the Euclidean distance were calculated by the software. The results showed that cluster 1 consist of 59 cluster members, cluster 2 consist of 38 cluster member, and cluster 3 consist of 3 cluster members. Table 2 gives the detail cluster profile.

**Table 1.** Davies Bouldin Index for K optimum

| Number of cluster (K) | DBI |
|---|---|
| 2 | 1,126 |
| **3** | **1,022** |
| 4 | 1,210 |
| 5 | 1,338 |

All of the member in cluster 1 are university student and dominated with customer with arrival frequency 3 until 5 times per month, come in weekend (Friday, Saturday, and Sunday), and the weight of the laundry is around 4 until more than 5 kg. Gender is not included as the attribute in this analysis but for additional information cluster 1 is dominated with female. Cluster 2 is also dominated with university student who come in weekdays (mostly on Monday, Wednesday, and Thursday) with weight of the laundry less than 3 kg. For cluster 2 there is no dominated arrival frequency. Half of the customer come 3 until 5 times per month, half of them come once until twice per month. It dominated with male. Meanwhile, cluster 3 is dominated with businessman, and civil servant. There is no dominated arrival frequency, day, and laundry weight for this cluster. In all cluster, there is no significant difference or dominated arrival time (morning, afternoon, evening). Thus, arrival time is not important attribute to be considered in this analysis.

The clustering result shows that the biggest customer segment (more than 50%) is university student who mostly come in weekend, with arrival frequency 3 until 5 times per month and weight of the laundry around 4 until 5 kg. This result is an important pattern than can be a useful information to create the business strategy.

### 4.2 Association rule mining

This step was conducted to reveal the customer pattern in using services offered by the laundry. To get the information, the customer was asked related with what kind of services that they used by filling the questioner. The data was transform into binary form, 0 for not using the service and 1 for using the service. Then the binary data was put into Rapid Miner software. Minimum support used was 5% and minimum confidence used was 5%. There were several rules generated, however only 8 rules had lift ratio bigger than one. Lift ratio measure the correlation between item set and it can use to measure the importance of the rule. Greater lift ratio indicates stronger associations, the highest lift which means highest correlation [23,24]. From the 8 rules, there are 2 rules which have high support and confidence value, as follows

bed sheet, shoes $\Rightarrow$ clothes (s:10%, c:91%)

clothes $\Rightarrow$ shoes (s:10%, c:90%)

Table 2. Cluster profile

| Cluster | Occupation | Arrival frequency | Arrival day | Arrival time | Laundry weight |
|---|---|---|---|---|---|
| Cluster 1 (59 member) | University Students (100%) | More than 3 times a month (71%) | Weekend: Friday, Saturday, Sunday (83%) | There is no domination | Minimum 4 kg (55%) |
| Cluster 2 (38 member) | University Students (100%) | less than 3 times a month (58%) | Weekdays: Monday, Wednesday, Thursday (53%) | There is no domination | 1 kg (32%) 2-3 kg (68%) |
| Cluster 3 (3 member) | Businessman and civil servant | There is no domination | There is no domination | There is no domination | |

The first rule means 10% of all transaction contain bed sheet, shoes and clothes and 91% of the transaction have the item bed sheet and shoes also contain clothes. While, the second rule means 10% of the transaction contain clothes and shoes, and 90% of the transaction have item clothes also contain shoes. This result shows that the tree items clothes, shoes, and bed sheet have strong relationship.

### 4.3 Service catalogue design

This step is conducted to design the content of laundry "A" catalogue service that contain an interesting information based on the result of the clustering and association rule analysis. Catalogue should inform the detail information about the service offered and also some interesting information which can attract the customer to use the service. This study is not designing the whole catalogue service, but only the attractive part of the catalogue that contain promos.

From the previous analysis, it can be known some important information. Not all the information is used on designing the contain of the catalogue, but only the most attractive one. Clustering analysis show that the biggest customer segment (more than 50%) is university student who mostly come on weekends with weight of the laundry 4 until 5 kg for each customer. The attractive information here is the weekend; it means that this laundry will be very busy on the weekend. To create an attractive business strategy and to reduce workload on weekend this laundry can give a discount for student who come on weekdays. By analysing the data, it is known that customer rarely come on Tuesday. So, this laundry can give discount for student who come on Tuesday with minimum laundry weight 4 kg.

From the association rule mining, it is known that clothes, shoes, and bed sheet has strong relationship. To create an attractive business strategy, by combining the clustering result, this laundry can make a packet promo for this three items. For example, free washing shoes for customer who have already came to wash the clothes 5 times in a month with minimum weight of the laundry 15 kg or free washing bed sheet for customer who wash clothes minimum 5 kg and one pair of shoes. The illustration of the catalogue service promo is shown in Figure 2.

## 5 Conclusion

From the clustering result it can be concluded that the biggest customer segment is university student who mostly come 3 until 5 times in a month on weekends and with weight of the laundry 4 until 5 kg. From the association rule it can be concluded that clothes, shoes, and bed sheet have strong relationship. These important results then used to create service catalogue promo as illustrated in Figure 2. For further research it is better to conduct preliminary study to determine the attributes in order to get a good clustering result. It is better to have more than 100 respondents so the information about customer preferences in using services or products will be better. This kind of analysis can be developed in other service industry or product and can help the owner of a business to create a suitable business strategy for the target market which can beat the competitors.

## References

1. W. J. Stanton, Prinsip Pemasaran, Erlangga, Jakarta, (2001)
2. J. Chance, 8 Keys to a Strong Marketing Strategy, Online access (July 28 2017), URL: https://www.businessknowhow.com/marketing/blocks.htm,(2017)
3. M. S. Kahreh, M. Tive, A. Babania, M. Hesan, Analyzing the applications of customer lifetime value (CLV) based on benefit segmentation for the banking sector. Procedia-Social and Behavioral Sciences, 109, 590-594, (2014)
4. P. Kotler, Marketing Management: Analysis, Planning, and Control, 4th edition, Prentice-Hall, Englewood Cliffs, N.J, 195, (1980)
5. P.N. Tan, M. Steinbach, V, Kumar, Introduction to Data Mining, Pearson Education, MA, (2006)
6. A.K. Jain, M.N. Murty, P.J. Flynn, Data clustering: a review, ACM Computer Survey, 31 (3), 264-323, (1999)
7. J. Han, M. Kamber, Data Mining: Concepts and Techniques, 2nd edition, Morgan Kaufmann, (2006)
8. A.K. Jain, Data clustering: 50 years beyond K-means. 19th International Conference in Pattern Recognition (ICPR), 31 (8), 651-666, (2010)

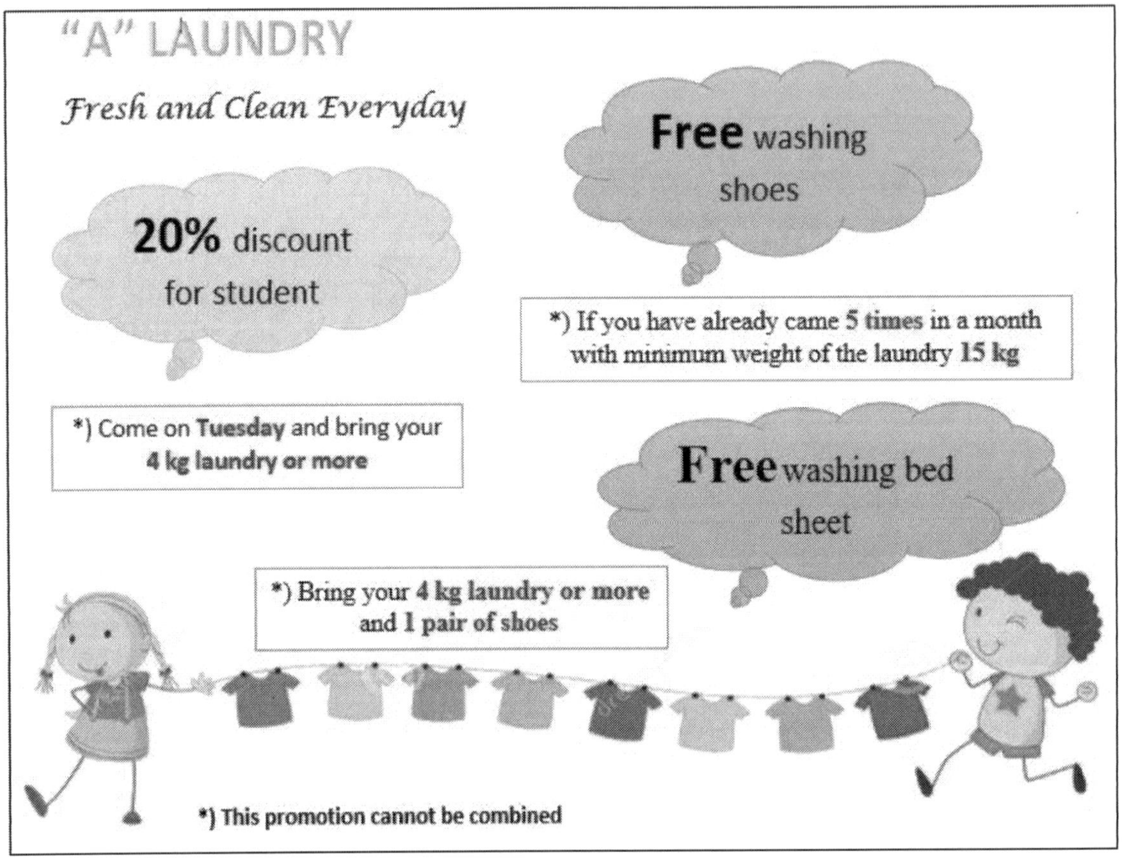


**Fig 2.** Service catalogue promo
(Source of the kids holding the tip of the wire:[25])

9. Z. Cai, W. Gong, C.X. Ling, H. Zhang, A clustering-based differential evolution for global optimization, Applied Soft Computing, 11, 1363-1379, (2010)
10. W. Kwedlo, A clustering method combining differential evolution with the K-means algorithm, Pattern Recognition Letters, 32, 1613-1621, (2011)
11. Z. Huang, Extensions to the k-Means Algorithm for Clustering Large Data Sets with Categorial Values, Data Mining and Knowledge Discovery, 2, 283-304, (1998)
12. S. Das, A. Abraham, A. Konar, Automatic Clustering Using an Improved Differential Evolution Algorithm. IEEE Transaction on System, Man, and Cybernetics, 38 (1), (2008)
13. S. M. S. Hosseini, A. Maleki, M. R. Gholamian, Cluster analysis using data mining approach to develop CRM methodology to assess the customer loyalty, Expert Systems with Applications, 37(7), 5259-5264, (2010)
14. T. Hong, E. Kim, Segmenting customers in online stores based on factors that affect the customer's intention to purchase, Expert Systems with Applications, 39(2), 2127-2131, (2012)
15. J. T. Wei, M. C. Lee, H. K. Chen, H. H. Wu, Customer relationship management in the hairdressing industry: An application of data mining techniques Expert Systems with Applications, 40(18), 7513-7518, (2013)
16. H. K. Han, H. S. Kim, S. Y. Sohn, Sequential association rules for forecasting failure patterns of aircrafts in Korean airforce. Expert systems with applications, 36(2), 1129-1133, (2009)
17. J. Hipp, U. Güntzer, G. Nakhaeizadeh, Algorithms for association rule mining—a general survey and comparison, ACM sigkdd explorations newsletter, 2(1), 58-64, (2000)
18. Y. Li, P. Ning, X. S. Wang, S. Jajodia, Discovering calendar-based temporal association rules, Data & Knowledge Engineering, 44(2), 193-218, (2003)
19. S. O. Abdulsalam, K. S. Adewole, A. G. Akintola, M. A. Hambali, Data Mining in Market Basket Transaction: An Association Rule Mining Approach. International Journal of Applied Information Systems (IJAIS), 7(10), 15-20, (2014)
20. A. Verma, S. D. Khan, J. Maiti, O. B. Krishna, Identifying patterns of safety related incidents in a

steel plant using association rule mining of incident investigation reports, Safety science, 70, 89-98, (2014)

21. G. Czibula, Z. Marian, I. G. Czibula, Software defect prediction using relational association rule mining, Information Sciences, 264, 260-278, (2014)
22. S. Kumar, D. Toshniwal, Analysing road accident data using association rule mining. In IEEE International Conference on Computing, Communication and Security (ICCCS), 1-6, (2015)
23. M. Hahsler, B. Grün, K. Hornik, Introduction to arules–mining association rules and frequent item sets. SIGKDD Explor, 2(4), (2007)
24. A. El-Halees, Mining students data to analyze e-Learning behavior: A Case Study. (2009).
25. Dannyphoto80, Illustration of the kids holding the tips of the wire for the hanging clothes on a white (royalty free), Online access (August 8th 2017), URL: https://www.dreamstime.com/royalty-free-stock-photos-kids-holding-tips-wire- hanging-clothes-illustration-white-background-image32709718, (2017)

# Quantitative analysis on the fluctuation of vegetable price in supermarket

*Suci* Miranda* and *Vembri* Noor Helia

Industrial Engineering Department, Faculty of Industrial Technology, Universitas Islam Indonesia, Indonesia

**Abstract.** In this paper we analyze the variables affecting the determination of vegetable sale price in supermarket and investigate each amount considered into the formulation. We use the supermarket pricing strategies literatures, cost management theory, and any information obtained from suppliers and supermarket to investigate the variables influenced the sale price gap between supplier sale price and supermarket sale price. In order to obtain the supermarket sale price, it is required to understand the concept of costs. Supermarket as a merchandising firm has two components of costs called purchase costs and operating expenses which are computed to be Cost of Goods Sold (COGS). We found that supplier sale price is the purchase costs and plays has the main role in determining the supermarket sale price. The operating expenses considered is the holding cost. The vegetable characteristics and consumer behaviour have driven the contribution of amount of variables into the sale price at supermarket. Finally, we present the variables satisfying the vegetable sale price formulation and how they are calculated becoming the supermarket sale price.

## 1 Introduction

Vegetable has a great economic value for agribusiness sector. As one of consumers' needs, they are not only sold in traditional markets, but also in modern markets such as supermarkets. In fact, most supermarkets have offered similar quality of Fresh Food and Vegetable (FFV) to wetmarkets but at higher prices on avegare 15%-60% above traditional retailers [4]. Competition in the supermarket industry is a complex phenomenon in which firms have to compete across the entire retail and marketing mix, enticing customers with an attractive set of products, competitive prices, convenient locations, and a host of other services, features, and promotional activities [3]. Firms choose a set of services and features that maximize profits, conditional on the types of consumers they expect to serve and their beliefs about the actions of their rivals.

A supermarket's pricing strategy is a main element in order to achieve its profit. Choosing a strategy is quite elaborate that managers must balance the preference of their consumers and their firm's own capabilities againts the expected actions of their rivals [7]. While according to data in Central American, the supermarket share in FFV retail is most four times smaller that in overall retail [4]. The clear advantage of supermarkets over traditional markets are convenience, safety, and cleanliness. Supermarkets must maintain consumer loyalty as well as to maximize profit.

From different point of view, some studies have been conducted on fluctuation in the prices of vegetables and also the supermarket pricing strategies. A detailed description of supply chain of vegetables has revealed the reason for great fluctuation in the prices in China and gives the recommendations for stabilizing the prices of vegetables [8]. A prediction of vegetable sale price has been established using a Markov Chain model in order to solve the problems of heavy price fluctuations and unbalanced reources allocation in the vegetable wholesale market caused by large daily demand and varied supply channels in China [10]. [7] observed the supermarket pricing strategy using a system of simultaneous dicrete choice models on the underlying features of the market, as well as each firm's beliefs regarding its rival's actions. [3] studied the inventory management decisions of retailer selling a single perishable good in a deterministics settings, formulate and analyze an EOQ model with demand decreasing in the age of the product for both fresh produce and packaged goods. [6] addressed the impact of consumer purchasing behaviour on the production planning of perishable food products for companies operating in the fast moving consumer good using direct store delivery. [5] modelled an EOQ for the inventory management of perishables in small medium-sized grocery stores using a regression approach from the product's holding, lifetime, and markdown policy. It recorded that managers frequently utilize markdown to stabilize demand when the products expiration date nears. While in this sale price analysis study, the markdown policy is not applied in supermarket observed. Yet, the lifetime will be considered as one of the variables influencing the sale price determination. It is along with another research that formulates and analyzes an EOQ

* Corresponding author: suci.miranda@uii.ac.id

model with demand decreasing in the age of the product [7].

Yogyakarta is one of special province (DIY) and is the second largest tourist destination in Indonesia with a small expatriate community living here reached 2,500 by the end of 2015 (Immigration, Yogyakarta). Due to the vegetable supply and demand, this has given a new opportunity in agribusiness providing vegetables for residents by establishing green groceries where are close to housing. These stores provides FFV with convenient places and cleanliness even though they are smaller than supermarkets. Competing with supermarkets and traditional markets, the green groceries need a pricing strategy in order to determine a sale price of vegetables. The main business purpose is to gain profit while these groceries are lacking of adequate information how to formulate the sale price [9]. Despite the supermarket pricing strategy, there has not been a representational framework of determining the vegetable sale price for a green grocery. Building on the supermarket pricing strategies literatures, production cost theory, and information obtained from suppliers, we analyze the major factors that drive vegetable pricing formulation in supermarkets particularly in Yogyakarta, and how much each factor contributes into the sale price formulation. We use a supply chain for investigating the sale price gap of every chain and the factors influenced, and cost plus pricing analysis considering the direct and indirect cost for the sale price formulation. Finally, we present the result of the variables and each percentage counted into the formulation.

The rest of the paper is organized as follows. Section 2 explains the reasons of the fluctuation in vegetable sale price in Indonesia. Section 3 describes the analysis of vegetable pricing method in determining the sale price at supermarket. Section 4 concludes and identifies opportunities for future research.

## 2 Analysis of Fluctuation in the Price

The objects of research are non-organic vegetable which have a great price fluctuation: red onion, garlic, red chilli pepper, big red pepper, and cayenne pepper. The high-valueable vegetables such as peppers, onions, and tomatoes are different with other vegetables such as kale, spinach, and green mustard. As the basic ingredients in most of Indonesian recipes such as rendang and curry, the supply and demand play a main role on the fluctuation. There are 93.1% and 80.7% of 462 respondents who consume onion and red pepper almost every day [1]. Total supply in the market will affect suppliers to meet demand from supermarkets particularly when the price of chili, onions, and tomatoes fluctuate sharply, it will influence the price offered by the supplier to supermarkets, then the higher sale price continues from supermarkets to the end consumer. These are the following reasons driving the fluctuation in Indonesia. First. The safety stock mechanism has triggered the increase if the supply are not sufficient for the demand. For instance, due to a shortage supply of onions, the rest onions will be imported for stabilizing the price. Meanwhile, the local onions are sold at a higher price instead. This condition can be caused by the high production costs applied (Bappenas, 2013). Second. Indonesia has 2 seasons, rainy season and dry season. Onions and pepper can not grow well during the rainy season which the production will decrease significantly. It is obvious that the price will be more expensive because the supply is lower than the demand. Third. The supply chain of vegetables. The vegetable industry is an industry with a strong concept of supply chain. The supply chain nodes can be analyzed one by one to subdivide the prices of vegetables into all nodes of supply chain of vegetables if the supply chain can be potrayed in detail in order to conduct effective analysis of great fluctuation in the prices [8].

This paper selects the main data on the prices of vegetables from a national supermarket with 8 branches in Yogyakarta and a main supplier of this supermarket. The sale price within 6 months observation are shown in Figure 1.

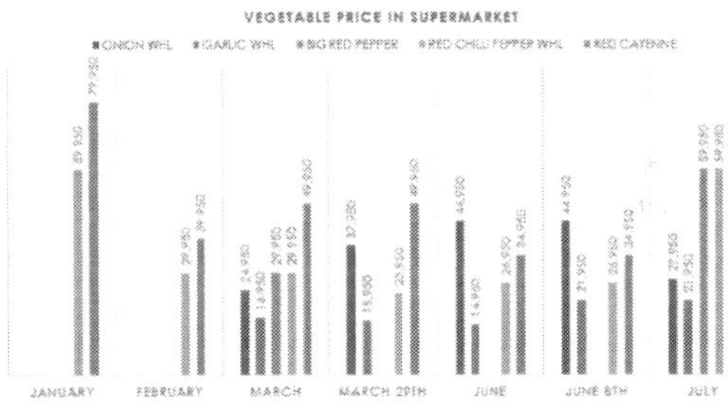


**Fig 1**. The fluctuation price of vegetables in supermarket (Rp)
Source: Data compiled from supplier and supermarket

As can be seen from Table 1, the sale price has risen almost every month especially for red cayenne and onions. Indonesia has lots of spicy food, so red cayenne and red chilli pepper are preferred into the main course, not only for home cooking, but also at restaurant like Padang restaurant. While garlic does not increace sharply because it is used little than onions in every recipe. Sambal or pepper sauce comprises 8 onions and 2 or 3 garlics, for example. Compared to the sale price from supplier to supermarket shown in Table 1, there is a gap sale price ($\Delta p$) that will be identified in further section.

**Table 1**. Supplier and supermarket sale price

| Item | Supplier (Rp) | Supermarket (Rp) | $\Delta p$ (Rp) |
|---|---|---|---|
| Garlic WHL | 15,500 | 16,950 | 1,450 |
| Onion WHL | 34,000 | 37,950 | 3,950 |
| Big red pepper WHL | 20,000 | 23,950 | 3,950 |
| Red chilli pepper WHL | 30,000 | 49,950 | 19,950 |
| Red cayenne WHL | 14,500 | 29,950 | 15,450 |

Source: Data compiled from supplier and supermarket

Table 1 shows the different sale price from supplier to supermarket and from supermarket to end consumer displayed at the shelves. The highest gap is red chilli pepper reaching up to Rp. 20,000 followed by red cayenne with Rp. 15,450 gap price which is 100% higher from the supplier sale price. This fluctuation price of vegetables is frequent in Indonesia, so we must face the abnormal price phenomenon.

## 3 Analysis of Vegetable Pricing Method

These vegetables observed are part of non-leafy vegetables with the aging group (a) is a+1 (they can last for more than one day) and their shelf life (t) is t+1 [9]. The non-leafy vegetables means the leaves are not consumed as the main course but their root or stalk. They will be placed at the shelf for more than one day (t+1) as their freshness can last more than one day (a+1) started from the day they are purchased. All of them are sold in kilogram without packaging and have a random lifetime (there is no expired date).

The supermarket is supported by more than 15 suppliers for FFV that every supplier has to follow the supermarket policy related to the vegetables demanded such as the packaging, the weight, barcode, and the size. When the vegetables are purchased a day before to each supplier, they are ready to be put into the shelves once they arrive at supermarket next morning, as the supply chain can be seen in Figure 2. Supermarket provides checkers as a quality controller of vegetable supply. The responsbilities of a checker are to check all vegetable from suppliers every morning, and to ensure the number and the physical condition of vegetables in order to meet the quality standards of supermarket. The checker also has to check the freshness and neatness of vegetables at their shelves regularly.

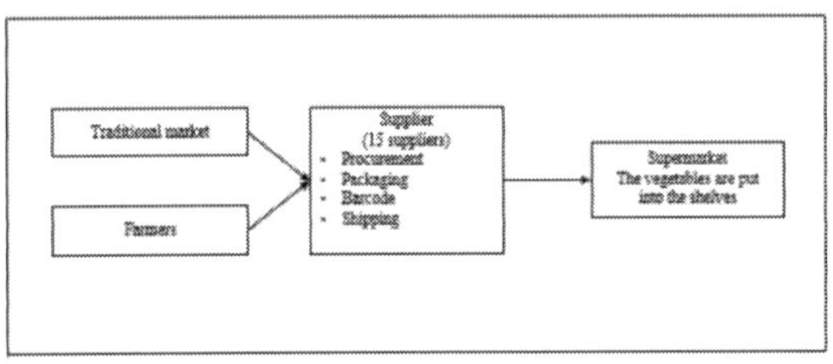


**Fig. 2** Vegetable supply chain

Before calculating the sale price of vegetable called the cost of goods sold (COGS), we need to define what kind of firm it is. There are three kinds of firms: manufacturing, merchandising, and service firm. Since supermarket does not manufacture inventory of vegetables, hence it is a merchandising firm who purchases inventory. It means that merchandising firm sells products but not make them. According to [2], the process is a merchandising firm is simpler than manufacturing firm which is purchases merhcandise and places it in the product inventory account. Once it is sold, it is transferred to the cost of goods sold (COGS) account.

All other costs for managing the firm and selling the product are not product costs such as advertising costs, data processing costs, facilities costs, and executive and staff salaries. The non product costs are called period costs. In merchandising firm, period costs are also sometimes referred to operating expenses or selling and administrative expenses. It has been explained before that these vegetables have the aging group (a) stated a+1 (they can last for more than one day) and their shelf life (t) defined t+1. They are purchased every day to suppliers, even though some of them are left in the shelves. Due to the stock on the shelf ( $t+1$) treated as inventory, so holding cost is considered as an operating expenses. The other operating expenses such as marketing cost, and general and administration such as the checker salary must be included in the calculation of the COGS. Unfortunately, we only know quite a bit of the information needed due to the confidential issues. The COGS formula is a way of determining the costs attributable to the products sold, after determinig how much of the inventory stocks are still on hand. The cost of goods sold (COGS) will be calculated as follows:

$$\text{BI} + \text{Purchase} - \text{EI} = \text{COGS} \quad (1)$$

Where:
BI = Beginning inventory
EI = Ending inventory
COGS = Cost of goods sold

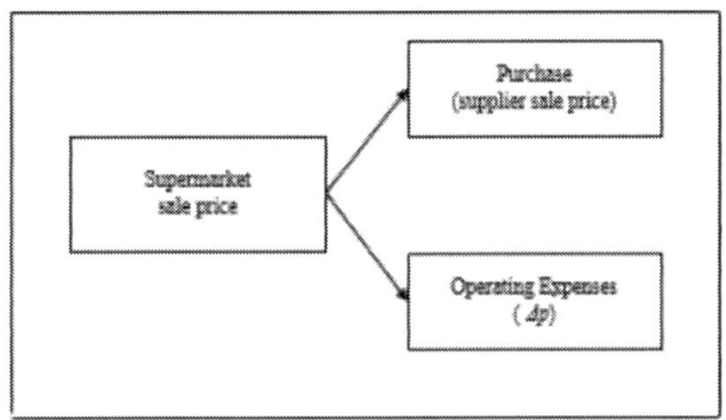


**Fig. 3** Conceptual nodel

Figure 3 explains the steps of computing supermarket sale price. According to the equation (1) above, purchase costs of vegetables is directly gained from the supplier sale price to supermarket. The price gap ($\Delta p$) called operating expenses will be broken down into variables considered and the amount of each. This cost will clearly be explained in Table 2 below.

The holding cost for perishable products ranges between 15% to 35% [3]. It is set fixed in every period $t + 1$ for each vegetable on the shelf [9]. For the holding cost of garlic, onion, and red chilli pepper, they are between 3% - 5% which is lower than the recommended holding cost around 15% - 35% [5]. While cayenne pepper and big red pepper are higher at 35% [3]. The difference treatment can be caused by the characteristics of red chili pepper, onion, and garlic which play a central role in Indonesian cuisine. Cayenne pepper is the most spicy among the peppers. Having a small bite of cayenne pepper will give a stronger taste instead. Meanwhile, the red chilli pepper is more spicy than the big red pepper. For red chilli pepper, it can be cooked to stir with thinly sliced so it feels more spicy. As for big red pepper is more often used to make a smoothed cooking spice. It can be concluded that red chilli pepper is consumed regularly compared to cayenne pepper and big red pepper. With a different number of demand, the supply of onion, garlic, and red chilli pepper into the shelf will be more frequent. Therefore, the red chilli pepper and big red pepper require a greater holding cost. The processing results are shown in Table 2.

**Table 2**. The cost of goods sold of vegetables in supermarket

| Item | Purchase (Rp) | Supermarket Sale Price (Rp) | $\Delta h$ (Rp) | Holding cost | (Rp) | (Rp) | GrossMargin (%) |
|---|---|---|---|---|---|---|---|
| | 1 | 2 | 3 | 4 | 5 = (1*4) | 6 = (3-5) | 7=(6/2) |
| GARLIC WHL | 15,500 | 16,950 | 1,450 | 3% | 465 | 985 | 6% |
| ONION WHL | 34,000 | 37,950 | 3,950 | 3% | 1,020 | 2,930 | 8% |
| RED CHILLI PEPPER | 20,000 | 23,950 | 3,950 | 5% | 1,000 | 2,950 | 12% |
| RED CAYENNE WHL | 30,000 | 49,950 | 19,950 | 35% | 10,500 | 9,450 | 19% |
| BIG RED PEPPER WHL | 14,500 | 29,950 | 15,450 | 30% | 5,075 | 10,375 | 35% |

| Item | GrossMargin (Rp) | Sale Price Predicted (Rp) | Price Gap (Rp) |
|---|---|---|---|
| | 8 = (7*2) | 9 | 10 = (9 - 2) |
| GARLIC WHL | 1,017 | 16,982 | 32 |
| ONION WHL | 3,036 | 38,056 | 106 |
| CURLY RED PEPPER WHL | 2,874 | 23,874 | 76 |
| RED CAYENNE WHL | 9,490.5 | 49,990.5 | 40.5 |
| RED PEPPER WHL | 10,482.5 | 30,0.57.5 | 107.5 |

The contribution margin rate (CMR) is suggested at 10% for retailers [5], but based on the calculation, each item is applied different margin rate. Red pepper is the highest margin rate which is equal to 35% of the total revenue. It may be caused by the utility of it in the main course recipes. Consumers prefer red cayenne and red chilli pepper that are more spicy than the big red pepper, thus it also affect its demand as well as the sales at supermarket. The margin can be set higher in order to gain a higher profit, so the sale price to the end consumer will be more expensive than the supplier sale price to supermarket. In calculating the merchandising firm gross margin, it is needed the sale data substract the cost of goods sold. However, the sales data can not be obtained from supermarket, so it is defined by using coloumn 6 divide by coloumn 2 in Table 3. As can be seen in Table 3, when the gap sale price between supplier and supermarket higher, the gross margin increases. For example, garlic's gap sale price is Rp. 1,450 or 9% higher than the supplier sale price, and gross margin is 6%. Whereas big red pepper is Rp. 15.450 or it is around 100% more expensive than the supplier sale price with gross margin of 35%.

Using this traditional pricing approach in merchandising firm, the predicted sale price is close to the actual sale price at supermarket showed in coloumn 9 in Table 3 above. Finally, the supermarket sale price consist of purchase cost, the holding cost, and gross margin.

## 4 Conclusions and Directions for Future Research

These vegetables have a lifespan of more than a day (a + 1) and the shelf life ($t$ + 1) with the random life (expiration date is not listed), and sold in kilogram. Since all vegetables are supplied by suppliers, so the supplier sale price to supermarket is treated as the purchase cost. It plays an important role in the formulatin of sale price at supermarket. The price gap ($\Delta h$) or operating expenses between supplier and supermarket sale price comprises selling expenses and administrative expenses. Price is a confidential issue so some informations of Beginning inventory, Ending inventory, sales, and other operating expenses are assumed 0. There is only one operating cost which can be identified, the holding cost, while the others expenses are set 0. For further research, the data needed could be obtained from more than one supplier and supermarket so it will provide a robust data for analyzing the variables. The sale price of vegetables from green groceries may also be added in order to gain the gap sale price between supermarkets and the green groceries. The qualitative analysis is important in identifying a characteristic of vegetables which may be considered in determining the sale price as well as the consumer behaviour in purchasing these vegetables. Finally, by gaining adequate informations related to factors influencing the determination of vegetable sale price both quantitatively and qualitatively, a better guide line of business growth can be provided for those green groceries.

## References

1. A. Witono (2011). *Faktor-faktor yang mempengaruhi perilaku dan keputusan konsumen untuk membeli kentang, bawang merah, dan cabai merah*. J.Hort. **21**, 3 (2011)
2. E. J. Blocher, D. E. Stout, and G. Cokins, *Cost Management*, fifth edition, *McGraw-Hill Irwin* (2010)
3. G. Dobson, E. J. Pinker, and O. Yildiz O, *An EOQ model for perishable goods with age-dependent demand rate*, Eur. J. of Operation Research., 1-5 (2016)
4. J. A. Berdeguè, J. A., F. Balsevich, L. Flores, and T. Reardon, *Central American Supermarkets' Private Standartd of Quality and Safety in Procurement of Fresh Fruits and Vegetables*. Food Policy, **30**, 254-269 (2005)
5. M. Ferguson, V. Jayaraman, and G. C. Souza, *Note: an application of the EOQ model with nonlinear holding cost to inventory management of perishables*. Eur J. of Operation Research, **18**, 485-490 (2007)
6. P. Amorim, A. M. Costa, and B. Almada-Lobo, *Influence of Consumer Purchasing Behavior on The Production Planning of Perishable Food*, OR Spectrum, (2013)
7. P. B. Ellickson, and S. Misra, *Supermarket Pricing Strategies*, Marketing Sciences, **27(5)**, 811-828 (2008)
8. S. Chen, and L. Hu, *Why Are There Great fluctuations in the Prices of Vegetables ?*, Asian Agri. Research, **5(3)**, 1-6 (2013)
9. S. Miranda., (2017). *A Qualitative and Quantitative Analysis of Vegetable Pricing in Supermarket*. Retrieved from http://www.researchgate.net/(2017)
10. X. Zhu and X. Xu, *Markov Chain Analysis and Prediction on The Fluctuation Cycle of Vegetable Price*, J. of Syst and Mngmnt Sciences, **2(3)**, 40-49 (2012)

# Modified technology acceptance model for hospital information system evaluation – a case study

*Vembri Noor* Helia*, *Vikha Indira* Asri, *Elisa* Kusrini, and *Suci* Miranda

Department of Industrial Engineering, Universitas Islam Indonesia, Jl.Kaliurang km 14.5, Yogyakarta, Indonesia

**Abstract.** Hospital Information Systems (HIS) has been widely used in Indonesia. This implementation in hospital needs comprehensive, integrated information system designed to manage all the aspects of a hospital's operation. One of the factors that currently plays an important role in the successful HIS's application is the user factor. The hospital can evaluate HIS from user factor, so that it becoming better, perfect and can support the objectives, vision and mission of the hospital. This study aims to use the Modified Technology Acceptance Model created in Structural Equation Modeling (SEM) and tested using the software Amos 20 to determine the variables that most influence in HIS's application using in several hospitals in Sleman, Daerah Istimewa Yogyakarta Province, Indonesia. Data collected by distributing questionnaires to employees of D grade hospitals in Sleman district. The results from this study are seven variables that may influence the user in using HIS (Subjective Norm, Perceived Usefulness, Perceived Ease of Use, User Satisfaction, Behavior Intention to Use, Attitude Toward Using, Actual System Usage) and significant or insignificant relationship among them.

## 1 Introduction

Hospital Information Systems (HIS) has been widely used in Indonesia. Having HIS is a must for an accredited hospital. In the DIY province, 82.21% of all hospitals have adopted HIS [1]. The hospital information system has an important role in clinical and administrative services. There is evidence that there are use of information technology and individual performance at a hospital in Yogyakarta [2]. In addition, information technology is able to save filtering and processing the data into information with the higher capacity than manual methods.

Rapid technological development, making the information system as one determinant in the success of an organization's profit and non-profit. The information system has a role in supporting the operations, and supporting management in making decisions. HIS evaluation needs to be done because it can quantify the benefits from the use of such information systems [3]. The evaluation results can be used as a benchmark to improve or enhance HIS. So that, it becoming better, perfect and can support the objectives, vision and mission of the organization [4].

HIS evaluation can be done in various ways. HIS performance evaluation measurement model by looking at the relationship between the human-technology-organization through HOT-Fit (Human-Technology-Organization-Net Benefit) [3, 4]. Before going into the organization, it needs evaluation from user acceptance. One way is using the Technology Acceptance Model (TAM) to determine the acceptance of hospital admissions, such as in Bangkinang hospital [5], TAM with modifications to the doctor's private clinic [6], the reception HIS in hospital in Iran with seven variable [7]. TAM can also applicable to find the adoption of software measure with two variable, which are perceived usefulness and perceived ease of use [8].

Although many models have been proposed to explain and predict the use of the system, the Technology Acceptance Model has become the most public attention Information Systems method [9]. Therefore, in this study using the Modified Technology Acceptance Model to determine the variables that most influence in HIS that is suitable for use in several hospitals in Sleman, Daerah Istimewa Yogyakarta Province, Indonesia. It is more relevant and widely used in the literature, especially for the application of information technology in hospitals [10].

## 2 Literature Review

### 2.1 Hospital Information System

Hospital Management Information System is a system of information and communications technology that processes and integrates the entire workflow process Hospital services in the form of network coordination, reporting, and administrative procedures to obtain information appropriately and accurately and is part of the Health Information System [11]. This implementation

* Corresponding author: vembri@uii.ac.id

needs comprehensive, integrated information system designed to manage all the aspects of a hospital's operation.

### 2.2 Technology Acceptance Model (TAM)

TAM was developed from a psychological theory that describes the behavior of computer users, which is based on belief, attitude, intention, and the relationship of user behavior (user behavior relationship).The purpose of this model is to explain the factors of user behavior on the acceptance of the use of technology.

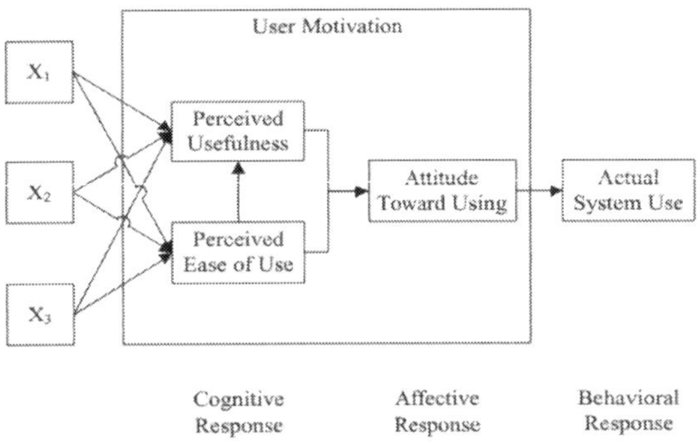


**Fig 1.** Technology Acceptance Model 1 [9]

Technology Acceptance Model is shown in Figure 1, with arrows represent causal relationships. Alternative systems are represented using a set of binary "design features" variable. According to the model, the overall attitude of potential users against using the system that is thought to be important determinants of whether he actually used it [9,12].

TAM has many limitations in explaining the reasons for which a person can see a useful system, so it is necessary to add a few variables from previous studies [12]. The first modification is done by the Technology Acceptance Model additional external factors such as shown in Figure 2.

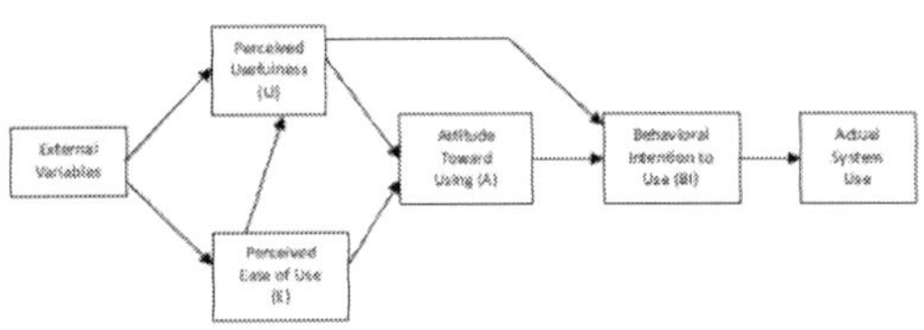


**Fig 2.** Technology Acceptance Model 2

There are studies which refer to the user and is used in hospitals as practiced in figure 3. Tabibi using a modified TAM with seven variables. The variables are *Perceived Ease of Use (PEU), Perceived Usefulness (PU), Subjective Norms (SN), User Satisfaction (SAT), User Behavioral Intention (BI), Mandatoriness of system use (Mand)* and *System Use* (USE).

## 3 Methodology

The research is to prove the hypothesis that built up with the approach of *Modified Technology Acceptance Model (TAM)* created in Structural Equation Modeling (SEM) and tested using the software Amos 20. Method of *Modified* TAM is able to analyze the factors that affect the user in using HIS at hospitals in Sleman. Data collected by distributing questionnaires to employees of D class hospital in Sleman district. The sample in this study was limited only to the medical nursing employee (outpatient, inpatient), medical non nursing (part laboratory, radiology, and pharmacy), and non-medical (part of medical records / registration, finance and accounting). Minimum number of samples in the study was 200 respondents.

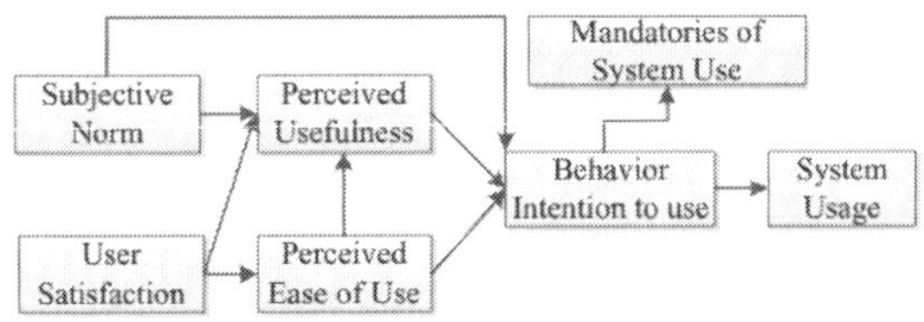


**Fig 3. Modified** Technology Acceptance Model for Hospital [7]

Questionnaires were administered to respondents referred to seven latent variables/constructs with 23 indicators. Variables Perceived Ease of Use (PEOU) has four indicators, which is easy to learn, easy to operate HIS, clear guidance on the use of HIS, supporting facilities provided by the hospital is adequate. Then, the variable Perceived Usefulness (PU) is to facilitate the task, the control for the work, the convenience to get information, and improve discipline. Subjective Norm (SN) variable has three indicator that influence the behavior in the use of systems and parties are important in the use of the system. Attitude Toward Using (ATU) variable has indicators, such as in the form of his attitude towards the implementation, in the spirit of the task, the variable of User Satisfaction (SAT) has covering his satisfaction, the work load becomes lighter, and improved performance in the task. Behavioral Intention (BI) has indicators, such as motivation to keep using, plan to keep using the future, motivating others to use, the motivation to provide inputs for use. Variable Actual Usage (AU) uses his intensity indicator use in completing the task, the accuracy of the information in the HIS, the reference of the information in the HIS, and the availability of a complete application in HIS.

## 4 Result and Discussion

### 4.1 The Development of the Theory-Based Model

In the study Tabibi et al. [7], Mandatories of system use (Mand) variable has no significance influence with other factors, so that this variable was not used. Referring to the previous studies, the factors that affect the TAM in the information system of hospital management to pay more attention to users. The variables are Subjective Norm,

User Satisfaction, Perceived Ease of Use, Perceived Usefulness, Behavioral Intention, and the Actual System Usage) [5,7,9,12 - 14].

From the scheme, it can be formulated hypothesis as follows:

- Hypothesis 1a: Subjective Norm affect Perceived Usefulness
- Hypothesis 1.b: Subjective Norm influence behavior Intention to Use
- Hypothesis 2.a.: User Satisfaction affect Perceived Ease of Use
- Hypothesis 2.b.: User Satisfaction affect Perceived Usefulness
- Hypothesis 3: Perceived Ease of Use affect Perceived Usefulness
- Hypothesis 4.a.: Perceived Usefulness affect Attitude Toward Using
- Hypothesis 4.b.: Perceived Ease of Use affect Attitude Toward Using
- Hypothesis 5 : UsingAttitude Toward affect Behavior Intention to Use
- Hypothesis 6 :Behavior Intention to Use affect Actual System Usage

Subjective Norm (SN) is a person's perception or assumptions about other people's expectations of certain behaviors that will do or will not do. Subjective norms also influenced by beliefs [15]. If an individual has a strong belief that the reference is important for them to do or not to do.

Perceived Ease of Use (PEOU) is something where the ease, freedom in the use of a system, in this case the computer. Ease can be seen from the intensity of technology use between user and the system. So it could be stated that the system is more often used to show the system is better known, is easier to operate and easier to use by users. Perception of usefulness (PU) is defined as a measure of where the use of a trustworthy technology will bring benefits to those who use it. Attitude Toward Using the TAM conceptualized as an attitude towards the use of the system in the form of acceptance or rejection as an impact when someone is using a technology called a jug in her job or user attitude towards the use of technology.

Behavioral intention refers to the behavior of individuals that have the intention or desire to do so. Behavioral Intention to Use (ITU) is the behavioral tendency to continue to use a technology. The interest can be seen from the use of a computer technology to the user so that attitudes, interests predictable.

Actual System Usage is the real conditions of use of the system. Such measurements as the amount of time spent using these technologies. A person will be satisfied to use the system if it believes that the system is easy to use and increases productivity.

### 4.2 Develop Path Diagram

The theoretical model on Modified Technology Acceptance Model which has been built in the first phase and then depicted in Path Diagram to do the estimation process by using AMOS. The theoretical model can be seen in Figure 5a.

### 4.3 Choose the Input Matrix for Data Analysis

Analysis of the data will be done by using a maximum likelihood estimation of the model gradually. The first is the estimated measurement models by using Confirmatory Factor Analysis. Next is Structural Equation Model with a full analysis of the model to test the suitability of the model and causality in models tested.

### 4.4 Model Identification and Criteria for Assessing Goodness of Fit

There are 235 questionnaires that were returned and analyzed decent outlier data found as many as 28 data, so it can be used 207 data. This amount is sufficient to be able to use the software AMOS in its analysis. Identification of the model is done by using test Confirmatory Factor Analysis for the overall model or full structural models. On early models show the results of the model does not fit the model so necessary modifications. After the modification of the model obtained the final model (Figure 5.b.) where Chi-square amounted to 482.999, the value of probability on this analysis shows the value of 0.00. Another value of Goodness of Fit shown by GFI (0842). It is approximately 0.90 within the tolerance value. Similar with the value of RMSEA (0.079) under 0.08 so the model is considered fit with existing data. So that the model can be accepted. A good model is influenced by the construct validity and reliability indicators, too. In this case all the data is valid and reliable.

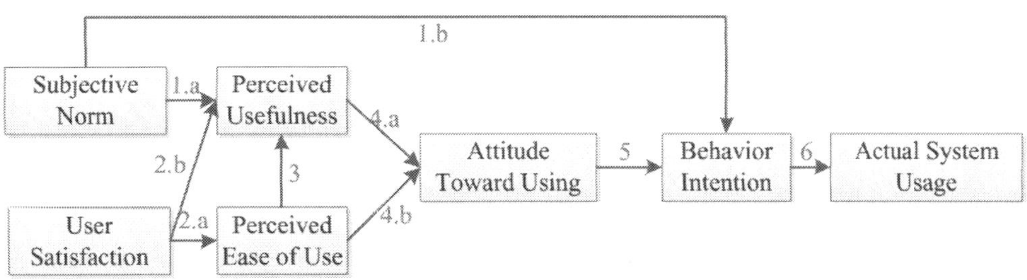


**Fig 4.** Modification Technology Acceptance Model Proposed on Selected Hospital in Sleman, Daerah Istimewa Yogyakarta, Indonesia

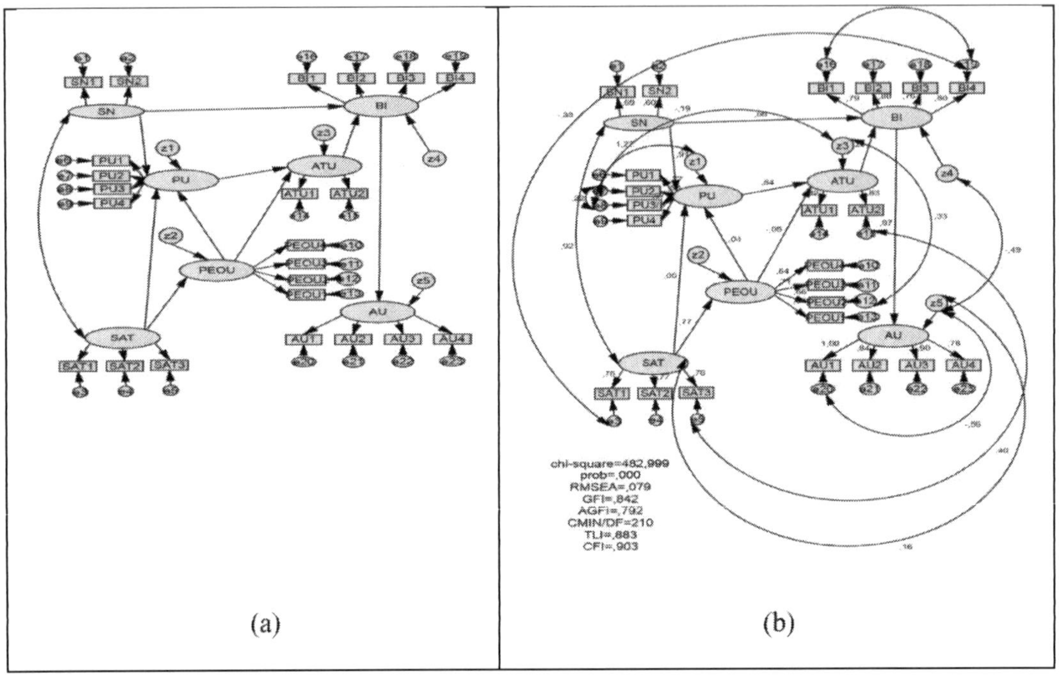


**Fig 5.a.** Early Path Diagram Model

**Fig 5.b.** Modified Path Diagram Model

### 4.5 Interpretation of the Model

After going through some testing, the last step is to conclude testing of hypotheses developed in this study. Value standard used which indicates that the hypothesis is accepted is by looking at the value t count> t-table. Hypothesis testing is done by analyzing the value of CR and the probability values (P), then compared with the required statistical limits, for the example above 1.96 for the value of CR and above 0.05 for the P value.

Hypothesis testing results show six accepted hypotheses, namely the hypothesis 1.a., the hypothesis 1.b., the hypothesis 2.a., the hypothesis 4.a., the hypothesis 5, and the hypothesis 6. The rejected hypotheses are hypothesis 2.b., hypothesis 3, and hypothesis 4.b. It means that the acceptance of technology Hospital Information System strongly influenced by perceived of usefullness. Thus the managers, developers, and software metric coordination should fully understand to the benefits of HIS's application. Thus they can encourage the user to adopt it [8]. Without support from all hospital stakeholders, the use of the application can not run optimally.

## 5 Conclusion

Modified Technology Acceptance Model (TAM) can be used to evaluate Hospital Information System (HIS). From this research, it is found that the factors that may influence the user in using HIS are Subjective Norm (SN), Perceived Usefulness (PU), Perceived Ease of Use (PEOU), User Satisfaction (SAT), Behavior Intention to Use (BI), Attitude Toward Using (ATU), and Actual System Usage (AU). Subjective Norm variable has a positive and significant relationship to Perceived Usefulness and Behavior Intention to Use. User Satisfaction variable has a positive and significant relationship to Perceived Ease of Use, but not with Perceived Usefulness. Perceived Ease of Use variable has an unfavorable and insignificant relationship to Perceived Usefulness and Attitude Toward Using.Good relationship occurs from Perceived Usefulness to Attitude Toward Using, followed by Attitude Toward Using to Behavior Intention to Use, and Behavior Intention to Use to Actual System Usage. Further research is to improve theories and models in other hospitals and in other fields, such as social, economics.

## References

1. E. Hariana, G.Y. Sanjaya, A.R. Rahmanti, B. Murtiningsih, E. Nugroho, Penggunaan sistem Informasi manajemen rumah sakit (SIMRS) di DIY, *Seminar Nasional Sistem Informasi Indonesia* (2013)
2. Sabihaini, Analisis Pemanfaatan Teknologi Informasi dan Kinerja Individual (Studi pada Rumah Sakit di Yogyakarta), *Jurnal widya manajemen dan akuntansi*, **vol.6** no.1, April 2006: 1-16 (2006)
3. E. Kristianto, Evaluasi Penerapan Sistem Informasi Manajemen Rumah Sakit di RSUP Dr. Sardjito Yogyakarta Dengan Menggunakan HOT-Fit, *Unpublished Thesis*, Magister Teknologi Informasi, UGM, Yogyakarta (2007)
4. A.B. Saputra, Evaluasi Faktor-Faktor Kesuksesan Implementasi Sistem Informasi Manajemen Rumah

Sakit Di RS PKU Muhammadiyah Sruweng Dengan Menggunakan Metode HOT-Fit, *Unpublished thesis*, Magister Teknik Informatika, UII, Yogyakarta (2013)

5. E. Saputra, Misfariyan, Analisis Penerimaan Sistem Informasi Manajemen Rumah Sakit Umum Daerah Bangkinang Menggunakan Metode Technology Acceptance Model (TAM), *Jurnal Sains, Teknologi dan Industri*, Vol.**10** No.2 (2013)
6. A.Ng.C. San, C.J.Yee, The Modified Technology Acccepetance Model for Private Clinical Physicians : A Case Study in Malaysia, Penang, *Int.j. of academic research in business and social sciences*, **vol 3** no.2, ISSN : 2222-6990 (2013)
7. S.J. Tabibi, A.A. Nasiripour, R.B. Kazemzadeh, A.A. Farhangi, P. Ebrahimi, Effective factors on Hospital Information System Acceptance : A confirmatory study in Iranian Hospitals; Middle-East *Journal of Scientific Research* **9** (1): 95-101, 2011, ISSN 1990-9233 ©IDIOSI Publications (2011)
8. L.G. Wallacae, S.D. Sheetz, The adoption of software measures : a technology acceptance model (TAM) perspective, *Int.J. of Information and Management* **51**, 245-259 (2014)
9. M.Chuttur, "Overview of the Technology Acceptance Model: Origins, Developments and Future Directions ," Indiana University, USA . *Sprouts: Working Papers on Information Systems,* **9**(37). http://sprouts.aisnet.org/9-37 (2009)
10. V.P. Anggelidis, P.D. Chatzoglou, Using a modified technology acceptance model in hospital, *Int. J. of medical informatics*, **vol.78**, issue 2, pp.115-126 (2009)
11. Menkes, Peraturan Menteri Kesehatan Republik Indonesia nomor 82 tahun 2013 tentang Sistem Informasi Manajemen Rumah Sakit, Jakarta (2013)
12. Han, Shengnan, Individual Adoption of Information Systems in Organisations: A Literature Review of Technology Acceptance Model, Turku Center for Computer Science Technical Report No 540 (2003)
13. B. Dharmarajan, K. Gangadharan., Applying Technology Acceptance (TAM) model to determine the acceptance of Nursing Information System (NIS) for Computer Generated Nursing Care Plan among nurses, *Int. J. of Computer Trends and Technology (IJCTT)*, Vol. **4** Issue 8 (2013)
14. F.I. Yunastiti, Z.Baridwan, Penerimaan Individu Terhadap Sistem Informasi Berbasis Komputer: Pendekatan Modified Technology Acceptance Model (TAM), *Jurnal Ilmiah Mahasiswa FEB Universitas Brawaijaya*, **Vol.2**. No.1 (2014).
15. Dreana, M.Tirza. Analisis Faktor-Faktor Yang Mempengaruhi Penerimaan Penggunaan SIMAWEB Di Fakultas Ekonomika Dan Bisnis Universitas Diponegoro: Integrasi TAM Dan TPB, *Unpublished Thesis*, Fakultas Ekonomika dan Bisnis, UNDIP, Semarang (2012)

# Risk control analysis of a furniture production activities using hazard identification and risk assessment method

*Sri* Indrawati, *Atyanti Dyah* Prabaswari*, and *M. Abdul* Fitriyanto

Industrial Engineering, Universitas Islam Indonesia, Yogyakarta, Indonesia

**Abstract.** The furniture industry in Indonesia is growing rapidly that leads to technological development. But the use of technology in the industry creates some risk to worker safety. The number of work accidents in Indonesia at the end of 2016 is relatively high. Therefore, this research is conducted to identify hazards and determine risk control using hazard identification and risk assessment method (HIRA). There are three main stages, i.e. identify and analyze hazards, classify hazards and risk control. The result shows that there are nine potential hazards in furniture production, 22% of it is categorized as high risk level, i.e. hand slashed and fingers cut by machine. Three types of risk control activities are determined to minimize the risk, i.e. administration and engineering control and personal protective equipment.

## 1 Introduction

In recent years, the furniture industry in Indonesia is growing rapidly so it gives a big influence to a rapid development of technology. This industry is one of the development priority sector in Indonesia because it is labor-intensive and export-oriented [1]. In fact, the use of technology in industry creates some risk to worker safety. Work accident is one of the factors that must be considered and minimized. Work accidents have three main elements, i.e. unexpected, causing material and non-material losses and disruption of work processes [2]. Some of these conditions are caused by poor worker safety and health management.

The number of work accidents in Indonesia at the end of 2016 is relatively high at 101,367 cases with 2.4% of deaths [3]. The highest accidents level in the world is wood processing industry that includes furniture industry [4]. Accidents that often occur in the majority industry are caused by human error and dangerous working environment conditions, i.e. manual handling, noise and hazard materials. Most workers are unable to properly identify the workplace potential hazards so that the risk of work becomes greater [5].

In the application of occupational safety and health systems, an industry must have risk management that is hazard identification, risk assessment and occupational risk management [6]. Hazard or the occurrence of things that can harm people, processes, environment around the industry should be well noticed because it is related to the condition of machines, work tools, the nature of work and the production process. Production processes that occur have a great potential hazards, the state of the environment that uses energy is one of the causes of workplace accidents in the work environment. Therefore, an evaluation is required to minimize the potential hazards and risks that occur based on the production process.

Measurement of hazards and risks can be done by choosing the appropriate method with the industry characteristics. An integrated occupational health and safety in plant layout designs is used to minimize hazards from facilities layout design [7]. A hazard identification and risk assessment method (HIRA) is ssuccesfuly used to identify some hazard control in chemistry laboratory [8]. HIRA is used to measure risk level of some potential hazards on the foundry production process [9]. HIRA also used to evaluate potential hazards in paper production. The result show that this method can identify 23 potential hazards with 10 of it categorized as high risks [10].

Hazard identification, risk assessment and determining control (HIRADC) is applied in manufacturing process to prevent accident in workplace [11]. A study is conducted to identify hazards in pharmaceutical manufacturing process using integrated HIRA and failure modes and effect analysis (FMEA) [12]. The integrated method has the ability to reduce the risk and create necessary recommendations to improve worker safety. 10 types of high risk hazard using HIRA method in spare part manufacturing are found [13]. HIRA also used to assess and prioritizing hazards in automotive industry [14]. There are five types of hazards that are identified and a recommendation is to minimize the risk level.

## 2 Basic Theory

An industry should identify all materials, tools and systems of production, to determine the potential hazards [15]. Because this identification is a process that can cause accidents [2]. As well as physical conditions that potentially cause losses such as accidents, equipment

* Corresponding author: atyanti.dyah@uii.ac.id

damage, damage to buildings [16]. The level of risk determination of work accidents are often called risk assessment [2]. The risk assessment can measure the likelihood of an activity whether it have a financially detrimental effect, health and safety over a period of time [17].

HIRA is a method for identifying and measuring potential hazard at a workplace that can cause a loss of company or individual. There are three ways to determine the risk, i.e.:

a. Qualitative analysis is a risk analysis method that uses tabulation based on descriptive (high, medium or low).
b. Quantitative analysis is a risk analysis method that uses numeric numbers to express impact and probability.
c. Semi-quantitative analysis, which is a risk analysis method that uses scale numbers for each qualitative category.

Assessment of occupational risks is performed using a scale of 1-5 as shown in Table 1.

**Table 1.** Risk Probability Level [18]

| Probability Level | Description |
|---|---|
| Level 5 (Almost Certain) | Incidence rate of 76% to 99,99% |
| Level 4 (Likely) | The probability level is 60% to 75,99% |
| Level 3 (Possible) | The probability rate is 26% to 50,99% |
| Level 2 (Unlikely) | The probability of occurrence is 1% to 25,99% |
| Level 1 (Rare) | Possible incidence rate up to 0,99% |

Once the risk value is identified, the process of job risk classification can be done using Table 2.

**Table 2.** Risk Cathegory [18]

| Rank | Level | Description |
|---|---|---|
| 17- 25 | Extreme High risk | Unacceptable risks with high priority mitigation |
| 10 - 16 | High Risk | The risk is not acceptable with medium priority mitigation |
| 5 - 9 | Medium Risk | Risks maybe received by monitoring |
| 1 - 4 | Low Risk | Risk maybe accepted without monitoring |

## 3 Research Method

This research is conducted in an Indonesian furniture industry located in West Java using HIRA method. There are four main stages, i.e. hazard identification, risk assessment, evaluation and risk control. Research begins with observing the work process in the furniture industry. At this stage, the identification of working facilities, work procedures and materials is done. The next stage is to conduct a survey for the identification and confirmation process of the identified hazard. The survey includes the type of hazard that occurs, who is at risk of hazard, when and how long the hazard exposure, the location of the hazard and how hazard exposure occurs. Furthermore, risk evaluation is done by qualitative approach. In-depth analysis is done to determine the priority of each type of hazard. Some hazard control and mitigation measures are developed based on three considerations, i.e. hazard source, exposure mode and exposed workers.

## 4 Result and Discussion

There are eight main production process in furniture industry, i.e. material preparation, cutting raw materials, drying, smoothing, cutting components, sizing and, assembly, finishing and packaging. Each type of process have 2-6 hazards. The highest number of hazards is in cutting components process. While the packaging process only have one type of hazards, i.e. slashed by sharp materials as seen in Appendix 1. There are nine types of hazards based on the identification process that have been done. Once the hazards and risks are identified, then the assessment of the risk of the work is done with the results shown in Table 3.

**Table 3.** Job Risk Assessment

| No | Hazards | Possible Weight Ratio | Weight Risk | Risk Value |
|---|---|---|---|---|
| 1 | Squashed wood | 2 | 1 | 2 |
| 2 | Hands slashed by machine | 2 | 3 | 6 |
| 3 | Wood fall | 3 | 4 | 12 |
| 4 | Eye irritation and respiratory disorders | 4 | 1 | 4 |
| 5 | Material collapse | 2 | 2 | 4 |
| 6 | Finger cut by sharp material | 2 | 3 | 6 |
| 7 | Kickback of the wood being cut | 3 | 3 | 9 |
| 8 | Exposure to paint | 4 | 1 | 4 |
| 9 | Fingers cut off by machine | 3 | 4 | 12 |

The next stage is risk classification to classified risk based on the total value of hazards. There are two types of hazard included in the category of high risk, i.e. hands slashed by machine and fingers cut of by machine as shown in Table 4.

Then the classification of risk level is converted to get the percentage of risk level as illustrated in Figure 1. Production process in Indonesian furniture industry is dominates with job hazards at low risk level around 45%. 22% of production process activities are categorized into high risk. Some action plans are developed to lower job hazards especially at high risk level. As a risk control priority that must be done by the industry consist of three types, i.e. administration control, engineering control and personal protective equipment as seen in Table 5.

**Table 4.** Risk Classification

| Risk Level | Hazards |
|---|---|
| Low Risk | a. Wooden tejepit<br>b. Exposure to powder (eye irritation and respiratory disorder)<br>c. Fall of material / at the foot<br>d. Exposure to paint (eye irritation and respiratory disorder) |
| Medium Risk | a. Wood fall<br>b. Finger cuts sharp objects<br>c. Kickback from the wood being cut |
| High Risk | a. Hands slashed by machine<br>b. Fingers cut of by machine |

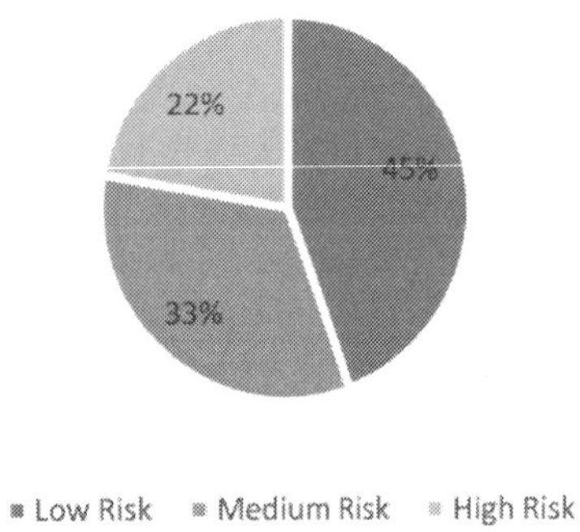


**Fig 1.** Risk Presentation Diagram

**Table 5.** Risk Handling

| Risk Level | Hazards | Risk Control |
|---|---|---|
| High Risk | Hands slashed by machine | 1. Administratiion control: simplification of work instruction and stick it to machine. |
| | | 2. Engineering control: Machine guarding and auxiliary tool. |
| | | 3. Personal protective equipment: gloves |

## 5 Conclussion

Based on the results of research that have been done, it can be concluded that:

a. There are nine job hazards in furniture production process, i.e. pinched fingers, slashed hands and kickback while cutting wood.
b. Job hazards is 22% at high risk level, 33% at medium risk level and 45% at low risk level.
c. Hazards that include high risk level are hands slashed by machine and fingers cut of by machine. Priority risk control plan is simplification and embedding of work instruction in working machines, using gloves, adding machine guarding and thrust tools.

Researchers thank to the support of Industrial Engineering Department, Faculty of Industrial Technology, Universitas Islam Indonesia.

## References

1 Hartarto, A., Kemenperin dan HIMKI Susun Strategi Pacu Kinerja Industri Mebel dan Kerajinan. Indonesian Furniture and Craft Industry Association, https://www.himki-indonesia.com/post/77/Kemenperin-dan-HIMKI-Susun-Strategi-Pacu-Kinerja-Industri-Mebel-dan-Kerajinan/id, Accessed on **7 April 2017**, (2017)

2 Tarwaka, Managemen dan Implementasi K3 di Tempat Kerja, Harapan Press, Surakarta, (2008)

3 Isafetynews, Kecelakaan Kerja Konstruksi 2017 Diprediksi Tetap Tinggi. http://isafetynews.com/2017/02/01/kecelakaan-kerja-konstruksi-2017-diprediksi-tetap-tinggi/, accessed on **1 Mei 2017**, (2017)

4 Vestlund, A.T., Health And Safety In The Furniture Industry, The Furniture Industry Research Association, (2013)

5 Bahn, S. T., Workplace Hazard Identification: What Do People Know And How Is It Done?. *Proceedings of Annual Conference of the Association of Industrial Relations Academics Australia and New Zealand.* **2012: 1-9**. (2012).

6 Suardi, R., Sistem Manajemen Keselamatan dan Kesehatan Kerja, PPM, (2008)

7 Kazerouni, A.M., Chinniah, Y., Agard, B, A Proposed Occupational Health And Safety Risk Estimation Tool For Manufacturing Systems, *International Journal of Production Research,* Vol. 53. Issue **15: 4459-4475**, (2015).

8 Legget, D.J., Lab-HIRA: Hazard Identification And Risk Analysis For The Chemical Research Laboratory: Risk Analysis Of Laboratory Operations, *Journal of Chemical Health and Safety*, Vol. 19. Issue **5: 25-36,** (2012)

9 SheikAllavudeen,S. & Sankar, S., Hazard Identification, Risk Assessment and Risk Control in Foundry, *SSRG International Journal of Industrial Engineering,* Volume 2. Issue **3: 1-4**, (2015)

10 Sari, R.M., Syahputri, K., Rizkya, I., Siregar, I., Identification of Potential Hazard using Hazard Identification and Risk Assessment, *IOP Conf. Series: Materials Science and Engineering*, Vol. **180. 012120**, (2017)

11 Shamsuddin, K.A., Norzaimi, C.M., Che-Ani, A.I., Ismail, A.K., Investigation the effective of the Hazard Identification, Risk Assessment and Determining Control (HIRADC) In Manufacturing Process. *International Journal of Innovative Research in Advanced Engineering*, Vol.2. Issue **8: 80-84**, (2015)

12 Swarnakar, M., Siddiqui, N.A., Baksi, S., Hazard Identification, Risk Analysis And Recommendations For Improving Safety In Pharmaceutical Industry,

*Journal of Management, Information Technology and Engineering,* Vol. 2, Issue **2: 23-4**, (2016).

13 Restuputri, D.P. & Fakhri, M., The Analysis Of Health And Safety Aspects By Using Hazard Identification And Risk Assessment (HIRA) Method, *Proceeding 8th International Seminar on Industrial Engineering and Management*, ER **37-ER44**, (2016)

14 Ramesh, R., Prabu, M., Magibalan, S., Senthilkumat, P., Hazard Identification and Risk Assessment in Automotive Industry, *International Journal of ChemTech Research*, Vol.10. No.**4: 352-358**, (2017)

15 Department of Occupatinal Safety and Health, Guidelines for Hazard Identification, Risk Assessment and Risk control, Malaysia, (2008)

16 Munawir, A., HAZOP, HAZID, VS JSA, Migas Indonesia, (2010)

17 Kolluru, Rao V., Risk Assesment and Management Handbook for Environmental, Health, and Safety Profesionanals, McGrawhill-Hill, United State Of America, (1996)

18 Global, *Risk Management Guidelines*, SAI Global. Australia, (2007)

# Ergonomics smoke machine for indigenous people in Indonesia

*Eko* Nurmianto*

Department of Industrial Eng., Faculty of Industrial Technology, Institute Teknologi Sepuluh Nopember, Surabaya 60111, Indonesia

**Abstract.** A participatory process is used in the design of a user-centered smoke machine, resulting in lasting benefits to the community. This program of IbM (Science and Technology for Society) has designed fish fumes that are mobile, portable and environmentally ergonomic, applying appropriate technology about fumigation so as to increase knowledge to the community. This program improve people skills especially in smoked fish production, increase employment in the field of fishery products processing business. With this activity can grow the interest of the community to entrepreneurship for those who have not started a business and will increase its business for those who have started the processing business of fishery processing. Benefits derived from IbM (Science and Technology for Society) program are, among others, reduced air pollution caused by open fish curing system so that air quality environment, especially around fish processing SMEs to be better, can support and strengthen fish processing industry in the coastal area that began to grow and especially support local SMEs, increase the quantity and quality of patents that can be generated from community service activities funded by Directorate General of Higher Education, Ministry of Technology Research and Higher Education, Republic of Indonesia.

## 1 Introduction

This study The fish processing industry in several country have produced musculo skeletal disorder, i.e: Swedia [1 - 4], United States [5, 6], Egypt [7], Malaysia [8], Canada [9], and Norway [10, 11]. Among those processing are small-scale industry [12].

The fish processing industry in SubDistrict Brondong Regency of Lamongan (one city of the East Java Province) has been around for a few years but it is still not well developed yet. Lack of capital makes them unable to increase production capacity and limited marketing. They still use traditional technology (Figure 1). In view of such circumstances, there is still an opportunity to increase efficiency and effectiveness through a Small Business Development Program in accordance with the situation and condition of such small-scale industry. The design of management, human resources, production, finance, and design will in turn improve the competitiveness of the small-scale industry.

The existing workforce in SME Fish Production has gained the skills of making fish from years of work experience and transmitted informally. Workers SME Fish Production are 6 people and still have a family relationship. The number of SME workers is only 5 people, so there is no job specialization. The administration system is not yet organized. They are still many concurrent jobs regardless of job specification and job description, so that management functions such as planning, organizing, directing and control are not clear and overlap.

At that time the tool used is a tool for roasting goat satay or chicken satay (Figure 1). The tool is insufficient since it can only burn approximately 20 to 30 fish in one process of fogging fish. Whereas in once the process of curing fish takes approximately one hour, so that less can meet the number of orders from consumers. Besides, this tool has a shortage during the processing of smoked fish, most of the smoke disturb neighbors or the surrounding community or less environmentally friendly (Figure 1).

As an ergonomist with a lifelong interest in indigenous stuff, I'd been fascinated for years by questions such as these: How can fishermen/fisherwomen create such smoke machine while constrained in a kneeling posture? Why do they kneel to smoke fish ? Are they in pain or polluted ? What do they love about the traditional way of working, and what would they change if they could ? Based on the above conditions then comes the idea of Ergo-Design of Fish Cultivation Tool (Figure 2-4).

The tool is a stuff to process fish fumes with a capacity of 120 in one process. This tool has a privilege that most of the smoke does not disturb the community (environmentally friendly) is marked by a chimney that is 3 meters high so that smoke can be wasted into the free air and does not disturb the community around SME. Besides, this tool is equipped with a door which can be opened at the front and rear so easy in the placement of fish before and after the process of fogging.

* Corresponding author: nurmieko@gmail.com

**Fig 1.** Tools for fish fumigation in previous situation [13].

**Fig 2.** Ergonomic smoke fish tools when first time founded by ergonomic designer [13].

The purpose of this activity are (a) Designing ergonomic smoke machine for fishermen that are mobile, portable and environmentally friendly; (b) Implementing the science of health and safety so as to increase knowledge to the community; (c) Increasing employment in fishery products processing business; (d) With this activity can foster the interest of the community to entrepreneurship for those who have not started a business and will increase its business for those who have started the processing business of fishery processing.

Benefits of this research are: (a) Reduced air pollution caused by open fish curing system so the quality of the air environment especially around the fish processing SMEs for the better, (b) Can support and strengthen fish processing industry that began to grow and especially support local SMEs, (c) Increase the quantity and quality of patents that can be generated from community service activities.

## 2 Literature review

At present, more than 90% of fishery production in Indonesia is produced by fishermen and 50% of the yield is traditionally obtained. Increased production of fishery products are generally oriented to economic activities, both increased production in fishery business people and industrial fisheries. Most of the fishery products processing industry still use simple technology and generally still household industry, while large scale industry is generally limited to canning and freezing of fish.

There are various preservation fish [8], among others by salting, drying, fermenting, fermenting and cooling fish. The benefits of eating fish are known to many people, as in Malaysia are the main foods in everyday dishes that provide youthful effects and higher life expectancy than other countries. Processing fish in various ways and flavors causes people to consume more fish.

In conditions where fish seasons reach peak fishermen have difficulty in marketing considering the selling price becomes low. One alternative that can be developed by the fishermen community and their families is to develop a business of local fish-based jerky fish.

## 3 Methods of Developing Appropriate Technology

The success of this program has come from combining appropriate technology with a respectful social process. It illustrates that a systems approach can have an impact well beyond enabling an ergonomist to design a user-centered product. As demonstrated in this article, participation in the systems assessment stimulates end-users to become more aware of themselves and their culture and helps empower them to create meaningful change in their own lives.

The ergonomic smoke machine (ESM) that we designed demonstrates that good ergonomic design can be both simple and human centre design. Indeed, this tool can minimised smoke to be inhaled. It can simply fume the fish untill it cooked. The one knew the fish cooked by obseving via window. The height and thw width is anthropometrically suit with the people in Indonesian costal area [14]. Good design can come from embracing a people's anthropometry cultural, environmental, and economic priorities while applying universal principles to the specific context in which they reside.

***Conceptualization.*** The concept for the ESM evolved from a participatory process that was facilitated by a variety of nonprofit organizations in Lamongan. In two villages we held exploratory meetings with women in their fisherwomen cooperatives and in their homes.

Fumigation Method. Smoking can be done in two ways, namely cold and hot fumigation :

A. Smoking cold (cold smoking). Is the process of fumigation by putting the fish that will be smoked somewhat away from the source of smoke (wood burning), with a temperature of about 40-500C with a long process of smoking several days to two weeks

B. Smoking hot (hot smoking). Smoking hot (hot smoking) is the process of fumigation where it will be smoked placed close enough to the source asap.Suhu around 70 - 1000C, the duration of fumigation 2 - 4 hours

## 4 Results

Ergonomic Smoke Machine (ESM) for fish curing tool. In the manufacture of Fish Curing Tool required working drawings, 3D Auto CAD drawings and photos of the products that have been made, as shown below.

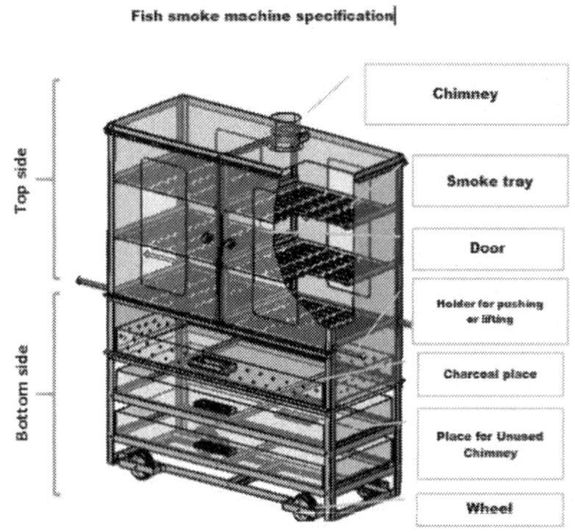


**Fig 3.** 3-D Smoke Machine

**Fig 4.** Ergonomic, mobile and portable fish smoke machine.

How to use the fish fumes:

1. Remove the smoke tray and place it on the ground.
2. Place the coconut shell on top of the smoke tray.
3. Burn the coconut shell so that it emits smoke.
4. While waiting for coconut shell charcoal to smoke a bit more, the fish will be smoked, washed, removed the contents of his stomach, and spiced then hung in a tool that has been available smoke.
5. Drawers that have issued a lot of smoke placed in the fish smoke machine so that the smoke can make the fish ripe.
6. Fish maturity level can be observed through glass windows.
7. If the fish is cooked then the smoke is thrown through the chimney by opening the valve.
8. After all the smoke comes out through the chimney then the fish can be taken for the next served

## 4.2 Fish smoking training

This fish smoking training begins with the introduction of tools and how to use it. The training materials comprises of:

- How to use ESM (Ergonomic Smoke Machine) and to process of fishery products.
- Manufacture of smoke fish machine.
- Finance and business management
- Organoleptic (Testing the quality of smoked fish products)
- Packaging and labelling.
- Marketing of smoked fish.
- Entrepreneurship.

## 4.3 Appropriate Technology for Society

Appropriate Technology of Fish Curing Tool is an effective tool to process fish into smoked fish. By using the tool, the Business Group Development of Smoked Fish in Sub District Brondong in Lamongan can run more optimally. From these tools can be smoked various types of fish, chicken, duck and other types of smoke ingredients. And can help people to recognize the smoke fish technology as an alternative to self-employment. Improving community skills is a fish-smoking skill that can be practically demonstrated by the following figures (Figures 5-9):

**Fig 5.** Fishermen test the prototype during the IbM fish cultivation business in Sub-District Brondong in Lamongan city

Two main themes emerged from this research: (1) To be culturally accepted, the ESM (Ergonomic Smoke Machine) would need to be simple and its use intuitively obvious. (2) Because we were designing for people who live in economic poverty, the design would need to balance high quality with low cost.

There were five criteria for economic value [8, 15, 16]:

a. Robustness, in relation to function;
b. Ergonomic fit, with particular regard to adjustability;

https://doi.org/10.1051/matecconf/201815401103

c. Durability, to withstand local environmental conditions such as dust and humidity with little or no maintenance;
d. Sustainability, to allow local people to use it from locally available materials, fostering community self-reliance and self-sufficiency; and
e. Affordability, so that as many fishermen as possible could acquire the technology without relying on charity.

**Fig 6.** People processing curing fish

**Fig 7.** (a) Fish before it is smoked

**Fig 7.** (b) Fish after it is smoked

**Fig 8.** The smoked fish is ready to be served

**Fig 9.** Participants of the program

The ESM (Ergonomics Smoke Machine) supports fisherwomen as a cottage and coastal industry and decreases the likelihood that women will be forced to seek jobs outside their community. Furthermore, the grassroots approach benefits the entire community. By adding the ESM to their product line, welder can increase their incomes and gain satisfaction from making and selling a product that helps their own people. Another benefit is that the welder can use their increased awareness of ergonomics to improve their own working methods.

The more obvious outcomes in health, productivity, and fish quality are accompanied by cultural and personal benefits (Figure 10). The women describe feeling enhanced pride in their identity as Brondong fisherwomen. The ESM enhances self-esteem and helps the indigenous people preserve their traditions.

The tools has already designed in proper way in order to have better understanding on ergonomics tools used, moreover the correlation between ergonomics approach, product design and community development.

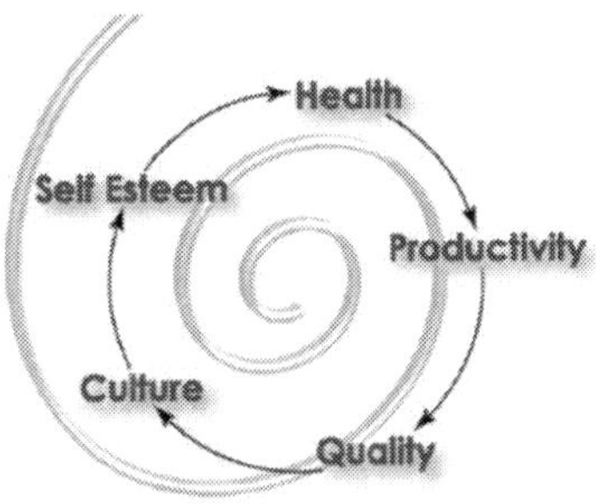


**Fig 10.** The elements of this ergonomics cycle of benefits apply universally, bringing all products an evolving sense of self-confidence, a connection with their cultural heritage, and development of their unique creative gifts of local wisdom products [15].

## 5 Conclusions

From the results of community service programs, the following conclusions can be drawn:

a. Has implemented the implementation of Community Service in the form of skills training on Fish Curing Method with the following materials: Development of entrepreneurial spirit, Financial management, Creative packaging design, Implementation of mobile fishing tool and portable to improve fish processing so it can increase competitiveness
b. Increasing community self-reliance and providing community income through entrepreneurial activities.
c. Skilled in processing smoked fish.
d. Functioning as a means of trying for the community.

### Award Recognition

This work on development of the ESM ergonomic smoke machine for fishermen has been recognized with an Indikatama Award for Innovating Technology Benefiting Humanity.

### Contribution and Collaboration of Research

The ESM of this study contribute to society especially the people in the coastal area along the Indonesian coast. Since Indonesia possess the second longest coastal in the world while the Canada is longest one.

Author is actively developing ergonomics resources across the spectrum of art media, art cultures and technology. We invite and welcome with individuals and organizations that share a passion for exploring the link between ergonomics and human creativity, and for bringing ergonomics to cultures and economies where it has been underrecognized and undervalued.

Authors acknowledge funding support from Directorate General of Higher Education, Ministry of Technology Research and Higher Education, Republic of Indonesia

## 6 References

1. M. Tbrner, G. Blide, H. Eriksson, R. Kadefors, R. Karlsson, and I. Petersen, "Musculo-skeletal symptoms as related to working conditions among Swedish professional fishermen," no. September, pp. 191–201, 1988.
2. M. Tisrner, R. Karlsson, S. Harald, and R. Kadefors, "Analysis of serious occupational accidents in Swedish fishery," vol. 21, 1995.
3. K. Kazmierczak, S. Erik, M. Forsman, and J. Winkel, "An integrated analysis of ergonomics and time consumption in Swedish ' craft-type ' car disassembly," vol. 36, no. 2005, pp. 263–273, 2006.
4. A.-C. Falck and M. Rosenqvist, "What are the obstacles and needs of proactive ergonomics measures at early product development stages? – An interview study in five Swedish companies," Int. J. Ind. Ergon., vol. 42, no. 5, pp. 406–415, Sep. 2012.
5. S. Fulmer, "Ergonomic Exposure Case Studies in Massachusetts Fishing Vessels," vol. 18, pp. 10–18, 2002.
6. J. M. Lincoln, M. B. O. Connor, K. D. Retzer, R. D. Hill, T. D. Teske, C. C. Woodward, D. L. Lucas, P. D. Somervell, J. T. Burton, N. A. Mode, B. J. Husberg, and G. A. Conway, "Occupational Fatalities in Alaska : Two Decades of Progress , 1990-1999 and 2000-2009," J. Safety Res., vol. 44, pp. 105–110, 2013.
7. M. A. Zytoon, "Occupational injuries and health problems in the Egyptian Mediterranean fisheries," Saf. Sci., vol. 50, no. 1, pp. 113–122, 2012.
8. et al Shaffril, "Adapting towards climate change impacts: Strategies for small-scale fishermen in Malaysia," Mar. Policy, vol. 81, no. December 2017, pp. 196–201, 2017.
9. B. Jackson, B. Neis, A. Canning, and S. Mackinnon, "Safety on Newfoundland ' s fishing wharves," vol. 60, pp. 1–12, 2013.
10. E. Mcguinness, H. L. Aasjord, I. B. Utne, and I. Marie, "Safety Scien ce Injuries in the commercial fishing fleet of Norway 2000 – 2011," Saf. Sci., vol. 57, pp. 82–99, 2013.
11. T. Thorvaldsen, "The importance of common sense : How Norwegian coastal fishermen deal with occupational risk," Mar. Policy, vol. 42, pp. 85–90, 2013.
12. N. M. Smith, S. Ali, C. Bo, and N. Collins, "Human health and safety in artisanal and small-scale mining : an integrated approach to risk mitigation," vol. 129, 2016.
13. E. Nurmianto, Ergonomi : Konsep Dasar dan Aplikasinya, 2nd ed. Surabaya: Guna Widya, 2004.
14. I. Z. Sutalaksana and A. Widyanti, "Anthropometry approach in workplace redesign in Indonesian Sundanese roof tile industries," Int. J. Ind. Ergon., vol. 53, pp. 299–305, 2016.
15. B. Y. K. Piegorsch, "An Ergonomic Bench for Indigenous Weavers," Ergon. Des., 2009.
16. G. A. Mirka, X. Ning, S. Jin, O. Haddad, and K. L. Kucera, "International Journal of Industrial Ergonomics Ergonomic interventions for commercial crab fi shermen," vol. 41, 2011.

# Survey of customer expectation and satisfaction: preliminary research of a modular product design approach for sheep cage design

*Yuli Agusti* Rochman[1,*] and *Dwi Ana Ratna* Wati[2]

[1]Industrial Engineering Department, Universitas Islam Indonesia, Indonesia
[2]Electrical Engineering Department, Universitas Islam Indonesia, Indonesia

**Abstract.** This study aims to produce a modular sheep cage design. Feeding, maintenance and integrated waste management are taking into account in the design. The modular and integrated product design provides advantages such as (1) easy way of making and modifying the product, (2) enabling the utilization of existing facilities such as shade, (3) giving the user a sense of convenience as the waste can be processed and not being a source of pollution. The Modular Quality Function Deployment (QFD) method is applied to identify modules associated with the customer needs of the product. A survey about customer expectation and satisfaction was conducted in order to evaluate the gap of both. The highest value of gap between customer expectation and satisfaction of products indicates that the customer needs are the focus of the problems. The most important customer needs are considered in modification and improvement of design. These are (1) the cage is not easy to fall down, (2) the sheep cage can last for long time, (3) it uses a strong frame, (4) the roof section materials is not easy to leak, and (5) It can be used even if one part of the cage is damaged.

## 1 Introduction

The development of technology is currently growing rapidly and precisely in creating innovative products in the future [7]. Technology is a special combination of knowledge, productive resources, input, and systematic services to produce the desired output [6]. The use of technology can help improve the efficiency and effectiveness of resource management. Quality function deployment (QFD) process is generally used as customer-oriented quality management and product development methodology in the manufacturing industry [1].

For rural communities livestock such as buffaloes, beef cattle, dairy cows, goats, sheep, ducks or domestic poultry have a strategic role because the animals can be used as life savings as well as labor source such as buffaloes and beef cattle. Livestock can also be used as a producer of organic fertilizer which is very good for increasing agricultural production. In addition, having many cattle will improve the social status. Problems related to livestock are public complaint about the adverse effects of livestock business activities because most farmers ignore the handling of waste from their business, some even dump their business waste into the river, resulting in environmental pollution. Livestock waste generated by livestock activities such as feces, urine, and feed residues cause pollution that trigger protests from nearby resident.

To overcome these problems, the role of technology used to make sheep cage. This study proposes a method of designing a modular sheep cage based on QFD. The design takes into account the integrated feeding, maintenance and waste management.

## 2 Research methods

This work is in the context of modular product design. In order to collect information about customer need dealing with sheep cage design, a survey was conducted. The survey includes several stages: (1) define the modularity of the sheep cage as the basis for establishing questions of the survey, (2) making and retrieving the questionnaire data, (3) processing the data with the QFD methodology. The object of this research is sheep cage. The number of respondents is 29 people. Research data was obtained from the survey to farmers in Turi, Sleman, Yogyakarta, Indonesia. The purpose of this survey is to collect information about customer expectation and satisfaction.

## 3 Results and discussion

### 3.1 Product modularity

The division of the product into an independent component raises the modularity. This independence allows companies to create variations of products and standardize components. Product modularity is frequently stated as a goal of good design practice, yet has not been as carefully studied as some other current design practices such as design-for-assembly [8]. Modularity is the focus

* Corresponding author: gusti@uii.ac.id

of the company's attention along with improvement efforts to rationalize the product line and to provide increasing varieties at a lower cost. Modularity arises from the division of a product into independent components, allowing companies to standardized components and create product variety. Product modularity is defined using two characteristics; (1) Similarity between the physical and functional architecture of the design and (2) Minimization of the degree of interaction between physical components [3].

Modular design is built on the basis of functional analysis of a product with the same function or with different functions, but has different specifications and performance. [4]. This is achieved by dividing and designing a series of general function modules, select and combine these modules to meet the requirements of the customers. Sheep cage is divided into 4 modules: sheep's house, cage roof, feeding basket, and waste basket. sheep's house is function to manage the sheep. Cage roof is function as protection from sun light and rain. Feeding basket is function as module for feeding and drinking of sheep. Waste basket is function as waste collector from the sheep such as feces and urine.

### 3.2 Quality Function Deployment (QFD)

Quality Function Deployment (QFD) is a method used for the planning and development of structured products that enable the development team to clearly determine the needs and wants of consumers, and evaluate each desired product or service capacity systematically provided to meet the wants and needs Consumers [2]. In general, QFD as a planning tool that is used to meet the sounds of consumers in the form of desires (expectations) and consumer needs. QFD will be used to translate the voice of consumers from the form of specific needs into the direction then engineering actions are disseminated through product planning, part, process planning, production planning, and service [5]. Customer voice are obtained by surveying the customers [4]. We measured the severity of customer's perception items/questions using likert scale (1 to 5). The customer voice of Sheep Cage are presented in Table 1. These are used to identify the customer expectation and satisfaction which then collected by survey. Result of the survey are shown by Table 2. The table has been sorted by the highest gap values. The largest gap is given by the question "The cage is not easy to fall" with a value of 0.83 while smallest gap is "The roof of the cage is made flexible to be removed or installed as needed". From this survey result, we leverage five most important customer need to be consider in the modular design of sheep cage. These are (1) the cage is not easy to fall down, (2) the sheep cage can last for long time, (3) it uses a strong frame, (4) the roof section of the cage uses materials that is not easy to leak, and (5) the cage still can be used even if one part of the cage is damaged.

**Table 1.** Customer voice

| No | Customer requirements |
|---|---|
| 1 | Easy to maintain |
| 2 | Clean |
| 3 | Durable |
| 4 | Wide |
| 5 | No smell |
| 6 | Can be dismantled pairs |
| 7 | Safe |
| 8 | Strong |
| 9 | Easy to access |
| 10 | Sufficient amount of light |
| 11 | Smooth air circulation |

**Table 2.** Survey result of customer expectations and satisfaction

| No | Survey questions | Expect. | Satisfac. | Gap |
|---|---|---|---|---|
| 1 | The cage is not easy to fall down | 4.59 | 3.76 | 0.83 |
| 2 | The sheep cage can last for long time | 3.90 | 3.17 | 0.72 |
| 3 | Use a strong frame | 4.34 | 3.72 | 0.62 |
| 4 | The roof section of the cage uses a material that is not easy to leak | 4.28 | 3.69 | 0.59 |
| 5 | the cage still can be used even if one part of the cage is damaged | 3.69 | 3.21 | 0.48 |
| 6 | The cage is made light | 2.83 | 3.24 | 0.41 |
| 7 | The roof section of the cage uses a lot of material to absorb light | 3.62 | 3.21 | 0.41 |
| 8 | The cage is easy to clean | 4.17 | 3.79 | 0.38 |
| 9 | Easy when picking up solid or liquid waste | 4.21 | 3.83 | 0.38 |
| 10 | Solid and liquid waste can be separated automatically | 4.41 | 4.07 | 0.34 |
| 11 | It is easy to move (portable) | 2.48 | 2.83 | 0.34 |
| 12 | The roof section of the cage is made tilted | 4.07 | 3.79 | 0.28 |
| 13 | It is easy to unloaded as needed | 2.62 | 2.86 | 0.24 |
| 14 | It has good air circulation | 3.97 | 3.72 | 0.24 |
| 15 | It is easy to supplay water for sheep | 3.24 | 3.00 | 0.24 |
| 16 | The cage floor is arranged to keep the sheep from falling easily | 4.00 | 3.76 | 0.24 |
| 17 | The cage is made on stage | 4.45 | 4.24 | 0.21 |
| 18 | Material price is cheap | 2.90 | 2.72 | 0.17 |

| | | | | |
|---|---|---|---|---|
| 19 | The size of the sheep cage can contain at least 4 sheeps | 2.83 | 2.72 | 0.10 |
| 20 | Where sheep food can be removed when they want to be cleaned | 2.83 | 2.90 | 0.07 |
| 21 | The roof of the cage is made flexible to be removed or installed as needed | 2.69 | 2.69 | 0.00 |

## 4 Conclusion

In this paper, modular QFD is apply to developing sheep cage and to know the design parameters that are important to the customer. A sheep cage is divided into 4 modules: sheep's house, cage roof, feeding place, and waste place. The five most important customer needs are considered into modification and improvement of the design. These parameters are the cage is not easy to fall, the sheep cage can last a long time, use a strong frame, the roof section of the cage uses materials that are not easy to leak, and the cage still can be used even if one part of the cage is damaged.

The authors would like to thank DP2M Dikti who has provided financial support for this research project and the reviewers for their great work and their suggestions and recommendations to improve this paper.

## References

1. Z. Baran, M.S. Yıldız, Quality Function Deployment and Application on a Fast Food Restaurant, International Journal of Business and Social Science, 122–131 (2015)
2. L. Cohen, Quality Function Deployment: How to Make QFD Work for You. Addison-Wesley. (1995)
3. G. Erixon, Modular function deployment: a method for product modularisation. Royal Inst. of Technology, Department of Manufacturing Systems, Assembly Systems Division. (1998)
4. Y. Liu, S.K. Ong, A.Y.C. Nee, Modular Design of Machine Tools to Facilitate Design for Disassembly, *Procedia CIRP*, 15, 443–448. https://doi.org/10.1016/j.procir.2014.06.085. (2014)
5. A. Rahman, H. Supomo, Analisa Kepuasan Pelanggan pada Pekerjaan Reparasi Kapal dengan Metode Quality Function Deployment (QFD). Jurnal Teknik ITS, 1(1), G297-G302, (2012)
6. Reijntjes, C., Haverkort, B., & Waters-Bayer, A.. Pertanian masa depan: pengantar untuk pertanian berkelanjutan dengan input luar rendah. Kanisius. (1999)
7. P. Tursch, C. Goldmann, R. Woll, Integration Of Triz Into Quality Function Deployment. Management and Production Engineering Review, 6(2), 56–62. https://doi.org/10.1515/mper-2015-0017, (2015)
8. K. Ulrich, Fundamentals of product modularity. In Management of Design, (pp. 219–231), Springer, (1994)

# Die design optimization on sheet metal forming with considering the phenomenon of springback to improve product quality

*Agung Setyo* Darmawan[1,*], *Agus Dwi* Anggono[1], and *Abdul* Hamid[2]

[1]Department of Mechanical Engineering, Faculty of Engineering, Universitas Muhammadiyah Surakarta, Jl. Ahmad Yani, Tromol Pos 1 Pabelan, Surakarta 57162, Indonesia
[2]Department of Mechanical Engineering, Faculty of Engineering, Universitas Batam, 29464 Batam, Indonesia

**Abstract.** The process of sheet metal forming is one of the very important processes in manufacture of products mainly in the automotive field. In sheet metal forming, it is added a certain size at the die to tolerate a result of the elasticity restoration of material. Therefore, when the product is removed from the die then the process elastic recovery will end within the allowable tolerance size. Extra size of the die is one method to compensate for springback. The aim of this research is to optimize the die by entering a springback value in die design to improve product quality that is associated with accuracy the final size of the product. Simulation processes using AutoForm software are conducted to determine the optimal parameters to be used in the forming process. Variations the Blank Holder Force of 77 N, 97 N, and 117 N are applied to the plate material. The Blank Holder Force application higher than 97 N cannot be conducted because the Forming Limit Diagram indicates the risk of tearing. Then the Blank Holder Force of 37 N, 57 N and 77 N are selected and applied in cup drawing process. Even though a few of wrinkling are appear, however there is no significant deviation of dimension between the product and the design of cup.

## 1 Introduction

The forming process commonly called stamping plate in the industrialized world is the forming process for automotive components [1]. Its main basic material is in the form of sheet metal, this can be mild steel, aluminium alloy or titanium alloys [2-4]. One of the forming processes is deep drawing. Deep drawing is a compression-tension forming process with the greatest range of applications involving rigid tooling, draw punches, a blank holder and a female die. In this process the blank is generally constrained over the draw punch into the die to give required shape of cavity [5].

During the process, the blank sheet is likely to develop defects if the process parameters are not selected properly. Therefore, it is important to optimize the process parameters to avoid defects in the parts and to minimize production cost. Optimization of the process parameters for instance die radius, blank holder force, coefficient of friction, etc. can be concluded according to their degree of importance on the sheet metal forming characteristic [6].

At the end of the bending process, after the object removed from the die, the material's elastic energy will be received by the object that resulted objects trying to get back in shape. It was referred to as elastic recovery, which in the world of stamping referred to as springback. In this case, springback angle is defined as the addition of the product bent on stamping after being removed from the die.

The aim of this research is to optimize the die by entering a springback value in die design to improve product quality that is associated with accuracy the final size of the product.

## 2 Springback

Springback phenomenon always occurs in the process of sheet metal forming (SMF), because of the nature of the material elasticity. Thus, springback cannot be eliminated, but can be minimized or compensated in the die. Therefore, the handling of springback is very important in the design of the die to produce a product with a very high accuracy measurement [7]. Illustration of the springback phenomenon can be seen in Fig. 1, where $R_a$ is radius after springback and $R_b$ is radius before springback.

Various methods are used to reduce the springback. The traditional method of experiments on die requires a very large cost, because the wrong die is disuse anymore, while the manufacturing cost is very expensive. Then the more modern method is come by using computer equipment and software to simulate springback process. After that, it is made a design change in the die before it is produced. As it is known that the development of computers so rapidly so that computer become a tool to perform computation in a very large scale. Finite Element Method (FEM) is one of the popular way in solving the problems of static, dynamic and vibration in the design of a component.

*Corresponding author: Agung.Darmawan@ums.ac.id

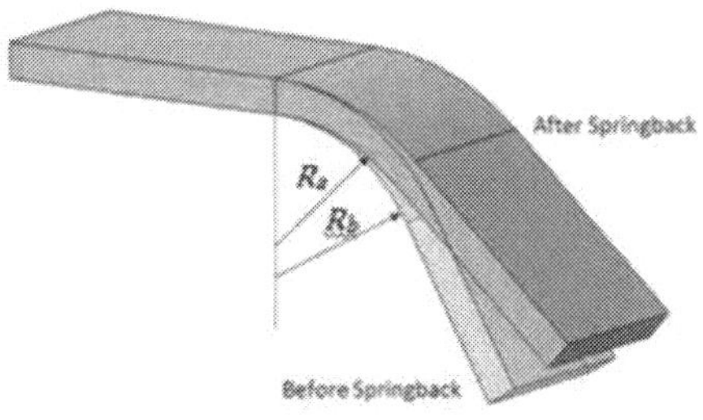


**Fig 1.** The phenomenon of springback in bending process.

FEM for numerical analysis and simulation has become an important tool in predicting material behaviour, collapse, cracking, fracture and fatigue [8, 9]. Furthermore, developing analytical models sometimes can be assisted by finite element results [10, 11].

## 3 Cup Drawing

Cup Drawing is a series of processes plat shaped like a bowl or a hat with a metal stamping which is often called deep drawing. Schematic of cup drawing can be seen in Fig. 2. On the Cup Drawing process, it is found many defects in the workmanship such as fracture, wrinkling, and thinning [12].

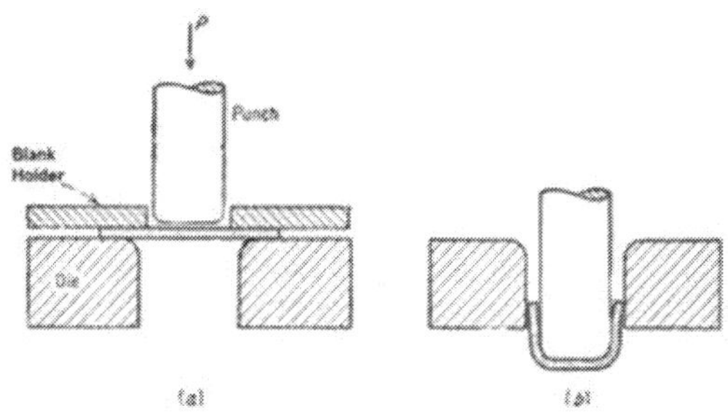


**Fig 2.** Cup drawing process (a) before process (b) after process.

The forms of these defects should be prevented or even avoided as much as possible in order to reduce production cost. In general, to avoid these problems, manufacturers require trial and error experimentation steps that would cost quite large. Therefore simulations need to avoid the occurrence of defects in the process Cup Drawing. Among them is to avoid defects wrinkling and thinning.

Wrinkling is undesirable defects that occur in the process Cup Drawing [13]. This defect occurs at the edges (flange) and the side wall of the cup.

Effects thinning or thickness variation in the cup drawing process cannot be avoided, because the thinning occurs as a result of push by the punch so that the plate which is clamped between the blank holder and dies will be pulled and will cause different thickness in some parts.

The main reason of the disability of wrinkling is the inability to resist Blank Holder Force (BHF) [14]. The optimization of BHF can be conducted by using FEM to find the BHF which minimum of wrinkling and thinning [15].

Appropriate BHF evolved through a process results in controlling the thickness variations in a deep drawn part and thus the quality of the part [16, 17].

## 4 Materials and Methods

The material used in this study is the aluminium alloy plate. Tensile testing in accordance with ASTM E8 standard performed on this material.

Design drawing is a step performed prior to the manufacture of workpiece machining. It is intended to minimize errors during the process of manufacturing or production. Solid Work 2013 is used to draw a design of a cup die. Die design refers to the standard Numisheet 2014 Benchmark Model 4. The schematic of tools and the dimension of die, punch, blank holder and blank sheet of cup drawing that will be used in the experiment are described in Fig. 3. All models are created in surface design and then save as IGES files. The files are common use to change ability.

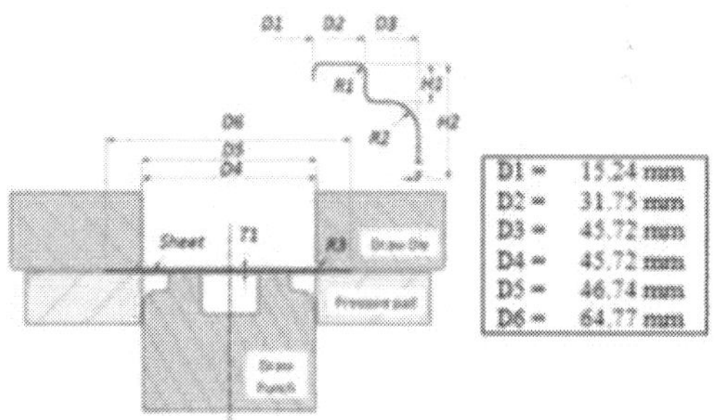


**Fig 3.** Schematic view of tools for the drawing process.

Manufacture of die sets consisting of die, blank holder and punch, is conducted by using the ST37 steel material in the form of a cylinder with a diameter of 150 mm. The material is processed with turning machine or CNC lathe in accordance with the size of the design drawings that have been made.

Aluminium alloy is selected for the blank sheet. Aluminium alloy sheet with the thickness 0.32 mm are considered in the investigation. Four equally space bolts were prepared for BHF. The machine has hydraulic press of 1500 kN capacity.

To determine the optimal parameters to be used in the forming process, then simulation processes using Auto Form software are conducted. Variations BHF of 77 N, 97 N and 117 N as well as the variation compensation factor of 0.0, 0.3, 0.5, 1.0 and 1.5 are applied to the plate material. The Forming Limit Diagram (FLD) is used to predict whether the plate will tear or not.

Finally, after the simulation process, the Cup Drawing experiment is conducted by considering the result from simulation process.

## 5 Results and Discussions

Based on the data that has been obtained from simulations and experiments that have been done before, data processing and graphically statistic analysis of cup drawing testing will be performed.

Springback is taken into consideration by providing compensation factor to the die. Fig. 4 shows that the compensation factor of 0.5 provides the lowest springback of 1.5 mm while the compensation factor of 0.0 delivers the biggest springback of 2 mm. Therefore, the compensation factor of 0.5 is applied in the experimental process.

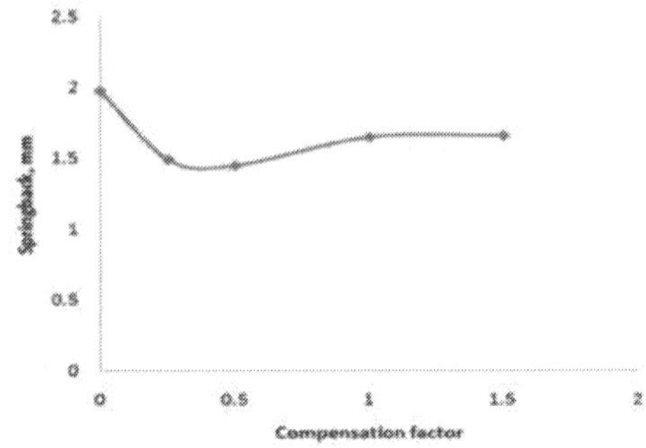


**Fig 4.** The Compensation factor versus Springback Curve.

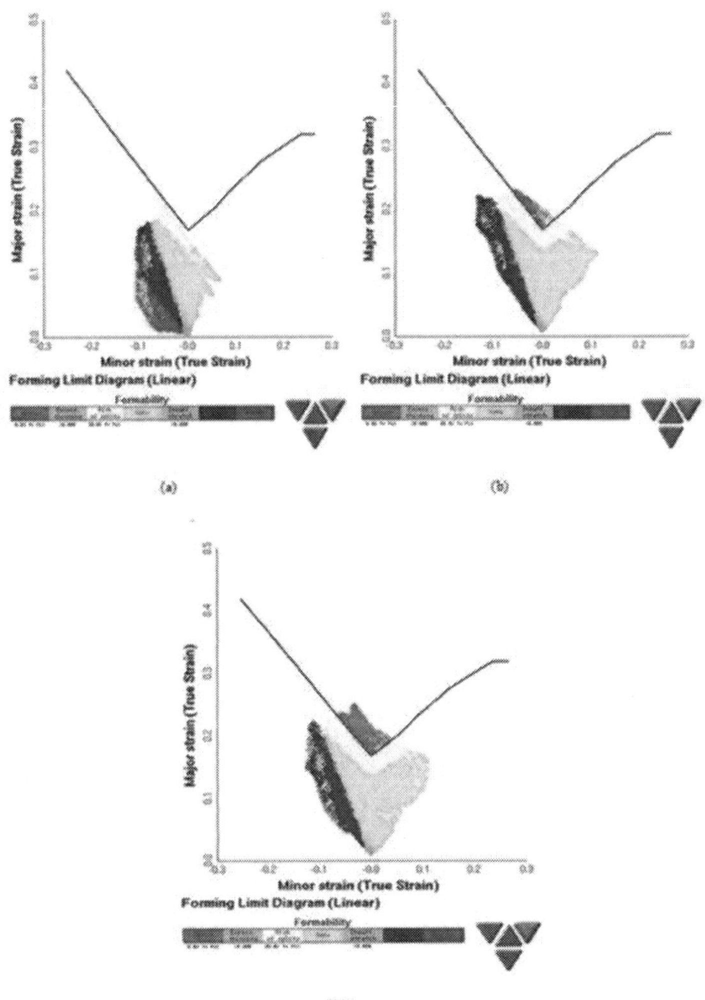


**Fig 5.** Forming limit diagram for BHF of (a) 77 N, (b) 97 N, and (c) 117 N.

The Forming Limit Diagrams which are resulted by using Auto Form software show that the process of cup drawing with BHF higher than 97 N will cause the plates to tear (see Fig. 5). Therefore, the BHF value under 97 is selected.

Experimental process of the cup drawing is conducted by applied the BHF of 77 N, BHF of 57 N, and BHF of 37 N with the compensation factor of 0.5. The study on the influence of the blank holding force over wrinkling has been made using dies that work according to the Table 1. Table 1 shows that the greater the BHF, the smaller the wrinkling that occurs. However it is still need awareness of the ultimate tensile strength of the material. Over this value the cracking and braking of the material occur.

In this Cup Drawing process, even though there is no tearing appears on the product, wrinkling defects in small size can be detected on the product of the cup drawing as can be seen in Fig. 6. It also can be seen clearly that the wrinkling are arise in the dome area and side wall.

Table 2 shows deviation of diameter of parameter D2 and D3. For Cup Drawing product, the deviation of diameter of parameter D2 and D3 are 0.78 % and 1.50 % respectively. These are considered as small deviation.

**Table 1.** Wrinkling on product of cup drawing process.

| No. | Blank Holder Force (N) | Wrinkling (mm) |
|---|---|---|
| 1 | 77 | 0.92 |
| 2 | 57 | 0.98 |
| 3 | 37 | 1 |

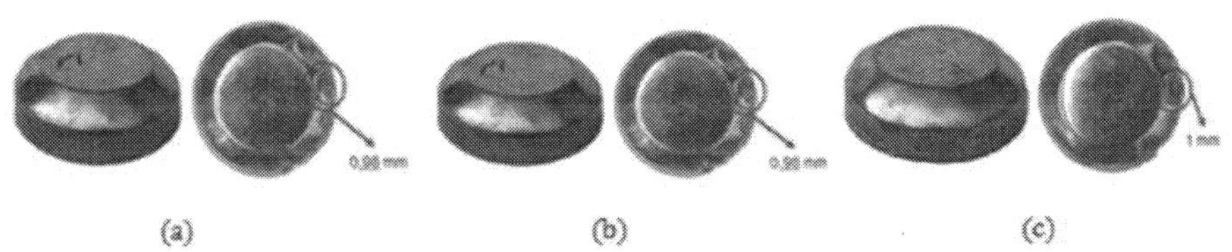


**Fig 6.** The Cup Drawing product of aluminium alloy with (a) BHF of 77 N, (b) BHF of 57 N, and (c) BHF of 37 N.

**Table 2.** Deviation of product diameter

| Parameter | Design (mm) | Cup Drawing Product (mm) | Deviation of Cup Drawing Product (mm) |
|---|---|---|---|
| D2 | 31.75 | 32.53 | 0.78 |
| D3 | 45.72 | 47.22 | 1.50 |

## 6 Conclusions

The following conclusions can be drawn from the research, by considering the springback phenomenon:

1. There is no tearing occur in the cup drawing test
2. There is no significant difference of cup dimension between the product and the design.

The authors would like to acknowledge Lembaga Penelitian dan Pengabdian Masyarakat (LPPM) Universitas Muhammadiyah Surakarta and Direktorat Jenderal Penguatan Riset dan Pengembangan, Ministry of Research Technology and Higher Education, Republic of Indonesia for the funding support of the research project [Project Number=186.21/A.3-III/LPPM/V/2016].

## References

1. J. Hirsch, Trans. Nonferrous Met. Soc. China (English Ed.), **24 (7)**, 1995 (2014).
2. N.B.K. Babu, M.J. Davidson, A.N. Rao, K. Balasubramanian, M. Govindaraju, Mater. Des., **55**, 35 (2014).
3. A. Khan, V.V.N.S. Suresh, S.P. Regalla, Procedia Mater. Sci., **6**, 401 (2014).
4. A.S. Darmawan, W.A. Siswanto, T. Sujitno, Applied Mechanics and Materials, **315**, 700 (2013).
5. G.M. Kakandikar, V.M. Nandedkar, Journal of Computational Design and Engineering, **3**, 63 (2015).
6. T. Altinbalik, A. Tonka. International Journal of Modern Manufacturing Technologies. **4 (2)**, 9 (2012).
7. A.D. Anggono, W.A. Siswanto, B. Omar, Res. J. Appl. Sci. Eng. Technol., **4 (14)**, 2036 (2012).
8. W.A. Siswanto, A.S. Darmawan, Research Journal of Applied Sciences, Engineering and Technology, **4 (10)**, 1277 (2012).
9. A.D. Anggono, T.W.B. Riyadi, W.A Siswanto, Appl. Mech. Mater., **660**, 337 (2014).
10. N. Liu, H. Yang, H. Li, M. Zhan, Z. Tao, X. Hu, Procedia Engineering, **81**, 2171 (2014).
11. L. Chen, H. Chen, Q. Wang, Z. Li, Materials & Design, **65**, 505 (2015).
12. M. Banu, M. Takamura, T. Hama, O. Naidim, C. Teodosiu, A. Makinouchi, Journal of Materials Processing Technology. **173 (2)**, 178 (2006).
13. X. Wang, J. Cao, J. of Manu. Proc., **2 (2)**, 100 (2000).
14. R.V. Reddy, T.A.J. Reddy, G.C.M. Reddy, International Journal of Engineering Trends and Technology, **3 (1)**, 53 (2012).
15. A. Agrawal, N.V. Reddy, P.M. Dixit, J. of Mat. Pro. Tech., **191**, 51 (2007).
16. H. Gharib, A.S. Wifi, M. Younan, A. Nassef, Journal of achievements in materials and manufacturing engineering, **18 (1-2)**, 291 (2006).
17. W. Zhang, R. Shivpuri, Reliability Engineering & System Safety, **94 (2)**, 152 (2009).

# Comparison of Corner-Butt 45 (Cb-45) and Corner-Lap (Cl) joints in friction stir welding

*Widia* Setiawan[1,*], *Djarot B.* Darmadi[2], *Wahyono* Suprapto[2], and *Rudy* Soenoko[2]

[1]Lecturer at Sekolah Vokasi University of Gadjah Mada, he is taking Doctorate at University of Brawijaya
[2] Senior Reseacher at University of Brawijaya

**Abstract.** The 10 mm thick Aluminum 6061 plates have been corner joined using varied design and those were 45° Corner Butt and Corner Lap Joints (CB-45 & CL). Friction tool was hardened EMS 45. True experimental method was used with independent parameters is feed rate which varied at 10 mm/min, 15 mm/min and 30 mm/min respectively. Other parameter such as rotating speed was kept constant. Experiment results show that, CB-45 yields better properties than CL. The tensile strength of CB-45 reaches 163.7 MPa for 10 mm/min feed rate. Whilst CL produces joint with tensile strength equal 120 MPa for equal parameters. Microstructure observation showed that CB-45 produces fine and homogenous appearance of MgO compared to CL. This phenomenon is caused by the pin of CB-45 joint which fully penetrates the nugget zone which is not found in CL design. This microstructure in turn promotes higher tensile strength of CB-45.

## 1. Introduction

Friction Stir Welding (FSW) offers some benefit that can overcome the drawback of fusion welding when applied to aluminum. Problems such as porosity, the existence of $Al_2O_3$ film, hot cracking and strength loss is exhibited in fusion welding beside others such as residual stress (Darmadi *et al.*, 2014). Mainly the problem of fusion welding is caused by phase transformation, from liquid to solid states (Darmadi *et al.*, 2016), which is not performed in FSW.

As it has been implicitly mention in previous paragraph, FSW is a solid phase joining technique on fabrication industry. Good quality single sided and double sided butt, "T", and lap joints. Invented in 1991, and was originally used to produce butt joints of aluminum alloys (Thomas *et al.*, 1991).

The Friction Stir Welding (FSW) is a solid-state, hot-shear joining process which a rotating tool with a shoulder and terminating in a threaded pin. It moves along the butt surfaces of two rigidly clamped plates placed on a backing plate as shown in Fig. 1.

The shoulder position is firm contact with the top surface of the work-piece. The heat generated by friction on the shoulder and a lesser extent on the pin surface, softens the material being welded. During FSW, heat is generated by friction between the tool and the work piece through via plastic deformation. The fraction of the plastic deformation energy is stored within the thermo mechanically processed region in the form defect densities increment (Nandan et al., 2008)

Friction stir welding (FSW) is a friction welding, requires no added ingredients. The heat generated from friction between probe and workpiece. Probe spin at a certain speed then placed on a connection that has gripped material. Friction of both objects generate heat 0.8 point liquid workpiece (Tang et al., 1998). Then the pin is pressed and moved in the direction of the weld line. Probe melting point and hardness are used in friction stir welding to be greater than the material being welded. Welding by the method of friction stir welding (FSW) is used for welding the same material (similar metal) or a material that is not the same (dissimilar metal) such as steel, stainless steel, aluminum with brass and possible to weld a combination of other materials which can not be welded by other welding method.

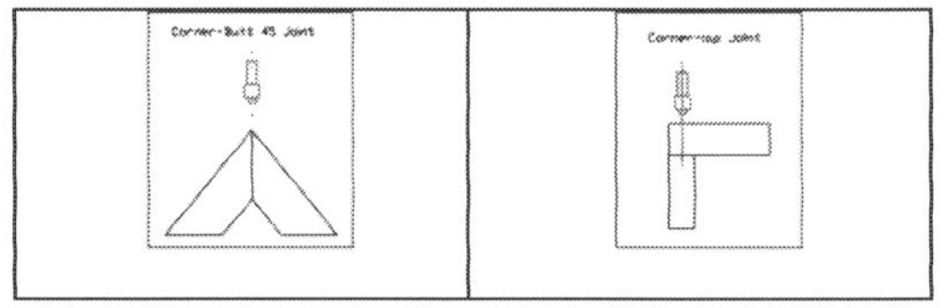

**Fig 1.** Corner-Butt 45 and Corner-Lap Joint on friction stir welding

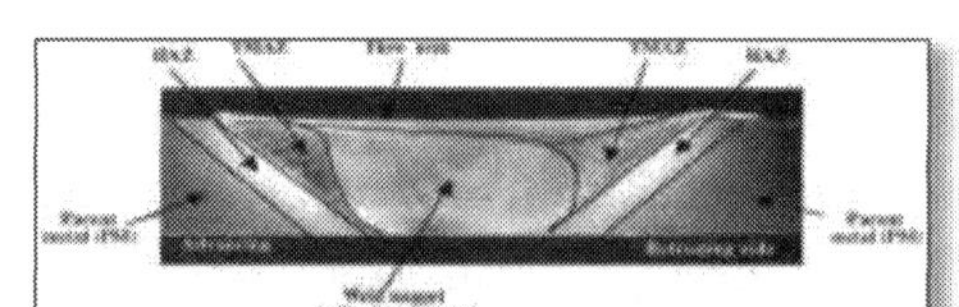

**Fig 2**. Heat zone pada friction stir welding (Mishra et al., 2007)

FSW process produces a typical microstructure. The

*Corresponding author: widia_s@.ugm.ac.id

resulting microstructure can be divided into four zones [4] as in figure 2.

The FSW has become an efficient option of welding method for the same or dissimilar aluminum alloys, especially those which are difficult or impossible to be welded by the conventional fusion welding without any hot crackings, blowholes or distortions (Kazi and Murr, 2001),( McNelley et al., 2008). The welding corner welds need to have a filled in order for the weld piece to achieve high enough rigidity. To be able to create this fillet additional material needs to be supplied into the weld. The additional material then needs to be mixed with the welds and shaped into a fillet. Because of the 90 degrees angle of the weld piece and the shaping of the fillet, it is not possible to have a rotating Shoulder. This is a stationary Shoulder is needed to be able to create a corner weld. When trying to create a corner weld with normal FSW without wire feeding the Shoulder will cut into the weld piece creating a non-rigid weld as seen Fig.3 (Alexander and Marcus, 2014).

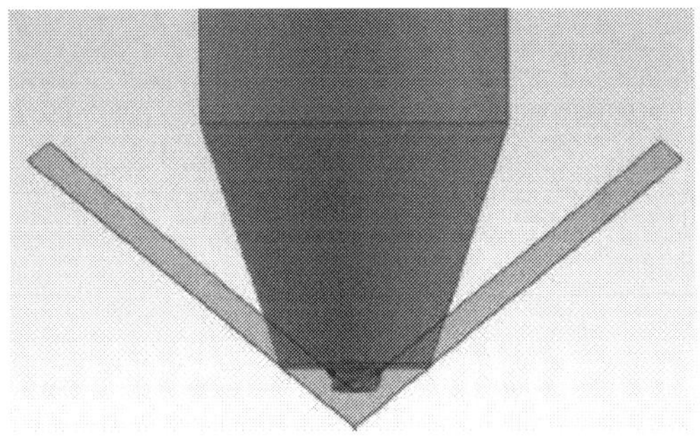

**Fig 3**. Corner jonit (Alexander and Marcus, 2014).

Friction Stir Welding is a new technique in the manufacturing industry, especially aluminum alloys. This non-consumable electrode technique makes use the heat generated by the rotating tool. Its deformation at the welding zone thereby affects the joint formation material into the solid state. The FSW has become an efficient option of welding method for the same or dissimilar aluminum alloys, especially those which are difficult or impossible to be welded by the conventional fusion welding without any hot crackings, blowholes or distortions (Padgett *et al*., 2003),( Ghosh et al., 2010).

In Journal Friction Stir Welded T-joint optimization [10]. One of the conclusions says at 1000 rpm, a depth of 3.90 mm diameter probe and the ratio of the shoulder/investigations 2.5 will be obtained an optimum connection.

**Table 1**. Level of the level parameters (Ana *et al*., 2013)

levels of the selected parameters.

| | Parameters | Unit | Level 1 | Level 2 | Level 3 |
|---|---|---|---|---|---|
| A | Tool rotational speed | rpm | 490 | 1000 | 1500 |
| B | Welding speed | mm/min | 76 | 216 | 360 |
| C | Shoulder/probe diameters ratio (D/d) | - | 2 (12/6) | 2.5 (15/6) | 3 (18/6) |
| D | Probe depth | mm | 3.50 | 3.70 | 3.90 |

Conducted experiment with two combinations modes T-lap and T-Butt-lap joints, special clamping joint was used. As shown in fig 3. (Hou et al., 2013)

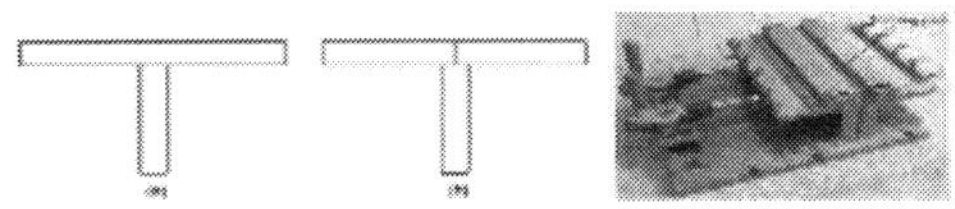

**Fig 4**. T-lap, T-Butt-lap joints, and special clamping fixture (Hou et al., 2013)

The same experiment with three combinations modes T-lap, T-butt-lap, and, T-double Butt joints is listed in fig.4 (Hou et al., 2013)

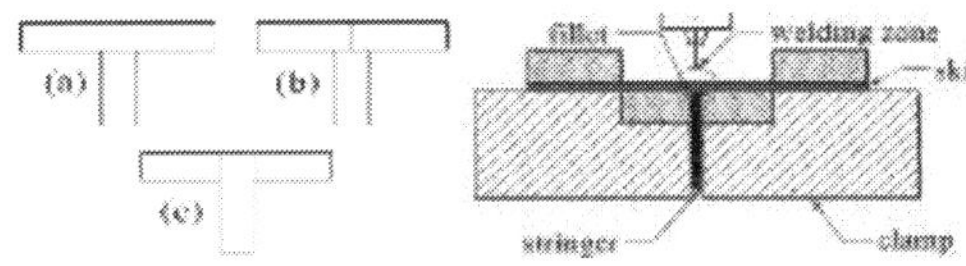


**Fig 5**. T-lap, T-butt-lap,T-double butt, and FSW process (Hou et al., 2013)

The Fig. 3 shows the effect of different rotational speeds on the recorded welding thermal cycles. The thermocouples were placed on the mid-plane of the plate on the advancing side (AS). It is 10 mm away from the weld center. It can be seen the peak temperature increases from 252 to 330 °C with the rotational speed increasing from 400 rpm to 1000 rpm. Moreover, the typically elevated-temperature over 150 °C exposure time 150 chronological increases from 57 to 100 s as the rotational speed increases from 400 rpm to 1000 rpm (Wang *et al*., 2015).

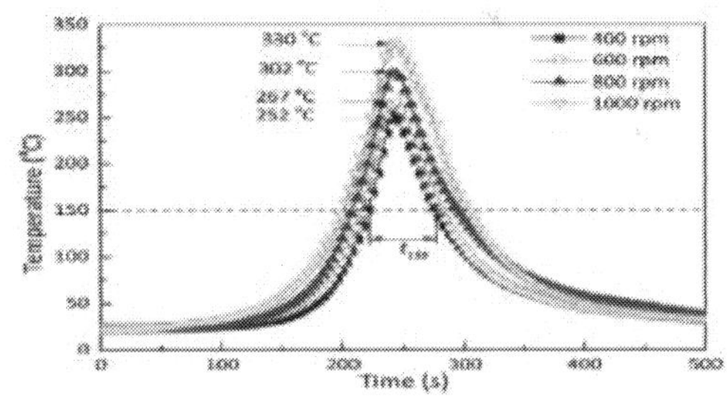


**Fig 6.** The welding thermal cycle of the joint produce (Wang *et al*., 2015).

## 2. Experimental procedures

In this study, the type 6061 aluminum material with dimensions of 150 mm x 70 mm x 10 mm. 6061 is a blend of pure aluminum with a Silicon (Si) and magnesium (Mg). Tool / probe used in the research process are made from medium carbon steel material EMS 45 which has undergone a hardening process. The function of the tool / probe is to make the material into plastic and stir the plastic material in the connection area.

**Tabel 1.** Chemical composition of 6061 aluminum alloy (wt.%). ASM Metal Handbook, vol 9, 2004

| Compositions | Mg | Si | Cu | Mn | Fe | Cr | Ti | Zn | Al |
|---|---|---|---|---|---|---|---|---|---|
| Content | 1.5 | 1.8 | 0.25 | 0.086 | 0.18 | 0.1 | 0.192 | 0.01 | Bal. |

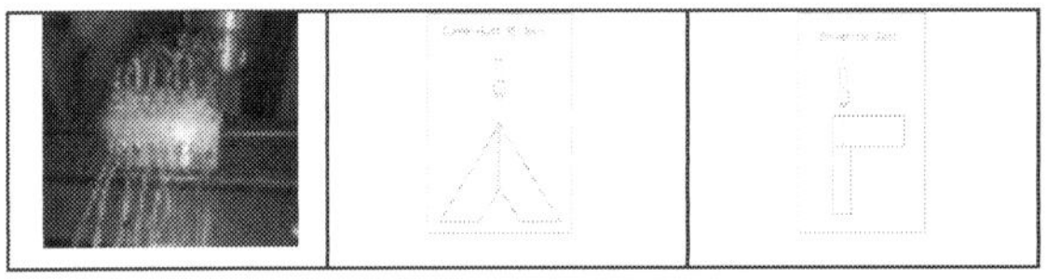

**Fig 7.** Installation of the workpiece in the jig fixture, thermocouple and design-Butt Joint Corner connection 45 (CB-45) and Corner-Lap (CL)

Design probes that will be used in the process by the method of friction stir welding on 6061 aluminum material, thickness 10 mm, the type of corner-Butt joint connection 45 (CB-45) and Corner-lap (CL) is as follows:

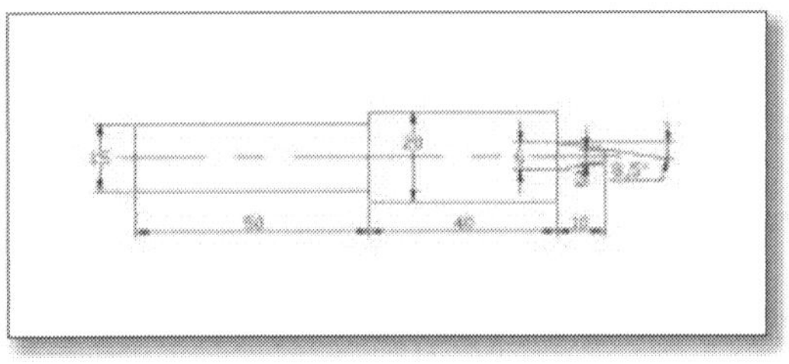

**Fig 8**. Design tool / probe

## 3 Result and Discussion

### 3.1 Comparison Visually

The result of the friction stir welding process of joining the 6061 design material connection CB-45 and CL, use a 2000 rpm spindle rotation and variation feedrate 10 mm / min, 15 mm / min and 30 mm / min is as follows:

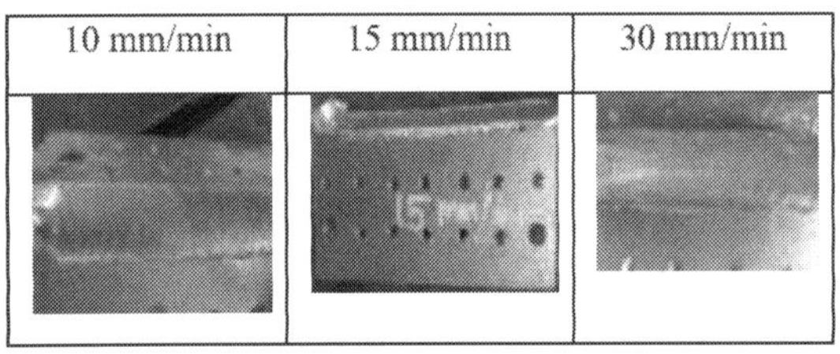


**Fig 9.** CB-45 Macro photos feedrate 10 mm/min, 15 mm/min, and 30 mm/min FSW

The first experiments carried out with feedrate 10 mm/min and 2000 rpm spindle rotation obtained good results seen visually, wherein the upper surface of the weld looks neat and almost all surfaces exposed to friction. This occurs because the time spent in a very long which is 15 minutes, resulting in plastic conditions which formed more homogeneous. Then the plastic material will be stirred by the PIN evenly and will cause the connection to be better. The plastic material that occurs will be stirred evenly by the PIN, making the joints tidy and flat for C-B 45 and C-L types. Si (silica) as an alloying element in aluminum 6061 also makes good flow, thus making the plastic material surface smooth for CB-45 and CL types. See figures 9 and 10 on a feedrate of 10 mm / min.

In the second experiment performed with a feedrate of 15 mm / min and spindle rotation of 2000 rpm, the results obtained were less good, compared to the first experiment with a feedrate of 10 mm / min visually seen. The welding results of the top surface look unkempt, because the temperature distribution is lower, then the time used for the connection process is shorter. So that plastic conditions are not well achieved (less homogeneous), this will result in the stirring PIN cannot do perfect mortar, resulting in rough macro photo surface conditions. This will cause poor connection quality. This will result in the stirring PIN is not able to do mortar material with less plastic conditions. Then it will cause poor connection quality for CB-45 and CL connection type at feedrate 15 mm / min. Can be seen in Fig. 9 and 10 feedrate 15 mm / min.

In the third experiment conducted with a feedrate of 30 mm / min and spindle rotation of 2000 rpm, obtained macro photo results look rough, because the plastic condition is clearly not achieved and not flat on each shift probe that occurred. Visually this illustrates the pastis condition of a material stirred by an imperfect pin, causing a rough surface texture of the surface. Because the time to warm up is less, about 5 minutes to make the connection process. Causing the resulting low temperature distribution. Will result in connection quality happening happened blowhole, porosity, Rough on the nugget area For the CB-45 connection type. (See fig. 9 at feedrate 30 mm / min). While for the CL connection type of heat is not reached, then plastic conditions are also not achieved. This results in cracks in joints along the line of connection. Can be seen fig. 10 at feedrate 30 mm/min.

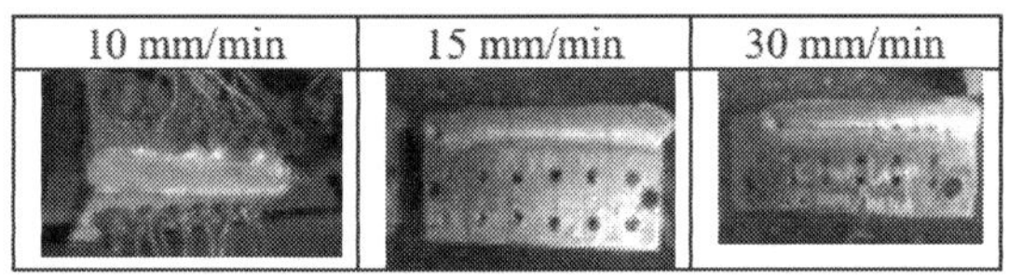


**Fig 10.** Macro photos CL feedrate 10 mm/min, 15 mm/min, and 30 mm/min FSW

### 3.2 Weld Thermal Cycles

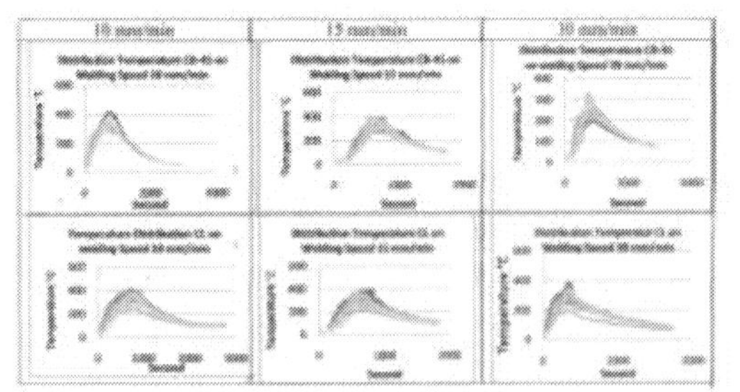

**Fig 11.** Distribution welding temperature of CB-45 and CL with feedrate 10 mm min, 15 mm/min, and 30 mm/min.

Measurement of temperature distribution that happened at aluminum connection 6061 with friction stirs welding method using data logger. Then the time required for feedrate10 mm / min is 900 seconds, while the time required for feedrate15 mm / min is 600 seconds, and the time to feed rate 30 mm / min is 300 seconds. The temperature graphs generated in the process are as follows:

Figure 11. shows six temperature measurements on the welding friction stir welding process. In the first graph, the temperature distribution is generated with peak temperature reaching 434 $^0$C in 747 seconds. At this temperature plastic material will be easily achieved. So the PIN will stir the plastic material perfectly. This will cause the connection quality to occur either, not Blowhole. Then visually macro photos more neat, smooth, with better connection quality. While at the third temperature graph distribution is 220 $^0$C the temperature measurement occurs on the parent metal or 30 mm of the shoulder (fig. 9).

In the measurement of temperature distribution friction stir welding is done with feedrate 15 mm/min position of CB-45 material Al 6061. In graph seen the first position indicates the distribution of temperature evenly in accordance with time reached. Peak temperature is 390 $^0$C in 545 seconds. This is enough to make plastic material on aluminum, but plastic material that formed less homogeneous, so that PIN will be difficult to do the mortar plastic material. Poor connection quality, can also blowhole material as result of low heat produced. Due to the FSW process will be good when the temperature reached is 0.8 liquid temperature (Tang et al., 1998). (fig. 9 and fig. 10).

Figure 10. Before the process of grafting friction stir welding is done, the temperature of material Al 6061 is 30 $^0$C. Then the probe/ tool begins to enter (plunge) and swipes the aluminum material 6061 (weld process), the temperature of the aluminum material 6061 began to rise slowly at about 500 seconds reaching the temperature of 311 $^0$C. The time used is 5 minutes, the material length is 150 mm, so the low temperature does not reach 0.8 Tm. This plastic condition is not achieved, so the probe PIN is not able to stir the semi-solid material. The results shows that defects is found, not fully melted interface, porosity and not good visual texture (Fig.10 and Fig. 11).

## 3.3 Temperature Distribution.

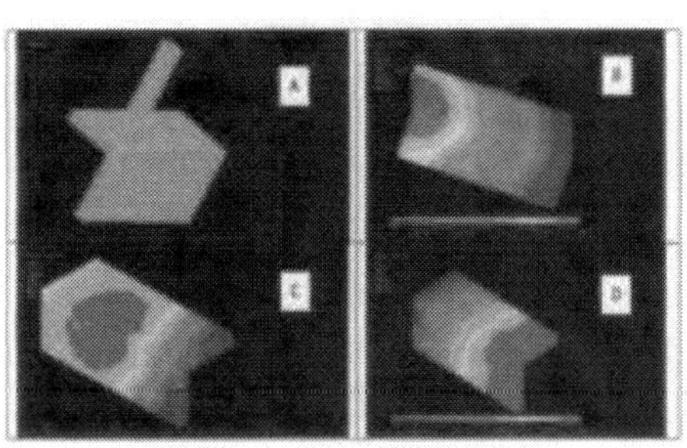

**Fig 12.** Mesh (A) and temperature distribution in 3 minutes (B), 8 minute (C) and 15 minutes (D)

Fig. 12 explains In the process of friction for 15 minutes, the heat will be centered on the shoulder probe, temperature about 400 $^0$C, this occurs at a temperature difference thermocouple measurements. It can occur because of the position of the thermocouple within 20 mm from the center of the heat (Shoulder Probe). Fig. 11.A shows the heat that occurs at the time of 8 minutes, the temperature distribution almost all temperatures approaching 350 $^0$C. This position close to the perfect plastic material would be causing the connection quality is also good. FSW process 14 minutes later the temperature distribution at 400 $^0$C nearly all plastic materials will be formed. So this will make the quality of the connection becomes stronger. (Fig 12.B)

## 3.4 Micro Structure

In Fig. 13, the 10 mm / min feed rate shows a very clear line of connections, which indicates good connection quality (supported fig. 11). Then the microstructure that occurred was obtained by a homogeneous plastic condition compared to the microstructure at the other feed rate because the temperature reached at 10 mm / min of high FSW process was 434 $^0$C. And the result of the plastic material PIN mix is more perfect, so it will obviously result in the mortar that occurs in the better microstructure photos, tensile stress will be high later.

| **Microstructure Corner Butt 45** | | |
|---|---|---|
| 10 mm/min | 15 mm/min | 30 mm/min |
| 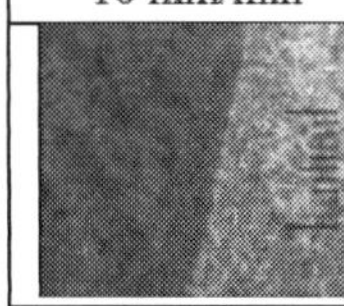 | 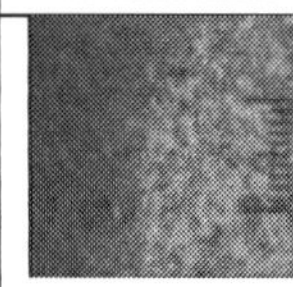 | 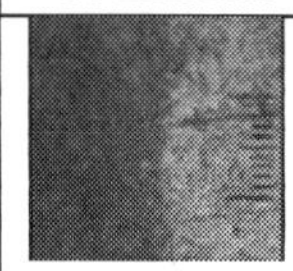 |
| **Microstructure Corner Lap** | | |
| 10 mm/min | 15 mm/min | 30 mm/min |
| 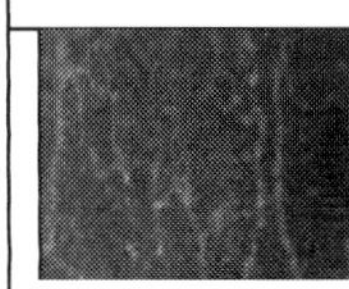 | 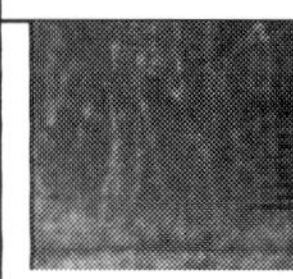 | 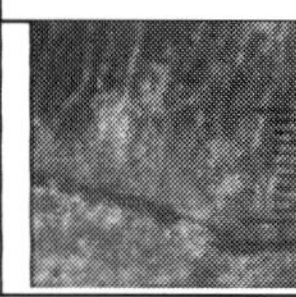 |

**Fig 13.** Microstructure corner butt 45 and corner lap Friction stir welding

The microstructure occurring is dominated by the β-$Mg_2Si$ phase, eventually forming a magnesium oxide (MgO) compound at a low temperature of 150 $^0$C. With reduced Magnesium in Al-Mg-Si will result in reduced mechanical properties of aluminum 6061. White color on Micro structure is a silica alloy (αAl phase) shows the mortar on the plastic material. The properties of silica provide a good flow of aluminum. Plastic material in direct contact with the tool/probe will form a finer homogeneous and black microstructure. Because at a temperature of 200 $^0$C, the magnesium-silicon alloy ($Mg_5Si_6$) is more stable and does not change the crystal structure. But if formed above the temperature of 200 $^0$C will form $Mg_5Si_6$ which dissolves dynamic recristallisation. This will decrease the mechanical properties and form a black precipitate. In the thermo-mechanically affected zone (TMAZ) region or area occurring on both sides of the stir zone, the structure looks coarser with a more dominant color of white. The black microstructure is the $\beta$-$Mg_2Si$ phase mixed with the Fe (iron) element of the probe/ tool while the white color is the $\alpha$-Al phase.

At Feed rate 15 mm / min, and 30 mm / min, the microstructure is formed brighter. This is as a result of high feed rate speed, MgO oxide compound that formed less due to the processing time of 10 minutes and 5 minutes, and oxygen that reacts to form a small magnesium oxide compound.

### 3.5 Micro Vickers

The hardness test is aimed to know the hardness of welding friction stir welding. This hardness testing is done by using micro Vickers method with 100 gf or 0.1 kgf indenter load.

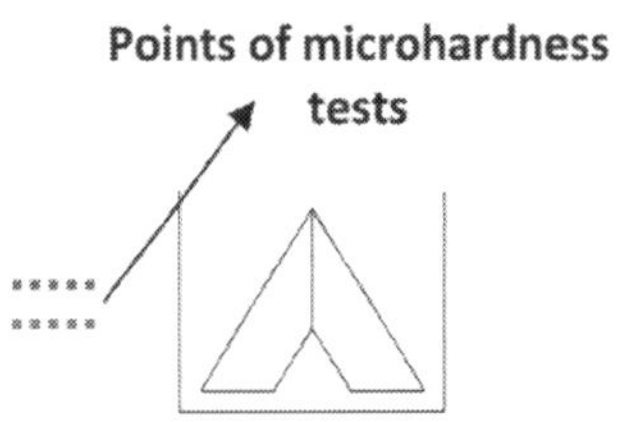


**Fig 14.** Points of microhardness vickers tests

Hardness testing is done on welding friction welding stirrer area with

Indent point as much as twelve points to the right and left of the center of the welded joint.

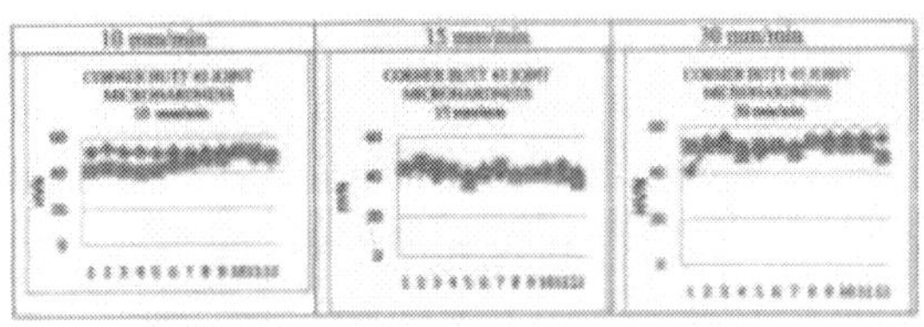

**Fig 15**. MicroVickers graph Corner Butt CB-45

From Figure 15,17 can be seen that a feed rate of 10 mm/min, and 15 mm/min of mean hardness is higher than federate 30 mm/min. This can happen because the resulting temperature is low, then generated high temperature that is 306 $^0$C, and 243 $^0$C this will cause the content of Mg (Magnesium) on Aluminum to form oxide compounds that MgO will increase the hardness, as well as mechanical properties. When viewed from the X-Ray test occurs Peak, it shows a lot of magnesium content to form oxide compounds. With the reduction of magnesium in alloys 6061 will cause the hardness to fall, but magnesium will bind to the oxide to form a magnesium oxide (MgO) compound that causes the mechanical properties to rise. So from the theory and research data on the FSW process at high temperatures close to 0.8 TM at Al 6061 will result in increased hardness (Hasan et al. 2015).

Figure 19, XR-D shows the magnesium content to form the oxide compound (MgO) more at a rate of 10 mm / min, this can happen because at a low O2 will bind more Mg about 150 °C. Consequently, the reduction of Mg in Aluminum and binding to the oxide causes the mechanical properties to rise. At a higher of 15 mm/min the oxide compound to form MgO will be lower.

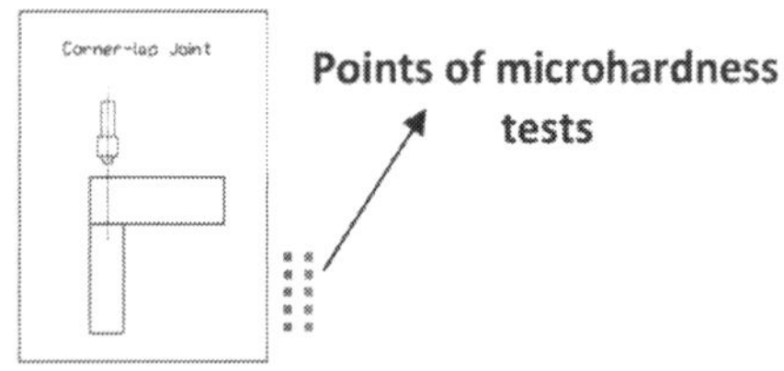


**Fig 16.** Points of microhardness vickers tests

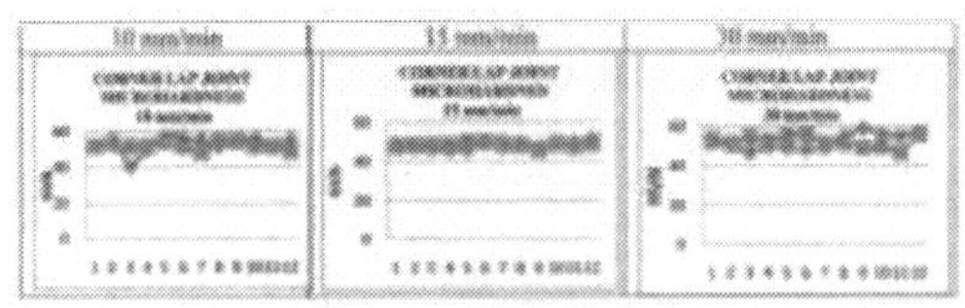

**Fig 17.** MicroVickers graph Corner Lap

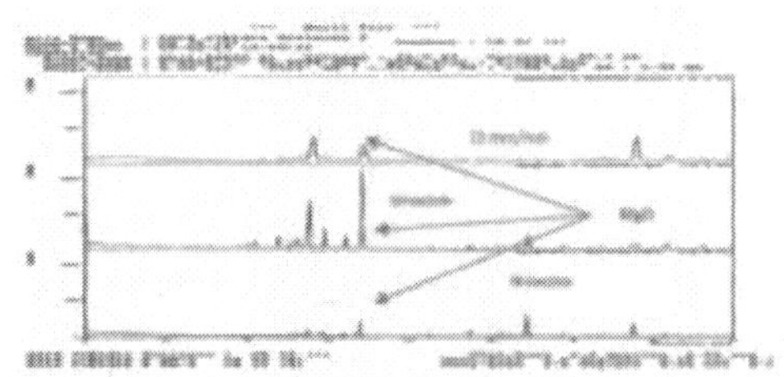

**Fig 18.** XR-D Corner-Butt 45 (CB-45) friction stir welding

### 3.6 Tensile

Tensile test of welding corner butt 45 corner and corner joint at aluminum 6061 friction stir welding method is used jig fixture fig. 19 as shown below:

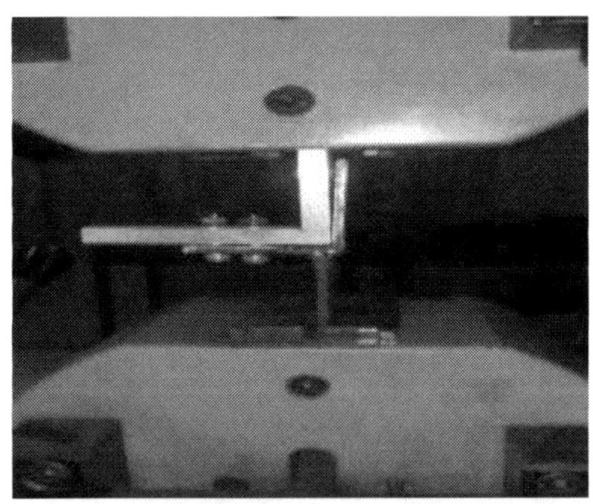

**Fig 19.** Tensile testing process with jig fixture

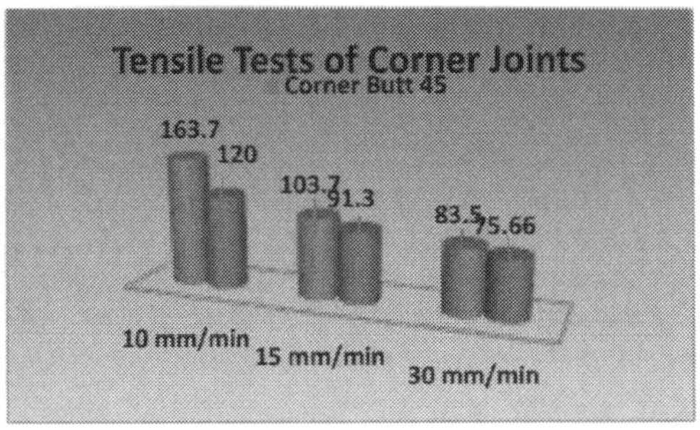


**Fig 20.** The average tensile stress graph of the FSW result

The highest tensile test value of 163.7 MPa at feedrate 10 mm/min could happen:

At low Feedrate will be generated high temperature distribution, resulting in a more widespread homogeneous plastic area. This large plastic area will affect the Tensile Test results.

In the Corner Lap joint, the Pin probe height which is rubbing against the line of connection is not in full contact with the plate thickness, so that the probe pin-stirred material is only partially. ut this not-so-thick corner connection design will make the plastic material more homogeneous, so that the mortar will be better than federate 15 mm/min, or 30 mm/min, and the tensile strength of the material is better.

Corner Butt 45 joint, high pin probe will stir the plastic material as thick as the material to be joined. With this mortar the connection quality will be stronger than the Lap Corner. Also the magnesium bonded to the oxide will be more evenly thicker as material.

## 4 Conclusions

The Welding material of aluminum 6061 comparison corner Butt joint 45 and corner lap using friction stir welding method can be concluded as follows:

1. Corner butt 45 joint with feedrate 10 mm/min has better result visually, micro structure, and temperature of 434 $^0$C. And the result of the connection has a smoother texture Compared to the corner lap joint.
2. From the SEM image, the microstructure, and the temparature distribution can be drawn to a red thread that is, the feedrate 10 mm/min microstructure shows a good connection, supported by the SEM test, this is due to the high peak temperature distribution.
3. Welding with feedrate 10 mm/min, and 15 mm/min average hardness higher, compared with 30 mm/min. Because Magnesium oxide is formed, and the peak temperature is high for corner butt 45 and corner lap.
4. The highest tensile stress is 154.67 MPa at Feedrate 10 mm/min. Because the plastic material is more homogeneous, compared to feedrate 15 mm/min or 30 mm/min.

## References

1. Ana C.F. Silva , Daniel F.O. Braga , M.A.V. de Figueiredo b, P.M.G.P. Moreira,2013, Friction stir welded T-joints optimization, Instituto de Engenharia Mecânica e Gestão Industrial (INEGI), R. Dr. Roberto Frias, 400 4200-465 Porto, Portugal Faculdade de Engenharia da Universidade do Porto (FEUP), R. Dr. Roberto Frias, s/n 4200-465 Porto, Portugal.
2. Beckham A., Sundstom M., 2014, Development of stationary for friction stir welding, Departement of product development, Chalmers University Of Technology, Gothenberg, Sweden.
3. Darmadi, D.B.; Kiet-Tieu, A. and Norrish, J. (2014), A validated thermo mechanical FEM model of bead-on-plate welding, International Journal of Materials and Product Technology, vol. 48, nos. 1-4, pp. 146-166.
4. Darmadi, D.B. (2016), Evaluation of the effects of melting phenomenon on the residual stress formation in ferritic pipeline multi pass girt-weld joints, International Journal of Engineering System Modeling and Simulation, vol. 48, no.3, pp. 205-217.
5. Dawood H. I. , Mohammed K., S. , Rahmat A. , Uday M.B.,2015, The influence of the surface roughness on the microstructures and mechanical properties of 6061 aluminium alloy using friction stir welding, Universiti Malaysia Perlis, School of Materials Engineering, Taman Muhibah-Jejawi-Arau, 02600 Perlis, Malaysia UTM — Centre for Low Carbon Transport in cooperation with Imperial College London, Institute for Vehicle Systems and Engineering, Universiti Teknologi Malaysia, 81310 Skudai, Johor, Malaysia
6. Ghosh M. , Kumar K., Kailas S.V., Ray A.K., 2010,Optimization of friction stir welding parameters for dissimilar aluminum alloys, Materials Science and Technology Division, National Metallurgical Laboratory (CSIR), Jamshedpur 831 007, India Department of Mechanical Engineering, Indian Institute of Science, Bangalore 560 012, India
7. Hou X, Xinqi Yang , Lei Cui, Guang Zhou,2013, Influences of joint T-joints, Tianjin Key Laborat[ory of Advanced Joining Technology, School of Materials Science and Engineering, Tianjin University, Tianjin 300072, China.
8. Hou X, Xinqi Yang , Lei Cui, Guang Zhou Influences of joint geometry on defects and mechanical properties of friction stir welded AA6061-T4 T-joints, Tianjin Key Laboratory of Advanced Joining Technology, School of Materials Science and Engineering, Tianjin University, Tianjin 300072, China,2013
9. Kazi SH, Murr LE. Complex flow phenomena associated with friction-stir welding of aluminum alloys. In: Mishra RS, et al., editors. Friction stir welding and processing, Warrendale (PA): TMS; 2001. p. 139–151.
10. McNelley TR, Swaminathan S, Su JQ. Recrystallization mechanisms during friction stir welding/processing of aluminum alloys. Scripta Mater 2008;58:349–54

11. Mishra, Rajiv S dan Mahoney, Murray W. 2007. *Friction Stir Welding and Processing.* ASM Internasional.
12. Nandan R., DebRoy T. , Bhadeshia H.K.D.H. Recent advances in friction-stir welding – Process,weldment structure and properties, Department of Materials Science and Engineering, Pennsylvania State University, University Park, PA 16802, USA Department of Materials Science and Metallurgy, University of Cambridge, Cambridge CB2 3QZ, UK, 2008.
13. Padgett PN, Paglia C, Buchheit RG. Characterization of corrosion behaviour in friction stir weld Al–Li–Cu AF/C458 Alloy. Friction stir welding and processing II. Warrendale (PA, USA): TMS; 2003.
14. Tang W., Guo X., McClure J.C., and, MURR. L.E., Metallurgicall, and Materials Engineering Departement, University of Texas at El Paso, El Paso TX 79968 USA, 1998
15. Thomas W.M., Nicholas E.D., Needham J.C., Murch M.G., P. Templesmith, C.J. Dawes, G.B Patent Application No. 9125978.8 (December 1991).
16. Wang F.F. , Li W.Y., Shen J. , Hu S.Y., dos Santos J.F., Effect of tool rotational speed on the microstructure and mechanicalproperties of bobbin tool friction stir welding of Al–Li alloy, State Key Laboratory of Solidification Processing, Shaanxi Key Laboratory of FrictionWelding Technologies, Northwestern Polytechnical University, Xi'an 710072, Shaanxi, PR China Helmholtz-ZentrumGeesthacht, Institute of Materials Research, Materials Mechanics, Geesthacht 21502, Germany, 2015

# The effect of pressure and frequency on the dynamic behavior and evaporation time of successive water droplets impacting onto hot surface

*Teguh* Wibowo[1,3*], *Arif* Widyatama[1], *Samsul* Kamal[1,3], Indarto[1], and Deendarlianto[1,2]

[1]Department of Mechanical & Industrial Engineering, Universitas Gadjah Mada, Jl. Grafika No.2 Yogyakarta, Indonesia.
[2]Center for Energy Studies, Universitas Gadjah Mada, Sekip Blok K 1A, Kampus UGM, Bulaksumur, Yogyakarta, Indonesia
[3]Department of Mechanical Engineering, Sekolah Tinggi Teknologi Adisutjipto, Jl. Janti Blok R. Lanud, Yogyakarta, Indonesia.

**Abstract.** Multiple droplets are the drop of water which continuously dropped on a surface in certain time difference. The phenomenon droplets impacting a solid surface can be found in a various field, one of them is the cooling process in the metal industry. The cooling process is carried out by spraying a number of water droplets to hot surface until it reaches the desired condition. When droplets impinge on the surface, the dynamics of droplets, such as the spreading and recoil phenomena, depend on some properties. In this study, the effect of pressure and drop frequency to the droplet characteristics as well as cooling effectiveness will be investigated. Visualization process is used to find maximum spreading which can indicate the effectiveness of hot surface cooling. The experiment is performed by setting the distance of water tank and nozzle into 100cm, 150cm and 200cm while the frequency of droplets is set at 245, 386, and 623 drops/minute. The material used in the present work is Stainless Steel with the temperature ranged from 100°C - 220°C. In addition, image processing technique is applied to gather the quantitative data from the images that successfully taken by using high-speed camera. Based on the study, it is found that the pressure of water tank affects the frequency of generated droplet. Furthermore, the frequency of droplet impacting onto the hot surface also influences the evaporation time.

## 1 Introduction

The application of multiple droplets on numerous industrial field has grown massively nowadays. For instance, in the material industry, spray cooling method is usually utilized in quenching process of metal since this method is able to distribute the heat better than the conventional method. As a result, it provides some benefits such as reducing the production cost and improving the accurateness as well as the efficiency of cooling process on high-quality metal forming which will significantly affect the profitability of the final product [1].

To obtain in depth understanding related to those phenomena, comprehensive study is essentially needed. Based on several previous research, it is noted that characteristics of spray cooling, such as the bubble dynamics and transient heat transfer, can be investigated by successive multiple droplet approaches. In addition, [2] has revealed that the deformation behavior of a single droplet impacting on a hot solid surface is affected by various factors, such as impact velocity and pre-impact diameter of droplet, liquid viscosity, surface tension, wettability between liquid and solid surface, liquid and solid temperatures, roughness of solid surface, and impact angle. Increasing impact velocity enhances heat flux from the substrate by only a small amount [3].
Kandiklar et al. [4] have investigated the effect of Webber number on the characteristics of droplet impinging on the solid surface. This research focused on the observation during droplet levitation and has classified that phenomena into four types. Furthermore, the phenomena of hydrodynamics and boiling phenomena of water droplet have been studied by Fujimoto et al [5]. It covers various condition (impact velocities, droplet diameters, and temperatures) and used high speed camera to observe the behavior of droplet and the presence of secondary droplet. However, the previous research has focused on the phenomena of single droplet while in real situations, the droplets usually impact upon a hot solid surface in the form of successive multiple droplets. Therefore, the comprehensive study related to multiple droplets should be conducted.

The objective of the present study is to investigate the effect of pressure and frequency to the dynamic behavior and heat transfer phenomena of successive water droplets impacting on to hot surface. The material used in this present study is stainless steel. The visualization by using

*Corresponding author: teguhwibowo76@yahoo.co.id

high speed video camera, supported by image processing technique, is applied in order to reveal the specific phenomena comprehensively.

## 2. Experimental apparatus and procedure

Figure 1 depicts the schematic of the experimental apparatus to observe the dynamic behavior of multiple successive droplets on a hot solid surface. It is placed in Fluid Mechanics Laboratory of the Department of Mechanical and Industrial Engineering, Universitas Gadjah Mada and generally consists of droplet generator, hot surface, an illumination system, a high-speed video camera, and personal computer.

The droplet generator consists of two main parts, the water tank, as a water storage, and droplet injector. The position of the water tank can be adjusted in vertical axis to observe support the aim of this study. In the present work, the height of tank is set on 100 cm, 150 cm, and 200 cm. Furthermore, the droplet diameter and the distance of droplet injector from hot surface are 2.8 mm and 70 mm, respectively. To observe the phenomena of multiple successive droplets clearly, a high-speed video camera is utilized and set on 1000 fps and very low shutter speed. The produce bright and optimal visualization, a set of LED lamp, is also installed in particular position.

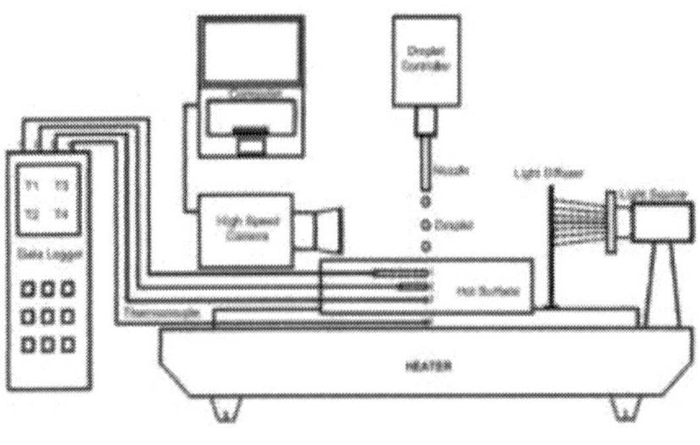


**Fig 1.** The schematic drawing of experimental apparatus

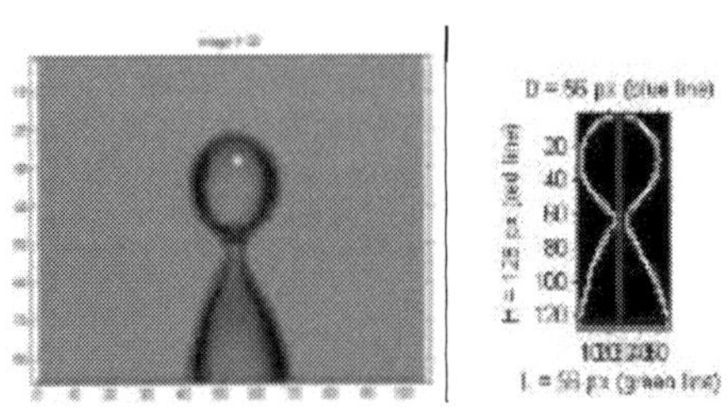

**Fig 2.** The example of image processing result

In the present study, stainless steel is used as experiment specimen. It is placed on the induction heater to achieve the desired hot solid surface condition which is set in 100°C, 140°C, 180°C, and 220°C. The benefit of this heater is the ability to adjust the temperature easily and can reach the experimental temperature fast. In general, the experimental condition used in this study is briefly shown in Table 1.

After the experimental data is recorded, the multiple figures of multiple droplets are processed through image processing technique. In this method, sequence steps are applied to raw images to produce binary image. The binary image is an image which only consists of two color, black and white, representing the background and the droplet. Therefore, the dynamics of the droplet can be identified. Figure 2 shows the final result of image processing technique utilized in present work. Based on the figure, it is noted that the boundary of the bubble can be clearly described. As a result, the height and the spreading diameter of the bubble on the particular time can be calculated. Those parameters will be compared to the initial diameter ($d_o$) and arranged in time series diagram.

**Table 1.** Experiment condition.

| High Water Tank (cm) | Weber Number | Temperature (°C) |
|---|---|---|
| 100 | 61 | 100-220 |
| 150 | 61 | 100-220 |
| 200 | 61 | 100-220 |

## 3 Result and discussion

In the present work, the water tank is configured on three different height so the frequency will also change, too. Figure 3. Shows the different behavior which occurs on three different configurations.

When the water tank is set at 100 cm from the nozzle, first droplet and second droplet impact the hot surface with time difference at 195.5 ms and the gap increased consistently until the fourteenth droplet which reaches 260 ms. On the fifteenth droplet, the frequency of droplet is almost stable since the time difference is quite similar.

Next, the similar behavior occurs when the water tank is set 150 cm from the nozzle. First, the time difference increase before it is stable on the fourteenth droplet. On the eighteenth droplet, there is a sudden increase of time difference. This phenomenon is caused by when the droplet valve has been closed, the potential energy of the water is still able to generate the droplet from the nozzle although it occurs in very slow frequency.

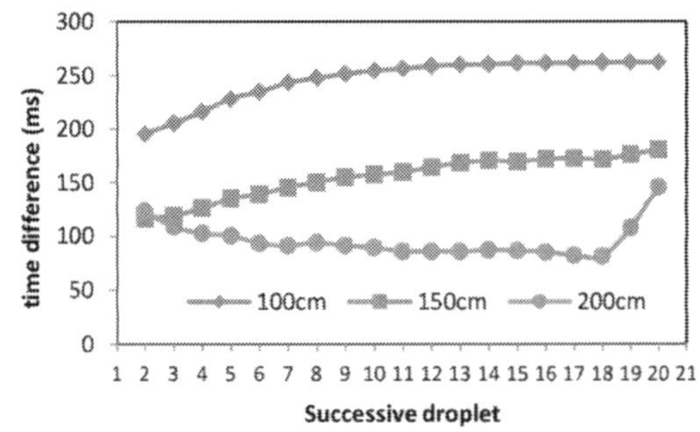


**Fig 3.** The effect of the position of water tank on the time difference between successive droplets

As the water tank is placed higher, on 200 cm from the nozzle, the droplet generator shows a different trend. After the first droplet hit the hot surface, the time difference between two consecutive droplets become shorter and reach the stable condition after the eleventh droplet is generated. At the end, it shows the same tendency with the droplet generated from the water tank which is set 150 cm from the nozzle due to the potential energy. Finally, based on the previous analysis, it is known that the frequency of droplet generated from the water tank placed on 100 cm, 150 cm, and 200 cm, are 4 dps (droplet per second), 6.4 dps, and 10.4 dps, respectively.

Figure 4 illustrates the time evolution of spreading ratio during the successive multiple droplets impacting on a hot surface where the water tank is placed 100 cm from the nozzle. Each line in the figure represents the surface temperature of 100, 140, 180, and 220°C. On 100°C, it is apparent after the first droplet impacting the hot surface, the spreading ratio suddenly increase followed by some recoil and fluctuation before it starts to reach a stable condition. Furthermore, the presence of trailing droplet is able to raise the spreading ratio again. On 140° C the same trend is also found regarding the dynamic behavior of successive multiple droplets. However, the spreading ratio of 140°C is slightly higher than that of on 100°C. Close observation on the figure also reveals that on 180°C and 220°C, the fluctuation of spreading ratio appears clearly compared to the lower temperature.

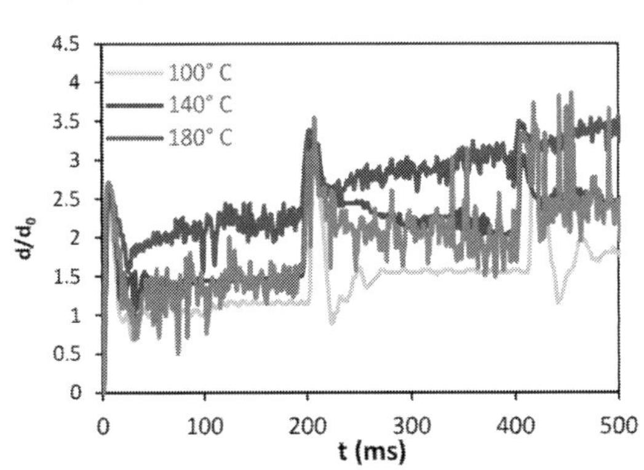


**Fig 4.** The effect of surface temperature on the evolution of spreading ratio of successive multiple droplets (height of water tank = 100 cm)

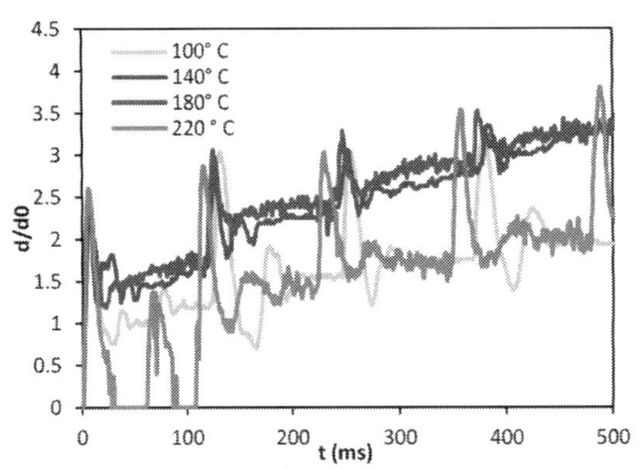


**Fig 5.** The effect of surface temperature on the evolution of spreading ratio of successive multiple droplets (height of water tank = 150 cm)

Figure 5 shows the time evolution of spreading ratio during the successive multiple droplets impacting on a hot surface where the water tank is placed 150cm from the nozzle. In general, compared to Figure 4, the sudden increase of spreading ratio occurs more frequent as a result of the increase in the number of successive multiple droplets impacting the hot surface during the same time. The dynamic behavior of successive multiple droplets is similar to the previous case except on the 220°C where the bouncing phenomena occur several times. The result of this study shows good agreement with that of Bai and Gosman [6] which revealed the effect of Webber number on bouncing phenomena

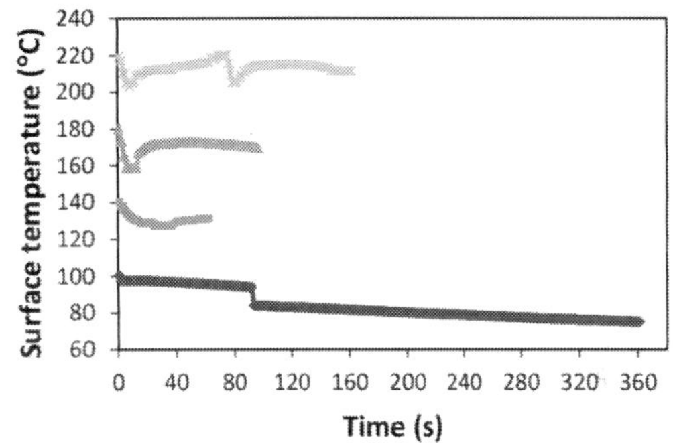


(a)

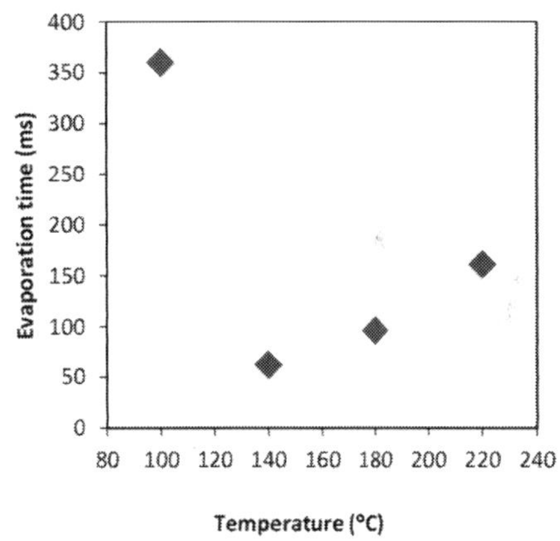


(b)

**Fig 6.** (a) Temperature-time history of a surface, (b) Evaporation time

Figure 6. (a) displays the temperature time history of stainless steel surface affecting by the presence of droplet on various surface temperature while the Figure 6. (b) shows the comparison of evaporation time after the successive multiple droplets impacting the hot surface on various temperature. The configuration of water tank placed 100 cm on the nozzle is chosen as an example. Based on the figure, it is observed that on 100°C, it need around 360s for droplet to perfectly evaporate. In addition, the fastest time occurs when the hot surface is set to 140°C. This phenomenon is similar to the result of an experiment conducted Nukiyama [7], who found that the maximum heat flux occurs around 140°C. On 180°C and 220°C, it is also observed that the droplet generates small bubbles which affect the bubble shape. In addition, the numerous secondary droplet also occurs and separate them from the main droplet

## 4 Conclusion

In the present work, the effect of pressure on water tank on the dynamics of successive multiple droplets is investigated experimentally. An image processing technique is utilized to observe the droplet behavior. Based on the study, it is found that the pressure and frequency affect the behavior of droplet impacting the hot surface. The presence of multiple droplets can effectively enhance the wettability performance which is proved by the increase of spreading ratio of the droplet. The further study can be carried out in wide range area of frequency parameter. It is also possible to improve the presence of high-quality database by conducting the experiment in different material.

This present work was carried out within a research project funded by Directorate General of Higher Education, Ministry of Education and Culture, Republic of Indonesia

## References

1. Deendarlianto, Yasuki, T., Sumitomo, H., Indarto, et al., *Effect of Static Contact Angle on The Droplet Dynamics During The Evaporation of a Water Droplet on The Hot Walls*, Int J Heat Mass Transf **71**, pp.691 - 705, (2014)
2. Fujimoto H., A.Y. Tong, H. Takuda, *Interaction phenomena of two water droplets successively impacting onto a solid surface.* Int. J. Therm. Sci. **47**, 229–236, (2008)
3. Pasandideh-F, S.D. Aziz, S. Chandra, J. Mostaghimi, Cooling effectiveness of a water drop impinging on a hot surface, Int. J. Heat Fluid Fl. **22**, 201-210, (2001).
4. Kandlikar, S. G., Steinke, M. E., & Singh, A.. *Effects of Weber Number and Surface Temperature on the Boiling and Spreading Characteristics of Impinging Water Droplets*. **35th** National Heat Transfer Conference, (2001).
5. Fujimoto. H, Oku. Y, Ogihara. T, Takuda. H., *Hydrodynamics and boiling phenomena of water droplets impinging on hot solid,* Int J. Multiph. Flow **36**, 620-642, (2010)
6. Bai, C., Gosman, A.D., *Development of methodology for spray impingement simulation*. Society Automot. Eng. 950283, 69–87, (1995).
7. Frank P. Incropera and David P. DeWitt, , *Fundamentals of Heat and Mass Transfer*, 3rd ed. This material is used by permission of John Wiley & Sons, Inc, (1990)

# Drag reduction due to recirculating bubble control using plasma actuator on a squareback model

Budiarso[1], Harinaldi[1,*], *Riza Farrash* Karim[2], and *James* Julian[2]

[1]Departement of Mechanical Engineering, Faculty of Engineering Universitas Indonesia. Kampus Baru UI Depok 16424, Indonesia
[2]Fluid Mechanics Laboratory, Faculty of Engineering Universitas Indonesia. Kampus Baru UI Depok 16424, Indonesia

**Abstract.** Flow control on a squareback object which resembles many engineering related objects is believed to be highly beneficial. One of the flow characteristics behind the object, recirculating bubble, is known to play significant role in pressure distribution. Meanwhile, plasma actuator implementation on such object is still underdeveloped in application basis. This paper focuses on acquiring a deeper understanding of plasma actuator effect on flow phenomenon behind a squareback object, especially on its application to recirculating bubble control in order to reduce drag. The experiment was divided into drag measurement experiment and visualization experiment. The drag measurement result shows that plasma actuator succeeded on reducing drag up to 15.36% in the lowest Reynolds number. Meanwhile, the visualization experiment shows that plasma actuator has shifted the recirculating bubble position to be closer to the object's wall.

## 1 Introduction

The squareback object has a lot shape similarity on its application in the real world such as structures like towers and bridges supports or vehicles like trains and truck trailers. This proves how important the knowledge of the flow at the squareback object really is and thus will provide great benefits. The squareback objects, resembles blunt objects which known to have form drag. This form drag is caused by the flow which seperated in the corner of the object squareback and formed the wake and ultimately give the effect of a large pressure difference on the object [1]. The effects later felt by the objects, such as excessive use of the fuel and the greater stress on structural objects. This condition is what makes an understanding in flow control on squareback object which focuses on its application in drag reduction becomes highly beneficial.

This later confirmed with a limited amount of flow control research on this particular objects. Even though several researches has shown using varied orifice area which correlates to the length of separation zone [2], using pulsed jets [3], and even suction only method [4], the amount of researches which connect the relation between flow control on this object to drag reduction is still difficult to find.

In order to understand what are the factors which contributes to drag force on squareback object, an understanding in flow phenomenon in squareback object is recommended. With the approach of the flow passes through the Backward Facing Step, it was noted there are three areas that affect the flow [5]. The three areas are the zone of reattachment, recirculation, and shear layer. It was also noted that in the area of shear layer, there are several factors which affected it, i.e. mechanism of vortex pairing and rolling, mechanism of vortex shedding, and the coherent structure. This coherent structure which is known to have long lifetime [6] and was made from roll up phase of the unstable separated layer [7], is also known to have correlation in reattachment length reduction. It is believed that coherent structure caused the increasing fluid entrainment in recirculating bubble [8]. This confirmed the importance of recirculating bubble region area to flow control in such object.

Flow control method to which the flow is being controlled is also important. One of the methods for active flow control is Dielectric Barrier Discharge or what is best known as plasma actuator. Plasma actuator in the last decade has gained interest of researchers in flow control field for its advantages in compare with other active flow control methods. Plasma actuator is known for its advantages in having fast responds, no moving components, and low energy consumption [9]. The application of plasma actuator itself on squareback object is not a new thing. Control on the vortex shedding has been done using plasma actuator which resulted in transverse actuator configuration as the best one [10] while another research suggested the placement to be at the upstream of separation point [11].

However, the research on application of plasma actuator as an active flow control on bluff body is still underdeveloped, especially on its application in reducing drag. This is predicted due to the focus of the research is still focusing in the mechanism and the phenomenon of the plasma actuator itself. Therefore, the research development on plasma actuator application is still quite wide and promising.

* Corresponding author: harinald@eng.ui.ac.id

## 2 Methodology

The main method which was used in this research is experimental method. The experiment itself was divided to two experiments: visualization experiment and drag measurement experiment. Visualization experiment was focused on observing the flow behind squareback object in order to get the recirculating bubble area data. Meanwhile, the drag measurement experiment was objected to acquire the plasma actuator effect on drag reduction as the integrated result of the actuation process. The model which was being used in the experiment was 0.366 m long, 0.078 m high, and 0.073 wide. The plasma actuators were placed on the back of the object, given the fact that the freestream came from the front of the object. The electrodes were placed on the top and the bottom of the back side of the object.

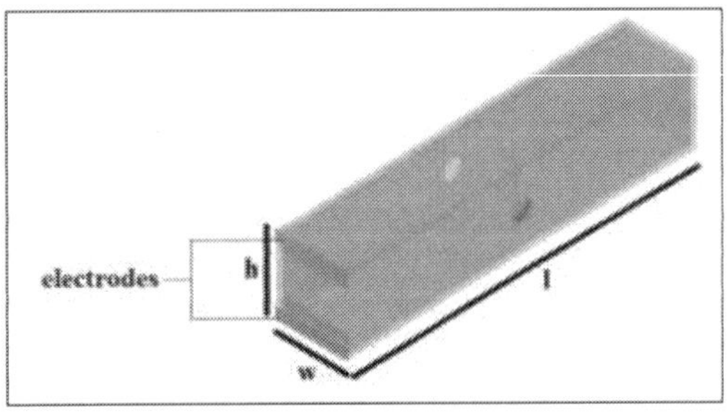


**Fig 1.** Experiment model

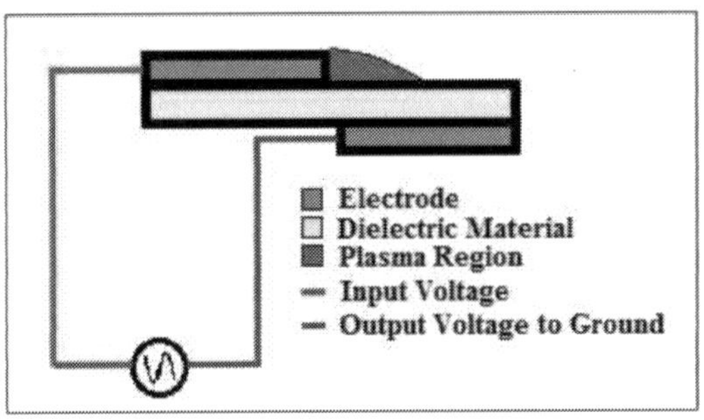


**Fig 2.** Plasma actuator configuration

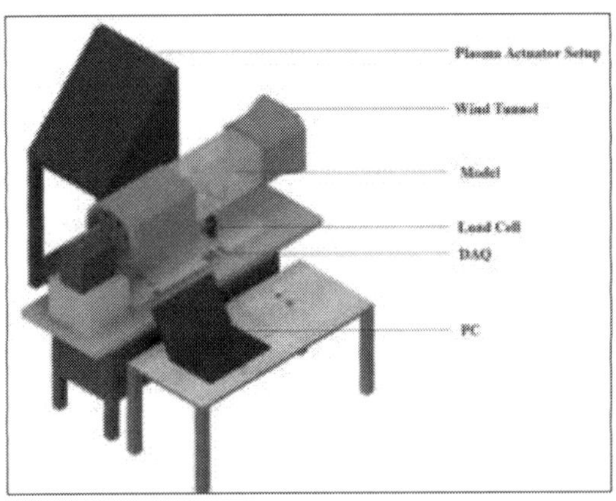


**Fig 3.** Drag measurement experiment setup

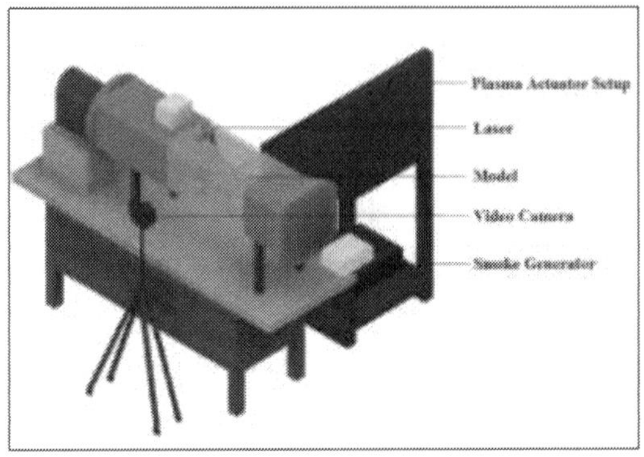


**Fig 4.** Visualization experiment setup

The plasma configuration is shown on fig. 1. The configuration consists of two electrodes and one dielectric material which made of copper and acrylic consecutively. These electrodes have a thickness of 10 mm and a width of 50μm, while the dielectric has a thickness of 3 mm. To generate plasma with the configuration, it takes a high voltage transformer with the type of step up that is able to generate the voltage of 8kVp-p with a current of 0.03 A. Plasma generation is done at a frequency of 10 kHz triangle wave type, with the voltage and the current which kept constant so that it was obtained the power of 120 W.

The drag measurement process was conducted using load cell which connected to DAQ then to the computer to be read. The velocity of the freestream ranged from $Re_D = 1.3x10^4 - 3.0x10^4$ (2.5 m/s – 5.6 m/s) with the model height being the D. Meanwhile, the visualization technique which was used in this research is smoke visualization. The object was placed within the wind tunnel, and later then being exposed with laser sheet. The smoke was generated with smoke generator which then supplemented to the inlet of the wind tunnel. Video camera later was used to capture the smoke particles which passed through the object and then scattered by the laser sheet. The video later processed frame by frame to acquire the measurement of flow properties which required to calculate the recirculating bubble area.

## 3 Results and Discussion

The first result which was shown in the fig. 5 shows the measurement of drag force which was acquired by the load cell. Due to higher velocity, it is shown that the drag force which was affecting the object to be higher than the lower velocity. The figure shows drag force reduction which was presented due to plasma actuator. On the first velocity variation (2.5 m/s) it was clear that the reduction was quite noticeable unlike the second velocity variation (5.6 m/s). This drag measurement indeed consisted of form drag and friction drag but due to the plasma configuration which was placed perpendicular to the freestream, it is safe to assume that the drag force reduction result happened due to plasma actuator effect on the form drag alone. It is also noticeable that due to plasma actuator, the acquired data appeared to fluctuate.

This result can be caused by the plasma actuator perturbation on the vortex shedding which affects the vibration on the load cell.

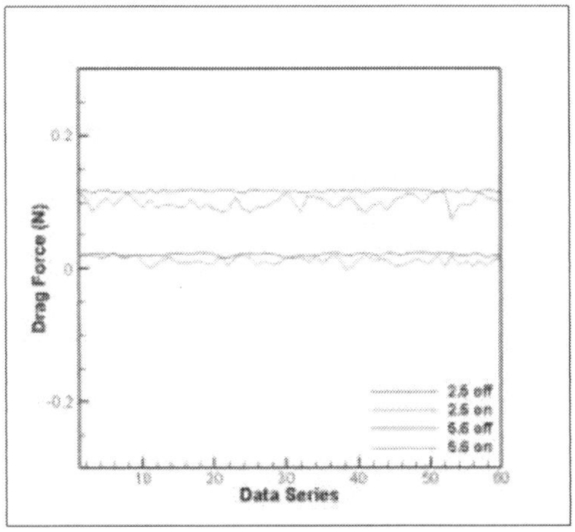


**Fig 5.** Drag force measurement result

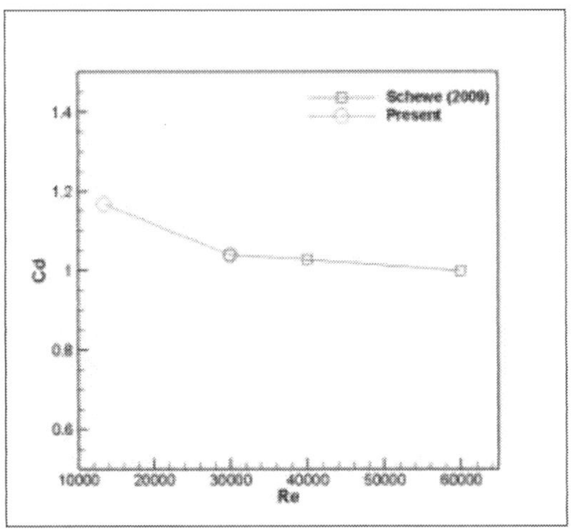


**Fig 6.** Drag coefficient on varied Reynolds number

Meanwhile, fig. 6 shows the drag coefficient which is the processed data version of drag force measurement. In order to assure that the acquired data was true, the figure also shows the literature data [12]. It is noticeable that the literature data value isn't too farfetched with the acquired data. It also noticeable the descending trend of drag coefficient which appeared both on acquired data and the literature. This can be caused by the domination of the shape of the object to the flow itself which caused by the quadratic ascending of velocity aspect in drag coefficient formula.

The drag reduction percentage graph which was shown on fig. 7 shows the descending trend of drag reduction as the Reynolds number increases. This phenomenon can be caused by the principal work of the plasma actuator itself which depends on the ionizing process of the surrounding gas. As this process needs time, the higher the Reynolds number means the faster the flow velocity, which means the shorter the time that plasma actuator uses to ionize the surrounding gas and later resulted in the descending trend of drag reduction percentage. The analogy of this phenomenon can be illustrated by the bypass factor phenomenon in heat transfer on different tube configurations. The best drag reduction percentage was achieved at the lowest Reynolds number with 15.36%. This trend was also achieved by another experiment [13].

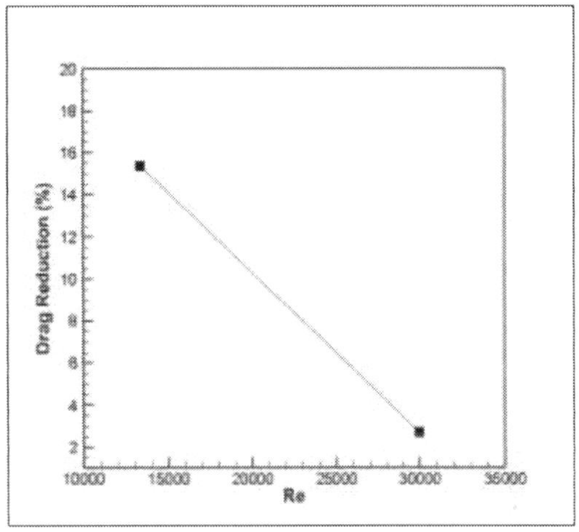


**Fig 7.** Drag reduction percentage

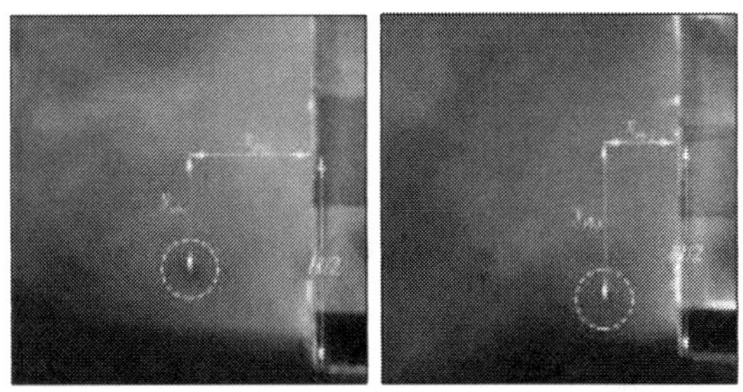

**Fig 8.** Recirculating bubble assessment method for visualization experiment

Visualization method was initiated by determining the flow assessment method. Fig. 8 shows the assessment method which recognizes the flow. Recirculating bubble is characterized by the center of the circulation which shown by the black hollow structure in the picture. This structure later processed frame by frame to show the mean location of the recirculating bubble both before and after plasma actuator contribution.

The result on fig. 9 shows that the recirculating bubble position shifted due to plasma actuator effect. X=0 point was located on the surface of the object's wall. This means that the recirculating bubble has shifted towards the objects wall. This can be caused by the plasma actuator effect on separated flow which passed the edge of the squareback object. This separated flow later perturbed by plasma actuator, then affected the coherent structure. Coherent structure, which was the product of the separated shear layer, is characterized by the large momentum on itself. By the perturbation of the plasma actuator, this property later became even larger, so that the coherent structure became even more influential than before. This later influenced the surrounding flow by its ability to entrain. The surrounding flow later influenced by the vorticity of the coherent structure which then formed into recirculating bubble. Due to plasma actuator, this location of recirculating bubble forming later shifted to the object's wall. Later, the area of the recirculating bubble itself shrunk and thus increased the pressure distribution behind the wall. This phenomenon later suggested to be the cause of the drag reduction effect by plasma actuator on the squareback objects

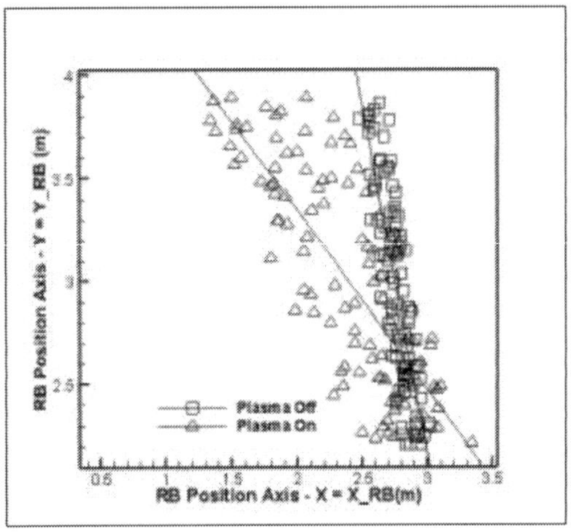


**Fig 9.** Recirculating bubble position due to plasma actuator

## 4 Conclusion

Plasma actuator has proven to be able to reduce drag on squareback object up to 15.36% on low Reynolds number. This result is correlated by the visualization result which shows that recirculating bubble position has shifted due to plasma actuator. Recirculating bubble which corresponds to the pressure distribution on the object, thus resulting on drag coefficient of the object. It is also suggested that future research on recirculating bubble area to be done in order to assure other recirculating bubble properties on drag reduction.

## References

1. B. R. Munson, D. F. Young, and T. H. Okiishi, *Fundamentals of Fluid Mechanics*. John Wiley, (1998).
2. V. Uruba, P. Jonáš, and O. Mazur, "Control of a channel-flow behind a backward-facing step by suction/blowing," *Int. J. Heat Fluid Flow*, vol. **28**, no. 4, pp. 665–672, (2007).
3. S. Chaligné, T. Castelain, M. Michard, and D. Juvé, "Active control of the flow behind a two-dimensional bluff body in ground proximity," *Comptes Rendus Mécanique*, vol. **341**, no. 3, pp. 289–297, (2013).
4. M. Sano, I. Suzuki, and K. Sakuraba, "Control of turbulent channel flow over a backward-facing step by suction," *J. fluid Sci. Technol.*, vol. **4**, no. 1, pp. 188–199, (2009).
5. J. Rajasekaran, "On the flow characteristics behind a backward-facing step and the design of a new axisymmetric model for their study." University of Toronto, (2011).
6. A. S. Monin, "Coherent structure definition," in *Akademiia Nauk SSSR Doklady*, vol. **318**, pp. 853–856, (1991).
7. S. C. C. Bailey, R. J. Martinuzzi, and G. A. Kopp, "The effects of wall proximity on vortex shedding from a square cylinder: three-dimensional effects," *Phys. Fluids*, vol. **14**, no. 12, pp. 4160–4177, (2002).
8. K.-B. Chun and H. J. Sung, "Control of turbulent separated flow over a backward-facing step by local forcing," *Exp. Fluids*, vol. **21**, no. 6, pp. 417–426, (1996).
9. T. C. Corke, C. L. Enloe, and S. P. Wilkinson, "Dielectric barrier discharge plasma actuators for flow control*," *Annu. Rev. Fluid Mech.*, vol. **42**, pp. 505–529, (2010).
10. G. Nati, M. Kotsonis, S. Ghaemi, and F. Scarano, "Control of vortex shedding from a blunt trailing edge using plasma actuators," *Exp. Therm. Fluid Sci.*, vol. **46**, pp. 199–210, (2013).
11. S. G. Pouryoussefi, M. Mirzaei, and M. Hajipour, "Experimental study of separation bubble control behind a backward-facing step using plasma actuators," *Acta Mech.*, vol. **226**, no. 4, pp. 1153–1165, (2015).
12. G. Schewe, "Reynolds Number Effects in Flow Around a Rectangular Cylinder With Aspect Ratio 1: 5," in *Proceedings of the Fifth European and African Conference on Wind Engineering, Florence, Italy*, (2009).
13. Harinaldi, Budiarso, and J. Julian, "The effect of plasma actuator on the depreciation of the aerodynamic drag on box model," in *AIP Conference Proceedings*, vol. **1737**, no. 1, p. 40004, (2016).

# The application of ceramic stove cover and loading distance to the performance of water boiling system

*Agung Sugeng* Widodo*

The Department of Mechanical Engineering, Brawijaya University, Indonesia

**Abstract.** Conventional gas stove (CGS) used in the water boiling system (WBS) is quite inefficient if the mechanism is not considered. In the present research, an assessment has been performed to a CGS by applying a ceramic stove cover (CSC) and also investigated the optimum distance between CSC and the loading to obtain very best efficiency of a CGS. Several parameters used to investigate the CGS's performance such as water temperature, heating value of the fuel and temperature of CSC have been used in reference to the definition of efficiency. In this study the distance between CSC and the loading is varied by 1 mm to 7 mm by increment of 1 mm. The results show that at a distance of 4 mm, the performance of a CGS in a certain fuel rate of 12.5 cc/s increase significantly indicated by the best efficiency of 46.4 %. It is because of an optimum condition is accomplished simultaneously in heat transfer processes such convection and radiation in the WBS. Although the fuel rate itself have an influence to the flame structure heating the loading, however, the used of ceramic cover have an effect to the exchange between fresh air and flue gas to the combustion process.

## 1 Introduction

The main fuel for households in Indonesia was subsidy kerosene or decades. The subsidy of oil for domestic gas stove was hundred billion only in years of 2006 – 2008. This subsidy was overwhelming to the PDB of Indonesia. For this reason, the government of Republic of Indonesia furthermore applied oils to gas conversion program start in year 2007 [1]. The use of LPG also has advantages in relatively clean combustion products. The use of LPG as domestic heating appliances shows cost reduction about 60% cheaper compared to those of using oil [2]. However, if the CGS mechanism is observed there are opportunities to improve its efficiency.

Many efforts have been performed to increase CGS's efficiency and to reduce emission as in Hou [3, 4, 5]. To improve efficiency, modification, and implementation some devices by Gohil [6] into a CGS yield an increasing efficiency of 66% from previously 48%. An addition of a heat collector device or a pyramidal stove cover has also increased by 10 % of the efficiency of a CGS respectively compared to a CGS without cover [7].

The influence of oblique angle and heating on an impinging laminar jet flame applied in domestic gas burner was investigated by Hou [8]. The results show that the oblique angle and heating height have a great influence to temperature field, flame structure, and efficiency of domestic gas burner. A maximum efficiency is obtained at the beginning and then decreased, as the heating height is increased for a fixed oblique angel.
Materials and CGS's design are also have a contribution to develop efficiency. Khan [9] conducted an experiment related to material and CGS's design. By replacing a head burner made from brass instead of cast iron, the efficiency of a CGS increase about 4 %. The efficiency of a CGS also increased around 8 % as a design of a CGS substituted by flat face burner instead of flower brass burner. Commercial CGS which is used in Indonesia is a CGS standard without any modification. Widiandra [10] investigated the used of stove cover and varied the material of stove cover and furthermore applied to a CGS. An indication has also been identified that the application of a CSC to CGS able to prevent heat release from combustion chamber at initially process of heating system. At the initial time of boiling process, CSC have a good efficiency compared to other stove covers due to CSC have a low thermal conductivity compares to others. Thus the CSC will prevent radiation heat transfer from combustion chamber into surrounding as shown in figure 1.

Figure 1(a) is total efficiency per unit time of a CGS without and using stove cover on various materials. From a CGS with and without cover, the time obtained to boil water is equal to 553 seconds without cover, 521 seconds for CSC, 530 seconds aluminum stove cover and 552 seconds for iron stove cover respectively. It indicates that CSC has the best efficiency compared to other stove cover with material of iron or aluminum. The higher efficiency of CGS reaches 46.4%. The value of efficiency for this case is relatively different with the efficiency obtains by Gohil [6], a maximum efficiency of 66% as a result because of different design of the stove cover and how to define the term of efficiency. Other information obtains from figure 1(a) is that the used of stove cover made from

* Corresponding author: agung_sw@ub.ac.id

iron have insignificant influence on the efficiency of the CGS. Aluminum stove cover relatively has a good efficiency compared to iron stove cover with maximum efficiency of 45.8%.

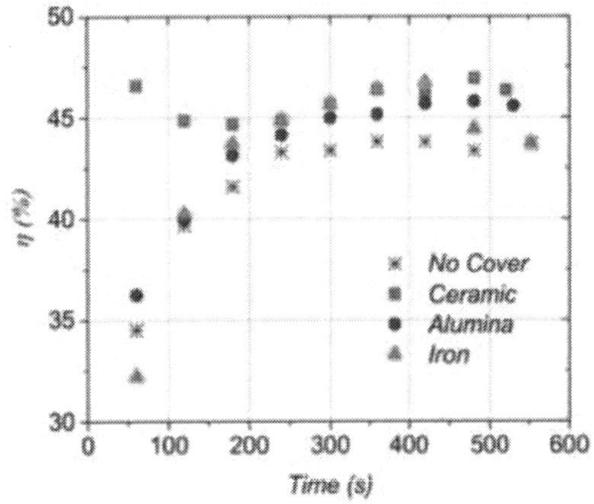


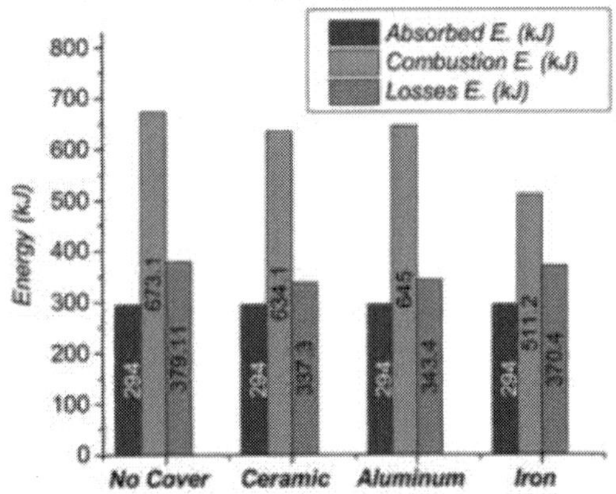


**Fig 1.** (a) Efficiency of a CGS per unit time with and without stove cover; (b) Energy distribution without & using a stove cover

Figure 1(b) also shows a distribution of energy absorbed by water, energy from fuel and losses energy during boiling process. At equally energy absorbed by the water, it is shown that a CGS without cover use more energy compared to a CGS using a CSC. Losses energy coming from a CGS without cover also show a larger value compared to a CGS using CSC.

Effects of heating height on flame appearance, temperature field and efficiency in a single impinging laminar jet flame used in domestic gas stoves without covering the flame inside the combustion chamber also have been investigated by Hou [5]. Hou categories flames into three different type of flames impinging the loading. It means there will be three different height of heating flame. The result showed that the highest efficiency is by using a C-type of flame which inner premixed and outer diffusion flame were opened and diverged.

An indication has been identified that as the distance between the stove cover and the loading is too close, the efficiency of the CGS tends to decrease. As conduction, convection and radiation heat transfer are elaborate, further investigation is needed. Thereby, the aim of this research is to investigate the optimum distance between loading and stove cover to improve efficiency of a CGS in a certain flow rate of fuel. In addition, in this research, the CSC is made as simple as possible for applicable and manufacture reason.

### 1.1 Efficiency of a CGS

The term efficiency in a CGS is defined by fraction of the heat input coming from combustion process produced by fuel which is converted to desired output of energy absorbed by water with ignoring the heat absorbed by the loading material as shown in the equation (1).

$$\eta = \frac{m_w C_{p,w} \Delta T}{m_f . LHV_f} x\ 100\% \qquad (1)$$

where is efficiency, mw is mass of water, $C_{p,w}$ is specific heat constant of water, $\Delta T$ is initial and end temperature differences, $Q_f$ is energy produced by fuel, $m_f$ is mass of fuel, $LHV_f$ is lower heating value of fuel.

Definition of efficiency in (1) is used for its relationship between energy losses in the heating system. The energy losses in the system are relatively complicated to identify because of the complexity of geometry of the heating system and heat transfer processes mechanism. The easiest way to identify one of the energy losses is radiation heat transfer from the CSC.

### 1.2 Energy of Heating Systems

Energy conservation in a WBS can be seen in equation (2). Identification of heat transfer processes in this research such as conduction and convection are difficult to be performed due to the complexity of the geometry of installation. It means that only radiation energy is explicated in the energy conservation.

$$E_C = E_{ab} + E_{r,cov} + E_l \qquad (2)$$

where $E_c$ is energy contained in the fuel, LPG, $E_{ab}$ is energy, which is absorbed by water, $E_{r,cov}$ is radiation energy that is emitted by stove cover and $E_l$ is energy losses instead of radiation energy. Efficiency of WBS is determined using definition of input and output in (1). Efficiency is calculated per unit time until the entire water start boiling and plotted with variations of the distance.

## 2 Data Acquisitions

To obtain data in this research, variables as water temperature, radiation heat transfer, and fuel rate of LPG were set up in an experimental installation as illustrated in figure 3.

A CGS is a stove Quantum type QGC-101R. A standard loading with diameter of 24-cm is hanged over a supporter. The loading is equipped with a glass lid for preventing vapor of water to escape. The glass lid is chosen as an observation window as the water start to boil. At the top of the lid, an Alcohol thermometer type Al-30100-010 with range of scale of -10°C – 110°C is used to measure temperature of the water. The loading is equipped with a stirrer driven by a motor to ensure uniform temperature distribution obtained inside the loading so that the reading of Alcohol thermometer represent a precisely temperature of the water. Mass fuel rate are measured using a flow meter Omega FL-1501A. Surface temperatures of stove cover are obtained using k-type thermocouple that is plugged to a thermometer-set KRISBOW KW06-278 Single Input Digital Thermometer.

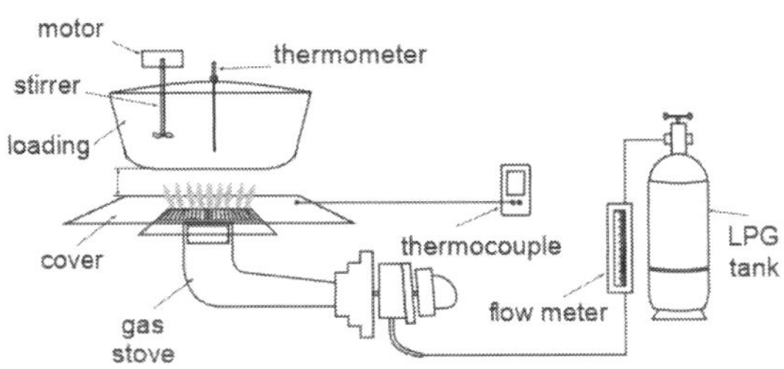


**Fig 3.** Illustration of experiment installation

## 3 Experimental Conditions

Many variables have to be collected to develop efficiency of WBS, furthermore experimental conditions are set up as seen in table 1.

**Table 1.** Experiment conditions

| Conditions | Value |
|---|---|
| water mass [kg] | 1 |
| initial temperature of water [°C] | 25 |
| end temperature of water [°C] | 95 |
| $C_{p\ water}$ [J/kg.°C] | 4196 |
| $Q$ LPG [L/m] | 0.75 |
| $\rho$ LPG [kg/L] | 0.0021 |
| *LHV* propane [MJ/kg] | 46.1 |
| *LHV* butane [MJ/kg] | 46.5 |

The LPG used in experiment was commercial LPG in Indonesia with a LHV of 46.44 MJ/kg. This value was obtained from proportionally composition of the fuel 30 % propane and 70% butane by volume. Average atmosphere (room) temperature was 25°C. The end temperature of water started to boil was 95°C correspond to the location of experiment in ± 476 from sea level. Reading of thermometer was recorded when the bottom solid interface was covering with bubbles and it rising to the top surface. Specific heat constant of water was obtained in an average operational temperature of boiling system. In every data acquisition, experiments were restarting in room temperature.

## 4 Results and Discussions

Efficiency of CGS is measured per unit time in various distances compared to a CGS without cover as shown in figure 4. It shows that efficiency tend to lower at the initial heating process because of energy produced from combustion process is absorbed by materials in the heating system including stove cover and loading. However, efficiency tends to be constant at time of 250 seconds exceeding a standard CGS (dash line). Actually, the distance of 5 mm has a good characteristic for covering heat release from combustion chamber. It is shown by the efficiency, which tends to increase at the initial of heating process. However, for after 300 seconds, the distance of 4 mm has a higher tendency characteristic compared to the distance of 5 mm.

Distance of 4 mm shows a maximum efficiency of 46.4 % differ 1 % compare to the distance of 5 mm. This value was identic with 116 secondds time saving equal to 141.2 kJ refer to the WBS without a CGS. The highest efficiency is reached because of the optimum condition is simultaneous achieved in convection and radiation heat transfer processes of the heating system.

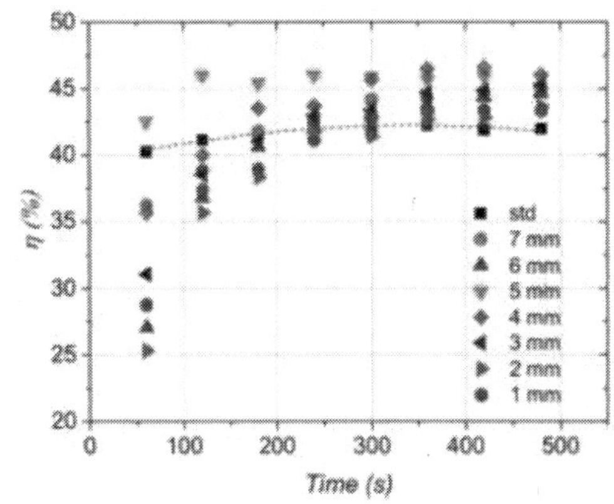


**Fig 4.** Efficiency of a CGS per unit time with and without stove cover

In a laminar jet flame, maximum temperature is located at the top of flame tip because of buoyancy effect. When the flowrate of the jet is increasing, the flame structure change causing combustion and heat transfer process in a WBS. As the distance between stove cover and loading is too close, there will be conditions as mention above. However, relating with fresh and flue gas exchange, there are two possibilities conditions happened. The first is the shape of the flame is not perfectly developed as a free flame because of the flame is hitting the loading. In such condition, stretch phenomenon is likely appear. Stretch affects the propagation rate by changing the diffusive flux that has significant effect to the flame temperature. Once the flame temperature decrease, it will influence the reaction rate and the burning velocity.

Secondly, the distance between CSC and the loading is a channel of convection heat transfer and also as a channel of the flow of air into fuel to make mixture in a combustion chamber. If the distance is too close, one of thus conditions may happened. In such condition, maximum temperature in combustion chamber is likely decrease and has an impact to radiation heat transfer. Conversely, as the distance is too far, convection and radiation heat energy easily escape from combustion chamber.

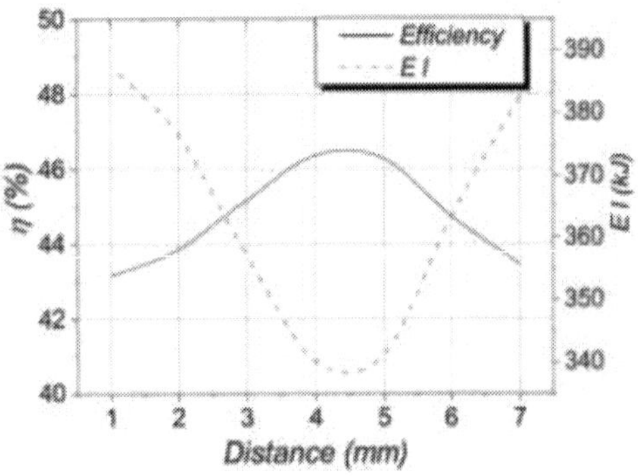


**Fig. 5**. Efficiency and energy losses;

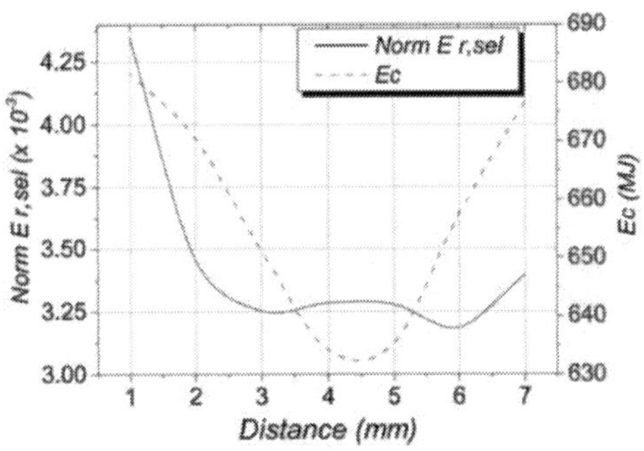


**Fig 6.** Norm. $E_{r,cov}$ and energy from fuel combustion in various distance

Figure 5 shows polynomial line for efficiency and energy losses in various distances. Maximum efficiency occurs in distance of 4 and 5 mm. The rest of distances indicate a decreasing efficiency as shown by values of $E_l$ that have a minimum value as efficiency reach a maximum value. $E_l$ of 0.34 MJ is the smallest energy losses at the distance of 4 mm. Otherwise, maximum $E_l$ is 0.48 MJ happened in a standard CGS.

To analyze the performance of the WBS, polynomial line also plotted to combustion and radiation energy from the stove cover, $E_{r,cov}$ as shown in figure 6. Term Norm. $E_{r,cov}$ is normalized $E_{r,cov}$ by $E_l$ in various distances. As can be seen in figure 6, distance of 1 mm indicates the highest Norm. $E_{r,cov}$ of $4.31 \times 10^{-3}$. It means that although radiation energy has a large value (indicated by surface temperature of 323.3°C), energy losses coming from convection heat transfer at a distance of 1 mm is relatively small compared to other distances. Distance of 6 mm indicated a Norm $E_{r,cov}$ of $3.17 \times 10^{-3}$ is the smallest one however this value is not referring to the smallest efficiency of the distance of 7 mm. Energy polynomial line which is produced by combustion process, $E_C$, (dash line in the figure 6) is showing a decreasing trend which is contrast with efficiency line. The highest value is obtained at a distance of 1 mm, meaning that the distance requires higher energy to boil water compare to others. This phenomenon is contrast with combustion energy at a distance of 4 mm.

## 5 Conclusions

In the present study, efficiency of a CGS has been assessed by varying distance between loading and stove cover in WBS. The used of CSC indicated a good characteristic for covering heat release from combustion chamber. From the discussion it is concluded that distance between pan and cover stove influences efficiency of a CGS. In a certain fuel rate of 12.5 cc/s, maximum efficiency is obtained in a distance of 4 mm. The distance of 4 mm indicates the highest efficiency of 46.4% of a CGS because of the optimum condition is simultaneous achieved in convection and radiation heat transfer processes of the WBS.

## References

1. Lontoh L., Beaton C., and Clarke K. 2015, Indonesia Energy Subsidy Review, International Institute for Sustainable Development, Issue 1. Volume 2. March 2015.
2. Anonymous. (2012). Conversion Oil to Gas, Reduce Subsidy for Long Term, Ministry and Mineral Resources of Republic of Indonesia.
3. Hou, S.S. (2007). Efficiency and Emissions of a New Domestic Gas Burner with a Swirling Flame, Energy Conversion and Management 48; 1401–1410.
4. Jugjai, S. et al. (2001) Thermal Efficiency Improvement of an LPG Gas Cooker by a Swirling Central Flame. Int. J. Energy Res; 25:657–74.
5. Hou, S.S., and Chou C.H. (2013). Parametric Study of High-Efficiency and Low-Emission Gas Burners, Advances in Materials Science and Engineering, vol. 2013, Article ID 154957, 7 pages.
6. Gohil, P.P., and Channiwala, S.A. (2011). Investigation of performance of Conventional LPG cooking stove. Fund. J. Therm. Sci. & Eng., Vol. 1, Issue 1, 2011, Pages 25-34.
7. Syarial, M. (2012). Performance of Biogas Stove Using Reflector to Improve Efficiency, Thesis Unpublished. Institut Teknologi Sepuluh November, Surabaya.
8. Hou, S.S., and Ko, Y.C. (2005). Influence of oblique angle and heating height on flame structure, temperature field and efficiency of an impinging laminar jet flame. Energy Conversion & Management; 46(6):941–58.
9. Khan, M.Y., and Saxena A. (2013). Performance of LPG Cooking Stove Using Different Design of

Burner Heads, International Journal of Engineering Research & Technology (IJERT), ISSN: 2278-0181, Vol. 2 Issue 7.

10. Widiandra, P., & Widodo, A.S. (2014), Effect of Materials Stove Cover to the Efficiency of Gas Stove Water Boiling System, Jurnal Rekayasa Mesin, Vol.5, No.3, 291-295, Brawijaya University.

11. Hou, S.S. and Ko, Y.C. (2004). Effects of heating height on flame appearance, temperature field and efficiency of an impinging laminar jet flame used in domestic gas stoves. Energy Conversion & Management; 45(9–10): 1583–1895.

# Numerical study on the effect of width of single curtain on the performance of Savonius wind turbine

*Triyogi* Yuwono*, *Aip Abdul* Latip, *Nabila Prastiya* Putri, *Ubaidillah* Muhammad, *Erik Noer* Mazhilna, *Citro* Ariyanto, *Ulfa* Andaryani, and *Anaz* Fauzi

Fluid Mechanics and Fluid Machineries Laboratory, Department of Mechanical Engineering Institut Teknologi Sepuluh Nopember Kampus ITS – Keputih – Surabaya, Indonesia

**Abstract.** This is a preliminary results of the flow around the Savonius wind turbine with installing curtain plate in front of the returning blade turbine. It was investigated numerically in a uniform flow at Reynolds number of 30,000 and 90,000. The velocity vector and pressure distribution around the turbine were simulate by varying the width of curtain plate relative to the diameter of rotor blade (S/D) of = 1.00, 1.02, 1.03, 1.15, 1.41, and 2.00, using STAR CCM++ Software. The k-ε realizable as turbulence model was used to visualize the flow phenomena occurred around the turbine, and where in this simulation, the rotor turbine was set static. The results show that it seems the width of the curtain installed in front of the returning blade of the turbine plays an important role in the performance of the turbine. In general, the installing of the curtain in front of the returning blade of the turbine is more effective to improve the turbine performance. This is not necessarily, but depends on the width of the curtain and the number of Reynolds (Re). For the width of the large curtain of S/D = 2 at Re = 90,000, the performance of the turbine is estimated lower than when the turbine without the curtain.

## 1 Introduction

Wind energy is a freely available natural resource that is abundant and available over the years. But unfortunately the wind speed in Indonesia is relatively low, which is about 4-7 m/s. The appropriate turbine for the low wind speed is the Savonius turbine. This turbine is a vertical pivot turbine type, Vertical Axis wind Turbine (VAWT). The Savonius turbine is one example of a drag type turbine. The phenomenon of flow across the Savonius turbine is the use of drag force to improve the performance of the turbine. The flow will cross the 2 (two) side of the advancing blade and the returning blade of the tubine. The difference of drag force between the returning blade and the advancing blade will result in a total drag to rotate the Savonius turbine, and finally to generate power to drive the generator producing an electrical energy. The advantages of this type of drag turbine are the self-starting ability with a small fluid velocity, so there is no need for external push support; it is very suitable for the region of Indonesia which has a relatively the low average speed.

However, this type of the wind turbine has the lowest performance compared to others types of the wind turbine. That is why various studies have been done to improve the performance of the turbine Savonius. Among others, by varying the number of bucket, the authors [1] concluded that the two-bucket configurations have better aerodynamic performance than the three-bucket configurations. The Authors [2] has conducted research on optimizing the design of Vertical Axis Wind Turbine (VAWT) Darrius-Savonius type. One is done to get a VAWT design that has a strong structure is by varying the length of the shaft top-side and bottom-side support. The authors [3] conducted a study on the Savonius turbine, by varying the overlapping ratio of 0, 0.1 and 0.2. They have observed that for varied wind speed the maximum torque has obtained at the overlap ratio of 0.2. The authors [4] studied experimentally on improvement of a Savonius rotor performance with curtaining. They used two curtain blades, composed of flat sheet plates, to directing the wind toward the advancing blade where one curtain placed in front of returning blade and one others positioned beside of advancing blade. They have proved that the arrangement of Savonius wind rotor with three different curtains have the power coefficient higher than one without curtain. The authors [5] conducted a study on the use Savonius turbine deflector on the front side of the advancing blade sideways, where the configuration can improve turbine performance. The phenomenon of flow across the advancing blade in the form of drag force will be increased by using two deflectors.

Based on the previous study, there is no research about the size of curtain plate, concerning its influence on the performance of Savonius turbine. So, it is interesting to study numerically the effect of width of single curtain, where the curtain in varied width is installed in the front of the returning blade, to improve the performance of Savonius turbine.

*Corresponding author: triyogi@me.its.ac.id

# 2 Numerical Study

## 2.1 Simulation Method

STAR CCM++ Software has used to simulate the flow characteristics around the Savonius turbine. The k-ε realizable as turbulence model was used to visualize the flow phenomena occurred around the rotor turbine. In this simulation, the rotor turbine was set static or not moving. This simulations is performed at Reynolds number (Re) of 30,000 and 90,000 based on the characteristic length of d = 2D-b and free stream velocity (U). There are 7 sizes of curtain plate were tested in this simulation, such as for the width of curtain plate relative to the diameter of rotor blade (S/D) of = 1.00, 1.02, 1.03, 1.15, 1.41, and 2.00. Figure 1 shows the scheme of modeling domain of the present study. As boundary conditions of the problem, it has been set at the inlet of the domain is velocity inlet of U = 2,48 m/s for Re = 30,000 and U = 7.45 m/s for Re = 90,000, at the outlet is pressure outlet, and the symmetry plane (walls) at the upper and lower side of the domain.

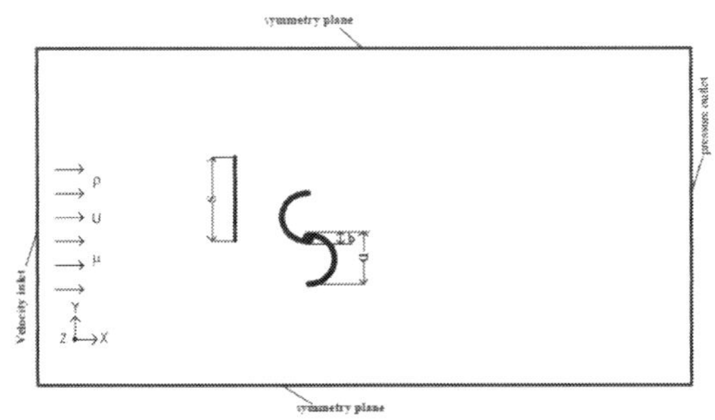


**Fig 1.** Scheme of the modeling domain.

## 2.2 Results and Discussions

Figure 2 shows the results of the simulation for flow arround the Savonius rotor turbine without curtain, for Re = 30,000 (figure 2a and 2b) and Re = 90,000 (figure 2c and 2d), where in this simulation the rotor turbine was set static or not moving.

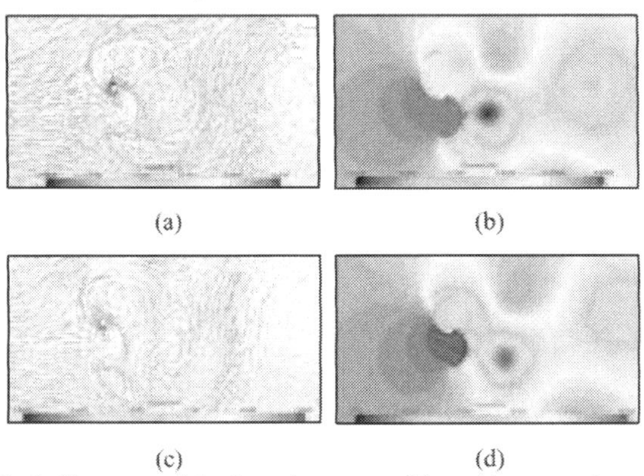


**Fig 2.** Flow around the Savonius rotor turbine without curtain, for Re = 30,000: (a) velocity vector, (b) pressure countur, and for Re = 90,000: (c) velocity vector and (d) pressure countur

Figure 2 shows that there is generally a different pressure between the pressure in front of the returning blade and the rear, where the pressure in the front of the returning blade higher than the rear, and this is also the same case for the flow around advancing blade. It causes the pressure drag in the front side is higher than the rear side. It seems clearly enough that the pressure drag acting on the advancing blade is higher than on the returning blade. It results a positive static torque and will cause the rotor turbine turning in counter clockwise. When compared between the flow with Re = 30,000 and 90,000, it seems that the rotor turbine being flowed with Re = 90,000 will have a static torque higher than with Re = 30,000. It is clearly indicated by the pressure in the rear of the advancing blade for Re = 30,000 (blue colour, figure 2d) lower than for Re = 90,000 (green colour, figure 2b). So, the negative torque on the returning blade for Re = 30,000 higher than the one for Re = 90,000.

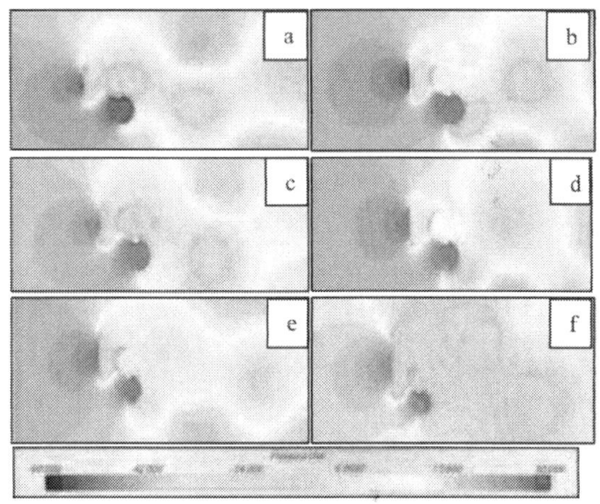


**Fig 3.** Pressure countur flow around the Savonius rotor turbine with curtain, for Re = 30,000: (a) S/D = 1,00, (b) S/D = 1.02, (c) S/D = 1.03, (d) S/D = 1.15, (e). 1.41 and (f) S/D = 2.00

Figure 3 shows the results of the simulation for flow arround the Savonius rotor turbine with curtain varied of S/D = 1,00; 1.02, 1.03, 1.15, 1.41 and 2.00, for Re = 30,000. As seen in the figure that due to the installing of curtain plate in front of returning blade rotor, the pressure in the rear of curtain plate or in front of returning blade area is decrease, meanwhile the pressure in the front of advacing blade is relatively not changing compared to the Savonius rotor turbine without curtain (see figure 2b). It causes the pressure drag acting on the returning blade decreased and as result the positive static torque increased. So, the installation of curtain plate in front of returning blade of the Savonius turbine can be expected to increase the performance of turbine.

Figure 4 shows the results of the simulation for flow arround the Savonius rotor turbine with curtain varied of S/D = 1,00; 1.02, 1.03, 1.15, 1.41 and 2.00, for Re = 90,000. Except for S/D = 2, for Re = 90,000 the phenomenon of the flow around the turbine rotor is almost exactly as for Re = 30,000.

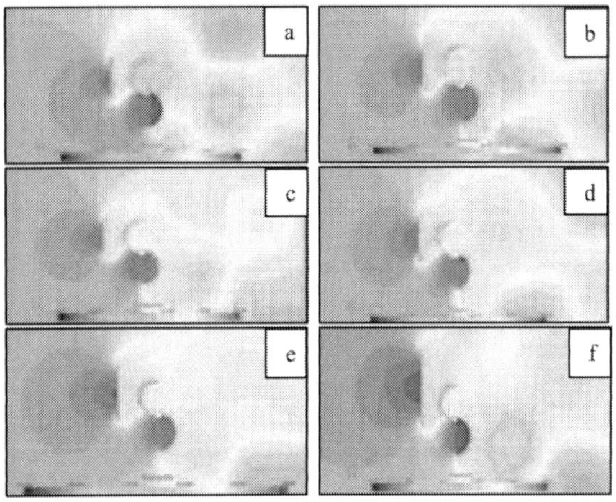


**Fig 4.** Pressure countur flow around the Savonius rotor turbine with curtain, for Re = 90,000: (a) S/D = 1,00, (b) S/D = 1.02, (c) S/D = 1.03, (d) S/D = 1.15, (e). 1.41 and (f) S/D = 2.00

For SD = 2, the curtain plate installing in front of the returning blade is predicted to decrease turbine performance. It is seemly due to the region with lower pressure in front of returning blade is larger than the one for S/D < 2, it causes the flow tends to pass in this region instead of push the advancing blade to produce the positive torque. As a result a decrease in turbine peformance was obtained. It is clearly indicated by the velocity vector around the turbine with the curtain (figure 5), when compared to the velocity vector around the turbine without the curtain (figure 1c). So, the installing of curtain plate which the size of S/D = 2 in front of returning blade of the Savonius turbine for Re = 90,000 is not efective to increase turbine performance. It is as well as ensures that the size of the curtain width, placing in front of the blades, has an important role in affecting the performance of the turbine. But this is required a justification, either through simulation with the turbine rotor in a rotating state, or by an experimental study.

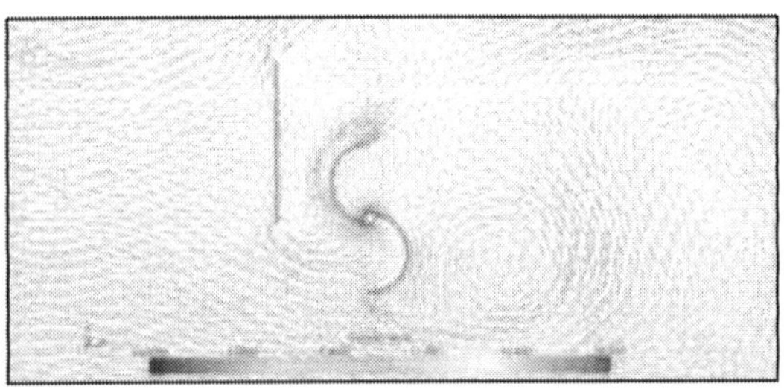

**Fig 5.** Velocity vector around the Savonius rotor turbine with the curtain of S/D = 2.0 for Re = 90,000.

## 2.3 Conclusions

The size of a single curtain installed in front of the returning blade of the Savonius wind turbine plays an important role in the performance of the turbine rotor. In general, the installing of the curtain in front of the returning blade of the Savonius wind turbine is more effective for improving turbine performance. This is not necessarily, but depends on the width of the curtain and the number of Reynolds (Re). For the width of the large curtain (S/D = 2) at Re = 90,000, the performance of the Savonius turbine is estimated lower than when the turbine is without the curtain. This is required a justification, either through a simulation with the turbine rotor in a rotating state, or an experimental study.

The research was funded by the University - Institut Teknologi Sepuluh Nopember (ITS) through a research laboratory' scheme, in 2017.

# References

1. R.E. Sheldahl, L.V. Feltz, and B.F. Blackwell, J. Energy **2** **(3)**, 160-164 (1978). (https://doi.org/10.2514/3.47966)
2. B.A. Dwiyantoro, T.Y. Yuwono , V. Suphandani, *ARPN JEAS*, **11 (2)**, 1073-1077 (2016).
3. C.R. Patel, V.K. Patel, S.V. Prabhu, T.I. Eldho, (2013). IJSCE **3(2)**, 379-383 (2013).
4. B.D. Altan, M. Atilgan, Renew. Energy, 35 (4), 821-829 (2009).
5. G. Kailash, T.I. Eldho, S.V. Prabhu, IJRM **2012** ID 679247, 12 pages, 2012. (http://dx.doi.org/10.1155/2012/679247)

# Flow-driven simulation on variation diameter of counter rotating wind turbines rotor

Y. *Fredrika* Littik[1], Y. *Heru* Irawan[2], and M. *Agung* Bramantya[3*]

[1,3]Department of Mechanical & Industrial Engineering, Universitas Gadjah Mada, Yogyakarta, Indonesia
[2]Mechanical Engineering, STTNAS Yogyakarta, Indonesia

**Abstract.** Wind turbines model in this paper developed from horizontal axis wind turbine propeller with single rotor (HAWT). This research aims to investigating the influence of front rotor diameter variation ($D_1$) with rear rotor ($D_2$) to the angular velocity optimal (ω) and tip speed ratio (TSR) on counter rotating wind turbines (CRWT). The method used transient 3D simulation with computational fluid dynamics (CFD) to perform the aerodynamics characteristic of rotor wind turbines. The counter rotating wind turbines (CRWT) is designed with front rotor diameter of 0.23 m and rear rotor diameter of 0.40 m. In this research, the wind velocity is 4.2 m/s and variation ratio between front rotor and rear rotor ($D_1/D_2$) are 0.65; 0.80; 1.20; 1.40; and 1.60 with axial distance ($Z/D_2$) 0.20 m. The result of this research indicated that the variation diameter on front rotor influence the aerodynamics performance of counter rotating wind turbines.

## 1 Introduction

Application of wind energy as an alternative energy using wind turbines have been carried out in Indonesia. Wind turbines are used as a power generation system that converts kinetic energy of wind into mechanical energy to drive an electric generator.

Horizontal axis wind turbine propeller type with single rotor is a wind turbine models commonly used in Indonesia. Single rotor wind turbine according to the blade diameter is divided into micro (≤ 0.1 m), small scale (0.1 m < diameter ≤ 1 m), medium scale (1m < diameter ≤ 5 m), and large scale (> 5 m).

The ideal maximum power coefficient of wind turbine dual rotor through actuator disk theory is 64%. Assuming a ratio of the axial velocity on the first rotor is 0.2 and 0.6 on the second rotor. Kumar et al [4], based on the theory of momentum (Betz limit), the maximum energy conversion efficiency of single rotor wind turbine about 59% while the counter rotating wind turbine with the same swept area increased to 64%.

Various research has been conducted to investigating the influence of diameter rotor using computational fluid dynamics (CFD) aims to analysis aerodynamics performance in wind turbine rotor, both numerically and experimentally demonstrated that ratio of the diameter influenced the performance of counter rotating wind turbine. However, variations in the diameter of the rotor used is still partially in a generally simulation methods.
In this research uses the fixed axial distance between front rotor and rear rotor, k-ω turbulence models (SST), the solution method Coupled-Scheme and simulation Flow-Driven for analyze aerodynamics performance of rotor wind turbine.

## 2 Methodology

### 2.1 Computational Fluid Dynamics (CFD)

Computational fluid dynamics (CFD) used to indicate the numerical solution of the different governing equations of fluid flows. It aims to calculate and analyze a problem that exists. In this case the computer is used as a medium for doing calculations in the form of iteration. CFD consists of three equations set the basis of the calculation process performed computer. The basic principle of a CFD is discretized set into the algebra equations that can be solved by a computer. Equation set in CFD consists of conservation of mass that show in equation (1) and conservation of momentum in equation (2), (3), (4), can be written in Cartesian coordinates as follows:

$$\frac{\partial \rho}{\partial t} + \nabla.(\rho\vec{V}) \tag{1}$$

$$\frac{\partial(\rho u)}{\partial t} + \nabla.(\rho u\vec{V}) = -\frac{\partial p}{\partial x} + \frac{\partial \tau_{xx}}{\partial x} + \frac{\partial \tau_{yx}}{\partial y} + \frac{\partial \tau_{zx}}{\partial z} + \rho f_x \tag{2}$$

$$\frac{\partial(\rho v)}{\partial t} + \nabla.(\rho v\vec{V}) = -\frac{\partial p}{\partial y} + \frac{\partial \tau_{xy}}{\partial x} + \frac{\partial \tau_{yy}}{\partial y} + \frac{\partial \tau_{zy}}{\partial z} + \rho f_y \tag{3}$$

$$\frac{\partial(\rho w)}{\partial t} + \nabla.(\rho w\vec{V}) = -\frac{\partial p}{\partial z} + \frac{\partial \tau_{xz}}{\partial x} + \frac{\partial \tau_{yz}}{\partial y} + \frac{\partial \tau_{zz}}{\partial z} + \rho f_z \tag{4}$$

Where t is the time, ρ is the density of the fluid, u is the velocity of a fluid particle, ∇ is the divergence, V is volume, τ is the viscous stress tensor, and *f* is represents of the inertial forces.

Three phases of the problem analysis in CFD consist of preprocessor, solver, and post-processing. The simulation conducted using k-ω turbulence model. This

* Corresponding author: bramantya@ugm.ac.id

model performs well with free shear flows, flat plate boundary layer flows, complicated adverse pressure gradient flows and separated flows.

## 2.2 Counter rotating wind turbines

Lee et al [5], a counter rotating wind turbines (CRWT) is the development of a dual rotor horizontal axle wind turbine, having two rotors rotating in opposite direction on the same axis. The Counter rotating wind turbine model design used in the research can be seen from Table 1.

The independent variables used in this study are variations on wind speed (u) of 4.2 m/s with the variation in the rotor diameter of the front and rotor ratios ($D_1/D_2$) 0.65; 0.80; 1.20; 1.40 and 1.60. The controlled variable is the ratio of axial distance between two rotors ($Z/D_1$) is 0.20 m.

**Table 1.** Properties of counter rotating wind turbines rotor.

| Rotor | Front | Rear |
|---|---|---|
| Diameter (mm) | 230 | 400 |
| Position | Upwind | Upwind |
| Airfoil | NACA 0012 | NACA 0012 |
| Rotation | Clockwise | Counter clockwise |
| Material | Balsa wood | Balsa wood |

## 2.3 Flow-Driven Simulation

Flow-driven simulation is a numerical approximation method used by the researcher to calculate the optimal angular velocity of both wind turbine rotor. Calculations are performed every time step until it reaches convergent and stable. This simulation uses the help of a six degree of freedom (6DOF) solver contained in ANSYS Fluent, to determine the direction of rotation of the wind turbine rotor. Using the force and moment imposed in this case the rotor to calculate the translational motion and rotational motion where its center of gravity has been defined.

Equation (5) is an integral form of the conservation equations contained in Flow-Driven simulation:

$$\frac{d}{dt}\int_V \rho\phi\, dV + \int_{\partial V} \rho\phi\,(\bar{u} - \bar{u}_g).d\bar{A} = \int_{\partial V} \Gamma\nabla\phi\,.d\bar{A} + \int_V S_\phi\, dV \quad (5)$$

Where $\rho$ is the fluid density, $\bar{u}$ is the flow velocity vector, $\bar{u}_g$ is the mesh velocity of the moving mesh, $\Gamma$ is the diffusion coefficient, and $S_\phi$ is the source term of $\phi$.

## 2.4 Setup and Solution Solver Simulation

The general setup used in this 3D simulation is a Transient Pressure-Based type, with turbulence model of SST k–ω As for the solution method used Coupled-Scheme with second order implicit and time step size 0.01. For mesh controls use advanced size function on Proximity and curvature with high smoothing sizing mesh. Quality inspection mesh should be taken to ensure no errors during the iterative process, but it also affects the quality of the mesh generated numerical solutions. The steps in this simulation can be seen in the following fig. 1.

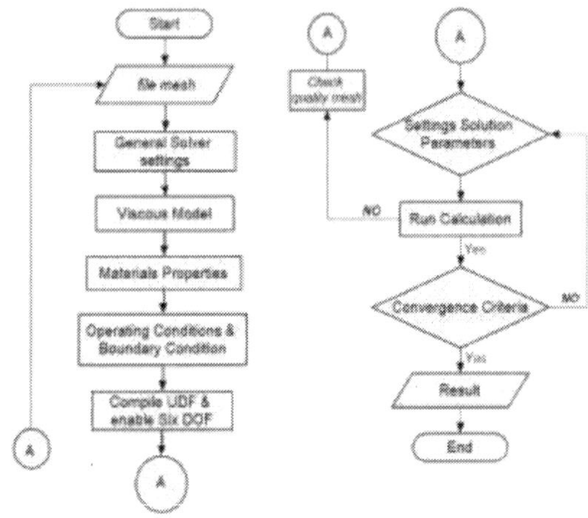


**Fig 1.** Simulation steps

# 3 Result and Discussion

The 3D simulation results of counter rotating wind turbines show that varying the rotor diameter ratio of front rotor to rear rotor with an incoming wind speed of 4.2 m/s, has an effect on the aerodynamics performance of the rotors.

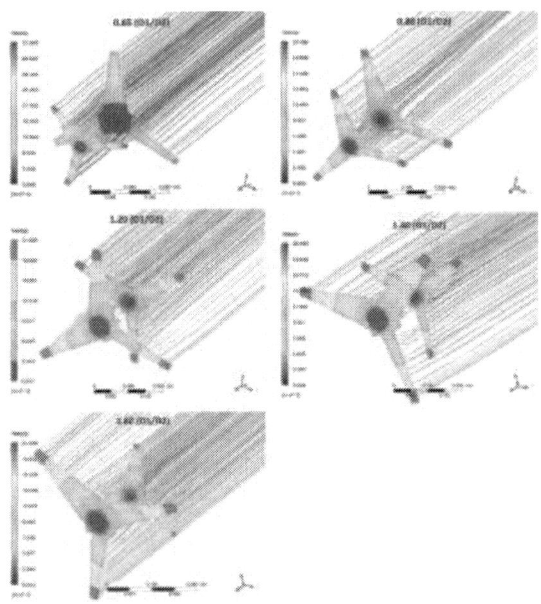

**Fig 2.** Result of the simulation in 3D contour of velocity

As shown in fig. 2, the wind velocity that occurs in the front rotor and rotor rear with a diameter ratio of 0.65 increases almost reaching 33.64 m/s (front) and 29.44 m/s (rear) in stable condition and optimal rotation speed. While the ratio of diameter 0.80 decreased wind speed on both rotor by 19.91 m/s, the ratio of diameter 1.20 wind velocity that occurred on the rotor increased by 21.03 m /

s due to the larger diameter of the rotor, the ratio of diameters 1.40 and 1.60 has a speed the highest wind is almost the same on both rotor pieces that is equal to 18.64 m/s and 18.67 m/s.

The highest wind speed occurs at the tip of rotor, either on the front or rear rotor at optimum rotation conditions. As for the lowest speed is in the hub of rotor, especially on the rear rotor.

**Table 2.** Comparison two model wind turbines.

| Wind Turbines Model | Front Rotor | | Rear Rotor | |
|---|---|---|---|---|
| | ω (rad/s) | TSR | ω (rad/s) | TSR |
| SRWT | - | - | 120.30 | 5.73 |
| CRWT (0.65 D1/D2) | 290.62 | 9.00 | 110.03 | 5.24 |
| CRWT (0.80 D1/D2) | 139.71 | 5.32 | 108.18 | 5.15 |
| CRWT (1.20 D1/D2) | 87.29 | 4.99 | 104.91 | 5.00 |
| CRWT (1.40 D1/D2) | 74.74 | 4.98 | 103.70 | 4.94 |
| CRWT (1.60 D1/D2) | 65.53 | 4.99 | 98.06 | 4.67 |

Table. 2 shows the performance of wind turbines with single rotor and dual rotor. The angular velocity with a single rotor using rotor rear as a comparison with counter rotating wind turbines.

The simulation results obtained by a single rotor wind turbine (SRWT) has an angular velocity of 120.30 rad / s and TSR 5.73, this is more or less consistent with the assumption (TSR = 5-6) for a three-blade wind turbine. The counter rotating wind turbine with a diameter ratio of 0.65 has an angular velocity of the front rotor of 290.62 rad / s and TSR 9, this is because the rotational speed of the rotor is too high. As for the ratio of diameters from 0.80 to 1.60 TSR values ranged from 4 to 5, this is influenced by decreased angular velocity on the front rotor and rear rotor.

## 4 Conclusions

Counter rotating wind turbines with variations in diameter only the front rotor effect on wind turbine aerodynamic performance, judging by the wind velocity occurring at the rotor end of the front and rear following the change in diameter size from the front rotor to the same axial distance. At a ratio of diameter of 1.20 be it the front rotor and the rear rotor, has a high wind speed at the tip of rotor is 21.03 m / s. This shows the stability of the rotational speed at this diameter ratio better than the ratio of the other diameter, so also when viewed from the results of angular velocity and its TSR. The greater the number of blades on a wind turbine rotor, the greater the angular velocity and the decreasing TSR value.

## References

1. C.J. Bai, W.C. Wang, Review of Computational and Experimental Approaches to Analysis of Aerodynamic Performance in Horizontal-Axis Wind Turbines (HAWT's), *Renewable and Sustainable Energy Reviews*, vol. **63**, pp. 506-519, (2016)
2. A.D. Hoang, C.J. Yang, Design and Performance Evaluation of a 10 kW Scale Counter Rotating Wind Turbine Rotor, *Journal of The Korean Society of Marine Environment and Safety*, vol. **20**, pp. 104-112, (2014)
3. S.N. Jung, T.S. No, K.W. Ryu, Aerodynamic Performance Prediction of a 30 kW Counter Rotating Wind Turbine System, Jeonju : ResearchGate, (2004)
4. P.S. Kumar, A. Abraham, R.J. Bensingh, and S. Ilangoyan, Computational and Experimental Analysis of a Counter Rotating Wind Turbine System, *Journal of Scientific and Industrial Research*, vol. **72**, pp. 300-306, (2013)
5. S. Lee, H. Kim, E. Son, S. Lee, Effects of Design Parameters on Aerodynamic Performance of a Counter Rotating Wind Turbine, *Renewable Energy*, vol. **42**, pp.140-144, (2012)
6. M.A. Bramantya, L. Al Huda, An Experimental Study on The Mechanics Power of Counter Rotating Wind Turbines Model Related with Axial Distance Between Two Rotors, *6th International Annual Engineering Seminar (InAES), Yogyakarta, Indonesia*, (2016)
7. L. Oggiano, CFD simulations on the NTNU wind turbine rotor and comparison with experiments, *Renewable Energy Research Conference, Energy Procedia* vol. **58**, pp. 111 – 116, (2014)
8. W.Z. Shen, V.A.K Zakkam, J.N. Sorensen, K. Appa, Analysis of Counter-Rotating Wind Turbines, *Journal of Physics: Conference Series* **75**, pp. 012003, (2007)
9. Y.H. Irawan, M.A. Bramatya, Numerical Simulation of The Effect of Axial Distance Between Two Rotors in Counter-Rotating Wind Turbines, *2nd International Conference on Science and Technology Computer (ICST), Yogyakarta, Indonesia*, (2016)

# Bolt connection behaviour of the cold-formed steel joint

*Sabril* Haris*, *Hazmal* Herman, Zaidir, and Rendy Thamrin

Civil Engineering Department, Andalas University, Kampus Unand Limau Manis, Padang, Indonesia

**Abstract.** This paper presents an experimental study of bolt connection behaviour of the cold-formed steel joint. A 5-mm bolt was used to combine two channel sections. The size of channel section was 75 mm height, 35 mm width and 0.75 mm thickness. Tension test was conducted to obtain material characteristics of the channel section. Two type of connections were observed: Connection Type-1 that back-to-back channel sections were joined on their web directly, and Connection Type-2 that two channel sections were joined on their flanges through a same-size channel connector. The specimen was pulled by an increasing monotonically load. It was found that ultimate condition of the joint occurred when the bolt was split due to shear failure and the bolt was rotated because the hole size increased; these two failure modes are covered in the Indonesian and Australian/New Zealand cold-formed steel codes. The ultimate loads from the test were compared to analytical calculations using the formulae provided in the codes. The analytical calculation predicted the resistances conservatively within the range of 4 – 22 %.

## 1 Introduction

Cold-formed materials have been applied in various areas including automotive industries and building constructions. Nowadays, cold-formed steel has become an alternative material to replace timber for the short span roof on the residential house or even substitute conventional steel material for the longer span roof on the industrial building. According to Yu in [1], cold-formed steel material has several advantages in light weight, high strength and stiffness, uniform quality and non-combustibility.

Most of the cold-formed steel structural elements are assembled by using screws because of simplicity and fast installation process; use of bolt as fastener is found rarely. Although use of screws is more favourable due to method and time consideration than bolt, some potential failures may limit bolt's capacity to resist the load. Unlike bolt, screw has no nut which can prevent excessive rotation of the bolt and separation of the connection due to increasing of the hole size such as reported in [2 – 4]. Experimental results of screw connection which improper installation was presented in [5]. A gap between two layer elements on the screw connection generated additional bending and torque moments which could decrease its resistance.

Application of bolt connection on the simple moment frame was investigated in [6]. The bolts of 12 mm diameter were used at the beam-column joints and the supports. Indentation around the hole was observed without failure of the bolts. Kulkarni and Vaghe [7] conducted experimental tests of bolt connections on the cold-formed channel section using packing plate; diameter of bolt was 12 mm. Failure only occurred on the packing plate and the cold-formed section. The resistance of 12 mm bolt was presumed to be much stronger than the resistance of the channel section, so that the bolts did not failure during the tests.

In this paper, experimental tests were conducted for connection of light gauge steel using 5 mm bolt(s). Two type of connections were observed, i.e.: Connection Type-1 which joined on their webs directly and Connection Type-2 which joined on their flanges through a same-size channel connector. Experimental tests were conducted by subjecting the specimens to axial load with a constant speed until failure occurred. Type of ultimate conditions when the maximum load reached were observed. The experimental results were compared to the analytical calculation using formulae in the codes.

## 2 Experimental Study

Experimental tests were carried out for the channel section joint connected by bolts. Nominal size of the channel section was 75 mm, 35 mm, and 0.75 mm in height, width, and thickness, respectively. Two type of connections were observed, i.e.: Connection Type-1 and Connection Type-2 (Figure 1). There were three similar specimens for each connection type. Connection Type-1 was two channel sections which were set in back-to-back position and joined on their webs directly using one bolt. In the second type, two channel sections were joined by two bolts on their flanges using a same-size connector channel. In total, four bolts were used in this connection type.

* Corresponding author: sabril_haris_hg@ft.unand.ac.id

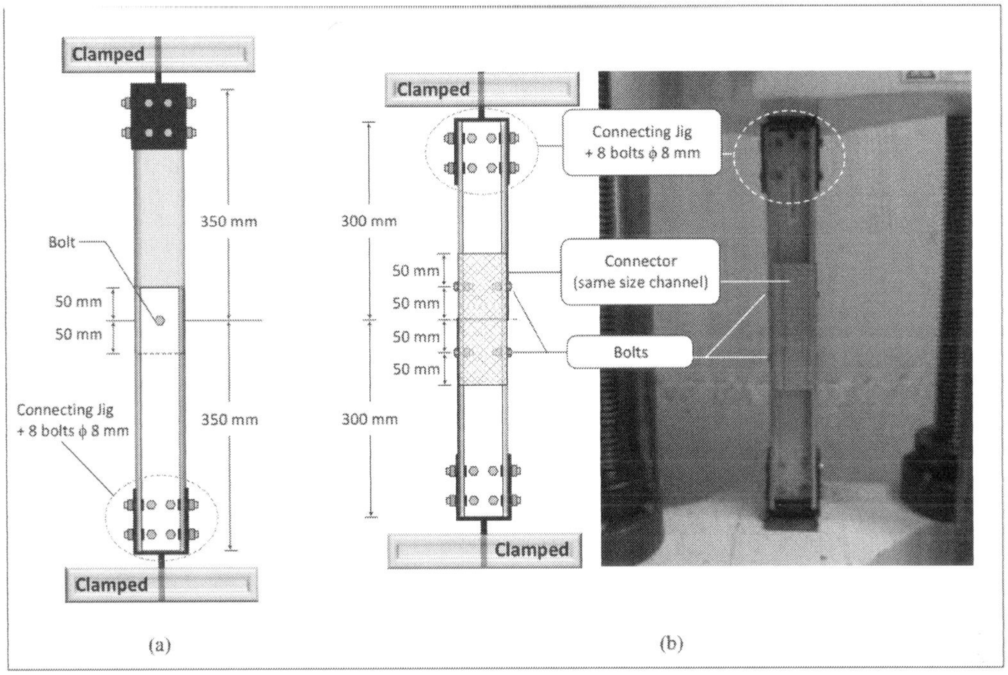


**Fig 1.** Illustration for two type of connections: (a) Connection Type-1, (b) Connection Type-2 and photo of the specimen

Each end of the specimen was clamped to the Universal Testing Material through a connecting jig. Eight bolts of 8 mm in diameter were used to fasten securely the ends of specimen and the connecting jigs; failures did not occur on this transition area. The specimen was subjected by axial tension load which increased monotonically until failure. The speed of loading was 2 mm per minute. The load and displacement data during the test was displayed on the screen, so that load-displacement curve could be observed directly.

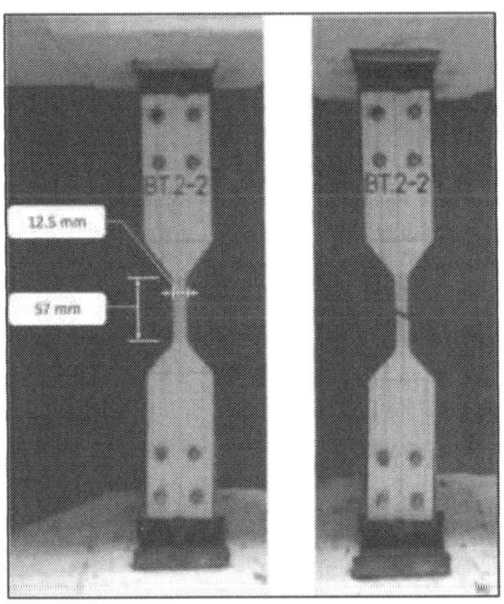


**Fig 2.** The specimen before and after tensile test

Cold-formed steel used in this study was a typical material which available and applied widely in residential construction in Indonesia. Properties of material were obtained through tensile test which refer to the standard for metals test [8]. Specimen for tensile test was taken from section web of the channel section; the channel was cut to form a profile of specimen as shown in Figure 2. The observed tensile zone defined in the reference was 57 mm in length and 12.5 mm in width. Both ends of the specimen were prolonged and linked to connecting jig with 4 bolts. The tensile specimen was pulled by the Universal Testing Machine with a constant speed of 2 mm per minute.

Typical ultimate condition was found on the tensile specimen that diagonal fracture occurred suddenly at the observed tensile zone, i.e. at the specimen width of 12 mm. In general, the ultimate stress was slightly higher than the yield stress with the maximum rupture strain was 3.51 % (see Table 1). Modulus of elasticity was assumed to be E = 200,000 MPa, a standard modulus for a typical cold-formed steel material. Diameter of the bolts used at the joint for both connection types was 5 mm; the bolt's grade was A307 with minimal tensile strength of 420 MPa [9].

**Table 1.** Tensile test results

| Connection Type | Yield Stress (fy, MPa) | Ultimate Stress (fu, MPa) | Ultimate Strain |
|---|---|---|---|
| 1 | 578.9 | 581.9 | 2.63 % |
| 2 | 589.9 | 589.9 | 3.51 % |

https://doi.org/10.1051/matecconf/201815401112

## 3 Connection Test Results and Discussion

### 3.1 Ultimate Condition

Ultimate condition for Connection Type-1 which two channel sections (CS-1 and CS-2) were joined with one bolt is shown in Figure 3. It was observed that the channel sections displaced away, and centre of the holes were no longer coincide at the ultimate state. The holes were dented by the bolt and their shape became oval. Indentation around the hole and displacement of the specimen indicated that the bolt was rotated and tilted before it failed. Failure at the connection occurred when the bolt was split at the shear plane, i.e. contact plane between two channel sections.

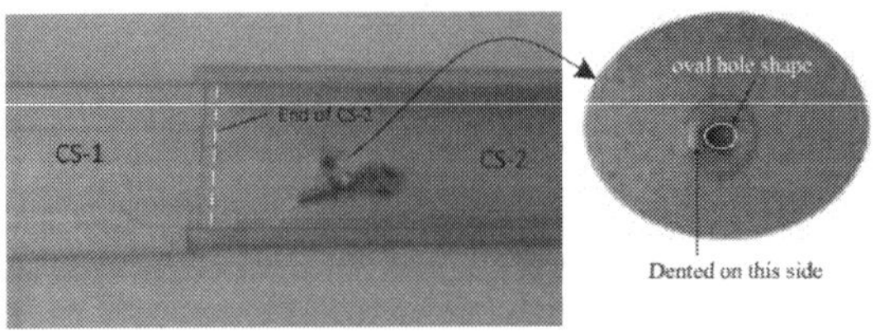


**Fig 3.** Ultimate state for Connection Type-1

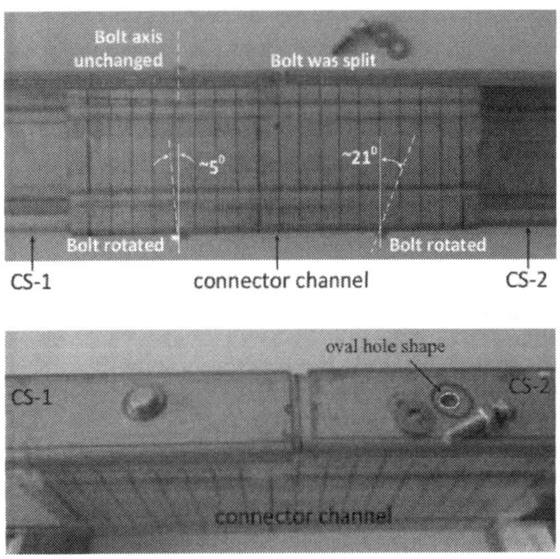


**Fig 4.** Ultimate state for Connection Type-2

Ultimate condition for Connection Type-2 is illustrated in Figure 4. There was four bolts in the connection system for which two bolts kept each channel section joining to the connector channel. However, all four bolts determined failure of the connection independently; failure on one bolt might cause failure of the connection system. It was observed from three specimens of Connection Type-2 that the ultimate condition occurred when one bolt was split, while other three bolts were not failure: two bolts were rotated and one bolt's axis was unchanged. Condition of these three bolt's axis indicated the stages of bolt before failure was initiated by rotation of the bolt. Regarding shape of the hole, similar condition was found. The hole became oval due to indentation.

### 3.2 Load-Displacement Curve

Response of the specimens to the subjected load are shown in Figure 5. It was noted that the loads used in the curve for Connection Type-2 (Figure 5.b) was a half of the total loads, in order to compare them to Connection Type-1 curves. In general, all load-displacement ($P$-$\delta$) curves showed similar tendency that the load increased almost linearly and dropped quickly after maximum load reached. The average ultimate loads ($P_u$) for Connection Type-1 and Type-2 were 4.77 kN and 5.50 kN, respectively. Other data related to load and displacement are collected in Table 2.

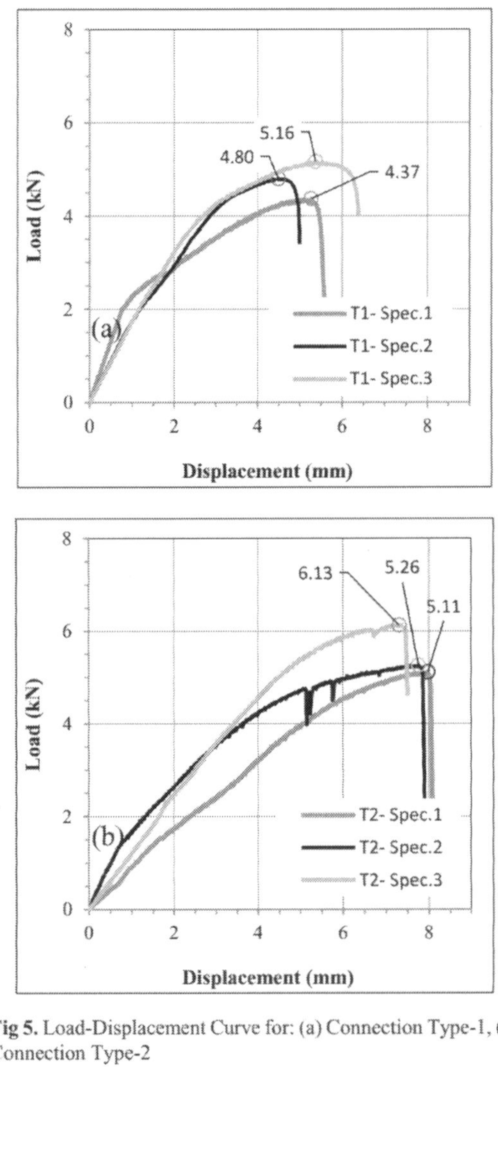


**Fig 5.** Load-Displacement Curve for: (a) Connection Type-1, (b) Connection Type-2

**Table 2.** Load and displacement data

| Type | Specimen Id | Experimental | | | Calculated $P_u$ | |
|---|---|---|---|---|---|---|
| | | $P_u$ (kN) | $\delta_u^*$ (mm) | $P_{u\text{-avrg}}$ (kN) | $P_{u\text{-bearing}}$ (kN) | $P_{u\text{-shear}}$ (kN) |
| 1 | T1-Spec.1 | 4.37 | 5.27 | 4.77 | 3.92 | 6.55 |
| | T1-Spec.2 | 4.80 | 4.52 | | | |
| | T1-Spec.3 | 5.16 | 5.37 | | | |
| 2 | T1-Spec.1 | 5.11 | 7.98 | 5.50 | 3.92 | 6.55 |
| | T1-Spec.2 | 5.26 | 7.76 | | | |
| | T1-Spec.3 | 6.13 | 7.31 | | | |

*displacement at $P_u$

### 3.3 Analytical Calculation

From the experimental tests, two conditions were identified when the ultimate condition occurred on the connection system, i.e.: indentation on the hole and shear failure of the bolt. Indonesian National Standard for Light Gauge Steel [10] and Australian/New Zealand Standard for Cold-formed Steel Structures [11] provide formula to calculate connection resistance for the state of indentation on the hole (bearing) and shear failure of the bolt, i.e.:

Bearing : $V_b = a\, C\, d_f\, t\, f_u$ (1)

Shear failure : $V_{fv} = 0.62 f_{uf} (n_n A_c + n_x A_o)$ (2)

where:

$a$ = modification factor for type of bearing connection
$C$ = bearing factor
$d_f$ = nominal bolt diameter
$t$ = base metal thickness
$f_{uf}$ = minimum tensile strength of a bolt
$n_n$ = number of shear planes with threads intercepting the shear plane
$A_c$ = minor diameter area of a bolt, as specified in AS 1275
$n_x$ = number of shear planes without threads intercepting the shear plane
$A_o$ = plain shank area of a bolt

Using appropriate value for constants and parameters in Eq.1 and Eq.2, resistances due to indentation around the hole and bolt shear were calculated, and it yielded the resistances of 6.55 kN and 3.92 kN, respectively. These resistances indicated that the connection will be dissemble at the minimum load of 3.92 kN because of the shear failure on the bolt.

### 3.4 Discussion

The shear failure on the bolt determined the ultimate condition of the connection because its resistance lower than the resistance due to indentation around the hole. It means the connection will be failed because of the bolt splitting, and thus indentation around the hole should not occur. However, test results showed that indentation around the hole still occurred although the ultimate condition was determined by shear failure on the bolt. The hole on the channel section was still prone to be indented. It caused change of shape and dimension of the hole. The hole shape became bigger and changed to oval form. The bolt was then rotated and tilted following direction of the applied load as shown in Figure 4.

The ultimate loads from the experimental test and analytical calculation were compared through following steps. For Connection Type-1, each ultimate experimental load in Table 2 was compared to analytical bearing resistance. It is because one bolt involved in this connection. For Connection Type-2, for which two bolts involved, the ultimate load in Table 2 was multiplied by two. The result was then compared to the sum of bearing and shear failure resistances which met failure condition of the bolts on the connection: one bolt was split and one bolt indented the hole.

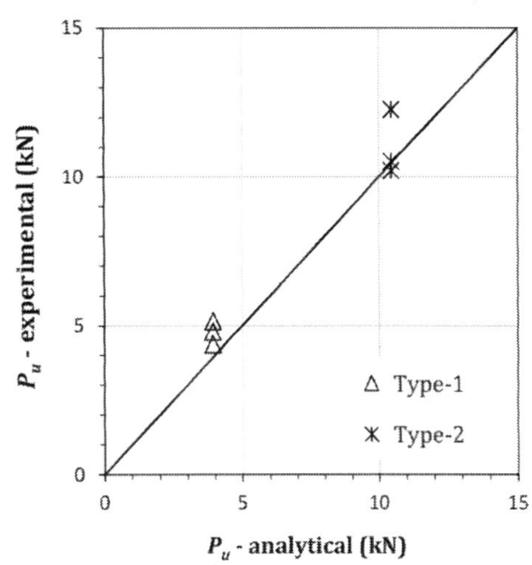


**Fig 6.** Comparison between experimental results and analytical calculations

Comparison between experimental results and analytical calculations is presented in Figure 6. It shows that the analytical formula conservatively predict the experimental results. In average, the bolt resistances from the test were 22% and 4% greater than analytic for Connection Type-1 and Type-2, respectively.

## 4. Conclusion

Experimental results of bolt connection behaviour on the cold-formed steel joints has been conducted in this study. For two different connection type, similar ultimate conditions on the specimens loading were found, i.e.: shear failure of the bolt and indentation on the hole. Analytical formula defined greater resistance due to indentation than resistance due to shear failure which indicated that indentation on the hole will not initiate if shear failure on the bolt occur. However, indentation on the hole still occurred and the hole shape was changed become oval. This condition caused the centre of the holes became no longer coincide and the bolt was rotated. Comparing to the experimental results, analytical calculation predict lower resistances within the range of 4 – 22 %.

## References

1. W.W. Yu. *Cold-Formed Steel Design*. (John Wiley and Sons Inc, New York, 2000)
2. S. Haris, H. Herman. Experimental study of connection behavior using screws on light gauge structural elements (in Indonesian). *Proceeding of Annual Civil Engineering Seminar 2015*. Pekanbaru, Indonesia, pp 390-396, 2015.
3. S.S. Fairuz, L.H. Ho. Influence of Screw Spacing on the Strength of Self-Drilling Screw Connection for the High Strength Cold-Formed Steel. Adv. Mat. Research **712-715**, pp 1054-1057. 2013.
4. R.M. Francka, R.A. LaBoube. Screw Connections Subject to Tension Pull-Out and Shear Forces. *Proceeding of Twentieth International Specialty Conference on Cold-Formed Steel Structures*. St. Louis, Missouri, pp 635-651. 2010.
5. Bambach, M.R. & Rasmussen, K.J.R. Behavior of Self-Drilling Screws in Light-Gauge Steel Construction. J Str Eng **133**, pp 895 – 898. 2007.
6. B.A. Ali, S. Saad, M.H. Osman. (2010). Cold-Formed Steel Frame with Bolted Moment Connections. Int J Civil and Struct. Eng. **1** (3), pp 534-544. 2010
7. R.B. Kulkarni, V.M. Vaghe. Experimental study of bolted connections using light gauge channel sections and packing plates at the joints. Int. J Adv Struct Eng **6** (4), pp 105–119. 2014
8. ASTM. *Annual Books of ASTM Standards, Section 3: Metals Test Methods and Analytical Procedures*. (American Society for Testing and Materials, Philadelphia, 1991).
9. C.G. Salmon, J.E. Johnson, F.S. Malhas. *Steel Structures: Design and Behavior*. 5th Ed. (Prentice Hall Indiana, US, 2008).
10. *Indonesian National Standard for Cold-formed Steel Structures* (in Indonesian), SNI 7971, 2013.
11. *Australian/New Zealand Standard for the Design of Cold-formed Steel Structures*, AS/NZS 4600, 2005.

# Point-to-point practical controller performance in an experimental two-axes belt driven mechanism

Purtoj* and Mulfiyandi

Mechanical Engineering Dept., Faculty of Industrial Engineering, Islamic University of Indonesia, Jl. Kaliurang km 14.5 Sleman, Yogyakarta, Indonesia

**Abstract.** High accuracy, fast response with no or small overshoot, and robust to object uncertainties are such performance required in a point-to-point positioning system. In order to achieve those performances, a good controller is required. A nominal characteristic trajectory following controller is one type of practical controllers which already has shown it good point-to-point positioning performance. Its structure consists of a nominal characteristic trajectory and a compensator. The design of the controller is based on a simple open loop experiment to construct a nominal characteristic trajectory. Information from nominal characteristic trajectory was then used to design the compensator. This practical approach was implemented in several different controlled objects, but limited to, system with a high stiffness mechanism. In this work, a nominal characteristic trajectory following controller was implemented on a low stiffness mechanism. An experimental two-axes belt driven mechanism was used as a controlled object. The actuator to drive the mechanism was electric motors with integrated encoders. Both axes with different characteristics identified through a simple open loop experiments and nominal characteristic trajectory constructed using speed and position data. A proportional integral compensator is selected in this work and was designed according to information from a nominal characteristic trajectory. In order to evaluate controller performance, point to point operation was conducted through simulation and experiment.

## 1 Introduction

Point-to-point (PTP) operation is widely used in industrial application. Pick and place operation, NC drill presses and spot welding operation, to name only few example applications of PTP operation. PTP operation as part of positioning systems requires a controller to satisfy such requirements as high accuracy, fast response with no or small overshoot, and robust to object uncertainties. Many kinds of controllers have been proposed and evaluated for positioning systems. The PID controller performance still in investigation dealing with positioning system, in the example, the cascaded PID controller was proposed to reduce nonlinearities factors such as friction and power saturation [1]. The PID controller designed with dominant pole placement method provides one more parameter than the conventional PID controller to get better performances [2]. The sliding mode controller also provides good tracking performance and robustness even in the presence of the parameter uncertainties [3]. Practical controller, nominal characteristic trajectory controller (NCTF) which was one of controller developed to handle point to point operation [4] provides good performance despite its simple design and its procedure. The NCTF controller has a simple structure. It consists of a nominal characteristic trajectory (NCT) and a compensator. Construction of the NCT is simply based on open loop responses of the controlled object. Moreover, the design of the compensator is based on NCT information. Its performance dealing with mechanism with small elasticity effectively compensated vibration cause by mechanical resonance [5]. In this paper, an implementation of nominal characteristic trajectory controller on an experimental two-axes belt driven mechanism which has low stiffness mechanism will be investigated.

## 2 Introduction

The structure of the NCTF control system is shown in Fig. 1. The controller consists of a nominal characteristic trajectory (NCT) and a compensator. The controller output is signal $u_p$, this signal is used to drive the object. The controller inputs are error, $e$ , and object motion, &. In principle, the controller compares the object motion input, &, with the error-rate provided by predetermined NCT, &, at a certain error. The difference between the actual error-rate of the object motion and that of the NCT is signal $u_p$, which is the output of the NCT. If the object motion perfectly follows the NCT, the value of signal $u_p$ is zero. Thus, no action is performed by the compensator. When the signal $u_p$ is not zero, the compensator forces the object motion so that the value of signal $u_p$ is zero. The objective of the compensator is to make the object motion

* Corresponding author: purtojo@uii.ac.id

follow the NCT, and finishes at the origin of the phase-plane $(e = 0, \dot{e} = 0)$.

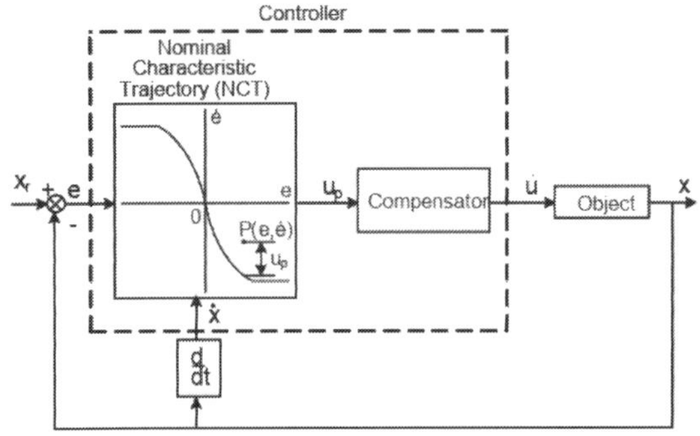


**Fig 1.** Structure of the NCTF control system

In order to determine the NCT, a simple open-loop experiment has to be conducted. In the experiment, stepwise inputs are applied to the object, and displacement and velocity responses of the object are measured. Fig. 2 (a) shows the stepwise input, and the resulting velocity and displacement responses, normalized respective to its input heights. The rated input to the actuator $u_r$ is used as a value of the stepwise inputs.

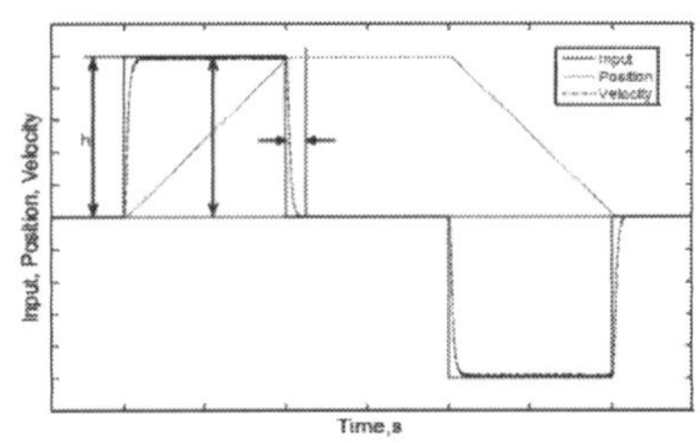


(a)

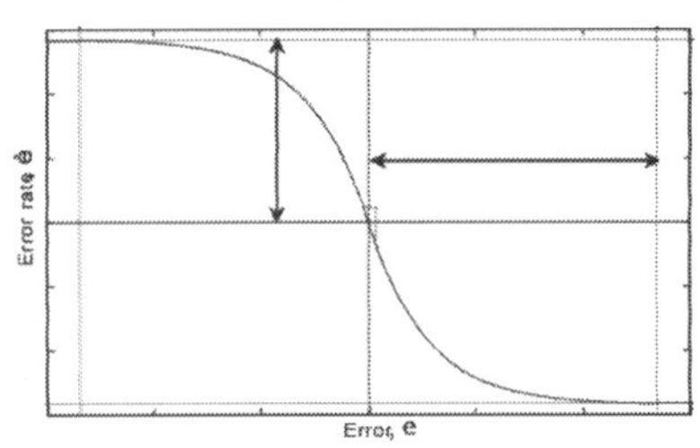


(b)

**Fig 2.** (a) Open-loop responses (b) NCT

The velocity and displacement responses are used to determine the NCT. The phase plane of the NCT has a horizontal axis of error, $e$, and a vertical axis of error-rate, $\dot{e}$, as shown in Fig. 2 (b). The horizontal axis shows the error from the desired position. Therefore, the values are taken from the displacement data of the open-loop responses. The error-rate is the velocity data within the deceleration process. In Fig. 2 (a), h is the maximum velocity, which depends on the input step height and gradually decreasing to zero. Thus the pair data of error and error-rate construct the NCT.

## 3 Experimental Setup

An experimental two-axes belt driven mechanism as a controlled object is shown in Fig. 3. It represents an industrial CNC machine. The mechanism provides three axes perpendicular motion, however discussion is limited to X and Y direction only. The table mounted on linear bearing slides along a circular rod. DC motors with integrated rotary encoders are used as actuators. Those are received signal from motor drivers connected to an Arduino Mega 256. The control signal is sent from Scilab Xcos model build in a personal computer.

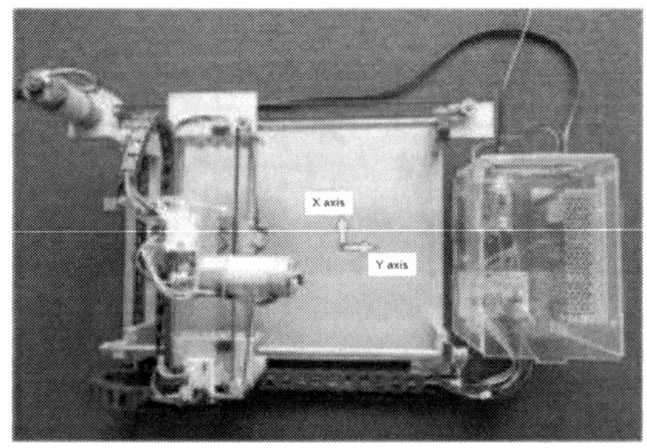


**Fig 3.** Experimental two-axes belt driven positioning system

The open-loop experiment using stepwise input had been conducted in both axes of the mechanism. Maximum input voltage was $u_r$ = 12 volts. Fig. 4 shows the velocity and displacement responses on both-axes.

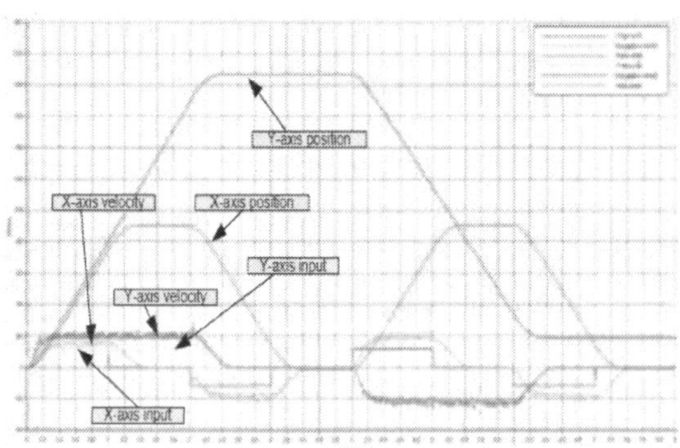


**Fig 4.** System responses to a stepwise inputs

### 3.1 NCT

Velocity and position data of system responses are then used to construct the NCT of both axes. The NCT of x-axis is shown Figure 5 (a) and Figure 5 (b) shows the NCT of Y-axis. Two important parameters in the NCT are the maximum error rate indicated by h and the near-zero slope of the NCT indicated by m. These parameters are required to design the compensator. X-axis NCT parameters are hx=25.31 mm/s and mx=-124.5, and Y-axis NCT parameters are hy=25.71 mm/s and my=-25.06.

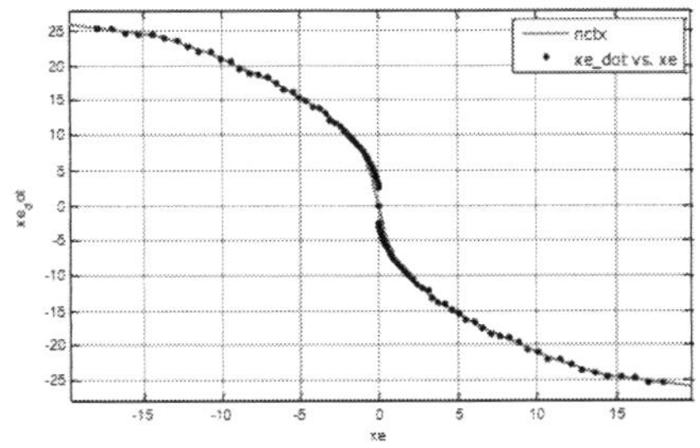


(a)

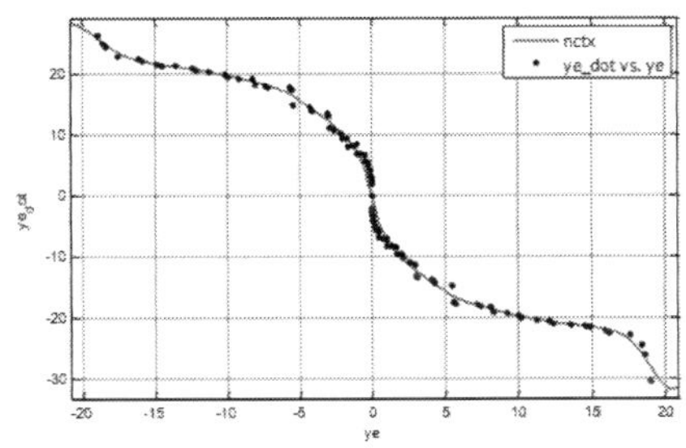


(b)

**Fig 5.** (a) NCT of X-axis, and (b) NCT of Y-axis

### 3.2 Compensator Design

In this study PI controller is used as a compensator to control the movement of the controlled object.

$$G_C(s) = K_P + \frac{K_i}{s} \tag{1}$$

PI compensator can be expressed as a function of the natural frequency and damping ratio as follows (Wahyudi, Sato, & Shimokohbe, 2005):

$$K_p = \frac{2\zeta\omega_n}{\alpha K} = \frac{2\zeta\omega_n u_r}{mh} \tag{2}$$

$$K_i = \frac{\omega_n^2}{\alpha K} = \frac{\omega_n^2 u_r}{mh} \tag{3}$$

## 4 Positioning Performances

Performance of the designed NCTF controller evaluated through simulation and experiment. The evaluation was conducted by driving the controlled object to a several inputs, 10 mm to 50 mm input. Figure 6 shows positioning performance of NCTF controller to a 10 mm input, Figure 6 (a) for X-axis and Figure 6 (b) for Y-axis respectively. In this paper only 10 mm input presented since the other responses to different input is similar.

Designed NCTF controller had been evaluated for position control using the PI controller as a compensator. This investigation shows that the maximum steady state error achieved on the X axis is 8.73% for a 10 mm input, and lower to higher inputs. Steady state condition is achieved with settling time of 0.28 seconds. On the Y axis the maximum steady state errors also present for a 10 mm input, which is 8.56%, and lower to higher inputs. The settling time is 0.2 seconds. Since the input larger 20 mm is out of NCT, the settling time obviously is slower, however the oscillation is reduced. Thus, because the system has a constant linear velocity, so the further the set point value is given, then the longer the system reaches the desired value.

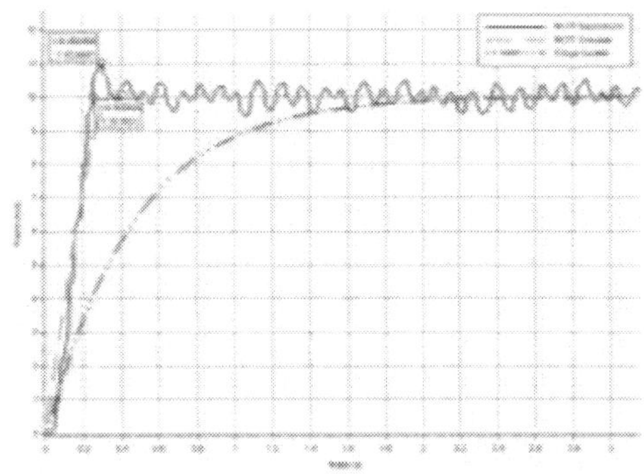

(a)

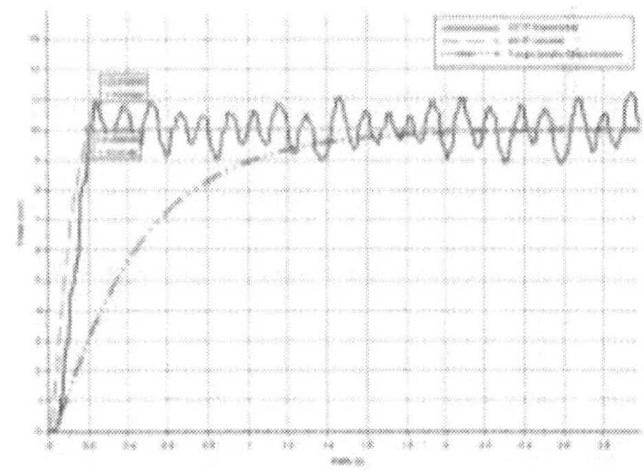

(b)

**Fig 5.** Positioning performance to a 10 mm input (a) X-axis, and (b) Y-axis.

Oscillation about desired position was occurring on both axes. The source of oscillation was the hardware selection problem, which is the belt to drive the mechanism. The stiffness of the belt is very low and not sufficient to withstand the inertia of the system. As shown in the X-axis, the oscillation is smaller compared to oscillation in Y-axis. This is related to the mass of components moved along each axis. Since the belt has a high elasticity compared to other more precise drive transmissions such as ball screw, it cannot withstand the inertia of the moving components. Overall the position control system has been designed to run in accordance with the expectation. The system response drive by the NCTF controller was able to accelerate the rise time on both the X-axis and the Y-axis. Although the system has a sufficiently large steady state error value, but it's still within the allowed tolerance of 2-10% of the steady state error.

## 4 Positioning Performances

The NCTF controller design procedure had been successfully applied to an experimental two-axes positioning system with belt driven mechanism. The open loop experiment provided necessary information to construct NCT and consecutively to design the PI compensators. Controller design is simple and successfully implemented to the controlled object. Performance evaluation through simulation and experiment show that the designed NCTF controller able to accelerate the rise time. However, controller positioning performance deteriorated by the very low stiffness of the belt. Oscillation about the steady state position occurred constantly. Thus, selection of sufficient hardware properties is a must prior to implementation of the controller.

## References

1. K.H. Ng, C.F. Yeong, E.L.M. Su, L.X. Wong. Procedia Engineering **41**, 244–250 (2012).
2. P.D. Mandić, T.B. Šekara, M.P. Lazarević, M. Bošković. ISA Transactions **67**, 76–86 (2017).
3. M.C. Pai. J of the Franklin Institute **347**(10), 1837–1849 (2010).
4. Wahyudi, K. Sato, A. Shimokohbe. Precision E. **27**(2), 157–169 (2003).
5. M.F.M. Yakub, W. Martono, R. Akmeliawati. *The IEEE Control & System Graduate Research Colloquium* pp. 61–67 (2010).
6. Wahyudi, K. Sato, A. Shimokohbe. Robotics and Autonomous Systems **52**(2–3), 247–256 (2005).

# Study on the effect of Weber Number to heat transfer of multiple droplets on hot stainless steel surface

*Aria* Riswanda, *Indro* Pranoto, Deendarlianto, Indarto, *and Teguh* Wibowo

Department of Mechanical Engineering, Faculty of Engineering, Universitas Gadjah Mada, Jl. Grafika No. 2, Yogyakarta, 55281, Indonesia·

**Abstract.** Multiple droplets are drops of water that continuously dropped onto a surface. Spray cooling is an application of the use of droplet on a cooling system. Spray cooling is usually used in a cooling system of electronic devices, and material quenching. In this study, correlations between Weber number and surface temperature decrease rate during multiple droplets impingement are investigated and analyzed. Visualization process is used to help determine the evaporation time of droplets impingement by using high speed camera. Induction stove is used to maintain a stainless steel surface temperature at 120°C, 140°C, and 160°C. The Weber number was varied at 15, and 52.5 to simulate low and medium Weber number. The result of this study shows that increase in Weber number does not increase the temperature decrease rate noticeably. Whereas the Weber number decrease the time required for surface temperature to reach its lowest surface temperature. It was also found that for low and medium Weber number, Weber number affect the evaporation time of multiple droplets after impingement.

## 1 Introduction

Multiple droplets are a series of droplet released with certain interval continuously onto a surface. Spray cooling is one the applications of multiple droplets. Widespread use of spray cooling in industrial application such as fire suppression by sprinkler system, cooling of turbine blades and electrical components. The rapid development of electrical components leads to smaller and more complex electronic component, but also accompanied with increase in power density. Making the conventional heat dissipation methods such as forced convection, pool boiling, jet impingement, are difficult to meet desired thermal control required. Spray cooling is known for having numerous advantages over conventional heat dissipation such as higher heat dissipation capacity, less flow rate demand, no temperature overshoot [1]. Spray cooling are also used in metal quenching process to increase the quality [2].

One of the factors that influences the heat transfer of droplet during impact onto a heated solid surface is the Weber number (We). This number is defined as the ratio of the inertia energy to the surface tension energy. Bernadin et al. [3] categorized Weber number into three regions (1) low Weber number, We < 30, (b) medium Weber number, 30 < We < 80, (c) high Weber number, We > 80. The correlation between Weber number and temperature of impinged surface to droplet spreading's characteristic has been analyzed and resulting in droplet characteristics after impingement which is levitation, disintegration, and bouncing [4]. The effects of the wettability of a droplet impingement onto a hot solid surface under medium Weber number had been studied experimentally using varying surface coating. Resulting in the higher the wettability, the larger the droplet spreading on the hot surface [5]. Analytical model approach to determine maximum spreading on low Weber and Reynold number had been made [6] with acceptable result. Research on low Weber number also had been conducted before. The result concludes that for low Weber number, the higher the static contact angle, the lower the wetting limit temperature [7].

Research on droplet evaporation on hot stainless steel plate had been conducted before [8]. With surface temperature varied between 63-605 to cover entire spectrum of heat transfer and droplet size ranging from 0.07 to 1.8 mm to simulate spray cooling diameter and also liquid included water and carbon fuel. The research resulting in droplet lifetime curve as a function of a wall temperature is similar for all liquid and all drop size. And the maximum heat transfer rate is independent of drop size.

The aim of present research is to study the effect of Weber number to heat transfer phenomena on multiple droplets impinging hot stainless steel surface, mainly the surface temperature decrease rate and droplets evaporation time. Visualization was used to help observation of droplet evaporation time after impingement.

* Corresponding author: ariarwd@gmail.com

## 2 Experimental Apparatus and Procedure

The detail of apparatus used in this study is shown in Figure 1. The apparatus consists of droplet injector, droplet frequency controller, thermocouple, data logger, high speed video camera, illumination system, and personal computer for acquisition of the surface temperature data.

The fluid used is deionized water with droplet size was kept constant at 2.75 mm by using nozzle with inner diameter 0.8 mm and dropped by switching open/close the solenoid valve inside the system. Distance between the droplet injector and specimen is set to 20, and 70 mm resulting in 15, 52.5 Weber number. Droplet temperature is kept constant at 27 °C. Frequency of droplets used are 6.7 drops/second and opened only for 3.0 seconds resulting in 20 droplets impinged onto stainless steel surface.

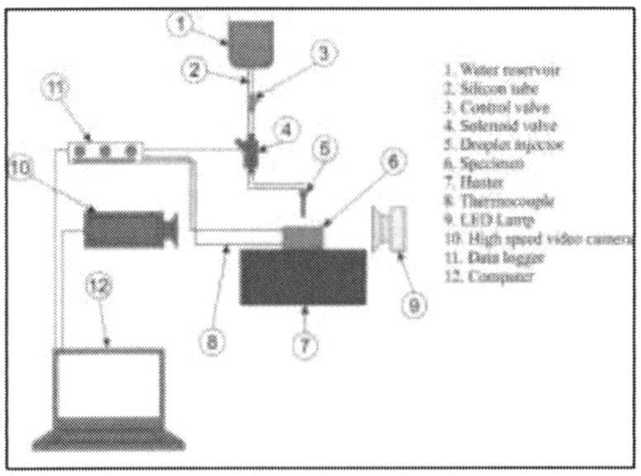


**Fig 1.** Schematic diagram of experiment apparatus

The specimen used is stainless steel with diameter of 50 mm and 0.06 μm hardness. Surface temperature are measured using interpolation of three thermocouples planted inside specimen in 5, 10, and 15 mm range in height. The heater used is made from induction stove, surface heating was performed until desired temperature reached. The temperature of specimen varied at 120 °C, 140 °C, and 160 °C. Surface temperature decrease data then obtained by using thermocouple connected to data logger.

In the present study, a macro lens attached to a high speed video camera is used to observe the droplet evaporation time after impingement. The video images were taken at 1000 frame per second with resolution of 640 x 480 pixels.

The initial diameter of a droplet was measured from the droplet image was not perfectly spherical. Both horizontal and vertical diameter of droplet was measured and using equation made by Šikalo [9] the equivalent diameter is

$$D = (D_h{}^2 D_v)^{1/3} \quad (1)$$

Where $D_h$ is the horizontal diameter and $D_v$ is vertical diameter shown in Figure 2.

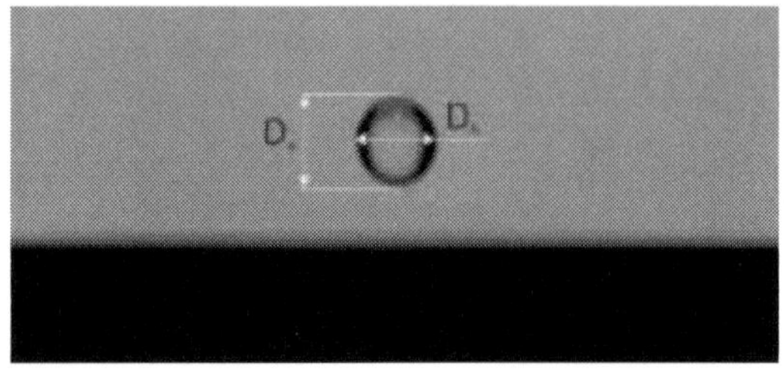

**Fig 2.** Droplet diameter measurement

In this experiments Weber numbers are 15, and 52.5. This Weber number can be achieved from the expressions as follows:

$$We = (\rho v^2 D)/\sigma \quad (2)$$

Where $\rho$, $v$, $\sigma$ are density, relative velocity of liquid, and surface tension respectively.

The rate of decrease in surface temperature can be determined by looking at changes in surface temperature impinged by the droplet. The rate of decrease in surface temperature can be calculated by

$$S = \frac{T_{S1} - T_{S2}}{t_2 - t_1} \quad (3)$$

Where $S$, $T_{s1}$, $T_{s2}$, $t_1$, $t_2$ are initial surface temperature, final surface temperature, initial time, final time respectively.

## 3 Result and Discussion

This present work is mainly focused on the effect of Weber number and surface temperature to temperature decrease rate during droplet impingement. Figure 3. Shows the temperature decrease over time value of stainless steel across all Weber numbers and surface temperatures. It is shown that in all surface temperature variation We = 52.5 has lower lowest temperature after impingement than We = 15 albeit the difference is not considerable. It is also shown that the higher the Weber number the less the time required to reach the lowest surface temperature after impingement. At surface temperature of 120 °C, the time required for surface temperature to rise again is longer than at 140 °C and 160 °C, this is due to heat transfer occurred between droplets and surface at 120 °C is lower than its counterparts. At surface temperature of 160 °C, the differences in lowest temperature and time required for temperature to rise again are notably large compared to the lower surface temperatures. This is due to surface temperature at 160 °C started to gain more increase in heat transfer rate between droplets and surface than previous surface temperature, in accordance to the droplets starting to reach its Critical Heat Flux (CHF) temperature

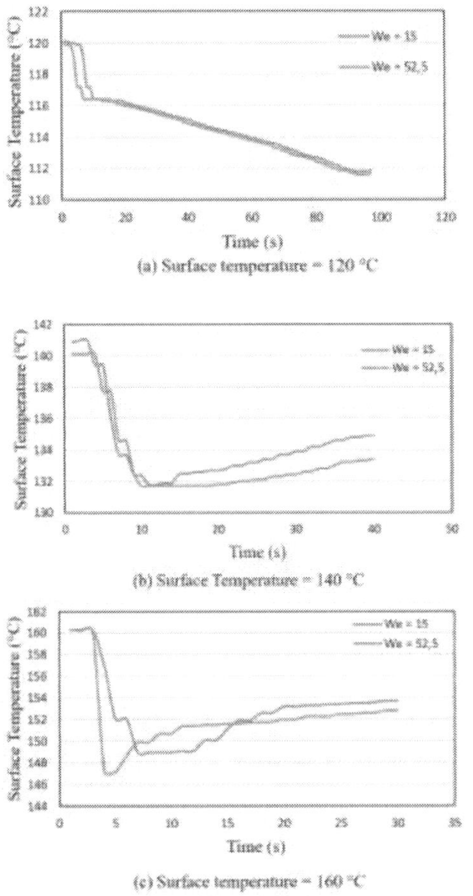


**Fig 3.** Surface temperature decrease over time

Temperature decrease rate achieved by measuring the initial and final surface temperature and using equation (3). Figure 4 shows the temperature decrease rate impinged by multiple droplets. It is shown that by increasing the Weber number, the temperature decrease rate is also increased, the increment in temperature decrease rate is increasing as surface temperature increased. During surface temperature at 160 °C there is significant rise to temperature decrease rate for both Weber numbers. This is mainly due to surface temperature at 160 °C gain significant increase in heat transfer rate.

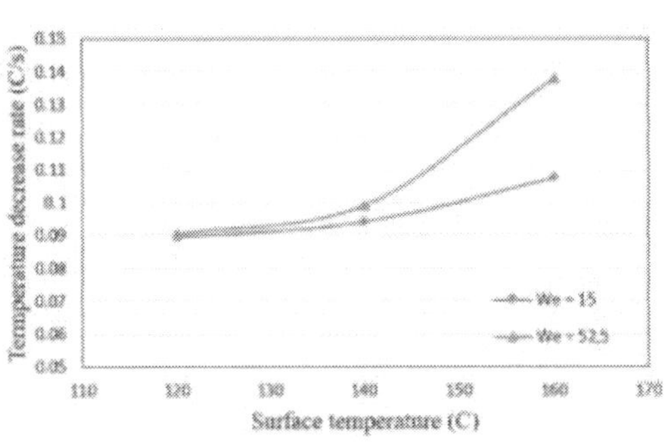


**Fig 4.** Surface temperature decrease rate

Figure 5 shows the evaporation time of multiple droplets impinging onto hot stainless steel surface. The evaporation time defined as time required for droplet to evaporate completely after impinging solid surface. From Figure 5 it can be seen that effect of Weber number on the evaporation time is apparent, the evaporation time decrease as the Weber number increases. The difference in evaporation time between We = 15 and We = 52.5 is getting smaller as the surface temperature is increased. The higher Weber number, the droplet spread diameter is larger, thus the contact area between droplet and solid surface is also higher, resulting a shorter evaporation time.

## 4 Conclusion

A study on the effect of Weber number of multiple droplets impinging hot stainless steel surface had been conducted. The surface temperature was varied from 120, 140, and 160 °C. Their effects on the heat transfer phenomena were studied, the results are summarized as follows:

1. The Weber number has not a significant effect on the lowest surface temperature after impingement. Increasing Weber number from low Weber number to medium Weber number does not change the lowest surface temperature significantly but reduce the time required to reach the lowest surface temperature.
2. At the surface temperature 120 °C, increasing Weber number does not change the rate of temperature decrease noticeably. By increasing surface temperature to 140 and 160 °C will increase the rate of temperature decrease more considerable, notably at surface temperature of 160 °C.
3. Weber number has apparent effect on the evaporation time. Which is the higher the Weber number the lower the evaporation time occurred. At lower temperature the difference in evaporation time between We = 15 and We = 52.5 is considerable, although the difference in evaporation time getting smaller as surface temperature increased. The higher the surface temperature is also making the evaporation time shorter.

## References

1. Cheng, W.L., Zhang, W., Chen, H., Hu, L, Spray Cooling and Flash Evaporation Cooling: The Current Development and Application, Renewable and Sustainable Energy Reviews, **55**, pp. 614-628 (2015)
2. Fujimoto, H., Watanabe, S., Okamoto, T., Hama, T., Takuda, H., Photographic Study of Hydrodynamics of Drops of Aqueous Polymer Solution Impinging on Hot Solid, Experimental Thermal and Fluid Science, **60**, pp. 66-74 (2015)
3. Bernadin, J.D., Stebbins, C.J., Mudawar, I., Mapping of Impact and Heat Transfer Regimes of Water Drops Impinging on a Polished Surface, International Journal of Heat and Mass Transfer, **40**, pp. 247-267 (1997)

4. Kandlikar, S.G., Steinke, M.E. & Singh, A., Effects of Weber Number and Surface Temperatur on the Boiling, Proceeding of 35th National Heat Transfer Conference, pp.1–10 (2001)
5. Mitrakusuma, W.H., Kamal, S., Indarto, Deendarlianto, et al., The Dynamics of the Water Droplet Impacting onto Hot Solid Surfaces at Medium Weber Numbers, Journal Of Heat and Mass Transfer, **53**, pp. 1-13 (2017)
6. Li, R., Ashgriz, N., Chandra, S., Maximum Spread of Droplet on Solid Surface: Low Reynold and Weber Numbers, Journal of Fluids Engineering, **48**, pp. 1025-1035 (2010)
7. Deendarlianto, Yasuki, T., Sumitorno, H., Indarto, et al., Effect of Static Contact Angle on The Droplet Dynamics During the Evaporation of a Water Droplet on The Hot Walls, International Journal of Heat and Mass Transfer ,**71**, pp. 691-705 (2014)
8. Xiong, T.Y, Yuen, M.C., Evaporation of a liquid droplet on a hot surface, International Journal of Heat Mass Transfer, **34**, pp.1881-1894 (1991)
9. Šikalo, M. Marengo, C. Tropea, E.N. Ganić, Analysis of Impact of Droplets on Horizontal Surfaces, Exp. Therm. Fluid Sci. **25**, pp. 503–510 (2002)

# A comparative pressure analysis of air flow between horizontal and V-Tail of UAV MALE of NACA0012H with speed variation

*Rahmat* Riza*, *Dicky* Kurniawan, and *Arif Budi* Wicaksono

Universitas Islam Indonesia, Yogyakarta, Indonesia

**Abstract.** NACA0012H is an airfoil type that could be used for Unmanned Aerial Vehicle Medium Altitude Long Endurance. This experiment was used to analyze stress in the surface of Tail of UAV MALE that was caused by air flow. The experiment was conducted using Computational Fluid Dynamics Software. Two designs of tail, horizontal and *V*-tail, were considered to simulate pressure occurred on the surface of leading edge, chamber and trailing edge. The simulation was developed varying the speed of the UAV MALE. The results showed that pressure occurred on the surface of horizontal tail higher than pressure on the *V*-tail.

## 1 Introduction

Unmanned Aerial Vehicles (UAV) simply could categorize as plane which remotely control outside of its plane [1]. Hence, the design of UAV would be identic with the actual aircraft where the ability of UAV to fly determined by the design of its aerofoil. This could be understood that the effectivity of the wing of the aircraft or UAV to produce aerodynamic force was determined by the design of air foil [2]. This force is important to obtain optimum process for the aircraft to fly in the air. The aerofoil referred to the profile of the wings of the aircraft which obtain by cutting section that was perpendicular to the wing [3]. Therefore, *tail* or *empennage* on a UAV plane has the same function as a full-scale aircraft, which is to provide both longitudinal (pitch) and directional stability (yaw).

The selection of tail geometries is based on mission requirements, as the example for take-off and landing conditions, manoeuvres, payload laying, propeller configurations, drive systems (servo) to aesthetic aspects. The tail design should consider the lightest and strongest material and construction and not disturbance the engine exhaust. The tail could be designed as small as possible to obtain minimum weight.

One variation of the tail has *V*-shape, known as *V*-Tail. This is generated combining the vertical fin and stabilizer into a pair of controls. The theoretical advantage of this design is lighter than *T*-tail and to reduce the interference drag associated with two surfaces [1]. This could be established by simulating the pressure resulted in the design. *Computational Fluid Dynamic* (CFD) could be used to simulate the pressure change especially in the change of the speed of air flow in the tail itself.

In this study, NACA0012H was applied to construct aerofoil profile in order to explore the pressure change of horizontal and *V*-tail design. This was caused that air foil design from NACA widely used in the aerofoil design for aircraft or UAV [3], especially for NACA0012 [4, 5].

## 2 Simulation Methods

In this work, two design of tails, horizontal and *V*-tails, were simulated using CFD software to analyse the change of its pressure resulted by variation of air flow through it surface. The model simulation was conducted to investigate the change of the pressure in three important region of the tail for both designs. The important areas of the tail are given on Fig.1 [3, 6].

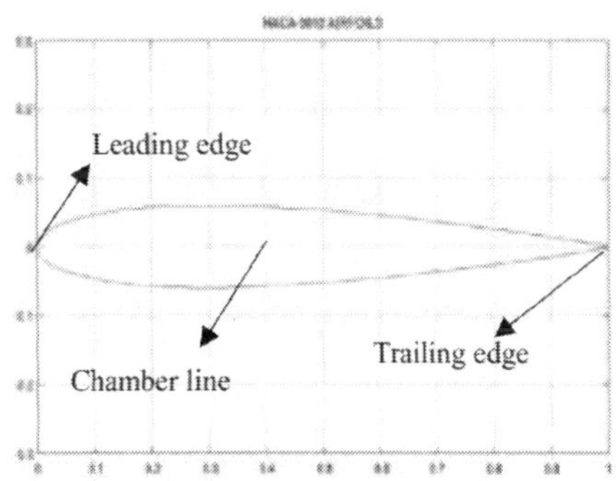


**Fig 1.** The geometry aerofoil of NACA 0012

Pressure which acted in all of the areas had been established as variation speed of the aircraft or UAV. The design of the tail was expected to reach speed of 240 km/h and to carry in till 400 kg payload. This could be categorized as UAV with specification for Medium Altitude Long Endurance (MALE). Four speed variations were selected in order to simulate the change of pressure in the important region of the tail, namely 165 km/h, 195 km/h, 210 km/h and 240 km/h. The results were used to analyse the advantages of either horizontal or *V*-tail.

* Corresponding author: rahmatriza@uii.ac.id

https://doi.org/10.1051/matecconf/201815401115

## 3 Results and Analysis

The accomplished CFD simulation had generated two results which would be analysed in this paper. The first result was a general description of the pressure acting in tail of UAV MALE in horizontal and *V*-tail design. This result would be discussed in Section 3.1 for the horizontal tail and Section 3.2 for the *V*-tail design. The second result was the pressure variation in the three specific region of tail either in the horizontal and *V*-tail design. This result would be discussed in Section 3.3.

### 3.1 Simulation Results on Horizontal Tail UAV MALE

The simulation of air flow in the area of the horizontal tail was conducted in the seven variation speed, namely 150 km/h, 165 km/h, 195 km/h, 210 km/h, 240 km/h. The overall results are given in Table 1.

**Table 1.** Simulation results for pressure of the horizontal tail

| Velocity (Km/h) | Pressure (Pa) | | Pressure Range (Pa) |
|---|---|---|---|
| | Maximum | Minimum | |
| 150 | 113815.75 | 93627.89 | 20187.86 |
| 165 | 116595.62 | 92041.64 | 24553.98 |
| 180 | 120745.51 | 89587.19 | 31158.32 |
| 195 | 124566.19 | 87941.76 | 36624.43 |
| 210 | 128310.97 | 85795.96 | 42515.05 |
| 225 | 132404.45 | 82952.14 | 49452.31 |
| 240 | 136866.99 | 79445.62 | 57421.37 |

Table 1 showed the simulation results for every speed variation. The values of Table 1 showed that the increase of aircraft speed had resulted parallel value with the increase of the maximum pressure. In other hand, the increase of aircraft speed looked contrary with minimum pressure value acting on the surface of the tail. This data showed that the effect of the increase of the speed had different impact between the maximum and minimum working pressure.

The entire acting pressure on the surface of the tail is given in Fig. 2 and Fig. 3. Fig. 2 showed the simulation results for the pressure acting on the surface of the tail with the speed of 165 km/h where Fig. 3 referred to the simulation for speed of 210 km/h.

Fig. 2 and Fig. 3 show pressure variation acting in the horizontal tail. This result was generated from the pressure acting caused by air flow with assumption that the tail is in the speed of 165 km/h and 210 km/h. Based on the legend, we could conclude that the tail had the highest pressure in the front side of the tail. They were concentrated in some points. Simulation results for variation speed of 165 km/h and 210 km/h overall had similar region of the pressure acting on the surface of the tail. However, simulation results for the speed of 210 km//h had generated lower pressure in the region which the minimum pressure act.

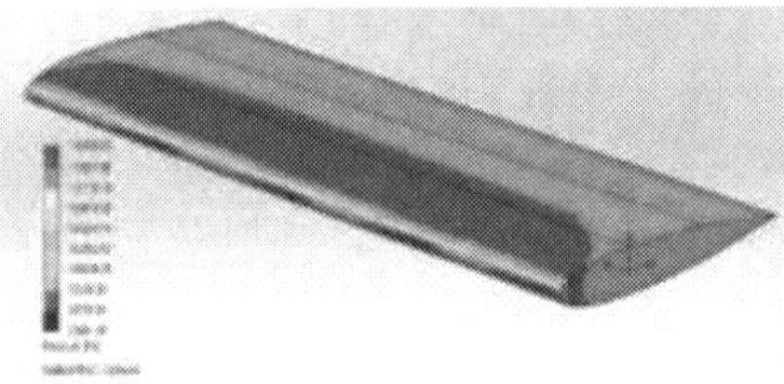

**Fig 2.** Pressure Surface Aerofoil with speed of 165 km/h

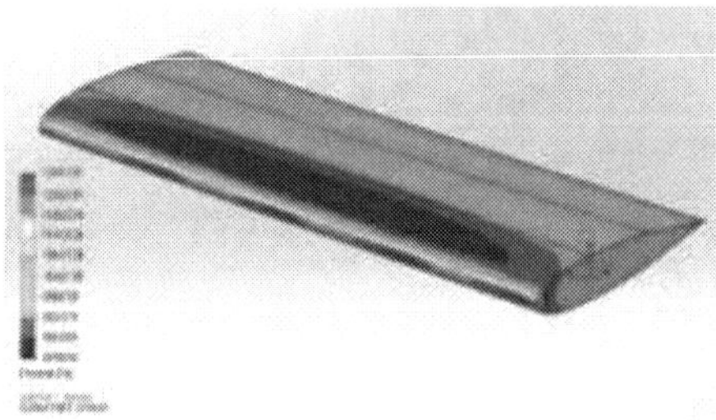

**Fig 3.** Pressure Surface Aerofoil with speed of 210 km/h

However, the pressure caused by the air flow acted in the end of the tail not too much different compare to the pressure acting in the front side. The area of the tail had received the lowest pressure located near to the area that had the highest pressure. This could be ensued due to the pressure of the air flow highly concentrate in some points.

The flow of the air through the tail are given in Fig. 3.

**Fig 4.** Flow Pressure Surface Aerofoil with speed of 165 km/h

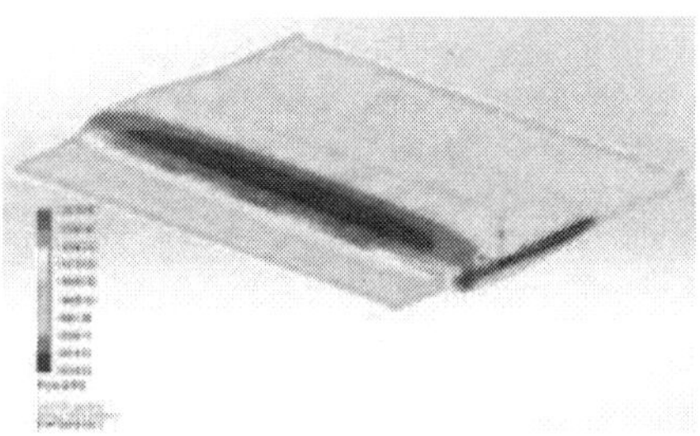

**Fig 5.** Flow Pressure Surface Aerofoil with speed of 210 km/h

In Fig. 4 and Fig. 5, the air flow was overall described. This results also showed where the area of air around the tail effected by the acting pressure on the surface of the tail caused by the speed of the aircraft. Those, However, the pressure caused by the air flow acted in the end of the tail not too much different compare to the pressure acting in the front side. The area of the tail had received the lowest pressure located near to the area that had the highest pressure. This could be ensued due to the pressure of the air flow highly concentrate in some points.

The samples contour of aerofoil are given in Fig. 6 for speed of 165 km/h and Fig. 7 for speed of 210 km/h. The pattern drawn of both simulations had quite different especially in low pressure. There is change the area of light blue pattern and the both area of green pattern. This showed the increase of pressure that was parallel with the increase the speed of the tail.

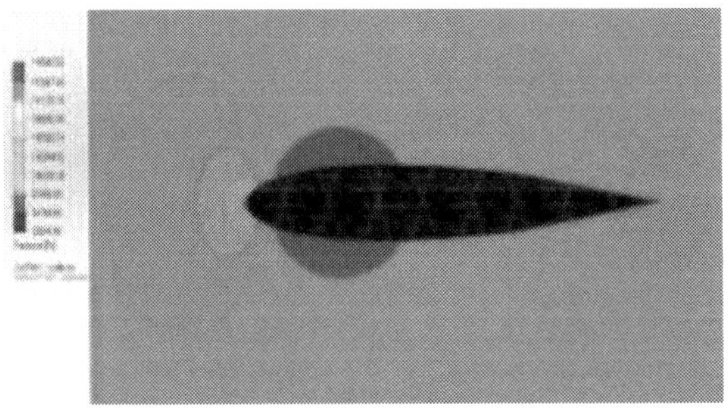

**Fig 6.** Contour Pressure Surface Aerofoil with speed of 165 km/h

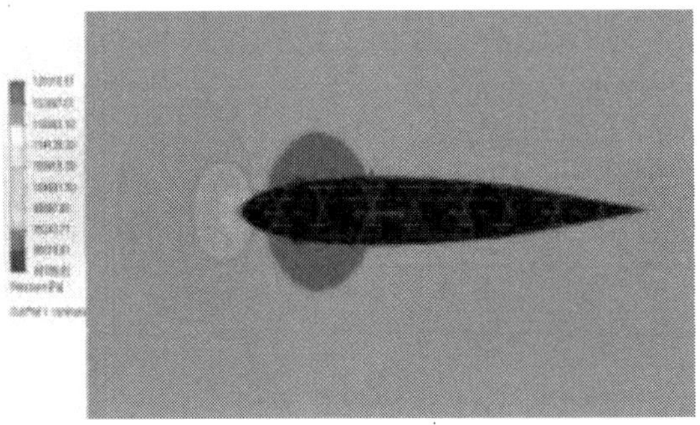

**Fig 7.** Contour Pressure Surface Aerofoil with speed of 210 km/h

The other area of the tail had received the pressure evenly distributed. This was shown by the area of each colour pattern. The results of both speed variations in this pressure range had been drawn within similar pattern.

### 3.2 Simulation results on *V*-tail UAV MALE

Besides the horizontal design, the tail of UAV had been proposed using *V*-shape or *V*-configuration [Babbu..]. Therefore, the analysis of NACA00012 aerofoil had been simulated in the *V*-configuration in this experiment. The overall results *V*-tail are given on Table 2.

The simulation results of the *V*-tail as given on the Table 2 showed the similar pattern as the results of the horizontal tail. The increase of the aircraft speed is in line with the increase of maximum pressure and the range pressure acting on the tail. However, this ensued till the speed 210 km/h. After this speed, there is a reduction of the maximum pressure acting on the aerofoil. The maximum pressure looked decreasing while the speed of aircraft rising from 210 km/h to 240 km/h. The increase speed, in fact, had caused the reduction of the minimum pressure acting on the aerofoil in this configuration.

**Table 2.** Simulation results for pressure of the *V*-tail

| Velocity (Km/h) | Pressure (Pa) | | Pressure Range (Pa) |
|---|---|---|---|
| | Maximum | Minimum | |
| 150 | 132984.51 | 91896.15 | 41088.36 |
| 165 | 170740.01 | 83655.90 | 87084.11 |
| 180 | 162794.92 | 79706.77 | 83088.15 |
| 195 | 188658.64 | 75497.31 | 113161.33 |
| 210 | 211780.13 | 78837.81 | 132942.32 |
| 225 | 187643.06 | 78571.59 | 109071.47 |
| 240 | 186030.17 | 71377.89 | 114652.28 |

Description of the working pressure on the tail with specific speed had been simulated as shown on Fig. 8 and Fig.9. Fig. 8 refers to the simulation results for the pressure action on the tail at speed 165 km/h while Fig. 9 refers to the speed of 210 km/s. The pattern resulted from the CFD simulation for both speed variations looked significantly different.

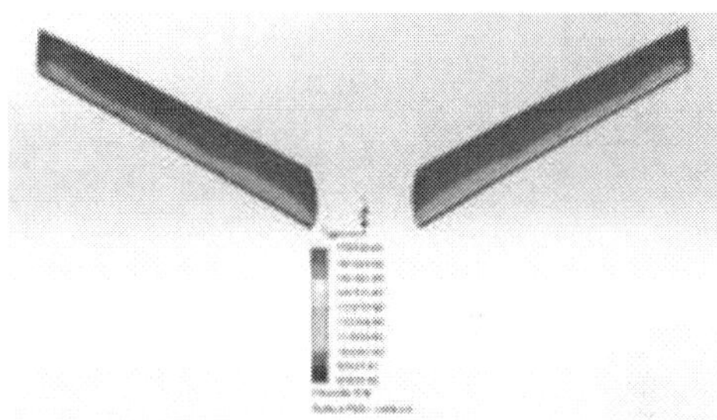

**Fig 8.** Pressure Surface Aerofoil with speed of 165 km/h

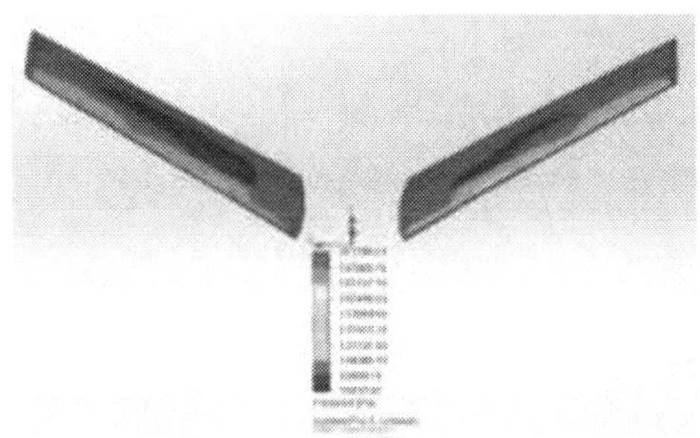

**Fig 9.** Pressure Surface Aerofoil with speed of 210 km/h

The pattern of simulation results for the speed of 210 km/h had dark blue colour. This colour refers to the lowest pressure acting on the tail. This is not found in the result for speed variation of 165 km/h. The maximum pressure working on the tail also had few desperate in the simulation results as described in Fig. 8 and Fig. 9. The speed variation of 210 km/h had yellow colour. This colour is pointing to the highest value of the pattern in both Fig. 8 and Fig. 9. This is consistent with the results given on the Table 2.

The description of the value of Table 2 was able to be illustrated with the comparison of the air flow pressure between specific speed value. In this case, simulation results for the speed of 165 km/h and 210 km/h were selected to compare the pattern. This is given on Fig.10 and Fig.11. Both figures visualized the results of lower speed and higher speed. Fig. 10 illustrated the working pressure of the air flow within speed of 165 km/h while Fig. 11 represented the pressure in the speed of 210 km/h.

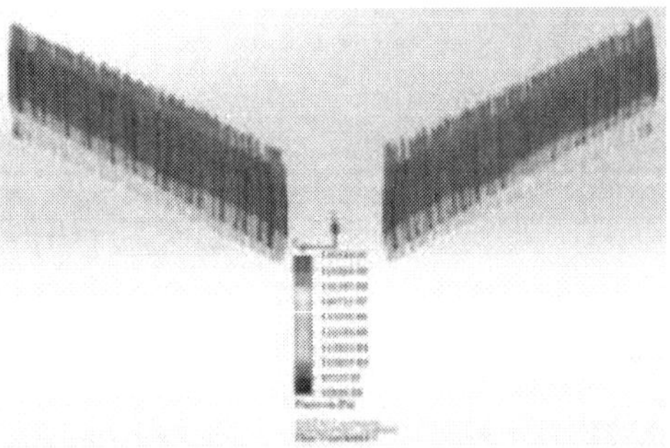

**Fig 10.** Flow Pressure Surface Aerofoil with speed of 165 km/h

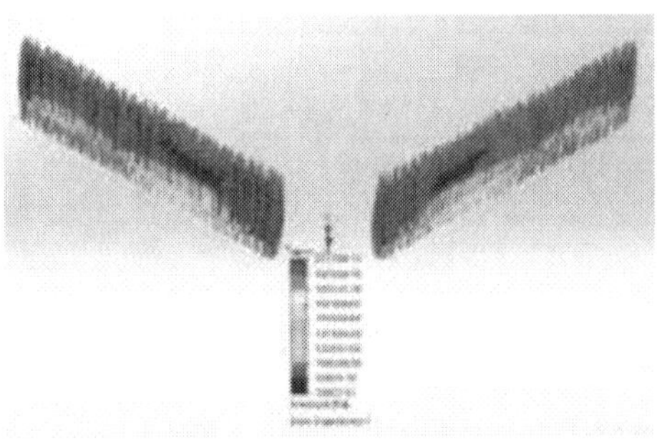

**Fig 11.** Flow Pressure Surface Aerofoil with speed of 210 km/h

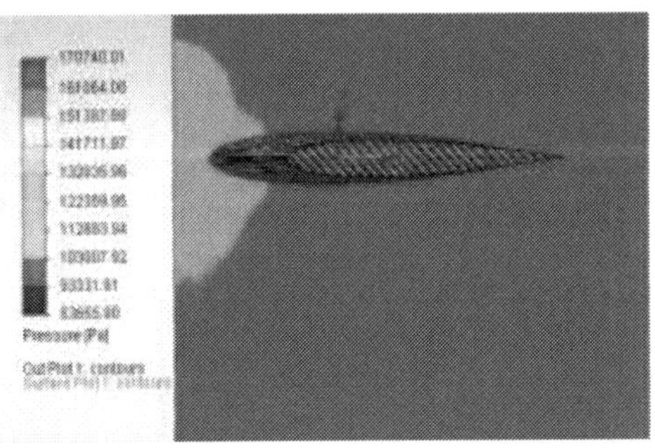

**Fig 12.** Contour Pressure Surface Aerofoil with speed of 165 km/h

At a glance, the pattern of flow pressure surface aerofoil with speed of 210 km/h is more complicated compare than the pattern for the speed of 165 km/h. This was supported by the contour pressure surface simulated by CFD as given on Fig. 12 and Fig. 13. Fig. 12 represent the contour pressure surface produced by air flow with speed of 165 km/h and Fig. 13 as the result of speed of 210 km/h.

The contour effected by air flow with speed of 210 km/h had more pattern compared than the area generated by air flow with speed of 165 km/h. This means that the pressure generated by air flow with speed of 210 had higher pressure range than air flow with speed of 165 km/h. The results at speed of 210 km had area with dark blue colour which is refer to the lowest pressure among the pattern in the Fig. 12 and Fig. 13.

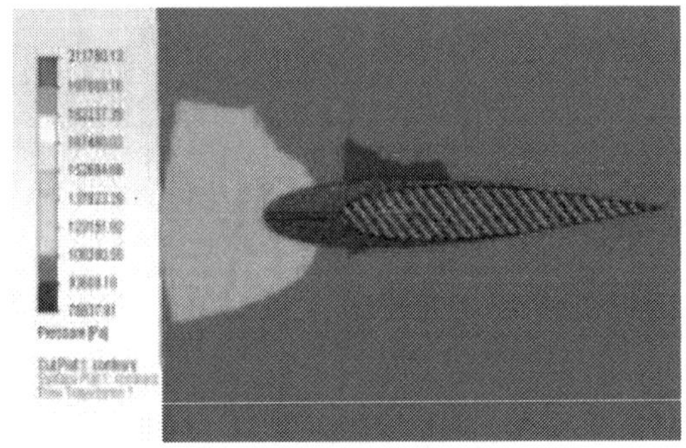

**Fig 13.** Contour Pressure Surface Aerofoil with speed of 210 km/h

At a glance, the pattern of flow pressure surface aerofoil with speed of 210 km/h is more complicated compare than the pattern for the speed of 165 km/h. This was supported by the contour pressure surface simulated by CFD as given on Fig. 12 and Fig. 13. Fig. 12 represent the contour pressure surface produced by air flow with speed of 165 km/h and Fig. 13 as the result of speed of 210 km/h.

## 3.3 Pressure comparison of the tail

The simulation had also performed to analyse pressure acting in the three important area of the tail. This will be discussed in the form of comparison among the horizontal and *V*-tail.

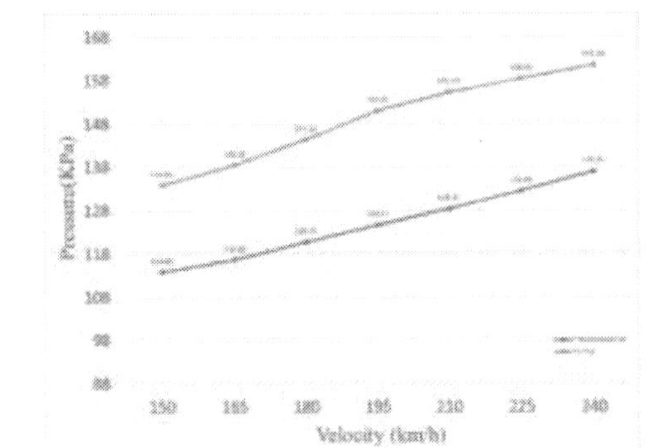


**Fig 14.** Pressure comparison in the area of maximum Leading Edge

### *3.3.1 Pressure comparison on the Leading Edge*

Based on the computer simulation, the pressure values on the horizontal and the *V*-Tail in the area of *Leading Edge* are presented in graphical format shown in Fig. 14 and Fig. 15.

https://doi.org/10.1051/matecconf/201815401115

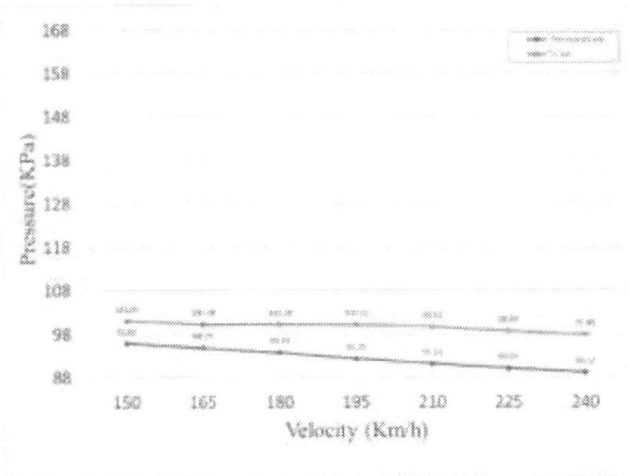


**Fig 15.** Pressure comparison in the area of minimum Leading Edge

Fig. 14 and Fig. 15 has shown the comparative graph of the pressure values occurring in the Leading Edge region of each wing model. Both graphs shown a relatively similar pattern. It appears that the maximum and minimum pressure values occurring on the V-Tail wing model are larger than the horizontal wing model. The pattern for the maximum *leading edge* region looked to have a tendency to decrease with the increasing speed of UAV MALE. In the other side, the trend of the minimum *leading edge* is in contrary tendency.

### *3.3.2 Pressure comparison on the Chamber*

Results of computer simulation on the *chamber* area are given on Fig. 16 and Fig. 17.

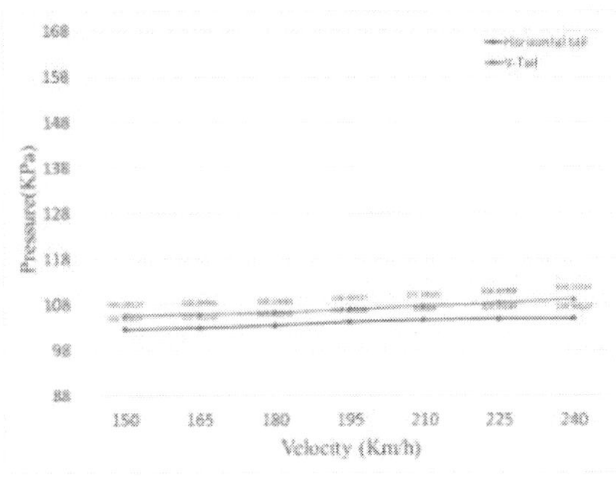


**Fig 16.** Pressure comparison in the area of maximum Chamber

Where Fig. 16 shows the pressure occurred in the maximum *chamber* region and Fig. 17 refers to the minimum *chamber* region. The trend of the pressure change for both horizontal and *V*-tail have similar tendency in the region of the maximum and minimum *chamber*. However, the pressure trend in the maximum *chamber* area has downward inclination. This is different with the pressure trend in the minimum *chamber* region.

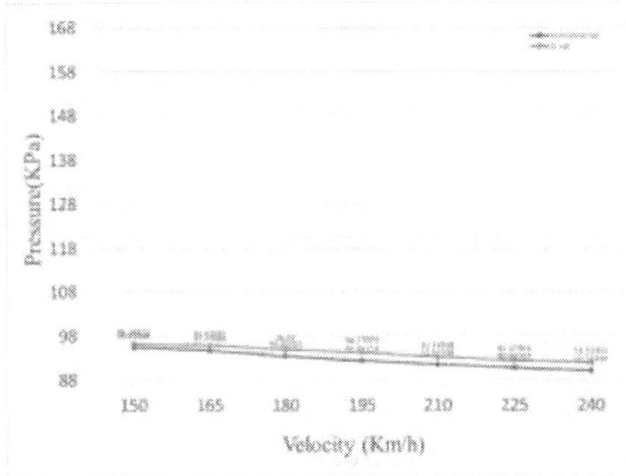


**Fig 17.** Pressure comparison in the area of minimum Chamber

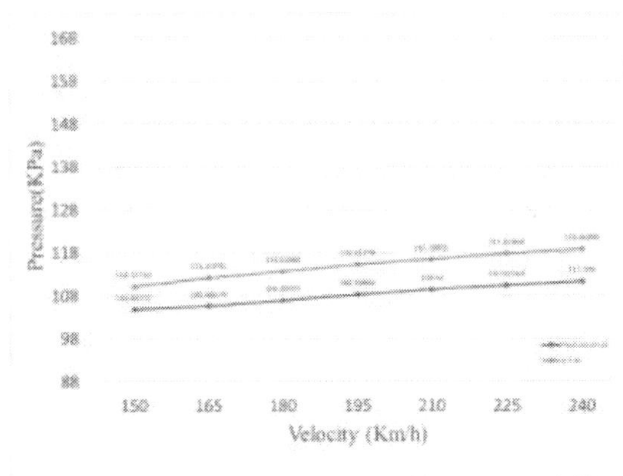


**Fig 18.** Pressure comparison in the area of maximum Trailing Edge

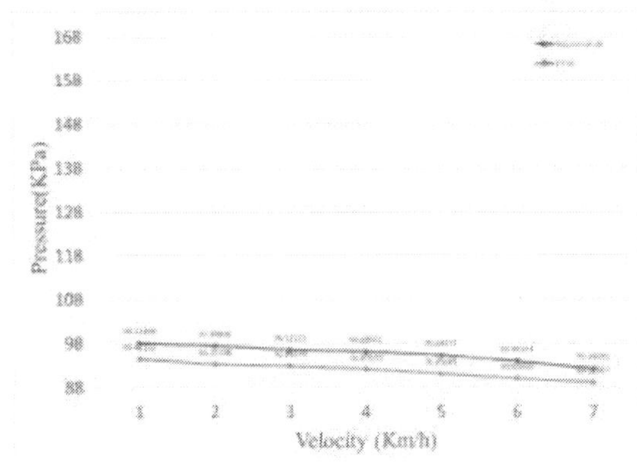


**Fig. 19** Pressure comparison in the area of minimum Trailing Edge

### *3.3.3 Pressure comparison on the Trailing Edge*

Results of the computer simulation on the *trailing edge* are given on Fig. 18 and Fig. 19.

In the region of *trailing edge*, the simulation results also demonstrated the adverse profile between maximum and minimum region of the *trailing edge*. This is in line with the illustration given by the value of Table 1 and Table 2.

## 5 Conclusion

From the CFD simulation results, it can be seen that the maximum pressure arising from the air flow rate is directly proportional to the speed. The faster the speed of the plane, the resulting pressure is also greater, however, the minimum pressure had a contrast condition for both horizontal and V-tail. This condition could also found in the specific geometry for both horizontal and V- tail.

Furthermore, the pressure range will be greater if the rate of aircraft getting faster for both horizontal and V-tail. The pressure range of V-tail was higher than range for the horizontal tail. The highest range occurred in the V-tail with value of 114652,28 Pa. This condition had taken place caused by the minimum pressure acting in aerofoil is hand in hand with its maximum values. However, the highest pressure acting still on the horizontal tail namely the working pressure in the speed of 210 km/h with value of 136866.99 Pa.

The author would like to thank to Mechanical Engineering Department to fully support for this work and all the people involved in this work.

## References

1. Babu, P. Hussain, Rodriguez Arthurs S.A, S. Mahaboob Basha, Sadiq. P., *Design and Analysis of V-Tail Unmanned Air Vehicle (UAV) for Surveillance Application*, International Journal of Engineering Trends and Technology (IJETT), **32**, Number 3 (2016)
2. Azeez VP, Abdul, John Paul *CFD Analysis of NACA 63-018 Airfoil at Different Reynolds-Number*, International Journal of Engineering Trends and Technology (IJETT), **12**, No. 5 (2014)
3. Mustak, Rubiat, Md. Habib Ullah Khan, Md. Harun-Or-Rashid Molla, *Design and Construction of NACA-4415 Airfoil with Various Shaped Surface Modification*, Asia Pacific Journal of Engineering Science and Technology, **3**, No. 1, PP. 28-38 (2017)
4. Garg, Pankaj, Neelesh Soni, *Aerodynamic Investigation of Flow Parameter over NACA 4415 Airfoil by Computational Fluid Dynamics*, International Journal of Recent Scientific Research, **7**, Issue 5, PP. 10737-10741, (2016)
5. Ahmadi-Baloutaki, Mojtaba, Ahmad Sedaghat, Mohsen Saghafian, Mohammad Ali Badri, *A computational study on robust prediction of transition point over NACA0012 aerofoil surface from laminar to turbulent flows*, Theoretical & Applied Mechanics Letter, **3**, 042004 (2013)
6. Ahmed, Tousif, Md. Tanjin Amin, S.M. Rafiul Islam, Shabbir Ahmed, *Computational study of flow around a NACA 0012 wing flapped at different flap angles with varying mach numbers*, **13**, 4 (2013)

# Ring Irrigation System (RIS) design through customer preference representation

*Infandra I.Z.* Ridwan, *Suchada* Rianmora *, and *Siwat* Werawatganon

School of Manufacturing Systems and Mechanical Engineering (MSME), Sirindhorn International Institute of Technology, Thammasat University, 12121 Pathumthani, Thailand

**Abstract.** In agricultural field, irrigation is one of the most interesting considerations affecting the rate of plant growth and development. Micro-irrigation as the dripping or sprinkle method is one of the irrigation types that applies the small amount of water for fulfilling the humidity requirement. The most important factors affecting the demand of water for plants are soil conditions and effect of climatic factors. With less human labour required, to improve the irrigation method from the recent days, analyzing water used or water permeation automatically through the soil moisture has been raised as the interesting topic. Proposed in this research is the ring irrigation system (RIS) which is introduced as an alternative channel for emitters that drip water directly onto the soil at the plant's root zone where the soil conditions before and after watering can be quickly detected by the sensors. This RIS can be used for the potted plant, green house, or other small farm fields. Product design and development (PDD) is applied in this research for assisting the designer to understand and create the RIS prototype properly according to the customer's requirements where the suggested functions obtained will be added and tested.

## 1 Introduction

During the recent decades, manufactures invest the amount of money to upgrade their product to achieve business competitiveness. An improvement sometimes just focuses on technical requirements in the explicit form. To guide the improvement, the comprehensive design takes attention on the customer perception and preference. In addition, technical merit with the marketplace analysis can minimize gaps between practice and theory. Instead, qualitative and subjective factors like usability, appearance, and comfort become imperative [1]. To quickly achieve their targets, the concept of product design development (PDD) has been applied. PDD is the innovation that decides the added value of the existing product to the specific product design [2]. Considering product development as knowledge-based capabilities, these processes help the firm to build its competencies and to learn about new technologies that can be exploited in order to serve market demands [3]. Two stages of PDD are concept development and system level design. In *concept development* has many activities which involve creating, adding, testing, modifying, and refining physical characteristics of a new product item. This stage has been applied for assisting the engineering design to fulfill the customer needs or support current market trends. In the second stage is *system level design*. This stage is performed where the designers and engineers develop the product architecture in detail, and manufacturers determine which components should be made or purchased, and identify the necessary suppliers.

Currently, PDD has been applied in various agricultural fields for obtaining high efficiency of farming works. World water assessment program explains that the future agricultural product consumption increases 70% based on the growth of the global population 2-3 billion people by 2050 and require more than 50% water consumption than was needed in 2012 [4]. Based on that, there were approximately 847,163 people in *Central Region of Thailand* decide to be farmers. The characteristics of farm operator households and their farm businesses have been analyzed and considered mainly on water resources [5]. It means that some experiments in the agricultural field which has already built in the laboratory should be implemented in the real farm further. Moreover, the product which gives the automation can provide the customers not only for the farm field but for the customers who have less place for planting also customers who cannot irrigate the plant in scheduling referred to timing and amount to make most efficient water. Especially, the product knowledge explores to the customer for making the easiest way to get information when product launch in the market.

Utilization the water resources as the traditional irrigation still popular to be used by the people in the past until in this era to saturate water in the agricultural field. This irrigation maximizes water resource near the planting field. Farmers build the small canal to flow the water into the field then they divide the water for all nearest farmers. The other way, farmers use diesel pump to flow water into the field. This method is suitable to use if the water resource is abundant and the location of the

* Corresponding author: suchada@siit.tu.ac.th

planting field is near the water resource like weir or river to supply anytime they want. Otherwise, farmers mostly use underground soil water if their field is not included in that condition. The significant change of environment in Thailand drives people to anticipate in every condition. Explained by Wichitarapongsakun [6], water security and drought become critical issues in Thailand and the greatest damage was in 2005 covering 71 provinces. He also explained that the Sakae Krang River, one of the main river basin, could supply only 62.51% of demand [6].

In order to support the water unpredictable condition, the people have built and used the irrigation system in their planting fields; the most popular one is micro irrigation system. For this system, the gravity flow or conventional pump from the water tank is replaced by dripping or sprinkler irrigation. The sprinkler device is designed for almost every garden size and shape, it is placed on the topmost surface layer of the soil as shown in Fig. 1 [7]. Using this method can save up to 50% on the water bill.

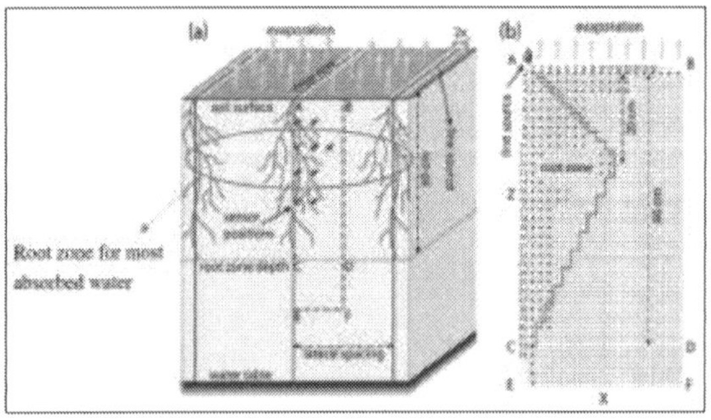


**Fig 1.** Dimensional physical model of surface micro irrigation system [7].

However, this irrigation faces the problem of providing insufficient amount of water since the irrigated water is uncontrollably; water that accidentally overflows. In active roots, there is a zone with root hairs. This zone is actively absorbing water and mineral nutrients. For the potted plants, small root hair, sometimes, can absorb massive quantities of water which might have the direct effects on overhydration. The ways to provide water of adequate quantity and quality have been studied and researched for supporting various styles of gardening.

## 2 Research Background

### 2.1 Customer Knowledge for Product Design

Customer may have the knowledge for the product development in the market. There is no limitation for the customer to get the new product knowledge which is based on generating and integrating customer information throughout the organization because the processes of generating and integrating customer knowledge are embedded in organizational cognitive activities and are not observed readily from outside. Customer knowledge was conceptualized by the impact of market knowledge on product advantage. To get that knowledge, the customers have information process refers to the set of behavior activities that generate customer knowledge pertaining to customers' current and potential needs for products and services [8]. In the market also exist several products to support every activity. The customers can choose the preferable product that has a different type of product as their own requirement. People also have the basic need for their life. When the people want to buy the product, they will consider many things to satisfy their feeling [9].

### 2.2 Development of Irrigation System Design

Irrigation plays a vital role in increasing plant yields and stabilizing production. Irrigation may be defined as the application of water to soil for the purpose of supplying the moisture essential for plant growth. In dry and semi-dry regions, irrigation is essential for economically feasible agriculture, whereas in semi-humid and humid areas, it is often required on a supplementary basis. Some of the most important factors affecting the demand of water for field-grown plants are: *nature of plants*; *nature of soil*; *effect of climatic factors*; *effect of nature of irrigation*; and *attack of pests and diseases* [10 – 12].

The method for water management consists of periodic flooding or sprinkling and dryness prediction during the soil water scarcity [13]. Advance technology in the agricultural generates the development of irrigation system. Several solutions have been proposed the requirements for plant growth. Temperature is a primary factor affecting the rate of plant development [14 - 15]. Farmers mostly used the semi-automatic equipment to watering their plant. In practices, they know the automatic irrigation is the device that used the timer to control the valve open and close and the setting time. In other case, irrigation has already supported with sensor shield for applying the smart farming. Sensors help to collect the information about the environmental attributes, give the feedback and accumulate information to have the control situation [16]. Sensors explained by Abbasi [16], are most widely used for indicating the conditions of soil where temperature, moisture, water flow, and salinity can be measured. To make more efficient used of water, the sensors are applying with the micro irrigation system like the sprinkler or dripping irrigation.

The basic of micro irrigation is used principles of hydraulic of pressuring flow from a pipe. Micro irrigation not only minimizes the problem in traditional irrigation but also provides higher water productivity [17]. However, this method well-known with the effectiveness, it still depends on the users how to create and make the combination of all devices correctly with the environment. The micro irrigation is determined by water requirement for given irrigation interval on the planted area. The affected of micro irrigation is from basic variation, hydraulic design, emitter grouping, emitter spacing and plugging [18]. The researchers tried to reveal the customer requirement for the purposed design of automatic irrigation with used the ring emitter. This device is called ring irrigation system (RIS).

### 2.3 Customer Perceived Utility of the Design

Before starting to design ring irrigation system (RIS), the researchers have tried to reveal the basic requirements and hidden needs from 150 respondents who were interviewed and discussed about their opinions (via self-administered questionnaires). The questions will be focused about the urban people who need to grow the plant, the problems occurred, and the ways to solve the problems.

***Question 1***: How to grow the plant?
***Question 2***: How do you manage to irrigate your plant, garden or farming land?
***Question 3***: What is your irrigation problem during the cultivation?

**Table 1.** How to grow the plant

| How | % of respondents |
|---|---|
| Home farming or garden | 41.18 |
| Potted plant | 27.45 |
| Farm field | 25.49 |
| Vertical planting | 3.92 |
| Greenhouse | 1.96 |

Table 1 presented the results of *Question 1* which was about "*the way to grow the plant*". The highest percentage was shown though "*growing plant in home farming or garden*" because they thought that the space required for this planting was quite small and might be easy to manage, whereas the greenhouse style was required larger space and the proper infrastructure was the main issue for well planting.

**Table 2.** The way to manage the irrigate your plant

| How | % of respondents |
|---|---|
| Manual irrigation | 37% |
| Automated irrigation system | 27% |
| Used water pump system | 25% |
| Used the water flow from the river of out source | 11% |

**Table 3.** Problems

| Problem | % of respondents |
|---|---|
| Not scheduled time of irrigation | 44% |
| Limited water | 31% |
| Watering in unnecessary area | 16% |
| Over flow of water | 5% |
| Others | 4% |

The results of *Question 2* and 3 were shown in Table 2 and 3. All interviewed data and analysis were interpreted into the rough design where the main function of the developed machine will be created according the potential direction of how to manage the irrigation of flowers or plants in the urban areas. 37% of the respondents decided to apply "*the manual irrigation*" where the valve was closed and opened manually by human. The soaker hoses, sprinkler hoses (water mist systems) and water conservation for the home were mentioned and discussed.

Watering of the plants like this manual method was suffering where the inadequate amounts of water might cause the plant death by overhydration (i.e., excessive water intake). In general, the comments were shown about watering plants in the cool of the evening or the very early morning; less water is lost.

The irrigation problems were discussed and grouped. The majority problem was shown through "*Not scheduled time of irrigation*" answer which was approximately 44%. This interviewed data could be considered as the issue for selecting the "*the automated irrigation system*" (27% of the respondents) where the automatic battery-operated garden irrigation system with home watering timer device, and electromagnetic valve timer controller were applied. This system could operate to open the valve in seconds, minutes or hours, and the irrigation frequency (e.g., 2 times/day or 1 time/30 days) could be set.

25% of the respondents chose "*the water pump system*" for increasing the pressure of the water for a long distance during watering the plant. The water from natural resource as river, groundwater or pond was still in consideration for 11% of the respondents for saving cost.

However, using the timer to close or open the valve automatically cannot minimize the risk of overhydration since the soil's conditions are not taken into considerations. Optimizing water for home farming or gardening properly by detecting the specific soil's conditions should be focused.

## 3 Conceptual and System Level Design

### 3.1 Conceptual Design

In current years, automatic irrigation devices have widespread in the market. The way to irrigate this device is using timer. The automatic irrigation will flow the water at the specified time which is already set by the user. Providing timer irrigation can save time during watering plants and less handling from the user. However, it always flows the water at the time setting even the soil humidity is still high or while raining.

Commonly, micro irrigation devices apply the sprinkle to water the plant in the large field with a large amount of water. Sometimes the water drops in the unnecessary areas which cannot be absorbed by the root of the plant. Overflow of water can make inefficiency of water used and makes oversaturated for the plant root.

The proposed ring irrigation system (RIS) will be developed by using the integration concepts among micro irrigation, submerged irrigation and watering potted plant where the water is emitted as the ring-like spraying subsurface direction, and the soil condition can be identified by the sensors for properly preparing the cultivation. Water is permeated from the circle-shape emitter with the small discharge; only a small amount of water sprayed. The emitter is embedded in the soil subsurface around the plant root.

In conclusion, "the circle-shape water emitter" is the main component of the system to distribute water from storage tank to the planted area according to the soil conditions. This system can be applied to various agricultural fields, especially in the drought condition or the place that has limited water resources. For these systems can be applied in the small holder agricultural areas. This innovation is derived for addressing in the potted plant to substantiate the sustenance of agricultural practiced by a small and marginal farmer. The concept of watering design is presented in Fig. 2.

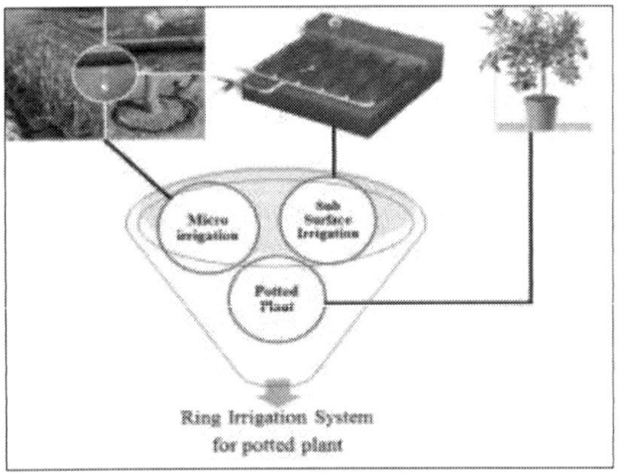


**Fig 2**. The concept of watering design

Machine embodies the following themes:

*Design:* It is designed to combine to support for the people who want to grow the plant with some obligation. Not just automatic flowing water like timer irrigation, the automation of ring irrigation system (RIS) combined with the ring emitter can support the people who would like to grow the plant with less handling and using optimum water during the cultivation. For supporting the one who has less time to water and take care of the plant, using the proposed RIS can flow the water with a suitable amount automatically by itself based on the soil condition.

*Function*: Providing the 3 sensors as temperature, moisture and light sensors can indicate and check the soil condition to flow the water properly to the plant. These sensors and components are controlled by microcontroller for supporting the automation work. Ring emitter placed in the subsurface of the plant will disperse a small amount of water directly to the root. This ring emitter also transfers the nutrient during the irrigation.

*Protection:* When either one of these three sensors are broken; the soil conditions cannot be detected properly and even less misleading measures may have the direct effects to the plant for surviving. Therefore, the controller will switch itself to apply the "*time constraint*" mode immediately for releasing the water in pattern.

Based on the climate condition and water resource in Thailand, the researchers have tried to study and analyze the condition of the soil for creating some constraints and boundary for the amount of water used. Since the ring emitter is to be submerged under the soil for a long time; the material property of the ring should be well considered. The drafted idea of the proposed RIS is presented in Fig. 3.

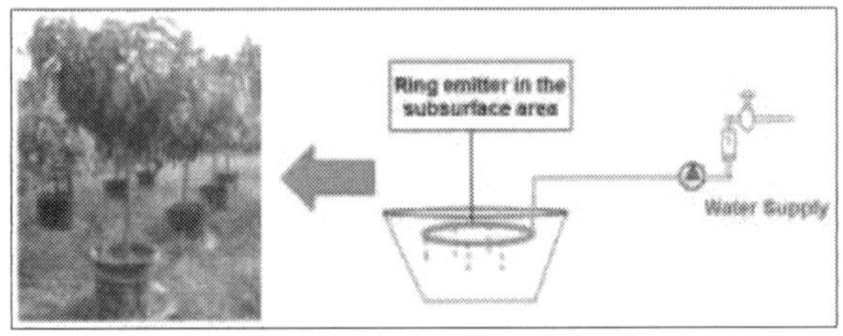


**Fig 3.** Drafted design of the proposed ring irrigation system (RIS)

### 3.2. System Level Design of Ring Irrigation

Soil water balance is the main factor for irrigation scheduling. In this RIS, the 3 sensors are added to form the automatic system to control the valve and pump for stopping or releasing the water from the source. The ring emitter applied in this RIS is made from the elastic material and can maintain its shape during varied conditions; steady shape like rubber hoses, shown in Fig. 4 and Table 4 show the description of the component. The size of hose is 5/8 inch (internal diameter 16 mm) and making the circle with 15-cm diameter.

## 4 Discussion

### 4.1 Device Performance

Fig. 5 shows the soil moisture reading for the condition of humidity in the soil. This result shown the device work to read of soil condition. The sensor is put into the center of root area and analyze the water content in the soil during the cultivation. The sensor detects the humidity and saved in the SD card module that is connected with microcontroller. It is clearly repeated in 24 hours in the suns time and in the cools time at the night. The plant analyzed with this RIS was "pak-choi" (collards leaves vegetables) that was analyzed in the small green house to control the water content from the rain. But the researchers cannot control the moisture from the dew in the morning.

The irrigation activity would be started when the sensors detected the condition as dryness, immediately, the water was released from the tank to the root area via the ring emitter. The observations had been performed during 19-28 August 2017, this month is still in the rainy season and it rains almost everyday where the dew always shows in the morning and the evaporation rate reduces [19].

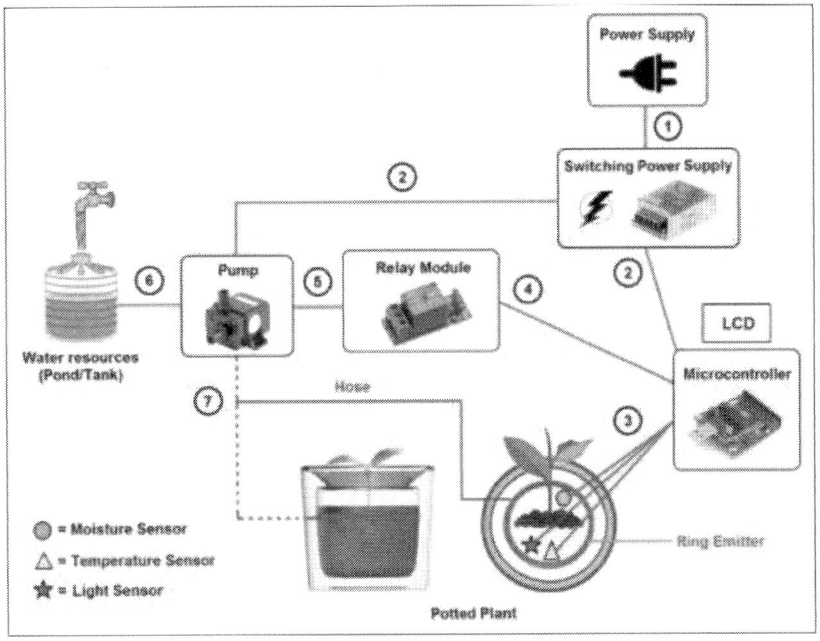


**Fig 4.** Schematic diagram of the ring irrigation

**Table 4.** Description path of the ring irrigation diagram

| Path | Related equipment | Descriptions |
|---|---|---|
| 1 | Power supply | The electrical energy is transferred to switching power supply. This power supply is the only main power for the RIS device |
| 2 | Switching power supply | This component has output DC 5 Volts; 35 Watts; 7 Amperes to control the conversion of electrical power. |
| 3 | Sensors | • *Moisture sensor:* soil moisture sensor module (LM393)<br>• *Temperature Sensor:* NTC thermistor temperature module<br>• *Light Sensor:* LDR photo resistor module |
| 4 | Relay module | This component uses 2 channels with 5 Volts and 10 Amperes |
| 5 | Pump | The maximum flow is 240 *l/hr* with 4.8 Watts and it can support feeding water downwardly or upwardly through the tube with 3 m-apart from the ground level. |
| 6 | Water resources | RIS device does not require the process to place water tank on an elevated height. Pond is also accepted. |
| 7 | Hose | 5 mm-diameter hose is required to flow water from the water resources to the ring emitter. |

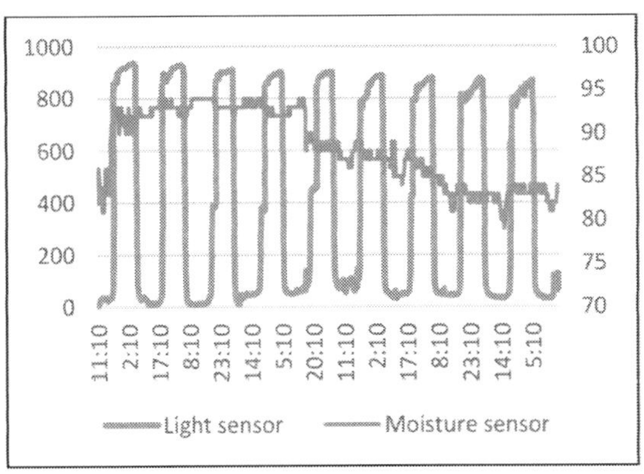


**Fig 5.** Water content in the soil using moisture sensor

The potted plant was selected because this was the most convenient gardening methods for analyzing data. From the graph, the value detected by the moisture sensor kept decreasing slightly in the middle, it caused the device to indicate the dryness of the soil and read the water content in the soil. Comparing the data obtained between two sensors; moisture, and light sensors, they provided the additional knowledge that, under the high-intensity sunlight, the light sensor presented the low level and decreasing in the level of soil humidity. The highest level of the soil humidity was at the beginning after the device released the water into the soil (95 % of the soil humidity). The lowest level was 79% of soil humidity.

### 4.2 Customer Preference on the Proposed Design Utility

The researchers have tried to figure out the ways for revealing the hidden requirements; two sets of self-administered questionnaires were raised at the concept development stage, and the detailed design stage, respectively. After accomplishing the conceptual design of the ring irrigation system (RIS), 150 respondents experienced in planting were asked and represented as the customers. Six issues presented in Fig. 6 were asked as the key considerations for the second set of questionnaires. A question was answered with a simple "Yes" or "No". Saying "Yes" on which particular function implied a chance that the designed utility might be survived in the real application was in the high level. 102 respondents who answered "YES" were analyzed about their opinions, ideas and acceptances on the designed functions.

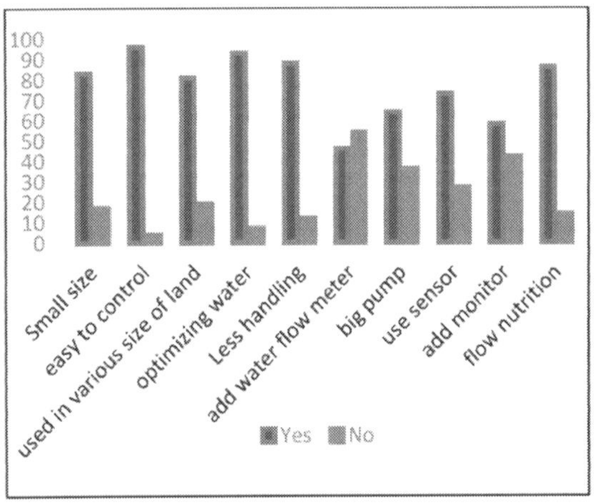


**Fig 6.** Percentage of customer preference on the proposed engineering issues

The first issue was focused on "*the size of the device*". 83% of the customers who answered "YES" preferred the smaller size since they could place it everywhere even in limited area. The second issue is about "*easy to control and use*" and 96% of customers needed to apply the user-friendly interface and applications. The control buttons and switches should be put at obvious and reachable positions, and 58% of customers required LCD screen for monitoring the real-time condition data of soil moisture, temperature and light measurement from sensors. For the connected components, either identical or disjoint, they should let the users get plugged in easily; this can make the device more convenient.

The third issue was about "*garden space and size*", the customers wondered what the area of garden is to apply one unit of the developed device. 81% of customers expected to have the multifunction device that could be applied in various sizes of land; the size of the pump was taken into consideration for supporting this requirement; 64% of the customers required a big pump for supporting a large scale of the planting field.

The forth issue was mentioned about "*optimizing water used*" and 93% of customers required the extra function for measuring the amount of water used. In this era, they have already had the knowledge about the efficiency of using water resource. The effect of some environmental conditions or factors (e.g., during the drought season) is significant; a plant stress usually reflects some sudden change in environmental condition. The irrigable areas through supplemental irrigation for different plants are provided to ensure optimal plant growth.

The fifth issue was about "*less handling and less human labour required*" where 88% of customers needed the automation system where the developed device should function by itself after being adjusted and assigned the watering parameters by the user. It would save time and reduce the number of watering activities rather than using the irrigation manually. The flow meter (46% of customers) and sensors (73% of customers) were mentioned and required for the new design. Moreover, applying sensors could help the users to easily recognize and indicate the real-time condition of the soil and planting environment comparing to the timer-valve application. Once people start to become aware of how much water they use, they would be able to start saving water by considering the water flow meter since it would measure the flow rate or quantity of water. The last issue was about "*flowing nutrients*". 86% of customers required the specific function of the watering device that could transfer the nutrients with water directly to the root zone.

However, 48 respondents who answered "NO" (they disagreed with researchers) even before knowing the whole explanation of the provided functions. 90% of those respondents explained briefly about their reasons, the interesting opinion was that the advance functions and devices would have direct effects on the price (49% of the respondents). Whereas another 10% had no comments or ideas of the watering design; all components might be necessary and required as shown in Fig. 7.

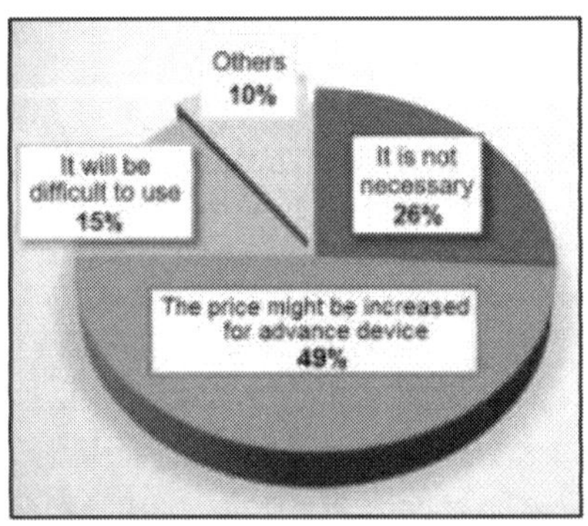


**Fig 7.** Other opinions of the customers on the developed design

The comments from Fig.7 can be applied for modifying the new version of irrigation system where the material selection and automatic functions will be the priority consideration to minimize the cost while maintaining high accuracy and reliability of the system.

For the contribution of the proposed system, the manufacturer or designer can define and identify the suitable device or equipment to support the customers who live and grow the plants in the small space/area by using "ring irrigation system-RIS". The new generation of automatic filling water can be modified and adapted for supporting the customer's requirements where the chance for producing waste materials or lack of design pattern can be eliminated before starting the conceptual design phase.

In term of marketing point of view, the designer or manufacturer should make and provide a prototype or effective final demo presentation about product's descriptions to the customers directly and clearly for increasing the selling/buying volumes. The clean and clear advertisement can be recorded as the good perception for purchasing decision.

## 5 Conclusions

Ring irrigation system (RIS) is the development of micro irrigation to support an automated watering system where the amount of water applied onto the area of interest can be optimized based on the environmental conditions. The specification of RIS has been considered and created according to the customer preference representation. Customer preference refers to their expectations and behaviors on a product where its function can be revealed and extracted into the specific characteristics by the concept of product design and development (PDD). Applying the proposed technique can support the manufacturer or designer to achieve the optimal design of the automated watering system where the number of trial-and-error activities can be minimized.

Three main issues about the function of RIS identified from the target customers' preferences (these might have the direct effect on purchasing decisions) are; *less handling and labour required*, *water optimization*, and *easy to operate the device*.

Moreover, as expected, most of the interviewed women and men reside in rural areas, price is the most consideration for buying decision. Therefore, to build the second prototype without ignoring the main function, the extra devices "*water flow meter*" and "*real-time monitor or digital screen*" are not recommended to apply.

For the future work, the obtained prototype will be tested refined more in various soil/weather/moisture conditions for supporting different types of environment and the potted plants can be grown up easily and quickly. This contribution can be applied in the large fields or farms where the design of the effective ring irrigation might be changed and modified while minimizing operational cost.

## References

1. Y. Wang, and M.M. Tseng, Integrating comprehensive customer requirements into product design. *CIRP Annals-Manufacturing Technology*, ***60*(1)**, 175-178. (2011)
2. L. P. Khoo, C. Chen, W. Yan An investigating on a prototype customer oriented information system for product concept development. *Computers in Industry*. **49**, 157-174. (2002)
3. O. C. Racela, Customer Orientation, Innovation Competencies, and Firm Performance: A Proposed Conceptual Model. *Procedia-Social and Behavioral Sciences*, **148**, 16-23. (2014)
4. WWAP (World Water Assesment Programme). Managing Water Under Uncertainity and Risk-United Nations World Water Assessment Programme United Nations World Water Development Report 4. The United Nations Educational, Scientific and Cultural Organization, Paris, France. Available: https://goo.gl/gm6igd (2012)
5. National Statistic of Thailand. Number of holdings by agricultural registration of holder and size of total area of holding in 2013. Accessed on 23 August 2017 [online]. Available: https://goo.gl/786kvv (2013)
6. P. Wichitarapongsakun, C. Sarin, P. Klomjek, and S. Chuenchooklin, Rainfall prediction and meteorological drought analysis in the Sakae Krang river basin of Thailand. *Agriculture and Natural Resources*, **50(6)**, 490-498. (2016)
7. K. X. Soulis, S. Elmaloglou, and N. Dercas, Investigating the effects of soil moisture sensors positioning and accuracy on soil moisture based drip irrigation scheduling systems. *Agricultural Water Management*, **148**, 258-268. (2015)
8. A. J. Campbell, Creating customer knowledge competence: managing customer relationship management programs strategically. *Industrial Marketing Management*, ***32*(5)**, 375-383. (2003)
9. B. Poston, An Exercise in Personal Exploration: Maslow's Hierarchy of Needs. Association of Surgical Technologists. (2009)
10. Natural Resources Management and Environment Department (NRMED), Water treatment and use in agricultural. Food and Agriculture Organization corporate document repository. Accessed on 26 August 2017 [online]. Available : https://goo.gl/qG9V4n (1992)
11. Department of Primary Industries, Chapter D9: Irrigation scheduling. NSW Government. Accessed on 28 August 2017 [online]. Available: https://goo.gl/qgSxig. (2007)
12. Raghav, 5 Important factors affecting the demand of water for crops. Accessed on 23 August 2017 [online]. Available: https://goo.gl/gXwDRj. (2017)
13. F. Olszewski, P. Jeranyama, C.D. Kennedy, and C.J. Demoranville, Automated cycled sprinkler irrigation for spring frost protection of cranberries. *Agricultural Water Management*, **189**, 19-26. (2017)
14. T. Foster, N. Brozovic, and A.P. Butler, Why well yield matters for managing agricultural drought risk. *Weather and Climate Extremes*. **10**, 11–19. (2015)

15. L.H. Jerry, and H.P. John, Temperature extremes: Effect on plant growth and development. *Weather and Climate Extremes*. **10**, 4-10. (2015)
16. A. Z. Abbasi, N. Islam, and Z. A. Shaikh, A review of wireless sensors and networks' applications in agriculture. *Computer Standards and Interfaces*, ***36*(2)**, 263-270. (2014)
17. M. Kumar, N. Kumar, K.P. Singh, P. Kumar, K. Srinivas, and A.K. Srivastva, Integrating water harvesting and gravity-fed micro-irrigation system for efficient water management in terraced land for growing vegetables. *Biosystems Engineering*, **102**(1), 106-113. (2009)
18. J. Barragan, L. Cots, J. Monserrat, R. Lopez, and I. P. Wu, Water distribution uniformity and scheduling in micro-irrigation systems for water saving and environmental protection. *Biosystems Engineering*, ***107*(3)**, 202-211. (2010)
19. Thai Meteorological Department. Weather forecasting in central Thailand. Accessed on 30 August 2017. Available: https://goo.gl/sSNarB. (2017)

# Analysis ballistic impact modeling of multicomponent fabrics with jalaba structure

*Asmanto* Subagyo

Department of Chemical Engineering, Concentration Textile Engineering, Faculty of Industrial Technology, Islamic University of Indonesia

**Abstract.** Various textile materials have been employed in the past for protection against the ballistic with only limited satisfaction without realizing in underlying mechanism of ballistic impact. The multicomponent fabrics was prepare by nylon yarn with plain constraction with thickness of 0.18 mm. The multicomponent fabrics were made 35 layers (laminated) with various thickness. A computational constitutive model has been developed to characterize the progressive failure behaviours of multicomponent fabrics with jalaba structure as reinforcement under high velocity impact conditions. Test result indicates that multicomponent fabric with jalaba structure can reach second level of International Standar of NIJ-010104 for hand gun (revolver) at shoot distance 5 meters. If compared with panel from biotextile composite the multicomponent fabrics have advantages in structure stability, flexibility, lightly and confortable. The integrated modeling was sucessful utilized to predict the damage and ballistic bihavior of multicomponent fabrics to various ballistic impact conditions.

## 1 Introduction

Ballistic protection is a class of protective cloting which aims at protecting the individuals from the bullets and steel fraghment from hand held weapons and exploding munitions. The use of armour to protect the personal has a long history and dates back to time immemorial. As the man developed for better clothing for protection. The list of body armour materials has included every conceivable material including the traditional material like steel and less traditional ones like silk and leather. The best known method of protecting the human body in old days from all kinds of missile was to use a hard rigid material which resisted the penetration and dissipated the load of impact. With the advent of synthetic textiles better ballistic protective systems have been developed. Although the basic idea of spreading the load over a large area is still applied, because it is better to dissipate the energy of impact by deformation and breaking the protective material (Aboot, 1999). Nylon fabric, which has high work of rupture was thought to be ideal for ballistoc protection. However, it became very clear that simple rupture is not the only creterion for ballistic protection. Due to the fact that ballistic protection process is a complex phenomenon involving the transverse velocity, tensile properties and the fine structure, however this article a brief account of multicomponent fabrics used for the desinging of ballistic protective clothing with JaLaBa structure.The analysis penetration and ballistic modeling of multicomponent fabrics has been also discussed.

### 1.1 Ballistic Requirements

The basic objective of the body armour research is to manage the conflicting goals of producing light hight yet confortable ballistic protection system with superior performance and low cost. On the other hand, for user the comfort degree mobility and maximum protection against injury are important, while for technologist, the level of protection required and energy absorption characteristics of the material are important.

According to Capilli and Rothulzen (1999), that energy absorption characteristics of the body armour are very important consist of some types namely; kinetic energy of the fabric moving towards the point of initial impact, in original fabric between the pyramid and the longtudinal wave form, strain energy accumulated in the yarns of the pyramid, energy dissipated as heat by friction between fiber and fiber or fiber and projectile.

The bullet resistant clothing has to stop bullet from penetrating and absorb its kinetic energy converting it into work deformation (Pushpa, 2009). Therefore, the primary factor which influence the performance of bulletproof material are strength and modulus and elongation at break, deformability of the projectile and the velocity of transverse shock wave in the fiber.

* Corresponding author: asmanto.subagyo@uii.ac.id

## 1.2 Nylon 6

Nylon fibers were material of choice for most of the ballistic applications in early times. The advantage of nylon ussually has absorbs twice the amount of energy. The transverse wave volocity is almost 3 until 4 times higher to that in nylon. Hence stress propagation is more efficient with aramids which is clear, as showed Figure 1. The elongation of the yarn around the point the point of impact was measured approximately 9.8 μs after the shot was fire at a speed 398 m/s. It is evident that the strain nylon fiber fabric is spread over much large area and the elongation is much lower. Moreover, nylon creeps under such high strain rates at which the ballistic operate and sometimes melting and fusion at the interlacement points (Prosser R A, 1988).

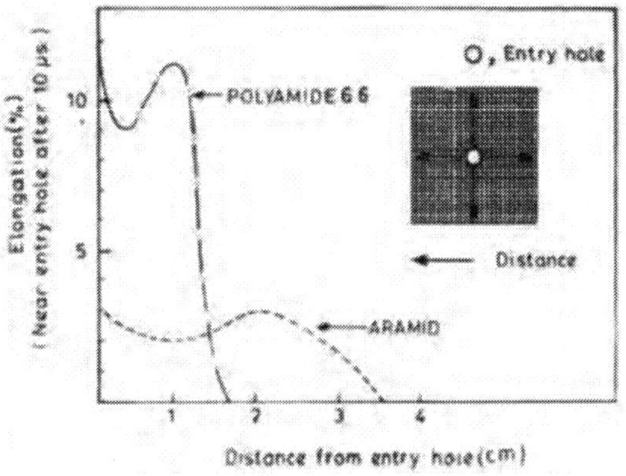


**Fig 1.** Comparison of ballistic between nylon and aramid fabrics

## 1.3 Fabric Design for Ballistic Protection

According to Cunniff. 1999, that to large extent the ballistic response of fabrics appears to be closely related to the ballistic response of the single yarns. In woven fabric since there is an extensive interaction among the yarns due to interlacements, it was found that tranverse deflection cause loading of the crossover yarns and up to 50% of the total energy absorption may accur in the secondary yarns. Therefore fabric construction particulars have a major role in the final ballistic protection.On the other hand, if the weave is too tight or the fabric is too stiff, deflection will be restricted, cuasing shear failure due to concentration of stress at the impact point. Therefore, it is the balance force distribution with maximum interlacements in a given unit area which makes plain weave contruction most suitable for ballistic protection.

## 1.4 Desinging of Ballistic Protective Clothing

Principally, soft body armour is constracted from multi layers of woven fabric with jalaba structure sewn together with crosswise seam. The calibre to be stopped depend on the yarn count, number of multicomponent fabric layer in making a bullet proof vest. Various fabric layer or multicomponent fabric are stacked with jalaba structure which harbours the risk that bullets impacting in succesion at not too great a distance from each other will damage the same warp or weft yarn all the multicomponent layers.

## 1.5 Modeling of Multicomponent Fabrics with Jalaba Structure

In this work, to enhance the modeling capability of the failure behaviour of multicomponent fabrics laminates due to transverse impact, a multicomponent fabrics lamina model based on the 3 Dimens stress field. The failure model can be used to effectively simulate fiber failure, jalaba stucture damage and behaviour under all conditions and sliding of failure surfaces. The major problem is that the whole process typically last only for few microseconds and it is very difficult to get specific information of all the processes that take place and then use them to validate a particular model. Furthermore, the failure modeling approach is advantageous as it enables one to predict delamination when locations of delamination sites can not be anticipated, namely locations of potential delamination initiation is caculted wuthout a prioridefinition of an inter laminar crack surface. Failure models based on the strains in a multicomponent fabrics layer with jalaba structure and improved progrssive failure modeling capability are established for a nylon plain weave fabric layer which consist of 35 layers. It can be used to effectively simulate the fiber failure and delamination behavior under high strain- rate and high pressure ballistic impact condistions. The multicomponent fabric failure criteria are expressed in term of ply level enineering strains with x, y and z denoting the in plane wrap and out of plain directions, respectively. On the other hand, the associated elastic moduli are $E_x$, $E_y$, $E_z$ and $G_{xy}$, $G_{yx}$, $G_{zx}$. The fiber on multicomponent failure criteria for a unidirectional layer are generalized to characterize the fiber damage in term of strain components for a plain weave multilayer. The fill and warp fiber tensile damage are given by the quadratic inter-action between the associated axial and through the thickness tesile strains as follow;

$$f_1 - r_1^2 = \left(\frac{E_x\langle\varepsilon_x\rangle}{S_{xT}}\right)^2 + \left(\frac{G_{xz}\varepsilon_{xz}}{S_{xFS}}\right)^2 - r_1^2 = 0 \quad (1)$$

$$f_2 - r_2^2 = \left(\frac{E_y\langle\varepsilon_y\rangle}{S_{yT}}\right)^2 + \left(\frac{G_{yz}\varepsilon_{yz}}{S_{yFS}}\right)^2 - r_2^2 = 0$$

Where $S_{xT}$ and $S_{yT}$ are axial tensile strength in the fill and wrap directions and $S_{xFS}$ and $S_{yFS}$ are layer multicomponent fabric shear strengths due to fiber shear failure in the fill and wrap directions. These multicomponent failure creteria are applicable when the associated $\varepsilon_x$ and $\varepsilon_y$ is positive. In equation (1) $r_1$ and $r_2$ are the damage thresholds which are equal to one without damage.

When $\varepsilon_x$ and $\varepsilon_y$ are compressiive, if assumed that damage in fill and warp directions are given by the maximum strain so the equations;

$$f_3 - r_3^2 = \left[\frac{E_x\langle\varepsilon_x'\rangle}{S_{xC}}\right]^2 - r_3^2 = 0, \quad \varepsilon_x' = -\varepsilon_x - \langle-\varepsilon_z\rangle\frac{E_z}{E_x}$$
$$f_4 - r_4^2 = \left[\frac{E_y\langle\varepsilon_y'\rangle}{S_{yC}}\right]^2 - r_4^2 = 0, \quad \varepsilon_y' = -\varepsilon_y - \langle-\varepsilon_z\rangle\frac{E_z}{E_y} \quad (2)$$

Note that the effect of through the thickness compressive strain on the multicomponent fabrics damage taken into account in equation (2), where $S_{xC}$ and $S_{yC}$ are the axial compressive strengths in the fill and warp directions respectively and $r_3$ and $r_4$ are the coressponding damage thresholds.

When a multicomponent with jalaba structure subjected to the transverse impact by a projectile, high compressive stresses will generally accour in the impact area with high shear stresses in the surrounding area between the projectile and the target material. Mean while the fiber shear punch damage due to the high shear stresses can be analysis for by equation (1) the crush damage due to the high through the thickness compressive pressure is modeled using the following criterion namely (Davies, 1995);

$$f_5 - r_5^2 = \left(\frac{E_z\langle\varepsilon_z\rangle}{S_{FC}}\right)^2 - r_5^2 = 0 \quad (3)$$

Where $S_{FC}$ is the fiber crush strengths and $r_5$ is associated damage thresholds.

A multicomponent fabrics weaver layer can damage under shear stressing without accorance of fiber breakage, shear stresses model is given by:

$$f_6 - r_6^2 = \left(\frac{G_{xy}\varepsilon_{xy}}{S_{xy}}\right)^2 - r_6^2 = 0 \quad (4)$$

Where $S_{xy}$ is the layer shear strength due to shear failure and $r_6$ is damage threshold.

Another failure analysis, which is due to the quadratic interaction between the thickness stresses is expected to be mainly a jalaba structure failure. This through the thickness multicomponent jalaba structure failure criterion is assumed to have the following form:

$$f_7 - r_7^2 = S^2\left\{\left(\frac{E_z\langle\varepsilon_z\rangle}{S_{zT}}\right)^2 + \left(\frac{G_{yz}\varepsilon_{yz}}{S_{yz0}+S_{SR}}\right)^2 + \left(\frac{G_{zx}\varepsilon_{zx}}{S_{zx0}+S_{SR}}\right)^2\right\} - r_7^2 = 0 \quad (5)$$

Where $r_7$ is damage threshold, and $S_{zT}$ is the through the thikcness tensile strength and $S_{yz0}$ and $S_{zx0}$ are the shear strength for tensile $\varepsilon_z$. The damage surface due to equation (5) is paralel to the multicomponent fabrics layering. When compressive $\varepsilon_z < 0$ the damage surface is considered to be closed and the damage strengths are assumed to depend on the compressive normal strain $\varepsilon_z$ similar to the Coulomb-Mohr theory:

$$S_{SR} = E_z \tan\varphi\langle-\varepsilon_z\rangle \quad (6)$$

Where φ is the Coulomb's friction

When damage predicted by this criterion accours within elements that are adjacent to the ply interface, the failure multicomponent is expected to be parallel to the layering multicomponent, can be referred to as delamination mode.Note that a scale factor S is introduced to provide better correlation of delamination area with experiments.

The scale factor S can be determined by fitting the analytical prediction to experimental data for the delamination area.

## 2 Analysis Damage Criteria

A set of damage variables ω with i value = 1, 2, .....6 are introduced to relate the onset and growth of damage to stiffness losses in the material. The compliance jalaba structure S is related to the damage variables namely;

$$[S] = \begin{bmatrix} \frac{1}{(1-\varpi_1)E_x} & \frac{-v_{yx}}{E_y} & \frac{-v_{zx}}{E_z} & 0 & 0 & 0 \\ \frac{-v_{xy}}{E_x} & \frac{1}{(1-\varpi_2)E_y} & \frac{-v_{zy}}{E_z} & 0 & 0 & 0 \\ \frac{-v_{xz}}{E_x} & \frac{-v_{yz}}{E_y} & \frac{1}{(1-\varpi_3)E_z} & 0 & 0 & 0 \\ 0 & 0 & 0 & \frac{1}{(1-\varpi_4)G_{xy}} & 0 & 0 \\ 0 & 0 & 0 & 0 & \frac{1}{(1-\varpi_5)G_{yz}} & 0 \\ 0 & 0 & 0 & 0 & 0 & \frac{1}{(1-\varpi_6)G_{zx}} \end{bmatrix} \quad (7)$$

The stiffness of jalaba structure C is obtained by inverting the compliance jalaba stucture,$[C] = [S]^{-1}$.

The growth rate damage as variables $\varpi_i$ is governed by damage rule of the form

$$\dot{\varpi}_i = \sum_j \phi_j q_{ij} \quad (8)$$

Where the scalar functions $\phi_i$ and I = 1, 2,.....6 control amount of growth and the vector-valued functions and $q_{ij}$ also where I = 1, 2, .....6 and j = 1, 2,....7 provide the coupling between in individual damage variables (i) in the various damage (j). The damage criteria $f_i - r_i^2 = 0$ of equations from 1 until 5 provide the danage surfaces in strain space. Damage growth in this case that $\phi_i > 0$, will occure when the strain path crosses the updated damage surface $f_i - r_i^2 = 0$ and the strain increment has a non-zero component in the direction of the normal to damage surface and combined with a damage growth function

$$\sum_j \frac{\partial f_i}{\partial \varepsilon_j}\dot{\varepsilon}_j > 0$$

and where γi (εj,ϖj) and j = 1, 2, ....6, $\phi_i$ is assumed to have the form equations (9)

$$\phi_i = \sum_j \gamma_i \frac{\partial f_i}{\partial \varepsilon_j}\dot{\varepsilon}_j \quad (\textit{no sumation over } i) \quad (9)$$

Choosing

$$\gamma_i = \frac{1}{2}(1-\varpi_i) f_i^{\frac{m}{2}-1} \quad (10)$$

$$\sum_j \frac{\partial f_i}{\partial \varepsilon_j}\dot{\varepsilon}_j = \dot{f}_i \quad (11)$$

and then for the quadratic functions of equations lead to

$$\phi_i = \frac{1}{2}(1-\varpi_i) f_i^{\frac{m}{2}-1}\dot{f}_i \ \text{(no summation over } i\text{)} \quad (12)$$

where ϖ'I is damage variable associated with the ith failure mode, and m is a material constant.

When fiber tensile damage is predicted in a layer consist of 35 layers fabrics ( one laminat) by equation (1), the load carrying capacity capacity of that layer in the associated direction is reduced to zero according to damage variable equation.

$$\varpi_i = 1 - e^{\frac{1}{m}(1-r_j^m)}, \quad r_j \geq 1 \quad (13)$$

This provides an initial elastic region bounded by the damage functions in strain space. The nonlinear response is modeled by loading on the damage surface to cause damage growth with increasing damage thresholds and values of damage variables. After damage initiated the progressive damage model assumes linear elastic responese within the part of strain space bounded by the updated damage thresholds. Figur 2 shows typical axial tensile and compressive stress – strain curve obtained from damage model analysis under monotonic loading. Figure 2 a, shows that the model provides an initial elastic region up to strength value and softening region after damage initiated. Figure 2 b, shows stress-strain curves due to cyclic loading along the axial direction x.

Note that the unloading and reloading stress-strain curves within the damage surface follow the reduced elastic moduls, which is given by the updated damage parameter for the axial fiber damage analysis.

For through the thicness multicomponent failure given by equation (5) the load carrying capacity within the element is assume to elastic. The load carrying behavior in through the thickness multicomponent fabrics direction is assumed to depend on the opening or closing of the jalaba structure damage. For tensile analysis when $\varepsilon_z>0$ the through the thickness stress multicomponents are softened and reduced to zero due to the damage criteria described above. For compressive analysis when $\varepsilon_z<0$ the damage surface is considered to be closed, and thus $\sigma_z$ is assumed to be elastic while $\tau_{yz}$ and $\tau_{zx}$ are allowed to reduce to a sliding friction stress.

## 3 Analysis Ballistic Mechanism in Fibers

Model analysis of ballistic impact of multicomponent were conducted by taking account for the strain-rate sensitivity properties. The analysis ballistic problem were 35 multicomponent fabric subjected to 0.5 caliber impact. The ballistic analyses were conducted using integrated with the proposed multicomponent fabric failure model.

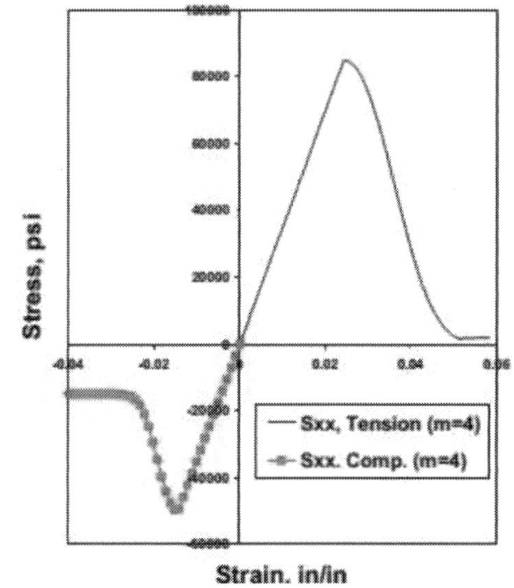


(a)

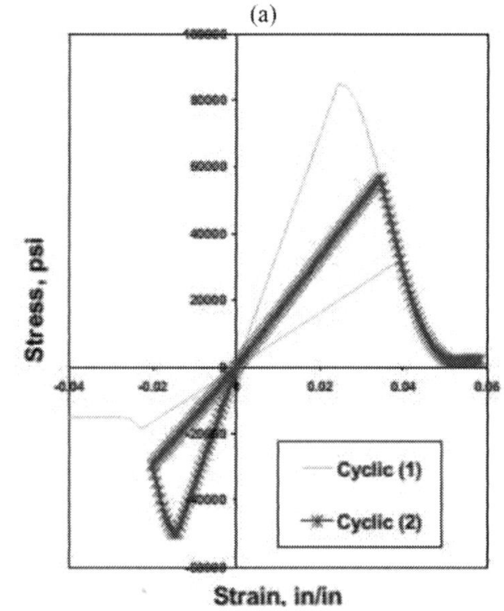


(b)

**Fig 2.** Stress-Strain curves for damage model under Monotonic and Cyclic loading

The finite element models for the multicomponent fabric with jalaba structure is shown in Figure 3. In these cases only one quadrant of the impact system was modeled due to the geometric and material symetry. There were 35 layers multicomponent fabrics of element through the tickness.

The basic idea of ballistic protection with fibrous materials is essentially based on the conversion of the kinetic energy into work of deformation. Hence, the principle influencing factors would be the tensile properties of the protective material, the deformability of the projectile and the armour.

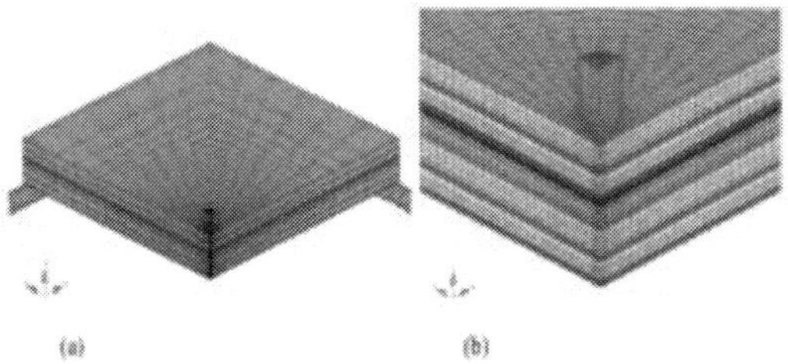


**Fig 3.** Finite element model for a multicomponent fabrics with jalaba structure and impactor, (a) Full view and (b) closed-up of impact area.

Figure 4. the mechanism of deformation when single yarn is subjected to impact by a projectile, with the longitudinal axis of the fiber perpendicular to the path of projectile. The impact leads to longitudinal wave propagation velocity C is nothing but speed of the sound in the material which can be expressed in term of density and modulus as under: $C = (E/\rho)^{1/2}$ where C is speed of longitudinal wave velocity, E = initial moduls and $\rho$ = density.

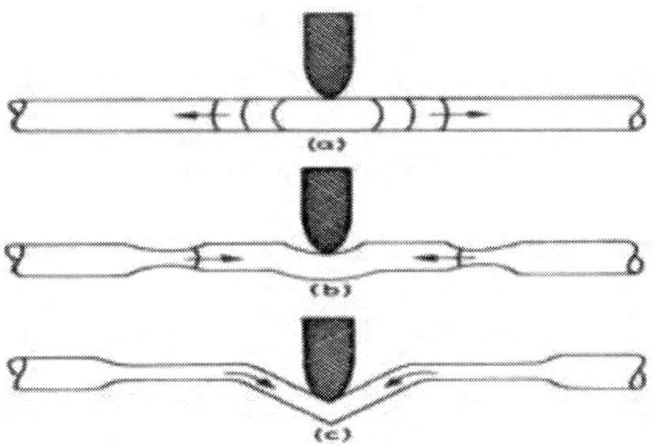


**Fig 4.** Failure mechanism of single fiber under ballistic impact. (a) Propagation of langitudinal compressive waves (b) Reflection as tensile wave (c) Elongation of fiber

The wave reaches the end of the fiber and is reflected back as a tensile wave. The reflection of the tensle wave is necessary so as to satisfy the boundary condition of zero stress at the fiber ends. The moment of tensile wave back towards the point of impact causes the flow of material in the same direction. Ultimately, a tensile strain is experienced by the fiber in contact with the the projectile and the magnitude of which is dictated by the impact velocity of the projectile. Continued wave reflections intensify the tensile strain with energy absorbed by the fiber being proportional to this strain (Subagyo. A. 2014). At the same time, a second and slower wave propagates along the transverse axis and cross the diameter of the fiber, moving paralel to the projectile and at the same velocity. Because of this transverse wave, the fiber starts moving in the same direction as the projectile path. The fiber continues to absorb energy and deflects until projectile is decelerated and stopped or another words the fiber strains past the dynamic yield point and breaks.

## 4 Conclusions

To predict the performance of analysis ballistics impact on multicomponent fabric used jalaba structure has succeeded in the accounting for all complexities that are involved with the penetration mechanism of the ballistic protection. The major problem is that the whole process typically last only for few microseconds and it very difficult to get the specific information of all processes that take place and then use them to validate a particular model.

This analysis model multicomponent fabric used jalaba structure used a technique which requires complex experimental to augment the numerical calculations. The success of this model is impeded maily due to the simplification of assumptions that are made ans axcessive dependence on the experimental data.

The model took care of the essential physics of impact mechanism along with impact mechanics, still keeping the solution reasonably tractable. The main assumption is that a multicomponent fabric can be thought as a single laminated and the impact is approximated as the point impact. The model successfully indicated that the majority of the energy absorbed by the fabric under ballistic impact was due to the strain and the kinetic energy in the yarns directly in contact with the projectile.

The authors gratefully acknowledge the financial support of the specialized Research Fund for the Directorate of Research and Community Service, Islamic University of Indonesia Yogyakarta.

## References

1 Abott T.A., *Protective Clothing*, Shirley Publication, Manchester. UK (1999).

2 Davies, G A. *Impact Damage Prediction in Carbon Composite Structure,* International Conference on Material. Washington, USA (1995)

3 Capilli K and Rothulzen., *High Performance Textile*, Textile Research Journal, Prince, South Carolina, USA (1992)

4 Cunniff. P.M, *Characteristic of Body Armour,* Textile Research Journal, Prince, USA (1999)

5 Pushpa Bajaj, *Ballistic Protective Cloting,* Indian Journal of Fiber and Research. New Delhi, India (2009)

6 Subagyo. A. *Model Impact Resistance of Bio-Textile Fabrics Composite from Agave and Pineapple Leaf Fibers.* Full Paper Proceeding ETAR-2014, Vol. **1**, 320-328 ISBN: 978-969-9948-23-7 . Filipine (2014)

MATEC Web of Conferences 154, 01118 (2018) https://doi.org/10.1051/matecconf/201815401118
ICET4SD 2017

# The effect of microwax composition on the staining quality of Klowong Batik Wax

*Abdul* Malik[1,*], *Fety Ilma* Rahmillah[2], *Bayu Dwi* Atmaja[3], and *Bilal Fatihul* Ihsan[4]

[1,3,4]Department of Chemical Engineering, Islamic University of Indonesia, Jl. Kaliurang Km 14.5, Yogyakarta, Indonesia
[2]Department of Industrial Engineering, Islamic University of Indonesia, Jl. Kaliurang Km 14.5, Yogyakarta, Indonesia

**Abstract.** Batik is one of the most highly developed Indonesian art forms that had been designated by UNESCO as a Masterpiece of Oral and Intangible Heritage of Humanity. Batik wax is considered as the main material to make Batik, especially for handmade Batik in which the pattern and carve applied directly by the artisans. Melted wax (Malam) is made from a mixture of melted damar mata kucing, gondorukem, beeswax, and paraffin. This study aimed to substitute the beeswax which was difficult to find into microwax made from paraffin which is available in huge amount as well as easily searchable. Microwax varied as many as four samples with the equal variable. The samples were then tested used Spectrophotometer to test the colour difference, colour ageing test, rub test (wet, dry, and endurance soap). The results showed that microwax can be used to substitute the function of beeswax with appropriate composition. It is highly expected that the increasing of melted wax quality can increase the quality of Batik itself as well as the preservation of Indonesian Batik as a cultural heritage.

## 1 Introduction

In a piece of batik cloth, there are social, cultural, as well as economic values which promote the honour of Indonesia [1]. In addition, Batik had been appointed as the masterpiece of an oral and intangible heritage of humanity by UNESCO [2]. The batik method dyeing also has been applied to tanned leather [3].

Batik wax is considered as the main material to make Batik, especially for handmade Batik in which the pattern and carve applied directly by the artisans. Batik wax is the material used to cover parts of cloth that are not to be dyed. Batik wax is not only composed of one material, but a mixture of several basic wax materials such as residue of pine-gum distillation (gondorukem), Damar Mata kucing (Shorea javanica), paraffin (white and yellow), microwax, animal fat (Kendal, animal lard), coconut oil, beeswax, and lancing wax. Beeswax is used for its malleability, paraffin is for friability, resins can be added to increase the adhesiveness, while the animal fats can make greater liquidity [8].

All of the ingredients have different prices. Beeswax is the most expensive component which is usually used in the formula to produce the finest quality of batik. Beeswax or lancing wax has been used since the beginning of creating Batik motive process. These animal wax usually called as "malam", therefore it is common to call "malam Batik". At the earlier times, javanese people tend to use beehive for batik, because there is a hexagonal structure made from a kind of wax which composed of ester fatty acid and long-chain alcohol compounds [8].

Experiences made people mix the pure malam or beeswax with plants such as Gondorukem and Damar Mata kucing. Then, animal lard or palm oil can be added to smooth and lower the melting point. In the last process of making batik, the entire wax is removed with hot water added, so that the batik wax off and after water used to remove wax becomes cold, it will be frozen and then it can be taken back. Wax derived from removal called used wax or black wax, because the colour is black. The former wax mixed again in the manufacture of a mixture of new wax. Principally, the work process of malam and dye in making batik use two opposite materials like oil and water. While the oil contains wax, the dye contains water. Certain parts waxed automatically cannot be penetrated by the dye [8].

Research about Batik wax is rarely conducted. Scholar [4] identify the effect of natural resin composition on temperature of wax removing for batik natural dye, while [5] determine the composition of shorea javanica required on klowong batik wax, and [6] research about good quality wax measured by ten testing items that become the standard criterion which has to be done by batik artisans who act as the raters. Study [6] conducted by using a many-facet Rasch model (MFRM) which at the same time measures the quality of the batik wax, the quality of the testing items, and the severity level of the rates. Scholar [7] use another material to make wax, so called soy wax in which the number of material used and the ratio is varied depending on the use and the experiences of the artisan.

According to the literature, no study tries to substitute the beeswax which was difficult to find into microwax

* Corresponding author: 815210102@uii.ac.id

made from paraffin which is available in huge amount as well as easily searchable. Therefore, this paper aims to know the potency of microwax as a substitute of beeswax and to know the influence of microwax composition towards a staining quality of Klowong Batik wax.

## 2 Method

The tools used for the experiment are:

a. To make batik patterns: pencil and mori cloth commonly used for batik.
b. To make the sample for experiments: wok (to heat the wax), stove, stirring tool, and mould to relocate the wax samples.

Four samples with different composition of microwax (A-10 gr, B-15 gr, C-20 gr, and D-25 gr) were used. The variations can be used to know its influence on the quality of wax. The experimental steps are making the sample and pattern; the process of making Batik; colouring process; removal wax process; and observation process.

The colouring process for four samples was done by the same treatment with small doses of the same naphthol, soaking the same time and the same drying time. So, from four samples were obtained the results of the same colour and the same relative. After dyeing finished sample of size 5x5 will be tested by firing spectrophotometer to test how well the wax covers the staining of a pattern. Then, some tests were done to the batik cloth such as dry and wet rub test, colour fastness test, colour ageing tests, and colour different test. The questionnaire was used as a qualitative analysis involving eight respondents and three questions related to the quality of Malam, Batik patterns, and staining results.

## 3 Results and Discussion

This study uses microwax as the material to make Batik wax. Microwax is derived from refined paraffin (wax) which is from petroleum refineries. The properties of microwax resemble beeswax flexible/lithe/ductile and easy to be removed. Its flexibility similar with beeswax (kote), so that it can be used to substitute or reduce the use of beeswax. Microwax properties are melting at 70 degrees Celsius, taking a long time to melt it; easy off in a water bath, hard to penetrate the fabric, and resistant to alkaline solution. Of these properties, microwax was fit to be the main material in this study which is to know the effect of its composition on Batik staining. Here is the results of the tests conducted toward four samples.

### 3.1 Dry and Wet Rub Test

According to the analysis of colour fastness test shown in Figure 1, all the varied experiments showed good index and the different concentration of microwax does not give impact to the colour quality.

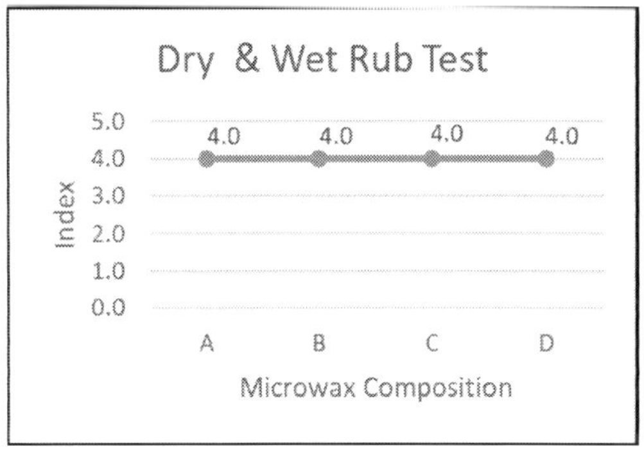


**Fig 1.** Dry and wet rub test

### 3.2 Color Fastness Test

This test plays an important role because it assesses the fastness of the cloth dyeing as well as it can be used as a decision-making tool in dye selection. Compared to samples A and B, samples C and D get a higher value of 4.5 which is just differ 0.5.

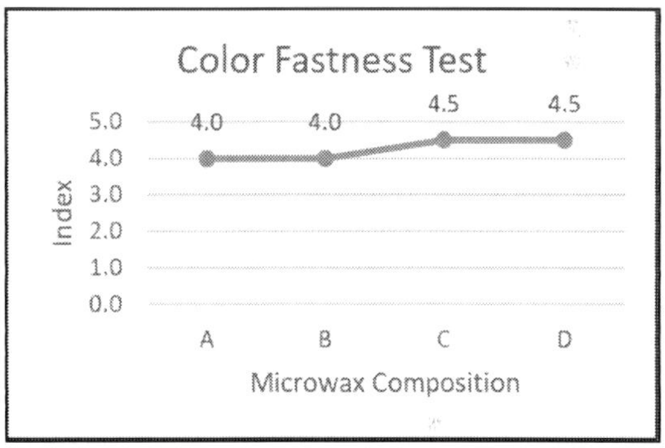


**Fig 2.** Color fastness test

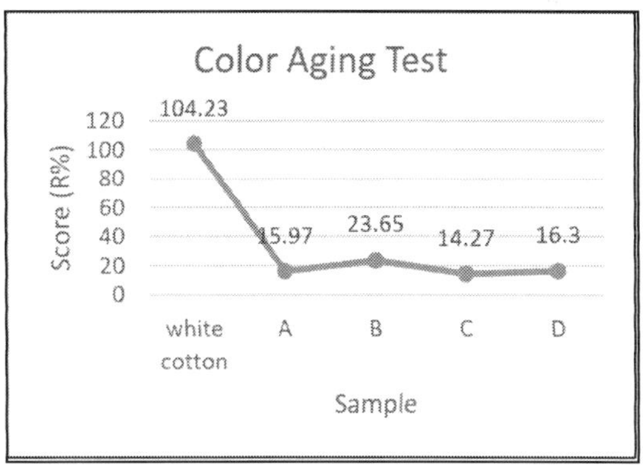


**Fig 3.** Color aging test

### 3.3 Color Aging Test

The colour ageing test with various microwax variations can be seen in Figure 3. The white cotton value 104.23 is used as a standard to be compared with the samples. The graph indicates that the smaller R% value, the darker the fabric fibres, inversely, the lighter the fibres lead to white

https://doi.org/10.1051/matecconf/201815401118

colour in which the standard score of white cotton is 104,23 or called as "blanko". The up and down of the score was due to the differing time of cloth dyeing.

### 3.4 Color Different Test (R%)

Colour different test with microwax variations can be seen in Table 1. The value of L* gets down from 96.19 until 68.97 as the smaller the microwax amount. However, the value of * increases from 24.37 into 43.67; and the value of b* goes up from 22.02 to 41.02; as well as the value of dE*ab which ascends from 33.16 into 67.79.

**Table 1.** Color different test (Microwax) (R%)

| Sample Code | Color Different Test (L*a*b*dE*ab) | | | |
|---|---|---|---|---|
| | L* | a* | b* | dE*ab |
| | 100,67 | -0,01 | -0,01 | 0 |
| A | 96,19 | 24,37 | 22,02 | 33,16 |
| B | 87,68 | 30,06 | 28,15 | 43,19 |
| C | 65,54 | 42,12 | 40,39 | 68,12 |
| D | 68,97 | 43,67 | 41,02 | 67,79 |

Table 2 shows the data of colour different test without microwax. There is an unstable value of L*; a*; b*; and dE*ab which show up and down. Graph 4 shows the relationship between wax from microwax and nonmicrowax.

**Table 2.** Color different test (NonMicrowax) (R%)

| Sample Code | Color Different Test (L*a*b*dE*ab) | | | |
|---|---|---|---|---|
| | L* | a* | b* | dE*ab |
| | 100,67 | -0,01 | -0,01 | 0 |
| A | 64,36 | 56,15 | 25,57 | 71,6 |
| B | 60,56 | 58,81 | 27,42 | 76,29 |
| C | 61,89 | 56,58 | 25,13 | 73,06 |
| D | 57,04 | 60,56 | 29,03 | 80,1 |

The graph illustrates the colour different test result on the various use of microwax composition ranging from zero to hundred. Overall, the more microwax used, the darker the colour and the lower the score of the different colour test. However, sample C and D have almost similar results which only differ 3.5 points for five grammes differ composition of microwax.

### 3.5 Batik Quality Test from Questionnaire

Three questions were asked to eight respondents. The questions as follows: how is the quality of Batik wax in colouring the sample cloth; how is the quality of batik pattern visually on the sample cloth; and how is the quality of colouring visually on the sample cloth.

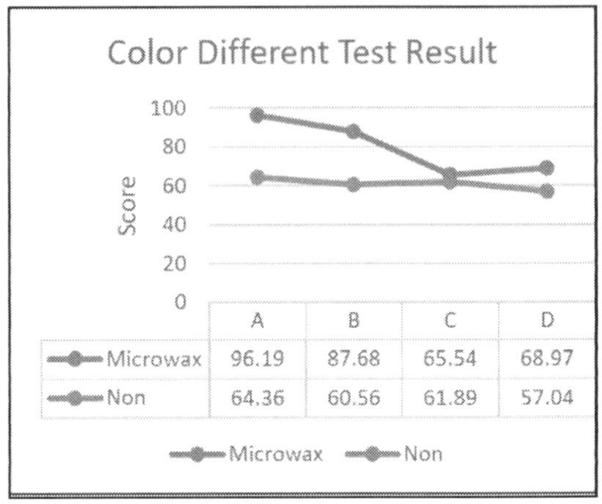


**Fig 4.** Color different test

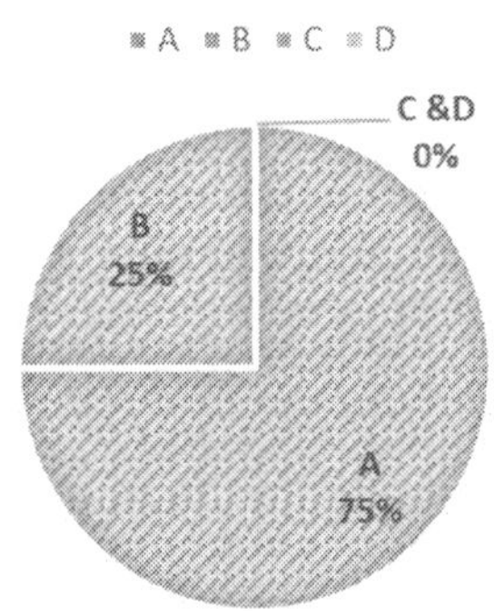


**Fig 5.** Quality of malam

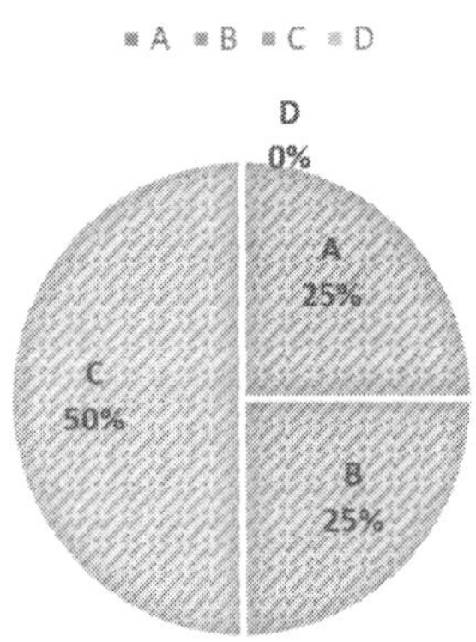


**Fig 6.** Batik pattern

Figure 5 shows that 75% respondents prefer sample A as the best malam while the rest of 25% choose sample B. Figure 5 shows that 50% respondents vote sample C as the best batik pattern, followed by sample B and C as much as 25%. While the best staining result is sample C (50%), sample B (38%), and sample A (12%) consecutively.

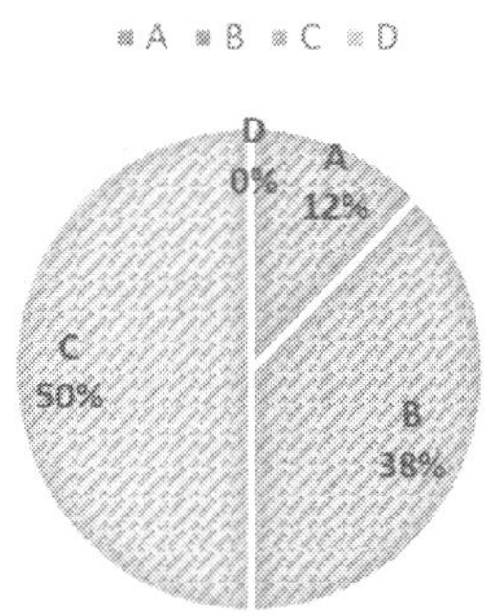


**Fig 7.** Staining result

In the process of making Batik wax and patterns, the steps were done carefully, especially when the process of Canting made. The researcher created patterns that look good but easily be tested during the process. In the colouring process, red dye naphthol was used so that the colour obtained has a strong colour and clear to make visible differences in colour with a cloth that cover the wax. In the process of removing wax with boiled water and caustic soda, the wax can be separated perfectly without leaving a melted wax around the fabric. The experiment resulted from different Batik wax in terms of colour and texture. The more Microwax composition, the brighter and clear. In terms of the texture, the more Microwax composition makes the cloth softer and easier to dry. While for the closure of the pattern is relatively the same. All samples can cover the pattern as well as resistant to colouring fabric.

## 4 Conclusion

The conclusion can be drawn as follows:

a. The more microwax composition is used, the lower the melting point during the process of melting the raw materials into a wax as well as the clearer and softer of the wax, visually.
c. The more microwax composition does not make the better quality of the wax. The best wax quality is obtained on a sample with the smallest amount of microwax.
d. The composition of microwax must be balanced with other ingredients in order to produce the best quality.
e. Microwax has the same function of beeswax, so that it can reduce the use of beeswax.

## References

1. E. Steelyana, *Binus Bussiness Rev.* (2012)
2. Unesco, Indonesian Batik - intangible heritage - Culture Sector - *UNESCO* (2009)
3. W. Pancapalaga, P. Bintoro, Y. B. Pramono, S. Triatmojo, Int. J. Appl. Sci. Tech. **4** (2), 236–242 (2014)
4. V. Atika, A. Haerudin, *Din. Kerajinan dan Batik* **30**(1), 23–30 (2013)
5. A. Malik, Nadia, Khairunnisa, Teknoin **22**(4), 252–260 (2016)
6. K. Kudiya, B. Sumintono, S. Sabana, and A. Sachari, *Pacific Rim Objective Measurement Society symposium* (2016)
7. D.B. Bowen, Surface Design J. **32**(1), 22-25 (2007)
8. Susanto, S. *Indonesian Batik and Handicraft Research Institute* (1980)

# Recycled fiber from straw waste: effect of take-up speed and spinneret diameter to linear density and tenacity

*Asril S.* Soekoco*, *Dody* Mustafa, Rosmawat, and Jane

Politeknik STTT Bandung, Textile Engineering Dept., 40272, Bandung, Indonesia

**Abstract.** Recycled fiber has a great potential as a solution to meet overgrowing synthetic fiber demand and reduce plastic waste in the same time. In the previous research, bottle cap waste is used as the material of the recycled fiber. The result shows that fiber tenacity of recyled fiber from bottle cap waste has low tenacity, different polymer. This phenomena may have been caused by the present of pigment molecule. This research focused on producing recycled fiber from straw waste which has low pigment content. Straw waste was washed and cleaned before the cutting process. Then, the waste was processed in the experiment melt spinning machine with plunger system and single hole orifice in various diameter, 4 mm, 10 mm and 15 mm. The processing temperature was 140° C temperature in three take-up speed, 3.00 m/minutes, 10.30 m/ minutes and 19.16 m/ minutes. The diameter and cross section shape of recycled polyethylene fiber were obtained by using electric microscope with software assistance. The linear density of the recycled fiber was analysed by calculating it with denier and the mechanical strength of the fiber was measured in accordance with the ASTM D 3379-75 standard. High take-up speed leads to higher linear density and take-up speed value is proportionally linear with the tenacity of fiber produced. Moreover, the spinneret diameter is proportionally linear with linear density but it is inversely linear with the tenacity of the fiber produced.

## 1 Introduction

Currently, plastic is the one of the most important materials in industrial world due to its low production cost, production process and wide range product properties. On the other hand, plastic material needs long time to degrade naturally, and this condition has lead to environmental problems. The demand of plastic product including synthetic fiber climbs up as the growth of clothing demand and populations. Indonesia places as the second rank for mismanaging plastic waste country in the world after China [1]. An effort to fulfil synthetic fiber demand and reduce plastic waste in the same time is one of solution to answer this challenge (Leonas, 2017).

Many research of producing recycled fiber using plastic waste has been developed, and one of the biggest challenge in using waste as material is to yield new the products which have minimum quality deference with the virgin material product. In a previous research, bottle cap waste used as the material of the recycled fiber. The result shows that fiber tenacity of recycled fiber from bottle cap waste has low tenacity, different polymer. This phenomena may have been caused by the present of pigment molecule [2].

This research focuses to produce recycled fiber from waste with straw waste which has low pigment content (see figure 2). Melt spinning method was chosen due to its low process cost and more environmental friendly compared to wet and dry spinning [5]. The melt spinning principle is heating materials to its melting temperature and then extruding the melted materials through small orifice of a spinning nozzle to produce fiber [3]. The fiber produced is then pulled and winded on take-up roll. Fiber take-up velocity can influencing mechanical properties and diameter of the fiber [4]. This research used various spinneret diameter and take-up speed to obtain the optimum fiber property.

## 2 Figures and Tables

### 2.1 Material

Transparent straw waste selected was due to its minimum pigment content. In order to reach homogenous material properties, this research only used transparent straw waste which has 110 mm length, 3 mm diameter and 0.24 gram weight (see figure 1). Transparent straw waste washed and cleaned prior to cutting process in 5 mm length.

Fourier Transform Infra-Red (FTIR) analysis was run on Shimadzu to identify the material polymer. FTIR spectra shows that all these plastic bags waste were polyethylene according to their finger prints. This result confirms to the fact that most of straw in the market is commonly made from polypropylene polymer (see figure 2).

* Corresponding author: asril-s@kemenperin.go.id

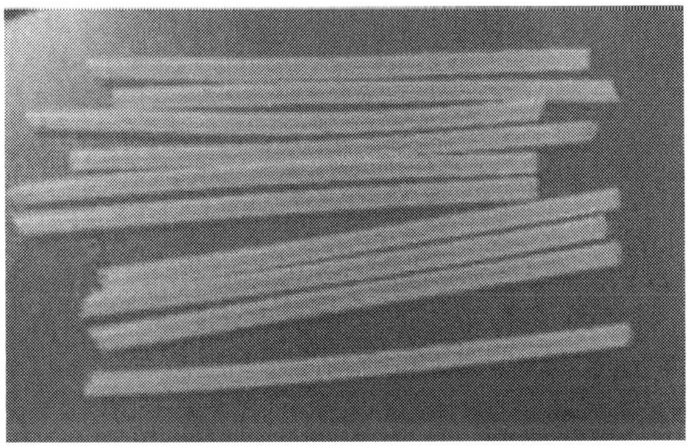

**Fig 1.** Straw waste

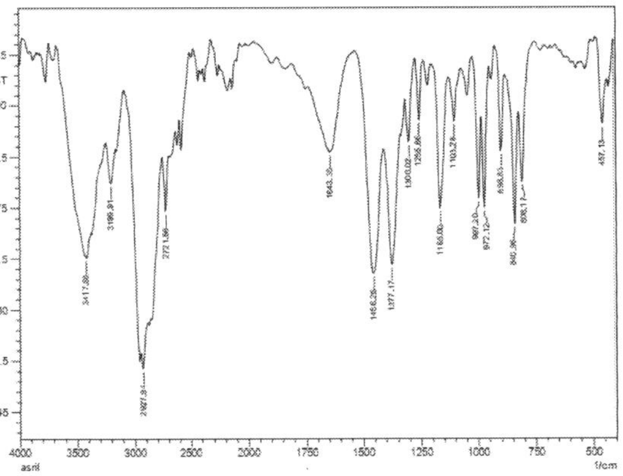


**Fig 2.** Polypropylene FTIR spectra

### 2.2 Melt Spinning Machine

The melt spinning machine used in this experiment was built using plunger system and utilized by electric motor with speed controller to push molten polymer pass out through spinneret with some adjustment for lab-scale used. The spinnerets have single hole each with 4 mm, 10 mm and 15 mm of diameter. The temperature controller was installed in this spinning machine in order to provide precise temperature management. This melt spinning machine has Fotek solid state relay with current load 40 Ampere and barrel with 15 mm of diameter. The maximum temperature can be reached with this melt spinning machine is 350 °C. The winding unit has various take-up speed, 3.00 m/minutes, 10.30 m/ minutes and 19,16 m/minutes and measure using tachometer.

### 2.3 Fiber Making

The fiber production was performed at 130 ± 2 °C and plunger movement at 0.25 m/minutes. After reaching the desired temperature, the heating temperature was maintained for 12 minutes to ensure the uniformity of processing temperature. At the first stage of experiment, 3 variations of take-up speed (3.00 m/minutes, 10.30 m/ minutes and 19.16 m/ minutes) applied in 15 mm of spinneret diameter. At the second stage of the experiment, 3 variations of spinneret (4 mm, 10 mm and 15 mm) were applied in same take-up speed, 10.3 m/minutes. All these adjustments were conducted to discover the most suitable setting to produce a fiber with the highest mechanical properties. The molten polymer was than quenched in open air at 24-27 °C temperature.

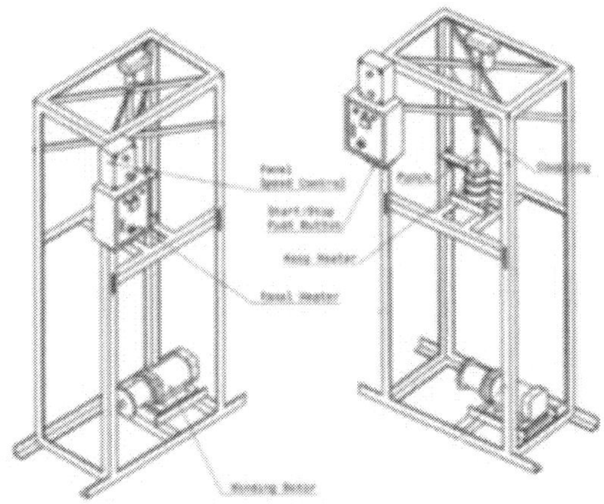

**Fig 3.** Laboratory scale melt spinning machine

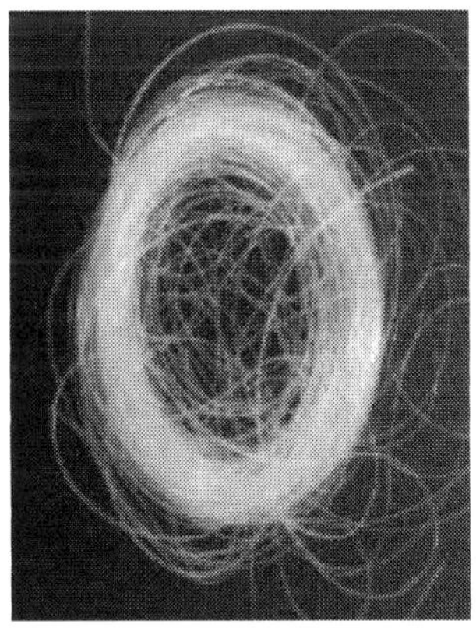

**Fig 4.** Recycled polypropylene fiber

### 2.4 Morphology and Tensile Strength Test

The mechanical strength of the fibers was measured in accordance with the ASTM standard (D 3379-75) with Textechno Favigraph 100 cN load cell. Fiber diameters were examined using a calibrated optical microscope with software assistance. The fiber examination result unit was pixel, than was converted to micrometre.

## 3 Result and Discussion

### 3.1 Fiber Diameter

The cross section of the fiber was circular as the shape of orifice in spinneret, because the melted polymer was spread by surface stress from take up motion [6]. The fiber diameter of these fibers was varied due to take-up speed and orifice diameter (see figure 5).

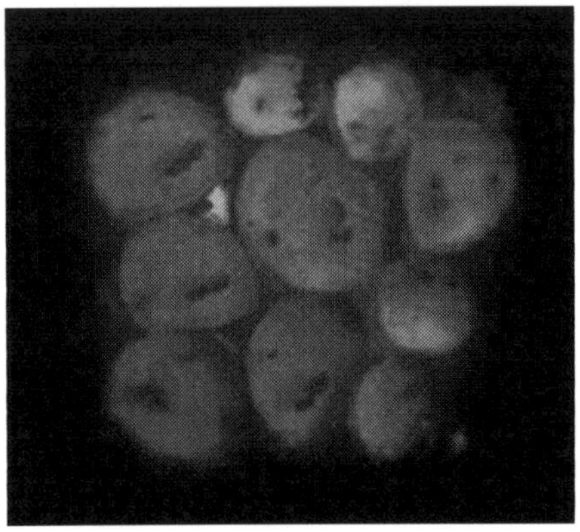

**Fig 5.** Cross section of recycled polypropylene fiber

### 3.2 Linear Density of Fiber

The linear density of the fiber was influenced by the take up speed and spinneret diameter. High take-up speed and small spinneret diameter lead to fine fiber produced. Comparison of take-up speed to melted polymer throughput is known as draw ratio, which was inversely proportional to linear density, the higher the take-up speed (see figure 6), the higher draw ratio and the finer fibers are produced [7].

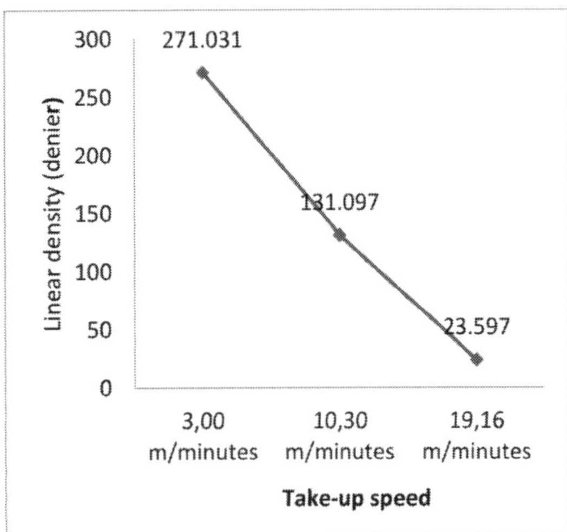


**Fig 6.** Relationship between take-up speed and linear density

The fiber produced from the highest take-up speed has the smallest linear density and the fiber produced from the lowest take-up speed has the smallest linear density. This phenomena was also influenced by the strain rate which occurred along the spin line. Higher take-up speed leads to higher strain rate. The fiber diameter will decrease as the increment of the distance of fiber from spinneret (see figure 7).

The linear density of the fiber produced was proportionally linear with the spinneret diameter. Bigger spinneret diameter leads to higher molten polymer throughput. Since the take-up speed in this session was fixed in 10.30 m/s, the highest molten polymer throughput received low strain rate and produced fiber with highest linear density (see figure 8).

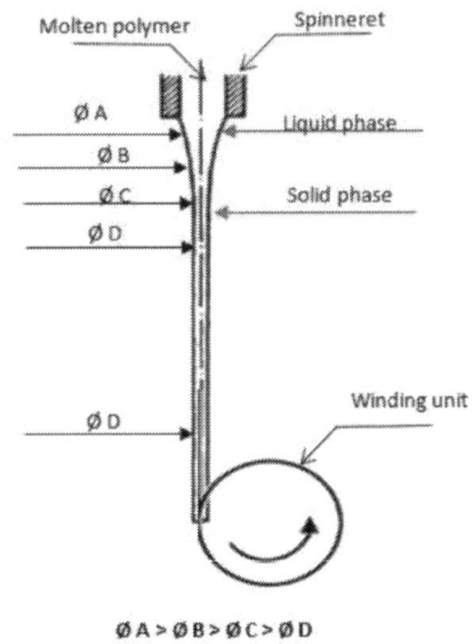


**Fig 7.** Relationship between take-up speed and fiber finest

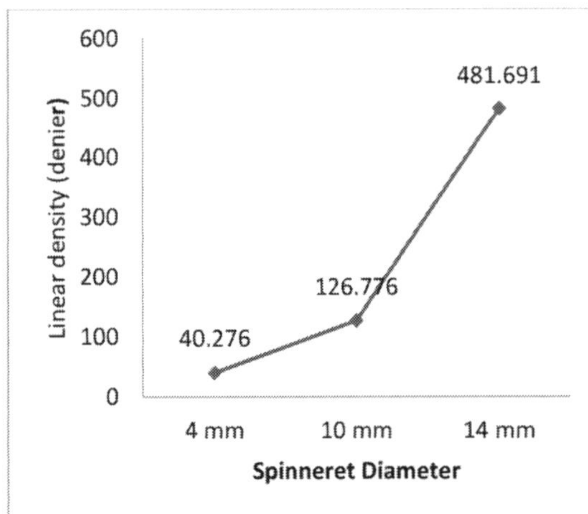


**Fig 8.** Relationship between spinneret diameter and linear density

### 3.3 Tenacity of Fiber

Tenacity of fiber produced from the highest take-up speed was 9.993 g/denier and tenacity of fiber from the lowest take-up speed was 0.666 g/denier. Take-up speed value is proportionally linear with the tenacity of fiber resulted. This condition is totally different with the second experiment, where spinneret diameter value is inversely linear with the tenacity of fiber resulted. The smallest spinneret diameter produces 3.161 g/denier of and the biggest spinneret diameter produces 0.787 g/denier of fiber tenacity (see figure 9 and 10).

The take-up speed is proportional to the draft resulted in the spinning line. Higher draft resulted leads to the increment of polymer molecular orientation degree [7].

The increment of polymer molecular orientation improves fiber tenacity (see figure 11).

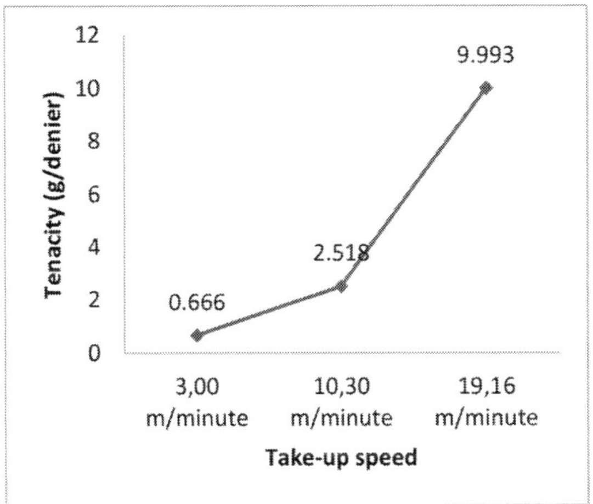


**Fig 9.** Relationship between take-up speed and tenacity

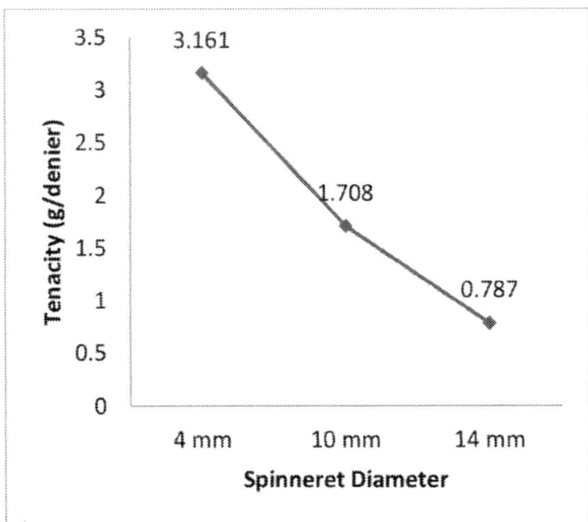


**Fig 10.** Relationship between spinneret diameter and tenacity

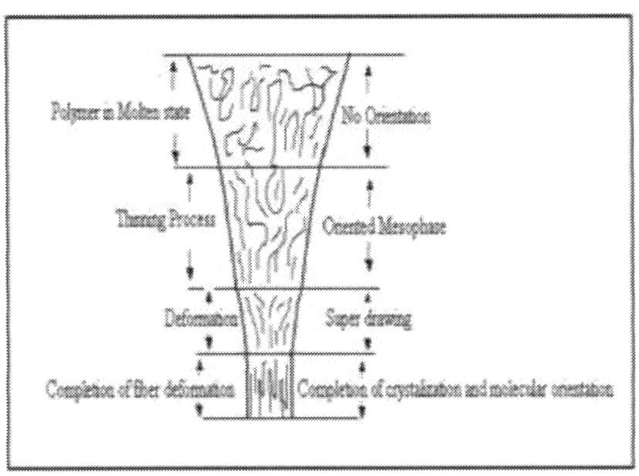


**Fig 11.** Molecular orientation during drafting

## 4 Conclusion

Transparent straw waste is a potential materials for synthetic fiber by melting spinning method due to its spinability. High take-up speed leads to fine fiber produced and take-up speed value is proportionally linear with the tenacity of fiber resulted. The linear density of the fiber produced from is proportionally linear with the spinneret diameter but inversely linear with the tenacity of fiber resulted.

The authors would like to acknowledge Politeknik STTT Bandung for supporting this research in funding.

## References

1. J. R. Jambeck, M. Perryman, R. Geyer, A. Andrady, C. Wilcox, T. R. Siegler, R. Narayan, K. Lavender. Law Plastic Waste Inputs from Land into the ocean. (2015)
2. A.S. Soekoco, Eco-friendly Fibre from Recycled Polypropylene of Bottle Cap Waste and Lignin, *IOP C. S.*, **105**, (2016).
3. F. Luca, I. D, R. Ceccato, Melt Spinning and Drawing of Polyethylene Nanocomposite Fiberwith Organically Modified Hydrotactile, *J. Appl. Polym. Sci.*, (2014)
4. T Nakajima., *Advanced Fiber Spinning Technology*, (2009).
5. Y. P. Jeon, *Simulation of Multifilament Semicrystalline Polymer Fiber Melt-Spinning*. (2009).
6. S.J. Eichhorn, W.S. Hearle, M. Jaffe, T. Kikutani. *Handbook of Textile Fibre Structure*, Woodhead Publishing in Textile New Delhi. (2009)
7. R. R. Hegde, M. G. Kamath, A. Dahiya, *Polymer Crystallinity*.

# Enhancing biogas quality of indigofera plant waste through co-digestion with cow dung

*Finda* Sihta[a,*], Suyitno[b], *Atmanto Heru* Wibowo[c], *and Retno* Tanding[d]

[a]Chemistry Department, Institut Pertanian Bogor, Jl. Raya Dramaga, Bogor, Indonesia
[b]Mechanical Engineering Department, Universitas Sebelas Maret, Jl. Ir. Sutami 36 A, Surakarta, Indonesia
[c]Chemisty Department, Universitas Sebelas Maret, Jl. Ir. Sutami 36 A, Surakarta, Indonesia
[d]Economic and Business Faculty, Universitas Sebelas Maret, Jl. Ir. Sutami 36 A, Surakarta, Indonesia

**Abstract.** The production process of indigofera plants yields 10% of blue natural dyes and wastes are 90%. The biomass wastes from the extraction of indigofera plants are classified as difficult to process in an Anaerobic Digester (AD) because of the large content of hemicellulose in the Indigofera stems. Therefore, the study reports the anaerobic digestion of Indigofera plant wastes co-digested with cow dung for enhancing the biogas quality. The digestion process was conducted in a vessel of 160 L with various weight ratios between Indigofera biomass and cow dung by 0%, 25%, 50%, 75%, and 100%. The results show that the AD reactor temperature kept above 29.5°C with a pH between 6-7 were the key factors for enhancing the biogas quality. The co-digestion between 50% indigofera wastes with 50% cow dung yielded biogas with a CH4 content of about 42% and H2S concentration below 40 ppm showing that the biogas has the potential to be further improved in quality and can be used to fuel the stove and power generation.

## 1 Introduction

Anaerobic digestion (AD) is an old technology which gains a renewed interest, especially in a new route for the process of biomass wastes into biogas. Raw materials commonly digested into biogas are cow dung, chicken manure, pig manure, palm oil mill effluent (POME), and sugar cane processing wastes. In the AD process, the quality of biogas produced is influenced by biological oxygen demand (BOD), chemical oxygen demand (COD), and total suspended solid (TSS) from raw materials. BOD, COD, and TSS of POME are about 10 g/L and 50 g/L, and 31 g/L respectively [6], with pH between 4 and 5. The methane yield of co-digestion between POME and empty fruit bunches (EFB) was 276-340 L/kg [11]. Meanwhile, BOD, COD, and pH of sugar cane processing wastes were 0.8-2.5 g/L, 1.6-6.8 g/L, and 10.2 [13, 4]. The methane yield of 72.1 L/kg was achieved with peroxide pretreatment digestion of bagasse [14]. It can be seen that BOD and COD strongly affected the composition and quality of biogas. Raw materials with low COD indicate the small amount of organic material that can be processed into biogas. Low COD also shows that during the AD process, bacteria will consume more acid.

Besides BOD, COD, and TSS, the following factors that affect to the composition and quality of biogas are the C/N ratio, pH, and digester temperature. In addition, agricultural wastes are generally rich in C content, but deficiency N with a ratio close to 200 [17]. In contrast, farm wastes are rich in N content, but lack of C [16], with the ratio ranging from 8 to 50. Ideal C/N ratio for AD is 30. Moreover, pH of agricultural and farm wastes is generally ranging from 5 to 7. The ideal pH is between 6,5 and 7,6 [15], with the ideal temperature for meshophilic bacteria is between 35 and 40°C [17], for the best growth of fermenter bacteria in AD.

AD is a complex process that is affected by many variables. Every raw material has a unique character, thus requiring different treatment. To the best of our knowledge, there is no report about AD with indigofera plant wastes from the extraction process of blue natural dyes. This success is important from practical point of view, because the amount of wastes in the extraction process of blue natural dyes can reach 90% by weight. Unfortunately, the most indigofera wastes are stems containing much hemicellulose and lignocellulose. The subtrates with high content of cellulose and lignin that interlink each other [10], becomes difficult to hydrolyze, so that the yield of biogas is low. To overcome this, some researchers have proposed some pretreatment processes before processing in AD, such as thermo-chemical [12], thermal [2, 1], mechanical-chemical-enzymatic [3], alkaline [7], and biological [9] pretreatments. The methods are successful for enhancing the yield of biogas, but the pretreatment processes are less practical, as they require additional equipment and also sometimes use additional chemicals that may add an operational cost.

Therefore, this study aims to enhance the quality of biogas from wastes of indigofera biomass that is co–digested with cow dung. Various variables such as

https://doi.org/10.1051/matecconf/201815402001

temperature and pH were observed real time in the AD placed above the ground.

## 2 Methods

Raw materials used were the waste of biomass from the extraction process of blue natural dyes, indigofera tinctoria. The main wastes contain substantial amount of stems but less amount of leaves. Another substrate used was cow dung. The bacteria that used in AD were the EM4 (effective microorganism).

There were two AD reactors with size of 160 L each. The first reactor was filled the substrate with a ratio of 17.5:17.5:35 kg and the second was with a ratio of 35:35:70 kg. Therefore, it could be observed the influence of the number of packing in the AD reactor. The AD reactor schematic can be seen in Figure 1. To investigate the effect of co-digestion, the indigo content in substrate was varied at 100%, 75%, 50%, 25%, and 0%.

Before feeding into the AD reactor, indigofera wastes were mechanically sliced to produce small pieces of biomass thickness of ± 1 cm. Then the results of sliced biomass were mixed with cow dung and water in accordance with its predetermined ratio. The amount of biogas produced, the temperature of the reactor, and the pH of AD were measured periodically. $CH_4$ and $H_2S$ contents in biogas were measured daily and taken the average.

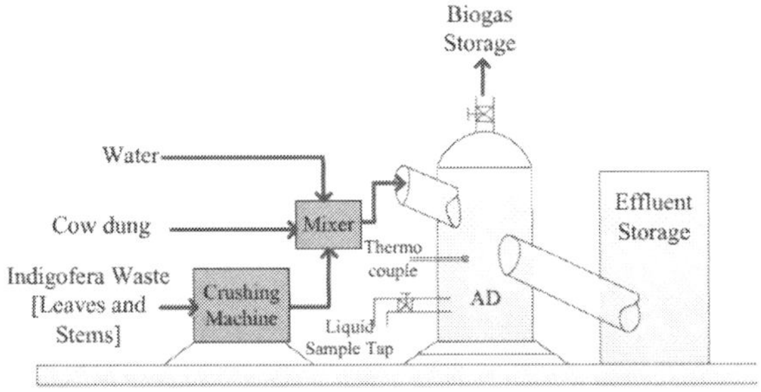


**Fig. 1**. Scheme of AD reactor

## 3 Results and Discussion

Figure 2 shows biogas output and temperature of the first and second AD reactor. It appears that the reactor temperature fluctuated and was influenced by the surrounding. The AD reactor was made of high-density polyethylene (HDPE) without insulation and placed on the ground in order the effect of the environmental temperature on AD reactor can be studied. The weather variable influencing to the temperature of AD was the rainy conditions especially at night and occurred on several days (days 12 to 15). When the rain occurred at night, the temperature of the reactor dropped dramatically even below 27°C. When the rain occurred during the day with medium intensity during 2-4 hours, the reactor temperature has decreased slightly but still able to stay above 29°C.

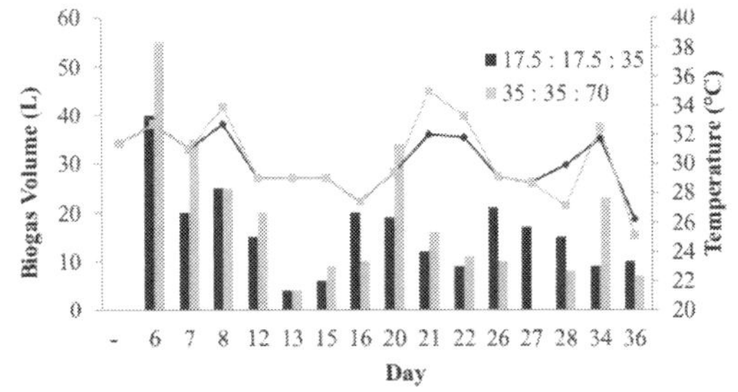


**Fig. 2**. Biogas yield and AD temperature.

The dramatically decrease of the AD temperature was also followed by a decrease of pH in the AD reactor as can be seen in Fig. 3 The condition of acidity in the first AD reactor is more sensitive to the effect of temperature than the second AD reactor. The response of increasing pH with an increase in temperature of AD was slow. This indicates that at low temperatures, anaerobic digestion conditions undergo acidogenesis.

It has been widely expressed by previous researchers that temperature is an important parameter for microorganisms to grow based on optimum requirements and to increase biogas production. Ranging temperature is an important technique for distinguishing the digestive process. There are three types of operation that can be used in anaerobic digesters: psychophilia (25°C), meshophilic (about 35°C) and thermophilic (about 55°C). Optimum growth of microorganism is in meshophilic and thermophilic temperature. AD microorganisms are very sensitive to temperature changing that affect the production of hydrogen and methane, and decomposition of organic matter [5]. Low temperatures cause the low production of volatile fatty acids (VFA) from lignocellulose thereby decreasing the production of biogas. In general, the pH in AD reactor of the co-digestion between indigofera biomass waste and cow dung was 6.54 and 6.32 at first and second reactor, respectively.

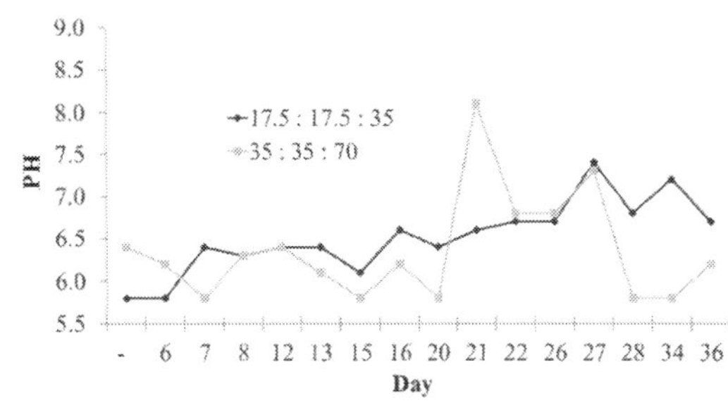


**Fig. 3**. pH conditions in AD reactor.

The pH conditions and temperature of the AD reactor were very influential on the yield of biogas. At temperatures higher than 29.5°C with a pH of about 6 to 7, biogas can be produced in large quantities. At temperatures less than 29.5°C the digestion reaction tends to be slow. The pH level at 6.0 causes the highest hydrolytic enzyme activity and therefore the content of VFA is also high. Methanogenesis is most efficient at pH 6.5–8.2, and the optimum pH is 7.0 at the common substrates. In the AD process, methanogenesis microorganisms are very sensitive to pH variations and prefer a pH around 7.0. Acidogenesis microorganisms are relatively less sensitive to pH and are tolerable to the range of 4.0–8.5. However, the optimum pH for hydrolysis and acidogenesis is between 5.5 and 6.5 (Mao *et al.*, 2015)

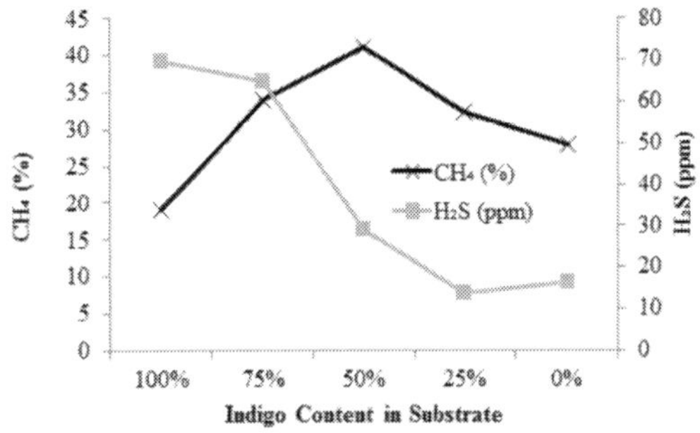


**Fig. 4**. $CH_4$ and $H_2S$ content from AD reactor with various content of indigofera in substrate

In the AD process with 100% indigofera, the biogas produces only contains 19% of $CH_4$ making it very difficult to burn. Addition cow dung up to 50% while maintaining a mass ratio of 1:1 between water and substrate can significantly increase the content of $CH_4$ about 42% as can be seen in Figure 4. Although still low, the biogas containing with 42% of $CH_4$ can be easily burned. Meanwhile, the concentration of $H_2S$ is below 40 ppm. The addition of cow dung could improve the process of methanogenesis because could increase the C/N ratio of indigofera wastes from 21.

## 4 Conclusions

The study has successfully demonstrated the co-digestion process between wastes of indigofera plant and cow dung. It is very important to keep the AD reactor temperature above 29.5 ° C with a pH between 6 and 7. With a $CH_4$ content of about 42% and $H_2S$ concentration below 40 ppm, the biogas produced by co-digestion between indigofera wastes and cow dung has the potential to be further improved in quality and can be used to fuel the stove and internal combustion engine.

This work is fully supported by the grant of INSINAS from the Ministry of Research, Technology, and Higher Education, the Republic of Indonesia with contract number 453/UN27.21/LT/2016 FY 2016 and 2017.

## References

1. Cuetos, M. J., Gómez, X., Otero, M. and Morán, A. (2010). Anaerobic digestion and co-digestion of slaughterhouse waste (SHW): Influence of heat and pressure pre-treatment in biogas yield. *Waste Management, 30*(10), 1780-9.
2. Ferrer, I., Ponsá, S., Vázquez, F. and Font, X. (2008). Increasing biogas production by thermal (70°C) sludge pre-treatment prior to thermophilic anaerobic digestion. *Biochemical Engineering Journal, 42*(2), 186-92.
3. Frigon, J.-C., Mehta, P. and Guiot, S. R. (2012). Impact of mechanical, chemical and enzymatic pre-treatments on the methane yield from the anaerobic digestion of switchgrass. *Biomass and Bioenergy, 36*(1-11.
4. Ghodale, M. D. and Kankal, S. B. (2014). Investigation of optimum operating parameters for BOD & COD removal using activated carbon. *International Journal of Recent Development in Engineering and Technology, 2*(4).
5. Hagos, K., Zong, J., Li, D., Liu, C. and Lu, X. Anaerobic co-digestion process for biogas production: Progress, challenges and perspectives. *Renewable and Sustainable Energy Reviews*.
6. Ibrahim, A. H., Dahlan, I., Adlan, M. N. and Dasti, A. F. (2012). Comparative study on characterization of Malaysian palm oil mill effluent. *Research Journal of Chemical Sciences, 2*(12).
7. Liew, L. N., Shi, J. and Li, Y. (2011). Enhancing the solid-state anaerobic digestion of fallen leaves through simultaneous alkaline treatment. *Bioresource Technology, 102*(19), 8828-34.
8. Mao, C., Feng, Y., Wang, X. and Ren, G. (2015). Review on research achievements of biogas from anaerobic digestion. *Renewable and Sustainable Energy Reviews, 45*(540-55.
9. Müller, H. W. and Trösch, W. (1986). Screening of white-rot fungi for biological pretreatment of wheat straw for biogas production. *Applied Microbiology and Biotechnology, 24*(2), 180-5.
10. Noike, T., Endo, G., Chang, J. E., Yaguchi, J. and Matsumoto, J. (1985). Characteristics of carbohydrate degradation and the rate-limiting step in anaerobic digestion. *Biotechnol Bioeng, 27*(10), 1482-9.
11. O-Thong, S., Boe, K. and Angelidaki, I. (2012). Thermophilic anaerobic co-digestion of oil palm empty fruit bunches with palm oil mill effluent for efficient biogas production. *Applied Energy, 93*(648-54.
12. Penaud, V., Delgenes, J. and Moletta, R. (1999). Thermo-chemical pretreatment of a microbial biomass: influence of sodium hydroxide addition on solubilization and anaerobic biodegradability. *Enzyme and microbial technology, 25*(3), 258-63.
13. Prasad, D. Y. (1991). Biogas generation from bagasse-based paper mills by anaerobic digestion. *Journal of Chemical Technology & Biotechnology, 51*(4), 515-24.
14. Rabelo, S. C., Carrere, H., Maciel Filho, R. and Costa, A. C. (2011). Production of bioethanol, methane and heat from sugarcane bagasse in a biorefinery concept. *Bioresource Technology, 102*(17), 7887-95.
15. Rittmann, B. E. and McCarty, P. L. (2001), *Environmental biotechnology: Principles and*

*applications,* McGraw-Hill.

16. Saputra, T., Triatmojo, S. and Pertiwiningrum, A. (2010). Produksi biogas dari campuran feses sapi dan ampas tebu (bagasse) dengan rasio C/N yang berbeda. *Buletin Peternakan, 34*(2).
17. Shuler, M. L. and Kargi, F. (2001), *Bioprocess engineering: Basic concepts,* Prentice Hall.

https://doi.org/10.1051/matecconf/201815402002

# Chromophoric and liability quantification of organic matters in the polluted rivers of Bandung watershed, Indonesia

*Mohamad Rangga* Sururi*, *Dwina* Roosmini, and *Suprihanto* Notodarmojo

Institut Teknologi Bandung, Environmental Engineering Department, Ganesha 10 Bandung, Indonesia

**Abstract.** The environmental pressure on raw water intakes has increased and affected the characteristic of Natural Organic Matters (NOMs) in many rivers in Indonesia. The purpose of this study is to determine the characteristic of NOMs and to understand the relationship among NOMs parameters in the polluted river intakes. This research was conducted in the raw water intakes of Water Supply Company in Bandung Indonesia. Statistical and spatial analysis were applied to obtain the most polluted water among the intakes. Then, water samplings from a selected intake were taken to measure three different terminologies of NOMs, which are (i) NOM which are related to lability (BOD and COD), (ii) Chromophoric NOM ($UV_{254}$, $UV_{355}$, and UV ¾), (iii) and fluorescent NOM (Humification Index (HIX)). In this study, the Fluorescent Excitation-Emission Matrix (F-EEM) was then measured to determine the origins of NOMs. Also, further results reveal that the characteristic of NOMs from the selected water intake is predominantly influenced by anthropogenic activities, as shown by weak relationship among lability NOM and Chromophoric NOM.

## 1 Introduction

The presence of Natural Organic Matter (NOM) is an important parameter to describe the quality of raw water. The problem related to NOM became a crucial since the anthropogenic activities give the pressure to the rivers as the water sources of Water Company. Human activities along the watershed have degraded the quality of water, thus it is important to determine the effects of land use on the organic composition in the river watershed [1].

The NOM composition in the raw water can be different among the watersheds due to the dissimilar of land use patterns [2]. Moreover, NOM characteristic can be affected by a variety of environmental factors, such as the geomorphology, hydrology, aquatic light intensity and microbial activity within the watershed [2, 3].

According to Yamashita, Maie, Briceño and Jaffé [3], the biogeochemistry of Dissolved Organic Matter (DOM) in typical tropical rivers, such as Indonesia, might be quite sensitive to climatic and land use change. The problems related to liability of NOM had enforced the challenges to the water treatment facilities in terms of operational optimization and proper process control [4, 5]. An additional problem related to NOM arose due to wide use of chlorination in water treatment plant. NOM that does not remove by the treatment process reacts with chlor to form Disinfection by Products (DBPs), such as Trihalomethanes (THMs).

NOM in the water is very heterogeneous, and consists of complex mixture. Different terminologies and parameters have been used to describe it in the water [6, 7]. Inappropriate measurements of NOM will lead to ambiguous perception.

In developing countries such as Indonesia, the organic content in the drinking raw water is measured as Biochemical Oxygen Demand (BOD) and Chemical Oxygen Demand (COD). Both acronyms were used in the study related NOM when the labiality criteria are involved [6]. BOD and COD worked in the best on waste-water effluents and they had the widest application in measuring wastes from the treatment plants [8]. Nevertheless, there are no strong relationship among BOD,COD and THMs [9, 10]. Meanwhile, some experts have concluded that Dissolved organic carbon, Chromophoric NOM (CNOM), SUVA (Specific Ultraviolet Absorbance) correlated with THMs, thus it served as surrogate parameter for THMs [11-16].

In Indonesia, one of the greatest challenges is the limited information regarding the characteristic and quantity CNOM ($UV_{254}$, $UV_{355}$ and UV3/4); and fluorescent NOM (HIX); at raw water sources. Both terminologies are important surrogate parameter of THMs forming potential in the raw water.

The objective of this study was to determine the characteristic of NOM and understanding relationship among NOM parameters on the polluted raw water. This study also determines the origin of NOM by the fluorescent excitation-emission matrix (F-EEM).

* Corresponding author: rangsoer@gmail.com

# 2 Material and Methods

## 2.1 The study area

Bandung is the capital of West Java Province in Indonesia. The water company who supplies drinking water for Bandung City is PDAM Tirtawening. They had generated the drinking water of 1,904.86 L/sec from surface water and less than 115 L/sec from others sources. In Water production region 2, there are three main rivers, providing raw water for treatment plant. These rivers are Cikapundung; Cipanjalu and Cibeureum, which part of the Citarum Catchment [17].

The study area is shown in Figure 1. The PDAM Tirtawening has measured BOD and COD every 3 months in period of 2013-2016. The data were analysed statistically by ANOVA using SPSS 19 to determine the most polluted intake which would be sampling location. In order to strengthen the statistic results, land use analysis was used to determine the natural area and anthropogenic areas which were affected by human activities in the catchment area of four intakes.

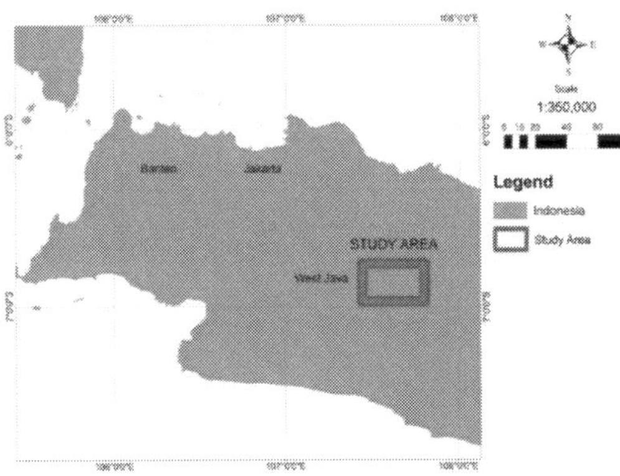


**Fig.1.** Study area

## 2.2 Water sample collection

The sampling on selected intake location was conducted on 9 days of measurements. The sampling was taken after first sedimentation before coagulation processes thus the composition of DOC has not changed.

## 2.3 NOM measurements

Table 1 shows, the list of NOM parameters which were measured and the standard of measurement. Spectrophotometer UV-VIS CAMSPEC M500 series was used to measure COC. $UV_{355}$ is describes the CDOM derived from terrestrial [18]. Comparison of UV300 and UV400, describes the humification in the sample, the lower of ratio, the higher humification and aromatization [19].

Spectrophotometer fluorescence Shimadzu RF5301PC was used to measure HIX, the ratio H/L, where H is the fluorescence intensity recorded at excitation wavelength 254 nm for emission spectrum integrated from 435 to 480 nm, and L the fluorescence intensity recorded at excitation wavelength 254 nm for emission spectrum integrated from 300 to 345 nm [20, 21]. The greater value of HIX, the higher value of aromatization [20, 21].

**Table 1.** Method of measurements

| No | Parameter | Measurement method |
|---|---|---|
| 1 | BOD | Standard methods 5210 B |
| 2 | COD | Standard methods 5220 C |
| 3 | DOC | Standard methods 3510 B |
| 4 | $UV_{254}$ | Standard methods 3510 B |

## 2.4 Correlation among NOM parameters

In this research, a correlation is a single number that describes the relationship between NOM parameters. The data were 2nalysed statistically by SPSS 19.

## 2.5 The sources of NOM determination

NOM from watershed which are affected by human activities dominated by protein like and others which are affected by natural activities dominated by humic like [1]. In this part of research, the samples were divided to filtered and unfiltered sample. The samples were filtered by Advantech membrane with diameter of 0.45 mμ to know dissolved organic sources.

Spectrophotometer fluorescence Shimadzu RF5301PC was used to measure fluorescent excitation-emission matrix (F-EEM). Table 2 Shows the F-EEM of fluorescence intensity peaks selected based on the F-EEMs contour plots. The maximum intensities of the two humic-like peaks are much higher than of the protein-like peaks.

**Table 2.** F-EEM of fluorescence intensity peaks

| Excitation, nm | | Emission, nm | | Fluorophore |
|---|---|---|---|---|
| (Coble, 1996) | (Baghoth, 2012) | (Coble, 1996) | (Baghoth, 2012) | |
| 275 | 280 | 310 | 320 | Tyrosine, protein like |
| 275 | 280 | 340 | 350 | Tryptophan, protein-like |
| 350 | 330 | 420-480 | 450 | Humic like |
| 312 | 310 | 380-420 | 410 | Humic, marine humic |

# 3 Result and discussion

## 3.1 Location determination

The most polluted water sample was determined by using series BOD and COD data from PDAM Tirtawening. The measurement result for BOD during period 2013-2016 can be seen at Figure 2, meanwhile COD can be seen at Figure 3.

Figures 2-3 show, during period of sample, BOD and COD concentration were fluctuated at every sampling point. The highest BOD and COD concentration were recorded at Bantar Awi intake with concentration of 85 mg/L and 262,78 mg/L respectively. These values are far above Indonesia stream standard for both parameters which are 2 mg/L and 10 mg/L for BOD and COD, respectively.

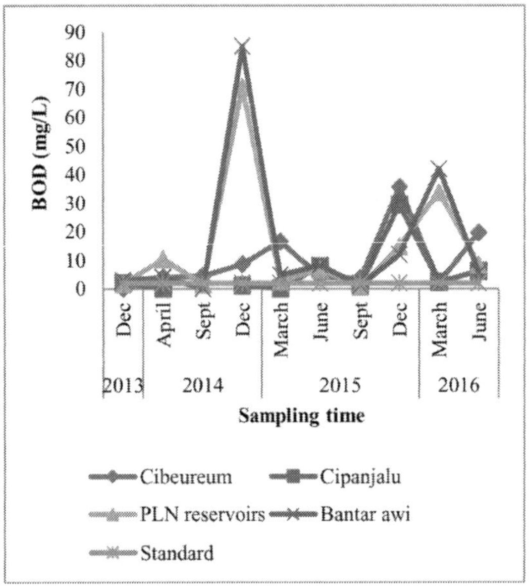


**Fig.2.** BOD concentration at intake between 2013-2016 (source: PDAM Tirtawening,2017)

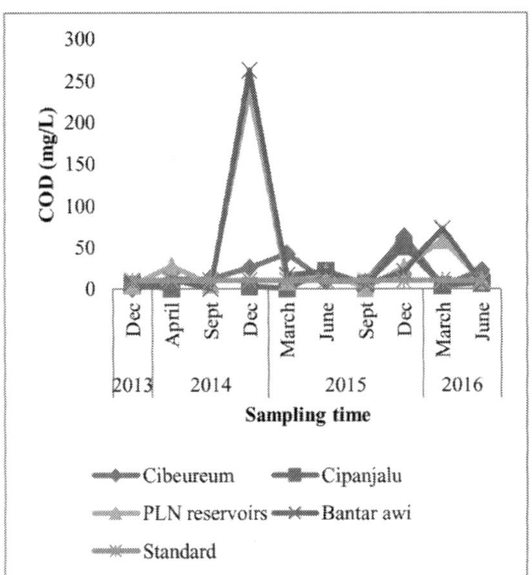


**Fig.3.** COD concentration at intake between 2013-2016 (source: PDAM Tirtawening,2017)

SPSS 19 was used for Homogeneities and Anova test. Homogeneities test results shows significant values for both parameters were 0.097; these were higher than the sig, so that Ho was accepted, and it means that the variance of BOD and COD on all four sources are homogen. Anova test result was showed there were no significant differences of mean value between all four sources, these were because the concentration of BOD and COD at every sampling points commonly exceed than standards. However, BOD and COD concentration at Bantar Awi intake have recorded the worst among the intake.

Figure 4 shows the catchment area all of intakes, although there were several type of land use, due to research purposes, these were grouped into two grouped namely:

1. Natural area (green), there were no significant human activities in this area,
2. Developed area (yellow), there were significant human activities in these area.

Figure 4 shows, the two largest developed area were found in Bantar Awi and PLN Reservoir, but at PLN reservoir intake, the water had been stored at the particular time at reservoir, so the qualities of the water at the intake was better compare to Bantar Awi intake.

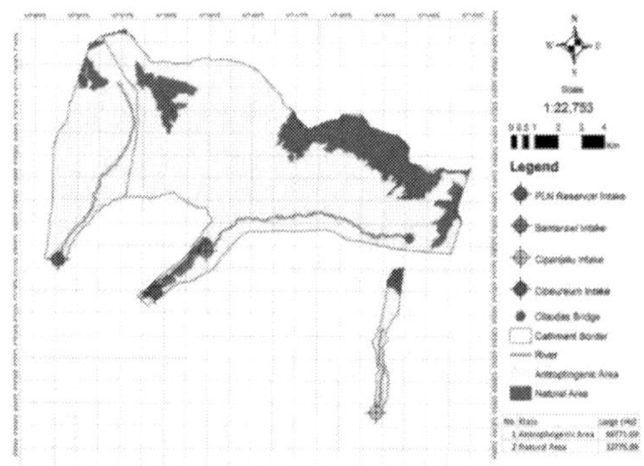


Fig.4. Land use pattern in catchment area of the intake

According statistical and spatial analysis, the worst organic concentration at raw water found in Bantar Awi intake. These intakes provide ±600 L/sec of raw water for Dago Pakar treatment plant.

## 3.2 NOM measurements

Table 3 shows the organic characteristic of the sample and Table 4 shows the correlation among organic parameters. Table 3 shows, the measurement result BOD/COD of the samples were between 0.17-0.46. These values indicated that non-biodegradable organic parts were far higher than biodegradable organic part. BOD of the sample was between 1.5 mg/L-13.2 mg/L and COD of the sample around 5 mg/L-57.6 mg/L. Table 4 shows; there was strong correlation between BOD and COD (sig 0.05 and p 0.05). However, it was found the relationships of both parameters with BOD/COD were weak.

**Table 3** Organic characteristic

| **Parameters** | **Value** | | | |
|---|---|---|---|---|
| BOD | 1.5 | - | 13.2 | mg/L |
| COD | 5 | - | 57.6 | mg/L |
| BOD/COD | 0.17 | - | 0.46 | - |
| UV $_{254}$ | 0.189 | - | 0.596 | (cm-1) |
| UV $_{355}$ | 0.076 | - | 0.447 | (cm-1) |
| A3/A4 | 1.49 | - | 2.65 | - |
| HIX | 0.71 | - | 1.79 | - |

Organic aromatic as $UV_{254}$ were between 0.189 $cm^{-1}$ and 0.596 $cm^{-1}$. Hua, Reckhow and Abusallout [22] reported that the $UVA_{254}$ value at three intakes in USA, wherein the values were between 0,114 $cm^{-1}$ and 0,313 $cm^{-1}$. That study also reported that the highest value of $UV_{254}$ recorded at the watershed of the Assomption River, was highly influenced by agricultural activities. The $UV_{254}$ values of Bantar Awi intake were higher than that of Assomption River because it was influenced by heterogeneous activities around the catchment.

$UV_{355}$ is used to determine organic aromatic from terrestrial (allochthonous NOM) [19]. The $UV_{355}$ values were between 0.076 $cm^{-1}$ and 0.447 $cm^{-1}$, these were lower than that of $UV_{254}$. This condition happened as in the water sample, there were others aromatic sources besides from terrestrial. These aromatic sources might from others allochthonus NOM or from secondary NOM (autochthonous NOM). However, based on the values of UV $_{254}$ and $UV_{355}$, organic aromatic in the water sample 60% (average) were from terrestrial, and these values were higher, compared to $UV_{355}$ at Yangtze estuary in China, where the effect of dilution is significant [23]. This finding shows that the allochthonus sources contributed significantly to organic aromatic composition in the water sample. This phenomenon was strengthen by strong correlation between $UV_{355}$ and $UV_{254}$ (sig 0,01) (Table 3).

**Table 4.** Correlation among BOD,COD and CTOC

| | | **BOD** | **COD** | **$UV_{254}$** | **A3/4** | **HIX** | **BOD/COD** | **$UV_{355}$** |
|---|---|---|---|---|---|---|---|---|
| BOD | Pearson Correlation | 1 | | | | | | |
| | Sig. (2-tailed) | | | | | | | |
| COD | Pearson Correlation | .736* | 1 | | | | | |
| | Sig. (2-tailed) | .024 | | | | | | |
| $UV_{254}$ | Pearson Correlation | -.478 | -.304 | 1 | | | | |
| | Sig. (2-tailed) | .193 | .427 | | | | | |
| A3/A4 | Pearson Correlation | -.016 | .325 | -.546 | 1 | | | |
| | Sig. (2-tailed) | .968 | .394 | .128 | | | | |
| HIX | Pearson Correlation | .209 | .064 | -.400 | .167 | 1 | | |
| | Sig. (2-tailed) | .590 | .870 | .286 | .667 | | | |
| BOD/COD | Pearson Correlation | .464 | -.243 | -.329 | -.462 | .130 | 1 | |
| | Sig. (2-tailed) | .208 | .529 | .387 | .211 | .738 | | |
| $UV_{355}$ | Pearson Correlation | -.318 | -.339 | .922** | -.750* | -.257 | -.033 | 1 |
| | Sig. (2-tailed) | .404 | .372 | .000 | .020 | .504 | .933 | |

* Correlation is significant at the 0.05 level (2-tailed).
** Correlation is significant at the 0.01 level (2-tailed).

A 3/4 describes the humification in the sample. Table 3 noted, the ratios were between 1.49-2.19. These showed that all the samples values were less than 5, and these indicated that the sample were dominated by humic acid material, compared to fulfic acid. Meanwhile, humification can be also related to an increase in aromatic character [24]. With regard these two terms, it was found that there was a negative correlation between A3/4 and $UV_{254}$ although it was not significant. Another term of humic can be explained by the composition and structure of humic depending on its sources. The structure made humic is categorized as Refractory Organic Substance (ROS) [25]. Thus, this why is, in this study, the ratio of BOD/COD ratio were low. Further result shows that ROS can increase not only due to the waste water effluent as discussed in Frimmel [26], but also due to direct discharge of anthropogenic activities as found during the sampling period of this study.

Table 3 also presents that there is no a significant correlation between COD and $UV_{254}$. According to Mrkva [27], a strong correlation between COD and $UV_{254}$ on polluted water was resulted in lignin contamination, wherein lignin can be impurities from nature. Furthermore, Zhang, Yin, Feng, Zhu, Shi, Liu and Zhang [28] also found that a strong correlation between the two parameters can occur mainly on humic-dominated samples. While, there was low correlation between the two parameters if the samples were derived from protein [29]. From this information, it is expected that the sample in this study was dominated by protein which will be discussed in final section of this paper.

The FDOM value was represented by HIX as presented in Table 3. The HIX of sample were 1,1-1,39. HIX values were <4 corresponding to biological material [20, 21]. In contrast, the value of BOD/COD was dominated by non-biodegradable organic as presented in Table 3. The HIX value also shows different interpretation with A3/4 showing the sample was dominated by humic material, and there was negative correlation between HIX and UV $_{355}$ and $UV_{254}$ (Table 4). This strengthen Chen, Zheng, Song and Qin [24]'s finding which reported that ultraviolet–visible absorption is negatively correlated with HIX.

### 3.3 The source of the NOM

The characteristic of the NOM is dominated by protein, as shown in Table 5 discussing the F-EEM of fluorescence intensity by single peaks using excitation's wavelength of 280 nm. It was found that the peak of emission was between 311 nm and 349 nm. These confirmed that the samples were dominated by protein-like humic both tyrosine-like and tryptophan-like. The land use around the catchment was dominated by built area such as traditional market, residential area, farms and agricultural. Yang, Hong, Guo, Huang, Li and Yu [1] discussed that protein like fluorophore dominated at watershed areas which are affected by human activities such as residential area.

**Table 5.** F-EEM of fluorescence intensity peaks of sample

| Sample | Excitation,nm | Emission,nm |
|---|---|---|
| D-1 | 280 | 311 |
| D+1 | 280 | 342 |
| D-2 | 280 | 312 |
| D+2 | 280 | 349 |
| D-3 | 280 | 312 |
| D+3 | 280 | 311 |
| D-4 | 280 | 311 |
| D+4 | 280 | 311 |
| D-5 | 280 | 311 |
| D+5 | 280 | 334 |

## 4 Conclusion

Land use pattern is affected the characteristic of organic matter in the water intakes. It was found significant correlations were found between (i) BOD and COD, (ii) $UV_{254}$ and $UV_{355}$ and (iii) A3/4 and $UV_{355}$, while the relationship between lability NOM and Chromophoric NOM was weak. This results reveal that the characteristic of NOMs from the selected water intake is predominantly influenced by anthropogenic activities which dominantly comprise of protein.

## References

1. L. Yang, H. Hong, W. Guo, J. Huang, Q. Li, and X. Yu, Regional Environmental Change, **12**, 145-151(2012)
2. C.A.Stedmon and S. Markager, Limnology and Oceanography, **50**, 686-697(2005)
3. Y.Yamashita, , N. Maie, H. Briceño, and R. Jaffé, Journal of Geophysical Research, **115**, (2010)
4. O Gibert, B. Lefèvre, A. Teuler, X. Bernat, and J. Tobella, Journal of Water Process Engineering, **6**, 64-71(2015).
5. R.Fabris, , C.W. Chow, M. Drikas, and B. Eikebrokk, Water research,**42**, 4188-4196 (2008)
6. M.Filella, Environmental chemistry letters, **7**, 21-35 (2009)
7. T.Pagano, M. Bida, and J.E. Kenny, Water, **6**, 2862-2897(2014)
8. E.M.Thurman, , Organic geochemistry of natural waters. Vol. 2. Springer Science & Business Media (1985)
9. M.A.El-Shafy and A. Grünwald, Water Research, **34**, 3453-3459 (2000)
10. M.Haddad, L. McNeil, and N. Omar, Arabian Journal for Science and Engineering,**39**,5883-5893(2014).
11. J.Awad, , J. van Leeuwen, C. Chow, M. Drikas, R.J. Smernik, D.J. Chittleborough, and E. Bestland,. Journal of Hazardous Materials,**308**, 430-439(2016).
12. D.A.Reckhow, P.C. Singer, and R.L. Malcolm,Environmental science & technology, **24**, 1655-1664(1990).
13. N Ates,., S.S. Kaplan, E. Sahinkaya, M. Kitis, F.B. Dilek, and U. Yetis, Journal of hazardous materials, **142**, 526-534(2007)
14. A.D.Pifer and J.L.Fairey,Environmental engineering science, **31**, 117-126(2014).
15. D.Ma, B. Peng, Y. Zhang, B. Gao, Y. Wang, Q. Yue, and Q. Li, Bioresource technology, **165**, 81-87(2014).
16. Z.-Y.Zhao, J.-D. Gu, X.-J. Fan, and H.-B. Li, Journal of hazardous materials, **134**, 60-66(2006).
17. I.Juwana, , B. Perera, and N. Muttil,"Conceptual framework for the development of West Java water sustainability index". in 18th World IMACS Congress and MODSIM09 International Congress on Modelling and Simulation, 3343-3349(2009)
18. P.G.Dainard, and C. Guéguen, Marine Chemistry, **157**, 216-223(2013).
19. R.Artinger, G. Buckau, S. Geyer, P. Fritz, M. Wolf, and J. Kim, Applied Geochemistry, **15**, 97-116(2000)
20. C.Guéguen, D.C. Burns, A. McDonald, and B. Ring, Chemosphere, **87**, 932-937(2012).
21. A.Huguet, L. Vacher, S. Relexans, S. Saubusse, J.-M. Froidefond, and E. Parlanti, Organic Geochemistry, **40**, 706-719(2009).
22. G.Hua, D.A. Reckhow, and I. Abusallout, Chemosphere,**130**, 82-89(2015).
23. Y.Wang, , D. Zhang, Z. Shen, J. Chen, and C. Feng, Chemosphere, **95**, 353-362(2014).
24. H.Chen, , B. Zheng, Y. Song, and Y. Qin, Aquatic Sciences,**73**, 103-112(2011).
25. H.-R.Schulten, B. Plage, and M. Schnitzer, Naturwissenschaften, **78**, 311-312(1991).
26. F.H.Frimmel, Biopolymers Online,**1**, (2005).
27. M.Mrkva, Water research, **17**, 231-235(1983)

https://doi.org/10.1051/matecconf/201815402002

28. Y.Zhang, Y. Yin, L. Feng, G. Zhu, Z. Shi, X. Liu, and Y. Zhang, water research, **45**, 5110-5122(2011).
29. J.Zhang, J. Yu, W. An, J. Liu, Y. Wang, Y. Chen, J. Tai, and M. Yang, Journal of Environmental Sciences, **23**, 183-188(2011).

# The concept of "Loop Cycle" in landfill management (Case study at Piyungan landfill, Yogyakarta, Indonesia)

*Hijrah Purnama* Putra[1,*], *Enri* Damanhuri[2], and *Akhmad* Marzuko[3]

[1]Department of Environmental Engineering, FTSP Universitas Islam Indonesia, Jalan Kaliurang km 14,5 Sleman, Yogyakarta 55584, Indonesia
[2]Department of Environmental Engineering, FTSL Institut Teknologi Bandung Jalan Ganesa 10, Bandung 40132, Indonesia
[3]Department of Civil Engineering, FTSP Universitas Islam Indonesia, Jalan Kaliurang km 14,5 Sleman, Yogyakarta 55584, Indonesia

**Abstract.** The amount of waste continues to increase from year to year, one of which is due to the population increase. With the target of 100% service level by 2020, Indonesia must prepare the land that will be used as a landfill location in order to accommodate the waste that continues to be produced. Apparently, the problem is not only limited to the provision of land, but operational challenges become more severe. One of its is experienced by Piyungan Landfill, Bantul, Yogyakarta which has been designed to expire by 2015. The government of Yogyakarta is optimizing for landfill can still operate until 2018. One solution that can be given in operation for the loop cycle or closed cycle concepts is landfill mining method, which is utilizing degraded waste into other designations so that the land can still be used to accommodate other waste. Sampling and analysis results show that the waste contained in 1st zone Piyungan landfill aged 15-20 years, with the highest composition is soil (59%) dominate other types of waste. The soil obtained has the potential to be utilized as cover soil and compost, but for compost is necessary to further study the modification of the design of the zoning zone, so that the waste not too long is in the soil so that the nutrient content is still high.

## 1 Introduction

Access to national waste management services in 2013 reached 79.80% consisting of 87% of services in urban areas (resource management: 41% and final management: 46%) and 72.60% in rural areas (source management: 69.2% and final management: 3.40%). Until now, the available final processing area (landfill) in Indonesia has been 87Ha, so that the target of service can be reached 100% by 2020, it is projected to need a landfill area of 875Ha in all parts of Indonesia with the estimated cost of 13,120 trillion rupiah [1]. With a great value, surely this is not an easy job because the operation and maintenance are much difficult than the construction at the beginning.

The study conducted by InSwa [2] related to the operation of the existing landfill in Indonesia, shows the results of 10 landfills in the metropolitan city of Indonesia only 10% of which do the waste, closing with the soil once a week, while the other closes fortnightly (40%), a month up to once a year (50%). Similar conditions are also obtained from 14 landfills in major cities of Indonesia, 58% of landfill closure from a period of a month to once a year, even 14% did not close at all during the operational period. The condition is due to many factors, including the unavailability of cover soil and high operational cost. Not only the closure of waste using the soil cover, the control of leachate and gas produced is still minimal management, so that newly constructed or rehabilitated landfill will be damaged again and can not work optimally and bring up various problems such as disease vector growth, air pollution, odor and smell, burning smoke, noise, social impact and pollution due to leachate.

Landfill management is a serious concern because of the increase of waste generation generated every time period. One of the causes is the increasing number of people in a region will affect the quantity of waste generated. Special Region of Yogyakarta Province one of 34 provinces owned by Indonesia, geographically located between $70^0$33'LS - $8^0$12'LS and $110^0$00'BT - $110^0$50'BT. It has an area of 3,185.8 $km^2$, consisting of 4 districts/cities, namely Sleman, Bantul, Gunungkidul, Kulonprogo, and Yogyakarta and is divided into 78 districts and 438 sub-districts/villages. The total population of Yogyakarta in 2015 reached 3,691,196 people, with the population still dominated in Sleman Regency by 31.78%, Bantul Regency 26.58% and the rest spread evenly in three other areas. The highest population density is owned by the city of Yogyakarta, which reached 12,699 people/$km^2$ [3].

Based on National Standard of Indonesia (SNI) 19-3983-1995 of waste generation specification in Indonesia with potential waste generation reaching 3.25 L/person/day or equivalent to 0.8 Kg/person/day,

* Corresponding author: hijrah@uii.ac.id

this province has a potential waste production of 11,996 m³/day or 2,953 tons/day [4]. Not all waste generated will end up in a landfill, some of which end up in unsustainable ways. Yogyakarta City has 85% waste level service, followed by Sleman Regency 30.71%, Gunungkidul 17%, Bantul 7.49% and Kulonprogo 7.2% [4-6].

Yogyakarta City, Sleman and Bantul Regency have a final handling system for waste in the same location, namely Piyungan Regional Landfill located in Sitimulyo Village, Piyungan District, Bantul Regency, Yogyakarta. This landfill has an area of 12.5 ha and started its operation since 1995, the management has changed several times from the Cipta Karya of Public Works of Yogyakarta Province (1995-2000), continued management done by the Joint Secretariat (Sekber Kartamantul) (2000-2014) is currently managed under the Sanitation and Urban Water Supply Infrastructure Management Center (Balai PISAMP), Public Works Department and Housing of Special Province of Yogyakarta [4].

Piyungan landfill has a storage capacity of 2.7 million m³ of waste divided into 3 work zones. In planning the life of the TPA 15-20 years, should be in 2015, Piyungan landfill has expired. However, until 2017 it is still operated by optimizing 1st zone which had previously been closed/full. However, the operation is expected to last only 1-2 years, so that in 2018 the Government of the Special Province of Yogyakarta must have a development plan for the landfill as a place to accommodate the waste that continues to be produced [7]. The reality in the field to find a new landfill location is very difficult, in addition to limited land, there is also a potential for high social conflicts in the community.

One of the methods that can be used in order to realize a sustainable landfill is a concept of "closed cycle" or "loop cycle" in landfill operations. The most feasible way to apply is landfill mining, an effort to recover beneficial ingredients from waste dumps in the landfill [8-11]. [12] Argue that the landfill mining method was first introduced in Tel Aviv, Israel in 1953 as a way of getting fertilizer for agricultural activities. Continued in India in 1989 as a pilot project of utilization of degraded organic matter as compost [8]. The Environmental Protection Agency (EPA) reported in 1993 that the first landfill mining activity was carried out with complete processing carried out at Collier Country Florida in 1986 [13]. This method continued to grow in the 1990s, as evidenced by more articles writing about this theme, some 39 papers, derived from landfill mining studies in the US (50%), Europe (30%) and Asia (20%) [11]. So the landfill mining method is one of the promising strategies to solve the problem of increasing the quantity of waste in a city, the limited land of the landfill, and minimizing the large funding for operational and post-operational activities of a landfill [14-15].

## 2 Research Methodology

In the landfill mining method, data on the composition of degraded waste according to the depth of the landfill is desirable. Sampling was conducted at 4 (four) points in 1st zone area of Piyungan Landfill, Bantul. Samples were taken by drilling using drill rig spindle type drill rig through spindle stroke with 7cm diameter drilled per one meter up to 13 meters. The sample was analyzed for the composition, in order to identify the type of waste at each depth. Further testing of nutrients and heavy metals contents in the soil from various variations of landfill drilling depth was performed. Later, the results were compared with SNI 19-7030-2004 for compost specification from domestic organic waste and Regulation of Minister of Agriculture 70/2011 for organic fertilizer, biological fertilizer, and soil enhancers.

## 3 Results and Discussion

The unavailability of data related to the age of waste contained in each depth of waste pile, it takes a simple approach so that the waste can be known age. Using some data as follows: (1) the waste data goes to the landfill annually, (2) the total area of 1st zone reaches 4 Ha (since 1st zone is used as the sampling site); (3) the current waste density is estimated to reach 800kg/m³, (4) degradation factor occurring by 30%, and (5) covering land used 3% (for sanitary landfill have 15-20% cover soil ratio, 3% figure is predicted because management of Piyungan landfill at 1995-2000 still not in accordance with applicable procedures). With that data, it is estimated that the depth of landfill in accordance with the operational time period (Table 1).

**Table 1**. Simple modeling of depth of waste dump in 1st Zone Piyungan landfill

| Year | Total (ton/ year) | Total (m³/ year) | Degra-dation (30%) | Cover Soil (3%) |
|---|---|---|---|---|
| 1995 | 87.500 | 109.375 | 32.813 | 3.281 |
| 1996 | 90.200 | 112.750 | 33.825 | 3.383 |
| 1997 | 93.000 | 116.250 | 34.875 | 3.488 |
| 1998 | 95.900 | 119.875 | 35.963 | 3.596 |
| 1999 | 98.800 | 123.500 | 37.050 | 3.705 |
| 2000 | 102.000 | 127.500 | 38.250 | 3.825 |
| Total Waste | Total (m³/ month) | Total (m³/ day) | Height (m/ year) | Depth Total (m) |
| 79.844 | 6.654 | 222 | 2,0 | 12,9 |
| 82.308 | 6.859 | 229 | 2,1 | 10,9 |
| 84.863 | 7.072 | 236 | 2,1 | 8,9 |
| 87.509 | 7.292 | 243 | 2,2 | 6,8 |
| 90.155 | 7.513 | 250 | 2,3 | 4,6 |
| 93.075 | 7.756 | 259 | 2,3 | 2,3 |

Source : [15]

The result of simple modeling in Table 1 shows that waste from 15-16 years ago (1999-2000) is 4 meters deep from the surface of the soil, the depth of 4-8 meters is waste with predicted age of 17-18 years (1997- 1998), while the depth> 8m to 13 meters is waste with predicted age ±19-20 years (1995-1996). Next per meter depth has separated the composition of the waste manually (maximizing the vision). Brief look samples comprise the majority of soil blackened, made possible because of the influence a long time in the soil litter. But after being parsed one by one, the more visible the difference between plastic, glass, cloth, metal, wood, bone, hair, paper and rubber.

In accordance with the understanding, the composition of waste is a representation of each component contained in waste and its distribution. The composition of waste is expressed in percent by weight of each component by the following equation:

$$\% \, weight = \frac{Weight \, of \, waste \, per \, composition \, (kg)}{Total \, weight \, of \, waste} \, x \, 100\% \quad (1)$$

**Table 2.** Composition of waste in each sampling depth

| Depth (m) | Weight Percentage (%) | | | | |
|---|---|---|---|---|---|
| | Soil | Plas-tic | Glass | Tex-tile | Me-tal |
| 1 | 44,27 | 34,16 | 1,37 | 12,48 | 1,24 |
| 2 | 56,86 | 31,02 | 0,85 | 1,83 | 0,93 |
| 3 | 70,87 | 22,07 | 0,21 | 1,51 | 0,08 |
| 4 | 69,57 | 24,39 | 0,51 | 0,15 | 0,02 |
| 5 | 67,97 | 26,19 | 0,61 | 0,66 | 0,38 |
| 6 | 70,87 | 22,71 | 1,40 | 1,06 | 0,08 |
| 7 | 57,44 | 30,28 | 1,11 | 2,22 | 0,87 |
| 8 | 67,51 | 25,17 | 0,65 | 0,62 | 0,24 |
| 9 | 49,56 | 32,91 | 0,79 | 1,94 | 3,51 |
| 10 | 54,29 | 37,69 | 0,84 | 1,47 | 0,62 |
| 11 | 48,51 | 43,47 | 0,48 | 3,80 | 0,12 |
| 12 | 49,81 | 35,81 | 1,56 | 11,89 | 0,00 |
| 13 | 59,52 | 32,47 | 0,75 | 2,22 | 0,60 |
| Average | 59,00 | 30,64 | 0,86 | 3,22 | 0,67 |

| Depth (m) | Weight Percentage (%) | | | |
|---|---|---|---|---|
| | Wood | Orga-nic | Paper | Rubber |
| 1 | 6,40 | 0,00 | 0,00 | 0,08 |
| 2 | 5,03 | 2,40 | 0,67 | 0,41 |
| 3 | 2,41 | 2,59 | 0,19 | 0,06 |
| 4 | 4,36 | 0,00 | 0,46 | 0,54 |
| 5 | 4,15 | 0,00 | 0,00 | 0,04 |
| 6 | 3,87 | 7,20 | 0,00 | 0,01 |
| 7 | 4,83 | 3,02 | 0,08 | 0,09 |
| 8 | 5,76 | 0,00 | 0,00 | 0,06 |
| 9 | 10,28 | 0,02 | 0,00 | 0,99 |
| 10 | 4,39 | 0,26 | 0,00 | 0,44 |
| 11 | 3,26 | 0,00 | 0,00 | 0,35 |
| 12 | 0,69 | 0,21 | 0,00 | 0,03 |
| 13 | 4,00 | 0,00 | 0,00 | 0,43 |
| Average | 4,57 | 1,21 | 0,11 | 0,27 |

The waste composition per depth is shown in Table 2, the soil dominates the results compared to other waste compositions, averaging 59%; followed by plastic 30.64%; wood 4.57%; 3.22% fabric; And the rest are glass, metal, organic bones and hair, and paper.

High soil components are derived from the degradation of organic waste, based on data of waste composition entering the Piyungan landfill dominated by organic waste, in the form of food, garden and garden waste (64,41-68,52%), plastic, glass and metal (0.56-11.19%), fabrics (3.49%), wood (3.24-6.11%) and paper (2.08-10.24%) [7, 16]. With the composition of organic waste high so it is possible that the resulting soil is also high, besides the soil is derived from the soil cover used in the landfill operation [17].

Potential of soil is 59% from each sampling point, with a storage capacity of 2.7 million $m^3$, the soil that can be obtained from the degradation of waste amounts to 1.5 million $m^3$. Or if taken into account in more detail than the 4 sampling points based on the height of the soil (6.28 m), the soil volume of 175,950.47 $m^3$ [17] is obtained. Of course, both approaches in determining the volume of soil produce large quantities, the soil is potentially useful for use as compost and as a cover for landfill applications under both controlled and sanitary landfill categories. Compost from the degradation of organic waste is very good for the plant, but because the composition of waste into the landfill is very mixed, it is necessary to test the quality of compost and contaminant content in it. Similarly, soil cover becomes an important issue, because so far the cost of soil cover becomes an obstacle from the operation of a landfill.

Research from [7] showed the results of testing on organic material owned by waste degradation land in Piyungan landfill has very low content. Only two samples have the appropriate organic material required by SNI 19-7030-2004, ie at sampling point 1 depth 4 m and sampling point 4 depth 8 m. In contrast to the C/N ratio, most of it fulfill the SNI requirement, which has at least a ratio of 10, while only 3 samples whose C/N ratio meet the requirements of Regulation of the Minister of Agriculture 70/2012. In general, the C/N ratio shows relatively low results compared to the requirements of organic compost.

The sample age will affect the quality of the nutrients it contains, with the sample age of 15-20 years, the organic material possessed by the waste in the form of carbon values has been depleted, causing the nutrients contained to be very low. If this soil will be utilized as organic fertilizer/compost it is necessary preliminary processing in order to increase the nutrients before applied to the plant. If the landfill will use the landfill mining concept, the operating zone's lifetime should be reduced so that the nutrient content of the degraded waste becomes higher.

While for metal parameters in soil analyzed include Zn, Cr, Cu, Cd, Ni, Pb, Hg, Mn and Fe. All metals parameters meet the requirements as organic compost, except Lead (Pb) which still exceeds the quality standard. The mixing of domestic urban waste with hazardous waste from domestic activities causes waste to potentially contain heavy metals, especially high Pb [7].

## 4 Conclusion

Based on sampling activities in order to plan a sustainable landfill, among the utilization of degraded waste that can be done as a compost and soil cover. Here are some points that can be concluded from this research:

a. The composition of degraded waste with aged 15-20 years of age is dominated by soil (59%). It is estimated that 1st zone of Piyungan Landfill can produce 175,000 until 1.5 million $m^3$ of soil.
b. The land obtained can be used as soil cover for landfill operations in other zones, while the mined zone can be reused as a new zone and ready to accommodate other waste.
c. The potential of soil for compost still needs to be studied because of the very low nutrients and the potential of heavy metals contained therein. Low nutrient content can be anticipated by improvements in the design of the zone of exposure especially regarding aspects of life. Age of waste will affect the nutrients contained

## References

1. Direktorat Jenderal Cipta Karya, *Buku pedoman penyelenggaraan pembangunan bidang Cipta Karya*, **Volume 4** : Pengembangan pengelolaan persampahan, Kementerian Pekerjaan Umum dan Perumahan Rakyat, Indonesia (2015)
2. Inswa, *Indonesia Solid Waste Newsletter : Untuk Indonesia yang lebih bersih*, **Edisi 2** Maret 2013, Jakarta (2013)
3. BPS Yogyakarta, *Data Jumlah Penduduk Propinsi Daerah Istimewa Yogyakarta* (2016)
4. H.P. Putra, E. Damanhuri, Performance and Operational of Landfill Piyungan as The Regional Landfill in Yogyakarta Special Region, Indonesia. In Proceding of *The 9th Asia-Pacific Landfill Symposium*. University of Hongkong. Hongkong (2016)
5. BLH DIY, *Rencana Aksi Sampah Daerah DIY*. Yogyakarta, Indonesia (2014)
6. S.A. Mulasari, A.H. Husodo, N. Muhadjir, Kebijakan pemerintah dalam pengelolaan sampah domestik. Kesmas, *Jurnal Kesehatan Masyarakat Nasional*, **Vol 8 No 8**, 404-410 (2014)
7. H.P. Putra, A. Marzuko, K. Sari, T. Septhiani, F. Rahmadani, Identification of compost potential on degraded solid waste in Piyungan landfill, Bantul, Yogyakarta as a step of landfill mining management optimization by using landfill mining methode, In Proceding of *4th International Conference on Sustainable Built Environment*, Faculty of Civil Engineering and Planning, Universitas Islam Indonesia, Yogyakarta (2016)
8. C. Barlow, W. Iqbal, S. Ashton, Landfill mining, Publication of University of Cambridge, Institute for Manufacturing, Departement of Engineering, download from https://www.srcf.ucam.org/awtbi/documents/Landfillmining.pdf. (2002).
9. Direktorat PPLP. *Materi Bidang Sampah : Diseminasi dan Sosialisasi Keteknikan Bidang Penyehatan Lingkungan Permukiman*, Direktorat Jenderal Cipta Karya, Kementerian Pekerjaan Umum, Indonesia (2012)
10. R. Schreiber, C. Russell, D. Abeln, Regulatory requiremen, Paper 5 in *Procedings of Global Landfill associated with landfill mining in the United States*, Paper 4 in Procedings of Global Landfill Mining Conference and Exhibition, PRo Publication International Ltd, UK (2010)
11. J. Krook, N. Svensson, M. Eklund, Landfill Mining: A Critical Review of Two Decades of Research. Waste Management Journal **32**, pp.513-520 (2012)
12. G.M. Savage, C.G. Golueke, E.L. von Stein, Landfill mining: past and present, Biocycle Journal **34**, 58–61 (1993)
13. R. Goeschl, System, technology and experience of 17Mt of landfill mining projects, Paper 5 in *Procedings of Global Landfill Mining Conference and Exhibition*, PRo Publication International Ltd, UK (2010)
14. W. Dickinson, Landfill mining comes of age. Solid Waste Technologies, **9**, 42–47 (1995)
15. H.P. Putra, D. Afrillah, A. Marzuko, Analisis karakteristik dan potensi pemanfaatan sampah plastik dari hasil penambangan sampah di Tempat Pemrosesan Akhir (TPA) (studi kasus TPA Piyungan, Bantul, Yogyakarta), in *Proceding National Conference Menuju Masyarakat Madani dan Lestari*, DPPM Universitas Islam Indonesia, Yogyakarta (2015)
16. K.P. Adidarma, L.M. Al Rosyid, H.P. Putra, and A.U. Farahdiba, Gas emissions inventory of methane (CH4) with First Order Decay (FOD) method in Piyungan Landfill, Bantul, DIY, In: *Proceedings The 3rd International Conference on Sustainable Built Environment (ICSBE)*, Faculty of Civil Engineering and Planning, Universitas Islam Indonesia, Yogyakarta, Indonesia (2014)
17. H.P. Putra, E. Damanhuri, A. Marzuko. Landfill Mining Prospect in Indonesia. *In Proceeding of 3rd Symposium of the Asian Regional Branch of International Waste Working Group*. Seoul National University. Seoul (2017)

# Evaluation of solid waste management at campus using the "Zero Waste Index": The case on campus of Islamic University of Indonesia

Kasam*, *Fajri Mulya* Iresha, and *Satrio Ajie* Prasojo

Department of Environmental Engineering, Faculty of Civil Engineering and Planning, Islamic University of Indonesia, Yogyakarta Indonesia, 55584

**Abstract.** Faculty of Civil Engineering and Planning (FCEP) Islamic University of Indonesia (UII) after doing zero waste program which has been going on September 2016. Zero waste concept are needed to measure how far the ability virgin material substitution to balance with the system of zero waste. The aims of this research is to calculate zero waste index (ZWI) value, to know the impact of zero waste index, and to give the solution for zero waste program. The location was doing in FCEP UII Campus. The method of sampling was using for this research is refers to SNI 19-3964-1994 about to calculate waste generation of residential. The result shows value of zero waste index at FCEP UII are 0,26. FCEP UII reuse 134,19 kg waste of total 516,37 kg waste that has been produced. The waste that has been reuse is organic waste 87,93 kg, plastic waste 21,49 kg, and paper waste 24,77 kg. 80,10% FCEP peoples already know about zero waste program at FCEP UII. 98,90% of FCEP peoples, was agree with waste segregate, and 57,50% FCEP peoples are don't know that waste from FCEP have been manage. Although as many as 29% of element campus do not understand about the zero waste program but the majority of them support the program.
**Keywords** : Zero Waste, Zero Waste Index, Waste Generation

## 1 Introduction

Rapid urbanization, population growth, migration to urban areas, lack of sufficient funds and affordable services often force city authorities to offer unreliable and inefficient waste management services [1]. Increasingly people move from rural to urban environments due to the economic activities and quality of life provided to inhabitants, causing cities to expand. Over-crowded cities are compromising the quality of urban life due to their rapid growth and ever-increasing generation of waste. One of the important mandates in the Law No. 18/2008 in the implementation of waste separation, reduction, reuse and recycle. Waste management carried out by the community has been started and is developing in many cities. Activities related to community involvement consist of separation, composting (at source and communal), and recycling. The municipality is the responsibility for the transportation of the residue to disposal sites [2]. Worldwide, the higher education (HE) sector has expanded phenomenally; for example, since the 1960s, the United Kingdom (UK) HE system has expanded sixfold to >2.4 million students. As a consequence, the overall production of waste at HEIs throughout the world is very large and presents significant challenges as the associated legislative, economic and environmental pressures can be difficult to control and manage [3].

In traditional waste management systems, waste is considered an 'end-of-life' product, produced in the last phase of the product-consumption process. Zero waste challenges the traditional definition of waste by recognising that waste is a transformation of resources which happens in the intermediate phase of the resource consumption process. The resources that are transformed into waste as a result of our consumption activities should therefore be redirected into the production process through holistic zero waste management systems. Thus zero waste management is a combination of integrated design and waste management philosophies. The zero waste index is a tool to measure the potentiality of virgin materials to be offset by zero waste management systems. One of the important goals of the zero waste concept is zero depletion of natural resources. Therefore, measuring the performance of the zerowaste city would eventually measure the resources that are extracted, consumed, wasted, recycled, recovered and finally substituted for virgin materials and offset resource extraction by the waste management systems [4].

Within the zero waste philosophy, product design and waste management principles are considered simultaneously to eliminate potential threats to the environment caused by human consumption and unsustainable behaviour. Zero waste product design ensures that the discarded product is easily reused and/or repaired to extend the product's lifespan. A zero waste

* Corresponding author: kasam@uii.ac.id

product is created by applying cradle-to-cradle design principles which eliminate waste from a product's life cycle. A zero waste product eliminates the 'waste phase' from the traditional product life cycle because after the end-of-life phase the zero waste product can either be reused, repaired or remanufactured to produce a secondary product. Zero waste management processes ensure that the discarded waste be recycled, recovered or easily nourished through natural processes, without polluting our natural environment. Hence the concept of zero waste safeguards the optimum utilization of natural resources with minimal environmental degradation [5].
The past decade has seen many researchers motivated to develop new approaches to waste management systems based on zero waste concepts. The major work in zero waste research has been done by [6] shown that the campus of University of Tabriz produced 2.5 metric tons of wastes per day that more than 80% of it could be diverted through waste reduction, recycling and composting activities. Compostable organic material was the most significant waste type.

The major work in zero waste research has been done by [7 - 10]. This research has identified the most vital areas of zero waste studies as consumption of resources, individuals' consumption behaviour, and product design based on cradle-to-cradle principles (ecoeffective product and system design), maximum waste diversion from landfill and optimum resource recovery. Waste diversion rates (determined by calculating the amount of waste diverted from landfill) are one of the performance assessment indicators for waste management systems [11]. As a performance indicator, waste diversion has different socio-economic and environmental issues [12] and has received criticism [4] from researchers due to its limited forecasting capacity. [13] proposed the ZWI as an alternative performance assessment tool. The higher ZWI value represents higher substitution of virgin materials, energy and water saving i.e. higher achievement. The study compared the waste management performance of Adelaide, San Francisco and Stockholm based on the virgin material substitution factor, energy, GHG and water savings from their respective waste management systems. The study found that San Francisco's municipal solidwaste management system has the highest virgin material substitution potentiality and that, in 2010, its Zero Waste Index was 0,51 with a 72% diversion rate followed by Adelaide (ZWI = 0,23) and Stockholm (ZWI = 0,17).

The aim of this study is determine the performance of waste management systems on campus by calculating the amount of solid waste diverted from landfill are one of assessment indicators.

# 2 Material and Methods

This study was conducted during the 2015-2016 academic year on the main campus of the FCEP UII. The faculty had, approximately, 6500 students and 150 staff work in the year of study. The research consisted of five main stages as below:

- Estimation of amount of daily generated solid wastes
- Waste characterization, analysis of the amounts and types of wastes.
- Determining the ZWI based on strategy for solid waste management within faculty

## 2.1 Estimation of Solid Waste Generation

In order to determine the solid wastes generation has been carried out sampling for 8 days in FCEP UII campus. Prior to sampling, waste disposal spots were identified and campus was divided into nine subareas according to the type of activity. These subareas included academic departments of the university (faculties, labs and workshops), administration buildings, health center, sport center, dormitories, staff residential area, conference halls, canteen and cafeterias and university gardens. Sampling methods and procedures for characterization of the wastes were derived through using Standard Test Method for Determination of the Composition of Unprocessed Municipal Solid Waste (SNI 19-3964-1994). Furthermore the reviews of studies performed by different universities and countries were used. According to ASTM recommendations, samples weighed 200 to 300 pound (91 to 136 kg) can be chosen and a single sample of 100 kg is a good representative of the total waste characteristics (ASTM, 2001). All samples were taken during on working days to ensure that the sampling results represent the normal university operations during the academic year. A total of nine sampling loads, one for each subarea, were scheduled over one week during each month of year. Samples were weighted by manual fish scale (accuracy ± 0,1 kg) and kept in plastic bags for the analyses.

## 2.2 Waste Characterization

In the separation site, all samples were hand sorted into 11 primary waste categories and weighted. In order to assess the recycling potential of the wastes, each category was further subdivided into subcategories. Waste characterization of categories and subcategories were adapted from various waste characterization methodologies, mainly from SNI 19-3964-1994 standards. Waste categories included paper and paper products, plastics, organic (compostable) materials, glass, metals, textiles, construction and demolition wastes, wood, hazardous wastes, electronic wastes and others. The percentage for each category was calculated using the following equation:

$$PC = \left(\frac{PL}{PT}\right)x100 \quad (1)$$

where PC is the percentage of each category, PL is the amount of category present in kg, and PT is the total weight of the sample in kg.

## 2.3 Determining the ZWI

The waste diversion rate does not indicate the virgin material replacement efficiency of the waste management system, which is very important in conservation of global natural resources. Thus, the zero waste index is a cutting-edge tool to measure virgin material substitution bywaste management systems. By introducing the zero waste index globally, we could measure the virgin material offset potentiality and the potential depletion of natural resources. The ZWI is also a useful tool to compare different waste management systems in different cities and it gives a broader picture of the potential demand for virgin materials, energy, carbon pollution and water in a city. The ZWI is thus a performance indicator to assess the overall performance of waste management systems [5].

$$ZWI = \frac{\sum_1^n WMSi * SFi}{\sum_1^n GWS} \quad (2)$$

WMSi = amount of waste managed by system i (i.e. i =1, 2, 3... n = amount of waste avoided, recycled, treated, etc.)

SFi = Substitution factor for different waste management systems based on their virgin material replacement efficiency

GWS = Total amount of waste generated (tonnes of all waste streams).

The Z is based on the value of material that can potentially replace the virgin material inputs. The substitution of energy, water and greenhouse gas emissions is also considered with the material substitutions. Substitution values for material, energy, water and greenhouse gases (GHG) emissions have been extracted from the life cycle database of different life cycle assessment tools and database sources. The amount of materials and resources substituted is positively related to the advancement of technology used in the material recovery process; therefore, the substitution value varies for different materials and for different waste management systems. Even though,waste prevention is one of the core components in the zero waste concept, but quantitative measurement of waste prevention by behaviour change has not been considered in this research due to limited scientific quantitative measurement data. Table 1 shows the substitution values for waste streams for different waste management systems.

Table 1. Substitution values of the resource from waste management systems for the Zero Waste Index

| Waste management systems | Waste category | Virgin material substitutions efficiency (tonnes) | Energy substitutions efficiency (GJLHV/tonne) | GHG emissions reduction (Tonne/$CO_{2e}$) | Water saving (kL/tonne) |
|---|---|---|---|---|---|
| Recycling | Paper | 0,84-1,00 | 6,33-10,76 | 0,60-3,20 | 2,91 |
| | Glass | 0,90-1,00 | 6,07-6,85 | 0,18-0,62 | 2,3 |
| | Metal | 0,79-0,96 | 36,09-191,42 | 1,40-17,8 | 5,97- 81,77 |
| | Plastic | 0,90-0,97 | 38,81-64,08 | 0,95-1,88 | -11,37 |
| | Mixed | 0,25-0,45 | 5,00-15,0 | 1,15 | 2,0-10 |
| Composting | Organic | 0,60-0,65 | 0,18-0,47 | 0,25-0,75 | 0,44 |
| Landfill | Mixed MW | 0 | 0,00-0,84 | (-)0,42-1,2 | 0 |

References; [5]

# 3 Results and Discussion

## 3.1 Solid Waste Generation and Composiion in FCEP UII

Universitas Islam Indonesia (UII) was founded on 8th of July 1945, one month before the proclamation of Indonesian independence. Located in the northern outskirt of Yogyakarta. In 2016, UII has 24,202 students and 692 faculty members, distributed in 8 faculties (Faculty of Civil Engineering & Planning, Economics, Law, Mathematics & Natural Sciences, Psychology & Socio-Cultural Sciences, Islamic Studies, Industrial Technology and Faculty of Medicine). The Faculty of Civil Engineering and Planning (FCEP) was founded in 12th of October 1964. It has 3 undergraduate programs (Environmental Engineering, Civil Engineering and Architecture), 1 master program of civil engineering and 1 architect professional program with 2841 students and faculty member are 112.

Based on sampling shown that the average weight of solid waste generation in FCEP UII was 64,55 kg per day. Applying Equation (1) resulted the average composition of solid waste in FCEP UII Campus is organic (28,38%), paper (5,70%), plastic (4,62%), and mixed (61,28%, mostly construction and demolition waste). Fig. 1 shows the composition of solid waste generation in FCSP UII Campus. Conducting a waste composition study is the first step if an effective waste management program development of campus are to be attained. It is estimated that the FCEP Campuss generate approximately 64,55 kg of waste per day, of which more than 35% may possibly be dealt with through waste reduction, recycling and composting activities. Compostable organic material was the most significant waste type. Therefore some strategies should be taken to manage these organic wastes. In addition, to manage other components of generated wastes in the FCEP UII Campus, there should be some plans to reuse and/or recycle them, included: reducing waste generation, reuse of generated wastes, composting of organic wastes, recycling of recyclable wastes, and disposal (landfill) of non-recyclable wastes.

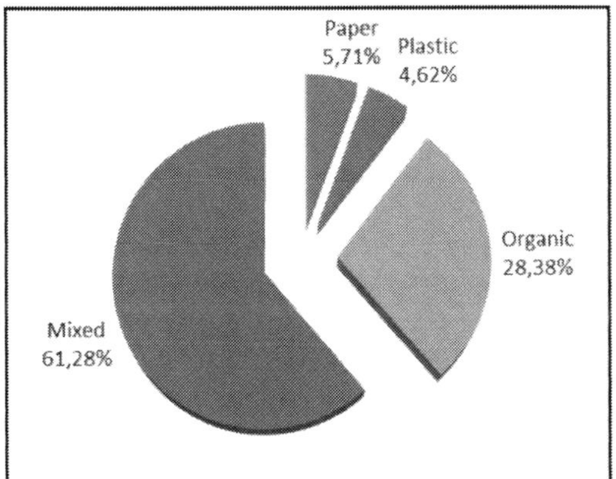


**Fig. 1.** Solid waste composition in FCEP UII Campus

### 3.2 The ZWI in FCEP UII

Base on Equation (2) and Table 2, the zero waste index in FCEP UII Campus is 0,26. That means around 26% of solid waste generation were recovered from the waste management systems. [4] study shows that in 2010 San Francisco achieved a 72% waste diversion rate with a ZWI of 0.51. FCEP UII generated around 516,37 kg of waste every week and the resources recovered and potentially substituted for virgin material was 134,19 kg. Waste management systems in FCEP UII potentially substitute the energy demand of 1139,64 megajoules (MJ). GHG emissions substituted were -55,90 kg CO2e and total water savings from the waste management systems was -121,16 L per week.

The zero waste program in FCEP has not succeeded in reducing GHG emissions by producing 55.9 kg $CO_2$ per week. Where garbage was disposed to landfill which accounts for the largest GHC that is 132,91 kg of $CO_2$, but for waste paper, plastics, and organic managed to reduce GHG, although not too large. Shown that paper, plastic and oranic waste can reduce of GHG amounted to 17.69 kg, 22.68 kg, 36.64 kg $CO_2$ of GHG per week respectively. Based on water saving, the zero waste program in FCEP UII also still not been successful because it become concume of water up to 121.16 L. Plastic waste became the largest contributor in the water that is equal to 271.5 L. The value is great because of the use of plastic bottles and plastic cups are relatively many in FCEP UII. But it helps in saving paper waste water, which can save 85.82 L and organic waste at 64.48 L

Table 2. Potential substitution of resources in the zero waste index

| Waste management systems | Waste category | Total waste managed in the campus (kg) | Potential total virgin material substituted (kg) | Total energy substituted (MJ) | Total GHG emissions reduction (kg/$CO_{2e}$) | Total water saving (L) | *Zero Waste Index* |
|---|---|---|---|---|---|---|---|
| Recycling | Paper | 29,49 | 24,77 | 186,67 | 17,69 | 85,82 | 0,26 |
| | Plastic | 23,88 | 21,49 | 926,59 | 22,68 | -271,5 | |
| Composting | Organic | 146,55 | 87,93 | 26,38 | 36,64 | 64,482 | |
| Landfill | Mixed | 316,46 | 0 | 0 | -132,91 | 0 | |
| Total | | 516,375 | 134,19 | 1139,64 | -55,9 | -121,16 | |

### 3.3 Participation of Faculty Member and Student

The zero waste program in FCEP UII involves all elements in the campus consists of faculty leaders, lecturers, academic staff and students. FCEP UII has committed to resource conservation, sustainability and environmental management and will take the initiative to apply material management methods and progressive zero waste practices. The form of activities through the reduction of consumption, the use of better materials, resource conservation and resource recovery on waste generated.

Faculty leaders have established policies on the implemntation of zero waste program in FCEP. Lecturers act as supervisors and educate all the elements in the campus to participate in the success of the zero waste program. While academic staff acts as a companion and field implementers such as placement and taking up to each container location. The student element that is the largest number of campus residents is to be very strategic for participation. Zero waste program in FCEP has established a task force of students with a total of 30 people, it aims to monitor and invite directly to all students to participate in handling waste in the campus environment.

The socialization activities are routinely conducted by the zero waste team to all FCEP elements. It intended to know the response and invite all campus residents to zero waste program. The form of socialization is done through posters, social media, and quesionnair. Based on the questionnaire, it is known that some campus residents have not understood about zero waste program in FCEP that is 29%, while 99% of campus support and agree with the program.

## 4 Conclusions

The study is a time series measurement of waste management performance in FCEP UII based on the ZeroWaste Index tool. From the assessment analysis, the study concludes that despite a potentially higher diversion rate (134,19 kg from 516,37 kg of waste generated) in 2016. FCEP UII may not be significantly advanced in virgin materials substitution and resource recovery by then and the projected ZWI would be 0.26. Participation of all faculty members and student in FCEP on zero waste program are very good, although the understanding of zero waste program still needs to be improved, because only 29% already understand.

This study is part of an on-going research project studying strategies for zero waste material flows conducted at the Zero Waste Program at the UII. The authors would like to give a special thanks to Mr. Dr Ing. Widodo Brontowiyono, dean of FCEP UII that has supported the study of the evaluation of zero waste programs in FCEP.

## References

1. Wilson, D., Velis, Costas, Cheeseman, Chris, 2006. Role of informal sector recycling in waste management in developing countries. Habitat Int. 30, 797-808.
2. Damanhuri, E., Handoko, W., Padmi, T., 2010. Municipal Solid Waste Management in Indonesia, Faculty of Civil and Environemntal Engineering, ITB, Bandung-Indonesia
3. Zhang, N., Williams, I.D., Kemp, S., Smith, N.F., 2011. Greening academia: Developing sustainable waste management at Higher Education Institutions, Journal of Waste Management 31 (2011) 1606-1616.
4. Zaman, A.U., Lehmann, S., 2013. The zero waste index: a performance measurement tool for waste management systems in a 'zero waste city', Journal of Cleaner Production, 50 (2013) 123-132.
5. Zaman, A.U., 2014. Measuring waste management performance using the 'Zero Waste Index': the case of Adelaide, Australia, Journal of Cleaner Production 66 (2014) 407– 419.
6. Taghizadeh, S., Ghassemzadeh, H.R, Vahed, M.M., Fellegari, R., 2012. Solid waste characterization and management within university campuses case study: university of Tabriz, Elixir Pollution 43 (2012) 6650-6654.
7. Kinuthia, J.M., Nidzam, R.M., 2011. Towards Zero Industrial Waste: Utilisation of Brick Dust Waste in Sustainable Construction. Waste Management 31 (2011), 1867-1878.
8. Phillips, P.S., Tudor, T., Bird, H, Bates, M., 2011. A Critical Review of a key Waste Strategy Initiative in England: Zero Waste Places Projects, Resources, Conservation and Recycling, 55, (2011), 335-343.
9. Zaman, A.U, Lehmann, S., 2011. Urban Growth and Waste Management Optimization Towards 'Zero Waste City'. City Cult. Soc, 2 (2011), 177-187.
10. Curran, T.,Williams, I.D., 2012. A zero Waste Vision for Industrial Networks in Europe, Journal of Hazardous Materials, 208 (2012), 3-7.
11. Yoshida, H., Gable, J.J., Park, J.K., 2012. Evaluation of organic waste diversion alternatives for greenhouse gas reduction. Resour. Conserv. Recycl. 60, 1- 9.
12. Mazzanti, Massimiliano, Montini, Anna, Nicolli, Francesco, 2009. The dynamics of Landfill Diversion: Economic Drivers, Policy Factors and Spatial Issues: Evidence from Italy Using Provincial Panel Data, Resources Conservation and Recycling 54 (2009), 53-61.
13. Mueller, W., 2013. The effectiveness of Recycling Policy Options: Waste Diversion or Just Diversions, Waste Managemrnt, 33 (2013), 508-518.

# Ambient air monitoring of nitrogen dioxide at Kalimati, Tirtomartani, Kalasan, Sleman, Yogyakarta

*Supriyanto*[1,*] and Indah Suci *Ramadhani*[2]

[1]Environmental Engineering Department, Faculty of Civil Engineering and Planning, Universitas Islam Indonesia , Jl. Kaliurang 14,5 Yogyakarta 55584
[2]Study Program of Environmental Engineering, Faculty of Civil Engineering and Planning, Universitas Islam Indonesia , Jl. Kaliurang 14,5 Yogyakarta 55584

**Abstract.** Nitrogen dioxide ($NO_2$) is one of parameters in air quality according to Indonesia Government No. 41/1999. $NO_2$ has influence in respiratory problem of human being if it exceeds threshold level (400 $\mu gNO_2/Nm^3$). This study is air monitoring of $NO_2$ concentration in ambient air and it uses a Griess Saltzman method using the spectrophotometer (Indonesia Standard SNI 19-7119.2-2005) near of Plywood Industry and Main Road. The sampling location has two points at Kalimati Village, Tirtomartani, Kalasan, Sleman, D.I Yogyakarta. On December 2016, the results show which $NO_2$ concentration is below threshold level with 0,469 $\mu g$ $NO_2/Nm^3$ at first point and 0,234 $\mu g$ $NO_2/Nm^3$ at second point.
**Keywords:** Ambient Air Monitoring, Nitrogen Oxide, Kalimati, Griess-Saltzzman

## 1 Introduction

Nitrogen Oxide ($NO_2$) is outdoor air pollutant [1] and it causes environmental problem such as health problem in respiratory problem and acid rain[2]–[4]. $NO_2$ come from combustion process with use oxygen excess [3]. $NO_2$ is also come from reaction between Nitric Oxide, NO and Oxygen, $O_2$ in internal combustion engine. Therefore, $NO_2$ is resulted from transportation sector both gasoline engine and diesel engine. This study objective is to monitor ambient air near Stack of Plywood industry at Kalimati Village, Purwomartani, Kalasan, Yogyakarta (LS 7°44'26.67" BT 110°27'45.47") and Main Road.

## 2 Methodology

### 2.1 Sampling Location

The air monitoring used two sample locations. The first sample location has agreement with the wind direction and the north of industrial plywood stack [5]. Table 1 shows the coordinate sampling point and table 2 presents the sampling scheduling on December 2016

**Table 1.** Coordinate of Sampling Location

| Location | Coordinate Point |
|---|---|
| First | LS 7°44'25.7028" BT 110°27'50.6016" |
| Second | LS 7°44'20.598" BT 110°27'47.1456" |

**Table 2.** Sampling Shcedule

| Day | Time |
|---|---|
| Monday/Tuesday Afternoon | 13.00 - 14.00 |
| Thusday Morning | 09.00 - 10.00 |

### 2.2 Sampling and Analysis

$NO_2$ sampling used single impinger with flow rate 0.4 lpm (liter per minute) and NO2 analysis used Griess-Saltzman Method with spectrophotometer 550 nm. The analysis concept is NO2 absorb by solution Griess-Saltzman to form azo-dye substance with pink in 15 minute [6].

## 3 Result and Discussion

### 3.1 Sampling Location

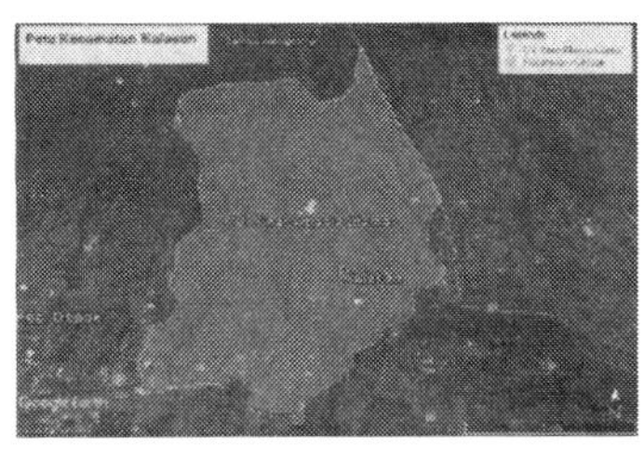

**Fig. 1.** Map of Kalasan, Sleman, Yogyakarta

*Corresponding author: supri.yanto@uii.ac.id

**Fig. 2.** Map of Sampling Location

**(a)** **(b)**

**Fig. 3.** Sampling location at first point (a) and second point (b)

### 3.2 Sampling and Analysis

$NO_2$ sampling used single impinger with flow rate 0.4 lpm (liter per minute) and $NO_2$ analysis used Griess-Saltzman Method with spectrophotometer 550 nm. The analysis concept is NO2 absorb by solution Griess-Saltzman to form azo-dye substance with pink in 15 minute.

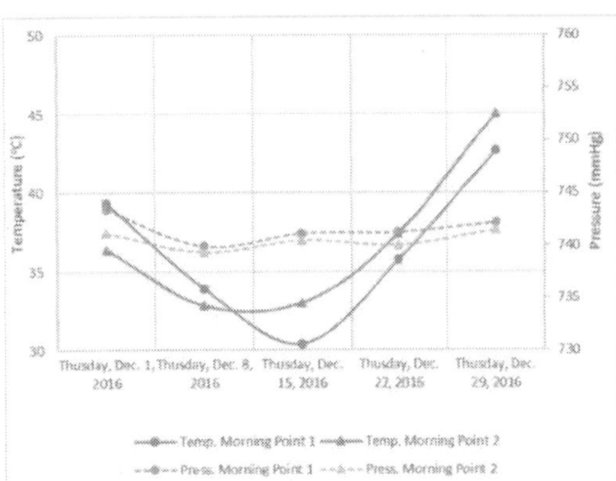


**Fig. 4.** Temperature and Pressure for Morning Day

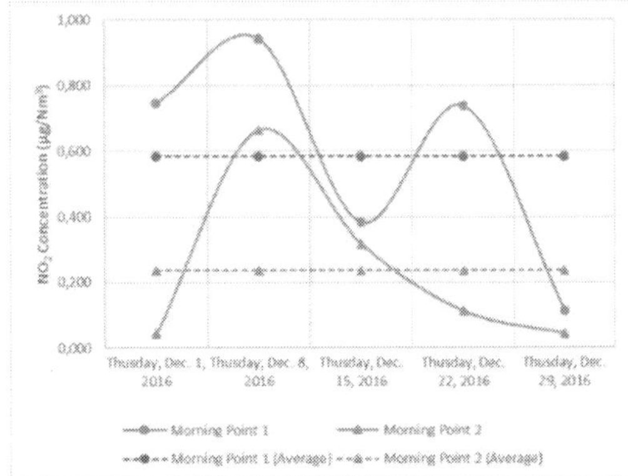


**Fig. 5.** $NO_2$ Concentration for Morning Day

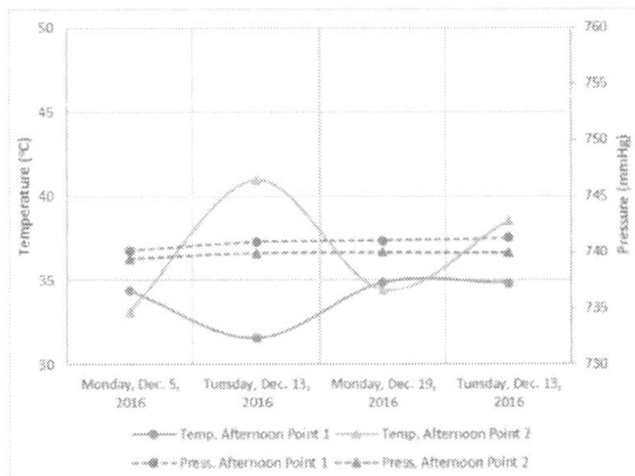


**Fig. 6**. Temperature and Pressure for Afternoon Day

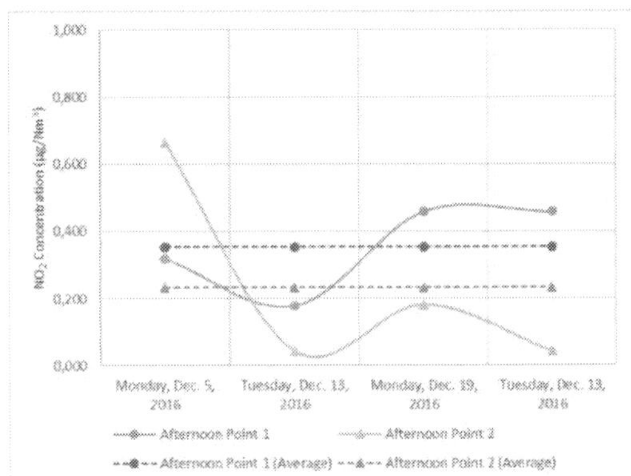


**Fig. 7.** $NO_2$ Concentration for Afternoon Day

At the first point in figure 4 Morning temperature is between 30-45°C, Pressure is around 740 mmHg, $NO_2$ concentration is slightly below 0.600 µg/Nm3 in average. During afternoon in figure 6, the temperature between 32 to 42°C, Pressure is around 740 mmHg. $NO_2$ concentration is slightly above 0.35 µg/Nm$^3$. Temperature has influence during ambient air monitoring [6]. Temperature from morning to afternoon 37oC in average, and the reduce concentration two times. At the second point in figure 5, Morning temperature is between 30-

45°C, Pressure is around 740 mmHg. $NO_2$ concentration is slightly below 0.600 μg/Nm$^3$ in average. During afternoon, the temperature between 32 to 42°C, Pressure is around 740 mmHg. Figure 7 shows which $NO_2$ concentration is somewhat above 0.35 μg/Nm$^3$. So temperature from morning to afternoon 37°C in average, and the reduce concentration two times.

## 4 Conclusions

The ambient air monitoring has conducted at Kalimati Village, Tirtomartani, Kalasan, Sleman, Yogyakarta. On December 2016. Sampling location of $NO_2$ used two points and analysis of $NO_2$ used Griess-Saltzman method. The results show which $NO_2$ concentration is below threshold level with 0,469 μg $NO_2$/Nm$^3$ at first point and 0,234 μg $NO_2$/Nm$^3$ at second point.

Authors thanks to Environmental Engineering Department, Universitas Islam Indonesia for Research and Conference Grant

## Reference

1. G. Hoek *et al.*, "Satellite NO2 data improve national land use regression models for ambient NO2 in a small densely populated country," *Atmos. Environ.*, vol. 105, pp. 173–180, 2015.
2. H. J. Lee and P. Koutrakis, "Daily ambient NO2 concentration predictions using satellite ozone monitoring instrument NO2 data and land use regression," *Environ. Sci. Technol.*, vol. 48, no. 4, pp. 2305–2311, 2014.
3. L. N. Lamsal, R. V. Martin, D. D. Parrish, and N. A. Krotkov, "Scaling relationship for NO2 pollution and urban population size: A satellite perspective," *Environ. Sci. Technol.*, vol. 47, no. 14, pp. 7855–7861, 2013.
4. United States Environmental Protection Agency, "Basic Information About NO2," *Nitrogen Dioxide (NO2) Pollution*, 2016. [Online]. Available: https://www.epa.gov/no2-pollution/basic-information-about-no2#What is NO2.
5. S. R. A. Rahman, S. N. S. Ismail, M. F. Ramli, M. T. Latif, E. Z. Abidin, and S. M. Praveena, "The Assessment of Ambient Air Pollution Trend in Klang Valley, Malaysia," *World Environ.*, vol. 5, no. 1, pp. 1–11, 2015.
6. Badan Standarisasi Nasional. SNI 7119-2.2005. Udara ambien - Bagian 2: Cara uji kadar nitrogen dioksida ($NO_2$) dengan metode *Griess Saltzman* menggunakan spektrofotometer, 2005.

# The development of green infrastructure through optimization of water resources carrying capacity in Yogyakarta urban area

*Widodo* Brontowiyono[1,*], *Dhandhun* Wacano[1], and *Ribut* Lupiyanto[1]

[1]Environmental Engineering Program, Universitas Islam Indonesia, Jln. Kaliurang KM 14.5, 55584 Yogyakarta, Indonesia
[2]Center for Environmental Study, Universitas Islam Indonesia, Jln. Kaliurang KM 14.5, 55584 Yogyakarta, Indonesia

**Abstract.** The development of Yogyakarta Urban Area has spread to its surroundings. As a result, water resources carrying capacity and environmental problems are worsening. Rainwater harvesting (RWH) development as part of green infrastructure is one of the important methods to increase water resources carrying capacity. Relating to the problem, this research was conducted in Yogyakarta Urban Area. The study was a combination of field research and studio analysis. The field study measured the settlement's physical parameters and surveyed the community's socio-economic and cultural aspects. According to the physical and economic analysis based on CBR, the recommended models of water resource conservation infrastructure were absorption well/1.10; conservation pond/4.53; open space/7.47; and green space/3.59. Meanwhile, rainwater reservoir was not suggested because the CBR value was only 0.18.

## 1 Introduction

The United Nations has predicted that three out of five people in the world would live in urban areas by 2030 [3]. Meanwhile, in Yogyakarta, the urbanization rate has increased significantly. The very limited land extent has forced the urbanization in Yogyakarta to spread to the area of Sleman Regency and Bantul Regency.

In Sleman, for example, the growing number of housing complexes is closely related to Yogyakarta's status as a study destination that is supported by the development of related business sectors. There are about 60% of university campuses in Sleman, and 63% of the students predictably come from other areas outside Yogyakarta Province. As a result, the demand for housing is increasing. In fact, the construction of housing property in Yogyakarta Province has reached 71% - the highest percentage of all - while 26% of the construction is used for commercial property and the remaining comprises industrial and conference construction [1].

Meanwhile, [13] forecasted that in 2030s the agricultural land in Sleman, Yogyakarta, and Bantul would be used up for settlement if there was no control. In addition, the development of settlement has caused such problem as a decrease in water resources carrying capacity. Therefore, this paper was aimed to recommend the models and strategies for settlement planning and management using the concept of green settlement to strengthen the water resources carrying capacity.

## 2 Research method

The research was conducted in Yogyakarta Urban Area. This area was chosen based on the complexity of area dynamics and potential degradation of environmental carrying capacity, particularly the water resources. The study was conducted by combining field research and studio analysis. The field research was done to measure the settlements' physical parameters and to survey the community's social, economic, and cultural aspects.

The social and economic data were gathered in a primary as well as secondary way. The secondary data came from reports or official sources from related boards while the primary data were collected through semi-structured interview using questionnaire. The research population was the households living in Yogyakarta Urban Area. The respondents were the householders that were determined using proportional cluster random sampling. The settlement cluster was divided into settlement areas in elite estate, medium estate, small estate, hamlet, and a mix of hamlet and estate. The total number of samples was 100 householders [4,8].

The settlements' physical data were collected from field survey, semi-structured interview using questionnaire, as well as in-depth interview. The field survey and questionnaire were conducted according to the settlement clusters that consist of elite estate, medium estate, small estate, hamlet, and a mix of hamlet and estate.

## 3 Results and discussion

### 3.1. Economic feasibility of green infrastructure models

* Corresponding author: widodo.bronto@uii.ac.id

Green settlement infrastructure (hereinafter referred to as green infrastructure) has several technical models in its implementation that can be an alternative. As recommended in the first year's report, the models include rainwater absorption well, rainwater reservoir, polder, open space, conservation pond, and vegetation. In the context of Yogyakarta Urban Area, all these types of infrastructure can be implemented.

According to the critical level of water resources, Yogyakarta Urban Area is categorized as an anthropogenic critical area. It is an area that has critical water resources because the high population growth has exceeded the water supply, and it is found mostly in urban areas with high population density and limited land availability.

The following analysis described the economic feasibility of each implementation model. The analysis was conducted for a period of 20 years, which was estimated as the average age of storage model with 16% discount level.

The first model is rainwater absorption well. The shape of absorption well is similar to that of ordinary well, yet the function is the opposite because absorption well is used for collecting rainwater from the roof or absorbing runoff into the ground. Various shapes and dimensions of absorption well have been widely used because there is no standard size. Technically, an accurate calculation should be conducted in order to reach maximum absorption with minimum construction cost. The calculation of construction cost in this study was based on the concept model from [10] with a diameter of 100 cm and a depth of 50 cm above the groundwater table. The prediction of construction cost is presented in Table 1.

**Table 1.** Estimation of construction cost for rainwater absorption well based on existing market price

| Component | Quantity and unit | Price/ unit IDR $10^3$ | Total cost IDR $10^3$ |
|---|---|---|---|
| Betonbuizen | 5 piece | 85 | 425 |
| Cement | 2 sack | 50 | 100 |
| Sand | 1/4 pickup truck | 120 | 120 |
| Gutter | 20 m | 8 | 160 |
| Pipe | 4 length | 80 | 320 |
| Worker | 2 days | 60 | 240 |
| Total | | | 1,365 |

According to Table 1, the total cost of absorption well construction was Rp 1,365,000.00. The cost included the expense for equipment, materials, and workers, and the value was based on the average market price in the research area. Besides the construction cost, the land price and maintenance cost were also considered. To make a realistic consideration, the minimum need for land was calculated based on the size of well section of 1 m2 that equaled Rp 500,000.00. Meanwhile, the maintenance was predicted to be done every 5 years of usage in the form of well emptying to normalize the absorption capability; the emptying process required Rp 150,000.00.

The following calculation was conducted for the benefit, which was merely based on the effect of absorption well on the increase in groundwater table. Other benefit variables were difficult to predict economically because of the extensive land size and time required. [6] showed that the existence of absorption well in Yogyakarta has raised the level of groundwater table by 0.282 meter per year. Since the research area was the same, the aforementioned value was used to estimate the benefit of absorption well in this study. For this research, the benefit value was calculated only by comparing the increase at the beginning and end of the study for three months. Therefore, the total volume increase was the volume of water use during the three months added by the difference value at the end of the research.

The economic value was based on the average piped-water (PDAM) bill for every cubic meter, which was Rp 3,000 per cubic meter in the research area. The average use of water per capita for domestic need reached 100 liters per day, in which the average number of household members in the research area was 4 people. Therefore, the total volume of the increase in groundwater table was 4 people x 0.1 $m^3$/day x 90 days + (3.14 x 0.5 x 0.5) x 0.282 = 36.221 $m^3$. On the other hand, the ability to absorb rainwater also depends on the rainfall.

The calculated value would be reached when the average rainfall was 800.66 mm. Since the average rainfall in the research area was 2133.33 mm [9], it was predicted that the total volume of the increase in groundwater table was (2133.33/800.6) x 36.221 x 3 = 144.59 $m^3$ or as much as Rp 433,770.00. Table 2 illustrates the result of cost-benefit ratio analysis based on the aforementioned values.

**Table 2.** Analysis of cost-benefit ratio for absorption well construction

| Parameter | Total value IDR $10^3$ |
|---|---|
| Net Cost (15%) | 1,917.344 |
| Net Benefit (15%) | 2,546.317 |
| Net Present Value (NPV) | 628.973 |
| Cost-Benefit Ratio | 1.33 |

The obtained net benefit was found higher than the net cost and the difference as much as Rp 628,973.00 was considered as the value of net benefit based on the present value (NPV). This positive NPV indicated that the cost-benefit ratio would reach more than 1, which was proven by the value of in the table. Therefore, it is concluded that, economically, absorption well was feasible to implement in the research area. The next analysis was to determine the payback period.

Graphically, this value could be determined using Fig. 1, in which the payback period is located on the intersection between the net cost and net benefit (break-even point). The graph shows that the payback period of absorption well would occur in the $7^{th}$ year of the project. Therefore, starting from the $7^{th}$ year, the community would enjoy the net benefit of the use of absorption well. The payback period would occur before the project period ended; this means that absorption well was feasible.

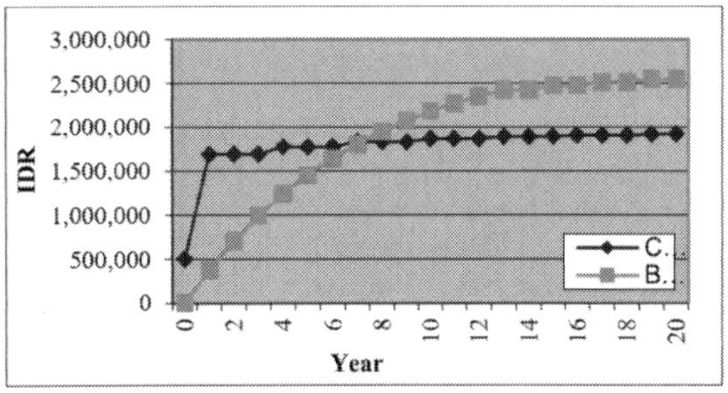


**Fig.1.** Graph of payback period for absorption well

The second model is rainwater reservoir (PAH) whose function is limited to storage only. The position can be above the building, on the ground, or under the ground. It is easier to use if the reservoir is located above the building, and it is more suitable for urban area with limited land, but the construction cost is high. It is more efficient when installed on the ground, but it requires land availability. The last position does not depend on land availability, but the construction cost and pumping cost are higher. This research analyzed on-the-ground reservoir because the land availability in the research area made it possible, and it was predicted that the cost would be more efficient.

The shape of rainwater reservoir can be cylindrical, rectangular, globular, or others. Considering the construction effectiveness and efficiency, this research chose the rectangular shape to be analyzed. Similar to that of absorption well, the cost component comprised the construction, land, and maintenance cost. The reservoir construction consisted of the cost of materials, workers, and equipment for water treatment. The collected rainwater can go through a certain treatment to be used as drinking water. In this study, a filter was chosen to keep water clean and to anticipate turbidity, which was more efficient than the regular use of alum. Table 3 illustrates the cost details.

**Table 3.** Estimation of Construction Cost for Rainwater Reservoir Based on Existing Market Price

| Component | Quantity and unit | Price IDR $10^3$ | Total IDR $10^3$ |
|---|---|---|---|
| Construction: | | | |
| • cement | 20 sack | 50 | 1,000 |
| • brick | 4000 piece | 0.425 | 1,700 |
| • sand | 1 pickup truck | 660 | 660 |
| • faucet | 2 piece | 15 | 40 |
| • iron | 20 length | 28 | 560 |
| • filter | 20 $m^2$ | 7.5 | 150 |
| Worker | 4 days | 25 | 400 |
| Total | | | 4,100 |

During the operation, the cost would be added by the maintenance cost to empty the reservoir in order to keep it clean at the end of the dry season. The emptying cost per year was Rp 50,000.00. The total cost was then compared to the benefit based on the economic value of rainwater that could be used during the dry season. It was assumed that in the dry season only rainwater reservoir was used.

In this study, the capacity of rainwater reservoir was based on the need during the dry season, which was 221 days x 4 people x 0.1 $m^3$ = 88.4 $m^3$. Therefore, the benefit was 88.4 x Rp 3,000 = Rp 265,320.00. This value was similar to the affordable price of piped water in the research area.

The result of cost and benefit calculation is presented in Table 4. It can be seen that the cost value far exceeded the benefit, so the NPV was negative and the CBR was below 1. This indicated that in the research area the construction of rainwater reservoir was not feasible. This was influenced by the limitation of capacity benefit that only works during the dry season, the need for large reservoir, and the high construction cost due to the size and land value. For houses that could allow the construction of reservoir on the top floor, the cost could be minimized, especially for the land value and ease of use.

**Table 4.** Result of cost-benefit ratio analysis for rainwater reservoir

| Parameter | Total value IDR $10^3$ |
|---|---|
| Net Cost (15%) | 4,326.295 |
| Net Benefit (15%) | 1,660.232 |
| Net Present Value (NPV) | -2,666.263 |
| Cost-Benefit Ratio | 0.38 |

The unfeasibility was also supported by the result of payback period and IRR analysis. Fig. 2 shows that the payback period was far above the project age, even unlimited, while when the trial-and-error process using up to 0% discount was conducted, the NPV still could not give positive value.

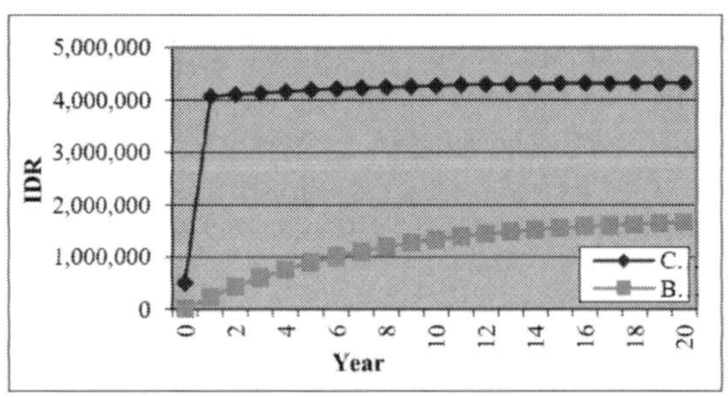


**Fig. 2.** Graph of payback period for rainwater reservoir

The next rainwater harvesting infrastructure that can be managed is conservation pond. In general, it is similar to a fish pond, but it is designed to have a frame on the surroundings while the bottom is only soil to ensure water absorption. The analysis of economic feasibility in this study was conducted for each square meter of the pond because there was no standard size. The construction cost included the dredging and making of surrounding frame that were estimated to spend Rp 480,000.00 with the land value of Rp 500,000.00 in Table 5.

**Table 5.** Estimation of initial investment value for conservation pond

| Component | Quantity and unit | Price IDR $10^3$ | Total IDR $10^3$ |
|---|---|---|---|
| Worker | 4 in 2 days | 60 | 480 |
| Land | 1 $m^2$ | 500 | 500 |
| Total | | | 980 |

The maintenance would be conducted by dredging the bottom to reduce sedimentation every 3 years for Rp 50,000.00. The measurable benefit was similar to that of absorption well, which was the effect on groundwater table. [12] described that the pitch, which is closest in model to this type of pond, had the ability to absorb 70% rainfall for every square meter of the pond. By comparing with the benefit of absorption well, the economic benefit of conservation pond was Rp 303,639.00 per year.

This benefit value far exceeded the cost, which is shown in Table 6. The NPV value was positive and the CBR value was 1.33. This value could increase if the pond was also used as a fish pond.

**Table 6.** result of cost-benefit ratio analysis for conservation pond

| Parameter | Total value IDR $10^3$ |
|---|---|
| Net Cost (15%) | 994.118 |
| Net Benefit (15%) | 1,900.871 |
| Net Present Value (NPV) | 906.683 |
| Cost-Benefit Ratio | 1.91 |

Based on this value, it was found that the payback period occurred in the 9th year in Fig 3. All the analysis results showed that the conservation pond was feasible to manage.

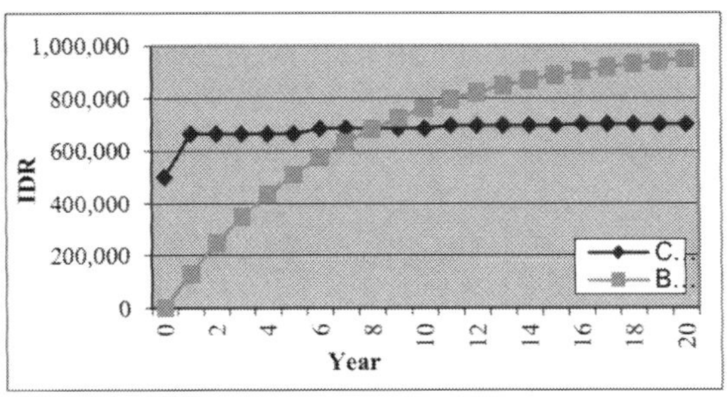


**Fig. 3.** Payback period for the construction of conservation pond

Another infrastructure to consider in rainwater harvesting is open space. The cost for its construction was very low, only Rp 500,000.00 per square meter of land and Rp 188,500.00 for building the foundation surrounding the land that showed in Table 7.

**Table 7.** Estimation of construction cost for open space per square meter

| Component | Quantity and unit | Price IDR $10^3$ | Total IDR $10^3$ |
|---|---|---|---|
| Construction : | | | |
| • cement | 1 sack | 50 | 50 |
| • brick | 100 piece | 0.425 | 42.5 |
| • sand | 0.1 pickup truck | 660 | 66 |
| Worker | 1 in 0.5 day | 60 | 30.5 |
| Total | | | 188.5 |

The calculation of benefit was the same as that of conservation pond, which was based on the absorption ability of the open space and the influence of absorption well on the groundwater table. [2] noted that an open space with sandy soil in the research area would result in a maximum of 15% runoff that could reduce approximately 50% evaporation and absorb around 35% of water. The value gave an economic benefit as much as Rp 151,820.00 (1 cubic meter of water equaled the price of piped water as much as Rp. 3,000.00). The analysis of cost-benefit resulted in positive NPV and the value of CBR reached 7.47, which was far above 1 can see in Table 8.

**Table 8.** Result of cost-benefit ratio analysis for open space construction (per $m^2$)

| Parameter | Total value IDR $10^3$ |
|---|---|
| Net Cost (15%) | 701.625 |
| Net Benefit (15%) | 950.439 |
| Net Present Value (NPV) | 248.814 |
| Cost-Benefit Ratio | 1.36 |

The value of CBR was quite high and even resulted in a payback period in the 8th year in Fig. 4.

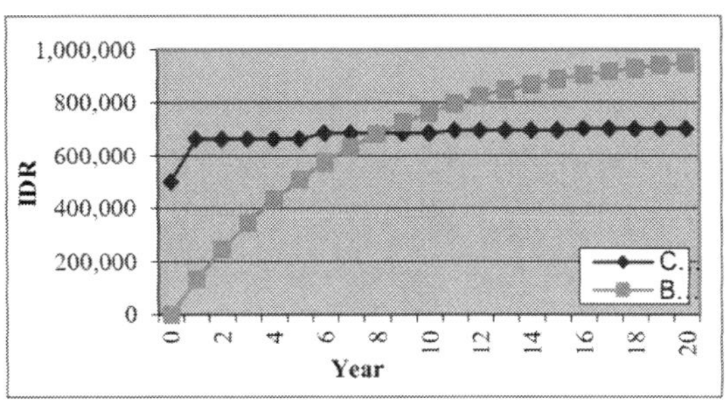


**Fig. 4.** Payback period for open space management

The last infrastructure to be economically analyzed is green space (RTH). The analysis was conducted for one plant, especially productive plant. Table 9 shows the estimation of investment value to manage vegetation; one plant needs 4 square meter of land. U.S. Forest Services [2] explained that productive vegetation could absorb maximum 25% runoff.

**Table 9.** Estimation of initial investment value for green space construction per m$^2$

| Component | Quantity and unit | Price IDR $10^3$ | Total IDR $10^3$ |
|---|---|---|---|
| Seedling | 1 plant | 25 | 25 |
| Fertilizer | 2 kg | 10 | 10 |
| Land | 1 m$^2$ | 500 | 500 |
| Total | | | 535 |

Therefore, when subtracted by 50% evaporation level, the 25% absorption capability would be equal to Rp 108,443.00. Another benefit was the plant's product that could be estimated to worth Rp 50,000.00 per plant, and this could be obtained after 3 years. Based on this value, the NPV was positive and CBP was above 1 in Table 10.

**Table 10.** Result of cost-benefit ratio analysis for green space construction per plant

| Parameter | Total value IDR $10^3$ |
|---|---|
| Net Cost (15%) | 530.436 |
| Net Benefit (15%) | 755.666 |
| Net Present Value (NPV) | 225.230 |
| Cost-Benefit Ratio | 1.43 |

The payback period was obtained in the 8th year of the project that can seen in Fig. 5.

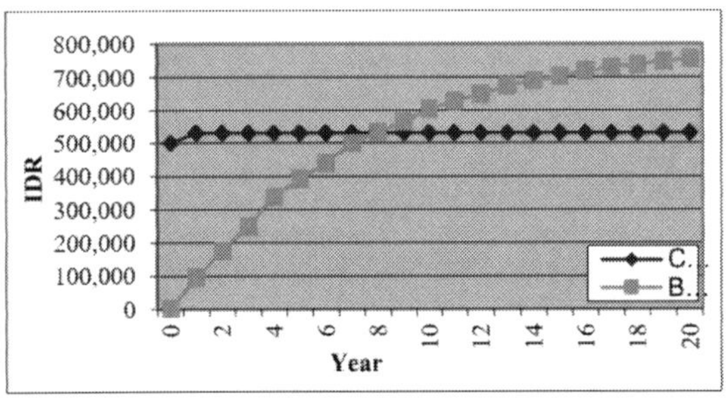


**Fig. 5.** Payback period for green space construction

Based on the analysis result, a recommendation priority could be determined for the construction of green infrastructure in Table 11. The priority showed the cost effectiveness of each infrastructure type. All the types of infrastructure were prioritized because the benefit value used in the calculation was very limited, so it had not yet involved indirect benefit and benefit for other areas.

In order of priority, the types of infrastructure that could be conservation pond, absorption well, greend space, open space, and rainwater reservoir respectively. Besides the aforementioned general assumption, the specific assumption was about the generalization of research area characteristics, such as the same physiographical condition, economic value of water, rainfall, land availability, and economic condition. This assumption did not apply to specific areas, such as naturally critical area; in this type of area, the priority could not be implemented.

**Table 11.** Priority of green infrastructure construction

| TI | NPV IDR $10^3$ | CBR | PBP (year) | EF | P |
|---|---|---|---|---|---|
| Aw | 628.973 | 1.33 | 7 | F | II |
| Rr | -2,666.263 | 0.38 | > 20 | Un-F | V |
| Cp | 906.683 | 1.91 | 5 | F | I |
| Os | 248.814 | 1.36 | 8 | F | IV |
| Gs | 225.230 | 1.43 | 8 | F | III |

Note:
TI : type of infrastructure; PBP: pay back period; EF: Economic Feasibility; F: feasible; Aw : Absorption Well; Rr : Rainwater Reservoir; Cp: Conservation Pond; Os: Open Space; Gs: Green Space

## 3.2 Social feasibility of the green infrastructure models

The construction of green infrastructure is required even at the level of households. Therefore, a feasibility that is merely based on profit-loss analysis is not enough to make a participative policy. The community's ability to pay and willingness to manage also need to be considered. The community's ability can be analyzed using ATP analysis while their willingness can be determined using WTP analysis [11]. That can be seen in Table 12.

**Table 12.** Respondent's ownership of green infrastructure

| Ownership | Quantity | Percentage |
|---|---|---|
| Own (at least 1 type) | 32 | 40 |
| Not own at all | 48 | 60 |

The critical area of research was in Yogyakarta Urban Area with dense settlement land use. This condition was strongly assumed as influencing the level of green infrastructure ownership. Table 13 illustrates that only 40% of the respondents in the surplus area owned green infrastructure. Besides the need for water, the environment's physical condition, especially the land, also became a factor that influenced the ownership level and type of green infrastructure as shown in Table 13.

**Table 13.** Respondents' ownership of each type of green infrastructure

| TI | Own | | Not own | |
|---|---|---|---|---|
| | quantity | % | quantity | % |
| Aw | 4 | 5 | 76 | 95 |
| Rr | 0 | 0 | 80 | 100 |
| Cp | 0 | 0 | 80 | 100 |
| Gs | 16 | 20 | 64 | 80 |
| Os | 16 | 20 | 64 | 80 |

Note:
TI : type of infrastructure; Aw : Absorption Well; Rr : Rainwater Reservoir; Cp: Conservation Pond; Os: Open Space; Gs: Green Space

In the research area, none of the types of green infrastructure was dominant although there were only three types, which were 5% ownership of absorption well and 20% ownership of green space or open space. Green space and open space can naturally appear although ideally they still need management and almost always exist if a household has a yard. This is supported by the data in Table 13 showing the types of green infrastructure that required better management. They were absorption well, rainwater reservoir, and conservation pond that was more varied and less owned. Table 14 shows other evidence of the encouragement of respondents who owned those types of green infrastructure.

**Table 14.** Respondents' support to manage green infrastructure

| Source of support | Quantity | % |
|---|---|---|
| Awareness | 20 | 62.5 |
| Government | 12 | 37.5 |
| Inadvertence | 0 | 0 |

Awareness had dominated the support to manage green infrastructure by 62.5%. This is directly proportional to the level of water demand. Meanwhile, none of the respondents in the anthropogenic critical area had green infrastructure without support. This was because of the limited land availability for natural vegetation and open space as well as due to the better level of education normally owned by urban community.

The management of green infrastructure is influenced not only by the physical environment as previously explained but also by the economic condition, especially the ability to manage. The respondents' economic conditions in the three areas were similarly good although there were a variety of levels. The easiest method to identify the economic level is by using poverty indicators. According to [7], one of the methods to determine the poverty level is based on the level of equivalent-to-rice household expense, in which the minimum limit is equivalent to 320 kg/capita/year (standard of Central Bureau of Statistics).

Based on the macro-economic condition of the three areas, which was also similar, the average value of poverty limit equivalent to the price of rice as much as Rp 4,000/kg for household expense was Rp 1,280,000/capita/year. On the average, the households in the research area consisted of 4 members, so the poverty limit for household expense was Rp 5,120,000/year or approximately Rp. 427,000/month.

In the research area, the average economic condition was good because it was above the poverty line in Table 15. Only 20% of the respondents were poor because the research area had non-agricultural activities that were more developed to earn a living.

**Table 15.** Respondent's Economic Condition

| Expense/ month IDR $10^3$ | Quantity | % | Poverty category | Quantity | % |
|---|---|---|---|---|---|
| ≤ 427 | 3 | 15 | Poor | 4 | 20 |
| 427 - 800 | 4 | 20 | Not poor | 16 | 80 |
| 800 - 1,200 | 10 | 50 | | | |
| ≥ 1,200 | 3 | 15 | | | |

In general, the condition indicated that the respondents had the ability to manage green infrastructure because most of the cost was only for initial construction. Table 15 illustrates the quantitative value of respondents' ability (ATP) to manage green infrastructure. The ATP in the research area was Rp. 1,500,000 due to the good economic condition and life demand. Based on this value, the types of green infrastructure that could be immediately managed by considering the cost were absorption well, conservation pond, green space, or open space.

Apart from the ATP analysis result, a WTP analysis was also conducted to strengthen the social feasibility analysis see Table 16. WTP analysis was carried out to identify the perception of green infrastructure implementation in the form of respondents' willingness level. The benefit of rainwater for primary need is mostly obtained indirectly, and there is a negative character of rainwater. Therefore, it should be considered that besides the economic feasibility, WTP also influenced respondents' perception of rainwater.

**Table 16.** respondents' perception of rainwater

| Type of perception | Quantity | Percentage |
|---|---|---|
| Should be collected | 48 | 60 |
| Should be overlooked | 12 | 15 |
| Should be discarded | 20 | 25 |

Amount 60% of the community in the research area perceived that rainwater should be harvested, 15% thought to overlook rainwater, and 25% believed that it should be disposed of. The perception to collect rainwater is affected by water demand as well as education level. Meanwhile, the opinion that rainwater should be ignored or immediately discarded was supported by the thought that rainwater is usually a source of disease and disaster without considering that the real cause is management. Even if this was already understood, some showed skepticism about whether the management had been optimum. That can be seen in Table 17.

**Table 17.** Respondents' willingness to manage green infrastructure

| Willingness | Quantity | Percentage |
|---|---|---|
| Willing | 15 | 75 |
| Not willing | 5 | 25 |

These levels of perception were used to direct the following question about the respondents' willingness to manage green infrastructure. The 75% WTP in the research area showed a positive condition for green infrastructure management in Table 17. It indicated that some respondents perceived to neglect rainwater while some others were willing to collect it because they were aware of its benefit. Therefore, the effort to convince the community becomes the key factor to achieve wider acceptance see Table 18.

**Table 18.** Socio-economic feasibility of green infrastructure

| TI | Cost benefit analysis | ATP | WTP | Socio-economic feasibility |
|---|---|---|---|---|
| Aw | + | + | + | + |
| Rr* | - | - | + | - |
| Cp | + | + | + | + |
| Gs | + | + | + | + |
| Os | + | + | + | + |

Note:
+ : feasible; * : not a priority of recommendation; - : unfeasible
TI : type of infrastructure; Aw : Absorption Well: Rr : Rainwater Reservoir; Cp: Conservation Pond; Os: Open Space; Gs: Green Space

## 4 Conclusion

The result of Cost-Benefit, ATP, and WTP analyses could be used to identify the socio-economic feasibility of green infrastructure management. All the types of green infrastructure in the research area were feasible, except the rainwater reservoir due to the high investment cost and limited land availability. In addition, physically, the research area had good absorption ability, so it would be more effective and proven as feasible to manage absorption well, conservation pond, green space, or open space. This area had limited land availability and good economic condition; therefore, the management of green infrastructure would be more effective if absorption well was prioritized because of its flexibility towards land availability and due to its higher benefit value.

## 5 References

1. Z. Arifin, Identification of factors relating to the growth of housing business in Sleman-*in bahasa*, LOGIKA R.J., **4**, (2007).
2. C. Asdak, *Hydrology and watershed management*, (2007).
3. M. Baiquni, *Unsustainable development: a critical reflection of Indonesian development*, Trans Media, (2002).
4. S. Faisal, *Social research formats: foundations and applications*, (Rajawali Press 1995).
5. Kadariah, *Evaluation of economic analysis projects, first edition*, (LPFE-UI, 1999).
6. R. Lupiyanto, 2009, Designing a green settlement along the city river: a case of Tegal Panggung, (BKPRN, 2009).
7. L. Muta'ali, Regional Analysis Engineering, (FGE-UGM, 2000).
8. M. Singarimbun, Survey research methods, (LP3ES, 1995)
9. B. Sulistiono, Bambang, Selecting the best rainfall distribution in a designed flood analysis using synthetic unit hydrograph of gamma I, Teknisia J. **10**, (2005).
10. Sunjoto, *Optimization of rainwater absorption well as one of the efforts to prevent seawater intrusion*, hydarulica and Hidrologi, (1988).
11. O. Tamim, Evaluation of public transport fares and analysis of Ability to Pay and Willingness to Pay in DKI Jakarta, Jurnal Transportasi Jurusan Teknik Sipil-ITB **1**, (1999).
12. D. K. Todd, Groundwater Hydrology 2nd edition, (John Wiley, 1980).
13. W. Brontowiyono, *Vulnerability Water Resources Mapping*, International SURED Integrated Water Management in Urban Environment, DAAD-Uni Karlsruhe-UKI, (2005).

# Integration of formal and informal sector (waste bank) in waste management system in Yogyakarta, Indonesia

*Hijrah Purnama Putra*[1,*], *Enri Damanhuri*[2], and *Emenda Sembiring*[2]

[1]Department of Environmental Engineering, FTSP Universitas Islam Indonesia, Jalan Kaliurang km 14,5 Sleman, Yogyakarta 55584, Indonesia
Doctoral Student in Environmental Engineering Department, FTSL, Institut Teknologi Bandung
[2]Department of Environmental Engineering, FTSL Institut Teknologi Bandung, Jalan Ganesa 10, Bandung 40132, Indonesia

**Abstract.** The change of waste management paradigm becomes an important thing to do, as a step adaptation to the increasing rate of waste generation every year in Indonesia. 100% management target has been divided into two parts, namely the reduction (30%) and waste handling (70%). Reductions focus on source limitation and 3R program optimization, whereas handling involves collecting and final processing activities. However, the current level of waste reduction is still very low (12%), the government made various efforts to increase it, one of its with the waste bank program. DIY province as a pioneer in the concept of waste bank continues to develop to increase the participation of the community, from 166 locations in 2013, increased to 792 locations in 2017 and 495 of its as the waste bank (62.5%). Average waste bank with 43 customers, able to manage the waste up to 2,078,064 kg/month, with the data can be estimated the amount of waste that can be managed in the city of Yogyakarta, Sleman and Bantul Regency. The city of Yogyakarta has 433 units of the waste bank, capable of managing waste up to 899,801.8 kg/month, Sleman Regency has 34 units of the waste bank (78.966,4 kg/month) and Bantul has 24 units of the waste bank (49.873,5 kg/month ). The integration of formal and informal sectors through waste banks can increase the percentage of waste management services. The level of service in Yogyakarta City increased from 85% to 95.5%, Sleman District from 30.71 to 31%, and Bantul Regency from 7.49 to 7.7%

## 1 Introduction

The enactment of Law 18/2008 on Waste Management, a turning point in Indonesia's waste management system, returns waste into a resource that must be managed to minimize negative impacts on the environment. Changes in the concept of waste management from Collect-Transport-Dispose to Collect-Transport-Manage accompanied by minimization on waste generation, sorting, reuse, and recycling or known as the 3R program (Reduce, Reuse and Recycle) through intelligent, efficient and programmed [1].

Based on the 2008 waste statistics, Indonesia has a waste generation rate of 38.5 million tons/year [2], this condition continues to increase until 2017, the waste generation has reached 65.8 million tons/year and is projected by 2025 to 71.3 million tons/year [3]. The waste management target is divided into 2 (two) major sections, namely, waste reduction and handling. Reduction through various activities of source limitation and the 3R program optimization, while handling is the activity of service of collecting, transporting to final processing done by the related institution, generally service area is in the urban area. Both must be consistent in order to achieve 100% service target by 2025, reaching 30% (± 21 million tons/year) through reduction activities and 70% (± 50 million tons/year) through handling activities.

The level of waste reduction in 2017 is still below 12% (7.82 million tons/year) that can be handled, while handling activities reaches 71% (46 million tons/year), the rest is still not handled and even handled by the community in a way that not environmentally friendly, through dumping, open burning, throw in the river, even ending up in the sea [4]. The main obstacle in increasing the level of waste reduction in the source is the mindset of the community to waste, especially in the understanding and low level of awareness in the sorting. Though the key to successful waste management at the source is sorting [5]. Disaggregated waste will have a higher value, both the selling value and the value of processing.

In order to change people's mindset about waste, the Ministry of the Environment implements various strategies to improve the community's role in the 3R program. One of its programs is the Waste Bank, a strategy to convert waste into financial benefits for the communities involved. Based on data in February 2012 the number of waste banks in Indonesia reached 471 units with 47,125 customers and the amount of waste can be managed to reach 755,600 kg/month or equivalent to $ 174,000/month. This number continues to increase, for example, within 3 months later then the number of waste

* Corresponding author: hijrah@uii.ac.id

bank reaches 886 units, with 84,623 customers and 2,001,788 kg of waste/month and potentially generate $ 283,000/month [1, 6]. Currently, Indonesia already has 4,820 units of waste banks spread in 30 provinces and 206 districts/cities [3].

Special Province of Yogyakarta (DIY) as a province of initiator in the concept of waste bank, precisely by Mr. Bambang Suwerda through Environmental Health Workshop which pioneered in the middle of 2006, now becomes "Gemah Ripah" Waste Bank, Bantul Regency of DIY Province continues to grow to increase the role of communities are involved in the management of waste banks

DIY Province has the potential to produce waste about 11,996 $m^3$/day or 2,953 tons/day (SNI 19-3983-1995, standard 3.25 L/person/day or equivalent to 0.8 kg/person/day) [7]. The potential of this waste is influenced by the population, DIY Province has more than 3.6 million people, with the spread of population dominated in Sleman Regency of 31.78%, Bantul Regency 26.58%, and the rest spread evenly in three other areas. The highest population density is owned by the City of Yogyakarta, which reached 12,699 people/$km^2$, and the lowest is Gunungkidul Regency 482 people/$km^2$ [8]. The level of waste service in DIY Province varies greatly between districts and other districts (Figure 1). This significant difference is caused by a system that is not always able to reach the whole region, minimal funding allocation, less professional management, limited infrastructure and human resources and access difficulties become the main obstacle [9-13].

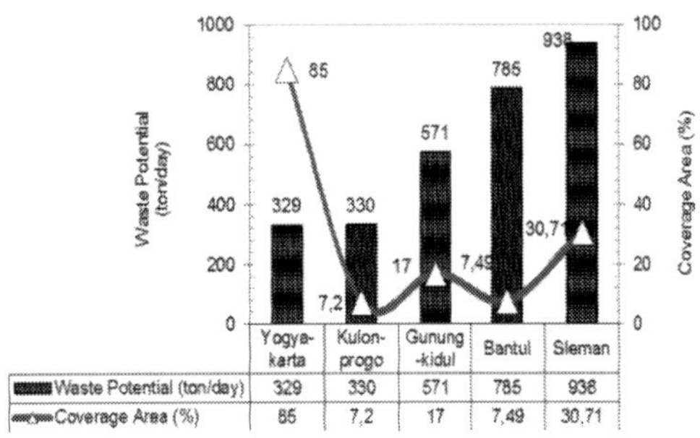


**Fig. 1.** Amount of waste and percentage of urban waste service (Source: Modify data from [14,15,4]).

The imbalance between the quantity of waste generated by the service capability raises the community's initiative to conduct waste management independently. From 2003 to 2011 in Yogyakarta Province, 117 groups of independent waste managers have been established [16], this data has increased to 155 groups by 2015 [17], the various groups are able to reduce waste up to 70% [18].

The role of the sector is so large (informal), but not yet considered in the planning of urban waste management system [19-21]. Strategic and comprehensive integration between the formal and the informal sectors is required, although it is a challenge in many developing countries [21, 22, 13].

This study aims to identify informal sector activities (garbage bank), among others, in Yogyakarta, the ability of garbage banks to accept waste and increasing percentage of waste management services in Yogyakarta.

## 2 Research Methodology

This research was conducted in several phases, starting with secondary data searching the number and location of the waste bank as representative of the informal sector in DIY Province, but limited only to Yogyakarta City, Sleman and Bantul Regency. Based on the results of data collection, the waste bank of Yogyakarta City has had a recapitulation of the number of customers and waste collected. Based on the data, projected number of customers and waste can be managed in 2 (two) other regencies. The amount of waste that can be managed is linked to the potential waste generated and the waste that is transported to the landfill in order to determine the percentage of integrated waste management services to the community.

The data required in this study are as follows:

a. The number of waste banks in the 3 research areas (Yogyakarta city, Sleman and Bantul regency);
b. The number of customers of each waste bank;
c. Total type of waste that can be received by the waste bank

The data is linked to garbage entering the landfill each month, with the level of service that has been owned by each region. So it can be known the contribution of waste banks reduce waste into landfills.

## 3 Results and Discussion

Waste Bank is a sorting and collection area for recyclable/reusable waste that has economic value and is one of the 3R (reduce, reuse and recycle) waste management strategies in community-level waste management. Economic value becomes the trigger of active participation from the community to join in the program. Before discussing further waste bank, the following is the development of community participation through various locations of community-based waste management in the study area.

Community-based solid waste management in the study area until 2017 amounted to 792 units, of which 495 units have status as the waste bank (62.5%) (Fig. 2). Technically the construction, operation, and management of waste banks are regulated in the Minister of Environment Regulation No. 13/2012, although the reality in the field of minimum requirements in this regulation has not been fulfilled. The distribution of waste bank in Yogyakarta City is quite evenly distributed, all districts have the waste bank. Umbulharjo district has 77 units of the waste bank and is the highest district of the number of waste banks owned, while the lowest is 13 waste banks owned by Gedongtengen district (Fig. 3).

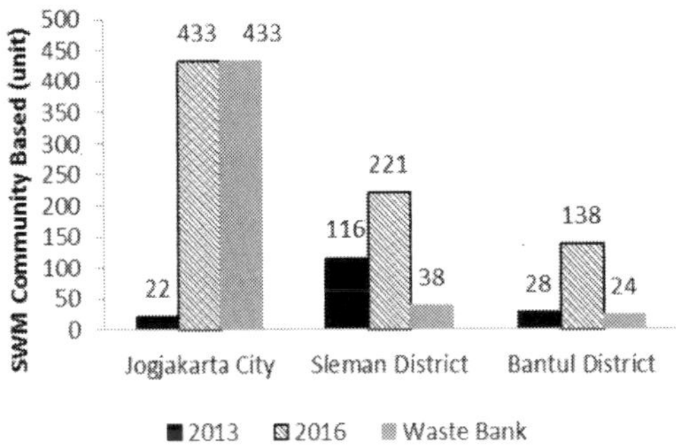


**Fig. 2.** Development of community-based waste management in the study area

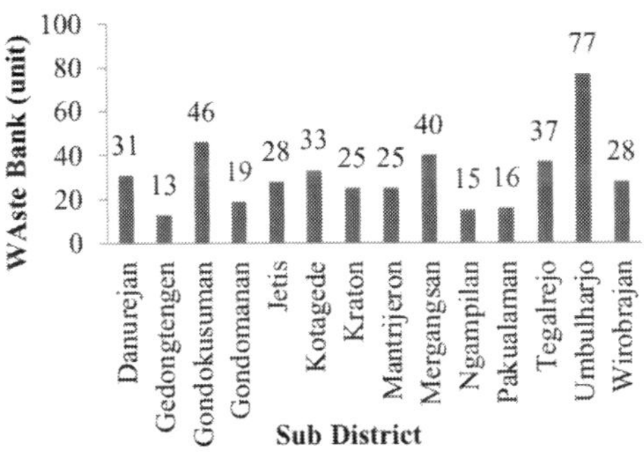


**Fig. 3.** Distribution of waste banks in Yogyakarta City

As for the other two regions, Sleman and Bantul regencies have districts that do not have a waste bank and in terms of numbers, the city of Yogyakarta has a larger number of waste banks. There are several factors that influence it, the main is the limited access to buyers of waste (collectors) because of the distance, in addition to the extent of the land owned yard, still allows people to burn or bury the waste because it is considered more practical without sorting.

Actually how the role of waste banks in reducing the amount of waste that goes into the formal system (urban service) or the amount of waste managed not environmentally friendly. The waste bank becomes a collection point for the communities involved, although not all waste generated is acceptable. In general, waste banks can receive paper, plastic, glass, metal and some other types of waste. However, there is also a waste bank that is more creative in sorting out different types of paper and plastic so as to obtain different types of paper and plastic. With 433 units of waste bank owned by Yogyakarta City, has managed to manage the waste worth recycling as much as 899,801,8 kg/month. There is the recapitulation of waste bank management in Yogyakarta City (Table 1):

**Table 1.** Recapitulation of waste-based on waste bank management in Yogyakarta City

| No | Specification | | Value |
|---|---|---|---|
| 1 | Customers | | |
| | a. Number of customers 433 units waste bank | | 18.434 customers |
| | b. Average customers per a waste bank | | 43 custumers/waste bank |
| 2 | The amount of waste by type | | |
| | a. Plastic | 417.088,9 kg/month | |
| | | 963,254 kg/month/waste bank | |
| | b. Paper | 465.860,7 kg/month | |
| | | 1.075,891 kg/month/waste bank | |
| | c. Glass | 8.989,8 kg/month | |
| | | 20,762 kg/month/waste bank | |
| | d. Metal | 5.601,9 kg/month | |
| | | 12,937 kg/month/waste bank | |
| | e. Others | 2.260,7 kg/month | |
| | | 5,221 kg/month/waste bank | |
| | d. Total | 899.801,8 kg/month | |
| | | 2.078,064 kg/month/waste bank | |

The average amount of waste that can be managed reaches 2,078,064 kg/month per waste bank, without considering other factors, with the data can be projected the amount of waste that can be received by waste banks in Sleman and Bantul Regency. With 34 waste bank units in Sleman Regency and 24 units in Bantul, the waste that can be managed through waste bank program in Sleman Regency reaches 78,966.4 kg/month and 49,873,5 kg/month in Bantul Regency.

The percentage of waste management services in the study area may increase (Fig. 1), because the service is also done by the waste bank. By connecting the potential waste of each region, the waste entering the landfill, the current percentage of services, and the amount of waste that can be managed by the waste bank, it can be calculated the percentage of new level of services with the integration of the waste bank system with a formal management system conducted by the relevant institutions.

The percentage of waste services has increased after being integrated with the waste bank in waste management. The integration pattern becomes important to improve the service to the community so that 100% service access can be achieved immediately.

**Table 2**. Integration of percentage of waste management services between formal and informal sector (waste bank) in the study area

| City/ Regency | Waste Potential (tons/ month) | Landfill Entering Waste (tons/ month) | Level of Service (%) |
|---|---|---|---|
| Yogyakarta | 8.558 | 7.023 | 85 |
| Sleman | 24.398 | 4.488 | 30,71 |
| Bantul | 20.408 | 1.706 | 7,49 |
| **City/ Regency** | **Waste Potential in Waste Bank (tons/ month)** | **Level of Service in Waste Bank (%)** | **New Level of Service (%)** |
| Yogyakarta | 899,8 | 10,5 | 95,5 |
| Sleman | 79,0 | 0,3 | 31,0 |
| Bantul | 49,9 | 0,2 | 7,7 |

## 4 Conclusion

a. The uneven level of formal sector waste management services in the study area is caused by factors such as limited infrastructure, human resources, and funding.
b. A significant increase in community participation in various community-based waste management activities until 2017 amounted to 792 locations and 492 of it as the waste bank (62.5%).
c. The waste bank has an average customer number of 43 customers, with the dominance of paper and plastic waste that is put into the waste bank. As per waste bank can manage as much as 2,078,064 kg of waste/month.
d. The number of the waste bank in Yogyakarta City 433 units can manage waste up to 899,801,8 kg/month, Sleman Regency 34 waste bank units (78.966,4 kg/month) and Bantul Regency 24 waste bank units (49.873,5 kg/month).
e. Integration of percentage of waste management services increased in Yogyakarta from 85% to 95,5%, Sleman Regency from 30,71 to 31%, and Bantul Regency from 7.49 to 7.7%.

## References

1. Asisten Deputi Pengelolaan Sampah, *Buku Profil Bank Sampah Indonesia 2012*. Kementerian Negara Lingkungan Hidup, Jakarta (2012)
2. Deputi Bagian PPL, *Statistik Persampahan Domestik Indonesia Tahun 2008*. Kementerian Negara Lingkungan Hidup, Jakarta (2008)
3. Ditjen Pengelolaan Sampah dan B3, *Kebijakan Nasional tentang Bank Sampah Induk*. Disampaikan di Rakornas Pengelolaan Sampah dan Raker Pengelolaan Sampah, Limbah dan B3. Kementerian Lingkungan Hidup dan Kehutanan. Palembang (2017)
4. H.P. Putra, E. Damanhuri, A. Marzuko, Landfill Mining Prospect in Indonesia. *Proceeding of 3rd Symposium of the Asian Regional Branch of International Waste Working Group*. Seoul National University. Seoul (2017)
5. H.P. Putra, S. E. Saputra, Bank Sampah sebagai Alternatif Model Pengelolaan Sampah Berbasis Masyarakat. *Proceeding National Conference II Forum Wahana Teknologi*, 253-263 (2015)
6. S. Raharjo, T. Matsumoto, T. Ihsan, I. Rachman, Community-based solid waste bank program for municipal solid waste management improvement in Indonesia:a case study of Padang city. Journal of Material Cycles and Waste Management, DOI 10.1007/s10163-015-0401-z (2015)
7. H.P. Putra, E. Damanhuri, Performance and Operational of Landfill Piyungan as The Regional Landfill in Yogyakarta Special Region, Indonesia. *Proceeding of The 9th Asia-Pacific Landfill Symposium*. University of Hongkong. Hongkong (2016)
8. BPS Yogyakarta. *Data Jumlah Penduduk Propinsi Daerah Istimewa Yogyakarta* (2016)
9. S.A. Ahmed, M. Ali, Partnerships for solid waste management in developing countries: linking theories to realities. Habitat International, **28**, 467–479 (2004)
10. E. Damanhuri, A future prospect of municipal solid waste management in Indonesia. Keynote Lecture *5th Asian-Pasific Landfill Symposium di Sapporo*, Jepang (2008)
11. Dirjen Cipta Karya, *Buku pedoman penyelenggaraan pembangunan bidang Cipta Karya*, **Volume 4** : Pengembangan pengelolaan persampahan, Kementeriaan Pekerjaan Umum dan Perumahan Rakyat, Indonesia (2015)
12. C. Meidiana, T. Gamse, Development of Waste Management Practices in Indonesia. *European* Journal of Scientific Research, **Vol.40 No.2**, 199-210 (2010)
13. W.L. Filho, L. Brandli, H. Moora, Kruopiene, A. Stenmarck, Benchmarking approaches and methods in the field of urban waste management. Journal of Cleaner Production, **112**, 4377-4386 (2016)
14. BLH DIY, *Rencana Aksi Sampah Daerah DIY*. Yogyakarta, Indonesia (2014)
15. S.A. Mulasari, A.H. Husodo, N. Muhadjir, Kebijakan pemerintah dalam pengelolaan sampah domestik. *Kesmas, Jurnal Kesehatan Masyarakat Nasional*, **Vol 8 No 8**, 404-410 (2014)
16. Merti Boemi Lestari, *Profil jejaring pengelola sampah mandiri Provinsi DIY*. Badan Lingkungan Hidup Propinsi DIY, Yogyakarta (2011)
17. BLH DIY, *Data pengelola sampah mandiri di Propinsi Yogyakarta Tahun 2015*. Yogyakarta (2015)

18. Faizah, Pengelolaan sampah rumah tangga berbasis masyarakat : studi kasus di Kota Yogyakarta. *Tesis Magister Ilmu Lingkungan*. Universitas Diponegoro. Semarang (2008)
19. M. Medina, Scavenger cooperatives in Asia and Latin America. Resources, Conservation and Recycling, **31**, 51–69(2000)
20. CA. Velis, D.C. Wilson, O. Rocca, S.R. Smith, A. Mavropoulos, C.R. Cheeseman, An analytical framework and tool ('InteRa') for integrating the informal recycling sector in waste and resource management systems in developing countries. Waste Management and Research, **30**, 43-66 (2012)
21. F. Fei, L. Qu, Z. Wen, Y. Xue, H. Zhang, (2016). How to integrate the informal recycling system into municipal solid waste management in developing countries: Based on a China's case in Suzhou urban area. Resources, Conservation and Recycling, **110**, 74-86 (2016)
22. E Sembiring, V. Nitivattananon, Sustainable solid waste management toward an inclusive society: Integration of the informal sector. Resources, Conservation and Recycling Journal, **54**, 802–809 (2010)
23. J.G. Paul, J. Arce-Jaque, N. Ravena, S.P. Villamor, Integration of the informal sector into municipal solid waste management in the Philippines – What does it need?, Waste Management Journal, **32**, 2018-2028 (2012)

# Study of voice data communication using network streams on dataflow programming

*Barlian Henryranu* Prasetio*, *Dahnial* Syauqy, *Rizal* Maulana, and *Gembong Edhi* Setyawan

System and Robotics Lab, Faculty of Computer Science, University of Brawijaya, Indonesia

**Abstract.** In this paper, we implement a voice data communications between sender and receiver through Wi-Fi. The system was designed and implemented using network stream protocol which was developed by National Instruments on Lab-VIEW between two devices. The device which acts as the sender was My-RIO board and the receiver device was Personal Computer. The voice data communication system was tested by observing the effect of various buffer sizes on the throughput, latency and CPU usage. The result showed that the larger the buffer size being used, the throughput values were also getting bigger. However, particular buffer size made the throughput became saturated. The change on the buffer size also directly affects CPU usage. It seemed that the buffer size did not affect latency significantly. It showed that the latency was stable with an average of 13 milliseconds on voice data transmission.

## 1 Introduction

Authentication process is a required part in a smart home system to ensure the safeness of its control system [1]. Set of codes implemented as authentication such as Personal Identification Number (PIN), smart cards and passwords methods are shown to have weaknesses [2]. Authentication system using human voice as a password code replacement has become new alternative in Smart Home [3-4].

Generally, human voice which is used as biometric authentication is captured by microphone embedded into node device. Since node device normally does not have high specs hardware, the captured voice will be sent to processing device which can perform faster computation. Therefore, methods to send the voice data become interesting [5].

The effectiveness of a data communication system depends on four fundamental characteristics, namely packet loss, throughput, delay and jitter [6]. In order to support real time communications between nodes, other than it requires lossless type of communication, it also requires type of communication that has the characteristics of a high throughput. In addition, for a command based system, low latency is also preferable to achieve dependable system [7].

In our design, there are 2 nodes which acts as sender and receiver. The sender nodes captures any voice command and directly send it to the receiver nodes that recognize the command. Since both nodes have already had complex function, CPU utilization in both nodes are also need to be monitored to ensure al function works properly [8].

There are three models of communication that have special characteristics. Network communication model characteristics are shown in Table 1. Table 1 can be used as a reference to determine the model and the network protocol used in the application configuration and system requirements [9]. By analyzing the system requirement, the appropriate type of voice communication is using streamed/ buffered model type.

**Table 1.** Communication model characteristics [9]

| Communication Type | Configuration | Characteristics | Requirement |
|---|---|---|---|
| Message Base | 1:N | Even driven, command | Low Latency, Guaranteed Delivery |
| Process Data | N:1 | Single Point, Current Value | Latest Value Instead of Guaranteed Delivery |
| Streaming/Buffered | 1:1 | Continuous Data Transfer | High Throughput, Guaranteed Delivery |

The data types in communication such as TCP and UDP is formatted in string. It means that the sender must flatten all type of data into a string and the receiver must unflatten the received data from string. Therefore, additional layers need to be added to the complexity of the data transmission of non-string. In this paper, we used Network Stream Protocol in Lab-VIEW which was developed by National Instruments because it can transfer most data types including clusters and arrays. But before implementing on a real machine, we need to test the system performance to see the influence of varying buffer size to the throughput, latency and CPU usage.

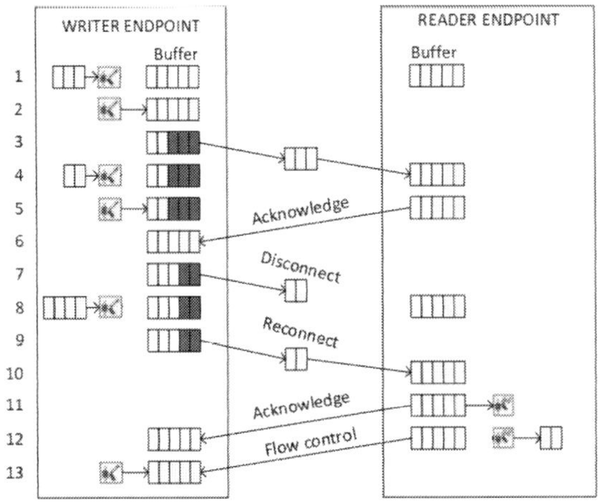


**Fig 1.** Sequence message diagram lossless data communication

## 2 Algorithms

Network stream communication is a point-to-point where the data are maintained until all data is sent or until the stream is disconnected. The data transmissions are obtained through the use of FIFO buffer on the endpoint and through the acknowledgment and flow control on message protocol. The FIFO buffers keep the application that contains the writer endpoint of the data overwrites if the data writing process is faster than the network transfer rate. Message acknowledgment serves to prevent loss of data from the network termination, and message flow control prevents the application that contains the writer endpoint from overwriting the data if the data is generated faster than the data reception by the application that contains endpoint reader. Figure 1 shows a sequence message diagram lossless data communication [10].

## 3 Research Methods

The Lossless communication by using network stream protocol requires two devices that will be used to transmit data to each other, which in this case are My-RIO and a PC. My-RIO is an evaluation board which is made by National Instruments (NI). NI My-RIO provides FPGA integrated with processor which can be programmed using Lab-VIEW. This evaluation board is also equipped with onboard accelerometer, analog and digital I/O, USB port and also audio jack [11].

The voice data is sent from the microphone that is connected to a 3.5mm port on My-RIO board. The voice data input will be perceived as analog data on the single writer section to write in the network streams on My-RIO. Before single writer transmit the voice data, first it requires creating endpoint at the initialization block on each side of the network stream. The initialization was done by setting parameters including the size of the buffer, sender/recipient ID and sender address/recipient data while creating the endpoint. After the initialization is done, the data can be sent from a single write My-RIO towards single read PC using Wi-Fi communications.

The process is done by sending the data directly from single write block using Wi-Fi communication to the single read contained in My-RIO. Once the data is received single read My-RIO, the destroy endpoint blocks will remove all endpoints in My-RIO. To perform the latency test, the received data on the PC need to be sent back to the My-RIO to determine the round trip time two-way communication. Therefore, we need to create Single Write and Single Read block in both devices. Block diagram of network stream system can be seen in Figure 2

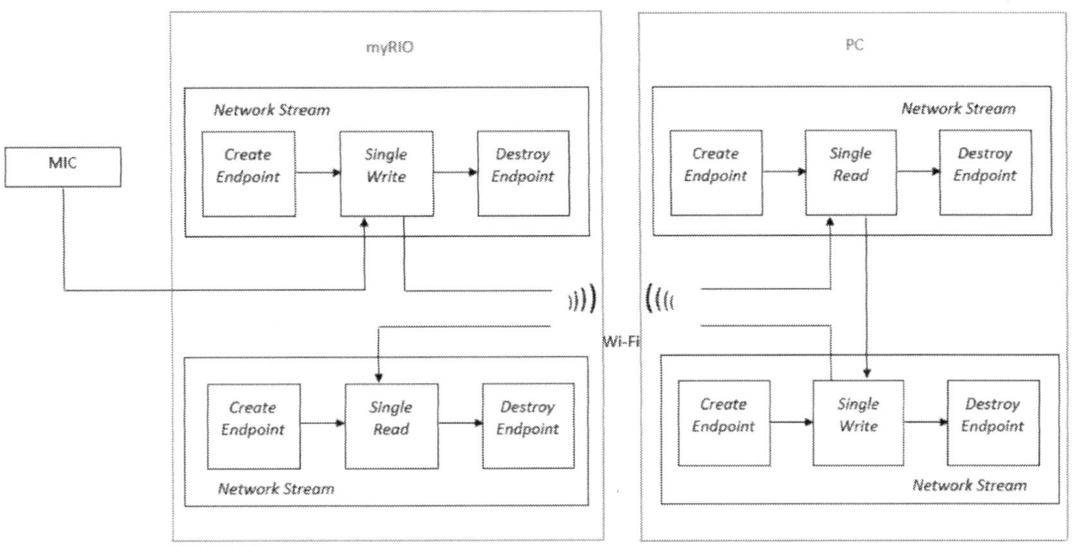


**Fig 2.** Network stream system block diagram

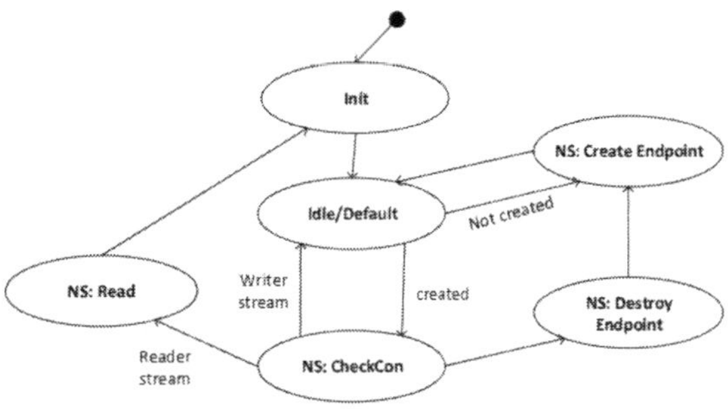


**Fig 3.** Network stream state diagram

The process of connection creation from one host to another is quite complex and involves multiple exchange of packets between two endpoints. Therefore, it requires management connection between endpoints. The state diagram for a network connection process stream can be seen in Figure 3.

The process is started from system initialization. In its default state, the system had an idle status. Then, in idle status, network stream makes connections to the endpoint. If the endpoint has not been established, the network stream will make the first endpoint. After that, the system back to the default status idles. Then, the system will try to make a connection because the endpoint has been made so that network stream check the condition of the destination endpoint. There are 3 possible conditions, i.e. write, read and destroy. When the endpoint is ready to write, the system performs a write stream. When the endpoint destination is ready, then system will read the stream and when the connection is timed out, the system will destroy the endpoint. The endpoint uses a FIFO buffer to receive and transmit data.

The system was implemented in Lab-VIEW 2014. Before implementing on real voice data, first we simulate the system using sensor data generated by onboard sensor. After getting appropriate parameters, we simulate the system using real voice data. Overall, the system was simulated in two phases:

• Simulation of Data Sensor (data z axis accelerometer sensor)
• Simulation of Voice Data (voice from microphone)

## 4 Result and Discussions

Network stream is a lossless, unidirectional, one-to-one communication consisting of a writer and a reader endpoint.

### 4.1 System Modeling

The system block diagram modeling at the sender which was implemented in My-RIO device can be seen in Figure 4. The Figure shows that the system will perform a loop for 60 seconds. This duration will be used to count how many data can be passed within each second. Before the looping process it initialize parameters including buffer size, sender ID and sender address in the endpoint creation block. During the process of looping occurs, the system retrieves the data on the z-axis accelerometer and then writes in a single write block. The data are then transmitted to the Reader Endpoint. The data retrieval process from the accelerometer and data writing on a single write block rely mainly on the processor speed which is being used without implement any delay. After the looping process is complete, the system will flush and destroy the buffer stream.

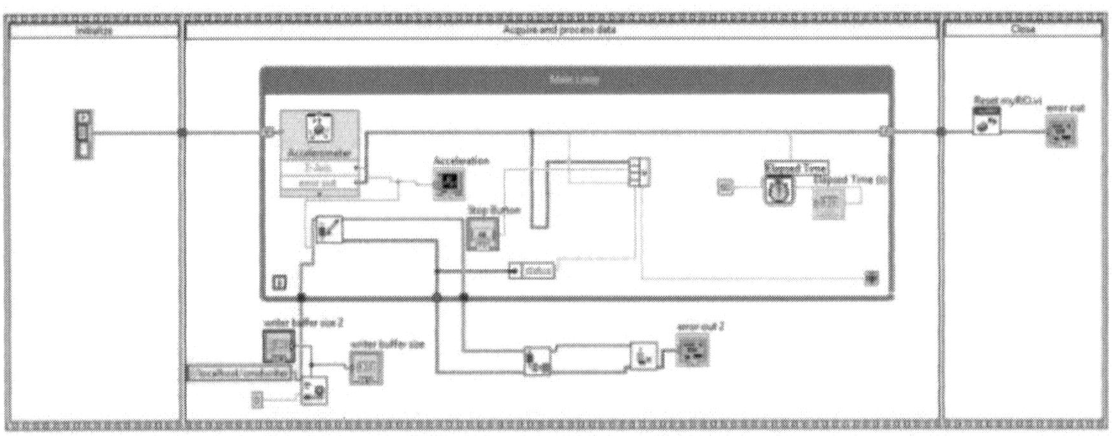

**Fig 4.** Blok diagram modeling of the system at endpoint writer

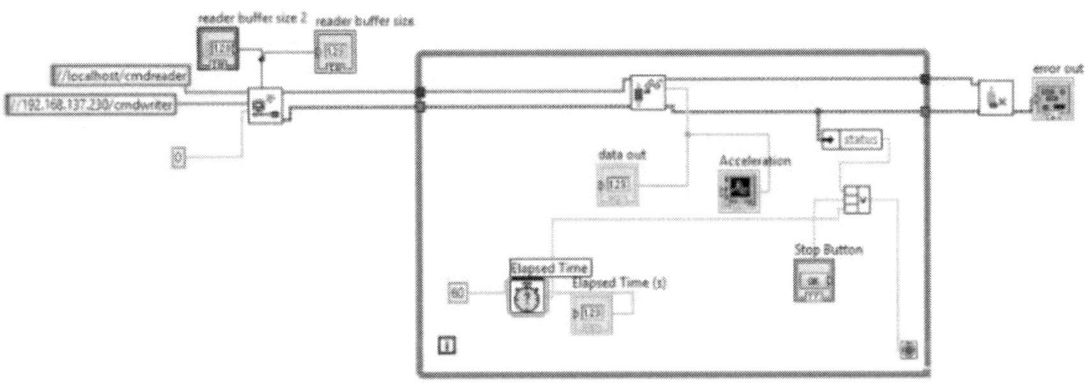


**Fig 5.** The block diagram modeling of the system at endpoint reader

The block diagram modeling of the system at reader endpoint can be seen in Figure 5. The Reader Endpoint initializes the buffer size, sender ID and the sender address. The buffer size of Endpoint Reader must be equal to the size of the buffer in the Endpoint Writer. Similar to the Writer Endpoint, the system will loop for 60 seconds. During the loop run, the system will receive the network stream data from the Writer Endpoint. After the loop is finished, the system will flush and destroy the buffer stream.

## 4.2 Testing Using Sensor Data

The testing focused on throughput, latency and CPU usage while varying the buffer size parameters which is used. In the first phase, we use sensor data as the input, and in the second phase we use real voice data as the input.

The network stream throughput testing aims to determine how much data can be sent in the system every second. The Z-axis accelerometer sensor data contained in My-RIO were sent continuously to the PC for 60 seconds. The z-axis accelerometer data on My-RIO were sent without delay. The data were sent as elements in which one element refers to DBL data type (64 bits). The test was done by giving variation in the buffer size that is used, and observe the effect on the throughput. The throughput testing result can be seen in Figure 6.

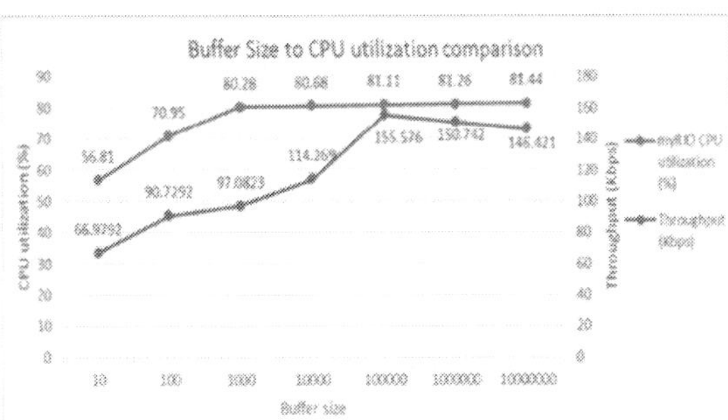


**Fig 6.** The throughput and CPU utilization testing result

Figure 6 shows that the throughput is linearly influenced by the size of the buffer. However, when the size of the buffer is above 100000, the throughput will be saturated despite the enlarged size of the buffer. In addition, it should be considered that by increasing the size of the buffer, and then the CPU usage on My-RIO also will increase.

The Latency test aims to determine the processing time required to transmit a data from My-RIO to the PC. The data packets from My-RIO were sent to a PC, and then after the data were received by the PC, the data were directly sent back to the My-RIO. The data transmission was done as many as 1000 iterations. The round trip transmission was assumed to be similar therefore the result will be divided by 2. The latency testing result can be seen in Figure 7.

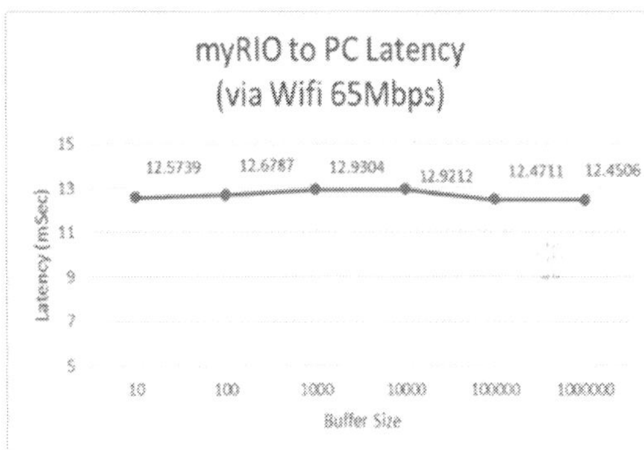


**Fig 7.** The latency testing result

## 4.3 Testing Using Voice Data

The voice was captured by a microphone which was connected to a 3.5 mm port on My-RIO. The voice duration is only set at 3 seconds since the voice is only used as short command. In this model, the buffer size is made permanent with value of 100000 as the results from previous throughput sensor data testing.

Figure 8 shows the front panel for testing the voice data. The voice data recording can be performed after the light turns on. The user is given only 3 seconds to record sound through a microphone. The top graph signal is a voice signal which has been recorded and is ready to be sent My-RIO to the PC. While bottom charts signal is a sound signal that has been sent back from the PC to My-RIO.

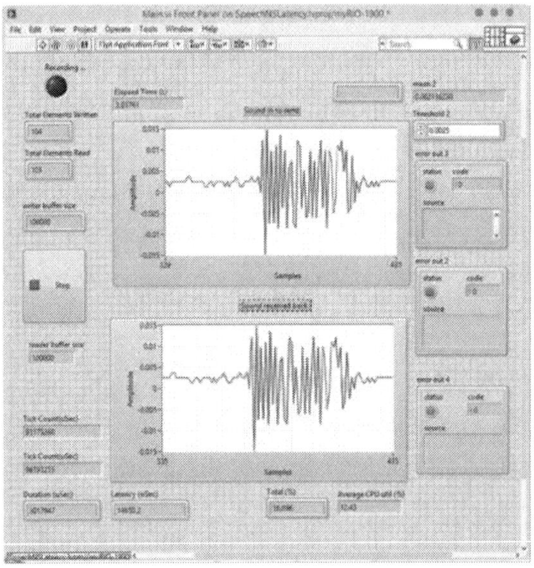

**Fig 8.** The front panel display using voice data

Figure 9 shows that the voice data latency testing is done by using four different commands, with each command is repeated three times. From the results, although there are differences in command, but the value of latency is not much different. The average latency value of this system is 13 milliseconds.

## 5 Conclusion

Network stream protocol supports transmission of all data types without normalizing the data to the intermediate data type. Network streams are designed and used for lossless communication, where data is maintained until all data sent and certainly there is no data transmission failure.

The throughput test is done by sending the data on My-RIO heads to PC while varying the size of the buffer. From the test results, the larger size of the buffer was being used the throughput values are also getting bigger. However, in certain circumstances, although buffer size is enlarged, the throughput is saturated.

The change on the buffer size also directly affects CPU usage. The larger the buffer size, the CPU usage on My-RIO will also increase. So the selection of the buffer size must be adapted to the complexity of the system to be designed.

In Latency test using the sensor data, the data were sent as many as 1000 iterations while varying the size of the buffer used. From the test results, the size of the buffer does not affect the speed of data transmission process. The latency is always stable with an average of 12 milliseconds even though the size of the buffer increases. In the latency test using sound data, the test is done by sending 3 seconds voice duration data using four different short commands and with buffer size of 100000. The result shows that the latency is stable with an average of 13 milliseconds.

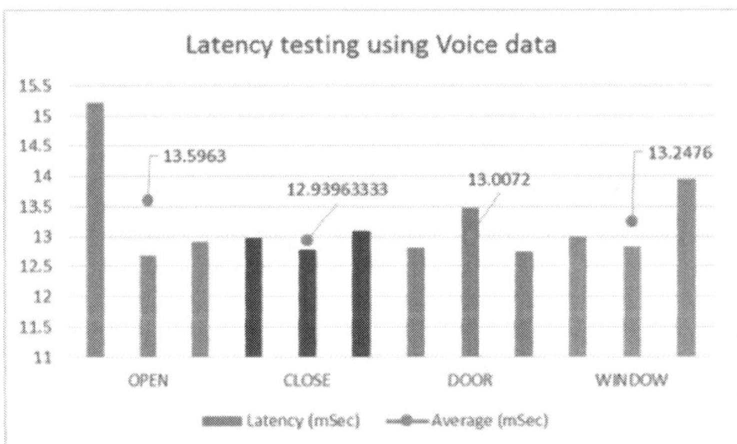


**Fig 9.** The latency voice data testing result

## References

1. S. Chitnis, N. Deshpande, A. Shaligram, An Investigative Study for Smart Home Security: Issues, Challenges and Countermeasures, Wireless Sensor Network, 8, 61-68. (2016)
2. Lawton, Biometrics: A New Era in Security, IEEE Computer, pp.16-18. (1998)
3. O. Joymala, N. Khare, Securing a Smart Home Network using Voice Biometric, International Journal of Application or Innovation in Engineering & Management (IJAIEM), 5:2. (2016)
4. S. Amrutha, S. Aravind, A. Mathew, S. Sugathan, R. Rajasree, and S. Priyalakshmi,Voice Controlled Smart Home, International Journal of Emerging Technology and Advanced Engineering, 5:1. (2015)
5. T. Obaid, H. Rashed, A.A. El Nour, M. Rehan, M.M. Saleh, and M. Tarique, Zigbee based Voice Controlled Wireless Smarthome System, International Journal of Wireless & Mobile Networks (IJWMN), 6:1. (2014)
6. A.F. Behrouz, Data Communication and Networking, USA: McGraw Hill. (2007)
7. T. Hori, S. Araki, T. Yoshioka, M. Fujimoto, S.

Watanabe, T. Oba, A. Ogawa, K. Otsuka, D. Mikami, K. Kinoshita, T. Nakatani, A. Nakamura and J. Yamato, Low-Latency Real-Time Meeting Recognition and Understanding Using Distant Microphones and Omni-Directional Camera, IEEE transactions on audio, speech, and language processing, vol. 20, no. 2, February (2012)

8. S. Nejati, S. Di Alesio, M. Sabetzadeh and L. Briand, Modeling and Analysis of CPU Usage in Safety-Critical Embedded Systems to Support Stress Testing, Model Driven Engineering Languages and Systems pp 759-775 (2012)
9. National Instruments, Using the Right Networking Protocol, http://www.ni.com/white-paper/12079/en/, access on April 15th 2016. (2013)
10. National Instruments, Network Streams Functions, http://zone.ni.com/reference/en-XX/help/371361H-01/lvcomm/ns_networkstreamspalette/, access on April 15th 2016. (2011)
11. National Instruments, NI MyRIO-1900 user guide and specifications, http://www.ni.com/pdf/ manuals/ 376047a.pdf. (2013)

# Disturbance rejection using feed-forward control system on self balancing robot

*Barlian Henryranu* Prasetio* and *Wijaya* Kurniawan

Computer Engineering and Robotics Laboratory-Faculty of Computer Science-University of Brawijaya, Indonesia

**Abstract.** This research implements self-balancing robot using 3 algorithms. There are PID Controller, Ensemble Kalman Filter and Feed-Forward Control system. The PID controller function is as a robot equilibrium control system. The Kalman Ensemble algorithm is used to reduce noise measurement of accelerometer and gyroscope sensors. The PID controller and Ensemble Kalman filter were implemented on self-balancing robot in previous research. In this paper, we added the Feed-Forward controller that serves to detect disturbance derived from the unevenness of the ground. Disturbance is detected using 2 proximity sensors. Base on test results that the system can detect disturbance with an average delay of 2.15 seconds at Kff optimal value is 2.92. Feed-Forward effects result in self-balancing robots increasing power so that the pitch of the robot changes to anticipation of disturbance.

## 1 Introduction

Self-Balancing Robot is an electric self-balancing vehicle with two wheels. Self-Balancing robot is implementing the inverted pendulum algorithm [1]. Self-Balancing Robot uses accelerometer [2] and gyro sensor in the reading position of the tilt [3]. This robot maintains its equilibrium by moving the motors which is connected to the wheels. Self-Balancing Robot is using a controller to control the motor speed and rotary direction. Controller PID is used to maintain the stability of the Self Balancing Robot [4]. In addition, the control system will be given additional filter. The filter function is to reduce the noise measurements from the sensors [5].

In our previous research entitled "Ensemble Kalman Filter and PID Controller Implementation on Self Balancing Robot", we have implemented a Self-Balancing Robot which can be burdened by a human. The system uses a PID controller and ensemble Kalman filter (EnKF) to harmonize and run the robot. Block diagram of the robot system and the design of the robot structure are shown in Figures 1 and 2.

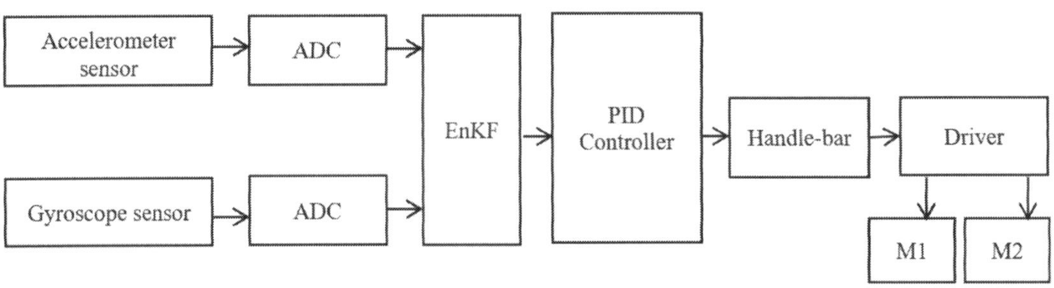


**Fig 1.** Block diagram of the robot system

From the test results, it showed that by varying the process noise covariance matrix, system overshoot can be reduced by regulating the process noise covariance matrix. The resulting data will provide the output for various levels of the slope and speed of the changes experienced by the robot chassis. It is then compared with each cycle of response time and a wide range of power applied to the motor. The expected result is an effective stability in the robot will be easily achieved. If the motor is not operated in a row, there will be a reverse bias on the robot chassis. Incorrect detection by the sensor will also affect the position of the robot. This will cause a delay in responding to the correct position and further enhance the computing time of the robot. The specific parameters which must be observed from the stability of the robot includes a response at various levels of forward and backward acceleration of the robot analyses were used to assess the recorded inertial sensor from time to time in connection with a motor output signal.

* Corresponding author: barlian@ub.ac.id

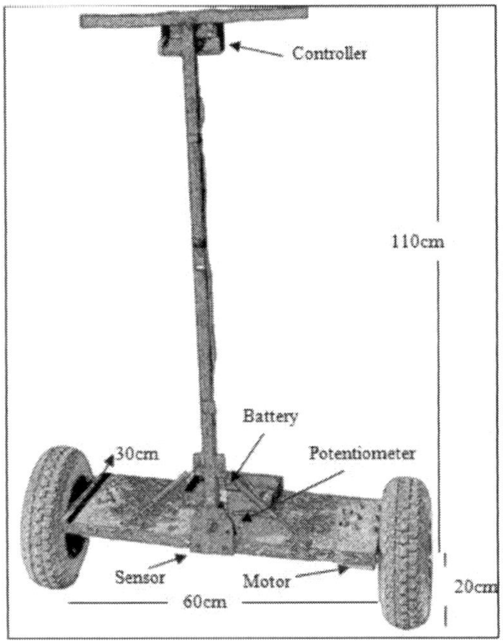


**Fig 2.** The robot structure

From Self Balancing Robot system testing in previous research, the robot cannot predict the disturbance caused by the instability of the area impassable. The instability of this area comes from the street sloping up or downhill. So from the disclosed problems we need a system that can detect the disturbance and make compensation before it led to a change in the value of the controlled variable.

## 2 Feed-Forward Controllers

The Feed-Forward controller is made to cope with the disturbance [6]. Feed-Forward controller is used to overcome the interference from the load. Block diagram of Feed-Back control that comes with Feed-Forward can be seen in Figure 3.

In Figure 3, it can be seen that the disturbance signal is detected by the sensor. The detection result will then be processed by the Feed-Forward computation element to generate the Feed-Forward control signal to be added to the Feed-Back control signal. Total control signal from the Feed-Forward and Feed-Back will then drive the actuator so that the process variable maintained equal to the set point [8]. The transfer function between the process output y and the measured disturbance from Figure 3 is

$$y(s) = \frac{d_c}{1 + PID * p} = \frac{(P_d - pq_{ff})d}{1 + PID * p} \quad (1)$$

Where: $d$ is disturbance

$$pd - pqff = 0 \quad (2)$$

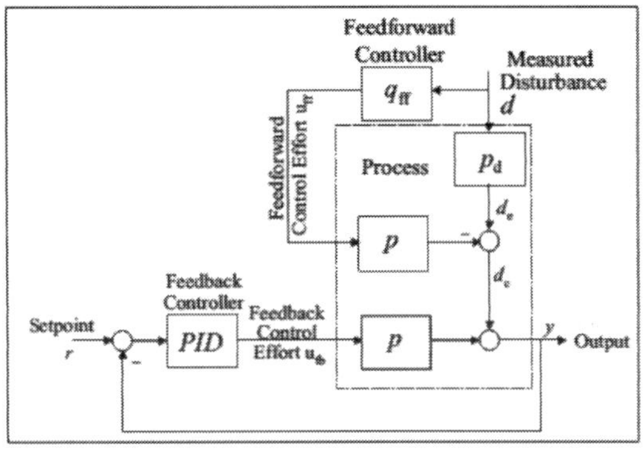


**Fig 3.** Block diagram of Feed-Back/Feed-Forward [7]

## 3 Research Methods

Disturbance signal may result in changes in the value of the process variable that must be processed to get the highest predictive control signal before entering the plant. Disturbance caused by unevenness of the ground is detected by the proximity sensor to observe changes in surface area [9]. Proximity sensor detected values are calculated to determine [10] how much additional value to be added to the self-balancing robot motor power [11] in order to remain balanced when going past the unevenness area in front of it. Block diagram of Feed-Forward on self-balancing robot system is shown in Figure 4

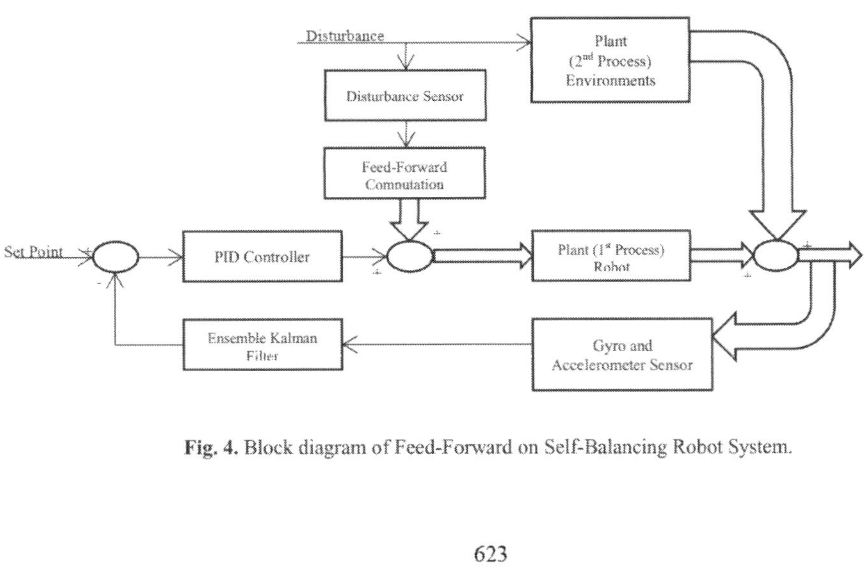


**Fig. 4.** Block diagram of Feed-Forward on Self-Balancing Robot System.

https://doi.org/10.1051/matecconf/201815403002

### 3.1 Feed-Forward Computation

Feed-Forward computation consists of two parts: Dead Time Function (Ttd), Gain and Lead-lag (Kff and Tld, Tlg) [12]. General transfer function of Feed-Forward computation can be seen in Figure 5.

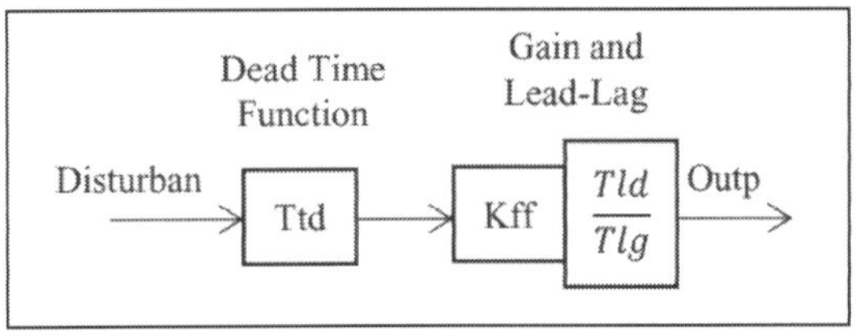


**Fig. 5.** General transfer fFunction Feed-Forward Controller.

Dead Time Function (Ttd) is the delay time between the disturbances with the output controller to dampen the disturbance. Kff is the value of gain of the Feed-Forward controller. This value will be tuned to look for a suitable value in the real system. Lead Time (Tld) is the delay time between output controllers with a process variable output of the plant. Lag Time (Tlg) is the delay time between emergence disturbances with the process variable output of the plant.

In the implementation of Self-Balancing Robot, the system is added with distance sensors where the sensors are installed to detect the unevenness of the ground in the path of the robot trajectory. Then, the experimental measurements were Ttd, Tld and Tlg in the following manner:

1. The Ttd is the time since the detected unevenness of the ground with a Self-Balancing Robot when going up on the uneven ground.
2. The Tld is the time since the output signal from the control until the onset of changes in the balance on a Self-Balancing Robot for the output signal of the control.
3. The Tlg is the time since the change of the balance of the Self-Balancing Robot to the rise of the output signal from the control to respond to changes in the balance.

After the value of Ttd, Tld and Tlg is obtained, then the next one will look for the value of the Kff suitable/appropriate for its real system. The Kff obtained from the tuning process of the Feed-Forward controller.

### 3.2 Gain Tuning Process

The tuning process is processes that aim to know how much Kff value. It is done by trial and error as follows:

1. Setting the Kff value with an initial value, Kff = 1.
2. Once the system is run and sees how much disturbance as well as how much the output value of the controller.
3. If the value of the disturbance is greater than the value of the output of the controller (a> b) as in figure 6a, the value of Kff raised.
4. If the value of the disturbance is less than the output of the controller (a <b) as in Figure 6b, the value of Kff minimized.
5. If the value of the disturbance equal to the output of the controller (a = b) as in Figure 6c, the value is used for Kff.

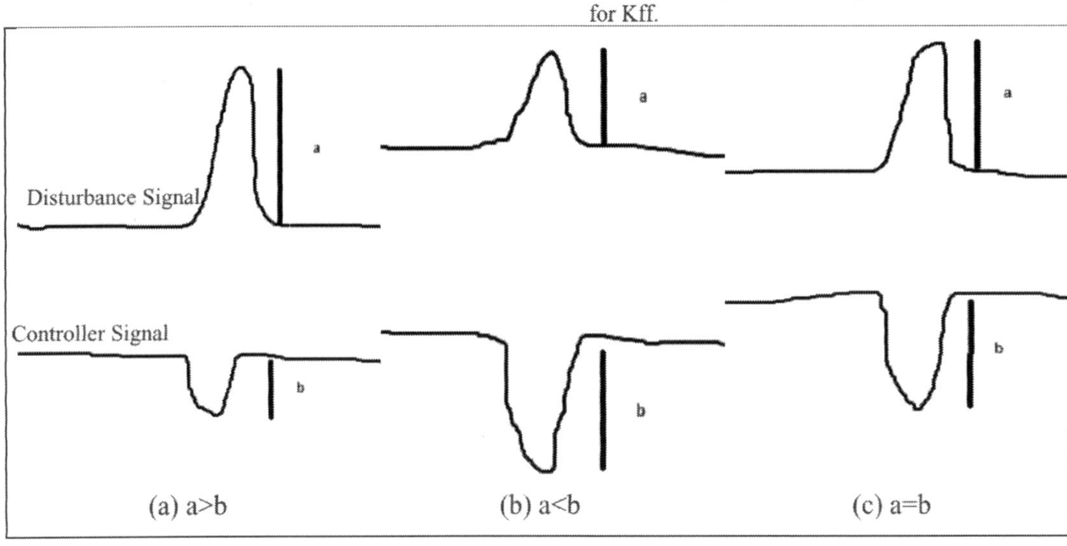


**Fig 6.** The comparisons between disturbance signal and control signal

## 4 Results and Discussion

The system was built using three algorithms. PID controller algorithm is used to maintain the stability of the robot. In addition, the system coupled with Ensemble Kalman Filter (EnKF) algorithm. EnKF is used to eliminate noise measurement of accelerometer sensor and reduce the deviation measurement of a gyro sensor. Feed-Forward controller algorithm used to predict the disturbance in the area of the robot [13].

### 4.1 Algorithm

The proximity sensor function is to detect unevenness of the ground to be traversed by the robot. The sensor detection results processed by the Feed-Forward

https://doi.org/10.1051/matecconf/201815403002

computation element to generate a Feed-Forward control signal and added it to the Feed-Back control signal. The tuning process for Kff value on Feed-Forward flowchart is shows in Figure 7.

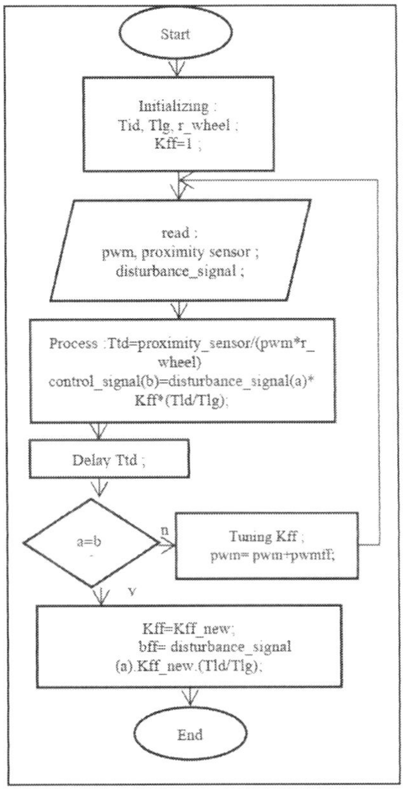


**Fig. 7.** The tuning process for Kff value on Feed-Forward flowchart

## 4.2 Experimental

Self-balancing robot tested in two scenarios. The first scenario, the robot is ascending to the slope of angle $\alpha$. The second Scenario, the robot descending with an inclination of angle $-\alpha$. Illustration of the experimental can be seen in Figure 8.

The disturbance is the angle of the ground. We use two proximity sensors with distance 10cm. When the ground is flat, the proximity sensors detect at 150cm distance. Whereas when the ground is ascending, the proximity sensors detect changes reduced. If the ground is descending, the proximity sensor will detects major changes suddenly. The Feed Forward control signal computation (a) flowchart can be seen in Figure 9.

The test results show that, when disturbance is detected, the speed of the robot changes, which results in the angle of the robot also changing according to the disturbance level. In the experiment, the Kff value is changed until it reaches the optimal value. The optimal of Kff value is 2.92. The relationship between the disturbance signal and the robot pitch can be seen in Figure 10.

Based on figures 10, the self-balancing robot increases the robot's power which is the effect of the Feed-Forward resulting in the pitch of the robot changing as it anticipates disturbance before disturbance occurs. The time delay between the disturbance detected and the robot velocity response is an average of 2.15 seconds.

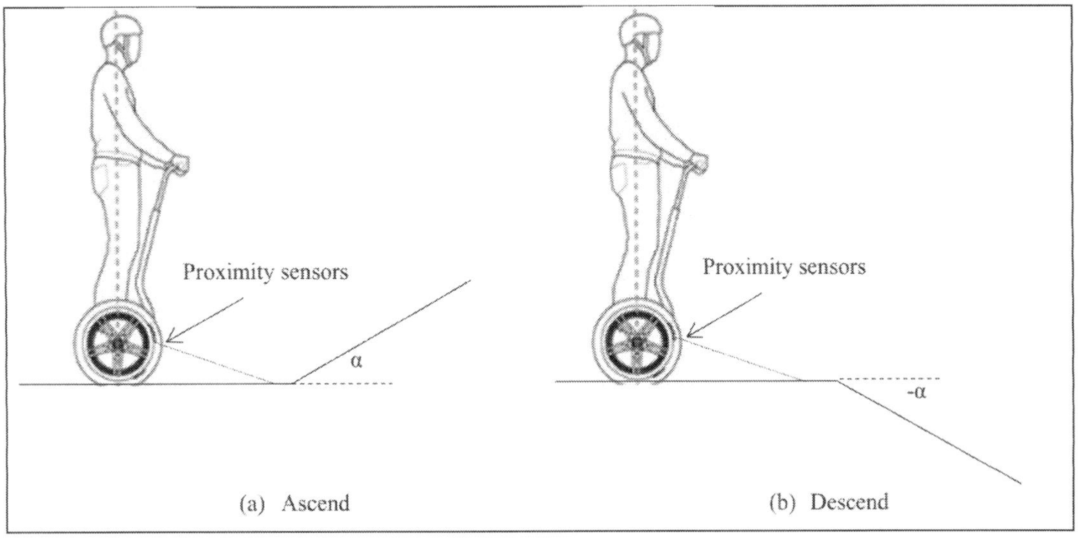


**Fig. 8.** The Experimental Illustration.

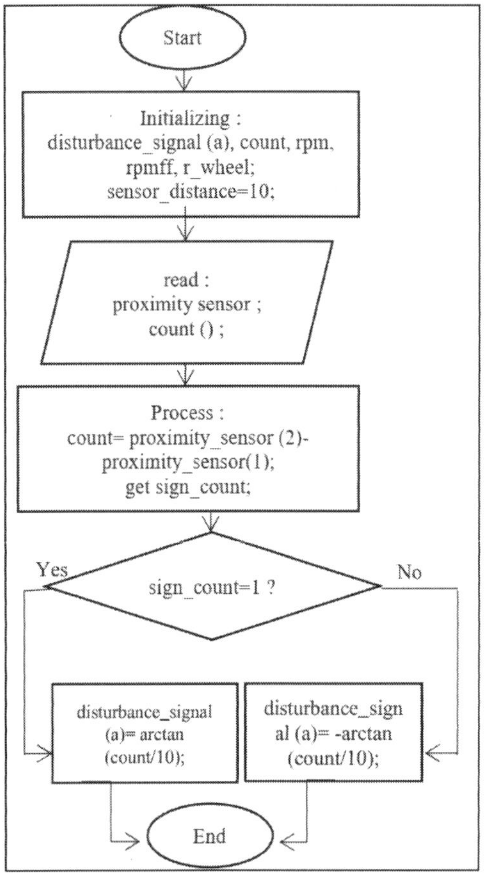


**Fig. 9.** The Feed-Forward control signal computation
(a) Flowchart

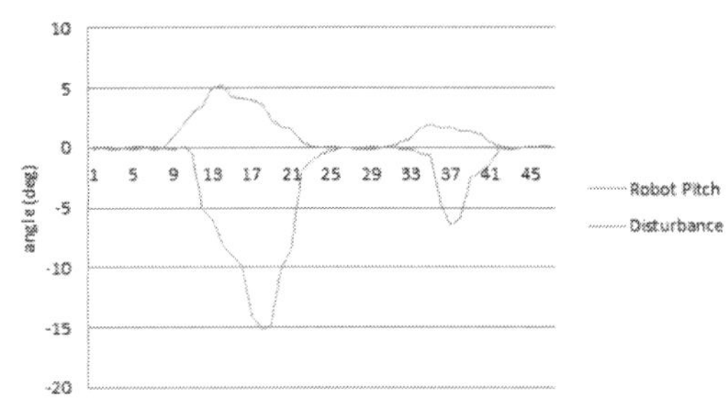


(a) Ascend

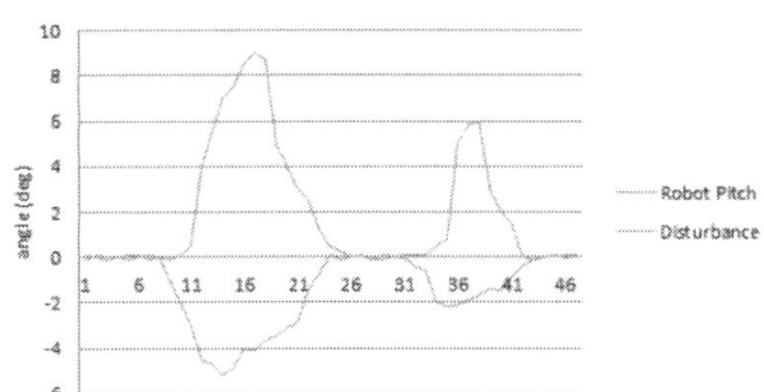


(b) Descend

**Fig 10.** The relationship between disturbance signal and robot pitch

## 5 Conclusions

Self-balancing robot system was using Feed-Forward controller to detect the disturbance. Disturbance is detected using 2 proximity sensors. The disturbance sensor detects the unevenness of the ground by comparing the distance of the two sensors. Testing is done by initializing Kff=1. Base on the test results show the self-balancing robot system can maintain equilibrium by adding or reducing speed with an average delay of 2.15 seconds. The optimal of Kff value is 2.92.

In the control system, the Feed-Forward control is used to remove disturbance in the control process. The self-balancing robot is a system with high disturbance level so it takes a lot of research to eliminate disturbance on this system.

All research activities were performed in Computer Systems and Robotics Laboratory, University of Brawijaya, Indonesia.

## References


1. K. Prakash, K. Thomas, *Study of controllers for a two wheeled self-balancing robot*, IEEE International Conference on Next Generation Intelligent Systems (ICNGIS), 1-7 (2016)
2. N. Guedelha, N. Kuppuswamy, S. Traversano, F. Nori, *Self-calibration of joint offsets for humanoid robots using accelerometer measurements*, IEEE-RAS 16th International Conference on Humanoid Robots (Humanoids), 1233-1238 (2002)
3. A.P. Abraham, S. Ashok, *Gyro-accelerometric SAG analysis and online-monitoring of transmission lines using Line Recon Robot*, Annual IEEE India Conference (INDICON), 1036-1040 (2012)
4. A.M.T. Rahman, S. Ahmad, *Performance comparison between PD-fuzzy and PID controller towards the stability of the extendable double-link two-wheeled mobile robot*, IEEE 10th Asian Control Conference (ASCC), 1-6 (2015)
5. J. Zhang, B. Zhou, S. Wei, Y. Song, *Study on sliding mode trajectory tracking control of mobile robot based on the Kalman filter*, IEEE International Conference on Information and Automation (ICIA), 1195-1199 (2016)
6. H. Hokal, P. Colaneri, L. Del Re, M. Schmidt, M. Paulweber, *Feed forward disturbance rejection by a multiple FIFO approach for transient operation of an engine test bench*, IEEE International Conference on Control Applications (CCA), 65-70 (2013)
7. C. Brosilow, B. Joseph, Techniques of Model-base Control. Prentice Hall PTR. Upper Saddle River, New Jersey, (2002)
8. C. Sagert, M. Walter, O. Sawodny, *DC/DC converter control for voltage ripple reduction in electric vehicles*, IEEE International Conference on Advanced Intelligent Mechatronics (AIM), 1442-1447 (2016)
9. K. Tanaka, H. Ishii, Y. Okamoto, etal, *Novel method of estimating surface condition for tiny mobile robot to improve locomotion performance*, IEEE/RSJ International Conference on Intelligent Robots and Systems (IROS), 6515-6520 (2015)
10. M.A. Marwah, M.E. Khaled, M.A. Abrar, *Development of efficient obstacle avoidance and line following mobile robot with the integration of fuzzy logic system in static and dynamic environments*, IEEE Long Island Systems, Applications and Technology Conference (LISAT), 1-6 (2016)
11. M. Barzegaran, H. Zargarzadeh, O.A. Mohammad, Wireless Power Transfer for Electric Vehicle using an Adaptive Robot, IEEE Transactions on Magnetics, **53**, 6 (2017)
12. A.A. Carlos, Automated Countinous Process Control, A Wiley-Interscience Publication. John Wiley & Son, inc, (2002)
13. A.A. Wardana, T. Takaki, T. Aoyama, I. Ishii, *Development of a Control System for a Stair-Climbing Inverted Pendulum Robot*, IEEE International Conference on Instrumentation, Control and Automation (ICA) Institut Teknologi Bandung(ITB), Bandung, Indonesia, 1-5 (2016)

# Interaction design of automatic faucet for standard hand-wash

*Hurriyatul* Fitriyah*, *Edita Rosana* Widasari, *Eko* Setiawan, and *Brian Angga* Kusuma

Dept. of Computer Engineering, Faculty of Computer Science, University of Brawijaya, Indonesia

**Abstract.** In hand-washing practice, people tend to forget to use soap and scrub for 20 seconds as stated in the standard. This paper attempts to develop an automatic faucet that is user-friendly and easy to plug in regular water pipe for standard hand-washing routine. An Interaction Design Process that aims to maximize product usability was applied during the development. The faucet utilized IR proximity sensor to detect presence of hands that would automatically commence the hand-washing process. First, it exited water and soap simultaneously to force users to use soap. The scrubbing duration was marked by buzzer's sound and LED's light. The validation test showed health practitioner agreed the faucet facilitates a standard hand-washing. The user test in 30 participants showed users used soap and scrubbed in the exact duration. The usability questionnaire filled by participants showed they were strongly agreed for its usefulness and agreed for its satisfaction and easiness.

## 1 Introduction

Hand hygiene is a major requirement for human health. Many infectious diseases can be emerged if proper hand hygiene procedures are not implemented [1, 2]. The infectious diseases, which are caused by hand non-hygiene, are viruses (hepatitis A, polio, diarrhoea), bacteria (typhoid, fever, dysentery), and parasites (amoeba dysentery, enterobijaza) [3]. Furthermore, the impact of hand non-hygiene can increase pandemic public health such as severe acute respiratory syndrome and avian influenza in the community [4]. Many studies have also been reported an association between improvements in hand hygiene and reductions in rates of infectious illness. The Medline database was searched from January 1980 to June 2001 shows the relation between hygiene practices and relative reduction in risk of illness is greater than 20% [5, 6].

Hand washing is the simplest, important and cost-effective way to improve hand hygiene in health care and support the prevention of infectious disease. The WHO standard obligates people for hand washing with non-antibacterial soap and water. The duration ranged on average as short as 20 seconds to 30 seconds, including rub the backs of hands, wrists, between fingers and under fingernails [1, 7]. Hand washing with soap and water is more effective for the removal of bacteria than washing with water alone to 23% [8].

Studies show that the compliance of hand-washing procedures by healthcare workers, householder and officer are only about 40%, and then barely reaches 50% even after various educational trainings or interventions [9]. A study also showed that the average duration of scrubbing was less than 15 seconds in the community [10].

Interaction design process aims to maximize usability of a product. It focuses not only on functionality, but user-friendliness of a product as well. It consists of 5 steps which are: (1) What is wanted, (2) Analysis, (3) Design, (4) Prototype, (6) Implementation and Deployment [11]. Analysis-Design-Prototype is iterative process, where the product kept redesigned until the product requirements are met. The Interaction design process had been applied widely not only to design a software application but computer hardware as well [12, 13].

## 2 Related Work

In market, automatic faucets that do not necessitate users to turn the faucet's lever to exit water are common. It uses proximity sensor that mostly based on infrared (IR) light to detect presence of hands. Several automatic faucets to exit soap also have used the similar usage of proximity sensor in detecting hands.

In term of designing automatic system for hand-washing standard procedure, [14] has developed a machine that accommodate wetting, lathering, scrubbing, rinsing and drying in automatic mode. The machine consisted of a sink and 3 faucets of water, soap, air blower that was controlled electronically. The hand-wash process used timer to exit soap, then water and then air blow. The design also included mechanical component of water piping, pump and housing. Other patented design entitled Automatic Hand Washing and Drying Machine were also consisted of water-soap-dryer [15]. The machine design included electronic components and mechanical piping and pump.

Both designs were in a form of integrated automatic basin. Both machines were required to be assembled as a

* Corresponding author: hfitriyah@ub.ac.id

complete basin system. In market, such automatic basins were available as well.

Instead of developing a complete basin for standard hand-wash, this paper aim to develop an automatic hand-wash faucet, which can be used simply by replacing the standard faucet with the developed one. It was designed for simpler solution of performing hand-wash standard procedure instead of buying the whole housing of hand-wash machine.

## 3 Methodology

Development of the automatic faucet used Interaction Design process to carefully consider the usability of the product that meet the standard hand-wash procedure as well.

### 3.1 What is Wanted

The very first step in interaction design is "What is Wanted". The developer must first enlist the exact definition or information of the product. The definition consists of product's functionality, environment, users or any expectation that should be met by the product.

- Definition of Product: (1) The faucet is equipped with soap, (2) It is a free-hand operation, (3) It must obey the hand-washing standard procedure.
- Definition of Environment: A wet environment, hence electronically components must be protected.
- Definition of Users: General users

There are two important notes from the hand-washing procedure that must be obeyed in designing the automatic faucet, which are:

#### *3.1.1 The automatic faucet force users to use soap*

Washing hands with soap is compulsory. Although it is not necessarily an antibacterial soap, a regular soap is required since the surfactants in soap could lift soil and microbes from skin. Using antibacterial soap might even lead to development of bacteria that are resistant to the product's antimicrobial agents [16]. Due to low percentage of users that use soap in their hand-washing routine, hence the automatic faucet must be designed to force the users to use soap with no exception.

#### *3.1.2 The automatic faucet force must guide users for exact 20 seconds duration of scrubbing*

The 20 seconds duration of scrubbing is important. CDC mentioned that the duration was necessarily required to kill bacteria in hands. Hence, the automatic faucet must be designed to necessitate the users to scrub in the exact duration.

### 3.2 Analysis

The "Analysis" step is a formulation of the "What is Wanted". It comprises structured Scenario and Task Analysis. Scenario is a story that contains visual description of how users would interact with the developed product. Task Analysis is a structured task decomposition of how users would perform the task using the developed product in a chronological order.

#### *3.2.1 Scenario*

The story of how a user would wash their hands using the developed automatic faucet is descripted below:

"Adam was about to having a dinner. His hand was dirty since he has spent his afternoon for gardening in his backyard. He then walk into his fully automatic hand-washing faucet and put his hand beneath. The presence of his hand automatically makes water ran from the end of the faucet along with soap. Realising there is soap in his hand, he has no choice than to wash his hands using the soap, the thing he usually forgot to use. He then lathered his hand and scrubs it few seconds until the faucet released water automatically for once more. He rinsed the soap from his hand and dried his hand with towel"

#### *3.2.2 Task analysis*

From the description in the "Scenario", the task of washing hands using the automatic faucet was structured in hierarchy. The hierarchy is named Hierarchical Task Analysis (HTA) that contains task decomposition in chronological orders. The HTA of the developed automatic faucet is composed below:

0. Washing hand with soap using the automatic faucet
1. Put hands under the faucet's end to exit water and soap simultaneously
2. Lather and scrub the hands with soap until the water run once more
3. Rinse the hands thoroughly

**Plan 0:** do 1 – 2 – 3 in that order

### 3.3 Design

The "Design" process consists of designing the product's System and Applying Design Principle. In this paper, the System includes Structure, Hardware Design and Housing Design. The Design principle was applied as a checklist of design characteristic that is advised to be fulfilled to maximize product's usability.

#### *3.3.1 Structure*

To perform the task as mentioned in the HTA above, the system of automatic faucet was designed to run in chronological order as well. The product's structures are: (1) Exit water & soap, (2) Stop water & soap, (3) Wait 20 seconds for scrubbing, (4) Exit water for rinsing, (5) Stop water.

#### *3.3.2 Hardware design*

The product's hardware was designed according to the 5 structures of the system. It was controlled electronically. Its operation was initiated by the presence of hands.

Proximity sensors are common sensor used for the role. An infrared-based proximity sensor was used in this paper since it would strictly detect the motion of human's hands. Once the hands are detected, the sensor sent signal to a microcontroller to start the opening of valves by activating a motor. At first, the valve would open two inlets, which are water's inlet and soap's inlet. This opening would make the faucet exited water and soap simultaneously. The water and soap were programmed to be exited for 3 seconds, a sufficient duration to wet the hands and lather the soap thoroughly.

The product then proceeds for scrubbing phase. The microcontroller instructed the motor to close the inlets of water and soap to stop the water and soap from running down the faucet. The closing is programmed for 20 seconds, the exact duration necessarily required for scrubbing according to the hand-washing procedure. A buzzer and LED lamp was activated during this process to make users intuitively start the scrubbing. It programmed to beep and flashed in accelerated intervals giving users guidance when the duration was about to end.

The product then proceeds for rinsing phase. The microcontroller instructed the motor to open the water's inlet only. The opening was programmed for 10 seconds, a sufficient duration for cleaning the soap form hands thoroughly. The Flow Diagram of the automatic faucet is shown in Figure 1.

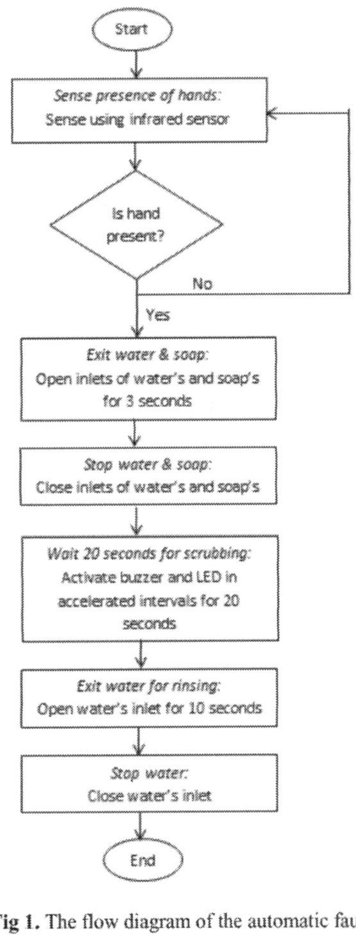


**Fig 1.** The flow diagram of the automatic faucet

From the Flow Diagram in Figure 1, the product's hardware consisted of IR proximity sensor as input, a microcontroller as controller, buzzer and motor as outputs. The block diagram is shown in Figure 2.

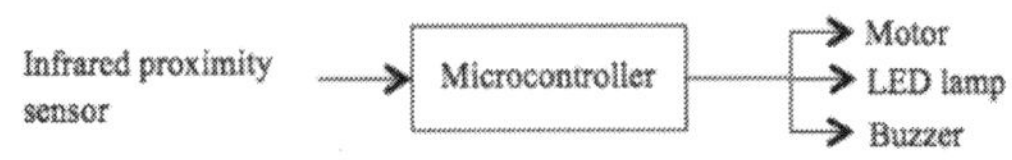


**Fig 2.** Block diagram of the automatic faucet's hardware

### *3.3.3 Housing design*

From the Hardware Design, the automatic faucet required 6 components of Housing, which are: (1) Soap container, (2), Water inlet, (3) Box for Microprocessor, (4) Box for motor, (5) Sliding lid to cover water and soap's inlet, (6) Water Outlet. Each component followed below requirements:

- Soap container: Tube shape. Soap outlet at the bottom. Refillable
- Water inlet: Standard connection size and type with water pipe
- Box for microprocessor: Waterproof to protect the electronics component
- Box for motor: Waterproof to protect the electronics component
- Sliding lid: Leak-proof material to cover 2 inlets, water and soap. The lid was full-opened to exit water and soap and half-opened to exit water only
- Water outlet: Standard pipe that run water or soap from top to bottom of the tip.

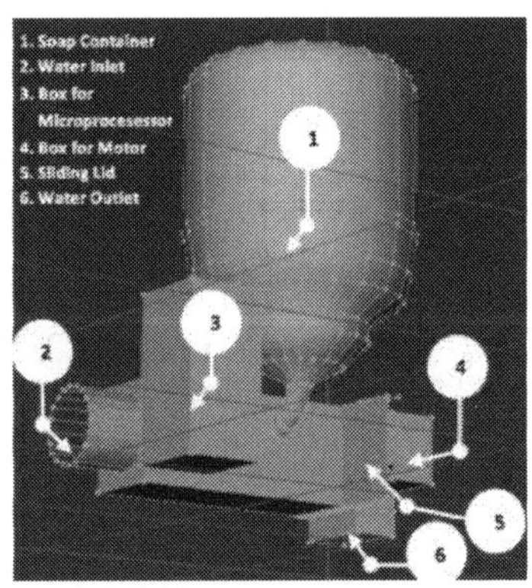


**Fig 3.** The housing design of automatic faucet

The components were set in a configuration as shown in Figure 3. The motor was placed close to the sliding lid, where it would open and close the passage of water and soap. The microcontroller was hidden in the back.

### *3.3.4 Applying design principle*

The Design Principles consist of Learnability, Flexibility and Robustness. Learnability, which consists of Predictability, Synthesizability, Familiarity, Generality, Consistency, is about designing an easy and intuitive operation of the product. Flexibility, which consists of Dialogue Initiative, Task Migratability, Multi-Threading, Subtitutivity and Customizability, is about designing a friendly and satisfied experience when using the product. Robustness, which consists of Observability, Responsiveness, Task Conformance, Recoverability, is about designing a useful product that also able to handle human's error. These principles are applied in designing the interaction of developed automatic faucet.

*a) Learnability*

- Predictability: The developed faucet did not equipped with lever similar to other automatic faucet; hence users intuitively put their hands under the faucet's tip. An upturned hands icon was also placed on the tip for guiding users to put their hands beneath it to activate the automatic faucet. Once the hand had been detected, the faucet exited water and soap simultaneously. Seeing soap in their hands, users could easily predict the next action, which was scrubbing their hands with it.
- Synthesizability: The water and soap was designed to come out once the hands were detected; hence users able to correlate the faucet would commence the hand-washing process by placing their hands beneath the faucet's tip.
- Familiarity: The developed faucet had icon of upturned hands on its tip as guidance for users to intuitively put their hands beneath it. This simple hands symbol in upturned position is easily understandable by users. The hands icons were also accompanied by water-drop and bubble icon for users to expect the water and soap to come out of the tip. The symbol is shown in Figure 4.

**Fig 4.** The icon of water drop and bubble

- Generality: The developed faucet shared general shape with common faucets. It had tip where users expect waters to come out from. It used standard water-pipe connector. The soap container shape was also a common tube where the refill access was put upside down.
- Consistency: Every hand-washing process in the developed faucet worked consistently based on timer. No manual operation was needed in the process.

*b) Flexibility*

- Dialogue Initiative: The developed faucet was designed for fully automatic and standardized operation; hence the faucet led the interaction. Users only initiate the interaction by placing their hands beneath the faucet's tip. This dialogue arrangement was designed for operation simplicity.
- Task Migratability: The developed faucet was designed for fully automatic operation; hence users do not have to perform a task of turning a lever to run the water and pressing the soap pump to run the soap such in common faucet's operation. Users also does not have to count for exact 20 second or singing "Happy Birthday" song as suggested in scrubbing phase, since the faucet provided 20-seconds-waiting for the process accompanied with buzzer's beep and LED's flash light.
- Multi-threading: The developed faucet was designed for simple hand-washing in standard procedure, hence no multi-threading is needed.
- Subtitutivity: The developed faucet works on common behaviour similar to other automatic faucets. The input is simply from the presence of hands only.
- Customizability: The developed faucet was designed for standardized hand washing. It is also utilized for shared-using. A customization based on users' preference was not necessary.

*c) Robustness*

- Observability: During the scrubbing process, users may observe the timing by listening to buzzer's beep that was programmed to sounds in accelerated fashion. An LED lamp was also programmed to flash synchronously to the buzzer's beep. Both buzzer's beep and LED's flash was applied to intuitively guide the users to scrub their hands in exactly 20 seconds. This feedback was also a guarantee to users that the faucet was working a progress and not just silently broken in the 20 seconds duration.
- Responsiveness: For fast response, the developed faucet's program used timer. Developer determined the 3 seconds duration of wetting-leathering and 10 seconds duration of rinsing based on trials.
- Task Conformance: The developed faucets worked as intended, which to assist users to perform a hand washing in standardize and suggested procedure.
- Recoverability: Although a reset button was a good tool for a user to re-do the operation when previous user abandoned the faucet in the middle of process, it was not provided in the faucet for two reasons. First reason was that the process only last for less than 1 minute. Second reason was that adding a reset button would burden the responsiveness due to hardware and software complexity.

*d) Prototype*

A prototype of the faucet was developed using an Adjustable IR Reflection Sensor as input to detect presence of hands; Microcontroller AT8353 as controller; DC motor, LED lamp and 6V round electrical buzzer as outputs. These components were connected as shown in Figure 5. Once the hands were detected by IR sensor, it sent digital signal 1 to the microcontroller. The microcontroller then started the program to control motor, LED lamp and buzzer according to the designed Block Diagram in Figure 1.

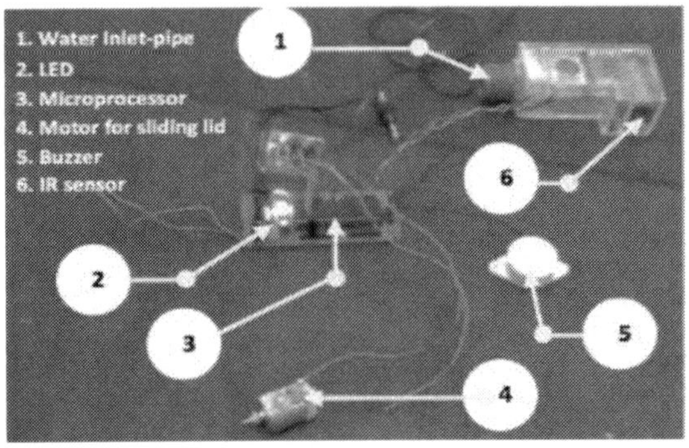


**Fig 5.** Prototype of the automatic faucet

The electrical circuit were then integrated with the designed Housing in Figure 3 to result an automatic faucet as shown in Figure 6. The case was made using acrylic as it is cheap and easy material to cut and glued. The water's inlet used a standard connector of PVC pipe. The soap container used a plastic bottle. The sliding lid was also made using acrylic, which was sealed by rubber for leak-proof guard.

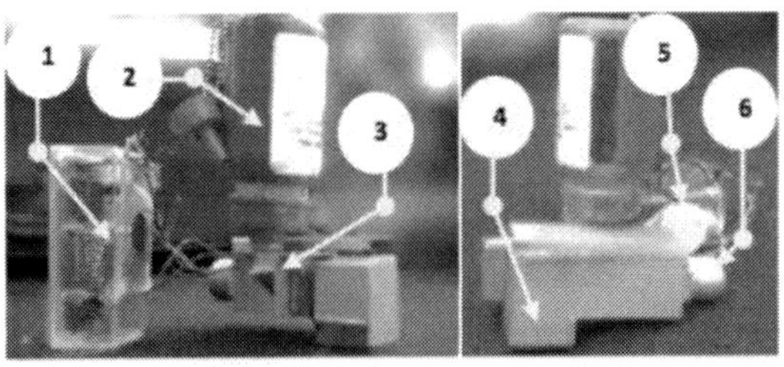


1. Microprocessor Box; 2. Soap Container; 3. Sliding Lid
4. Outlet pipe; 5. Buzzer; 6. Water-inlet pipe

**Fig 6.** The integrated case of an automatic faucet

*e) Implementation and development*

The Implementation and Deployment step includes designing Product Documentation and User Support. In this paper, both were bundled in a form of User's Manual Book. It consisted of:

- Installation Guide: Instruction on assembling the water inlet to water pipe
- Device Layout: Diagrammatic layout of device's part
- Operation Manual: Step by step instruction accompanied with pictures on how to perform hand-wash using the developed automatic faucet.
- Maintenance: Advice on caring the water inlet and soap outlet
- Troubleshooting: What-to-do instruction on handling jammed motor and microcontroller

## 4 Testing Results and Discussion

Three tests are performed on the developed automatic faucet, which are Verification, Validation and User-Participation Test.

### 4.1 Verification on faucet functionality

The verification on faucet functionality was performed to find the performance of: (1) Distance of adjustable IR reflection sensor. The result in Table 1 shows the minimum distance that can be detected by the sensor was 3 cm and the maximum distance was 80 cm. This distance was reasonable distance to detect presence of human hands beneath the automatic faucet's tip. (2) DC Motor connectivity. The result in Table 2 shows the DC motor could drive the sliding lid to fully open, half open and fully closed the inlets of water and soap. (3) Microcontroller to buzzer and LED lamp connectivity. The Microcontroller was able to function the Buzzer and LED in accelerated fashion within 20 seconds.

**Table 1.** Distance of adjustable IR refection sensor.

| Distance (cm) | LED Lamp (as Indicator) |
|---|---|
| 3 | ON |
| >80 | OFF |

**Table 2.** DC motor connectivity.

| Variable | Movement | Result |
|---|---|---|
| Full Open | DC motor drived the sliding lid in full rotation counter-clockwise | The automatic faucet exited water and soap simultaneously |
| Half Open | DC motor drived the sliding lid in half rotation counter-clockwise | The automatic faucet exited water only |
| Full Close | DC motor drived the sliding lid in full rotation clockwise | The automatic faucet stop water and soap from running |

### 4.2 Validation by health practitioner

In Validation Test, Health practitioner was asked to perform hand-wash using the developed automatic faucet. The result showed that it was conformed the hand-wash standard procedure which emphasizes usage of soap and scrubbing duration of 20 seconds. Figure 7 shows a Health Practitioner used the faucet for hand-washing.

### 4.3 User-participation test

The User-Participation Test consisted of two test, which were: (1) Observation on users' execution time of hand-washing using conventional faucet with separated soap and the developed automatic faucet; (2) Usability Evaluation by users using USE Questionnaire. Total 30 respondents range from teenager to elderly people was gathered in correlation with the definition of users in "What is Wanted".

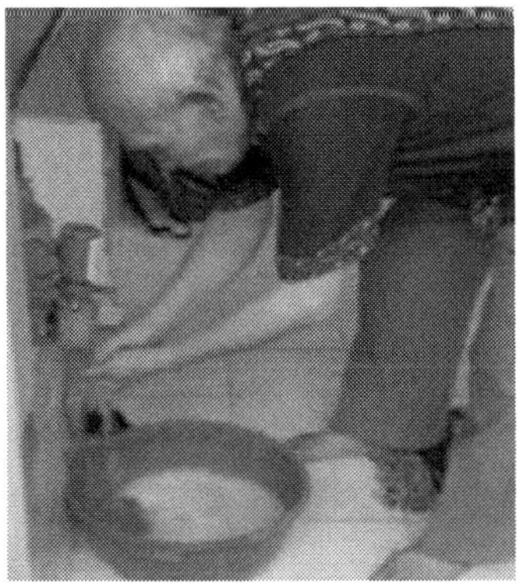

**Fig 7.** A healthcare practitioner used the automatic faucet for hand-washing

#### *4.3.1 Observation on user's execution time*

In the test, users where asked to wash their hands using a conventional faucet and then the automatic faucet. Since there are no guidance for exact duration of scrubbing in the conventional faucet, users scrubbed their hands varied in 4 to 44 seconds (average 15.4 seconds). This duration is less than the standard. Whilst, the automatic faucet guided the users to scrub their hands in exactly 20 seconds in accordance with the standard rules. The automatic faucet were also successfully forced the users to wash their hands with soap. They also directly scrubbed their hands with the soap intuitively.

#### *4.3.2 Usability evaluation using USE questioner*

USE questionnaire is generally used to evaluate user attitudes towards a variety of consumer products using three dimensions: Usefulness, Satisfaction, and Easiness (Ease of Use and Ease of Learn) [17]. It consists of 30 questions where users answer using Likert Scale of 1 to 7. This scale is often found in the questionnaire to facilitate the respondents to give answers in structured and limited options [18]. The scale used in this test is a scale of 1 (strongly disagree), 2 (disagree), 3 (rather disagree), 4 (doubtful), 5 (rather agree), 6 (agree), and 7 (strongly agree). A statistical parameter of mode is commonly used to analyse categorical data from Likert scale as used in the USE questionnaire.

The result shows a mode of grade 7 for Usefulness, grades 6 for Satisfaction, grades 6 for Easy to Use, and grade 6 for Easy to Learn. The mode of all usability criteria is grade 6 which mean users "agree" that the developed automatic faucet has maximized its usability.

## 5 Conclusion

In this paper, a fully automatic faucet for hand-washing has been developed using Interaction Design process to maximize its usability. The analysis on Verification, Validation and User-Participation test has shown that it functions well to force users to use soap in their hand-washing routine and to scrub in 20 seconds as well as excel the usability in terms of Usefulness, Satisfaction and Easiness. Further development should focused on optimizing the hardware, housing and electrical source. An addition of dryer should also be designed to meet the hand-washing standard that includes drying activity.

## References

1. D. Pittlet, WHO Guidelines on Hand Hygiene in Helath Care: a Summary. World Health Organization Patient Safety: University of Geneva Hospitals (2009)
2. A. E. Aiello, R. B. Coulborn, V. Perez, and E. L. Larson, *Effect oh Hand Hygiene on Infectious Disease Risk in the Community Setting: A-Meta-Analysis*, in American Journal of Public Health, **98**(8), 1372-1381 (2008)
3. M. Lovely, Disease of Dirty Hands. Medicine Adviser: your health (2013)
4. E. N. Perencevich, M. T. Wong, and A. D. Harris, *National and Regional Assessmnt of the Antibacterial Soap Market: A Step Toward Determining in the Impact of Prevalent Antibacterial Soap*, In American Journal of Infection Control, **29**(5), 281-283, (2001)
5. A. E. Aiello and E. L. Larson, *What is the Evidence for a Causal Link between Hygiene and Infections*, In Lancet Infectious Diseases, **2**(2), 103-110, (2002).
6. K. B. Kamm, D. R. Feikin, G. M. Bigogo, G. Aol, M. M. Shah, J. Yu, R. F. Breiman, and P. K. Ram, *Associations between Presence of Handwashing Stations and Soap in the Home and Diarrhoea and Resiratory Illness, in Children Less than Five Years Old in Rural Western Kenya*, In Tropical Medicine and International Health Journal, **19**(4), 398-406, (2014)
7. Mayo Clinic Staff, Hand-washing: Do's and Don'ts. Healthy Lifestyle Adulth Health (2014)
8. M. Burton, E. Cobb, P. Donachie, G. Judah, V. Curtis and W. P. Schmidt, *The Effect of Handwashing with Water or Soap on Bacterial Contamination of hands*, In International Journal of Environmental Research and Public Health, **8**(1), 97-104 (2011)
9. V. Boscart, K. McGilton, A. Levchenko, G. Hufton, P. Holliday, and G. Fernie, *Acceptability of a wearable hand hygiene device with monitoring capabilities*, Journal of Hospital Infection, **70**, 216-222 (2008)
10. J. M. Boyce and D. Pittet, Guideline for Hand Hygiene in Health-Care Settings: Recommendaions of the Healthcare Infection Control Practies Advisory Comitte and the HICPAC/SHEA/APIC/IDSA Hand Hygiene Task Force. Department of Health and Human Services: Morbidity and Mortality Weekly Report, USA (2002)
11. A. Dix, F. Janet, D. Gregory . B. A. Rusell, Human Computer Interaction, 3rd ed., Haddigton: Scotprint (2004)
12. J. H. Goldberg and X. P. Kotval, *Computer Interface Evaluation using Eye Movement: Methods and Constructs*, in International Journal of Industrial Ergonomics, **24**(6), 631-645, (1999)

13. A. Dix, Beyond Interaction Pushing Boundaries with Incidental Interaction: Lancaster University, UK (2002)
14. A. Bianchi, D. M, WHO Guidelines on Hand Hygiene in Helath Care: a Summary. World Health Organization Patient Safety: University of Geneva Hospitals (2009)
15. E. Stanley and Sr. Flowers, *Automatic Hand Washing and Drying Machine*, U.S. Patent US5924148A (1999)
16. C. P. Borchgrevink, J. Cha and S. Kim, *Hand Washing Practices in a Collage Town Environment*, Journal of Environ Health, **75**, 18-24 (2013)
17. A. M. Lund, *Measuring Usability with the USE Questionnaire*, STC Usability SIG Newsletter, 2-8, (2001)
18. R. A. Likert, *A Technique for The Measurement of Attitudes*, Archives of Psychology, **140**, 1-55 (1993)

# Enterprise firewall virtualization design

*Ari* Sujarwo[1,*] and *Jefferson* Tan[2]

[1] *Universitas Islam Indonesia, Jalan Kaliurang Km.14,5, Sleman, Indonesia*
[2] *Monash University, 900 Dandenong Rd, Caulfield East VIC 3145, Australia*

**Abstract.** Grid computing could consists of thousand of computers across the world with the role as the resource provider or Grid users. This conditions make a challenge for engineer to keep the Grid secure but at the same the Grid must be easily joined by clients and providers. This paper aims to propose a method to control enterprise firewall which is basically a static engine to be a dynamic engine to cope with the Grid needs. At the end of the paper, the results shows that this method is works and was tested using a real enterprise firewall.

## 1 Introduction

This paper proposes the design of *enterprise firewall virtualization* for Grid. The enterprise firewall virtualization is a composite system that is designed to manage a firewall via a firewall agent for Grid computing. The Grid needs a more advanced security system to filter the incoming traffic than the current security system. Currently, the Grid computing system is secured by a basic certificate-based authentication system and a static firewall. This static firewall possibly raises an open pinhole problem caused by several ports, which is configured to remain open. Globus recommended many ports to remain open when the Grid in operation: as much as 10 for each connecting client [1]. These open ports could be the way attackers compromise the security of Grid.

Several research projects have been conducted to propose an idea to solve such a problem for example: Port knocking approach [2], transparent firewall virtualization [3], and a firewall virtualization that works based on RADIUS [4].

The concept of firewall virtualization itself is the basis for the work Tan in the Romulus project [5]. Before Romulus, Tan [6] had worked on Remus, which is a firewall traversal method based on tunneling over SSH protocol. The project found that tunneling over SSH introduced performance degradation compared to the direct connection caused by the encapsulations overhead during tunneling process [6]. The latency was increased about 25% of the direct connection latency. Based on that fact over Remus, this paper try to propose a technique that might be another approach than Remus. Here in this project, another model based on Romulus is proposed to solve the problem by managing an enterprise firewall based on the client's details that can be identified during the authentication, authorization, and accounting (AAA) processes.

Several components are involved in this model. The firewall agent which manages the firewall consists of an authentication server and some other components. This authentication server validates the credentials of each incoming user. Moreover, the user details are extracted from the AAA logs. Another component formulates and pushes the appropriate firewall rules out to the firewall.

## 2 Firewall Virtualization System Architecture

As briefly explained above, this system consists of several cooperating components. Each of them has a responsibility to support the entire system in meeting its aim. Fig 1 shows the steps on how the system works, while Fig 2 shows the network architecture.

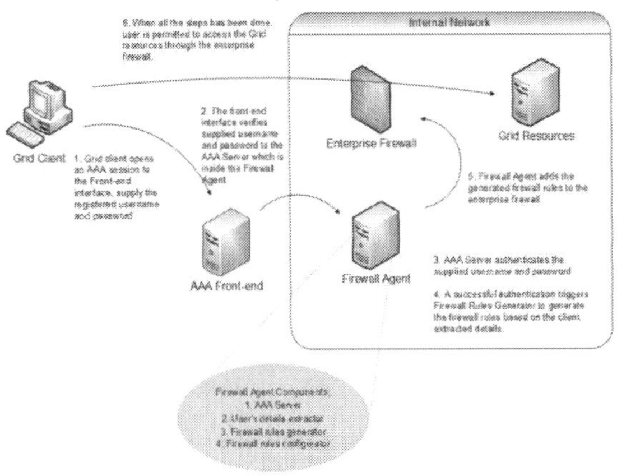


**Fig 1.** Firewall virtualization diagram

* Corresponding author: ari.sujarwo@uii.ac.id

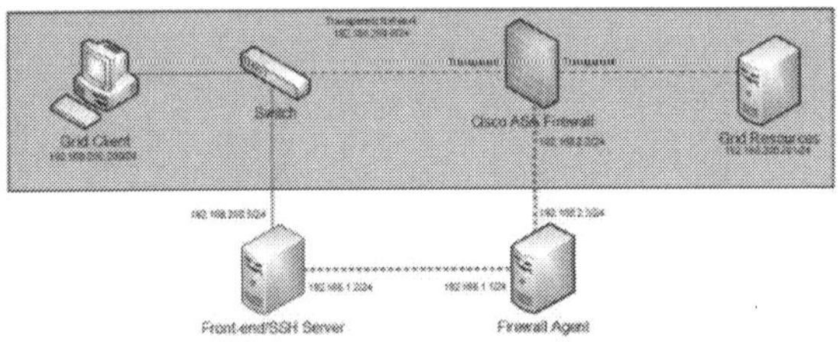


**Fig 2.** Network architecture

As shown on the figure, the authentication process has a role to validate the incoming user. The incoming user which provides a credential must be matched with the known user credentials to ensure that this is a valid credential holder [7]. It is essential to identify Grid users. Since the firewall is dropping all incoming data to the Grid by default, a *pinhole* is needed to enable the authorized client to access the Grid resources. The firewall rule that is used to open the pinhole is generated based on the client's details. This situation makes successful user identification compulsory. Once a client is authenticated, the following phases are authorization and accounting. The client's details from the AAA phase are then extracted to build the pinhole opening rules. After the firewall rule is added to the firewall, the Grid client is allowed to access the Grid resources. The authentication system explained above is used to match the credential. However, this authentication system needs a front-end to let user enter the username and password. Among authentication protocols, there are some *front-end* components available based on the *web interface* and *command line interface (CLI)*.

Selecting an appropriate front-end depends on the features of the authentication engine itself. Some AAA protocols have many plug-ins developed to work with, for example, a combination of secure web interface and AAA as implemented in the previous research [4]. Other AAA systems may use simpler applications as the front-end. This kind of AAA can use a common operating system console as front-end for users to supply the credentials. The console-based front-end applies to a server which provides command line-based remote access for a Grid user. The basic function of this server is to enable users to get in to a server in secure mode then this occasion written to the accounting logs.

Logs are crucial. They record almost all activities happening within the system. There are so many logs in an operating system that each of them tells of specific events regarding the configured services. The most highlighted log in this project is accounting log. This log records when a Grid user starts and stops the session after being authorized. This log provides each user's IP Address and username to be used to generate firewall rules. The agent manages the firewall by configuring firewall rules based on user identity so that the identification of the user is the most crucial step. First, firewall agent must be able to get the user details from the log. As explained above, the log contains the essential information relating to authenticated users such as IP address and username. Those details are extracted and then used to generate the firewall rules which are the second function of this component.

To extract the user details from the log, the generator will search for keywords that contain relevant information. The user's source IP address is used as the source address parameter in the firewall rule, and then the username is used as the firewall rule's name. However, since the Grid resource has a fixed address, the destination address in the firewall generation would use a single IP address.

## 3 Enterprise Firewall

An enterprise firewall is typically used to protect a large production system. Such a large organization needs a powerful security device which protect their system professionally and is strong enough to filter any incoming or outgoing data. Grid infrastructure today is no longer a prototype system. It has been long considered a production system which provides users substantial computing resources. It is no longer a simple system which serves a few users, but thousands of users are now connected to the Grid, wherever it is. A simple one layer security is no longer suitable to Grid infrastructure needs.

The firewall performs traffic checks during its operation. The firewall drops all traffic except the packets which are sent by an authenticated user. The traffic which comes from an authenticated user match with a rule or a policy implemented by the firewall. The firewall rules contain platform-specific syntax to allow the authenticated client's traffic to pass the firewall. Up to now, the manufactured firewalls work based on statically configured rules. An administrator must get into the firewall, and then put some rules to implement what is desired.

However, to make a firewall more dynamic to allow traffic based on user needs, the firewall must be managed by the agent. The firewall is configured to enable client's traffic to pass the firewall. The firewall rules that the firewall agent prepares needs to be configured and activated on the firewall. Once the firewall is configured with the new rules, the matched traffic are permitted to pass the firewall.

The firewall virtualization is designed to be fault tolerant. As normal equipment, the Grid components always have risks of system failures for various reasons. The recovery strategy is focus on the management of the enterprise firewall. When the firewall goes down, the current firewall configuration might be lost. Since the firewall usually works in a dynamic way, almost all access control lists rules are configured remotely by the firewall agent. The firewall has only the basic pre-configuration that contains a basic communication configuration that enables the firewall to be accessed remotely by the firewall agent, and basic and static access control lists rules that drop all incoming packets unless allowed by the agent.

### 3.1 Firewall Pinhole Opening Steps

In the authentication procedure, Client sends a request to open firewall. To do so, the client must run two steps in his client machine: open an AAA session to the front-end interface, and supply the username and password, and then open the Grid client application. The AAA front-end allows the user to enter his credential, at the time, when the firewall is still dropping the packets (Fig 3 Step 1). In the background, the supplied username and password (Step 2) is authenticated by the AAA server inside the firewall agent host (Step 3). The AAA server validates the incoming credentials with the database. If the authenticating user is identified as a valid user, the AAA server grants the user specific rights as an authenticated user. Otherwise, the AAA server rejects the authentication request and writes the incident to the log file.

A successful authentication process brings the user to the authorization process, which is a granting step to give the user rights for a further step. The next step is accounting. The username and the IP address from which users are logged-in from, are recorded in the accounting log. This accounting event triggers the user's detail extractor to obtain all the details and runs the firewall rules generator to create the firewall rules. After the firewall rules have been generated, the agent opens the management session to the firewall. The firewall agent adds the generated firewall rules to the enterprise firewall to achieve the opening of the desired firewall ports (Step 4). Step 5 is the step when users are able to access the Grid resources.

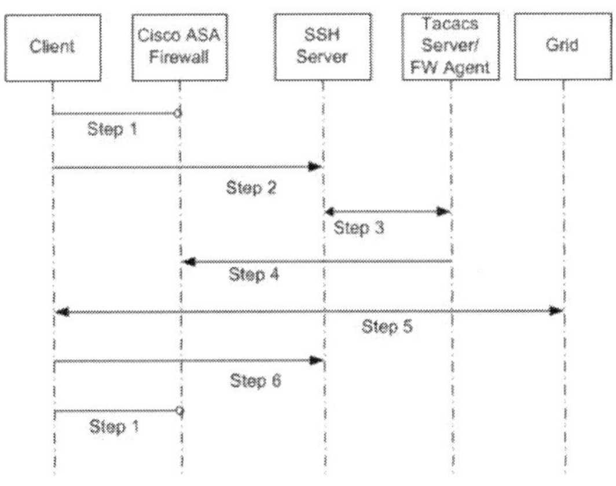


**Fig 3.** Pinhole opening steps

### 3.2 Ending Firewall Pinhole Opening Session

The pinhole is open during the Grid user activity and can close at any time when the Grid user wants. This firewall virtualization has a simple design. It makes it convenient for the Grid user to authenticate himself to the system. The way to end the session is designed in a simple way as well. The user just needs to type **exit** on the opened AAA session. This command sends the message to the AAA server to close the session for this user (Fig 3 Step 6). The AAA server will then record the timestamp of when the session closed, and triggers the firewall agent to remove the user-related firewall rules from the enterprise firewall.

### 3.3 Cisco ASA with Clogin

This section proposes the design and implementation details of enterprise firewall virtualization which manages a Cisco ASA enterprise firewall unit by a firewall agent. Several prepared components work to build such a firewall agent. The firewall agent consist of an authentication server which validates each of incoming user's credential, couple of scripts which work to fetch user details from accounting log file, and a mechanism to write the firewall rules to the firewall engine.

The Cisco ASA firewall is placed as a security check point between client and the Grid system which prevents unauthorized access by client. A client will need to pass the firewall through the open pinholes before it is able to send any data to the Grid. The pinhole itself opened by Clogin. Clogin is an application that enables an application or a system communicate with a Cisco device interactively. Clogin installed within the firewall agent machine, and the only established communications are with the Cisco ASA firewall and the SSH Server.

Clogin has a mechanism to configure a Cisco device remotely from a Linux operating system where it is installed. Based on the Ubuntu manual, Clogin is explained to be an *Expect script*, which is a script that automates a logging activity and configure a device. Clogin works in two steps: logging in to the Cisco device and executing the written script that contains new device instructions. To be able to log into a Cisco device, the Clogin reads a configuration file which contains credentials to enter the Cisco device. This file is a hidden file, which is secretly secured from unauthorized access other than Clogin. Several preprocessing are applied to this file before the Clogin ready to be used.

### 3.4 Firewall Rules for Grid

Based on the Globus toolkit document [1], a Grid needs at least 10 ports to be open per simultaneous user to let the functions of Grid work. Considering this condition, the default firewall settings which are mostly static will either conflict with Globus recommendations or make the firewall always to open on those ports.

In the Globus-toolkit, among those 10 recommended opened ports, there are three ports which open for Grid basic services: *Grid resources Allocation and Management (GRAM)* which work over port 2119 and 8443, and *GridFTP* which needs port 2811. Those ports will be facilitated by firewall agent to be opened.

## 4 Performance Evaluation

The performance of firewall virtualization was measured by two different tests. The first test was measuring the consumed time when a Grid user was requesting a pinhole opening. The second test was measuring the performance of the clogin script to generate the firewall rules and reconfigure the firewall.

There are only two steps of test which shows that a hundred percent of service has been delivered: when there is only one user requests and a maximum of a couple users request at a time. This means that before any further investigation conducted in the future to resolve the problem, these tests limit maximum of two user can send requests at a time to get the system works properly. Results of the tests can be seen in Fig 4 and Fig 5 below.

Figure 4 shows the performance results of the new approach. The graphics shows that the system works with some notes:

1. The approach send firewall rules by Agent to the enterprise firewall. Since the rules were not preprocessed (Figure 5), and delivered as is, it makes packet data massively moved by Agent when user increased.
2. The massive amount of rules generated by Agent reduce the preformance of system delivery.
3. Figure 6 and 7 shows how the system tested in security test environment to show pinhole were successfully opened and closed. It allows data moved while open and reject data while port closed.

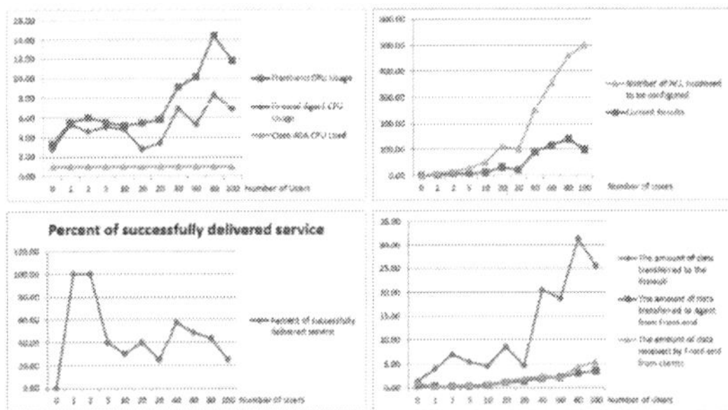


**Fig 4.** Stress test results

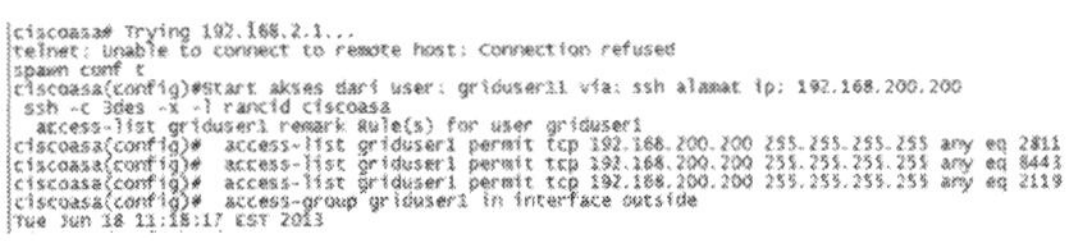


**Fig 5.** Clogin controls the firewall

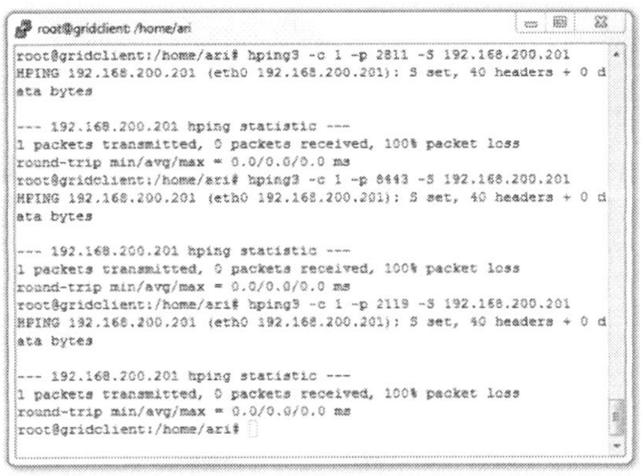


**Fig 6.** Port detected as closed before pinhole opening

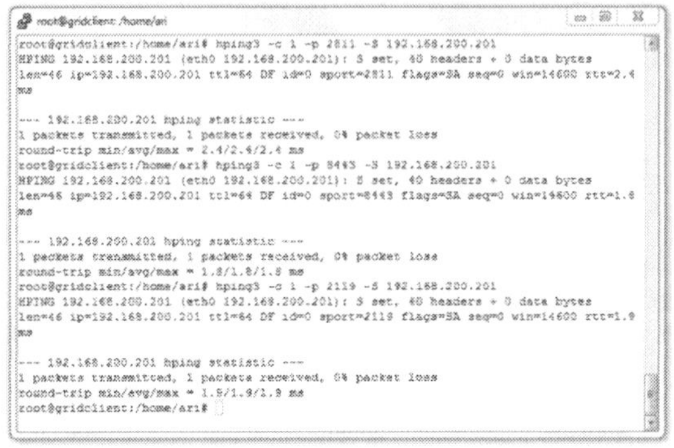


**Fig 7.** Port detected as open after pinhole opening

## 5 Conclusion

This paper shows the enterprise firewall virtualization model which worked as a group of security elements. The firewall virtualization infrastructure contained an enterprise firewall device which separated the network of the Grid resources from the outside, a firewall agent which offered firewall management based on Authentication, Authorization, and Accounting (AAA), and a front-end machine which accepted credentials which were supplied by Grid clients. Firewall virtualization was proposed as an advanced security solution for Grid infrastructure, and offered a controllable firewall which is located in front of the Grid.

## References

1. V. Welch and O. Mulmo, Usg. Glbs. Tlkit. Frwlls (2006)
2. W. Lian, Prt. Knckng Auth. Mechm. Grd. Svc. Mon. Uni (2011)
3. G. Vilcans, Transp. Frwll. Virt. Grd. Cls. Mon. Uni (2011)
4. Mahjoubi, P. RAD. Frwll. Virt. Grds. Mon. Uni (2010)
5. J. Tan, D. Abramson, and C. Enticott, A virt. Con. lyr grds. E. Sc (2009).
6. J. Tan, D. Abramson, and C. Enticott. REMUS: a rer. and mult. Sys. Grd. Con. Acr. Frwlls. J. Grd Com., vol. 7, no. 1, pp. 25–50, (2009)
7. C. Metz. AAA prot. Auth. Autho. Acc. Int. IEEE Internet Comput., vol. 3, no. 6, pp. 75–79 (1999)

# Web-based audiovisual phonetic table program application as e-learning of pronunciation practice in undergraduate degree program

*Ayu Bandu* Retnomurti* and *Rr. Astri Indriana* Octavita H.

English Education Department, Indraprasta PGRI University, Jakarta, Indonesia

**Abstract.** Verbal-based learning such as English pronunciation practice requires the existence of an effective e-learning because if it is directly given without any learning media, inaccuracies in pronunciation, spelling, repetition will usually occur in the spoken language. Therefore, this study aims to develop e-learning to be used in the Pronunciation Practice class, Indraprasta PGRI University. This research belongs to Research and Development are: requires an analysis, develops syllabus and teaching materials, creates and develops e-learning, tries and revises the media. Consequently, there is a need to develop module in the classroom into a versatile technology web-based module in the form of Phonetic Table Program. The result is carried out in pronunciation practice classes to find more details on some parts that may still not be detected by the researchers. Thus, the use of technology has become a necessity to assist students in achieving the learning objectives. Therefore, the process of communication in learning will attract more students' interest and provide facilities to understand the sound system of English as it is equipped with buttons to practice presented by nonnative speakers. Non-native speakers' selection are based on the consideration that they quickly adapt helping other students who are less fluent in English.

## 1 Introduction

Language learning is essentially a process of learning a language. Learning a language is certainly not infallible. Khansir (2012) states everyone is not infallible in learning a language. Keep in mind that error is a source of inspiration to adjust. The study of error and its relationship to language teaching should be encouraged because it is through the study of error, various things related to language error made by students or learners can be expressed. Pronunciation Practice class in the English Department is a media which can be used to develop verbal language skills, listening and speaking. Without adequate mastery of pronunciation, it is impossible a person can speak English well. Error in pronunciation can lead to misunderstanding. In order to make someone understand us in English, we must be able to pronounce English words correctly and in order to understand English, one must be able to capture and understand the pronunciation of others. New technology, especially multimedia has an increasingly important role in learning process. Many people believe that multimedia will be able to bring us to the situation to replace learning with effort to learning with fun. Thus, fun, creative and interesting learning process will be the right choice for educators. Conventional learning system that has been applied so far (faculty of teaching) is strongly connected with instructional concept and is considered to be less suitable to the enthusiasm of rapid development of science and technology. Conventional learning system is less flexible in accommodating the competency of materials development because educators should intensively adjust the teaching materials with the latest technological developments. In fact, at present, Indonesia is entering the information age, an age which is characterized with rapid increase of information medium. The spread of information will always use media, either electronic or non-electronic one. The explanation above shows that the presence of the media has influenced the whole aspects of life including our education system, although in a varying degree. Students should have the skills to read the sound symbols used in the dictionaries so that if once they find a new word whose pronunciation is unknown they will be able to consult the dictionary. But the reality is not always so. On many occasions, even in the undergraduate thesis examination, it is quite common to find students who mispronounced words, even words that are often used in everyday communication and learning. Even worse, if they are asked to check into a dictionary, they will not be able to read the symbols with the right sound. This fact shows that learning Pronunciation Practice has not been managed optimally and will leave negative impact on the mastery of English as a whole, especially in spoken language. Students are the major components in determining the learning outcomes because they are the subject of learning activities. Students with proper prior ability are quite likely to be easier to accept and process teaching materials and training compared to students with

* Corresponding author: ayubandu@gmail.com

lower prior ability. Initial capability is also influential in other learning components, such as the determination of the syllabus, the level of difficulty of teaching materials, teaching techniques and a variety of exercises (Riswanto & Endang, 2012: 9).

Based on previous observation, proper learning outcomes of pronunciation practice can only be achieved by students who have had adequate prior knowledge enough, while students with lower initial ability result in less satisfactory results (Nilsen, 2002: 89). The role of lecturers in Pronunciation Practice learning is as a manager, as a motivator, a facilitator and a model. Pronunciation Practice lecturers should be able to devise proper teaching and learning activities, select appropriate materials, assist students in training, monitor exercises and learning progress of students, and should also be an example or model of how to pronounce English properly (Yulia and Ouda: 2004). The lecture that has been implemented so far is centered on the lecturer. Student activities in class seem to be monotonous and unattractive. Learning activities are focused only on the student handbook. Students are only required to imitate lecturers individually after finding or writing the proper pronunciation through the dictionary. Consequently, in pronunciation practice, students are still unconfident to apply the correct pronunciation like a native in an English conversation. The correct pronunciation will sound strange and funny when practiced by students who are reluctant to apply the aspects of correct pronunciation.

Based on the background of the problems described above, it expects the presence of design concept development through PROTATIK audio-visual (PROTATIK means in Indonesian *Program Tabel Fonetik)* web-based e-learning as solution to effective Pronunciation Practice in college, so it gives good results on pronunciation skills of the students.

## 2 Method

To produce learning media suitable for use and requirement, a research approach and development need to be taken. Research and development are methods to develop and test a product. According to Sadiman (2005: 64) in his book entitled "Learning Media for Education" which is in the field of education, research and production/development can be used to develop books, modules, instructional media, evaluation instruments, models of curriculum, learning, evaluation and other - other. According to him, there are three steps of research and development, namely: 1) conduct a preliminary study to examine the theory and supervise the product or activity, 2) develop a product or a new activity program, and 3) test or validate a product or a new activity program. The first step has been carried out, a preliminary study by reviewing the theories which are included in the literature review about the creation of media and media content that will be developed. Whereas, the development of Pronunciation Practice learning media takes some steps: 1) conduct an analysis on needs, 2) develop syllabus, lessons and training, 3) create a learning media in the form of web-based audio-visual Phonetic Table Program, and 4) conduct a test and improve learning media. Research data are taken from lecturers and students of the Pronunciation Practice classes of Indraprasta University PGRI Jakarta. Data are in the form of qualitative data, on opinions and suggestions from both lecturers and students which are collected in peer discussion, class of observation, and questionnaires. According to Sadiman (2005: 68) there are three steps of research and development in general, i.e.: : 1) conduct a preliminary study to examine the theory and supervise the product or activity, 2) develop a product or a new activity program, and 3) test or validate a product or a new activity program. The first step has been carried out, a preliminary study by reviewing the theories which are included in the literature review about the creation of media and media content that will be developed. Whereas, the development of Pronunciation Practice learning media takes some steps: 1) conduct an analysis on needs, 2) develop syllabus, lessons and training, 3) create a learning media in the form of web-based audio-visual Phonetic Table Program, and 4) conduct a test and improve learning media. The instruments in this research include: records of implementation and questionnaire. Next, data are discussed with colleagues and students, classroom observation, and questionnaires which are analyzed in qualitative descriptive manner and are directly applied to the development and improvement of learning media in the Pronunciation Practice course.

## 3 Results and output achieved

The result of this research is an application of web- based Pronunciation Practice audiovisual Phonetic Table Program, e-learning media that contains a table presenting the pronunciation of consonants and vowels in English which can be heard and are equipped with buttons exercises to practice. The followings are the process of the design and development process of the learning media. The development of audiovisual Phonetic Table Program, a learning media for Pronunciation Practice takes the following steps: 1) the analysis of needs, 2) the development of the syllabus and the preparation of materials and training materials, 3) the creation of the media, 4) the test and repair on the media (Sadiman, 2005: 64).

### 3.1 Analysis of Needs

In order to recognize the suitable media needed in learning Pronunciation Practice in English Education study program, Faculty of Language and Art, UNINDRA PGRI, the research team deliberates on pronunciation practice class learning that has been practiced. The results of this deliberation are discussed in teams of lecturers who teach Pronunciation Practice classes to get extra practice and validation of data on current issues in the Pronunciation Practice class and the media needed to overcome the existing problems. Pronunciation Practice teaching so far has not given any satisfactory result as shown by many facts. First, student results on exams are less satisfying. Students can imitate any given sounds or words by the lecturer in class, but they immediately forget about it. The research team realizes that they need the latest pronunciation practice model that they can listen to and imitate whenever or wherever they need. At present, manual of pronunciation practice model is provided directly by the lecturers while teaching in the classroom without being accompanied by recordings. Lecturers have to shout repeating the same words, phrases, and sentences from class to class. To solve this problem, a good lecture module is indispensable. Class teaching material of Pronunciation Practice in English Department is a handbook lecture entitled Teaching to Mastery: PRONUNCIATION PRACTICE: (Purwanto Andri, Retnomurti Ayu & Mutiara Dewi, 2016) This module/UNINDRA handbook contains all the lecture materials and is accompanied by adequate practice materials. Unfortunately, this module is not accompanied by recording of speech models or practices that are required by both lecturers and students. From this point of view, it is acceptable to say that we need a lecture module that is accompanied by a recording of acceptable speech models. Making recording materials to assist the existing module is feasible to improve the quality of learning Pronunciation Practice, so that the medium of instruction not only in the form of modules but also tables of phonetic sounds in English and Phonetic Table Program, a recording material for practice or exercise that will facilitate the work of lecturers in delivering speech model to be used by students to learn at home. However, when considering the progress of IT today, teaching to use the manual module perhaps is not so interesting anymore. It is time the researchers develop existing manual module to be an effective technology-based module, a Phonetic Table Program in the form of cutting-edge media to make Pronunciation Practice teaching more attractive as it uses web-based technology. Results of observation on the Pronunciation Practice teaching in English Department is taken to the discussion forum of Pronunciation Practice lecturers to generate comments, insights and clarity on media that is actually needed. Here are the results of these discussions: The lecturers generally face similar problems in the teaching of Pronunciation Practice, so all agreed that what is needed is a lecture material that has been arranged systematically accompanied by a number of exercises, recording materials, exercises or practices that can be used in Pronunciation Practice teaching in the classroom. A lecturer confirms that it doesn't have to be a sophisticated media, modules and cassettes shall be enough. But there is also opinions saying we should keep up with technology, so there is no harm if this media is made to be computer-based to optimize the learning process. The materials must be included in this media covers a phonetic table of English sounds. While the materials of exercises or practices should cover loose the words, groups of words, texts, and sentences of various forms, dialogs, or paragraphs.

## 3.2 Syllabus, Lecture Material and Practice or Exercise Materials

Syllabus or Unit Course Events is based on course descriptions of Pronunciation Practice Curriculum in English Education, Faculty of Language and Art, University of Indraprasta PGRI 2015/2016, the purpose of learning and teaching material scope of UNINDRA Pronunciation Practice Handbook (2016) and Hancock (2003): English Pronunciation in Use; Self Study and Classroom Use and the results of the discussion among colleagues who teach Pronunciation Practice on need analysis. Practice or training materials are developed in accordance with the lecture material added with other exercises for the development of the required skill which include the pronunciation of loose words, groups of words, texts, sentences, paragraphs and dialogues. This course syllabus includes four components: 1) The topic of learning, 2) learning objectives, 3) learning materials, and 4) practices or exercises.

**Table 1**: Syllabus / Pronunciation Practice SAP

| Topic | Purpose | Material Description | Practice or Exercise |
|---|---|---|---|
| 1.Introduction | Students Have an overall picture of the contents of the Pronunciation Practice lecture | Definition of Pronunciation<br>Table of English sounds<br>Letters and sounds | Loose words (regular, irregular verbs, plural forms), phrases, texts, sentences of various forms, paragraphs, and dialogues |
| 2. Vowels 1. /i/ & /i:/<br>2. /ʊ/ & /u:/ 3. /a:/ & /Λ/<br>4. /ɒ/ & /ɔ:/<br>5. /e/ & /æ/<br>6. /3:/ & /ə/<br>7. /iə/ & /eə/<br>8. /ei/, /ai/ & /ɔi/<br>9. /əʊ/ & /aʊ/ | 1, Students can pronounce English vowel sounds<br>2. Students can produce English vowel sounds | The Vowel sounds | Loose words of all English vowel sounds considered difficult for the students. |
| 3. Consonants: 10. /p/ & /b/<br>11. /t/ & /d/<br>12. /k/ & /g<br>13. /v/ & /f/<br>14. /Θ/ & /ð/<br>15. /s/ & /z/<br>16. /ʃ/ & /Ʒ/<br>17. /tʃ/ & /dƷ/ | 1. Students have an overview of the English consonant sounds<br>2. Students can pronounce examples of consonant sounds in words correctly without consulting the dictionary | The consonant sound and a summary in table form Consonant sounds of words with words | Loose words with consonant sounds |

| | | | |
|---|---|---|---|
| 18. /m/, /n/ & /ŋ/<br>19. /h/, /l/ & /r/<br>20. /w/ and /j/ | | | |
| 4. -Syllables<br>-Words | 1. The student can distinguish the differences between syllables and words<br>2. Students can pronounce syllables and words with different stress<br>3. The students can determine the number of syllables of each word in a phrase or a sentence<br>4. The students can pronounce syllables, words, phrases or sentences correctly | Introduction to syllables<br>Consonants at the start of syllables<br>Consonants at the end of syllables<br>Syllables: plural and other -s endings<br>Syllables: adding past tense endings<br>Introduction to word stress<br>Stress in two-syllable words<br>Stress in compound words<br>Stress in longer words 1<br>Stress in longer words 2 | Loose words: verbs and nouns, words begins with a consonant in the initial of syllables, final of syllables, plural syllable, suffix -s, past tense suffixes, word stress, plural words, and complex sentences and compound sentences. |
| 5. *Sentences (part I)*<br>*Sentences (part II)*<br>*Sentences (part III)* | Students can pronounce phrases, sentences, texts and dialogues properly and smoothly | Introduction to sentence stress<br>Sentences with all the words stressed<br>Unstressed words<br>Pronouns and contractions<br>Pronouncing the verb be<br>*Auxiliary verbs*<br>Pronouncing short words (a, of, or)<br>Joining words 1<br>Joining words 2<br>Joining words 3 | Examples of sentences that contain elements of stress / unstressed in the sentence, pronouns, auxiliary verbs, and article |

### 3.3 Media Making

Lecture and practice or exercise materials by speaker/non-native speaker models which have been organized are made into an application form (web-based audio-visual Phonetic Table Program). This e-learning is created using the PHP program, Java Script, Jquery and CSS. Macromedia Dreamweaver and Adobe Photoshop are also used. The application development is divided into several stages: analysis, program design, testing and evaluation. The materials developed in the previous step are analyzed, designed and typed in Microsoft Word program because if the materials are typed directly on the program PHP, Java Script, Jquery or CSS, the materials would be too long and very difficult to edit. After all the analyzed material is completed, the next step is to copy the material into the PHP, Java Script, Jquery and CSS program. Next, each of PHP page is equipped with the desired command, namely, forward or return to previous page, play and stop the sound, to the other files, and call the sound symbol table. The next process is recording the non-native speakers. Sound recordings are used to fill the key icon practice contained in this Phonetic Table Program application. The sound is recorded from the models, lecturers from English Department Faculty who pass the selection. Sound is recorded using Macromedia Dreamweaver and Adobe Photoshop with Windows Media Player from Microsoft. After recording the voice, the voice file is then stored in WAV format, MP3, and then edited to suit low high, volume, and to eliminate existing noise. Edited sound file is then connected to a PHP file that has been inserted with commands. After each page is linked to a sound file, the next processes are conducting a trial and then checking each page. This process is done to look at the suitability of a given command with the desired results. This trial is necessary because after checking, it turns out there are many commands that do not run or do not correspond to the desired result. Once the file is tested, then the researchers repair the application. In this mater, the research team uses non-native speakers sound for teaching and learning in the classroom according to the agreement between the lecturers of Pronunciation Practice class. In order to achieve the teaching target in the classroom, the exercises of the pronunciation practice use non-native speakers from their colleagues who have passed the selection. The reason behind the selection is that "Native speakers are usually monolingual and are not versatile enough in adjusting themselves to many different languages" as stated by Jenkins, a professor of global language-English at the University of Southampton in England. She adds, it turns out those who are non-native speakers tend to speak more regularly and carefully, typical of someone who speaks a second or third language. On the other hand, English speakers / native speakers, often talk too fast to be followed by others, and use jokes, slang language and refer to specific things of their own culture. Non-native-speakers typically use a more limited vocabulary and phrases that are simpler, without figurative language or slang. Therefore, they understand each other through facial gestures. After finishing the media making, the last step is tests and repairs on the media.

### 3.4 Tests and repairs on the media

Web-based Phonetic Table Program e-learning media application that has been developed is then tested for learning in the classroom. The test is done in pronunciation practice classes to find out more details on some parts that may still not be detected by the research team. This process is also very necessary to get suggestions from students. The test involves Pronunciation Practice classes (R2A and R2B) such as they are required to give comments on Phonetic Table Program that they have just used. Their comments include three variables, the advantages and disadvantages of the media as well as suggestions for improvement.

**Table 2.** Comment on test media students in the first year

| No | Comments | Frequency of students |
|---|---|---|
| 1 | Excess of the media: | |
| | 1. Very useful for students in learning Pronunciation Practice in the classroom | 17 |
| | 2. Easy to operate by user | 11 |
| | 3. Relatively clear English symbol sounds | 6 |
| | 4. Contains relatively complete exercises and practices | 5 |
| | 5. Help students in individual practices | 2 |
| | 6.Flexibel, students can choose their preferred materials/exercises | 2 |
| | 7. Attractive design | 2 |
| | 8. non-native speakers by lecturer sounds clear | 1 |
| | 9. Recording sound of PROTATIK is clear | 1 |
| 2 | Weaknesses of the media: | |
| | 1 Packaging is not attractive enough | 5 |
| | 2 Cannot be used if the students do not have computers | 2 |
| | 3. Less practical, because students must click many icons | 1 |
| | 4. Lack of exercise in the form of poems or texts of songs in order to check whether our vocabulary is already correct | 4 |
| 3 | suggestions for improvement: | |
| | 1. Fix existing errors | 4 |
| | 2. Add animation | 3 |
| | 3. Make slower pace | 2 |
| | 4. Add music in the middle of the program | 1 |
| | 5. Add more examples | 1 |
| | 6. soften the color | 1 |
| | 7. brighten the colors | 1 |
| | 8. make the web-based software application PROTATIK free for students | 1 |

**Fig 1.** Documentation in R2A & R2B classrooms

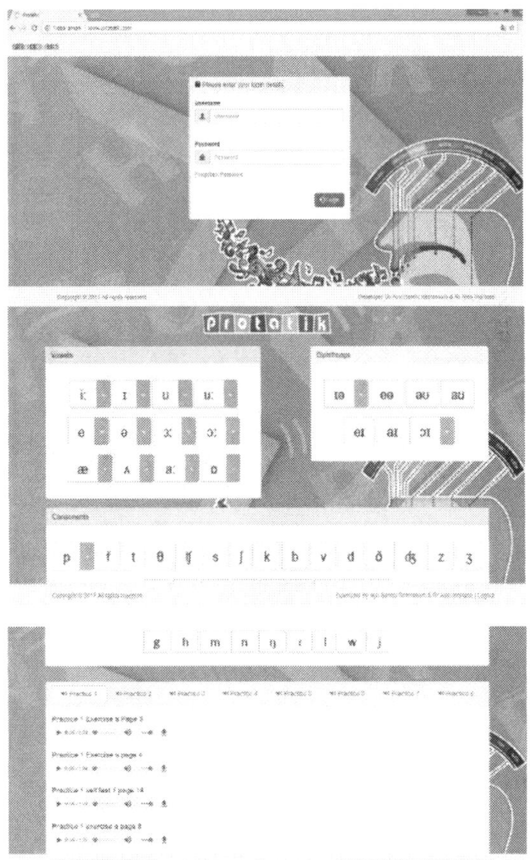

**Fig 2.** Display of web-based PROTATIK e-learning in http://www.protatik.com/

## 4 Conclusion

After developing the web-based Phonetic Table Program learning media and conducting development tests in the first year of use of the media in learning Pronunciation Practice in the class by giving comments include three variables, the advantages and disadvantages of the media as well as suggestions for improvement, the researchers can conclude several things, namely:

1. The process of development of the learning media is an interdisciplinary field. Here synergic cooperation between lecturers who master the field of science and teaching, and those skilled in the technical field of making the media, the master computer programs or information technology and also understand the teaching field to be developed is greatly required. In addition, the research team uses the voice of non-native speakers / lecturers of Pronunciation Practice for the voice, in this case the person reading lessons. Peer selection is made between lecturers who do not only have good voice, but also have the ability to say near-perfect English.

2. The contents of the web-based Phonetic Table Program learning media are taken from available sources, both text and voice recording. These sources belonging to other people, thus this Phonetic Table Program learning media which is still developed is not for sell both to students and the research institutions, before its intellectual property right being formally certified. This web-based Phonetic Table Program learning media is temporary only used for a limited group, namely teaching Pronunciation Practice in English Department, Language and Art faculty, Indraprasta PGRI Jakarta University in the first year. This media which is still in its development process is in the form of a web-based application Program of Phonetics Table. Therefore, in the teaching process, the application needs to be assisted with specific facility, for example the language laboratory, computer laboratory, or classrooms equipped with computers, LCD, and speakers. For students who want to use Flash Disk or network for independent study at home, they should have at least a computer / laptop / notebook at home. Although the use of the media requires certain conditions, namely their computer / laptop / notebook, the media is quite effective in facilitating Pronunciation Practice lecturers and students in the learning process. Lecturer in the class gets the ease of delivering the material and gives the correct speech model.

The researchers' team would like to thank profusely to the Indonesia Government through Appreciation and gratitude to Directorate of Research and Community Service, Directorate General of Research and Development, Ministry of Research, Technology and Higher Education who has financed national competitive research activities Applicable Product Research (PPT) Year 2017 entitled: Rancang Bangun melalui Audiovisual TATIK (Program Tabel Fonetik) Berbasis Website sebagai Media Pembelajaran Pronunciation Practice pada Program Sarjana ". Thanks also to Kopertis Wilayah III Jakarta and Institute for Research and Community Service of Indraprasta PGRI University who has assisted this research activity KEMENRISTEKDIKTI as the organizer of funds at the expense of Applied Research Products, LP2M UNINDRA and Indraprasta PGRI Jakarta University as institutions that facilitate and support the progress and success of this research through Research Contract 0428/ K3/KM/2017, Date May 24, 2017. And UNINDRA Research Agreement/Contract Number: 0552/SKP.LT/LPPM/UNINDRA/VI/2017, Date June 05, 2017. Finally, the researchers' team also would like to thank to the second ICET4SD Yogyakarta & MATEC Web Conferences for our publication committee.

## References

1. A, A, Khansir, *Error analysis and second language acquisition, Theory and practice in language studies*, 2(5), 1027 (2012)
2. Andri, Purwanto, Ayu B, Retnomurti. Dewi, Mutiara, Indah A, *Teaching to Master Pronunciation Practice-A Pronunciation Teaching Learning Tool-An Effective Pronunciation-Teaching Learning Strategy*, UNINDRA. (2016)
3. A, Sadiman, *Media Pendidikan*, Jakarta. PT Raja Grafindo Perkasa (2005)
4. Don, L.F Nilsen. Pace, Nilsen Alleen. *Pronunciation Contrast in English*, Pearson Education Inc, Waveland Illinois USA. (2002)
5. Mark, Hancock, *English Pronunciation in Use; Self Study and Classroom Use*, Cambridge: Cambridge University Press. A. 80 (2003)
6. M.F, Yulia, Ena Teda Ouda, *"Pronunciation Problems of Indonesians EFL Teachers"*, the Proceeding of the 9th Eng. in South. Asia Conf. USD. Yogyakarta (2004)
7. Riswanto, Endang Haryanto, *"Improving Students' Pronunciation through Communicative Drilling Technique at Senior High School (SMA) 07 South Bengkulu, Indonesia"*, International Journal of Humanities and Social Science, **2 (21)**, 10 (2012)

# Decision making model design for antivirus software selection using Factor Analysis and Analytical Hierarchy Process

*Ai* Nurhayati[1,*], *Aditya* Gautama[2], and *Muchammad* Naseer[3]

[1,3] Sekolah Tinggi Teknologi Bandung, Bandung, Indonesia
[2]Politeknik Negeri Batam, Batam, Indonesia

**Abstract.** Virus spread increase significantly through the internet in 2017. One of the protection method is using antivirus software. The wide variety of antivirus software in the market tends to creating confusion among consumer. Selecting the right antivirus according to their needs has become difficult. This is the reason we conduct our research. We formulate a decision making model for antivirus software consumer. The model is constructed by using factor analysis and AHP method. First we spread questionnaires to consumer, then from those questionnaires we identified 16 variables that needs to be considered on selecting antivirus software. This 16 variables then divided into 5 factors by using factor analysis method in SPSS software. These five factors are security, performance, internal, time and capacity. To rank those factors we spread questionnaires to 6 IT expert then the data is analyzed using AHP method. The result is that performance factors gained the highest rank from all of the other factors. Thus, consumer can select antivirus software by judging the variables in the performance factors. Those variables are software loading speed, user friendly, no excessive memory use, thorough scanning, and scanning virus fast and accurately.

## 1 Introduction

Nowadays, almost every person is connected to the internet. It has a lot of impact in daily lives that can be positive or negative. One of the negative impact is the infection of computer virus, spreading faster through the internet.

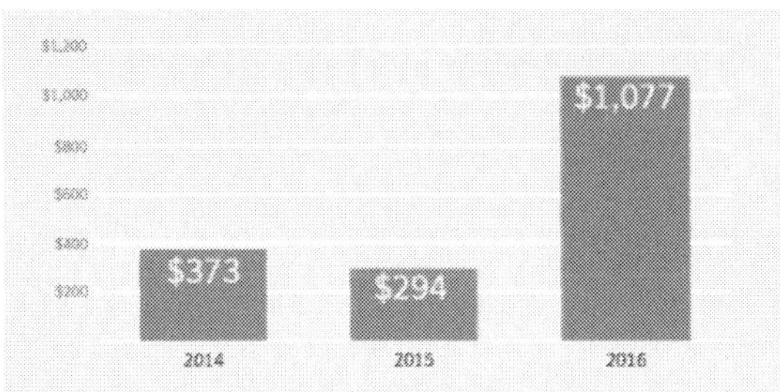


**Fig 1.** Internet Security Threat Report. (Source: Symantec 2017 Internet Security Threat Report)

This problem have made the amount of antivirus software increased in 2017. People who usually connected to the internet, such as students, businessman, and teacher, need to anticipate virus infection by using one of the antivirus software that are available in the market. The variety of antivirus software brand have made it difficult for them to choose which one is the best for their computer system. In this research, we will identify which variables that needs to be considered in choosing antivirus software. The methods that we used are through survey, questioner and direct interview with the user. The variables are identified from those methods. The next step is using factor analysis to group those variables into 5 factors. In order to see the importance rank from these factors we will use AHP. The user can choose which antivirus is best for them based on the factors analysis and AHP method. This research will help common people in choosing the right antivirus for them.

### 1.1 Research Objective

1. Identifying which factors that need to be considered in choosing antivirus software by using factor analysis
2. Ranking the identified factors using AHP

### 1.2 Research Benefits

1. Helping the common people in deciding which antivirus software for them.
2. As an additional information for the IT business industry.

## 2 Literature Review

### 2.1. Factor Analysis

Factor analysis including both principal component analysis and common factor analysis is a statistical approach that can be used to analyzed interrelationships among a large number of variables and to explain these

* Corresponding author: ai.nurhayati@sttbandung.ac.id

variables in terms of their common underlying dimensions (actors) [6]. The objective is to find a way of condensing the information contained in a number of original variables into a smaller set of variates (factors) with a minimal loss of information [6]. For example, it is possible that variations in six observed variables mainly reflect the variations in two unobserved variables. Factor analysis searches for such joint variations in response to unobserved latent variables. Factor analysis aims to find independent latent variables. Followers of factor analytic methods believe that the information gained about the interdependencies between observed variables can be used later to reduce the set of variables in a dataset. The factor analysis model is scale invariant. The loadings are not unique (only up to multiplication by an orthogonal matrix) [18].

### 2.2 Analytic Hierarchy Process

Saaty [16] describes the seven pillars of the AHP as follows:

1. Ratio scales, proportionality and normalised ratio scales.
2. Reciprocal paired comparisons.
3. The sensitivity of the principal right eigenvector.
4. Clustering and using pivots to extend the scale.
5. Synthesis to create a one-dimensional ratio scale for representing the overall outcome.
6. Rank preservation and reversal.
7. Integrating group judgements.

The AHP provides a means of decomposing the problem into a hierarchy of subproblems which can more easily be comprehended and subjectively evaluated [10]. The subjective evaluations are converted into numerical values and processed to rank each alternative on a numerical scale [10]. The methodology of the AHP can be explained in following steps [10] :

Step 1: The problem is decomposed into a hierarchy of goal, criteria, sub-criteria and alternatives.

Step 2: Data are collected from experts or decision-makers corresponding to the hierarchic structure, in the pairwise comparison of alternatives on a qualitative scale.

Step 3: The pairwise comparisons of various criteria generated at step 2 are organised into a square matrix.

Step 4: The principal eigenvalue and the corresponding normalised right eigenvector of the comparison matrix give the relative importance of the various criteria being compared.

Step 5: The consistency of the matrix of order $n$ is evaluated.

Step 6: The rating of each alternative is multiplied by the weights of the sub-criteria and aggregated to get local ratings with respect to each criterion.

## 3 Research Methodology

First we spread questionnaires to consumer, then from those questionnaires we identified 16 variables that needs to be considered on selecting antivirus software.

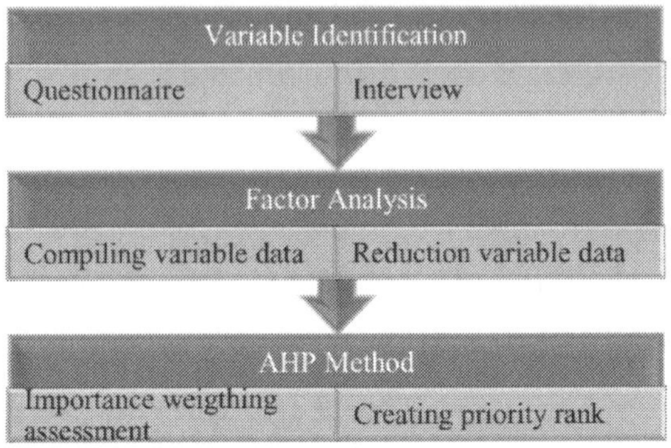


**Fig 2.** Flow chart of research

The 16 variables are anti-spyware, internet security, anti-trojan, anti-worm, anti-spam, online virus detection, software loading speed, user friendly, no excessive memory use, scanning virus fast and accurately, thorough scanning, affordable price, installing time, and data and identity protection. This variables then divided into factors by using factor analysis method in SPSS software. To rank those factors we spread questionnaires to 6 IT experts then the data is analyzed using AHP method. They all gained expertise in Anti Viruses for two years or certified.

## 4 Result and Review

First we spread questionnaires to consumer, then from those questionnaires we identified 16 variables that needs to be considered on selecting antivirus software.

### 4.1 Result from Factor Analysis

Output result from SPSS software program using Factor Analysis:

**Table 1.** KMO and Bartlett's Test

| | | |
|---|---|---|
| Kaiser-Meyer-Olkin Measure of Sampling Adequacy. | | .712 |
| Bartlett's Test of Sphericity | Approx. Chi-Square | 384.008 |
| | Df | 120 |
| | Sig. | .000 |

**Table 2.** Rotated component Matrix

| | Component | | | | |
|---|---|---|---|---|---|
| | 1 | 2 | 3 | 4 | 5 |
| Software loading speed | .181 | .744 | .178 | .020 | .108 |
| Software harddisk space | .118 | .127 | .032 | .157 | .839 |
| Anti spyware | .673 | .361 | .193 | .171 | -.105 |
| Internet security | .664 | -.058 | .188 | .054 | .236 |
| Anti Trojan | .830 | .300 | -.008 | .080 | -.021 |
| Anti worm | .879 | -.011 | -.103 | -.057 | .149 |
| Installing time | .340 | -.106 | .077 | .803 | .162 |
| Anti spam | .717 | .063 | -.062 | .272 | .323 |
| User friendly | .188 | .641 | .138 | -.205 | .097 |
| No excessive memory use | -.155 | .871 | .114 | .206 | -.045 |
| Scanning virus fast and accurately | .241 | .680 | .024 | -.282 | .180 |
| Thorough scanning | .179 | .589 | .180 | -.487 | .260 |
| Virus scanning before download | .249 | .345 | .374 | -.313 | .607 |
| Affordable Price | .110 | .128 | .819 | .334 | .126 |
| Data and identity protection | .015 | .240 | .842 | -.221 | .047 |
| Online virus detection | .614 | .248 | .362 | -.179 | -.036 |

The 1st Factor variables are: Anti spyware, Internet security, Anti-trojan, Anti-worm, Anti-spam, Online virus detection. If we named this factor then it will be called Security Factor. This factor indicate that there are a group of consumer who bought antivirus software based on security factor. The 2nd Factor variables are: Software loading speed, User friendly, No excessive memory use, Scanning virus fast and accurately, Thorough scanning. If we named this factor then it will be called Performance Factor. The 3rd Factor variables are: Affordable price, Data and identity protection. This factor will be named Internal Factor. The 4th Factor variables are: Installing time. This factor will be named Time Factor. The 5th Factor variables are: Software harddisk space, Virus identification before download. This factor will be named Capacity Factor.

## 4.2 Result from AHP Calculations

The 16 variables then divided into 5 factors by using factor analysis. These five factors are security, performance, internal, time and capacity.

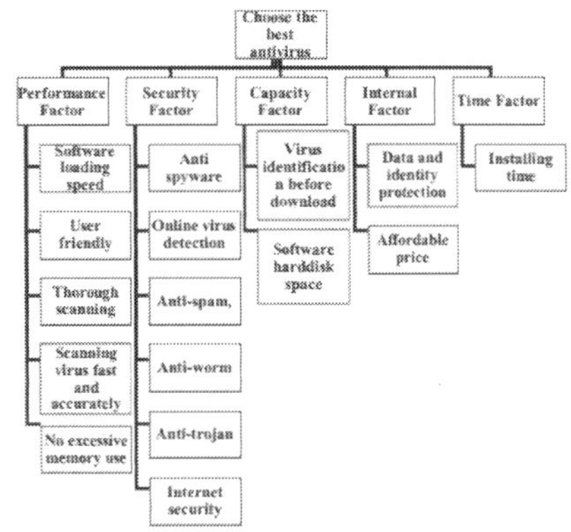


**Fig 3.** Hierarchy criteria

To rank those factors we spread questionnaires to 6 IT expert then the data is analyzed using AHP method.

**Table 3.** Expert 1

| Expert 1 | Security | Performance | Internal | Time | Capacity |
|---|---|---|---|---|---|
| Security | 1 | 4 | 1/5 | 5 | 5 |
| Performance | ¼ | 1 | 5 | 5 | 5 |
| Internal | 5 | 1/5 | 1 | 4 | 5 |
| Time | 1/5 | 1/5 | 1/4 | 1 | 3 |
| Capacity | 1/5 | 1/5 | 1/5 | 1/3 | 1 |

**Table 4.** Expert 2

| Expert 2 | Security | Performance | Internal | Time | Capacity |
|---|---|---|---|---|---|
| Security | 1 | 1/9 | 5 | 5 | 5 |
| Performance | 9 | 1 | 9 | 9 | 9 |
| Internal | 1/5 | 1/9 | 1 | 1/5 | 1/5 |
| Time | 1/5 | 1/9 | 5 | 1 | 1/5 |
| Capacity | 1/5 | 1/9 | 5 | 5 | 1 |

**Table 5.** Expert 3

| Expert 3 | Security | Performance | Internal | Time | Capacity |
|---|---|---|---|---|---|
| Security | 1 | 1/5 | 1/6 | 7 | 1/6 |
| Performance | 5 | 1 | 1/6 | 6 | 7 |
| Internal | 6 | 6 | 1 | 5 | 5 |
| Time | 1/7 | 1/6 | 1/5 | 1 | 1/5 |
| Capacity | 6 | 1/7 | 1/5 | 5 | 1 |

**Table 6.** Expert 4

| Expert 4 | Security | Performance | Internal | Time | Capacity |
|---|---|---|---|---|---|
| Security | 1 | 1/3 | 1/3 | 1/3 | 1/4 |
| Performance | 3 | 1 | ¼ | 1/4 | 1/3 |
| Internal | 3 | 4 | 1 | 1/3 | 1/3 |
| Time | 3 | 4 | 3 | 1 | 1/3 |
| Capacity | 4 | 3 | 3 | 3 | 1 |

**Table 7.** Expert 5

| Expert 5 | Security | Performance | Internal | Time | Capacity |
|---|---|---|---|---|---|
| Security | 1 | 1/4 | ½ | 3 | 1/5 |
| Performance | 4 | 1 | 4 | 5 | 1/6 |
| Internal | 2 | 1/4 | 1 | 1/5 | 1/5 |
| Time | 1/3 | 1/5 | 5 | 1 | 1/6 |
| Capacity | 5 | 6 | 5 | 6 | 1 |

**Table 8.** Expert 6

| Expert 6 | Security | Performance | Internal | Time | Capacity |
|---|---|---|---|---|---|
| Security | 1 | 7 | 1/5 | 1/3 | 1/8 |
| Performance | 1/7 | 1 | 8 | 8 | 1/6 |
| Internal | 5 | 1/8 | 1 | 9 | 7 |
| Time | 3 | 1/8 | 1/9 | 1 | 3 |
| Capacity | 8 | 1/6 | 1/7 | 1/3 | 1 |

Average assessment from 6 Experts:

$$GM_{12} = \sqrt[6]{4 \times \frac{1}{9} \times \frac{1}{5} \times \frac{1}{3} \times \frac{1}{4} \times 7} = 0{,}611$$

$$GM_{13} = \sqrt[6]{\frac{1}{5} \times 5 \times \frac{1}{6} \times \frac{1}{3} \times \frac{1}{2} \times \frac{1}{5}} = 0{,}421$$

$$GM_{14} = \sqrt[6]{5 \times 5 \times 7 \times \frac{1}{3} \times 3 \times \frac{1}{3}} = 1{,}969$$

$$GM_{15} = \sqrt[6]{5 \times 5 \times \frac{1}{6} \times \frac{1}{4} \times \frac{1}{5} \times \frac{1}{8}} = 0{,}544$$

$$GM_{21} = \sqrt[6]{\frac{1}{4} \times 9 \times 5 \times 3 \times 4 \times \frac{1}{7}} = 1{,}637$$

$$GM_{23} = \sqrt[6]{5 \times 9 \times \frac{1}{6} \times \frac{1}{4} \times 4 \times 8} = 1{,}979$$

$$GM_{24} = \sqrt[6]{5 \times 9 \times 6 \times \frac{1}{4} \times 5 \times 8} = 3{,}732$$

$$GM_{25} = \sqrt[6]{5 \times 9 \times 7 \times \frac{1}{3} \times \frac{1}{6} \times \frac{1}{6}} = 1{,}195$$

$$GM_{31} = \sqrt[6]{5 \times \frac{1}{5} \times 6 \times 3 \times 2 \times 5} = 2{,}376$$

$$GM_{32} = \sqrt[6]{\frac{1}{5} \times \frac{1}{9} \times 6 \times 4 \times \frac{1}{4} \times \frac{1}{8}} = 0{,}505$$

$$GM_{34} = \sqrt[6]{4 \times \frac{1}{5} \times 5 \times \frac{1}{3} \times \frac{1}{5} \times 9} = 1{,}157$$

$$GM_{35} = \sqrt[6]{5 \times \frac{1}{5} \times 5 \times \frac{1}{3} \times \frac{1}{5} \times 7} = 1{,}152$$

$$GM_{41} = \sqrt[6]{\frac{1}{5} \times \frac{1}{5} \times \frac{1}{7} \times 3 \times \frac{1}{3} \times 3} = 0{,}508$$

$$GM_{42} = \sqrt[6]{\frac{1}{5} \times \frac{1}{9} \times \frac{1}{6} \times 4 \times \frac{1}{5} \times \frac{1}{8}} = 0{,}268$$

$$GM_{43} = \sqrt[6]{\frac{1}{4} \times 5 \times \frac{1}{5} \times 3 \times 5 \times \frac{1}{9}} = 0{,}864$$

$$GM_{45} = \sqrt[6]{3 \times \frac{1}{5} \times \frac{1}{5} \times \frac{1}{3} \times \frac{1}{6} \times 3} = 0{,}521$$

$$GM_{51} = \sqrt[6]{\frac{1}{5} \times \frac{1}{5} \times 6 \times 4 \times 5 \times 8} = 1{,}837$$

$$GM_{52} = \sqrt[6]{\frac{1}{5} \times \frac{1}{9} \times \frac{1}{7} \times 3 \times 6 \times \frac{1}{6}} = 0{,}46$$

$$GM_{53} = \sqrt[6]{\frac{1}{5} \times 5 \times \frac{1}{5} \times 3 \times 5 \times \frac{1}{7}} = 0{,}868$$

$$GM_{54} = \sqrt[6]{\frac{1}{3} \times 5 \times 5 \times 3 \times 6 \times \frac{1}{3}} = 1{,}919$$

**Table 9**

| | Security | Performance | Internal | Time | Capacity |
|---|---|---|---|---|---|
| Security | 1 | 0,611 | 0,421 | 1,969 | 0,544 |
| Performance | 1,637 | 1 | 1,979 | 3,732 | 1,195 |
| Internal | 2,376 | 0,505 | 1 | 1,157 | 1,152 |
| Time | 0,508 | 0,268 | 0,864 | 1 | 0,521 |
| Capacity | 1,837 | 0,46 | 0,868 | 1,919 | 1 |
| Total | 7,358 | 2,844 | 5,132 | 9,777 | 4,412 |

**Table 10**

| | Security | Performance | Internal | Time | Capacity | Total | Priority Weight |
|---|---|---|---|---|---|---|---|
| Security | 0,136 | 0,215 | 0,082 | 0,201 | 0,123 | 0,757 | 0,151 |
| Performance | 0,222 | 0,352 | 0,386 | 0,382 | 0,271 | 1,613 | 0,323 |
| Internal | 0,323 | 0,178 | 0,195 | 0,118 | 0,261 | 1,075 | 0,215 |
| Time | 0,069 | 0,094 | 0,168 | 0,102 | 0,118 | 0,551 | 0,11 |
| Capacity | 0,25 | 0,162 | 0,169 | 0,196 | 0,227 | 1,004 | 0,201 |
| Total | 1 | 1 | 1 | 1 | 1 | 5 | 1 |

Calculation of matrix consistency test:

a. Calculate value [**A**]

$$[\mathbf{A}] = \text{matrix} \times \text{priority weights} \tag{1}$$

$$\mathbf{A} = \begin{vmatrix} 1 & 0{,}611 & 0{,}421 & 1{,}969 & 0{,}544 \\ 1{,}637 & 1 & 1{,}979 & 3{,}732 & 1{,}195 \\ 2{,}376 & 0{,}505 & 1 & 1{,}157 & 1{,}152 \\ 0{,}508 & 0{,}268 & 0{,}864 & 1 & 0{,}521 \\ 1{,}837 & 0{,}46 & 0{,}868 & 1{,}919 & 1 \end{vmatrix} \times \begin{vmatrix} 0{,}151 \\ 0{,}323 \\ 0{,}215 \\ 0{,}11 \\ 0{,}201 \end{vmatrix}$$

$$= \begin{vmatrix} 0{,}765 \\ 1{,}646 \\ 1{,}096 \\ 0{,}564 \\ 1{,}025 \end{vmatrix}$$

b. Calculate vector $\mathbf{B} = \frac{\text{Vector [A]}}{\text{Priority weights}}$ (2)

$$\mathbf{B} = \begin{vmatrix} \frac{0{,}765}{0{,}151} & \frac{1{,}646}{0{,}323} & \frac{1{,}096}{0{,}215} & \frac{0{,}564}{0{,}11} & \frac{1{,}025}{0{,}201} \end{vmatrix}$$

$$= |5{,}066 \quad 5{,}096 \quad 5{,}098 \quad 5{,}127 \quad 5{,}099|$$

c. Calculate Maximum Eigen Value

$$\text{MEV} = \frac{\text{Sum element on Matrix B}}{\text{N}} \tag{3}$$

$$\text{MEV} = \frac{5{,}066+5{,}096+5{,}098+5{,}127+5{,}099}{5} = 5{,}097$$

d. Calculate Consistency Index $= \frac{\text{M.E.V}-\text{N}}{\text{N}-1}$ (4)

$$\text{Consistency Index or CI} = \frac{5{,}097-5}{5-1} = 0{,}024$$

e. Search for random index value from table. If N = 5, then index random value = 1,12

f. Calculating and checking the Consistency Ratio.

$$\text{Consistency Ratio} = \frac{\text{CI}}{\text{RI}} \tag{5}$$

$$\text{Consistency Ratio} = \frac{0{,}024}{1{,}12} = \text{CR} = 0{,}021$$

From AHP calculations, we can conclude that the calculation result are consistent since CR = 0.021 (smaller than 0.1). AHP method also give us the importance rank from each of the factors. The most important factors are performance, followed by internal, capacity, security, and time.

The factors priority weight score are:
Performance = 0,323
Internal = 0,215
Capacity = 0,201
Security = 0,151
Time = 0,11

## 4 Conclusions

The conclusion of the results of this research:

1. The 16 research variables are grouped to 5 factors. This is the result from Factor Analysis.
2. The factors and variables are :
   a. Security factor are: Anti spyware, Internet security, Anti-trojan, Anti-worm, Anti-spam, Online virus detection.
   b. Performance factor are: Software loading speed, User friendly, No excessive memory use, Scanning virus fast and accurately, Thorough scanning.
   c. Internal factor are: Affordable price, Data and identity protection.
   d. Time factor are: Installing time.

   e. Capacity factor are: Software harddisk space, Virus identification before download.
3. The factors priority weight score are:
   a. Performance = 0,323
   b. Internal = 0,215
   c. Capacity = 0,201
   d. Security = 0,151
   e. Time = 0,11
4. Consumer can select antivirus software by judging the variables in the performance factors. Those variables are software loading speed, user friendly, no excessive memory use, thorough scanning, and scanning virus fast and accurately. Those variables can viewed at antivirus software.

## References

1. D. K. Devi, Dr. K. M. Kumar, *An Analysis of Various Anti-Virus Software Tools Based On Different Effective Parameters*. ijcstjournal, **4**(4) (2016)
2. F. Mamaghani, Evaluation and selection of an antivirus and content filtering software, Information Management & Computer Security, Vol. **10** Issue: 1, pp.28-32(2002)
3. F. Mata, J. L. García-Dorado, J. Aracil, J. E. López de Vergara, Factor analysis of Internet traffic destinations from similar source networks, Internet Research, Vol. **22** Issue: 1, pp.29-56, (2012)
4. H. Chong, R. M. Zin, Selection of dispute resolution methods: factor analysis approach", Engineering, Construction and Architectural Management, Vol. **19** Issue: 4(2012)
5. I. Khan, An introduction to computer viruses: problems and solutions, Library Hi Tech News, Vol. **29** Issue: 7, pp.8-12(2012)
6. J. F. Hair et al. *Multivariate Data Analysis*. Pearson (2010)
7. M. Brunelli, *Introduction to the Analytic Hierarchy Process*. Springer International Publishing, Finland (2015)
8. M. Hosseini, M. S. Owlia, Designing a model for measuring and analyzing the relational capital using factor analysis: Case study, Ansar bank, Journal of Intellectual Capital, Vol. **17** Issue: 4, pp.734-757(2016)
9. M. Peters, S. Zelewski, Pitfalls in the application of analytic hierarchy process to performance measurement, Management Decision, Vol. **46** Issue: 7, pp.1039-1051(2008)
10. N. Bhushan, & K. Rai, *Strategic Decision Making Applying the Analytic Hierarchy Process*. Bangalore: Springer (2004)
11. R.Sriramachandramurthy, S.K.Balasubramanian, M. A. Hodis, Spyware and Adware: How Do Internet Users Defend Themselves?, American Journal of Business, Vol. **24** Issue: 2, pp.41-52(2009)
12. S. Chen, Y. Wu, L. Yang, Application of the Analytic Hierarchy Process for the selection of wastewater reuse targets, Management Decision, Vol. **52** Issue: 7, pp.1222-1235(2014)
13. S. Lin, M. Lu, Characterizing disagreement and inconsistency in experts judgments in the analytic hierarchy process, Management Decision, Vol. **50** Issue: 7, pp.1252-1265(2012)
14. S. Sipahi, M. Timor, The analytic hierarchy process and analytic network process: an overview of applications, Management Decision, Vol. **48** Issue: 5, pp.775-808(2010)
15. T. Saaty. *Decision making with the Analytic Hierarchy Process*. Int. J. Services Sciences Int. J. Services Sciences (2008)
16. T.L.Saaty, L.G.Vargas, *Models,Methods, Concepts and Applications of the Analytic Hierarchy Process*. Kluwer,Dordrecht (2001)
17. U. Roy, *Vulnerability of Watersheds to Climate Change Assessed by Neural Network and Analytical Hierarchy Process*. India: Springer (2016)
18. W. Hardley, L. Simar, *Applied Multivariate Statistical Analysis*. Springer (2003)
19. Wendler, T., & Gröttrup, S. *Factor Analysis*. Switzerland: Springer International Publishing (2016)

# The using of readutainment as e-learning to improve students' reading comprehension skill

*Nina* Dwiastuty*, Susilawati, and *Muhammad* Sulhan

The University of Indraprasta PGRI, Jakarta, Indonesia

**Abstract**. Technology grows and has changed life styles in many aspects. It offers simplicity and practicality to human to switch the life easier from traditional to modern creativity. In the education context, the use of e-learning as a product of technology in the language learning is inevitable and has made language teachers find new teaching strategies. This action research aims to prove whether the innovation of Readutainment application as e-learning enables students to improve their reading comprehension skill. This application usage is to facilitate students to comprehend. English texts. The research object were private vocational students of grade XI taken from 6 of 12 vocational schools in the districts of Singaparna and Leuwisari, West Java. It uses a purposive random sampling by considering that all students are homogeneous in their cognitive levels. The method of this research is qualitative action based on students' studying in the cloud classrooms in two cycle designs. The result shows that the use of Readutainment can improve students' ability to comprehend English texts. It is proven that after being treated by using the Readutainment in two cycles, the average score increases from 68 to 70 in the first cycle and 75 in the second cycle.

## 1 Introduction

The concept of teaching and learning has been, these days, switching from conventional to more creative one as the impact of the fast growing of information technology. The conventional way of teaching is, in one hand, still mostly applied in which the activity of teaching and learning occur face-to-face in the classrooms. The more creative way of teaching in this modern-gadget era, on the other hand, requires educators to have no choice but consequently apply it anyway. The process of teaching and learning can be done through e-learning in the cyberspace as a new concept. In short, the teaching and learning process does not mainly depend on space and time due to the presence of internet that has been widely used in the form of e-learning.

The use of e-learning in education becomes very possible to be implemented in line with the fact that most people connect with internet for many purposes. Survey on internet usage done by the Indonesia Internet Service Provider Association or *Asosiasi Penyelenggara Jasa Internet Indonesia* found that in 2016, internet users reached up to 132 million people. How people access internets vary in some ways as such that 2 million of them or 50.7 percent access internet via smartphones and computers, 63.1 million people or 47.6 percent access internet via smartphones, and 2.2 million people or 1.7 percent access internet only via computers. (Kompas.com, Oct. 24, 2016). This indicates that teachers and students very possibly furnish themselves with smartphones which are not only used for communication but for accessing internet as well. Therefore, teacher-student interaction can be built via e-learning in the process of teaching and learning at anytime and anywhere.

E-learning is worth implementing as an alternative in which teachers, for example, have another way of delivering their teaching materials to students outside the classrooms (Mohammadi et al., 2011:464–468). Though lesson plans have been previously prepared, teachers might often face problems of time arrangement to discuss all teaching material as planned in one meeting in the classrooms. It was due to two different factors; practical factor and professional factors. The practical factor refers to the large number of students in class, class size, time allotment, lack of appropriate resources to support the implementation of the approach. While, professional factors may cover the teacher preparations, mastery of the discussed topics, and teaching-learning strategies (Marcellino, 2008). Hence, students have insufficient of time to deliver questions for unclear matters and teachers may not satisfy their explanation to the students due to the limited time. Another problem emerges in such a way that teachers lost opportunity to fully concentrate in the teaching activities. Teaching in a crowded classroom can cause failure to attract students' attention and involvement as some students may make noises and not focus in the learning process.

In the Indonesian context, e-learning has been arising as research subject and continuously developed in the teaching and learning practices. It was previously hard to find references and research findings on the concept of

* Corresponding author: ninadwiastuty@yahoo.com

learning in the virtual classrooms by using e-learning in Indonesia. Rather, most educators discussed mostly about how to teach and learn effectively in the real classrooms (Damayanti et. al., 2007). However, the rapid changing in the Information and Communication Technology (ICT) has encouraged teachers and lecturers to do research on the effectiveness of using e-learning in their related knowledge background. In short, the stakeholders in the education have realized that the teaching and learning do not, anymore, depend on face-to-face meeting in the real classrooms.

At least, two examples of research might represent to prove that e-learning affects significantly in the teaching and learning of certain subjects. First, Daud and Rahmadana (2015) conducted a research and development (R & D) to develop an educational product; that is, Biology Instructional Media Based E-learning for Excretion System Concepts at Class XI IPA3 SMAN 4 Makassar and to validate its usage. They implemented a software called *Joomla* and Wondershare Quiz Creator (WQR) for the evaluation tool. They applied 5 steps in this R & D; (1) Conducting first study, (2) Designing the product, (3) Developing the first product, (4) Trials and revision of the first product, and (5) Trials and revision of final product. The result shows that the biology instructional media based e-learning by using *Joomla* and WQC is considered valid, practical, and effective.

Another research done by Suharyanto and Mailangkay (2016) conclude that e-learning contributes a positive and significant impact on the quality of learning. The influence is strong confirming that the more intensive e-learning is used, the quality of learning will be increasing. These conclusions derive from their investigation and analysis of other researchers' findings on the using of e-learning in some educational institutions The data were taken from 5 (five) researchers; (1) Edy conducted a study in 2011 in Central Java, (2) Budi conducted a study in 2012 in the Directorate of Diploma Program of the Institute of Agriculture Bogor, (3) Agustina conducted a study in 2013 in the University of Bina Darma, (4) Tiara conducted a study in 2015 in SMA Negeri 4, Jember-East Java, and (5) Harahap conducted a study in 2015 in the Faculty of Economy in UMSU. It can be said that this is an indirect research as all data were taken from the researchers' findings.

Being inspired by the above previous studies, the researchers offer a product of e-learning called Readutainment application to be implemented in this current research. It is true that the application is not really new in the e-learning innovation product especially in language learning. In fact, the similar products can be easily found in the internet these days. However, the researchers have made some modification from the existing application to comply with Indonesian higher school students' levels of thinking.

The name Readutainment deals with entertaining reading activities while its users may also sharpen and enrich the other features of language such as vocabulary, part of speech, etc. Exercises and discussion which come after the texts may foster students to reach understanding. In short, it is expected that this application enables students to improve their EFL reading skill achievement. The research question would be, then, stated as: *Can the use of Readutainment as e-learning improve students' reading comprehension skill?*

The researchers focus on the students' reading skill improvement because it is considered difficult to the Indonesian EFL students. In most cases, EFL students do not have good English reading habit although they have formally learned English from elementary school (Iftanti, 2012). They are lack of reading strategy consequently and even fail to get the message of the English texts due to insufficient or limited vocabularies. It is perhaps because English teachers fail to guide students to have higher levels of thinking in the reading classrooms. Instead, the teaching or reading comprehension occur in the lower levels of cognition (Sulhan, 2015). This may lead to the failure on students' effort to catch the writer's message in the English reading texts.

As a matter of fact, reading is the construction of meaning from a printed or written message and material (Richards, 1998:12; Fox, 2015:12) and the meaning itself is always contextual, generated from the interaction between long-term memory traces and the momentary context existing in working memory (Kintsch and Mangalath, 2011). Moreover, the nature of reading arises in its multiple forms and purposes (Blachowi and Donna, 2008:13). It implies that written texts consist of words, phrases, clauses, and sentences to build meaning as the authors' idea to be delivered to the readers. In order to reach comprehension, the readers will automatically connect the information from the written message with the previous knowledge. However, the process of connecting the readers and their experience depends on the sociocultural context which mediates students' experiences, just as students' experiences influence the context (https://www.rand.orgcontentdamrandpubsmonograph_rep ortsMR1465MR1465.ch2.pdf).

If the text message is successfully caught through appropriate strategies, reading may empower, bewitch, enrich the experience because it appears to be superficially and parasitically, waving, tool making gathering, face and gender recognition and any others, whereby a flood of visual data-shapes, units, patterns, orientation, sequencing- is assessed at a glance (Fischer, 2003:7).

The flowchart below simply figures out the framework proposed in this research to show how e-learning as a creative teaching and learning model (a product of technology) in in the form of Readutainment application is expected to improve students' reading comprehension skill.

## 2 Research Method

The sample of this current classroom action research were totally 180 students in Singaparna and Leuwisari, West Java which were randomly taken from 6 (six) private vocational schools in these districts. It means that 30 (thirty) students involved as the representative participants of each school. The researchers used the Edmodo network as the artificial platform to accommodate students in the virtual classrooms *in the cloud*. There were 6 virtual classrooms established and all students were hoped to actively engage in the teaching and learning of reading comprehension by using Readutainment application through this network into cycles of teaching as the design of action research. The researchers shared the teaching materials which have been set in the application and let students access it at any convenient time and anywhere.

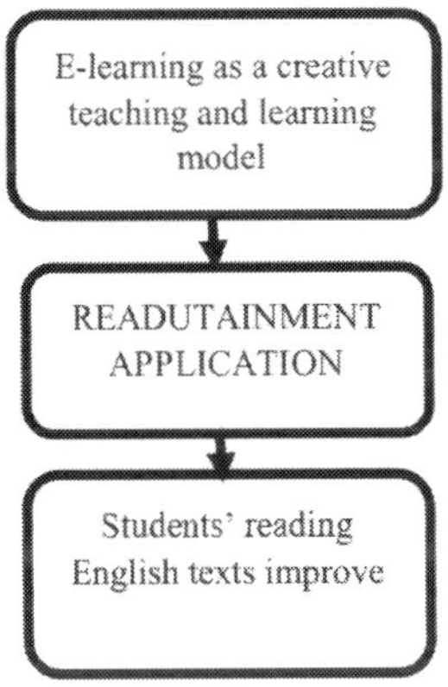


**Fig 1.** Flowchart of framework of thinking

The said application contains reading comprehension texts complete with exercises and discussion, quizzes and assignments. It provides mini dictionaries or glossaries of word classes following the texts. When students click VERB, for example, all verbs in the text highlight automatically showing its particular green color background and the Indonesian translations can be found beside it. The students may move to click the other word classes and the same results emerge spontaneously. In addition, the comprehension strategy offered in the Readutainment application is expected to strengthen the students' previous reading capability. The improvement scores can be measured as they were assigned to submit their works after completing all multiple choice exercises. Importantly, the cycle of teaching would have been stopped until the scores were averagely increasing.

## 3 Result and Discussion

| Score | Before treatment | Cycle I | Cycle II |
|---|---|---|---|
| Average score | 68 | 71 | 75 |
| Lowest score | 55 | 52 | 62 |
| Highest score | 77 | 80 | 89 |

Before the treatment, the researchers collected the students' average score of reading comprehension from the local English teachers of 6 vocational schools involved in this research. The scores were respectively noted 68 as the average score, 55 as the lowest score and 77 as the highest score. This was the starting point to decide how many cycles of treatment to be implemented for the next steps. Considerably, this action research was set to have 2 (two) cycles and in the end of each cycle, the students were given an objective test of reading comprehension. Each cycle consists of 4 (four) steps; planning, action, observation, and reflection.

In cycle I, students were still confused to cope with the Readutainment application (as it was a bit new thing for them) and faced problem with the internet connection. It was unfortunately less than 60% of students participating in cycle I. The achievement of reading test in the end of the first cycle proved to be unsatisfying reaching up to 71 for the average score; 52 for the lowest score (even decreasing from the previous score 55 before the treatment), and 80 for the highest score.

The researchers continued to do reflection by controlling and checking to get more students engaged in cycle II. Observation and reflection were done, previously, by analyzing all works executed in the cycle I. The observation was done indirectly through network to check students' participation to be present in the application. The researchers also visited directly to the real classrooms to gather students' learning problems in using e-learning. These all generated new plans and strategy, as a reflection, to apply in the next step. To continue the research to the cycle II, some texts needed to be changed with those within students' levels of knowledge and experience. For example, the texts of Indonesian prominent figures and national or international daily products were proven to enhance students' confidence in doing the exercise. It was worth doing since as an online course, Readutainment is offered as a distance education program. The content should be pre-constructed (Anderson, 2010:41) appropriate with students' capacity so that they were able to connect with their previous knowledge to build comprehension (Woolley, 2011).

To overcome the problem of internet connection, the researchers distributed internet package data from provider which was familiar and easier to be accessed in these area freely to the students. However, group works were suggested and considered more helpful for students to deal with the application. It was more practical as only the leader or the representative of the group required to connect the internet when they got together to do the assignments after the school time. There was less complaint of internet connection from students afterwards. It was noted that as many as 165 students or 91.7 % of the sample successfully submitted their works through the application in the cycle II.

Overall, the table below shows the students' reading comprehension achievements before treatment, in cycle I and in cycle II. It can be seen that in cycle II, the average score increases to reach 75, the lowest score is 62, and the highest score increases significantly to 89.

## 4 Conclusion and Suggestion

As the impact of the fast growing technology, e-learning applied in a more creative teaching and learning model is worth implementing. This has led to the production of, for example, the Readutainment application, as one kind of innovation in technology, and is proven to have helped students involved and engaged with fun in the reading classrooms. The research question formulated in the current research is finally answered saying that the use of Readutainment application as e-learning can improve the students' reading comprehension skill. Nevertheless, in order to obtain a more general result, the next research is suggested to be conducted in the other places and the content of the application should be adjustably modified within the range of students' knowledge and experience in the new area of the research.

We hereby express our thanks to all parties who have directly and indirectly supported and contributed from the preparation till the completion of this research. Our special gratitude shall be delivered to the vocational schools in Singaparna and Leuwisari, West Java; the headmasters for their permission and the students of XI graders for their participation in this action research. We also wish to thank the Ministry of Research, Technology, and Higher Education who facilitated us with the fund to conduct the research, our beloved Prof. Dr. H. Sumaryoto, the rector of the University of Indraprasta-PGRI, Indonesia for his support and facilitation on this research. Finally, we wish to extend a special word of thanks to IEC4SD who facilitated us to join the conference.

## References

1. N. Mohammadi, V. Ghorbani, F. Hamidi, Eff. of e-learning on LL, *Procedia Computer Science* **3**, 464–468 (2011)
2. M. Marcellino, ELT in Ind: A Cont. Chall. in Ed. and Cult. Diversity, *TEFLIN Journal,* **19**, 1 (2008)
3. T. Darmayanti, M. Y. Setiani, B. Oetojo, E-Learning pada Pendidikan Jarak Jauh, *Jurnal Pendidikan Terbuka dan Jarak Jauh,* **8**, 2 (2007)
4. F. Daud & A. Rahmadana. Pengembangan Media Pembelajaran Biologi Berbasis E-Learning, Jurnal Bionature, **16**, 1 (2015)
5. Suharyanto dan A. B. L.Mailangkay, Pen. E-Learning sbg Alat Bantu Mengjr Dlm Dunia Pend., *Jurnal Ilmiah Widya*, **3**, 4 (2016)
6. E. Iftanti, A Survey on the Eng. Reading Habits of EFL Students in Ind. *TEFLIN Journal,* **23**, 2 (2012)
7. M. Sulhan, Map. Out The Teaching of Read. Compr. to SHS Students: A Bloom's Tax. Persp., *Journal of Developing English and Language Teaching.* **05**, 3 (2015)
8. J. C. Richards, *Extensive Reading in The Second Language Classroom*, (Cambr. UP, UK, 1998)
9. R. Fox, Res-Based Strat. for Stud. with Learn. Disabil.: Focus on Phon. and Fluency. http://opus.ipfw.edu/grad_research/31
10. W. Kintsch and P. Mangalath, The Constr. of Meaning. *Topics in Cognitive Science,* **3**, 346–370.doi:10.1111/j.1756-8765.2010.01107.x
11. C. Blachowi & D. Ogle, *Reading Comprehension - Strategies for Independent Learners. Second Edition* (The Gulf. Press, NY, 2008)
12. Fischer, S. Roger *A History of Reading*, (React. Books, GB, 2003)
13. J. Anderson, *ICT Transforming Education*. (UNESCO, Bangkok, 2010)
14. G. Woolley, Reading Comprehension: Assisting Children with Learning Difficulties, DOI 10.1007/978-94-007-1174-7_2, © Springer Science +B. Media B.V. (2011)

# Integration of EA and IT service to improve performance at higher education organizations

*Agus* Hermanto* and Supangat

Universitas 17 Agustus 1945 Surabaya, Informatics Departement, Semolowaru 45, Surabaya, Indonesia

**Abstract.** One of the perceived benefits of the use of information technology is the increased accuracy and speed of information that greatly helped the operations of an institution or organization. Politeknik XYZ as one of the organizations / institutions in the field of education, now need to consider the strategic role of information technology in supporting academic activities and develop a wide range of academic services, especially for students. Accordingly, in this study has been carried out the planning and improvement of information technology services in accordance with EA and IT Service standards, based on analysis using ACMM maturity level. The end result of architectural planning and service enhancement developed with TOGAF and ITIL is to be used as a reference in the improvement the role of information systems and technology in supporting business process organizations at Polytechnic XYZ. This study demonstrates the importance of raising awareness of the interdependent relationship between architecture and quality of service management in the organization of information technology.

## 1 Introduction

One of the perceived benefits of the use of information technology is the increased accuracy and speed of information that greatly helped the operations of the institution or organization. Therefore, it can be said with the increasing role of information technology will be directly proportional to the investment made. This will require careful planning in the implementation of information technology investment in the future, thus their governance requires good information technology within an organization, which starts from planning to implementation. Information technology governance has a lot of tools, such as TOGAF-ADM (The Open Group Architecture Enterprise-Architecture Development Method) and ITIL (IT Infrastructure Library), which is used as a guide in conducting the audit [1].

The appropriate use of TOGAF and ITIL in information technology governance can be obtained by analyzing and measuring the current state of an institution or organization in advance, so as to align the organization's business strategy and technology strategy to deliver maximum results for the organization. As in this study, the analysis and the measurement conditions at the Politeknik XYZ using ACMM (Architecture Capability Maturity Model), which has six levels and nine characteristics in assessing the maturity level of an organization [2].

Politeknik XYZ which is part of the national education system, at this time need to pay attention to the strategic role of information technology in supporting academic activities and develop a wide range of services and capabilities that can provide an edge over competitive pressure in regional markets to acquire new students. Accordingly, in this study will be carried out a gap analysis and architectural planning the development of information technology has been used by the Politeknik XYZ in order to determine the level of maturity so that the unknown levels of performance and effectiveness of information technology which has been running at the Politeknik XYZ based on standard TOGAF and ITIL.

## 2 Theoretical Basis

There is so much understanding of IT governance including IT governance is the responsibility of top management and management executives of an organization. It also explains that IT governance is part of organizational management as a whole consisting of leadership and organizational structure and the existing process is to ensure the continuity of IT organization and development of strategy and goals of the organization [3] [4]. Furthermore, in another definition says information technology governance is an organizational action undertaken by IT management to control the formulation and implementation of IT strategy and how to believe in business and IT itself, but also carries risks for the organization, resulting in a complexity of complexity, no high focus and operating and maintenance costs [5-7].

As John Zachman said that the increasing complexity in the application of information systems, it is necessary to use the architecture for the determination and control of interfaces and integration of all system components [8]. To achieve the organization's mission through the optimal performance of business processes with the efficiency of the IT environment, the application of EA should be

* Corresponding author: hermanto_if@untag-sby.ac.id

 https://doi.org/10.1051/matecconf/201815403008

incorporated into the roadmap of the organization. There are five focus areas for information technology governance: strategic alignment, value delivery, risk management, resource management, and performance measurement. In this research, we will use two standards at the same time that is TOGAF and ITIL. TOGAF is a framework developed by The Open Group's Architecture Framework in 1995. TOGAF is widely used in areas such as banking, manufacturing and education industries. One of the advantages of using this TOGAF framework is because it is a flexible and open source. TOGAF views enterprise architecture into four categories [2], as shown in Figure 1 below.

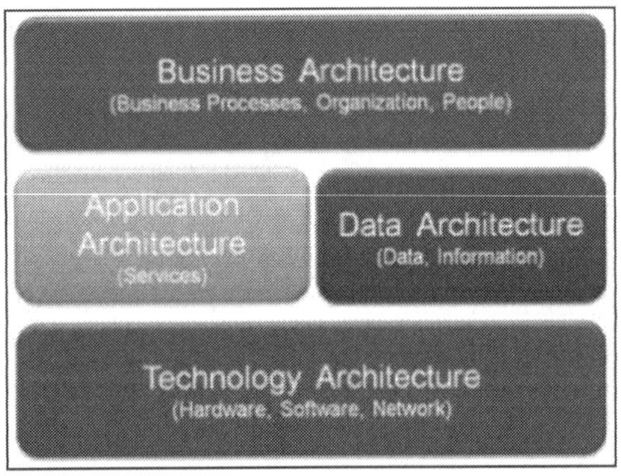


**Fig 1.** Categories Of Architrcture Domains In TOGAF

One form of deliverable from ADM is a recommendation of information technology-based service. Related to this, the framework used is ITIL which is the best practice of technology management information, and focuses on continuous measurement and improvement of quality information technology services, with several benefits, including: the first to adapt to business needs, the second is the efficiency in providing a well-defined service process, and the third is IT services that can be improved and measured [9-11].

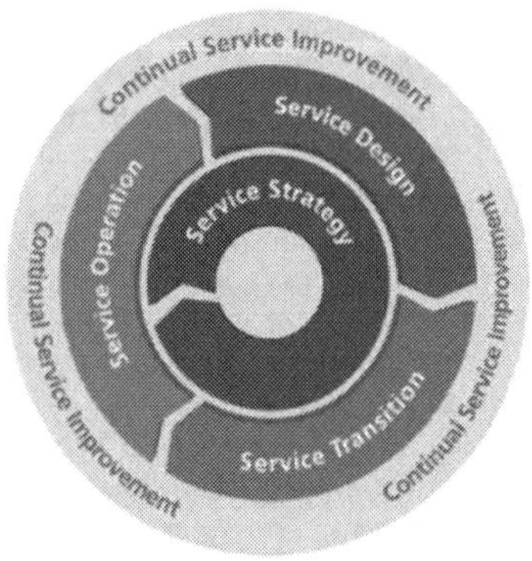


**Fig 2.** ITIL service life cycle diagram

## 3 Previuos Research

Based on the results of research conducted by Brian H. Cameron and Eric McMillan, it was concluded that the most popular framework used in information technology governance is TOGAF by comparing various attributes or criteria for respondents in choosing the EA Framework, including: process completeness, ADM, interoperability or flexibility in using the elements, availability of architectural knowledge, neutrality, and alignment with industry standards (Cameron & McMillan, 2013). In another study, various categories have been introduced for design patterns as a way to capture and re-analyze, design and apply well applied to the TOGAF framework, as well as provide motivational examples of how design patterns can be structured to create TOGAF framework architecture is generic and to improve the organization's performance in terms of IT services include the utilization of CSF that has been successfully identified in implementing ITIL [12-16].

## 4 Scope and Research Method

In this study, we conducted a literature study to get the relevant concepts on the problems encountered. We then conducted observations and interviews with stakeholders to get an overview of the conditions of the organization. We also developed questionnaires to conduct assessments to obtain technological maturity levels with reference to ACMM. Based on the ACMM analysis, we designed the information technology architecture using TOGAF, then integrated and mapped the architecture with ITIL. In Figure 3 below, we demonstrate the TOGAF and ITIL research and integration sequences, from key business processes to delivering and maintaining the IT services shown in Figure 2. The limitations of this study are the mapping of preliminary phase and phase A to service strategy, and phase A with requirement management to service design.

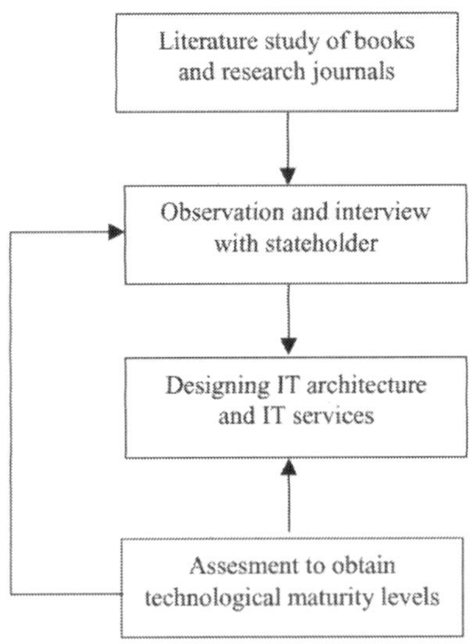


**Fig 3.** Overall research flow

The integration of TOGAF and ITIL is quite necessary in this case because the business architecture is supported by TOGAF, but not by ITIL and IT services are handled by ITIL but not by TOGAF. While other elements (information architecture, technology architecture and IT solutions) are covered in both frameworks, although the level of detail is different for each framework.

## 5 Result and Discussion

As explained in the research methodology section, we conducted an assessment by preparing a questionnaire drawn from ACMM and we assigned to various levels of XYZ Polytechnic management to answer questions in the questionnaire. From the results of the questionnaire can be known condition of maturity level as shown in Figure 5, it can be seen that the level of maturity of Polytechnic XYZ is very low, due to the large gap in all indicators. Based on CMM score card that has 12 indicators, there is only one indicator that reaches level one (initial), namely IT Security (indicator 10), while other indicators are still at level 0 (none) with maturity level measurement score is currently at 0.11 from 2.04 (expected maturity level score). The 12 indicators in ACMM are: Architecture process, Architecture development, Business linkage, Senior management involvement, Operating unit participation (two sub-indicators), Architecture communication (three sub-indicators), IT security, Architecture governance, IT investment and acquisition strategy. It shows that organizational management does not have awareness in information technology governance, so it can be said that business processes that occur in organizations have low efficiency and effectiveness.

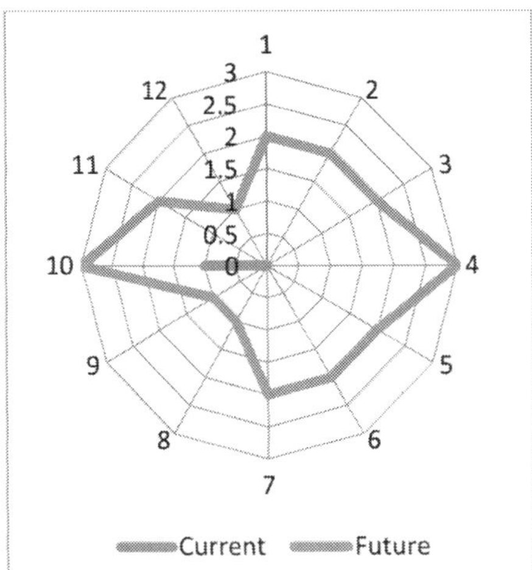


**Fig 4.** Assessment results in maturity level

In ITIL mentioned that IT Service Design is part of the whole process of business change. Services Design definition that refers to Figure 5, it can be even more clearly emphasize that ITIL and TOGAF can do integration (in a friendly way) in their respective areas (Sante & Ermers, 2009), to be: the design of the service is an appropriate and innovative design on the IT-based Services, including architectural design, development process, policy analysis, and documentation, in accordance with current and future agreements. The assessment and analysis results for each phase of TOGAF will be mapped to ITIL, with the following description:

### 5.1 The Preliminary and Phase A are Mapped to the Service Strategy

Preliminary phase is used to determine the scope of enterprise architecture (EA) to be developed, determine commitment with management in EA development and conduct assessment using EA-CMM to know current governance and expected governance, using value chain analysis, while the phase A (Architecture Vision) is creating uniformity of views on the importance of EA to achieve organizational goals formulated in the form of strategy and determine the scope of the architecture to be developed, by making the concept of the solution diagram. Furthermore, the value chain and the concept of solution will be input on the service strategy, as shown in table 1.

**Table 1.** Value chain diagram and the concept of solution mapped to the service strategy.

| Activities | | | | |
|---|---|---|---|---|
| PRIMARY | Inbound Logistics | Student Registration & transfer | E-Registration | Strategy Management for IT Services, Service Portfolio Management, Financial Management for IT Services, Demand Management, Business Relationship Management |
| | Operation | Lectures | E-Learning | |
| | | Exam | E-Learning | |
| | | Internship | E-Message | |
| | | Final project | E-Learning | |
| | Outbond Logistic | Yudisium & Graduation | E-Message | |
| | Marketing | Promotion | Web Portal | |
| | Services | Academic services | E-Scheduler & E-StudyPro | |
| | Firm infrastructure | Financial | E-Payments | |
| | | Libraries | E-library | |
| | Human Resource Management | Staffing (Lecturers & Staff) | E-Attendance | |
| | Technology Development | Computer lab | E-Labs | |
| | Procurement | Facilities & infrastructure | E-Procurement | |

### 5.2 Phase A and Requirement Management are Mapped to the Service Design

Requirement management aims to provide the process of managing architectural needs throughout the phase of the ADM cycle. This stage identifies the organization's core business, organizational issues and conceptualizes business solution concept solutions and information system solution concepts (SI) based on current organizational issues. Furthermore, the solution concept in phase A and the IS solution are mapped to the service design, as shown in Table 2.

Basically, the service strategy is to determine the strategy to serve the customer, with the first step is the assessment of customer and market needs, so that the explanation of the mapping process of value chain analysis and architectural vision, tailored to the main process contained in service strategy, while the requirement managament and architectural vision mapped to service design to develop effective and efficient service solutions that meet current business needs as well as IT needs, derived from service strategy. In Service Design, there are five aspects that must be integrated to ensure that the integrated enterprise architecture that is produced consists of a satisfactory set of standards, design and architecture, all the management and operational requirements of the service, as well as the functionality required by the business.

**Table 2.** The concept of solution and requirement management mapped to the service design.

| Concept of Solution | IS Solution | Service Design |
|---|---|---|
| E-Learning | Academic-IS | • Design Coordination<br>• Service Catalogue Management<br>• Service Level Management<br>• Availability Management<br>• Capacity Management<br>• IT Service Continuity Management<br>• Information Security Management<br>• Supplier Management |
| E-Scheduling | | |
| E-StudyPro | | |
| E-Labs | | |
| E-Message | Student-IS | |
| E-Registration | | |
| E-library | | |
| E-Payments | | |
| E-Budgeting | Administration-IS | |
| E-Attendance | | |
| Web Portal | | |
| E-Procurement | | |

Combining ITIL in the TOGAF context provides a toolkit to convey the vision that XYZ Polytechnic has an opportunity to align their services and achieve alignment of business and technology objectives. This is because ITIL provides a framework that guides organizations in delivering IT services. On the other hand, TOGAF provides methodologies and frameworks that guide Business and IT stakeholders to transform IT throughout the organization. As shown in Tables 1 and 2, the process of developing the concept of solutions can be done simultaneously using both TOGAF and ITIL frameworks. The report in this study is just the first step in integrating the two frameworks to reduce duplication of activities in IT development.

## 6 Conclusion

Combining ITIL in the TOGAF context provides a toolkit to convey the vision that XYZ Polytechnic has an opportunity to align their services and achieve alignment of business and technology objectives. This is because ITIL provides a framework that guides organizations in delivering IT services. On the other hand, TOGAF provides methodologies and frameworks that guide Business and IT stakeholders to transform IT throughout the organization. As shown in Tables 1 and 2, the process of developing the concept of solutions can be done simultaneously using both TOGAF and ITIL frameworks. The report in this study is just the first step in integrating the two frameworks to reduce duplication of activities in IT development.

However, this research has shown that by integrating EA and IT service management provides benefits to the organization in the form of alignment of vision, mission and objectives of the organization in setting up IT development to support the organization's progress and consideration in preparing its resources

## References

1. T. v. Sante and J. Ermers, Whitepaper, OGC Pub., (2009).
2. The Open Group, TOGAF, TOG Pub., (2011).
3. ITGI, IT Gov. Imp. Guide, ITGI, (2007).
4. P. Weill and J. W. Ross, IT Gov.., MIT, (2004).
5. W. Grembergen, Stra. for IT Gov., Idea Group, Inc., (2003).
6. W. V. Grembergen, S. D. Haes dan E. Guldentops, Struct., Proc. and Rel.Mech. For IT Gov, Idea Group Publishing, (2004).
7. R. R. Peterson, Integ. Strat. and Tactics for IT Gov, *Strat for IT Gov.*, Idea Group Inc. , **37** (2004)
8. J. Zachman, A Framework for Inf. Syst. Arc, *IBM Systems Journal,* (1987).
9. ITGI, ITIL for ITSM, 26 March 2017. [Online]. Available: https://www.itgovernance.co.uk/itil
10. Cab. Office, ITIL Best Practise Manag., TSO, 2011.
11. S. Kempter and D. A. Kempter, IT Process WIKI, 26 March 2017. [Online]. Available: http://wiki.en.it-processmaps.com/index.php/Main_Page
12. M. Taleb and O. Cherkaoui, Pattern-Oriented Approach for EA, *Journal of Software Engineering and Applications,* **45** (2012).
13. J. Lussem and H. Harrach, *AICCS),* (2013).
14. N. Ahmad and Z. M. Shamsudin, *Information Technology and Quantitative Management,* **237** (2013).
15. J. Iden and T. R. Eikebrokk, *International Journal of Information Management,* **512** (2013).
16. J. N. Cruz-Hinojosa dan J. A. Gutiérrez-de-Mesa, *Computer Standards & Interfaces,* **124** (2016).
17. TOGAF, "TOGAF Online > Part I: Introduction > Core Concepts," 10 March 2017. [Online]. Available: http://www.togaf.info/togaf9/chap02.html
18. B. H. Cameron and E. McMillan, *Journal of Enterprise Architecture,,* **60** (2013).

# The design of e-learning applications by considering aspects of the user's personality based on students take courses in human-computer interaction

Supangat[a,*], *Francisca Haryati* Chandar[b], and *Agus* Hermanto[c]

[a]Universitas 17 Agustus 1945, Jl. Semolowaru 45, Surabaya, Indonesia
[b]Sekolah Tinggi Teknik Surabaya, Surabaya, Indonesia
[b]Universitas 17 Agustus 1945, Jl. Semolowaru 45, Surabaya, Indonesia.

**Abstract.** The success of an E-Learning is determined from two elemen such as instructional design and user interface design. This study focuses on user interface design. To get a user interface design that suits the personality of students, with carried out research using data mining in e-learning participants. This study uses association rule mining to user design interface of an e-learning application. From the results of the process of training and testing of the 344 (three hundred and forthty-four) dataset consisting of 233 datasets training and 121 datasets testing, introverted personality types, shows that there are two combinations that come from these two processes when counting frequent itemset, namely: { Times New Roman, Blue, White} and {White, Black, Calibri}, whereas personality types extrovert, no combination of dominant favored by this type, and the results are identical to the results of the process of training data, with a choice of preferred color is Blue as foreground and background colors are White. Both of these colors appear in the results of the process of training and testing datasets, with selected font is Arial.

## 1 Introduction

Information technology and advanced rapidly in the last decade, especially the rapid proliferation in the use of the internet and the world wide web is changing the way of life, work, communicate, and human learning. Several studies have shown that a variety of technologies, interactive distance learning cause positive changes in teaching and learning process when compared to the previous distance learning system.

Distance learning is interactive in the current era is the E-learning. E-learning in its broadest sense to encompass learning is done in electronic media (internet) either formally or informally. E-learning refers to the use of Internet technology to transmit a series of solutions that improve the knowledge and skills. It's important to understand how this new technology can affect learning when used by different types of learners.

The visual part of the program E-learning is the appearance of colors, text size, layout, look and feel of the menu or function of the parameters in affecting user interaction with the learning media. As the researchers, the goal in designing and selecting colors, navigation, and buttons have to anticipate what learners need to do and how they will do it. In addition, it should be considered is the use of the method in designing the interface according to the personality of E-learning. So the E-learning is not just learning media alone but can be overcome and help participants E-learning in the plurality of their personality.

To analyze the design of the user interface as it suits the personality of students will require a study using data mining methods to explore the personality of E-learning with user interface design. Researchers used Association rule mining algorithms using a priori to explore each students's personality relevance with user interface design of an E-learning which already exist. From the findings of existing rules which were analyzed by the researchers then used a reference in making e-learning that will be tested on the same students and analyzed the results in order to determine the level of student satisfaction in the use of E-Learning and increase students' ability in understanding the content of its E- Learning.

## 2 Literature Review

More than twenty years ago, Perraton says that "distance education has been successful without theory" (Perraton, 1981). The same thing can be said today of E-Learning, though whether or not it has' succeeded very well may not be so accurate. However, of the various articles that have been published, indicate that institutional investment in practice and uptake of Web-based educational tools for the past decade has proven practice E-Learning has reached a momentum that will make it a central part of future education. However, most of the literature on E Learning is a practice-based and usually presented in a descriptive format.

* Corresponding author: supangat@untag-sby.ac.id

With the development of online learning terms, these terms provide an overview of education that occurs only through the Web, ie, does not consist of physical learning materials issued to learners or face to face. The essential pure of online learning to use the tools E-Learning in distance education mode via the Web as the only medium for all students' and contacts.

While the term mixed mode based learning resources, in turn, explains the educational approach that incorporates a face-to-face and distance education between the instructor or teacher to meet with students (either in person or through technology) and material resources basic content and learning activities available for learners. Learning Management System (LMS) is a learning management system that can be used as a platform for online courses or components built using a particular program.

The interface is an interactive E-Learning, there are two types, namely interactivity indicative and simulative. Interactivity indicative denoted by the use of rollover buttons and navigation. Clicking a button to start an animation or change the page is indicative interactivity. While the simulative interactivity is interactivity that allows students to learn from their own choices in a way that provides some form of feedback. The ability to choose between different Web pages is indicative interactivity; the ability to fly a virtual plane in a realistic virtual environment is simulative interactivity.

The next thing to note is to make sure the learning that occurs through e-learning programs, should be a priority in designing instruction for e-learning itself. Related to the above, that the theory of constructivism learning, which focuses on the construction of knowledge based on experience of previous participants, is suitable to be applied to e-learning because of the certainty of the learning process among students (Harman & Koohang, 2007; Hung & Nichani, 2001; Hung, 2001). The author of this study was used associate rule mining method that is used to categorize participants of E-Learning and getting relations elements in doing interface design in accordance with certain personality group as a participant E-Learning.

As has been explained that the associate rule mining is used in this research will discover certain patterns that associate the data with other data. To search for association rule from a data set, the first step to do is find frequent itemset first. A frequent itemset is a set of items that frequently appear simultaneously. After all patterns of a frequent itemset are found, then search for the associative rules or rules of relevance are eligible who have been determined (Arockiam & Selvaraj, 2013; Fatahi, et al., 2009).

E-learning is a relatively new discipline to grow in the field of education. Advances in technology and the information age is able to change the pedagogy and the way we learn. Many researchers have shown that E-Learning is one tool that can outperform traditional learning. Personalized learning and teaching can be regarded as the ultimate level of the instruction . E-learning process can be influenced by cognitive and affective factors. Among the effective factor are emotion, personality and individual differences, which can be an effective parameter on human activities such as learning. People with different personalities show different emotions in the face of an event. The differences in individual characteristics are reflected in their daily activities and their works. In the case of teaching and learning, personality differences among learners play an important role. Personality learners will determine the level of effectiveness in learning style.

An experiment conducted by Seto Mulyadi (2011) through the learning process approach in E-Learning centered on the development of personality, show that to produce graduates with intellectual talents were very good and creative personality can only be achieved if the learning is done with a process approach in the E-learning program. The relationship between cognitive style, personality, and perception of quality multimedia users explored by Stephen R. Gulliver, et al (2010) of the Informatics Research Center, UK, through experiments on cognitive style and personality. They showed that the type of personality and cognitive style affect user information assimilation.

## 3 Research Methodology

Based on the selection of correlational research method used in this study, the researchers focused his efforts on achieving information that can explain the complex phenomena through the relationship between variables. Thus, researchers can also conduct exploratory study via partial correlation techniques, where researchers eliminate one of the variables that influence can be seen the relationship between two variables that are considered important so that it can be seen a variety of variables which are estimated to have a significant role in achieving the learning process. The reasons for the researchers chose to use the correlation study are as follows:

1. There is a need for information that there is a relationship between the variables for the achievement of the correlation coefficient.
2. There is a need to measure several variables and relationships that exist in a realistic setting.
3. The need for researchers to achieve a prediction formula, which is the state that shows the relationship between variables assumptions.

Participants in this study are students who take courses in HCI. The number of students programmed the HCI courses in one semester as many as 344 students. Each student data field has the desired user interface design including background color, font type, font color, shortcut type, feedback type and simple error handling, obtained from questionnaire results. This study aims to determine the effect of personality on the performance of learning with e-learning, especially to know the difference between traditional e-learning and e-learning adaptation in terms of the MBTI personality types, especially extraversion vs. introvert, based on the assessment process performed by a psychologist

The algorithm used in this case study is the Apriori algorithm (Hongyan, et al., 2009). The rules of apriori algorithm consist of two parameters, support (support value) which is the percentage of the combination of those items in the database and confidence (value certainty) that the strong relationship between items. This analysis aims to determine all associative rules that meet the minimum support and minimum confidence.

## 4 Results and Discussion

The first step is to load a combination of 1 itemsets at each criteria and frequency of each feature support calculated according to the data in the table of personality types of participants in the database.

The results of the calculation process and the formation of frequent itemset for introvert and extrovert personality types using the minimum support value of 25% and a minimum of 75% confident, with the dataset used as training data as much as 223 datasets, will result in the formation of frequent itemset as follows:

a. Introverted Personality Types

```
{Blue, Green 1} = 4
{Arial, Green 1} = 5
{Blue, Arial, Black} = 10
{Blue, Arial, Red} = 5
{Blue, Black, Calibri} = 5
{Blue, Black, Times New Roman} = 36
{Blue, Times New Roman, Red} = 15
{Blue, Times New Roman, White} = 8
{Arial, Black, Red} = 9
{Arial, Black, White} = 8
{Black, Calibri, Green 1} = 7
{Black, Calibri, White} = 4
{Black, Times New Roman, Green 1}= 6
{Black, Times New Roman, Red} = 11
{Black, Times New Roman, White} = 13
{Times New Roman, Green 1, Red} = 4
```

b. Extrovert Personality Type

```
{Arial, Black} = 4
{Black, Green 1} = 3
{Black, Calibri} = 3
{Black, White} = 5
{Times New Roman, Green 1} = 3
{Times New Roman, Red} = 3
{Blue, Black, Times New Roman} = 3
```

Based on the outcomes obtained from the training process above, can be explained as follows:

1. In the introverted personality types, the order of the largest combination favored by this type is {Blue, Black, Times New Roman} with a value of 36 datasets, the next combination is {Black, Times New Roman, White} with a value of 13 datasets. This personality type like Times New Roman font, Calibri, and Arial. While the preferred colors are blue, black, green one, red dan white.
2. In the extroverted personality types, the order of the largest combination favored by this type is {Blue, Black, Times New Roman} with a value of 5 datasets, the next combination is {Arial, Black} with a value of 4 datasets. The above results indicate that this type is not a too complex combination of variables used in designing e-learning applications, compared introverted type who tend to be more complex combinations. The type of font types like this personality type is Times New Roman and Arial. While the colors are preferably the same as the introverted personality types: white black, green and red one.

While the results of the calculation process and the formation of frequent itemset for introvert and extrovert personality types using the minimum support value of 25% and a minimum of 75% confident, with the data set used as the test data as much as 121 datasets, will result in the formation of frequent itemset as follows:

a. Introverted Personality Types

```
{Red, Times New Roman, White} = 12
{Red, White, Calibri} = 6
{Times New Roman, Blue, White} = 23
{Times New Roman, White, Black} = 5
{Blue, Arial, White} = 10
{Blue, White, Calibri} = 10
{Arial, White, Green 1} = 3
{White, Black, Calibri} = 3
```

b. Extrovert Personality Type

```
{Blue, White} = 3
{Red, White} = 3
{White, Arial} = 3
```

The outcome of the application for the establishment of an association rule introvert and extrovert personality types with datasets that are used as training data as much as 223 datasets, can be seen below:

a. Association Rule Type Introvert

```
Calibri => Black = 80.95%
Blue, Red => Times New Roman = 75%
Arial, White => Black = 80% x
Calibri, Green 1 => Black = 87.5%
Green 1, Red => Times New Roman = 80%
```

b. Association Rule Type Extrovert

```
Arial => Black = 80%
White => Black = 100%
Blue, Black => Times New Roman = 75%
```

While the results of the process of formation of the association rules introvert and extrovert personality types using the minimum support value of 25% and a minimum of 75% confident, with the data set used as the test data as much as 121 datasets, will result in the formation of frequent itemset as follows:

a. Association Rule Type Introvert

```
Times New Roman => White = 91.3%
Red, Times New Roman => White = 75%
Calibri => White = 95%
Red, Calibri => White = 100%
Blue => White = 97.73%
Times New Roman, Blue => White = 100%
Black => White = 100%
Times New Roman, Black => White =
100%
Arial => White = 100%
Blue, Arial => White = 100%
Blue, Calibri => White = 90.91%
Green 1 => White = 100%
Arial, Green 1 => White = 100%
Black, Calibri => White = 100%
```

b. Association Rule Type Extrovert

```
Blue => White = 75%
Red => White = 75%
Arial => White = 100%
```

## 5 Conclusion

The use of Association rule with a priori algorithm can provide good results in determining combination parameters of the interface in e-learning applications in accordance with the user's personality with the number of participants for the training dataset 223 and 121 participants for the testing, the execution time below 1 second. Based on the results obtained during the training and testing, it can be concluded that the difference in the number of processed dataset affects the association rule frequent itemset formed or produced. For the introverted personality types, examination results between the training dataset and testing dataset shows that there are two combinations that come from these two processes during frequent itemset counting, namely: {Times New Roman, Blue, White} and {White, Black, Calibri}, so that it can be said that combination will be selected by the personality of this type. On extrovert personality type, no combination is favored by dominant for this type, and the result is identical to the process of training data. While preferred colors preferred by this personality are blue and white which can be selected as foreground or background colors. Both of these colors appear in the results of the process of training and testing dataset, while the selected font is Arial. Establishment of association rule during the process of training and testing, for the introverted personality types, can be concluded that, if the user selects the Arial font and Calibri, then the foreground color selected is White with a value of 100% confident. These results also emerge from the process of additional training data with background color options Black. For the extroverted personality types, there are differences in the results of formation of association rule during the process of training and testing.

## References

1. Arockiam, L. & Selvaraj, J. C., 2013. User Interface Design for Effective E-Learning based on Personality Traits. *International Journal of Computer Applications*, 61(14), pp. 28-32.
2. Fatahi, S., Kazemifard, M. & Ghasem-Aghaee, N., 2009. Design and Implementation of an E-Learning Model by Considering Learner's Personality and Emotions. In: *Advances in Electrical Engineering and Computational Science*. s.l.:Springer Science And Business Media B.V., p. 423.
3. Harman, K. & Koohang, A., 2007. *Learning Objects: Standards, Metadata, Repositories, and LCMS*. s.l.:Informing Science.
4. Hongyan, L., Zhong, W. & Cheng, L., 2009. *The Air Cargo Strategy Based on Apriori and ID3*. Beijing, s.n.
5. Hung, D., 2001. *Theories of learning and computer-mediated instructional technologies*. s.l.:Education Media International.
6. Hung, D. & Nichani, M., 2001. *Constructivism and e-learning: Balancing between the individual and social levels of cognition*. s.l.:Educational Technology.
7. Perraton, H., 1981. *A Theory For Distance Education*. s.l.:Prospects.

# Clustering on Twitter: case study Twitter account of higher education institution in Indonesia

*Almed* Hamzah* and *Ahmad Fathan* Hidayatullah

Department of Informatics, Universitas Islam Indonesia, Jalan Kaliurang KM 14,5 Yogyakarta, Indonesia

**Abstract.** Recently, higher education institutions have been using Twitter as one of tools to enhance their communication network. This paper aims to cluster Twitter data retrieved from the official Twitter account of higher education institutions in Indonesia. We expect to obtain a valuable information from the tweet posted. Furthermore, we use Twitter's hashtag as a basis of clustering. We collect data from n=10 institutions that have an official account on Twitter. The Affinity Propagation algorithm was employed to perform the clustering task. According to the clustering results, we conclude that higher education in Indonesia mostly utilize Twitter to post general information, news, agenda, announcement, information to the new students, and achievement.

## 1 Introduction

The popularity of social media is increasing most rapidly over the last decade. Social media has become a trend in people's daily life over the last decade. Social media has been utilized by many individual and institutions to share any information about various topics [1].

In the academic context, social media is the one of the tool that could help academic activities. Among the social networking sites, Twitter is the one of social media that has been widely used by academic institutions such as higher education. Previous research has reported that Twitter is the one of the most well-known social media in the world and has significant influence in the academic world compared to other social media including Facebook [2]. Therefore, there are some benefits that can be obtained by higher education from using Twitter for academic purposes.

This research aims to extract information from Twitter data which posted by higher education in Indonesia based on Twitter hashtags. In addition, Furthermore, the clustering task would be performed to identify the topic group the hashtags. Therefore, the result of the clustering task could illustrate the meaningful topic and what kind of topics that posted by higher education in Indonesia.

The rest of the paper is organized as follows. Section 2 will describe about review result of related works that have been done by other researcher. Section 3 will explain about methods that has been used in this study. Section 4 will deliver the research result and followed by relevant discussions. Section 5 will end this paper by concluding remarks.

## 2 Related Works

In this part, the related works which has been conducted before are discussed. Rosa, et al. [3] have investigated about tweet classification into some categories. The approach of the research inspired from Google News service provided by Google. They used LDA and K-Means to cluster the tweet data. Moreover, they also proposed new method to complement the lack of Twitter in annotating the topic using hashtag. The method to classify the tweet are unsupervised and supervised method. The unsupervised method has revealed that the tweet classification conducted based on the language or word similarity not based on the discussed topic. On the other hand, the supervised method has worked better on the short text and high noise.

Tang, et al. [4] have reported that there are two issues on tweet clustering process. Firstly, the issue about data sparsity caused by the short message on Twitter which limited no more than 140 characters. Secondly, the issue about the language ambiguity which frequently used by people in tweet messages. Accordingly, they proposed the concept from Wikipedia in representing tweet into vector concept.

Tripathy, et al. [5] proposed the clustering technique based on word frequency and Wikipedia topic taxonomy to find the discussed topic in the tweet. The research has revealed that the proposed algorithm has given better result than the algorithm which only involve word frequency.

Kim, et al. [6] has suggested the method called Core-Topic-based-Clustering (CTC) to extract meaningful topics then cluster the tweet based on the topic. As comparison, the research also conducted the clustering task using K-Means method. The experiments showed that CTC method can efficiently extract the meaningful topic. In addition, the CTC algorithm has better performance than K-Means.

The use of social media in the context of the

* Corresponding author: almed.hamzah@uii.ac.id

institutions has not been done. This is most visible from publications that there are currently only discusses the use of social media by individuals, although some publications discussed in the context of the academic world. This is similar to that expressed by [2]. However, a few publications are trying to reveal the role of social media at the university level.

Based on studies conducted by [2], there are two main interests that encourage the use of social networking media by the university, i.e. marketing activities and delivering information about the university to the students. These findings are based on the perception of the university; however, the findings are not supported by empirical data. Other studies revealed similar things, even supported by empirical data. This research was conducted by [7] found that, on Twitter, universities use it for marketing and public relations (50%) and communicating with students (50%).

## 3 Data Retrieval

The data in this research were retrieved from the Twitter account of top 10 universities and colleges in Indonesia based on the rank from the minister of research, technology and higher education in 2015. Before retrieving tweet data, we identified the official Twitter account of those top 10 higher education in Indonesia. The tweet datasets were retrieved using Twitter API v1.1. To access the API, we utilized Tweepy library in Python. Data retrieval was carried out by using GET statuses/user_timeline provided by Twitter. GET statuses/user_timeline method provides a set of tweets from a particular user with screen_name or user_id parameters.

The downloaded Twitter messages are the tweets which are not protected by the owner of the account. Furthermore, the tweet data obtained using the above method is similar to the tweet data viewed by the user profile on Twitter. However, this method can only access approximately 3240 latest tweets from a particular user timeline. We have obtained 31351 tweets with various amount of Tweets for each account being observed.

## 4 Topic Clustering Based on Hashtags using Affinity Propagation

In this section, the topic clustering task based on hashtags would be explained. This step aims to group the tweets based on their similarity of hashtag-based topics. In Twitter, users use hashtag (#) to represent the subject of their posts. Hence, we can extract the topic of the messages by identifying their hashtag only. Rosa et al. [3] stated that tweet clustering based on hashtag has given better result than content based clustering.

### 4.1 Affinity Propagation

The chosen algorithm to cluster the hashtags is Affinity Propagation (AP). This algorithm clusters the data by sending messages among the data points until convergence is reached [8]. Before running the algorithm, it is not necessary specify the number of clusters since it works by iteratively refining a randomly-chosen initial set of exemplars, which are recognized as those most representative of other samples.

The algorithm works through a number of iterations until the data has reached the convergence condition. Moreover, there are two messages passing categories in each iteration, responsibility and availability.

The responsibility $r(i,k)$ measures how well-suited $k$ is to serve as the exemplar for point $i$. Furthermore, the responsibility of a sample $k$ to be the exemplar of $i$ is updated by the function:

$$r(i,k) \leftarrow s(i,k) - max\left[a(i,\acute{k}) + s(i,\acute{k}) \forall\, \acute{k} \neq k\right] \quad (1)$$

The availability $a(i,k)$ shows how appropriate the point $i$ to be picked in sample $k$ as its exemplar. The availability calculation is based on the function:

$$a(i,k) \leftarrow min\left[0, r(k,\acute{k}) + \sum_{\acute{i}\, s.t.\, \acute{i} \notin \{i,k\}} r(\acute{i},k)\right] \quad (2)$$

Initially, all values for $r$ and $a$ are set to zero. After that, the calculation will run iteratively until reach the convergence condition.

### 4.2 Hashtags Analysis

From the Twitter dataset, we extract all of the hashtags. The extraction process of hashtags from Twitter was conducted by identifying words or phrases preceded by the hash (#) symbol. Overall, there are 12813 hashtags and 1825 unique hashtags from total of 31351 tweets. These hashtags will become raw data for clustering task. Figure 1 shows the top 20 most frequent hashtags in our dataset.

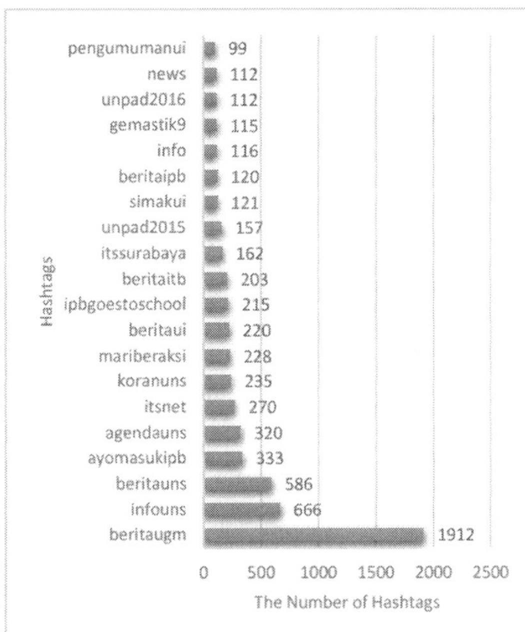


**Fig 1.** Top 20 most frequent hashtags

The top 20 most frequent hashtags illustrate the most topic that are frequently posted by the Twitter account of higher education in Indonesia. At a glance, the information topic posted are talking about news or general information about the college.

According to the figure, it can be seen that half of top 20 represents about news or information, for instance *beritaugm, infouns, beritauns, koranuns, beritaui, beritaitb, beritaipb, info, news,* and *pengumumanui.* The other hashtags indicate the agenda or activity such as *agendauns, ipbgoestoschool, simakui,* and *gemastik9.* The hashtags *itsnet, itssurabaya, unpad2015,* and *unpad2016* could be identified as the general information about the university itself. The rest of the hashtags such as *ayomasukipb* and *mariberaksi* illustrates persuasion to do specific action.

### 4.3 Clustering Result

The hashtags clustering was performed by using the Affinity Propagation algorithm. Before the clustering process, we compute the similarity matrix by multiplying negative and Levenshtein distance. In addition, this research works with precomputed distance measure by assigning *"precomputed"* value to affinity parameter. The convergence iteration is 10 which means the number of iterations with no change in the number of estimated clusters that stops the convergence after 10 iterations. Maximum number of iterations are 100 and damping factor is 0.5.

According to the clustering process, the estimated number of clusters is 274 as can be seen in figure 2. However, the silhouette coefficient of this result is -0.358. This low silhouette value indicates that the clustering configuration may have too many or too few clusters.

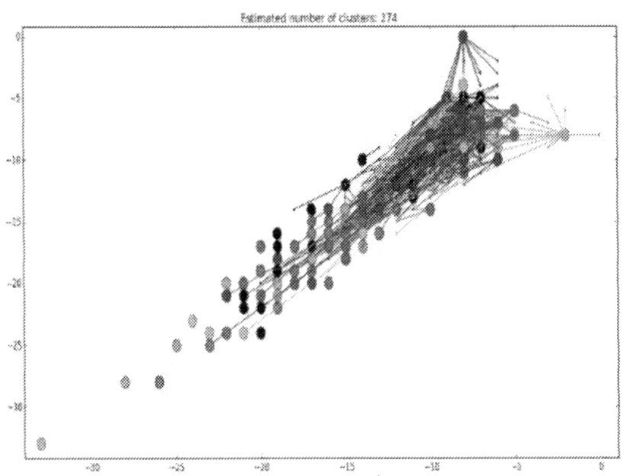


**Fig 2.** Clustering plot

Table 1 shows the top ten cluster results with the hashtags in the clusters. We tried to label each of the cluster by looking the most frequent hashtags in the cluster.

**Table 1.** Top ten hashtags clustering results

| Clusters | Hashtags | Topics |
|---|---|---|
| Cluster – 1 | 38thuns, cdcuns, expouns, fotouns, ikauns, info, infocdugm17, infogizi, infolomba, infonh, infopkm, infougm, infoui, infouns, internship, lingkungan, lnfouns, | Info |
| Cluster – 2 | agenda, agenda2, agendabandung, agendafh, agendaipb, agendanh, agendaugm, agendauns, agendaunsuns, | Agenda |
| Cluster – 3 | koranuns, civitasuns, congratssss, garansindo, gelorarun, gowesuns, hope4uns, kabaruns, karyapenuhmakna, kerjahumas, klipinguns, kontakuns, koranuns, koranunsuns, korauns, kotasolo | News about UNS (Universitas Negeri Sebelas Maret) |
| Cluster – 4 | beritafema, beritafh, beritafik, beritafkui, beritaipb, beritaitb, beritaskpm, beritaugm, beritaui, beritauns, berituns, berkasub | News |
| Cluster – 5 | econews , enews, ieo17, ieyc, junewish, nes, new, news, newyear, niche, nots, now, runners, sdgs, ubnews, unsnews | News |
| Cluster – 6 | pengumuman, pengumumandiib, pengumumanfik, pengumumanui, pengumumanuns, penundaanwisuda, | Announcement |
| Cluster – 7 | ub, ub51, ub53, ubdi, ubstlh, ubtv, ui, uktub, uns, uts | Specific information about colleges |
| Cluster – 8 | camabaub2013sila, camabaub2014, camabaub2103, mabaub2013 | New Students of Universitas Brawijaya |
| Cluster – 9 | award, budaya, calonjuara, charoen, eksyarjuara, itsjuara, iysca, jakarta, jgtc39, juara, juaraan, juarai, jumatberkah, mobiljuara | Achievement |
| Cluster – 10 | drpmupdate, fkhmdelegates, funnyquote, liveupdate, puprgtcw, quote, unstoppable, unsupdate, update, updates | Information update and quotes |

It can be seen that the hashtags in the cluster-1 are dominated by the word *"info"* which means that the topic in cluster-1 is general information about the college. The hashtags in cluster-2 denote the agenda of the higher education. Cluster-3 is more specific talking about the news regarding UNS (Universitas Negeri Sebelas Maret). Cluster-4 and 5 represent the news topic. The hashtags in cluster-6 are about the announcement. Cluster-7 shows the specific information about the colleges. In cluster-8, all the hashtags represent the information of new students of Universitas Brawijaya. Cluster-9 is talking about the achievement of the colleges. Cluster-10 shows the information update and quotes.

## 5 Discussion

In this section, we will discuss further about the clustering results. According to the hashtags analysis in the previous section, it is mentioned that a half of the top 20 hashtags depicts the news and general information regarding the colleges. The top 10 clustering results show that cluster-1, cluster-3, cluster-4, cluster-5, cluster-7, and cluster-10 also represent about news and general information on the higher education. Therefore, it can be concluded that most of higher education in Indonesia mainly utilizes Twitter to post general information and news.

Furthermore, there are more specific topic that represented by the cluster-2, cluster-6, cluster-8, and cluster-9. The topic of cluster-2 has the same topic with the previous hashtags analysis which illustrate about the agenda. From the cluster 2, 6, 8, and 9, there are some important topics that frequently posted via Twitter such as agenda, announcement, information to the new students, and achievement.

## 6 Concluding Remarks

Twitter is among the alternatives for higher education to held an effective communication with their stakeholders. It can be a means to share various information regarding to academic activity. Moreover, it allows the user to create specific topic by using Twitter's hashtag which in turn will increase reader's awareness about particular information. In this study, we use the hashtags to cluster the information that being posted by universities on their Twitter's account.

The clustering method we used in this research is Affinity Propagation Algorithm. According to the clustering results, we conclude that higher education in Indonesia mostly utilize Twitter to post general information, news, agenda, announcement, information to the new students, and achievement.

Overall, from the Twitter messages posted by the higher education in Indonesia could give the illustration about what kind of information that frequently discussed. The information posted could be give beneficial for the internal people in the colleges and external people outside the colleges.

The authors would like to thank Department of Informatics Universitas Islam Indonesia to fund this research in the scheme of Internal Research Grant.

## References

1. J. K. Sinclaire and C. E. Vogus, "Adoption of social networking sites: An exploratory adaptive structuration perspective for global organizations," *Inf. Technol. Manag.*, vol. 12, no. 4, pp. 293–314, 2011.
2. C. H. F. Davis III, R. Deil-Amen, C. Rios-Aguilar, and M. S. G. Canche, "Social Media in Higher Education: A literature review and research directions." 2012.
3. K. Dela Rosa, R. Shah, B. Lin, A. Gershman, and R. Frederking, "Topical clustering of tweets," *Proc. ACM SIGIR SWSM*, 2011.
4. G. Tang, Y. Xia, W. Wang, R. Lau, and F. Zheng, "Clustering tweets using Wikipedia concepts." in *LREC*, 2014, pp. 2262–2267.
5. R. M. Tripathy, S. Sharma, S. Joshi, S. Mehta, and A. Bagchi, "Theme Based Clustering of Tweets," in *Proceedings of the 1st IKDD Conference on Data Sciences*, 2014, pp. 1–5.
6. S. Kim, S. Jeon, J. Kim, Y.-H. Park, and H. Yu, "Finding core topics: Topic extraction with clustering on tweet," in *Cloud and Green Computing (CGC), 2012 Second International Conference on*, 2012, pp. 777–782.
7. R. Reuben, "The use of social media in higher education for marketing and communications: A guide for professionals in higher education." 2008. Retrieved from http://doteduguru.com/wp-content/uploads/2008/08/social-media-inhigher-education.pdf
8. B. J. Frey and D. Dueck, "Clustering by passing messages between data points," *Science (80-.)*, vol. 315, no. 5814, pp. 972–976, 2007.

# Determining e-learning success factor in higher education based on user perspective using Fuzzy AHP

*Rini* Anggrainingsih*, *Muhammad Zuhrul* Umam, and *Haryono* Setiadi

Informatics Dept, Sebelas Maret University, Jl Ir Sutami 36 A, Kentingan, Surakarta, Indonesia 57126

**Abstract.** Recently almost all universities in the world have implemented E-learning to support their academic system. Previous studies have been conducted to determine CSF using Analytic Hierarchy Process (AHP) method. However, AHP method cannot handle the uncertainty and vagueness of the human's opinion, so then it causes less appropriate decision. Some researcher has proposed to use fuzzy sets theory with AHP to increase the ability of AHP to deal problem regarding the uncertainty/fuzziness. This study aims to determine ranks of priorities of the multiple factors which influence the E-learning success using FAHP method. The respondents consist of ten e-learning's experts, 305 lecturers, and 4195 students at Sebelas Maret University. The result describes similar success factors ranking between both experienced and non-experienced user (lecturer and student). Then, the result shows that there are five most influencial success factors of e-learning at Sebelas Maret University based on the lectures perspective *Financial Policy, Regulatory Policy, Course quality, Relevant Content* and *Technical Support*. On the other hand, according to the student's point of view five most e-learning, critical success factors are *Quality of Course, Relevant of Content, Completeness of Content, Attitudes toward Student*, and *Flexibility in taking Course*. Therefore, this finding can be used by E-learning management of Sebelas Maret University to deteremine a strategy to to achieve successful implementation of e-learning at Sebelas Maret University with consider these factors.

## 1 Introduction

Almost all universities in the world have implemented e-learning to support their academic system today. There are various advantages of e-learning such as students can complete courses in anytime and anywhere, and enable to choose subjects according to his ability. Moreover, e-learning can reduce the costs related to instructor's salaries, classroom, student and lecture travel. Also, students can interact directly with lectures [1]. So, understanding various factors which influence the success of e-learning implementation is important. This study aims to identify the success factors of e-learning programs in Sebelas Maret University from Lecturer and student's perspective.

A literature review of previous research on CSF e-learning gives us information that five factors identified as success factors are University Involvement, Quality of Infrastructure and System, Quality of Design, Attitude of Students and Lecturer Toward E-learning.

Previous studies have been conducted to determine CSF using Analytic Hierarchy Process (AHP) method m [1 - 3]. However, AHP method has a drawbacks that AHP can only compute information which has an absolute value and improper to deal the uncertainty value of human subjective judgment cause less precision result of priority ranking [4]. To deal this issue, some researchers suggest to using Fuzzy theory to merge with AHP to improve the ability of AHP to handle uncertainty/Fuzzyness [4 - 8]. This article aims to determine E-learning CSF rankings based on student and lecturer's perception using Fuzzy-AHP.

## 2 Literature Review

### 2.1 E-learning Critical Success Factors

According to various previous studies, there are many factors influence success of e-learning in the universities as follow

#### *2.1.1 Universities Involvement*

The involvement of university's management has a significant role in influencing the success of E-learning. This factor consist of a university policies (Financial Policy) [2, 9, 10], Regulatory policies [2, 9, 10], Technical Support [2, 9 - 11], Held seminars and training [2, 10].

#### *2.1.2 Quality of Infrastructure and System*

Infrastructure and e-learning system is a major factor in supporting the success of E-learning. This element consists of Level portability products [2, 11], The standard of reliability product [2, 10, 11], Easy to understand and easy to use [2, 10, 11], Design and user interface system [2].

* Corresponding author: rini.anggrainingsih@staff.uns.ac.id

### 2.1.3 Quality of Design and Material

Design and good course materials can significantly increase user satisfaction E-learning. This factors consist of Course quality [2, 11, 12], relevant content [2, 11, 12], completeness of content [2, 11, 12], and Flexibility in taking course [2, 12].

### 2.1.4 Student's Characteristic

These factors consist of expertise and insight into the use of computers [2, 12], expertise and insight into the utilization of the internet [2, 12], Attitudes toward e-learning [2, 12], forum/discussion availability [12].

### 2.1.5 Lecture's Characteristic

Lecture characteristics that affect the success of E-learning, such as attitude towards students [2, 12], timely response [2, 12], liveliness lectures [2, 11, 12], and attitudes toward e-learning [2, 11, 12].

## 2.2 AHP and Fuzzy AHP

Analytic Hierarchy Process (AHP) is a multicriteria decision-making approach which arranges some relevant decision factors in a hierarchy structure [13]. The international scientific community has accepted AHP method as a robust and flexible multi-criteria decision-making tool for dealing complex decision problems [13].

However, there is a drawback of AHP method that AHP is just able to compute an absolute value in number only and does not has adequate to reveal precision decision on a subjective judgment of human [4]. A Fuzzy method is merged with AHP to increase the ability of AHP on uncertainty/Fuzzyness problem [4 - 8].

Fuzzy-AHP is an improvement of AHP method which combines fuzzy theory and AHP method. Fuzzy theory is a theory to overcome the confusion and uncertainty of a judgment [5]. The fuzzy-AHP method can deal vagueness of opinion which cannot be accommodated using AHP only. The uncertainty data can be represented in a fuzzy number, as known as Triangular Fuzzy Number (TFN). TFN denoted by $(l, m, u)$ where $l$ is the possibility of the smallest value, $m$ is the value of the most promising, and $u$ is the opportunity to the greatest value.

AHP method can combine ratings and preferences of individuals and groups who then called AHP Group or AHP Group Decision Making [14]. AHP Group or AHP Group Decision Making is the development of AHP method which is individual. The judgement is not only done by one person but can be several people at once. There are two techniques in the merging of personal preferences in the group [15], there is *Aggregation of Individual Judgements (AIJ)* and *Aggregation of Individual Priorities (AIP)*.

*Aggregation of Individual Judgements (AIJ)* method combines a collection of different choices in the group into a single person / single group. Some decision makers made pairwise comparisons are combined into one group pairwise comparisons using the formula (1).

$$A^{[G]} = \left(a_{ij}^{[G]}\right), where\ a_{ij}^{[G]} = \prod_{k=1}^{r}\left(a_{ij}^{[k]}\right)^{\beta_k}, i, i \in \{1, n\} \quad (1)$$

Where $A^{[G]}$ a new pairwise comparison matrix, $a_{ij}^{[G]}$ he scale of the new interest in the new comparative matrix, $r$ a number of decision-makers, $k$ a number of columns (factors) in the pairwise comparison matrix, $a_{ij}^{[k]}$ the scale of the interests of the individual comparison matrix column $k$, dan $\beta k$ is the weight of the decision maker. If the weight of the decision maker synonymous people who make decisions in the group have the same interests and strengths then $\beta k = 1/r$

*Aggregation of Individual Priorities (AIP)* is merging technique whereas each preference would be to have each priority alternative. So every person in a group decision-makers sought alternative weighting of pairwise they make, then the priority weight of each was combined into a priority group. To combine the preferences of each priority person group can use the formula (2).

$$w_i^{[G]} = \prod_{k=1}^{r}\left(w_i^{[k]}\right)^{\beta_k} \quad (2)$$

Where $w_i^{[G]}$ is a joint priority weight, $r$ a number of decision-makers, $k$ a number of columns (factors) in the priority weighting matrix, and $\beta k$ is the weight of the decision maker. If the weight of the decision maker synonymous people who make decisions in the group have the same interests and strengths then $\beta k = 1/r$.

Technique incorporation of preference on Fuzzy-AHP is calculated by combining pairwise made each person into a matrix of the pairwise group by using the formula (3) [16].

$$l_{ij} = \left(\prod_{k=1}^{K} l_{ijk}\right)^{1/K}, \quad m_{ij} = \left(\prod_{k=1}^{K} m_{ijk}\right)^{1/K}, \quad u_{ij} = \left(\prod_{k=1}^{K} u_{ijk}\right)^{1/K} \quad (3)$$

Where $l_{ij}$ is $l$ (lowest triangular fuzzy number) in a new row $I$ column $j$, $K$ is a number of decision-makers. Then, $l_{ijk}$ is $l$ (most inferior triangular fuzzy-number) policymakers to $k$ row $i$ column $j$, $m_{ij}$ is $m$ (average triangular fuzzy number) of new row $i$ column $j$, and $u_{ij}$ is $u$ (upper triangular fuzzy number) of new row $i$ column $j$.

## 3 Research Method

This study investigated e-learning CSFs using three sets of a questionnaire for expert, lecturer and student at Sebelas Maret University, Surakarta, Indonesia. The survey was designed as ratio type, whereas the questions consist of factors and subfactors were compared each other one by one. We used Likert scale ratio and weights respectively Strongly Agree (SS), Agree (S), Normal (N), Disagree (TS) and Strongly Disagree (STS ) for five (5), four (4), three (3), two (2) and one (1) respectively.

The questionnaire for experts developed using Google form while questionnaire for students and lecturers was developed using PHP laravel and PostgreSQL. We conducted an online survey on two stages. Firstly, we conducted survey for ten e-learning expert which consists of ten statements to obtain weight and rank of five factors that identified as e-learning success factors namely; University Involvement (A), Quality of Infrastructure and System (B), Quality of Design (C), Attitude of Students and Lecturer Toward E-learning (D & E). E-learning experts are both lecturer and technician of Sebelas Maret University who have experience in managing online class as an admin and who have experience research in e-learning topic.

**Table 2.** E-learning critical success factors & subfactors

| No | Sub-Factor | Code | Reference |
|---|---|---|---|
| **University Involvement (A)** | | | |
| 1 | University's (Financial Policy) | Sub F 1 | [2, 9, 10] |
| 2 | Universities Policies (Regulatory Policy) | Sub F 2 | [2, 9, 10] |
| 3 | Technical Support | Sub F 3 | [2, 9, 10, 11] |
| 4 | Seminars and Training Availability | Sub F 4 | [2, 10] |
| **Quality of Infrastructure and Systems (B)** | | | |
| 5 | The level of portability products | Sub F 5 | [2, 11] |
| 6 | The level of product reliability | Sub F 6 | [2, 10, 11] |
| 7 | Easy to understand & easy to use | Sub F 7 | [2, 10, 11] |
| 8 | Design and user interface system | Sub F 8 | [2] |
| **Quality of Design and Courses (C)** | | | |
| 9 | Course quality | Sub F 9 | [2, 11, 12] |
| 10 | Relevant Content | Sub F 10 | [2, 11, 12] |
| 11 | Completeness of Content | Sub F 11 | [2, 11, 12] |
| 12 | Flexibility in taking Course | Sub F 12 | [2, 12] |
| **Student's Characteristics (D)** | | | |
| 13 | Expertise and insight to use a computer | Sub F 13 | [2, 12] |
| 14 | Expertise and insight to use the internet | Sub F 14 | [2, 12] |
| 15 | Attitudes toward e-learning | Sub F 15 | [2, 12] |
| 16 | Forum / discussion availability | Sub F 16 | [12] |
| **Lecture's Characteristics (E)** | | | |
| 17 | Attitudes toward Student | Sub F 17 | [2, 12] |
| 18 | Timely response | Sub F 18 | [2, 12] |
| 19 | liveliness lectures | Sub F 19 | [2, 11, 12] |
| 20 | Attitudes toward e-learning | Sub F 20 | [2, 11, 12] |

Then, the second survey was conducted to find weight and rank of subfactors of five e-learning success factors as shown in Table 2 from lecturer and student point of view. Whereas, the lecturers gave 30 statements of questions while the students were given 18 statements of questions.

The application of survey was life online at http://survey.uns.ac.id/elearning. Then, the collected data was calculated using the Fuzzy-AHP application as following steps. First, make a pairwise comparison group using Aggregation of Individual Judgements (AIJ), then determined the fuzzy weight and defuzzification and normalised using formula (1) (2) and (3).

## 4 Result and Discussion

We ran Survey application from 31 October to 21 November 2016 and obtained 10 data from e-learning expert of Sebelas Maret University Surakarta and 305 lecturers and 4195 students who answered questionnaire completely (as shown in Table 3, 4, 5). The lectures consist of 93 experienced lecturers and 212 non-experienced on e-learning, while the students consist of 1370 students experienced and 2825 non-experienced students on e-learning.

**Table 3.** Survey result for e-learning expert at UNS

| Factors Compared | 1 | 2 | 3 | 4 | 5 |
|---|---|---|---|---|---|
| | STS | TS | N | S | SS |
| A more important than B | 0 | 1 | 1 | 4 | 4 |
| A more important than C | 0 | 3 | 1 | 6 | 0 |
| A more important than D | 0 | 1 | 2 | 7 | 0 |
| A more important than E | 0 | 0 | 5 | 5 | 0 |
| B more important than C | 1 | 5 | 3 | 1 | 0 |
| B more important than D | 0 | 2 | 3 | 3 | 2 |
| B more important than E | 0 | 6 | 0 | 3 | 1 |
| C more important than D | 0 | 2 | 3 | 1 | 4 |
| C more important than E | 0 | 2 | 2 | 4 | 2 |
| D more important than E | 0 | 4 | 2 | 4 | 0 |

**Table 4.** Survey result for lecturers

| Subfactor Compared | 1 | 2 | 3 | 4 | 5 |
|---|---|---|---|---|---|
| | STS | TS | N | S | SS |
| 1 more important than 3 | 22 | 41 | 78 | 131 | 33 |
| 1 more important than 4 | 22 | 42 | 88 | 124 | 29 |
| 2 more important than 3 | 21 | 49 | 94 | 113 | 28 |
| 2 more important than 4 | 17 | 40 | 102 | 116 | 30 |
| 3 more important than 4 | 14 | 39 | 120 | 107 | 25 |
| 5 more important than 6 | 17 | 36 | 104 | 115 | 33 |
| 5 more important than 7 | 11 | 17 | 84 | 139 | 54 |
| 5 more important than 8 | 9 | 17 | 98 | 128 | 53 |
| 6 more important than 7 | 8 | 18 | 105 | 128 | 46 |
| 6 more important than 8 | 7 | 13 | 93 | 135 | 57 |
| 7 more important than 8 | 7 | 12 | 113 | 131 | 42 |
| 9 more important than 10 | 9 | 5 | 96 | 144 | 51 |
| 9 more important than 11 | 19 | 39 | 130 | 97 | 20 |
| 9 more important than 12 | 12 | 31 | 122 | 109 | 31 |
| 10 more important than 11 | 15 | 61 | 134 | 81 | 14 |
| 10 more important than 12 | 11 | 26 | 126 | 109 | 33 |
| 11 more important than 12 | 15 | 44 | 133 | 96 | 17 |
| 13 more important than 14 | 15 | 20 | 136 | 105 | 29 |
| 13 more important than 15 | 8 | 26 | 107 | 132 | 32 |

| Subfactor Compared | 1 STS | 2 TS | 3 N | 4 S | 5 SS |
|---|---|---|---|---|---|
| 13 more important than 16 | 6 | 8 | 90 | 148 | 53 |
| 14 more important than 15 | 7 | 21 | 111 | 135 | 31 |
| 14 more important than 16 | 4 | 11 | 100 | 144 | 46 |
| 15 more important than 16 | 8 | 17 | 127 | 125 | 28 |
| 17 more important than 18 | 5 | 6 | 98 | 149 | 47 |
| 17 more important than 19 | 11 | 27 | 125 | 117 | 25 |
| 17 more important than 20 | 11 | 27 | 127 | 115 | 25 |
| 18 more important than 19 | 12 | 32 | 121 | 112 | 28 |
| 18 more important than 19 | 8 | 17 | 124 | 126 | 30 |
| 18 more important than 20 | 9 | 19 | 142 | 100 | 35 |
| 19 more important than 20 | 6 | 28 | 139 | 105 | 27 |

**Table 5.** Survey result for students

| Subfactor Compared | 1 STS | 2 TS | 3 N | 4 S | 5 SS |
|---|---|---|---|---|---|
| 5 more important than 7 | 211 | 260 | 1645 | 1590 | 489 |
| 5 more important than 8 | 201 | 265 | 1690 | 1537 | 502 |
| 6 more important than 7 | 192 | 174 | 1726 | 1532 | 571 |
| 6 more important than 8 | 209 | 182 | 1745 | 1523 | 536 |
| 7 more important than 8 | 190 | 161 | 1726 | 1584 | 534 |
| 9 more important than 10 | 186 | 249 | 1730 | 1498 | 532 |
| 9 more important than 11 | 191 | 448 | 1796 | 1305 | 455 |
| 9 more important than 12 | 202 | 359 | 1885 | 1311 | 438 |
| 10 more important than 11 | 173 | 340 | 1848 | 1377 | 457 |
| 10 more important than 12 | 175 | 265 | 1933 | 1396 | 426 |
| 11 more important than 12 | 169 | 220 | 1918 | 1428 | 460 |
| 17 more important than 18 | 235 | 164 | 1780 | 1430 | 586 |
| 17 more important than 19 | 231 | 171 | 1867 | 1428 | 498 |
| 17 more important than 20 | 219 | 218 | 1894 | 1341 | 523 |
| 18 more important than 19 | 221 | 155 | 1926 | 1388 | 505 |
| 18 more important than 20 | 217 | 162 | 1979 | 1367 | 470 |
| 19 more important than 20 | 246 | 251 | 2052 | 1205 | 441 |

Then, employed Fuzzy-AHP method to compute the data and rank the influence e-learning success factors. To make computation process easier, we designed an application program to apply Fuzzy-AHP algorithm using PHP laravel and PostgreSQL as a database. While weight for each factor was changed to the fuzzy-AHP ratio scale as seen in Table 6

**Table 6.** The Fuzzy-AHP proportion [5]

| Ratio | Triangular fuzzy Scale | Triangular fuzzy Scale (Reciprocal) |
|---|---|---|
| Normal (N) | (1, 1, 1) | (1/1, 1/1, 1/1) |
| Disagree (TS) | (1/7, 1/5, 1/3) | (3, 5, 7) |
| Strongly Disagree (STS) | (1/9, 1/7, 1/5) | (5, 7, 9) |
| Agree (S) | (3, 5, 7) | (1/7, 1/5, 1/3) |
| Strongly Agree (SS) | (5, 7, 9) | (1/9, 1/7, 1/5) |

The calculation result for expert point of view shown in Table 7. According to Table 7, we get information that the most important factor is 'University Involvement' then, the second is 'Quality of design and courses' following by 'Lecturer's Characteristic', 'Quality of Infrastructure and System' and 'Student's Characteristics'.

**Table 7.** E-learning success factors ranking based on expert perception

| Factor | Ranking |
|---|---|
| University Involvement | 1 |
| Quality Of Infrastructure and Systems | 4 |
| Quality Of Design and Courses | 2 |
| Student's Characteristics | 5 |
| Lecture's Characteristics | 3 |

Table 8 illustrates the calculation results of critical e-learning priority from a lecturer who has both e-learning experienced and non-e-learning experienced lecturers point of views. Table 8 shows that both experienced and non-experienced lecture have similar notion that there are five essential sub-factors for e-learning success 'Financial Policy', 'Regulation Policy', 'Technical Support', 'Course Quality' and 'Relevant Content'. Both experienced and non-experienced lecture agreed that the most important factor is university policy (financial policy and regulation policy).

**Table 8.** The Priority of e-learning success factor based on lecturers point of view

| Subfactor | Exp Lecture | Non-exp Lecture |
|---|---|---|
| University Policies (Financial Policy) | 1 | 1 |
| University Regulation (Regulation Policy) | 2 | 2 |
| Technical Support | 4 | 3 |
| Seminars and Training Availability | 6 | 6 |
| Portability products | 8 | 8 |
| Reliability product | 13 | 13 |
| Easy to understand and easy to use | 16 | 17 |
| Design and user interface system | 20 | 19 |
| Course quality | 3 | 4 |
| Relevant Content | 5 | 5 |
| Completeness of Content | 9 | 7 |
| Flexibility to taking Course | 10 | 10 |
| Expertise to use a computer | 11 | 9 |
| Expertise to use the internet | 14 | 14 |
| Attitudes toward e-learning | 18 | 18 |
| Forum / discussion availability | 19 | 20 |
| Attitudes toward Student | 7 | 11 |
| Respond time | 12 | 12 |
| liveliness lectures | 15 | 15 |
| Attitudes toward e-learning | 17 | 16 |

Then, calculations results of critical subfactors of e-learning priority according to students point of view as shown in Table 9.

**Table 9.** Calculations results of critical e-learning priority according to students point of view

| Subfactor | Exp Student | Non-Exp Student |
|---|---|---|
| The level of portability products | 6 | 6 |
| The level of product reliability | 8 | 8 |
| Easy to understand and easy to use | 10 | 11 |
| Design and user interface system | 12 | 12 |
| Course quality | 1 | 1 |

| Subfactor | Exp Student | Non-Exp Student |
|---|---|---|
| Relevant Content | 2 | 3 |
| Completeness of Content | 3 | 2 |
| Flexibility in taking Course | 4 | 4 |
| Attitudes toward Student | 5 | 5 |
| Timely response | 7 | 7 |
| liveliness lectures | 9 | 10 |
| Attitudes toward e-learning | 11 | 9 |

Table 9 describes that both experienced and not experienced students on e-learning have similar notion that the most five e-learning CSFs are subfactors 'Course quality', Relevant Content, Completeness of Content, Flexibility in taking Course, and Lecture's Attitudes toward Student. Then, both experienced and non-experienced student agreed that the most important e-learning success subfactor is 'Course Quality'.

## 5 Conclusion

Based on the research, we conclude that both experienced and non-experienced e-learning user have similar notion about e-learning critical success factor as follows

- Five factors that most influence the CSF E-learning UNS according to the lectures perception are *Financial Policy, Regulatory Policy, Course quality, Relevant Content and Technical Support*
- Five factors that most influence the CSF E-learning UNS based on students point of view are *Course quality, Relevant Content, Completeness of Content, Attitudes toward Student, and Flexibility in taking Course*

Therefore, this finding can be used by E-learning management of Sebelas Maret University to determine a strategy to achieve successful implementation of e-learning at Sebelas Maret University with considering these factors.

## References

1. Gautam, S. and M.K. Tiwari, *Components and benefits of E-learning system*. 2014.
2. Bhuasiri, W., et al., *Critical success factors for e-learning in developing countries: A comparative analysis between ICT experts and faculty*. Computers & Education, 2012. **58**(2): p. 843-855.
3. Anggrainingsih, Rini, et al. "Determining elearning critical success factor at sebelas maret university using Analytical Hierarchy Process (AHP)." *Technology of Information and Communication (ISemantic), International Seminar on Application for*. IEEE, 2016.
4. Ayhan, M.B., *A Fuzzy AHP approach for supplier selection problem: a case study in a gear motor company*. arXiv preprint arXiv:1311.2886, 2013.
5. Kabir, G. and A. Akhtar Hasin, *Evaluation of customer oriented success factors in mobile commerce using fuzzy AHP*. Journal of Industrial Engineering and Management, 2011. **4**(2): p. 361-386.
6. Kimiafar, K., et al., *Prioritizing factors influencing nurses' satisfaction with hospital information systems: a fuzzy analytic hierarchy process approach*. CIN: Computers, Informatics, Nursing, 2014. **32**(4): p. 174-181.
7. Zain, D. and B. Santoso, *Determination of Naval Based Locations: Strategy to Maximize Performance Monitoring of Defense and Security Systems in The Sea (Study on Maritime Security and Defense Systems in Indonesia)*. Jurnal Aplikasi Manajemen, 2011. **9**(1): p. 254-263.
8. Tiznado, M. and L. Pradenas, *LOCATION OF PREHOSPITAL CARE BASIS THROUGH COMBINED FUZZY AHP AND GIS METHOD*. 2014.
9. Biškupić, I.O., S. Lacković, and K. Jurina, *Successful and Proactive e-learning Environment Fostered by Teachers' Motivation in Technology Use*. Procedia-Social and Behavioral Sciences, 2015. **174**: p. 3656-3662.
10. McGill, T.J., J.E. Klobas, and S. Renzi, *Critical success factors for the continuation of e-learning initiatives*. The Internet and Higher Education, 2014. **22**: p. 24-36.
11. Lwoga, E.T., *Critical success factors for adoption of web-based learning management systems in Tanzania*. International Journal of Education and Development using Information and Communication Technology, 2014. **10**(1): p. 4.
12. Chow, W.S. and S. Shi, *Investigating students' satisfaction and continuance intention toward e-learning: An Extension of the expectation–confirmation model*. Procedia-Social and Behavioral Sciences, 2014. **141**: p. 1145-1149.
13. Srdevic, Z., B. Blagojevic, and B. Srdevic, *AHP based group decision-making in ranking loan applicants for purchasing irrigation equipment: a case study*. Bulgarian Journal of Agricultural Science, 2011. **17**(4): p. 531-543.
14. Huong, N.Q., *COMBINING AHP GROUP ANALYSIS AND GIS IN VULNERABILITY ASSESSMENT OF PROTECTED AREA IN VIETNAM*. 2014.
15. Ssebuggwawo, D.D., S.S. Hoppenbrouwers, and H.E. Proper, *Group Decision Making in Collaborative Modeling*. 2010.
16. Meixner, O., *Fuzzy AHP group decision analysis and its application for the evaluation of energy sources*. Institute of Marketing and Innovation. Vienna, Austria, 2009.
17. Abdullah, L., *Ranking of the factors associated with road accidents using correlation analysis and fuzzy TOPSIS*. 2010.
18. Govindan, K., A. Diabat, and K.M. Shankar, *Analyzing the drivers of green manufacturing with*

*fuzzy approach*. Journal of Cleaner Production, 2015. **96**: p. 182-193.

19. Astuty, W., *An Analysis on the Impact of Participatory Budgeting and Procedural Fairness Toward Manager's Commitment and Performance*. Global Journal of Management And Business Research, 2015. **14**(5).
20. Chakraborty, R., A. Ray, and P. Dan, *Multi criteria decision making methods for location selection of distribution centers*. International Journal of Industrial Engineering Computations, 2013. **4**(4): p. 491-504.
21. Hussain, N. and S. Subramoniam, *Ranking Critical Success Factors of Healthcare Management Information Systems using AHP*. 2014, ISAHP.

# Implementation of dataflow programming based Fuzzy Logic algorithm for gas concentration index in around of Sidoarjo mudflow, Indonesia

*Edita* Rosana Widasari*, *Barlian Henryranu* Prasetio, *Hurriyatul* Fitriyah, and *Reza* Hastuti

Dept. of Computer Engineering, Faculty of Computer Science, University of Brawijaya, Indonesia

**Abstract.** Sidoarjo mudflow or known as Lapindo mudflow erupted since 2006. The Sidoarjo mudflow is located in Sidoarjo City, East Java, Indonesia. The mudflow-affected area has high air pollution level and high health risk. Therefore, in this paper was implemented a system that can categorize the level of air pollution into several categories. The air quality index can be categorized using fuzzy logic algorithm based on the concentration of air pollutant parameters in the mudflow-affected area. Furthermore, Dataflow programming is used to process the fuzzy logic algorithm. Based on the result, the measurement accuracy of the air quality index in the mudflow-affected area has an accuracy rate of 93.92% in Siring Barat, 93.34% in Mindi, and 95.96% in Jatirejo. The methane concentration is passes the standard quality even though the air quality index is safe. Hence, the area is indicated into Hazardous level. In addition, Mindi has highest and stable methane concentration. It means that Mindi has high-risk air pollution.

## 1 Introduction

Air is a natural resource that played important role in the lives of human and other living things [1]. The air quality must be monitored for the maintenance of human health and protection of other living thing [2]. The air was around in everyday life is called ambient air [3]. One of the areas that monitored by ambient air pollution is the Sidoarjo mudflow-affected area [4]. The mudflow-affected area is an area that closes to hot mudflow accordance with the mapping of the Muds Area Control Map (Sidoarjo Mudflow Management).

The Sidaorjo mudflow or known as Lapindo mudflow erupted in Porong district, Sidoarjo city, East Java, Indonesia. The Sidoarjo mudflow is located at 900 km east Jakarta (the capital city of Indonesia) or 14 km south of Surabaya (the capital of East Java). The mudflow erupted because of drilling activities for gas exploration sine 2006. The mudflow has enormous volume and spread quickly. Hence, it covers surrounding area and cause air pollution [5].

Indonesia has six national standards for air quality criteria in ambient air, such as: carbon monoxide (CO), nitrogen oxide ($NO_x$), sulphur oxide ($SO_x$), hydro-carbon (HC), ozone ($O_3$), and dust particles [6]. Carbon monoxide (CO) and dust particles are the important gas parameter in the mudflow-affected area. Beside of that, methane ($CH_4$) is a dangerous air pollutant in the mudflow-affected area [7]. Moreover, The air pollution can cause respiratory disturbances. Respiratory disturbance is caused by high concentration of air pollutant parameters. It can make adverse effects on the human body and hamper the daily activities [8]. Therefore, the mudflow-affected area has high air pollution level and high health risk.

Numerous researchers have observed the concentration value of air pollution parameters based on Ambient Air Quality Standard and Unmoved Source Emission [9, 10]. However, that is less effective because there is not dangerous levelling while the concentration of air pollution crossing the standard. Therefore, in this paper was implemented a system that can categorize the level of air pollution into several categories. The air quality index can be categorized using fuzzy logic algorithm based on the concentration of air pollutant parameters in the mudflow-affected area. Furthermore, Dataflow programming is used to execute parallel execution and pipelining technique [11]. It is required in the processes of data acquisition of sensors, data processing of sensors, and fuzzy logic algorithm.

## 2 Air Quality Index

The air quality index is used to determine the air quality status. AirNOW is the international standard of air quality index used. US Environmental Protection Agency (EPA) publishes it [12]. The categories of air quality index those shown in Table 1.

## 3 System Overview

The system is consisting of hardware part and software part. The hardware part has three kinds of sensors: 1) MQ4 sensor, which is methane gas sensor, 2) MQ7 sensor, which is CO gas, and 3) Dust sensor SHARP

* Corresponding author: editarosanaw@gmail.co.id

GP2Y1010AU0F, which is a sensor to detect dust particles. All of sensors are connected to NI MyRio through analog input pin is shown in Figure 1. NI MyRio also used as microcontroller. Therefore, NI LabVIEW is used for software part. There are the function of NI LabVIEW in this system: 1) It used to display the value of gas parameters (methane, CO, and dust), 2) It is stored the data into *.csv* format file on USB flash-disk through USB port on MyRio, 3) It is displayed the result of fuzzy logic algorithm.

**Table 1.** Air quality index

| No | Range | Level |
|---|---|---|
| 1 | 0 - 50 | Good |
| 2 | 50 - 100 | Moderate |
| 3 | 100 - 150 | Unhealthy for sensitive group |
| 4 | 150 - 200 | Unhealthy |
| 5 | 200 - 300 | Very unhealthy |
| 6 | 300 - 500 | Hazardous |

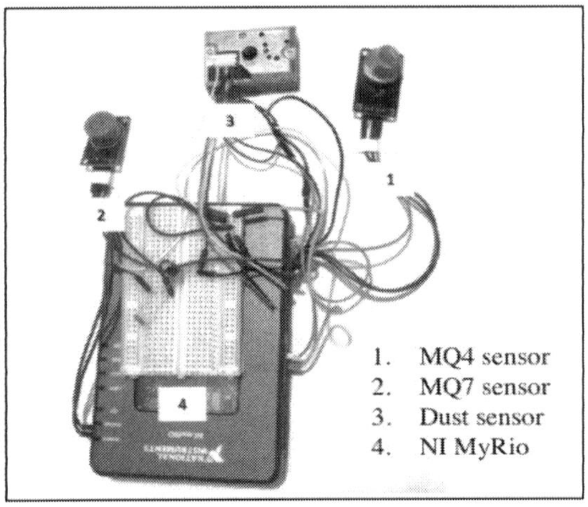


**Fig 1.** The hardware of the system

## 4 Methods

### 4.1 Linear Regression

Linear regression is used to determine the concentration of each gas parameters. The value of gas parameters is obtained from datasheet on each sensor. The concentration of methane gas, CO gas, and dust gas can be seen in Equation 1, Equation 2, and Equation 3, respectively. Where, $y$ is the gas concentration in units of part per million (ppm) and $x$ represents the shifting of ($R_s/R_o$) ratio. $R_s$ is resistance at 1000 ppm gas parameters in ambient air while $R_o$ represents sensor resistance in another gas.

$$y = 1021x^{-2.7887} \quad (1)$$

$$y = 96.311x^{-1.239} \quad (2)$$

$$y = 0.172x^{0.0999} \quad (3)$$

### 4.2. Fuzzy Logic Algorithm

The fuzzy logic algorithm consists of the following steps.

#### *4.2.1 Defining the variables*

The system uses three gas parameters as input variable (methane, CO, dust) to calculate air quality index. However, CO and dust were used as input fuzzy. It is based on air quality criteria [6]. While, air quality index is used as output fuzzy.

#### *4.2.2 Fuzzyfication*

The fuzzyfication is used to transform the crisp input into fuzzy input. The membership function must first be determined for each input and output variable. The membership function for CO, dust, and air quality was already adopted by international standard of air quality index as shown in Figure 2, Figure 3, and Figure 4.

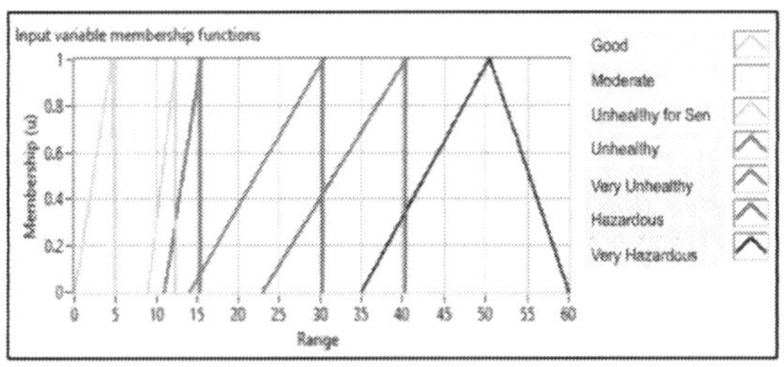


**Fig 2.** Membership function for CO (as input)

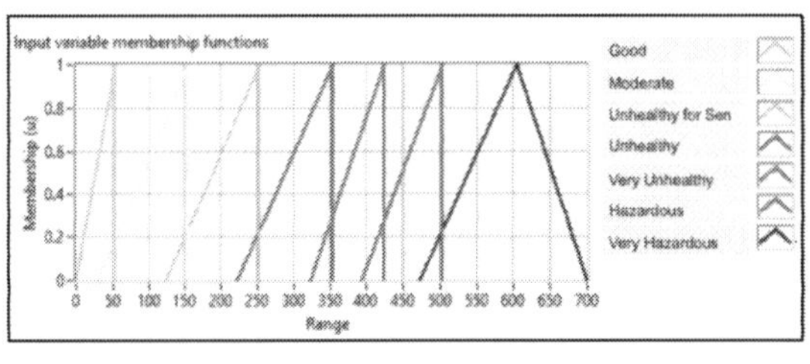


**Fig 3.** Membership function for dust (as input)

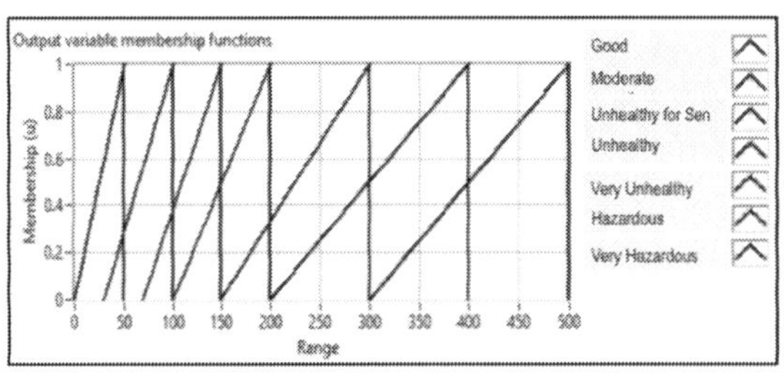


**Fig 4.** Membership function for air quality (as output)

#### *4.2.3 Fuzzy Interface Rules*

The rules are used to determine the input and output of fuzzy logic algorithm. The rules based 7x7 membership function those shown in Table 2. Fuzzy Mamdani or known as Min-Max Method is related to process the rules.

**Table 2.** Rule based 7x7 membership function

| Dust Particle/ CO | Very Hazardous | Hazardous | Very Unhealthy | Unhealthy | Unhealthy for sensitive group | Moderate | Good |
|---|---|---|---|---|---|---|---|
| Good | Very Hazardous | Hazardous | Very Unhealthy | Unhealthy | Unhealthy for sensitive group | Moderate | Good |
| Moderate | Very Hazardous | Hazardous | Very Unhealthy | Unhealthy | Unhealthy for sensitive group | Moderate | Moderate |
| Unhealthy for sensitive group | Very Hazardous | Hazardous | Very Unhealthy | Unhealthy | Unhealthy for sensitive group | Unhealthy for sensitive group | Unhealthy for sensitive group |
| Unhealthy | Very Hazardous | Hazardous | Very Unhealthy | Unhealthy | Unhealthy | Unhealthy | Unhealthy |
| Very Unhealthy | Very Hazardous | Hazardous | Very Unhealthy | Very Unhealthy | Very Unhealthy | Very Unhealthy | Very Unhealthy |
| Hazardous | Very Hazardous | Hazardous | Hazardous | Hazardous | Hazardous | Hazardous | Hazardous |
| Very Hazardous | Very Hazardous | Very Hazardous | Very Hazardous | Very Hazardous | Very Hazardous | Very Hazardous | Very Hazardous |

*4.2.4 Defuzzyfication*

The center of area (CoA) method is used in this step. It is used to get a crisp output. The method calculates the weighted average of a fuzzy set.

### 4.3 Linear Interpolation

The linear interpolation used to calculate air quality index manually by using Equation 4. Where, $I$ is air quality, $I_a$ is upper limit of air quality, $I_b$ is lower limit of air quality, $X_a$ is upper limit of ambient air, $X_b$ is lower limit of ambient air, and $X_x$ is air measurement results. Furthermore, the calculate result will be compared with the result of fuzzy logic algorithm.

$$I = \frac{I_a - I_b}{X_a - X_b}(X_x - X_b) + I_b \qquad (4)$$

## 4 The implementation and Testing Result

Figure 5 shows the implementation of all system on Dataflow programming. Firstly, USB flash drive folder is used to call directory path of fuzzy logic file (*.fs*) that is saved in USB flash-disk. Then, FL Load Fuzzy System will load and save the fuzzy logic file (*.fs*). Local variable will call the output data of MQ4 sensor, MQ7 sensor, and Dust sensor. Furthermore, FL Fuzzy Controller will process the data. There are fuzzycation process, fuzzy interface rules, and defuzzycation process in the FL Fuzzy Controller.

The process of MQ4 sensor, MQ7 sensor, and Dust sensor to determine the concentration of methane, CO, and dust are shown in Figure 6, Figure 7, and Figure 8. The linear regression method (Equations 1-3) is used in these processes. The value of each sensor output in ppm will becomes an input on FL Fuzzy Controller.

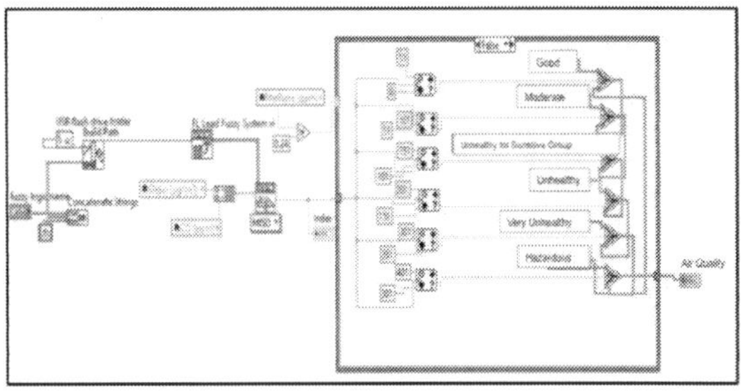

**Fig 5.** Implementation of all system on Dataflow programming

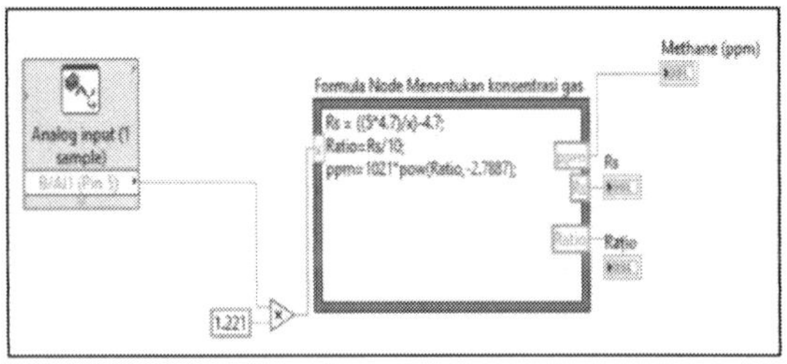


**Fig 6.** Dataflow programming to determine the concentration of methane

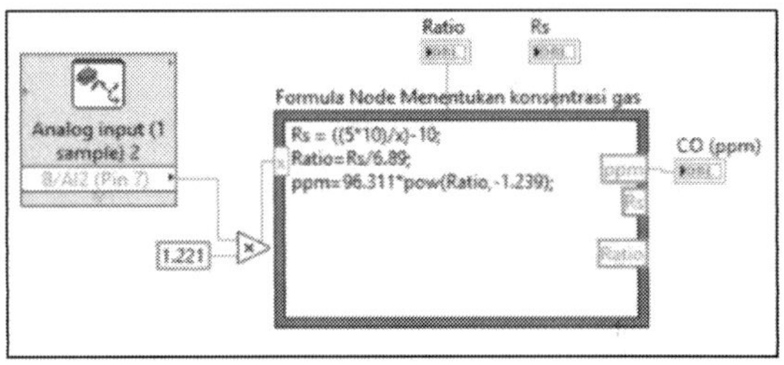


**Fig 7.** Dataflow programming to determine the concentration of CO

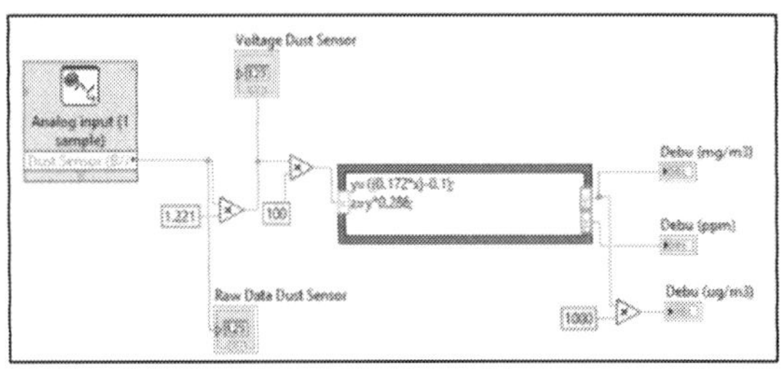


**Fig 8.** Dataflow programming to determine the concentration of dust

Figure 9 shows the implementation of fuzzy logic algorithm. The fuzzy interface rules is consist of 7x7 membership function that must be declared. The rules are obtained based on the combination of each input to fuzzy output.

The system was already implemented in three areas, which closed to Sidoarjo mudflow. The areas are: 1) Siring Barat, 2) Mindi, 3) Jatirejo. The result of air quality test in Siring Barat, Mindi, and Jatirejo are shown in Table 3, Table 4, and Table 5, respectively.

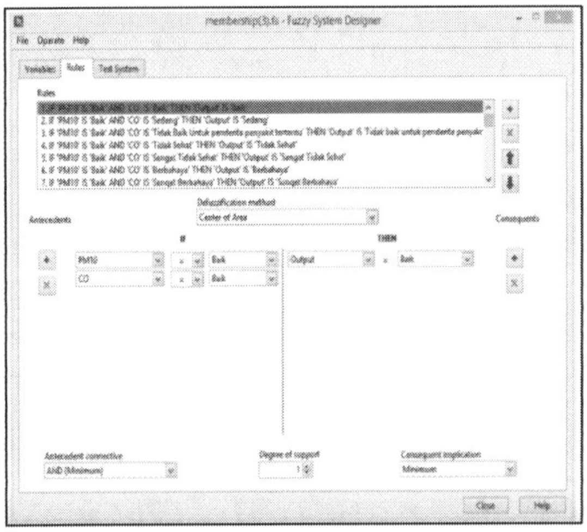

**Fig 9.** Fuzzy logic algorithm on Dataflow programming

**Table 3.** The result of air quality test in Siring Barat

| Methane (ppm) | Dust (ppm) | CO (ppm) | Air Quality (Linear Interpolation) | Air Quality (Fuzzy Logic Algorithm) | Error (%) | Category |
|---|---|---|---|---|---|---|
| 21.00 | 26.27 | 0.45 | 24.00 | 23.80 | 0.83 | Hazardous |
| 12.30 | 76.00 | 1.20 | 63.00 | 69.11 | 9.70 | Hazardous |
| 14.77 | 79.04 | 1.60 | 63.00 | 7.46 | 11.84 | Hazardous |
| 8.89 | 28.18 | 1.52 | 26.00 | 25.95 | 0.19 | Hazardous |
| 6.40 | 26.50 | 2.45 | 24.00 | 27.34 | 13.92 | Hazardous |
| 9.25 | 29.15 | 2.00 | 27.00 | 27.00 | 0.00 | Hazardous |
| Average | | | | | 6.08 | |

**Table 4.** The result of air quality test in Mindi

| Methane (ppm) | Dust (ppm) | CO (ppm) | Air Quality (Linear Interpolation) | Air Quality (Fuzzy Logic Algorithm) | Error (%) | Category |
|---|---|---|---|---|---|---|
| 1.88 | 27.05 | 0.78 | 25.00 | 23.80 | 4.80 | Hazardous |
| 2.49 | 80.05 | 3.00 | 63.00 | 71.04 | 12.76 | Hazardous |
| 2.49 | 79.45 | 1.49 | 63.00 | 69.88 | 10.92 | Hazardous |
| 2.49 | 53.82 | 1.49 | 49.00 | 47.09 | 3.90 | Hazardous |
| 1.11 | 40.00 | 1.04 | 37.00 | 39.61 | 7.05 | Hazardous |
| 2.29 | 45.30 | 1.49 | 42.00 | 42.23 | 0.55 | Hazardous |
| Average | | | | | 6.66 | |

## 5 Analysis

Based the testing result on Table 3, Table 4, and Table 5, all area is indicated into hazardous level. It is caused by methane concentration passed the standard quality. The standard quality of methane is 0.24 ppm [13]. In other words, all area has methane concentration more than 0.24 ppm.

**Table 5.** The result of air quality test in Jatirejo

| Methane (ppm) | Dust (ppm) | CO (ppm) | Air Quality (Linear Interpolation) | Air Quality (Fuzzy Logic Algorithm) | Error (%) | Category |
|---|---|---|---|---|---|---|
| 6.76 | 28.12 | 1.50 | 26.00 | 25.91 | 0.35 | Hazardous |
| 5.15 | 182.01 | 1.49 | 114.00 | 114.54 | 0.47 | Hazardous |
| 5.40 | 182.30 | 1.60 | 114.00 | 114.97 | 0.85 | Hazardous |
| 3.94 | 23.82 | 1.55 | 49.00 | 46.83 | 4.43 | Hazardous |
| 3.94 | 55.01 | 1.57 | 51.00 | 68.37 | 17.37 | Hazardous |
| 2.90 | 28.18 | 1.50 | 26.00 | 26.20 | 0.77 | Hazardous |
| Average | | | | | 4.04 | |

However, there is a change in each methane concentration result. It means that MQ4 sensor produces different methane concentration ($R_s$ value) in each area. Therefore, the comparison between methane concentration and ($R_s/R_o$) ratio is necessary to determine the influence of methane concentration in each area.

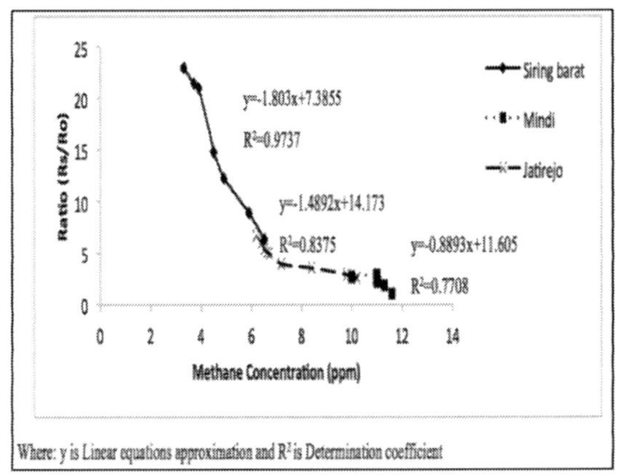


**Fig 10.** Comparison between methane concentration and ($R_s/R_o$) ratio

Figure 10 shows the linear regression of methane concentration and ($R_s/R_o$) ratio. 97.37% of methane concentration is influenced by ($R_s/R_o$) ratio while 2.63% of methane concentration is influenced by other variables in Siring Barat. 83.75% of methane concentration is influenced by ($R_s/R_o$) ratio while 16.25% of methane concentration is influenced by other variables in Jatirejo. 77.08% of methane concentration is influenced by ($R_s/R_o$) ratio while 22.93% of methane concentration is influenced by other variables in Mindi. The linear regression graph also shows that Mindi has the lowest gradient slope. Therefore, Mindi has highest and stable methane concentration.

## 6 Conclusions

The accuracy of the air quality index in the mudflow-affected area has an accuracy rate of 93.92% in Siring Barat, 93.34% in Mindi, and 95.96% in Jatirejo. The methane concentration is passes the standard quality even though the air quality index is safe. Hence, the area is indicated into Hazardous level. In addition, Mindi has highest and stable methane concentration. It means that Mindi has high-risk air pollution. A development real-time gas concentration index monitoring system in around of Sidoarjo mudflow is necessary for future work. The system is expected can give early warnings when the hazardous level comes.

## References

1. C. Chiu and Z. Zhang, *The Air Quality Evaluation Based on Gas Array*, IEEE Semiconductor Technology International Conferencem (2017)
2. World Health Organization (WHO), *Air Quality Guidelines*, (WHO Regional Office for Europe, Denmark, 2006)
3. World Health Organization (WHO), *Ambient Air Pollution: A Global Assessment of Exposure and Burden of Disease*, (WHO Document Production Services, 2016)
4. C. Poh, *Lapindo Brantas and The Mud Volcano, Sidoarjo, Indonesia*. Friends of the Earth International and Friends of the Earth Europe (2007)
5. L. Bayuaji et al, *Temporal Analysis of Land Deformation on Erupted Mud Volcano in Sidoarjo, Indonesia using DInSAR Technique*. IEEE International Geoscience and Remote Sensing Symposium (2011)
6. Peraturan Pemerintah, Peraturan Pemerintah Tentang Pengendalian Pencemaran Udara, Jakarta: Sekertariat Negara, **41** (1999)
7. National Agency for Disaster Management, *Dangerous Air Pollutant* (2010).
8. S. Rajasegarar et al, *High Resolution Spatio-Temporal Monitoring of Air Pollutants using Wireless Sensor Networks*, IEEE international Conference on Intelligent Sensors, Sensor Networks and Information Processing (2014).
9. X. Jiang et al, *The statistical distributions of SO2, NO2 and PM10 concentrations in Xi'an, China*, IEEE International Symposium on Water Resource and Environmental Protection (2011).
10. H. Sulistyarso, *Dynamic and Geological-Ecological Spatial Planning Approach in Hot Mud Volcano Affected Area in Porong-Sidoarjo*, The Journal for Technology and Science, **21**, 2-3 (2010)
11. National Instruments, F*uzzy Logic for G. Toolkit Reference Manual*, Austin: National Instruments (1997)
12. EPA, *AirNow*, US Environmental Protection Agency (2015)
13. Center for Earth and Disaster Studies, LPPM ITS (2009).

# Enhancing usability using Near Field Communication for mobile application

*Endar* Wihidayat[1,*], *Ovide D.W.* Ardi [2]

[1]Pendidikan Teknik Informatika dan Komputer, Universitas Sebelas Maret, Surakarta, Indonesia
[2]D3 Teknik Informatika, Universitas Sebelas Maret, Surakarta, Indonesia

**Abstract.** Near Field Communication (NFC) as relatively new wireless communication technology pushes new challenges to application developers to make their applications easier to use and simpler to operate. This point of view known as usability element. Usability is one of the elements for creating good quality applications. This study aims to analyse the usability of mobile-based application embeds with NFC. We also try to evaluate usability in applications used by children. We developed an application called Receptionist which has a primary function as a communication tool between students, teachers and parents at a middle school. To know the impact of the NFC, the Receptionist input system is designed with two methods, via conventional navigation (using buttons) and via NFC. To understand the usability of each method, we do user testing and questioners on students. The results show, using the NFC there is a significant increase in usability attributes: efficiency, effectiveness, and learnability. On the other hand, there is decreases of user satisfaction comparing to conventional method. In general, this study demonstrates the potential of new input device technologies that can improve the usability of mobile-based applications.

## 1 Introduction

Many studies discuss the use of NFC in mobile applications. Related to usability purposes, most research on NFC focuses on improving the effectiveness, efficiency, and ease of use of the applications. The use of NFC is aimed at improving effectiveness in product identification [1]. Furthermore, another study shows the application of NFC is more efficient than barcodes [2]. Within the study, it has been demonstrated that 95% of participants agreed that the system was easy to use and easy to learn, with a pleasant interface. Using technology with NFC-enabled help older adults to make the technology easy to use, useful, and enhancing the sense of power [3]. Using an integrated application with NFC for helping children with language impairment proved that the application was easy to understand and learn [4].

Most of the studies do not mention the significance of NFC compared the standard navigation like touch screen buttons or keyboard. When a device is embedded with the NFC, it will the only way to perform certain functions. That situation won't let us know whether NFC is better than other input methods, for example using the button for navigation, or input form using a keyboard for entering the password for authentication. Furthermore, we try to analyze the usability impact of NFC on children. We want to know, is it better or is it matter to make faster input process using NFC, especially for children? So, this study is trying to answer those questions. We build an application, called Receptionist, that provides two ways of navigation to compare its significance. Furthermore, this research will be tested on children as one of the application users.

This research is divided into several sections. The second section will discuss general usability concepts in software, usability in mobile applications, NFC as input devices, usability components and measurement, and usability designed for children apps. The third section will discuss the methods used. The fourth section will show results. And finally, the conclusions of this article are in Section 5.

## 2 Usability, Mobile Application and Children Usability Testing

### 2.1 Software usability

Usability is defined as the ease with which a user can learn to operate, prepare inputs for, and interpret the outputs of a system or component (IEEE Std 610.12-1990). Usability is one of key factor to create a high-quality software. International Organization for Standardization (ISO) standard divides software quality into six general categories of characteristics: functionalities, reliability, usability, effectiveness, maintainability and portability. Usability refers to some concepts such as execution time, performance, user satisfaction and ease of learning [5].

Based on the ISO 9126 (1991), usability is defined as a set of attributes in an application that is intended to meet the needs at the time of use and assessment in the process of use. These attributes are understandability, learnability,

* Corresponding author: endars@staff.uns.ac.id

operability, attractiveness and usability compliance. In the 2001 edition of ISO/IEC 9126-1 define usability as the capability of the software product to be understood, learned and liked by the user, when used under specified conditions. Furthermore, based on the ISO 9241 standard, software is usable when it allows the user to execute his task effectively, efficiently and with satisfaction in the specified context of use. So this standard has three attributes: effectiveness, efficiency, and satisfaction.

Regarding usability, ISO develops several different standards but can be broken down into two categories, product-oriented standards (ISO 9126, 2001; ISO 14598, 2001) and process-oriented standards (ISO 9241, 1992/2001; ISO 13407, 1999). Both are applied to complement each other.

## 2.2 Mobile Application Usability Challenge

The term of mobile application is commonly used refer to software which runs on the mobile devices operation system such as Android, Windows Mobile and iOS. Nowadays, they are evolving rapidly, making ubiquitous information access at anytime and anywhere [6]. Developing applications for mobile devices that pay attention to usability factors is a new challenge. These challenges arise related to the characteristic of mobile devices which has limited hardware capabilities (such as small screen size, limited memory, limited processor, virtual keyboard), unreliable wireless network connections, and many disruptive use environments. Mobile devices have some uniqueness compared to desktop computers as taken from [6] which explains at least five characteristics that are considered still relevant for now.

***Mobile Context:*** This term can be defined as an environmental condition when the users use their mobile devices. Environmental conditions are influenced by many variables such as user locations, the presence of objects such as vehicles, people. The weather, noise and lighting, etc. All these elements affect the usability of the application.

***Connectivity***: Mobile devices are connected to wireless networks that are not reliable compared to fixed networks (using LAN cable). Data transfer rates are affected by wireless networks used (normally, WIFI are more stable than cellular networks).

***Screen Size***: Screen size of touchscreen technology did not change significantly for a decade. And it does not seem to be a significant shift in the future, considering the portability and ergonomic factors of small screen size. What may be different in recent years is the use of an external or extended screen that has been enabled via wireless. However, the use of an extended screen is only possible in conditions where the user is not mobile and only works in a particular mobile context (for instance: indoor use).

***Power and Processing Capability***: Compared to a decade ago, processor capabilities have increased significantly, even multiplied. But the applications that are created also require an increasing resource. In general, mobile devices have lower specs than desktop devices.

***Data Entry Method***: Performing data entry on devices with small screen is a major challenge in application design. Virtual keyboard reduces speed and increases error. No precise mouse cursor also increases navigation errors. In the end, all lead to reduced convenience in the use of the application. In a study by [7], they simplified the application data entry method using NFC as a key feature. This study emphasizes the advantages of NFC which can represent the real-world actions (by tapping as if the user performs other physical activities such as pressing switches).

With those characteristics, in general, applications designed for mobile devices will emphasize simplicity by limiting the information displayed. The app developer will select the most necessary information to be presented.

## 2.3 Near Field Communication (NFC) as Shortcut

The NFC standards and protocols are based on RFID standards outlined in ISO / IEC 14443. The NFC is attached to a smartphone/tablet device that falls into the High-Frequency category at 13.56 MHz NFC works at close range, within a few centimetres range. It consists of an active device (has its own energy, for example, NFC Reader, smartphone or tablet), and passive devices (do not have their own energy, for example, NFC card-shaped tags, stickers, etc.in). Because it works at close range, then NFC tag needs to be taped or pasted onto the active device. NFC has three modes of communication: (1) reader/writer mode, (2) peer-to-peer mode and (3) card emulation mode. This new method of communication means new potential research in improving Quality of Service [8].

In 2006 Nokia issued the 6131-phone model which is the first smartphone device compatible with NFC. And in 2010 Samsung with Google released Nexus S, the first Android with NFC compatible. Within Android OS-based, the NFC function works by overriding or interrupting ongoing activities. This condition makes the event triggered by NFC can be a shortcut to trigger an action or command. This shortcut function is commonly used as an advantage or privilege of the NFC when this feature is first introduced by smartphone vendors. This study is trying to apply the advantages.

## 2.4 Usability Measurement

A study proposed five usability components with several measurement parameters as shown in Table 1 [5]. The component is a proposed improvement of two ISO usability standards of ISO 9241 and ISO 9126.

## 2.5 Usability Testing for Children

Usability testing refer to a process that employs people as testing participants who are representative of the target audience to evaluate the degree to which a product meets specific usability criteria [9]. Currently, mobile applications are commonly used by children in everyday life. But in contrast, usability study of mobile applications, especially applications for children is very little. Testing usability for children has several challenges such as child's tendency to remain silent during testing, span their attention during the test, and in the end making difficult to collect the data [10].

**Table 1.** Usability characteristic and measurement

| Characteristic | Measurement |
|---|---|
| Effectiveness | Percentage of task accomplished, Ratio on failure of handling, Percentage of task achieved per unit of time |
| Efficiency | Repetition number of failed command, documentation or help's use frequency error percentage, time spent on error, time to achieve one task, number of good and bad characteristics recalled by users, number of available commands not called upon |
| Satisfaction | percentage user favorable and unfavorable comments, number of times that user expresses his frustration, rating scale for user satisfaction with functions and characteristics |
| Security | access audibility, access controllability, data corruption prevention, data encryption |
| Learnability | time to learn |

Children are very few in providing feedback and only focus on running the applications. Different child ages require different test adaptations. Commonly, children age is divided into three ranges, preschool (2-5 years), elementary school (5-10 years) and middle school (10-14 years).

Pre-schoolers require the most extensive adaptations of usability testing because their attention span, their motivation to please adults, and their ability to adjust to strange surroundings and new people may change from one moment to the next [11]. [11] recommends seven things need to be prepared for usability testing in children: (1) making a comfortable and child-friendly lab and (2) using a familiar device input. (3) Use laboratory equipment as effectively as one unobtrusively as possible. A one-way mirror is not recommended because it keeps children out of a given task (grooming behaviour). (4) When planning a series of tasks, switch the order around for different children. (5) Screening the children who have at least some experience with the computer, (6) but, exclude the children who have computer expertise (they have advanced computer skill). (7) Not to use our own children or our colleague's children as usability testing participants.

## 2.6 Receptionist Main Feature

Receptionist has several actors and features as shown in Table 2.

**Table 2**. Main feature of *Receptionist*

| Main Functional: | Actor | Detail functional | NFC |
|---|---|---|---|
| SMS Broadcast | Teacher/Admin | SMS to a parent | Y |
| | Teacher/Admin | SMS broadcast | Y |
| Phone call | Teacher, Students | Phone call to parents | Y |
| SMS | Student | SMS to parent | Y |
| Student Call | Parents, Teachers | Students call via speaker | Y |
| Data Management | Teacher/Admin | Import and Manage Student Data | N |
| | Teacher/Admin | Manage user | N |
| | Teacher/Admin | Voice Setting | N |
| | Teacher/Admin | Web Setting | N |
| | Teacher/Admin | SMS Setting | N |
| | Teacher/Admin | NFC Setting | N |

As shown in Table 2, not all navigation supported by NFC. Using a NFC Tag card, only one command can be provided. So, the user has one command that can be assigned using a NFC tag. User testing will be performed on function with the student (children) as the actor and only function with available NFC navigation. In Fig. 1 we can see the user interface of Receptionist.

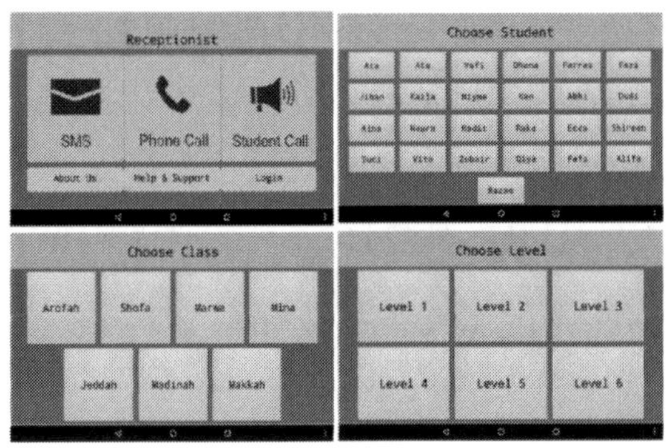


**Fig 1**. Receptionist user interface

## 2.7 Receptionist Architecture

The receptionist is installed in the school hall and administration room. Using acrylic case with extended antenna for NFC as shown in Fig. 2. The Receptionist connected to the WIFI network as a data path to the server and attached to an external sound system for student call function. The tablet has SMS and Call capability via cellular networks to perform SMS and phone call functions (Fig. 3).

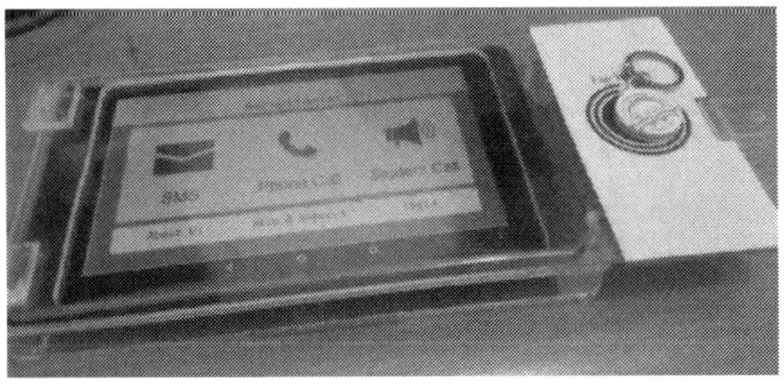

**Fig 2.** Receptionist installed tablet

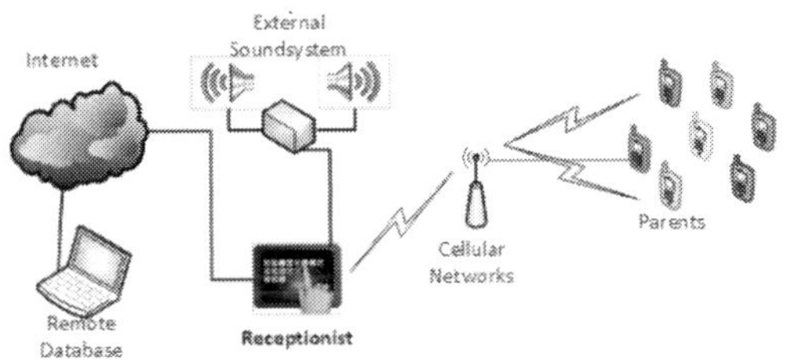


**Fig 3.** Receptionist architecture

## 3 Research Method

Two commonly used methodologies for testing usability are experimental testing within the lab and by field testing or real mobile context testing. Two commonly used methodologies for testing usability are experimental testing within the lab and field testing or real mobile context testing. The advantages of the lab experiment are: (1) full control of the experiment, (2) easy to measure usability attributes, at the same time exclude the irrelevant attributes, and (3) allowing the use of audio and video recording to capture user reactions during testing. [6]. Testing is conducted in a class (own participants class) which makes students feel comfortable. Testing is guided by a moderator. The moderator will guide the task to be done and record the time required to complete each task. Moderators will also be given a hidden video recorder, so hopefully does not affect the spontaneity of the attitude of the user. To help the process of data retrieval, then before usability testing the participants given a way of thinking aloud protocol.

### 3.1 Usability Test Participants

With 20 testing participants than at least 95% usability problems will be found [12]. As much as 28 children with the variation of age 12-13 years who plays as a student chosen taking the test. Participants were selected randomly from sixth-grade students of International Islamic Elementary School (SDII) Al Abidin (private elementary school at Central Java Province, Indonesia). Following the guidelines described in Section 2.6, all students have sufficient background using mobile devices and also familiar with touchscreen input devices.

### 3.2 Usability Components and Data Collection

Usability measurements were taken from [5] as we can see at Table 3. The parameters used in user testing are tailored to the research context that aims to compare the usability between conventional methods (navigation using touchscreen button) and NFC methods. Table 3 shows the chosen measurement parameters.

**Table 3.** NFC usability testing components

| Usability Component | Measurement | Data collection/capturing method |
|---|---|---|
| Effectiveness | Number of successful task<br>Number of error | Video record of testing process |
| Efficiency | Time to achieve one task | Video record of testing process |
| Satisfaction | Percentage of users favorable and unfavorable comments | Questioners |
| Learnability | Percentage of successful user to accomplish task on first trial | Video record of testing process |

For questioners, we adopted the questions from Software Usability Scale (SUS). SUS is known as "quick-and-dirty" usability scale [13], but for preliminary research, it is sufficient to demonstrate the effect of NFC on usability. Because of the mobile context and the user's context (children), it is necessary to adapt the questions that to be more easily to understand by children. There are 15 questions used as shown in Table 4.

**Table 4.** Questioners for usability testing

| No | Question |
|---|---|
| 1 | The application is running fast |
| 2 | No found error during operation |
| 3 | No trouble when first use |
| 4 | I'm very sure to determine the next step |
| 5 | It's no need extra time to learn |
| 7 | Sometimes I felt quite tense when using this app |
| 8 | I do not need an assistant for use this app |
| 9 | I enjoy my moments using this app |
| 10 | The application is fun |
| 11 | This app is very easy to understand |
| 12 | I want to use this app in everyday life |
| 13 | The design is quite attractive |
| 14 | It's easy and fast to move between pages |
| 15 | At the first start, it's easy to find the necessary navigation |

### 3.3 Testing scenario

This research focus to know the effect of NFC implementation for mobile app usability, specifically for children. So, the tested functions are only functions with alternative NFC navigation. Table 5 shows three tasks of function and the action flow for accomplishing the tasks, via button vs. via NFC.

**Table 5.** Action flow via button and via NFC

| Function | Action Flow | |
|---|---|---|
| | Via Button | Via NFC |
| SMS | Home →Pick SMS → Pick Level →Pick Class →Pick Student →**[SMS Sent]** | Home → **[NFC Tap]** →Pick SMS → **[SMS Sent]** |
| Phone call | Home →Pick Phone Call →Pick Level →Pick Class →Pick Student → **[Phone called]** | Home → **[NFC Tap]** →Pick phone call → **[Phone called]** |
| Students Call | Home →Students Call → Pick Level →Pick Class → Pick Student → **[Student Called]** | Home → **[NFC Tap]** → Pick student call→ **[Phone called]** |

## 4. Results and Analysis

From the questioner's results, it can be seen that children are more satisfied using the button (Fig 3), even there is no significant gap. Furthermore, noticeable differences such as ease of navigation to start the task show that children find it difficult to start using NFC.

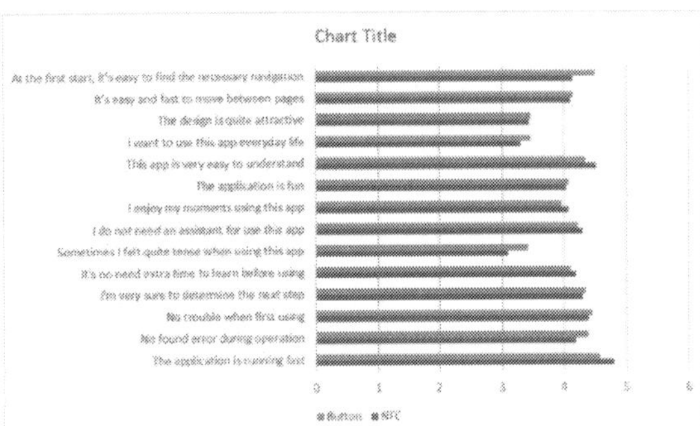


**Fig 3.** Questioner results

The use of NFC was also caused more tension than using a button which is understandable because NFC is not familiarly used as input devices among children. From the unfavorable comment gathered it is known that 12 participants criticize application user interface that less appealing. This result is matched with the questionnaire result (third question). Testing result shows that the NFC is more efficient to accomplish the tasks. The average time required for a single task is 3.5 seconds, a lot different when using a button that takes an average of 8 seconds. Using NFC, students can also complete tasks more effectively. Each method has a failure number while performing the task. If the NFC is due to the less sensitive of the sensor (this is ensured due to the use of the extended antenna that aims to increase the NFC range, originally laid back of the tablet but extend to the side as shown in Fig. 2). Button method failure comes from the interrupt timer which is activated. It happens when the student takes too long for choosing the class name or the student name.

## 5. Conclusion and Discussion

This research is in the early stages to know the role of NFC to increase usability, especially at mobile applications for children. From the preliminary analysis result, it is known unsurprisingly that efficiency, effectiveness and learnability factor are increased by NFC. But in contrast, user satisfaction was decreased.

During the observation, the authors' personal opinion predicted that the speed and outcome, for children, were not the only goal they want to. For children, a longer process to interact is more fun. So, the main feature NFC that speeds up the process actually reduces their fun. Another thing that may affect children's user experience is the fact that all participants are still unfamiliar with NFC, in contrast, they are already familiar with the touchscreen device. With the difference between the NFC and the buttons which is not significant, if they are familiar with both input methods, the results may be different. Suggestions for further research, for children, NFC needs to be applied not only to finish the task faster but also more fun. The developer got extra time by implementing the NFC (because of the faster process). Extra time can be used to make better feedback to users after finished the task, and also could be a more fun way such as displaying fun messages for the users.

## References

1. M. Khosravi, M. Karbasi, A. Shah, I. A. Brohi, and N. I. Ali, *An adoption of halal food recognition system using mobile Radio Frequency Identification (RFID) and Near Field Communication (NFC)*, Proc. - 6th Int. Conf. Inf. Commun. Technol. Muslim World, ICT4M 2016, pp. 70–75 (2017)
2. A. Landman et al., "Efficiency and usability of a near field communication-enabled tablet for medication administration," J. Med. Internet Res.,**16**, no. 6 (2014)
3. W. M. Lim, P. Teh, P. K. Ahmed, A. H. S. Chan, S. Cheong, and W. Yap, *Enhancing the Sense of Power and User Adoption in Gerontechnology: An Experimental Investigation of Near-Field Communication Lighting Systems*, Proc. 2016 IEEE IEEM, pp. 1–5 (2016)
4. M. Luisa Lorusso, E. Biffi, M. Molteni, and G. Reni, "Exploring the learnability and usability of a near field communication-based application for semantic enrichment in children with language disorders," Assist. Technol., vol. 0, no. 0, pp. 1–12 (2017)
5. A. Abran, A. Khelifi, W. Suryn, and A. Seffah, "Usability meanings and interpretations in ISO standards," Softw. Qual. J., vol. 11, no. 4, pp. 325–338 (2003)
6. D. Zhang and B. Adipat, "Challenges, Usability, Methodologies, and Issues in the Usability Testing of

Mobile Applications," Int. J. Human-Computer ..., **18**, no. 3, pp. 293–308 (2005)

7. S. L. Ghiron, S. Sposato, C. M. Medaglia, and A. Moroni, *NFC ticketing: A prototype and usability test of an NFC-based virtual ticketing application*, Proc. - 2009 1st Int. Work. Near F. Commun. NFC 2009, pp. 45–50 (2009)
8. V. Coskun, B. Ozdenizci, and K. Ok, A survey on near field communication (NFC) technology, Wirel. Pers. Commun., **71**, no. 3, pp. 2259–2294 (2013)
9. J. Rubin and D. Chisnel, *Handbook of Usability Testing, Second Edition: How to Plan, Design, and Conduct Effective Tests Published*, 2nd ed., no. 1. Indiana: Wiley Publishing, Inc. (2008)
10. A. Donker and P. Reitsma, *Usability Testing With Young Children*, Interact. Des. Child., pp. 43–48 (2004).
11. L. Hanna, K. Risden, and K. Alexander, "Guidelines for usability testing with children," Interactions, **4**, no. 5, pp. 9–14 (1997).
12. L. Faulkner, Beyond the five-user assumption: Benefits of increased sample sizes in usability testing, Behav. Res. Methods, Instruments Comput., **35**, no. 3, pp. 379–383 (2003).
13. J. Sauro and J. Lewis, Quantifying the User Experience Practical Statistics for User Research. Elsevier Inc. (2012).

# Simulation of trust-based attacks in Internet of Things

*Vera* Suryani[1,2,*], *Selo* Sulistyo[2,*], and *Widyawan* Widyawan[2]

[1]School of Computing, Telkom University, Jl. Telekomunikasi no. 1, Bandung, Indonesia
[2]Department of Electrical Engineering and Information Technology, Universitas Gadjah Mada, Jl. Grafika No. 2, Sleman, Yogyakarta, Indonesia

**Abstract.** Security issue in Internet of Things (IoT) is one of the most important aspects to be resolved. Some attacks that occurred in Internet technology are also penetrated IoT. A trust-based security system is a promising solution in IoT as it is suitable for the characteristic of IoT which is distributed, decentralized control and on-off connections. Knowing the behaviors of the attacks might help up to understand of how we may prevent the attacks and plan to mitigate the attacks. In this paper, we simulated trust-based attacks in IoT environment by giving the fake reputation values of an object. For this purpose, we utilized ConTrust model, a trust-based security model. Matlab was used to simulate the attacks, and the simulation result showed that ConTrust model was outperformed on mitigating a trust-based attack. The attack was detected and resolved correctly.

## 1 Introduction

Internet of Things (IoT) is a new paradigm that is born and developed from Internet technology. The Internet here is used to connect the 'things', which are the IoT forming components. Things can be in the form of physical objects such as vehicles, sensors, cameras, and people. In this paper, the term 'things' are called as objects. Objects can be connected to communicate and interact with each other remotely via the internet. Various internet technologies can be used in IoT, ranging from Radio Frequency Identification (RFID) to large battery powered systems [1]. These technologies and connected objects can be used to form applications on smart cities, smart transportation, smart home, smart logistics, and other various applications.

In smart transportation, smart cameras can be used to monitor the density of the roadways, and sensors can be used to make it easier for car drivers to find the empty parking slots, also speed detecting sensors can be mounted at certain spots to monitor overrunning speed vehicles, reducing potential accidents that may occur. Meanwhile, in the smart homes, sensors are useful for lighting automation, home appliance automation: smart refrigerator, automatic room temperature setting, or automatic window cover in the case of the homeowner forgot to close the window when leaving home. With the IoT, many conveniences are offered in people's everyday life.

However, behind the easiness offered by IoT, there is a security concern in IoT to be considered. Since the IoT was born from Internet technology, it causes various threats that might be appearing there, ranging from worm attacks to Distributed Denial of Service (DDoS). Preventive measures have been done before, from the physical aspects such as authentication models that are embedded in the chip of the sensor to the use of security framework that applies to all components in a system in IoT.

One possible method that can be used to improve security aspect in IoT is by utilizing the trust value among objects. Each object can assess the trust value of another object quantitatively, and the result of its assessment becomes a reference to determine whether the communication process will proceed or not. In our previous work, we have proposed ConTrust model [2] that used a mathematical model to compute the trust value of objects in the IoT environment. Preceded by the authentication process using the Diffie-Hellman algorithm, each object can calculate the trust value of another object independently or without involving a third party. The ConTrust model still requires proof to check how robust the model is against possible attacks, especially trust-based attacks. Being robust is essential for a security framework, especially concerning about attack resistance because the purpose of the framework is to protect the object from various possible vulnerabilities and attacks. In this paper, simulated trust-based attacks are performed to determine the robustness of the ConTrust model. The main contribution of the study is as follows. We outline that some security attacks might happen in the IoT network; one of them is a trust-based attack. We then present the alternative framework to prevent that attack, namely ConTrust model. Some simulations were made to find out the robustness of this model against trust-based attack. As a result, we discovered that this model was robust enough to prevent a trust-based attack.

This paper is organized as follows: Section II describes the related work specializing on security model in IoT, Section III describes the definition of security in

* Corresponding author: [1]vera.s3te14@mail.ugm.ac.id, [2]selo@ugm.ac.id

IoT, including our previous work ConTrust model, Section IV reports simulation result and discussion, and Section V presents the conclusion and future works.

## 2 Related Works

Security on the Internet of Things (IoT) is a very open research area since the security is a fundamental aspect to be managed for any technologies. One component in IoT is an object, and this object has a dynamic character concerning of joining and leaving the network. This dynamic character is a challenge for security aspect.

Generally, an authentication scheme is used to secure the environment in IoT. Several authentication schemes can be applied such as centralization, semi-centralization, and decentralization. The centralization scheme usually involves a third party to handle the authentication processes; one is Certificate Authority (CA). The object should contact a trusted third party in order to be classified as an authenticated object. Some researchers already used this scheme to secure their system [3]. The advantage of centralization scheme is the energy savings from the computing side, yet the disadvantage is difficulties to choose a third party that can be considered reliable. The selection of responsible third parties becomes unnecessary in the decentralization scheme. In this scheme, each object may authenticate each other without involving a third party. However, behind the ease offered by this scheme, there is a relatively high computation cost, considering a load of authentication process calculation is charged to each object [4]. The decentralized scheme is then developed to a semi-decentralized scheme, from which one of the objects in the decentralization scheme is selected as coordinator to handle the authentication process. Local security handled by the local coordinator is considered better than one which involves a third party that might not necessarily be trusted.

Both decentralized and semi-centralized schemes rely on quantitative models to improve the security aspect, such as the trust value model. A study conducted by [5] utilized trust management into three phases of trust model: neighbor discovery, service request, and trust-based computation. The model was able to detect and mitigate on-off attack from malicious objects in the network.

Meanwhile, other researchers used distributed trust schemes and centrality degrees to obtain trust among objects [4]. The proposed model was named TC-BAC, using the trust calculation of objects in the network through direct and indirect trust calculations. Besides, this model also used access control in its architecture.

These previous studies can indeed mitigate some attacks, one of which is an on-off attack. However, their research has not been proven yet to cope against a trust-based attack. In this paper, we conducted some simulations to mitigate the trust-based attacks. The model used in this study was compared to another model, namely TC-BAC to find out whether the model used here is reliable enough to resist a trust-based attack as well as TC-BAC model.

## 3 Security in Internet of Things

Security in IoT becomes interesting and challenging issue because there are always two parties involved in it, the parties who try to defend the system with various prevention efforts, and those who try to attack the system or attackers. No security system is 100% secure, and there are only preventive efforts to stem attacks with various methods, algorithms, models, and security framework.

### 3.1 ConTrust Model

The model used in this paper is ConTrust Model [2]. In this model there are 4 processes that involve: pre-processing, calculating the value of trust, giving the value of trust, and recommendations. The topology used in ConTrust model is Social Internet of Things (SIoT) which proposed by [6].

#### *3.1.1 Pre-processing Stage:*

Objects are defined in matrix form, consisting of three independent default matrices that contain the object IDs, the initial trust values, and the initial recommendation values. These matrices can be filled with any values ranging from zero to one. The fulfilment of these values depends on the assumptions used in the simulation. If all objects are considered trustable then they can be assigned value of one. Otherwise, the value of zero means all objects are assumed not trustable.

#### *3.1.2 Trust Assessment Stage:*

For objects located in same network or intra network then the calculation of trust value is defined as follows:

$$f_{ij}^{m}(t) = \frac{\frac{1}{N}\sum_{j=1,i\neq j}^{m} T_{[j]}(t) + T_{[j]}(t-1)}{2} \quad (1)$$

And trust calculation for objects in a different network can be seen in equation (2).

$$g_{jl}^{n}(t) = \frac{1}{N}\sum_{l=1,l\neq m}^{m} T_{[l]}(t) \quad (2)$$

$$h_{ijl}^{nm}(t) = \frac{f_{ij}^{n}(t) + \frac{1}{N}\sum_{l=1}^{m} g_{jl}^{m}(t)}{2} \quad (3)$$

where:

$f_{ij}^{m}(t)$ = trust function of the object $i$ to object $j$ at time $t$
$g_{jl}^{n}(t)$ = trust function of the object $j$ to object $l$ at time $t$
$h_{ijl}^{nm}(t)$ = function to calculate the average of trust value in the same and different community

*3.1.3 Trust Value Granting Stage:*

The trust value used in the ConTrust model is ranging from zero to one. The minimum value of zero is identified as an object that is not trusted at all, and the maximum value of one is considered as a trustable object.

a. Total Trust Value calculation:

$$T(t) = \propto . h_{ijl}^{nm}(t) + (1-\propto). R(t) \quad (4)$$

where:
$T(t)$ =total of trust value
$h_{ijl}^{nm}(t)$ = direct trust assessment of object *i* to object *j* at time *t*, in the same community *n*, and different community *l* to *m*
*R(t)*= object reputation value
$\alpha$ = given weight of history function [0,1]

b. Reputation Calculation. The equation for reputation value:

$$R(t) = \frac{T(t)}{1 + e^{-\beta t}} \quad (5)$$

with

$$\beta = T(t) - T(t-1) \quad (6)$$

where
$\beta$= historical trust value of the object which is a reduction from the current trust value and the previous trust value
*R(t)* is indirect trust value produced from object's reputation in certain time periods

*3.1.4 Recommendation Stage:*

After the calculation of the total trust value, then the resulting value categorized as follows:

$$K = \begin{cases} very\ trustable\ if\ 0.7 \le T(t) \le 1 \\ trustable\ if\ 0.5 < T(t) \le 0.7 \\ untrustable\ if\ 0.3 < T(t) \le 0.5 \\ very\ untrustable\ if\ 0 < T(t) \le 0.3 \end{cases} \quad (7)$$

The recommendation value is useful for assessing the trust level of an object. It is also useful for deciding whether the data communication process to proceed or not.

### 3.2. Trust-based Attack

Attacks may occur among objects in IoT, especially those relating to trust. The types of attacks on fake trust scores [7] include:

- Good-mouthing attacks: gives a false trust value of an object, that is exaggerated value
- Bad-mouthing attacks: gives a false trust value of an object, that is to vilify an object
- Ballot-stuffing attacks: provides a congregation false trust value of an object

In the ConTrust model, these attacks are prevented through an authentication process between objects that is performed before the trust value is calculated. Thus only authenticated objects can perform trust value calculations. The authentication process used in the ConTrust model is Diffie-Hellman key distribution [8].

## 4 Simulation and Discussion

In this section, we give simulation result of trust-based attack as an output of ConTrust model usage. The simulation was conducted using MatLab tool considering the ConTrust model as a mathematical model. We compared our work with TC-BAC method, as well as without any trust method to fulfil the performance evaluation. The scenario given was to deploy false reputation values randomly to an object, in which these fake values would imply the trust values. Figure 1 depicted the normal distribution of reputation value when there was no such attack happened in the simulation. As can be seen in Figure 1, reputation values in ConTrust model remained stable compared to TC-BAC and without any trust method. It means that ConTrust model was proven to be a stable trust assessment model.

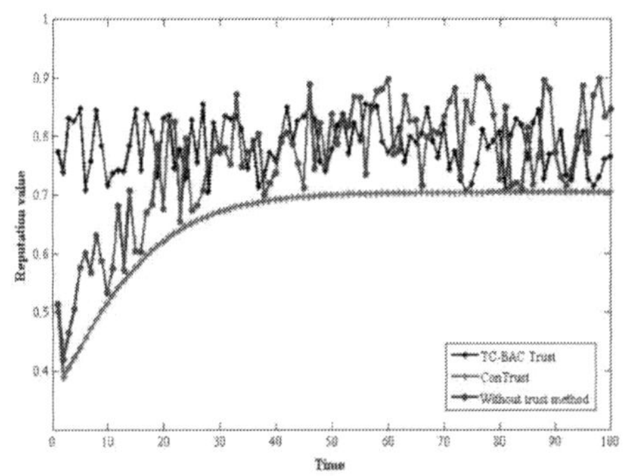


**Fig 1.** Normal distribution of reputation value

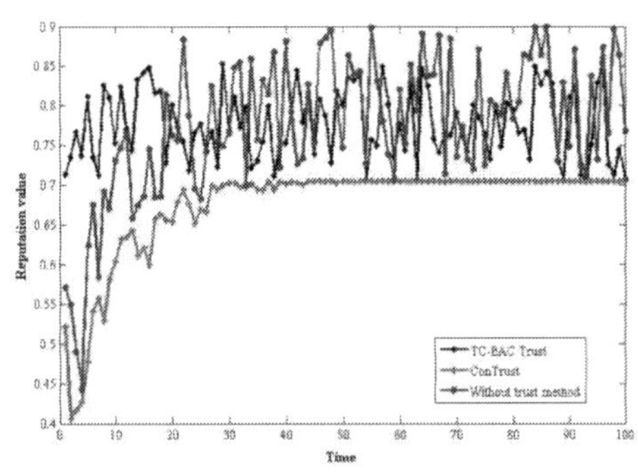


**Fig 2.** Distribution of reputation value affected by good mouthing attack

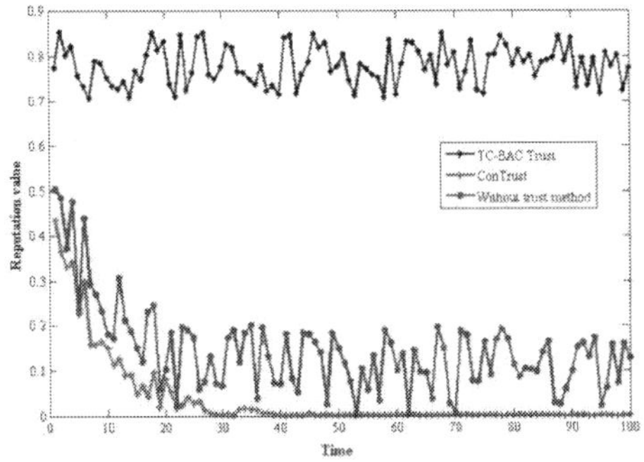


**Fig 3.** Distribution of reputation value affected by bad mouthing attack

Meanwhile, after a good mouthing attack simulation was conducted, it appears that the ConTrust model experiences a fluctuation of the trust value up to a certain time, and becomes stable again afterwards, as shown in Figure 2. The same result occurs for the type of bad mouthing attack simulation, it becomes stable again after fluctuating for a while, as shown in Figure 3. The simulation results of both good and bad mouthing attack are shown in Figure 2 and 3, respectively. ConTrust model was able to detect and dampened fluctuations of trust value that was changing because of the attack. This was due to the negative feedback mechanism as written in Eq. 5, where the current reputation value was influenced by previous input value. The changing of input value is expressed in Eq. 6, which can be analogous to the history parameter used in the ConTrust model.

## 5 Conclusion and Future Works

Our work presents simulations of trust-based attack that had been solved using ConTrust model, TC-BAC and without any trust assessment method. From simulation results, we can conclude that ConTrust model was proven to be a stable model compared to two other methods mentioned previously. However, ConTrust model still needs to be developed further in order to handle other attacks apart from trust-based attack. Authentication provided by Diffie-Hellman key distribution or a statistical approach could be deployed to mitigate some attacks that might be launched by attackers.

## References

1. S. Chakrabarty , D. W. Engels. *13th IEEE Annual Consumer Communications & Networking Conference (CCNC)* (2016)
2. V. Suryani, S. Sulistyo, W. Widyawan. *Int. J. Intell. Eng. Syst.*, **10,** 3 (2017)
3. X. Zou, Y. S. Dai, Y. Pan. *Trust and Security In Collaborative Computing* (World Scientific, Singapore, 2008)
4. J. Duan, D. Gao, C. H. Foh, H. Zhang. *Ad Hoc Networks.* **11**, 8 (2013)
5. C. V. L. Mendoza, J. H. Kleinschmidt. *Int. J. Distrib. Sens. Networks.* **2015** (2015)
6. L. Atzori, A. Iera, G. Morabito. *IEEE Commun. Lett.* **15**, 11(2011)
7. F. Bao, I.-R. Chen, J. Guo. *2013 IEEE Eleventh International Symposium on Autonomous Decentralized Systems (ISADS)* (2013)
8. E. Rescorla. *RFC 2631: Diffie-Hellman Key Agreement Method.* (RTFN, Virginia, 1999)

MATEC Web of Conferences **154**, 03015 (2018) https://doi.org/10.1051/matecconf/201815403015
*ICET4SD 2017*

# Syllables sound signal classification using multi-layer perceptron in varying number of hidden-layer and hidden-neuron

*Domy* Kristomo[1,2,*], *Risanuri* Hidayat[1], and *Indah* Soesanti[1]

[1]Department of Electrical Engineering and Information Technology, Universitas Gadjah Mada, Grafika Street No. 2, Yogyakarta, 55281 Indonesia
[2]Study Program of Computer Engineering, STMIK Akakom Yogyakarta, Raya Janti Street 143 Karang Jambe Yogyakarta, 55198 Indonesia

**Abstract.** The research on signal processing of syllables sound signal is still the challenging tasks, due to non-stationary, speaker-dependent, variable context, and dynamic nature factor of the signal. In the process of classification using multi-layer perceptron (MLP), the process of selecting a suitable parameter of hidden neuron and hidden layer is crucial for the optimal result of classification. This paper presents a speech signal classification method by using MLP with various numbers of hidden- layer and hidden-neuron for classifying the Indonesian Consonant-Vowel (CV) syllables signal. Five feature sets were generated by using Discrete Wavelet Transform (DWT), Renyi Entropy, Autoregressive Power Spectral Density (AR-PSD) and Statistical methods. Each syllable was segmented at a certain length to form a CV unit. The results show that the average recognition of WRPSDS with 1, 2, and 3 hidden layers were 74.17%, 69.17%, and 63.03%, respectively.

## 1 Introduction

One of main goal in speech recognition is to obtain the best accuracy for recognizing or classifying the speech signal uttered by speaker. The technology of speech recognition is currently used in many applications, such as smart phones, security systems, etc. However, these systems still have some difficulties in distinguishing syllables or word that sound similar. The common stages in speech recognition system are pre-processing, feature extraction and recognition stage. In the recognition stage, the process of selecting the suitable parameter in the classifier system is crucial for optimal result of classification.

Many classification methods were applied for classifying the speech signal by previous researchers. Several methods such as the Hidden Markov Model (HMM), Support Vector Machines (SVM), Gaussian Mixture Model (GMM), and Multi-layer Perceptron (MLP) or Artificial Neural Network (ANN) as classifiers [1 - 8]. MLP or ANN is a method when learning algorithm is performed and converged. It involves of weights and the ability of the underlying networks to implement desired function using sufficient number of hidden neuron. Generally, hidden layer is not needed in case of small samplés number in the data set, However, there is no proved or accepted theory in determining the numbers of neurons in hidden layer for function approximation. The number of neurons in hidden layer influenced the network.

Several previous studies has been done using MLP classifier [2, 3, 8, 9] for speech classification. In [5], three discrete wavelet families (db, sym, and coif) with a different number of coefficients was used and evaluated with two classifiers (GMM and MLP). The result in computation time showed that MLP has better performance than GMM [5]. In [8], MLP was used for classification of Hindi CV syllables. In [3], the MLP was used for classifying syllables by using combination of Discrete Wavelet Transform (DWT) and statistical features with variation of mother wavelet of Haar, Coiflet and Daubechies. The experiment result showed that Daubechies is the most effective mother wavelet compared to Haar and Coif. In [2], The MLP was used for classifying syllables sound by using DWT, Renyi Entropy (RE), and Autoregressive Power Spectral Density (AR-PSD) features.

This paper presents the classification of Indonesian CV syllables sound signal by using the MLP in varying number of hidden neuron, and the signal processing by using DWT, RE, AR-PSD, and statistical for generating features [10]. Five feature set are performed in this study. Feature Set 1 is the combination of the DWT and statistical (WS). The wavelet type used is db2 at the 7th level of decomposition [11]. Feature Set 2 is the RE features. Feature set 3 is the combination of AR-PSD and statistical features in frequency and time domain (PSDS) [2]. Feature Set 4 is the combination of AR-PSD and the RE features after selected by using Correlation-based feature selection method or CFS (RPSDS). Feature Set 5 is the combination of WS, RE, and PSDS (WRPSDS).

* Corresponding author: domykristomo@gmail.com

## 2 Methodology

A database of 360 CV syllable utterances taken from six different speakers was created. It is formed by three Indonesian consonants (/k, g, l, r/) followed by the three vowels (/a, i, u/). The consonant /g/ and /k/ represent articulation place of the velar as well as the part of the stop consonants [1, 12], while /l/ and /r/ represent articulation place of the alveolar. Frequency sampling used was 8 kHz and 16 bits mono per sample.

After the phase of recording, the next step was the cropping phase, which was basically a windowing phase to form a rectangular window of the signal. From the acoustic study by the previous researcher (Sharma et al. 2013), it was found that duration for all relevant acoustic parameters was about 60 ms [1]. Therefore, the duration manually taken for each CV unit in this study was about 60 ms, starting from release burst of the associate consonant to steady state of the following vowel. The significant events regions of the CV unit /ka/ are shown in Fig. 1a [13]. The next phase was the peak normalization. By applying peak normalization, the signal magnitude variation which is caused by the differences in the recording condition (such as speaker distance and loudness factor) can be avoided [14].

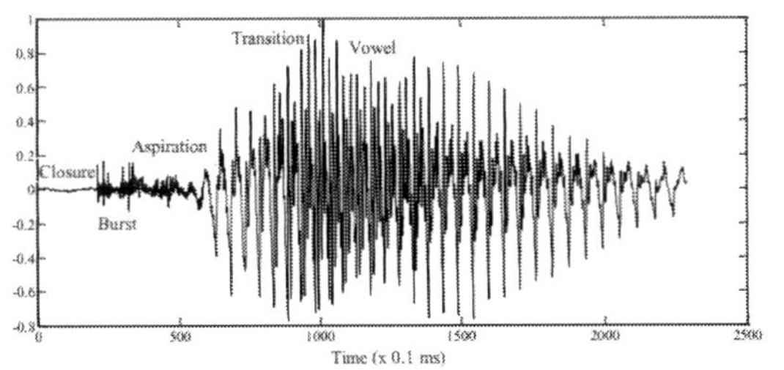


**Fig 1a.** Example of syllable /ka/ and its significant events regions

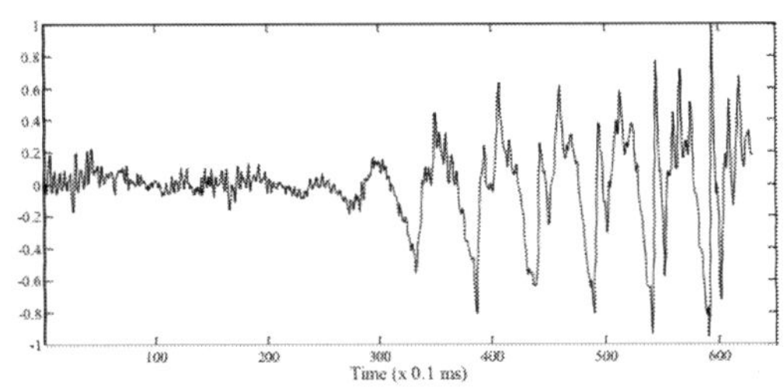


**Fig 1b.** The syllable /ka/ after windowing phase

### 2.1. Wavelet

The wavelet transform (WT) is a signal processing method which can decompose a signal into several bands using a low-pass filter and a high-pass filter. In this part, feature extraction using DWT at 7th level of decomposition was conducted. In the decomposition phase of DWT, only at a lower frequency band which is also called as approximation. By decomposing at 7th level, it gives the highest frequency band of 2000-4000 Hz and the lowest frequency band of 0–31.25 Hz. More level decomposition will be insignificant to improve recognition rate because a very low frequency band will not have discriminatory information [15].

In the DWT, the process of selecting the suitable mother wavelet is crucial for optimal result of classification. Based on previous research [1, 3, 11], it was found that Daubechies 2 (db2) was the one of the effective mother wavelet. The Daubechies wavelet of class D-2N can be written as:

$$\psi(x) := \sqrt{2}\sum_{k=0}^{2N-1}(-1)^k h_{2N-1-k}\varphi(2x-k) \tag{1}$$

Where h0,..., h2N-1 ∈ ℝ are the constant filter coefficients satisfying the condition, and φ is the (Daubechies) scaling function. After transformation process by using DWT, then the result was a signal in frequency domain. The moving average feature was calculated of each twenty sample of the signal magnitude until the maximum sample of the signal magnitude. As the additional feature, the signal in frequency domain was calculated using a statistical method [3].

### 2.2 Renyi Entropy

The Renyi entropy (RE) is a generalization of the Shannon entropy, the collision entropy, the Hartley entropy, and the min entropy. The function of generalized entropy for discrete variable X can be defined in Equation below.

$$H_\alpha(X) = \frac{1}{1-\alpha}\log\left(\sum_{i=1}^{n} P_i^\alpha\right) \tag{2}$$

Where pi is the probability of X belonging to possible outcome, $o_1$, $o_2$,..., $o_n$. The order of entropy, $\alpha$, has the constraint of ≠ 1. In special case of $\alpha = 1$, it converges to Shannon entropy [16], [17].

### 2.3 Autoregressive Power Spectral Density (AR-PSD)

In this study, PSD using Yule-Walker AR algorithm was performed. The AR model in P order can be defined in equation bellow:

$$x_{pp}(t) = -\sum_{k=1}^{p} a_k x_{pp}(t-k) + e(t) \tag{3}$$

Where
$a_k$ = AR's Coefficient

Then, using 256 point $x_{pp}(t)$ with Hamming's window, AR-PSD estimation can be formulated as follows:

 https://doi.org/10.1051/matecconf/201815403015

$$P_{AR}(f) = {T\sigma_W^2}\Big/{|1 + \sum_k^p a_k e^{-2\pi f k T}|^2} = T\sum_{m=1}^{C-1} r_{xx} e^{-2\pi f m k T} \quad (4)$$

Where $r_{xx}$ is extrapolation of data series autocorrelation bias estimation data from AR model, T is the period of sampling, and $\sigma_W^2$ is the variance of the drive noise input.

### 2.4 Single Hidden Layer in Multi-layer Perceptron (MLP)

In the classification process by using MLP, the process of selecting the suitable parameter and architecture is crucial for the optimal result of classification [18], [19]. The architecture used in this section consists of three layer, they are input layer, hidden layer, and output layer. The input layer represents the features of each feature extraction method (WS, RE, PSDS, RPSDS, or WRPSDS). The hidden layer consists of 1-20 hidden neurons. The output layer consist of twelve neurons, it represents the classification result of the syllables. To estimate the reliability of the classification results, the data verification was performed. The verification technique used on the test set was k-fold cross validation or the hold out method [20].

### 2.5 Hidden Layers in MLP

In this experiment, we used 2 layers in Hidden layer. Based on the previous experiment on single layer of hidden neuron, the optimal result for WRPSD was 55 nodes. Therefore in this part we used 55-55 nodes.

### 2.6 Hidden Layers in MLP

In this part of experiment, we used 3 layers. As the previous research recommendation [9], the nodes architecture for 3-hidden layer was 20-20-15 nodes.

## 3 Result and Discussion

In this study, the number of features generated by using WS, RE, PSDS, RPSDS, and WRPSDS were twenty-nine, twenty, thirteen, nineteen, and sixty-two features, respectively. After feature extraction, the next phase was classification that uses MLP-BP. The parameter of the learning rate and the momentum was 0.3 and 0.2, respectively. In the previous study [14], the experiment for the number of hidden neurons 0 to 20 has been done. In this study we continued the experiment for the number of hidden neurons 20-60 as shown in Fig. 2.

Figure 2 shows the percentage accuracy for WS, RE, PSDS RPSDS, and WRPSDS in varying number of hidden neuron. The result showed that WRPSDS has the highest score in average accuracy, but at ninth hidden neuron, the score of WS is higher than WRPSDS. It indicated that the number of hidden neuron can influence the classification result, but the increase of accuracy was not linear.

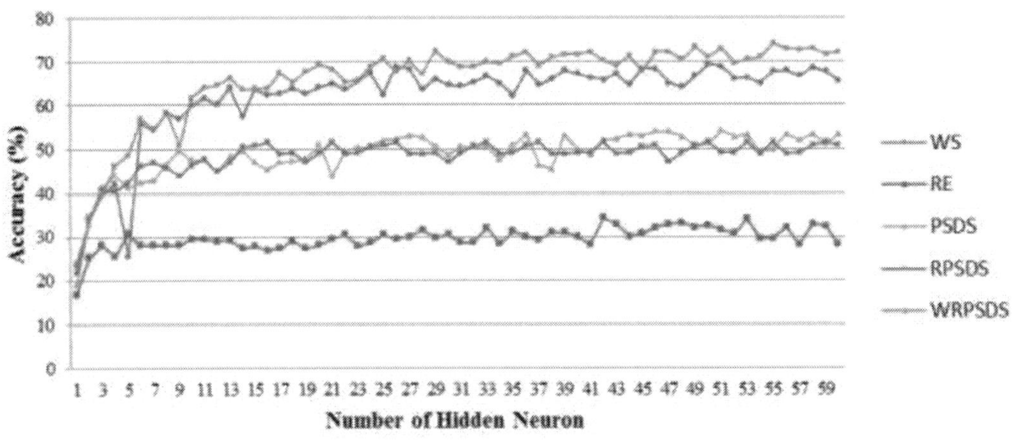


**Fig 2.** Accuracy for each feature extraction method with various number of hidden neuron in MLP

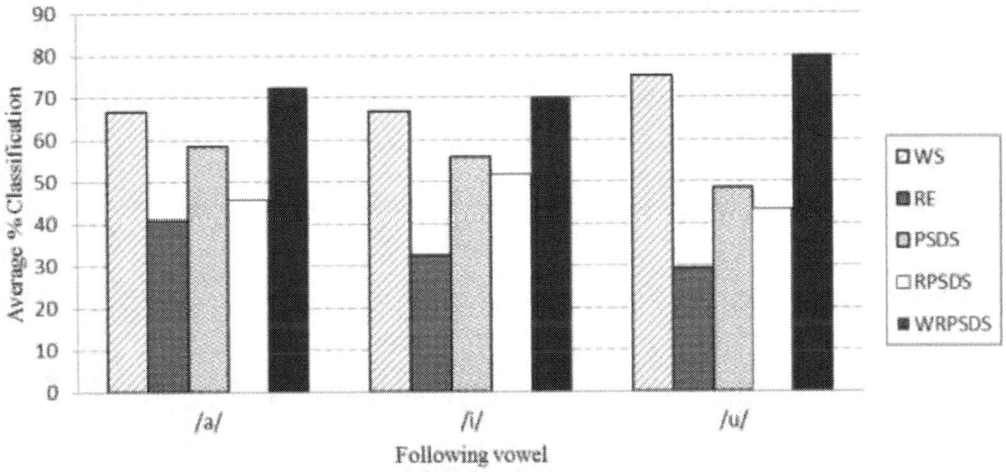


**Fig 3.** Graph of the average percentage classification at the optimal number of hidden neuron of each feature extraction method for classifying the Indonesian consonant following vowel contexts

**Table 1.** Accuracy for each feature extraction method with various hidden layer in MLP

| Feature extraction Method | 1-Hidden Layer (55 nodes) | | | | 2-Hidden Layer (55-55 nodes) | | | | 3-Hidden Layer (20-20-15 nodes) | | | |
|---|---|---|---|---|---|---|---|---|---|---|---|---|
| | Following vowel | | | Average % recognition | Following vowel | | | Average % recognition | Following vowel | | | Average % recognition |
| | /a/ | /i/ | /u/ | | /a/ | /i/ | /u/ | | /a/ | /i/ | /u/ | |
| WS | 65 | 61.67 | 76.67 | 67.78 | 59.17 | 60.83 | 68.33 | 62.78 | 55.83 | 59.17 | 6 | 58.33 |
| RE | 31.67 | 28.33 | 28.33 | 29.44 | 34.17 | 26.67 | 28.33 | 29.72 | 34.17 | 23.33 | 17.5 | 25 |
| PSDS | 51.67 | 52.5 | 45.83 | 50 | 55 | 56.67 | 51.67 | 54.44 | 45.83 | 4 | 39.17 | 41.67 |
| RPSDS | 45.8 | 51.65 | 43.32 | 46.92 | 49.17 | 53.33 | 46.67 | 49.72 | 48.33 | 43.33 | 38.33 | 43.33 |
| WRPSDS | **72.5** | **70** | **79.99** | **74.17** | 64.17 | 67.5 | 75.83 | 69.17 | 68.33 | 58.33 | 62.5 | 63.03 |

Table I shows the percentage classification scores using Ten-fold cross validation (Ten-FCV) for 1, 2 and 3 hidden layer. The experiment result showed that the average recognition for 1, 2, and 3 hidden layers that using WRPSDS were 74.17%, 69.17%, and 63.03% respectively. It indicated that the MLP architecture in 1-hidden Layer (55 nodes) gives better performance of classification compared to 2-Hidden Layer (55-55 nodes) and 3-Hidden Layer (20-20-15) nodes.

In case of /a/, the highest score was 72.5% by using WRPSDS features. (For the vowel /i/ the highest classification score was 70%. In case of /u/ the highest score was 79.99% which was the highest score among the other vowel. Some feature showed better performance in 2-Hidden Layer architecture (RE, PSDS, RPSDS), but overall 1 layer was still better.

## 4 Conclusion

In this paper, a classification of the Indonesian syllables sound using classifier of multi-layer perceptron in varying number hidden layer and hidden neuron fusing with Wavelet, Renyi entropy (RE), AR-PSD features was proposed and implemented. Based on the experimental result presented in this paper, it can be concluded that the MLP architecture in 1-hidden Layer (55 nodes) when fusing with WRPSDS gives a better performance of classification score compared to 2-Hidden Layer (55-55 nodes) and 3-Hidden Layer (20-20-15) nodes as shown by accuracy of 74.17%, 69.17%, and 63.03% respectively. Some feature such as RE, PSDS, RPSDS showed better performance in 2-Hidden Layer architecture, but overall 1 hidden layer architecture was still better. The future work recommended for this re-search is to use bigger syllable dataset, applied to the Indonesian stop consonant or the other place of articulation (such as labial, dental, etc.), to use different combination of feature extraction technique, and to use different testing procedure of classification process.

## References

1. R. P. Sharma, O. Farooq, and I. Khan, "Wavelet based sub-band parameters for classification of unaspirated Hindi stop consonants in initial position of CV syllables," *Int. J. Speech Technol.*, **16**, no. 3, pp. 323–332 (2013)
2. D. Kristomo, R. Hidayat, and I. Soesanti, "Classification of the Syllables Sound Using Wavelet , Renyi Entropy and AR-PSD Features," in *2017 IEEE 13th International Colloquium on Signal Processing & its Application (CSPA 2017)*, **13**, pp. 97–102 (2017)
3. D. Kristomo, R. Hidayat, and I. Soesanti, "Feature extraction and classification of the Indonesian syllables using Discrete Wavelet Transform and

statistical features," in *2016 2nd International Conference on Science and Technology-Computer (ICST)*, **2**, pp. 88–92 (2016)

4. S. Hidayat, R. Hidayat, and T. B. Adji, "Speech recognition of CV-patterned indonesian syllable using MFCC, wavelet and HMM," *J. Ilm. Kursor*, **8**, no. 2, pp. 67–78 (2015)
5. P. Král, "Discrete Wavelet Transform for automatic speaker recognition," *Image Signal Process. (CISP), 2010 3rd Int. Congr.*, **7**, pp. 3514–3518 (2010)
6. X. Zhao, Z. Wu, J. Xu, K. Wang, and J. Niu, "Speech Signal Feature Extraction Based on Wavelet Transform," *2011 Int. Conf. Intell. Comput. Bio-Medical Instrum.*, no. 1, pp. 179–182 (2011)
7. M. El Ayadi, M. S. Kamel, and F. Karray, "Survey on speech emotion recognition: Features, classification schemes, and databases," *Pattern Recognit.*, **44**, no. 3, pp. 572–587 (2011)
8. C. Chandra and B. Yegnanarayana, "A constraint satisfaction model for recognition of stop consonant-vowel (SCV) utterances," *IEEE Trans. Speech Audio Process.*, **10**, no. 7, pp. 472–480, (2002)
9. G. Dede and M. H. Sazlı, "Speech recognition with artificial neural networks," *Digit. Signal Process.*, **20**, no. 3, pp. 763–768 (2010)
10. S. Theodoridis and K. Koutroumbas, *Pattern Recognition*, 4th ed. United States ofAmerica (2009)
11. R. Hidayat, Priyatmadi, and W. Ikawijaya, "Wavelet based feature extraction for the vowel sound," in *2015 International Conference on Information Technology Systems and Innovation (ICITSI)*, pp. 1–4 (2015)
12. F. L. Hardjono and R. A. Fox, "Stop Consonant Characteristics: VOT and Voicing in American-Born-Indonesian Children's Stop Consonants," The Ohio State University (2011)
13. A. K. Vuppala, K. S. Rao, and S. Chakrabarti, "Spotting and recognition of consonant-vowel units from continuous speech using accurate detection of vowel onset points," *Circuits, Syst. Signal Process.*, **31**, no. 4, pp. 1459–1474 (2012)
14. D. Kristomo, R. Hidayat, I. Soesanti, and A. Kusjani, "Heart sound feature extraction and classification using autoregressive power spectral density (AR-PSD) and statistics features," in *AIP Conference Proceedings*, **1755**, pp. 90007-1-90007–7 (2016)
15. O. Farooq and S. Datta, "Phoneme recognition using wavelet based features," *Elsevier Inf. Sci.*, **150**, pp. 5–15 (2003)
16. A. Rényi, "On measures of entropy and information," in *Fourth Berkeley Symposium on Mathematical Statistics and Probability*, **1**, pp. 547–561 (1961)
17. C. Y. Kee, S. G. Ponnambalam, and C. K. Loo, "Binary and multi-class motor imagery using Renyi entropy for feature extraction," *Neural Comput. Appl.*, pp. 1–12 (2016)
18. N. M. Nawi *et al.*, "The Effect of Pre-Processing Techniques and Optimal Parameters selection on Back Propagation Neural Networks," **7**, no. 3, pp. 770–777 (2017)
19. A. Kuri-Morales, "The Best Neural Network Architecture," *Springer*, (2015)
20. R. Kohavi, "A Study of Cross-Validation and Bootstrap for Accuracy Estimation and Model Selection," *Int. Jt. Conf. Artif. Intell.*, **14**, no. 12, pp. 1137–1143 (1995)

# Color feature extraction of HER2 Score 2+ overexpression on breast cancer using Image Processing

*Izzati* Muhimmah[1,*], *Dadang* Heksaputra[2], and Indrayanti[3]

[1]Master of Informatics Engineering Universitas Islam Indonesia, Indonesia
[2]Master of Informatics Engineering Universitas Islam Indonesia, Indonesia
[3] Faculty of Medicine Universitas Muhammadiyah Yogyakarta, Indonesia

**Abstract.** One of the major challenges in the development of early diagnosis to assess HER2 status is recognized in the form of Gold Standard. The accuracy, validity and refraction of the Gold Standard HER2 methods are widely used in laboratory (Perez, et al., 2014). Method determining the status of HER2 (human epidermal growth factor receptor 2) is affected by reproductive problems and not reliable in predicting the benefit from anti-HER2 therapy (Nuciforo, et al., 2016). We extracted color features by methods adopting Statistics-based segmentation using a continuous-scale naïve Bayes approach. In this study, there were three parts of the main groups, namely image acquisition, image segmentation, and image testing. The stages of image acquisition consisted of image data collection and color deconvolution. The stages of image segmentation consisted of color features, classifier training, classifier prediction, and skeletonization. The stages of image testing were image testing, expert validation, and expert validation results. Area segmentation of the membrane is false positive and false negative. False positive and false negative from area are called the area of system failure. The failure of the system can be validated by experts that the results of segmentation region is not membrane HER2 (noise) and the segmentation of the cytoplasm region. The average from 40 data of HER2 score 2+ membrane images show that 75.13% of the area is successfully recognized by the system.

## 1 Introduction

The accuracy of the status Human Epidermal Growth Factor Receptor 2 (HER2) is very important to optimize the result of breast cancer treatmementPerez, et al., 2014). Human epidermal growth factor receptor 2 (HER2) overexpression in breast cancer is an indicator of poor prognosis. HER2 is a prerequisite for a treatment with tissue targeting the epidermal growth receptor (Cresti, et al., 2016). One of the major challenges in the development of early diagnosis to assess HER2 status is recognized in the form of Gold Standard. The accuracy, validity and refraction of the Gold Standard HER2 methods are widely used in laboratory (Perez, et al., 2014). Method determining the status of HER2 (human epidermal growth factor receptor 2) is affected by reproductive problems and not reliable in predicting the benefit from anti-HER2 therapy (Nuciforo, et al., 2016).

The computer-assisted research is carried out to improve the accuracy and reproducibility of breast tissue assessment. Breast tissue specimen image assessment has been done in recent years. Brugmann, et al. (2012) developed and validated a software which is capable of IHC scoring algorithms automatically based on membrane connectivity on each image. Keller, Chen, and Gavrielides (2012) developed a system which classifies IHC score based on color feature. The system is built using algorithms fuzzy c-means clustering in the HSV color space. Hall et al. (2008) resulted in a system which is able to give a score of IHC using multiple stages, such as selection and image capturing, color decomposition, and membrane isolation algorithm (MIA) with three features based on the average intensity of membrane area. Masmoudi et al. (2009) developed an automated assessment method of HER2 immunohistochemistry. Labellapansa, Muhimmah, and Indrayanti (2013) performed an image analysis by providing a classification HER2 score 1+ and HER2 score 3+ using Otsu's thresholding segmentation method.

The state of the art from the research domain in table 1 shows the gap to conduct this research. We extracted color features by methods adopting Statistics-based segmentation using a continuous-scale naïve Bayes approach.

* Corresponding author: izzati@uii.ac.id

**Table 1.** State of the art

| No | Name | Methods | | | | | | | | | | | | | |
|---|---|---|---|---|---|---|---|---|---|---|---|---|---|---|---|
| | | HER2-CONNECT™ | Fuzzy C-Means | Clustering | Linear regression | Transformasi watersheet | Minimum cluster distance | Eclipse approach | Images selection and capturing | Color decomposition | Brown's model | Color feature | Color deconvolution | Naïve bayes | Skeletonization |
| 1 | Bru¨gmann, et al (2012) | √ | | | | | | | | | | | | | |
| 2 | Keller, Chen, and Gavrielides (2012) | | √ | √ | | | | | | | | | | | |
| 3 | Masmoudi, et al (2009) | | | | √ | √ | √ | √ | | | | | | | |
| 4 | Hall, et al (2008) | | | | | | | | √ | √ | | | | | |
| 5 | Labellapansa, Muhimmah, and Indrayanti (2013) | | | | | | | | | | √ | | | | |
| 6 | Laursen, et al (2014) | | | | | | | | | | | √ | | √ | |
| 7 | proposed | | | | | | | | | | | √ | √ | √ | √ |

## 2 Methods

A. Color Deconvolution

Color deconvolution is needed in the application of membrane extraction method. The approach has been used in previous of this method to divide the allegations of cell membrane overexpression. Ruifrok and Johnston (2001) gave OD matrix for IHC staining H, E and DAB. Color decomposition was applied to the original image and resulted in grayscale and image with DAB stain with hematoxylin stain (Hall, et al., 2008).

B. Calculation of Color Feature

The pixel values were obtained from the clustering using color deconvolution method. The pixel values were used as criteria for location of the position coordinates. RGB values of the layer were taken to apply the calculation of color features. The chromaticities (r,g,b) were given by the raw color value divided with the sum of all the color total value (Laursen, et al., 2014).

$$(r,g,b)^T = \frac{(R,G,B)^T}{R+G+B} \quad (1)$$

There were also excess green (ExG) and excess red (ExR) (Meyer and Neto, 2008).

$$ExG = 2.G - R - B \quad (2)$$

$$ExR = 1.4R - G \quad (3)$$

C. Classifier Training

Flexible naïve Bayes was used in this section (Perez, Larranaga, and Inza, 2009). The classifier training consisted of estimating the probability density function for each object and background color feature. The probability density was determined from the annotated images. Numerical vectors were created, containing the values of the four raw colors, namely chromaticities, ExG and ExR values of each pixel. The probability function [$P(F_k = x|\mathbb{C})$] for each of the k was given input variables conditional on class $\mathbb{C}$. Kernel density estimation with a Gaussian kernel K(x) has a bandwidth h (Silverman, 1986), where L is the number of observations (Laursen, et al., 2014).

$$P(F_k = x|\mathbb{C}) = \frac{1}{L.h}\sum_{i-1}^{L} K\left(\frac{x-x_i}{h}\right) \quad (6)$$

$$K(x) = \frac{1}{\sqrt{2\pi}} e^{\frac{-x^2}{2}} \quad (7)$$

D. Classifier Prediction

The color feature extraction is based on the criteria diagnosed by a physician as HER2 membrane. ($\mathbb{S}$) is a group of background and ($\mathbb{V}$) is the object. Support groups $\mathbb{C}$ are symbolized by $P(\mathbb{C},\vec{F})$, where the vector $\vec{F}$ is a vector of $F_1, \ldots, F_M$ features M. $P(\mathbb{C})$ is the previous probability of group $\mathbb{C}$.

$$P(\mathbb{C},\vec{F}) = P(\mathbb{C}).P(F_1|\mathbb{C}).P(F_2|F_1,\mathbb{C}).P(F_3|F_1,F_2,\mathbb{C})\ldots \quad (8)$$

$$P(\mathbb{C},\vec{f}) = P(\mathbb{C}).P(F_1 = f_1|\mathbb{C}).P(F_2 = f_2|F_1 = f_1,\mathbb{C}).P(F_3 = f_3|F_1 = f_1,F_2 = f_2,\mathbb{C})\ldots \quad (9)$$

Characteristic assumptions of color features include probability functions. It can be written to equation (10).

$$P(\mathbb{C},\vec{F}) \simeq \prod_{m=1}^{M} P(F_k|\mathbb{C}) \quad (10)$$

Feature input is denoted as $F_m$, where $m \in \{1, \ldots, m\}$. Naïve Bayes classifier is chosen because it provides values in predicting the dependent input, and determining the increase in the ratio of the input that is independent (Hand dan Yu, 2001). A pixel is classified as an object if $P(\mathbb{V}|\vec{F}) \geq P(\mathbb{S},\vec{F})$ and if not then in the background (Laursen, et al., 2014).

$$P(\mathbb{V}|\vec{F}) = \frac{P(\mathbb{V},\vec{F})}{P(\mathbb{V},\vec{F})+P(\mathbb{S},\vec{F})} \quad (11)$$

E. Skeletonization

The extraction results of the membrane skeletonization technique were used to obtain a linear line to shape the object of the membrane extraction model. Skeleton object is an important topological description of the object of two-dimensional binary image. It has been used in a wide range of applications (Levine, 1985; Serra, 1982; Pitas & Venetanopoulos, 1990). The skeleton formula of Serra (1982) can be seen in equation (12).

$$SK(X) = \bigcup_{n=0}^{N} S_n(X) = \bigcup_{n=0}^{N}[(X \ominus nB^s) - (X \ominus nB^s)_B] \quad (12)$$

where Sn (X) shows a part of skeleton SK (X) corresponding to the maximum radius in disk.

## 3 Research Methodology

We extracted color features by methods adopting Statistics-based segmentation using a continuous-scale naïve Bayes approach. In this study, there were three parts of the main groups, namely image acquisition, image segmentation, and image testing. The stages of image acquisition were image data collection and color deconvolution. The stages of image segmentation consisted of color features, classifier training, classifier prediction, and skeletonization. The stages of image testing were image testing, expert validation, and expert validation results (see fig. 1).

The datasets of HER2 images were obtained from the Histology Laboratory of Pathology, Universitas Muhammadiyah Yogyakarta. The tool was sigma hd microscope camera, olympus. The images were taken from patient preparation with a microscope at 400x magnification. The staining of HER2 was done by using IHC which was less expensive than FISH and CISH (Dobson, et al., 2010). The datasets used 40 images. The data were a diagnosis of HER2 score 2+. The color deconvolution was the automatic annotation to separate the alleged membranes and cells in the image. The results of this annotation were used as criteria to object and background color feature. The calculation of color features was done by using equation (1), equation (2), and equation (3). The calculation values of the color features were processed by classifier training and classifier prediction. They were the eight-layer image namely red (R), green (G), blue (B), excess green (ExG), Chromaticities red (r), Chromaticities green (g) , Chromaticities blue (b), and excess red (EXR). The results of classifier training and classifier prediction were to perform skeletonization. The skeletonization was done by experts to justify the overexpression of the membrane criteria.

The researchers used the approach to the average value of the threshold pixel in the image to provide a solution to the justification of experts. The results of the final image underwent the validation process by experts to provide justification to determine the percentage of successful segmented membrane by the system model.

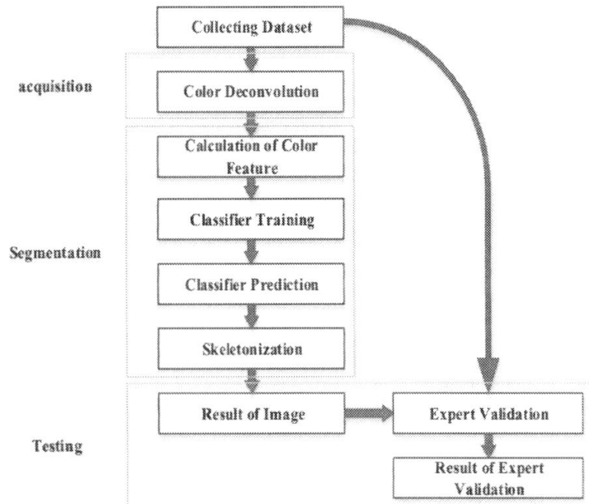


**Fig 1**. Research process

## 4 Results

This section shows the results of the proposed method. Figure 2 is an example of the implementation of the system model. The system model implementation was designed from the datasets. Ten samples from the image datasets were taken to the modeling system. The end results were validated by physicians. The validation was intended to determine the accuracy of the percentage rate by the membrane systems model. Figure 2 (a) is the original image without image processing. The initial image was the result of the image on hd microscope sigma camera, olympus. Figure 2 (b) shows the results of the color deconvolution, calculation of color features, classifier training, and classifier prediction. Figure 2 (c) is the result of the image skeletonization and application. Figure 2 (d) is the final image. The final image was validated by experts to provide an assessment of the membrane percentage. It is segmented by the system model.

Membrane features contain microscopic images of breast cancer cells using stain Immunuhistochemistry/IHC. They are contained on the distribution of RGB image layer. Sample distribution histogram value of the RGB layer on the membrane characteristics can be seen in figure 3. The x-axis is the range of variations in the intensity image pixel value ranging from 0 to 255 (grayscale). The y-axis represents the sum of the number of pixels contained in the image. Figure 3 is a sample of the RGB histogram distribution on HER2 score 2+. The results are compared with the variation in the intensity of membrane staining positive control specimen. The results of the distribution, HER2 score 2+ membrane values are in the middle of the range of values ranging from 100 to 200 of variations in the intensity image pixel values.

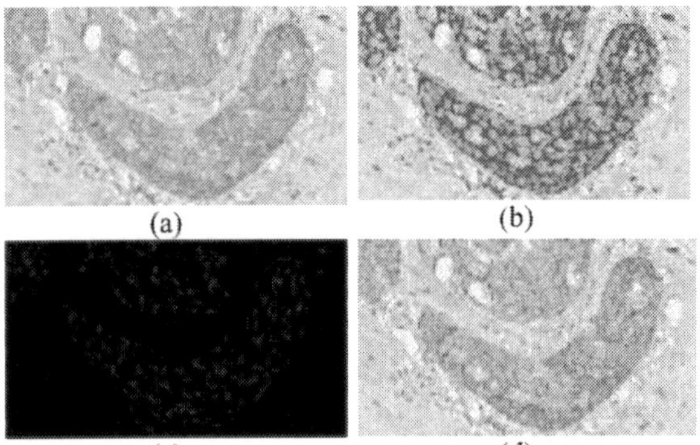

**Fig 2.** Example of the stages of the system model implementation. (a) Original image. (b) Image of the alleged membranes by color feature extraction. (c) Binary image skeletonization result. (d) Final result.

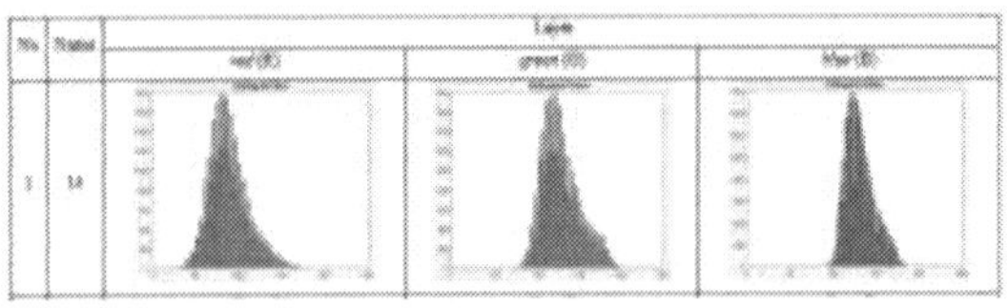

**Fig 3.** Histogram distribution sample on HER2 score 2+

## 5 Testing

The result testing showed the accuracy of the system model which was designed and implemented. The percentage value was derived from physician assessment of anatomic pathology (Table 2).

**Table 2.** Results of validation HER2 Score 2+ by physician

| No | File Name | Level of success | | | | |
|---|---|---|---|---|---|---|
| | | 60% | 70% | 80% | 90% | 95% |
| **1** | 14 | √ | | | | |
| **2** | 15 | | √ | | | |
| **3** | 16 | | | √ | | |
| **4** | 17 | | | | √ | |
| **5** | 18 | | | | √ | |
| **6** | 19 | | | | | √ |
| **7** | 20 | | | | | √ |
| **8** | 26 | | | √ | | |
| **9** | 27 | | | | √ | |
| **10** | 28 | | | | √ | |
| **11** | 29 | | √ | | | |
| **12** | 30 | | | | | √ |
| **13** | 31 | √ | | | | |
| **14** | 32 | √ | | | | |
| **15** | 33 | | √ | | | |
| **16** | 34 | | √ | | | |
| **17** | 35 | | √ | | | |
| **18** | 45 | √ | | | | |
| **19** | 46 | √ | | | | |
| **20** | 48 | | | √ | | |
| **21** | 49 | | √ | | | |
| **22** | 50 | | | √ | | |
| **23** | 60 | | √ | | | |
| **24** | 62 | | √ | | | |
| **25** | 66 | | √ | | | |
| **26** | 67 | √ | | | | |
| **27** | 68 | √ | | | | |
| **28** | 87 | | √ | | | |
| **29** | 88 | | √ | | | |
| **30** | 96 | | | √ | | |
| **31** | 97 | | | √ | | |
| **32** | 98 | | | √ | | |
| **33** | 102 | | | √ | | |
| **34** | 103 | | | √ | | |
| **35** | 104 | | | √ | | |
| **36** | 105 | | √ | | | |
| **37** | 106 | | √ | | | |
| **38** | 107 | | √ | | | |
| **39** | 117 | | √ | | | |
| **40** | 119 | | | | √ | |

Segmentation of the membrane can be false positive and false negative. Regional false positive and false negative are called to the area of system failure. The failure of the system can be validated by experts to have the results.

a. Segmentation of the areas is not membrane HER2 (noise).
b. Segmentation shows cytoplasmic region.

The average of the images in all the datasets shows membrane HER2 score 2+. 75.13% of the area is successfully recognized by the system.

## 6 Conclusion

Based on the discussion and testing of the system, it can be concluded that:

1. The results of the distribution, HER2 score 2+ membrane values are in the middle of the range of values ranging from 100 to 200 of variations in the intensity image pixel values.
2. The algorithm to extract membrane of breast cancer cells with image processing has been shown successfully to identify membrane in HER2 score 2+.

A research grant from Hibah PUPT from Ministry RISTEK-DIKTI of Indonesia for the period of 2015-2016, contract no. 001/HB-LIT/III/2015, is gratefully acknowledged.

## References

1. Bru¨gmann, A., et al. (2012). Digital image analysis of membrane connectivity is a robust measure of HER2 immunostains. *Breast Cancer Res Treat*, 132:41–49.
2. Dobson, L., et al. (2010). Image analysis as an adjunct to manual HER-2 immunohistochemical review : a diagnostic tool to standardize interpretation. *Blackwell Publishing Ltd, Histopathology*, *57*, 27-38.
3. Hall, B. H., et al. (2008). Computer-assisted assessment of the Human Epidermal Growth Factor Receptor 2 immunohistochemical assay in imaged histologic sections using a membrane isolation algorithm and quantitative analysis of positive

controls. *BMC Medical Imaging*, 1-13.

4. Hand , D. and Yu, K. (2001). Idiot's Bayes : Not to Stupid after All. *int. stat. Rev. 69*, 385-398.
5. Labellapansa, A., Muhimmah, I., and Indrayanti. (2013). Klasifikasi Citra Immunohistokimia Sel Kanker Payudara HER2 Skore 1+ dan 3+. *Seminar Nasional Informatika Medis (SNIMed)*, 24.
6. Laursen, M. S., et al. (2014). Statistics-based Segmentation using a Continuos Scale Naive Bayes Approch. *Elsevier Computer in Electronics in Algiculture*, 271-277.
7. Levine, M. D. (1985). *Vision in Main and Machine.* McGraw-Hill.
8. Masmoudi, H., et al. (2009). Automated quantitative assessment of HER-2/neu immunohistochemical expression in breast cancer. *IEEE transactions on medical imaging*, 28(6), 916-925.
9. Meyer, G. E. and Neto, J. C. (2008). Verification of Color Vegetation in this for automated Crop Imaging Applications. *Comput. Electron. Agr*, 282-293.
10. Perez, A., Larranaga, P., and Inza, I. n. (2009). Bayesian Classifiers Based on Kernel Density Estimation : Flexibel Classifiers. *IMT. J. Approx Reason 50*, 341-362.
11. Perez, E. A., et al. (2014). HER2 testing: Current status and future directions. *ELSEVIER*, Cancer Treatment Reviews 40 276–284.
12. Pitas, I., and Venetanopoulos, A. N. (1990). *Nonliniear Digital Filters : Principles and Applications.* Kluwer Academic.
13. Ruifrok, A. C. and Johnston, D. A. (2001). Quantification of histochemical staining by color deconvolution. *Anal Quant Histol*, 291-299.
14. Serra, J. (1982). *Image Analysis and Mathematical Morphology.* Academic Press.
15. Silverman, B. W. (1986). Density Estimation for Statistics for Data Analisys. *Appl. Stat. 37*, 120.

MATEC Web of Conferences **154**, 03017 (2018) https://doi.org/10.1051/matecconf/201815403017
*ICET4SD 2017*

# On the improvement of IT process maturity: assessment, recommendation and validation

*Teduh* Dirgahayu[1,*] and *Setiyowati*[2]

[1]Universitas Islam Indonesia, Jl. Kaliurang KM 14.5, Yogyakarta, Indonesia
[2]STMIK Sinar Nusantara, Jl. KH. Samanhudi 84-86, Surakarta, Indonesia

**Abstract.** The use of information technology (IT) in enterprises must be governed and managed appropriately using IT processes. The notion of IT process maturity is useful to measure the actual performance and to define the desired performance of IT processes. Improvements are necessary when there are gaps between the actual and desired performance. Most literatures focus on IT process maturity assessment. They do not address how to improve IT process maturity. This paper proposes an approach to enterprise IT process maturity improvement for COBIT processes. The approach consists of three activities, i.e. IT process maturity assessment, recommendation, and validation. Assessment is to recognise the process' control objectives maturity. From the assessment results, recommendation identifies control objectives that must be improved and then suggests improvement actions. The prescriptive nature of the control objectives facilitates in suggesting those actions. Recommendations for managements are defined by abstracting similar actions. Validation checks whether the recommendations match with the enterprise needs and capability. It includes a scale for validation, in which enterprise's capability is categorized into (i) not capable, (ii) capable with great efforts, and (iii) fully capable. The paper illustrates the approach with a case study.

## 1 Introduction

As enterprises now depend much on information technology (IT) to carry out their business, the use of IT in the enterprises must be governed and managed appropriately. IT governance and management is to make sure that IT aligns strategically with enterprises' business, IT value are delivered as promised, IT risks are managed properly, IT resources are managed effectively and efficiently, and IT performance is measured against its investment [1, 2].

Frameworks and standards have been established to give guidance for senior IT executives in conducting enterprise IT governance and management, e.g. COBIT [3, 4], ITIL [5, 6], ISO 20000 [7] and ISO 38500 [8]. Those frameworks define a set of generic IT processes. For example, COBIT 5 comprises 37 processes; each of which consists of a number of activities to achieve one or more IT-related goals [4].

The frameworks usually also provide maturity models to assess the status of IT processes. IT process maturity indicates the degree of reliability of an IT process. For example, COBIT 4.1 maturity models has five levels, i.e. non-existent, initial/ad hoc, repeatable but intuitive, defined, managed and measurable, and optimised [3]. Concepts of maturity models can also be found, among others, in the field of business process management [9, 10], information security [11, 12], and software development process [13].

The notion of process maturity is useful to measure the actual performance, to define a desired performance, and to determine necessary improvements to close gaps between the actual and desired performance [4]. Nonetheless, the aforementioned frameworks do not suggest any method to assess IT process maturity. Several researches hence proposed assessment methods for COBIT processes [14-17]. Those researches however do not address how to formulate recommendations to improve IT process maturity.

The objective of this paper is to propose an approach to enterprise IT process maturity improvement. It consists of three activities, i.e. assessment, recommendation and validation. The assessment is to recognise the maturity of an IT process. Recommendation is to suggest necessary actions to improve process maturity. Validation checks whether the suggested actions match with the enterprise needs.

This paper is further organised as follows. Section 2 presents the concepts of IT process maturity. Section 3 describes a case study conducted in our research that is used to illustrate our approach. Section 4 presents our assessment method for IT process maturity. Section 5 and 6 consecutively presents our recommendation and validation methods. Section 7 concludes this paper and identifies future work.

* Corresponding author: teduh.dirgahayu@uii.ac.id

## 2 IT Process Maturity

In this paper, we refer to COBIT 4.1 processes and maturity model. Although COBIT 5 has been available since 2012, actually it expands COBIT 4.1 with other IT governance frameworks and standards [4]. Thus COBIT 4.1 is still worth to be a reference in our research.

COBIT 4.1 consists of 34 processes that are grouped into four domains, i.e. plan and organise (PO), acquire and implement (AI), deliver and support (DS), and monitor and evaluate (ME). A process includes a number of control objectives, i.e. a set of high-level requirements for effective control of the process. Control objectives can be organisational structures, policies, and procedures [3].

All processes have the same attributes; each of which represents an aspect of a process. The attributes are (i) awareness and communication; (ii) policies, plans and procedures; (iii) tools and automation; (iv) skill and expertise; (v) responsibility and accountability; and (vi) goal setting and measurement. Those attributes are supposely to be used in assessing the maturity of a process; therefore, they are called *maturity attributes* [3].

**Table 1.** Maturity model of process DS7 (ITGI, 2007)

| Level | Status | Description |
|---|---|---|
| 0 | Non-existent | The organisation is in a complete lack of a training and education programme. |
| 1 | Initial/ad hoc | The organisation recognises the need for a training and education programme, but there are no standardised processes. |
| 2 | Repeatable but intuitive | The organisation is aware of the need for a training and education programme and for associated processes. |
| 3 | Defined | The organisation institutes and communicates a training and education programme. |
| 4 | Managed and measurable | The organisation conductes a comprehensive training and education programme that yields measurable results. |
| 5 | Optimised | The organisations benefits from a training and education programme that result in an improvement of individual performance. |

The attribute of *awareness and communication* (AC) of a process indicates whether the enterprise fully understands of the importance of that process and whether the management communicates this issue using mature communication techniques. The attribute of *policies, plans and procedures* (PPP) indicates whether the process is sound and complete, i.e. all related policies are approved, and procedures are adopted, followed and documented. The attribute of *tools and automation* (TA) indicates whether the enterprise implements standardised tools to automate and monitor the process. The attribute of *skill and expertise* (SE) indicates whether the enterprise updates skills required for the process, which includes training plans, certifications and knowledge sharing. The attribute of *responsibility and accountability* (RA) indicates whether process responsibility and accountability is defined clearly and whether a reward culture is employed to motivate positive actions towards the process. Lastly, the attribute of *goal setting and measurement* (GSM) indicates whether the enterprise measures the effectiveness and efficiency of the process and links the results to the enterprise's business strategic goals [3].

The maturity of a process evolves from a non-existent to an optimised process. COBIT provides a generic maturity model from which each process adapts it to define its specific maturity model. Table 1 shows the maturity model of process DS7 (educate and train users).

Several assessment methods for IT process maturity have been established [15-17]. Those methods however do not consider maturity attributes in a proper and comprehensive way. Hence, a method for assessing IT process maturity based on process attributes and control objective has later been proposed [14].

## 3 Case Study

Our proposed approach is presented in the next sections. We conducted a study to apply our approach in a real-world case. The case study was about process DS7 during the implementation of electronic identity (e-ID) card in a citizen registration agency in a district in the province of Central Java, Indonesia. For assessing process maturity, we distributed a questionnaire to persons responsible for e-ID card implementation, i.e. head of citizen registration division, head of IT division, and IT operators. We observed current practices and asked the desired process maturity to those persons. Based on the resulted maturity gap, we recommended actions to improve the maturity of process DS7 by referring COBIT to practices. We then validated the recommendations by confirming them to persons that have authority to run the recommendations, i.e. head of citizen registration division and head of IT division. If the agency was not capable to implement a recommendation with proper effort, we further asked about any constraints that might hinder the agency.

This paper focuses on our approach for IT process maturity assessment, recommendation, and validation. This paper does not intend to give full and detailed description on how we apply our approach in the case study.

## 4 Maturity Assessment

In COBIT, an IT process consists of control objectives and activities (ITGI, 2007). Control objectives are requirements that must be satisfied by the process; while activities are means to satisfy the control objectives (Dirgahayu et.al, 2015). When control objectives are activities must have been done. Therefore we use control objectives as the basis for IT process maturity assessment. We can calculate IT process maturity from the process'

**Table 2.** Maturity of process DS7.

| Control objectives | Maturity attributes | | | | | | Control objectives maturity | IT process maturity |
|---|---|---|---|---|---|---|---|---|
| | AC | PPP | TA | SE | RA | GSM | | |
| DS7.1 | 2.56 | 2.33 | 1.67 | 2.78 | 2.33 | 2.44 | 2.35 | |
| DS7.2 | 2.11 | 1.89 | 1.11 | 2.56 | 2.22 | 2.67 | 2.09 | 2.13 (level 2) |
| DS7.3 | 1.67 | 1.56 | 1.22 | 2.33 | 2.22 | 2.78 | 1.96 | |

**Table 3.** Recommendation for process DS7 – attribute TA.

| Control objectives | Current practices | Actions for improvement | Recommendation for management |
|---|---|---|---|
| DS7.1 | Tools are available, but their use are based on personal initiative. | Give instruction to use tools systematically. | Procedures to use tools systematically shall be defined. |
| DS7.2 | Tools are available but less used. | Give instruction to use tools. | |
| DS7.3 | Tools are available but less used. | Give instruction to use tools. | |

satisfied, we can assume that their corresponding control objectives maturity. A control objective maturity is calculated based on the maturity of all attributes in that control objective. This way allows us to identify which control objectives have to be addressed to improve IT process maturity.

For example, process DS7 has three control objectives, i.e. DS7.1 (*identification of education and training needs*), DS7.2 (*delivery of training and education*), and DS7.3 (*evaluation of training received*). Table 2 depicts the calculation of process maturity in the case study. IT process maturity is the average of its control objectives maturity; while a control objective maturity is the average of that control objective's attributes maturity. The score of a maturity attribute of a control objective is the average of the scores given by our respondents. The result indicates that the process maturity is at level 2 (*repeatable but intuitive*). This level was obtained by rouding the IT process maturity score to the nearest integer.

A control objective may be described in several actions, e.g. DS7.2 is described as (i) identify target groups, (ii) identify efficient delivery mechanisms, (iii) appoint trainers, (iv) organise training sessions, and so on (ITGI, 2007). This prescriptive nature of the control objectives shall facilitate us in formulating recommendations to improve the maturity of the corresponding IT process.

## 5 Recommendation

In the case study, the desired process maturity was at level 4 (*managed and measurable*). As there was a maturity gap, actions and recommendations to improve IT process maturity were necessary. We define two steps for suggesting such actions and recommendations.

First, we identified control objectives and their attributes that need improvement, i.e. their actual maturity are lower than the target level. Note that when a control objective maturity is already at the target level or higher, its control objective does not need urgent improvement, although the maturity of some of their attributes are below the target level.

In Table 2, all control objectives DS7.1, DS7.2 and DS7.3 need to be improved. Also, the maturity of their attributes must be improved. Due to space limitation, this paper addresses attribute TA only. In the case study, tools and automation refers to methods, forms and softwares that can help in identifying training needs, delivering training contents, and evaluating training received.

Second, we identified the current practices related to those attributes and suggested actions for improvement. Similar actions were then grouped and abstracted into a recommendation for senior management. Recommendations at a management level are necessary to ensure that actions can be carried out successfully. Table 3 depicts this step for attribute TA.

**Table 4.** Validating the recommendations for process DS7

| Recommendations | Not required | Not capable | | | Capable with great efforts | | | Fully capable | | |
|---|---|---|---|---|---|---|---|---|---|---|
| | 0 | 1 | 2 | 3 | 4 | 5 | 6 | 7 | 8 | 9 |
| Procedures to use tools systematically shall be defined. | | | | | | √ | | | | |
| Performance indicators shall be defined | | | √ | | | | | | | |

## 6 Validation

In the case study, we provided seven recommendations. We validated them using a range of scores as depicted in Table 4. This table lists two recommendations only due to space limitation. Since two authorative persons were asked for confirmation, their answers were averaged and the results are indicated as checkmarks (√) in the table. The validation was to check whether the agency was able to apply the recommendations (score is between 1 and 9). The agency may also indicate if the recommendations were not required (score = 0).

Since all answers were less than 7 (below the agency's minimum capability for implementing the recommendations with proper effort), we further asked about any constraint that might hinder the agency. The main constraint was lack of personnels with sufficient expertise in managing and conducting training programes; however, no sufficient budget was allocated for improving personnels' expertise.

## 7 Concluding Remarks

Many literatures focus on methods for IT process maturity assessment [14-17]. Many other researches present the application of those methods in case studies. In this paper we have presented an approach to IT process maturity improvement, that consists of activities of maturity assessment, recommendation, and validation.

The calculation in IT process maturity assessment is based on the maturity of the process' control objectives. The maturity of a control objective maturity is calculated from the maturity of the control objective's attributes. As illustrated in Table 2, this way of calculation is done horizontally as indicated with an arrow below the table. It differs from the calculation in [14] that is done vertically, i.e. IT process maturity is the average of its attributes maturity; while an attribute maturity is the average of that particular attribute maturity in all control objectives of that process

By calculating the process maturity horizontally, we can recommend actions to improve process maturity based on its control objectives improvement. This way of recommendation is enabled by the prescriptive nature of control objectives. Such improvement actions typically formulated at a technical level. In the other hand, recommendations at a management level is also necessary in order to ensure that the actions can be carried out successfully in a proper effort. Recommendations for management are formulated by grouping and abstracting similar actions. Grouping is done based on maturity attributes.

Recommendations are validated to make sure that they match with the enterprise needs and capability. For recommendations that are required by the enterprise, we have developed a scale for validation, in which enterprise's capability is defined in three categories, i.e. (i) not capable, (ii) capable with great efforts, and (iii) fully capable. Each category can be further divided for finer granularity. Recommendation and validation should be done iteratively until they results in recommendations that are required by and within the capability of the enterprise.

In the future, we will use our approach in more case studies in different domains. From those case studies, we expect to learn valuable lessons that are beneficial to improve our approach. Futhermore, we will integrate IT Balance Score Card [18] in the recommendation.

## References

1. ITGI. *Board briefing on IT governance 2nd edition*. IT Governance Institute (2003)
2. NGC. *IT governance: Developing a successful governance strategy*. National Computing Centre (2005)
3. ITGI. *COBIT 4.1*. IT Governance Institute (2007)
4. ISACA. *COBIT 5: A business framework for the governance and management of enterprise IT*. ISACA (2012).
5. OGC. *ITIL service strategy version 3*. Office of Government Commerce (2007)
6. TSO. *ITIL Service Strategy 2011 edition*. The Stationery Office (2011).
7. ISO. *ISO/IEC 20000-1: Information technology – Service management – Part 1: Service management system requirements*. International Organization for Standardization (2011)
8. ISO. *ISO/IEC 38500: Corporate governance of information technology*. International Organization for Standardization (2008)
9. M. Röglinger, J. Pöppelbuß, J. Becker. Maturity models in business process management. Business

Process Management Journal, **18**(2), pp. 328-346 (2012)

10. M. Rosemann, T. De Bruin. Application of a holistic model for determining BPM maturity. BP Trends, February 2005, pp. 1-21 (2005)
11. M. Siponen. Towards maturity of information security maturity criteria: six lessons learned from software maturity criteria. Information Management & Computer Security, **10**(5), pp. 210-224 (2002)
12. V.A. Canal. *ISM3 1.0. Information security management maturity model.* Institute for Security and Open Methodologies (2004)
13. SEI. *CMMI for development version 1.3* Technical report CMU/SEI-2010-TR-033. Software Engineering Institute (2010)
14. T. Dirgahayu, D. Ariyadi, Assessment to COBIT 4.1 maturity model based on process attributes and control objectives. *Proc. 2015 Int. Conf. Science in Information Technology*, pp. 343-347 (2015)
15. A. Pederiva. The COBIT maturity model in a vendor evaluation case. Information Systems Control Journal, **3**, pp. 26-29 (2003)
16. M. Simonsson, P. Johnson, H. Wijkström. Model-based IT governance maturity assessments with COBIT. *Proc. European Conf. Information Systems*, pp 1276-1287 (2007)
17. A. Walker, T. McBride, G. Basson, R. Oakley. ISO/IEC 15504: Measurement applied to COBIT process maturity. Benchmarking: An International Journal, **19**(2), 159-176 (2012)
18. W. Van Grembergen. The balanced scorecard and IT governance. *IRMA Conference*, pp. 1123-1124 (2000)

# Changes in folate characteristics and its identification in broccoli (*Brassica oleracea Italica*) extract fermented by Lactic Acid Bacteria Mixed Culture (LAB)

*Yati* Maryati, and *Agustine* Susilowati

[1]Research Center for Chemistry, Indonesian Institute of Sciences, Indonesia

**Abstract.** Broccoli (*Brassica oleracea Italica*) was fermented by cultures of lactic acid bacteria (LAB) as a potential source of natural folic acid. This study aimed to evalte characteristic changes and to identify folate compounds from broccoli extract, fermented by mixed LAB cultures (*L. bulgaricus, S. thermophulus, L.acidophilus, Bd. bifidum*). The formulation of broccoli extract was fermented with variation of LAB starter culture with concentrations of 10 and 20%(v/v), and the change of characteristic of folic acid compound during fermentation (0 to 48 hours) with an interval of 8 hours was evaluated. The results showed that the fermentation of broccoli extract with different concentration of LAB culture had an effect on the concentration of folic acid produced, as well as the change of concentration of folic acid during the fermentation time interval. The optimum condition was obtained based on the highest folic acid concentration of 6.74%, at culture concentration of 20% during 24 hour fermentation with the value of folic acid concentration of 72.11 μg/mL, pH value of 4.29, total sugars of 34.61%, total acids of 0, 97%, dissolved protein of 14.64 mg/mL and total LAB of log 13.02 + 0.05 cfu / ml.

## 1 Introduction

Vegetables from the *Cruciferous* family such as broccoli (*Brassica oleracea Italica*) are foods that are often consumed and can good for health. The benefits provided from broccoli secondary plants are linked to compounds that have bioactivity; Glucosinolates and phenolic acids [1]. Broccoli contains less fat. It is low in calories and rich in vitamins, inorganic substances, and fiber [2]. Broccoli is also rich in β-carotene, ascorbic acid, selenium, quercetin, and glutathione, potentially as antioxidants. In addition, it contains high enough folic acid (folate).
Humans cannot synthesize folate, hence, it is a need for human to intake folate to prevent the deficiency of this vitamin [3]. The recommended daily intake (RDI) of folate for adults is 200-400 μg [4, 5]. For pregnant women is recommended to be 400-600 μg. Although folate is always present in every food consumed by humans, folate deficiency is still common, even in developing countries [6]. Milk and fermented milk products are folate-derived foods. Folate has a very unstable nature. Therefore, to synthesize folate, culture of lactic acid bacteria in fermented milk products combined with broccoli vegetables can be used to increase folate content.

Lactic acid bacteria (LAB) include *lactobacilli* and *bifidobacteria* that are dominant in the gut and able to produce lactic acid through saccharide fermentation. LAB also acts as a live bacterial activator that can promote the growth of beneficial bacteria in the body, prevent various diseases, and regulate physiological activity by improving digestive function, inhibits cholesterol absorption, controlling immunity, enhances absorption, improves nutrient utilization and others [7,8] . It is well known that it is not only yogurt starter culture and *Lactobacillus lactis* which has the ability to produce folate, but other lactic acid bacteria (LAB) also possess the ability to produce folate. *L. acidophilus* is reported to be able to increase the concentration of folate in fermented milk [9]. It has also been observed that *L. plantarum* is capable of producing folate, even at low folate levels, when it is grown on a folate-free medium with a certain chemical composition [10]. In addition, recent research has shown that some probiotic microorganisms have the ability to synthesize folic acid such as *bifidobacteria* and propionibacteria. In this study we examined changes in the characteristics of folate and its identification in fermented broccoli beverages with the addition of lactic acid bacteria enriched with FOS as soluble fiber that is beneficial for the health of the body.

## 2 Experimental

### 2.1 Materials

* Corresponding author: maryati97@yahoo.com

The ingredients used in this study were broccoli vegetable concentrate, fructooligosaccharide (FOS), skim milk, mixed cultures LAB powder (*L. bulgaricus, S. thermophulus, L. acidophilus, Bifidum*), supplied by Hansel. Co. Ltd. MRSA, MRSB, 3-aminophenol, Hydrochloric (4M), Sodium Nitrite, Sulfamic Acid were supplied by Sigma-Aldrich (St. Louis, MO). Ethanol, Folin-Ciocalteau reagent, HCl, $KH_2PO_4$, potassium ferric cyanide, trichloroacetic acid (TCA), NaOH, Buffer Phosphate were obtained from E-Merck.

### 2.2 Fermentation process

Cleaned broccoli vegetables were blanched for 5 minutes at 80°C, followed by blending process of vegetable and water ratio of 1: 4, so that broccoli vegetable porridge was obtained. The next step of filtration was done with 80 mesh filters to obtain filtrate/broccoli vegetable extract. The broccoli extract is enriched by the addition of skim milk and FOS as a source of prebiotic oligosaccharides at 9% (w/v) for each. The result of the broccoli extract formulation was sterilized at 121 °C for 15 minutes. The sterilized formulated broccoli extract was rapidly cooled to a temperature of 45 °C, and then inoculated with 10% and 20% (v/v) of LAB starter mix culture (containing about $10^9$ cfu/mL) obtained from 1% (w/v) mixed LAB powder culture (*L Bulgaricus, S. thermophulus, L. acidophilus, B. bifidum*), inoculated against MRSB-based media and incubated at 37 °C for 24 hours. The obtained subculture was then inoculated with variations of 10% and 20% (v/v) on the sterile formula of 15% (w/v) of skim milk solution and 15% (w/v) of FOS, with incubation temperature of 37 °C for 0–48 hours, with an interval of 8 hours.

### 2.3 Cell number of LAB, pH, total sugar and Total acid (TA)

LAB fermented broccoli extract was analyzed for bacterial growth, determined by cell number of LAB calculation from several series dilution with 0,1 mL sample to 10 ml of sterile water, then using plating method with MRSA, which was done as much as 2 replicates (duplo), incubated at temperature 37 °C for 48 hours. The number of colonies grown in the medium was calculated by multiplying the average number of colonies by dilution factor [11]. Similarly, the pH value measurements of the samples were measured by digital pH meter and total sugar measurements were made by the phenol sulfate method [12] performed at 0–48 hours. The total acid was determined by titration of each sample with 0.1 M NaOH. The result was expressed as a percentage of lactic acid, determined according to standard procedure [13].

### 2.4 Analysis of folic acid

Folic acid analysis was performed using spectrophotometric method based on the reaction of diazotizing p-aminobenzoylglutamat acid produced after the reduction reaction of folic acid and 3-aminophenol to form a complex of yellow-orange. About 1 mL of standard folic acid or sample was added into 1 ml of 4 M HCl , 1 mL of 1% (w/v) sodium nitrite, 1 mL of 1% (w/v) sulfamic acid, and 1 mL of 1% (w/v) 3-aminophenol. After mixing, it formed a yellow-orange complex solution. Furthermore, the absorbance was measured using UV–VIS spectrophotometer at a wavelength of 460 nm [14].

### 2.5 Folic acid identification by LC-MS

After fermentation process, fermented brocoli extract were then filtered by microfiltration. Identification of folic acid and glutamate acid compounds in brocoli fermentation at the optimum condition. The obtanied permeate or purified brocoli extracts and folic acid standard. Oligomer analysis was performed by LC-MS using Mariner Biospectrometry integrated with Q-tof mass spectrometer (MS) through ESI (electrospray ionisation) system where the scan mode performed in the range of 100-1200 sqm at 140°C. C18 column Supelco (RP 18, 250 x 2 mm with a particle size of 5 µ) was used for the LC (Hitachi L 6200). Solvent was a mixture of water containing 0.3% acetic acid (A) and methanol containing 0.3% acetic acid (B) at a ratio of 80% methanol and 20% water with a flow rate of 1 mL/min. The injection volume 20 µL was used in this analysis [15].

## 3 Results and Discussion

### 3.1 Material Characterization

Fresh broccoli (*Brassica oleracea L.*), blanching broccoli at 80 °C for 5 minutes, and blend broccoli at ratio of broccoli and water (1:4) as broccoli pulp, and broccoli extract filtered through 100 mesh sieve showed a difference in chemical compositions, as tabulated in Table 1. This difference was showed at filtrate 100 mesh (broccoli extract) containing total solids 1.54% higher compared with total solids 0.79% and in broccoli pulp. This difference also seen at total polyphenol in broccoli pulp (0.4740%) and broccoli extract (0.1836%) were lower compared with fresh broccoli (0.0179%). but different from the content of folic acid in fresh broccoli, folic acid is lower than the blanching treatment.

* Corresponding author: maryati97@yahoo.co.

**Table 1.** Composition of broccoli as substrates in LAB fermentation

| Type of broccoli | Composition | | |
|---|---|---|---|
| | **Total solids (%)** | **Total polyphenol (%)** | **Folic acid (µg/mL)** |
| Fresh broccoli | 25.51 | 0.0179 | 8.76 |
| Blanched broccoli* | 0.5129 | 0.2704 | 32.11 |
| Broccoli pulp** | 0.79 | 0.4740 | 30.53 |
| Broccoli extract*** | 1.54 | 0.1836 | 37.72 |

**result of blanching 80 °C for 5 minutes*
***blending at a mixture of 1 part of broccoli and 4 parts of water*
****filtered via 100 mesh*

Blanching process decreased possibility components contained in broccoli, besides pulverizing and filtration. Broccoli extract showed that polyphenol was trapped in other components so that not significant difference in pulp and extract. Soluble component effect on total solids, although this matter was affected by solubility properties of polyphenol in water. It had been known that both components were eased to solve in water [16].

### 3.2 Cell number of LAB

Characteristics of fermentation of broccoli extract with mixed LAB culture to total LAB count continued to increase depending on the amount of LAB culture concentration added to broccoli extract medium. The main factor influencing the increase of LAB population was the ability of lactic acid bacteria as starter culture that serves as a trigger of a metabolic process, in this case in particular is fermentation. The nature of the LAB is its ability to ferment sugars to acid and other factors that were suspected to cause an increase in LAB population was the condition of the media environment that supports optimum LAB growth. The LAB population for the 10% concentration of culture reached a maximum of 11.26 log cfu/ml at 48 hours fermentation time and for a culture concentration of 20% reached a maximum of 13.68 log cfu/mL at 40 fermentation hours, as shown in Figure 1.

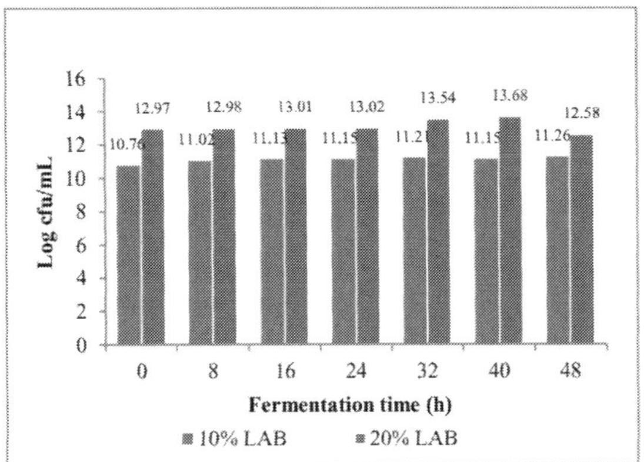


**Figure 1.** Changes in cell number of LAB (Log cfu/mL) from broccoli extract (*Brassica oleracea Italica*) fermented by mixed culture of Lactic acid bacteria (LAB) for 0-48 hours

### 3.3 pH and Total acid (%)

The concentration of free hydrogen ions is reflected by the pH value [17]. Changes in pH and total acid during fermentation of broccoli extract inoculated by mixtures of lactic acid bacteria and *bifidobacteria* were presented in Figure 2.

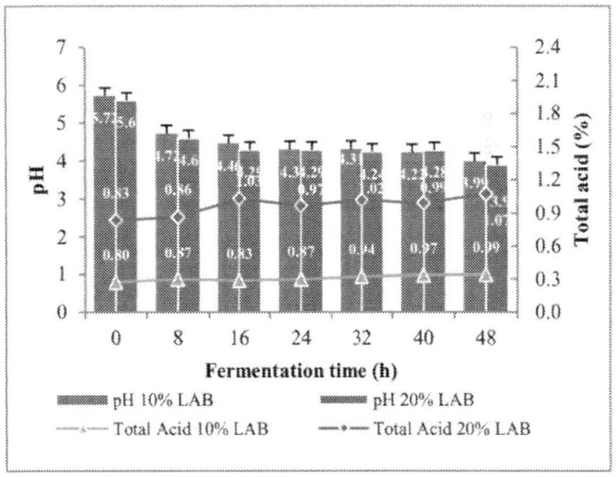


**Figure 2.** Relation of total acid and pH to characteristic changes of broccoli extract (*Brassica oleracea Italica*) is fermented by mixed culture of Lactic acid bacteria (LAB)

In general, broccoli extract during the fermentation time of 0 to 48 hours decreased the pH and increased the total acid. The concentration of lactic acid bacteria influenced the pH change. Low pH values and higher acid production were obtained from both 10% of LAB starch culture concentrations and 20% of fermented broccoli extracts.

The decrease of pH value after 48 hours incubation in broccoli extract was significantly different from each concentration of starter culture LAB added. In this study, the culture concentrations of 10% and 20% had a pH value of 5.72 and 5.60 at the beginning of fermentation (0 hours) and at the end of fermentation (48 hours) ranged from 3.98 and 3.90, while the total acid ranged from 0.83 and 0.80 up to 1.07% and 0.99% at the end of fermentation (48 hours). The decrease in pH occurred along with the production of various organic acids during the fermentation of broccoli extract by LAB. The total acid (TA) value after fermentation by LAB increased significantly depending on the concentration of LAB culture added to the broccoli extract medium. This was due to the measurement of TA that the measured acid component consists of dissociated and un-dissociated acids, whereas for pH only measures the dissociated acid component in the form of $H^+$ ions [17]. Similarly, the pattern of TA value change in treatment was consistent with the change in pH values (Figure 3). Similarly, the pattern of TA value changes in starter culture was consistent with the pattern of pH value changes in starter culture. Similar to pH values, the pattern of TA values change in general was consistent with the pattern of TA value changes in starter control.

### 3.4 Total Sugar and Reducing sugar

The change in total sugar is in line with the changes that occur in reducing sugars. The pattern of changes in total sugar and reducing sugars occurred in both culture concentration of 10% and 20% of LAB culture during fermentation 0 to 48 hours. Changes in starter culture concentration and fermentation time can be seen in Figure 3.

Changes in total sugar decrease occur at 20% concentration, this is because the greater the total content of sugar will accelerate bacterial activity in converting to acid [18]. In other words, it indicates that a single carbon source derived from prebiotics FOS (DP 2-8) in broccoli beverages is utilized by BAL for its growth, or an existing carbon source can be metabolized in β(2-1) fructans with a DP Higher than 2 (two). The total sugar increased during the fermentation process (0-48 h) at the culture concentration of 10% to 20%, the increase of total sugar and reducing sugar showed that too large BAL concentration resulted in inefficient culture in metabolizing FOS compounds in the time range until 48 hours at concentrations of 10% and 32 hours at a concentration of 20% of LAB cultures, this may be influenced by physical and chemical factors from the tested sample.

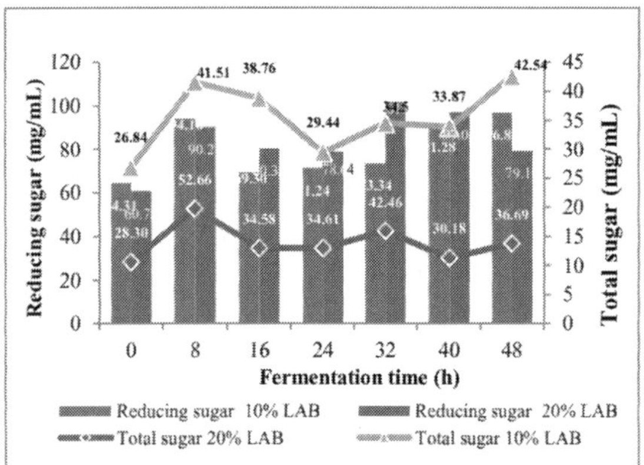


**Figure 3.** Relationship of total sugar and sugar reduction to the characteristic changes of broccoli extract (*Brassica oleracea italica*) fermented by mixed cultures Lactic acid bacteria (LAB)

### 3.5 Folic acid and Dissolved protein

The tendency of yield on the rate of fermentation to folic acid, the data showed during the fermentation time of broccoli extract gave the result of folic acid fluctuated, the data obtained was significantly different during the fermentation time.

Folic acid at optimum condition was obtained at initial fermentation (32 hours) at 10% LAB culture concentration while at optimum 20% culture concentration at 24 hours fermentation time, the concentration of folic acid in each starter culture was 71.20 and 72.11 μg / mL. Folic acid is a compound that is very sensitive to light, oxygen and temperature, therefore fermentation by using lactic acid bacteria culture can improve the recovery of natural folic acid because it is done under anaerobic conditions to minimize interaction with light and oxygen. In addition, microbial activity in the culture of lactic acid bacteria is expected to help the stability of natural folic acid during fermentation. The fermentation time will increase the chemical composition change from the fermentation of broccoli extract. Changes in dissolved protein levels during the fermentation time occur because of the role of the enzyme produced by the BAL during the fermentation process takes place. Protease is an enzyme capable of hydrolyzing peptide bonds in proteins. Lactic acid bacteria have the ability to produce proteolytic enzymes around cell walls, cytoplasmic membranes, or in cells [19]. Increased levels of dissolved protein in fermented broccoli extract may also occur due to increased growth and proliferation of BAL cells (*L. acidophilus, Bd. bifidum, L. bulgaricus and S. thermophilus*). This is due to the protein in the extract of broccoli undergoing hydrolysis process by proteolic enzyme produced by BAL during fermentation. Rahman [20] states that lactic acid-forming bacteria in fermentation produce proteolic enzymes that help protein degradation.

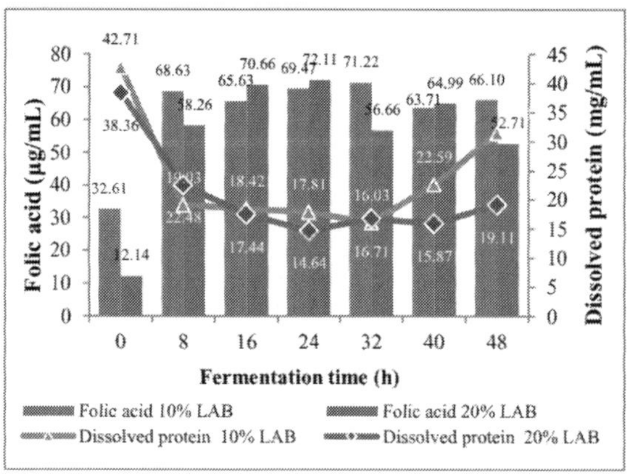


**Figure 4.** The relation of soluble protein and folic acid to changes in the characteristics of broccoli extract (*Brassica oleracea italica*) fermented by mixed cultures Lactic acid bacteria (LAB).

Folic acid at optimum condition was obtained at initial fermentation (32 hours) at 10% LAB culture concentration while at optimum 20% culture concentration at 24 hours fermentation time, the concentration of folic acid in each starter culture was 71.20 and 72.11 μg / mL. Folic acid is a compound that is very sensitive to light, oxygen and temperature, therefore fermentation by using lactic acid bacteria culture can improve the recovery of natural folic acid because it is done under anaerobic conditions to minimize interaction with light and oxygen. In addition, microbial activity in the culture of lactic acid bacteria is expected to help the stability of natural folic acid during fermentation. The fermentation time will increase the chemical composition change from the fermentation of broccoli extract. Changes in dissolved protein levels during the fermentation time occur because of the role of the enzyme produced by the BAL during the fermentation process takes place. Protease is an enzyme capable of hydrolyzing peptide bonds in proteins. Lactic acid bacteria have the ability to produce proteolytic enzymes around cell walls, cytoplasmic membranes, or in cells [19]. Increased levels of dissolved protein in fermented broccoli extract may also occur due to increased growth and proliferation of BAL cells (*L. acidophilus, Bd. bifidum, L. bulgaricus and S. thermophilus*). This is due to the protein in the extract of broccoli undergoing hydrolysis process by proteolic enzyme produced by BAL during fermentation. Rahman [20] states that lactic acid-forming bacteria in fermentation produce proteolic enzymes that help protein degradation.

### 3.6 Identification of folic acid with LCMS

Broccoli extract contains high enough folic acid and can be seen the changes during the fermentation process with the addition of lactic acid bacteria. Folic acid is a heterocyclic compound with a conjugated pentoric acid structure with one or more L-glutamates linked via an amino acid-carbonyl group. Folic acid has one L-glutamic residue with the name "pteroilglutamic acid". Folic acid can be reduced to H2 folate or tetrahydrofolate (H4 folate) which is an active form of coenzyme vitamin [21].

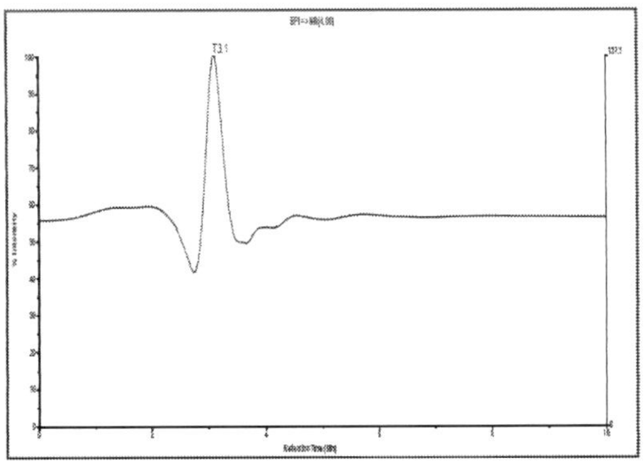

(a)

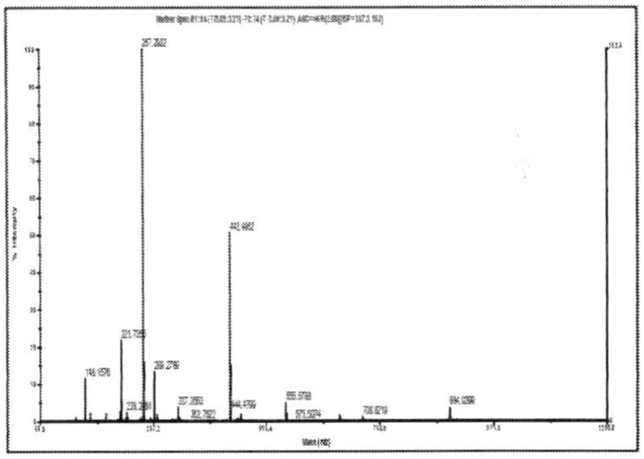

(b)

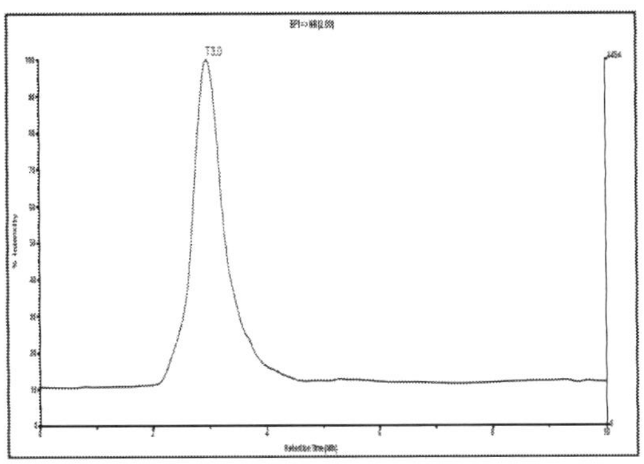

(c)

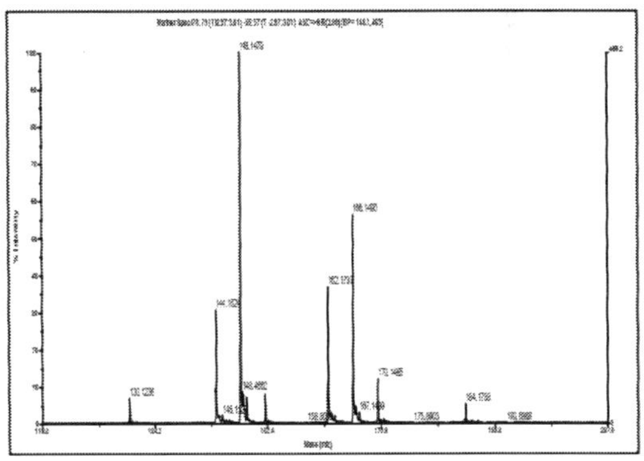

(d)

**Figure 5**. Standard solution of glutamate (c) and mass spectra from T 3.0 from chromatogram folic acid standard (b), Standard solution of glutamate (c) and mass spectra from T 3.0 from chromatogram glutam e standard (d)

Figures 5a and 5b show that the standard of folic acid is dominated by compounds with BM 442,5, 443,16, 443,51 and 444,48 Da, respectively. With the successive intensities of 50.55, 8.35, 15 and 2.44% in other words folic acid is possible on the BM. In Figs. 5c and 5d show that the standard of glutamic acid is dominated by compounds with BM 148.15, 148.47 and 149.14 Da respectively. With successive intensities of 100, 8.51 and 7% in other words glutamic acid is possible on the BM. The results of mass spectra of broccoli extract fermentation by mixed LAB cultures were obtained, at a culture concentration of 20% during 24-hour fermentation

(Fig. 6) than the standard mass spectra of folic acid and glutamate (Figs. 5a and 5c), identified eight compounds at T 3.7 where 2 (two) compounds are identical as folic acid with molecular weight 441.23, and 443.29, with the relative intensity respectively of 3.48% and 2.11%.

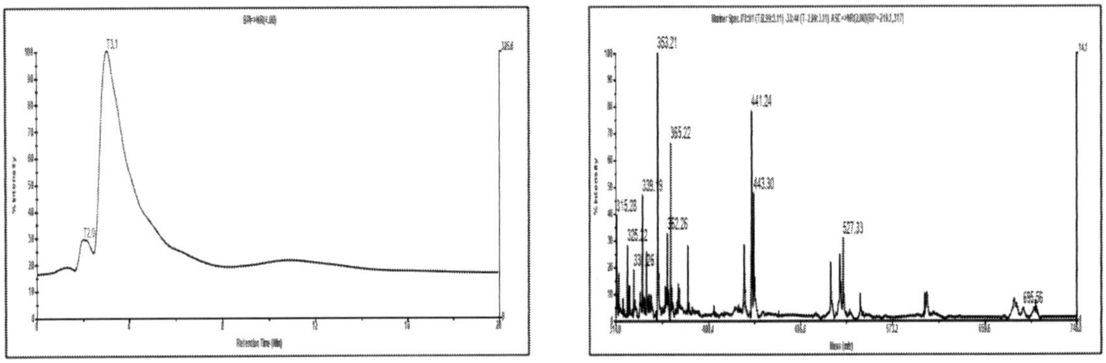

(a) (b)

**Figure 6**. Chromatogram of fermented broccoli (a) and mass spectra of T3.07 from chromatogram of fermented broccoli and optimum process conditions based on highest folic acid in a culture concentration of 20% during fermentation time of 24 h (b)

While the identical compound as glutamic acid is obtained 5 (five) compounds with molecular weight of 145.08, 159.046, 163.09, 175.10, and 191.08, with relative intensity of 0.86%, 12.15%, 1.58%, 2.95%, and 28% of 43 compounds respectively. The presence of folic acid in the fermentation of broccoli extract is presented in Table 2.

**Table 2**. Identification of folic acid and degraded folic acid compounds in fermented extract broccoli using lactic acid bacteria (LAB) mix culture by LC-MS

| No. | Centroid Mass | Relative Intensity (%) | Area | Represented Compounds |
|---|---|---|---|---|
| 1. | 145.156904 | 0.86 | 26.85 | 2-aminopentanedioateglutamate $C_5H_7NO_4^{-2}$ |
| 2. | 159.060508 | 12.15 | 265.59 | Lithium L-glutamate C5H8NNaO4 |
| 3. | 163.167017 | 1.58 | 30.97 | 2-amino-4-hydroxypteridine |
| 4. | 175.183459 | 2.95 | 59.96 | L-Glutamic acid 5-ethyl ester C7H13NO4 |
| 5. | 191.118699 | 28 | 653.26 | L-Glutamic acid disodium salt C5H7NNa2O4 |
| 6. | 192.073319 | 1.2 | 24.69 | Pterin-carboxylic acid C7H5N5O3 |
| 7. | 441.397255 | 3.48 | 101.33 | Isofolic Acid C19H19N7O6 |
| 8. | 443.417706 | 2.11 | 40.96 | (6S)-5,6,7,8-tetrahydrofolate dianion C19H21N7O6-2 |

## 4 Conclusion

Characteristics changes of the bioactive component of fermented broccoli extract mixed of starter cultures of LAB (*L. bulgaricus, S. thermophilus, L. acidophilus, Bd. bifidum*) with different concentrations of LAB cultures and different time of fermentation affect to folic acid, dissolved protein, total sugar, total acids, and total BAL. The optimum conditions obtained by the highest concentration folic acid at a concentration of 20% during the fermentation time of 24 h with a value of folic acid concentration of 72.11 µg/mL, pH value of 4.29, total sugars of 34.61%, total acids of 0, 97%, dissolved protein of 14.64 mg/mL and total LAB of log 13.02 + 0.05 cfu / ml. Identification of glutamate and folic acid through LCMS showed identified eight compounds at T 3.7. Where 2 (two) compounds are identical as folic acid and as glutamic acid is obtained 5 (five) compounds.

The authors thank Puspa Dewi Lotulung for assistance assistance in the use of the LC MS at the NPPC (Natural Products and Pharmachemical Chemistry, Research Center for Chemistry).

## References

1. Rebecca J. Robbins, Anna-Sigrid Keck, Gary Banuelos, and John W. Finley. (2005), Cultivation conditions and selenium fertilization alter the phenolic profile, glucosinolate, and sulforaphane content of broccoli. *Journal of Medicinal Food*. July 2005, 8(2): 204-214.
2. Fahey J. W., Zhang Y, & Talalay, P. (1997). Broccoli sprouts: an exceptionally rich source of inducers of enzymes that protect against chemical carcinogens. *Proc Natl Acad Sci* 94, 10367–72.
3. LeBlanc J. G., de Giori G. S., Smid E. J., Savoy G., Hugenholtz J., & Fernando Sesma. (2007). Folate production by lactic acid bacteria and other food-

grade microorganisms, Communicating Current Research and Educational Topics and Trends in Applied Microbiology A. Méndez-Vilas (Ed.).

4. Institute of Medicine. (1998). Dietary Reference Intakes for Thiamin, Riboflavin, Niacin, Vitamin B6, Folate, Vitamin B12, Panthothenic Acid, Biotin, and Choline, National Acad Pr; Washington, DC.
5. FAO/WHO. (2002). in: *Human vitamin and mineral requirements*, Bangkok, Thailand
6. Konings E. J., H. H. Roomans, E. Dorant, R. A. Goldbohm, W. H. Saris and P. A. van den Brandt. (2001). Folate intake of the Dutch population according to newly established liquid chromatography data for foods, *Am J Clin Nutr* 73. 4. 765
7. Kim M. J., & Kim G. R. (2006). In vitro evaluation of cholesterol reduction by lactic acid bacteria extracted from Kimchi. *Korean J Culinary* Res 12, 259–68.
8. Gao M. T., Kaneko M., Hirata M., Toorisaka E., and Hano T. (2008). Utilization of rice bran as nutrient source for fermentative. *Bioresour Technol* 99,3659–64
9. Lin, M. Y., and C. M. Young. (2000). Folate levels in cultures of lactic acid bacteria. *Int. Dairy J.* 10:409-414
10. Sybesma W. , Starrenburg, M., Tijsseling, L., Hoefnagel M. H. & Hugenholtz J. (2003). Effects of Cultivation Conditions on Folate Production by Lactic Acid Bacteria, *Appl Environ Microbiol* 69. 8. 4542
11. [BAM]. Bacteriological Analitical Manual. 2001. Aerobic plate count [Internet]. Retrieved from http://www.fda.gov/Food/FoodScienceResearch/Laboratory/Methods/ucm 063346. htm
12. Dubois, M., Gilles K. A., Hamilton J. K., Rebers, P., Smith, F. (1956). Colorimetric method for determination of sugars and related substances, *Anal. Chem.* 28 (3); 350–356.
13. AOAC. 1984. *Official Methods of Analysis* 14th edn. Washington, DC, Association of Official Analytical Chemists.
14. Ruengsitagoon, W. & Hattanat, N. (2012). *Simple spectrophotometric method for determination of folic acid,* The 4th Annual Northeast Pharmacy Research Conference, Thailand.
15. Eichhorn, P., and Knepper, T. P., 2001, Electrospray ionization mass spectrometric studies on the amphoteric surfactant cocamidopropylbetaine, *Journal of Masspectrometry,* Vol. 36, pp. 677 – 684.
16. Belitz H. D. & Grosch, W. (2002). *Food Chemistry*, 2nd Edition (Springer-Verlag, Berlin Heidelberg, Germany.
17. Nielsen S. S. (2003). *Food Analysis.* 3th Ed. Kluwer Academic/Plenum Publisher, New York.
18. Triyono, A. 2010. Mempelajari Pengaruh Maltodekstrin Dan Susu Skim Terhadap Karakteristik Yoghurt Kacang Hijau. Balai Besar Pengembangan Teknologi Tepat Guna, Teknik Kimia, Universitas Diponegoro Semarang.
19. Thomas, T. D., Pritchard, G. G., (1987), Proteolytic enzymes from dairy starter cultures. *Fed. Eur. Microbiol.Soc.Microbiol.* Rev. 46:245
20. Rahman, A. (1992). *Teknologi Fermentasi*. Penerbit Arcan. Jakarta
21. Eitenmiller, W. O., Landen Jr, L Ye, (2008), *Vitamin analysis for health and food science*, second edition, CRC Press.

# Burn drug made from ozonated vegetable oil mixture with white tumeric and cassava leaves extract

*Farah* Moulydia, *Nabila* Salsabila, *Ria Kusuma* Dewi, *Adinda* Nirmala, *Setijo* Bismo*

*Chemical Engineering Department, Universitas Indonesia, Kampus Baru UI Depok, Depok 16424, Indonesia*

**Abstract.** This research aims to create a burn treatment performed with ozonation process from a mixture of vegetable oil and added extracts of herbal ingredients. Ozonation on vegetable oils proven to kill bacteria and safe for the body. Ozonated vegetable oil produced from the ozone reactor batch process by doing a variety of extraction mixture to Oleozon® and vegetable oils. Then the results of ozonation is added extracts of herbal ingredients that cassava leaves and white turmeric to increase effectiveness in killing bacteria. Cassava leaves have anti-inflammatory agent, namely Vitamin C. While white turmeric Curcuma zedoaria have substance, which of the two compounds can inhibit and kill bacteria. The quality of ozonated oil (Oleozon®) analytically were tested by the method of iodine number, acid number, peroxide number, and FTIR. Ozonation increased the peroxide and acid values for both oils, the increase being higher for mixture of coconut oil and soybean oil. The results of such mixing is then tested in bacteria to determine their effectiveness in killing the bacteria. The best ozonation condition is in an increase of 386,85% acid value, peroxide value about 102,91 meq/kg oil, and decrease in iodine number up to 21%. The result showed that under these conditions, ozonized oil has an antiseptic effect against *Staphylococcus aureus*. The final results of this study are expected to be a new innovation in the healing of skin wounds caused by burns as an anti-inflammatory that is effective, safe, and environmentally friendly.

## 1 Introduction

Burns are damages to our body tissues caused by heat, chemicals, electricity, sunlight or radiation. Besides, burns can cause swelling, blistering, leaving scars and can cause serious shock. They can also cause infections because they damage the protective layer of the skin. Bacteria that are often the cause of infection in burns are: *Streptoccoccus pyogenes, Pseudomonas aeruginosa, Proteus sp, Klebsiella spp, Seratia* and *Staphylococcus aureus*.

The ozonated vegetable oils or Oleozon® have been shown to have antibacterial and antifungal capabilities in food applications, cosmetics and pharmaceutical industries. The reaction of ozone with vegetable oils produces *Criegee ozonide*, hydroperoxide, and aldehyde products [2]. The research on the ozonated vegetable oils used as medicine in humans does not indicate a secondary reaction.

From the results of research on the characteristics of the ozonated vegetable oil [3] it can be concluded that soybean oil and coconut oil also have the potential to be used as a vegetable oil due to its high unsaturated fat content. In addition, mixing vegetable oil with natural ingredients can also be done to increase the effectiveness of the oil. Selected herbal ingredients are white turmeric and cassava leaves. Both of these herbal ingredients proved to have substances that can be anti-inflammatory in skin wounds. Therefore, in this study used extracts of cassava leaves and white turmeric to be mixed into a mixture of coconut oil and soybean oil.

## 2 Methodology and Results

### 2.1 Methodology/Experimental

This study uses the main component of ozonator as an ozone-producing tool to applied in the ozonation process of vegetable oil mixture. The ozonation reaction of the oil will occur in a batch reactor equipped with a 4 L/min air flow rate and a coolant to keep the reaction temperature at 25°C. At this stage the ozonation of vegetable oil is done continuously for 72 hours to see the relationship between the length of time ozonation with the resulting anti-bacterial effects. The vegetable oil mixture will be ozoneated with concentrations of the oil to be varied.

After oil ozonation, herbal extraction is performed. This stage aims to produce herbal plant extracts that are cassava leaves and white turmeric. These two herbs will be macerated using IPA and Hexane solvents for 5 days and stirring every 4 hours. After obtaining the extract by evaporating, the extract will be mixed with the ozonated vegetable oil to add antibacterial properties.

The test method used is the test of iodine number, acid number, peroxide number, FT-IR, and stability test of Oleozon®. Iod number to know the amount of double bond in oil. Acid number to indicate the amount of free fatty acids present in the oil. Peroxide numbers to show the level of oxidation of ozone to fatty acids. Then for FT-IR to find out the differences in functional groups that increase or decrease during preoccupation

*Corresponding author: sbismo@che.ui.ac.id

and after the diozonation. As well as stability test Oleozon® for a month at room temperature. This test is performed to determine the success of ozonation, based on changes in the physical and chemical properties of each of the ozonated vegetable oils.

This stage is the stage of addition of vegetable oil into the culture of bacteria that have been prepared before then seen the ability of bacterial disinfection qualitatively in the form of visual observations with the ratio of bacterial culture conditions before and after addition. In this method also can see the mechanism of activity of the ozonated vegetable oil against destruction and inhibition of bacterial cell wall and become bacterial disinfectant.

## 2.2 Results and Discussions

Based on Table 1 & Figure 1, it can be seen that on all variations of pressure, the ozonated soybean oil (SBO) has decreased the iodine number so it can be inferred that the combined design ozonator succeeds in attacking and lowering the double bond on soybean oil.

The highest decrease in iodine number is in the ozonation process of pure coconut oil (CO) which changes up to 21% during ozonation process. This indicates that in this condition, the double bond reaction by ozone to produce ozonide is better than in both variations of other oil mixtures. So that pure coconut oil can be an alternative to soybean oil as well.

**Table 1.** Iodine Number of Ozonated Oil

| Parameter | Iodine Number (gram iod/100 gram) | | |
|---|---|---|---|
| | Coconut Oil (CO) | CO & SBO (85:15) | CO & SBO (75:25) |
| Before Ozonated (0 hours) | 2.06% | 4.30% | 5.01% |
| After Ozonated (72 hours) | 1.61% | 4.01% | 4.80% |
| % Change | 21% | 6.70% | 4.20% |

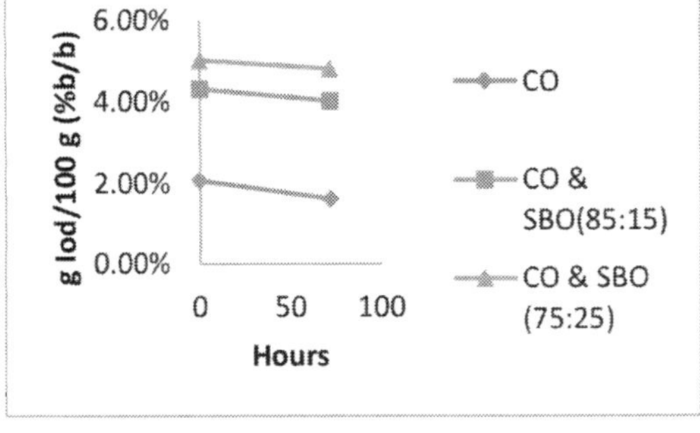


**Figure 1.** Iodine Number of Ozonated Oil

Based on the available data in Figure 2 & Table 2, it can be seen that the acid number of oil after the diozonation has increased, it seems that the ozonation in this research was successful. Increasing the acid number can indicate that ozonation is successful. The increase in acid numbers is triggered by increased peroxide decomposition and oxidation of aldehydes to carboxylic acids during the ozonation reaction

**Table 2.**Acid Number of Ozonated Oil

| Parameter | Acid Number (mg KOH/g oil) | | |
|---|---|---|---|
| | Coconut Oil (CO) | CO & SBO (85:15) | CO & SBO (75:25) |
| Before Ozonated (0 hours) | 0.2104 | 0.8377 | 0.3645 |
| After Ozonated (72 hours) | 0.8131 | 1.6423 | 0.7678 |
| % Change | 386.45% | 196.05% | 210.64% |

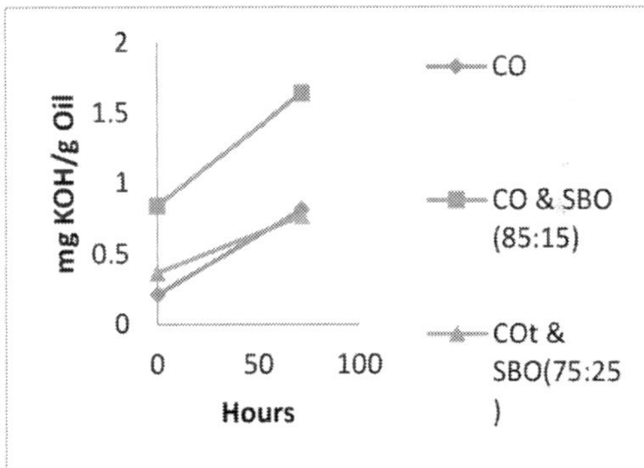


**Figure 2.** Acid Number of Ozonated Oil

The resulting peroxide acts as an oxidizing agent in lowering the growth factor of Staphylococcus aureus bacteria. The higher the reaction pressure, it allows rapid collision between ozone and soybean oil fatty acid molecules. This ozone will quickly break the double bonds in soybean oil fatty acids, so as to produce the desired ozonolysis product better. The speed of this reaction is also influenced by the stirring system using turbine stirrer, causing high turbulence in the ozonation reaction. Therefore, the pure coconut oil ozonation process is recommended because it has high peroxide numbers. The following Table 3 presents the results of peroxide numbers.

**Table 3.** Peroxide Number of Ozonated Oil

| Parameter | Peroxide Number (meq/Kg Oil) | | |
|---|---|---|---|
| | Coconut Oil (CO) | CO & SBO (85:15) | CO & SBO (75:25) |
| Before Ozonated (0 hours) | 0 | 0 | 0 |
| After Ozonated (72 hours) | 102.91 | 95.44 | 83.93 |

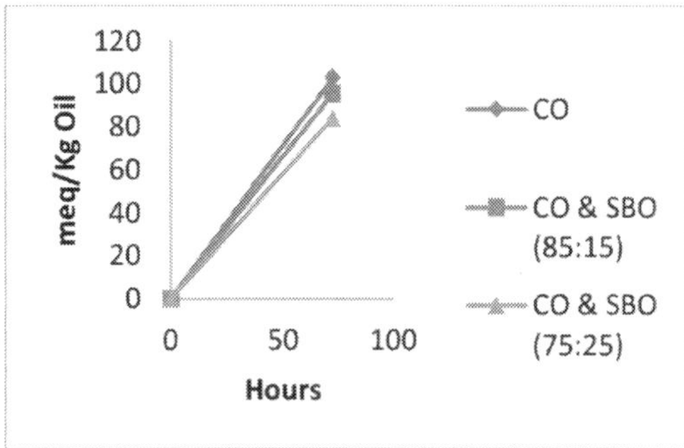


**Figure 3.** Peroxide Number of Ozonated Oil

Table 4 shows the results of FTIR data on vegetable oil before and after ozonated on a particular mixture of oils. In the data, it appears that there is an increase and there is some decrease in transmittance percent either in aromatic groups, alkene groups, or carbonyl groups. It shows that the concentration decreases in the bonds C = O, C = C, and = C-H when an increase in transmittance occurs. Increased transmittant percent indicates that there is a release of bonding on a large carboxylic acid, so the greater the percent increase in transmittance, the more acidic the pH resulting from ozonated soybean and the higher the acid number obtained.

Then for the bacterial disinfection test it is necessary to know how effective the mixed sample of the ozonated vegetable oil with herbal extract can inhibit the growth of Staphylococcus aureus bacteria.

In the samples of the ozonated vegetable oils indicated a magnitude of inhibit zone of 7.93 mm. That is, the length of time ozonation affects the effectiveness of the ozonolysis product as an antiseptic. 72 hours of ozonation time is a good time in the ozonolysis reaction of soybean oil.

**Table 4.** FTIR Analysis Results

| Sample | Alkena Bond (C=C) | | Aromatic Bond (=C-H) | | Karbonil (C=O) | |
|---|---|---|---|---|---|---|
| | %T | Wavelength ($cm^{-1}$) | %T | Wavelength ($cm^{-1}$) | %T | Wavelength ($cm^{-1}$) |
| CO (Blanco) | 86.298 | | 88.791 | | 57.243 | |
| Ozonated CO | 89.356 | | 90.794 | | 66.959 | |
| CO & SBO (85:15) Blanco | 86.376 | | 87.668 | | 57.962 | |
| Ozonated CO & SBO (85:15) | 86.31 | 1500 | 88.003 | 720 | 57.947 | 1700 |
| CO & SBO (75:25) Blanco | 86.356 | | 87.029 | | 58.331 | |
| Ozonated CO & SBO (75:25) | 86.246 | | 87.136 | | 58.467 | |

**Table 5.** FTIR Analysis Results

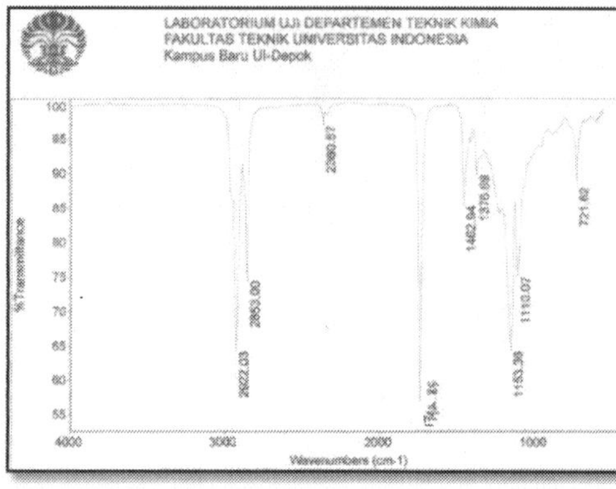


## Conclusion

The optimum temperature for ozonation of coconut oil was obtained at 25 ° C with optimum ozonation time for 72 hours and flow rate of 4 L/min. Oleozon® in the form of oil has increased peroxide number as much as 102,91 meq/Kg Oil, increase in acid number of 386,85%, and decrease the amount of iodine equal to 21%. The ability of Staphylococcus aureus bacterial inhibition against a mixture of Oleozon® with a positive 72 hour herbal extract can inhibit bacteria.

The authors acknowledge Universitas Indonesia and PITTA program for the financial support. We are also thankful to AKA Bogor and Fakultas Farmasi Universitas Pancasila for the instruments to characterize the Oleozon®.

## References

1. Almeida, Nathalia R de, Adilson Beatriz, Ana Camila Micheletti, dan Eduardo J. de Arruda. *Oozonized Vegetable Oils and Ttherapeutic Properties: A Review*. Campo Grande: *Orbital Eletronic Journal of Chemistry*. (2012).
2. Anonymous. *Top Types of Burn Wounds and How to Care For Them*, [Online]. Avaiable: http://www.advancedtissue.com/top-types-burn-wounds-care/ (Akses : 14 September 2016). (2013).
3. Bismo, Setijo. *Sintesis Biodiesel dengan Teknik Ozonasi: Ozonolisis Etil- Ester Minyak Sawit Sebagai Suatu Bahan Bakar Mesin Diesel. Alternatif.* Depok: Chemical Engineering UI. (2005).
4. Diaz, Maritza F., Dayan Gil, Barbara Manet, Magali Gomez, Eliane T. Sousa, Marcia C. Da C. Veloso, pedro Aa. De P. Pereira, Solange Pizzolatto, Antonio S. Mmangrich, Angelo C. Pinto, Jailson B. De Andrade. *Physicochemical Characterization of Ozonated Sunflower Oil Obtained by Different Procedures*. Cuba: Ozone Research Center. (2012).
5. Sadowska, Justyna et al. *Characterization of Ozonated Vegetable Oil by Spectroscopic and*

*Chromatographic Methods.* Ireland: Elsevier. (2007).
6. Sechi. *Antibacterial Activity of Ozonated Sunflower Oil (Oleozon®).* Journal Application Microbiology. 90. 279-284. (2001).
7. Tjitrosoepomo, Gembong., *Taksonomi Tumbuhan Obat-obatan*, Gadjah mada University Press, Yogyakarta. (2005),
8. USDA. *Soybean Research Advisory Institute*. United States: Departement of Aglicultur. (2015).

# Antiproliferative activity of natural coumarins from *Calophyllum incrasaptum* M.R Henderson-Wyatt Smith against human breast cancer cells MCF

*Jamilah* Abbas*, *Linar* Z Udin, and *Muhammad* Hanafi

Research Center for Chemistry, Indonesian Institute of Sciences (LIPI). PUSPIPTEK, Serpong, 15314, Indonesia

**Abstract.** Objective: to evaluated the antiproliferative activity of natural coumarin from Calophyllum incrasaptum M.R Henderson-Wytt Smith against human breast cancer cells MCF-7. Methode : Coumarin from ethyl acetate fraction of C. incrasaptum M.R Henderson- Wyatt Smith was isolated by coloumn chromatographyic and structure elucidated by using spectroscopic methods and isolate compound was evaluated for their antiproliferative activities in the alamar blue assay. Result: Coumarin have antiproliferative activity against MCF-7 cancer cell lines through alamar blue assay for 4 h after treatment. Conclusions: coumarin showed good activity against cancer cell lines with IC50 value of 2.23 µg/mL.
**Keywords:** antiproliferative, C. incrasaptum M.R Henderson- Wyatt Smith, MCF-7

## 1 Introduction

Cancer is a fatal disease standing next to the cardiovascular disease. Cancer, a diverse group of diseases characterized by uncontrolled growth and spread of abnormal cells, if the spread is not controlled, it can result in death or mortality, estimate 1,660,299 new cancer cases are expected to be diagnosed in 2013, and die about 580,350 in 2013 and almost 1,600 people die per day. New cases of breast cancer in the US during 2013 is 232,340 expected in women and 2,240 in men. Deaths : an estimated 40,030 breast cancer death (39,620 women, 410 men). Risk factors breast cancer for women is bigger than men, potentially risk factor after age 18, being overweight or obese (for postmenopausal breast cancer), used of menopausal hormone therapy (combined estrogen and progestin), risk is also increased by a family history of breast cancer [1]) and the number of deaths from cancer will continue to rise, with an estimated 13.1 million people dying in 2030 [2].

Emergency and rapid spread of cancer, represent a major problem for prophylaxis and treatment cancer which becomes more difficult, and the medicines used as treatment cancer have clear limitation and unfortunately cancer projected as the primary cause of death in the future [3]. Therefore, it is necessary to discovery anticancer drugs from traditional medicine plant for curing cancer.

Some experiments have been done to develop compound which it's having anticancer activity, such as: Cao et al 1998 have isolated some of the phenyl coumarins., which it's showed cytotoxic activity against mouse leukemia P388, mouse fibrosarcoma (WEH 1640), human monocytic leukemia (HTP-1) and human lymphoblastic leukemia (MOLT4) [4].: Guilet et al., 2001[5], and Reyes-chilpa et al., 2004 have isolated coumarin derivate, costatolide, soulattrolide, & calanolide F and *C. dispar* contains coumarin of the mammea type, some of them resulted highly cytotoxic against human epidermoid carcinoma cell (KB, PC3, K562, U251[6]; Li,Y-Z et al 2010 have also isolated triterpenoid from *C. inophyllum* and their growth inhibitory effects on human leukemia HL-60 cells [7], Mousavi, S.H.(2015) have evaluated and compered of cytotoxic effect of prenyloxycoumarin and herniarinon to MCF-7 cell line [8]. Other researcher (Khasman et.al 1992) have isolated calanolide A from *C. lanigerum* var austrocoriaceum as anti-HIV. [9].

Preventing of cancer by administrating a 6-substituted coumarin derivative was patented, The substituent at the 6-position may have five or more carbon atoms such as ten or preferable fifteen carbon atom [(C-5(6-prenyloxycoumarins), C-10(6-prenyloxycoumarins), C-15(6-prenyloxycoumarins)], [10]. Preventing of cancer by using coumarin/calcone derivative was described by (Carrico-Moniz). The structure of coumarin/calcone as anticancer was saw in figure 1, 2,3,4 [9]

The structure of coumarin derivate with substituted C-5, C10 and C15 at position C-6

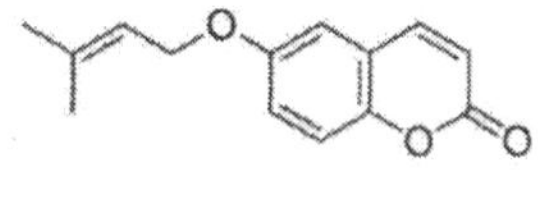


**Fig 1**. Structure of C-5(6-prenyloxycoumarins)

* Corresponding author: jamilahabbas@yahoo.com.

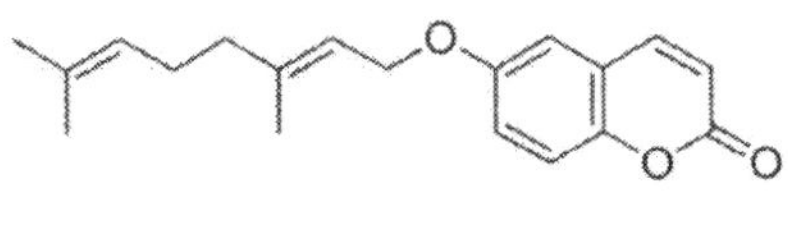

**Fig 2.** Structure of C-10(6-prenyloxycoumarins)

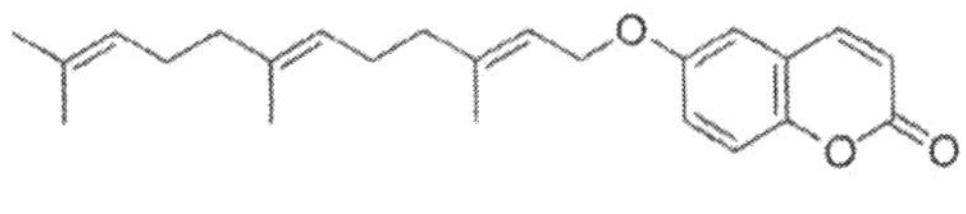

**Fig 3.** Structure of C-15(6-prenyloxycoumarins)

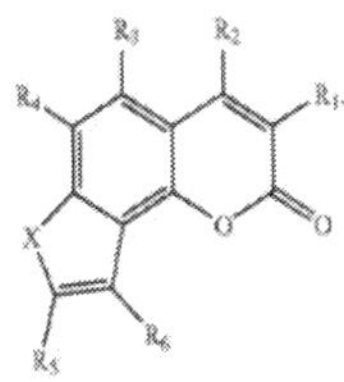

**Fig 4.** Structure of coumarin/chalcopne

Since coumarin showed interesting antiproliferative or anticancer activity, we decided to investigated *Calophyllum* species led to the isolation of one chemically distinct group of coumarin. In these study we are described the antiproliferative activity of the phenylcoumarin to MCF-7 cell lines from *C. incrasaptum*.

## 2 Materials and methods

### 2.1 Materials

Stem bark of *C. incrasaptum* M.R Henderson- Wyatt Smith was collected from Kalimantan Island in Indonesia. The sample was identified at the Herbarium Unit, Department of Botany Indonesian Institute of Sciences (LIPI). Solvents and chemicals used in this study are analytical grade and were purchased from Merck (Darmstadt, Germany).

### 2.2 General aspects

$^1H$ and $^{13}C$ NMR spectra were recorded on a JEOL NMR spectrometer at 500 and 125 MHz, respectively. The abbreviations s, d, and br used to describe $^1H$ spectra refer to singlet, doublet, and broad, respectively. NMR spectra were recorded in $CDCl_3$ solution on a Jeol JNM ECA 500 MHz instrument, using TMS as the internal standard. Silica gel (60–230 mesh and 230–400 mesh) were used for column chromatography and preparative TLC plates coated with silica gel (2000 micron thickness) was purchased from Merck (Darmstadt, Germany). Precoated Si gel plates SIL G/UV$_{254}$, 0.25 mm were used for TLC. The compounds were detected under UV light at 254 and 366 nm. Melting points were determined on an Electrothermal Fisher melting point apparatus (Scientific serial 903N0056) and are uncorrected. IR spectra were recorded on a FTIR Prestige 21 Shimadzu spectrophotometer (KBr dish) and UV spectra was taken on a Hitachi U-2000 spectrophotometer. LC-MS were recorded on Mariner Bio Spectrometric equipment with pneumatically assisted electronspray ionization (EIS) source.

## 3 Methods

### 3.1 Extraction procedure

The air-dried stems bark of *C. incrasaptum* (2.1 kg) were minced finely and shredded in an electric mill, macerated three time with MeOH (6L x72h weight of extract after solvent was removed : 400 gr), then was diluted in $H_2O$ and partitioned with between *n*-hexane and $H_2O$: EtOAc and $H_2O$ : and butanol – $H_20$ . The n-hexane, EtOAc and buthanol was removed under reduced pressure to give 65.8 g, 138.8 g and 609.4 g *n*-hexane, EtOAc and BuOH extract respectively.

EtOAc extract (120 gr) was separated by flash coloumn chromatography, It was eluted with a stepwise gradient of *n*-hexane followed by increasing amounts of EtOAc (20:1; 10:1; 5:1; 3:1; 1:1 ; EtOAc 100%) give 6 fractions (Fr. A1–A6), then were eluted by EtOAc:MEOH (20:1; 10:1; 5:1; 3:1 ,1:1) to give 5 fractions (Fr A7-A11). Fraction A5 was further purified several time by normal CC system on silica gel medium-pressure liquid chromatography (MPLC), eluting with *n*-hexane and 10% stepwise gradient of acetone to afford several fraction, fraction were collected and regrouped on the basis of TLC analysis. Sub fraction 68-74 was then applied to an silica column eluted with *n*-hexane – DCM (dichoromethane 50:50→0:100). Fraction 3 was futher purified by sephadex LH-20 , approximately 15 ml was collected for each fraction, collection sub fraction 33 - 47 were pooled after TLC profile analysis. We was able to pure crystal of the phenyl coumarin by recrystalyzing methode using *n*-hexane : acetone (3:1).

### 3.2 Preparation of Cancer Cells Line

Cancer cells line used in this study is breast carcinoma (MFC-7), and the cell line was cultured in DME Medium with FBS 10%. These cell was cultured at 37°C with moisture content of 95% and 5% $CO_2$ for 3 days until the cell cultures undergo confluence 60-70%. After that the old media removed, replaced with new medium and incubated again for 24 hours. Culture cells then washed with PBS 1-2 times as much and was suspended using trypsin-EDTA solution. Cells that have been suspended added with new.

### 3.3 In vtro antiproliferative assay

Antiproliferative assay was used by Alamar Blue method in triplicate against human breast carcinoma (MFC-7) cell lines. The MCF culture cells line of 100 μL was supplemented with 10 μL of pure test compound (05; 1 and 1.5 ppm). It was then incubated for 24 hours at a 37°C.

Colourring process was done by adding Alamar blue solution for 4 hours. The colour intensity of the cell line was measured by using ELISA plate reader at 560 nm (excitation) and 590 nm (emission) wavelength. Percent viability is calculated as in equation 1. While $IC_{50}$ of the active extract was calculated by linear regression analysis between percent survival showed in equation 2. Alamar Blue was used to color the sample, and the $IC_{50}$ of the pure compound (Alamar Blue Assay, U.S. Patent No.5,501,959). [12]

$$\%\ Viability = \frac{OD\ (cell+sample) - OD\ (negative\ control}{OD\ (cells) - OD\ (negative\ control)} \times 100$$

## 4 Result and Discussion

### 4.1 Results of Elucidation structure

Phenylcoumarin was isolated via column chromatography (CC) as an yellow amourphous from an EtOAC extract of steam bark of *C incrasaptum* M.R Henderson- Wytt Smith, mp 184-185°C. The UV absorbtion of phenylcoumarin characterized a 5,7,10 threeoxygenated coumarin, and give a positive test with a vaniline/$H_2SO_4$ reagent, an intense deep blue spot develops over 24 hour period. LC-MS revealed the molecular formula to be $C_{21}H_{18}O_5$ (*m/z* 351,40 $[M+1]^+$), by ESIMS analysis, i,e., the molecule had threeteen degrees of unsaturation. The UV spectrum exhibited two maxima at $\lambda_{max}$ 256 and 319 nm were indicative of a dimethylchromanone derivative, whilst the FTIR spectra showed important peaks explaining the bending, stretching, and double-bond absorptions of the coumarin. The C-H stretching of absorption for $CH_3$ occurred at wavelengths of 2937 $cm^{-1}$. One alkane peaks were observed showing the bending absorption for $CH_3$ ($v_{max}$ 1332 $cm^{-1}$). spectrum indicated absorption bands $v_{max}$ 1732 $cm^{-1}$ attributed to non-conjugated carbonyl group (C=O), and $v_{max}$ 1691 $cm^{-1}$ as conjugated C=O carbonyl groups, $v_{max}$ 1597-1332 (aromatic C=C) and $v_{max}$ 1153 $cm^{-1}$ (C-O), and mono substituted benzene ring at $v_{max}$ 767 and 702 $cm^{-1}$ (Figure 5).

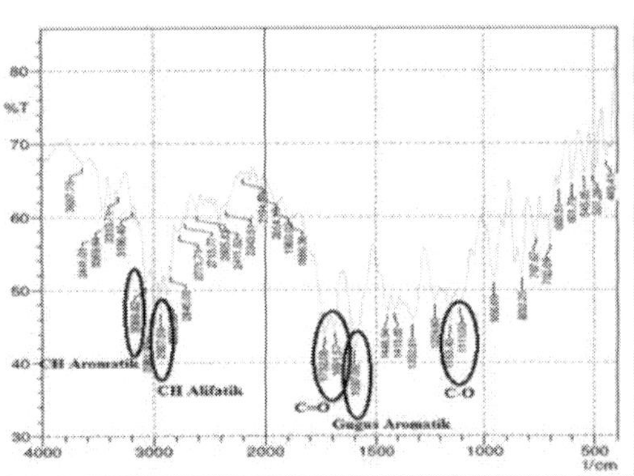


**Fig 5.** IR spectrum of isolate compound

Determination of structure was performed usinga NMR device. The 1-NMR spectrum of isolaed compound (phenyl coumarin) is evidenced by the presence of two proton singlet ($\delta_H$ 6.07, 1H,s, H-3) and δH 6.73, 1H,s,H-6) and group of signal belonging to five aromatic proton of the phenyl group [(δ7.34-7.40, *m* : $\delta_H$ = 7,34 (1H, *m*), 2 protons at $\delta_H$ 7.40 (2H, *m*), and 2 protons at $\delta_H$ 7.38 (2H, *m*)]. The presence of a singlet-shaped signal at δH 3.14 ppm (3H,s, H-13) shows the proton of one methoxy group. Addition signal included these of dimethylchromanone ring ($\delta_H$ 4.28, 1H, *m*, H-8, and $\delta_H$ 2.55, 1H, *m*,H-9 : $\delta_H$ 1.52, 3H, *d*, *J* = 6.5 Hz, H-11 and $\delta_H$ 1.18, 3H, *d*, *J* = 7.1 Hz, H-12) (Figure 6).

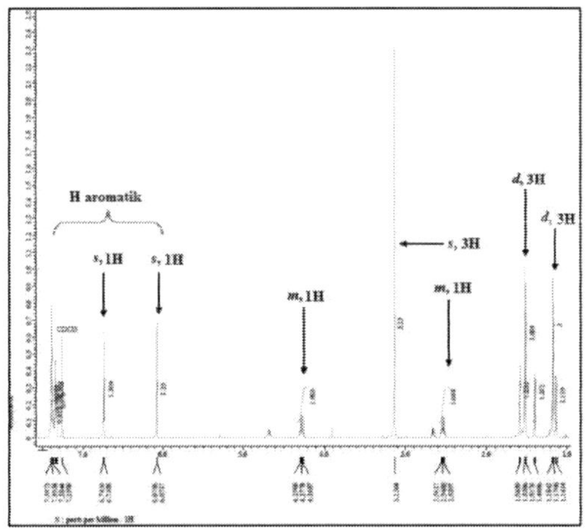


**Fig 6.** [1]H-NMR spectrum of isolate compound

The $^{13}C$-NMR data shows 21 peaks of carbon atoms, nine C quartener atoms (including two carbonyl downfield signals at δ 159.43 and 191.65), seven C-H aromatic, two aliphatic protons ($\delta_H$ 1.52, *m*, H-8 & $\delta_H$ 2.55, *m*, H-9), the presence of two methyl groups ($CH_3$) is supported by $^{13}C$-NMR data with a carbon chemical shift of 19.65 and 10.49 ppm and one methoxy group at 62.77 ppm (Figure 7).

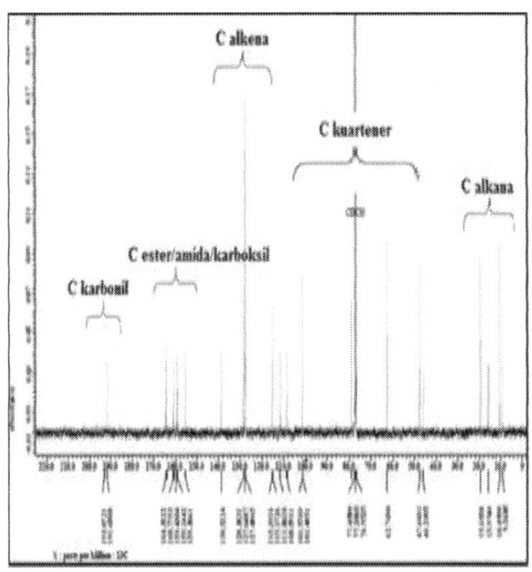


**Fig 7.** $^{13}C$-NMR spectrum of isolate compound

This phenyl coumarin compoun have a methoxy sustituted at C-5(singlet methyl signal at $\delta_H$ 3.14 ppm). The NMR signals observed were in accordance with the presence of coumarin nucleus, with the benzenoid ring ($\delta_C$ 159.43, 115.17, 155.38, 108.59, 160.77, 101,48, 164.35, 111.88 and 159.14) an unsubtituted phenyl ring (138.12, 128.38, 127.49 and 127.65) one a methoxyl group ($\delta_C$ = 62.77) and a cromanone ring ($\delta_C$ = 191.5, 78.99, 47.68, 19.65 and 10.49). A correlation from the phenyl C1' (δ 138.2) to H-3 confirmed that a phenyl group was attached to C-4 as in isolated compound.

Long-distance 3J and 2J (HMBC) proton H 6,07 ppm on C-3 has been proved by H-correlation at δH 6.07 with C-4a (δC 108,59), C-2 (δC 155,43) and C-1 '(δC 138.52), this proves the position of H 6.07 ppm at C-3. The presence of two doublet-shaped signals at δH 1.52 ppm (3H, *d J*, 6.5 Hz) and 1.18 ppm (3H, *d J* 7,1 Hz) show protons of two methyl groups. HMBC data for proton correlation of methoxy group (δH = 3.14 ppm) with carbon C-5 (δC 160,77) assured the position of methoxy on C-5.

Phenyl group in C-4 positions is evidenced by the correlation of proton aromatics $\delta_H$ 7.38 ppm with carbon at $\delta_C$ 155.38 ppm, and proton correlation $\delta_H$ 6.07 ppm (H3) with carbon at $\delta_C$ 138.52 ppm (Figure 8). HMBC correlation of isolate compound was explained in figure 8.

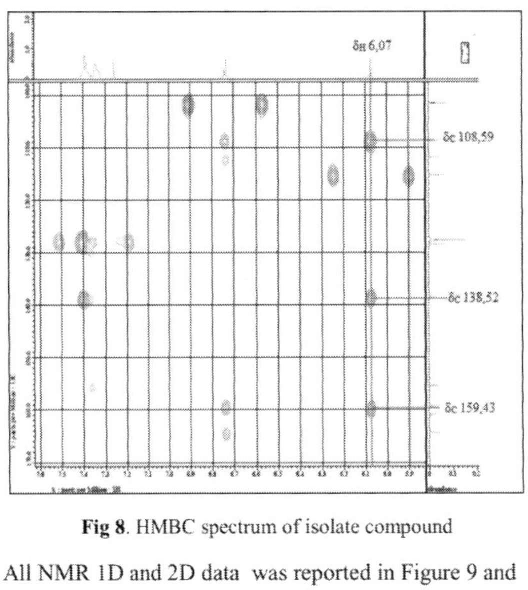


**Fig 8**. HMBC spectrum of isolate compound

All NMR 1D and 2D data was reported in Figure 9 and Table 1.

**Table 1.** Data NMR 1D and 2D and COSY chemical shift (ppm) of phenyl coumarin

| No | C | H-NMR (δH) ppm | Cosy |
|---|---|---|---|
| 1 | 0 | | |
| 2 | 159.43 | - | |
| 3 | 115.17 | 6.07 (1H,*s*) | - |
| 4 | 155.38 | - | - |
| 4a | 108.59 | - | - |
| 5 | 160.77 | - | - |
| 6 | 101.48 | 6.73 (1H,*s*) | - |
| 6a | 164.35 | - | - |
| 7 | 0 | - | - |
| 8 | 78,99 | 4.28 (1H, *m,J* =6.5 Hz) | H-8 - H9 |
| 9 | 47.68 | 2.55 (1H, *m,J* =6,5 Hz) | H-9- H-8 |
| 10 | 191.65 | - | - |
| 10a | 111.88 | - | |
| 10b | 159.14 | - | - |
| 11 | 19.65 | 1.52 (3H, *d, J*=6.5 Hz) | H11-H8 |
| 12 | 10.49 | 1.18 (3H,*d. J* = 7.1 Hz) | H12-H-9 |
| 1' | 138.2 | - | - |
| 2' | 128.38 | 7.38 (1H, *m*) | H-2'-H-3' |
| 3' | 127.65 | 7.40 (1H,*m*) | H-2'-H-4' |
| 4' | 127.49 | 7,34 (1H,*m*) | H-3'–H5' |
| 5' | 127.65 | 7.40 (1H,*m*) | H-4'. -6' |
| 6' | 128.38 | 7.38 (1H, *m*) | H-6'-H-5' |
| $OCH_3$ | 62.77 | 3.14 (3H, s) | |

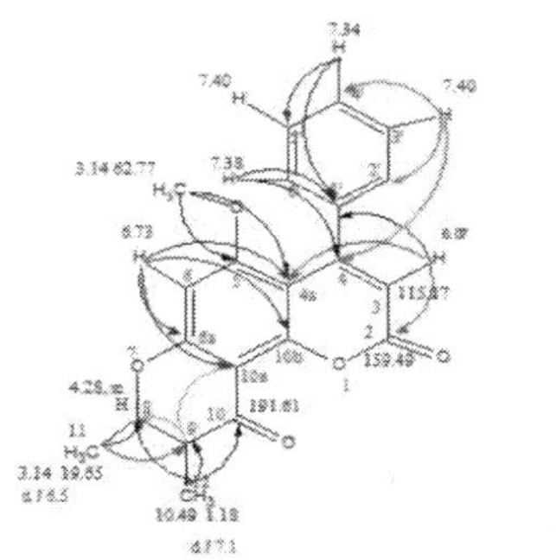

**Fig 9.** HMBC correlation of of isolate compound

This paper report the crystal structure of one isolated compound from *C. incrasaptum* as phenyl coumarin (Figure 10).

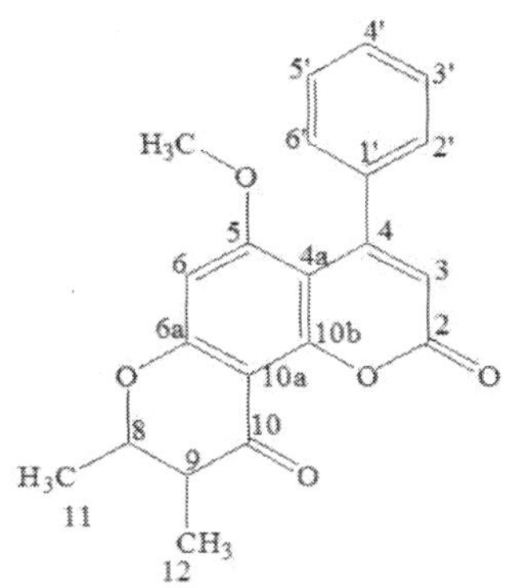

**Fig 10**. Chemical structure of phenyl coumarin

Hormon oestrogen has the crucial role in development of the breast cancer, the most frequent malignant disease in women, therefore many therapies are designed to block

https://doi.org/10.1051/matecconf/201815404003

this activity. Cinnamoyl-coumarin derivatives were especially effective in eostrogen-dependent cancer, such as breast (MCF7) cancer cell line. The phenylcoumarin in these paper have anticancer activity too, becauce coumarin are proven to posses a wide range of biological activitied , e. i as anticancer.

### 4.2 Results of cell viability

Inhibition of cell viability caused by prenyl coumarins including phenylcoumarin from *C. incrasaptum* was examined using Alamar Blue assay. The cytotoxic effect of pure test compound (phenyl coumarin) against MCF-7 breast cancer cells was determined with alamar blue assay at range of 0-1.5 ppm after 24 h of treatment period. The cell viability can be qualitatively assessed by the color intensity of alamar blue. The intensity of the alamar blue dye increased when the concentration of pure test compound decreased. The results are graphically represented in Figure 11`. Cell viability (%) was observed to be conversely proportioned to the concentration of the pure test compound, these results showed that these phenylcoumarin compound indicated decreased cell viability (that treatments showed growth inhibition) cel MCF7 responded in a dose-dependent. The pure test compound showed effective cytotoxic effect with the $IC_{50}$ value of 2,6 ppm in MCF-7 breast cancer cells

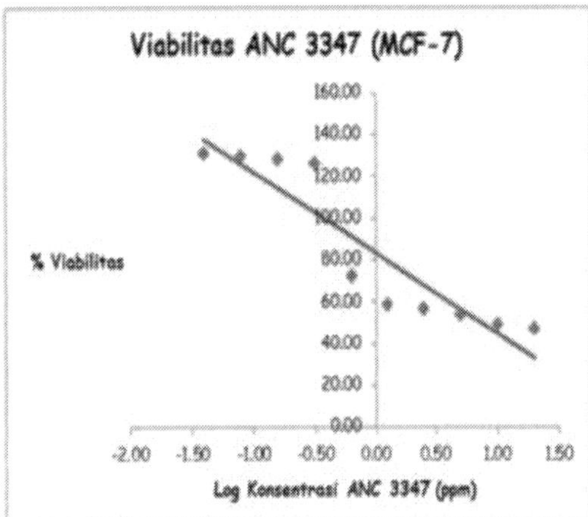


**Fig 11.** Cytotoxic effects of phenylcoumarin on MCF7 cells. Relationship between log concentration with % viabilirty

lines obtained from regression : Y = -54.04 Log X + 68.80 with R = 0.90 $IC_{50}$ = 2,23 µg/mL
$IC_{50}$ values in (µg/mL), $IC_{50}$ < 4 µg/mL is higest cytotoxic,
$IC_{50}$ 4-30 µg/mL considered cytotoxic and in active, $IC_{50}$=> 40 µg/mL

### 4.3 Discussion

According to the results, we treated the MCF7 cell line with different concentrations of phenylcoumarin for 24 h. The results indicated phenylcoumarin induced cell death in concentration –dependent manner.

Phenyl coumarin is a group of compounds which have shown to psses preventive and therapeutic effects on breast cancer.

## 5 Conclusions

Using a combination of NMR techniques ($^1H$ and $^{13}C$ NMR, HMQC, HMBC and COSY) and IR spectroscopy combined with DEPT calculations, it has been possible to determine the structure of phenyl coumarin (Figure 10)

Phenylcoumarin exerts antiproliferative effect in breast carcinoma cell line, and can be considered for furher mechanistic evaluation in human its biosafety and anticancer effect.

The results demonstrated that treatmen of MCF7 cells with phenylcoumarin results in a significant decrease in cell viabilty, $IC_{50}$ values against MCF7 is 2.23 µg/mL.

The authors would like to thank the LIPI (Indonesian Institute of Sciences) for financial support through Kompetitive project thanks are due to Mr. Ismail Rahman at the Herbarium staff in Research Centre for Biology- LIPI Cibinong for identifying the plant material.

## References

1. American Cancer Society. Cancer Facts & Figure. Atlanta American Cancer Society. (2013).
2. Klenkar J and Molnar M. Natural and synthetic coumarins as potential anticancer agents. J of Chemical and pharmaceutical Research, **7** (7) : 123-1238 (2015)
3. Sashidhara, K.V., Kumar, A., Kumar, M., Sarkar, J., Sinha, S.K Unites State Patent Application Publication, Pub No: US 2013/02/0210909 A1 (2013).
4. Cao, S. G., Wu, X. H., Sim, K. Y., Tan, B. H. K., Vittal, J. J., Perera, J. T and Goh, S-H. Minor coumarins from *Calophyllum teysmannii* var *inophylloide* and synthesis of cytotoxic calanone cerivates. Helvetica Chimica Acta **81**(5-8), 1404-1415. (1998).
5. Guilet, D., Seraphin, D., Rondeau. D, Richomme, P. and Bruneton, J. Cytotoxic coumarin from *Calophyllum dispar.* Phytochemistry, **58** (4), 571-575. (2001).
6. Reyes-chilpa,R., Estrada-Muniz, E., Apan, R. T., Amekraz, B., Aumelas., A. Jankowski, Christopher and Vazquez-Torres M. Cytotoxic Effects of Mammea type coumarins from *Calophyllum brasilienc. Life brasiliense. Life Science,* **75** (3) 1635-1647. (2004).
7. Li,Y., Li, Z-L., Yin, S-L., Shi, G., Liu M-S. Triterpenoid from Calophyllum inophyllum and their growth inhibitory effects on human leukemia HL-60 cells. (2010).
8. Mousavi, S.H., Davari, A-S., Iranshahi, M., Sabouri-Rad, S., Najaran, Z. T. Comparative analysis of the cytotoxic effect of 7-prenyloxycoumarin compounds

and herniarin on MCF-7 cell line. *Avicenne J of Phytomedicine (AJP),* **5** (6), 520-529. (2015).

9. Moniz-Corroco. United States Patent No US 9.403,791 B1, Aug, 2. (2016).
10. Noldin, V. F., Isaias, D. B., and Filho, V. D Calophyllum Genus: Chemical and Pharmacological importance, *Quim Nova*, **29** (3), 549-554. (2006).
11. Vital, J. J.,m Cao, S-H., Goh, S-H, Tan, G-K and Sim, K-Y. Three novel isomeric pyranocoumarins from Calophyllum teysmannii: Calanone, isocalanone and teysmanone A, *Acta Cryst C.* **54**, 1536-1540.(1998)
12. U.S. Patent No.**5**,501,959.

# In vitro antimalarial activity of *Calophyllum bicolor* and hemozoin crystals observed by Transmission Electron Microscope (TEM)

*Jamilah* Abbas*

Research Center for Chemistry, Indonesian Institute of Sciences (LIPI) PUSPIPTEK, Serpong, Indonesia

**Abstract.** *Objective* : In continuation of our antimalarial candidate drug discovery program on Indonesia medicinal plants especially from stem bark of *Calophyllum bicolor*. *Metode* : We extracted of bioactive crude extract with hexane, acetone and methanol from stem bark of *Calophyllum bicolor* and evaluated their antimalarial activity by using parasite *Plasmodium falciparum in vitro*. *Results:* Methanol fraction showed most active and potent antimalarial activity dose dependent in *in vitro* experiments with $IC_{50} = 0.52$ ppb, and hexane fraction showed moderate antimalarial activity, but dose not dependent, while acetone fraction have not antimalarial activity. TEM (Trasmission Electron Microscopy) analysis confirmed a remarkable reduction of hemozoin in the presence of the bioactive fraction. *Conclusion*: The results suggest that the antimalarial activity of *Calophyllum bicolor* is due, methanol fraction showed the best *in vitro* antimalarial activity then another fraction.
**Keywords:** *Calophyllum bicolor, Plasmodium falciparum, in-vitro*, TEM (Transmition Electron Mecroscope)

## 1 Introduction

Malaria is a serious parasitic disease transmitted by the bites of the female *Anopheles* mosquites , there are 4 species of plasmodium, e.i. *Plasmodium falciparum, P vivax, P malaria,* and *P ovale,* which *P falciparum* the most lethal [1, 2, 3]. Parasites belonging to *Phyllum* protozoa and genus *Plasmodium* are the most important causal pathogens and cause several human infection with globally massive impact[4]. Malaria remain one of the most significant world-wide public health problem, especially in tropical and subtropical regions [5]. Malaria is endemic to 90 counties world health and effects nearly 40% of the the world's population. Malaria is responsible for the death of 1-2 million each year with more than 90% of cases found in sub Saharan Africa and ≥1 case per 1000 population, and 80% of such cases are concentrated in 13 cauntires, over half in Negeria, Congo, Ethiopia, Tanzania and Kenya. [6, 7, 1]. Nigeria accounts for a quarter of all malaria cases in Africa. In the southern part of the country, transmission occurs all year round while in the north it is more seasonal [1].

One major virulence factor of this parasite is the highly variant *P. falciparum* erythrocyte membrane protein-1 (PfEMP) family. The malaria parasite in its blood stage is a voracious consumer of the globin component of hemoglobin, which causes the release of copious amounts (> 4000 mM) of toxic heme. Proteolysis of hemoglobin yields amino acid for protein synthesis as well as toxic heme. However, the parasite withstands heme toxicity by employing efficeinets mechanisms of heme detoxification. The conversion of heme to the relative benign heme polymer, hemozoin, is achieved by a process akin to template-mediated biomineralization. Templates for heme polymerization include performance hemozoin, histidin rich protein, and lipid or parasite cannot enzymatically cleave the porphyrin ring, heme is'detoxified' by conversion to an insolubly polymer, hemozoin [8 ,9].

The emergence and rapid spread of multidrug-resistant strains of *Plasmodium*, particularly *Plasmodium falciparum* has led to serious problem for prophylaxis and treatment malaria, which becomes more difficult and limits the ability to control this disease and also limits the choice of drugs used. This has been identified as the current primary cause of control failure, this highlights the need to develop quickly more effective and less toxic new antimalarial drugs with different mechanism of action [9. 10]. It has been suggested that hemozoin formation is inhibited by the extract *Calophyllum*.

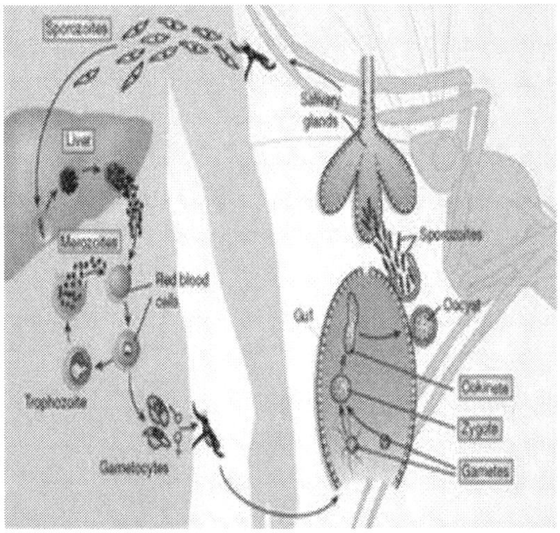


**Fig 1**. Live cycleof *P.falciparum* parasite

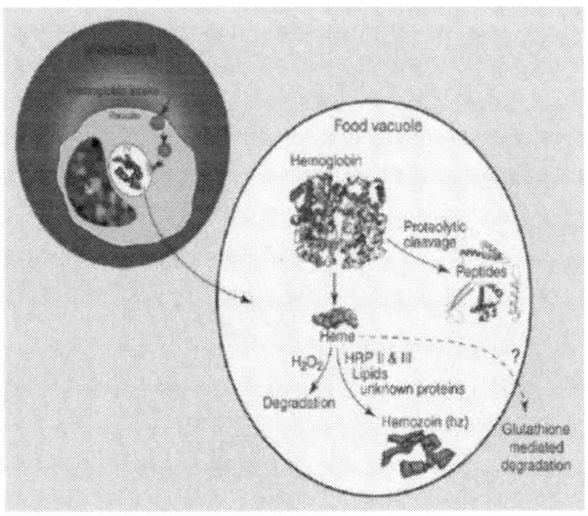


**Fig 2**. Heme degradation and hemozoin formatoion

Transmission electron microscope (TEM). TEM is one of type of electron microscope that has three essential systems: (1) an electron gun, generating the electron beam, and the condenser system, which focusing the beam onto the object, (2) the image-producing system, consisting of the objective lens, movable specimen stage, and intermediate and projector lenses, focusing the electrons passing through the specimen to form a real, highly magnified image, and (3) the image-recording system, converting the electron image toward some form perceptible to the human eye. The image-recording system usually covers of a fluorescent screen for showing and focusing the image and a digital camera for permanent records. Bisides,on, a vacuum system, consisting of pumps and their associated gauges and valves, and power supplies are needed.see Figure 4 [11]

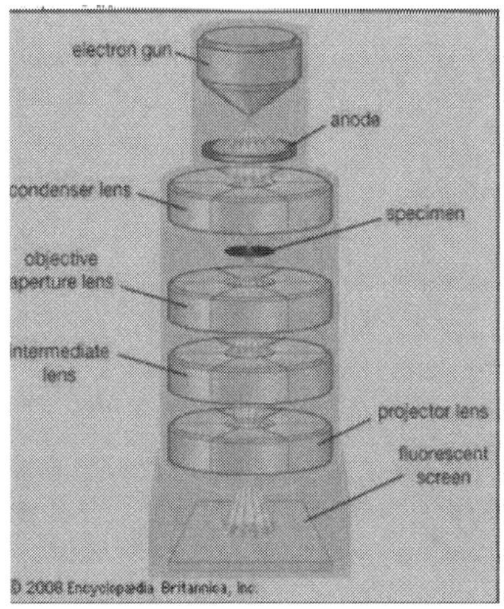


**Fig 3**. Transmission electron microscope (TEM)

# 2 Material and Methods

## 2.1 Material

### *2.1.1 General experimental procedures*

TEM (transmission electrone microscope), maserator, evaporator,UV lamp.

### *2.1.2 Sample*

*C. bicolor* P.F Stevent (2.858 kg) and *C. europhyllum* (1.555 kg) were collected in Bulungan, Kalimantan Island, in Indonesia. The plant were determined by staff from Herbarium Research Centre for Biology - LIPI and voucher specimen was deposited at the Herbarium Research Centre for Biology-LIPI. Indonesia.

### *2.1.3 Chemical material*

RPMI media and glutamin were obtained from Nissui, $NaHCO_3$ was purchased from E. Merck. Silica gel 200-300 mesh, TLC plates were purchased from E. Merck and all organic solvent (methanol, hexane, $CH_2Cl_2$, ethyl acetate) were purchased from local market. RPMI media contain 2,5 µg/mL gentamicin, 50 µg/mL hypoxanthin, 25 mM buffer HEPES (N-2-*hydroxyethylpiperazine-N'-2- ethanne sulfonic acid, (GibcoBRL)*, 25 mM natrium bikarbonat, 10% human serum AB+, 5 % hematokrit, human erythrocytes, and chloroquin was used as a positif control. DMSO were used to dilute of the samples

## 2.2 Methods

### *2.2.1 Extraction*

The air-dired *C. bicolor* P.F Stevent (2.858 kg) and *C. europhyllum* Lauter (2,555 kg) were successively extracted with *n*-hexane, acetone and methanol at room temperature for 4 day and such was repeated three times, the solvent were evaporated under reduced pressure to give hexane extract. The residue was extract successively with acetone (10 lt x 3) and methanol (10 lt x 3), then acetone and methanol were evaporated under reduced pressure to give a acetone and methanol extracts.

### *2.2.2 Antimalarial test*

Erythrocytes infected with *Plasmodium falciparum* strain 3D7, from cultures obtained using the method Trager and Janson up to density of parasites 2%, were suspended in complete culture medium at a hematocrit of 5%. The suspension parasite was distributed in 96-well micro titer plates (200 µl per well). To determine the antimalarial activity against *plasmodium falciparum*, each extract (5 mg) was first dissolved in dimethyl sulfoxide (100 µl DMSO mix with vortek until all samples dilution completely, and then diluted with completed medium (RPMI) to obtain the desired concentration (5 mg x $10^{-2}$ µl until x 5 mg x$10^{-9}$ µl), total volume in each well plate are 1 ml) and cloroquin was used as a positive control.

Compounds were tested in duplicate in 2% parasitemia cultures mostly at ring stage. For each assay, a parasite culture was incubated with the compound at 37 °C for 48 h in 5% $CO_2$ at 95% relative humidity and frozen until the biochemical assay could be run. After 48 hours of incubation ach well was harvested to a glass were then fixation by methanol, also added Geimsa reagent.

Concentration of extract required to inhibit 50% parasite growth ($IC_{50}$) was determined by computer. Parasit amount was monitoring by microscope, and computer by linear program Sigma-plot ($IC_{50}$) was used for calculated of $IC_{50}$. The antimalarial activity was expressed as the $IC_{50}$ value (ppm), which was defined as the concentration of extract required to inhibit 50% parasite growth after 48-h incubation at 38°C.

### *2.2.3 TEM (Transmission electron microscope)*

Brief, small amount of extract were placed on the grid (formvar carbon coated Cu messed 300), at the acceleration voltage of 80 kV, using Zeiss. Model EM 10c (Jeol Japan). In addition, the apparatus was used for imaging at two different scales 24 and 30 nm. *in-situ* and after isolation was used for detect hemozoin

### *2.2.4 Flow chart of methode*

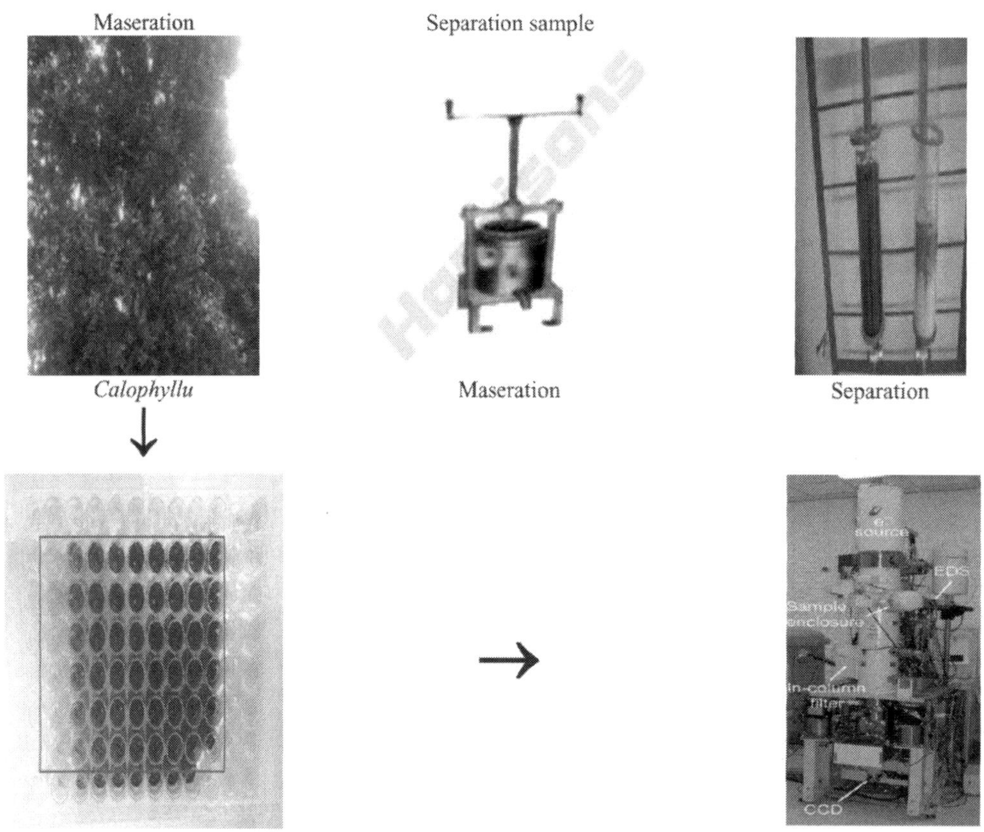


**Fig. 4** . Flow chart of isolation and TEM analysis

## 3 Results and Discussion

### 3.1 TEM result

#### *3.1.1 TEM analysis of the fraction containing bioactive material showed significan hemozoin reduction.*

In TEM, fixed-illumination mode, individual spectra are obtained from whichever area of the sample is illuminated. To control the sampling volume, the microscopist converges the beam to illuminate only the region of interest uses postspecimen apertures to select the divergence angle of electrons reaching the spectrometer. Because the sample and beam are fixed relative to each other, spectra, EDS and/or EELS, can be obtained in TEM mode

#### *3.1.2 TEM imaging- Producing system.*

The specimen grid is conducted in a small holder in amoved specimen stage,. The objective lens is usually of short focsal length 1-5 mm (0.04-0.2 inch) and results a real intermediate image that is furher magnified by the projector lenses. A single projector lens will provide a range of magnification of 5:1. For practical reasons of image stability and brightness, the microscope is operated to give a final magnification of 1,000-250,000 x on the sreen to show the hemozoin crystal.

Reduction of hemozoin crystal formation in the present of bioactive fraction had investigated by using TEM (Transmition Electrone Microscope) technich *in-situ* and after isolation. The identification of spesific area of hemozoin an image, or pixels with specified characteristic,

To furher characterize the antimalarial activity of *Calophyllum*, methanol extract (from the late sate of hemozoin formation) in the present sample after 1x24 hr incubation and in present sample after after 2x24 hr incubation were examined by TEM. There existed numerous hemozoin with diameter 1μm after 1x24 incubated, but there are not hemozoin after 2x24 hr incubation).

The transmission electron microscope (TEM) image in Figure 5 showed hemozoin crystal extracted from the sample previously used for the MO measurement. The typical length of the extracted hemozoin crystal range 500 nm. For comparison., a TEM image of synthetic hemozoin crystal in our MO study, is showen in panel A and B. The synthetic crystal have more elongated shape in typical length 500 nm. Panel A and B display TEM images of intact infected erythrocytes in the schizont stage. Distinct component of the parasite and erythrocyte can be observed including the erhythrocyte membrane (EM), parasite membrane (PM). Food vacuole membrane (FVM), merozoite (M), hemoglobin transport vesicles (HbTV) knons (9K) and hemozoin (Hz).Figure 5 and Figure 6.

#### *3.1.3 Figure from TEM results*

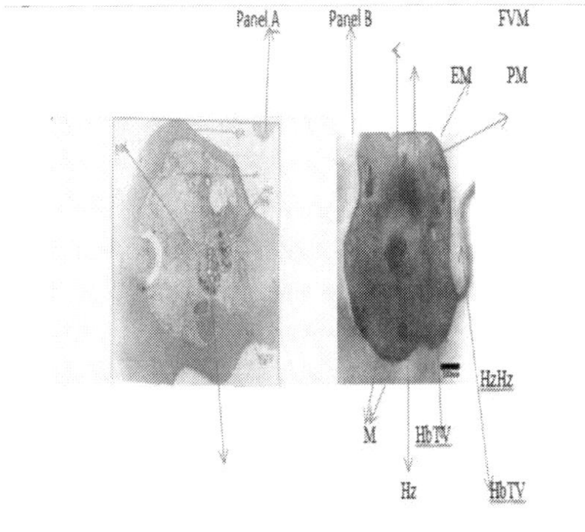


**Fig 5**. Hemozoin crystal showed in figure as white crystal after 1x24 incubattion

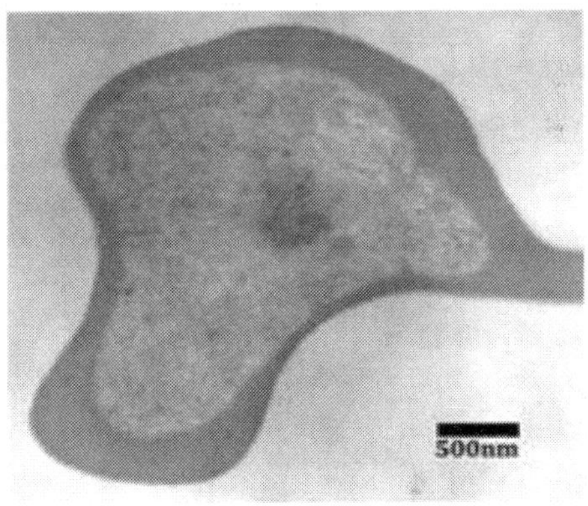


**Fig 6**. Hemozoin crystal have not showed in figure after 2x24 incubattion

### 3.2 Sepration results

The present investigation for antimalarial principles from *C. bicolor* has led to the finding of three fraction, e.i, hexane (208.29 g), acetone (413.62 g) and methanol 149.8 g.

### 3.3 Antimalarial activity results

According to the previous paper [12] which it's explained that antimalarial activity of methanol fraction of *C. bicolor* have better than hexane fraction, and hexane fraction have better than acetone fraction while the acetone fraction have not antimalarial activity, these data showed in Figure 7.

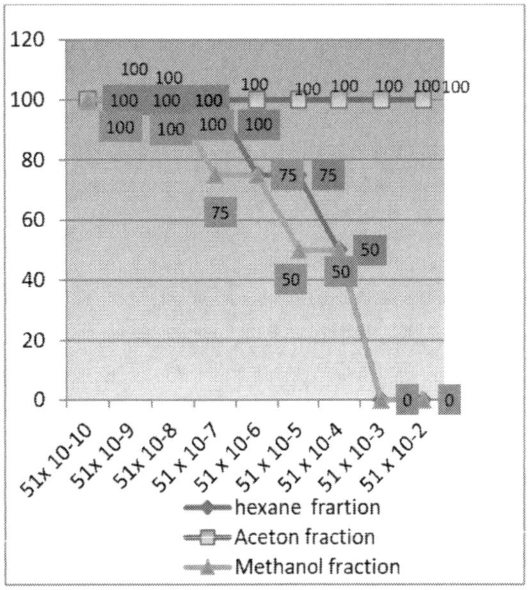


**Fig 7**. Grafik antimalarial activity of hexane,acetone and methanol fraction of C bicolor.

Antimalarial activity of the metanol fraction best compare hexane and acetone fraction, these methanol fraction are devided further into five sub fraction (e.i sub fraction 9 :1 ; 8 :2 ; 7 :3 ; 6 :4 ; 5 : 5. The data antimalrial activity explaned in Table 1

**Table 1.** Antimalarial activity of sub fraction methanol

| DOSA GE | C.Bi MeOH 9:1 | C.Bi MeOH 8:2 | C.Bi MeOH 7:3 | C.Bi MeOH 6:4 | C.Bi MeOH 5:5 |
|---|---|---|---|---|---|
| 1.00E-09 | 83 | 145 | 65.2 | 111 | 100 |
| 1.00E-08 | 92 | 118 | 43.75 | 100 | 100 |
| 1.00E-07 | 92 | 136 | 43.75 | 133 | 127 |
| 1.00E-06 | 117 | 127 | 37.5 | 122 | 127 |
| 1.00E-05 | 141 | 136 | 31.25 | 100 | 127 |
| 1.00E-04 | 75 | 136 | 75 | 88 | 118 |
| 1.00E-03 | 83 | 127 | 62.5 | 122 | 81 |
| 1.00E-02 | 83 | 118 | 37.5 | 55 | 9 |
| 1.00E-01 | 16 | 18 | 31.25 | 0 | 0 |

## 4 Discussion

The present investigation for antimalarial principal from methanol fraction of *C. bicolor* has led to the finding of five sub fraction e.i sub fration MeOH 9 :1 ; MeOH 8 :2 ; MeOH 7 :3 ; MeOH 6 :4 and MeOH 5 :5. Among the sub fraction which had weak-to-moderate antimalarial activities on *P. falcioparum* parasite. Antimalarial activities of sub fraction methanol e.i sub fraction MeOH 9 :1 ; MeOH 8 :2 ; MeOH 7 :3 ; MeOH 6 :4 and MeOH 5 :5 were elustrated in Figure 8.9

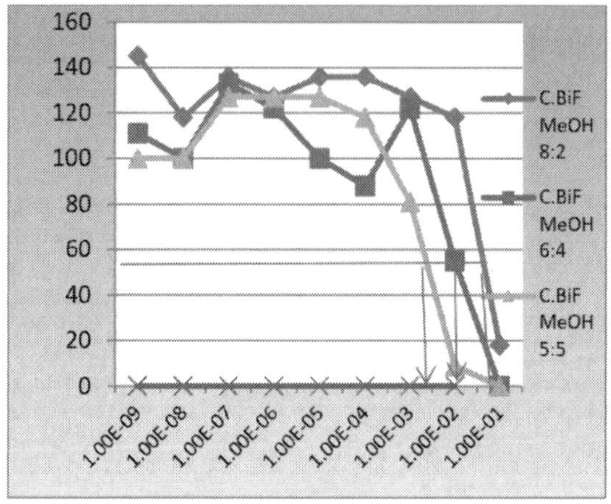


**Fig 8**. Antimalarial activity of sub fraction MeOH 8 :2 ; 6 :4 and 5 :5

Percentage of parasite *P. falciparum* growth versus sample concentration ($10^{-9}$ to $10^{-1}$ mg/mL) was observed. The increasing in sample concentration should decrease the parasite growth percentage. Increasing sample concentration from $10^{-9}$ to $10^{-8}$ mg/mL leads to a decrease the percentage of parasite growth from 145 to 111 : 65.2 to 43.75 and 111 to 100% for sub fraction 8 :2 ; 6 :4 and 5 :5 respectively.

Unfortunately,In the fact state that the increasing sample concentration from $10^{-8}$ to $10^{-7}$ mg /mL did not cause a decreasing percentage of parasite growth, this illustrates that the percentage decreasing parasite growth does not depend on the sample concentration (see Figure 8).

According to the Figure 8, we can calculated the $IC_{50}$ of sub faction MeOH 8 :2 ; 6 :4 and 5 :5 samples are 50 $x10^{-1}$mg/mL, 50 $x10^{-2}$ mg/mL and 50x $10^{-3}$ mg/mL respectively (equivalent 5000 ppm, 500 ppm and 50 ppm respectively).

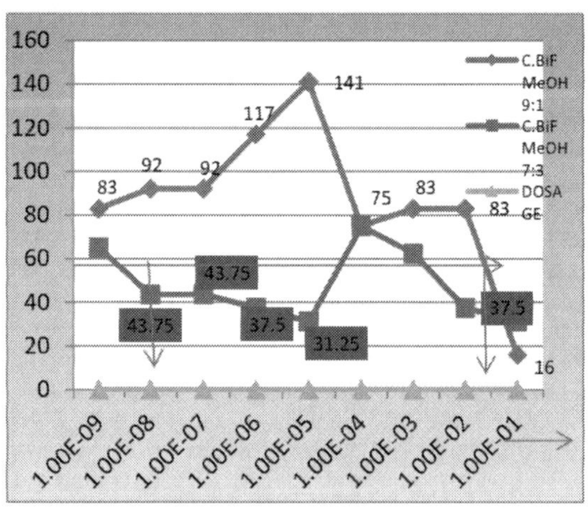


**Fig 9**. Antimalarial activity of sub fraction MeOH 9 :1 ; and 7 :3

The antimalarial activity of sub fraction 9: 1 and 7: 3 are described in Figure 9. In case of sub fraction 7 :3 sample, increasing the sample concentration from $10^{-9}$ to $10^{-8}$ ; $10^{-7}$ to $10^{6}$ ; $10^{-6}$ to $10^{-5}$ mg/mL leads to a decrease the percentage of parasite growth from 65.2 to 43.75 %: 43.75 to 37.5 and 37.5 to 31.24 % respectively. Unfortunately increasing sample concentration from $10^{-5}$ to $10^{-4}$ mg /mL did not cause a decreasing percentage of parasite growth.

According to the Figure 9, we can calculated the $IC_{50}$ of sub faction MeOH 7 :3 is 50 x $10^{-8}$ mg/mL (equivalent 0.005 ppm), and $IC_{50}$ of sub fraction 9 :1 sample is 50 x $10^{-2}$ mg/mL (equivalent 0.5 ppm).According to the Figure 8 and 9, conclution taht the sub fraction MeOH 7 :3 is better as antimalarial compare to any another sub fraction.

One pure compound have isolated from the sub fraction of *C. bicolor* and have antimalarial activity with $IC_{50}$ is 50 x $10^{-2}$ mg/mL (equivalent is 500 ppm) (see Table 2 and Figure 10

**Table 2.** Antimalarial activity of isolate compound

| | Conc mg/mL | Stage | | | T | x̄ | Growth rate (%) |
|---|---|---|---|---|---|---|---|
| No | Conc | R | Tr | Sc | | | |
| 1 | 50x $10^{-9}$ | 2 | 2 | 2 | 6 | | 116 |
| | | 3 | 3 | 2 | 8 | 7 | |
| 2 | 50 x $10^{-8}$ | 2 | 1 | 1 | 4 | | 67 |
| | | 2 | 1 | 1 | 4 | 4 | |
| 3 | 50x $10^{-7}$ | 3 | 1 | 1 | 5 | | 92 |
| | | 2 | 3 | 1 | 6 | 5.5 | |
| 4 | 50 x $10^{-6}$ | 4 | 4 | 1 | 9 | | 133 |
| | | 3 | 2 | 2 | 7 | 8 | |
| 5 | 50 x $10^{-5}$ | 3 | 2 | 1 | 6 | 7 | 117 |
| | | 5 | 2 | 1 | 8 | | |
| 6 | 50 x $10^{-4}$ | 4 | 1 | 1 | 6 | 5.5 | 92 |
| | | 3 | 1 | 1 | | | |
| 7 | 50 x $10^{-3}$ | 1 | 2 | 1 | 4 | 4 | 67 |
| | | 2 | 1 | 1 | 4 | | |
| 8 | 50 x $10^{-2}$ | 2 | 5 | 1 | 8 | 7 | 116 |
| | | 1 | 4 | 1 | 6 | | |
| 9 | 50x $10^{-1}$ | 1 | 0 | 0 | 1 | 1 | 16 |
| | | 1 | 0 | 2 | 1 | | |
| | Conc = Concentration, R = ring, Tr = Tropozoid, S= Schizont, T=Total, C= Conclusion. NA = Not active | | | | | | |

One pure compound have isolated as trepenoid from stem bark of *C. bicolor*. Triterpen as white powder. LC-MS (Figure 7) displayed a positive molecular ion paek at m/z 387,86 $[M+H]^+$ indicating a molecular formula $C_{27}H_{46}O$.

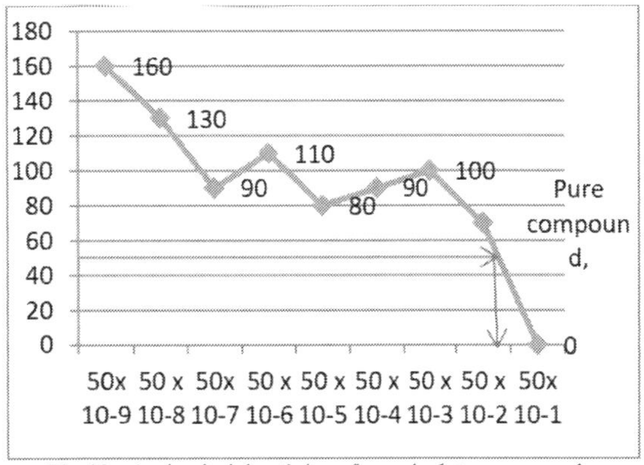


**Fig 10.** Antimalarial activity of pure isolate compound

Pure isolate compound have tested by using TLC. The TLC result saw single spot ( Figure 11).

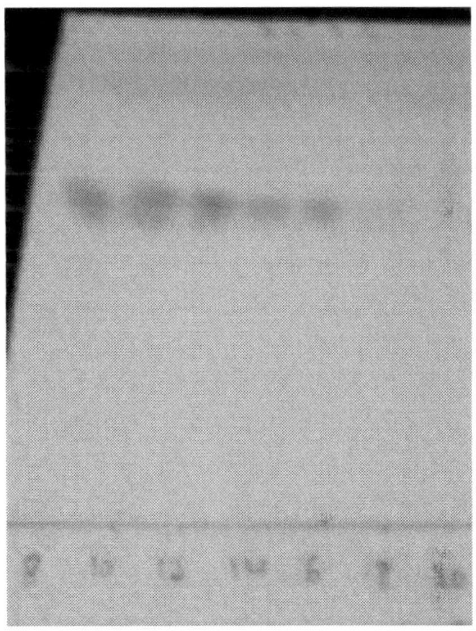

**Fig 11**. TLC spot of pure isolate compound

One compound have isolated as trepenoid from stem bark of *C. bicolor*. Triterpen as white powder. LC-MS (Figure 12) displayed a positive molecular ion paek at m/z 387,86 $[M+H]^+$ indicating a molecular formula $C_{27}H_{46}O$.

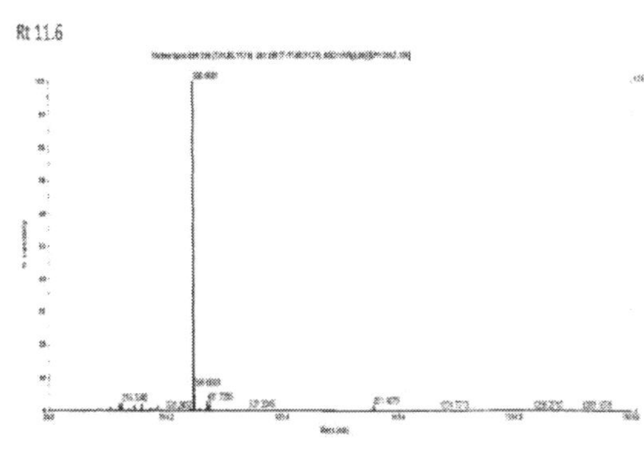


**Fig 12**. LC-MS spectrum of pure isolate compound

Isolated compound give typical triterpen IR absorption at 3210-3550 $cm^{-1}$ (OH) , 2930 (sp3/CH3 aliphatic and 2864 $cm^{-1}$ sp2/$CH_2$ aliphatic), 1681 (C=C) cm-1 and 1456 $cm^{-1}$ as CH bending (Figure 13)

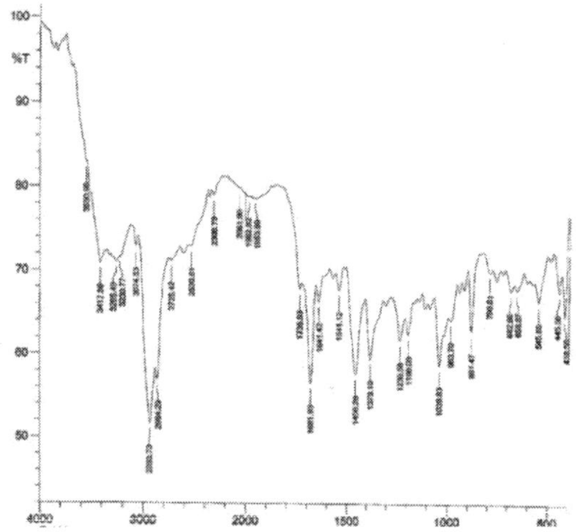

**Fig 13**. Infra Red spectrum of pure isolate compound

UV spectrum of pure isolate compoune was showed in figure 14. Maximum absorption were observed at wave length 233,5 nm as electron transition of $\pi$ to $\pi^*$ from C=C group and wave length 295 nm as electron transition of n- $\pi^*$ from OH group.

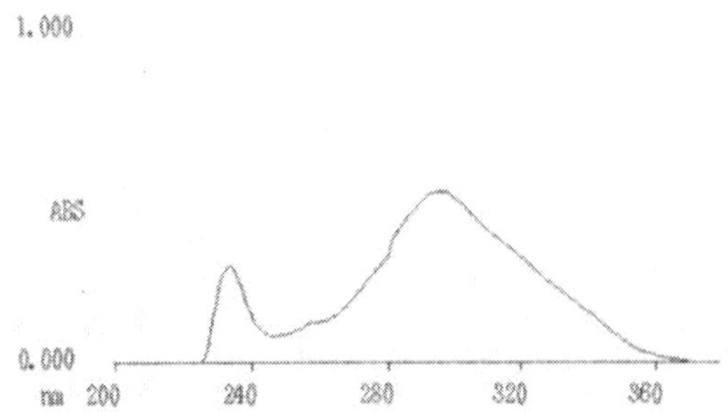


**Fig 14.** UV spectrum of pure isolate compound

The [1]H-NMR spectrum revealed the presence of one hydroxyl proton doublet at $\delta_H$ 12 ppm (1s, OH), one alyllic proton signal at $\delta_H$ 4.72 (dd, 2H, J = 2.6 and 4.56 Hz) for H-21 recpectively. There are 6 six methyls singlet signal at $\delta_H$ 0,75 ppm; 0,85 ppm; $\delta_H$ 0,95 ppm; $\delta_H$ 0,96 ppm; $\delta_H$ 1,01 ppm and $\delta_H$ 1,7 ppm (sp$_3$ CH) (Figure 15)

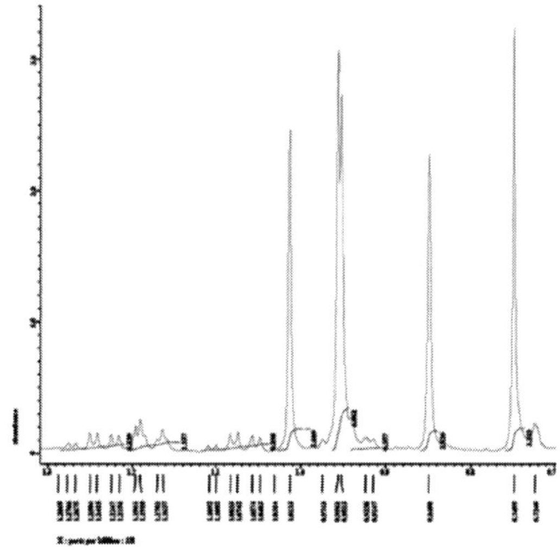

**Fig 15.** [1]H-NMRspectrum of pure isolate compound

There are ten metilen ($CH_2$) signal at $\delta_H$ 1,36 ppm; $\delta_H$ 1,56 ppm; $\delta_H$ 1,91 ppm; $\delta_H$ 1,30 ppm; $\delta_H$ 1,46 ppm; $\delta_H$ 1,43 ppm; $\delta_H$ 1,63 ppm; $\delta_H$ 1,51 ppm; $\delta_H$ 1,48 ppm; $\delta_H$ 3,05 ppm and $\delta_H$ 4,72 ppm (2H, *dd* J 2,6 & 4,92).

Signal at $\delta_H$ 3,12 ppm (1H, *dd* J 5,28 & 11,65) is signal of mehine (CH) at position C-3 of triterpenoid compound, Isolate compound also have one hydroxil group at C-3. Unfortunately isolate compound have not signal at 6-8 ppm, these indicated that isolate compound have not aromatic proton.

There are six methine (CH) signal at $\delta_H$, 3.12 ppm, 2,35 ppm (1H, *ddd* J 3,25; 2,95, $\delta_H$ 2,24 ppm, $\delta_H$ 1,18 ppm and $\delta_H$ 1,08 ppm, CH signal will be explained in figure 16

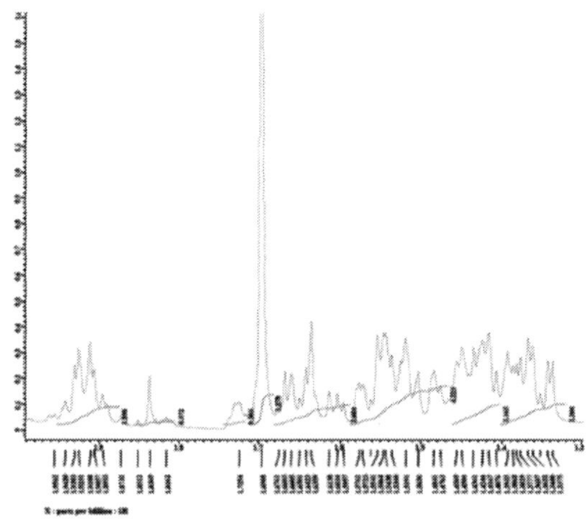

**Fig 16.** [1]H-NMR spectrum of pure isolate compound

The [13]C-NMR spectrum and DEPT experiment showed the presence 6 six methyl, ten methylene, six methine five quartener carbons at ∂ 39.6 (C-4), 56.4 (C-5), 39.1 (C-8), 39.7 (C-10) and 151.7 (C-21) respectively see Figure 17

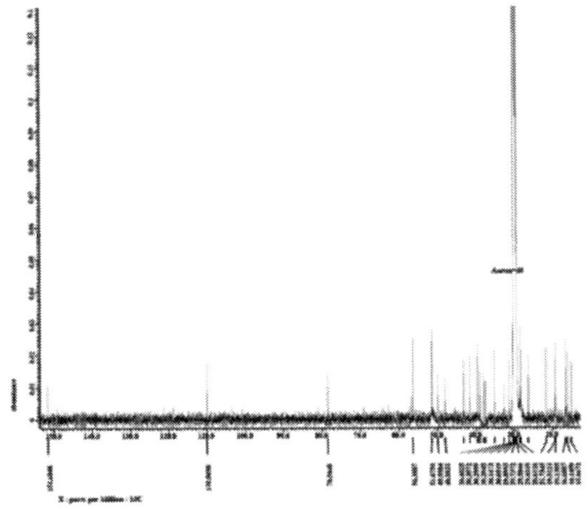

**Fig 17.** $^{13}$C-NMRspectrum of pure isolate compound

$^{1}$H-NMR and $^{13}$C-NMR data showed that the isolate compound have 27 carbon atom, fourty six hydrogen and one oxygen, with structure shown in figure 18

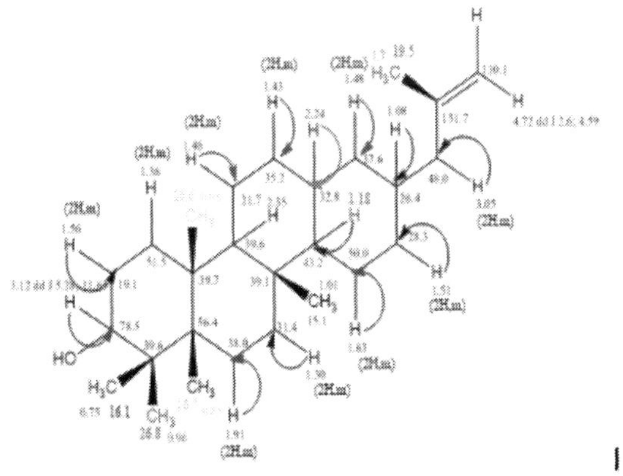

**Fig 18.** Structure terpenoid from C *bicolor* P.F Steven ($CDCl_3$) have antimalarial activity

## 5 Conclusions

1. These data showed the biological activity potency of Indonesian Clusiaceae plant (*C. bicolor*) as a source of antimalarial drugs and also showed that hexane and methanol fraction have antimalarial activity, but acetone fraction have not antimalarial activity, methanol fraction more active then hexane fraction,
2. According to the Figure 8 and 9, conclution that the sub fraction MeOH 7 :3 is better as antimalarial compare to any another sub fraction.
3. Hemozoin crystal were detected by TEM after 24 hours incubation of parasite in extract sample.
4. One triterpen compound have isolated and have antimalarial activity with $IC_{50}$ is 500 ppm

## 6 Equations and mathematics

$$\% \text{ parasetemia} = \frac{\text{Total eritrosit was infected by parasite}}{\text{total eritrosit}} \times 100\% \quad (1)$$

Percentase inhibition was obtained from parasetemia in control positive minus parasetemia in sampel /parasetemia in control x 100% :

$$\text{Inhibisi} = \frac{\text{Parasetemia ( in control - sample )}}{\text{Parasetemia in control positive}} \times 100 \quad (2)$$

The authors are thankful to Ministry of Research Technology and Higher Education of the Repoblic of Indonesia through the scheme of Insentive Research Program for the National Innovation System (04/INS-2/PPK/E/E4/2017). for financial support and thank to LIPI for recording the 2 D NMR, thanks are also due to the Eijkman Insitute to detect the hemozoin crystal, and Mr Ismail Rahman from Hermatium - Researh Centre for Botani - LIPI for identification of sample.

## References

1. Adebayo, J. O, and Krettl, A. U. Potential Antimalarials from Negerian plants: A reiview. *Journal ethnopharmacologically*, **113**, 289-302 (2011). Collins, W. J., Greenhouse, B., , Rosenthal, P. J, and Dorsey G. The used of genotyping in antimalarial clinical trials: a systematic review of published studies from Malaria J., *5*: 122, 1-8. (2011).
2. Silva, J. R de A., Ramos, A de S, Machado, M., Moura, D., Neto, Z., Cavalheiro., M. M. C., Fiqueiredo, V E do., Amaral, A. C F., Lopes, D. A Review of antimalarial plants used in traditional medicine in communities in Portuguese-Speaking countries: Brazil, Mozambique, Cape Verde, Guinea-Bissau, Sao Tome and Principe and Anggola. *Mem Inst Oswaldo Crus, Rio de Janeiro*, Vol **106**, 142-158, (2011).
3. Ojha, P. K, and Roy, K., Chemometric modeling, docking and *in silico* design of triazolopyrimidine-based dihydroorotate dehydrogenase, European J. of Medicinal Chemistry, ***45***, 4645-4556.
4. Duffy R., Wade Cristine. Discovery of anticancer drugs from antimalarial natural products: a Medline literature review, *Feature*, 1359-6446/06/s. www. Drugdiscoverytoday.com. ***Vol 17***, no 17/18, September (2012).
5. Li Jiazhong, Li. S., Bai, C., Liu, H., Gramatica, P. Structural requirements of 3-carboxyl-4(1H)-quinolones as potential antimalarials from 2D and 3D QSAR analysis, J. of Molecular Graphics and Modelling. **44**, 266-277. (2013).
6. Tames, A. P., Lantivt, D., Lim, E., Pezzuto, J. M., Chemosentizing action of cepharanthine against drug-resisitant human malaria, *Plasmodium falciparum*. J. of ethnopharmacoology, ***98***, 137-142. (2005).
7. Pouplin J N,Tran Hop, Tran Hung,Phan T A,Dolecek C, Farrar J, Tran T H, Caron P, Bodo B, Grellier P

(2007). Antimalarial and cytotoxic activities of ethnopharma cologically selected medicinal plants from South Vietnam. .J. of Ethnopharma cology **109**,p 417-427. (2010)

8. Sahal D.,Kannan and Chauhan, V.SApplying malaria parasite's heme detoxification system for screening potential antimalarial drugs. *Analytical Biochemisy.* 312, 258-260 (2003).
9. Ignatushchenko, M. V.. Winter, R.W., Bachinger, H. P., Hinrichs, D.V and Riscoe, M.K Xanthones as antimalarial agents : studies of a possible mode of action, *FEBS letter* ***409***, 67-73. (1997).
10. Geral, N., Mahajan B., and Kumar S., Mitosis in the Human malaria parasite Plasmodium falciparum, E*ukaryotic Cell, Vol 10*, no4, 474-481 (2011)
11. Brian. J, Ford C, J, Bradbury S, Transmission electron microscope (TEM). http://www.britania.com/technology/transmission-electron-microscope
12. A. Jamilah. M, Hanafi, M., Artanti. N., Asih.P,B,A., and Syafruddin D, Antiplasmodial activity of hexane extract of *Calophyllum Lowii.* . J. of Tropical Medicinal Plants Vol ***16*** (Jan-Dec 2016)

# Rheumatoid arthritis and the challenge of using nanoparticles for its treatment

*Lutfi* Chabib[1,2,*], *Zullies* Ikawati[1], *Ronny* Martien[1], *Hilda* Ismail[1], *Made Dwi Pradipta* Wahyudi[3], *Dewa Ayu* Arimurni[3], *Wildan Khairi* Muhtadi[1], and *Arif* Hidayat[4]

[1]Faculty of Pharmacy, Gadjah Mada University, Yogyakarta, 55281, Indonesia
[2]Department of Pharmacy, Islamic University of Indonesia, Yogyakarta, 55584, Indonesia
[3]School of Pharmacy Mahaganesha, Denpasar, Bali, 80226, Indonesia
[4]Chemical Engineering Department, Islamic University of Indonesia, Yogyakarta, 55584, Indonesia

**Abstract.** Rheumatoid arthritis is the most common autoimmune disease that affects the joints. The cause of the disease is unknown, many studies proposed hypothesis about the etiology of rheumatoid arthritis. The clinical manifestations of arthritis are different in each patients. In addition, the development of the medication is still continue to achieve the most effective role with less side effect. Nanoparticles may be the answer to this problem, since they have been widely used to improve the pharmacokinetic and pharmacodynamics of rheumatoid arthritis drugs. Using nanoparticles-tagged folate or PEG to deliver rheumatoid arthritis drugs may increase the specificity of the drugs to the target and consequently, may decrease the side effects of the drugs. The purpose of this review is to summarize the etiology, clinical manifestation and highlighting the use of nanoparticles in rheumatoid arthritis treatment.

## 1 Introduction

Rheumatoid arthritis is the most common autoimmune disease that affects the joints [1]. This disease associated with progressive disability, systemic complications, early death, and socioeconomic costs [2]. RA is characterised by inflammation of the synovium and increased synovial exudate, which result in thickening of the synovium and joint swelling [3]. The condition is characterised by a symmetrical small joint polyarthritis involving the hands and feet, but larger joints can also be affected. The overall world prevalence of rheumatoid arthritis is approximately 0.5% to 1%. Based on the data from United Nations World Population Database in 2004, there are over 21 million people worldwide suffering from Rheumatoid arthritis. It can affect people of all age especially in adult in the range 40-50 [4]. The highest incidence appears in Pima indians (5.3%) and chippewa indians (6.8%), and the lowest in people from china and Japan (0.2%-0.3%) [5]. These finding can be conclude that Rheumatoid Arthritis caused by multifactorial like gene, and environment. The cause of rheumatoid arthritis is still unknown but the leading hypothesis for this disease is the result of an environmental (hormon, dietary factor, infection, ciggarete smoke, and many more) exposure or "trigger" in a genetically susceptible individual [6].

Nowadays, to deliver the drugs to the target site directly is still a problem. Attaching drugs to individually design carriers, like nanoparticles, can make a cell-specific targeting becomes achievable. Many researchers in nanotechnology found that nanoparticles, which exhibit a unique physicochemical and biological property because of their size, is a potential drug carrier. Nanoparticles have shown that they have the ability to enhance reactive area as well as an ability to cross cell and tissue barriers. That will be the reason behind the widely usage of nanoparticles for biomedical applications [7].

## 2 Rheumatoid arthritis

### 2.1 Etiology

Even the cause of rheumatoid arthritis is still unknown, however many studies have explained that rheumatoid arthrtitis is a result from the interaction of genetics, environment, and the immune system [8].

### 2.2 Genetic

More than 80% of patients carry the epitope ofthe HLA-DRB1*04 cluster [9]. Patients expressing two HLA-DRB1*04 alleles have a greater effect on gain that lead to nodular disease, major organ involvement and surgery related to joint destruction [10]. Another gen that have role in RA is A1-B8-DR3 which is the haplotype located near the haplotype and HLA-DPB1 gene [1]. Other RA-associated loci are PTPN22, PADI4, STAT4, TRAF1-C5 and TNFAIP3. The various gen above are MHC family, but non-MHC also can lead to this disease only 3-5% possibility. Environmental factors, such as smoking and infection, may also interfere the development, rate of progression and severity of RA [11].

* Corresponding author: lutfi.chabib@uii.ac.id

### 2.3 Environment

Multiple environmental factors including hormones, dietary factors, infections and exposure tobacco smoke as well as associated with increased risk for rheumatoid arthritis (RA) [12].

#### *2.3.1 Smoking*

Multiple case-control and prospective cohort studies have demonstrated that cigarette smoking is the strongest environmental factor linked with RA, and it can increased the risk of rheumatoid arthritis until 25% [13]. Furthermore, several studies demonstrate a dose-response between heavier smoking and RA, some studies show that people who have the habit of smoking more than 20 years led to persistent RA [14]. However, the specific mechanism of how smoking can induce autoimmunity is still unknown.

#### *2.3.2 Reproductive and hormon*

The abundant study evidence shows there is relation between hormon and incident of RA. Women are 2 to 4 times more likely than men to develop RA [15]. women with irregular menstruation and menarche under 10 years, may increase the risk of RA [16].

#### *2.3.3 Infection*

The body's immune system consists of cellular and humoral. When there is an infection, the immune system raises inflammatory substances to eliminate infectious agents, this inflammation can occur in the joints. The best example to explain this concept is an infection of *porphyromonas gingivalis*. The causal direction remains uncertain, but recent discoveries suggest that inflammation and in particular infection by *Porphyromonas gingivalis* maybe important in the pathophysiology of RA development [17]. The current hypothesis is that *Porphyromonas gingivalis* mediated citrullination of human peptides might responsible for the initial breakdown in self-tolerance that leads to the development of RA-related autoimmunity [18].

#### *2.3.4 Dietary intake*

The type of food or nutrient which is having relation with RA are vitamin D and protein. Study show that vitamin D has pleotropic effects on the immune system, inhibiting pro-inflammatory cytokines, upregulating anti-inflammatory cytokines, and regulating the innate and adaptive immune system through the vitamin D receptor [19]. In addition, Vitamin D prevents the development of inflammatory arthritis in collagen-induced mouse models [20]. Foods contain high protein and Fe is red meat. Based on the study of Norfolk Arthritis Register (NOAR, Great Britain) found the relationship between protein from red meat with incident RA [21].

### 2.4 Clinical manifestation

Symptoms appear in patients with RA vary, according to Arthritis Research UK, the common symtomps are joint pain (three or more joint) and swelling, morning stiffnes (at least one hour), anemia flu-like symtomps (such as feeling generaly ill, feeling hot and sweating), fatigue, irritability, and depression [4]. Another less common symtomp are weight loss, eye inflamation (Episcleritis), rheumatoid nodule, inflamation in other body part such as lung, blood vessel (Vasculitis), entrapment neuropathies median nerve involvement sjogren's syndrome(deficient moisture production of the lacrimal, salivary and otherglands), lymphadenopathy (enlargement of the lymph nodes or vessels), and less common symtomp are felty's syndrome (hypersplenism) and pericarditis [4].

## 3 Therapeutic drug for rheumatoid arthritis

The goal of acute onset therapy for rheumatoid arthritis is to control the underlying inflammatory disease [22]. Attainment of this goal will alleviate pain, improving patients' quality of life, and preserve independence and ability to perform activities of daily living and vocational and avocational pursuits. Furthermore, major long-term goals of treatment are to prevent joint destruction and prevent comorbidities of disease and treatment, including heart disease and osteoporosis [23].

There are four main groups of drugs are used to treat RA. These are painkillers (analgesics), non steroidal anti-inflamatory drugs (NSAIDs), disease-modifying anti-rheumatic drugs (DMARDs), and steroids [4]. RA patients generally consume more than one type of drug, this is due to every drug has different mechanisms and effects.

### 3.1 Analgesics

As explained earlier, that one of the major symptoms of RA are joint pain. This condition will really disrupt the patients, because the joint is the main organ of the movement. Analgesics works as an painkillers. Among analgesics commonly used are type of opioid, paracetamol and aspirin, the other is co-codamol, co-proxamol, co-drydramol and tylex. Some analgesics containing codeine can cause constipation [4].

### 3.2 Steroid

Some steroid or corticosteroid produced in the body, for example cortisol. Currently, steroids are widely produced and used to reduce pain. This medication can be used intravenously or orally. As for the side effects can result from steroid medications are facial flushing, interference in menstrual cycle, thinning, and other change in the site of injection (atrophy). But oral tablet has more side effects including weight gain, osteoporosis, muscle weakness, cataract, a rise of blood sugar or blood pleasure, and increased of developing infections [4].

### 3.3 Non steroidal anti-inflamatory drugs (NSAIDs)

NSAID is used to relieve pain and swelling. NSAID works by inhibiting the enzyme siklooksigenase. There are two COX isoforms and non-selective NSAIDs inhibit both COX-1 and COX-2 in differing ratios. Some of NSAID are ibuprofen dan diclofenac sodium, cox II selective, meloxicam, Cox II specific, celecoxib. Opioids is also used to relieve moderate to severe pain, some of them are dihydrocodeine tartrate, tramadol hydrochloride. The form of NSAID commonly tablet, capsules, dan liquid. In addition NSAIDs can also be given as a patch or cream that can be used directly at the site of action. but the use of NSAIDs can cause many side effects depends on the dose and duration of therapy, the most severe is the gastrointestinal (GI) and cardiovascular side effects are of particular concern [24].

### 3.4 Disease-modifying anti-rheumatic drugs (DMARDs)

DMARDs mecanism are to reduce pain, swelling and stiffness over a period of weeks or months by slowing down the disease and its effects [4]. This drug is contraindicated for patients with moderate-to-severe heart failure, active tuberculosis. the DMARDs for use in RA include ciclosporin A, hydroxycholoroquine (HCQ), leflunomide (LEF), methotrexate (MTX), intramuscular gold, penicillamine and sulfasalazine (SASP). The efficacy of MTX, intramuscular gold, LEF, penicillamine and SASP, is similar. Intramuscular gold has the highest toxicity and therefore increased treatment drop-out rates compared to SASP, HCQ and MTX. Systematic review found LEF, MTX and SASP to have comparable efficacy. MTX has the most favourable efficacy/toxicity trade-off. SASP scored close to MTX and had more adverse events initially. HCQ had a relatively low rate of toxicity. Essentially, these are monoclonal antibodies (MAbs) directed against chemical mediators of inflammation,notably TNF-a, and interleukins IL-1 and IL-6. These MAbs compete withthe endogenous ligands at their receptor sites on cells, e.g. CD20 and CD22, some of which have no identified endogenous ligand [24].

## 4 Challenge

Commonly, RA was treated symptomatically with non-steroidal and steroidal anti-inflammatory drugs. However, use of these drugs encountered many obstacles. The most corrosive effects of nonsteroidal drugs such as aspirin on the gut and the serious effects of steroidal drugs is water retention and redistribution of body fat. To reduce these effect is used paracetamol, but it has a relatively low therapeutic index. Furthermore these treatments do not slow the progression of tissue damage and loss mobility. Several drugs, e.g. methotrexate, penicillamine, gold and azathioprine, commonly known as DMARDs, or disease-modifying anti-rheumatic drugs, have been used for several years. These may provide symptomatic relief and slow the progression of the disease, but are associated with serious adverse effects and relatively non-specific in their actions. The harmful effect is that MAbs suppress the immune response and may facilitate opportunistic infections. Therefore it is challenge that the other drug therapy effective in inhibiting the progression of RA but fewer side effects.

## 5 New molecules for rheumatoid arthritis

Recently it has been proved that biologic RA therapy can solve the problem, this substance works specifically targeted. The target specific therapies either act on the cytokines or its receptors; thus modify immune mediated damage during the "window of opportunity". The potential targets in RA include cytokines such as TNFa, IL1 and IL6, B cells, molecules that cause interaction between antigen presenting cells and T cells, RANK ligand (receptor activator of nuclear factor kB ligand), and small molecules that mediate intracellular signalling. TNFa, which is produced by activated macrophages, monocytes, fibroblasts, mast cells, and natural killer (NK) cells, is one of the main cytokines that plays a significant role in the pathophysiology of RA. TNFa possesses an ability to induce signaling pathways of the immune system that lead to the tissue inflammation [25]. Currently, There are nine biologics approved for the treatment of RA, including TNF inhibitors (infliximab, etanercept, adalimumab, certolizumabpegol, and golimumab), anakinra (IL-1 receptorantagonist), abatacept (CTLA4-Ig fusion protein), rituximab (antiCD20 antibody), and tocilizumab (anti-IL-6 receptor antibody). Rituximab and tocilizumab are currently approved for RA patients who have failed treatment with at least 1 TNF inhibitor. Denosumab is amonoclonal antibody against RANKL which is playing role in the differentiation of macrophages to osteoclasts as well as activation of osteoclasts to cause rheumatoid bone erosions. Blocking the action of RANKL with denosumab is therefore likely to be useful in preventing erosions. The type of biological molecule which is currently widely observed by researcher is kinase inhibitor (janus kinases or JAKs, spleen tyrosine kinase or SyK, and p38 mitogen-activated protein kinases or MAPK), which targeted intracellular molecules involved in signal transduction, are also currently in development. Other new targets are regulatory T cell activation agonist, granulocyte colonys timulating factor (GCSF) and granulocyte macrophage colony-stimulating factor (GMCSF) antagonists than can significantly role in early stage of development of RA [26]. In Addition, nowadays being develop nanoparticle delivery system, which can be combined with antiinflamation natural product from curcumin to eradicate inflamation in RA. This product is expected to be the best solution for RA treatment because the product is biodegradable with high bioavailibility [27].

## 6 Nanoparticle challenge as novel treatment of rheumatoid arthritis

Nanoparticles delivery system has been describe as particles formulation which been disperse in the scale of nanometer. NNI (National Nanotechnology Iniative) defined nanoparticles as structures of sizes ranging from 1 to 100 nm. This size can be easily characterize by visual observation. Usually, they have a slightly transparent appearance and have a lower precipitating time compare to another particle with size more than 100 nm [28].

There are 2 types of nanoparticles based upon their preparation, such as nanospheres and nanocapsules [29]. Nanospheres has a monolithic type structure, where the active compound is disperse actively in the surface or adsorbed into the surface of its carrier matrix. Nanocapsules form a membrane-like structure with the active compound trapped in to the "heart" of its structure or adsorbed on the membrane surface. However, it is hard to classify a nanoparticles complex in to nanospheres or nanocapsules type [30].

One of the simplest preparation methods of polymeric nanoparticles is via ionic gelation method. Ionic gelation concept make the usage of two biopolymers in one formulation system becomes possible. The two biopolymers should have a different charge, so they can form a flexible matrix to entrap various kind of drugs with different characteristics. For example, many researchers used a combination of chitosan-alginate [31] or chitosan-carrageenan [32] that can be used to physically entrap drugs. Grenha [32] reported that chitosan-carrageenan nanoparticles complex could become a carrier for ovalbumin with entrapment efficiency around 4-17%. On another research, Gazori [33] used a combination of chitosan-alginate to make nanoparticles. They reported that this combination could increase drug stability in the systemic fluid.

Some of the advantages of using nanoparticles are their small size enable them to penetrate intercellular spaces [34], cell walls compared to bigger particles, and flexibility to combine with other technologies [35]. Nanoparticles can be used to encapsulate, protect drugs from degradation, and improve targeted drug delivery. Nanoparticles can modify the drug release, and can be produce in large, reproducible scale [36]. Nanoparticles' wide variety of physical and chemical properties affect the encapsulated drug biomedical potential such as bioavailability and biodistribution [37]. The use of nanomaterials particulate as drug carrier, resulted accumulation of nanoparticles in mononuclear phagocytic system that present in spleen and liver, therefore liver targeting by nanoparticle may be favorable and could be a promising drug therapy for liver diseases. Surface modification of nanoparticles particularly with PEG, could prolong its residence time in the circulation, reduce in vitro toxicity, prevents agglomeration that lead to destabilization of nanoparticle suspension [38]. The other advantage is that the drugs in the form of nanoparticles possess good storage stability. The study examined the storage stability of self-nanoemulsifying drug delivery systems (SNEDDS), which one of the nanoparticle based drug deliveries, found that the SNEDDS indicated no phase separation and no significant changes on the particle size were found during several tests that had been conducted. The small particle size of the nanoparticle based dosage form prevents creaming or sedimentation due to a decrease of gravitation and Brownian motion. The small particle size also prevents the flocculation to occur, which allows the drugs to be stored for a long period [39].

Despite various benefits of nanoparticle as drug carrier, there are several disadvantages such as the possibility to generate toxicity. Due to small size of nanoparticle, it has wide biodistribution in the body and increase the potency of drug, which not only beneficial for therapy but also increase the possibilities of adverse effect of attached drug in the organs other than the main target of therapy. Nanoparticles also known to affect immune response to common allergens.

Nanoparticles have been widely used to improve the pharmacokinetic and pharmacodynamics of drug molecules. Most of the non-steroidal anti-inflammatory drugs that commonly used to treat RA have a low solubility in gastric fluid but have a good solubility in intestine fluid [40]. Many researchers have proved the ability of nanoparticles in improving the bioavailability of drugs with low solubility [41] while delivered via oral [42], transdermal [43], intravenous [44], and pulmonary [45,46]. On the other hand, the improvement of the amount of drugs in systemic will increase the risk of side effects occurrence until the possibility of achieving toxic level [47]. The compatibility between nanoparticles dosage form and target tissue is needed to achieve an optimum therapy results. The guarantee of target therapy achievement becomes an absolute requirement to introduce a new drug delivery system.

Many researchers have been proposed so many materials to deliver drugs for treating cancer. Those materials have been used to deliver rheumatoid arthritis therapeutic agents in recent years. Nanoparticles, like micelles and liposome, can be further improved its specificity and therapeutic efficacy by combining with an antibody, peptide, or polysaccharide whose specific for the receptors located within tissue affected by RA. By combining passive and active targeting strategies, the delivery of drugs can be optimized, so the toxicity and the unwanted side effect will be reduce, and in the end it will improve patient outcome [48].

Nanoparticle delivery systems of RA drugs primarily based upon polymers. For example, many researches have been done to explore the use of poly(D,L-lactic/glycolic acid) (PLGA) nanoparticles. The PLGA nanoparticles shows a promising result that it can prolong the circulation time and control the drug release from matrix [48]. Higaki [49] make a betamethasone nanoparticles using PLGA with the size of particles around 100-200 nm. On that research, they found that the PLGA-betamethasone system was more effective to reduce the inflammatory response on arthritic rats and mice while administered intravenously compare to free glucocorticoid agents. The modification of PLGA nanoparticles with PEG can improve the targeting ability and efficacy of betamethasone. That system is called "stealth nanosteroids" [50].

Nowadays, the development of drug delivery systems for rheumatoid arthritis treatment is based upon the principles of passive targeting. The improvement of drugs specificity can be achieved by incorporating an active targeting compound like folate receptor β, which is an active targeting compound of Rheumatoid Arthritis Synovial Macrophage (RASMs). The researches show that the system of cationic and anionic poly(amidoamine) dendrimers loaded with indomethacin exhibit better effect in treating arthritic rats when incorporated with folate [51-53]. Therefore, the combination of nanoparticles with folate or with folate-PEG may be potential to be used as drug delivery system for RA treatment in the future.

## 7 Conclusion

Rheumatoid arthritis is the most common autoimmune disease that Affects the joints. RA is characterized by inflammation of the synovium and synovial Increased exudate which result in thickening of the synovium and joint swelling. Cause of RA is unknown, but the most stronger hypothesis is as a result of the interaction between genes and the environment. The drugs used for RA are steroids, NSAIDs, and DMARDs. However the drugs effectiveness is still scanty and harmful side effects. Nowadays researchers are being developed RA drugs that tend to inhibit the progression of RA in early development. It is still a challenge for researchers to find precise formula which is effective and have minimal side effects. Thus, incorporating drugs in nanoparticles-tagged folate or PEG may become a promising delivery system for rheumatoid arthritis treatment.

We would like to acknowledge Curcumin Research Centre, Faculty of Pharmacy, Gadjah Mada University and Pharmacy Nanoscience, Pharmaceutical Technology Laboratory, Department of Pharmacy, Islamic University of Indonesia. We also would like to address a gratitude to the grant of Penelitian Disertasi Doktor Kemeristekdikti. Authors declare no conflict of interest.

## References

1. I. B. McInnes, G. Schett, The Pathogenesis of Rheumatoid Arthritis, N Engl J Med, **365**, 2205-2219, (2011).
2. G. S. Firestein, Evolving concepts of rheumatoid arthritis, Nature, **423**, 356-361, (2003).
3. S. Ryan, Continuing professional development, autoimmune disease, Royal College of Nursing, **16**, (2002).
4. UK AR, Condition Rheumatoid Arthritis United Kingdom, Arthritis Research UK, (2014).
5. A. Gibofsky, Overview of epidemiology, pathophysiology, and diagnosis of rheumatoid arthritis, Am J Manag Care, **18**, 295-302, (2012).
6. A. Gibofsky, R. J. Winchester, M. Patarroyo, M. Fotino, H. G. Kunkel. Disease associations of the Ia-like human alloantigens: contrasting patterns in rheumatoid arthritis and systemic lupus erythematosus, J Exp Med, **148**, 1728-1732, (1978).
7. A. Z. Wilczewska, K. Niemirowicz, K. H. Markiewicz, H. Car, Nanoparticles as drug delivery systems, Pharmacol Rep, **64**, 1020-1037, (2012).
8. D. S. Majka, K. D. Deane, L. A. Parrish, A. A. Lazar, A. E. Baron, C. W. Walker, M. V. Rubertone, W. R. Gilliland, J. M. Norris, V. M. Holers, The duration of preclinical rheumatoid arthritis-relatedautoantibody positivity increases in subjects with older age at time of disease diagnosis, Ann Rheum Dis, **67**, 801–807, (2008).
9. J. S. Smolen, G. Steiner, Therapeutic strategies for rheumatoid arthritis, Nat Rev Drug Discov, **2**, 473-488, (2003).
10. J. S. Smolen, D. Aletaha, M. Koeller, M. H. Weisman, P. Emery, New therapies for treatment of rheumatoid arthritis. Lancet, 370, 1861-1874, (2007).
11. M. T. Getts, S. D. Miller, 99th Dahlem conference on infection, inflammation and chronic inflammatory disorders: triggering of autoimmune diseases by infections, Clin Exp Immunol, **160**, 15-21, (2010).
12. E. W. Karlson, K. Deane, Environmental and gene-environment interactions and risk of rheumatoid arthritis. Rheum Dis Clin North Am, **38**, 405-426, (2012).
13. K. H. Costenbader, D. Feskanich, L. A. Mandl, E. W. Karlson, Smoking intensity, duration, and cessation, and the risk of rheumatoid arthritis in women, Am J Med, **119**, 503–511, (2006).
14. L. Klareskog, P. Stolt, K. Lundberg, H. Källberg, C. Bengtsson, J. Grunewald, J. Rönnelid, H. E. Harris, A. K. Ulfgren, S. Rantapää-Dahlqvist, A. Eklund, L. Padyukov, L. Alfredsson, A new model for an etiology of rheumatoid arthritis: Smoking may trigger HLA-DR (shared epitope)-restricted immune reactions to autoantigens modified by citrullination, Arthritis Rheum, **54**, 38-46, (2006).
15. D. P. Symmons, E. M. Barrett, C. R. Bankhead, D. G. Scott, A. J. Silman, The incidence of rheumatoid arthritis in the United Kingdom: results from the Norfolk Arthritis Register, Br J Rheumatol, **33**, 735-739, (1994).
16. E. W. Karlson, L. A. Mandl, S. E. Hankinson, F. Grodstein, Do breast-feeding and other reproductive factors influence future risk of rheumatoid arthritis? Results from the Nurses' Health Study, Arthritis Rheum, **50**, 3458–3467, (2004).
17. K. Lundberg, N. Wegner, T. Yucel-Lindberg, P. J. Venables, Periodontitis in RA-the citrullinated enolase connection, Nature Reviews Rheumatology, **6**, 727–730, (2010).
18. N. Wegner, R. Wait, A. Sroka, S. Eick, K. A. Nguyen, K. Lundberg, A. Kinloch, S. Culshaw, J. Potempa, P. J. Venables, Peptidylarginine deiminase from Porphyromonas gingivalis citrullinates human fibrinogen and alpha-enolase: implications for autoimmunity in rheumatoid arthritis, Arthritis Rheum, **62**, 2662–2672, (2010).
19. Y. Arnson, H. Amital, Y. Shoenfeld, Vitamin D and autoimmunity: new aetiological and therapeutic considerations, Ann Rheum Dis, **66**, 1137–1142, (2007).

20. M. T. Cantorna, C. E. Hayes, H. F. DeLuca, 1,25-Dihydroxycholecalciferol inhibits the progression of arthritis in murine models of human arthritis, J Nutr, **128**, 68–72, (1998).
21. D. J. Pattison, D. P. Symmons, M. Lunt, A. Welch, R. Luben, S. A. Bingham, K. T. Khaw, N. E. Day, A. J. Silman, Dietary risk factors for the development of inflammatory polyarthritis: evidence for a role of high level of red meat consumption, Arthritis Rheum, **50**, 3804–3812, (2004).
22. J. M. Davis, E. L. Matteson, My treatment approach to rheumatoid arthritis, Mayo Clin Proc, **87**, 659-673, (2012).
23. L. F. Callahan, D. S. Cordray, G. Wells, T. Pincus, Formal education and five-year mortality in rheumatoid arthritis: mediation by helplessness scale score, Arthritis Care Res, **9**, 463-472, (1996).
24. SIGN (Scottish Intercollegiate Guidelines Network), Management of early rheumatoid arthritis, Edinburgh, (2011). (SIGN publication no. 123). [cited February 2011]. Available from URL: http://www.sign.ac.uk
25. R. Awaluddin, W. K. Muhtadi, L. Chabib, Z. Ikawati, R. Martien, H. Ismail, Molecular docking and ADME-toxicity studies of potential compounds of medicinal plants grown in Indonesia as an anti-rheumatoid arthritis, AIP Conf Proc, **1823**, 020033, 1-9, (2017).
26. M. Joshi, Advances in the management of rheumatoid arthritis, J Mahatma Gandhi Inst Med Sci, 17, 1-7, (2012).
27. L. Chabib, Z. Ikawati, R. Martien, H. Ismail, Review of rheumatoid arthritis: the pharmacological therapy, the potential of curcumin and its analogues, and the development of nanoparticle system, Jurnal Pharmascience, 3, 10-18, (2016).
28. R. B. Gupta, U. B. Kompella, *Nanoparticle technology of drug delivery*, Taylor & Francis Grup, New York, 4-6, 13-16, (2006).
29. E. Alleman, R. Gurny, E. Doelker, Drug-loaded nanoparticles-preparation methods and drug targeting issues, Eur. J. Pharm. Biopharm., **39**, 173-191, (1993).
30. W. Tiyaboonchai, Chitosan nanoparticles: A promising system for drug delivery, Naresuan Univ. J., **11**, 51-66, (2003).
31. P. Li, Y. Dai, J. Zhang, A. Wang, Q. Wei, Chitosan-alginate nanoparticles as a novel drug delivery system for nifedipine, Int J Biomed Sci, **4**, 221-228, (2008).
32. A. Grenha, M. E. Gomes, M. Rodrigues, V. E. Santo, J. F. Mano, N. M. Neves, R. L. Reis RL, Development of new chitosan/carrageenan nanoparticles for drug delivery application, J Biomed Mater Res, **92**, 1265–1272, (2009).
33. T. Gazori, M. R. Khoshayand, E. Azizi, P. Yazdizade, A. Nomani, I. Haririan, Evaluation of alginate/chitosan nanoparticles as antisense delivery vector: formulation, optimization and in vitro characterization, Carb. Pol., **77**, 599-606, (2009).
34. C. Buzea, L. Pacheco, K. Robbie, Nanomaterials and nanoparticles: sources and toxicity, Biointerphases, **2**, 17-71, (2007).
35. Y. Kawashima, H. Yamamoto, H. Takeuchi, Y. Kuno, Mucoadhesive DL-lactide/glycolide copolymer nanoparticles coated with chitosan to improve oral delivery of elcatonin, Pharm Dev Technol, **5**, 77-85, (2000).
36. A. C. Anselmo, S. Mitragotri, Cell-mediated delivery of nanoparticles: taking advantage of circulatory cells to target nanoparticles, J Control Release, **190**, 531-541, (2014).
37. R. Stevenson, A. J. Hueber, A. Hutton, I. B. McInnes, D. Graham, Nanoparticles and inflammation, Sci World J, **11**, 1300–1312, (2011)
38. T. Niidome, M. Yamagata, Y. Okamoto, Y. Akiyama, H. Takahashi, T. Kawano, Y. Katayama, Y. Niidome, PEG-modified gold nanorods with stealth character for in vivo application, J Control Release, **114**, 343-347, (2006)
39. L. Chabib, W. K. Muhtadi, Z. Ikawati, R. Martien, H. Ismail, Stability study of gamavuton (GVT-0) self-nanoemulsifying drug delivery system (SNEDDS) with myritol as the oil phase, International Journal of Current Innovation Research, **3**, 590-594. (2017).
40. M. Yazdanian, K. Briggs, C. Jankovsky, A. Hawi, The "high solubility" definition of the current FDA guidance on biopharmaceutical classification system may be too strict for acidic drugs, Pharm Res, 21, 293-299, (2004).
41. A. Bhatia, P. Shard, D. Chopra, T. Mishra, Chitosan nanoparticles as carrier of immunorestoratory plant extract: synthesis, characterization and immunorestoratory efficacy, International Journal of Drug Delivery, **3**,381-385, (2011).
42. R. Martien, B. Loretz, A. Bernkop-Schnürch, Oral gene delivery: design of polymeric carrier systems shielding toward intestinal enzymatic attack, Biopolymers, **83**, 327-336, (2006).
43. R. Ravichandran, Nanoparticles in drug delivery: Potential green nanobiomedicine applications, Int J Green Nanotech Biomed, **1**, 108-130, (2009).
44. F. Li, J. Li, X. Wen, S. Zhou, X. Tong, P. Su, H. Li, D. Shi, Anti-tumor activity of paclitaxel-loaded chitosan nanoparticles: An in vitro study, Mater Sci Eng C, **29**, 2392-2397, (2009).
45. W. F. Tonnis, G. F. Kersten, H. W. Frijlink, W. L. J. Hinrichs, A. H. de Boer, J. P. Amorij, Pulmonary vaccine delivery: a realistic approach?, J Aerosol Med Pulm Drug Deliv, **25**, 249-260, (2012).
46. P. Muttil, C. Prego, L. Garcia-Contreras, B. Pulliam, J. K. Fallon, C. Wang, A. J. Hickey, D. Edwards, Immunization of guinea pigs with novel hepatitis B antigen as nanoparticle aggregate powders administered by the pulmonary route, AAPS J, **12**, 330–337, (2010).

47. K. Poelstra, J. Prakash, L. Beljaars, Drug targeting to the diseased liver, J Control Release, **161**, 188-197, (2012).
48. R. A. Bader, *The Development of Targeted Drug Delivery Systems for Rheumatoid Arthritis Treatment, Rheumatoid Arthritis - Treatment*, Dr. Andrew Lemmey (Ed.), ISBN: 978-953-307-850-2, InTech, (2012).
49. M. Higaki, T. Ishihara, N. Izumo, M. Takatsu, Y. Mizushima, Treatment of experimental arthritis with poly(D, L-lactic/glycolic acid) nanoparticles encapsulating betamethasone sodium phosphate, Ann Rheum Dis, **64**, 1132-1136, (2005).
50. T. Ishihara, T. Kubota, T. Choi, M. Higaki, Treatment of experimental arthritis with stealth-type polymeric nanoparticles encapsulating betamethasone phosphate, J Pharm Exp Ther, **329**, 412–417, (2009).
51. D. Chandrasekar, R. Sistla, F. J. Ahmad, R. K. Khar, P. V. Diwan, The development of folate-PAMAM dendrimer conjugates for targeted delivery of anti-arthritic drugs and their pharmacokinetics and biodistribution in arthritic rats, Biomaterials, **28**, 504-512, (2007).
52. D. Chandrasekar, R. Sistla, F. J. Ahmad, R. K. Khar, P. V. Diwan, Folate coupled poly(ethyleneglycol) conjugates of anionic poly(amidoamine) dendrimer for inflammatory tissue specific drug delivery, J Biomed Mater Res A, **82**, 92-103, (2007).
53. A. S. Chauhan, N. K. Jain, P. V. Diwan, A. J. Khopade, Solubility enhancement of indomethacin with poly(amidoamine) dendrimers and targeting to inflammatory regions of arthritic rats, J Drug Target, **12**, 575-583, (2004)

# Potential medicinal plants for improve the immune system from Borneo Island and the prospect to be developed as nanomedicine

*Lutfi* Chabib[1,2,*], *Wildan Khairi* Muhtadi[3], *Muhammad Ikhwan* Rizki[4], and *Rifazul Aulia* Rahman[4], *Mohamad Rahman* Suhendri[5], and *Arif* Hidayat[6]

[1]Department of Pharmacy, Islamic University of Indonesia, Jl. Kaliurang km. 14,5, Yogyakarta, 55584, Indonesia
[2]Professional Pharmacy Education Program, Islamic University of Indonesia, Jl. Kaliurang km. 14,5, Yogyakarta, 55584, Indonesia
[3]Faculty of Pharmacy, Gadjah Mada University, Bulaksumur, Yogyakarta, 555281, Indonesia
[4]Lambung Mangkurat University, Jl. A. Yani km 36, Banjarbaru, Indonesia
[5]Faculty of Medicine, Islamic University of Indonesia, Jl. Kaliurang km. 14,5, Yogyakarta, 55584, Indonesia
[6]Chemical Engineering Department, Islamic University of Indonesia, Yogyakarta, 55584, Indonesia

**Abstract.** The immune system is a combination of cells, molecules, and tissues that play a role in preventing, detecting and eliminating a pathogen that enters the body. Indonesian society, hereditary and traditionally used medicinal plants (herbs) to increase immunity or immune system that fight the bacterias and pathogens. This paper aims to review the potential plants for immunomodulator from Borneo Island and the prospect to be developed as nanomedicines. The review is made using online and offline literature. Based on the literature, the immune system can be improved by various compounds from plants such as bawang dayak (*Eleutherine palmifolia* (L) Merr), kelakai (*Stenochlaena palustris* (Burm.f) Bedd), pasak bumi (*Eurycoma longifolia* Jack), kasturi (*Mangifera casturi* Kosterm.), binjai (*Mangifera caesia*), ramania (*Bouea macrophylla* Griff), gerunggang (*Cratoxylum arborescens* (Vahl) Bl) and akar kuning (*Fibraurea chloroleuca* Miers). In general, the mechanism of the plants to improve the immune system is by facing the cause of the disease directly as an effector and also work by regulating immunity. The medicinal plants also have prospects to be developed as nanomedicines.

## 1 Introduction

The immune system is the combination of cells, molecules, and tissues that play a role or in preventing, detecting and eliminating a pathogen that enters the body [1]. When the immune condition is decreasing, the body's defense against pathogen's attack will also decrease and the body is very susceptible to illness [2]. There are two types of responses generated by the immune system: innate, which inhibits the entry of microbes that successfully enter the host tissues rapidly, and adaptive, which is activated by the repeated exposure of pathogens and will memorize the pathogen so that the responses generated in the next infection will be much stronger [3].

Many diseases caused by weak defense of immune systems against pathogens such as HIV/AIDS, microbial infections (bacteria or viruses), allergies, herpes, malaria and more. One of the most dangerous diseases is HIV/AIDS, where the disease affects the immune system of its host and causes weakening of the immune system and makes it susceptible to be attacked by dangerous pathogens. Based on data from the Ministry of Health of the Republic of Indonesia in 2016, the number of HIV infections as many as 41.250 people [4]. The other disease that attacks the other immune system is malaria, which is caused by the plasmodium virus. The incidence of malaria in Indonesia based on the diagnostic is 19 per 1,000 population [5]. In Southeast Asia, there are 28 million cases of malaria with a death rate of 38 thousand people where 95% of deaths occur in 3 countries namely Indonesia, India, and Myanmar [6]. Therefore, substances that can be used to improve the immune system are required.

The diseases that attack the immune system requires a serious treatment such as increasing body immunity. The immune system can be increased by taking medicinal plants (herbs) to reduce the resistance caused by bacteria or pathogens. Medicinal plants work in the immune system not only as an effector that directly face the cause of the disease but also by regulating immunity [7]. Certain herbs possess immunomodulatory contents that modulate the function and activate the immune system. Stimulation and activation of the immune system regulated by T lymphocytes by TH-1 (CD 4+) will cause an increase in the protective properties of antigens that produce cytokines and activate macrophages [8]. Macrophages are the cells that play an important role in an immune response, which conduct phagocytosis or act as the antigen presenting cells (APC). The ability of plants in improving the immune system is caused by its chemical content. There are many kinds of compounds that possess immunostimulation activity, such as alkaloid, terpenoids, quinones, simple phenolic compounds, polysaccharides, peptides, glycoproteins, and nucleotides [7].

The use of nanotechnology in medicine is spreading rapidly throughout the world. Many pharmaceutical scientists are developing nanomedicine using the active substances from synthetic materials and also natural products. The development of herbal nanomedicines is growing rapidly due to an increase of the use of natural products as the treatment of many diseases [9]. The ability to cross the cell membranes, including the blood-brain barrier, gain access to cells and also translocate around the body via blood and lymph, make nanoparticle a promising tool for delivering drugs to the site of action [10]. In this review, we provide various plants live in Borneo Island that contain compounds with activity as an immunomodulatory and potential to be developed as nanomedicines.

# 2 The Plants in Borneo with the Ability to Improve the Immune System

## 2.1 Bawang Dayak (*Eleutherine palmifolia*(L) Merr)

Results of phytochemical screening of the root of bawang dayak showed the existence of secondary metabolites such as alkaloids, glycosides, flavonoids, phenolic, quinones, steroids, astringent tannins and essential oils. The leaves and roots of this plant also contain flavonoids and polyphenolic compounds [11]. In another journal stated that the efficacy of the root ethanol extract as an antioxidant derived from secondary metabolites such as triterpenoids, flavonoids, phenolic, alkaloids and tannins. Flavonoids contained in this plants is proven to stimulate the immune system by increasing the activity of macrophages and T lymphocytes and also known that the ethanol extract of this plants showed the ability as an immunomodulator in mice and produce immunoglobulin such as immunoglobulin G (IgG) and immunoglobulin M (IgM) [12]. Flavonoids also showed immunostimulant activity by stimulating the human mononuclear cell to secrete cytokines such as IL-1β, IFN-γ, and TNF-α, which are able to induce the death and differentiation of U973 cells, which play a role in the progression of leukemia disease [13].

**Fig 1.** The root of bawang dayak (*Eleutherine palmifolia*(L) Merr) [14]

## 2.2 Kelakai (*Stenochlaena palustris* (Burm.f) Bedd)

Kelakai (*Stenochlaena palustris* (Burm.f) Bedd) is a typical wetland (swamp) plant in South Kalimantan and a favorite food for Dayak people in Central Kalimantan. The Kenyah Dayak tribe uses this plant empirically for the treatment of anemia, fever reliever, and skin diseases. In the extract kelakai found the main bioactive substances are quercetin flavonoids that have anti-inflammatory effect. The total flavonoid contained in aqueous water extract was 14.5 μg/ml. Other ingredients in plant kelakai such as alkaloids and steroids [6]. Research has shown that extracts of the kelakai in doses of 100 mg/kg play a role for the immunomodulator in order to modulate the production of IL-10 cytokines in *P. berghei* infection. The role of IL-10 is the response of T-helper cells in malaria to prevent Th1 cells proliferating, thus suppressing the production of IFN-γ, IL-6, TNF-α by T cells, which in turn will neutralize the pathology of macrophages by inhibiting IFN secretion -γ and TNF-α in cerebral malaria [15].

**Fig 2.** Kelakai (*Stenochlaena palustris* (Burm.f) Bedd) [16]

## 2.3 Pasak Bumi (*Eurycoma longifolia* Jack)

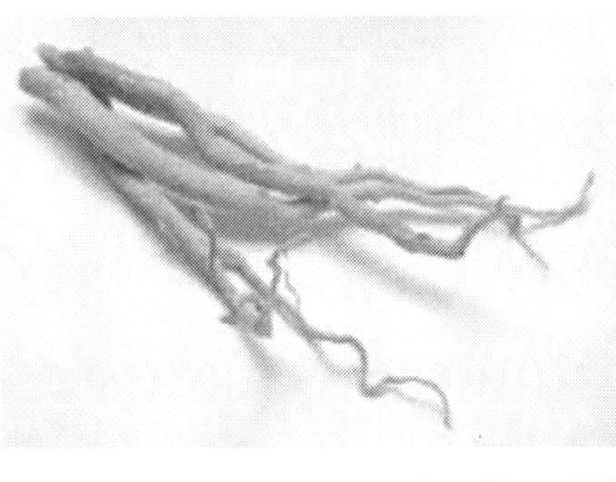

**Fig 3.** Root of pasak bumi (*Eurycoma longifolia* Jack) [19]

The bioactive contents of the root from pasak bumi possess anticancer, chemopreventive, and immunomodulatory effects. The bioactive compounds contained in this plant include quassinoids, flavonoids, and alkaloids (9-methoxycantin-6-on alkaloids and canthinone alkaloids). The mechanism of quassinoids as an immunomodulatory is by increasing the in vitro secretion of IL-12. IL-12 is a congenital immune response amplifier whose function activates macrophages against

https://doi.org/10.1051/matecconf/201815404006

microbial stimuli by developing the function of NK cell effector [17]. In the other research, quassin, which is the member of the quassinoids group, showed its ability to produce the derivatives of nitric oxide in order to kill the intracellular *Leishmania* parasites. Quassin was also found able to increase the release and mRNA expression of TNF-α, which is the inflammatory cytokine that plays a role in the eradication of *Leishmania* parasites. The abilities of quassin found in the study proven its potentiality as an immunomodulator to increase immunity for the treatment of chronic infectious diseases [18].

## 2.4 Kasturi (*Mangifera casturi* Kosterm.)

Kasturi is an endemic plant found in the South Kalimantan area which is commonly used by local people to be consumed due to its sweet taste and also has a distinctive aroma of fun. Phytochemical screening and preliminary tests on kasturi's tree trunks were reported to contain bioactive substances such as terpenoids, steroids, and saponins, and it was proposed that the methanol extract of kasturi also had bioactive components such as terpenoids / streroids and phenolics also contained flavonoids [20, 21]. The kasturi fraction also has an antibacterial content [22]. The isolated terpenoids from *Mangifera casturi* fruit with the concentration of 6,25 μg/mL and 12,5 μg/mL indicated immunomodulatory activity. The terpenoids enhanced the ability of macrophage to phagocytize the latex beads which indicated with the latex beads particles swallowed into the macrophage cells [23].

**Fig 4.** Kasturi (*Mangifera casturi* Kosterm.) [24]

## 2.5 Binjai (*Mangifera caesia*)

Binjai is including in manga type with oval leaf shape, thick, clay, rather large, absent bone, fruit weight between 192-316 g, thin-skinned, soft young flesh and soft when ripe, rough fibrous, and the smell of typical fruits sting. Large seeds with elongated round shape [25]. Binjai (*Mangifera caesia*) is one of the *Mangifera* genus, which is often consumed by the people of South Kalimantan, has a sour taste and a little sweet taste with a distinctive taste and commonly used as a mixture in sauce. The leaves of this plant contain bioactive substances such as flavonoids, which are compounds that have functions as anti-inflammatory, anti-cancer, anti-leukemia, anti malarial and can be used as HIV treatment [26]. Another study found that binjai plants contain a compound that is not found in other plant species that is alkenilfenol which has activity as an antioxidant [27]. The results of phytochemical screening performed on the roots and stems of binjai and kasturi plants contain saponins. Saponin is a secondary metabolite compound belonging to terpenoid groups, ie compounds containing isoprene skeletons $CH_2{=}C(CH_3){-}CH{=}CH_2$ [28]. Saponin has the ability as an immunostimulant by stimulating the response of innate immune, which makes saponin a promising compound to be used as vaccine adjuvant [29].

**Fig 5.** The tree and fruit of Binjai (*Mangifera caesia*) [30]

## 2.6 Ramania (*Bouea macrophylla* Griff)

Ramania (*Bouea macrophylla* Griff) is a typical plant of the Borneo region. In the leaves of ramania plants contain a large number of substances such as flavonoids, saponins, triterpenoids. The secondary compound of flavonoids acts as an antioxidant and accelerates wound healing by the mechanism of accelerating the proliferation of fibroblast cells and the production of collagen fibers so that it can treat wounds quickly and the cell turnover will accelerate and will surely enhance the body's immune system [31]. Another immunostimulant activity of flavonoids was found through its ability to inhibit the growth *Legionella pneumophila* in macrophages [32]. The other content of ramania is vitamin C, which is 172.4 mg in 100 grams of its fruits. Vitamin C possesses the ability to increase the immune system [33]. Vitamin A is also found in ramania fruits with amount of 1,020 IU that can be seen from the color of its orange fruit flesh [34].

**Fig 6.** The leaf and fruit of Ramania (*Bouea macrophylla* Griff) [35]

## 2.7 Gerunggang (*Cratoxylum arborescens* (Vahl) Bl)

This plant is commonly found in the wet areas (swamps) in Kalimantan. The study conducted by Yusro suggests that the bark extracts possess the ability as the treatment of headache and cancer. The phytochemical screening of the bark of gerunggang showed that it contains several compounds such as tannin, saponin, flavonoids, and quinone. Flavonoids have properties for anti-virus, anti-microbial, anti-inflammatory and healing subcutaneous capillary hemorrhage, while saponin has the function as an immunostimulant that stimulates the immune system in the body [36].

**Fig 7.** Gerunggang (*Cratoxylum arborescens* (Vahl) Bl) [37]

**Fig 8.** The leaves of akar kuning (*Fibraurea chloroleuca* Miers) [41]

### 2.8 Akar Kuning (*Fibraurea chloroleuca* Miers)

Dayak community in Kalimantan uses akar kuning as the treatment of abdominal pain, eye drops and yellow pain medication. There are secondary metabolite compounds contained in *F. chloroleuca* such as alkaloid compounds that are commonly found in the stems, roots and leaves. The other studies shown that alkaloids in the stems and roots of *F. chloroleuca* have similar species. The other secondary metabolite compounds are terpenoid compounds, both on stems, leaves and roots. Several other studies also suggested that the compounds contained in the *F. chloroleuca* plant are berberine, which is an alkaloid that has been used in a long time as a medicine. The type of bioactive compound that can be identified is 8-oxoprotoberberine [38]. Berberine possesses activity as an immunostimulant by increasing the ability of macrophages to kill bacteria swallowed in the body [39]. Another study found that berberine increased the production of interleukin-12, which stimulated the immunoregulatory activity by inhibiting the profile of T helper-2 cytokine in $CD4^+$ cells. The findings proven that medicinal plants containing berberine can be used as the therapy of allergic diseases, particularly the T helper type 2 cell-mediated immune disease [40].

## 2 The Prospects of Nanomedicine from Medicinal Plants as Immunostimulant

There are many medicinal plants containing compounds that possess immunomodulatory activity, either suppress or stimulate the immunity that can be used in the treatment of most disease. Many studies have been conducted to explore the most effective delivery methods to deliver the compounds to get the significant immunomodulatory effect in the human body. One of the methods that attract much attentions is to formulate them as the nanomedicine, which makes the drug in the form of nanoparticles (NPs). This method is being developed widely nowadays. NPs are very small forms of material substances, measuring from 1 to 100 nanometers, which make them possess large surface area. Due to the very small size and large surface area, NPs possess unique abilities. For example, NPs readily cross the cell membranes, including the blood-brain barrier, possibly to gain access to cells and also translocate around the body via blood and lymph. The ability to access the inside of the cells makes them a promising tool for delivering drugs to the site of action [10].

The study conducted by Seo found that *R. Coreanus* extract-encapsulated gelatin nanoparticles with an average size of 143 ± 18 nm that had been produced in the study possessed the possibility to be targeted to the particular tissues to achieve a higher local concentration compared to the ingestion of the crude extract only. The nanoparticles also showed a low toxicity and high biocompatibility. The immunomodulatory effects of *R. Coreanus* extract-encapsulated gelatin nanoparticles, that were determined in T-cells, B-cells, and natural killer (NK) cells, were found with a higher potency compared with the crude extract or ferulic acid as the control. There were two immune activities of the nanoparticles attracting attention in the study. First, the proliferation of B- and T-cells. It was found that the nano-encapsulated extracts induced a significantly higher proliferation in T- and B-cells, which was ~3.4 × control and ~2.9 × control respectively, while the extract only showed ~1.8 × control and ~1.9 × control in T- and B-cells respectively. Second, the proliferation of natural killer (NK) cells. The primary function of NK cells is to eliminate aberrant cells, including virally infected and tumorigenic cells. The result showed that the nano-encapsulated extract increases the proliferation of NK cells extremely compared to control and that the quite significant effect compared to the crude extract only. It was assumed that the result was caused by the ability of nanoparticle to induce cellular penetration and uptake, that lead to a higher local concentration [42].

Another study conducted by Rather found that green synthesis of silver nanoparticles (G-AgNPs) from *Azadirachta indica* (neem) possessed the highest phagocytic activity compared to the control (the fish which were used as the experimental animal were given no treatment) and neem extract only. It might be due the G-AgNP has activated an alternative pathway that only can be done by nanoscale particles that increase the immunomodulation activity. Nevertheless, the mechanism of immunomodulatory effects of G-AgNPs needs to be investigated further [43].

The result of the two studies indicates that there is a possibility to formulate nanomedicines of medicinal plants that possess immunomodulatory activity. It is assumed that the medicinal plants from Borneo Island that had been reviewed in this article also possible to be formulated as the nanomedicines.

## 3 Conclusion

There are many plants that grow in Borneo Island that can be used to improve the immune system such as bawang dayak (*Eleutherine palmifolia* (L) Merr), kelakai (*Stenochlaena palustris* (Burm.f) Bedd), pasak bumi (*Eurycoma longifolia* Jack), kasturi (*Mangifera casturi* Kosterm.), binjai (*Mangifera caesia*), ramania (*Bouea macrophylla* Griff), gerunggang (*Cratoxylum arborescens* (Vahl) Bl) and akar kuning (*Fibraurea chloroleuca* Miers). The medicinal plants are assumed to have the possibility to be developed as the immunomodulatory nanomedicines that possess many advantages.

We would like to acknowledge Pharmacy Nanoscience, Pharmaceutical Technology Laboratory, Department of Pharmacy, Islamic University of Indonesia. We also would like to address a gratitude to the grant of Penelitian Madya DPPM UII. Authors declare no conflict of interest.

## References

1. J. Kelly, *Understanding the Immune System How It Works*, NIAID Science Education, USA, (2007).
2. D. Mayasari, A. Pratiwi, Relation between immune respond and stress with the level of typhoid fever recurrence in the population of Puskesmas Colomadu Karanganyar region, Jurnal Berita Ilmu Keperawatan, **2**, 13-18, (2009).
3. P. J. Delves, I. M. Roitt, The Immune System. First of two parts, N Engl J Med, **343**, 37-49, (2000).
4. Ministry of Health of Republic of Indonesia, *Data and information of Indonesian health profile 2016*, Ministry of Health RI, (2017).
5. Ministry of Health of Repulic of Indonesia, *Health profile of Indonesia*, Ministry of Health RI, Jakarta, (2015).
6. D. P. N. H. Margono, E. Suhartono, H. Arwati, Potentiality of kelakai (*Stenochlaena palustris (Burm.f) Bedd)* extract against the level of tumor necrosis factor-alfa (TNF-α) in the BALB/c mice infected with *Plasmodium Berghe* Anka, Berkala Kedokteran, **12**, 77-85, (2016).
7. Y. P. Utami, Aliyah, R. Syukur, Immunostimulant effect combination syrup of *Carthamus tinctorius L.* and *Eleutherine palmifolia* extracts on mice *(Mus musculus)*, JST Kesehatan, **6**, 179-184, (2016).
8. A. N. Toemon, Effects of orally bawang dayak *(Eleutherine bulbosa (Mill.) Urb.)* root ethanol extract on Balb/c mice against the prevention of the decrease of IFN-ɤ-expressed cell amounts and the increase of CD 14-expressed cell amounts, Jurnal Biosains Pascasarjana. **17**, 1-14, (2015).
9. A. Pandey, G. Pandey, Usefulness of nanotechnology for herbal medicines, Plant Archives, **13**, 617-621, (2013).
10. K. S. Rao, A. Ghorpade, V. Labhasetwar, Targeting anti-HIV drugs to the CNS, Expert Opin Drug Deliv, **6**, 771-784, (2009).
11. R. Puspadewi, P. Adirestuti, R. Menawati, Efficacy of bawang dayak (*Eleutherine palmifolia* (L.) Merr.) root as the skin antimicrobial herbal, Kartika Jurnal Ilmiah Farmasi, **1**, 31-37, (2013).
12. A. B. Carmelita, Effects of orally bawang dayak (*Eleutherine palmifolia (L.) Merr.*) root extract on BALB/c mice against the prevention of the decrease Germinal Center parameters on the lymph gland and IgG serum concentration, Jurnal Biosains Pascasarjana, **18**, 1-13, (2016).
13. D. Y. Liao, Y. C. Chai, S. H. Wang, C. W. Chen, M. S. Tsai, Antioxidant activities and contents of flavonoids and phenolic acids of *Talinum triangulare* extracts and their immunomodulatory effects, J Food Drug Anal, **23**, 294-302, (2015).
14. Anonymous, https://id.wikipedia.org/wiki/Bawang_dayak, accessed 23 September 2017.
15. D. P. N. H. Margono, E. Suhartono, H. Arwati, Effects of kelakai (*Stenochlaena palustris (Burm.f) Bedd)* extract against the level of interleukin-10 (IL-10) on mice, Medical Laboratory Technology Journal, **2**, 31-36, (2016).
16. Anonymous, https://id.wikipedia.org/wiki/Lemidi, accessed 23 September 2017.
17. Arifah, A. N & Nurkhasanah, The effect of ethyl acetate fraction of ethanolic extract of pasak bumi (*Eurycoma longifolia*, Jack) on macrophages phagocytic activity in vitro, Pharmaciana, **4**, 9-14, (2104).
18. S. Bhattacharjee, G. Gupta, P. Bhattacharya, A. Mukherjee, S. B. Mujumdar, A. Pal, S. Majumdar, Quassin alters the immunological patterns of murine macrophages through generation of nitric oxide to exert antileishmanial activity, J Antimicrob Chemother, **63**, 317-324, (2009).
19. Anonymous, http://www.eurytest.com/science/, accessed 23 September 2017.
20. N. Fakhrudin, P. S. Putri, Sutomo, S. Wahyuono, Antiinflamatory activity of methanolic extract of

*Mangifera casturi* in thioglycollate-induced leukocyte migration on mice, Trad Med J, **18**, 151-156, (2013).

21. Sutomo, N. Agustina, Arnida, Fadilaturrahmah, Pharmacognostic study and nonspecific parameter test of the stem bark methanolic extract of kasturi (*Mangifera casturi* Kosterm.), Jurnal Pharmascience, **4**, 94-101, (2017).
22. K. Rosyidah, S. A. Nurmuhaimina, N. Komari, M. D. Astuti, Antibacterial activity of the saponin fraction of the stem bark of kasturi (*Mangifera casturi*), Alchemy, **1**, 53-103, (2010).
23. Sutomo, S. Wahyuono, E. P. Setyowati, A. Yuswanto, Activity of *Mangifera casturi* Kosterm. fruit isolates as an immunomodulator *in vitro*, Prosiding Seminar Nasional & Workshop "Perkembangan Terkini Sains Farmasi & Klinik 5", 260-266, (2015).
24. Anonymous, https://en.wikipedia.org/wiki/Mangifera_casturi, accessed 23 September 2017.
25. M. Noor, M. Saleh, H. Subagio, Potency of biodiversity of fruit plants in swamp land and its utilization, Prosiding Seminar Nasional Masyarakat Biodiversitas Indonesia, **1**, 1348-1358, (2015).
26. M. Syafarina, I. Taufiqurrahman, Edyson, Difference of total flavonoid between natural and artificial drying stage on binjai (*Mangifera caesia*) leaf extract, Jurnal Kedokteran Gigi, **1**, 84-88, (2017).
27. K. Rosyidah, Siska, M. D. Astuti, Isolation of the antioxidant compounds from the stem bark of binjai (*Mangifera Caesia*), Jurnal Sains dan Terapan, **5**, 8-14, (2011).
28. K. Mustikasari, D. Ariyani, The study potency of binjai (*Mangifera caesia*) and kasturi (*Mangifera casturi*) as antidiabetic by phytochemistry screening on roots and stem, Sains dan Terapan Kimia, **2**, 64-73, (2008).
29. H. Top, N. B. Sarikahya, A. Nalbantsoy, S. Kirmizigul, Immunomodulatory, hemolytic properties and cytotoxic activity potent of triterpenoid saponins from Cephalaria balansae, Phytochemistry, **137**, 139-147, (2017).
30. Anonymous, http://www.fruitipedia.com/binjai_mangifera_caesia.htm, accessed 23 September 2017.
31. A. Rahman, I. Taufiqurrahman, Edyson, Difference of the total flavonoids between the method of maseration and soxhletation on ramania (*Bouea macrophylla* Griff) leaves extract, Jurnal Kedokteran Gigi, **1**, 22-27, (2017).
32. M. Vajdy, Immunomodulatory properties of vitamins, flavonoids and plant oils and their potential as vaccine adjuvants and delivery systems, Expert Opin Biol Ther, **11**, 1501-1513, (2011).
33. S. Susi, Utilization potency of the nutrient of exotic fruits from South Kalimantan, Jurnal Ziraa'ah, **39**, 144-150, (2014).
34. S. S. Antarlina, Identification of physicochemical properties of local fruits of Kalimantan, Buletin Plasma Nutfah, **15**, 80-91, (2009).
35. Anonymous, https://en.wikipedia.org/wiki/Bouea_macrophylla, accessed 23 September 2017.
36. F. Yusro, Rendement of ethanol extract and the phytochemical test of three types of medicinal plants of West Kalimantan, Jurnal Tengkawang, **1**, 29-36, (2011).
37. Sabah Forestry Department, Serungan/geronggang tree species (*Cratoxylum arborescens*). Sabah, (2007).
38. M. S. R. Wahyuningsih, S. Wahyuono, D. Santosa, J. Setiadi, Soekotjo, S. M. Widiastuti, R. Rakhmawati, D. S. C. Wahyuni, Exploration of Central Kalimantan's forest plants as bioactive compound resources, Biodiversitas, **9**, 169-172, (2008).
39. C. Li, L. Yan, Y. Jing, L. Xu, Y. Liang, H. Wei, B. Hu, H. Pan, Q. Zha, D. Ouyang, X. He, Berberine augments ATP-induced inflammasome activation in macrophages by enhancing AMPK signaling, Oncotarget, **8**, 95-109, (2017).
40. T. S. Kim, B. Y. Kang, D. Cho, S. H. Kim, Induction of interleukin-12 production in mouse macrophages by berberine, a benzodioxoloquinolizine alkaloid, deviates CD4+ T cells from a Th2 to a Th1 response, Immunology, **109**, 407-414, (2003).
41. Anonymous, http://indiabiodiversity.org/species/show/268035, accessed 23 September 2017.
42. Y. C. Seo, W. Y. Choi, C. G. Lee, S. W. Cha, Y. O. Kim, J. Kim, G. P. C. Drummen, H. Y. Lee, Enhanced immunomodulatory activity of gelatin-encapsulated *Rubus coreanus* Miquel nanoparticles, Int J Mol Sci, **12**, 9031-9056, (2011).
43. M. A. Rather, I. A. Bhat, N. Sharma, A. Gora, P. A. Ganie, R. Sharma, Synthesis and characterization of *Azadirachta indica* constructed silver nanoparticles and their immunomodulatory activity in fish, Aquacult Res, **2016**, 1-13, (2016)

# AUTHOR INDEX

# AUTHOR INDEX

# AUTHOR INDEX

# AUTHOR INDEX

**EDP Sciences – Web of Conferences**
17, Avenue du Hoggar
Parc d'Activité de Courtabouf
BP 112
F-91944 Les Ulis Cedex A
France

ISBN 978-1-5108-5910-4

9 781510 859104

CURRAN ASSOCIATES INC.
proceedings
.com